Conversion between USCS and SI

To convert from USCS to SI: To convert the value of a variable from USCS units to equivalent SI units, ***multiply*** the value to be converted by the right-hand side of the corresponding equivalency statement in the Table of Equivalencies.

Example: Convert a length $L = 3.25$ in. to its equivalent value in millimeters.

Solution: The corresponding equivalency statement is: 1.0 in. = 25.4 mm

$$L = 3.25 \text{ in.} \times (25.4 \text{ mm/in.}) = \textbf{82.55 mm}$$

To convert from SI to USCS: To convert the value of a variable from SI units to equivalent USCS units, ***divide*** the value to be converted by the right-hand side of the corresponding equivalency statement in the Table of Equivalencies.

Example: Convert an area $A = 1000$ mm^2 to its equivalent in square inches.

Solution: The corresponding equivalency statement is: 1.0 in.^{2} = 645.16 mm^2

$$A = 1000 \text{ mm}^2 /(645.16 \text{ mm}^2/\text{in.}^2) = \textbf{1.55 in.}^2$$

FUNDAMENTALS OF MODERN MANUFACTURING

Materials, Processes, and Systems

Second Edition

Mikell P. Groover
Professor of Industrial and Manufacturing
  Systems Engineering
Lehigh University

JOHN WILEY & SONS, INC.

New York • Chichester • Weinheim • Brisbane • Singapore • Toronto

ACQUISITIONS EDITOR: Joseph Hayton
MARKETING MANAGER: Katherine Hepburn
SENIOR PRODUCTION EDITOR: Patricia McFadden
SENIOR DESIGNER: Dawn L. Stanley
ILLUSTRATION EDITOR: Sandra Rigby
PRODUCTION MANAGEMENT SERVICES: Ingrao Associates

Cover Photo: Courtesy of Kennametal Inc.

This book was set in Times Ten by TechBooks and printed and bound by Hamilton Printing. The cover was printed by Phoenix Color, Inc.

This book is printed on acid-free paper.

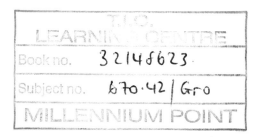
ISBN 0-471-40051-3

Printed in the United States of America

PREFACE

Fundamentals of Modern Manufacturing: Materials, Processes, and Systems is designed for a first-course or two-course sequence in manufacturing, mechanical engineering, industrial engineering, or manufacturing engineering. It may also be appropriate for technology programs related to these disciplines. Most of the book's content is on manufacturing processes (about 65% of the text), but it also provides substantial coverage of engineering materials and production systems. Materials, processes, and systems are the basic building blocks of manufacturing and the three broad subject areas covered in the book.

The author's objective in the first edition, to provide a treatment of its subject that is more ***modern*** and ***quantitative*** than competing books, also holds true for the second edition. Its claim to be "modern" is based on (1) its more balanced coverage of the three basic engineering materials (metals, ceramics, and polymers, as well as composites of these materials); (2) its inclusion of recently developed manufacturing processes in addition to the traditional processes that have been used and refined over many years; and (3) its more comprehensive coverage of electronics manufacturing technologies. Competing textbooks tend to emphasize metals and their processing at the expense of the other engineering materials, whose applications and methods of processing have grown significantly in the last several decades. For example, the volume of polymers processed commercially in the world today exceeds the volume of metals processed (the tonnage is still much less, simply because the average density of metals is several times greater than the average density of polymers). Also, competing books provide minimum coverage of electronics manufacturing. Yet the commercial importance of electronics products and their associated industries have increased substantially during recent decades. To illustrate, consider the 30 companies listed in the Dow Jones Industrials (DJI) Average. Five of these companies are in businesses related to electronics and computers (General Electric, Hewlett Packard, Intel, IBM, and Microsoft), compared to seven companies involved in traditional manufacturing (Alcoa, Allied Signal, Boeing, Caterpillar, Eastman Kodak, GM, United Technologies). The other 18 companies in the DJI are in service industries (e.g., banking, retail, communications) or manufacturing other than the processes covered in this book (e.g., food, beverage, oil, pharmaceuticals). However, when we look at market capitalization, the total stock market value of the five companies involved in electronics is approximately five times the market value of the seven companies in traditional manufacturing and fully one-third the market value of all 30 DJI companies. Is electronics manufacturing important? Even moreso today than when the first edition was released in 1995.

The text's claim to be more "quantitative" is based on its emphasis on manufacturing science and greater use of equations and quantitative (end-of-chapter) problems than any of its predecessors. In the case of some processes, it was the first manufacturing processes textbook to ever provide a quantitative engineering coverage of the topic.

FEATURES RETAINED FROM THE FIRST EDITION

(1) Sections on *Guide to Processing* in each of the four chapters on engineering materials.
(2) Sections on *Product Design Considerations* in many of the manufacturing process chapters.
(3) Multiple-choice quizzes at the end of all but one chapters.
(4) *Historical Notes* on many of the technologies covered.

The scope and content of the new edition are similar to the first edition. The author's objective of providing a modern and quantitative treatment remains the same.

NEW FEATURES IN THE SECOND EDITION

Readers who are familiar and comfortable with the first edition will not be overwhelmed by a completely different book. However, a number of new features have been added to keep the text current.

➤ A new chapter on *Rapid Prototyping* (Chapter 34).
➤ A new chapter on *Microfabrication* and *Nanofabrication* Technologies (Chapter 37).
➤ Expanded coverage on Cutting Tools used in machining processes (Chapter 23).
➤ More than 500 problems, including almost 150 new or revised problems compared to the previous edition.
➤ The principal engineering units have been changed to System International (metric), but both metric and U.S. Customary Units are used throughout the text.
➤ The organization of the book has been revised to be more logical.

INSTRUCTOR'S RESOURCES

A *Solutions Manual* can be obtained from the publisher by instructors who adopt the book as the text for their courses. The Solutions Manual contains answers to all end-of-chapter review questions, multiple-choice quizzes, and problems. Other support materials may be found at the Wiley web site www.wiley.com/college/groover. Individual questions or comments may be directed to me personally at Mikell.Groover@Lehigh.edu.

ACKNOWLEDGMENTS

The first edition was published in 1996 by Prentice Hall. In June 1999, the publication rights for the book were acquired by John Wiley & Sons, Inc., and the second edition is being published by Wiley.

I would like to express my appreciation to the following people who served as technical reviewers of individual sets of chapters for the first edition: Iftikhar Ahmad (George Mason University); J. T. Black (Auburn University); David Bourell (University of Texas at Austin); Paul Cotnoir (Worcester Polytechnic Institute); Robert E. Eppich (American Foundryman's Society); Osama Eyeda (Virginia Polytechnic Institute and State University); Wolter Fabricky (Virginia Polytechnic Institute and State University); Keith Gardiner (Lehigh University); R. Heikes (Georgia Institute of Technology); Jay R. Geddes (San Jose State University); Ralph Jaccodine (Lehigh University); Steven Liang (Georgia Institute of Technology); Harlan MacDowell (Michigan State University); Joe Mize (Oklahoma State University); Colin Moodie (Purdue University); Michael Philpott (University of Illinois at Champaign-Urbana); Corrado Poli (University of Massachusetts at Amherst); Chell Roberts (Arizona State University); Anil Saigal (Tufts University); G. Sathyanarayanan (Lehigh University); Malur Srinivasan (Texas A&M University); A. Brent Strong (Brigham Young University); Yonglai Tian (George Mason University); Gregory L. Tonkay (Lehigh University); Chester VanTyne (Colorado School of Mines); Robert Voigt (Pennsylvania State University); and Charles White (GMI Engineering and Management Institute).

I would like to thank the following individuals for their helpful reviews of certain chapters in the second edition: John T. Berry (Mississippi State University); Rajiv Shivpuri (Ohio State University); James B. Taylor (North Carolina State University); Joel Troxler (Montana State University); and Ampere A. Tseng (Arizona State University).

I would also like to thank several of my colleagues in the Department of Industrial and Manufacturing Systems Engineering at Lehigh for their encouragement during the several years it took me to write the manuscript for the first edition, for their comments on the book after its publication, and for their encouragement to prepare a second edition. Those colleagues include Keith Gardiner, Louis Martin-Vega, Nicholas Odrey, G. Sathyanarayanan (deceased), Marlin Thomas, Gregory Tonkay, David Wu, and Emory Zimmers.

In addition, it seems appropriate to acknowledge my Wiley colleagues who have been so helpful during preparation of this second edition. These include Joe Hayton, Steve Peterson, Patricia McFadden, and Ingrao Associates.

ABOUT THE AUTHOR

Mikell P. Groover is Professor of Industrial and Manufacturing Systems Engineering at Lehigh University, where he serves as Director of the George E. Kane Manufacturing Technology Laboratory. He received a B.A. (1961) in Arts and Science, B.S. (1962) in Mechanical Engineering, M.S. (1966) and Ph.D. (1969) in Industrial Engineering, all from Lehigh. He is a Registered Professional Engineer in Pennsylvania (since 1972). His industrial experience includes several years as a manufacturing engineer with Eastman Kodak Company. Since joining Lehigh, he has done consulting, research, and project work for a number of industrial companies, including Bethlehem Steel, Ingersoll-Rand, Air Products & Chemicals, and Hershey Foods.

His teaching and research areas include manufacturing processes, metal cutting theory, production systems, automation, robotics, material handling, facilities planning, and work systems. He has received a number of teaching awards at Lehigh University, as well as the *Albert G. Holzman Outstanding Educator Award* from the Institute of Industrial Engineers (1995) and the *SME Education Award* from the Society of Manufacturing Engineers (2001). His publications include over 75 technical articles and papers for *Industrial Engineering, IIE Transactions, ASME Transactions, IEEE Spectrum, International Journal of Production Systems, Encyclopaedia Britannica, SME Technical Papers,* and others. Professor Groover's avocation is writing textbooks on topics related to manufacturing and automation. His previous books are used throughout the world and have been translated into French, German, Spanish, Portuguese, Russian, Japanese, Korean, and Chinese. The first edition of the current book *Fundamentals of Modern Manufacturing* received the *IIE Joint Publishers Award* (1996) and the *M. Eugene Merchant Manufacturing Textbook Award* from the Society of Manufacturing Engineers (1996).

Dr. Groover is a member of the Institute of Industrial Engineers, American Society of Mechanical Engineers (ASME), the Society of Manufacturing Engineers (SME), the North American Manufacturing Research Institute (NAMRI), and ASM International. He is a Fellow of IIE (1987) and SME (1996).

Previous Books by the Author

Automation, Production Systems, and Computer-Aided Manufacturing, Prentice Hall, 1980.

CAD/CAM: Computer-Aided Design and Manufacturing, Prentice Hall, 1984 (co-authored with E. W. Zimmers, Jr.).

Industrial Robotics: Technology, Programming, and Applications McGraw-Hill Book Company, 1986 (co-authored with M. Weiss, R. Nagel, and N. Odrey).

Automation, Production Systems, and Computer Integrated Manufacturing, Prentice Hall, 1987.

Fundamentals of Modern Manufacturing: Materials, Processes, and Systems, First Edition, originally published by Prentice Hall, 1966; now published by John Wiley & Sons, Inc., 1999.

Automation, Production Systems, and Computer Integrated Manufacturing, Second Edition, Prentice Hall, 2001.

CONTENTS

1 INTRODUCTION 1

   1.1 What Is Manufacturing? 2
   1.2 Materials in Manufacturing 8
   1.3 Manufacturing Processes 10
   1.4 Production Systems 17
   1.5 Organization of the Book 20
   1.6 Images of Manufacturing 21

Part I Material Properties and Product Attributes

2 THE NATURE OF MATERIALS 24

   2.1 Atomic Structure and the Elements 24
   2.2 Bonding between Atoms and Molecules 26
   2.3 Crystalline Structures 29
   2.4 Noncrystalline (amorphous) Structures 34
   2.5 Engineering Materials 36

3 MECHANICAL PROPERTIES OF MATERIALS 39

   3.1 Stress–Strain Relationships 39
   3.2 Hardness 52
   3.3 Effect of Temperature on Properties 55
   3.4 Fluid Properties 57
   3.5 Viscoelastic Behavior of Polymers 60

4 PHYSICAL PROPERTIES OF MATERIALS 66

   4.1 Volumetric and Melting Properties 66
   4.2 Thermal Properties 69
   4.3 Mass Diffusion 71
   4.4 Electrical Properties 72
   4.5 Electrochemical Processes 74

5 DIMENSIONS, TOLERANCES, AND SURFACES 77

   5.1 Dimensions, Tolerances, and Related Attributes 77
   5.2 Surfaces 78
   5.3 Effect of Manufacturing Processes 85

Part II Engineering Materials

6 METALS 88

   6.1 Alloys and Phase Diagrams 89
   6.2 Ferrous Metals 93
   6.3 Nonferrous Metals 109
   6.4 Superalloys 120
   6.5 Guide to the Processing of Metals 122

7 CERAMICS 125

   7.1 Structure and Properties of Ceramics 127
   7.2 Traditional Ceramics 129
   7.3 New Ceramics 132
   7.4 Glass 134
   7.5 Some Important Elements Related to Ceramics 138
   7.6 Guide to Processing Ceramics 140

8 POLYMERS 143

   8.1 Fundamentals of Polymer Science and Technology 145
   8.2 Thermoplastic Polymers 155
   8.3 Thermosetting Polymers 162
   8.4 Elastomers 165
   8.5 Guide to the Processing of Polymers 172

9 COMPOSITE MATERIALS 175

   9.1 Technology and Classification of Composite Materials 176
   9.2 Metal Matrix Composites 184
   9.3 Ceramic Matrix Composites 186
   9.4 Polymer Matrix Composites 187
   9.5 Guide to Processing Composite Materials 190

Part III Solidification Processes

10 FUNDAMENTALS OF METAL CASTING 193

   10.1 Overview of Casting Technology 196
   10.2 Heating and Pouring 198
   10.3 Solidification and Cooling 202

11 METAL CASTING PROCESSES 213

   11.1 Sand Casting 214
   11.2 Other Expendable Mold Casting Processes 219
   11.3 Permanent Mold Casting Processes 225
   11.4 Foundry Practice 233
   11.5 Casting Quality 236
   11.6 Metals for Casting 239
   11.7 Product Design Considerations 240

12 GLASSWORKING 246

   12.1 Raw Materials Preparation and Melting 246
   12.2 Shaping Processes in Glassworking 247
   12.3 Heat Treatment and Finishing 253
   12.4 Product Design Considerations 254

13 SHAPING PROCESSES FOR PLASTICS 256

13.1 Properties of Polymer Melts 258
13.2 Extrusion 260
13.3 Production of Sheet and Film 270
13.4 Fiber and Filament Production (spinning) 272
13.5 Coating Processes 274
13.6 Injection Molding 275
13.7 Compression and Transfer Molding 285
13.8 Blow Molding and Rotational Molding 287
13.9 Thermoforming 292
13.10 Casting 296
13.11 Polymer Foam Processing and Forming 297
13.12 Product Design Considerations 299

14 RUBBER PROCESSING TECHNOLOGY 306

14.1 Rubber Processing and Shaping 306
14.2 Manufacture of Tires and Other Rubber Products 311
14.3 Product Design Considerations 315

15 SHAPING PROCESSES FOR POLYMER MATRIX COMPOSITES 317

15.1 Starting Materials for PMCs 319
15.2 Open Mold Processes 321
15.3 Closed Mold Processes 325
15.4 Filament Winding 327
15.5 Pultrusion Processes 329
15.6 Other PMC Shaping Processes 331

Part IV Particulate Processing of Metals and Ceramics

16 POWDER METALLURGY 334

16.1 Characterization of Engineering Powders 336
16.2 Production of Metallic Powders 340
16.3 Conventional Pressing and Sintering 342
16.4 Alternative Pressing and Sintering Techniques 348
16.5 Materials and Products for PM 351
16.6 Design Considerations in Powder Metallurgy 352

17 PROCESSING OF CERAMICS AND CERMETS 358

17.1 Processing of Traditional Ceramics 359
17.2 Processing of New Ceramics 366
17.3 Processing of Cermets 369
17.4 Product Design Considerations 371

Part V Metal Forming and Sheet Metalworking

18 FUNDAMENTALS OF METAL FORMING 374

18.1 Overview of Metal Forming 374
18.2 Material Behavior in Metal Forming 377
18.3 Temperature in Metal Forming 378
18.4 Strain Rate Sensitivity 380
18.5 Friction and Lubrication in Metal Forming 383

19 BULK DEFORMATION PROCESSES IN METAL WORKING 386

19.1 Rolling 387
19.2 Other Deformation Processes Related to Rolling 395
19.3 Forging 397
19.4 Other Deformation Processes Related to Forging 408
19.5 Extrusion 413
19.6 Wire and Bar Drawing 423

20 SHEET METAL WORKING 435

20.1 Cutting Operations 436
20.2 Bending Operations 442
20.3 Drawing 447
20.4 Other Sheet-Metal-Forming Operations 454
20.5 Dies and Presses for Sheet Metal Processes 457
20.6 Sheet-Metal Operations Not Performed on Presses 463
20.7 Bending of Tube Stock 469

Part VI Material Removal Processes

21 THEORY OF METAL MACHINING 475

21.1 Overview of Machining Technology 477
21.2 Theory of Chip Formation in Metal Machining 481
21.3 Force Relationships and the Merchant Equation 485
21.4 Power and Energy Relationships in Machining 490
21.5 Cutting Temperature 493

22 MACHINING OPERATIONS AND MACHINE TOOLS 499

22.1 Turning and Related Operations 502
22.2 Drilling and Related Operations 511
22.3 Milling 515
22.4 Machining Centers and Turning Centers 522
22.5 Other Machining Operations 524
22.6 High-Speed Machining 529

23 CUTTING TOOL TECHNOLOGY 534

23.1 Tool Life 534
23.2 Tool Materials 541
23.3 Tool Geometry 550
23.4 Cutting Fluids 558

24 ECONOMIC AND PRODUCT DESIGN CONSIDERATIONS IN MACHINING 565

24.1 Machinability 565
24.2 Tolerances and Surface Finish 568
24.3 Selection of Cutting Conditions 572
24.4 Product Design Considerations in Machining 578

25 GRINDING AND OTHER ABRASIVE PROCESSES 585

25.1 Grinding 586
25.2 Related Abrasive Process 603

26 NONTRADITIONAL MACHINING AND THERMAL CUTTING PROCESSES 610

26.1 Mechanical Energy Processes 611
26.2 Electrochemical Machining Processes 615
26.3 Thermal Energy Processes 619
26.4 Chemical Machining 627
26.5 Application Considerations 633

Part VII Property Enhancing and Surface Processing Operations

27 HEAT TREATMENT OF METALS 639

27.1 Annealing 640
27.2 Martensite Formation in Steel 640
27.3 Precipitation Hardening 644
27.4 Surface Hardening 645
27.5 Heat Treatment Methods and Facilities 647

28 CLEANING AND SURFACE TREATMENTS 651

28.1 Chemical Cleaning 651
28.2 Mechanical Cleaning and Surface Preparation 654
28.3 Diffusion and Ion Implantation 656

29 COATING AND DEPOSITION PROCESSES 659

29.1 Plating and Related Processes 660
29.2 Conversion Coatings 664
29.3 Physical Vapor Deposition 665
29.4 Chemical Vapor Deposition 668
29.5 Organic Coatings 671
29.6 Porcelain Enameling and Other Ceramic Coatings 674

29.7 Thermal and Mechanical Coating Processes 674

Part VIII Joining and Assembly Processes

30 FUNDAMENTALS OF WELDING 679

30.1 Overview of Welding Technology 681
30.2 The Weld Joint 683
30.3 Physics of Welding 686
30.4 Features of a Fusion-Welded Joint 690

31 WELDING PROCESSES 694

31.1 Arc Welding 695
31.2 Resistance Welding 705
31.3 Oxyfuel Gas Welding 712
31.4 Other Fusion-Welding Processes 716
31.5 Solid-State Welding 719
31.6 Weld Quality 724
31.7 Weldability 728
31.8 Design Considerations in Welding 729

32 BRAZING, SOLDERING, AND ADHESIVE BONDING 734

32.1 Brazing 735
32.2 Soldering 740
32.3 Adhesive Bonding 744

33 MECHANICAL ASSEMBLY 752

33.1 Threaded Fasteners 753
33.2 Rivets and Eyelets 759
33.3 Assembly Methods based on Interference Fits 760
33.4 Other Mechanical Fastening Methods 763
33.5 Molding Inserts and Integral Fasteners 765
33.6 Design for Assembly 766

Part IX Special Processing and Assembly Technologies

34 RAPID PROTOTYPING 772

34.1 Fundamentals of Rapid Prototyping 773
34.2 Rapid Prototyping Technologies 774
34.3 Applications Issues in Rapid Prototyping 782

35 PROCESSING OF INTEGRATED CIRCUITS 786

35.1 Overview of IC Processing 787
35.2 Silicon Processing 792
35.3 Lithography 796
35.4 Layer Processes Use in IC Fabrication 800

35.5 Integrating the Fabrication Steps 807
35.6 IC Packaging 808
35.7 Yields in IC Processing 813

36 ELECTRONICS ASSEMBLY AND PACKAGING 819

36.1 Electronics Packaging 819
36.2 Printed Circuit Boards 821
36.3 Printed Circuit Board Assembly 831
36.4 Surface Mount Technology 835
36.5 Electrical Connector Technology 839

37 MICROFABRICATION TECHNOLOGIES 845

37.1 Microsystem Products 845
37.2 Microfabrication Processes 850
37.3 Nanotechnology 857

Part X Manufacturing Systems

38 NUMERICAL CONTROL AND INDUSTRIAL ROBOTICS 860

38.1 Numerical Control 861
38.2 Industrial Robotics 873
38.3 Programmable Logic Controllers 878

39 GROUP TECHNOLOGY AND FLEXIBLE MANUFACTURING SYSTEMS 884

39.1 Group Technology 884
39.2 Flexible Manufacturing Systems 889

40 PRODUCTION LINES 896

40.1 Fundamentals of Production Lines 896
40.2 Manual Assembly Lines 900
40.3 Automated Production Lines 904

Part XI Manufacturing Support Systems

41 MANUFACTURING ENGINEERING 912

41.1 Process Planning 913
41.2 Problem Solving and Continuous Improvement 921
41.3 Concurrent Engineering and Design for Manufacturability 922

42 PRODUCTION PLANNING AND CONTROL 928

42.1 Aggregate Planning and the Master Production Schedule 930
42.2 Inventory Control 931
42.3 Material and Capacity Requirements Planning 935
42.4 Just-in-Time and Lean Production 939
42.5 Shop Floor Control 942

43 QUALITY CONTROL 947

43.1 What Is Quality? 947
43.2 Process Capability 949
43.3 Statistical Tolerancing 949
43.4 Taguchi Methods 953
43.5 Statistical Process Control 955

44 MEASUREMENT AND INSPECTION 965

44.1 Metrology 966
44.2 Inspection Principles 969
44.3 Conventional Measuring Instruments and Gages 971
44.4 Measurement of Surfaces 978
44.5 Advanced Measurement and Inspection Technologies 980

INDEX 989

1

INTRODUCTION

CHAPTER CONTENTS

1.1   What Is Manufacturing?
        1.1.1   Manufacturing Defined
        1.1.2   Manufacturing Industries and Products
        1.1.3   Manufacturing Capability
1.2   Materials in Manufacturing
        1.2.1   Metals
        1.2.2   Ceramics
        1.2.3   Polymers
        1.2.4   Composites
1.3   Manufacturing Processes
        1.3.1   Processing Operations
        1.3.2   Assembly Operations
        1.3.3   Production Machines and Tooling
1.4   Production Systems
        1.4.1   Production Facilities
        1.4.2   Manufacturing Support Systems
1.5   Organization of the Book
1.6   Images of Manufacturing
References

Manufacturing is important—technologically, economically, and historically. ***Technology*** can be defined as the application of science to provide society and its members with those things that are needed or desired. Technology affects our daily lives, either directly or indirectly, in many ways. Consider the list of products in Table 1.1. They represent various technologies that help our society and its members to live better. What do these products have in common? They are all manufactured. These technological wonders would not be available if they could not be produced. Manufacturing is the essential factor that makes technology possible.

Economically, manufacturing is an important means by which a nation creates material wealth. In the United States, the manufacturing industries account for about 20% of gross national product (GNP). A country's natural resources, such as agricultural lands, mineral deposits, and oil reserves, also create wealth. In the United States, agriculture, mining, and similar industries account for less than 5% of GNP. Construction and public utilities make up slightly more than 5%. The rest is service industries, which include

TABLE 1.1 Products representing various technologies, most of which affect nearly all of us.

Athletic shoes	High-density PC diskette	Photocopying machine
Automatic teller machine	Home security system	Pull-tab beverage cans
Automatic dishwasher	Incandescent light bulb	Quartz crystal wristwatch
Ballpoint pen	Industrial robot	Self-propelled mulching lawnmower
Cellular telephone	Ink-jet color printer	Sport utility vehicle (SUV) with all-
Compact disc (CD)	Integrated circuit	wheel drive, dual air bags, antilock
Compact disc player	Large-screen color television	brakes, cruise control, and AM–FM
Contact lenses	Magnetic resonance imaging (MRI)	radio with CD player
Digital camera	machine for medical diagnosis	Supersonic aircraft
Digital video disc (DVD)	Microwave oven	Tennis racket of composite materials
Digital video disc player	One-piece molded plastic patio chair	Videocassette recorder
Fax machine	Optical scanner	Video games
Hand-held electronic calculator	Personal computer (PC)	Washing machine and dryer

retail, transportation, banking, communication, education, and government. The service sector accounts for approximately 70% of U.S. GNP. Government alone accounts for about as much of GNP as the manufacturing sector, but government services do not create wealth. In the modern international economy, a nation must have a strong manufacturing base (or it must have significant natural resources) if it is to provide a strong economy and a high standard of living for its people.

Historically, the importance of manufacturing in the development of civilization is usually underestimated. But throughout history, human cultures that were better at making things were more successful. By making better tools, they had better crafts and weapons. Better crafts allowed them to live better. Better weapons allowed them to conquer neighboring cultures in times of conflict. In the American Civil War (1861–65), one of the great advantages of the North over the South was its industrial strength—its capacity to manufacture. In World War II (1939–45), the United States outproduced Germany and Japan—a decisive advantage in winning the war. To a significant degree, the history of civilization is the history of humans' ability to make things.

In this opening chapter, we consider some general topics about manufacturing. What is manufacturing? How is it organized in industry? What are the materials, processes, and systems by which production is accomplished? This chapter concludes with a collection of color plates depicting various manufactured products and manufacturing operations.

1.1 WHAT IS MANUFACTURING?

The word **manufacture** is derived from two Latin words **manus** (hand) and **factus** (make); the combination means *made by hand*. The English word *manufacture* is several centuries old, and *made by hand* accurately described the manual methods used when the word was first coined.[1] Most modern manufacturing is accomplished by automated and computer-controlled machinery that is manually supervised (Historical Note 1.1).

---

[1]As a noun, the word *manufacture* first appeared in English around 1567 A.D. As a verb it first appeared around 1683 A.D.

Historical Note 1.1
History of manufacturing

T he history of manufacturing can be separated into two subjects: (1) man's discovery and invention of materials and processes to make things, and (2) development of the systems of production. The materials and processes to make things predate the systems by several millenia. Some of the processes—casting, hammering (forging), and grinding—date back 6,000 years or more. The early fabrication of implements and weapons was accomplished more as crafts and trades than as manufacturing as we know it. The ancient Romans had what might be called factories to produce weapons, scrolls, pottery and glassware, and other products of the time, but the procedures were largely based on handcraft.

Let us examine the systems aspects of manufacturing here, and postpone materials and processes until Historical Note 1.2. **Systems of manufacturing** refers to the ways of organizing people and equipment so that production can be performed more efficiently. Several historical events and discoveries stand out as having had a major impact on the development of modern manufacturing systems.

Certainly one significant discovery was the principle of **division of labor**—dividing the total work into tasks and having individual workers each become a specialist at performing only one task. This principle had been practiced for centuries, but the economist Adam Smith (1723–90) is credited with first explaining its economic significance in **The Wealth of Nations.**

The **Industrial Revolution** (circa 1760–1830) had a major impact on production in several ways. It marked the change from an economy based on agriculture and handicraft to one based on industry and manufacturing. The change began in England, where a series of machines were invented and steam power replaced water, wind, and animal power. These advances gave British industry significant advantages over other nations, and England attempted to restrict export of the new technologies. However, the revolution eventually spread to other European countries and to the United States. Several Inventions of the Industrial Revolution greatly contributed to the development of manufacturing: 1. **Watt's steam engine**—a new power generating technology for industry; 2. **Machine tools,** starting with John Wilkinson's boring machine around 1775 (Historical Note 22.1); 3. The **spinning jenny, power loom,** and other machinery for the textile industry that permitted significant increases in productivity; and 4. The **factory system**—a new way of organizing large numbers of production workers based on division of labor.

While England was leading the Industrial Revolution, an important concept was being introduced in the United States: **interchangeable parts** manufacture. Much credit for this concept is given to Eli Whitney (1765–1825), although its importance had been recognized by others [6]. In 1797, Whitney negotiated a contract to produce 10,000 muskets for the U.S. government. The traditional way of making guns at the time was to custom-fabricate each part for a particular gun and then hand-fit the parts together by filing. Each musket was unique, and the time to make it was considerable. Whitney believed that the components could be made accurately enough to permit parts assembly without fitting. After several years of development in his Connecticut factory, he traveled to Washington in 1801 to demonstrate the principle. Before government officials, including Thomas Jefferson, he laid out components for 10 muskets and proceeded to select parts randomly to assemble the guns. No special filing or fitting was required, and all of the guns worked perfectly. The secret behind his achievement was the collection of special machines, fixtures, and gages that he had developed in his factory. Interchangeable parts manufacture required many years of development before becoming a practical reality, but it revolutionized methods of manufacturing. It is a prerequiste for mass production. Because its origins were in the United States, interchangeable parts production came to be known as the **American System** of manufacture.

The mid- and late 1800s witnessed the expansion of railroads, steam-powered ships, and other machines that created a growing need for iron and steel. New steel production methods were developed to meet this demand (Historical Note 6.1). Also during this period, several consumer products were developed, including the sewing machine, bicycle, and automobile. In order to meet the mass demand for these products, more eficient production methods were required. Some historians identify developments during this period as the **Second Industrial Revolution,** characterized in terms of its effects on manufacturing systems by the following: (1) mass production, (2) scientific management movement, (3) assembly lines, and (4) electrification of factories.

In the late 1800s, the **scientific management** movement was developing in the United States in response to the need to plan and control the activities of growing numbers of production workers. The movement was led by Frederick W. Taylor (1856–1915), Frank Gilbreath (1868–1924) and his wife Lilian (1878–1972), and others. Scientific management included several features: 1. *motion study,* aimed at finding the best method to perform a given task; 2. *time study* to establish work standards for a job; 3. Extensive use of industry *standards*; 4. The *piece rate system* and similar labor incentive plans; and 5. Use of data collection, record keeping, and cost accounting in factory operations.

Henry Ford (1863–1947) introduced the *assembly line* in 1913 at his Highland Park plant (Historical Note 40.1). The assembly line made mass production of complex consumer products possible. Use of assembly line methods permitted Ford to sell a Model T automobile for as little as $500, thus making ownership of cars feasible for a large segment of the American population.

In 1881, the first electric power generating station had been built in New York City, and soon electric motors were being used as a power source to operate factory machinery. This was a far more convenient power delivery system than steam engines, which required overhead belts to distribute power to the machines. By 1920, electricity had overtaken steam as the principal power source in U.S. factories. The twentieth century was a time of more technological advances than in all other centuries combined. Many of these developments resulted in the *automation* of manufacturing.

1.1.1 Manufacturing Defined

As a field of study in the modern context, manufacturing can be defined two ways, one technologic, the other economic. Technologically, *manufacturing* is the application of physical and chemical processes to alter the geometry, properties, and/or appearance of a given starting material to make parts or products. Manufacturing also includes assembly of multiple parts to make products. The processes to accomplish manufacturing involve a combination of machinery, tools, power, and manual labor, as depicted in Figure 1.1(a). Manufacturing is almost always carried out as a sequence of operations. Each operation brings the material closer to the desired final state.

Economically, *manufacturing* is the transformation of materials into items of greater value by means of one or more processing and/or assembly operations, as depicted in Figure 1.1(b). The key point is that manufacturing *adds value* to the material by changing its shape or properties, or by combining it with other materials that have been similarly altered. The material has been made more valuable through the manufacturing operations performed on it. When iron ore is converted into steel, value is added. When sand is transformed into glass, value is added. When petroleum is refined into plastic, value is added. And when plastic is molded into the complex geometry of a patio chair, it is made even more valuable.

The words *manufacturing* and *production* are often used interchangeably. The author's view is that production has a broader meaning than manufacturing. To illustrate, we might speak of "crude oil production," but the phrase "crude oil manufacturing" seems out of place. Yet when used in the context of products such as metal parts or automobiles, either word is acceptable.

1.1.2 Manufacturing Industries and Products

Manufacturing is an important activity, but it is not carried out simply for its own sake. It is performed as a commercial activity by companies that sell products to customers. The type of manufacturing done by a company depends on the kind of product it makes. Let us explore this relationship by first examining the types of industries in manufacturing, and then identifying the products they make.

Manufacturing Industries   Industry consists of enterprises and organizations that produce or supply goods and services. Industries can be classified as primary, secondary, or tertiary. *Primary industries* are those that cultivate and exploit natural resources,

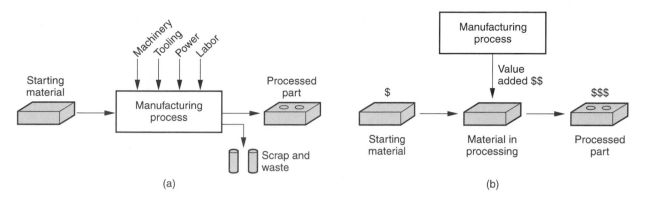

FIGURE 1.1   Two ways to define manufacturing: (a) as a technical process, and (b) as an economic process.

such as agriculture and mining. ***Secondary industries*** take the outputs of the primary industries and convert them into consumer and capital goods. Manufacturing is the principal activity in this category, but construction and power utilities are also included. ***Tertiary industries*** constitute the service sector of the economy. A list of specific industries in these categories is presented in Table 1.2.

In this book, we are concerned with the secondary industries in Table 1.2; these are the companies engaged in manufacturing. However, the International Standard Industrial Classification (ISIC) used to compile Table 1.2 includes several industries whose production technologies are not covered in this text (e.g., beverages, chemicals, and food processing). In our book, manufacturing means production of ***hardware,*** which ranges from nuts and bolts to digital computers and military weapons. We include plastic and ceramic products, but exclude apparel, chemicals, food, and software. Our short list of manufacturing industries appears in Table 1.3.

Manufactured Products   Final products made by the industries listed in Table 1.3 can be divided into two major classes: consumer goods and capital goods. ***Consumer goods*** are products purchased directly by consumers, such as cars, personal computers, TVs, and tennis rackets. ***Capital goods*** are those purchased by other companies to produce

TABLE 1.2   Specific industries in the primary, secondary, and tertiary categories.

Primary	Secondary		Tertiary (service)	
Agriculture	Aerospace	Food processing	Banking	Insurance
Forestry	Apparel	Glass, ceramics	Communications	Legal
Fishing	Automotive	Heavy machinery	Education	Real estate
Livestock	Basic metals	Paper	Entertainment	Repair and maintenance
Quarries	Beverages	Petroleum refining	Financial services	Restaurant
Mining	Building materials	Pharmaceuticals	Government	Retail trade
Petroleum	Chemicals	Plastics (shaping)	Health and medical	Tourism
	Computers	Power utilities	Hotel	Transportation
	Construction	Publishing	Information	Wholesale trade
	Consumer appliances	Textiles		
	Electronics	Tire and rubber		
	Equipment	Wood and furniture		
	Fabricated metals			

TABLE 1.3 Manufacturing industries whose materials, processes, and systems are likely to be covered in this book.

Industry	Typical Products	Industry	Typical Products
Aerospace	Commercial and military aircraft	Equipment	Industrial machinery, rail equipment
Automotive	Cars, trucks, buses, motorcycles	Fabricated metals	Machined parts, stampings, tools
Basic metals	Iron, steel, aluminum, copper, etc.	Glass, ceramics	Glass products, ceramic tools, pottery
Computers	Mainframe and personal computers	Heavy machinery	Machine tools, construction equipment
Consumer appliance	Large and small home appliances	Plastics (shaping)	Plastics moldings, extrusions
Electronics	TVs, VCRs, audio equipment	Tire and rubber	Tires, shoe soles, tennis balls

goods and supply services. Examples of capital goods include aircraft, mainframe computers, railroad equipment, machine tools, and construction equipment.

Other manufactured products include *materials, components,* and *supplies* used by the companies that make the final products. Examples of these items include sheet steel, bar stock, metal stampings, machined parts, plastic moldings and extrusions, cutting tools, dies, molds, and lubricants. Thus, the manufacturing industries consist of a complex infrastructure with various categories and layers of intermediate suppliers that the final consumer never deals with.

In this book we are generally concerned with *discrete items*—individual parts and assembled products rather than items produced by *continuous processes.* A metal stamping is a discrete item, but the sheet metal coil from which it is made is continuous (almost). Many discrete parts start out as continuous or semi-continuous products, such as extrusions and electrical wire. Long sections made in almost continuous lengths are cut to desired size. An oil refinery is a better example of a continuous process.

Production Quantity and Product Variety   The quantity of products made by a factory has an important influence on the way its people, facilities, and procedures are organized. Annual production quantities can be classified into three ranges: (1) *low* production, quantities in the range 1 to 100 units per year; (2) *medium* production, from 100 to 10,000 units annually; and (3) *high* production, 10,000 to millions of units. The boundaries between the three ranges are somewhat arbitrary (author's judgment). Depending on the kinds of products, these boundaries may shift by an order of magnitude or so.

Production quantity refers to the number of units produced annually of a particular product type. Some plants produce a variety of different product types, each type being made in low or medium quantities. Other plants specialize in high production of only one product type. It is instructive to identify product variety as a parameter distinct from production quantity. Product variety refers to different product designs or types that are produced in the plant. Different products have different shapes and sizes; they perform different functions; they are intended for different markets; some have more components than others; and so forth. The number of different product types made each year can be counted. When the number of product types made in the factory is high, this indicates high product variety.

There is an inverse correlation between product variety and production quantity in terms of factory operations. If a factory's product variety is high, then its production quantity is likely to be low; but if production quantity is high, then product variety will be low (Figure 1.2). Manufacturing plants tend to specialize in a combination of production quantity and product variety that lies somewhere inside the diagonal band in Figure 1.2.

Although we have identified product variety as a quantitative parameter (the number of different product types made by the plant or company), this parameter is much

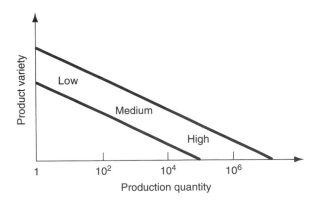

FIGURE 1.2   Relationship between product variety and production quantity in discrete product manufacturing.

less exact than production quantity because details on how much the designs differ is not captured simply by the number of different designs. Differences between an automobile and an air conditioner are far greater than between an air conditioner and a heat pump. And within each product type, there are differences between specific models.

The extent of product differences may be small or great, as illustrated in the automotive industry. Each of the U.S. automotive companies produces cars with two or three different nameplates in the same assembly plant, although the body styles and other design features are virtually the same. In different plants, the company builds heavy trucks. We might use the terms *soft* and *hard* to describe these differences in product variety. **Soft product variety** is when there are only small differences between products, such as the differences between car models made on the same production line. In an assembled product, soft variety is characterized by a high proportion of common parts among the models. **Hard product variety** is when the products differ substantially, and there are few common parts, if any. The difference between a car and a truck is hard.

1.1.3 Manufacturing Capability

A manufacturing plant consists of a set of **processes** and **systems** (and people, of course) designed to transform a certain limited range of **materials** into products of increased value. These three building blocks—materials, processes, and systems—constitute the subject of modern manufacturing. There is a strong interdependence among these factors. A company engaged in manufacturing cannot do everything. It must do only certain things, and it must do those things well. **Manufacturing capability** refers to the technical and physical limitations of a manufacturing firm and each of its plants. We can identify several dimensions of this capability: (1) technological processing capability, (2) physical size and weight of product, and (3) production capacity.

Technological Processing Capability   A plant's (or company's) **technological processing capability** is its available set of manufacturing processes. Certain plants perform machining operations, others roll steel billets into sheet stock, and others build automobiles. A machine shop cannot roll steel, and a rolling mill cannot build cars. The underlying feature that distinguishes these plants is the processes they can perform. Technological processing capability is closely related to material type. Certain manufacturing processes are suited to certain materials, while other processes are suited to other materials. By specializing in a certain process or group of processes, the plant is simultaneously specializing in certain material types. Technological processing capability includes not only the physical processes, but also the expertise possessed by plant personnel in these processing technologies. Companies must concentrate on the design and manufacture of products that are compatible with their technological processing capability.

Physical Product Limitations   A second aspect of manufacturing capability is imposed by the *physical product.* A plant with a given set of processes is restricted to certain size and weight limitations. Large, heavy products are difficult to move. To move these products about, the plant must be equipped with cranes of the required load capacity. Smaller parts and products made in large quantities can be moved by conveyor or other means. The limitation on product size and weight extends to the physical capacity of the manufacturing equipment as well. Production machines come in different sizes. Larger machines must be used to process larger parts. The set of production equipment, material handling, storage capability, and plant size must be planned for products that lie within a certain size and weight range.

Production Capacity   A third limitation on a plant's manufacturing capability is the production quantity that can be produced in a given time period (e.g., month or year). This quantity limitation is commonly called *plant capacity,* or *production capacity,* defined as the maximum rate of production that a plant can achieve under assumed operating conditions. The operating conditions refer to number of shifts per week, hours per shift, direct labor manning levels in the plant, and so on. These factors represent inputs to the manufacturing plant. Given these inputs, how much output can the factory produce?

Plant capacity is usually measured in terms of output units, such as annual tons of steel produced by a steel mill, or number of cars produced by a final assembly plant. In these cases, the outputs are homogeneous. In cases where the output units are not homogeneous, other factors may be more appropriate measures, such as available labor hours of productive capacity in a machine shop that produces a variety of parts.

Materials, processes, and systems are the basic building blocks of manufacturing and the three broad subject areas of this book. Let us provide an overview of these subjects.

1.2 MATERIALS IN MANUFACTURING

Most engineering materials can be classified into one of three basic categories: (1) *metals,* (2) *ceramics,* and (3) *polymers.* Their chemistries are different, their mechanical and physical properties are dissimilar, and these differences affect the manufacturing processes that can be used to produce products from them. In addition to the three basic categories, there are (4) *composites*—nonhomogeneous mixtures of the other three basic types rather than a unique category. The relationship of the four groups is pictured in Figure 1.3. In this section, we survey these materials. In Chapters 6 through 9, we cover the four material types in more detail.

1.2.1 Metals

Metals used in manufacturing are usually *alloys,* which are composed of two or more elements, at least one of which is a metallic element. Metals can be divided into two basic groups: (1) ferrous, and (2) nonferrous.

Ferrous Metals   *Ferrous metals* are based on iron; the group includes steel and cast iron. These metals constitute the most important group commercially—more than

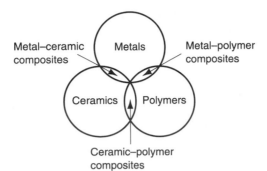

FIGURE 1.3   Venn diagram showing the three basic material types plus composites.

three-quarters of the metal tonnage throughout the world. Pure iron has limited commercial use, but when alloyed with carbon, iron has more uses and greater commercial value than any other metal. Alloys of iron and carbon form steel and cast iron.

Steel can be defined as an iron–carbon alloy containing 0.02 to 2.11% carbon. It is the most important category within the ferrous metal group. Its composition often includes other alloying elements as well, such as manganese, chromium, nickel, and molybdenum, to enhance the properties of the metal. Applications of steel include construction (e.g., bridges, I- beams, and nails), transportation (trucks, rails and rolling stock for railroads), and consumer products (automobiles and appliances).

Cast iron is an alloy of iron and carbon (2% to 4%) used in casting (primarily sand casting). Silicon is also present in the alloy (in amounts from 0.5% to 3%), and other elements are often added, too, to obtain desirable properties in the cast part. Cast iron is available in several different forms, of which gray cast iron is the most common; its applications include blocks and heads for internal combustion engines.

Nonferrous Metals   Nonferrous metals include the other metallic elements and their alloys. In almost all cases, the alloys are more important commercially than the pure metals. The nonferrous metals include the pure metals and alloys of aluminum, copper, gold, magnesium, nickel, silver, tin, titanium, zinc, and other metals.

1.2.2 Ceramics

A *ceramic* is defined as a compound containing metallic (or semimetallic) and nonmetallic elements. Typical nonmetallic elements are oxygen, nitrogen, and carbon. Ceramics include a variety of traditional and modern materials. Traditional ceramics, some of which have been used for thousands of years, include: *clay* (abundantly available, consisting of fine particles of hydrous aluminum silicates and other minerals used in making brick, tile, and pottery); *silica* (the basis for nearly all glass products); and *alumina* and *silicon carbide* (two abrasive materials used in grinding).

Modern ceramics include some of the preceding materials, such as *alumina,* whose properties are enhanced in various ways through modern processing methods. Newer ceramics include *carbides,* metal carbides such as tungsten carbide and titanium carbide, which are widely used as cutting tool materials; and *nitrides,* metal and semimetal nitrides like titanium nitride and boron nitride, used as cutting tools and grinding abrasives.

For processing purposes, ceramics can be divided into (1) crystalline ceramics and (2) glasses. Different methods of manufacturing are required for the two types. Crystalline ceramics are formed in various ways from powders and then sintered (heated to a temperature below the melting point to achieve bonding between the powders). The glass ceramics (namely, glass) can be melted and cast, and then formed in processes such as traditional glass blowing.

1.2.3 Polymers

A *polymer* is a compound formed of repeating structural units called *mers,* whose atoms share electrons to form very large molecules. Polymers usually consist of carbon plus one or more other elements such as hydrogen, nitrogen, oxygen, and chlorine. Polymers divide into three categories:

1. *Thermoplastic polymers.* These can be subjected to multiple heating and cooling cycles without substantially altering the molecular structure of the polymer. Common thermoplastics include polyethylene, polystyrene, polyvinylchloride, and nylon.

2. *Thermosetting polymers.* These molecules chemically transform (cure) into a rigid structure upon cooling from a heated plastic condition; hence the name *thermosetting.* Members of this type include phenolics, amino resins, and epoxies. Although the name *thermosetting* is used, some of these polymers cure by mechanisms other than heating.

3. *Elastomers.* These polymers exhibit significant elastic behavior; hence the name *elastomer.* Elastomers include natural rubber, neoprene, silicone, and polyurethane.

1.2.4 Composites

Composites do not really constitute a separate category of materials; they are mixtures of the other three types. A *composite* is a material consisting of two or more phases that are processed separately and then bonded together to achieve properties superior to those of its constituents. The term *phase* refers to a homogeneous mass of material, such as an aggregation of grains of identical unit cell structure in a solid metal. The usual structure of a composite consists of particles or fibers of one phase mixed in a second phase, called the *matrix.*

Composites are found in nature (e.g., wood), and they can be produced synthetically. The synthesized type are of greater interest here, and they include glass fibers in a polymer matrix, such as fiber-reinforced plastic; polymer fibers of one type in a matrix of a second polymer, such as an epoxy-Kevlar composite; and ceramic in a metal matrix, such as a tungsten carbide in a cobalt binder to form a cemented carbide cutting tool.

Properties of a composite depend on its components, the physical shapes of the components, and the way they are combined to form the final material. Some composites combine high strength with light weight and are suited to applications such as aircraft components, car bodies, boat hulls, tennis rackets, and fishing rods. Other composites are strong and hard, and capable of maintaining these properties at elevated temperatures; for example, cemented carbide cutting tools.

1.3 MANUFACTURING PROCESSES

Manufacturing processes can be divided into two basic types: processing operations and assembly operations. A *processing operation* transforms a work material from one state of completion to a more advanced state that is closer to the final desired product. It adds value by changing the geometry, properties, or appearance of the starting material. In general, processing operations are performed on discrete workparts, but some processing operations are also applicable to assembled items. An ***assembly operation*** joins two or more components in order to create a new entity called an assembly, subassembly, or some other term that refers to the joining process (e.g., a welded assembly is called a ***weldment***). A classification of manufacturing processes is presented in Figure 1.4. Some of the basic processes used in modern manufacturing date from antiquity (Historical Note 1.2).

Historical Note 1.2
Manufacturing materials and processes

Most of the historical developments that form the modern practice of manufacturing have occurred only during the last few centuries (Historical Note 1.1), but several of the basic processes of manufacturing date as far back as the Neolithic period (circa 8000–3000 B.C.). It was during this period that processes such as the following were developed: carving and other **woodworking**, hand forming and **firing** of clay pottery, **grinding** and **polishing** of stone, **spinning** and **weaving** of textiles, and **dyeing** of cloth.

Metallurgy and metalworking also began during the Neolithic, in Mesopotamia and other areas around the Mediterranean. It either spread to, or developed independently in, regions of Europe and Asia. Gold was found by early man in relatively pure form in nature; it could be **hammered** into shape. Copper was probably the first metal to be extracted from ores, thus requiring **smelting** as a processing technique. Copper could not readily be hammered because it strain hardened; instead, it was shaped by **casting** (Historical Note 10.1). Other metals used during this period were silver and tin. It was discovered that copper alloyed with tin produced a more workable metal than copper alone (casting and hammering could both be used). This heralded the important period known as the **Bronze Age** (circa 3500–1500 B.C.).

Iron was also first smelted during the Bronze Age. Meteorites may have been one source of the metal, but iron ore was also mined. Temperatures required to reduce iron ore to metal are significantly higher than for copper, which made furnace operations more difficult. Other processing methods were also more difficult for the same reason. Early blacksmiths learned that when certain irons (those containing small amounts of carbon) were sufficiently **heated** and then **quenched**, they became very hard. This permitted grinding a very sharp cutting edge on knives and weapons, but it also made the metal brittle.

Toughness could be increased by reheating at a lower temperature, a process known as **tempering**. What we have described is, of course, the **heat treatment** of steel. The superior properties of steel caused

it to succeed bronze in many applications (weaponry, agriculture, and mechanical devices). The period of its use has subsequently been named the **Iron Age** (starting around 1000 B.C.). It was not until much later, well into the nineteenth century, that the demand for steel grew significantly and more modern steelmaking techniques were developed (Historical Note 6.1).

The beginnings of machine tool technology occurred during the Industrial Revolution. During the period 1770–1850, machine tools were developed for most of the conventional **material removal processes**, such as **boring, turning, drilling, milling, shaping,** and **planing** (Historical Note 22.1). Many of the individual processes predate the machine tools by centuries; for example, drilling and sawing (of wood) date from ancient times, and turning (of wood) from around the time of Christ.

Assembly methods were used in ancient cultures to make ships, weapons, tools, farm implements, machinery, chariots and carts, furniture, and garments. The processes included **binding** with twine and rope, **riveting** and **nailing**, and **soldering.** By around the time of Christ, **forge welding** and **adhesive bonding** had been developed. Widespread use of screws, bolts, and nuts as fasteners—so common in today's assembly— required the development of machine tools (e.g., Maudsley's screw cutting lathe, 1800) that could accurately form the required helical shapes. It was not until around 1900 that **fusion welding** processes started to be developed as assembly techniques (Historical Note 30.1).

Natural rubber was the first polymer to be used in manufacturing (if we overlook wood, which is a polymer composite). The **vulcanization** process, discovered by Charles Goodyear in 1839, made rubber a useful engineering material (Historical Note 8.2). Subsequent developments included plastics such as cellulose nitrate in 1870, Bakelite in 1900, polyvinylchloride in 1927, polyethylene in 1932, and nylon in the late 1930s (Historical Note 8.1). Processing requirements for plastics have led to the development of *injection molding* (based on die casting, one of the metal casting processes) and other polymer shaping techniques.

Electronics products have imposed unusual demands on manufacturing in terms of miniaturization. The evolution of the technology has been to package more and more devices into a smaller area—in some cases a million transistors onto a flat piece of semi-conductor material that is only 6 mm (0.25 in.) on a side. The history of electronics processing and packaging dates from only a few decades (Historical Notes 35.1, 36.1, and 36.2).

1.3.1 Processing Operations

A processing operation uses energy to alter a workpart's shape, physical properties, or appearance in order to add value to the material. The forms of energy include mechanical, thermal, electrical, and chemical. The energy is applied in a controlled way by means of machinery and tooling. Human energy may also be required, but human workers are generally employed to control the machines, to oversee the operations, and to load and unload parts before and after each cycle of operation. A general model of a processing operation is illustrated in Figure 1.1(a). Material is fed into the process, energy is applied by the machinery and tooling to transform the material, and the completed workpart exits the process. Most production operations produce waste or scrap, either as a natural aspect of the process (e.g., removing material as in machining) or in the form of occasional

FIGURE 1.4   Classification of manufacturing processes.

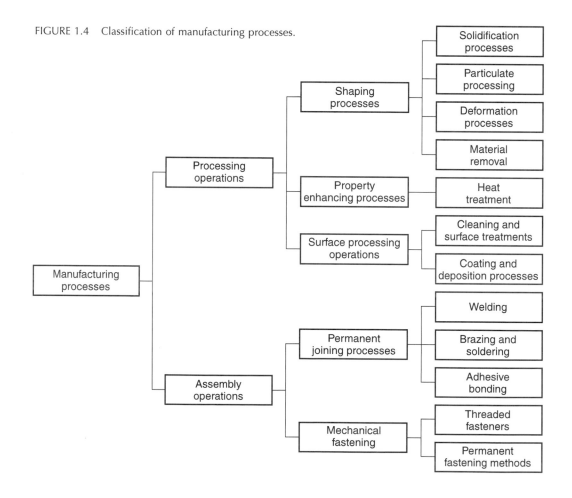

defective pieces. It is an important objective in manufacturing to reduce waste in either of these forms.

More than one processing operation is usually required to transform the starting material into final form. The operations are performed in the particular sequence required to achieve the geometry and condition defined by the design specification.

Three categories of processing operations are distinguished: (1) shaping operations, (2) property-enhancing operations, and (3) surface processing operations. ***Shaping operations*** alter the geometry of the starting work material by various methods. Common shaping processes include casting, forging, and machining. ***Property-enhancing operations*** add value to the material by improving its physical properties without changing its shape. Heat treatment is the most common example. ***Surface processing operations*** are performed to clean, treat, coat, or deposit material onto the exterior surface of the work. Common examples of coating are plating and painting. Shaping processes are covered in Parts III through VI, corresponding to the four main categories of shaping processes in Figure 1.4. Property-enhancing processes and surface processing operations are covered in Part VII. Our coverage of electronics manufacturing in Part IX includes several surface treatments and processes related to that technology.

Shaping Processes   Most shape-processing operations apply heat or mechanical force or a combination of these to effect a change in geometry of the work material. There are various ways to classify the shaping processes. The classification used in this book is based on the state of the starting material, by which we have four categories: (1) ***solidification processes,*** in which the starting material is a heated ***liquid*** or ***semifluid*** that cools and solidifies to form the part geometry. (2) ***particulate processing,*** in which the starting material is a ***powder,*** and the powders are formed and heated into the desired geometry. (3) ***deformation processes,*** where the starting material is a ***ductile solid*** (commonly metal) that is deformed to shape the part. (4) ***material removal processes,*** in which the starting material is a ***solid*** (ductile or brittle), from which material is removed so that the resulting part has the desired geometry.

In the first category, the starting material is heated sufficiently to transform it into a liquid or highly plastic (semifluid) state. Nearly all materials can be processed in this way. Metals, ceramic glasses, and plastics can all be heated to sufficiently high temperatures to convert them into liquids. With the material in a liquid or semifluid form, it can be poured or otherwise forced to flow into a mold cavity and allowed to solidify, thus taking a solid shape that is the same as the cavity. Most processes that operate this way are called casting or molding. ***Casting*** is the name used for metals, and ***molding*** is the common term used for plastics. This category of shaping process is depicted in Figure 1.5.

FIGURE 1.5   Casting and molding processes start with a work material heated to a fluid or semi-fluid state. The process consists of (1) pouring the fluid into a mold cavity and (2) allowing the fluid to solidify, after which the solid part is removed from the mold.

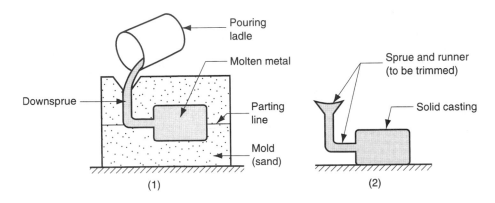

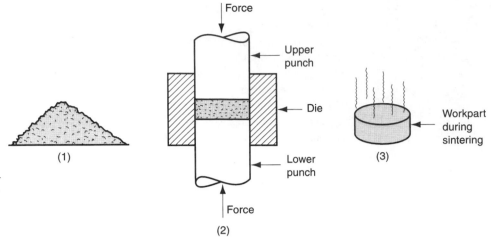

FIGURE 1.6 Particulate processing: (1) the starting material is powder; the usual process consists of (2) pressing and (3) sintering.

In *particulate processing,* the starting materials are powders of metals or ceramics. Although these two materials are quite different, the processes to shape them in particulate processing are quite similar. The common technique involves pressing and sintering, illustrated in Figure 1.6, in which the powders are first squeezed into a die cavity under high pressure and then heated to bond the individual particles together.

In *deformation processes,* the starting workpart is shaped by application of forces that exceed the yield strength of the material. For the material to be formed in this way, it must be sufficiently ductile to avoid fracture during deformation. To increase ductility (and for other reasons), the work material is often heated prior to forming to a temperature below the melting point. Deformation processes are associated most closely with metalworking and include operations such as *forging* and *extrusion,* shown in Figure 1.7.

Material removal processes are operations that remove excess material from the starting workpiece so that the resulting shape is the desired geometry. The most important processes in this category are *machining* operations such as *turning, drilling,* and *milling,* shown in Figure 1.8. These cutting operations are most commonly applied to solid metals, performed using cutting tools that are harder and stronger than the work metal. *Grinding* is another common process in this category. Other material removal processes are known as *nontraditional processes* because they use lasers, electron beams, chemical erosion, electric discharge, and electrochemical energy to remove material rather than cutting or grinding tools.

FIGURE 1.7 Some common deformation processes: (a) *forging*, in which two halves of a die squeeze the workpart, causing it to assume the shape of the die cavity; and (b) *extrusion*, in which a billet is forced to flow through a die orifice, thus taking the cross-section shape of the orifice.

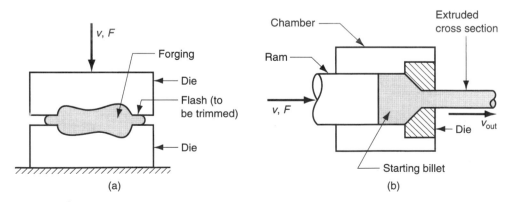

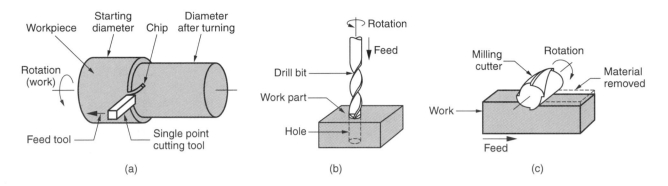

FIGURE 1.8   Common machining operations: (a) *turning,* in which a single-point cutting tool removes metal from a rotating workpiece to reduce its diameter; (b) *drilling,* in which a rotating drill bit is fed into the work to create a round hole; and (c) *milling,* in which a workpart is fed past a rotating cutter with multiple edges.

It is desirable to minimize waste and scrap in converting a starting workpart into its subsequent geometry. Certain shaping processes are more efficient than others in terms of material conservation. Material removal processes (e.g., machining) tend to be wasteful of material, simply by the way they work. The material removed from the starting shape is waste, at least in terms of the unit operation. Other processes, such as certain casting and molding operations, often convert close to 100% of the starting material into final product. Manufacturing processes that transform nearly all of the starting material into product and require no subsequent machining to achieve final part geometry are called *net shape processes.* Other processes require minimum machining to produce the final shape and are called *near net shape processes.*

Property-Enhancing Processes   The second major type of part processing is performed to improve mechanical or physical properties of the work material. These processes do not alter the shape of the part, except unintentionally in some cases. The most important property-enhancing processes involve *heat treatments,* which include various annealing and strengthening processes for metals and glasses. *Sintering* of powdered metals and ceramics, already mentioned, is also a heat treatment that strengthens a pressed powder metal workpart.

Surface Processing   Surface processing includes (1) cleaning, (2) surface treatments, and (3) coating and thin film deposition processes. *Cleaning* includes both chemical and mechanical processes to remove dirt, oil, and other contaminants from the surface. *Surface treatments* include mechanical working such as shot peening and sand blasting, and physical processes like diffusion and ion implantation. *Coating* and *thin film deposition* processes apply a coating of material to the exterior surface of the workpart. Common coating processes include *electroplating, anodizing* of aluminum, organic *coating* (call it *painting*), and porcelain enameling. Thin film deposition processes include *physical and chemical vapor deposition* to form extremely thin coatings of various substances.

Several surface-processing operations have been adapted to fabricate semiconductor materials into integrated circuits for microelectronics. These processes include chemical vapor deposition, physical vapor deposition, and oxidation. They are applied to very localized areas on the surface of a thin wafer of silicon (or other semiconductor material) to create the microscopic circuit.

1.3.2 Assembly Operations

The second basic type of manufacturing operation is assembly, in which two or more separate parts are joined to form a new entity. Components of the new entity are connected, either permanently or semipermanently. Permanent joining processes include *welding, brazing, soldering,* and *adhesive bonding.* They form a joint between components that cannot be easily disconnected. *Mechanical assembly* methods are available to fasten two (or more) parts together in a joint that can be conveniently disassembled. The use of screws, bolts, and other *threaded fasteners* are important traditional methods in this category. Other mechanical assembly techniques that form a more permanent connection include *rivets, press fitting,* and *expansion fits.*

Special assembly methods are used in electronics. Some are identical to or adaptations of the above techniques. For example, soldering is widely used in electronics assembly. Electronics assembly is concerned primarily with the assembly of components (e.g., integrated circuit packages) to printed circuit boards to produce the complex circuits used in so many of today's products.

Joining and assembly processes are discussed in Part VIII, and the specialized assembly techniques for electronics are described in Part IX.

1.3.3 Production Machines and Tooling

Manufacturing operations are accomplished using machinery and tooling (and people). The extensive use of machinery in manufacturing began with the Industrial Revolution. It was at that time that metal cutting machines started to be developed and widely used. These were called *machine tools*—power-driven machines used to operate cutting tools previously operated by hand. Modern machine tools are described by the same basic definition, except that the power is electrical rather than water or steam, and the level of precision and automation is much greater today. Machine tools are among the most versatile of all production machines. They are used to make not only parts for consumer products, but also components for other production machines. Both in a historic sense and in a reproductive sense, the machine tool is the mother of all machinery.

Other production machines include *presses* for stamping operations, *forge hammers* for forging, *rolling mills* for rolling sheet metal, *welding machines* for welding, *insertion machines* for inserting electronic components into printed circuit boards, and so forth. The name of the equipment usually follows from the name of the process.

Production equipment can be general purpose or special purpose. *General purpose equipment* is more flexible and adaptable to a variety of jobs. It is commercially available for any manufacturing company to invest in. *Special purpose equipment* is usually designed to produce a specific part or product in very large quantities. The economics of mass production justify large investments in special purpose machinery to achieve high efficiencies and short cycle times. This is not the only reason for special purpose equipment, but it is the dominant one. Another reason is because the process is unique and commercial equipment is not available. Some companies with unique processing requirements develop their own special purpose equipment.

Production machinery usually requires *tooling,* which customizes the equipment for the particular part or product. In many cases, the tooling must be designed specifically for the part or product configuration. When used with general purpose equipment, it is designed to be exchanged. For each workpart type, the tooling is fastened to the machine and the production run is made. When the run is completed, the tooling is changed for the next workpart type. When used with special purpose machines, the tooling is often

TABLE 1.4   Production equipment and tooling used for various manufacturing processes.

Process	Equipment	Special Tooling (Function)
Casting	[a]	Mold (cavity for molten metal)
Molding	Molding machine	Mold (cavity for hot polymer)
Rolling	Rolling mill	Roll (reduce work thickness)
Forging	Forge hammer or press	Die (squeeze work to shape)
Extrusion	Press	Extrusion die (reduce cross section)
Stamping	Press	Die (shearing, forming sheet metal)
Machining	Machine tool	Cutting tool (material removal)
		Fixture (hold workpart)
		Jig (hold part and guide tool)
Grinding	Grinding machine	Grinding wheel (material removal)
Welding	Welding machine	Electrode (fusion of work metal)
		Fixture (hold parts during welding)

[a]Various types of casting setups and equipment (Chapter 11).

designed as an integral part of the machine. Since the special purpose machine is likely being used for mass production, the tooling may never need changing except for replacement of worn components or for repair of worn surfaces.

The type of tooling depends on manufacturing process. In Table 1.4, we list examples of special tooling used in various operations. Details are provided in the chapters that discuss these processes.

1.4 PRODUCTION SYSTEMS

To operate effectively, a manufacturing firm must have systems that allow it to efficiently accomplish its type of production. Production systems consist of people, equipment, and procedures designed for the combination of materials and processes that constitute a firm's manufacturing operations. Production systems can be divided into two categories: (1) production facilities and (2) manufacturing support systems. ***Production facilities*** refer to the physical equipment and the arrangement of equipment in the factory. ***Manufacturing support systems*** are the procedures used by the company to manage production and solve the technical and logistics problems encountered in ordering materials, moving work through the factory, and ensuring that products meet quality standards. Both categories include people. People make these systems work. In general, direct labor people (blue collar workers) are responsible for operating the manufacturing equipment; and professional staff people (white collar workers) are responsible for manufacturing support.

1.4.1 Production Facilities

Production facilities consist of the factory, production equipment, and material handling equipment. The equipment comes in direct physical contact with the parts and/or assemblies as they are being made. The facilities "touch" the product. Facilities also include the way the equipment is arranged in the factory—the ***plant layout.*** The equipment is usually organized into logical groupings, let us call them ***manufacturing systems,*** such as an automated production line, or a machine cell consisting of an industrial robot and two machine tools.

A manufacturing company attempts to design its manufacturing systems and organize its factories to serve the particular mission of each plant in the most efficient way.

Over the years, certain types of production facilities have come to be recognized as the most appropriate way to organize for a given *type* of manufacturing—the combination of product variety and production quantity discussed in Section 1.1.2. Different facilities are required for each of the three quantity ranges.

Low-Quantity Production   In the low-quantity range (1 to 100 units/year), the term ***job shop*** is often used to describe the type of production facility. A job shop makes low quantities of specialized and customized products. The products are typically complex, such as space capsules, prototype aircraft, and special machinery. The equipment in a job shop is general purpose and the labor force is highly skilled.

A job shop must be designed for maximum flexibility in order to deal with the wide product variations encountered (hard product variety). If the product is large and heavy, and therefore difficult to move, it typically remains in a single location during its fabrication or assembly. Workers and processing equipment are brought to the product, rather than moving the product to the equipment. This type of layout is referred to as a ***fixed-position layout***, shown in Figure 1.9(a). In the pure situation, the product remains in a single location during its entire production. Examples of such products include ships, aircraft, locomotives, and heavy machinery. In actual practice, these items are usually built in large modules at single locations, and then the completed modules are brought together for final assembly using large-capacity cranes.

The individual components of these large products are often made in factories in which the equipment is arranged according to function or type. This arrangement is called a ***process layout***. The lathes are in one department, the milling machines are in another department, and so on, as in Figure 1.9(b). Different parts, each requiring a different op-

FIGURE 1.9   Various types of plant layout: (a) fixed-position layout, (b) process layout, (c) cellular layout, and (d) product layout.

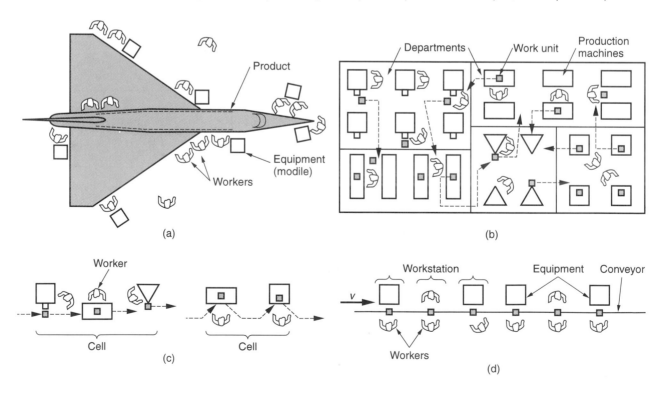

eration sequence, are routed through the departments in the particular order needed for their processing, usually in batches. The process layout is noted for its flexibility; it can accommodate a great variety of operation sequences for different part configurations. Its disadvantage is that the machinery and methods to produce a part are not designed for high efficiency.

Medium-Quantity Production   In the medium-quantity range (100 to 10,000 units annually), we distinguish two different types of facility, depending on product variety. When product variety is hard, the usual approach is ***batch production,*** in which a batch of one product is made, after which the manufacturing system is changed over to produce a batch of the next product, and so on. The production rate of the equipment is greater than the demand rate for any single product type, and so the same equipment can be shared among multiple products. The changeover between production runs takes time—time to change tooling and set up the machinery. This setup time is lost production time, and this is a disadvantage of batch manufacturing. Batch production is commonly used for make-to-stock situations, in which items are manufactured to replenish inventory that has been gradually depleted by demand. The equipment is usually arranged in a process layout, Figure 1.9(b).

An alternative approach to medium-range production is possible if product variety is soft. In this case, extensive changeovers between one product style and the next may not be necessary. It is often possible to configure the manufacturing system so that groups of similar products can be made on the same equipment without significant lost time due to setup. The processing or assembly of different parts or products is accomplished in cells consisting of several workstations or machines. The term ***cellular manufacturing*** is often associated with this type of production. Each cell is designed to produce a limited variety of part configurations; that is, the cell specializes in the production of a given set of similar parts, according to the principles of ***group technology*** (Section 39.1). The layout is called a ***cellular layout*** (the term ***group technology layout*** is also common), depicted in Figure 1.9(c).

High Production   The high-quantity range (10,000 to millions of units per year) is referred to as ***mass production.*** The situation is characterized by a high demand rate for the product, and the manufacturing system is dedicated to the production of that single item. Two categories of mass production can be distinguished: quantity production and flow line production. ***Quantity production*** involves the mass production of single parts on single pieces of equipment. It typically involves standard machines (such as stamping presses) equipped with special tooling (e.g., dies and material handling devices), in effect dedicating the equipment to the production of one part type. Typical layouts used in quantity production are the process layout and cellular layout, Figure 1.9(b) and (c).

Flow line production involves multiple pieces of equipment or workstations arranged in sequence, and the work units are physically moved through the sequence to complete the product. The workstations and equipment are designed specifically for the product to maximize efficiency. The layout is called a ***product layout***, and the workstations are arranged into one long line, as in Figure 1.9(d); or into a series of connected line segments. The work is usually moved between stations by mechanized conveyor. At each station, a small amount of the total work is completed on each unit of product.

The most familiar example of flow line production is the assembly line, associated with products such as cars and household appliances. The pure case of flow line production is where there is no variation in the products made on the line. Every product is identical, and the line is referred to as a ***single-model production line***. In order to successfully market a given product, it is often useful to introduce feature and model variations so that individual customers can choose the exact merchandise that appeals to them. From a production viewpoint, the feature differences represent a case of soft product variety. The term *mixed-model production line* applies to these situations where there is soft variety in the products made on the line. Modern automobile assembly is an example. Cars coming off the assembly line have variations in options and trim representing different models and in many cases different nameplates of the same basic car design.

1.4.2 Manufacturing Support Systems

To operate the facilities efficiently, a company must organize itself to design the processes and equipment, plan and control the production orders, and satisfy product quality requirements. These functions are accomplished by manufacturing support systems—people and procedures by which a company manages its production operations. Most of these support systems do not directly contact the product, but they plan and control its progress through the factory. Manufacturing support functions are often carried out in the firm by people organized into departments such as the following:

> ➤ *Manufacturing engineering.* The manufacturing engineering department is responsible for planning the manufacturing processes—deciding which processes should be used to make the parts and assemble the products. This department is also involved in designing and ordering the machine tools and other equipment used by the operating departments to accomplish processing and assembly.
> ➤ *Production planning and control.* This department is responsible for solving the logistics problem in manufacturing—ordering materials and purchased parts, scheduling production, and making sure that the operating departments have the necessary capacity to meet the production schedules.
> ➤ *Quality control.* Producing high-quality products should be a top priority of any manufacturing firm in today's competitive environment. It means designing and building products that conform to specifications and satisfy or exceed customer expectations. Much of this effort is the responsibility of the QC department.

1.5 ORGANIZATION OF THE BOOK

The previous three sections provide a preview and an overview of our book. The remaining 43 chapters are organized into 11 parts. The block diagram in Figure 1.10 summarizes the major topics that are covered. It shows the production system (outlined in dashed lines) with engineering materials entering from the left and finished products exiting at the right. Part I, titled Material Properties and Product Attributes, consists of four chapters that describe the important characteristics and specifications of materials and the products made from them. Part II discusses the four basic engineering materials: metals, ceramics, polymers, and composites.

The largest block in Figure 1.10 is labeled "Manufacturing processes and assembly operations." The processes and operations included in our text are those identified in Figure 1.4. Part III begins our coverage of the four categories of shaping processes. Part III consists of six chapters on the solidification processes: casting of metals (two chapters), glassworking, polymer shaping, rubber processing technology, and shaping processes for

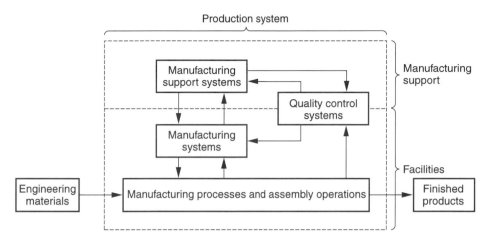

FIGURE 1.10   Overview of major topics covered in the book.

polymer-based composite materials. In Part IV, the particulate processing of metals and ceramics is covered in two chapters. Part V deals with metal deformation processes such as rolling, forging, extrusion, and sheet metalworking. Finally, Part VI discusses the material removal processes. Four chapters are used for machining, and two additional chapters cover the abrasive processes (e.g., grinding) and nontraditional material removal technologies.

The other types of processing operations, property enhancing and surface processing, are covered in Part VII. It consists of three chapters, covering heat treatment, cleaning and surface treatments, and coating and deposition processes.

Joining and assembly processes are considered in Part VIII, which is organized into four chapters on welding, brazing, soldering, adhesive bonding, and mechanical assembly.

Several unique processes that do not neatly fit into our classification scheme of Figure 1.4 are covered in Part IX, titled Special Processing and Assembly Technologies. Its four chapters are concerned with rapid prototyping, processing of integrated circuits, electronics assembly and packaging, and microfabrication.

We next turn our attention to the remaining blocks in Figure 1.10, which are concerned with the systems of production. Part X is titled "Manufacturing Systems" (note its position in the block diagram). It covers the major systems technologies and equipment groupings located in the factory: numerical control, industrial robotics, group technology, cellular manufacturing, flexible manufacturing systems, and production lines. Finally, Part XI, "Manufacturing Support Systems," deals with manufacturing engineering, production planning and control, quality control, and inspection.

1.6 IMAGES OF MANUFACTURING

In this section we describe the series of color plates that show various aspects of manufacturing and manufactured products. One of the largest manufactured products is an airplane. Boeing's 777 is one of the latest and most advanced commercial aircraft, shown in Color Plate 1. Commercial airplanes are made in annual production quantities of a few hundred at most. They are built of lightweight alloys of aluminum (Section 6.3.1), magnesium (Section 6.3.2), and titanium (Section 6.3.5). Modern aircraft construction also makes increasing use of fiber-reinforced polymer composites (Section 9.4.1). Turbines and other components for the 777's two jet engines are made of high-strength superalloys (Section 6.4) that can operate at elevated temperatures for maximum thrust and efficiency.

Contrasting in size and production quantity with the Boeing 777 is an integrated circuit (IC), many of which are used in modern aircraft. Shown in Color Plate 2 is one of Intel's Pentium microprocessors—the heart of many of today's personal computers. The Pentium is fabricated on a thin "chip" of high-purity silicon (Section 7.6.2), measuring only 25 mm (1 in.) on a side, yet it contains several million transistors. Although the Pentium is a single monolithic piece, its processing sequence is one of the most complex of all manufactured products, consisting of hundreds of individual processing steps (Chapter 35). ICs such as the Pentium are often produced in millions of units under "clean room" conditions, as suggested by Color Plate 3, which shows a robot handling silicon wafers.

Integrated circuit chips are packaged into modules that are assembled to printed circuit boards (PCBs). One PCB may hold hundreds of individual chips, and the complete electronic assembly may consist of many assembled PCBs. The PCBs must also be manufactured, which involves fabrication of the particular circuit to accomplish the electronic application specified by the design engineer. Color Plate 4 shows one of the operations in PCB assembly.

Many of the components in airplanes and most other engineered products are made of metal. Metals are the most important engineering materials, and steel is the most important metal. One of the spectacular scenes in steelmaking (Section 6.2.2) is the charging of a basic oxygen furnace (BOF), shown in Color Plate 5. Molten pig iron produced in a blast furnace is poured into the BOF, where it is heated for about 20 minutes to burn off impurities and make steel of a specified chemistry. Temperatures in the basic oxygen furnace are around 1650°C (3000°F).

Steels and other metals are shaped by many different manufacturing processes. One of the most widely used processes is machining, in which material is removed from a starting workpiece to create the desired part geometry. An important machining process is turning (Section 22.1), illustrated in Color Plate 6. In our figure, the cutting tool is the gold-colored object marked KC990 (tradename), shown removing metal in the form of "chips" from a steel workpiece. Color Plate 7 shows a photomicrograph of the cross-section of a KC990 cutting tool. The substrate is made of cemented carbide (Section 23.2.3) on which has been deposited a series of very thin coatings of aluminum oxide (black-colored) and titanium nitride (gold colored). The thicker grey-colored layer between the substrate and the series of thin layers is titanium carbonitride. The coatings provide wear and heat resistance to allow the tool to withstand the harsh cutting environment. They are applied by physical vapor deposition (Section 29.3) and chemical vapor deposition (Section 29.4).

Machining operations in modern factories are performed on highly automated machine tools, like the one shown in Color Plate 8. This is a horizontal machining center (Section 22.4) equipped with automatic tool changer and parts carousel. The worker to the right is loading a large workpart onto the carousel while machining proceeds on another part. The carousel can be loaded with up to six different part styles, and the machine tool automatically indexes them into position for machining, thus permitting untended operation for extended periods. Machines like this operate under computer numerical control (Section 38.1).

Machining is a conventional material removal process, dating from the Industrial Revolution (Historical Note 1.2). Other material removal processes are based on modern technologies such as lasers (Section 26.3.3). Color Plate 9 shows a laser-cutting process, used to cut a part outline from sheet metal. The cutting path is again controlled by computer numerical control.

Welding is an important family of manufacturing processes, used to assemble multiple components into a single fabricated unit. Two welding operations are shown in Color Plates 10 and 11, illustrating some of the wide differences among these processes. Continuous arc welding, Color Plate 10, is often performed manually. Considerable skill and concentration are required by the welder under conditions that are uncomfortable and potentially unsafe. Operations like this are generally used for low- and medium-quantity production, as well as construction work. Color Plate 11 shows a robotic spot-welding cell in an automobile assembly plant—high production. Sparks fly as inner and outer door panels are welded together.

The assembly line is a symbol of mass production, sometimes reveled for its efficiency and reviled for its subjugation of human labor. The automobile final assembly plant is the ultimate assembly line. A portion of an automobile assembly line is pictured in Color Plate 12.

REFERENCES

[1] DeGarmo, E. P., Black, J. T., and Kohser, R. A. *Materials and Processes in Manufacturing.* 8th ed. Macmillan, New York, 1997.

[2] Emerson, H. P., Naehring, D. C. E. *Origins of Industrial Engineering.* Industrial Engineering & Management Press, Institute of Industrial Engineers, 1988.

[3] Flinn, R. A., and Trojan, P. K. *Engineering Materials and Their Applications.* Houghton Mifflin, Boston, Mass., 1990.

[4] Garrison, E. *A History of Engineering and Technology.* CRC Press, Inc., Boca Raton, Fla., 1991.

[5] Groover, M. P. *Automation, Production Systems, and Computer Integrated Manufacturing.* 2nd ed., Prentice Hall, Upper Saddle River, N. J., 2001.

[6] Hounshell, D. A. *From the American System to Mass Production, 1800-1932.* The Johns Hopkins University Press, Baltimore, Md., 1984.

2 THE NATURE OF MATERIALS

CHAPTER CONTENTS

2.1   Atomic Structure and the Elements
2.2   Bonding between Atoms and Molecules
2.3   Crystalline Structures
        2.3.1   Types of Crystal Structures
        2.3.2   Imperfections in Crystals
        2.3.3   Deformation in Metallic Crystals
        2.3.4   Grains and Grain Boundaries in Metals
2.4   Noncrystalline (Amorphous) Structures
2.5   Engineering Materials

An understanding of materials is fundamental in the study of manufacturing processes. In Chapter 1, manufacturing was defined as a transformation process. It is the material that is transformed; and it is the behavior of the material when subjected to the particular forces, temperatures, and other physical parameters of the process that determines the success of the operation. We find that certain materials respond well to certain types of manufacturing processes and poorly, or not at all, to others. What are the characteristics and properties of materials that determine their capacity to be transformed by the different processes?

In this chapter, we consider the atomic structure of matter and the bonding between atoms and molecules. We also show how atoms and molecules in engineering materials organize themselves into two structural forms: crystalline and noncrystalline. It turns out that the basic engineering materials—metals, ceramics, and polymers—can exist in either form, although there is usually a preference for a particular form exhibited by a given material. Metals, for example, almost always exist as crystals in their solid state. Glass (e.g., window glass), a ceramic, assumes a noncrystalline form.

2.1 ATOMIC STRUCTURE AND THE ELEMENTS

The basic structural unit of matter is the atom. Each atom is composed of a positively charged nucleus, surrounded by a sufficient number of negatively charged electrons so that the charges are balanced. The number of electrons identifies the atomic number and the element of the atom. There are slightly more than 100 elements (not counting a few extras that have been artificially synthesized), and these elements are the chemical building blocks of all matter.

Just as there are differences between the elements, there are also similarities. The elements can be grouped into families and relationships established between and within the families by means of the Periodic Table, shown in Figure 2.1. In the horizontal direction there is a certain repetition, or periodicity, in the arrangement of elements. Metallic elements occupy the left and center portions of the chart, and nonmetals are located to the right. Between them, along a diagonal, is a transition zone containing elements called ***metalloids*** or ***semimetals.*** In principle, each of the elements can exist as a solid, liquid, or gas, depending on temperature and pressure. At room temperature and atmospheric pressure, they each have a natural phase; for example, iron (Fe) is a solid, mercury (Hg) is a liquid, and nitrogen (N) is a gas.

In the table, the elements are arranged into vertical columns and horizontal rows in such a way that similarities exist between elements in the same columns. For example, in the extreme right column are the ***noble gases*** (helium, neon, argon, krypton, xenon, and radon), all of which exhibit great chemical stability and low reaction rates. The ***halogens*** (fluorine, chlorine, bromine, iodine, and astatine) in column VIIA share similar properties (hydrogen is not included among the halogens). And the ***noble metals*** (copper, silver, and gold) in column IB have similar properties. Generally, there are correlations in properties among elements within a given column, whereas differences exist between elements in different columns.

Many of the similarities and differences among the elements can be explained by their respective atomic structures. The simplest model of atomic structure, called the planetary model, shows the electrons of the atom orbiting around the nucleus at certain fixed distances, called shells, as shown in Figure 2.2. The hydrogen atom (atomic num-

FIGURE 2.1 Periodic Table of Elements. The atomic number and symbol are listed for the 103 elements.

Metals — Transition Zone — Nonmetals

Period	IA	IIA	IIIB	IVB	VB	VIB	VIIB	VIIIB	VIIIB	VIIIB	IB	IIB	IIIA	IVA	VA	VIA	VIIA	VIIIA
1	1 H																1 H	2 He
2	3 Li	4 Be											5 B	6 C	7 N	8 O	9 F	10 Ne
3	11 Na	12 Mg											13 Al	14 Si	15 P	16 S	17 Cl	18 Ar
4	19 K	20 Ca	21 Sc	22 Ti	23 V	24 Cr	25 Mn	26 Fe	27 Co	28 Ni	29 Cu	30 Zn	31 Ga	32 Ge	33 As	34 Se	35 Br	36 Kr
5	37 Rb	38 Sr	39 Y	40 Zr	41 Nb	42 Mo	43 Tc	44 Ru	45 Rh	46 Pd	47 Ag	48 Cd	49 In	50 Sn	51 Sb	52 Te	53 I	54 Xe
6	55 Cs	56 Ba	57 La	72 Hf	73 Ta	74 W	75 Re	76 Os	77 Ir	78 Pt	79 Au	80 Hg	81 Tl	82 Pb	83 Bi	84 Po	85 At	86 Rn
7	87 Fr	88 Ra	89 Ac															

58 Ce	59 Pr	60 Nd	61 Pm	62 Sm	63 Eu	64 Gd	65 Tb	66 Dy	67 Ho	68 Er	69 Tm	70 Yb	71 Lu
90 Th	91 Pa	92 U	93 Np	94 Pu	95 Am	96 Cm	97 Bk	98 Cf	99 Es	100 Fm	101 Md	102 No	103 Lw

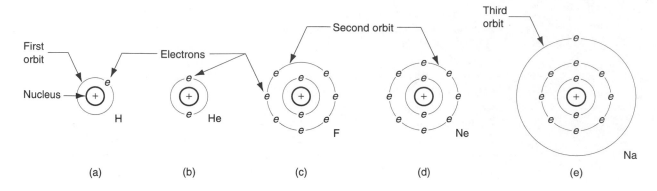

FIGURE 2.2   Simple model of atomic structure for several elements: (a) hydrogen, (b) helium, (c) fluorine, (d) neon, and (e) sodium.

ber 1) has one electron in the orbit closest to the nucleus. Helium (atomic number 2) has two. Also shown in the figure are the atomic structures for fluorine (atomic number 9), neon (atomic number 10), and sodium (atomic number 11). One might infer from these models that there is a maximum number of electrons that can be contained in a given orbit. This turns out to be correct, and the maximum is defined by

$$\text{Maximum number of electrons in an orbit} = 2n^2 \qquad (2.1)$$

where n identifies the orbit, with $n = 1$ closest to the nucleus.

The number of electrons in the outermost shell, relative to the maximum number allowed, determines to a large extent the atom's chemical affinity for other atoms. These outer-shell electrons are called **valence electrons.** For example, since a hydrogen atom has only one electron in its single orbit, it readily combines with another hydrogen atom to form a hydrogen molecule H_2. For the same reason, hydrogen also reacts readily with various other elements (e.g., to form H_2O). In the helium atom, the two electrons in its only orbit are the maximum allowed ($2n^2 = 2(1)^2 = 2$), and so helium is very stable. Neon is stable for the same reason. Its outermost orbit ($n = 2$) has eight electrons (the maximum allowed), so neon is an inert gas.

In contrast to neon, fluorine has one fewer electron in its outer shell ($n = 2$) than the maximum allowed and is readily attracted to other elements that might share an electron to make a more stable set. The sodium atom seems divinely made for the situation, with one electron in its outermost orbit. It reacts strongly with fluorine to form the compound sodium fluoride, as pictured in Figure 2.3.

At the low atomic numbers considered here, the prediction of the number of electrons in the outer orbit is straightforward. As the atomic number increases to higher levels, the allocation of electrons to the different orbits becomes somewhat more complicated. There are rules and guidelines, based on quantum mechanics, that can be used to predict the positions of the electrons among the various orbits and to explain their characteristics. A discussion of these rules is somewhat beyond the scope of our coverage of materials for manufacturing.

2.2 BONDING BETWEEN ATOMS AND MOLECULES

Atoms are held together in molecules by various types of bonds that depend on the valence electrons. By comparison, molecules are attracted to each other by weaker bonds, which generally result from the electron configuration in the individual molecules. Thus,

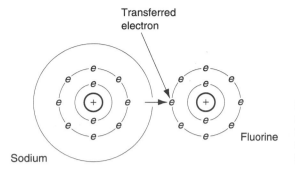

FIGURE 2.3   The sodium fluoride molecule, formed by the transfer of the "extra" electron of the sodium atom to complete the outer orbit of the fluorine atom.

we have two types of bonding: (1) primary bonds, generally associated with the formation of molecules; and (2) secondary bonds, generally associated with attraction between molecules. Primary bonds are much stronger than secondary bonds.

Primary Bonds   Primary bonds are characterized by strong atom-to-atom attractions that involve the exchange of valence electrons. Primary bonds include the following forms: (a) ionic, (b) covalent, and (c) metallic, as illustrated in Figure 2.4. Ionic and covalent bonds are called *intra*molecular bonds because they involve attractive forces between atoms within the molecule.

In the *ionic bond,* the atoms of one element give up their outer electron(s), which are in turn attracted to the atoms of some other element to increase their electron count in the outermost shell to eight. In general, eight electrons in the outer shell is the most stable atomic configuration (except for the very light atoms), and nature provides a very strong bond between atoms that achieve this configuration. Our previous example of the reaction of sodium and fluorine to form sodium fluoride (Figure 2.3) illustrates this form of atomic bond. Sodium chloride (table salt) is a more common example. Because of the transfer of electrons between the atoms, sodium and fluorine (or sodium and chlorine) *ions* are formed, from which this bonding derives its name. Properties of solid materials with ionic bonding include low electrical conductivity and poor ductility.

The *covalent bond* is one in which electrons are shared (as opposed to transferred) between atoms in their outermost shells to achieve a stable set of eight. Fluorine and diamond are two examples of covalent bonds. In fluorine, one electron from each of two atoms is shared to form F_2 gas, as shown in Figure 2.5(a). In the case of diamond, which is carbon (atomic number 6), each atom has four neighbors with which it shares electrons. This produces a very rigid three-dimensional structure, not adequately represented in Figure 2.5(b), and accounts for the extreme high hardness of this material. Other forms

FIGURE 2.4   Three forms of primary bonding: (a) ionic, (b) covalent, and (c) metallic.

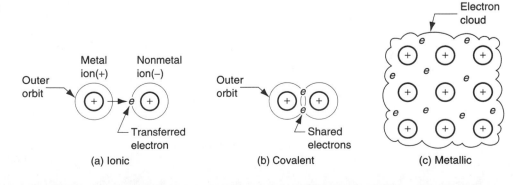

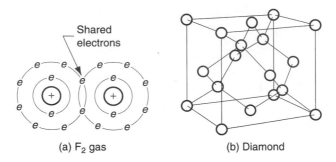

FIGURE 2.5 Two examples of covalent bonding: (a) fluorine gas F_2, and (b) diamond.

of carbon (e.g., graphite) do not exhibit this rigid atomic structure. Solids with covalent bonding generally possess high hardness and low electrical conductivity.

The metallic bond is, of course, the atomic bonding mechanism in pure metals and metal alloys. Atoms of the metallic elements generally possess too few electrons in their outermost orbits to complete the outer shells for all of the atoms in, say, a given block of metal. Accordingly, instead of sharing on an atom-to-atom basis, **metallic bonding** involves the sharing of outer-shell electrons by all atoms to form a general electron cloud that permeates the entire block. This cloud provides the attractive forces to hold the atoms together and form a strong, rigid structure in most cases. Because of the general sharing of electrons, and their freedom to move within the metal, metallic bonding provides for good electrical conductivity. By contrast, the other types of primary bonds involve local sharing of electrons between neighboring atoms only, so that these materials are poor electrical conductors. Other typical properties of materials characterized by metallic bonding include good conduction of heat and good ductility. (Although we have yet to define some of these terms, we rely on the reader's general understanding of material properties.)

Secondary Bonds   Whereas primary bonds involve atom-to-atom attractive forces, secondary bonds involve attraction forces between molecules, or **inter**molecular forces. There is no transfer or sharing of electrons in secondary bonding, and these bonds are therefore weaker than primary bonds. There are three forms of secondary bonding: (a) dipole forces, (b) London forces, and (c) hydrogen bonding, illustrated in Figure 2.6. Types (a) and (b) are often referred to as **van der Waals** forces, after the scientist who first studied and quantified them.

Dipole forces arise in a molecule comprising two atoms that have equal and opposite electrical charges. Each molecule therefore forms a dipole, as shown in Figure 2.6(a) for hydrogen chloride. Although the material is electrically neutral in its aggregate form, on a molecular scale the individual dipoles attract each other, given the proper orientation of positive and negative ends of the molecules. These dipole forces provide a net intermolecular bonding within the material.

London forces involve attractive forces between nonpolar molecules; that is, the atoms in the molecule do not form dipoles in the sense of the preceding paragraph. However, owing to the rapid motion of the electrons in orbit around the molecule, temporary dipoles form when more electrons happen to be on one side of the molecule than the other, as suggested by Figure 2.6(b). These instantaneous dipoles provide a force of attraction between molecules in the material.

Finally, *hydrogen bonding* occurs in molecules containing hydrogen atoms that are covalently bonded to another atom (e.g., oxygen in H_2O). Since the electrons needed to complete the shell of the hydrogen atom are aligned on one side of its nucleus, the opposite side has a net positive charge that attracts the electrons of atoms in neighboring molecules. Hydrogen bonding is illustrated in Figure 2.6(c) for water, and is generally a

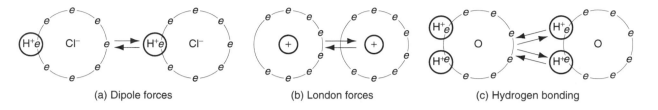

(a) Dipole forces   (b) London forces   (c) Hydrogen bonding

FIGURE 2.6   Types of secondary bonding: (a) dipole forces, (b) London forces, and (c) hydrogen bonding.

stronger intermolecular bonding mechanism than the other two forms of secondary bonding. It is important in the formation of many polymers.

2.3 CRYSTALLINE STRUCTURES

Atoms and molecules are used as building blocks for the more macroscopic structure of matter that we consider here and in the following section. When materials solidify from the molten state, they tend to close ranks and pack tightly, in many cases arranging themselves into a very orderly structure, and in other cases, not quite so orderly. Two fundamentally different material structures can be distinguished: crystalline and noncrystalline. Crystalline structures are examined in this section, and noncrystalline in the next.

Many materials form into crystals upon solidification from the molten or liquid state. It is characteristic of virtually all metals, as well as many ceramics and polymers. A ***crystalline structure*** is one in which the atoms are located at regular and recurring positions in three dimensions. The pattern may be replicated millions of times within a given crystal. The structure can be viewed in the form of a ***unit cell,*** which is the basic geometric grouping of atoms that is repeated. To illustrate, consider the unit cell for the body-centered cubic (BCC) crystal structure shown in Figure 2.7, one of the common structures found in metals. The simplest model of the BCC unit cell is illustrated in Figure 2.7(a). Although this model clearly depicts the locations of the atoms within the cell, it does not indicate the close packing of the atoms that occurs in the real crystal, as in Figure 2.7(b). Figure 2.7(c) shows the repeating nature of the unit cell within the crystal.

2.3.1 Types of Crystal Structures

In metals, three lattice structures are common: body-centered cubic (BCC), face-centered cubic (FCC), and hexagonal close-packed (HCP), illustrated in Figure 2.8. Crystal structures for the common metals are presented in Table 2.1. We should note that some met-

FIGURE 2.7   Body-centered cubic (BCC) crystal structure: (a) unit cell, with atoms indicated as point locations in a three-dimensional axis system; (b) unit cell model showing closely packed atoms (sometimes called the hard-ball model); and (c) repeated pattern of the BCC structure.

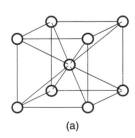

(a)

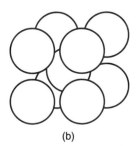

(b)

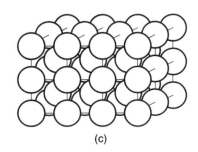

(c)

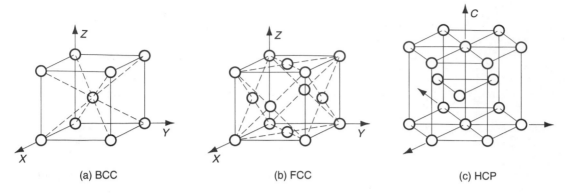

FIGURE 2.8   Three types of crystal structures in metals: (a) body-centered cubic, (b) face-centered cubic, and (c) hexagonal close-packed.

als undergo a change of structure at different temperatures. Iron, for example, is BCC at room temperature; it changes to FCC above 912°C (1674°F) and back to BCC at temperatures above 1400°C (2550°F). When a metal (or other material) changes structure like this, it is referred to as being ***allotropic.***

2.3.2 Imperfections in Crystals

Thus far, we have discussed crystal structures as if they were perfect—the unit cell repeated in the material over and over in all directions. A perfect crystal is sometimes desirable to satisfy aesthetic or engineering purposes. For instance, a perfect diamond that contains no flaws is more valuable than one containing imperfections. In the production of electronic chips, large single crystals of silicon possess desirable processing characteristics for forming the microscopic details of the circuit pattern.

However, there are various reasons why a crystal's lattice structure may not be perfect. The imperfections often arise naturally due to the inability of the solidifying material to continue the replication of the unit cell indefinitely without interruption. Grain boundaries in metals are an example. In other cases, the imperfections are introduced purposely during the manufacturing process; for example, the addition of an alloying ingredient in a metal to increase its strength.

The various imperfections in crystalline solids are also called defects. Either term, ***imperfection*** or ***defect,*** refers to deviations in the regular pattern of the crystalline lattice structure. They can be catalogued as: (1) point defects, (2) line defects, and (3) surface defects.

Point defects are imperfections in the crystal structure involving either a single atom or a few number of atoms. The defects can take various forms including, as shown

TABLE 2.1   Crystal structures for the common metals (at room temperature).

Body-Centered Cubic (BCC)	Face-Centered Cubic (FCC)	Hexagonal Close-Packed (HCP)
Chromium (Cr)	Aluminum (Al)	Magnesium (Mg)
Iron (Fe)	Copper (Cu)	Titanium (Ti)
Molybdenum (Mo)	Gold (Au)	Zinc (Zn)
Tantalum (Ta)	Lead (Pb)	
Tungsten (W)	Silver (Ag)	
	Nickel (Ni)	

▲ PLATE 1
■

Boeing 777 commercial airplane
(photo courtesy of Boeing Company)

PLATE 2 ▶
■

An Intel microprocessor, a
high-density integrated circuit
(photo courtesy of Intel Corporation)

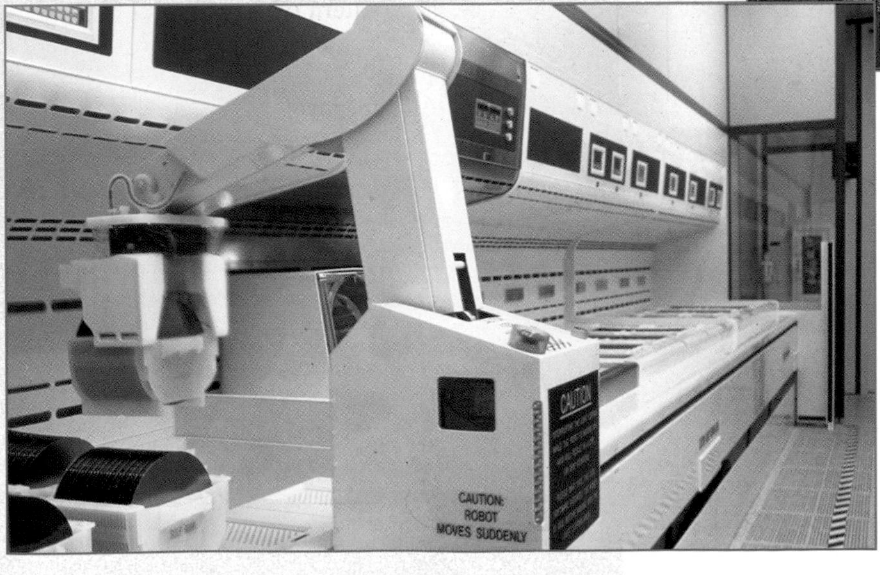

◀ PLATE 3
■

An industrial robot handling
silicon wafers during integrated
circuit manufacturing.
(photo courtesy of Intel Corporation)

◄ PLATE 4

Close-up view of electronic components being assembled to a printed circuit board by an automated insertion machine
(photo courtesy of Universal Instruments Corporation)

▼ PLATE 5

Charging a basic oxygen furnace to make steel
(photo courtesy of Bethlehem Steel Corporation)

◄ PLATE 6

Close-up view of a turning operation in which metal is removed from a rotating workpiece by a coated cemented carbide cutting tool
(photo courtesy of Kennametal Inc.)

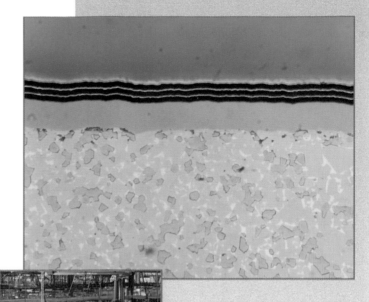

PLATE 7 ▶

Photomicrograph of a cross section
of the cutting tool in Plate 6
showing the multiple coatings
of wear-resistant materials
(photo courtesy of Kennametal Inc.)

◀ PLATE 8

Horizontal machining center
capable of automatically changing
its own cutting tools; parts are
loaded onto a carousel (the large
table) and machined in sequence
(photo courtesy of Cincinnati Milacron)

PLATE 9 ▶

Cutting operation performed
by a laser beam
(photo courtesy of PRC Corporation)

◄ Plate 10

Manual arc welding operation
(photo courtesy of Lincoln Electric Company)

▼ Plate 11

Automatic spot welding operation performed by an industrial robot
(photo courtesy of Ford Motor Company)

Plate 12 ▼

A portion of an automobile final assembly line. Cars are produced on such a line at the rate of about one every minute
(photo courtesy of Chrysler Corporation)

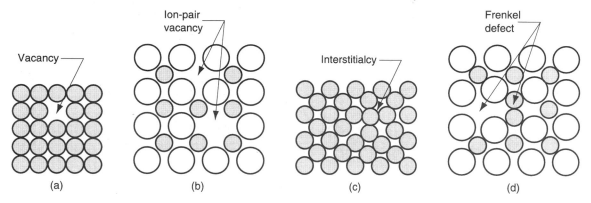

FIGURE 2.9   Point defects: (a) vacancy, (b) ion-pair vacancy, (c) interstitialcy, and (d) displaced ion.

in Figure 2.9: (a) *vacancy*—the simplest defect, involving a missing atom within the lattice structure; (b) *ion-pair vacancy,* also called a ***Schottky defect***—involves a missing pair of ions of opposite charge in a compound that has an overall charge balance; (c) *interstitialcy*—a lattice distortion produced by the presence of an extra atom in the structure; and (d) *displaced ion*—known as a ***Frenkel defect,*** which occurs when an ion becomes removed from a regular position in the lattice structure and inserted into an interstitial position not normally occupied by such an ion.

A ***line defect*** is a connected group of point defects that forms a line in the lattice structure. The most important line defect is the ***dislocation,*** which can take two forms: (a) edge dislocation and (b) screw dislocation. An ***edge dislocation*** is the edge of an extra plane of atoms that exists in the lattice, as illustrated in Figure 2.10(a). A ***screw dislocation,*** Figure 2.10(b), is a spiral within the lattice structure wrapped around an imperfection line, like a screw is wrapped around its axis. Both types of dislocations can arise in the crystal structure during solidification (e.g., casting), or they can be initiated during a deformation process (e.g., metal forming) performed on the solid material. Dislocations are useful in explaining certain aspects of mechanical behavior in metals.

Surface defects are imperfections that extend in two directions to form a boundary. The most obvious example is the external surface of a crystalline object that defines its shape. The surface is an interruption in the lattice structure. Surface boundaries can also lie inside the material. Grain boundaries are the best example of these internal surface interruptions. We discuss metallic grains in a moment, but first let us consider how

FIGURE 2.10   Line defects: (a) edge dislocation and (b) screw dislocation.

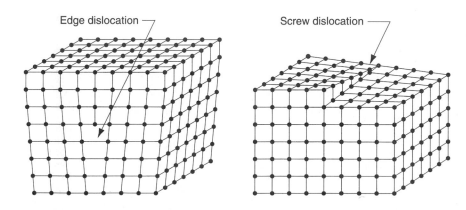

deformation occurs in a crystal lattice, and how the process is aided by the presence of dislocations.

2.3.3 Deformation in Metallic Crystals

When a crystal is subjected to a gradually increasing mechanical stress, its initial response is to deform *elastically*. This can be likened to a tilting of the lattice structure without any changes of position among the atoms in the lattice, in the manner depicted in Figure 2.11(a) and (b). If the force is removed, the lattice structure (and therefore the crystal) returns to its original shape. If the stress reaches a high value relative to the electrostatic forces holding the atoms in their lattice positions, a permanent shape change occurs, called *plastic deformation.* What has happened is that the atoms in the lattice have permanently moved from their previous locations, and a new equilibrium lattice has been formed, as suggested by Figure 2.11(c).

The lattice deformation shown in (c) of the figure is one possible mechanism, called slip, by which plastic deformation can occur in a crystalline structure. The other is called twinning.

Slip involves the relative movement of atoms on the opposite sides of a plane in the lattice, called the *slip plane.* The slip plane must be somehow aligned with the lattice structure (as indicated in our sketch), and so there are certain preferred directions along which slip is more likely to occur. The number of these *slip directions* depends on the lattice type. The three common metal crystal structures are somewhat more complicated, especially in three dimensions, than the square lattice depicted in Figure 2.11. It turns out that HCP has the fewest slip directions, BCC the most, and FCC falls in between. HCP metals show poor ductility and are generally difficult to deform at room temperature. Metals with BCC structure would figure to have the highest ductility, if the number of slip directions were the only criterion. However, nature is not so simple. These metals are generally stronger than the others, which complicates the issue, and the BCC metals usually require higher stresses to cause slip. In fact, some of the BCC metals exhibit poor ductility. Low carbon steel is a notable exception; although relatively strong, it is widely used with great commercial success in sheetmetal-forming operations, where it exhibits good ductility. The FCC metals are generally the most ductile of the three crystal structures, combining a good number of slip directions with (usually) relatively low to moderate strength. All three of these metal structures become more ductile at elevated temperatures, and this fact is often exploited in shaping them.

Dislocations play an important role in facilitating slip in metals. When a lattice structure containing an edge dislocation is subjected to a shear stress, the material deforms much more readily than in a perfect structure. This is explained by the fact that the dislocation is put into motion within the crystal lattice in the presence of the stress, as shown in the series of sketches in Figure 2.12. Why is it easier to move a dislocation through the

FIGURE 2.11 Deformation of a crystal structure: (a) original lattice; (b) elastic deformation, with no permanent change in positions of atoms; and (c) plastic deformation, in which atoms in the lattice are forced to move to new "homes."

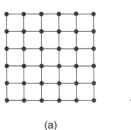

(a)

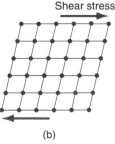

Shear stress

(b)

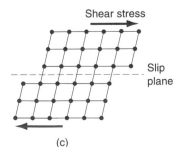

Shear stress

Slip plane

(c)

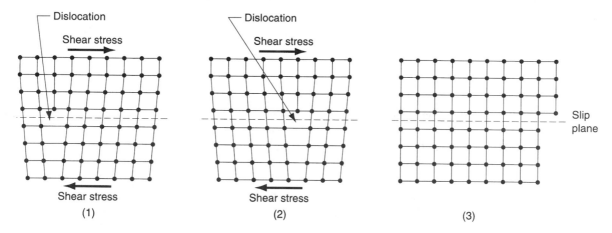

FIGURE 2.12   Effect of dislocations in the lattice structure under stress. In the series of diagrams, the movement of the dislocation allows deformation to occur under a lower stress than in a perfect lattice.

lattice than it is to deform the lattice itself? The answer is that the atoms at the edge dislocation require a smaller displacement within the distorted lattice structure in order to reach a new equilibrium position. Thus, a lower energy level is needed to realign the atoms into the new positions than if the lattice were missing the dislocation. A lower stress level is therefore required to effect the deformation. Since the new position manifests a similar distorted lattice, movement of atoms at the dislocation continues at the lower stress level.

The slip phenomenon and the influence of dislocations have been explained here on a very microscopic basis. On a larger scale, slip occurs many times over throughout the metal when subjected to a deforming load, thus causing it to exhibit the macroscopic behavior with which we are familiar. Dislocations represent a good-news-bad-news situation. Because of dislocations, the metal is more ductile and yields more readily to plastic deformation (forming) during manufacturing. However, from a design viewpoint, the metal is not nearly as strong as it would be in the absence of dislocations.

Twinning is a second way in which metal crystals plastically deform. ***Twinning*** can be defined as a mechanism of plastic deformation in which atoms on one side of a plane (called the twinning plane) are shifted to form a mirror image of the other side of the plane. It is illustrated in Figure 2.13. The mechanism is important in HCP metals (e.g., magnesium, zinc) because they do not slip readily. Besides structure, another factor in twinning is the rate of deformation. The slip mechanism requires more time than twinning, which can occur almost instantaneously. Thus, in situations where the deformation

FIGURE 2.13   Twinning involves the formation of an atomic mirror image (i.e., a "twin") on the opposite side of the twinning plane: (a) before, and (b) after twinning.

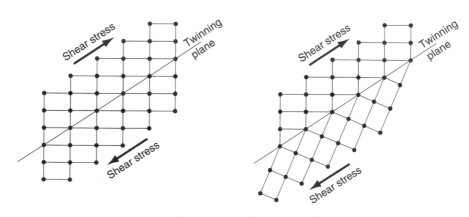

rate is very high, metals twin that would otherwise slip. Low carbon steel is an example that illustrates this rate sensitivity; when subjected to high strain rates it twins, while at moderate rates it deforms by slip.

2.3.4 Grains and Grain Boundaries in Metals

A given block of metal may contain millions of individual crystals, called **grains.** Each grain has its own unique lattice orientation; but collectively, the grains are randomly oriented within the block. We refer to such a structure as **polycrystalline.** It is easy to understand how such a structure is the natural state of the material. When the block is cooled from the molten state and begins to solidify, nucleation of individual crystals occurs at random positions and orientations within the liquid. As these crystals grow they finally interfere with each other, forming at their interface a surface defect—a **grain boundary.** The grain boundary consists of a transition zone, perhaps only a few atoms thick, in which the atoms are not aligned with either grain.

The grain size in the metal block is determined by the number of nucleation sites in the molten material, and by the cooling rate of the mass, among other factors. In a casting process, the nucleation sites are often created by the relatively cold walls of the mold, which motivates a somewhat preferred grain orientation at these walls.

Grain size is inversely related to cooling rate: faster cooling promotes smaller grain size, whereas slow cooling has the opposite effect. It is important in metals because it affects mechanical properties. Smaller grain size is generally preferable from a design viewpoint because it means higher strength and hardness. It is also desirable in certain manufacturing operations (e.g., metal forming) because it means higher ductility during deformation and a better surface on the finished product.

Another factor influencing mechanical properties is the presence of grain boundaries in the metal. They represent imperfections in the crystalline structure of the metal, which interrupt the continued movement of dislocations. This helps to explain why smaller grain size—therefore more grains and more grain boundaries—increases the strength of the metal. By interfering with dislocation movement, grain boundaries also contribute to the characteristic property of a metal to become stronger as it is deformed. The property we are referring to is **strain hardening,** and we examine it more closely in our discussion of mechanical properties in Chapter 3.

2.4 NONCRYSTALLINE (AMORPHOUS) STRUCTURES

Many important materials are noncrystalline: liquids and gases, for example. Water and air have noncrystal structures. A metal loses its crystalline structure when it is melted. Mercury is a liquid metal at room temperature, with its melting point of −38°C (−37°F). Important classes of engineering materials have a noncrystalline form in their solid state; the term **amorphous** is often used to describe these materials. Glass, many plastics, and rubber are materials that fall into this category. Many important plastics are mixtures of crystalline and noncrystalline forms. Even metals can be amorphous rather than crystalline, given that the cooling rate during transformation from liquid to solid is fast enough to inhibit the atoms from arranging themselves into their preferred regular patterns. This can happen, for instance, if the molten metal is poured between cold, closely spaced rotating rolls.

Two closely related features differentiate noncrystalline from crystalline materials: (1) absence of a long-range order in the molecular structure of a noncrystalline material, and (2) differences in melting and thermal expansion characteristics.

The difference in molecular structure can be visualized with reference to Figure 2.14. The closely packed and repeating pattern of the crystal structure is shown on the left, and the less dense and random arrangement of atoms in the noncrystalline material on the right. The difference is demonstrated by a metal when it melts. One of the effects is that the more loosely packed atoms in the molten metal show an increase in volume (reduction in density) compared to the material's solid crystalline state. This effect is characteristic of most materials when melted (a notable exception is ice; liquid water is denser than solid ice). It is a general characteristic of liquids and solid amorphous materials that they are absent of long-range order as on the right in our figure.

Let us examine the melting phenomenon in more detail, and in doing so, define the second important difference between crystalline and noncrystalline structures. As already indicated, a metal experiences an increase in volume when it melts from the solid to the liquid state. For a pure metal, this volumetric change occurs rather abruptly, at a constant temperature (i.e., the melting temperature T_m), as indicated in Figure 2.15. The change represents a discontinuity from the slopes on either side in the plot. The gradual slopes characterize the metal's ***thermal expansion***—the change in volume as a function of temperature, which is usually different in the solid and liquid states. Associated with the sudden volume increase as the metal transforms from solid to liquid at the melting point is the addition of a certain quantity of heat, called the ***heat of fusion,*** which causes the atoms to lose the dense, regular arrangement of the crystalline structure. The process is reversible; it operates in both directions. If the molten metal is cooled through its melting temperature, the same abrupt change in volume occurs (except that it is a decrease), and the same quantity of heat is given off by the metal.

An amorphous material exhibits quite a different behavior than that of a pure metal when it changes from solid to liquid, or vice versa, shown in Figure 2.15. The process is again reversible, but let us observe the behavior of the amorphous material during cooling from the liquid state, rather than during melting from the solid, as before. We will use glass (silica, SiO_2) to illustrate. At high temperatures, glass is a true liquid, and the molecules are free to move about as in the usual definition of a liquid. As the glass cools, it transforms into the solid state, going through a transition phase, called a ***supercooled liquid,*** before finally becoming rigid. It does not show the sudden volumetric change that is characteristic of crystalline materials; instead it passes through its melting temperature T_m without a change in its thermal expansion slope. In this supercooled liquid region, the material becomes increasingly viscous as the temperature continues to decrease. As it cools further, a point is reached at which the supercooled liquid converts to a solid. This is the ***glass-transition temperature*** T_g. At this point, there is a change in the thermal expansion slope (it might be more precise to refer to it as the thermal contraction slope; however, the slope is the same for expansion and contraction). The rate of thermal expansion is lower for the solid material than for the supercooled liquid.

The difference in behavior between crystalline and noncrystalline materials can be traced to the response of their respective atomic structures to changes in temperature. When a pure metal solidifies from the molten state, the atoms arrange themselves into

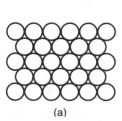
(a)

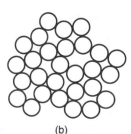
(b)

FIGURE 2.14   Illustration of difference in structure between (a) crystalline and (b) noncrystalline materials. The crystal structure is regular, repeating, and denser, while the noncrystalline structure is more loosely packed and random.

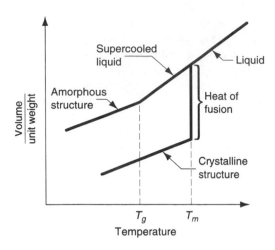

FIGURE 2.15 Characteristic change in volume for a pure metal (a crystalline structure), compared to the same volumetric changes in glass (a noncrystalline structure).

a regular and recurring structure. This crystal structure is much more compact than the random and loosely packed liquid from which it formed. Thus, the process of solidification produces the abrupt volumetric contraction observed in Figure 2.15 for the crystalline material. By contrast, amorphous materials do not achieve this repeating and closely packed structure at low temperatures. The atomic structure is the same random arrangement as in the liquid state; thus, there is no abrupt volumetric change as these materials transition from liquid to solid.

2.5 ENGINEERING MATERIALS

Let us summarize how atomic structure, bonding, and crystal structure (or absence thereof) are related to the type of engineering material—metals, ceramics, and polymers.

Metals Metals have crystalline structures in the solid state, almost without exception. The unit cells of these crystal structures are almost always BCC, FCC, or HCP. The atoms of the metals are held together by metallic bonding, which means that their valence electrons can move about with relative freedom (compared with the other types of atomic and molecular bonding). These structures and bonding generally make the metals strong and hard. Many of the metals are quite ductile (capable of being deformed, which is useful in manufacturing), especially the FCC metals. Other general properties of metals related to structure and bonding include: high electrical and thermal conductivity, opaqueness (impervious to light rays), and reflectivity (capacity to reflect light rays).

Ceramics Ceramic molecules are characterized by ionic or covalent bonding, or both. The metallic atoms release or share their outermost electrons to the nonmetallic atoms, and a strong attractive force exists within the molecules. The general properties that result from these bonding mechanisms include high hardness and stiffness (even at elevated temperatures), brittleness (no ductility), electrically insulating (nonconducting), refractory (thermally resistant), and chemically inert.

Ceramics possess either a crystalline or noncrystalline structure. Most ceramics have a crystal structure, while glasses based on silica (SiO_2) are amorphous. In certain cases, either structure can exist in the same ceramic material. For example, silica occurs in nature as crystalline quartz. When this mineral is melted and then cooled, it solidifies to form fused silica, which has a noncrystalline structure.

Polymers   A polymer molecule consists of many repeating *mers* to form very large molecules held together by covalent bonding. Elements in polymers are usually carbon plus one or more other elements such as hydrogen, nitrogen, oxygen, and chlorine. Secondary bonding (van der Waals) holds the molecules together within the aggregate material (intermolecular bonding). Polymers have either a glassy structure or mixture of glassy and crystalline. There are differences among the three polymer types. In ***thermoplastic polymers,*** the molecules consist of long chains of mers in a linear structure. These materials can be heated and cooled without substantially altering their linear structure. In ***thermosetting polymers,*** the molecules transform into a rigid three-dimensional structure on cooling from a heated plastic condition. If thermosetting polymers are reheated they degrade chemically rather than soften. ***Elastomers*** have large molecules with coiled structures. The uncoiling and recoiling of the molecules when subjected to stress cycles motivates the aggregate material to exhibit its characteristic elastic behavior.

The molecular structure and bonding of polymers provide them with the following typical properties: low density, high electrical resistivity (some polymers are used as insulating materials), and low thermal conductivity. Strength and stiffness of polymers vary widely. Some are strong and rigid (although not matching the strength and stiffness of metals or ceramics), while others exhibit highly elastic behavior.

REFERENCES

[1]   Dieter, G. E. *Mechanical Metallurgy.* 3rd ed. McGraw-Hill, New York, 1986.

[2]   Flinn, R. A., and Trojan, P. K. *Engineering Materials and Their Applications.* Houghton Mifflin, Boston, 1990.

[3]   Guy, A. G., and Hren, J. J. *Elements of Physical Metallurgy.* 3rd ed. Addison-Wesley, Reading, Mass., 1974.

[4]   Van Vlack, L. H. *Elements of Materials Science and Engineering.* 6th ed. Addison-Wesley, Reading, Mass., 1989.

REVIEW QUESTIONS

2.1. The elements listed in the Periodic Table can be divided into three categories. What are these categories? Give an example of each.

2.2. Which elements are the noble metals?

2.3. What is the difference between primary and secondary bonding in the structure of materials?

2.4. Describe how ionic bonding works.

2.5. What is the difference between crystalline and noncrystalline structures in materials?

2.6. What are some common point defects in a crystal lattice structure?

2.7. Define the difference between elastic and plastic deformation in terms of the effect on the crystal lattice structure.

2.8. How do grain boundaries contribute to the strain-hardening phenomenon in metals?

2.9. Identify some materials that have a crystalline structure.

2.10. Identify some materials that possess a noncrystalline (amorphous) structure.

2.11. What is the basic difference in the solidification (or melting) process between crystalline and noncrystalline (amorphous) structures?

MULTIPLE CHOICE QUIZ

There is a total of 20 correct answers in the following multiple choice questions (some questions have multiple answers that are correct). To attain a perfect score on the quiz, all correct answers must be given, since each correct answer is worth 1 point. For each question, each omitted answer or wrong answer reduces the score by 1 point, and each additional answer beyond the number of answers required reduces the score by 1 point. Percentage score on the quiz is based on the total number of correct answers.

2.1. The basic structural unit of matter is which one of the following? (a) atom, (b) electron, (c) element, (d) molecule, or (e) nucleus.

2.2. Approximately how many different elements have been identified (one answer)? (a) 10, (b) 50, (c) 100, (d) 200, or (e) 500.

2.3. In the Periodic Table, the elements can be divided into which of the following categories (more than one)? (a) ceramics, (b) gases, (c) liquids, (d) metals, (e) non-metals, (f) polymers, (g) semi-metals, and (h) solids.

2.4. The element with the lowest density and smallest atomic weight is which one of the following? (a) aluminum, (b) argon, (c) helium, (d) hydrogen, or (e) magnesium.

2.5. Which of the following bond types are classified as primary bonds (more than one)? (a) covalent bonding, (b) hydrogen bonding, (c) ionic bonding, (d) metallic bonding, and (e) van der Waals forces.

2.6. How many atoms are there in the face-centered cubic (FCC) unit cell (one answer)? (a) 8, (b) 9, (c) 10, (d) 12, or (e) 14.

2.7. Which of the following are point defects in a crystal lattice structure (more than one)? (a) edge dislocation, (b) interstitialcy, (c) Schottky defect, or (d) vacancy.

2.8. Which one of the following crystal structures has the fewest slip directions, and therefore the metals with this structure are generally more difficult to deform at room temperature? (a) BCC, (b) FCC, or (c) HCP.

2.9. Grain boundaries are an example of which one of the following types of crystal structure defects? (a) dislocation, (b) Frenkel defect, (c) line defects, (d) point defects, or (e) surface defects.

2.10. Twinning is which of the following (more than one)? (a) elastic deformation, (b) mechanism of plastic deformation, (c) more likely at high deformation rates, (d) more likely in metals with HCP structure, (e) slip mechanism, and (f) type of dislocation.

2.11. Polymers are characterized by which of the following bonding types (more than one answer)? (a) adhesive, (b) covalent, (c) hydrogen, (d) ionic, (e) metallic, and (f) van der Waals.

3 MECHANICAL PROPERTIES OF MATERIALS

CHAPTER CONTENTS

3.1   Stress–Strain Relationships
        3.1.1   Tensile Properties
        3.1.2   Compression Properties
        3.1.3   Bending and Testing of Brittle Materials
        3.1.4   Shear Properties
3.2   Hardness
        3.2.1   Hardness Tests
        3.2.2   Hardness of Various Materials
3.3   Effect of Temperature on Properties
3.4   Fluid Properties
3.5   Viscoelastic Behavior of Polymers

Mechanical properties of a material determine its behavior when subjected to mechanical stresses. These properties include elastic modulus, ductility, hardness, and various measures of strength. Mechanical properties are important in design because the function and performance of a product depend on its capacity to resist deformation under the stresses encountered in service. In design, the usual objective is for the product and its components to withstand these stresses without significant change in geometry. This capability depends on properties such as elastic modulus and yield strength. In manufacturing, the objective is just the opposite. Here, we want to apply stresses that exceed the yield strength of the material in order to alter its shape. Mechanical processes such as forming and machining succeed by developing forces that exceed the material's resistance to deformation. Thus, we have the following dilemma: Mechanical properties that are desirable to the designer, such as high strength, usually make the manufacture of the product more difficult. It is helpful for the manufacturing engineer to appreciate the design objective and for the designer to be aware of the manufacturing objective.

In this chapter, we examine the mechanical properties of materials. Limitations of scope and space force us to consider only those properties that are most relevant in manufacturing.

3.1 STRESS–STRAIN RELATIONSHIPS

There are three types of static stresses to which materials can be subjected: tensile, compressive, and shear. Tensile stresses tend to stretch the material, compressive stresses tend to squeeze it, and shear involves stresses that tend to cause adjacent portions of the material to slide against each other. The stress-strain curve is the basic relationship that describes the mechanical properties of materials for all three types.

3.1.1 Tensile Properties

The tensile test is the most common procedure for studying the stress–strain relationship, particularly for metals. In the test, a force is applied that pulls the material, tending to elongate it and reduce its diameter, as shown in Figure 3.1(a). Standards by ASTM (American Society for Testing and Materials) specify the preparation of the test specimen and the conduct of the test itself. The typical specimen and general setup of the tensile test is illustrated in Figure 3.1(b) and (c), respectively.

The starting test specimen has an original length L_o and area A_o. The length is measured as the distance between the gage marks, and the area is measured as the (usually round) cross-section of the specimen. During the testing of a metal, the specimen stretches, then necks, and finally fractures, as shown in Figure 3.2. The load and the change in length of the specimen are recorded as testing proceeds, to provide the data required to determine the stress–strain relationship. There are two different types of stress–strain curves: (1) engineering stress–strain and (2) true stress–strain. The first is more important in design, and the second is more important in manufacturing.

Engineering Stress–Strain   The engineering stress and strain in a tensile test are defined relative to the original area and length of the test specimen. These values are of interest in design because the designer expects that the strains experienced by any component of the product will not significantly change its shape. The components are designed to withstand the anticipated stresses encountered in service.

A typical engineering stress–strain curve from a tensile test of a metallic specimen is illustrated in Figure 3.3. The *engineering stress* at any point on the curve is defined as the force divided by the original area:

$$\sigma_e = \frac{F}{A_o} \tag{3.1}$$

FIGURE 3.1   Tensile test: (a) tensile force applied in (1), and (2) resulting elongation of material; (b) typical test specimen; and (c) setup of the tensile test.

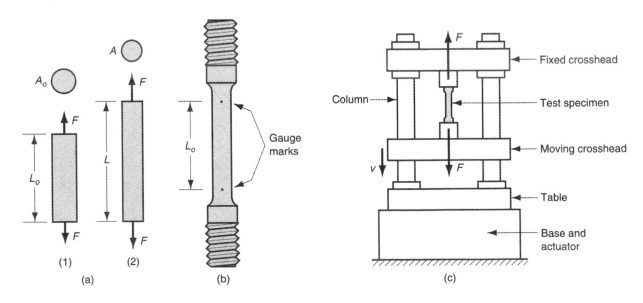

FIGURE 3.2   Typical progress of a tensile test: (1) beginning of test, no load; (2) uniform elongation and reduction of cross-sectional area; (3) continued elongation, maximum load reached; (4) necking begins, load begins to decrease; and (5) fracture. If pieces are put back together as in (6), final length can be measured.

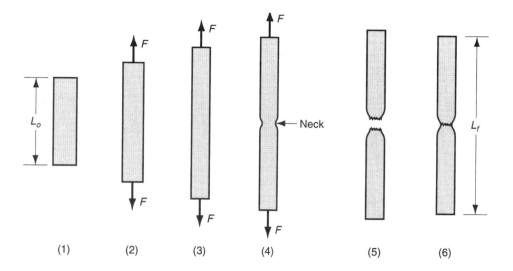

where σ_e = engineering stress, MPa (lb/in.^{2}), F = applied force in the test, N (lb), and A_o = original area of the test specimen, mm^2 (in.^{2}). The ***engineering strain*** at any point in the test is given by

$$e = \frac{L - L_o}{L_o} \tag{3.2}$$

where e = engineering strain, mm/mm (in./in.); L = length at any point during the elongation, mm (in.); and L_o = original gauge length, mm (in.). The units of engineering strain are given as mm/mm (in./in.), but we can think of it as representing elongation per unit length, without units.

The stress–strain relationship in Figure 3.3 has two regions, indicating two distinct forms of behavior: elastic and plastic. In the elastic region, the relationship between stress and strain is linear, and the material exhibits elastic behavior by returning to its original length when the load (stress) is released. The relationship is defined by ***Hooke's Law***:

$$\sigma_e = E e \tag{3.3}$$

FIGURE 3.3   Typical engineering stress–strain plot in a tensile test of a metal.

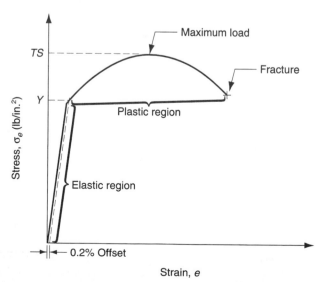

TABLE 3.1 Elastic modulus for selected materials.

Metals	Modulus of Elasticity		Ceramics and Polymers	Modulus of Elasticity	
	MPa	(lb/in.^{2})		MPa	(lb/in.^{2})
Aluminum and alloys	69×10^3	(10×10^6)	Alumina	345×10^3	(50×10^6)
Cast iron	138×10^3	(20×10^6)	Diamond[a]	1035×10^3	(150×10^6)
Copper and alloys	110×10^3	(16×10^6)	Plate glass	69×10^3	(10×10^6)
Iron	209×10^3	(30×10^6)	Silicon carbide	448×10^3	(65×10^6)
Lead	21×10^3	(3×10^6)	Tungsten carbide	552×10^3	(80×10^6)
Magnesium	48×10^3	(7×10^6)	Nylon	3.0×10^3	(0.40×10^6)
Nickel	209×10^3	(30×10^6)	Phenol formaldehyde	7.0×10^3	(1.00×10^6)
Steel	209×10^3	(30×10^6)	Polyethylene (low density)	0.2×10^3	(0.03×10^6)
Titanium	117×10^3	(17×10^6)	Polyethylene (high density)	0.7×10^3	(0.10×10^6)
Tungsten	407×10^3	(59×10^6)	Polystyrene	3.0×10^3	(0.40×10^6)

Compiled from [8], [10], [11], [14], [15], and other sources.

[a]Although diamond is not a ceramic, it is often compared with the ceramic materials.

where E = **modulus of elasticity,** MPa (lb/in.^{2}). E is a measure of the inherent stiffness of a material. It is a constant of proportionality whose value is different for different materials. Table 3.1 presents typical values for several materials, metals and nonmetals.

As stress increases, some point in the linear relationship is finally reached at which the material begins to yield. This **yield point, Y,** of the material can be identified in the figure by the change in slope at the end of the linear region. Because the start of yielding is usually difficult to see in a plot of test data (it does not usually occur as an abrupt change in slope), Y is typically defined as the stress at which a strain offset of 0.2% from the straight line has occurred. The yield point is a strength characteristic of the material, and is therefore often referred to as the **yield strength** (other names include **yield stress** and **elastic limit**).

The yield point marks the transition to the plastic region and the start of plastic deformation of the material. The relationship between stress and strain is no longer guided by Hooke's Law. As the load is increased beyond the yield point, elongation of the specimen proceeds, but at a much faster rate than before, causing the slope of the curve to change dramatically, as shown in Figure 3.3. Elongation is accompanied by a uniform reduction in cross-sectional area, consistent with maintaining constant volume. Finally, the applied load F reaches a maximum value, and the engineering stress calculated at this point is called the **tensile strength** or **ultimate tensile strength** of the material. We denote it as TS where $TS = F_{max}/A_o$. TS and Y are important strength properties in design calculations (we also use them in manufacturing calculations). Some typical values of yield strength and tensile strength are listed in Table 3.2 for selected metals. Conventional tensile testing of ceramics is difficult, and an alternative test is used to measure the strength of these brittle materials (Section 3.1.3). Polymers differ in their strength properties from metals and ceramics due to viscoelasticity (Section 3.5).

To the right of the tensile strength on the stress–strain curve, the load begins to decline, and the test specimen typically begins a process of localized elongation known as **necking.** Instead of continuing to strain uniformly throughout its length, straining becomes concentrated in one small section of the specimen. The area of that section narrows down (necks) significantly until failure occurs. The stress calculated immediately before failure is known as the **fracture stress.**

The amount of strain that the material can endure before failure is also a mechanical property of interest in many manufacturing processes. The common measure of this property is **ductility,** the ability of a material to plastically strain without fracture. This measure can be taken as either elongation or area reduction. Elongation is defined as

TABLE 3.2   Yield strength and tensile strength for selected metals.

Metal	Yield Strength MPa	(lb/in.^{2})	Tensile Strength MPa	(lb/in.^{2})	Metal	Yield Strength MPa	(lb/in.^{2})	Tensile Strength MPa	(lb/in.^{2})
Aluminum, annealed	28	(4,000)	69	(10,000)	Nickel, annealed	150	(22,000)	450	(65,000)
Aluminum, CW[a]	105	(15,000)	125	(18,000)	Steel, low C[a]	175	(25,000)	300	(45,000)
Aluminum alloys[a]	175	(25,000)	350	(50,000)	Steel, high C[a]	400	(60,000)	600	(90,000)
Cast iron[a]	275	(40,000)	275	(40,000)	Steel, alloy[a]	500	(75,000)	700	(100,000)
Copper, annealed	70	(10,000)	205	(30,000)	Steel, stainless[a]	275	(40,000)	650	(95,000)
Copper alloys[a]	205	(30,000)	410	(60,000)	Titanium, pure	350	(50,000)	515	(75,000)
Magnesium alloys[a]	175	(25,000)	275	(40,000)	Titanium alloy	800	(120,000)	900	(130,000)

Compiled from [8], [10], [11], [15], and other sources.

[a] Values given are typical. For alloys, there is a wide range in strength values depending on composition and treatment (e.g., heat treatment, work hardening).

$$EL = \frac{L_f - L_o}{L_o} \tag{3.4}$$

where EL = elongation, often expressed as a percent; L_f = specimen length at fracture, mm (in.), measured as the distance between gage marks after the two parts of the specimen have been put back together; and L_o = original specimen length, mm (in.). Area reduction is defined as

$$AR = \frac{A_o - A_f}{A_o} \tag{3.5}$$

where AR = area reduction, often expressed as a percent; A_f = area of the cross-section at the point of fracture, mm^2 (in.^{2}); and A_o = original area, mm^2 (in.^{2}). There are problems with both of these ductility measures because of necking that occurs in metallic test specimens and the associated nonuniform effect on elongation and area reduction. Despite these difficulties, percent elongation and percent area reduction are the most commonly used measures of ductility in engineering practice. Some typical values of percent elongation for various materials (mostly metals) are listed in Table 3.3.

TABLE 3.3   Ductility as percent elongation (typical values) for various selected materials

Material	% elongation	Material	% elongation
Metals		***Metals,*** *continued*	
Aluminum, annealed	40	Steel, low C[a]	30
Aluminum, cold worked	8	Steel, high C[a]	10
Aluminum alloys, annealed[a]	20	Steel, alloy[a]	20
Aluminum alloys, heat treated[a]	8	Steel, stainless, austenitic[a]	55
Aluminum alloys, cast[a]	4	Titanium, nearly pure	20
Cast iron, gray[a]	0.6	Zinc alloy	10
Copper, annealed	45	***Ceramics***	0[b]
Copper, cold worked	10	***Polymers***	
Copper alloy: brass, annealed	60	Thermoplastic polymers	100
Magnesium alloys[a]	10	Thermosetting polymers	1
Nickel, annealed	45	Elastomers (e.g., rubber)	1[c]

Compiled from [8], [10], [11], [15], and other sources.

[a] Values given are typical. For alloys, there is a range of ductility that depends on composition and treatment (e.g., heat treatment, degree of work hardening).

[b] Ceramic materials are brittle; they withstand elastic strain but virtually no plastic strain.

[c] Elastomers endure significant elastic strain, but their plastic strain is very limited, only around 1% being typical.

True Stress–Strain   Thoughtful readers may be troubled by the use of the original area of the test specimen to calculate engineering stress, rather than the actual (instantaneous) area that becomes increasingly smaller as the test proceeds. If the actual area were used, the calculated stress value would be higher. The stress value obtained by dividing the instantaneous value of area into the applied load is defined as the ***true stress***:

$$\sigma = \frac{F}{A} \tag{3.6}$$

where σ = true stress, MPa (lb/in.^{2}); F = force, N (lb); and A = actual (instantaneous) area resisting the load, mm^2 (in.^{2}).

Similarly, ***true strain*** provides a more realistic assessment of the "instantaneous" elongation per unit length of the material. The value of true strain in a tensile test can be estimated by dividing the total elongation into small increments, calculating the engineering strain for each increment on the basis of its starting length, and then adding up the strain values. In the limit, true strain is defined as

$$\epsilon = \int_{L_o}^{L} \frac{dL}{L} = \ln \frac{L}{L_o} \tag{3.7}$$

where L = instantaneous length at any moment during elongation. At the end of the test (or other deformation) the final strain value can be calculated using $L = L_f$.

If the engineering stress–strain curve in Figure 3.3 were plotted using the true stress and strain values, the resulting curve would appear as in Figure 3.4. In the elastic region, the plot is virtually the same as before. Strain values are small, and true strain is nearly equal to engineering strain for most metals of interest. The respective stress values are also very close to each other. The reason for these near equalities is that the cross-sectional area of the test specimen is not significantly reduced in the elastic region. Thus, Hooke's Law can be used to relate true stress to true strain: $\sigma = E\epsilon$.

The difference between the true stress-strain curve and its engineering counterpart lies in the plastic region. The stress values are higher in the plastic region because the instantaneous cross-sectional area of the specimen, which has been continuously reduced during elongation, is now used in the computation. As in the previous curve, a downturn finally occurs as a result of necking. A dashed line is used in the figure to indicate the projected continuation of the true stress-strain plot if necking had not occurred.

FIGURE 3.4   True stress–strain curve for the previous engineering stress–strain plot in Figure 3.3.

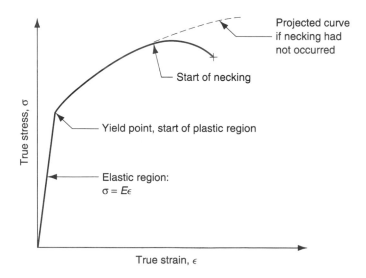

Projected curve if necking had not occurred

Start of necking

True stress, σ

Yield point, start of plastic region

Elastic region:
$\sigma = E\epsilon$

True strain, ϵ

As strain becomes significant in the plastic region, the values of true strain and engineering strain diverge. True strain can be related to the corresponding engineering strain by

$$\epsilon = \ln(1 + e) \tag{3.8}$$

Similarly, true stress and engineering stress can be related by the expression

$$\sigma = \sigma_e(1 + e) \tag{3.9}$$

In Figure 3.4, we should note that stress increases continuously in the plastic region until necking begins. When this happened in the engineering stress-strain curve, its significance was lost because an admittedly erroneous area value was used to calculate stress. Now when the true stress also increases, we cannot dismiss it so lightly. What it means is that the metal is becoming stronger as strain increases. This is the property called ***strain hardening*** that we had mentioned in the previous chapter in our discussion of metallic crystal structures, and it is a property that most metals exhibit to a greater or lesser degree.

Strain hardening, or ***work hardening*** as it is often called, is an important factor in certain manufacturing processes, particularly metal forming. Let us examine the behavior of a metal as it is affected by this property. If the portion of the true stress–strain curve representing the plastic region were plotted on a log-log scale, the result would be a linear relationship, as shown in Figure 3.5. Because it is a straight line in this transformation of the data, the relationship between true stress and true strain in the plastic region can be expressed as

$$\sigma = K \epsilon^n \tag{3.10}$$

This equation is called the ***flow curve,*** and it captures a good approximation of the behavior of metals in the plastic region, including their capacity for strain hardening. The constant K is called the ***strength coefficient,*** MPa (lb/in.^{2}); and it equals the value of true stress at a true strain value equal to one. The parameter n is called the ***strain hardening exponent,*** and it is the slope of the line in Figure 3.5. Its value is directly related to a metal's tendency to work harden. Typical values of K and n for selected metals are given in Table 3.4.

Necking in a tensile test and in metal forming operations that stretch the workpart is closely related to strain hardening. Let us examine this relationship as it can be observed during a tensile test. As the test specimen is elongated during the initial part of the test (before necking begins), uniform straining occurs throughout the length because if any element in the specimen becomes strained more than the surrounding metal, its strength increases due to work hardening, thus making it more resistant to additional strain until the surrounding metal has been strained an equal amount. Finally, the strain

FIGURE 3.5 True stress-strain curve plotted on log-log scale.

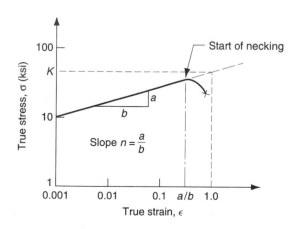

becomes so large that uniform straining cannot be sustained. A weak point in the length develops (due to build-up of dislocations at grain boundaries, impurities in the metal, or other factors), and necking is initiated, leading to failure. Empirical evidence reveals that necking begins for a particular metal when the true strain reaches a value equal to the strain hardening exponent n. Therefore, a higher n value means that the metal can be strained further before the onset of necking during tensile loading.

Types of Stress–Strain Relationships   Much information about elastic–plastic behavior is provided by the true stress–strain curve. As we have indicated, Hooke's Law ($\sigma = E\epsilon$) governs the metal's behavior in the elastic region, and the flow curve ($\sigma = K\epsilon^n$) determines the behavior in the plastic region. Three basic forms of stress–strain relationship describe the behavior of nearly all types of solid materials, shown in Figure 3.6:

(a) **Perfectly elastic.**   The behavior of this material is defined completely by its stiffness, indicated by the modulus of elasticity E. It fractures rather than yielding to plastic flow. Brittle materials such as ceramics, many cast irons, and thermosetting polymers possess stress–strain curves that fall into this category. These materials are not good candidates for forming operations.

(b) **Elastic and perfectly plastic.**   This material has a stiffness defined by E. Once the yield strength Y is reached, the material deforms plastically at the same stress level. The flow curve is given by $K = Y$ and $n = 0$. Metals behave in this fashion when they have been heated to sufficiently high temperatures that they recrystallize rather than strain harden during deformation. Lead exhibits this behavior at room temperature because room temperature is above the recrystallization point for lead.

(c) **Elastic and strain hardening.**   This material obeys Hooke's Law in the elastic region. It begins to flow at its yield strength Y. Continued deformation requires an ever-increasing stress, given by a flow curve whose strength coefficient K is greater than Y and whose strain hardening exponent n is greater than zero. The flow curve is generally represented as a linear function on a natural logarithmic plot. Most ductile metals behave this way when cold worked.

Manufacturing processes that deform materials through the application of tensile stresses include wire and bar drawing (Section 19.6) and stretch forming (Section 20.6.1).

TABLE 3.4   Typical values of strength coefficient K and strain hardening exponent n for selected metals.

Material	Strength Coefficient, K		Strain Hardening Exponent, n
	MPa	(lb/in.^{2})	
Aluminum, pure, annealed	175	(25,000)	0.20
Aluminum alloy, annealed[a]	240	(35,000)	0.15
Aluminum alloy, heat treated	400	(60,000)	0.10
Copper, pure, annealed	300	(45,000)	0.50
Copper alloy: brass[a]	700	(100,000)	0.35
Steel, low C, annealed[a]	500	(75,000)	0.25
Steel, high C, annealed[a]	850	(125,000)	0.15
Steel, alloy, annealed[a]	700	(100,000)	0.15
Steel, stainless, austenitic, annealed	1200	(175,000)	0.40

Compiled from [9], [10], [11], and other sources.

[a] Values of K and n will vary according to composition, heat treatment, and work hardening.

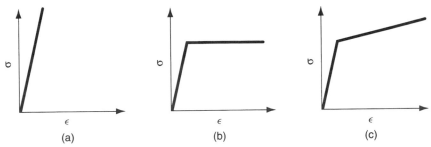

FIGURE 3.6 Three categories of stress–strain relationship: (a) perfectly elastic, (b) elastic and perfectly plastic, (c) elastic and strain hardening.

3.1.2 Compression Properties

A compression test applies a load that squeezes a cylindrical specimen between two platens, as illustrated in Figure 3.7. As the specimen is compressed, its height is reduced and its cross-sectional area is increased. Engineering stress is defined as

$$\sigma_e = \frac{F}{A_o} \tag{3.11}$$

where A_o = original area of the specimen. This is the same definition of engineering stress used in the tensile test. The engineering strain is defined as

$$e = \frac{h - h_o}{h_o} \tag{3.12}$$

where h = height of the specimen at a particular moment into the test, mm (in.); and h_o = starting height, mm (in.). Since the height is decreased during compression, the value of e will be negative. The negative sign is usually ignored when expressing values of compression strain.

When engineering stress is plotted against engineering strain in a compression test, the results appear as in Figure 3.8. The curve is divided into elastic and plastic regions, as before, but the shape of the plastic portion of the curve is different from its tensile test

FIGURE 3.7 Compression test: (a) compression force applied to test piece in (1), and (2) resulting change in height; and (b) setup for the test, with size of test specimen exaggerated.

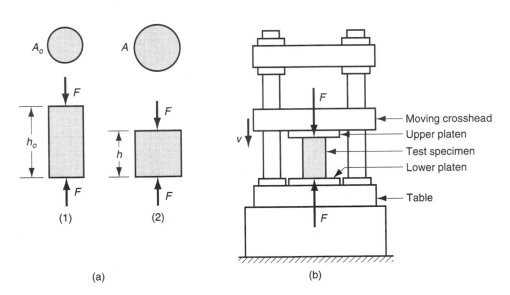

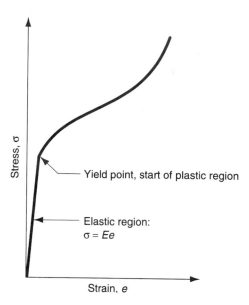

FIGURE 3.8   Typical engineering stress–strain curve for a compression test.

complement. Since compression causes the cross section to increase (rather than decrease as in the tensile test), the load increases more rapidly than previously. This results in a higher value of calculated engineering stress.

Something else happens in the compression test that contributes to the increase in stress. As the cylindrical specimen is squeezed, friction at the surfaces in contact with the platens tends to prevent the ends of the cylinder from spreading. Additional energy is consumed by this friction during the test, and this results in a higher applied force. It also shows up as an increase in the computed engineering stress. Hence, owing to the increase in cross-sectional area and friction between the specimen and the platens, we obtain the characteristic engineering stress–strain curve in a compression test as seen in our figure.

Another consequence of the friction between the surfaces is that the material near the middle of the specimen is permitted to increase in area much more than at the ends. This results in the characteristic **barreling** of the specimen, as seen in Figure 3.9.

Although differences exist between the engineering stress–strain curves in tension and compression, when the respective data are plotted as true stress–strain, the relationships are nearly identical (for almost all materials). Since tensile test results are more abundant in the literature, we can derive values of the flow curve parameters (K and n)

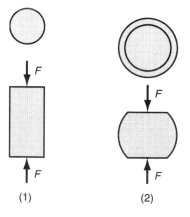

FIGURE 3.9   Barreling effect in a compression test: (1) start of test; and (2) after considerable compression has occurred.

from tensile test data and apply them with equal validity to a compression operation. What must be done in using the tensile test results for a compression operation is to ignore the effect of necking, a phenomenon that is peculiar to straining induced by tensile stresses. In compression, there is no corresponding collapse of the work. (One might argue that buckling of long, thin sections in compression could be considered the counterpart to necking. However, buckling is a failure mode that involves bending of the specimen, so the stress is no longer limited solely to compression. We consider bending stresses in the next section.) In our previous plots of tensile stress–strain curves, we have extended the data beyond the point of necking by means of the dashed lines. The dashed lines better represent the behavior of the material in compression than the actual tensile test data.

Compression operations in metal forming are much more common than stretching operations. Important compression processes in industry include rolling, forging, and extrusion (Chapter 19).

3.1.3 Bending and Testing of Brittle Materials

Bending operations are used to form metal plates and sheets. As shown in Figure 3.10, the process of bending a rectangular cross-section subjects the material to tensile stresses (and strains) in the outer half of the bent section and compressive stresses (and strains) in the inner half. If the material does not fracture, it becomes permanently (plastically) bent as shown in (3) of Figure 3.10.

Hard brittle materials (e.g., ceramics), which possess elasticity but little or no plasticity, are often tested by a method that subjects the specimen to a bending load. These materials do not respond well to traditional tensile testing because of problems in preparing the test specimens and possible misalignment of the press jaws that hold the specimen. The **bending test** (also known as the **flexure test**) can test the strength of these materials, using a setup illustrated in the first diagram in Figure 3.10. In this procedure, a specimen of rectangular cross section is positioned between two supports, and a load is applied at its center. In this configuration, the test is called a three-point bending test. A four-point configuration is also sometimes used. These brittle materials do not flex to the exaggerated extent shown in Figure 3.10; instead they deform elastically until immediately before fracture. Failure usually occurs because the ultimate tensile strength of the outer fibers of the specimen has been exceeded. This results in **cleavage,** a failure mode associated with ceramics and metals operating at low service temperatures, in which separation rather than slip occurs along certain crystallographic planes. The strength value

FIGURE 3.10   Bending of a rectangular cross section results in both tensile and compressive stresses in the material: (1) initial loading; (2) highly stressed and strained specimen; and (3) bent part.

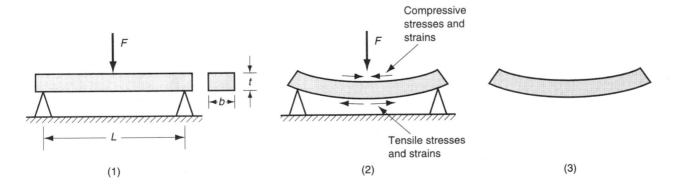

derived from this test is called the **transverse rupture strength,** calculated from this formula:

$$TRS = \frac{1.5\,FL}{bt^2} \qquad (3.13)$$

where TRS = transverse rupture strength, MPa (lb/in.²); F = applied load at fracture, N (lb); L = length of the specimen between supports, mm (in.); and b and t are the dimensions of the cross section of the specimen as shown in the figure, mm (in.).

The flexure test is also utilized for certain nonbrittle materials such as thermoplastic polymers. In this case, since the material is likely to deform rather than fracture, TRS cannot be determined based on failure of the specimen. Instead, either of two measures are used: (1) the load recorded at a given level of deflection, or (2) the deflection observed at a given load.

3.1.4 Shear Properties

Shear involves application of stresses in opposite directions on either side of a thin element to deflect it, as shown in Figure 3.11. The shear stress is defined as

$$\tau = \frac{F}{A} \qquad (3.14)$$

where τ = shear stress, MPa (lb/in.²); F = applied force, N (lb); and A = area over which the force is applied, mm² (in.²). Shear strain can be defined as

$$\gamma = \frac{\delta}{b} \qquad (3.15)$$

where γ = shear strain, mm/mm (in./in.); δ = the deflection of the element, mm (in.); and b = the orthogonal distance over which deflection occurs, mm (in.).

Shear stress and strain are commonly tested in a **torsion test,** in which a thin-walled tubular specimen is subjected to a torque as shown in Figure 3.12. As torque is increased, the tube deflects by twisting, which is a shear strain for this geometry.

The shear stress can be determined in the test by the equation

$$\tau = \frac{T}{2\pi R^2 t} \qquad (3.16)$$

where T = applied torque, N-mm (lb-in.); R = radius of the tube measured to the neutral axis of the wall, mm (in.); and t = wall thickness, mm (in.). The shear strain can be determined by measuring the amount of angular deflection of the tube, converting this into a distance deflected, and dividing by the gauge length L. Reducing this to a simple expression,

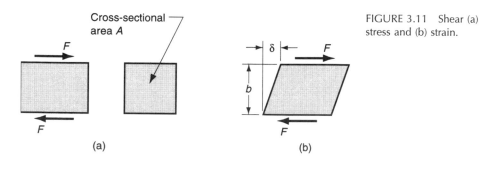

FIGURE 3.11   Shear (a) stress and (b) strain.

(a)   (b)

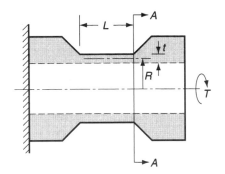
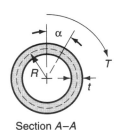

FIGURE 3.12   Torsion test setup.

Section A–A

$$\gamma = \frac{R\alpha}{L} \qquad (3.17)$$

where α = the angular deflection (radians).

A typical shear stress–strain curve is shown in Figure 3.13. In the elastic region, the relationship is defined by

$$\tau = G\gamma \qquad (3.18)$$

where G = the **shear modulus,** or **shear modulus of elasticity,** MPa (lb/in.^{2}). For most materials, the shear modulus can be approximated by $G = 0.4E$, where E is the conventional elastic modulus.

In the plastic region of the shear stress–strain curve, the material strain hardens to cause the applied torque to continue to increase until fracture finally occurs. The relationship in this region is similar to the flow curve. The shear stress at fracture can be calculated, and this is used as the **shear strength** S of the material. Shear strength can be estimated from tensile strength data by the approximation $S = 0.7(TS)$.

Since the cross-sectional area of the test specimen in the torsion test does not change as it does in both the tensile and compression tests, the engineering stress-strain curve for shear derived from the torsion test is virtually the same as the true stress–strain curve.

Shear processes are common in industry. Shearing action is used to cut sheet metal in blanking, punching, and other cutting operations (Section 20.1). In machining, the material is removed by the mechanism of shear deformation (Section 21.2).

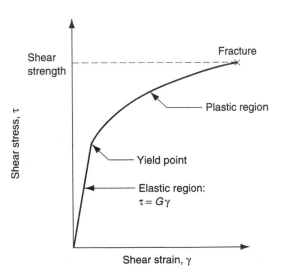

FIGURE 3.13   Typical shear stress–strain curve from a torsion test.

3.2 HARDNESS

The *hardness* of a material is defined as its resistance to permanent indentation. Good hardness generally means that the material is resistant to scratching and wear. For many engineering applications, including most of the tooling used in manufacturing, scratch and wear resistance are important characteristics. As we shall see later in this section, there is a strong correlation between hardness and strength.

3.2.1 Hardness Tests

Hardness tests are commonly used for assessing material properties because they are quick and convenient. However, a variety of testing methods are appropriate due to differences in hardness among different materials. The most well-known hardness tests are Brinell and Rockwell.

Brinell Hardness Test   The Brinell hardness test is widely used for testing metals and nonmetals of low to medium hardness. It is named after the Swedish engineer who developed it around 1900. In the test, a hardened steel (or cemented carbide) ball of 10 mm diameter is pressed into the surface of a specimen using a load of 500, 1,500, or 3,000 kg. The load is then divided into the indentation area to obtain the Brinell Hardness Number (BHN). In equation form,

$$HB = \frac{2F}{\pi D_b\left(D_b - \sqrt{D_b^2 - D_i^2}\right)} \tag{3.19}$$

where HB = Brinell Hardness Number (BHN), F = indentation load (kg), D_b = diameter of the ball, mm, and D_i = diameter of the indentation on the surface, mm. These dimensions are indicated in Figure 3.14(a). The resulting BHN has units of kg/mm², but the units

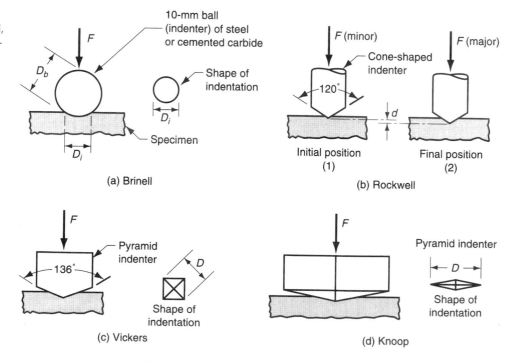

FIGURE 3.14   Hardness testing methods: (a) Brinell, (b) Rockwell: (1) initial minor load and (2) major load, (c) Vickers, and (d) Knoop.

are usually omitted in expressing the number. For harder materials (above 500 BHN), the cemented carbide ball is used, since the steel ball experiences elastic deformation that compromises the accuracy of the reading. Also, higher loads (1,500 and 3,000 kg) are typically used for harder materials. Because of differences in results under different loads, it is considered good practice to indicate the load used in the test when reporting *HB* readings.

Rockwell Hardness Test   This is another widely used test, named after the metallurgist who developed it in the early 1920s. It is convenient to use, and several enhancements over the years have made the test adaptable to a variety of materials.

In the Rockwell Hardness Test, a cone-shaped indenter or small diameter ball, with diameter = 1.6 or 3.2 mm (1/16 or 1/8 in.) is pressed into the specimen using a minor load of 10 kg, thus seating the indenter in the material. Then, a major load of 150 kg (or other value) is applied, causing the indenter to penetrate into the specimen a certain distance beyond its initial position. This additional penetration distance *d* is converted into a Rockwell hardness reading by the testing machine. The sequence is depicted in Figure 3.14(b). Differences in load and indenter geometry provide various Rockwell scales for different materials, shown in Table 3.5.

Vickers Hardness Test   This test, also developed in the early 1920s, uses a pyramid-shaped indenter made of diamond. It is based on the principle that impressions made by this indenter are geometrically similar regardless of load. Accordingly, loads of various size are applied, depending on the hardness of the material to be measured. The Vickers Hardness (*HV*) is then determined from the formula

$$HV = \frac{1.854 \, F}{D^2} \tag{3.20}$$

where *F* = applied load, kg; and *D* = the diagonal of the impression made by the indenter, mm, as indicated in Figure 3.14(c). The Vickers test can be used for all metals and has one of the widest scales among hardness tests.

Knoop Hardness Test   The Knoop test, developed in 1939, uses a pyramid-shaped diamond indenter, but the pyramid has a length-to-width ratio of about 7 : 1, as indicated in Figure 3.14(d), and the applied loads are generally lighter than in the Vickers test. It is a microhardness test, meaning that it is suitable for measuring small, thin specimens or hard materials that might fracture if a heavier load were applied. The indenter shape facilitates reading of the impression under the lighter loads used in this test. The Knoop hardness value (*HK*) is determined according to the formula:

$$HK = 14.2 \, \frac{F}{D^2} \tag{3.21}$$

where *F* = load, kg; and *D* = the long diagonal of the indentor, mm. Because the impression made in this test is generally very small, considerable care must be taken in preparing the surface to be measured.

TABLE 3.5   Common Rockwell Hardness Scales.

Rockwell Scale	Hardness Symbol	Indenter	Load (kg)	Typical Materials Tested
A	HRA	Cone	60	Carbides, ceramics
B	HRB	1.6 mm ball	100	Nonferrous metals
C	HRC	Cone	150	Ferrous metals, tool steels

Scleroscope   The previous tests base their hardness measurements on either the ratio of applied load divided by the resulting impression area (Brinell, Vickers, and Knoop) or by the depth of the impression (Rockwell). The Scleroscope is an instrument that measures the rebound height of a "hammer" dropped from a certain distance above the surface of the material to be tested. The hammer consists of a weight with diamond indenter attached to it. The Scleroscope therefore measures the mechanical energy absorbed by the material when the indenter strikes the surface. The energy absorbed gives an indication of resistance to penetration, which matches our definition of hardness. If more energy is absorbed, the rebound will be less, meaning a softer material. If less energy is absorbed, the rebound will be higher—thus a harder material. The primary use of the Scleroscope seems to be in measuring the hardness of large parts of steel and other ferrous metals.

Durometer   The previous tests are all based on resistance to permanent or plastic deformation (indentation). The durometer is a device that measures the elastic deformation of rubber and similar flexible materials by pressing an indenter into the surface of the object. The resistance to penetration is an indication of hardness, as the term is applied to these types of materials.

3.2.2 Hardness of Various Materials

In this section, we compare the hardness values of some common materials in the three engineering material classes: metals, ceramics, and polymers.

Metals   The Brinell and Rockwell hardness tests were developed at a time when metals were the principal engineering materials. A significant amount of data has been collected using these tests on metals. Table 3.6 lists hardness values for selected metals.

    For most metals, hardness is closely related to strength. Since the method of testing for hardness is usually based on resistance to indentation, which is a form of compression, one would expect a good correlation between hardness and strength properties determined in a compression test. But strength properties in a compression test are nearly the same as those from a tension test, after allowances for changes in cross-sectional area of the respective test specimens; so the correlation with tensile properties should also be good.

TABLE 3.6   Typical hardness of selected metals.

Metal	Brinell Hardness, HB	Rockwell Hardness, HR[a]	Metal	Brinell Hardness, HB	Rockwell Hardness, HR[a]
Aluminum, annealed	20		Magnesium alloys, hardened[b]	70	35B
Aluminum, cold worked	35		Nickel, annealed	75	40B
Aluminum alloys, annealed[b]	40		Steel, low C, hot rolled[b]	100	60B
Aluminum alloys, hardened[b]	90	52B	Steel, high C, hot rolled[b]	200	95B, 15C
Aluminum alloys, cast[b]	80	44B	Steel, alloy, annealed[b]	175	90B, 10C
Cast iron, gray, as cast[b]	175	10C	Steel, alloy, heat treated[b]	300	33C
Copper, annealed	45		Steel, stainless, austenitic[b]	150	85B
Copper alloy: brass, annealed	100	60B	Titanium, nearly pure	200	95B
Lead	4		Zinc	30	

Compiled from [10], [11], [15], and other sources.

[a]*HR* values are given in the B or C scale as indicated by the letter designation. Missing values indicate that the hardness is too low for Rockwell scales.

[b]*HB* values given are typical. Hardness values will vary according to composition, heat treatment, and degree of work hardening.

TABLE 3.7   Hardness of selected ceramics and other hard materials, arranged in ascending order of hardness.

Material	Vickers Hardness, HV	Knoop Hardness, HK	Material	Vickers Hardness, HV	Knoop Hardness, HK
Hardened tool steel[a]	800	850	Titanium nitride, TiN	3000	2300
Cemented carbide (WC-Co)[a]	2000	1400	Titanium carbide, TiC	3200	2500
Alumina, Al_2O_3	2200	1500	Cubic boron nitride, BN	6000	4000
Tungsten carbide, WC	2600	1900	Diamond, sintered polycrystal	7000	5000
Silicon carbide, SiC	2600	1900	Diamond, natural	10,000	8000

Compiled from [13], [15], and other sources.

[a] Hardened tool steel and cemented carbide are the two materials commonly used in the Brinell Hardness Test.

TABLE 3.8   Hardness of selected polymers.

Polymer	Brinell Hardness, HB	Polymer	Brinell Hardness, HB
Nylon	12	Polypropylene	7
Phenol formaldehyde	50	Polystyrene	20
Polyethylene, low density	2	Polyvinyl-chloride	10
Polyethylene, high density	4		

Compiled from [4], [8], and other sources.

Brinell hardness HB exhibits a close correlation with the ultimate tensile strength TS of steels, leading to the relationship [9], [14]:

$$TS = K_h(HB) \tag{3.22}$$

where K_h is a constant of proportionality. If TS is expressed in MPa, then $K_h = 3.45$; and if TS is in lb/in.^{2}, then $K_h = 500$.

Ceramics   The Brinell Hardness Test is not appropriate for ceramics because the materials being tested are often harder than the indenter ball. The Vickers and Knoop Hardness Tests are used to test these hard materials. Table 3.7 lists hardness values for several ceramics and hard materials. For comparison, the Rockwell C hardness for hardened tool steel is 65 HRC. The HRC scale does not extend high enough to be used for the harder materials.

Polymers   Polymers have the lowest hardness among the three types of engineering materials. Table 3.8 lists several of the polymers on the Brinell hardness scale, although this testing method is not normally used for these materials. It does, however, allow comparison with the hardness of metals.

3.3 EFFECT OF TEMPERATURE ON PROPERTIES

Temperature has a significant effect on nearly all properties of a material. It is important for the designer to know the material properties at the operating temperatures of the product when in service. It is also important to know how temperature affects mechanical properties in manufacturing. At elevated temperatures, materials are lower in strength

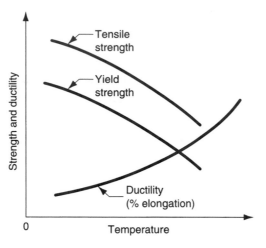

FIGURE 3.15   General effect of temperature on strength and ductility.

and higher in ductility. The general relationships for metals are depicted in Figure 3.15. Thus, most metals can be formed more easily at elevated temperatures than when they are cold.

Hot Hardness   A property often used to characterize strength and hardness at elevated temperatures is hot hardness. ***Hot hardness*** is simply the ability of a material to retain hardness at elevated temperatures; it is usually presented as either a listing of hardness values at different temperatures, or as a plot of hardness versus temperature, as in Figure 3.16. Steels can be alloyed to achieve significant improvements in hot hardness, as shown in the figure. Ceramics exhibit superior properties at elevated temperatures. These materials are often selected for high temperature applications, such as turbine parts, cutting tools, and refractory applications. The outside skins of the shuttle spacecrafts are lined with ceramic tiles to withstand the friction heat of high-speed re-entry into the atmosphere.

Good hot hardness is also desirable in the tooling materials used in many manufacturing operations. Significant amounts of heat energy are generated in most metalworking processes, and the tools must be capable of withstanding the high temperatures involved.

FIGURE 3.16   Hot hardness—typical hardness as a function of temperature for several materials.

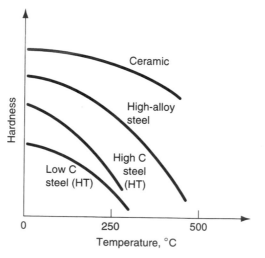

Recrystallization Temperature   Most metals behave at room temperature according to the flow curve in the plastic region. As the metal is strained, it increases in strength due to strain hardening (the strain hardening exponent $n > 0$). But if the metal is heated to a sufficiently elevated temperature and then deformed, strain hardening does not occur. Instead, new grains are formed that are free of strain, and the metal behaves as a perfectly plastic material; that is, with a strain hardening exponent $n = 0$. The formation of new strain-free grains is a process called recrystallization, and the temperature at which it happens is about one-half the melting point ($0.5\,T_m$) as measured on the absolute scale (R or K) — called the **recrystallization temperature.** Recrystallization takes time. The recrystallization temperature for a particular metal is usually specified as the temperature at which complete formation of new grains requires about one hour.

Recrystallization is a temperature-dependent characteristic of metals that we can exploit in manufacturing. By heating the metal to the recrystallization temperature prior to deformation, the amount of straining that the metal can endure is substantially increased, and the forces and power required to carry out the process are significantly reduced. Forming metals at temperatures above the recrystallization temperature is called **hot working** (Section 18.3).

3.4 FLUID PROPERTIES

Fluids behave differently than solids. Fluids flow; they take the shape of the container that holds them. Solids do not flow; they possess a geometric form that is independent of their surroundings. Fluids include liquids and gases; our interest in this section is on the former. Many manufacturing processes are accomplished on materials that have been converted from solid to liquid state by heating. Metals are cast in the molten state; glass is formed in a heated and highly fluid state; and polymers are almost always shaped as thick fluids.

Viscosity   Although flow is a defining characteristic of fluids, the tendency to flow varies for different fluids. Viscosity is the property that determines fluid flow. Roughly, **viscosity** can be defined as the resistance to flow that is characteristic of a fluid. It is a measure of the internal friction that arises when velocity gradients are present in the fluid—the more viscous the fluid is, the higher the internal friction and the greater the resistance to flow. The reciprocal of viscosity is **fluidity**—the ease with which a fluid flows.

Viscosity can be defined more precisely with respect to the setup in Figure 3.17, in which two parallel plates are separated by a distance d. One of the plates is stationary

FIGURE 3.17   Fluid flow between two parallel plates, one stationary and the other moving at velocity v.

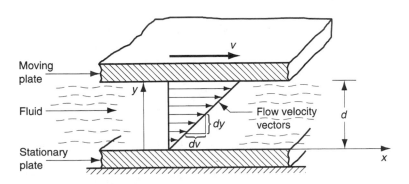

while the other is moving at a velocity v, and the space between the plates is occupied by a fluid. Orienting these parameters relative to an axis system, d is in the y-axis direction and v is in the x-axis direction. The motion of the upper plate is resisted by force F that results from the shear viscous action of the fluid. This force can be reduced to a shear stress by dividing F by the plate area A:

$$\tau = \frac{F}{A} \tag{3.23}$$

where τ = shear stress, N/m^2 or Pa (lb/in.^{2}). This shear stress is related to the rate of shear, which is defined as the change in velocity dv relative to dy. That is,

$$\dot{\gamma} = \frac{dv}{dy} \tag{3.24}$$

where $\dot{\gamma}$ = shear rate, 1/s; dv = incremental change in velocity, m/s (in./sec); and dy = incremental change in distance y, m (in.). The shear viscosity is the fluid property that defines the relationship between F/A and dv/dy; that is,

$$\frac{F}{A} = \eta \frac{dv}{dy} \text{ or } \tau = \eta \dot{\gamma} \tag{3.25}$$

where η = a constant of proportionality called the coefficient of viscosity, Pa-s (lb-sec/in.^{2}). Rearranging Eq. (3.25), we can express the coefficient of viscosity as follows:

$$\eta = \frac{\tau}{\dot{\gamma}} \tag{3.26}$$

Thus, the viscosity of a fluid can be defined as the ratio of shear stress to shear rate during flow; where shear stress is the frictional force exerted by the fluid per unit area, and shear rate is the velocity gradient perpendicular to the flow direction. The viscous characteristics of fluids defined by Eq. (3.26) were first stated by Newton. He observed that viscosity was a constant property of a given fluid, and such a fluid is referred to as a **_Newtonian fluid._**

The units of coefficient of viscosity require explanation. In the International System of units (SI), since shear stress is expressed in N/m^2 or Pascals and shear rate in $1/s$, it follows that η has units of N-s/m^2 or Pascal-seconds, abbreviated Pa-s. In the U.S. customary units, the corresponding units are lb/in.^{2} and 1/sec, so that the units for coefficient of viscosity are lb-sec/in.^{2} Other units sometimes given for viscosity are poise, which equals dyne-sec/cm^2 (10 poise = 1 Pas and 6895 Pas = 1 lb-sec/in.^{2}). Some typical values of coefficient of viscosity for various fluids are given in Table 3.9. One can observe in several of the materials listed that viscosity varies with temperature.

Viscosity in Manufacturing Processes   For many metals, the viscosity in the molten state compares to that of water at room temperature. Certain manufacturing processes, notably casting and welding, are performed on metals in their molten state, and success in these operations requires low viscosity so that the molten metal fills the mold cavity or weld seam before solidifying. In other operations, such as metal forming and machining, lubricants and coolants are used in the process, and again the success of these fluids depends to some extent on their viscosities.

Glass ceramics exhibit a gradual transition from solid to liquid states as temperature is increased; they do not suddenly melt as pure metals do. The effect is illustrated by the viscosity values for glass at different temperatures in Table 3.9. At room temperature, glass is solid and brittle, exhibiting no tendency to flow; for all practical purposes,

TABLE 3.9 Viscosity values for selected fluids.

Material	Coefficient of Viscosity		Material	Coefficient of Viscosity	
	Pa-s	(lb-sec/in.^{2})		Pa-s	(lb-sec/in.^{2})
Glass[b], 540 C (1000 F)	10^{12}	(10^8)	Pancake syrup (room temp)	50	(73×10^{-4})
Glass[b], 815 C (1500 F)	10^5	(14)	Polymer[a], 151 C (300 F)	115	(167×10^{-4})
Glass[b], 1095 C (2000 F)	10^3	(0.14)	Polymer[a], 205 C (400 F)	55	(80×10^{-4})
Glass[b], 1370 C (2500 F)	15	(22×10^{-4})	Polymer[a], 260 C (500 F)	28	(41×10^{-4})
Mercury, 20 C (70 F)	0.0016	(0.23×10^{-6})	Water, 20 C (70 F)	0.001	(0.15×10^{-6})
Machine oil (room temp)	0.1	(0.14×10^{-4})	Water, 100 C (212 F)	0.0003	(0.04×10^{-6})

Compiled from various sources.

[a]Low-density polyethylene is used as the polymer example here; most other polymers have slightly higher viscosities.

[b]Glass composition is mostly SiO_2; compositions and viscosities vary; values given are representative.

its viscosity is infinite. As glass is heated, it gradually softens, becoming less and less viscous (more and more fluid), until it can finally be formed by blowing or molding at around 1100°C (2000°F).

Most polymer-shaping processes are performed at elevated temperatures, where the material is in a liquid or highly plastic condition. Thermoplastic polymers represent the most straightforward case, and they are also the most common polymers. At low temperatures, thermoplastic polymers are solid; as temperature is increased, they typically transform first into a soft rubbery material, and then into a thick fluid. As temperature continues to rise, viscosity decreases gradually, as in Table 3.9 for polyethylene, the most widely used thermoplastic polymer. However, with polymers the relationship is complicated by other factors. For example, viscosity is affected by flow rate. The viscosity of a thermoplastic polymer is not a constant. A polymer melt does not behave in a Newtonian fashion. Its relationship between shear stress and shear rate can be seen in Figure 3.18. A fluid that exhibits this decreasing viscosity with increasing shear rate is called ***pseudoplastic.*** This behavior complicates the analysis of polymer shaping.

FIGURE 3.18 Viscous behaviors of Newtonian and pseudoplastic fluids. Polymer melts exhibit pseudoplastic behavior. For comparison, the behavior of a plastic solid material is shown.

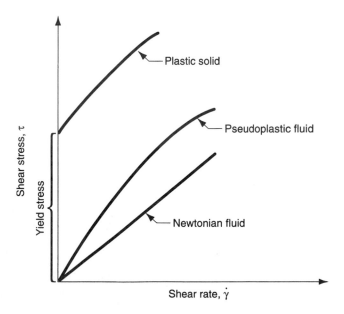

3.5 VISCOELASTIC BEHAVIOR OF POLYMERS

Another property that is characteristic of polymers is viscoelasticity. ***Viscoelasticity*** is the property of a material that determines the strain it experiences when subjected to combinations of stress and temperature over time. As the name suggests, it is a combination of viscosity and elasticity. Let us explain viscoelasticity with reference to Figure 3.19. The two parts of the figure show the typical response of two materials to an applied stress below the yield point during some time period. The material in (a) exhibits perfect elasticity; when the stress is removed, the material returns to its original shape. By contrast, the material in (b) shows viscoelastic behavior. The amount of strain gradually increases over time under the applied stress. When stress is removed, the material does not immediately return to its original shape; instead, the strain decays gradually. If the stress had been applied and then immediately removed, the material would have returned immediately to its starting shape. However, time has entered the picture and played a role in affecting the behavior of the material.

A simple model of viscoelasticity can be developed using the definition of elasticity as a starting point. Elasticity is concisely expressed by Hooke's Law: $\sigma = E\epsilon$, which simply relates stress to strain through a constant of proportionality. In a viscoelastic solid, the relationship between stress and strain is time-dependent; it can be expressed:

$$\sigma(t) = f(t)\epsilon \tag{3.27}$$

The time function $f(t)$ can be conceptualized as a modulus of elasticity that depends on time. We might write it $E(t)$ and refer to it as a viscoelastic modulus. The form of this time function can be complex, sometimes including strain as a factor. Without getting into the mathematical expressions for it, we can nevertheless explore the effect of the time dependency. One common effect can be seen in Figure 3.20, which shows the stress-strain behavior of a thermoplastic polymer under different strain rates. At low strain rate, the material exhibits significant viscous flow. At high strain rate, it behaves in a much more brittle fashion.

Temperature is a factor in viscoelasticity. As temperature increases, the viscous behavior becomes more and more prominent relative to elastic behavior. The material be-

FIGURE 3.19   Comparison of elastic and viscoelastic properties: (a) perfectly elastic response of material to stress applied over time; and (b) response of a viscoelastic material under the same conditions. The material in (b) takes a strain that is a function of time and temperature.

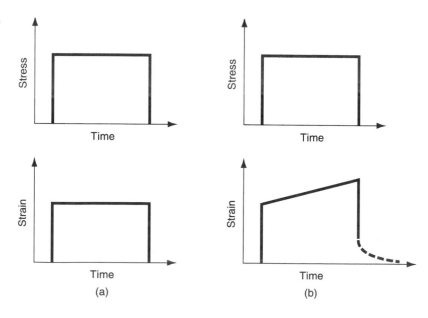

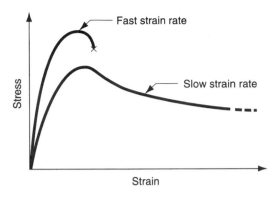

FIGURE 3.20   Stress–strain curve of a viscoelastic material (thermoplastic polymer) at high and low strain rates.

comes more like a fluid. Figure 3.21 illustrates this temperature dependence for a thermoplastic polymer. At low temperatures, the polymer shows elastic behavior. As T increases above the glass transition temperature T_g, the polymer becomes viscoelastic. As temperature increases further, it becomes soft and rubbery. And at still higher temperatures, it exhibits viscous characteristics. The temperatures at which these modes of behavior are observed vary, depending on the plastic. Also, the shapes of the modulus versus temperature curve differ according to the proportions of crystalline and amorphous structures in the thermoplastic. Thermosetting polymers and elastomers behave differently than shown in our figure; after curing, these polymers do not soften as thermoplastics do at elevated temperatures. Instead, they degrade (char) at high temperatures.

Viscoelastic behavior manifests itself in polymer melts in the form of shape memory. As the thick polymer melt is transformed during processing from one shape to another, it "remembers" its previous shape and attempts to return to that geometry. For example, a common problem in extrusion of polymers is die swell, in which the profile of the extruded material grows in size, reflecting its tendency to return to its larger cross section in the extruder barrel immediately before being squeezed through the smaller die opening. We examine the properties of viscosity and viscoelasticity in more detail in our discussion of plastic shaping (Chapter 13).

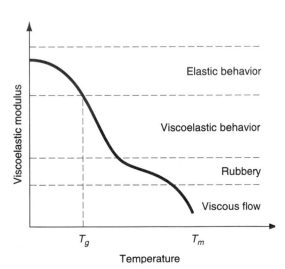

FIGURE 3.21   Viscoelastic modulus as a function of temperature for a thermoplastic polymer.

REFERENCES

[1] Avallone, E. A., and Baumeister III, T. (eds.). *Mark's Standard Handbook for Mechanical Engineers.* 10th ed. McGraw-Hill, New York, 1996.

[2] Beer, F. P., and Russell, J. E. *Vector Mechanics for Engineers.* 6th ed. McGraw-Hill, New York, 1999.

[3] Budynas, R. G. *Advanced Strength and Applied Stress Analysis.* 2nd ed. McGraw-Hill, New York, 1998.

[4] Chandra, M., and Roy, S. K. *Plastics Technology Handbook.* 3rd ed. Marcel Dekker, Inc., New York, 1998.

[5] Dieter, G. E. *Mechanical Metallurgy.* 3rd ed. McGraw-Hill, New York, 1986.

[6] DeGarmo, E. P., Black, J. T., and Kohser, R.A., *Materials and Processes in Manufacturing.* 8th ed. Wiley, New York, 1997.

[7] *Engineering Plastics.* Engineered Materials Handbook, Vol. 2, ASM International, Metals Park, Ohio, 1987.

[8] Flinn, R. A., and Trojan, P. K. *Engineering Materials and Their Applications.* 4th ed., Houghton Mifflin, Boston, 1990.

[9] Kalpakjian, S., and Schmid, S. R. *Manufacturing Processes for Engineering Materials.* 4th ed. Prentice Hall, Upper Saddle River, N.J., 2001.

[10] *Metals Handbook.* 10th ed. Volume 1, Properties and Selection: Iron, Steels, and High Performance Alloys, ASM International, Metals Park, Oh., 1990.

[11] *Metals Handbook.* 10th ed. Volume 2, Properties and Selection: Nonferrous Alloys and Special Purpose Materials, ASM International, Metals Park, Oh., 1991.

[12] Morton-Jones, D. H. *Polymer Processing.* Chapman and Hall, London, 1989.

[13] Schey, J. A. *Introduction to Manufacturing Processes.* 3rd ed. McGraw-Hill, New York, 2000.

[14] Van Vlack, L. H. *Elements of Materials Science and Engineering.* 6th ed. Addison-Wesley, Reading, Mass., 1991.

[15] Wick, C., and Veilleux, R. F. (eds). *Tool and Manufacturing Engineers Handbook.* 4th ed. Volume 3—Materials, Finishing, and Coating, Society of Manufacturing Engineers, Dearborn, Mich., 1985.

REVIEW QUESTIONS

3.1. What is the dilemma between design and manufacturing in terms of mechanical properties?

3.2. What are the three types of static stresses to which materials are subjected?

3.3. State Hooke's Law.

3.4. What is the difference between engineering stress and true stress in a tensile test?

3.5. Define *tensile strength* of a material.

3.6. Define *yield strength* of a material.

3.7. Why cannot a direct conversion be made between the ductility measures of elongation and reduction in area using the assumption of constant volume?

3.8. What is *work hardening*?

3.9. In what case does the strength coefficient in the flow curve equation have the same value as the yield strength?

3.10. How does the change in cross-sectional area of a test specimen in a compression test differ from its counterpart in a tensile test specimen?

3.11. What is the complicating factor that occurs in a compression test?

3.12. Tensile testing is not appropriate for hard brittle materials such as ceramics. What is the test commonly used to determine the strength properties of such materials?

3.13. How is the shear modulus of elasticity G related to the tensile modulus of elasticity E, on average?

3.14. How is shear strength S related to tensile strength TS, on average?

3.15. What is *hardness* and how is it generally tested?

3.16. Why are different hardness tests and scales required?

3.17. Define the *recrystallization temperature* for a metal.

3.18. Define *viscosity* of a fluid.

3.19. What is the defining characteristic of a *Newtonian fluid?*

3.20. What is *viscoelasticity,* as a material property?

MULTIPLE CHOICE QUIZ

There is a total of 18 correct answers in the following multiple choice questions (some questions have multiple answers that are correct). To attain a perfect score on the quiz, all correct answers must be given, since each correct answer is worth 1 point. For each question, each omitted answer or wrong answer reduces the score by 1 point, and each additional answer beyond the number of answers required reduces the score by 1 point. Percentage score on the quiz is based on the total number of correct answers.

3.1. Which of the following are the three basic types of static stresses to which a material can be subjected (three answers)? (a) compression, (b) hardness, (c) reduction in area, (d) shear, (e) tensile, (f) true stress, and (g) yield.

3.2. Which one of the following is the correct definition of ultimate tensile strength, as derived from the results of a tensile test on a metal specimen? (a) the stress encountered when the stress–strain curve transforms from elastic to plastic behavior, (b) the maximum load divided by the final area of the specimen, (c) the maximum load divided by the original area of the specimen, or (d) the stress observed when the specimen finally fails.

3.3. If stress values were measured during a tensile test, which of the following would have the higher value? (a) engineering stress, or (b) true stress.

3.4. If strain measurements were made during a tensile test, which of the following would have the higher value? (a) engineering strain, or (b) true strain.

3.5. The plastic region of the stress–strain curve for a metal is characterized by a proportional relationship between stress and strain: (a) true or (b) false.

3.6. Which one of the following types of stress–strain relationship best describes the behavior of brittle materials such as ceramics and thermosetting plastics: (a) elastic and perfectly plastic, (b) elastic and strain hardening, or (c) perfectly elastic.

3.7. Which one of the following types of stress–strain relationship best describes the behavior of most metals at room temperature: (a) elastic and perfectly plastic, (b) elastic and strain hardening, or (c) perfectly elastic.

3.8. Which one of the following types of stress–strain relationship best describes the behavior of metals at temperatures above their respective recrystallization points: (a) elastic and perfectly plastic, (b) elastic and strain hardening, or (c) perfectly elastic.

3.9. Which one of the following materials has the highest modulus of elasticity? (a) aluminum, (b) diamond, (c) steel, (d) titanium, or (e) tungsten.

3.10. The shear strength of a metal is usually (a) greater than, or (b) less than its tensile strength.

3.11. Most hardness tests involve pressing a hard object into the surface of a test specimen and measuring the indentation (or its effect) that results: (a) true or (b) false.

3.12. Which one of the following materials has the highest hardness? (a) alumina ceramic, (b) gray cast iron, (c) hardened tool steel, (d) high carbon steel, or (e) polystyrene.

3.13. Viscosity can be defined as the ease with which a fluid flows: (a) true or (b) false.

3.14. Viscoelasticity has features of which of the following more traditional material properties (more than one)? (a) elasticity, (b) plasticity, (c) viscosity.

PROBLEMS

Strength and Ductility in Tension

3.1. A tensile test uses a test specimen that has a gauge length of 50 mm and an area = 200 mm^2. During the test the specimen yields under a load of 98,000 N. The corresponding gage length = 50.23 mm. This is the 0.2 percent yield point. The maximum load = 168,000 N is reached at a gauge length = 64.2 mm. Determine: (a) yield strength Y, (b) modulus of elasticity E, and (c) tensile strength TS.

3.2. A test specimen in a tensile test has a gauge length of 2.0 in. and an area = 0.5 in.^{2} During the test the specimen yields under a load of 32,000 lb. The corresponding gauge length = 2.0083 in. This is the 0.2 percent yield point. The maximum load = 60,000 lb is reached at a gage length = 2.60 in. Determine: (a) yield strength Y, (b) modulus of elasticity E, and (c) tensile strength TS.

3.3. In Problem 3.1, (a) determine the percent elongation. (b) If the specimen necked to an area = 92 mm^2, determine the percent reduction in area.

3.4. In Problem 3.2, (a) determine the percent elongation. (b) If the specimen necked to an area = 0.25 in^2, determine the percent reduction in area.

3.5. The following data are collected during a tensile test in which the starting gauge length = 125.0 mm and the cross-sectional area = 62.5 mm^2:

Load (N)	0	17,793	23,042	27,579	28,913	27,578	20,462
Length (mm)	0	125.23	131.25	140.05	147.01	153.00	160.10

The maximum load is 28,913 N, and the final data point occurred immediately prior to failure. (a) Plot the engineering stress strain curve. Determine: (b) yield strength Y, (c) modulus of elasticity E, (d) tensile strength TS.

Flow Curve

3.6. In Problem 3.5, determine the strength coefficient and the strain hardening exponent. Be sure not to use data after the point at which necking occurred.

3.7. In a tensile test on a metal specimen, true strain = 0.08 at a stress = 265 MPa. When the true stress = 325 MPa, the true strain = 0.27. Determine the flow curve parameters n and K.

3.8. During a tensile test, a metal has a true strain = 0.10 at a true stress = 37,000 lb/in.^{2}. Later, at a true stress = 55,000 lb/in.^{2}, the true strain = 0.25. Determine the flow curve parameters n and K.

3.9. In a tensile test a metal begins to neck at a true strain = 0.28 with a corresponding true stress = 345.0 MPa. Without knowing any more about the test, can you estimate the flow curve parameters n and K?

3.10. A tensile test for a certain metal provides flow curve parameters: $n = 0.3$ and $K = 600$ MPa. Determine (a) the flow stress at a true strain = 1.0, and (b) true strain at a flow stress = 600 MPa.

3.11. The flow curve for a certain metal has parameters: $n = 0.22$ and $K = 54,000$ lb/in.^{2}. Determine (a) the flow stress at a true strain = 0.45, and (b) the true strain at a flow stress = 40,000 lb/in.^{2}.

3.12. A metal is deformed in a tension test into its plastic region. The starting specimen had a gauge length = 2.0 in. and an area = 0.50 in.^{2} At one point in the tensile test, the gauge length = 2.5 in. and the corresponding engineering stress = 24,000 lb/in.^{2}; and at another point in the test prior to necking, the gauge length = 3.2 in. and the corresponding engineering stress = 28,000 lb/in.^{2} Determine the strength coefficient and the strain hardening exponent for this metal.

3.13. A tensile test specimen has a starting gauge length = 75.0 mm. It is elongated during the test to a length = 110.0 mm before necking occurs. (a) Determine the engineering strain. (b) Determine the true strain.

(c) Compute and sum the engineering strains as the specimen elongates from (1) 75.0 to 80.0 mm, (2) 80.0 to 85.0 mm, (3) 85.0 to 90.0 mm, (4) 90.0 to 95.0 mm, (5) 95.0 to 100.0 mm, (6) 100.0 to 105.0 mm, and (7) 105.0 to 110.0 mm. (d) Is the result closer to the answer to part (a) or part (b)? Does this help to show what is meant by the term *true strain*?

3.14. A tensile specimen is elongated to twice its original length. Determine the engineering strain and true strain for this test. If the metal had been strained in compression, determine the final compressed length of the specimen such that (a) the engineering strain is equal to the same value as in tension (it will be negative value because of compression), and (b) the true strain would be equal to the same value as in tension (again, it will be negative value because of compression). Note that the answer to part (a) is an impossible result. True strain is therefore a better measure of strain during plastic deformation.

3.15. Derive an expression for true strain as a function of D and D_o for a tensile test specimen of round cross section.

3.16. Show that true strain = $\ln(1 + e)$.

3.17. Based on results of a tensile test, the flow curve has parameters calculated as $n = 0.40$ and $K = 551.6$ MPa. Based on this information, calculate the (engineering) tensile strength for the metal.

3.18. A copper wire of diameter 0.80 mm fails at an engineering stress = 248.2 MPa. Its ductility is measured as 75% reduction of area. Determine the true stress and true strain at failure.

3.19. A steel tensile specimen with starting gauge length = 2.0 in. and cross-sectional area = 0.5 in.^{2} reaches a maximum load of 37,000 lb. Its elongation at this point is 24%. Determine the true stress and true strain at this maximum load.

Compression

3.20. A metal alloy has been tested in a tensile test to determine the following flow curve parameters: $K = 620.5$ MPa and $n = 0.26$. The same metal is now tested in a compression test in which the starting height of the specimen = 62.5 mm and its diameter = 25 mm. Assuming that the cross section increases uniformly, determine the load required to compress the specimen to a height of (a) 50 mm and (b) 37.5 mm.

3.21. The flow curve parameters for a certain stainless steel are $K = 1100$ MPa and $n = 0.35$. A cylindrical specimen of starting cross-sectional area = 1000 mm^2 and

height = 75 mm is compressed to a height of 58 mm. Determine the force required to achieve this compression, assuming that the cross section increases uniformly.

3.22. A steel test specimen ($E = 30 \times 10^6$ lb/in.^{2}) in a compression test has a starting height = 2.0 in. and diameter = 1.5 in. The metal yields (0.2% offset) at a load = 140,000 lb. At a load of 260,000 lb, the height has been reduced to 1.6 in. Determine: (a) yield strength Y, (b) flow curve parameters K and n. Assume that the cross-sectional area increases uniformly during the test.

Bending and Shear

3.23. A bend test is used for a certain hard material. If the transverse rupture strength of the material is known to be 1000 MPa, what is the anticipated load at which the

specimen is likely to fail, given that its dimensions are: $b = 15$ mm, $h = 10$ mm, and $L = 60$ mm?

3.24. A special ceramic specimen is tested in a bend test. Its cross-sectional dimensions are $b = 0.50$ in. and $h = 0.25$ in. The length of the specimen between supports $= 2.0$ in. Determine the transverse rupture strength if failure occurs at a load $= 1700$ lb.

3.25. A piece of metal is deformed in shear to an angle of 42° as shown in Figure P3.25. Determine the shear strain for this situation.

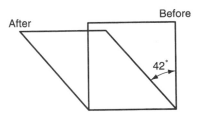

FIGURE P3.25

Hardness

3.30. In a Brinell hardness test, a 1500 kg load is pressed into a specimen using a 10 mm diameter hardened steel ball. The resulting indentation has a diameter $= 3.2$ mm. Determine the BHN for the metal.

3.31. One of the inspectors in the quality control department has frequently used the Brinell and Rockwell hardness tests, for which equipment is available in the company. He claims that all hardness tests are based on the same principle as the Brinell test, which is that hardness is always measured as the applied load divided by the area of the impressions made by an indentor. (a) Is he correct? (b) If not, what are some of the other principles

Viscosity of Fluids

3.34. Two flat plates, separated by a space of 4 mm, are moving relative to each other at a velocity of 5 m/sec. The space between them is occupied by a fluid of unknown viscosity. The motion of the plates is resisted by a shear stress of 10 Pa due to the viscosity of the fluid. Assuming that the velocity gradient of the fluid is constant, determine the coefficient of viscosity of the fluid.

3.35. Two parallel surfaces, separated by a space of 0.5 in. that is occupied by a fluid, are moving relative to each other at a velocity of 25 in./sec. The motion is resisted by a shear stress of 0.3 lb/in.[2] due to the viscosity of the

3.26. A torsion test specimen has a radius $= 25$ mm, wall thickness $= 3$ mm, and gauge length $= 50$ mm. In testing, a torque of 900 N-m results in an angular deflection $= 0.3°$. Determine (a) the shear stress, (b) shear strain, and (c) shear modulus, assuming the specimen had not yet yielded.

3.27. In a torsion test, a torque of 5000 ft-lb is applied that causes an angular deflection $= 1°$ on a thin-walled tubular specimen whose radius $= 1.5$ in., wall thickness $= 0.10$ in., and gage length $= 2.0$ in. Determine (a) the shear stress, (b) shear strain, and (c) shear modulus, assuming the specimen had not yet yielded.

3.28. In Problem 3.26, failure of the specimen occurs at a torque $= 1200$ N-m and a corresponding angular deflection $= 10°$. What is the shear strength of the metal?

3.29. In Problem 3.27, the specimen fails at a torque $= 8000$ ft-lb and an angular deflection $= 23°$. Calculate the shear strength of the metal.

involved in hardness testing, and what are the associated tests?

3.32. Suppose in Problem 3.30 that the specimen is steel. Based on the BHN determined in that problem, estimate the tensile strength of the steel.

3.33. A batch of annealed steel has just been received from the vendor. It is supposed to have a tensile strength in the range 60,000 to 70,000 lb/in.[2] A Brinell hardness test in the receiving department yields a value of BHN $= 118$. (a) Does the steel meet the specification on tensile strength? (b) Estimate the yield strength of the material.

fluid. If the velocity gradient in the space between the surfaces is constant, determine the viscosity of the fluid.

3.36. A 125.0 mm diameter shaft rotates inside a stationary bushing whose inside diameter $= 125.6$ mm and length $= 50.0$ mm. In the clearance between the shaft and the bushing is contained a lubricating oil whose viscosity $= 0.14$ Pas. The shaft rotates at a velocity of 400 rev/min; this speed and the action of the oil are sufficient to keep the shaft centered inside the bushing. Determine the magnitude of the torque due to viscosity that acts to resist the rotation of the shaft.

4 PHYSICAL PROPERTIES OF MATERIALS

CHAPTER CONTENTS

4.1   Volumetric and Melting Properties
    4.1.1   Density
    4.1.2   Thermal Expansion
    4.1.3   Melting Characteristics
4.2   Thermal Properties
    4.2.1   Specific Heat and Thermal Conductivity
    4.2.2   Thermal Properties in Manufacturing
4.3   Mass Diffusion
4.4   Electrical Properties
    4.4.1   Resistivity and Conductivity
    4.4.2   Classes of Materials by Electrical Properties
4.5   Electrochemical Processes

Physical properties, as we are using the term, define the behavior of materials in response to physical forces other than mechanical. They include volumetric, thermal, electrical, and electrochemical properties. Components in a product must do more than simply withstand mechanical stresses. They must conduct electricity (or prevent its conduction), allow heat to escape, transmit light, and satisfy a myriad of other functions.

Physical properties are important in manufacturing because they often influence the performance of the process. For example, thermal properties of the work material in machining determine the cutting temperature, which affects how long the tool can be used before it fails. In microelectronics, electrical properties of silicon and the way in which these properties can be altered by various chemical and physical processes is the basis of semiconductor manufacturing.

In this chapter, we discuss the physical properties that are most important in manufacturing—properties that we encounter in subsequent chapters of the book. We divide them into major categories such as volumetric, thermal, electrical, and so on. We also relate these properties to manufacturing, as we did in the previous chapter on mechanical properties.

4.1 VOLUMETRIC AND MELTING PROPERTIES

These properties are related to the volume of solids and how they are affected by temperature. The properties include density, thermal expansion, and melting point. They are explained below, and a listing of typical values for selected engineering materials is presented in Table 4.1.

4.1.1 Density

In engineering, the density of a material is its weight per unit volume. Its symbol is ρ, and typical units are g/cm^3 $(lb/in.^3)$. The density of an element is determined by its atomic number and other factors such as atomic radius and atomic packing. The term *specific gravity* expresses the density of a material relative to the density of water and is therefore a ratio with no units.

Density is an important consideration in the selection of a material for a given application, but it is generally not the only property of interest. Strength is also important, and the two properties are often related in a *strength-to-weight ratio,* which is the tensile strength of the material divided by its density. The ratio is useful in comparing materials for structural applications in aircraft, automobiles, and other products where weight and energy are of concern.

TABLE 4.1   Volumetric properties in U.S. customary units for selected engineering materials.

Material	Density, ρ		Coefficient of Thermal Expansion, α		Melting Point, T_m	
	g/cm^3	$(lb/in.^3)$	$°C^{-1} \times 10^{-6}$	$(°F^{-1} \times 10^{-6})$	$°C$	$°(F)$
Metals						
Aluminum	2.70	(0.098)	24	(13.3)	660	(1220)
Copper	8.97	(0.324)	17	(9.4)	1083	(1981)
Iron	7.87	(0.284)	12.1	(6.7)	1539	(2802)
Lead	11.35	(0.410)	29	(16.1)	327	(621)
Magnesium	1.74	(0.063)	26	(14.4)	650	(1202)
Nickel	8.92	(0.322)	13.3	(7.4)	1455	(2651)
Steel	7.87	(0.284)	12	(6.7)	a	a
Tin	7.31	(0.264)	23	(12.7)	232	(449)
Tungsten	19.30	(0.697)	4.0	(2.2)	3410	(6170)
Zinc	7.15	(0.258)	40	(22.2)	420	(787)
Ceramics						
Glass	2.5	(0.090)	1.8–9.0	(1.0–5.0)	b	b
Alumina	3.8	(0.137)	9.0	(5.0)	NA	NA
Silica	2.66	(0.096)	NA	NA	b	b
Polymers						
Phenol resins	1.3	(0.047)	60	(33)	c	c
Nylon	1.16	(0.042)	100	(55)	b	b
Teflon	2.2	(0.079)	100	(55)	b	b
Natural rubber	1.2	(0.043)	80	(45)	b	b
Polyethylene:						
Low density	0.92	(0.033)	180	(100)	b	b
High density	0.96	(0.035)	120	(66)	b	b
Polystyrene	1.05	(0.038)	60	(33)	b	b

Compiled from [2], [4], [5], [6], and other sources.

[a] Melting characteristics of steel depend on composition.

[b] Softens at elevated temperatures and does not have a well-defined melting point.

[c] Chemically degrades at high temperatures.

NA = Not available; value of property for this material could not be obtained.

4.1.2 Thermal Expansion

The density of a material is a function of temperature. The general relationship is that density decreases with increasing temperature. Put another way, the volume per unit weight increases with temperature. Thermal expansion is the name given to this effect that temperature has on density. It is usually expressed as the **coefficient of thermal expansion,** which measures the change in length per degree of temperature, as mm/mm/°C (in./in./°F). It is a length ratio rather than a volume ratio because this is easier to measure and apply. It is consistent with the usual design situation in which dimensional changes are of greater interest than volumetric changes. The change in length corresponding to a given temperature change is given by

$$L_2 - L_1 = \alpha L_1 \, (T_2 - T_1) \tag{4.1}$$

where α = coefficient of thermal expansion, $°C^{-1}$ ($°F^{-1}$); and L_1 and L_2 are lengths, mm (in.), corresponding, respectively, to temperatures T_1 and T_2, °C (°F).

Values of coefficient of thermal expansion given in Table 4.1 suggest that it has a linear relationship with temperature. This is only an approximation. Not only is length affected by temperature, but the thermal expansion coefficient itself is also affected. For some materials it increases with temperature; for other materials it decreases. These changes are usually not significant enough to be of much concern, and values like those in the table are quite useful in design calculations for the range of temperatures contemplated in service. Changes in the coefficient are more substantial when the metal undergoes a phase transformation, such as from solid to liquid, or from one crystal structure to another.

In manufacturing operations, thermal expansion is put to good use in shrink fit and expansion fit assemblies (Section 33.3.2), in which a part is heated to increase its size or cooled to decrease its size in order to permit insertion into some other part. When the part returns to ambient temperature, a tightly fitted assembly is obtained. Thermal expansion can be a problem in heat treatment (Chapter 27) and welding (Section 31.6.1) due to thermal stresses that develop in the material during these processes.

4.1.3 Melting Characteristics

For a pure element, the **melting point** T_m is the temperature at which the material transforms from solid to liquid state. The reverse transformation, from liquid to solid, occurs at the same temperature and is called the **freezing point.** For crystalline elements, such as metals, the melting and freezing temperatures are the same. A certain amount of heat energy, called the **heat of fusion,** is required at this temperature in order to accomplish the transformation from solid to liquid.

Melting of a metal element at a specific temperature, as we have described it, assumes equilibrium conditions. Exceptions occur in nature; for example, when a molten metal is cooled, it may remain in the liquid state below its freezing point if nucleation of crystals does not initiate immediately. When this happens, the liquid is said to be **supercooled.**

There are other variations in the melting process—differences in the way melting occurs in different materials. For example, unlike pure metals, most metal alloys do not have a single melting point. Instead, melting begins at a certain temperature, called the **solidus,** and continues as the temperature increases until finally converting completely to the liquid state at a temperature called the **liquidus.** Between the two temperatures, the alloy is a mixture of solid and molten metals, the amounts of each being inversely proportional to their relative distances from the liquidus and solidus. Exceptions include eutectic alloys, which melt (and freeze) at a single temperature. We examine these issues in our discussion of phase diagrams in Chapter 6.

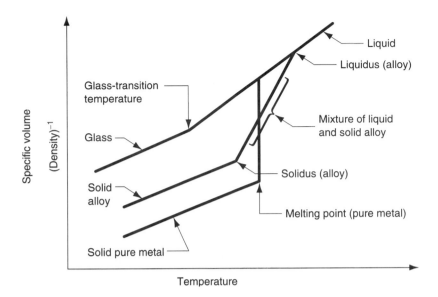

FIGURE 4.1 Changes in volume per unit weight (1/density) as a function of temperature for a hypothetical pure metal, alloy, and glass; all exhibiting similar thermal expansion and melting characteristics.

Another difference in melting occurs with noncrystalline materials (glasses). In these materials, there is a gradual transition from solid to liquid states. The solid material gradually softens as temperature increases, finally becoming liquid at the melting point. During softening, the material has a consistency of increasing plasticity (increasingly like a fluid) as it gets closer to the melting point.

These differences in melting characteristics between pure metals, alloys, and glass are portrayed in Figure 4.1. The plots show changes in density as a function of temperature for three hypothetical materials: a pure metal, alloy, and glass. Plotted in the figure is the volumetric change, which is the reciprocal of density.

The importance of melting in manufacturing is obvious. In metal casting (Chapters 10 and 11), the metal is melted and then poured into a mold cavity. Metals with lower melting points are generally easier to cast, but if the melting temperature is too low, the metal loses its applicability as an engineering material. Melting characteristics of polymers are important in plastic molding and other polymer shaping processes (Chapter 13). Sintering of powdered metals and ceramics requires knowledge of melting points. Sintering does not melt the materials, but the temperatures used in the process must approach the melting point in order to achieve the required bonding of the powders.

4.2 THERMAL PROPERTIES

Much of the previous section is concerned with the effects of temperature on volumetric properties of materials. Certainly, thermal expansion, melting, and heat of fusion are thermal properties because temperature determines the thermal energy level of the atoms, leading to the changes in the materials. In the current section we examine several additional thermal properties—ones that relate to the storage and flow of heat within a substance. The usual properties of interest are specific heat and thermal conductivity, values of which are compiled for selected materials in Table 4.2.

4.2.1 Specific Heat and Thermal Conductivity

The **specific heat** C of a material is defined as the quantity of heat energy required to increase the temperature of a unit mass of the material by one degree. Some typical values are listed in Table 4.2. To determine the amount of energy needed to heat a certain weight of a metal in a furnace to a given elevated temperature, the following equation can be used:

$$H = C W (T_2 - T_1) \tag{4.2}$$

where H = amount of heat energy, J (Btu); C = specific heat of the material, J/kg °C (Btu/lb °F); W = its weight, kg (lb); and $(T_2 - T_1)$ = change in temperature, °C (°F).

The volumetric heat storage capacity of a material is often of interest. This is simply density multiplied by specific heat ρC. Thus, **volumetric specific heat** is the heat energy required to raise the temperature of a unit volume of material by one degree, J/mm³ °C (Btu/in.³ °F).

Conduction is one of the fundamental heat-transfer processes. It involves transfer of thermal energy within a material from molecule to molecule by purely thermal motions; no transfer of mass occurs. The thermal conductivity of a substance is therefore its capability to transfer heat through itself by this physical mechanism. It is measured by the **coefficient of thermal conductivity** k, which has typical units of J/s mm °C (Btu/in. hr °F). The coefficient of thermal conductivity is generally high in metals, low in ceramics and plastics.

The ratio of thermal conductivity to volumetric specific heat is frequently encountered in heat transfer analysis. It is called the **thermal diffusivity** K and is determined as

$$K = \frac{k}{\rho C} \tag{4.3}$$

We make use of it to calculate cutting temperatures in machining (Section 21.5.1).

4.2.2 Thermal Properties in Manufacturing

Thermal properties play an important role in manufacturing because heat generation is common in so many processes. In some operations heat is the energy that accomplishes the process; in others, heat is generated as a consequence of the process.

Specific heat is of interest for several reasons. In processes that require heating of the material (e.g., casting, heat treating, hot metal forming), specific heat determines the amount of heat energy needed to raise the temperature to a desired level, according to Eq. (4.2).

TABLE 4.2 Values of common thermal properties for selected materials. Values are at room temperature, and these values change for different temperatures.

Material	Specific Heat Cal/g °C[a] or (Btu/lbm °F)	Thermal Conductivity J/s mm °C	(Btu/hr in °F)	Material	Specific Heat Cal/g °C[a] or (Btu/lbm °F)	Thermal Conductivity J/s mm °C	(Btu/hr in °F)
Metals				*Ceramics*			
Aluminum	0.21	0.22	(9.75)	Alumina	0.18	0.029	(1.4)
Cast iron	0.11	0.06	(2.7)	Concrete	0.2	0.012	(0.6)
Copper	0.092	0.40	(18.7)				
Iron	0.11	0.072	(2.98)	*Polymers*			
Lead	0.031	0.033	(1.68)	Phenolics	0.4	0.00016	(0.0077)
Magnesium	0.25	0.16	(7.58)	Polyethylene	0.5	0.00034	(0.016)
Nickel	0.105	0.070	(2.88)	Teflon	0.25	0.00020	(0.0096)
Steel	0.11	0.046	(2.20)	Natural rubber	0.48	0.00012	(0.006)
Stainless steel[b]	0.11	0.014	(0.67)				
Tin	0.054	0.062	(3.0)	*Other*			
Zinc	0.091	0.112	(5.41)	Water (liquid)	1.00	0.0006	(0.029)
				Ice	0.46	0.0023	(0.11)

Compiled from [2], [3], [6], and other sources.

[a] Specific heat has the same numerical value in Btu/lbm-F or Cal/g-C. 1.0 Calory = 4.186 Joule.

[b] Austenitic (18-8) stainless steel.

In many processes carried out at ambient temperature, the mechanical energy to perform the operation is converted into heat, which raises the temperature of the workpart. This is common in machining and cold forming of metals. The temperature rise is a function of the metal's specific heat. Coolants are often used in machining to reduce these temperatures, and here the fluid's heat capacity is critical. Water is almost always employed as the base for these fluids because of its high heat-carrying capacity.

Thermal conductivity functions to dissipate heat in manufacturing processes, sometimes beneficially, sometimes not. In mechanical processes, such as metal forming and machining, much of the power required to operate the process is converted to heat. The ability of the work material and tooling to conduct heat away from its source is highly desirable in these processes.

On the other hand, high thermal conductivity of the work metal is undesirable in fusion welding processes such as arc welding and electric resistance welding. In these operations, the heat input must be concentrated at the joint location so that the metal can be melted. To illustrate, copper is generally difficult to weld because of its high thermal conductivity; the heat is rapidly conducted into the work from the energy source.

4.3 MASS DIFFUSION

In addition to heat transfer in a material, there is also mass transfer. *Mass diffusion* involves movement of atoms or molecules within a material or across a boundary between two materials in contact. It is perhaps more appealing to one's intuition that such a phenomenon occurs in liquids and gases, but it also occurs in solids. It occurs in pure metals, in alloys, and between materials that share a common interface. Because of thermal agitation of the atoms in a material (solid, liquid, or gas), atoms are continuously moving about. In liquids and gases, where the level of thermal agitation is high, it is a free-roaming movement. In solids (metals in particular), the atomic motion is facilitated by vacancies and other imperfections in the crystal structure.

Diffusion can be illustrated by the series of sketches in Figure 4.2 for the case of two metals suddenly brought into intimate contact with each other. At the start, both metals have their own atomic structure, but with time there is an exchange of atoms, not only across the boundary, but within the separate pieces. Given enough time, the assembly of two pieces will finally reach a uniform composition throughout.

Temperature is an important factor in diffusion. At higher temperatures, thermal agitation is greater and the atoms can move about more freely. Another factor is the concentration gradient dc/dx, which indicates the concentration of the two types of atoms in a direction of interest defined by x. The concentration gradient is plotted in Figure 4.2(b) to correspond to the instantaneous distribution of atoms in the assembly. The relationship often used to describe mass diffusion is *Fick's first law*:

$$dm = -D\left(\frac{dc}{dt}\right) A \, dt \tag{4.4}$$

where dm = small amount of material transferred, D = diffusion coefficient of the metal which increases rapidly with temperature, dc/dx = concentration gradient, A = area of the boundary, and dt represents a small time increment. An alternative expression of Eq. (4.4) gives the mass diffusion rate:

$$\frac{dm}{dt} = -D\left(\frac{dc}{dt}\right) A \tag{4.5}$$

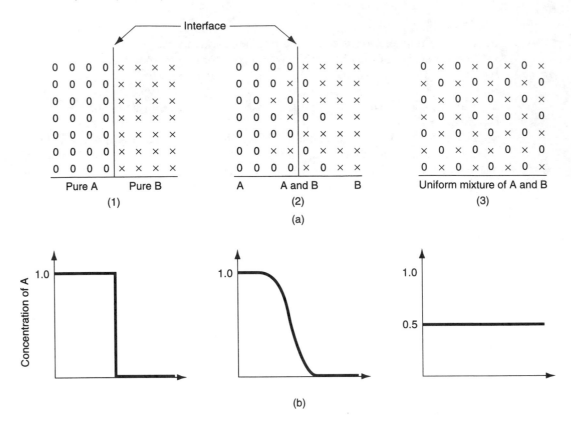

FIGURE 4.2 Mass diffusion (a) model of atoms in two solid blocks in contact: (1) at the start when two pieces are brought together, they each have their individual compositions; (2) after some time, an exchange of atoms has occurred; and (3) eventually, a condition of uniform concentration occurs. The concentration gradient dc/dx for metal A is plotted in (b) of the figure.

Although these equations are difficult to use in calculations because of the problem of assessing D, they are helpful in understanding diffusion and the variables on which it depends.

Mass diffusion is utilized in several processes. A number of surface-hardening treatments are based on diffusion (Section 27.4), including carburizing and nitriding. Among the welding processes, diffusion welding (Section 31.5.2) is used to join two components by pressing them together and allowing diffusion to occur across the boundary to create a permanent bond. Diffusion is also used in electronics manufacturing to alter the surface chemistry of a semiconductor chip in very localized regions to create circuit details (Section 35.4.3).

4.4 ELECTRICAL PROPERTIES

Engineering materials exhibit a great variation in their capability to conduct electricity. In this section, we define the physical properties by which this capability is measured.

4.4.1 Resistivity and Conductivity

Flow of electrical current involves movement of *charge carriers*—infinitesimally small particles possessing an electrical charge. In solids, these charge carriers are electrons. In a liquid solution, charge carriers are positive and negative ions. The movement of charge carriers is driven by the presence of electric voltage, and resisted by the inherent characteristics of the material, such as atomic structure and bonding between atoms and molecules. This is the familiar relationship defined by Ohm's law:

$$I = \frac{E}{R} \qquad (4.6)$$

where I = current, A; E = voltage, V; and R = electrical resistance, Ω. The resistance in a uniform section of material (e.g., a wire) depends on its length L, cross-sectional area A, and the resistivity of the material r; thus,

$$R = r\frac{L}{A} \text{ or } r = R\frac{A}{L} \qquad (4.7)$$

where resistivity has units of Ω-m²/m or Ω-m (Ω-in.). **Resistivity** is the basic property that defines a material's capability to resist current flow. Table 4.3 lists resistivity for selected materials. Resistivity is not a constant; instead it varies, as do so many other properties, with temperature. For metals, it increases with temperature.

It is often more convenient to consider a material as conducting electrical current rather than resisting its flow. The **conductivity** of a material is simply the reciprocal of resistivity:

$$\text{Electrical conductivity} = \frac{1}{r} \qquad (4.8)$$

where conductivity has units of $(\Omega\text{-m})^{-1}((\Omega\text{-in.})^{-1})$.

4.4.2 Classes of Materials by Electrical Properties

Metals are the best **conductors** of electricity, because of their metallic bonding. They have the lowest resistivity (Table 4.3). Most ceramics and polymers, whose electrons are tightly bound by covalent and/or ionic bonding, are poor conductors. Many of these materials are used as **insulators** because they possess high resistivities.

An insulator is sometimes referred to as a dielectric, since the term **dielectric** means nonconductor of direct current. It is a material that can be placed between two electrodes

TABLE 4.3 Resistivity of selected materials.

Material	Resistivity - Ω-m[a]	Material	Resistivity - Ω-m[a]
Conductors	$10^{-6} - 10^{-8}$	*Conductors (cont'd.)*	
Aluminum	2.8×10^{-8}	Tin	11.5×10^{-8}
Aluminum alloys	4.0×10^{-8} [b]	Zinc	6.0×10^{-8}
Cast iron	65.0×10^{-8} [b]	Carbon	5000×10^{-8} (approximate)
Copper	1.7×10^{-8}		
Gold	2.4×10^{-8}	*Semiconductors*	$10^1 - 10^5$
Iron	9.5×10^{-8}	Silicon	1.0×10^3
Lead	20.6×10^{-8}		
Magnesium	4.5×10^{-8}	*Insulators*	$10^{12} - 10^{15}$
Nickel	6.8×10^{-8}	Natural rubber	1.0×10^{12} (approximate)
Silver	1.6×10^{-8}	Polyethylene	100×10^{12} (approximate)
Steel, low C	17.0×10^{-8}		

Compiled from various standard sources.

[a] To convert to ohm-in., multiply ohm-m by 39.4.

[b] Value varies with alloy composition.

without conducting current between them. However, if the voltage is high enough, the current will suddenly pass through the material, for example, in the form of an arc. The *dielectric strength* of an insulating material, then, is the electrical potential required to break down the insulator per unit thickness. Appropriate units are volts/m (volts/in.).

A *superconductor* is a material that exhibits zero resistivity. It is a phenomenon that has been observed in certain metals and ceramics at very low temperatures—approaching absolute zero. We might expect the existence of this phenomenon, due to the significant effect that temperature has on resistivity. That these superconducting materials exist is of great scientific interest. If materials could be developed that exhibit this property at more normal temperatures, there would be significant practical implications in power transmission, electronic switching speeds, and magnetic field applications.

Semiconductors have already proven their practical worth, their applications ranging from mainframe computers to household appliances and automotive engine controllers. As one would guess, a *semiconductor* is a material whose resistivity lies between insulators and conductors. The typical range is shown in Table 4.3. The most commonly used semiconductor material today is silicon (Section 7.5.2), largely because of its abundance in nature, relative low cost, and ease of processing. What makes semiconductors unique is the capacity to significantly alter conductivities in their surface chemistries in very localized areas to fabricate integrated circuits (Chapter 35).

Electrical properties play an important role in various manufacturing processes. Some of the nontraditional processes use electrical energy to remove material. Electric discharge machining (Section 26.3.1) uses the heat generated by electrical energy in the form of sparks to remove material from metals. Most of the important welding processes (Chapter 31) use electrical energy to melt the joint metal. And as we have mentioned, the capacity to alter the electrical properties of semiconductor materials is the basis for microelectronics manufacturing.

4.5 ELECTROCHEMICAL PROCESSES

Electrochemistry is a field of science concerned with the relationship between electricity and chemical changes, and with the conversion of electrical and chemical energy.

In a water solution, the molecules of an acid, base, or salt are dissociated into positively and negatively charged ions. These ions are the charge carriers in the solution—they allow electric current to be conducted, playing the same role that electrons play in metallic conduction. The ionized solution is called an *electrolyte*; and electrolytic conduction requires that current enter and leave the solution at *electrodes*. The positive electrode is called the *anode* and the negative electrode is the *cathode*. The whole arrangement is called an *electrolytic cell*. At each electrode, some chemical reaction occurs, such as the deposition or dissolution of material, or the decomposition of gas from the solution. *Electrolysis* is the name given to these chemical changes occurring in the solution.

Consider a specific case of electrolysis: decomposition of water, illustrated in Figure 4.3. To accelerate the process, dilute sulfuric acid (H_2SO_4) is used as the electrolyte, and platinum and carbon (both chemically inert) are used as electrodes. The electrolyte dissociates in the ions H^+ and SO_4^-. The H^+ ions are attracted to the negatively charged cathode; upon reaching it they acquire an electron and combine into molecules of hydrogen gas:

$$2H^+ + 2e \rightarrow H_2(\text{gas}) \tag{4.9a}$$

The SO_4^- ions are attracted to the anode, transferring electrons to it to form additional sulfuric acid and liberate oxygen:

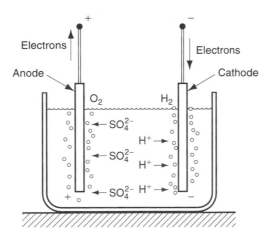

FIGURE 4.3 Example of electrolysis: decomposition of water.

$$2SO_4^- - 4e + 2H_2O \rightarrow 2H_2SO_4 + O_2 \qquad (4.9b)$$

The product H_2SO_4 is dissociated into ions of H^+ and SO_4^- again and so the process continues.

In addition to production of hydrogen and oxygen gases, as illustrated by our example above, electrolysis is also used in several other industrial processes. Two examples are (1) *electroplating* (Section 29.1.1)—an operation that adds a thin coating of one metal (e.g., chromium) to the surface of a second metal (e.g., steel) for decorative or other purposes; and (2) *electrochemical machining* (Section 26.2)—a process in which material is removed from the surface of a metal part. Both of these operations rely on electrolysis to either add or remove material from the surface of a metal part. In electroplating, the workpart is set up in the electrolytic circuit as the cathode, so that the positive ions of the coating metal are attracted to the negatively charged part. In electrochemical machining, the workpart is the anode, and a tool with the desired shape is the cathode. The action of electrolysis in this setup is to remove metal from the part surface in regions determined by the shape of the tool as it slowly penetrates (feeds) into the work.

The two physical laws that determine the amount of material deposited or removed from a metallic surface were first stated by the British scientist Michael Faraday: (1) The mass of a substance liberated in an electrolytic cell is proportional to the quantity of electricity passing through the cell. (2) When the same quantity of electricity is passed through different electrolytic cells, the masses of the substances liberated are proportional to their chemical equivalents. We make use of Faraday's laws in our subsequent coverage of electroplating and electrochemical machining.

REFERENCES

[1] Guy, A. G., and Hren, J. J. *Elements of Physical Metallurgy.* 3rd ed. Addison-Wesley, Reading, Mass., 1974.

[2] Flinn, R. A., and Trojan, P. K. *Engineering Materials and Their Applications.* 4th ed. Houghton Mifflin, Boston, 1990.

[3] Kreith, F. *Principles of Heat Transfer.* 5th ed. Wadsworth, Belmont, Calif., 1996.

[4] *Metals Handbook.* 10th ed. Vol. 1, Properties and Selection: Iron, Steel, and High Performance Alloys, ASM International, Metals Park, Oh., 1990.

[5] *Metals Handbook.* 10th ed. Vol. 2, Properties and Selection: Nonferrous Alloys and Special Purpose Materials, ASM International, Metals Park, Oh., 1990.

[6] Van Vlack, L. H. *Elements of Materials Science and Engineering.* 6th ed. Addison-Wesley, Reading, Mass., 1989.

REVIEW QUESTIONS

4.1. Define the property *density* of a material.

4.2. What is the difference in melting characteristics between a pure metal element and an alloy?

4.3. Describe the melting characteristics of a noncrystalline material such as glass.

4.4. Define the *specific heat* property of a material.

4.5. What is the thermal conductivity of a material?

4.6. Define *thermal diffusivity.*

4.7. What are the important variables that affect mass diffusion?

4.8. Define the *resistivity* of a material.

4.9. Why are metals better conductors of electricity than ceramics and polymers?

4.10. What is the *dielectric strength* of a material?

4.11. What is an *electrolyte?*

MULTIPLE CHOICE QUIZ

There is a total of 12 correct answers in the following multiple choice questions (some questions have multiple answers that are correct). To attain a perfect score on the quiz, all correct answers must be given, since each correct answer is worth 1 point. For each question, each omitted answer or wrong answer reduces the score by 1 point, and each additional answer beyond the number of answers required reduces the score by 1 point. Percentage score on the quiz is based on the total number of correct answers.

4.1. Which one of the following metals has the lowest density? (a) aluminum, (b) copper, (c) magnesium, or (d) tin.

4.2. Polymers typically exhibit greater thermal expansion properties than metals: (a) true, or (b) false.

4.3. In the heating of most metal alloys, melting begins at a certain temperature and concludes at a higher temperature. In these cases, which of the following temperatures marks the beginning of melting? (a) liquidus, (b) solidus.

4.4. Which of the following materials has the highest specific heat? (a) aluminum, (b) concrete, (c) polyethylene, or (d) water.

4.5. Copper is generally considered easy to weld, because of its high thermal conductivity: (a) true, or (b) false.

4.6. The mass diffusion rate dm/dt across a boundary between two different metals is a function of which of the following variables (more than one?): (a) concentration gradient dc/dx, (b) contact area, (c) density, (d) melting point, (e) temperature, or (f) time.

4.7. Which of the following pure metals is the best conductor of electricity? (a) aluminum, (b) copper, (c) gold, or (d) silver.

4.8. A superconductor is characterized by which of the following (choose one best answer): (a) very low resistivity, (b) zero conductivity, or (c) resistivity properties between those of conductors and semiconductors?

4.9. In an electrolytic cell, the anode is the electrode which is (a) positive, or (b) negative.

PROBLEMS

4.1. The starting diameter of a shaft is 25.00 mm. This shaft is to be inserted into a hole in an expansion fit assembly operation. To be readily inserted, the shaft must be reduced in diameter by cooling. Determine the temperature to which the shaft must be reduced from room temperature (20°C) in order to reduce its diameter to 24.98 mm. Refer to Table 4.1.

4.2. Aluminum has a density of 2.70 g/cm³ at room temperature (20°C). Determine its density at 650°C, using data in Table 4.1 as a reference.

4.3. With reference to Table 4.1, determine the increase in length of a steel bar whose length is 10.0 in. if the bar is heated from room temperature (70°F) to 500°F.

4.4. With reference to Table 4.2, determine the quantity of heat required to increase the temperature of an aluminum block that is 10 cm × 10 cm × 10 cm from room temperature (21°C) to 300°C.

4.5. What is the resistance R of a length of copper wire whose length is 10 m and whose diameter = 0.10 mm? Use Table 4.3 as a reference.

5 DIMENSIONS, TOLERANCES, AND SURFACES

CHAPTER CONTENTS
5.1   Dimensions, Tolerances, and Related Attributes
        5.1.1   Dimensions and Tolerances
        5.1.2   Other Geometric Attributes
5.2   Surfaces
        5.2.1   Characteristics of Surfaces
        5.2.2   Surface Texture
        5.2.3   Surface Integrity
5.3   Effect of Manufacturing Processes
        5.3.1   Tolerances and Manufacturing Processes
        5.3.2   Surfaces and Manufacturing Processes

In addition to mechanical and physical properties of materials, other factors that determine the performance of a manufactured product include the dimensions and surfaces of its components. ***Dimensions*** are the linear or angular sizes of a component specified on the part drawing. Dimensions are important because they determine how well the components of a product fit together during assembly. When fabricating a given component, it is nearly impossible and very costly to make the part to the exact dimension given on the drawing. Instead we allow a limited variation from the dimension, and that allowable variation is called a ***tolerance.***

The surfaces of a component are also important. They affect product performance, assembly fit, and aesthetic appeal that a potential customer might have for the product. A ***surface*** is the exterior boundary of an object with its surroundings, which may be another object, a fluid, or space; or combinations of these. The surface encloses the object's bulk mechanical and physical properties.

In this chapter we discuss dimensions, tolerances, and surfaces—three attributes specified by the product designer. Their physical realizations are determined largely by the manufacturing processes used to make the parts and products. In Chapter 44, we consider how these attributes are measured and inspected.

5.1 DIMENSIONS, TOLERANCES, AND RELATED ATTRIBUTES

In this first section, we define the basic parameters used by design engineers to specify sizes of geometric features on a part drawing. The parameters include dimensions and tolerances, flatness, roundness, and angularity.

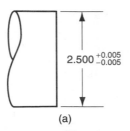 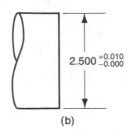 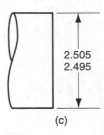

$2.500 \, ^{+0.005}_{-0.005}$ $2.500 \, ^{+0.010}_{-0.000}$ 2.505 / 2.495

FIGURE 5.1 Three ways to specify tolerance limits for a nominal dimension of 2.500: (a) bilateral, (b) unilateral, and (c) limit dimensions.

(a)   (b)   (c)

5.1.1 Dimensions and Tolerances

ANSI [3] defines a **dimension** as "a numerical value expressed in appropriate units of measure and indicated on a drawing and in other documents along with lines, symbols, and notes to define the size or geometric characteristic, or both, of a part or part feature." Dimensions on part drawings represent nominal or basic sizes of the part and its features. These are the values that the designer would like the part size to be, if the part could be made to an exact size with no errors or variations in the fabrication process. However, there are variations in the manufacturing process, which are manifested as variations in the part size. Tolerances are used to define the limits of the allowed variation. Quoting again from the ANSI standard [3], a **tolerance** is "the total amount by which a specific dimension is permitted to vary. The tolerance is the difference between the maximum and minimum limits."

Tolerances can be specified in several ways, illustrated in Figure 5.1. Probably most common is the **bilateral tolerance,** in which the variation is permitted in both positive and negative directions from the nominal dimension. For example, in Figure 5.1(a), the nominal dimension = 2.500 linear units (e.g., mm, in.), with an allowable variation of 0.005 units in either direction. Parts outside these limits are unacceptable. It is possible for a bilateral tolerance to be unbalanced; for example, 2.500 +0.010, −0.005 dimensional units. A **unilateral tolerance** is one in which the variation from the specified dimension is permitted in only one direction, either positive or negative. Figure 5.1(b) illustrates a positive unilateral tolerance.

An alternative method to specify the permissible variation in a part feature size is **limit dimensioning,** which consists of the maximum and minimum dimensions allowed, as in Figure 5.1(c).

5.1.2 Other Geometric Attributes

Dimensions and tolerances are normally expressed as linear (length) values. Other geometric attributes of parts are also important, such as flatness of a surface, roundness of a shaft or hole, parallelism between two surfaces, and so on. Definitions of these terms are listed in Table 5.1.

5.2 SURFACES

A surface is what we touch when we hold an object such as a manufactured part. The designer specifies the part dimensions, relating the various surfaces to each other. These **nominal surfaces,** representing the intended surface contour of the part, are defined by

TABLE 5.1 Definitions of geometric attributes of parts.

Angularity—The extent to which a part feature such as a surface or axis is at a specified angle relative to a reference surface. If the angle = 90°, then the attribute is called perpendicularity or squareness.

Circularity—For a surface of revolution such as a cylinder, circular hole, or cone, circularity is the degree to which all points on the intersection of the surface and a plane perpendicular to the axis of revolution are equidistant from the axis. For a sphere, circularity is the degree to which all points on the intersection of the surface and a plane passing through the center are equidistant from the center.

Concentricity—The degree to which any two (or more) part features (such as a cylindrical surface and a circular hole) have a common axis.

Cylindricity—The degree to which all points on a surface of revolution such as a cylinder are equidistant from the axis of revolution.

Flatness—The extent to which all points on a surface lie in a single plane.

Parallelism—The degree to which all points on a part feature such as a surface, line, or axis are equidistant from a reference plane or line or axis.

Perpendicularity—The degree to which all points on a part feature such as a surface, line, or axis are 90° from a reference plane or line or axis.

Roundness—Same as circularity.

Squareness—Same as perpendicularity.

Straightness—The degree to which a part feature such as a line or axis is a straight line.

lines in the engineering drawing. The nominal surfaces appear as absolutely straight lines, ideal circles, round holes, and other edges and surfaces that are geometrically perfect. The actual surfaces of a manufactured part are determined by the processes used to make it. The variety of processes available in manufacturing result in wide variations in surface characteristics, and it is important for engineers to understand the technology of surfaces.

Surfaces are commercially and technologically important for a number of reasons; there are different reasons for different product applications: (1) Aesthetic reasons—surfaces that are smooth and free of scratches and blemishes are more likely to give a favorable impression to the customer. (2) Surfaces affect safety. (3) Friction and wear depend on surface characteristics. (4) Surfaces affect mechanical and physical properties—for example, surface flaws can be points of stress concentration. (5) Assembly of parts is affected by their surfaces—for example, the strength of adhesively bonded joints (Section 32.3) is increased when the surfaces are slightly rough. (6) Smooth surfaces make better electrical contacts.

Surface technology is concerned with (1) defining the characteristics of a surface, (2) surface texture, (3) surface integrity, and (4) the relationship between manufacturing processes and the characteristics of the resulting surface. The first three topics are covered in this section; the final topic is presented in Section 5.3.

5.2.1 Characteristics of Surfaces

A microscopic view of a part's surface would reveal that it is less than perfect. The features of a typical surface are illustrated in the highly magnified cross section of the surface of a metal part in Figure 5.2. Although our discussion here is focused on metallic surfaces, our comments apply to ceramics and polymers, with modifications due to differences in struc-

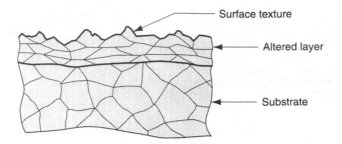

FIGURE 5.2   A magnified cross section of a typical metallic part surface.

ture of these materials. The bulk of the part, referred to as the ***substrate,*** has a grain structure that depends on previous processing of the metal; for example, the metal's substrate structure is affected by its chemical composition, the casting process originally used on the metal, and any deformation operations and heat treatments performed on the casting.

The exterior of the part is a surface whose topography is anything but straight and smooth. In this highly magnified cross section, the surface has roughness, waviness, and flaws. Although not shown here, it also possesses a pattern and/or direction resulting from the mechanical process that produced it. All of these geometric features are included in the term ***surface texture.***

Just below the surface is a layer of metal whose structure differs from that of the substrate. This might be called the ***altered layer,*** and it is a manifestation of the actions that have been visited upon the surface during its creation and afterward. Manufacturing processes involve energy, usually in large amounts, which operate on the part against its surface. The altered layer may result from work hardening (mechanical energy), heating (thermal energy), chemical treatment, or even electrical energy. The metal in this layer is affected by the application of energy, and its microstructure is altered accordingly. This altered layer falls within the scope of ***surface integrity,*** which is concerned with the definition, specification, and control of the surface layers of a material (most commonly metals) in manufacturing and subsequent performance in service. The scope of surface integrity is usually interpreted to include surface texture as well as the altered layer beneath.

In addition, most metal surfaces are also coated with an ***oxide film,*** given sufficient time after processing for the film to form. Aluminum forms a hard, dense, thin film of Al_2O_3 on its surface (that serves to protect the substrate from corrosion), and iron forms oxides of several chemistries on its surface (rust, which provides virtually no protection at all). There is also likely to be moisture, dirt, oil, adsorbed gases, and other contaminants on the part's surface.

5.2.2 Surface Texture

Surface texture consists of the repetitive and/or random deviations from the nominal surface of an object; it is defined by four elements: roughness, waviness, lay, and flaws, shown in Figure 5.3. ***Roughness*** refers to the small, finely spaced deviations from the nominal surface that are determined by the material characteristics and the process that formed the surface. ***Waviness*** is defined as the deviations of much larger spacing; they occur due to work deflection, vibration, heat treatment, and similar factors. Roughness is superimposed on waviness. ***Lay*** is the predominant direction or pattern of the surface texture. It is determined by the manufacturing method used to create the surface, usually from the

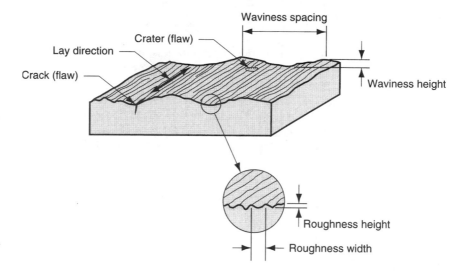

FIGURE 5.3 Surface texture features.

action of a cutting tool. Figure 5.4 presents most of the possible lays a surface can take, together with the symbol used by a designer to specify them. Finally, *flaws* are irregularities that occur occasionally on the surface; these include cracks, scratches, inclusions, and similar defects in the surface. Although some of the flaws relate to surface texture, they also affect surface integrity (Section 5.2.3).

FIGURE 5.4 Possible lays of a surface (source: [1]).

Lay symbol	Surface pattern	Description
=		Lay is parallel to line representing surface to which symbol is applied.
⊥		Lay is perpendicular to line representing surface to which symbol is applied.
X		Lay is angular in both directions to line representing surface to which symbol is applied.
M		Lay is multidirectional.
C		Lay is circular relative to center of surface to which symbol is applied.
R		Lay is approximately radial relative to the center of the surface to which symbol is applied.
P		Lay is particulate, nondirectional, or protuberant.

Surface Roughness and Surface Finish   Surface roughness is a measurable characteristic based on the roughness deviations as defined above. ***Surface finish*** is a more subjective term denoting smoothness and general quality of a surface. In popular usage, surface finish is often used as a synonym for surface roughness.

The most commonly used measure of surface texture is surface roughness. With respect to Figure 5.5, ***surface roughness*** can be defined as the average of the vertical deviations from the nominal surface over a specified surface length. An arithmetic average (AA) is generally used, based on the absolute values of the deviations, and this roughness value is referred to by the name ***average roughness***:

$$R_a = \int_0^{L_m} \frac{|y|}{L_m}\, dx \tag{5.1}$$

where R_a = arithmetic mean value of roughness, m (in.); y = the vertical deviation from nominal surface (converted to absolute value), m (in.); and L_m = the specified distance over which the surface deviations are measured. An approximation of Eq. (5.1), perhaps easier to comprehend, is given by

$$R_a = \sum_{i=1}^{n} \frac{|y_i|}{n} \tag{5.2}$$

where R_a has the same meaning as above; y_i = vertical deviations (converted to absolute value) identified by the subscript i, m (in.); and n = the number of deviations included in L_m. We have indicated that the units in these equations are m (in.). In fact, the scale of the deviations is very small, so more appropriate units are μm which = m $\times$ 10^{-6} or mm $\times$ 10^{-3} (μ-in. which = inch $\times$ 10^{-6}). These are the units commonly used to express surface roughness.

The AA method is the most widely used averaging method for surface roughness today. An alternative, sometimes used in the United States, is the ***root-mean-square*** (RMS) average, which is the square root of the mean of the squared deviations over the measuring length. RMS surface roughness values will almost always be greater than the AA values. This is because the larger deviations will figure more prominently in the calculation of the RMS value.

Surface roughness suffers the same kinds of deficiencies of any single measure used to assess a complex physical attribute. For example, it fails to account for the lay of the surface pattern; thus, surface roughness may vary significantly, depending on the direction in which it is measured.

FIGURE 5.5   Deviations from nominal surface used in the two definitions of surface roughness.

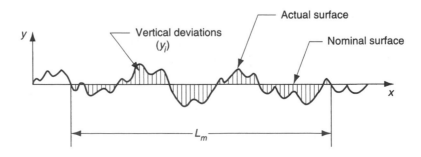

Another deficiency is that waviness can be included in the R_a computation. To deal with this problem, a parameter called the ***cutoff length*** is used as a filter that separates the waviness in a measured surface from the roughness deviations. In effect, the cutoff length is a sampling distance along the surface. A sampling distance shorter than the waviness width will eliminate the vertical deviations associated with waviness and only include those associated with roughness. The most common cutoff length used in practice is 0.8 mm (0.030 in.). The measuring length L_m is normally set at about five times the cutoff length.

Symbols for Surface Texture   Designers specify surface texture on an engineering drawing by means of symbols as in Figure 5.6. The symbol designating surface texture parameters is a check mark (looks like a square root sign), with entries as indicated for average roughness, waviness, cutoff, lay, and maximum roughness spacing. The symbols for lay are from Figure 5.4.

5.2.3 Surface Integrity

Surface texture alone does not completely describe a surface. There may be metallurgical or other changes in the altered layer beneath the surface that can have a significant effect on the material's mechanical properties. ***Surface integrity*** is the study and control of this subsurface layer and the changes in it that occur during processing that may influence the performance of the finished part or product.

There are many possible changes in the subsurface layer of the material that can result from a manufacturing process. They range from clearly observable cracks in the surface to subtle transformations in the metallic structure below. We present a list of the possible alterations and injuries to the surface layer in Table 5.2. The surface changes are caused by the application of various forms of energy during processing—mechanical, thermal, chemical, and electrical. Mechanical energy is the most common form used in manufacturing; it is applied against the work material in operations such as metal forming (e.g., forging, extrusion), pressworking, and machining. Although its primary function in these processes is to change the geometry of the workpart, mechanical energy can also cause residual stresses, work hardening, and cracks in the surface layers. Table 5.3 indicates the various types of surface and subsurface alterations that are attributable to the different forms of energy applied in manufacturing. Most of the alterations in our table refer to metals, for which surface integrity has been most intensively studied.

FIGURE 5.6   Surface texture symbols in engineering drawings: (a) the symbol, and (b) symbol with identification labels. Values of R_a are given in microinches; units for other measures are given in inches. Designers do not always specify all of the parameters on engineering drawings.

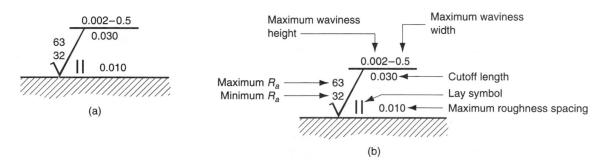

TABLE 5.2   Surface and subsurface alterations that define surface integrity.

Absorption—Impurities are absorbed and retained in surface layers of the base material, possibly leading to embrittlement or other property changes.

Alloy depletion—Critical alloying elements are lost from the surface layers, with possible loss of properties in the metal.

Cracks—Narrow ruptures or separations either at or below the surface that alter the continuity of the material. Cracks are characterized by sharp edges and length-to-width ratios of 4:1 or more. They are classified as macroscopic (can be observed with magnification of 10X or less) and microscopic (requires magnification of more than 10X).

Craters—Rough surface depressions left in the surface by short circuit discharges; associated with electrical processing methods such as electric discharge machining and electrochemical machining (Chapter 26).

Hardness changes—Hardness differences at or near the surface.

Heat affected zone—Regions of the metal that are affected by the application of thermal energy; the regions are not melted but are sufficiently heated that they undergo metallurgical changes that affect properties. Abbreviated HAZ, the effect is most prominent in fusion welding operations (Chapter 31).

Inclusions—Small particles of material incorporated into the surface layers during processing; they are a discontinuity in the base material. Their composition usually differs from the base material.

Intergranular attack—This refers to various forms of chemical reaction at the surface, including intergranular corrosion and oxidation.

Laps, folds, seams—Irregularities and defects in the surface caused by plastic working of overlapping surfaces.

Pits—Shallow depressions with rounded edges formed by any of several mechanisms, including selective etching or corrosion; removal of surface inclusions; mechanically formed dents; or electrochemical action.

Plastic deformation—Microstructural changes from deforming the metal at the surface; it results in strain hardening.

Recrystallization—Formation of new grains in strain hardened metals; associated with heating of metal parts that have been deformed.

Redeposited metal—Metal that is removed from the surface in the molten state and then reattached prior to solidification.

Resolidified metal—This is a portion of the surface that is melted during processing and then solidified without detaching from the surface. The name *remelted metal* is also used for resolidified metal. *Recast metal* is a term that includes both redeposited and resolidified metal.

Residual stresses—Stresses remaining in the material after processing.

Selective etch—A form of chemical attack that concentrates on certain components in the base material.

Compiled from [2].

TABLE 5.3   Forms of energy applied in manufacturing and the resulting possible surface and subsurface alterations that can occur.

Energy Form	Possible Alterations and Damages
Mechanical	Residual stresses in subsurface layer Cracks—microscopic and macroscopic Plastic deformation Laps, folds, or seams Voids or inclusions introduced mechanically Hardness variations (e.g., work hardening)
Thermal	Metallurgical changes (recrystallization, grain size changes, phase changes at surface) Redeposited or resolidified material Heat-affected zone (includes some of the metallurgical changes listed above) Hardness changes
Chemical	Intergranular attack Chemical contamination Absorption of certain elements such as H and Cl in the metal surface Corrosion, pitting, and etching Stress corrosion Dissolving of microconstituents Alloy depletion and resulting hardness changes
Electrical	Changes in conductivity and/or magnetism Craters resulting from short circuits during certain electrical processing techniques

Based on [2].

5.3 EFFECT OF MANUFACTURING PROCESSES

The ability to achieve a certain tolerance or surface is a function of the manufacturing process. In this section, we describe the general capabilities of various processes in terms of tolerance and surface roughness and surface integrity.

5.3.1 Tolerances and Manufacturing Processes

Some manufacturing processes are inherently more accurate than others. Most machining processes are quite accurate, capable of tolerances of ±0.05 mm (±0.002 in.) or better. By contrast, sand castings are generally inaccurate, and tolerances of 10 to 20 times those used for machined parts should be specified. In Table 5.4, we list a variety of manufacturing processes and indicate the typical tolerances for each process. Our tolerances are based on the process capability for the particular manufacturing operation, as defined in Section 43.2. The tolerance that should be specified is a function of part size; larger parts require more generous tolerances. Our table lists tolerance for moderate-sized parts in each processing category.

5.3.2 Surfaces and Manufacturing Processes

The manufacturing process determines surface finish and surface integrity. Some processes are inherently capable of producing better surfaces than others. In general, processing cost increases with improvement in surface finish. This is because additional operations and more time are usually required to obtain increasingly better surfaces. Processes noted for providing superior finishes include honing, lapping, polishing, and superfinishing (Chapter 26). Table 5.5 indicates the usual surface roughness that can be expected from various manufacturing processes.

TABLE 5.4 Typical tolerance limits, based on process capability (Section 43.2), for various manufacturing processes.

Process	Typical Tolerance Limits mm	(inches)	Process	Typical Tolerance Limits mm	(inches)
Sand casting:			Abrasive processes:		
Cast iron	±1.3	(±0.050)	Grinding	±0.008	(±0.0003)
Steel	±1.5	(±0.060)	Lapping	±0.005	(±0.0002)
Aluminum	±0.5	(±0.020)	Honing	±0.005	(±0.0002)
Die casting	±0.12	(±0.005)	Nontraditional processes:		
Plastic molding:			Chemical machining	±0.08	(±0.003)
Polyethylene	±0.3	(±0.010)	Electric discharge	±0.025	(±0.001)
Polystyrene	±0.15	(±0.006)	Electrochem. grind	±0.025	(±0.001)
Machining:			Electrochem. machine	±0.05	(±0.002)
Drilling, diameter:			Electron beam cutting	±0.08	(±0.003)
6 mm (0.250 in.)	+0.08, −0.03	(+0.003, −0.001)	Laser beam cutting	±0.08	(±0.003)
25 mm (1.000 in.)	+0.13, −0.05	(+0.006, −0.002)	Plasma arc cutting	±1.3	(±0.050)
Milling	±0.08	(±0.003)			
Turning	±0.05	(±0.002)			

Compiled from [4], [5], and other sources.

TABLE 5.5 Surface roughness values produced by the various manufacturing processes.

Process	Typical Surface Finish	Range of Roughness[a]	Process	Typical Surface Finish	Range of Roughness[a]
Casting:			Abrasive:		
Die casting	Good	1–2 (30–65)	Grinding	Very good	0.1–2 (5–75)
Investment	Good	1.5–3 (50–100)	Honing	Very good	0.1–1 (4–30)
Sand casting	Poor	12–25 (500–1000)	Lapping	Excellent	0.05–0.5 (2–15)
Metal forming:			Polishing	Excellent	0.1–0.5 (5–15)
Cold rolling	Good	1–3 (25–125)	Superfinish	Excellent	0.02–0.3 (1–10)
Sheet metal draw	Good	1–3 (25–125)	Nontraditional:		
Cold extrusion	Good	1–4 (30–150)	Chemical milling	Medium	1.5–5 (50–200)
Hot rolling	Poor	12–25 (500–1000)	Electrochemical	Good	0.2–2 (10–100)
Machining:			Electric discharge	Medium	1.5–15 (50–500)
Boring	Good	0.5–6 (15–250)	Electron beam	Medium	1.5–15 (50–500)
Drilling	Medium	1.5–6 (60–250)	Laser beam	Medium	1.5–15 (50–500)
Milling	Good	1–6 (30–250)	Thermal:		
Planing	Medium	1.5–12 (60–500)	Arc welding	Poor	5–25 (250–1000)
Reaming	Good	1–3 (30–125)	Flame cutting	Poor	12–25 (500–1000)
Shaping	Medium	1.5–12 (60–500)	Plasma arc cutting	Poor	12–25 (500–1000)
Sawing	Poor	3–25 (100–1000)			
Turning	Good	0.5–6 (15–250)			

Compiled from [1], [2], and other sources.

[a]Subjective description and typical range of surface roughness values are given, μm (μ-in). Roughness can vary significantly for a given process, depending on process parameters.

REFERENCES

[1] American National Standards Institute, Inc. *Surface Texture.* ANSI B46.1-1978, American Society of Mechanical Engineers, New York, 1978.

[2] American National Standards Institute, Inc. *Surface Integrity.* ANSI B211.1-1986, Society of Manufacturing Engineers, Dearborn, Mich., 1986.

[3] American National Standards Institute, Inc. *Dimensioning and Tolerancing.* ANSI Y14.5M-1982, American Society of Mechanical Engineers, New York, 1982.

[4] Bakerjian, R., and P. Mitchell. *Tool and Manufacturing Engineers Handbook.* 4th ed. Vol. VI: *Design for Manufacturability,* Society of Manufacturing Engineers, Dearborn, Mich., 1992.

[5] Drozda, T. J., and C. Wick. *Tool and Manufacturing Engineers Handbook.* 4th ed. Vol. I: *Machining,* Society of Manufacturing Engineers, Dearborn, Michigan, 1983, Chap. 1.

[6] *Machining Data Handbook.* 3rd ed. Vol. Two. Machinability Data Center, Cincinnati, Oh., 1980, Ch 18.

[7] Mummery, L. *Surface Texture Analysis—The Handbook.* Hommelwerke Gmbh, Germany, 1990.

[8] Oberg, E., F. D. Jones, H. L. Horton, and H. Ryffel. *Machinery's Handbook.* 26th ed. Industrial Press Inc., New York, 2000.

[9] Schaffer, G. H. "The Many Faces of Surface Texture." Special Report 801, *American Machinist and Automated Manufacturing.* June 1988, p 61–68.

[10] Sheffield Measurement, a Cross & Trecker Company. *Surface Texture and Roundness Measurement Handbook.* Dayton, Oh., 1991.

[11] Wick, C., and R. F. Veilleux. *Tool and Manufacturing Engineers Handbook.* 4th ed. Vol. IV: *Quality Control and Assembly,* Society of Manufacturing Engineers, Dearborn, Mich., 1987, Section 1.

REVIEW QUESTIONS

5.1. What is a *tolerance?*

5.2. What are some of the reasons why surfaces are important?

5.3. Define *nominal surface.*

5.4. Define *surface texture.*

5.5. How is surface texture distinguished from surface integrity?

5.6. Within the scope of surface texture, how is roughness distinguished from waviness?

5.7. Surface roughness is a measurable aspect of surface texture; what does *surface roughness* mean?

5.8. Indicate some of the limitations of using surface roughness as a measure of surface texture.

5.9. Identify some of the changes and injuries that can occur at or immediately below the surface of a metal.

5.10. What causes the various types of change that occur in the altered layer just beneath the surface?

5.11. Name and describe some of the tests used to assess surface integrity.

5.12. Name some manufacturing processes that produce very poor surface finishes.

5.13. Name some manufacturing processes that produce very good or excellent surface finishes.

MULTIPLE CHOICE QUIZ

There is a total of 19 correct answers in the following multiple choice questions (some questions have multiple answers that are correct). To attain a perfect score on the quiz, all correct answers must be given, since each correct answer is worth 1 point. For each question, each omitted answer or wrong answer reduces the score by 1 point, and each additional answer beyond the number of answers required reduces the score by 1 point. Percentage score on the quiz is based on the total number of correct answers.

5.1. A tolerance is which one of the following? (a) clearance between a shaft and a mating hole, (b) measurement error, (c) total permissible variation from a specified dimension, or (d) variation in manufacturing.

5.2. Which of the following two geometric terms have the same meaning? (a) circularity, (b) concentricity, (c) cylindricity, and (d) roundness.

5.3. Surface texture includes which of the following characteristics of a surface (more than one)? (a) deviations from the nominal surface, (b) feed marks of the tool that produced the surface, (c) hardness variations, (d) oil films, and (e) surface cracks.

5.4. Which averaging method generally yields the higher value of surface roughness, (a) AA or (b) RMS?

5.5. Surface texture is included within the scope of surface integrity: (a) true or (b) false.

5.6. Thermal energy is normally associated with which of the following changes in the altered layer? (a) cracks, (b) hardness variations, (c) heat affected zone, (d) plastic deformation, (e) recrystallization, or (f) voids.

5.7. A better finish (lower roughness value) will tend to have which of the following effects on the fatigue strength of a metal surface? (a) increase, (b) decrease, or (c) no effect.

5.8. Which of the following are included within the scope of surface integrity? (a) chemical absorption, (b) microstructure near the surface, (c) microcracks beneath the surface, (d) substrate microstructure, (e) surface roughness, or (f) variation in tensile strength near the surface.

5.9. Which one of the following manufacturing processes will likely result in the best surface finish? (a) arc welding, (b) grinding, (c) machining, (d) sand casting, or (e) sawing.

5.10. Which one of the following manufacturing processes will likely result in the worst surface finish? (a) cold rolling, (b) grinding, (c) machining, (d) sand casting, or (e) sawing.

6 METALS

CHAPTER CONTENTS

6.1   Alloys and Phase Diagrams
        6.1.1   Alloys
        6.1.2   Phase Diagrams
6.2   Ferrous Metals
        6.2.1   The Iron–Carbon Phase Diagram
        6.2.2   Iron and Steel Production
        6.2.3   Steels
        6.2.4   Cast Irons
6.3   Nonferrous Metals
        6.3.1   Aluminum and Its Alloys
        6.3.2   Magnesium and Its Alloys
        6.3.3   Copper and Its Alloys
        6.3.4   Nickel and Its Alloys
        6.3.5   Titanium and Its Alloys
        6.3.6   Zinc and Its Alloys
        6.3.7   Lead and Tin
        6.3.8   Refractory Metals
        6.3.9   Precious Metals
6.4   Superalloys
6.5   Guide to the Processing of Metals

In Part II, we discuss the four types of engineering materials: (1) metals, (2) ceramics, (3) polymers, and (4) composites. Metals are the most important engineering materials today. They have properties that satisfy a wide variety of design requirements. The manufacturing processes by which they are shaped into products have been developed and refined over many years; indeed, some of the processes date from ancient times (Historical Note 1.2). Engineers understand metals. The properties and other technical issues related to metals are considered in this chapter.

The technological and commercial importance of metals is due to the following general properties possessed by virtually all of the common metals:

➤ *High stiffness and strength.* Metals can be alloyed for high rigidity, strength, and hardness; thus, they are used to provide the structural framework for most engineered products.

> *Toughness.*   Metals have the capacity to absorb energy better than other classes of materials.
> *Good electrical conductivity.*   Metals are conductors because of their metallic bonding that permits the free movement of electrons as charge carriers.
> *Good thermal conductivity.*   Metallic bonding also explains why metals generally conduct heat better than ceramics or polymers.

In addition, certain metals have specific properties that make them attractive for specialized applications. Many common metals are available at relatively low cost per unit weight and are often the material of choice simply because of their low relative cost.

Metals are converted into parts and products using a variety of manufacturing processes. The starting form of the metal differs, depending on the process. The major categories are: (1) *cast metal,* in which the initial form is a casting; (2) *wrought metal,* in which the metal has been worked or can be worked (e.g., rolled or otherwise formed) after casting; better mechanical properties are generally associated with wrought metals compared to cast metals; and (3) *powdered metal,* in which the metal is purchased in the form of very small powders for conversion into parts using powder metallurgy techniques. Most metals are available in all three forms. In this chapter, our discussion will focus on categories (1) and (2), which are of greatest commercial and engineering interest. Powder metallurgy techniques are examined in Chapter 16.

Metals are classified into two major groups: (1) *ferrous*—those based on iron; and (2) *nonferrous*—all other metals. The ferrous group can be further subdivided into steels and cast irons. Most of our discussion will be organized around this classification, but let us first examine the general topic of alloys and phase diagrams.

6.1 ALLOYS AND PHASE DIAGRAMS

Although some metals are important as pure elements (e.g., gold, silver, copper), most engineering applications require the improved properties obtained by alloying. Through alloying, it is possible to enhance strength, hardness, and other properties compared to pure metals. In this section, we define and classify alloys; we then discuss phase diagrams which indicate the phases of an alloy system as a function of composition and temperature.

6.1.1 Alloys

An *alloy* is a metal composed of two or more elements, at least one of which is metallic. The two main categories of alloys are (1) solid solutions and (2) intermediate phases.

Solid Solutions   A *solid solution* is an alloy in which one element is dissolved in another to form a single-phase structure. The term *phase* describes any homogeneous mass of material, such as a metal in which the grains all have the same crystal lattice structure. In a solid solution, the solvent or base element is metallic, and the dissolved element can be either metallic or nonmetal. Solid solutions come in two forms, shown in Figure 6.1. The first is a *substitutional solid solution,* in which atoms of the solvent element are replaced in its unit cell by the dissolved element. Brass is an example, in which zinc is dissolved in copper. To make the substitution, several rules must be satisfied [1], [3], [4]: (1) the atomic radii of the two elements must be similar, usually within 15%; (2) their lattice types must be the same; (3) if the elements have different valences,

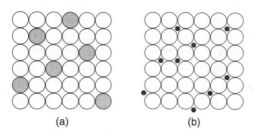

FIGURE 6.1 Two forms of solid solutions: (a) substitutional solid solution, and (b) interstitial solid solution.

the lower valence metal is more likely to be the solvent; and (4) if the elements have high chemical affinity for each other, they are less likely to form a solid solution and more likely to form a compound.

The second type of solid solution is an *interstitial solid solution,* in which atoms of the dissolving element fit into the vacant spaces between base metal atoms in the lattice structure. It follows that the atoms fitting into these interstices must be small compared to those of the solvent metal; for example, hydrogen, carbon, nitrogen, and boron. The most important example of this second type is carbon dissolved in iron to form steel.

In both forms of solid solution, the alloy structure is generally stronger and harder than either of the component elements.

Intermediate Phases There are usually limits to the solubility of one element in another. When the amount of the dissolving element in the alloy exceeds the solid solubility limit of the base metal, a second phase forms in the alloy. The term *intermediate phase* is used to describe it because its chemical composition is intermediate between the two pure elements. Its crystalline structure is also different from those of the pure metals. Depending on composition, and recognizing that many alloys consist of more than two elements, these intermediate phases can be of several types, including (1) metallic compounds consisting of a metal and nonmetal such as Fe_3C; and (2) intermetallic compounds—two metals that form a compound, such as Mg_2Pb. The composition of the alloy is often such that the intermediate phase is mixed with the primary solid solution to form a two-phase structure, one phase dispersed throughout the second. These two-phase alloys are important because they can be formulated and heat treated for significantly higher strength than solid solutions.

6.1.2 Phase Diagrams

As we shall use the term in this text, a *phase diagram* is a graphical means of representing the phases of a metal alloy system as a function of composition and temperature. Our discussion of the diagram will be limited to alloy systems consisting of two elements at atmospheric pressures. This type of diagram is called a *binary phase diagram.* Other forms of phase diagrams are discussed in texts on materials science, such as [3].

The Copper–Nickel Alloy System The best way to describe the phase diagram and its use is by example. Figure 6.2 presents one of the simplest cases, the Cu–Ni alloy system. Composition is plotted on the horizontal axis and temperature on the vertical axis. Thus, any point in the diagram indicates the overall composition and the phase or phases present at the given temperature. Pure copper melts at 1083°C (1981°F), and pure nickel at 1455°C (2651°F). Alloy compositions between these extremes exhibit gradual melting that commences at the solidus and concludes at the liquidus as temperature is increased.

The copper–nickel system is a solid solution alloy throughout its entire range of compositions. Anywhere in the region below the solidus line, the alloy is a solid solution;

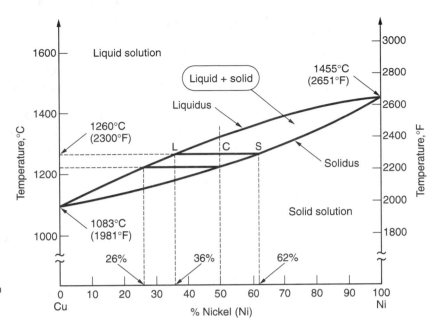

FIGURE 6.2   Phase diagram for the copper–nickel alloy system.

there are no intermediate solid phases in this system. However, there is a mixture of phases in the region bounded by the solidus and liquidus. Recall from Chapter 4 that the solidus is the temperature at which the solid metal begins to melt as temperature is increased, and the liquidus is the temperature at which melting is completed. We now see from the phase diagram that these temperatures vary with composition. Between the solidus and liquidus, the metal is a solid–liquid mix.

Determining Chemical Compositions of Phases   Although the overall composition of the alloy is given by its position along the horizontal axis, the compositions of the liquid and solid phases are not the same. It is possible to determine these compositions from the phase diagram by drawing a horizontal line at the temperature of interest. The points of intersection between the horizontal line and the solidus and liquidus indicate the compositions of the solid and liquid phases present, respectively. We simply construct the vertical projections from the intersection points to the x-axis and read the corresponding compositions.

**EXAMPLE 6.1
Determining
Compositions from
the Phase Diagram**

To illustrate the procedure, suppose we want to analyze the compositions of the liquid and solid phases present in the copper–nickel system at an aggregate composition of 50% nickel and a temperature of 1260°C (2300°F).

Solution: A horizontal line is drawn at the given temperature level, as shown in Figure 6.2. The line intersects the solidus at a composition of 62% nickel, thus indicating the composition of the solid phase. The intersection with the liquidus occurs at a composition of 36%, corresponding to the analysis of the liquid phase.

As the temperature of the 50–50 Cu–Ni alloy is reduced, we reach the solidus line at about 1221°C (2230°F). Applying the same procedure used in the example, we find the composition of the solid metal is 50% nickel, and the composition of the last remaining liquid to freeze is about 26% nickel. How is it, the reader might ask, that the last ounce of molten metal has a composition so different from the solid metal into which it freezes? The answer is that the phase diagram assumes that equilibrium conditions are allowed to prevail. In fact, the binary phase diagram is sometimes called an equilibrium diagram due to this assumption. What it means is that sufficient time is permitted for diffusion to occur and for the solid metal to gradually change its composition to achieve that which is indicated by the intersection point along the liquidus. In practice, when an alloy freezes (e.g., a casting), *segregation* occurs in the solid mass because of nonequilibrium conditions. The first liquid to solidify has a composition that is rich in the metal element with the higher melting point. Then as additional metal solidifies, its composition is different from that of the first metal to freeze. As the nucleation sites grow into a solid mass, compositions within the mass are distributed, depending on the temperature and time in the process at which freezing occurred. The overall composition is the average of the distribution.

Determining Amounts of Each Phase   We can also determine the amounts of each phase present at a given temperature from the phase diagram. This is done by the *inverse lever rule*: (1) using the same horizontal line as before that indicates the overall composition at a given temperature, measure the distances between the aggregate composition and the intersection points with the liquidus and solidus, identifying the distances as CL and CS, respectively (refer back to Figure 6.2); (2) the proportion of liquid phase present is given by

$$L \text{ phase proportion} = \frac{CS}{(CS + CL)} \qquad (6.1)$$

(3) the proportion of solid phase present is given by

$$S \text{ phase proportion} = \frac{CL}{(CS + CL)} \qquad (6.2)$$

The proportions given by Eqs. (6.1) and (6.2) are by weight, same as the phase diagram percentages. Note that the proportions are based on the distance on the opposite side of the phase of interest; hence the name inverse lever rule. One can see the logic in this by taking the extreme case when, say, $CS = 0$; at that point, the proportion of the liquid phase is zero because we have reached the solidus and the alloy is therefore completely solidified.

The methods for determining chemical compositions of phases and the amounts of each phase are applicable to the solid region of the phase diagram as well as the liquidus–solidus region. Wherever there are regions in the phase diagram in which two phases are present, these methods can be utilized. When only one phase is present (in Figure 6.2, this is in the entire solid region), the composition of the phase is its aggregate composition under equilibrium conditions; and the inverse lever rule does not apply since there is only one phase.

The Tin–Lead Alloy System   A more complicated phase diagram is the Sn–Pb system, shown in Figure 6.3. Tin–lead alloys are widely used in soldering (Section 32.2) for making electrical connections. The phase diagram exhibits several features not included in the previous Cu–Ni system. One feature is the presence of two solid phases, alpha (α) and beta (β). The α phase is a solid solution of tin in lead at the left side of the diagram, and the β phase is a solid solution of lead in tin that occurs only at elevated temperatures around 200°C (375°F) at the right side of the diagram. Between these solid solutions lies a mixture of the two solid phases, $\alpha + \beta$.

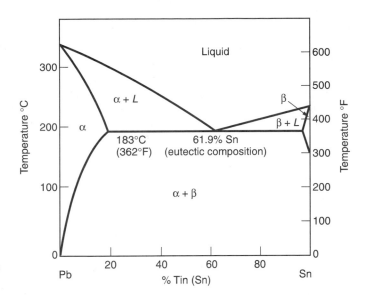

FIGURE 6.3   Phase diagram for the tin–lead alloy system.

Another feature of interest in the tin–lead system is how melting differs for different compositions. Pure tin melts at 232°C (449°F) and pure lead melts at 327°C (621°F). Alloys of these elements melt at lower temperatures. The diagram shows two liquidus lines that begin at the melting points of the pure metals and meet at a composition of 61.9% Sn. This is the eutectic composition for the tin–lead system. In general, a *eutectic alloy* is a particular composition in an alloy system for which the solidus and liquidus are at the same temperature. The corresponding *eutectic temperature,* the melting point of the eutectic composition, is 183°C (362°F) in the present case. The eutectic temperature is always the lowest melting point for an alloy system (eutectic is derived from the Greek word *eutektos,* meaning easily melted).

Methods for determining the chemical analysis of the phases and the proportions of phases present can be readily applied to the Sn–Pb system just as it was used in the Cu–Ni system. In fact, these methods are applicable in any region containing two phases, including two solid phases. Most alloy systems are characterized by the existence of multiple solid phases and eutectic compositions, and so the phase diagrams of these systems are often similar to the tin–lead diagram. Of course, many alloy systems are considerably more complex. We examine one of these as we consider next the alloys of iron and carbon.

6.2 FERROUS METALS

The ferrous metals are based on iron, one of the oldest metals known to man (Historical Note 6.1). The properties and other data relating to iron are itemized in Table 6.1(a). The ferrous metals of engineering importance are alloys of iron and carbon. These alloys divide into two major groups: steel and cast iron. Together, they constitute approximately 85% of the metal tonnage in the United States [3]. Let us begin our discussion of the ferrous metals by examining the iron–carbon phase diagram.

6.2.1 The Iron–Carbon Phase Diagram

The iron–carbon phase diagram is shown in Figure 6.4. Pure iron melts at 1539°C (2802°F). During the rise in temperature from ambient, it undergoes several solid phase transfor-

Historical Note 6.1   *Iron and steel ([2] and other sources).*

Iron was discovered sometime during the Bronze Age. It was probably uncovered from ashes of fires built near iron ore deposits. Use of the metal grew, finally surpassing bronze in importance. The Iron Age is usually dated from about 1200 B.C., although artifacts made of iron have been found in the Great Pyramid of Giza in Egypt, which dates to 2900 B.C. Iron-smelting furnaces have been discovered in Israel dating to 1300 B.C. Iron chariots, swords, and tools were made in ancient Assyria (northern Iraq) around 1000 B.C. The Romans inherited ironworking from their provinces, mainly Greece, and they developed the technology to new heights, spreading it throughout Europe. The ancient civilizations learned that iron was harder than bronze and that it took a sharper, stronger edge.

During the Middle Ages in Europe, the invention of the cannon created the first real demand for iron; only then did it finally exceed copper and bronze in usage. Also, the cast iron stove, the appliance of the seventeenth and eighteenth centuries, significantly increased demand for iron (Historical Note 11.3).

In the nineteenth century, industries such as railroads, shipbuilding, construction, machinery, and the military created a dramatic growth in the demand for iron and steel in Europe and the United States. Although large quantities of (crude) **pig iron** could be produced by **blast furnaces,** the subsequent processes for producing wrought iron and steel were slow. The necessity to improve productivity of these vital metals

was the "mother of invention." Henry Bessemer in England developed the process of blowing air up through the molten iron that led to the **Bessemer converter** (patented in 1856). Pierre and Emile Martin in France built the first **open hearth furnace** in 1864. These methods permitted up to 15 tons of steel to be produced in a single batch (heat), a substantial increase from previous methods.

In the United States, expansion of the railroads after the Civil War created a huge demand for steel. In the 1880s and 1890s, steel beams were first used in significant quantities in construction. Skyscrapers came to rely on these steel frames.

When electricity became available in abundant amounts in the late 1800s, this energy source was used for steelmaking. The first commercial **electric furnace** for production of steel was operated in France in 1899. By 1920, this had become the principal process for making alloy steels.

The use of pure oxygen in steelmaking was initiated just prior to World War II in several European countries and the United States. Work in Austria after the war culminated in the development of the **basic oxygen furnace** (BOF). This has become the leading modern technology for producing steel, surpassing the open hearth method around 1970. The Bessemer converter had been surpassed by the open hearth method around 1920 and ceased to be a commercial steelmaking process in 1971.

mations as indicated in the diagram. Starting at room temperature the phase is alpha (α), also called *ferrite.* At 912°C (1674°F), ferrite transforms to gamma (γ), called *austenite.* This, in turn, transforms at 1394°C (2541°F) to delta (δ), which remains until melting occurs. The three phases are distinct; alpha and delta have BCC structures, and between them, gamma is FCC.

Iron as a commercial product is available at various levels of purity. *Electrolytic iron* is the most pure, at about 99.99%, for research and other purposes where the pure metal is required. *Ingot iron,* containing about 0.1% impurities (including about 0.01% carbon), is used in applications where high ductility or corrosion resistance are needed. *Wrought iron* contains about 3% slag but very little carbon, and is easily shaped in hot forming operations such as forging.

TABLE 6.1   Basic data on the metallic elements: (a) iron.

Symbol:	Fe	Principal ore:	**Hematite** (Fe_2O_3)
Atomic number:	26	Alloying elements:	Carbon; also chromium, manganese, nickel, molybdenum, vanadium, and silicon.
Specific gravity:	7.87		
Crystal structure:	BCC		
Melting temperature:	1539°C (2802°F)	Typical applications:	Construction, machinery, automotive, railway tracks and equipment.
Elastic modulus:	209,000 MPa (30×10^6 lb/in^2)		

Compiled from [3], [7], [8], and other references.

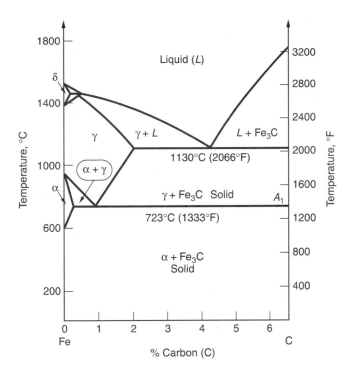

FIGURE 6.4   Phase diagram for iron–carbon system, up to about 6% carbon.

Solubility limits of carbon in iron are low in the ferrite phase—only about 0.022% at 723°C (1333°F). Austenite can dissolve up to about 2.1% carbon at a temperature of 1130°C (2066°F). This difference in solubility between alpha and gamma leads to opportunities for strengthening by heat treatment, but let us leave that for later. Even without heat treatment, the strength of iron increases dramatically as carbon content increases, and we enter the region in which the metal is called steel. More precisely, *steel* is defined as an iron–carbon alloy containing from 0.02% to 2.1% carbon; of course, steels can also contain other alloying elements as well.

We see in the diagram a eutectic composition at 4.3% carbon. There is a similar feature in the solid region of the diagram at 0.77% carbon and 723°C (1333°F). This is called the *eutectoid composition.* Steels below this carbon level are known as *hypoeutectoid steels,* and above this carbon level, from 0.77 to 2.1%, they are called *hypereutectoid steels.*

In addition to the phases mentioned, one other phase is prominent in the iron–carbon alloy system. This is Fe_3C, also known as *cementite,* an intermediate phase: a metallic compound of iron and carbon that is hard and brittle. At room temperature under equilibrium conditions, iron–carbon alloys form a two-phase system at carbon levels even slightly above zero. The carbon content in steel ranges between these very low levels and about 2.1%C. Above 2.1%C, up to about 4% or 5%, the alloy is defined as *cast iron.*

6.2.2 Iron and Steel Production

Our coverage of iron and steel production begins with the iron ores and other raw materials required. We then discuss ironmaking, in which iron is reduced from the ores; and steelmaking, in which it is then refined to obtain the desired purity and composition (alloying). We then consider the casting processes that are accomplished at the steel mill.

Iron Ores and Other Raw Materials   The principal ore used in the production of iron and steel is **hematite** (Fe_2O_3). Other iron ores include **magnetite** (Fe_3O_4), **siderite** ($FeCO_3$), and **limonite** ($Fe_2O_3-xH_2O$, where x is typically around 1.5). Iron ores contain from 50% to around 70% iron, depending on grade (hematite is almost 70% iron). In addition, scrap iron and steel are widely used today as raw materials in iron- and steelmaking.

Other raw materials needed to reduce iron from the ores are coke and limestone. **Coke** is a high carbon fuel produced by heating bituminous coal in a limited oxygen atmosphere for several hours, followed by water spraying in special quenching towers. Coke serves two functions in the reduction process: (1) it is a fuel that supplies heat for the chemical reactions; and (2) it produces carbon monoxide (CO) to reduce the iron ore. **Limestone** is a rock containing high proportions of calcium carbonate ($CaCO_3$). The limestone is used in the process as a flux to react with and remove impurities in the molten iron as slag.

Iron making   To produce iron, a charge of ore, coke, and limestone are dropped into the top of a blast furnace. A **blast furnace** is a refractory-lined chamber with a diameter of about 9 to 11 m (30 to 35 ft) at its widest and a height of 40 m (125 ft), in which hot gases are forced into the lower part of the chamber at high rates to accomplish combustion and reduction of the iron. A typical blast furnace and some of its technical details are illustrated in Figures 6.5 and 6.6. The charge slowly descends from the

FIGURE 6.5   Cross section of iron-making blast furnace showing major components.

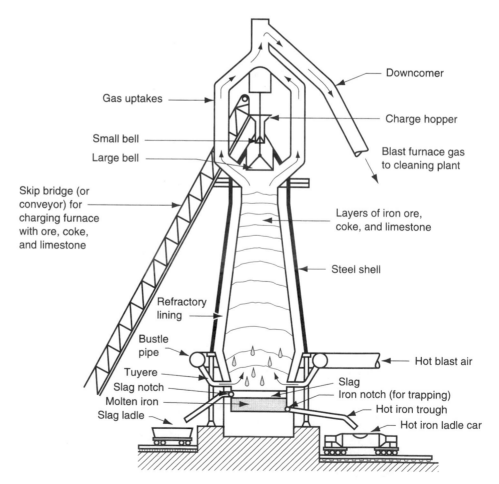

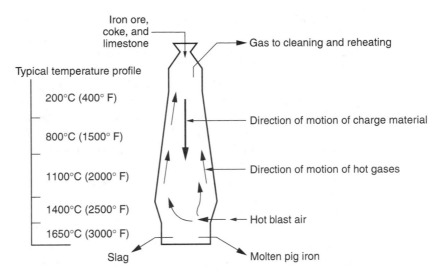

Iron ore, coke, and limestone

Gas to cleaning and reheating

Typical temperature profile

200°C (400° F)

800°C (1500° F)

Direction of motion of charge material

1100°C (2000° F)

Direction of motion of hot gases

1400°C (2500° F)

1650°C (3000° F)

Hot blast air

Slag

Molten pig iron

FIGURE 6.6 Schematic diagram indicating details of the blast furnace operation.

top of the furnace toward the base and is heated to temperatures around 1650°C (3000°F). Hot gases (CO, H_2, CO_2, H_2O, N_2, O_2, and fuels) pass upward through the layers of charge material to burn the coke. Carbon monoxide is supplied as hot gas, and is also formed from the combustion of coke. It has a reducing effect on the iron ore; the reaction (simplified) can be written as follows (using hematite as the starting ore):

$$Fe_2O_3 + CO \rightarrow 2FeO + CO_2 \tag{6.3a}$$

Carbon dioxide reacts with coke to form more carbon monoxide:

$$CO_2 + C \text{ (coke)} \rightarrow 2CO \tag{6.3b}$$

which then accomplishes the final reduction of FeO to iron:

$$FeO + CO \rightarrow Fe + CO_2 \tag{6.3c}$$

The molten iron drips downward, collecting at the base of the blast furnace. This is periodically tapped into hot iron ladle cars for transfer to subsequent steelmaking operations.

The role played by limestone can be summarized as follows. First, the limestone is reduced to lime (CaO) by heating, as follows:

$$CaCO_3 \rightarrow CaO + CO_2 \tag{6.4}$$

The lime combines with impurities such as silica (SiO_2), sulfur (S), and alumina (Al_2O_3) in reactions that produce a molten slag that floats on top of the iron.

It is instructive to note that approximately 7 tons of raw materials are required to produce 1 ton of iron. The ingredients are proportioned about as follows: 2.0 tons of iron ore, 1.0 ton of coke, 0.5 ton of limestone, and (here's the amazing statistic) 3.5 tons of gases. A significant proportion of the byproducts are recycled.

The iron tapped from the base of the blast furnace (called *pig iron*) contains more than 4% C, plus other impurities: 0.3–1.3% Si, 0.5–2.0% Mn, 0.1–1.0% P, and 0.02–0.08% S [7]. Further refinement of the metal is required for both cast iron and steel. A furnace called a *cupola* (Section 11.4.1) is commonly used for converting pig iron into gray cast iron. For steel, compositions must be more closely controlled and impurities brought to much lower levels.

Steelmaking   Since the mid-1800s, a number of processes have been developed for refining pig iron into steel. Today, the two most important processes are the basic oxygen furnace (BOF) and the electric furnace. Both are used to produce carbon and alloy steels.

The ***basic oxygen furnace*** accounts for about 70% of U.S. steel production. The BOF is an adaptation of the Bessemer converter. Whereas the Bessemer process used air blown up through the molten pig iron to burn off impurities, the basic oxygen process uses pure oxygen. A diagram of the conventional BOF during the middle of a heat is illustrated in Figure 6.7. The typical BOF vessel is about 5 m (16 ft) inside diameter and can process 150 to 200 tons in a heat.

The BOF steelmaking sequence is shown in Figure 6.8. Integrated steel mills transfer the molten pig iron from the blast furnace to the BOF in railway cars called hot-iron ladle cars. In modern practice, steel scrap is added to the pig iron, accounting for about 30% of a typical BOF charge. Lime (CaO) is also added. Color Plate 5 shows the furnace during charging. After charging, the lance is inserted into the vessel so that its tip is about 1.5 m (5 ft) above the surface of the molten iron. Pure O_2 is blown at high velocity through the lance, causing combustion and heating at the surface of the molten pool. Carbon dissolved in the iron and other impurities such as silicon, manganese, and phosphorous are oxidized. The reactions are:

$$2C + O_2 \rightarrow 2CO \ (CO_2 \text{ is also produced}) \tag{6.5a}$$

$$Si + O_2 \rightarrow SiO_2 \tag{6.5b}$$

$$2Mn + O_2 \rightarrow 2MnO \tag{6.5c}$$

$$4P + 5O_2 \rightarrow 2P_2O_5 \tag{6.5d}$$

The CO and CO_2 gases produced in the first reaction escape through the mouth of the BOF vessel and are collected by the fume hood; the products of the other three reactions are removed as slag, using the lime as a fluxing agent. The C content in the iron decreases almost linearly with time during the process, thus permitting fairly predictable control

FIGURE 6.7   Basic oxygen furnace showing BOF vessel during processing of a heat.

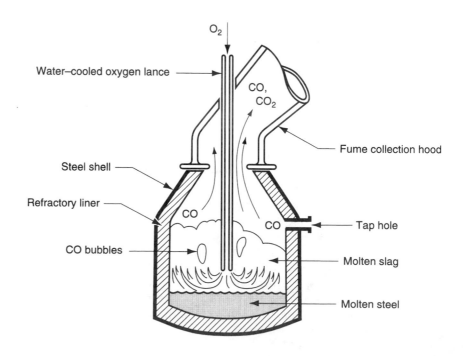

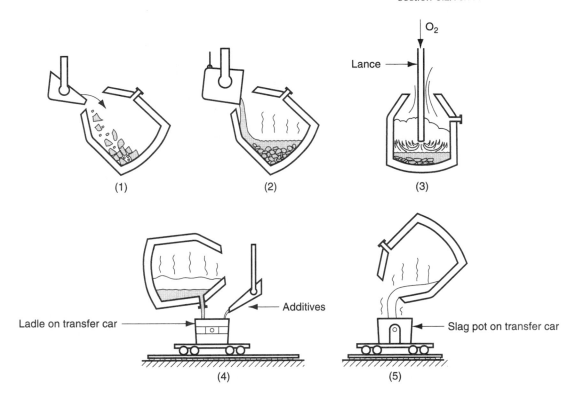

FIGURE 6.8   BOF sequence during processing cycle: (1) charging of scrap; and (2) pig iron; (3) blowing (Figure 6.7); (4) tapping the molten steel; and (5) pouring off the slag.

over carbon levels in the steel. After refining to the desired level, the molten steel is tapped; alloying ingredients and other additives are poured into the heat; then the slag is poured. A 200-ton heat of steel can be processed in about 20 minutes, although the entire cycle time (tap-to-tap time) takes about 45 minutes.

Recent advances in the technology of the basic oxygen process include the use of nozzles in the bottom of the vessel through which oxygen is injected into the molten iron. This allows better mixing than the conventional BOF lance, resulting in shorter processing times (a reduction of about 3 min), lower carbon contents, and higher yields.

The *electric arc furnace* accounts for about 30% of U.S. steel production. Although pig iron was originally used as the charge in this type of furnace, scrap iron and scrap steel are the primary raw materials today. Electric arc furnaces are available in several designs; the direct arc type shown in Figure 6.9 is currently the most economical type. These furnaces have removable roofs for charging from above; tapping is accomplished by tilting the entire furnace. Scrap iron and steel selected for their compositions, together with alloying ingredients and limestone (flux), are charged into the furnace and heated by an electric arc that flows between large electrodes and the charge metal. Complete melting requires about two hours; tap-to-tap time is four hours. Capacities of electric furnaces commonly range between 25 and 100 tons per heat. Electric arc furnaces are noted for better quality steel but higher cost per ton, compared to the BOF. The electric arc furnace is generally associated with production of alloy steels, tool steels, and stainless steels.

Casting of Ingots   Steels produced by BOF or electric furnace are solidified for subsequent processing either as cast ingots or by continuous casting. Steel *ingots* are large

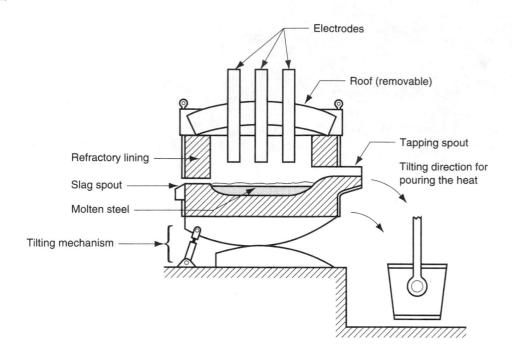

FIGURE 6.9 Electric arc furnace for steelmaking.

discrete castings weighing from less than one ton up to around 300 tons (the weight of an entire heat). Ingot molds are made of high carbon iron and are tapered at the top or bottom for removal of the solid casting. A **big-end-down mold** is illustrated in Figure 6.10. The cross section may be square, rectangular, or round, and the perimeter is usually corrugated to increase surface area for faster cooling. The mold is placed on a platform called a **stool**; after solidification the mold is lifted, leaving the casting on the stool.

The solidification process for ingots as well as other castings is described in our chapter on casting principles (Chapter 10). Because ingots are such large castings, the time required for solidification and the associated shrinkage are significant. Porosity caused by the reaction of carbon and oxygen to form CO during cooling and solidification is a problem that must be addressed in ingot casting. These gases are liberated from the molten steel due to their reduced solubility with decreasing temperature. Cast steels are often treated to limit or prevent CO gas evolution during solidification. The treatment involves adding elements such as Si and Al that react with the oxygen dissolved in the molten steel, so it is not available for CO reaction. The structure of the solid steel is thus free of pores and other defects caused by gas formation.

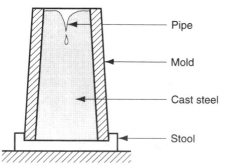

FIGURE 6.10 A big-end-down ingot mold typical of type used in steelmaking.

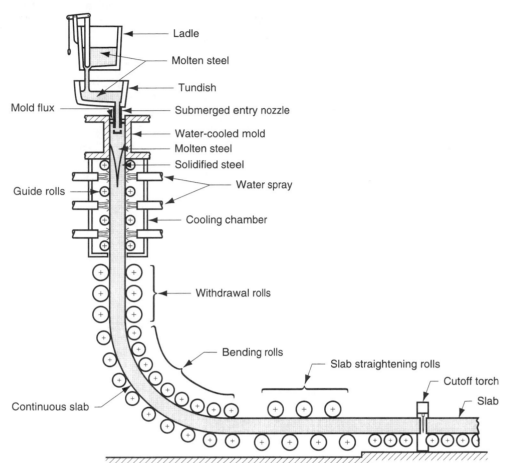

FIGURE 6.11 Continuous casting; steel is poured into tundish and distributed to a water-cooled continuous casting mold; it solidifies as it travels down through the mold. Thickness of slab is exaggerated for clarity.

Continuous Casting Continuous casting is widely applied in aluminum and copper production, but its most noteworthy application is in steelmaking. The process is replacing ingot casting due to dramatic increases in productivity. Ingot casting is a discrete process. Since the molds are relatively large, solidification time is significant. For a large steel ingot, it may take 10 to 12 hours for the casting to solidify. The use of continuous casting reduces solidification time by an order of magnitude.

The continuous casting process, also called *strand casting,* is illustrated in Figure 6.11. Molten steel is poured from a ladle into a temporary container called a *tundish,* which dispenses the metal to one or more continuous casting molds. The steel begins to solidify at the outer regions as it travels down through the water-cooled mold. Water sprays accelerate the cooling process. While still hot and plastic, the metal is bent from vertical to horizontal orientation. It is then cut into sections or fed continuously into a rolling mill (Section 21.1) in which it is formed into plate or sheet stock or other cross-sections.

6.2.3 Steels

Steel is an alloy of iron that contains carbon ranging by weight between 0.02% and 2.11%. It often includes other alloying ingredients as well: manganese, chromium, nickel, and molybdenum; but it is the carbon content that turns iron into steel. There are hundreds of compositions of steel available commercially. For purposes of organization here, they can be grouped into the following categories: (1) plain carbon steels, (2) low alloy steels, (3) stainless steels, and (4) tool steels.

Plain Carbon Steels These steels contain carbon as the principal alloying element, with only small amounts of other elements (about 0.5% manganese is normal). The strength of plain carbon steels increases with carbon content; a typical plot of the relationship is illustrated in Figure 6.12.

According to a designation scheme developed by the American Iron and Steel Institute (AISI) and the Society of Automotive Engineers (SAE), plain carbon steels are specified by a four-digit number system: 10XX, where 10 indicates that the steel is plain carbon, and XX indicates the percent of carbon in hundredths of percentage points. For example, 1020 steel contains 0.20% C. The plain carbon steels are typically classified into three groups according to their carbon content:

(1) *Low carbon steels* contain less than 0.20% C and are by far the most widely used steels. Typical applications are automobile sheetmetal parts, plate steel for fabrication, and railroad rails. These steels are relatively easy to form, which accounts for their popularity where high strength is not required. Steel castings usually fall into this carbon range, also.

(2) *Medium carbon steels* range in carbon between 0.20% and 0.50% and are specified for applications requiring higher strength than the low-C steels. Applications include machinery components and engine parts such as crankshafts and connecting rods.

(3) *High carbon steels* contain carbon in amounts greater than 0.50%. They are specified for still higher strength applications and where stiffness and hardness are needed. Springs, cutting tools and blades, and wear-resistant parts are examples.

Increasing carbon content strengthens and hardens the steel, but its ductility is reduced. Also, high carbon steels can be heat treated to form martensite, making the steel very hard and strong (Section 27.2).

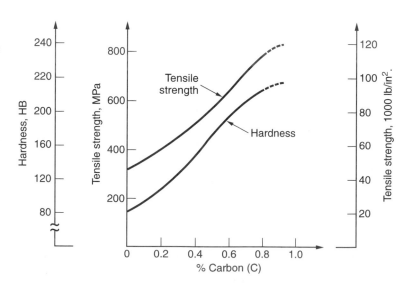

FIGURE 6.12 Tensile strength and hardness as a function of carbon content in plain carbon steel (hot rolled).

Low Alloy Steels Low alloy steels are iron–carbon alloys that contain additional alloying elements in amounts totaling less than about 5% by weight. Owing to these additions, low alloy steels have mechanical properties that are superior to those of the plain carbon steels for given applications. Superior properties usually mean higher strength, hardness, hot hardness, wear resistance, toughness, and more desirable combinations of these properties. Heat treatment is often required to achieve these improved properties.

Common alloying elements added to steel are chromium, manganese, molybdenum, nickel, and vanadium, sometimes individually but usually in combinations. These elements typically form solid solutions with iron and metallic compounds with carbon (carbides), assuming sufficient carbon is present to support a reaction. We can summarize the effects of the principal alloying ingredients as follows:

> *Chromium* (Cr) improves strength, hardness, wear resistance, and hot hardness. It is one of the most effective alloying ingredients for increasing hardenability (Section 27.2.3). In significant proportions, Cr improves corrosion resistance.

> *Manganese* (Mn) improves the strength and hardness of steel. When the steel is heat treated, hardenability is improved with increased manganese. Because of these benefits, manganese is a widely used alloying ingredient in steel.

> *Molybdenum* (Mo) increases toughness and hot hardness. It also improves hardenability and forms carbides for wear resistance.

> *Nickel* (Ni) improves strength and toughness. It increases hardenability but not as much as some of the other alloying elements in steel. In significant amounts it improves corrosion resistance and is the other major ingredient (besides chromium) in certain types of stainless steel.

> *Vanadium* (V) inhibits grain growth during elevated temperature processing and heat treatment, which enhances strength and toughness of steel. It also forms carbides that increase wear resistance.

The AISI–SAE designations of many of the low alloy steels are presented in Table 6.2, which indicates nominal chemical analysis. As before, carbon content is specified by XX in 1/100% of carbon. For completeness, we include plain carbon steels (10XX). To obtain an idea of the properties possessed by some of these steels, we have compiled Table 6.3, which lists the treatment to which the steel is subjected for strengthening and its strength and ductility.

Low alloy steels are not easily welded, especially at medium and high carbon levels. Since the 1960s, research has been directed at developing low carbon, low alloy steels that have better strength-to-weight ratios than plain carbon steels but are more weldable than low alloy steels. The products developed out of these efforts are called *high-strength low-alloy* (HSLA) steels. They generally have low carbon contents (in the range 0.10 to 0.30% C) plus relatively small amounts of alloying ingredients (usually only about 3% total of elements such as Mn, Cu, Ni, and Cr). HSLA steels are hot-rolled under controlled conditions designed to provide improved strength compared to plain C steels, yet with no sacrifice in formability or weldability. Strengthening is by solid solution alloying; heat treatment is not feasible due to low carbon content. Table 6.3 lists one HSLA steel, together with properties (chemistry is: 0.12 C, 0.60 Mn, 1.1 Ni, 1.1 Cr, 0.35 Mo, and 0.4 Si).

Stainless Steels *Stainless steels* are a group of highly alloyed steels designed to provide high corrosion resistance. The principal alloying element in stainless steel is

TABLE 6.2   AISI–SAE designations of steels.

Type	Name of Steel	Nominal Chemical Analysis, %							
		Cr	Mn	Mo	Ni	V	P	S	Si
10XX	Plain carbon		0.4				0.04	0.05	
11XX	Resulfurized		0.9				0.01	0.12	0.01
12XX	Resulfurized, rephosphorized		0.9				0.10	0.22	0.01
13XX	Manganese		1.7				0.04	0.04	0.3
20XX	Nickel steels		0.5		0.6		0.04	0.04	0.2
31XX	Nickel–chrome	0.6			1.0		0.04	0.04	0.3
40XX	Molybdenum		0.8	0.25			0.04	0.04	0.2
41XX	Chrome–molybdenum	1.0	0.8	0.2			0.04	0.04	0.3
43XX	Ni–Cr–Mo	0.8	0.7	0.25	1.8		0.04	0.04	0.2
46XX	Nickel–molybdenum		0.6	0.25	1.8		0.04	0.04	0.3
47XX	Ni–Cr–Mo	0.4	0.6	0.2	1.0		0.04	0.04	0.3
48XX	Nickel–molybdenum		0.6	0.25	3.5		0.04	0.04	0.3
50XX	Chromium	0.4	0.4				0.04	0.04	0.3
52XX	Chromium	1.4	0.4				0.02	0.02	0.3
61XX	Cr–Vanadium	0.8	0.8			0.1	0.04	0.04	0.3
81XX	Ni–Cr–Mo	0.4	0.8	0.1	0.3		0.04	0.04	0.3
86XX	Ni–Cr–Mo	0.5	0.8	0.2	0.5		0.04	0.04	0.3
88XX	Ni–Cr–Mo	0.5	0.8	0.35	0.5		0.04	0.04	0.3
92XX	Silicon		0.8				0.04	0.04	2.0
93XX	Ni–Cr–Mo	1.2	0.6	0.1	3.0		0.02	0.02	0.3
98XX	Ni–Cr–Mo	0.8	0.8	0.25	1.0		0.04	0.04	0.3

Source: [7].

chromium, usually above 15%. The chromium in the alloy forms a thin, impervious oxide film in an oxidizing atmosphere, which protects the surface from corrosion. Nickel is another alloying ingredient used in certain stainless steels to increase corrosion protection. Carbon is used to strengthen and harden the metal; however, increasing the carbon content has the effect of reducing corrosion protection because chromium carbide forms to reduce the amount of free Cr available in the alloy.

TABLE 6.3   Treatments and mechanical properties of selected steels.

AISI Code	Treatment[a]	Tensile Strength		Elongation, %
		MPa	(lb/in.^{2})	
1010	HR	304	(44,000)	47
1010	CD	366	(53,000)	12
1020	HR	380	(55,000)	28
1020	CD	421	(61,000)	15
1040	HR	517	(75,000)	20
1040	CD	587	(85,000)	10
1055	HT	897	(130,000)	16
1315	None	545	(79,000)	34
2030	None	566	(82,000)	32
3130	HT	697	(101,000)	28
4130	HT	890	(129,000)	17
4140	HT	918	(133,000)	16
4815	HT	635	(92,000)	27
9260	HT	994	(144,000)	18
HSLA	None	586	(85,000)	20

Compiled from [3], [7], and other sources.

[a] HR = hot-rolled; CD = cold-drawn; HT = heat treatment involving heating and quenching, followed by tempering to produce tempered martensite (Section 27.2).

In addition to corrosion resistance, stainless steels are noted for their combination of strength and ductility. Although these properties are desirable in many applications, they generally make these alloys difficult to work in manufacturing. Also, stainless steels are significantly more expensive than plain C or low alloy steels.

Stainless steels are traditionally divided into three groups, named for the predominant phase present in the alloy at ambient temperature:

(1) *Austenitic stainless* These steels have a typical composition of 18% Cr and 8% Ni, and are the most corrosion resistant of the three groups. Owing to this composition, they are sometimes identified as 18-8 stainless. They are nonmagnetic and very ductile; but they show significant work hardening. The nickel has the effect of enlarging the austenite region in the iron–carbon phase diagram, making it stable at room temperature. Austenitic stainless steels are used to fabricate chemical and food processing equipment, as well as machinery parts requiring high corrosion resistance.

(2) *Ferritic stainless* These steels have around 15% to 20% chromium, low carbon, and no nickel. This provides a ferrite phase at room temperature. Ferritic stainless steels are magnetic and are less ductile and corrosion resistant than the austenitics. Parts made of ferritic stainless range from kitchen utensils to jet engine components.

(3) *Martensitic stainless* These steels have a higher carbon content than ferritic stainlesses, thus permitting them to be strengthened by heat treatment (Section 27.2). They have as much as 18% Cr but no Ni. They are strong, hard, and fatigue resistant, but not generally as corrosion resistant as the other two groups. Typical products include cutlery and surgical instruments.

Most stainless steels are designated by a three-digit AISI numbering scheme. The first digit indicates the general type, and the last two digits give the specific grade within the type. Table 6.4 lists the common stainless steels with typical compositions and mechanical properties.

The traditional stainless steels were developed in the early 1900s. Since then, several additional high alloy steels have been developed that have good corrosion resistance and other desirable properties. These are also classified as stainless steels; continuing our list:

(4) *Precipitation hardening stainless* A typical composition is 17% Cr and 7% Ni, with additional small amounts of alloying elements such as aluminum, copper, titanium, and molybdenum. Their distinguishing feature among stainlesses is that they can be strengthened by precipitation hardening (Section 27.3). Strength and corrosion resistance are maintained at elevated temperatures, which suits these alloys to aerospace applications.

(5) *Duplex stainless* These steels possess a structure that is a mixture of austenite and ferrite in roughly equal amounts. Their corrosion resistance is similar to the austenitic grades, and they show improved resistance to stress-corrosion cracking. Applications include heat exchangers, pumps, and wastewater treatment plants.

Tool Steels *Tool steels* are a class of (usually) highly alloyed steels designed for use as industrial cutting tools, dies, and molds. To perform in these applications, they must possess high strength, hardness, hot hardness, wear resistance, and toughness under impact. To obtain these properties, tool steels are heat treated. Principal reasons for the

TABLE 6.4 Compositions and mechanical properties of selected stainless steels.

Type	Chemical Analysis, %						Tensile Strength		Elongation, %
	Fe	Cr	Ni	C	Mn	Other	MPa	(lb/in.^{2})	
Austenitic									
301	73	17	7	0.15	2		620	(90,000)	40
302	71	18	8	0.15	2		515	(75,000)	40
304	69	19	9	0.08	2		515	(75,000)	40
309	61	23	13	0.20	2		515	(75,000)	40
316	65	17	12	0.08	2	2.5 Mo	515	(75,000)	40
Ferritic									
405	85	13	—	0.08	1		415	(60,000)	20
430	81	17	—	0.12	1		415	(60,000)	20
Martensitic									
403	86	12	—	0.15	1		485	(70,000)	20
403[b]	86	12	—	0.15	1		825	(120,000)	12
416	85	13	—	0.15	1		485	(70,000)	20
416[b]	85	13	—	0.15	1		965	(140,000)	10
440	81	17	—	0.65	1		725	(105,000)	20
440[b]	81	17	—	0.65	1		1790	(260,000)	5

Compiled from [7]).

[a] All of the grades in the table contain about 1% (or less) Si plus small amounts (well below 1%) of phosphorous and sulfur and other elements such as aluminum.

[b] Heat treated.

high levels of alloying elements are (1) improved hardenability, (2) reduced distortion during heat treatment, (3) hot hardness, (4) formation of hard metallic carbides for abrasion resistance, and (5) enhanced toughness.

The tool steels divide into major types, according to application and composition. The AISI uses a classification scheme that includes a prefix letter to identify the tool steel. In the following listing of tool steel types, we identify the prefix, and present some typical compositions in Table 6.5:

T, M   ***High-speed tool steels***—These are used as cutting tools in machining processes (Section 23.2.2). They are formulated for high wear resistance and hot hardness. The original high-speed steels (HSS) were developed around 1900. They permitted dramatic increases in cutting speed compared to previously used

TABLE 6.5 Tool steels by AISI prefix identification, with examples of composition and typical hardness values.

AISI	Example	Chemical Analysis, %[a]							Hardness, HRC
		C	Cr	Mn	Mo	Ni	V	W	
T	T1	0.7	4.0				1.0	18.0	65
M	M2	0.8	4.0		5.0		2.0	6.0	65
H	H11	0.4	5.0		1.5		0.4		55
D	D1	1.0	12.0		1.0				60
A	A2	1.0	5.0		1.0				60
O	O1	0.9	0.5	1.0				0.5	61
W	W1	1.0							63
S	S1	0.5	1.5					2.5	50
P	P20	0.4	1.7		0.4				40[b]
L	L6	0.7	0.8		0.2	1.5			45[b]

[a]% composition rounded to nearest tenth.

[b] Hardness estimated.

tools; hence their name. The two AISI designations indicate the principal alloying element: T for tungsten and M for molybdenum.

H   *Hot-working tool steels*—These are intended for hot-working dies for forging, extrusion, and diecasting.

D   *Cold-work tool steels*—These die steels are used for cold working operations such as sheetmetal pressworking, cold extrusion, and certain forging operations. The designation D stands for die. Closely related AISI designations are A and O. A and O stand for air- and oil-hardening. They all provide good wear resistance and low distortion.

W   *Water-hardening tool steels*—These tool steels have high carbon with little or no other alloying elements. They can only be hardened by fast quenching in water. They are widely used because of low cost, but they are limited to low temperature applications. Cold heading dies are a typical application.

S   *Shock-resistant tool steels*—These tool steels are intended for use in applications where high toughness is required, as in many sheetmetal shearing, punching, and bending operations.

P   *Mold steels*—These are used to make molds for molding plastics and rubber.

L   *Low alloy tool steels*—These are generally reserved for special applications.

Tool steels are not the only tool materials. Plain carbon, low alloy, and stainless steels are used for many tool and die applications. Cast irons and certain nonferrous alloys are also suitable for tooling applications. In addition, several of the ceramic materials are seeing increased service as high-speed cutting inserts, abrasives, and other tools.

6.2.4 Cast Irons

Cast iron is an iron alloy containing from 2.1% to about 4% carbon and from 1% to 3% silicon. Its composition makes it highly suitable as a casting metal. In fact, the tonnage of cast iron castings is several times that of all other cast metal parts combined (excluding cast ingots made during steelmaking that are subsequently rolled into bars, plates, and similar stock). The overall tonnage of cast iron is second only to steel among metals.

There are several types of cast iron, the most important being gray cast iron. Other types include ductile iron, white cast iron, malleable iron, and various alloy cast irons. Typical chemical compositions of gray and white cast irons are shown in Figure 6.13, indicating their relationship with cast steel. Ductile and malleable irons possess chemistries similar to the gray and white cast irons, respectively, but result from special treatments to be described below. Table 6.6 presents a listing of chemistries for the principal types together with mechanical properties.

Gray Cast Iron   Gray cast iron accounts for the largest tonnage among the cast irons. It has a composition in the range 2.5% to 4% carbon and 1% to 3% silicon. This chemistry results in the formation of graphite (carbon) flakes distributed throughout the cast product upon solidification. The structure causes the surface of the metal to have a gray color when fractured; hence the name gray cast iron. The dispersion of graphite flakes accounts for two attractive properties: (1) good vibration damping, which is desirable in engines and other machinery; and (2) internal lubricating qualities, which makes the cast metal machinable.

The strength of gray cast iron spans a significant range. The American Society for Testing of Materials (ASTM) uses a classification method for gray cast iron that is in-

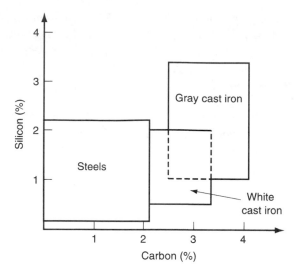

FIGURE 6.13 Carbon and silicon compositions for cast irons, with comparison to steels (most steels have relatively low silicon contents—cast steels have the higher Si content). Ductile iron is formed by special melting and pouring treatment of gray cast iron, and malleable iron is formed by heat treatment of white cast iron.

tended to provide a minimum tensile strength specification for the various classes: Class 20 gray cast iron has a *TS* of 20,000 lb/in.^{2}, Class 30 has a *TS* of 30,000 lb/in.^{2}, and so forth, up to around 70,000 lb/in.^{2} (see Table 6.6 for equivalent *TS* in metric units). The compressive strength of gray cast iron is significantly greater than its tensile strength. Properties of the casting can be controlled to some extent by heat treatment. Ductility of gray cast iron is very low; it is a relatively brittle material. Products made from gray cast iron include automotive engine blocks and heads, motor housings, and machine tool bases.

Ductile Iron This is an iron with the composition of gray iron in which the molten metal is chemically treated before pouring to cause the formation of graphite spheroids rather than flakes. This results in a stronger and more ductile iron—hence the name *ductile iron.* Applications include machinery components requiring high strength and good wear resistance.

TABLE 6.6 Compositions and mechanical properties of selected cast irons.

Type	Typical Composition, %					Tensile Strength		Elongation, %
	Fe	C	Si	Mn	Other	MPa	(lb/in.^{2})	
Gray cast irons								
Class 20	93.0	3.5	2.5	0.65		138	(20,000)	0.6
Class 30	93.6	3.2	2.1	0.75		207	(30,000)	0.6
Class 40	93.8	3.1	1.9	0.85		276	(40,000)	0.6
Class 50	93.5	3.0	1.6	1.0	0.67 Mo	345	(50,000)	0.6
Ductile irons								
ASTM A395	94.4	3.0	2.5			414	(60,000)	18
ASTM A476	93.8	3.0	3.0			552	(80,000)	3
White cast iron								
Low-C	92.5	2.5	1.3	0.4	1.5 Ni, 1 Cr, 0.5 Mo	276	(40,000)	0
Malleable irons								
Ferritic	95.3	2.6	1.4	0.4		345	(50,000)	10
Pearlitic	95.1	2.4	1.4	0.8		414	(60,000)	10

Compiled from [7].

Cast irons are identified by various systems. We have attempted to indicate the particular cast iron grade using the most common identification for each type.

[a] Cast irons also contain phosphorous and sulfur usually totaling less than 0.3%.

White Cast Iron This cast iron has less carbon and silicon than gray cast iron. It is formed by more rapid cooling of the molten metal after pouring, thus causing the carbon to remain chemically combined with iron in the form of cementite (Fe_3C), rather than precipitating out of solution in the form of flakes. When fractured, the surface has a white crystalline appearance that gives the iron its name. Owing to the cementite, white cast iron is hard and brittle, and its wear resistance is excellent. Strength is good, with *TS* of 276 MPa (40,000 lb/in.^{2}) being typical. These properties makes white cast iron suitable for applications where wear resistance is required. Railway brake shoes are an example.

Malleable Iron When castings of white cast iron are heat treated to separate the carbon out of solution and form graphite aggregates, the resulting metal is called malleable iron. The new microstructure can possess substantial ductility (up to 20% elongation)— a significant difference from the metal out of which it was transformed. Typical products made of malleable cast iron include pipe fittings and flanges, certain machine components, and railroad equipment parts.

Alloy Cast Irons Cast irons can be alloyed for special properties and applications. These alloy cast irons are classified as follows: (1) heat-treatable types that can be hardened by martensite formation; (2) corrosion-resistant types, whose alloying elements include nickel and chromium; and (3) heat-resistant types containing high proportions of nickel for hot hardness and resistance to high temperature oxidation.

6.3 NONFERROUS METALS

The nonferrous metals include metal elements and alloys not based on iron. The most important engineering metals in the nonferrous group are aluminum, copper, magnesium, nickel, titanium, and zinc, and their alloys.

Although the nonferrous metals as a group cannot match the strength of the steels, certain nonferrous alloys have corrosion resistance and/or strength-to-weight ratios that make them competitive with steels in moderate-to-high stress applications. In addition, many of the nonferrous metals have properties other than mechanical that make them ideal for applications in which steel would be quite unsuitable. For example, copper has one of the lowest electrical resistivities among metals and is widely used for electrical wire. Aluminum is an excellent thermal conductor, and its applications include heat exchangers and cooking pans. It is also one of the most readily formed metals, and is valued for that reason also. Zinc has a relatively low melting point, so zinc is widely used in die-casting operations. The common nonferrous metals have their own combination of properties that make them attractive in a variety of applications. In the following nine sections, we discuss the nonferrous metals that are the most commercially and technologically important.

6.3.1 Aluminum and Its Alloys

Aluminum and magnesium are light metals, and they are often specified in engineering applications for this feature. Both elements are abundant on Earth—aluminum on land and magnesium in the sea, although neither is easily extracted from the states in which they are found naturally.

TABLE 6.1 (continued): (b) Aluminum.

Symbol:	Al		Principal ore:	Bauxite [impure mix of Al_2O_3 and $Al(OH)_3$]
Atomic number:	13			
Specific gravity:	2.7		Alloying elements:	Copper, magnesium, manganese, silicon, and zinc
Crystal structure:	FCC			
Melting temperature:	660°C (1220°F)		Typical applications:	Containers (aluminum cans), wrapping foil, electrical conductors, pots and pans, parts for construction, aerospace, automotive, and other uses where light weight is important
Elastic modulus:	69,000 MPa (10×10^6 lb/in.^{2})			

Properties and other data on aluminum are listed in Table 6.1(b). Among the major metals, it is a relative newcomer, dating only to the late 1800s (Historical Note 6.2). Our coverage in this section includes (1) a brief description of how aluminum is produced and (2) a discussion of the properties and the designation system for the metal and its alloys.

Historical Note 6.2 *Aluminum* [2].

In 1807, the English chemist Humphrey Davy, believing that the mineral **alumina** (Al_2O_3) had a metallic base, attempted to extract the metal. He did not succeed, but was sufficiently convinced that he proceeded to name the metal anyway: **alumium,** later changing the name to **aluminum.** In 1825, the Danish physicist/chemist Hans Orsted finally succeeded in separating the metal. He noted that it "resembles tin." In 1845, the German physicist Friedrich Wohler was the first to determine the specific gravity, ductility, and various other properties of aluminum.

The modern electrolytic process for producing aluminum was based on the concurrent but independent work of Charles Hall in the United States and Paul Heroult in France around 1886. In 1888, Hall and a group of businessmen started the Pittsburgh Reduction Co. The first ingot of aluminum was produced by the electrolytic smelting process that same year. Demand for aluminum grew. The need for large amounts of electricity in the production process led the company to relocate in Niagara Falls in 1895 where hydroelectric power was becoming available at very low cost. In 1907, the company changed its name to the Aluminum Company of America (Alcoa). It was the sole producer of aluminum in the United States until World War II.

Aluminum Production The principal aluminum ore is *bauxite,* which consists largely of hydrated aluminum oxide (Al_2O_3–H_2O) and other oxides. Extraction of the aluminum from bauxite can be summarized in three steps: (1) washing and crushing the ore into fine powders; (2) the Bayer process, in which the bauxite is converted to pure alumina (Al_2O_3); and (3) electrolysis, in which the alumina is separated into aluminum and oxygen gas (O_2). The *Bayer process,* named after the German chemist who developed it, involves solution of bauxite powders in aqueous caustic soda (NaOH) under pressure, followed by precipitation of pure Al_2O_3 from solution. Alumina is commercially important in its own right as an engineering ceramic (Chapter 7).

Electrolysis to separate Al_2O_3 into its constituent elements requires dissolving the precipitate in a molten bath of cryolite (Na_3AlF_6) and subjecting the solution to direct-current between the plates of an electrolytic furnace. The electrolyte dissociates to form aluminum at the cathode and oxygen gas at the anode.

Properties and Designation Scheme Aluminum has high electrical and thermal conductivity, and its resistance to corrosion is excellent due to the formation of a hard thin oxide surface film. It is a very ductile metal and is noted for its formability. Pure alu-

TABLE 6.7(a)   Designations of wrought and cast aluminum alloys.

Alloy Group	Wrought Code	Cast Code
Aluminum, 99.0% or higher purity	1XXX	1XX.X
Aluminum alloys, by major element(s):		
Copper	2XXX	2XX.X
Manganese	3XXX	
Silicon + copper and/or magnesium		3XX.X
Silicon	4XXX	4XX.X
Magnesium	5XXX	5XX.X
Magnesium and silicon	6XXX	
Zinc	7XXX	7XX.X
Tin		8XX.X
Other	8XXX	9XX.X

minum is relatively low in strength, but it can be alloyed and heat treated to compete with some of the steels, especially when weight is taken into consideration.

The designation system for aluminum alloys is a four-digit code number. The system has two parts, one for wrought aluminums and the other for cast aluminums. The difference is that a decimal point is used following the third digit for cast aluminums. The designations are presented in Table 6.7(a).

Because properties of aluminum alloys are so influenced by work hardening and heat treatment, the temper (strengthening treatment, if any) must be designated in addition to the composition code. The principal temper designations are presented in Table 6.7(b). This designation is attached to the preceding four-digit number, separated from it by a hyphen, to indicate the treatment or absence thereof; for example, 1060-F. Of course, temper treatments that specify strain hardening do not apply to the cast alloys. Some ex-

TABLE 6.7(b)   Temper designations for aluminum alloys.

Temper	Description
F	As fabricated—no special treatment.
H	Strain hardened (wrought aluminums). H is followed by two digits, the first indicating a heat treatment, if any; and the second indicating the degree of work hardening remaining; for example:
	H1X No heat treatment after strain hardening, and X = 1 to 9, indicating degree of work hardening.
	H2X Partially annealed, and X = degree of work hardening remaining in product.
	H3X Stabilized, and X = degree of work hardening remaining. *Stabilized* means heating to slightly above service temperature anticipated.
O	Annealed to relieve strain hardening and improve ductility; reduces strength to lowest level.
T	Thermal treatment to produce stable tempers other than F, H, or O. It is followed by a digit to indicate specific treatments; for example:
	T1 = cooled from elevated temperature, naturally aged.
	T2 = cooled from elevated temperature, cold worked, naturally aged.
	T3 = solution heat treated, cold worked, naturally aged.
	T4 = solution heat treated and naturally aged.
	T5 = cooled from elevated temperature, artificially aged.
	T6 = solution heat treated and artificially aged.
	T7 = solution heat treated and overaged or stabilized.
	T8 = solution heat treated, cold worked, artificially aged.
	T9 = solution heat treated, artificially aged, and cold worked.
	T10 = cooled from elevated temperature, cold worked, and artificially aged.
W	Solution heat treatment, applied to alloys that age harden in service; it is an unstable temper.

TABLE 6.8   Compositions and mechanical properties of selected aluminum alloys.

Code	Typical Composition, %*						Temper	Tensile Strength		Elongation, %
	Al	Cu	Fe	Mg	Mn	Si		MPa	(lb/in.²)	
1050	99.5		0.4			0.3	O	76	(11,000)	39
							H18	159	(23,000)	7
1100	99.0		0.6			0.3	O	90	(13,000)	40
							H18	165	(24,000)	10
2024	93.5	4.4	0.5	1.5	0.6	0.5	O	185	(27,000)	20
							T3	485	(70,000)	18
3004	96.5	0.3	0.7	1.0	1.2	0.3	O	180	(26,000)	22
							H36	260	(38,000)	7
4043	93.5	0.3	0.8			5.2	O	130	(19,000)	25
							H18	285	(41,000)	1
5050	96.9	0.2	0.7	1.4	0.1	0.4	O	125	(18,000)	18
							H38	200	(29,000)	3
6063	98.5		0.3	0.7		0.4	O	90	(13,000)	25
							T4	172	(25,000)	20

Compiled from [8].

[a] In addition to elements listed, alloys may contain trace amounts of other elements such as copper, magnesium, manganese, vanadium, and zinc.

amples of the remarkable differences in the mechanical properties of aluminum alloys that result from the different treatments are presented in Table 6.8.

6.3.2 Magnesium and Its Alloys

Magnesium (Mg) is the lightest of the structural metals. Its specific gravity and other basic data are presented in Table 6.1(c). Magnesium and its alloys are available in both wrought and cast forms. It is relatively easy to machine. However, in all processing of magnesium, small particles of the metal (such as small metal cutting chips) oxidize rapidly, and care must be taken to avoid fire hazards.

Magnesium Production   Sea water contains about 0.13% $MgCl_2$, and this is the source of most commercially produced magnesium. To extract Mg, a batch of sea water is mixed with milk of lime—calcium hydroxide ($Ca(OH)_2$). The resulting reaction precipitates magnesium hydroxide ($Mg(OH)_2$) that settles and is removed as a slurry. The slurry is then filtered to increase ($Mg(OH)_2$) content. This mixture is next mixed with hydrochloric acid (HCl), which reacts with the hydroxide to form concentrated $MgCl_2$—much more concentrated than the original sea water. Electrolysis is used to decompose the salt into magnesium (Mg) and chlorine gas (Cl_2). The magnesium is then cast into ingots for subsequent processing. The chlorine is recycled to form more $MgCl_2$.

Properties and Designation Scheme   As a pure metal, magnesium is relative soft and lacks sufficient strength for most engineering applications. However, it can be al-

TABLE 6.1   (continued): (c) Magnesium.

Symbol:	Mg	Extracted from:	$MgCl_2$ in sea water by electrolysis
Atomic number:	12	Alloying elements:	See Table 6.9
Specific gravity:	1.74	Typical applications:	Aerospace, missiles, bicycles, chain
Crystal structure:	HCP		saw housings, luggage, and other
Melting temperature:	650°C (1202°F)		applications where light weight is a
Elastic modulus:	48,000 MPa (7×10^6 lb/in.²)		primary requirement

TABLE 6.9   Code letters used to identify alloying elements in magnesium alloys.

A	aluminum (Al)	H	thorium (Th)	M	manganese (Mn)	Q	silver (Ag)	T	tin (Sn)
E	rare earth metals	K	zirconium (Zr)	P	lead (Pb)	S	silicon (Si)	Z	zinc (Zn)

TABLE 6.10   Compositions and mechanical properties of selected magnesium alloys.

Code	Typical Composition, %						Process	Tensile Strength		Elongation, %
	Mg	Al	Mn	Si	Zn	Other		MPa	(lb/in.^{2})	
AZ10A	98.0	1.3	0.2	0.1	0.4		Wrought	240	(35,000)	10
AZ80A	91.0	8.5			0.5		Forged	330	(48,000)	11
HM31A	95.8		1.2			3.0 Th	Wrought	283	(41,000)	10
ZK21A	97.1				2.3	6 Zr	Wrought	260	(38,000)	4
AM60	92.8	6.0	0.1	0.5	0.2	0.3 Cu	Cast	220	(32,000)	6
AZ63A	91.0	6.0			3.0		Cast	200	(29,000)	6

Compiled from [8].

loyed and heat treated to achieve strengths comparable to aluminum alloys. In particular, its strength-to-weight ratio is an advantage in aircraft and missile components.

The designation scheme for magnesium alloys uses a three-to-five character alphanumeric code. The first two characters are letters that identify the principal alloying elements (up to two elements can be specified in the code, in order of decreasing percentages, or alphabetically if equal percentages). These code letters are listed in Table 6.9. The letters are followed by a two-digit number that indicates, respectively, the amounts of the two alloying ingredients to the nearest%. Finally, the last symbol is a letter that indicates some variation in composition, or simply the chronological order in which it was standardized for commercial availability. Magnesium alloys also require specification of a temper, and the same basic scheme presented in Table 6.7(b) for aluminum is used for magnesium alloys.

Some examples of magnesium alloys, illustrating the designation scheme and indicating tensile strength and ductility of these alloys, are presented in Table 6.10.

6.3.3 Copper and Its Alloys

Copper (Cu) is one of the oldest metals known (Historical Note 6.3). Basic data on the element copper are presented in Table 6.1(d).

Historical Note 6.3   *Copper* [2].

Copper was one of the first metals used by human cultures (gold was the other). Discovery of the metal was probably around 6000 B.C. At that time, copper was found in the free metallic state. Ancient peoples fashioned implements and weapons out of it by hitting the metal (cold forging). Pounding copper made it harder (strain hardening); this and its attractive reddish color made it valuable in early civilizations.

Around 4000 B.C., it was discovered that copper could be melted and cast into useful shapes. It was later found that copper mixed with tin could be more readily cast and worked than the pure metal. This led to the widespread use of bronze and the subsequent naming of the Bronze Age, dated from about 2000 B.C. to the time of Christ.

To the ancient Romans, the island of Cyprus was almost the only source of copper. They called the metal *aes cyprium* (ore of Cyprus). This was shortened to **Cyprium** and subsequently renamed **Cuprium.** From this derives the chemical symbol Cu.

TABLE 6.1 (continued): (d) Copper.

Symbol:	Cu	Ore extracted from:	Several: e.g., chalcopyrite ($CuFeS^2$).
Atomic number:	29	Alloying elements:	Tin (bronze), zinc (brass), aluminum,
Specific gravity:	8.96		silicon, nickel, and beryllium
Crystal structure:	FCC	Typical applications:	Electrical conductors and components,
Melting temperature:	1083°C (1981°F)		ammunition (brass), pots and pans, jewelry,
Elastic modulus:	110,000 MPa (16×10^6 lb/in.^{2})		plumbing, marine applications, heat exchangers,
			springs (Be–Cu)

Copper Production In ancient times, copper was available in nature as a free element. Today these natural deposits are more difficult to find, and copper is now extracted from ores that are mostly sulfides, such as *chalcopyrite* ($CuFeS_2$). The ore is crushed (Section 17.1.1), concentrated by flotation, and then *smelted* (melted or fused, often with an associated chemical reaction to separate a metal from its ore). The resulting copper is called *blister copper,* which is between 98% and 99% pure. Electrolysis is used to obtain higher purity levels suitable for commercial use.

Properties and Designation Scheme Pure copper has a distinctive reddish-pink color, but its most distinguishing engineering property is its low electrical resistivity—one of the lowest of all elements. Because of this property, and its relative abundance in nature, commercially pure copper is widely used as an electrical conductor (we should note here that the conductivity of copper decreases significantly as alloying elements are added). Cu is also an excellent thermal conductor. Copper is one of the noble metals (gold and silver are also noble metals), so it is corrosion resistant. All of these properties combine to make copper one of the most important metals.

On the downside, strength and hardness of copper are relatively low, especially when weight is taken into account. Accordingly, to improve strength (as well as for other reasons), copper is frequently alloyed. *Bronze* is an alloy of copper and tin (typically, about 90% Cu and 10% Sn), still widely used today despite its ancient ancestry. Additional bronze alloys have been developed, based on other elements than tin; these include aluminum bronzes, and silicon bronzes. *Brass* is another familiar copper alloy, composed of copper and zinc (typically around 65% Cu and 35% Zn). The highest strength alloy of copper is beryllium–copper (only about 2% Be). It can be heat treated to tensile strengths of 1035 MPa (150,000 lb/in.^{2}). Be–Cu alloys are used for springs.

TABLE 6.11 Compositions and mechanical properties of selected copper alloys.

	Typical Composition, %					Tensile Strength		
Code	Cu	Be	Ni	Sn	Zn	MPa	(lb/in.^{2})	Elongation, %
C10100	99.99					235	(34,000)	45
C11000	99.95					220	(32,000)	45
C17000	98.0	1.7	[a]			500	(70,000)	45
C24000	80.0				20.0	290	(42,000)	52
C26000	70.0				30.0	300	(44,000)	68
C52100	92.0			8.0		380	(55,000)	70
C71500	70.0		30.0			380	(55,000)	45
C71500[b]	70.0		30.0			580	(84,000)	3

Compiled from [8].

[a] Small amounts of Ni and Fe + 0.3 Co.

[b] Heat treated for high strength.

The designation of copper alloys is based on the Unified Numbering System for Metals and Alloys (UNS), which uses a five-digit number preceded by the letter C (C for copper). The alloys are processed in wrought and cast forms, and the designation system includes both. Some copper alloys with compositions and mechanical properties are presented in Table 6.11.

6.3.4 Nickel and Its Alloys

Nickel (Ni) is similar to iron in many respects; see Table 6.1(e). It is magnetic, and its modulus of elasticity is virtually the same as that of iron and steel. However, it is much more corrosion resistant, and the high temperature properties of its alloys are generally superior. Because of its corrosion-resistant characteristics, it is widely used as an alloying element in steel, such as stainless steel, and as a plating metal on other metals such as plain carbon steel.

Nickel Production   The most important ore of nickel is **pentlandite** $((Ni, Fe)_9S_8)$. To extract the nickel, the ore is first crushed and ground with water. Flotation techniques are used to separate the sulfides from other minerals mixed with the ore. The nickel sulfide is then heated to burn off some of the sulfur, followed by smelting to remove iron and silicon. Further refinement is accomplished in a Bessemer-style converter to yield high-concentration nickel sulfide (NiS). Electrolysis is then utilized to recover high-purity nickel from the compound. Ores of nickel are sometimes mixed with copper ores, and the recovery technique described here also yields copper in these cases.

Nickel Alloys   Alloys of nickel are commercially important in their own right and are noted for corrosion resistance and high temperature performance. Composition, tensile strength, and ductility of some of the nickel alloys are given in Table 6.12. In addition, a number of superalloys are based on nickel (Section 6.4).

TABLE 6.1   (continued): (e) Nickel.

Symbol:	Ni	Ore extracted from:	Pentlandite $((Fe, Ni)_9S_8)$
Atomic number:	28	Alloying elements:	Copper, chromium, iron, aluminum
Specific gravity:	8.90	Typical applications:	Stainless steel alloying ingredient,
Crystal structure:	FCC		plating metal for steel, applications
Melting temperature:	1453°C (2647°F)		requiring high temperature and
Elastic Modulus:	209,000 MPa (30×10^6 lb/in.^{2})		corrosion resistance

TABLE 6.12   Compositions and mechanical properties of selected nickel alloys.

Code	Typical Composition, %							Tensile Strength		Elongation, %
	Ni	Cr	Cu	Fe	Mn	Si	Other	MPa	(lb/in^2)	
270	99.9		a	a				345	(50,000)	50
200	99.0		0.2	0.3	0.2	0.2	C, S	462	(67,000)	47
400	66.8		30.0	2.5	0.2	0.5	C	550	(80,000)	40
600	74.0	16.0	0.5	8.0	1.0	0.5		655	(95,000)	40
230	52.8	22.0		3.0	0.4	0.4	b	860	(125,000)	47

Compiled from [8].

[a]Trace amounts

[b]Other alloying ingredients in Grade 230: 5% Co, 2% Mo, 14% W, .3% Al, .1% C.

TABLE 6.1 (continued): (f) Titanium.

Symbol:	Ti	Ores extracted from:	Rutile (TiO$_2$) and Ilmenite (FeTiO$_3$)
Atomic number:	22	Alloying elements:	Aluminum, tin, vanadium, copper, and
Specific gravity:	4.51		magnesium.
Crystal structure:	HCP	Typical applications:	Jet engine components, other
Melting temperature:	1668°C (3034°F)		aerospace applications, prosthetic
Elastic modulus:	117,000 MPa (17×10^6 lb/in.^{2})		implants.

6.3.5 Titanium and Its Alloys

Titanium (Ti) is fairly abundant in nature, constituting about 1% of Earth's crust (aluminum, the most abundant, is about 8%). The density of Ti is between aluminum and iron; this and other data are presented in Table 6.1(f). Its importance has grown in recent decades due to its aerospace applications where its light weight and good strength-to-weight ratio are exploited.

Titanium Production   The principal ores of titanium are *rutile,* which is 98 to 99% TiO$_2$, and *ilmenite,* which is a combination of FeO and TiO$_2$. Rutile is preferred as an ore due to its higher Ti content. In recovery of the metal from its ores, the TiO$_2$ is converted to titanium tetrachloride (TiCl$_4$) by reacting the compound with chlorine gas. This is followed by a sequence of distillation steps to remove impurities. The highly concentrated TiCl$_4$ is then reduced to metallic titanium by reaction with magnesium; this is known as the *Kroll process.* Sodium can also be used as a reducing agent. In either case, an inert atmosphere must be maintained to prevent O$_2$, N$_2$ or H$_2$ from contaminating the Ti, owing to its chemical affinity for these gases. The resulting metal is used to cast ingots of titanium and its alloys.

Properties of Titanium   Ti's coefficient of thermal expansion is relatively low among metals. It is stiffer and stronger than aluminum, and it retains good strength at elevated temperatures. Pure titanium is reactive, which presents problems in processing, especially in the molten state. However, at room temperature it forms a thin adherent oxide coating (TiO$_2$) that provides excellent corrosion resistance.

These properties give rise to two principal application areas for titanium: (1) in the commercially pure state, Ti is used for corrosion resistant components, such as marine components and prosthetic implants; and (2) titanium alloys are used as high strength components in temperatures ranging from ambient to above 550°C (1000°F), especially

TABLE 6.13 Compositions and mechanical properties of selected titanium alloys.

Code[a]	Typical Composition, %						Tensile Strength		Elongation, %
	Ti	Al	Cu	Fe	V	Other	MPa	(lb/in.^{2})	
R50250	99.8			0.2			240	(35,000)	24
R56400	89.6	6.0		0.3	4.0	[b]	1000	(145,000)	12
R54810	90.0	8.0			1.0	1 Mo,[b]	985	(143,000)	15
R56620	84.3	6.0	0.8	0.8	6.0	2 Sn,[b]			

Compiled from [8].

[a]United Numbering System (UNS).

[b]Traces of C, H, O.

where its excellent strength-to-weight ratio is exploited. These latter applications include aircraft and missile components. Some of the alloying elements used with titanium include aluminum, manganese, tin, and vanadium. Some compositions and mechanical properties for several alloys are presented in Table 6.13.

6.3.6 Zinc and Its Alloys

Table 6.1(g) lists basic data on zinc. Its low melting point makes it attractive as a casting metal. It also provides corrosion protection when coated onto steel or iron; *galvanized steel* is steel that has been coated with zinc.

Production of Zinc   Zinc blende or *sphalerite* is the principal ore of zinc; it contains zinc sulfide (ZnS). Other important ores of zinc include *smithsonite,* which is zinc carbonate ($ZnCO_3$), and *hemimorphate,* which is hydrous zinc silicate ($Zn_4Si_2O_7OH-H_2O$),

Sphalerite must be concentrated (*beneficiated,* as it is called) due to the small fraction of zinc sulfide present in the ore. This is accomplished by first crushing the ore, then grinding with water in a ball mill (Section 17.1.1) to create a slurry. In the presence of a frothing agent, the slurry is agitated so that the mineral particles float to the top and can be skimmed off (separated from the lower grade minerals). The more concentrated zinc sulfide is then roasted at around 1260°C (2300°F), so that zinc oxide (ZnO) is formed from the reaction.

There are various thermochemical processes for recovering zinc from this oxide, all of which reduce zinc oxide by means of carbon. The carbon combines with oxygen in ZnO to form CO and/or CO_2, thus freeing Zn in the form of vapor that is condensed to yield the desired metal.

An electrolytic process is also widely used, accounting for about half the world's production of zinc. This process also begins with the preparation of ZnO, which is mixed with dilute sulfuric acid (H_2SO_4), followed by electrolysis to separate the resulting zinc sulfate ($ZnSO_4$) solution to yield the pure metal.

Zinc Alloys and Applications   Zinc alloys are widely used in die casting to mass produce components for the automotive and appliance industries. Another major application of zinc is in galvanized steel. As the name suggests, a galvanic cell is created in galvanized steel (Zn is the anode and steel is the cathode) that protects the steel from corrosive attack. Finally, a third important use of zinc is in brass. As previously indicated, this alloy consists of the two metals copper and zinc, in the ratio of about 2/3 Cu to 1/3 Zn. Brass was covered in our discussion of copper. Several alloys of zinc are listed in Table 6.14, with data on composition, tensile strength, and applications.

TABLE 6.1   (continued): (g) zinc.

Symbol:	Zn	Ore extracted from:	Sphalerite (ZnS)
Atomic number:	30	Alloying elements:	Aluminum, magnesium, copper
Specific gravity:	7.13	Typical applications:	Galvanized steel and iron, die
Crystal structure:	HCP		castings, alloying element in brass
Melting temperature:	419°C (786°F)		
Elastic modulus:	90,000 MPa (13×10^6 lb/in.^{2})[a]		

[a]Zinc creeps, which makes it difficult to measure modulus of elasticity; some tables of properties omit E for zinc for this reason.

TABLE 6.14   Compositions, tensile strength, and applications of selected zinc alloys.

Code[a]	Typical Composition, %					Tensile Strength		Application
	Zn	Al	Cu	Mg	Fe	MPa	(lb/in.^{2})	
Z33520	95.6	4.0	0.25	0.04	0.1	283	(41,000)	Die casting
Z35540	93.4	4.0	2.5	0.04	0.1	359	(52,000)	Die casting
Z35635	91.0	8.0	1.0	0.02	0.06	374	(54,000)	Foundry alloy
Z35840	70.9	27.0	2.0	0.02	0.07	425	(62,000)	Foundry alloy
Z45330	98.9		1.0	0.01		227	(33,000)	Rolled alloy

Compiled from [8].
[a]UNS-Unified Numbering System for metals.

6.3.7 Lead and Tin

Lead (Pb) and tin (Sn) are often considered together because of their low melting temperatures, and because they form the soldering alloys used in making electrical connections. The phase diagram for the tin–lead alloy system is depicted in Figure 6.3. Basic data for lead and tin are presented in Table 6.1(h).

TABLE 6.1   (continued): (h) lead and tin.

	Lead	Tin
Symbol:	Pb	Sn
Atomic number:	82	50
Specific gravity:	11.35	7.30
Crystal structure:	FCC	HCP
Melting temperature:	327°C (621°F)	232°C (449°F)
Modulus of elasticity:	21,000 MPa (3×10^6 lb/in.^{2})	42,000 MPa (6×10^6 lb/in.^{2})
Ore from which extracted:	Galena (PbS)	Cassiterite (SnO$_2$)
Typical alloying elements:	Tin, antimony	Lead, copper
Typical applications:	See text	Bronze, solder, tin cans

Lead is a dense metal with a low melting point; other properties include low strength, low hardness (the word *soft* is appropriate), high ductility, and good corrosion resistance. In addition to its use in solder, applications of lead and its alloys include: pipes for plumbing, bearings, ammunition, type metals, x-ray shielding, storage batteries, and vibration damping. It has also been widely used in chemicals and paints. Principal alloying elements with lead are tin and antimony.

Tin has an even lower melting point than lead; other properties include low strength, low hardness, and good ductility. The earliest use of tin was in bronze, the alloy consisting of copper and tin developed around 3000 B.C. in Mesopotamia and Egypt. Bronze is still an important commercial alloy (although its relative importance has declined during 5,000 years). Other uses of tin include tin-coated sheet steel containers ("tin cans") for storing food and, of course, solder metal.

6.3.8 Refractory Metals

The **refractory metals** are metals capable of enduring high temperatures. The most important metals in this group are molybdenum and tungsten; see Table 6.1(i). Other re-

TABLE 6.1   (continued): (i) refractory metals.

	Molybdenum	Tungsten
Symbol:	Mo	W
Atomic number:	42	74
Specific gravity:	10.2	19.3
Crystal structure:	BCC	BCC
Melting point:	2619°C (4730°F)	3400°C (6150°F)
Elastic modulus:	324,000 MPa (47×10^6 lb/in.^{2})	407,000 MPa (59×10^6 lb/in.^{3})
Principal ores:	Molybdenite (MoS_2)	Scheelite ($CaWO_4$), Wolframite (($Fe,Mn)WO_4$)
Alloying elements:	See text	[a]
Applications:	See text	Light filaments, rocket engine parts, WC tools

[a]Tungsten is used as a pure metal and as an alloying ingredient, but few alloys are based on W.

fractory metals are columbium (Cb) and tantalum (Ta). In general, these metals and their alloys are capable of maintaining high strength and hardness at elevated temperatures.

Molybdenum has a high melting point and is relatively dense, stiff, and strong. It is used both as a pure metal (99.9 + % Mo) and as an alloy. The principal alloy is TZM, which contains small amounts of titanium and zirconium (less than 1% total). Mo and its alloys possess good high temperature strength, and this accounts for many of its applications, which include heat shields, heating elements, electrodes for resistance welding, dies for high temperature work (e.g., die casting molds), and parts for rocket and jet engines. In addition to these applications, molybdenum is also widely used as an alloying ingredient in other metals, such as steels and superalloys.

Tungsten (W) has the highest melting point among metals and is one of the densest. It is also the stiffest and hardest of all pure metals. Its most familiar application is filament wire in incandescent light bulbs. Applications of tungsten are typically characterized by high operating temperatures, such as parts for rocket and jet engines and electrodes for arc welding. W is also widely used as an element in tool steels, heat resistant alloys, and tungsten carbide (Section 7.3.2).

A major disadvantage of both Mo and W is their propensity to oxidize at high temperatures, above about 600°C (1000°F), thus detracting from their high temperature properties. To overcome this deficiency, either protective coatings must be used on these metals in high temperature applications or the metal parts must operate in a vacuum. For example, the tungsten filament must be energized in a vacuum inside the glass light bulb.

6.3.9 Precious Metals

The precious metals, also called the ***noble metals*** because they are chemically inactive, include gold, platinum, and silver. They are attractive metals, available in limited supply, and have been used throughout civilized history for coinage and to underwrite paper currency. They are also widely used in jewelry and similar applications that exploit their high value. As a group, the precious metals possess high density, good ductility, high electrical conductivity and corrosion resistance, and moderate melting temperatures; see Table 6.1(j).

Gold (Au) is one of the heaviest metals; it is soft and easily formed, and possesses a distinctive yellow color that adds to its value. In addition to currency and jewelry, its applications include electrical contacts (owing to its good electrical conductivity and cor-

Table 6.1 (continued): (j) the precious metals.

	Gold	Platinum	Silver
Symbol:	Au	Pt	Ag
Atomic number:	79	78	47
Specific gravity:	19.3	21.5	10.5
Crystal structure:	FCC	FCC	FCC
Melting temperature:	1063°C (1945°F)	1769°C (3216°F)	961°C (1762°F)
Principal ores:	a	a	a
Alloying elements:	b	b	b
Applications:	See text	See text	See text

[a]All three precious metals are mined from deposits in which the pure metal is mixed with other ores and metals. Silver is also mined from the ore **Argentite** (Ag_2S).

[b]The precious metals are generally not alloyed.

rosion resistance), dental work, and plating onto other metals for decorative purposes. **Platinum** (Pt) is the only metal (among the common metals) whose density is greater than that of gold. Although not as widely used as gold, its applications are diverse and include jewelry, thermocouples, electrical contacts, and catalytic pollution control equipment for automobiles.

Silver (Ag) is less expensive per unit weight than gold or platinum. Nevertheless, its attractive "silvery" luster makes it a highly valued metal in coins, jewelry, and tableware (which even assumes the name of the metal: *silverware*). It is also used for fillings in dental work. Silver has the highest electrical conductivity of any metal, which makes it useful for contacts in electronics applications. Finally, it should be mentioned that light-sensitive silver chloride and other silver halides are the basis for photography.

6.4 SUPERALLOYS

Superalloys constitute a category that straddles the ferrous and nonferrous metals. Some of them are based on iron, while others are based on nickel and cobalt. In fact, many of the superalloys contain substantial amounts of three or more metals, rather than consisting of one base metal plus alloying elements. Although the tonnage of these metals is not significant compared to most of the other metals we have discussed in this chapter, they are nevertheless commercially important because they are very expensive; and they are technologically important because of what they can do.

The **superalloys** are a group of high-performance alloys designed to meet very demanding requirements for strength and resistance to surface degradation (corrosion and oxidation) at high service temperatures. Conventional room temperature strength is usually not the important criterion for these metals, and most of them possess room temperature strength properties that are good but not outstanding. Their high temperature performance is what distinguishes them; tensile strength, hot hardness, creep resistance, and corrosion resistance at very elevated temperatures are the mechanical properties of interest. Operating temperatures are often in the vicinity of 1100°C (2000°F). These metals are widely used in gas turbines—jet and rocket engines, steam turbines, and nuclear power plants—systems in which operating efficiency increases with higher temperatures.

The superalloys are usually divided into three groups, according to their principal constituent: iron, nickel, or cobalt:

TABLE 6.15   Some typical superalloy compositions.

Superalloy	Chemical Analysis, %[a]								
	Fe	Ni	Co	Cr	Mo	W	Nb	Ti	Other[b]
Iron-based									
Incoloy 802	46	32		21				<1	<1
Haynes 556	29	20	20	22	3				6
Nickel-based									
Incoloy 807	25	40	8	21		5			1
Inconel 718	18	53		19	3		5		1
Rene 41		55	11	19	1			3	2
Hastelloy S	1	67		16	15				1
Nimonic 75	3	76		20				<1	<1
Cobalt-based									
Stellite 6B	3	3	53	30	2	5			4
Haynes 188	3	22	39	22		14			
L-605		10	53	20		15			2

Compiled from [8].

[a]Compositions given to the nearest%; percentages less than 1 are indicated by <1.

[b]These other elements include: carbon, tungsten, manganese, and silicon.

> *Iron-based alloys.*   These alloys have iron as the main ingredient, although in some cases the iron is less than 50% of the total composition.

> *Nickel-based alloys.*   These alloys generally have better high temperature strength than alloy steels. Nickel is the base metal. The principal alloying elements are chromium and cobalt; lesser elements include aluminum, titanium, molybdenum, niobium (Nb), and iron. Some familiar names in this group include Inconel, Hastelloy, and Rene 41.

> *Cobalt-based alloys.*   The main elements in these alloys are cobalt (around 40%) and chromium (perhaps 20%); other alloying elements include nickel, molybdenum, and tungsten.

TABLE 6.16   Strength properties of the superalloys at room temperature and 870°C (1600°F).

Superalloy	Tensile Strength at Room Temperature		Tensile Strength at 870°C (1600°F)	
	MPa	(lb/in.^{2})	MPa	(lb/in.^{2})
Iron-based				
Incoloy 802	690	(100,000)	195	(28,000)
Haynes 556	815	(118,000)	330	(48,000)
Nickel-based				
Incoloy 807	655	(95,000)	220	(32,000)
Inconel 718	1435	(208,000)	340	(49,000)
Rene 41	1420	(206,000)	620	(90,000)
Hastelloy S	845	(130,000)	340	(50,000)
Nimonic 75	745	(108,000)	150	(22,000)
Cobalt-based				
Stellite 6B	1010	(146,000)	385	(56,000)
Haynes 188	960	(139,000)	420	(61,000)
L-605	1005	(146,000)	325	(47,000)

Compiled from [7].

In virtually all of the superalloys, including those based on iron, strengthening is accomplished by precipitation hardening. The iron-based superalloys do not use martensite formation for strengthening. Typical compositions for some of the alloys are presented in Table 6.15, while strength properties for the same alloys at room temperature and elevated temperature are displayed in Table 6.16.

6.5 GUIDE TO THE PROCESSING OF METALS

A wide variety of manufacturing processes are available to shape metals, enhance their properties, assemble them, and finish them for appearance and protection.

Shaping, Assembly, and Finishing Processes   Metals are shaped by all of the basic processes, including casting, powder metallurgy, deformation processes, and material removal. In addition, metal parts are joined to form assemblies by processes such as welding, brazing and soldering, and mechanical fastening. Heat treating is performed to enhance properties. And finishing processes are commonly used to improve the appearance of metal parts or to provide corrosion protection. These finishing operations include electroplating and painting. Table 6.17 provides a road map to the many metal processing technologies described in this book.

Enhancement of Mechanical Properties in Metals   Mechanical properties of metals can be altered by a number of techniques. We have referred to some of these techniques in our discussion of the various metals. Methods for enhancing mechanical properties of metals can be grouped into three categories: (1) alloying, (2) cold working, and (3) heat treatment. Some of these methods are used at the very beginning of the manufacturing sequence, while others are applied at the very end. *Alloying* has been discussed throughout this chapter and is an important technique for strengthening metals.

Cold working has previously been referred to as strain hardening; its effect is to increase strength and reduce ductility. The degree to which these mechanical properties are affected depends on the amount of strain and the strain hardening exponent in the flow curve, Eq. (3.10). Cold working can be used on both pure metals and alloys. It is accomplished during deformation of the workpart by one of the shape forming processes, such as rolling, forging, or extrusion. Strengthening of the metal therefore occurs as a byproduct of the shaping operation.

Heat treatment refers to several types of heating and cooling cycles performed on a metal to beneficially change its properties. They operate by altering the basic mi-

TABLE 6.17   Processes for metals in manufacturing.

Process	Chapter	Process	Chapter
Casting	10 and 11	Heat treatment	27
Powder metallurgy	16	Cleaning and surface treatments	28
Forming processes:		Coating processes	29
Bulk deformation	19	Assembly processes:	
Sheetmetal working	20	Welding	30 and 31
Material removal processes:		Brazing and soldering	32
Conventional machining	21 and 22	Mechanical assembly	33
Abrasive processes	25		
Nontraditional processes	26		

crostructure of the metal, which in turn determines mechanical properties. Some heat treatment operations are applicable only to certain types of metals; for example, the heat treatment of steel to form martensite is somewhat specialized since martensite is unique to steel. Heat treatments are described in Chapter 27.

REFERENCES

[1] Brick, R. M., Pense, A. W., and Gordon, R. B. *Structure and Properties of Engineering Materials.* 4th ed. McGraw-Hill, New York, 1977.

[2] *Encyclopaedia Britannica,* Vol. 21, *Macropaedia,* Encyclopaedia Britannica, Inc., Chicago, 1990, under section: Industries, Extraction and Processing.

[3] Flinn, R. A., and Trojan, P. K. *Engineering Materials and Their Applications.* 4th ed. Houghton Mifflin, Boston, 1990.

[4] Guy, A. G., and Hren, J. J. *Elements of Physical Metallurgy.* 3rd ed. Addison-Wesley, Reading, Mass., 1974.

[5] Hume-Rothery, W., Smallman, R. E., and Haworth, C. W. *The Structure of Metals and Alloys.* Institute of Materials, London, England, 1988.

[6] Lankford, W. T., Jr., Samways, N. L., Craven, R. F., and McGannon, H. E. *The Making, Shaping, and Treating of Steel.* 10th ed. United States Steel Co., Pittsburgh, Penn., 1985.

[7] *Metals Handbook.* 10th ed. Volume 1, *Properties and Selection: Iron, Steels, and High Performance Alloys,* ASM International, Metals Park, Oh., 1990.

[8] *Metals Handbook.* 10th ed. Volume 2, *Properties and Selection: Nonferrous Alloys and Special Purpose Materials.* ASM International, Metals Park, Ohio, 1990.

[9] Moore, C., and Marshall, R. I. *Steelmaking.* The Institute for Metals, The Bourne Press, Ltd., Bournemouth, U.K., 1991.

[10] Wick, C., and Veilleux, R. F. (eds.), *Tool and Manufacturing Engineers Handbook.* 4th ed. Volume 3—*Materials, Finishing, and Coating,* Society of Manufacturing Engineers, Dearborn, Mich., 1985.

REVIEW QUESTIONS

6.1. What are some of the general properties that distinguish metals from ceramics and polymers?

6.2. What are the two major groups of metals? Define them

6.3. What is the definition of an *alloy*?

6.4. What is a solid solution in the context of alloys?

6.5. Distinguish between a substitutional solid solution and an interstitial solid solution.

6.6. What is an intermediate phase in the context of alloys?

6.7. The copper–nickel system is a simple alloy system, as indicated by its phase diagram. Why is it so simple?

6.8. What is the range of carbon percentages that defines an iron–carbon alloy as a steel?

6.9. What is the range of carbon percentages that defines an iron–carbon alloy as cast iron?

6.10. Identify some of the common alloying elements other than carbon in low alloy steels.

6.11. What are some of the mechanisms by which the alloying elements other than carbon strengthen steel.

6.12. What is the mechanism by which carbon strengthens steel in the absence of heat treatment?

6.13. What is the predominant alloying element in all of the stainless steels?

6.14. Why is austenitic stainless steel called by that name?

6.15. Besides high carbon content, what other alloying element is characteristic of the cast irons?

6.16. Identify some of the properties for which aluminum is noted?

6.17. What are some of the noteworthy properties of magnesium?

6.18. What is the most important engineering property of copper that determines most of its applications?

6.19. What elements are traditionally alloyed with copper to form (a) bronze and (b) brass?

6.20. What are some of the important applications of nickel?

6.21. What are the noteworthy properties of titanium?

6.22. Identify some of the important applications of zinc.

6.23. What important alloy is formed from lead and tin?

6.24. Name the important refractory metals. What does the term refractory mean?

6.25. Name the four principal noble metals. Why are they called noble metals?

6.26. The superalloys divide into three basic groups, according to the base metal used in the alloy. Name the three groups.

6.27. What is so special about the superalloys? What distinguishes them from other alloys?

6.28. What are the three basic methods by which metals can be strengthened?

MULTIPLE CHOICE QUIZ

There is a total of 23 correct answers in the following multiple choice questions (some questions have multiple answers that are correct). To attain a perfect score on the quiz, all correct answers must be given, since each correct answer is worth 1 point. For each question, each omitted answer or wrong answer reduces the score by 1 point, and each additional answer beyond the number of answers required reduces the score by 1 point. percentage score on the quiz is based on the total number of correct answers.

6.1. Which of the following properties or characteristics are inconsistent with the metals (more than one)? (a) good thermal conductivity, (b) high strength, (c) high electrical resistivity, (d) high stiffness, or (e) ionic bonding.

6.2. Which of the metallic elements is the most abundant on Earth? (a) aluminum, (b) copper, (c) iron, (d) magnesium, or (e) silicon.

6.3. The predominant phase in the iron-carbon alloy system for a composition with 99% Fe at room temperature is which of the following? (a) austenite, (b) cementite, (c) delta, (d) ferrite, or (e) gamma.

6.4. A steel with 1.0% carbon is known as which of the following: (a) eutectoid, (b) hypoeutectoid, (c) hypereutectoid, or (d) wrought iron.

6.5. The strength and hardness of steel increase as carbon conntent increases: (a) true or (b) false.

6.6. Plain carbon steels are designated in the AISI code system by which of the following? (a) 01XX, (b) 10XX, (c) 11XX, (d) 12XX, or (e) 30XX.

6.7. Which of the following elements is the most important alloying ingredient in steel? (a) carbon, (b) chromium, (c) nickel, (d) molybdenum, or (e) vanadium.

6.8. Which of the following is not a common alloying ingredient in steel? (a) chromium, (b) manganese, (c) nickel, (d) vanadium, or (e) zinc.

6.9. Solid solution alloying is the principal strengthening mechanism in high-strength low-alloy (HSLA) steels: (a) true or (b) false.

6.10. Which of the following alloying elements are most commonly associated with stainless steel (name two)? (a) chromium, (b) manganese, (c) molybdenum, (d) nickel, and (e) tungsten.

6.11. Which of the following is the most important cast iron commercially? (a) ductile cast iron, (b) gray cast iron, (c) malleable iron, or (d) white cast iron.

6.12. Which of the following metals has the lowest density? (a) aluminum, (b) magnesium, (c) tin, or (d) titanium.

6.13. Which of the following metals has the highest density? (a) gold, (b) lead, (c) platinum, (d) silver, or (e) tungsten.

6.14. From which of the following ores is aluminum derived? (a) alumina, (b) bauxite, (c) cementite, (d) hematite, or (e) scheelite.

6.15. Which of the following metals possess good electrical conductivity (more than one)? (a) aluminum, (b) copper, (c) gold, (d) silver, or (e) tungsten.

6.16. Traditional brass is an alloy of which of the following metallic elements? (a) aluminum, (b) copper, (c) gold, (d) tin, or (e) zinc.

6.17. Which of the following has the lowest melting point? (a) aluminum, (b) lead, (c) magnesium, (d) tin, or (e) zinc.

PROBLEMS

6.1. For the copper–nickel phase diagram in Figure 6.2, find the compositions of the liquid and solid phases for a nominal composition of 70% Ni and 30% Cu at 1371°C (2500°F).

6.2. For the preceding problem, use the inverse lever rule to determine the proportions of liquid and solid phases present in the alloy.

6.3. For the lead–tin phase diagram of Figure 6.3, is it possible to design a solder (lead-tin alloy) with a melting point of 260°C (500°F)? If so, what would be its nominal composition?

6.4. Using the lead–tin phase diagram in Figure 6.3, determine the liquid and solid phase compositions for a nominal composition of 40% Sn and 60% Pb at 204°C (400°F).

6.5. For the preceding problem, use the inverse lever rule to determine the proportions of liquid and solid phases present in the alloy.

6.6. Using the lead–tin phase diagram in Figure 6.3, determine the liquid and solid phase compositions for a nominal composition of 90% Sn and 10% Pb at 204°C (400°F).

6.7. For the preceding problem, use the inverse lever rule to determine the proportions of liquid and solid phases present in the alloy.

6.8. In the iron–iron carbide phase diagram of Figure 6.4, identify the phase or phases present at the following temperatures and nominal compositions: (a) 650°C (1200°F) and 2% Fe_3C, (b) 760°C (1400°F) and 2% Fe_3C, and (c) 1095°C (2000°F) and 1% Fe_3C.

7

CERAMICS

CHAPTER CONTENTS

7.1 Structure and Properties of Ceramics
   7.1.1 Mechanical Properties
   7.1.2 Physical Properties
7.2 Traditional Ceramics
   7.2.1 Raw Materials
   7.2.2 Traditional Ceramic Products
7.3 New Ceramics
   7.3.1 Oxide Ceramics
   7.3.2 Carbides
   7.3.3 Nitrides
7.4 Glass
   7.4.1 Chemistry and Properties of Glass
   7.4.2 Glass Products
   7.4.3 Glass Ceramics
7.5 Some Important Elements Related to Ceramics
   7.5.1 Carbon
   7.5.2 Silicon
   7.5.3 Boron
7.6 Guide to Processing Ceramics

Engineers traditionally consider metals to be the most important class of engineering materials. However, it is interesting to note that ceramic materials are actually more abundant and widely used. Included in this category are clay products (bricks, tile, pottery, and chinaware), glass, cement, and concrete. (Concrete is a composite material, but its two components are both ceramic.) Also included are modern ceramic materials such as tungsten carbide and cubic boron nitride.

The importance of ceramics as engineering materials is based on their abundance in nature and their mechanical and physical properties, which are quite different from those of metals. A *ceramic* material is an inorganic compound consisting of a metal (or semimetal) and one or more nonmetals. Important examples of ceramic materials are *silica,* or silicon dioxide (SiO_2), the main ingredient in most glass products; *alumina,* or aluminum oxide (Al_2O_3), used in applications ranging from abrasives to artificial bones; and more complex compounds such as hydrous aluminum silicate ($Al_2Si_2O_5(OH)_4$), known as *kaolinite,* the principal ingredient in most clay products. The elements in these compounds are the most common in Earth's crust; see Table 7.1. The group includes many additional compounds, some of which occur naturally, while others are manufactured.

Ceramic raw materials are generally formed into solid products by the action of heat, such as the firing of clay or the heating of glass for blowing or molding. The word *ceramic* traces from the Greek *keramos,* meaning potter's clay or wares made from fired clay. Thus the modern word describes both the material itself and many of the products made of it.

The general properties that make ceramics useful in engineered products are high hardness, good electrical and thermal insulating characteristics, chemical stability, and high melting temperatures. Some ceramics are translucent, window glass being the clearest example. They are also brittle and possess virtually no ductility, which can cause problems in both processing and performance of the ceramic products.

The commercial and technological importance of ceramics is best demonstrated by the variety of products and applications that are based on this class of material. The list includes the following:

> *Clay construction products,* such as bricks, clay pipe, and building tile
> *Refractory ceramics*—ceramics capable of high temperature applications such as furnace walls, crucibles, and molds
> *Cement* used in *concrete,* used for construction and roads
> *Whiteware products,* including pottery, stoneware, fine china, porcelain, and other tableware, based on mixtures of clay and other minerals
> *Glass*—bottles, glasses, lenses, window panes, and light bulbs
> *Glass fibers* for thermal insulating wool, reinforced plastics (fiberglass), and fiber optics communications lines
> *Abrasives,* such as aluminum oxide and silicon carbide
> *Cutting tool materials,* including tungsten carbide, aluminum oxide, and cubic boron nitride
> *Ceramic insulators*—applications include electrical transmission components, spark plugs, and microelectronic chip substrates
> *Magnetic ceramics,* for example in computer memories
> *Nuclear fuels* based on uranium oxide (UO_2)
> *Bioceramics,* such as artificial teeth and bones

For purposes of organization, we classify ceramic materials into three basic types: (1) *traditional ceramics*—silicates used for clay products such as pottery and bricks, common abrasives, and cement; (2) *new ceramics*—more recently developed ceramics based on nonsilicates such as oxides and carbides, and generally possessing mechanical or physical properties that are superior or unique compared to traditional ceramics; and (3) *glasses*—based primarily on silica and distinguished from the other ceramics by their noncrystalline structure. In addition to the three basic types, we have *glass ceramics*—glasses that have been transformed into a largely crystalline structure by heat treatment.

TABLE 7.1   Most common elements in the earth's crust, with approximate percentages.

Oxygen	Silicon	Aluminum	Iron	Calcium	Sodium	Potassium	Magnesium
50%	26%	7.6%	4.7%	3.5%	2.7%	2.6%	2.0%

Compiled from [5].

Also included in this chapter is coverage of several elements that are related to the ceramics because they are used in similar applications and are often competitive materials. These elemental materials are carbon, silicon, and boron.

7.1 STRUCTURE AND PROPERTIES OF CERAMICS

Ceramic compounds are characterized by covalent and ionic bonding. These bonds are stronger than metallic bonding in metals, which accounts for the high hardness and stiffness but low ductility of ceramic materials. Just as the presence of free electrons in the metallic bond explains why metals are good conductors of heat and electricity, the presence of tightly held electrons in ceramic molecules explains why these materials are poor conductors. The strong bonding also provides these materials with high melting temperatures; in fact, some ceramics decompose rather than melt at high temperature.

Ceramics usually take a crystalline structure. The structures are generally more complex than those of most metals. There are several reasons for this. First, ceramic molecules usually consist of atoms that are significantly different in size. Second, the ion charges are often different, as in many of the common ceramics such as SiO_2 and Al_2O_3. Both of these factors tend to force a more complicated physical arrangement of the atoms in the molecule and in the resulting crystal structure. In addition, many ceramic materials consist of more than two elements, such as $(Al_2Si_2O_5(OH)_4)$, also leading to further complexity in the molecular structure. Crystalline ceramics can be single crystals or polycrystalline substances. In the more common second form, mechanical and physical properties are affected by grain size; higher strength and toughness are achieved in the finer-grained materials.

Some ceramic materials tend to assume an amorphous structure or *glassy phase,* rather than a crystalline form. The most familiar example is, of course, glass. Chemically, most glasses consist of fused silica. Variations in properties and colors are obtained by adding other glassy ceramic materials such as oxides of aluminum, boron, calcium, and magnesium. In addition to these pure glasses, many ceramics possessing a crystal structure use the glassy phase as a binder for their crystalline phase.

7.1.1 Mechanical Properties

Basic mechanical properties of ceramics are presented in Chapter 3. Ceramic materials are rigid and brittle, exhibiting a stress-strain behavior best characterized as perfectly elastic (see Figure 3.6(a)). As seen in Table 7.2, hardness and elastic modulus for many of the new ceramics are greater than those of metals (compare Tables 3.1, 3.6, and 3.7). Stiffness and hardness of traditional ceramics and glasses are significantly less than for new ceramics.

Theoretically, the strength of ceramics should be higher than that of metals because of their atomic bonding. The covalent and ionic bonding types are stronger than metallic bonding. However, metallic bonding has the advantage that it allows for slip, the basic mechanism by which metals deform plastically when subjected to high stresses. Bonding in ceramics is more rigid and does not permit slip under stress. The inability to slip makes it much more difficult for ceramics to absorb stresses. Yet ceramics contain the same imperfections in their crystal structure as metals—vacancies, interstitialcies, displaced atoms, and microscopic cracks. These internal flaws tend to concentrate the stresses, especially when a tensile, bending, or impact loading is involved. As a result of these factors, ceramics fail by brittle fracture under applied stress much more

TABLE 7.2 Selected mechanical properties of ceramic materials.

Material	Hardness	Elastic Modulus, E	
		GPa	(lb/in.^{2})
Traditional ceramics			
Brick-fireclay	N.A.	95	(14×10^6)
Cement, Portland	N.A.	50	(7×10^6)
Silicon carbide (SiC)	2600 HV	460	(68×10^6)
New Ceramics			
Alumina (Al_2O_3)	2200 HV	345	(50×10^6)
Boron nitride, cubic	6000 HV	N.A.	N.A.
Titanium carbide (TiC)	3200 HV	300	(45×10^6)
Tungsten carbide (WC)	2600 HV	700	(100×10^6)
Glass			
Silica glass (SiO_2)	500 HV	69	(10×10^6)

Compiled from [2], [3], [4], [5], [8], [9], and other sources.
N.A. = Not available.

readily than metals. Their tensile strength and toughness are relatively low. Also, their performance is much less predictable due to the random nature of the imperfections and the influence of processing variations, especially in products made of traditional ceramics.

The frailties that limit the tensile strength of ceramic materials are not nearly so operative when compressive stresses are applied. Ceramics are substantially stronger in compression than in tension. For engineering and structural applications, designers have learned to use ceramic components so that they are loaded in compression rather than tension or bending.

Various methods have been developed to strengthen ceramics, nearly all of which have as their fundamental approach the minimization of surface and internal flaws and their effects. These methods include (1) making the starting materials more uniform; (2) decreasing grain size in polycrystalline ceramic products; (3) minimizing porosity; (4) introducing compressive surface stresses; for example, through application of glazes with low thermal expansions, so that the body of the product contracts after firing more than the glaze, thus putting the glaze in compression; (5) using fiber reinforcement; and (6) heat treatments, such as quenching alumina from temperatures in the slightly plastic region to strengthen it (see [6]).

7.1.2 Physical Properties

Several of the physical properties of ceramics are presented in Table 7.3. Most ceramic materials are lighter than metals and heavier than polymers (refer to Table 4.1). Melting temperatures are higher than for most metals, some ceramics preferring to decompose rather than melt.

Electrical and thermal conductivities of most ceramics are lower than for metals; but the range of values is greater, permitting some ceramics to be used as insulators while others are electrical conductors. Thermal expansion coefficients are somewhat less than for the metals, but the effects are more damaging in ceramics because of their brittleness. Ceramic materials with relatively high thermal expansions and low thermal conductivities are especially susceptible to failures of this type, which result from significant temperature gradients and associated volumetric changes in different regions of the same part. The terms *thermal shock* and *thermal cracking* are used in connection with such failures. Certain glasses (for example, those containing high proportions of SiO_2) and

TABLE 7.3 Selected physical properties of ceramic materials, as well as graphite and diamond.

Material	Specific Gravity	Melting Temperature	
		°C	(°F)
Traditional ceramics			
Alumina (Al$_2$O$_3$)	3.8	2054	(3729)
Brick, building	2.3	N.A.	N.A.
Cement, Portland	2.4	N.A.	N.A.
Kaolinite (Al$_2$Si$_2$O$_5$(OH)$_4$)	2.6	N.A.	N.A.
Silicon carbide (SiC)	3.2	2700[a]	(4892)[a]
New Ceramics			
Alumina (Al$_2$O$_3$)	3.8	2054	(3729)
Cubic boron nitride (BN)	2.3	3000[a]	(5430)[a]
Silicon nitride (SiN)	3.2	1900[a]	(3450)[a]
Titanium carbide (TiC)	4.9	3250	(5880)
Tungsten carbide (WC)	15.7	N.A.	N.A.
Glass			
Silica glass (SiO$_2$)	2.2	[b]	[b]

Compiled from [2], [3], [4], [5], and other sources.

[a]The ceramic material chemically dissociates or, in the case of diamond and graphite, sublimes (vaporizes), rather than melts.

[b]Glass, being noncrystalline, does not melt at a specific melting point. Instead, it gradually exhibits fluid properties with increasing temperature. It becomes liquid at around 1400°C (2500°F).

N.A. = Not applicable or not available.

glass ceramics are noted for their low thermal expansion and are particularly resistant to these thermal failures. (***Pyrex*** is a familiar example.)

7.2 TRADITIONAL CERAMICS

These materials are based on mineral silicates, silica, and mineral oxides. The primary products are fired clay (pottery, tableware, brick, and tile), cement, and natural abrasives such as alumina. These products, and the processes used to make them, date back thousands of years (see Historical Note 7.1). Glass is also a silicate ceramic material, and is often included within the traditional ceramics group [3], [5]. We are covering glass in a later section because it is distinguished from the above crystalline materials by its amorphous or vitreous structure (the term ***vitreous*** means glassy, or possessing the characteristics of glass).

***Historical* Note 7.1** *Ancient pottery ceramics.*

Making pottery has been an art since the earliest civilizations. Archeologist examine ancient pottery and similar artifacts to study the cultures of the ancient world. Ceramic pottery does not corrode or disintegrate with age nearly as rapidly as artifacts made of wood, metal, or cloth.

Somehow, early tribes discovered that clay is transformed into a hard solid when placed near an open fire.

Burnt clay articles have been found in the Middle East that date back nearly 10,000 years. Earthenware pots and similar products became an established commercial trade in Egypt by around 4000 B.C.

The greatest advances in pottery making were made in China, where fine white stoneware was first crafted as early as 1400 B.C. By the Ninth century, the Chinese were making articles of porcelain, which was fired at higher temperatures than earthenware or stoneware to partially vitrify the more complex mixture of raw materials and produce translucency in the final product. Dinnerware made of Chinese porcelain was highly valued in Europe; it was called *china*. It contributed significantly to trade between China and Europe and influenced the development of European culture.

Raw Materials Mineral silicates, such as clays of various compositions, and silica, such as quartz, are among the most abundant substances in nature and constitute the principal raw materials for traditional ceramics. These solid crystalline compounds have been formed and mixed in the Earth's crust over billions of years by complex geological processes.

The clays are the raw materials used most widely in ceramics. They consist of fine particles of hydrous aluminum silicate that become a plastic substance that is formable and moldable when mixed with water. The most common clays are based on the mineral *kaolinite,* $(Al_2Si_2O_5(OH)_4^-)$. Other clay minerals vary in composition, both in terms of proportions of the basic ingredients and through additions of other elements such as magnesium, sodium, and potassium.

Besides its plasticity when mixed with water, a second characteristic of clay that makes it so useful is that it fuses into a dense, strong material when heated to a sufficiently elevated temperature. The heat treatment is known as *firing.* Suitable firing temperatures depend on clay composition. Thus, clay can be shaped while wet and soft, and then fired to obtain the final hard ceramic product.

Silica (SiO_2) is another major raw material for the traditional ceramics. It is the principal component in glass, and an important ingredient in other ceramic products including whiteware, refractories, and abrasives. Silica is available naturally in various forms, the most important of which is *quartz.* The main source of quartz is *sandstone.* The abundance of sandstone and its relative ease of processing means that silica is low in cost; it is also hard and chemically stable. These features account for its widespread use in ceramic products. It is generally mixed in various proportions with clay and other minerals to achieve the appropriate characteristics in the final product. Feldspar is one of the other minerals often used. *Feldspar* refers to any of several crystalline minerals that consist of aluminum silicate combined with either potassium, sodium, calcium, or barium. The potassium blend, for example, has the chemical composition $KAlSi_3O_8$. Mixtures of clay, silica, and feldspar are used to make stoneware, china, and other tableware.

Another important raw material for traditional ceramics is *alumina.* Most alumina is processed from the mineral *bauxite,* which is an impure mixture of hydrous aluminum oxide and aluminum hydroxide plus similar compounds of iron or manganese. Bauxite is also the principal source of metallic aluminum. A more pure but less common form of Al_2O_3 is the mineral *corundum,* which contains alumina in massive amounts. Slightly impure forms of corundum crystals are the colored gemstones sapphire and ruby. Alumina ceramic is used as an abrasive in grinding wheels and as a refractory brick in furnaces.

Silicon carbide, another ceramic used widely as an abrasive, does not occur as a mineral. Instead, it is produced by heating mixtures of sand (source of silicon) and coke (carbon) to a temperature of around 2200°C (3900°F), so that the resulting chemical reaction forms SiC and carbon monoxide.

7.2.2 Traditional Ceramic Products

The minerals discussed above are the ingredients for a variety of ceramic products. We organize our coverage here by major categories of traditional ceramic products. A summary of these products, and the raw materials and ceramics out of which they are made, is presented in Table 7.4. We limit our coverage to materials commonly used with manufactured products, thus omitting certain important commercial ceramics such as cement.

Pottery and Tableware   This category is one of the oldest, dating back thousands of years; yet it is today one of the most important. It includes tableware products that we all use: earthenware, stoneware, and china. The raw materials for these products are clay usually combined with other minerals such as silica and feldspar. The wetted mixture is shaped and then fired to produce the finished piece.

Earthenware is the least refined of the group; it includes pottery and similar articles made in ancient times. Earthenware is relatively porous and is often glazed. *Glazing* involves application of a surface coating, usually a mixture of oxides such as silica and alumina, to make the product less pervious to moisture and more attractive to the eye. *Stoneware* has lower porosity than earthenware, resulting from closer control of ingredients and higher firing temperatures. *China* is fired at even higher temperatures, which produces the translucence in the finished pieces that characterize their fine quality. The reason for this is that much of the ceramic material has been converted to the glassy (vitrified) phase, which is relatively transparent compared to the polycrystalline form. Modern *porcelain* is nearly the same as china and is produced by firing the components, mainly clay, silica, and feldspar, at still higher temperatures to achieve a very hard, dense, glassy material. Porcelain is used in a variety of products, ranging from electrical insulation to bathtub coatings.

Brick and Tile   Building brick, clay pipe, unglazed roof tile, and drain tile are made from various low-cost clays containing silica and gritty matter widely available in natural deposits. These products are shaped by pressing (molding) and firing at relatively low temperatures.

Refractories   Refractory ceramics, often in the form of bricks, are critical in many industrial processes that require furnaces and crucibles to heat and/or melt materials. The useful properties of refractory materials are high temperature resistance, thermal insulation, and resistance to chemical reaction with the materials (usually molten metals) being heated. As we have mentioned, alumina is often used as a refractory ceramic, together with silica. Other refractory materials include magnesium oxide (MgO) and calcium oxide (CaO). The refractory lining often contains two layers, the outside layer being more porous because this increases the insulation properties.

Abrasives   Traditional ceramics used for abrasive products, such as grinding wheels and sandpaper, are alumina and silicon carbide. Although SiC is the harder material (hardness of SiC is 2600 HV versus 2200 HV for alumina), the majority of grinding wheels are based on Al_2O_3 because it gives better results when grinding steel, the most

TABLE 7.4   Summary of traditional ceramic products.

Product	Principal Chemistry	Minerals and Raw Materials
Pottery, tableware	$Al_2Si_2O_5(OH)_4$, SiO_2, $KAlSi_3O_8$	Clay + silica + feldspar
Porcelain	$Al_2Si_2O_5(OH)_4$, SiO_2, $KAlSi_3O_8$	Clay + silica + feldspar
Brick, tile	$Al_2Si_2O_5(OH)_4$, SiO_2 plus fine stones	Clay + silica + other
Refractory	Al_2O_3, SiO_2 Others: MgO, CaO	Alumina and silica
Abrasive: silicon carbide	SiC	Silica + coke
Abrasive: aluminum oxide	Al_2O_3	Bauxite or alumina

widely used metal. The abrasive particles (grains of ceramic) are distributed in the wheel using a bonding material such as a shellac, polymer resin, or rubber. The use of abrasives in industry involves material removal, and the technology of grinding wheels and other abrasive methods to remove material is presented in Chapter 25.

7.3 NEW CERAMICS

The term *new ceramics* refers to ceramic materials that have been developed synthetically over the last several decades and to improvements in processing techniques that have provided greater control over the structures and properties of ceramic materials. In general, new ceramics are based on compounds other than variations of aluminum silicate (which form the bulk of the traditional ceramic materials). New ceramics are usually simpler chemically than traditional ceramics; for example, oxides, carbides, nitrides, and borides. The dividing line between traditional and new ceramics is sometimes fuzzy, because aluminum oxide and silicon carbide are included among the traditional ceramics. The distinction in these cases is based more on methods of processing than chemical composition.

We organize the new ceramics into chemical compound categories: oxides, carbides, and nitrides, discussed in the following sections. More complete coverage of new ceramics is presented in several of our references ([2], [4], and [7]).

7.3.1 Oxide Ceramics

The most important oxide new ceramic is *alumina.* Although also discussed in the context of traditional ceramics, alumina is today produced synthetically from bauxite, using an electric furnace method. Through control of particle size and impurities, refinements in processing methods, and blending with small amounts of other ceramic ingredients, strength and toughness of alumina have been improved substantially compared to its natural counterpart. Alumina also has good hot hardness, low thermal conductivity, and good corrosion resistance. This is a combination of properties that promote a wide variety of applications, including abrasives (grinding wheel grit), bioceramics (artificial bones and teeth), electrical insulators, electronic components, alloying ingredients in glass, refractory brick, cutting tool inserts (Section 23.2.5), spark plug barrels, and engineering components (see Figure 7.1).

FIGURE 7.1   Alumina ceramic components (photo courtesy of Insaco Inc.).

7.3.2 Carbides

The carbide ceramics include silicon carbide (SiC), tungsten carbide (WC), titanium carbide (TiC), tantalum carbide (TaC), and chromium carbide (Cr_3C_2). Silicon carbide was discussed previously. Although it is a man-made ceramic, the methods for its production were developed a century ago, and therefore it is generally included in the traditional ceramics group. In addition to its use as an abrasive, other SiC applications include resistance heating elements and additives in steelmaking.

WC, TiC, and TaC are valued for their hardness and wear resistance in cutting tools and other applications requiring these properties. ***Tungsten carbide*** was the first to be developed (Historical Note 7.2) and is the most important and widely used material in the group. WC is typically produced by carburizing tungsten powders that have been reduced from tungsten ores such as ***wolframite*** ($FeMnWO_4$) and ***scheelite*** ($CaWO_4$). ***Titanium carbide*** is produced by carburizing the minerals ***rutile*** (TiO_2) or ***ilmenite*** ($FeTiO_3$). And ***tantalum carbide*** is made by carburizing either pure tantalum powders or tantalum pentoxide (Ta_2O_5) [10]. ***Chromium carbide*** is more suited to applications where chemical stability and oxidation resistance are important. Cr_3C_2 is prepared by carburizing chromium oxide (Cr_2O_3) as the starting compound. Carbon black is the usual source of carbon in all of these reactions.

Historical Note 7.2   *Tungsten carbide* [10]

The compound WC does not occur in nature. It was first fabricated in the late 1890s by the Frenchman Henri Moissan. However, the technological and commercial importance of the development was not recognized for two decades.

Tungsten became an important metal for incandescent lamp filaments in the early 1900s. Wire drawing was required to produce the filaments. The traditional tool steel draw dies of the period were unsatisfactory for drawing tungsten wire due to excessive wear. There was a need for a much harder material. The compound WC was known to possess such hardness. In 1914 in Germany, H. Voigtlander and H. Lohmann developed a fabrication process for hard carbide draw dies by sintering parts pressed from powders of tungsten carbide and/or molybdenum carbide. Lohmann is credited with the first commercial production of sintered carbides.

The breakthrough leading to the modern technology of cemented carbides is linked to the work of K. Schroter in Germany in the early and mid 1920s. He used WC powders mixed with about 10% of a metal from the iron group, finally settling on cobalt as the best binder, and sintering the mixture at a temperature close to the melting point of the metal. The hard material was first marketed in Germany as "Widia" in 1926. The Schroter patents were assigned to the General Electric Company under the trade name "Carboloy"— first produced in the United States around 1928.

Widia and Carboloy were used as cutting tool materials, with cobalt content in the range 4% to 13%. They were effective in the machining of cast iron and many nonferrous metals, but not in the cutting of steel. When steel was machined, the tools would wear rapidly by cratering. In the early 1930s, carbide cutting tool grades with WC and TiC were developed for steel cutting. In 1931, the German firm Krupp started production of Widia X, which had a composition 84% WC, 10% TiC, and 6% Co. And Carboloy Grade 831 was introduced in the U.S. in 1932; it contained 69% WC, 21% TiC, and 10% Co.

Except for SiC, each of the carbides discussed here must be combined with a metallic binder such as cobalt or nickel in order to fabricate a useful solid product. In effect, the carbide powders bonded in a metal framework creates what is known as a ***cemented carbide***—a composite material, specifically a ***cermet*** (reduced from ***cer***amic and ***met***al). We examine cemented carbides and other cermets in Section 11.2.1. The carbides have little engineering value except as constituents in a composite system.

7.3.3 Nitrides

The important nitride ceramics are silicon nitride (Si_3N_4), boron nitride (BN), and titanium nitride (TiN). As a group, the nitride ceramics are hard and brittle, and they melt at high temperatures (but not generally as high as the carbides). They are usually electrically insulating, TiN being an exception.

Silicon nitride shows promise in high temperature structural applications. Si_3N_4 oxidizes at about 1200°C (2125°F) and chemically decomposes at around 1900°C (3400°F). It has low thermal expansion, good resistance to thermal shock and creep, and resists corrosion by molten nonferrous metals. These properties have provided applications for this ceramic in gas turbines, rocket engines, and melting crucibles.

Boron nitride exists in several structures, similar to carbon. The important forms of BN are (1) hexagonal, similar to graphite; and (2) cubic, same as diamond; in fact, its hardness is comparable to that of diamond. This latter structure goes by the names *cubic boron nitride* and *borazon,* symbolized cBN, and is produced by heating hexagonal BN under very high pressures. Owing to its extreme hardness, the principal applications of cBN are in cutting tools (Section 23.2.6) and abrasive wheels (25.1.1). Interestingly, it does not compete with diamond cutting tools and grinding wheels. Diamond is suited to non-steel machining and grinding, while cBN is appropriate for steel.

Titanium nitride has properties similar to those of other nitrides in this group, except for its electrical conductivity; it is a conductor. TiN has high hardness, good wear resistance, and a low coefficient of friction with the ferrous metals. This combination of properties makes TiN an ideal material as a surface coating on cutting tools. The coating is only around 0.006 mm (0.0003 in.) thick, so the amounts of this material used are low.

A new ceramic material related to the nitride group, and also to the oxides, is the oxynitride ceramic called *sialon.* It consists of the elements silicon, aluminum, oxygen, and nitrogen; and its name derives from these ingredients: Si–Al–O–N. Its chemical composition is variable, a typical composition being $Si_4Al_2O_2N_6$. Properties of sialon are similar to those of silicon nitride, but it has better resistance to oxidation at high temperatures than Si_3N_4. Its principal application is for cutting tools, but its properties may make it suitable for other high temperature applications in the future.

7.4 GLASS

The term *glass* is somewhat confusing because it describes a state of matter as well as a type of ceramic. As a state of matter, the term refers to an amorphous, or noncrystalline, structure of a solid material. The glassy state occurs in a material when insufficient time has been allowed during cooling from the molten condition for the crystalline structure to form. It turns out that all three categories of engineering materials (metals, ceramics, and polymers) can assume the glassy state, although the circumstances for metals to do so are quite rare.

As a type of material, *glass* is an inorganic, nonmetallic compound (or mixture of compounds) that cools to a rigid condition without crystallizing; it is a ceramic that is in the glassy state as a solid material. This is the material we shall discuss in this section—a material that dates back more than 4,000 years (Historical Note 7.3).

Historical Note 7.3   *History of glass.*

The oldest glass specimens, dating from around 2500 B.C., are glass beads and other simple shapes found in Mesopotamia and ancient Egypt. These were made by painstakingly sculpturing glass solids, rather than by molding or shaping molten glass. It was a thousand years before the ancient cultures exploited the fluid properties of hot glass, by pouring it in successive layers over a sand core until sufficient thickness and rigidity had been attained in the product, a cup-shaped vessel. This pouring technique was used until around 200 B.C., when a simple tool was developed that revolutionized glassworking—the blowpipe.

   Glass blowing was probably first accomplished in Babylon and later by the Romans. It was performed using an iron tube several feet long, with a mouthpiece on one end and a fixture for holding the molten glass on the other. A blob of hot glass in the required initial shape and viscosity was attached to the end of the iron tube, and then blown into shape by an artisan either freely in air or into a mold cavity. Other simple tools were utilized to add the stem and/or base to the object.

   The ancient Romans showed great skill in their use of various metallic oxides to color glass. Their technology is evident in the stained glass windows of cathedrals and churches of the middle ages in Italy and the rest of Europe. The art of glassblowing is still practiced today for certain consumer glassware; and automated versions of glassblowing are used for mass-produced glass products such as bottles and light bulbs (Chapter 12).

7.4.1 Chemistry and Properties of Glass

The principal ingredient in virtually all glasses is *silica* (SiO_2), most commonly found as the mineral quartz in sandstone and silica sand. Quartz occurs naturally as a crystalline substance; but when melted and then cooled, it forms vitreous silica. Silica glass has a very low thermal expansion coefficient and is therefore quite resistant to thermal shock. These properties are ideal for elevated temperature applications; accordingly, Pyrex and chemical glassware designed for heating are made with high proportions of silica glass.

In order to reduce the melting point of glass for easier processing, and to control properties, the composition of most commercial glasses includes other oxides as well as silica. Silica remains as the main component in these glass products, usually making up 50% to 75% of total chemistry. The reason why SiO_2 is used so widely in these compo-

TABLE 7.5   Typical or average compositions of selected glass products.

Product	Chemical Composition (by weight to nearest %)								
	SiO_2	Na_2O	CaO	Al_2O_3	MgO	K_2O	PbO	B_2O_3	Other
Soda-lime glass	71	14	13	2					
Window glass	72	15	8	1	4				
Container glass	72	13	10	2[a]	2	1			
Light bulb glass	73	17	5	1	4				
Laboratory glass:									
Vycor	96			1				3	
Pyrex	81	4		2				13	
E-glass (fibers)	54	1	17	15	4			9	
S-glass (fibers)	64			26	10				
Optical glasses:									
Crown glass	67	8				12		12	ZnO
Flint glass	46	3				6	45		

Compiled from [3], [4], and [9], and other sources.

[a]May include Fe_2O_3 with Al_2O_3.

sitions is because it is the best *glass former.* It naturally transforms into a glassy state upon cooling from the liquid, whereas most ceramics crystallize upon solidification. Table 7.5 lists typical chemistries for some common glasses.

The additional ingredients are contained in a solid solution with SiO_2, and each has a function: (1) acting as flux (promoting fusion) during heating; (2) increasing fluidity in the molten glass for processing; (3) retarding *devitrification*—the tendency to crystallize from the glassy state; (4) reducing thermal expansion in the final product; (5) improving the chemical resistance against attack by acids, basic substances, or water; (6) adding color to the glass; and (7) altering the index of refraction for optical applications (e.g., lenses).

7.4.2 Glass Products

Following is a list of the major categories of glass products. We examine the roles played by the different ingredients in Table 7.5 as we discuss these products.

Window Glass   This glass is represented by two chemistries in Table 7.5: (1) soda-lime glass and (2) window glass. The soda-lime formula dates back to the glass-blowing industry of the 1800s and before. It was (and is) made by mixing soda (Na_2O) and lime (CaO) with silica (SiO_2) as the major ingredient. The blending of ingredients has evolved empirically to achieve a balance between avoiding crystallization during cooling and achieving chemical durability of the final product. Modern window glass and the techniques for making it have required slight adjustments in composition and closer control over its variation. Magnesia (MgO) has been added to help reduce devitrification.

Containers   In previous times, the same basic soda-lime composition was used for manual glass-blowing to make bottles and other containers. Modern processes for shaping glass containers cool the glass more rapidly than older methods. Also, the importance of chemical stability in container glass is better understood today. Resulting changes in composition have attempted to optimize the proportions of lime (CaO) and soda (Na_2O_3). Lime promotes fluidity. It also increases devitrification, but since cooling is more rapid, this effect is not as important as in prior processing techniques with slower cooling rates. Reducing soda reduces chemical instability and solubility of the container glass.

Light Bulb Glass   Glass used in light bulbs and other thin glass items (e.g., drinking glasses, Christmas ornaments) is high in soda and low in lime; it also contains small amounts of magnesia and alumina. The chemistry is dictated largely by the economics of large volumes involved in light bulb manufacture. The raw materials are inexpensive and suited to the continuous melting furnaces used today.

Laboratory Glassware   These products include containers for chemicals (e.g., flasks, beakers, glass tubing). The glass must be resistant to chemical attack and thermal shock. Glass that is high in silica is suitable because of its low thermal expansion. The trade name "Vicor" is used for this high-silica glass. This product is very insoluble in water and acids. Additions of boric oxide also produce a glass with low coefficient of thermal expansion, so some glass for laboratory ware contains B_2O_3 in amounts of around 13%. The trade name "Pyrex" is used for the borosilicate glass developed by the Corning Glass Works. Both Vicor and Pyrex are included in our listing as examples of this product category.

Glass Fibers   Glass fibers are manufactured for a number of important applications, including fiberglass reinforced plastics, insulation wool, and fiber optics. The compositions vary according to function. The most commonly used glass reinforcing fibers in plastics are E-glass. It is high in CaO and Al_2O_3 content, is economical, and it possesses good tensile strength in fiber form. Another glass fiber material is S-glass, which has higher strength but is not as economical as E-glass. Compositions are indicated in our table.

Insulating fiberglass wool can be manufactured from regular soda-lime-silica glasses. The glass product for fiber optics consists of a long, continuous core of glass with high refractive index surrounded by a sheath of lower refractive glass. The inside glass must have a very high transmittance for light in order to accomplish long distance communication.

Optical Glasses   Applications for these glasses include lenses for eyeglasses and optical instruments such as cameras, microscopes, and telescopes. To achieve their function, the glasses must have different refractive indices, but each lens must be homogenous in composition. Optical glasses are generally divided into: crowns and flints. *Crown glass* has a low index of refraction, while *flint glass* contains lead oxide (PbO) that gives it a high index of refraction.

7.4.3 Glass-Ceramics

Glass-ceramics comprise a class of ceramic material produced by conversion of glass into a polycrystalline structure through heat treatment. The proportion of crystalline phase in the final product typically ranges between 90% and 98%, with the remainder being unconverted vitreous material. Grain size is usually between 0.1 and 1.0 μm (4 and 40 μ-in.), significantly smaller than the grain size of conventional ceramics. This fine crystal microstructure makes glass-ceramics much stronger than the glasses from which they are derived. Also, due to their crystal structure, glass-ceramics are opaque (usually grey or white) rather than clear.

The processing sequence for glass-ceramics is as follows: (1) The first step involves heating and forming operations used in glassworking (Section 12.2) to create the desired product geometry. Glass-shaping methods are generally more economical than pressing and sintering to shape traditional and new ceramics made from powders. (2) The product is cooled. (3) The glass is reheated to a temperature sufficient to cause a dense network of crystal nuclei to form throughout the material. It is the high density of nucleation sites that inhibits grain growth of individual crystals, thus leading ultimately to the fine grain size in the glass-ceramic material. The key to the propensity for nucleation is the presence of small amounts of nucleating agents in the glass composition. Common nucleating agents are TiO_2, P_2O_5, and ZrO_2. (4) Once nucleation is initiated, the heat treatment is continued at a higher temperature to cause growth of the crystalline phases.

Several examples of glass-ceramic systems and typical compositions are listed in Table 7.6. The Li_2O–Al_2O_3–SiO_2 system is the most important commercially; it includes Corning Ware (Pyroceram), the familiar product of the Corning Glass Works.

There are significant advantages of glass-ceramics: (1) Efficiency of processing in the glassy state; (2) close dimensional control over the final product shape; and (3) good mechanical and physical properties. Properties include high strength (stronger than glass), absence of porosity, low coefficient of thermal expansion, and high resistance to thermal shock. These properties have resulted in applications in cooking ware, heat exchangers, and missile radomes. Certain systems (e.g., MgO–Al_2O_3–SiO_2 system) are also characterized by high electrical resistance, suitable for electrical and electronics applications.

TABLE 7.6   Several glass-ceramic systems.

Glass-Ceramic System	Typical Composition (to nearest %)						
	Li_2O	MgO	Na_2O	BaO	Al_2O_3	SiO_2	TiO_2
$Li_2O–Al_2O_3–SiO_2$	3				18	70	5
$MgO–Al_2O_3–SiO_2$		13			30	47	10
$Na_2O–BaO–Al_2O_3–SiO_2$			13	9	29	41	7

Compiled from [4], [5], and [9].

7.5 SOME IMPORTANT ELEMENTS RELATED TO CERAMICS

In this section, several elements of engineering importance are discussed: carbon, silicon, and boron. We encounter these materials on occasion in subsequent chapters. Although they are not ceramic materials according to our definition, they sometimes compete for applications with ceramics. And they have important applications of their own. Basic data on these elements are presented in Table 7.7.

7.5.1 Carbon

Carbon occurs in two alternative forms of engineering and commercial importance: graphite and diamond. They compete with ceramics in various applications: graphite in situations where its refractory properties are important, and diamond in industrial applications where hardness is the critical factor (such as cutting and grinding tools).

Graphite   Graphite has a high content of crystalline carbon in the form of layers. Bonding between atoms in the layers is covalent and therefore strong, but the parallel layers are bonded to each other by weak van der Waals forces. This structure makes graphite quite anisotropic; strength and other properties vary significantly with direction. It explains why graphite can be used both as a lubricant and as a fiber in advanced composite materials. In powder form, graphite possesses low frictional characteristics due to the ease with which it shears between the layers; in this form, graphite is valued as a lubricant. In fiber form, graphite is oriented in the hexagonal planar direction to produce a filament material of very high strength and elastic modulus. These graphite fibers are used in structural composites ranging from tennis rackets to fighter aircraft components.

TABLE 7.7   Some basic data and properties of carbon, silicon, and boron.

	Carbon	Silicon	Boron
Symbol	C	Si	B
Atomic number	6	14	5
Specific gravity	2.25	2.42	2.34
Melting temperature	3727°C[a] (6740°F)	1410°C (2570°F)	2030°C (3686°F)
Elastic modulus, GPa (lb/in.^{2})	240[b] (35×10^6)[b]	N.A.	393 (57×10^6)
	1035[c] (150×10^6)[c]		
Hardness (Mohs scale)	1[b], 10[c]	7	9.3

[a] Carbon sublimes (vaporizes) rather than melt.

[b] Carbon in the form of graphite (typical value given).

[c] Carbon in the form of diamond.

N.A. = not available.

Graphite exhibits certain high temperature properties that are both useful and unusual. It is resistant to thermal shock and its strength actually increases with temperature. Tensile strength at room temperature is about 100 MPa (15,000 lb/in.^{2}), but increases to about twice this value at 2500°C (4530°F) [4]. Theoretical density of carbon is 2.22 gm/cm^3, but apparent density of bulk graphite is lower due to porosity (around 1.7 gm/cm^3). This is increased through compacting and heating. It is electrically conductive, but its conductivity is not as high as most metals. A disadvantage of graphite is that it oxidizes in air above around 500°C (900°F). In a reducing atmosphere it can be used up to around 3000°C (5400°F), not far below its sublimation point of 3727°C (6740°F).

The traditional form of graphite is polycrystalline with a certain amount of amorphous carbon in the mixture. Graphite crystals are often oriented (to a limited degree) in the commercial production process to enhance properties in a preferred direction for the application. Also, strength is improved by reducing grain size (similar to ceramics). Graphite in this form is used for crucibles and other refractory applications, electrodes, resistance heating elements, antifriction materials, and fibers in composite materials. Thus, graphite is very versatile. As a powder it is a lubricant. In traditional solid form it is a refractory. When formed into graphite fibers, it is a high strength structural material.

Diamond   Diamond is carbon that possesses a cubic crystalline structure with covalent bonding between atoms, as shown in Figure 2.5(b). This structure is three-dimensional rather than layered, as in graphite carbon, and this accounts for the very high hardness of diamond. Single crystal natural diamonds (mined in South Africa) have a hardness of 10,000 HV, while the hardness of an industrial diamond (polycrystalline) is around 7000 HV. The high hardness accounts for most of the applications of industrial diamond. It is used in cutting tools and grinding wheels for machining hard, brittle materials, or materials that are very abrasive. For example, diamond tools and wheels are used to cut ceramics, fiberglass, and hardened metals

FIGURE 7.2   Synthetically produced diamond powders (photo courtesy GE Superabrasives, General Electric Company).

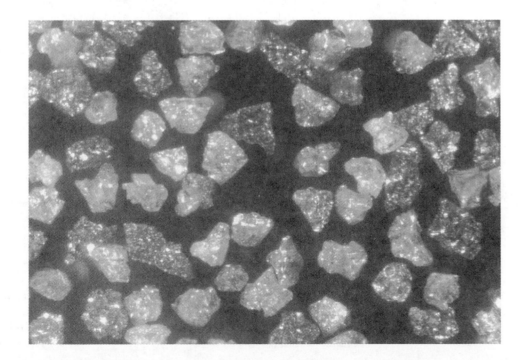

other than steels. Diamond is also used in dressing tools to sharpen grinding wheels that consist of other abrasives such as alumina and silicon carbide. Similar to graphite, diamond has a propensity to oxidize (decompose) in air at temperatures above about 650°C (1200°F).

Industrial or synthetic diamonds date back to the 1950s and are fabricated by heating graphite to around 3000°C (5400°F) under very high pressures, Figure 7.2. This process approximates the geological conditions by which natural diamonds were formed millions of years ago.

7.5.2 Silicon

Silicon is a semimetallic element in the same group in the periodic table as carbon, Figure 2.1. Silicon is one of the most abundant elements in the Earth's crust, comprising about 26% by weight (Table 7.1). It occurs naturally only as a chemical compound—in rocks, sand, clay, and soil—either as silicon dioxide or as more complex silicate compounds. As an element it has the same crystalline structure as diamond, but its hardness is lower. It is hard but brittle, lightweight, chemically inactive at room temperature, and is classified as a semiconductor.

The greatest amounts of silicon in manufacturing are in ceramic compounds (SiO_2 in glass and silicates in clays) and alloying elements in steel, aluminum, and copper alloys. It is also used as a reducing agent in certain metallurgical processes. Of significant technological importance is pure silicon as the base material in semiconductor manufacturing in electronics. The vast majority of integrated circuits produced today are made from silicon (Chapter 35).

7.5.3 Boron

Boron is a semimetallic element in the same periodic group as aluminum. It is only about 0.001% of the Earth's crust by weight, commonly occurring as the minerals *borax* ($Na_2B_4O_7$–$10H_2O$) and *kernite* ($Na_2B_4O_7$–$4H_2O$). Boron is lightweight, semiconducting in electrical properties (conductivity varies with temperature; it is an insulator at low temperatures but a conductor at high temperatures), and very stiff (high modulus of elasticity) in fiber form.

As a material of industrial significance, boron is usually found in compound form. As such, it is used as a solution in nickel electroplating operations, an ingredient (B_2O_3) in certain glass compositions, a catalyst in organic chemical reactions, and as a nitride (cubic boron nitride) for cutting tools. In nearly pure form it is used as a fiber in composite materials (Sections 9.4.1 and 15.1.2).

7.6 GUIDE TO PROCESSING CERAMICS

The processing of ceramics can be divided into two basic categories: molten ceramics and particulate ceramics. The major category of molten ceramics is glassworking (Chapter 12). Particulate ceramics include traditional and new ceramics; their processing methods constitute most of the rest of the shaping technologies for ceramics (Chapter 17). Cermets, such as cemented carbides, are a special case since they are metal matrix composites (Section 17.3). Table 7.8 provides a guide to the processing of ceramic materials.

TABLE 7.8   Guide to the processing of ceramic materials.

Material	Chapter or Section	Material	Chapter or Section
Glass	Chapter 12	Synthetic diamonds	Section 23.2.6
Glass fibers	Section 12.2.3	Silicon	Section 35.2
Particulate ceramics	Chapter 17	Carbon fibers	Section 15.1.2
Cermets	Section 17.3	Boron fibers	Section 15.1.2

REFERENCES

[1] Chiang, Y-M., D. P. Birnie III, and W. D. Kingery. *Physical Ceramics.* Wiley, New York, 1997.

[2] *Engineered Materials Handbook*: Volume 4, *Ceramics and Glasses.* ASM International, 1991.

[3] Flinn, R. A., and Trojan, P. K. *Engineering Materials and Their Applications.* 4th ed. Houghton Mifflin, Boston, 1990.

[4] Hlavac, J. *The Technology of Glass and Ceramics.* Elsevier Scientific Publishing Company, New York, 1983.

[5] Kingery, W. D., Bowen, H. K., and Uhlmann, D. R. *Introduction to Ceramics.* 2nd ed. Wiley, New York, 1995.

[6] Kirchner, H. P. *Strengthening of Ceramics.* Marcel Dekker, Inc., New York, 1979.

[7] Richerson, D. W. *Ceramics—Applications in Manufacturing.* Society of Manufacturing Engineers, Dearborn, Mich., 1989.

[8] Richerson, D. W. *Modern Ceramic Engineering.* 2nd ed. Marcel Dekker, Inc., New York, 1992.

[9] Scholes, S. R., and Greene, C. H. *Modern Glass Practice.* 7th ed. CBI Publishing Company, Boston, Mass., 1993.

[10] Schwarzkopf, P., and Kieffer, R. *Cemented Carbides.* MacMillan, New York, 1960.

[11] Singer, F., and Singer, S. S. *Industrial Ceramics.* Chemical Publishing Company, New York, 1963.

[12] Somiya, S. (ed.). *Advanced Technical Ceramics.* Academic Press, Inc., San Diego, Calif., 1989.

REVIEW QUESTIONS

7.1. What is a ceramic?

7.2. What are the four most common elements in the Earth's crust?

7.3. What is the difference between the traditional ceramics and the new ceramics?

7.4. What distinguishes glass from the traditional and new ceramics?

7.5. Why are graphite and diamond not classified as ceramics?

7.6. What are the general mechanical properties of ceramic materials?

7.7. What are the general physical properties of ceramic materials?

7.8. What type of atomic bonding characterizes the ceramics?

7.9. What do bauxite and corundum have in common?

7.10. What is clay, used in making ceramic products?

7.11. What is glazing, as applied to ceramics?

7.12. What does the term refractory mean?

7.13. What are some of the principal applications of the cemented carbides, such as WC-Co?

7.14. What is one of the important applications of titanium nitride, as mentioned in the text?

7.15. What elements comprise the ceramic material Sialon?

7.16. Define glass.

7.17. What is the primary mineral in glass products?

7.18. What are some of the functions of the ingredients that are added to glass?

7.19. What does the term devitrification mean?

7.20. What is graphite?

MULTIPLE CHOICE QUIZ

There is a total of 18 correct answers in the following multiple choice questions (some questions have multiple answers that are correct). To attain a perfect score on the quiz, all correct answers must be given, since each correct answer is worth 1 point. For each question, each omitted answer or wrong answer reduces the score by 1 point, and each additional answer beyond the number of answers required reduces the score by 1 point. Percentage score on the quiz is based on the total number of correct answers.

7.1. Which one of the following is the most common element in the Earth's crust? (a) aluminum, (b) calcium, (c) iron, (d) oxygen, or (e) silicon.

7.2. Glass products are based primarily on which one of the following minerals? (a) alumina, (b) corundum, (c) feldspar, (d) kaolinite, or (e) silica.

7.3. Which of the following contains significant amounts of aluminum oxide (more than one)? (a) alumina, (b) bauxite, (c) corundum, (d) quartz, or (e) sandstone.

7.4. Which of the following ceramics are commonly used as abrasives in grinding wheels (two best answers)? (a) aluminum oxide, (b) calcium oxide, (c) carbon monoxide, (d) silicon carbide, or (e) silicon dioxide.

7.5. Which one of the following is generally the most porous of the clay-based pottery ware? (a) china, (b) earthenware, (c) porcelain, or (d) stoneware.

7.6. Which of the following is fired at the highest temperatures? (a) china, (b) earthenware, (c) porcelain, or (d) stoneware.

7.7. Which one of the following comes closest to expressing the chemical composition of clay? (a) Al_2O_3, (b) $Al_2(Si_2O_5)(OH)_4$, (c) $3AL_2O_3-2SiO_2$, (d) MgO, or (e) SiO_2.

7.8. Glass ceramics are polycrystalline ceramic structures that have been transformed into the glassy state: (a) true or (b) false.

7.9. Which one of the following materials is closest to diamond in hardness? (a) aluminum oxide, (b) carbon dioxide, (c) cubic boron nitride, (d) silicon dioxide, or (e) tungsten carbide.

7.10. Which of the following best characterizes the structure of glass-ceramics? (a) 95% polycrystalline, (b) 95% vitreous, or (b) 50% polycrystalline.

7.11. Properties and characteristics of the glass-ceramics include which of the following (more than one)? (a) efficiency in processing, (b) electrical conductor, (c) high thermal expansion, or (d) strong, relative to other ceramics.

7.12. Diamond is the hardest material known: (a) true or (b) false.

7.13. The specific gravity of graphite is closest to the following: (a) 1.0 (b) 2.0, (c) 4.0, (d) 8.0, or (e) 16.0.

7.14. Synthetic diamonds date to (a) ancient times, (b) 1800s, (c) 1950s, or (d) 1980.

8
POLYMERS

CHAPTER CONTENTS

8.1   Fundamentals of Polymer Technology
    8.1.1   Polymerization
    8.1.2   Polymer Structures and Copolymers
    8.1.3   Crystallinity
    8.1.4   Thermal Behavior of Polymers
    8.1.5   Additives
8.2   Thermoplastic Polymers
    8.2.1   Properties of Thermoplastic Polymers
    8.2.2   Important Commercial Thermoplastics
8.3   Thermosetting Polymers
    8.3.1   General Properties and Characteristics
    8.3.2   Important Thermosetting Polymers
8.4   Elastomers
    8.4.1   Characteristics of Elastomers
    8.4.2   Natural Rubber
    8.4.3   Synthetic Rubbers
8.5   Guide to the Processing of Polymers

Of the three basic types of materials, polymers are the newest and at the same time the oldest known to man. A ***polymer*** is a compound consisting of long-chain molecules, each molecule made up of repeating units connected together. There may be thousands, even millions of units in a single polymer molecule. The word is derived from the Greek words ***poly,*** meaning many, and ***meros*** (reduced to ***mer***), meaning part. Most polymers are based on carbon and are therefore considered organic chemicals. However, the group also includes a number of inorganic polymers.

Polymers form the living organisms and vital processes of all life on Earth. To ancient man, biological polymers were the source of food, shelter, and many of his implements. However, our interest in this chapter is in materials other than biological polymers. With the exception of natural rubber, nearly all of the polymeric materials used in engineering are synthetic—they are made by chemical processing.

Polymers can be separated into ***plastics*** and ***rubbers.*** As engineering materials, they are relatively new compared to metals and ceramics, dating only from around the mid-1800s (Historical Notes 8.1 on plastics, and 8.2 and 8.3 on rubbers). For our purposes in covering polymers as a technical subject, it is appropriate to divide them into the following three categories, where (1) and (2) are the plastics and (3) are the rubbers:

Historical Note 8.1   *Plastics [13], [16], and other sources.*

Certainly one of the milestones in the history of polymers was Charles Goodyear's discovery of vulcanization of rubber in 1839 (Historical Note 8.2). In 1851, his brother Nelson patented hard rubber, called *ebonite*, which in reality is a thermosetting polymer. It was used for many years for combs, battery cases, and dental prostheses.

At the 1862 International Exhibition in London, an English chemist Alexander Parkes demonstrated the possibilities of the first thermoplastic, a form of *cellulose nitrate* (cellulose is a natural polymer in wood and cotton). He called it *Parkesine* and described it as a replacement for ivory and tortoiseshell. The material became commercially important due to the efforts of an American John Hyatt, who combined cellulose nitrate and camphor (which acts as a plasticizer) together with heat and pressure to form the product he called *Celluloid.* His patent was issued in 1870. Celluloid plastic was transparent, and the applications subsequently developed for it included photographic and motion picture film and windshields for carriages and early motorcars.

Several additional products based on cellulose were developed around the turn of the century. Cellulose fibers, called *Rayon,* were first produced around 1890. Packaging film, called *Cellophane,* was first marketed around 1910. *Cellulose acetate* was adopted as the base for photographic film around the same time. This material was to become an important thermoplastic for injection molding during the next several decades.

The first synthetic plastic was developed in the early 1900s by the Belgian-born American chemist L. H.

Baekeland. It involved the reaction and polymerization of phenol and formaldehyde to form what its inventor called **Bakelite.** This thermosetting resin is still commercially important today. It was followed by other similar polymers: urea-formaldehyde in 1918 and melamine-formaldehyde in 1939.

The late 1920s and 1930s saw the development of a number of thermoplastics of major importance today. A Russian I. Ostromislensky had patented **polyvinylchloride** in 1912, but it was first commercialized in 1927 as a wall covering. Around the same time, **polystyrene** was first produced in Germany. In England, fundamental research was started in 1932 that led to the synthesis of **polyethylene;** the first production plant came on line just before the outbreak of World War II. This was low density polyethylene. Finally, a major research program initiated in 1928 under the direction of W. Carothers at Du Pont in the United States led to the synthesis of the polyamide **nylon;** it was later commercialized in the late 1930s. Its initial use was in ladies' hosiery; subsequent applications during the war included low friction bearings and wire insulation. Similar efforts in Germany provided an alternative form of nylon in 1939.

Several important special-purpose polymers were developed in the 1940s: **Fluorocarbons (Teflon), silicones,** and **polyurethanes** in 1943; **epoxy** resins in 1947, and **acrylonitrile-butadiene-styrene** copolymer (ABS) in 1948. During the 1950s: **polyester** fibers in 1950; and **polypropylene, polycarbonate,** and **high density polyethylene** in 1957. **Thermoplastic elastomers** were first developed in the 1960s. The ensuing years have witnessed a tremendous growth in the use of plastics.

---

(1) ***Thermoplastic polymers,*** or ***Thermoplastics*** (TP), as they are often called, are solid materials at room temperature, but they become viscous liquids when heated to temperatures of only a few hundred degrees. This characteristic allows them to be easily and economically shaped into products. They can be subjected to this heating and cooling cycle repeatedly without significant degradation of the polymer.

(2) ***Thermosetting polymers,*** or Thermosets (TS) cannot tolerate repeated heating cycles as thermoplastics can. When initially heated, they soften and flow for molding; but the elevated temperatures also produce a chemical reaction that hardens the material into an infusible solid. If reheated, thermosetting polymers degrade and char rather than soften.

(3) ***Elastomers.*** These are the rubbers. Elastomers (E) are polymers that exhibit extreme elastic extensibility when subjected to relatively low mechanical stress. Some elastomers can be stretched by a factor of 10 and yet completely recover to their original shape. Although their properties are quite different from thermosets, they share a similar molecular structure that is different from the thermoplastics.

Thermoplastics are commercially the most important of the three types, constituting around 70% of the tonnage of all synthetic polymers produced. Thermosets and elastomers share the remaining 30% about evenly, with a slight edge for the former. Common TP polymers include polyethylene, polyvinylchloride, polypropylene, polystyrene, and nylon. Examples of TS polymers are phenolics, epoxies, and certain polyesters. The most common example given for elastomers is natural (vulcanized) rubber; however, synthetic rubbers exceed the tonnage of natural rubber.

Although the classification of polymers into TP, TS, and E categories will suit our purposes quite adequately for organizing the topic in this chapter, we should note that the three types sometimes overlap: certain polymers that are normally thermoplastic can be made into thermosets; some polymers can be either thermosets or elastomers (we indicated that their molecular structures are similar); and some elastomers are thermoplastic. However, these are exceptions to the general classification scheme.

The growth in applications of synthetic polymers is truly impressive. On a volumetric basis, current annual usage of polymers exceeds that of metals. There are several reasons for the commercial and technological importance of polymers:

> ⟩ Plastics can be formed by molding into intricate part geometries, usually with no further processing required. They are very compatible with **net shape** processing.
> ⟩ Plastics possess an attractive list of properties for many engineering applications where strength is not a factor: (1) low density relative to metals and ceramics; (2) good strength-to-weight ratios for certain (but not all) polymers; (3) high corrosion resistance; and (4) low electrical and thermal conductivity.
> ⟩ On a volumetric basis, polymers are cost competitive with metals.
> ⟩ On a volumetric basis, polymers generally require less energy to produce than metals. This is generally true because the temperatures for working these materials are much lower than for metals.
> ⟩ Certain plastics are translucent and/or transparent, which makes them competitive with glass in some applications.
> ⟩ Polymers are widely used in composite materials (Chapter 9).

On the negative side, polymers in general have the following limitations: (1) strength is low, relative to metals and ceramics; (2) modulus of elasticity or stiffness is also low—in the case of elastomers, of course, this may be a desirable characteristic; (3) service temperatures are limited to only a few hundred degrees because of the softening of thermoplastic polymers or degradation of thermosetting polymers; (4) some polymers degrade when subjected to sunlight and other forms of radiation; and (5) plastics exhibit viscoelastic properties (Section 3.5), which can be a distinct limitation in load bearing applications.

In this chapter we examine the technology of polymeric materials. The first section is devoted to an introductory discussion of polymer science and technology. Subsequent sections survey the three basic categories of polymers: thermoplastics, thermosets, and elastomers.

8.1 FUNDAMENTALS OF POLYMER SCIENCE AND TECHNOLOGY

Polymers are synthesized by joining many small molecules together into very large molecules, called **macromolecules,** that possess a chain-like structure. The small units, called **monomers,** are generally simple unsaturated organic molecules such as ethylene C_2H_4.

The atoms in these molecules are held together by covalent bonds; and when joined to form the polymer, the same covalent bonding holds the links of the chain together. Thus, each large molecule is characterized by strong primary bonding. Synthesis of the polyethylene molecule is depicted in Figure 8.1. As we have described its structure here, polyethylene is a linear polymer; its mers form one long chain.

A mass of polymer material consists of many macromolecules; the analogy of a bowl of just-cooked spaghetti (without sauce) is sometimes used to visualize the relationship of the individual molecules to the bulk material. Entanglement among the long strands helps to hold the mass together, but atomic bonding is more significant. The bonding between macromolecules in the mass is due to van der Waals and other secondary bonding types. Thus, the aggregate polymer material is held together by forces that are substantially weaker than the primary bonds holding the molecules together. This explains why plastics in general are not nearly as stiff and strong as metals or ceramics.

When a thermoplastic polymer is heated, it softens. The heat energy causes the macromolecules to become thermally agitated, exciting them to move relative to each other within the polymer mass (here, the bowl of spaghetti analogy loses its appeal). The material begins to behave like a viscous liquid, viscosity decreasing (fluidity increasing) with rising temperature.

Let us expand on these opening remarks, tracing how polymers are synthesized and examining the characteristics of the materials that result from the synthesis.

8.1.1 Polymerization

As a chemical process, the synthesis of polymers can occur by either of two methods: (1) addition polymerization and (2) step polymerization. Production of a given polymer is generally associated with one method or the other.

Addition Polymerization   In this process, exemplified by polyethylene, the double bonds between carbon atoms in the ethylene monomers are induced to open up so that they join with other monomer molecules. The connections occur on both ends of the expanding macromolecule, developing long chains of repeating mers. Because of the way the molecules are formed, the process is also known as ***chain polymerization.*** It is initiated using a chemical catalyst (called an ***initiator***) to open the carbon double bond in some of the monomers. These monomers, which are now highly reactive because of their unpaired electrons, then capture other monomers to begin forming chains that are reactive. The chains propagate by capturing still other monomers, one at a time, until large molecules have been produced and the reaction is terminated. The process proceeds as indicated in Figure 8.2. The entire polymerization reaction takes only seconds for any given macromolecule. However, in the industrial process, it may take many minutes or even hours to complete the polymerization of a given batch, since all of the chain reactions do not occur simultaneously in the mixture.

Other polymers typically formed by addition polymerization are presented in Figure 8.3, along with the starting monomer and the repeating mer. Note that the chem-

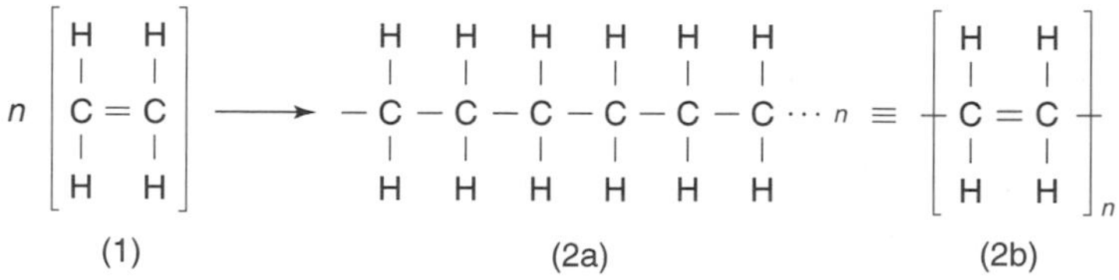

Figure 8.1   Synthesis of polyethylene from ethylene monomers: (1) n ethylene monomers yields (2a) polyethylene of chain length n; (2b) concise notation for depicting the polymer structure of chain length n.

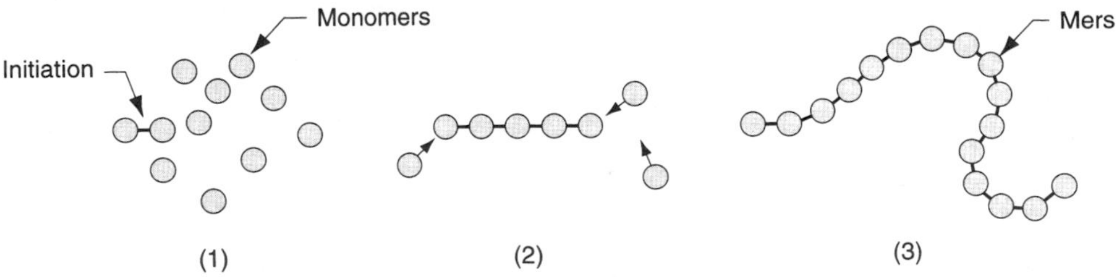

FIGURE 8.2 Model of addition (chain) polymerization: (1) initiation, (2) rapid addition of monomers, and (3) resulting long chain polymer molecule with n mers at termination of reaction.

ical formula for the monomer is the same as that of the mer in the polymer. This is a characteristic of this method of polymerization. Note also that many of the common polymers involve substitution of some alternative atom or molecule in place of one of the H atoms in polyethylene. Polypropylene, polyvinylchloride, and polystyrene are examples of this substitution. Polytetrafluoroethylene replaces all four H atoms in the structure with atoms of fluorine (F). Most addition polymers are thermoplastics. The exception in Figure 8.3 is polyisoprene, the polymer of natural rubber. Although formed by addition polymerization, it is an elastomer.

Step Polymerization In this form of polymerization, two reacting monomers are brought together to form a new molecule of the desired compound. In most (but not all) step polymerization processes, a byproduct of the reaction is also produced. The

FIGURE 8.3 Some typical polymers formed by addition (chain) polymerization.

Polymer	Monomer	Repeating mer	Chemical formula
Polypropylene	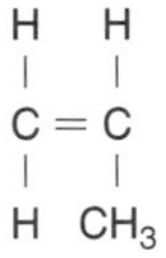	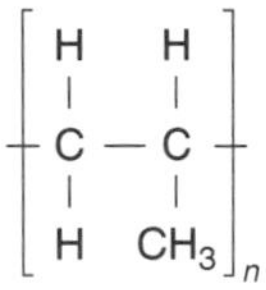	$(C_3H_6)_n$
Polyvinyl chloride		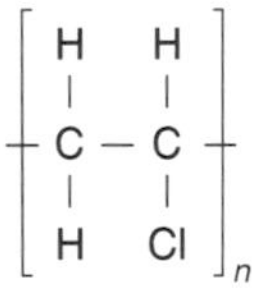	$(C_2H_3Cl)_n$
Polystyrene	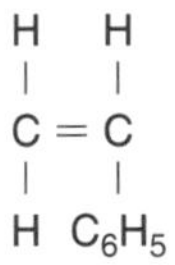	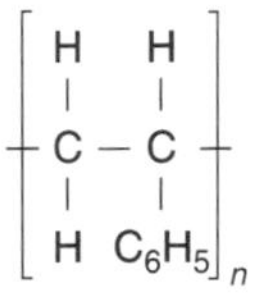	$(C_8H_8)_n$
Polytetrafluoroethylene (Teflon)	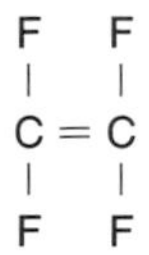	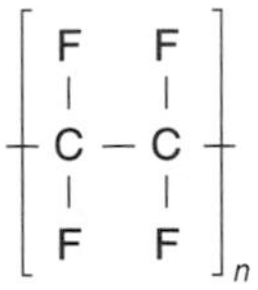	$(C_2F_4)_n$
Polyisoprene (natural rubber)	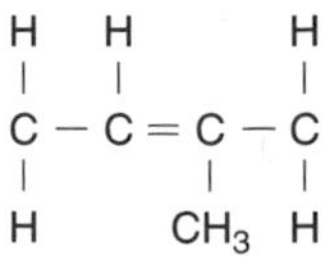	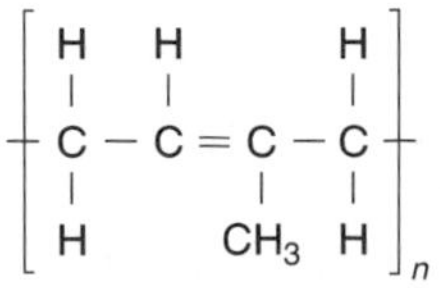	$(C_5H_8)_n$

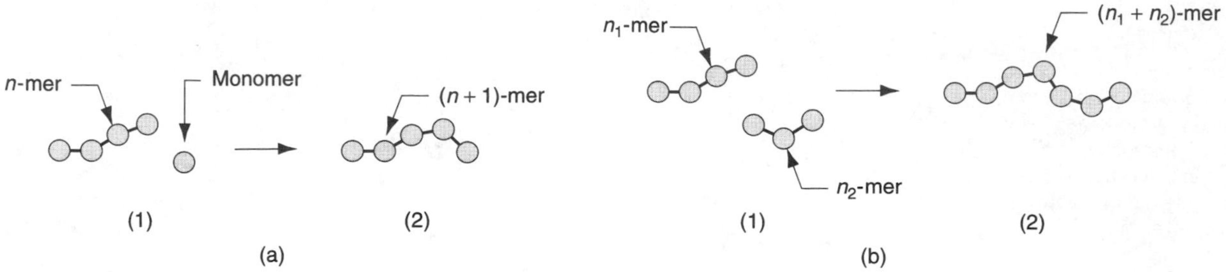

FIGURE 8.4 Model of step polymerization showing the two types of reactions occurring: (a) n-mer attaching a single monomer to form a $(n + 1)$-mer; and (b) n_1-mer combining with n_2-mer to form a $(n_1 + n_2)$-mer. Sequence is shown by (1) and (2).

byproduct is typically water, which condenses; hence, the term **condensation polymerization** is often used for processes that yield the condensate. As the reaction continues, more molecules of the reactants combine with the molecules first synthesized to form polymers of length $n = 2$, then polymers of length $n = 3$, and so on. Polymers of increasing n are created in a slow, stepwise fashion. In addition to this gradual elongation of the molecules, intermediate polymers of length n_1 and n_2 also combine to form molecules of length $n = n_1 + n_2$, so that two types of reactions are proceeding simultaneously once the process is under way, as illustrated in Figure 8.4. Accordingly, at any point in the process, the batch contains polymers of various lengths. Only after sufficient time has elapsed are molecules of adequate length formed.

FIGURE 8.5 Some typical polymers formed by step (condensation) polymerization (simplified expression of structure and formula; ends of polymer chain are not shown).

Polymer	Repeating unit	Chemical formula	Condensate
Nylon-6, 6	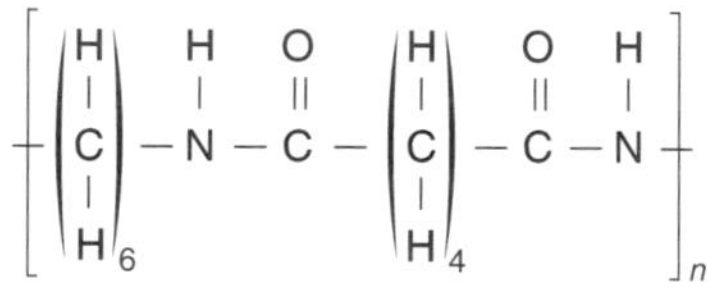	$[(CH_2)_6 (CONH)_2 (CH_2)_4]_n$	H_2O
Polycarbonate	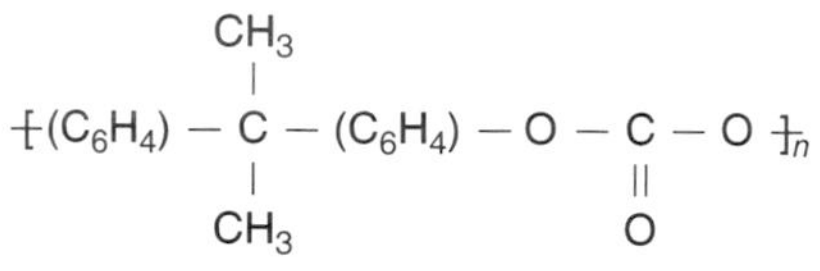	$(C_3H_6 (C_6H_4)_2CO_3)_n$	HCl
Phenol formaldehyde	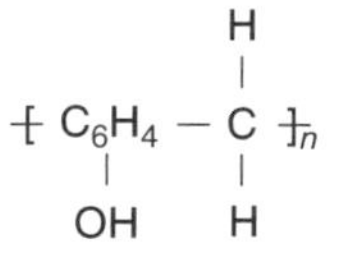	$[(C_6H_4)CH_2OH]_n$	H_2O
Urea formaldehyde	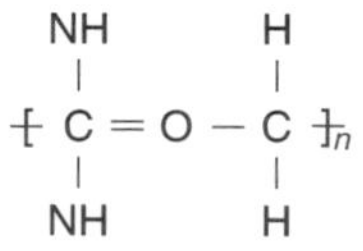	$(CO(NH)_2 CH_2)_n$	H_2O

It should be noted that water is not always the byproduct of the reaction; for example, ammonia (NH_3) is another simple compound produced in some reactions. Nevertheless, the term condensation polymerization is still used. It should also be noted that although most step polymerization processes involve condensation of a byproduct, some do not. Examples of commercial polymers produced by step (condensation) polymerization are given in Figure 8.5. Both thermoplastic and thermosetting polymers are synthesized by this method; nylon-6,6 and polycarbonate are TP polymers, while phenol formaldehyde and urea formaldehyde are TS polymers.

Degree of Polymerization, Molecular Weight, and Structure   A macromolecule produced by polymerization consists of *n* repeating mers. Since molecules in a given batch of polymerized material vary in length, *n* for the batch is an average; its statistical distribution is normal. The mean value of *n* is called the *degree of polymerization* (DP) for the batch. The degree of polymerization affects the properties of the polymer: higher DP increases mechanical strength but also increases viscosity in the fluid state, which makes processing more difficult.

The *molecular weight* (MW) of a polymer is the sum of the molecular weights of the mers in the molecule; it is *n* times the molecular weight of each repeating unit. Since *n* varies for different molecules in a batch, the molecule weight must be interpreted as an average. Typical values of DP and MW for selected polymers are presented in Table 8.1.

8.1.2 Polymer Structures and Copolymers

There are structural differences among polymer molecules, even molecules of the same polymer. In this section we examine three aspects of molecular structure: (1) stereoregularity, (2) branching and cross-linking, and (3) copolymers.

Stereoregularity   *Stereoregularity* is concerned with the spatial arrangement of the atoms and groups of atoms in the repeating units of the polymer molecule. An important aspect of stereoregularity is the way the atom groups are located along the chain for a polymer that has one of the H atoms in its mers replaced by some other atom or atom group. Polypropylene is an example; it is similar to polyethylene except that CH_3 is substituted for one of the four H atoms in the mer. Three tactic arrangements are possible, illustrated in Figure 8.6: (a) *isotactic,* in which the odd atom groups are all on the same side; (b) *syndiotactic,* in which the atom groups alternate on opposite sides; and (c) *atactic,* in which the groups are randomly along either side.

The tactic structure is important in determining the properties of the polymer. It also influences the tendency of a polymer to crystallize (Section 8.1.3). Continuing with our polypropylene example, this polymer can be synthesized in any of the three tactic structures. In its isotactic form, it is strong and melts at 175°C (347°F); the syndiotactic

TABLE 8.1   Typical values of degree of polymerization and molecular weight for selected thermoplastic polymers.

Polymer	Degree of Polymerization (n)	Molecular Weight
Polyethylene	10,000	300,000
Polystyrene	3,000	300,000
Polyvinylchloride	1,500	100,000
Nylon	120	15,000
Polycarbonate	200	40,000

Compiled from [7].

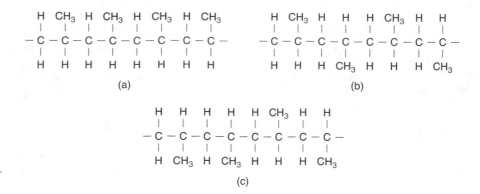

FIGURE 8.6 Possible arrangement of atom groups in polypropylene: (a) isotactic, (b) syndiotactic, and (c) atactic.

structure is also strong, but melts at 131°C (268°F); but atactic polypropylene is soft and melts at around 75°C (165°F) and has little commercial use [6], [9].

Linear, Branched, and Cross-Linked Polymers   We have described the polymerization process as yielding macromolecules of a chain-like structure, called a ***linear polymer.*** This is the characteristic structure of a thermoplastic polymer. Other structures are possible, as portrayed in Figure 8.7. One possibility is for side branches to form along the chain, resulting in the ***branched polymer*** shown in Figure 8.7(b). In polyethylene, this occurs because hydrogen atoms are replaced by carbon atoms at random points along the chain, initiating the growth of a branch chain at each location. For certain polymers, primary bonding occurs between branches and other molecules at certain con-

FIGURE 8.7   Various structures of polymer molecules: (a) linear, characteristic of thermoplastics; (b) branched, (c) loosely cross-linked as in an elastomer, and (d) tightly cross-linked or networked structure as in a thermoset.

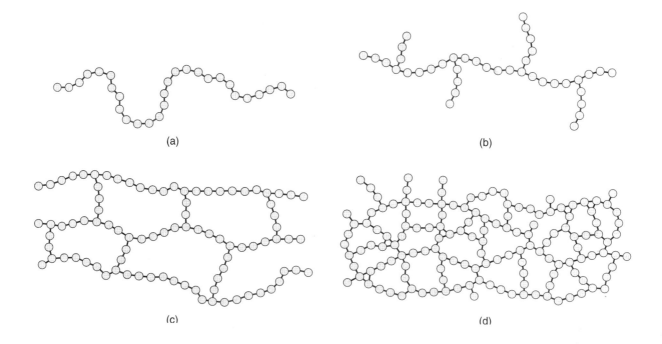

nection points to form ***cross-linked polymers*** as pictured in Figure 8.7(c) and (d). Cross-linking occurs because a certain proportion of the monomers used to form the polymer are capable of bonding to adjacent monomers on more than two sides, thus allowing branches from other molecules to attach. Lightly cross-linked structures are characteristic of elastomers. When the polymer is highly cross-linked we refer to it as having a ***network structure,*** as in (d); in effect, the entire mass is one gigantic macromolecule. Thermosetting plastics take this structure after curing.

The presence of branching and cross-linking in polymers has a significant effect on properties. It is the basis of the difference between the three categories of polymers: TP, TS, and E. Thermoplastic polymers always possess linear or branched structures, or a mixture of the two. Branching increases entanglement among the molecules, usually making the polymer stronger in the solid state and more viscous at a given temperature in the plastic or liquid state. Thermosetting plastics and elastomers are cross-linked polymers. Cross-linking causes the polymer to become chemically set; the reaction cannot be reversed. The effect is to permanently change the structure of the polymer; upon heating, it degrades or burns rather than melts. Thermosets possess a high degree of cross-linking, while elastomers possess a low degree of cross-linking. Thermosets are hard and brittle, while elastomers are elastic and resilient.

Copolymers   Polyethylene is a ***homopolymer;*** so is polypropylene, polystyrene, and many other common plastics; their molecules consist of repeating mers that are all the same type. ***Copolymers*** are polymers whose molecules are made of repeating units of two different types. An example is the copolymer synthesized from ethylene and propylene to produce a copolymer with elastomeric properties. The ethylene-propylene copolymer can be represented as follows:

$$—(C_2H_4)n(C_3H_6)m—$$

where n and m range between 10 and 20, and the proportions of the two constituents are around 50% each. We find in Section 8.4.3 that the combination of polyethylene and polypropylene with small amounts of diene is an important synthetic rubber.

Copolymers can possess different arrangements of their constituent mers. The possibilities are shown in Figure 8.8: (a) ***alternating copolymer,*** in which the mers repeat every other place; (b) ***random,*** in which the mers are in random order, the frequency depending on the relative proportions of the starting monomers; (c) ***block,*** in which mers of the same type tend to group themselves into long segments along the chain; and (d) ***graft,*** in which mers of one type are attached as branches to a main backbone of mers of the other type. The ethylene-propylene diene rubber, mentioned previously, is a block type.

FIGURE 8.8 Various structures of copolymers: (a) alternating, (b) random, (c) block, and (d) graft.

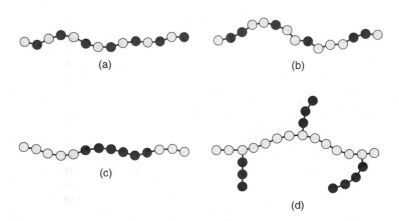

Synthesis of copolymers is analogous to alloying of metals to form solid solutions. As with metallic alloys, differences in the ingredients and structure of copolymers can have a substantial effect on properties. An example is the polyethylene–polypropylene mixture we have been discussing. Each of these polymers alone is fairly stiff; yet a 50–50 mixture forms a copolymer of random structure that is rubbery.

It is also possible to synthesize **ternary polymers,** or **terpolymers,** which consist of mers of three different types. An example is the plastic ABS (acrylonitrile–butadiene–styrene—no wonder they call it ABS).

8.1.3 Crystallinity

Both amorphous and crystalline structures are possible with polymers, although the tendency to crystallize is much less than for metals or nonglass ceramics. Not all polymers can form crystals. For those that can, the ***degree of crystallinity*** (the proportion of crystallized material in the mass) is always less than 100%. As crystallinity is increased in a polymer, so does (1) density, (2) stiffness, strength, and toughness, and (3) heat resistance. In addition, (4) if the polymer is transparent in the amorphous state, it becomes opaque when partially crystallized. Many polymers are transparent, but only in the amorphous (glassy) state. Some of these effects can be illustrated by the differences between low-density and high-density polyethylene, presented in Table 8.2. The underlying reason for the property differences between these materials is the degree of crystallinity.

Linear polymers consist of long molecules with thousands of repeated mers. Crystallization in these polymers involves the folding back and forth of the long chains upon themselves to achieve a very regular arrangement of the mers, as pictured in Figure 8.9(a). The crystallized regions are called ***crystallites.*** Owing to the tremendous length of a single molecule (on an atomic scale), it may participate in more than one crystallite. Also, more than one molecule may be combined in a single crystal region. The crystallites take the form of lamellae, as pictured in Figure 8.9(b), that are randomly mixed in with the amorphous material. Thus, a polymer that crystallizes is a two-phase system—crystallites interspersed throughout an amorphous matrix.

A number of factors determine the capacity and/or tendency of a polymer to form crystalline regions within the material. The factors can be summarized as follows: (1) as a general rule, only linear polymers can form crystals; (2) stereoregularity of the molecule is critical [13]: isotactic polymers always form crystals; syndiotactic polymers sometimes form crystals; atactic polymers never form crystals; (3) copolymers, due to their molecular irregularity, rarely form crystals; (4) slower cooling promotes crystal formation and growth, as it does in metals and ceramics; (5) mechanical deformation, as in the stretching of a heated thermoplastic, tends to align the structure and increase crystallization; (6) plasticizers (chemicals added to a polymer to soften it) reduce the degree of crystallinity.

TABLE 8.2 Comparison of low-density polyethylene and high-density polyethylene.

Polyethylene type	Low Density	High Density
Degree of crystallinity	55%	92%
Specific gravity	0.92	0.96
Modulus of elasticity	140 MPa (20,000 lb/in.^{2})	700 MPa (100,000 lb/in.^{2})
Melting temperature	115°C (239°F)	135°C (275°F)

Compiled from [6]. Values given are typical.

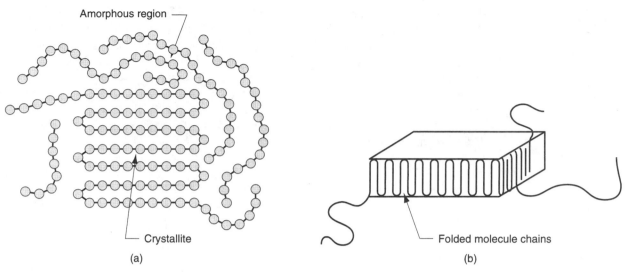

FIGURE 8.9   Crystallized regions in a polymer: (a) long molecules forming crystals randomly mixed in with the amorphous material; and (b) folded chain lamella, the typical form of a crystallized region.

8.1.4 Thermal Behavior of Polymers

The thermal behavior of polymers with crystalline structures is different from that of amorphous polymers (Section 2.4). The effect of structure can be observed on a plot of specific volume (reciprocal of density) as a function of temperature, as plotted in Figure 8.10. A highly crystalline polymer has a melting point T_m at which its volume undergoes an abrupt change. Also, at temperatures above T_m, the thermal expansion of the molten material is greater than for the solid material below T_m. An amorphous polymer does not undergo the same abrupt changes at T_m. As it is cooled from the liquid, its coefficient of thermal expansion continues to decline along the same trajectory as when it was molten, and it becomes increasingly viscous with decreasing temperature. During cooling below T_m, the polymer changes from liquid to rubbery. As temperature continues to drop, a point is finally reached at which the thermal expansion of the amor-

FIGURE 8.10   Behavior of polymers as a function of temperature.

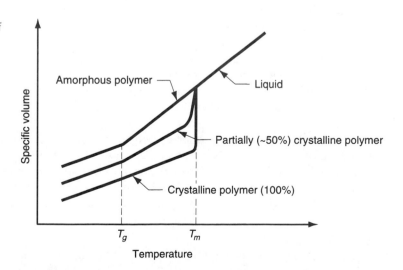

phous polymer suddenly becomes lower. This is the ***glass-transition temperature***, T_g (Section 3.5), seen as the change in slope. Below T_g, the material is hard and brittle.

A partially crystallized polymer lies between these two extremes, as indicated in Figure 8.10. It is an average of the amorphous and crystalline states, the average depending on the degree of crystallinity. Above T_m it exhibits the viscous characteristics of a liquid; between T_m and T_g it has viscoelastic properties; and below T_g it has the conventional elastic properties of a solid.

What we have described in this section applies to thermoplastic materials, which can move up and down the curve of Figure 8.10 multiple times. The manner in which they are heated and cooled may change the path that is followed. For example, fast cooling rates may inhibit crystal formation and increase the glass-transition temperature. Thermosets and elastomers cooled from the liquid state behave like an amorphous polymer until cross-linking occurs. Their molecular structure restricts the formation of crystals. Once their molecules are cross-linked, they cannot be reheated to the molten state.

8.1.5 Additives

The properties of a polymer can often be beneficially changed by combining them with additives. Additives either alter the molecular structure of the polymer or add a second phase to the plastic, in effect transforming a polymer into a composite material. Additives can be classified by function as (1) fillers, (2) plasticizers, (3) colorants, (4) lubricants, (5) flame retardants, (6) cross-linking agents, (7) ultraviolet light absorbers, and (8) antioxidants.

Filler   *Fillers* are solid materials added to a polymer usually in particulate or fibrous form to alter its mechanical properties or to simply reduce material cost. Other reasons for using fillers are to improve dimensional and thermal stability. Examples of fillers used in polymers include cellulosic fibers and powders (e.g., cotton fibers and wood flour, respectively); powders of silica (SiO_2), calcium carbonate ($CaCO_3$), and clay (hydrous aluminum silicate); and fibers of glass, metal, carbon, asbestos, or other polymers. Fillers that improve mechanical properties are called ***reinforcing agents***, and composites thus created are referred to as ***reinforced plastics***; they have higher stiffness, strength, hardness, and toughness than the original polymer. Fibers provide the greatest strengthening effect.

Plasticizers   *Plasticizers* are chemicals added to a polymer to make it softer and more flexible, and to improve its flow characteristics during forming. The plasticizer works by reducing the glass transition temperature to below room temperature. Whereas the polymer is hard and brittle below T_g, it is soft and tough above it. Addition of a plasticizer[1] to polyvinylchloride (PVC) is a good example; depending on the proportion of plasticizer in the mix, PVC can be obtained in a range of properties, from rigid and brittle to flexible and rubbery.

Colorants   An advantage of many polymers over metals or ceramics is that the material itself can be obtained in most any color. This eliminates the need for secondary coating operations. Colorants for polymers are of two types: pigments and dyes. *Pigments* are finely powdered materials that are insoluble in and must be uniformly distributed throughout the polymer in very low concentrations, usually less that 1%. They

---

[1]The common plasticizer in PVC is dioctyl phthalate, a phthalate ester.

often add opacity as well as color to the plastic. ***Dyes*** are chemicals, usually supplied in liquid form, that are generally soluble in the polymer. They are normally used to color transparent plastics such as styrene and acrylics.

Other Additives   *Lubricants* are sometimes added to the polymer to reduce friction and promote flow at the mold interface. Lubricants are also helpful in releasing the part from the mold in injection molding. Mold release agents, sprayed onto the mold surface, are often used for the same purpose.

Nearly all polymers burn if the required heat and oxygen are supplied. Some polymers are more combustible than others. ***Flame retardants*** are chemicals added to polymers to reduce flammability by any or a combination of the following mechanisms: (1) interfering with flame propogation, (2) producing large amounts of incombustible gases, and/or (3) increasing the combustion temperature of the material. The chemicals may also function to (4) reduce the emission of noxious or toxic gases generated during combustion.

We should include amongst the additives those that cause cross-linking to occur in thermosetting polymers and elastomers. The term ***cross-linking agent*** refers to a variety of ingredients that cause a cross-linking reaction or act as a catalyst to promote such a reaction. Important commercial examples are (1) sulfur in vulcanization of natural rubber, (2) formaldehyde for phenolics to form phenolic thermosetting plastics, and (3) peroxides for polyesters.

Many polymers are susceptible to degradation by ultraviolet light (e.g., from sunlight) and oxidation. The degradation manifests itself as the breaking of links in the long chain molecules. Polyethylene, for example, is vulnerable to both types of degradation, which lead to a loss of mechanical strength. ***Ultraviolet light absorbers*** and ***antioxidants*** are additives that reduce the susceptibility of the polymer to these forms of attack.

8.2   THERMOPLASTIC POLYMERS

In this section, we discuss the properties of the thermoplastic polymer group and then survey its important members.

8.2.1   Properties of Thermoplastic Polymers

The defining characteristic of a thermoplastic polymer is that it can be repeatedly heated from a solid state to a viscous liquid state and then cooled back down to solid before significantly degrading. The reason for this characteristic is that TP polymers consist of linear (and/or branched) macromolecules that do not cross-link upon heating. By contrast, thermosets and elastomers undergo a chemical change when heated, which cross-links their molecules and permanently sets these polymers.

In truth, thermoplastics do deteriorate chemically with repeated heating and cooling. In plastic molding, a distinction is made between new or ***virgin*** material, and plastic that has been previously molded (e.g., sprues, defective parts) and therefore has experienced thermal cycling. For some applications, only virgin material is acceptable. Thermoplastic polymers also degrade gradually when subjected to continuous elevated temperatures below T_m. This long-term effect is called ***thermal aging*** and involves slow chemical deterioration. Some TP polymers are more susceptible to thermal aging than others, and for a given material the rate of deterioration depends on temperature.

Mechanical Properties   In our discussion of mechanical properties in Chapter 3, we compared polymers to metals and ceramics. The typical thermoplastic at room temperature is characterized by the following: (1) much lower stiffness, the modulus of elasticity being two (in some cases, three) orders of magnitude lower than metals and ceramics; (2) lower tensile strength, about 10% of the metals; (3) much lower hardness; and (4) greater ductility on average, but there is a tremendous range of values, from 1% elongation for polystyrene to 500% or more for polypropylene.

Mechanical properties of thermoplastics depend on temperature. The functional relationships must be discussed in the context of amorphous and crystalline structures. Amorphous thermoplastics are rigid and glass-like below their glass transition temperature T_g and flexible or rubber-like just above it. As temperature increases above T_g, the polymer becomes increasingly soft, finally becoming a viscous fluid (it never becomes a thin liquid due to its high molecular weight). The effect on mechanical behavior can be portrayed as in Figure 8.11, in which mechanical behavior is defined as deformation resistance. This is analogous to modulus of elasticity but it allows us to observe the effect of temperature on the amorphous polymer as it transitions from solid to liquid. Below T_g, the material is elastic and strong. At T_g, a rather sudden drop in deformation resistance is observed as the material transforms into its rubbery phase; its behavior is viscoelastic in this region. As temperature increases, it gradually becomes more fluid-like.

A theoretical thermoplastic with 100% crystallinity would have a distinct melting point T_m at which it transforms from solid to liquid, but would show no perceptible T_g point. Of course, real polymers have less than 100% crystallinity. For partially crystallized polymers, the resistance to deformation is characterized by the curve that lies between the two extremes, its position determined by the relative proportions of the two phases. The partially crystallized polymer exhibits features of both amorphous and fully crystallized plastics. Below T_g, it is elastic with deformation resistance sloping downward with rising temperatures. Above T_g, the amorphous portions of the polymer soften, while the crystalline portions remain intact. The bulk material exhibits properties that are generally viscoelastic. As T_m is reached, the crystals now melt, giving the polymer a liquid consistency; resistance to deformation is now due to the fluid's viscous properties. The degree to which the polymer assumes liquid characteristics at and above T_m depends on molecular weight and degree of polymerization. Higher DP and MW reduces flow of the polymer, making it more difficult to process by molding and similar shaping methods. This is a dilemma faced by those who select these materials because higher MW and DP mean higher strength.

Physical Properties Physical properties of materials are discussed in Chapter 4. In general, thermoplastic polymers have the following characteristics: (1) lower densities

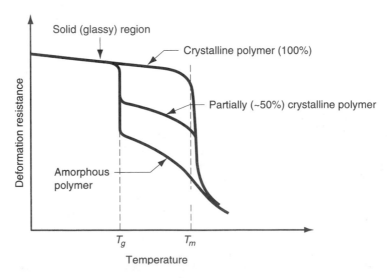

FIGURE 8.11 Relationship of mechanical properties, portrayed as deformation resistance, as a function of temperature for an amorphous thermoplastic, a 100% crystalline (theoretical) thermoplastic, and a partially crystallized thermoplastic.

than metals or ceramics—typical specific gravities for polymers are around 1.2, for ceramics around 2.5, and for metals around 7.0; (2) much higher coefficient of thermal expansion—roughly 5 times the value for metals and 10 times the value for ceramics; (3) much lower melting temperatures; (4) specific heats that are two to four times those of metals and ceramics; (5) thermal conductivities that are about three orders of magnitude lower than those of metals; and (6) insulating electrical properties.

8.2.2 Important Commercial Thermoplastics

Thermoplastic products include molded and extruded items, fibers, films, and sheets, packaging materials, and paints and varnishes. They are normally supplied to the fabricator in the form of powders or pellets in bags, drums, or larger loads by truck or rail car. The most important TP polymers are discussed in alphabetical order in this section. For each plastic, a table is presented listing chemical formula and selected properties. Approximate market share is given relative to all plastics (thermoplastic and thermosetting).

Acetals *Acetal* [Table 8.3(a)] is the popular name given to *polyoxymethylene,* an engineering polymer prepared from formaldehyde (CH_2O) with high stiffness, strength, toughness, and wear resistance. In addition, it has a high melting point, low moisture absorption, and is insoluble in common solvents at ambient temperatures. Because of this combination of properties, acetal resins are competitive with certain metals (e.g., brass and zinc) in automotive components such as door handles, pump housings, and similar parts; appliance hardware; and machinery components.

TABLE 8.3   Important commercial thermoplastic polymers: (a) acetal.

Polymer:	Polyoxymethylene, also known as polyacetal $(OCH_2)_n$		
Symbol:	POM	Elongation:	25%–75%
Polymerization method:	Step (condensation)	Specific gravity:	1.42
Degree of crystallinity:	75% typical	Glass transition temperature:	−80°C (−112°F)
Modulus of elasticity:	3500 MPa (500,000 lb/in.^{2})	Melting temperature:	180°C (356°F)
Tensile strength:	70 MPa (10,000 lb/in.^{2})	Approximate market share:	Much less than 1%

Compiled from [2], [4], [6], [7], [9], and [14].

Acrylics The acrylics are polymers derived from acrylic acid ($C_3H_4O_2$) and compounds originating from it. The most important thermoplastic in the acrylics group is *poly-methylmethacrylate* (PMMA), or Plexiglas (Rohm & Haas's trade name for PMMA). Data on PMMA are listed in Table 8.3(b). It is an amorphous linear polymer. Its outstanding property is excellent transparency, which makes it competitive with glass in optical applications. Examples include automotive tail-light lenses, optical instruments, and aircraft windows. Its limitation when compared with glass is a much lower scratch resistance. Other uses of PMMA include floor waxes and emulsion latex paints. Another important use of acrylics is in fibers for textiles; *polyacrylonitrile* (PAN) is an example that goes by the more familiar trade names Orlon (Du Pont) and Acrilan (Monsanto).

TABLE 8.3   (continued): (b) acrylics (thermoplastic).

Representative polymer:	Polymethylmethacrylate $(C_5H_8O_2)_n$		
Symbol:	PMMA	Elongation:	5
Polymerization method:	Addition	Specific gravity:	1.2
Degree of crystallinity:	None (amorphous)	Glass transition temperature:	105°C(221°F)
Modulus of elasticity:	2800 MPa (400,000 lb/in.^{2})	Melting temperature:	200°C (392°F)
Tensile strength:	55 MPa (8,000 lb/in.^{2})	Approximate market share:	About 1%

Acrylonitrile-Butadiene-Styrene ABS is called an engineering plastic due to its excellent combination of mechanical properties, some of which are listed in Table 8.3(c). ABS is a two-phase terpolymer, one phase being the hard copolymer styrene-acrylonitrile, while the other phase is styrene-butadiene copolymer that is rubbery. The name of the plastic is derived from the three starting monomers, which may be mixed in various proportions. Typical applications include components for automotive, appliances, business machines; and pipes and fittings.

TABLE 8.3 (continued): (c) acrylonitrile-butadiene-styrene.

Polymer:	Terpolymer of acrylonitrile (C_3H_3N), butadiene (C_4H_6), and styrene (C_8H_8).		
Symbol:	ABS	Tensile strength:	50 MPa (7,000 lb/in.^{2})
Polymerization method:	Addition	Elongation:	10% to 30%
Degree of crystallinity:	None (amorphous)	Specific gravity:	1.06
Modulus of elasticity:	2100 MPa (300,000 lb/in.^{2})	Approximate market share:	About 3%

Cellulosics *Cellulose* ($C_6H_{10}O_5$) is a carbohydrate polymer commonly occurring in nature. Wood and cotton fibers, the chief industrial sources of cellulose, contain about 50% and 95% of the polymer, respectively. When cellulose is dissolved and reprecipitated during chemical processing, the resulting polymer is called *regenerated cellulose.* When this is produced as a fiber for apparel it is known as *rayon* (of course, cotton itself is a widely used fiber for apparel). When it is produced as a thin film, it is *cellophane,* a widely used packaging material. Cellulose itself cannot be used as a thermoplastic because it decomposes before melting when its temperature is increased. However, it can be combined with various compounds to form several plastics of commercial importance; examples are *cellulose acetate* (CA) and *cellulose acetate–butyrate* (CAB). CA, data for which are given in Table 8.3(d), is produced in the form of sheets (for wrapping), film (for photography), and molded parts. CAB is a better molding material than CA and has greater impact strength, lower moisture absorption, and better compatibility with plasticizers. The cellulosic thermoplastics share about 1% of the market.

TABLE 8.3 (continued): (d) cellulosics.

Representative polymer:	Cellulose acetate ($C_6H_9O_5$–$COCH_3$)$_n$		
Symbol:	CA	Elongation:	10% to 50%
Polymerization method:	Step (condensation)	Specific gravity:	1.3
Degree of crystallinity:	Amorphous	Glass transition temperature:	105°C (221°F)
Modulus of elasticity:	2800 MPa (400,000 lb/in.^{2})	Melting temperature:	306°C (583°F)
Tensile strength:	30 MPa (4,000 lb/in.^{2})	Approximate market share:	Less than 1%

Fluoropolymers *Polytetrafluorethylene* (PTFE), commonly known as *Teflon,* accounts for about 85% of the family of polymers called *fluoropolymers* [see Table 8.3(e)] in which F atoms replace H atoms in the hydrocarbon chain. PTFE is extremely resistant to chemical and environmental attack, is unaffected by water, possesses good electrical properties, good heat resistance, and very low coefficient of friction. These latter two properties have promoted its use in nonstick household cookware. Other applications that rely on the same property include nonlubricating bearings and similar components. PTFE also finds applications in chemical equipment and food processing.

TABLE 8.3   (continued): (e) fluoropolymers.

Representative polymer:	Polytetrafluorethylene $(C_2F_4)_n$		
Symbol:	PTFE	Elongation:	100% to 300%
Polymerization method:	Addition	Specific gravity:	2.2
Degree of crystallinity:	About 95% crystalline	Glass transition temperature:	127°C (260°F)
Modulus of elasticity:	425 MPa (60,000 lb/in.^{2})	Melting temperature:	327°C (620°F)
Tensile strength:	20 MPa (2,500 lb/in.^{2})	Approximate market share:	Less than 1%

Polyamides An important polymer family that forms characteristic amide linkages (CO–NH) during polymerization is called the polyamides (PA). The most important members of the PA family are *nylons,* of which the two principal grades are nylon–6 and nylon 6,6 (the numbers are codes that indicate the number of carbon atoms in the monomer). The data given in Table 8.3(f) are for nylon–6,6, which was developed at Du Pont in the 1930s. Properties of nylon–6, developed in Germany are similar. Nylon is strong, highly elastic, tough, abrasion resistant, and self-lubricating. It retains good mechanical properties at temperatures up to about 125°C (250°F). One shortcoming is that it absorbs water with an accompanying degradation in properties. The majority of applications of nylon (about 90%) are in fibers for carpets, apparel, and tire cord. The remainder (10%) are in engineering components; nylon is commonly a good substitute for metals in bearings, gears, and similar parts where strength and low friction are needed.

TABLE 8.3   (continued): (f) polyamides.

Representative Polymer:	Nylon–6,6 $((CH_2)_6(CONH)_2(CH_2)_4)_n$		
Symbol:	PA–6,6	Elongation:	300%
Polymerization method:	Step (condensation)	Specific gravity:	1.14
Degree of crystallinity:	Highly crystalline	Glass transition temperature:	50°C (122°F)
Modulus of elasticity:	700 MPa (100,000 lb/in.^{2})	Melting temperature:	260°C (500°F)
Tensile strength:	70 MPa (10,000 lb/in.^{2})	Approximate market share:	1% for all polyamides

A second group of polyamides is the *aramids* (aromatic polyamides) of which *Kevlar* (Du Pont trade name) is gaining in importance as a fiber in reinforced plastics. The reason for the interest in Kevlar is that its strength is the same as steel at 20% of the weight.

Polycarbonate Polycarbonate (PC) [Table 8.3(g)] is noted for its generally excellent mechanical properties, which include high toughness and good creep resistance. It is one of the best thermoplastics for heat resistance—it can be used to temperatures around 125°C (250°F). In addition, it is transparent and fire resistant. Applications include molded machinery parts, housings for business machines, pump impellers, safety helmets, and compact disks (e.g., audio, video, and computer). It is also widely used in glazing (window and windshield) applications.

TABLE 8.3   (continued): (g) polycarbonate

Polymer:	Polycarbonate $(C_3H_6(C_6H_4)_2CO_3)_n$		
Symbol:	PC	Elongation:	110%
Polymerization method:	Step (condensation)	Specific gravity:	1.2
Degree of crystallinity:	Amorphous	Glass transition temperature:	150°C (302°F)
Modulus of elasticity:	2500 MPa (350,000 lb/in.^{2})	Melting temperature:	230°C (446°F)
Tensile strength:	65 MPa (9,500 lb/in.^{2})	Approximate market share:	Less than 1%

Polyesters   The polyesters form a family of polymers made up of the characteristic ester linkages (CO–O). They can be either thermoplastic or thermosetting, depending on whether cross-linking occurs. Of the thermoplastic polyesters, a representative example is *polyethylene terephthalate* (PET), data for which are compiled in [Table 8.3(h)]. It can be either amorphous or partially crystallized (up to about 30%), depending on how it is cooled after shaping. Fast cooling favors the amorphous state, which is highly transparent. Significant applications include blow-molded beverage containers, photographic films, and magnetic recording tape. In addition, PET is widely used as fibers in apparel. Polyester fibers have low moisture absorption and good deformation recovery, both of which make them ideal for "wash and wear" garments that resist wrinkling. The PET fibers are almost always blended with cotton or wool. Familiar trade names for polyester fibers include Dacron (Du Pont), Fortrel (Celanese), and Kodel (Eastman Kodak).

TABLE 8.3   (continued): (h) polyesters (thermoplastic).

Representative polymer:	Polyethylene terephthalate $(C_2H_4–C_8H_4O_4)_n$		
Symbol:	PET	Elongation:	200%
Polymerization method:	Step (condensation)	Specific gravity:	1.3
Degree of crystallinity:	Amorphous to 30% crystalline	Glass transition temperature:	70°C (158°F)
Modulus of elasticity:	2300 MPa (325,000 lb/in.^{2})	Melting temperature:	265°C (509°F)
Tensile strength:	55 MPa (8,000 lb/in.^{2})	Approximate market share:	About 2%

Polyethylene   Polyethylene (PE) was first synthesized in the 1930s, and today it accounts for the largest volume of all plastics. The features that make PE attractive as an engineering material are low cost, chemical inertness, and easy processing. Polyethylene is available in several grades, the most common of which are *low-density polyethylene* (LDPE) and *high-density polyethylene* (HDPE). The low-density grade is a highly branched polymer with lower crystallinity and density. Applications include sheets, film, and wire insulation. HDPE has a more linear structure, with higher crystallinity and density. These differences make HDPE stiffer and stronger and give it a higher melting temperature. HDPE is used to produce bottles, pipes, and housewares. Both grades can be processed by most polymer shaping methods (Chapter 13). Properties for the two grades are given in Table 8.3(i).

TABLE 8.3   (continued): (i) polyethylene.

Polyethylene:	$(C_2H_4)_n$ (low density)	$(C_2H_4)_n$ (high density)
Symbol:	LDPE	HDPE
Polymerization method:	Addition	Addition
Degree of crystallinity:	55% typical	92% typical
Modulus of elasticity:	140 MPa (20,000 lb/in.^{2})	700 MPa (100,000 lb/in.^{2})
Tensile strength:	15 MPa (2,000 lb/in.^{2})	30 MPa (4,000 lb/in.^{2})
Elongation:	100–500%	20–100%
Specific gravity:	0.92	0.96
Glass transition temperature:	−100°C (−148°F)	−115°C (−175°F)
Melting temperature:	115°C (240°F)	135°C (275°F)
Approximate market share:	About 20%	About 15%

Polypropylene   Polypropylene (PP) has become a major plastic, especially for injection molding, since its introduction in the late 1950s. PP can be synthesized in isotactic, syndiotactic, or atactic structures, the first of these being the most important and for which the characteristics are given in [Table 8.3(j)]. It is the lightest of the plastics and its strength-to-weight ratio is high. PP is frequently compared with HDPE because its cost and many of its properties are similar. However, the high melting point of polypropylene allows certain applications that preclude use of polyethylene; for example, components that must be sterilized. Other applications are injection molded parts for automotive and houseware, and fiber products for carpeting. A special application suited to polypropylene is one-piece hinges that can be subjected to a high number of flexing cycles without failure.

TABLE 8.3   (continued): (j) polypropylene.

Polymer:	Polypropylene $(C_3H_6)_n$		
Symbol:	PP	Elongation:	10%–500%[a]
Polymerization method:	Addition	Specific gravity:	0.90
Degree of crystallinity:	High, varies with processing	Glass transition temperature:	$-20°C$ $(-4°F)$
Modulus of elasticity:	1400 MPa (200,000 lb/in.^{2})	Melting temperature:	176°C (249°F)
Tensile strength:	35 MPa (5,000 lb/in.^{2})	Approximate market share:	About 13%

[a]Elongation depends on additives.

Polystyrene   There are several polymers, copolymers, and terpolymers based on the monomer styrene (C_8H_8), of which polystyrene (PS) is used in the highest volume [Table 8.3(k)]. It is a linear homopolymer with amorphous structure that is generally noted for its brittleness. PS is transparent, easily colored, and readily molded, but degrades at elevated temperatures and dissolves in various solvents. Because of its brittleness, some PS grades contain 5% to 15% rubber, and the term ***high-impact polystyrene*** (HIPS) is used for these types. They have higher toughness, but transparency and tensile strength are reduced. In addition to injection molding applications (e.g., molded toys, housewares), polystyrene also finds uses in packaging in the form of PS foams.

TABLE 8.3   (continued): (k) polystyrene.

Polymer:	Polystyrene $(C_8H_8)_n$		
Symbol:	PS	Elongation:	1%
Polymerization method:	Addition	Specific gravity:	1.05
Degree of crystallinity:	None (amorphous)	Glass transition temperature:	100°C (212°F)
Modulus of elasticity:	3200 MPa (450,000 lb/in.^{2})	Melting temperature:	240°C (464°F)
Tensile strength:	50 MPa (7,000 lb/in.^{2})	Approximate market share:	About 10%

Polyvinylchloride   Polyvinylchloride (PVC) [Table 8.3(l)] is a widely used plastic whose properties can be varied by combining additives with the polymer. In particular, plasticizers are used to achieve thermoplastics ranging from rigid PVC (no plasticizers) to flexible PVC (high proportions of plasticizer). The range of properties makes PVC a versatile polymer, with applications that include rigid pipe (used in construction, water and sewer systems, irrigation), fittings, wire and cable insulation, film, sheets, food packaging, flooring, and toys. PVC by itself is relatively unstable to heat and light, and stabilizers must be added to improve its resistance to these environmental conditions. Care must be taken in the production and handling of the vinyl chloride monomer used to polymerize PVC, due to its carcinogenic nature.

TABLE 8.3 (continued): (l) polyvinylchloride.

Polymer:	Polyvinylchloride $(C_2H_3Cl)_n$		Elongation:	2% with no plasticizer
Symbol:	PVC		Specific gravity:	1.40
Polymerization method:	Addition		Glass transition temperature:	81°C (178°F)[a]
Degree of crystallinity:	None (amorphous structure)		Melting temperature:	212°C (414°F)
Modulus of elasticity:	2800 MPa (400,000 lb/in.^{2})[a]		Approximate market share:	About 16%
Tensile strength:	40 MPa (6,000 lb/in.^{2})			

[a] With no plasticizer.

8.3 THERMOSETTING POLYMERS

Thermosetting (TS) polymers are distinguished by their highly cross-linked three-dimensional structure. In effect, the formed part (e.g., the pot handle or electrical switch cover) becomes one large macromolecule. Thermosets are always amorphous and exhibit no glass transition temperature. In this section, we examine the general characteristics of the TS plastics and identify the important materials in this category.

8.3.1 General Properties and Characteristics

Owing to differences in chemistry and molecular structure, properties of thermosetting plastics are different from those of thermoplastics. In general, thermosets are (1) more rigid—modulus of elasticity is two to three times greater; (2) brittle—they possess virtually no ductility; (3) less soluble in common solvents; (4) capable of higher service temperatures; and (5) not capable of being remelted—instead they degrade or burn.

The differences in properties of the TS plastics are attributable to cross-linking, which forms a thermally stable, three-dimensional, covalently bonded structure within the molecule. Cross-linking is accomplished in three ways [7]:

1. *Temperature-activated systems*—In the most common systems, the changes are caused by heat supplied during the part shaping operation (e.g., molding). The starting material is a linear polymer in granular form supplied by the chemical plant. As heat is added, the material softens for molding; continued heating results in cross-linking of the polymer. The term *thermosetting* is most aptly applied to these polymers.
2. *Catalyst-activated systems*—Cross-linking in these systems occurs when small amounts of a catalyst are added to the polymer, which is in liquid form. Without the catalyst, the polymer remains stable; once combined with the catalyst it changes into solid form.
3. *Mixing-activated systems*—Most epoxies are examples of these systems. The mixing of two chemicals results in a reaction that forms a cross-linked solid polymer. Elevated temperatures are sometimes used to accelerate the reactions.

The chemical reactions associated with cross-linking are called **curing** or **setting.** Curing is done at the fabrication plants that shape the parts rather than the chemical plants that supply the starting materials to the fabricator.

8.3.2 Important Thermosetting Polymers

Thermosetting plastics are not as widely used as the thermoplastics, perhaps because of the added processing complications involved in curing the TS polymers. The largest vol-

TABLE 8.4   Important commercial thermosetting polymers: (a) amino resins.

Representative polymer:	Melamine-formaldehyde		
Monomers:	Melamine ($C_3H_6N_6$) and formaldehyde (CH_2O)		
Polymerization method:	Step (condensation)	Elongation:	Less than 1%
Modulus of elasticity:	9000 MPa (1,300,000 lb/in.^{2})	Specific gravity:	1.5
Tensile strength:	50 MPa (7,000 lb/in.^{2})	Approx. market share:	About 4% for urea-formaldehyde and melamine-formaldehyde.

Compiled from [2], [4], [6], [7], [9], and [14].

ume thermoset is phenolic resins, whose annual volume is about 6% of the total plastics market—significantly less than the leading thermoplastic, polyethylene, which has about 35% of the market. Technical data for these materials are given in Table 8.4. Market share data refers to total plastics (TP plus TS).

Amino Resins   Amino plastics [Table 8.4(a)] characterized by the amino group (NH_2), consist of two thermosetting polymers, urea-formaldehyde and melamine-formaldehyde, which are produced by the reaction of formaldehyde (CH_2O) with either urea ($CO(NH_2)_2$) or melamine ($C_3H_6N_6$), respectively. In commercial importance, the amino resins rank just below the other formaldehyde resin, phenol-formaldehyde, discussed below. *Urea-formaldehyde* is competitive with the phenols in certain applications, particularly as a plywood and particle-board adhesive. The resins are also used as a molding compound. It is slightly more expensive than the phenol material. *Melamine-formaldehyde* plastic is water resistant and useful for dishware and as a coating in laminated table and countertops (Formica, trade name of Cyanamid Co.). When used as molding materials, amino plastics usually contain significant proportions of fillers, such as cellulose.

Epoxies   Epoxy resins [Table 8.4(b)] are based on a chemical group called the *epoxides.* The simplest formulation of epoxide is ethylene oxide (C_2H_3O). Epichlorohydrin (C_3H_5OCl) is a much more widely used epoxide for producing epoxy resins. Uncured, epoxides have a low degree of polymerization. To increase molecular weight and to cross-link the epoxide, a curing agent must be used. Possible curing agents include polyamines and acid anhydrides. Cured epoxies are noted for strength, adhesion, and heat and chemical resistance. Applications include surface coatings, industrial flooring, glass fiber-reinforced composites, and adhesives. Insulating properties of epoxy thermosets make them useful in various electronic applications, such as encapsulation of integrated circuits and lamination of printed circuit boards.

TABLE 8.4   (continued): (b) epoxy.

Example chemistry:	Epichlorohydrin (C_3H_5OCl) plus curing agent such as triethylamine ($C_6H_5–CH_2N–(CH_3)_2$)		
Polymerization method:	Condensation	Elongation:	0%
Modulus of elasticity:	7000 MPa (1,000,000 lb/in.^{2})	Specific gravity:	1.1
Tensile strength:	70 MPa (10,000 lb/in.^{2})	Approx. market share:	About 1%

Phenolics   Phenol (C_6H_5OH) [Table 8.4(c)] is an acidic compound that can be reacted with aldehydes (dehydrogenated alcohols), formaldehyde (CH_2O) being the most reactive. *Phenol-formaldehyde* is the most important of the phenolic polymers; it was first commercialized around 1900 under the trade name *Bakelite.* It is almost always combined with fillers such as wood flour, cellulose fibers, and minerals when used as a molding material. It is brittle and possesses good thermal, chemical, and dimensional stability. Its capacity to accept colorants is limited—it is available only in dark colors. Molded products constitute only about 10% of total phenolics use. Other applications include adhesives for plywood, printed circuit boards, counter tops, and bonding material for brake linings and abrasive wheels.

TABLE 8.4 (continued): (c) phenol formaldehyde.

Monomer ingredients:	Phenol (C_6H_5OH) and formaldehyde (CH_2O)		
Polymerization method:	Step (condensation)	Elongation:	Less than 1%
Modulus of elasticity:	7000 MPa (1,000,000 lb/in.^{2})	Specific gravity:	1.4
Tensile strength:	70 MPa (10,000 lb/in.^{2})	Approx. market share:	6%

Polyesters Polyesters [Table 8.4(d)], which contain the characteristic ester linkages (CO–O), can be thermosetting as well as thermoplastic (Section 8.2). Thermosetting polyesters are used largely in reinforced plastics (composites) to fabricate large items such as pipes, tanks, boat hulls, auto body parts, and construction panels. They can also be used in various molding processes to produce smaller parts. Synthesis of the starting polymer involves reaction of an acid or anhydride such as maleic anhydride ($C_4H_2O_3$) with a glycol such as ethylene glycol ($C_2H_6O_2$). This produces an ***unsaturated polyester*** of relatively low molecular weight (MW = 1000 to 3000). This ingredient is mixed with a monomer capable of polymerizing and cross-linking with the polyester. Styrene (C_8H_8) is commonly used for this purpose, in proportions of 30% to 50%. A third component, called an inhibitor, is added to prevent premature cross-linking. This mixture forms the polyester resin system that is supplied to the fabricator. Polyesters are cured either by heat (temperature-activated systems), or by means of a catalyst added to the polyester resin (catalyst-activated systems). Curing is done at the time of fabrication (molding or other forming process) and results in cross-linking of the polymer.

TABLE 8.4 (continued): (d) unsaturated polyester.

Example chemistry:	Maleic anhydride ($C_4H_2O_3$) and ethylene glycol ($C_2H_6O_2$) plus styrene (C_8H_8)		
Polymerization method:	Step (condensation)	Elongation:	0%
Modulus of elasticity:	7000 MPa (1,000,000 lb/in.^{2})	Specific gravity:	1.1
Tensile strength:	30 MPa (4,000 lb/in.^{2})	Approx. market share:	3%

An important class of polyesters are the ***alkyd*** resins (the name derived by abbreviating and combining the words ***al***cohol and a***cid*** and changing a few letters). They are used primarily as bases for paints, varnishes, and lacquers. Alkyd molding compounds are also available, but their applications are limited.

Polyurethanes This includes a large family of polymers [Table 8.4(e)], all characterized by the urethane group (NHCOO) in their structure. The chemistry of the polyurethanes is complex and there are many chemical varieties in the family. The characteristic feature is the reaction of a ***polyol,*** whose molecules contain hydroxyl (OH) groups, such as butylene ether glycol ($C_4H_{10}O_2$); and an ***isocyanate,*** such as diphenylmethane diisocyanate ($C_{15}H_{10}O_2N_2$). Through variations in chemistry, cross-linking, and processing, polyurethanes can be thermoplastic, thermosetting, or elastomeric materials, the latter two being the most important commercially. The largest application of polyurethane is in foams. These can range between elastomeric and rigid, the latter being more highly cross-linked. Rigid foams are used as a filler material in hollow construction panels and refrigerator walls. In these types of applications, the material provides excellent thermal insulation, adds rigidity to the structure, and does not absorb water in significant amounts. Many paints, varnishes, and similar coating materials are based on urethane systems. We discuss polyurethane elastomers in Section 8.4.

TABLE 8.4   (continued): (e) polyurethane.

Polymer:	Polyurethane is formed by the reaction of a polyol and an isocyanate. Chemistry varies significantly.			
Polymerization method:	Step (condensation)		Elongation:	Depends on cross-linking.
Modulus of elasticity:	Depends on chemistry and processing.		Specific gravity:	1.2
Tensile strength:	30 MPa (4,000 lb/in.^{2})^a		Approx. market share:	About 4%, including elastomers

^{a}Typical for highly cross-linked polyurethane.

Silicones   Silicones are inorganic and semi-inorganic polymers, distinguished by the presence of the repeating siloxane link (–Si–O–) in their molecular structure. A typical formulation combines the methyl radical (CH$_3$) with (SiO) in various proportions to obtain the repeating unit –((CH$_3$)$_m$–SiO)–, where m establishes the proportionality. By variations in composition and processing, polysiloxanes can be produced in three forms: (1) fluids, (2) elastomers, and (3) thermosetting resins. Fluids (1) are low molecular weight polymers used for lubricants, polishes, waxes, and other liquids—not really polymers in the sense of this chapter, but important commercial products nevertheless. Silicone elastomers (2), covered in Section 8.4, and thermosetting silicones (3), treated here, are cross-linked. When highly cross-linked, polysiloxanes form rigid resin systems used for paints, varnishes, and other coatings; and laminates such as printed circuit boards. They are also used as molding materials for electrical parts. Curing is accomplished by heating or by allowing the solvents containing the polymers to evaporate. Silicones are noted for their good heat resistance and water repellence, but their mechanical strength is not as great as other cross-linked polymers. Data in Table 8.4(f) are for a typical silicone thermosetting polymer.

TABLE 8.4   (continued): (f) silicone thermosetting resins.

Example chemistry:	(CH$_3$)$_6$–SiO)$_n$			
Polymerization method:	Step (condensation), usually		Elongation:	0%
Tensile strength:	30 MPa (4,000 lb/in.^{2})		Specific gravity:	1.65
			Approx. market share:	Less than 1%

8.4 ELASTOMERS

Elastomers are polymers capable of large elastic deformation when subjected to relatively low stresses. Some elastomers can withstand extensions of 500% or more and still return to their original shape. The more popular term for elastomer is, of course, rubber. We can divide rubbers into two categories: (1) natural rubber, derived from certain biological plants; and (2) synthetic polymers, produced by polymerization processes similar to those used for thermoplastic and thermosetting polymers. Before discussing natural and synthetic rubbers, let us consider the general characteristics of elastomers.

8.4.1 Characteristics of Elastomers

Elastomers consist of long-chain molecules that are cross-linked (like thermosetting polymers). They owe their impressive elastic properties to the combination of two features: (1) the long molecules are tightly kinked when unstretched, and (2) the degree of cross-linking is substantially below that of the thermosets. These features are illustrated in the model of Figure 8.12(a), which shows a tightly kinked cross-linked molecule under no stress.

When the material is stretched, the molecules are forced to uncoil and straighten as shown in Figure 8.12(b). The molecules' natural resistance to uncoiling provides the initial elastic modulus of the aggregate material. As further strain is experienced, the covalent bonds of the cross-linked molecules begin to play an increasing role in the modulus, and the stiffness increases as illustrated in Figure 8.13. With greater cross-linking, the elastomer becomes stiffer and its modulus of elasticity is more linear. These characteristics are shown in the figure by the stress-strain curves for three grades of rubber: natural crude rubber, whose cross-linking is very low; cured (vulcanized) rubber with low to medium cross-linking; and hard rubber (ebonite), whose high degree of cross-linking transforms it into a thermosetting plastic.

For a polymer to exhibit elastomeric properties, it must be amorphous in the unstretched condition, and its temperature must be above T_g. If below the glass transition temperature, the material is hard and brittle; above T_g the polymer is in the "rubbery" state. Any amorphous thermoplastic polymer will exhibit elastomeric properties above T_g for a short time, because its linear molecules are always coiled to some extent, thus allowing for elastic extension. It is the absence of cross-linking in TP polymers that prevents them from being truly elastic; instead they exhibit viscoelastic behavior.

Curing is required to effect cross-linking in most of the common elastomers today. The term for curing used in the context of natural rubber (and certain synthetic rubbers) is ***vulcanization,*** which involves the formation of chemical cross-links between the polymer chains. Typical cross-linking in rubber is one to ten links per hundred carbon atoms in the linear polymer chain, depending on the degree of stiffness desired in the material. This is considerably less than the degree of cross-linking in thermosets.

An alternative method of curing involves the use of starting chemicals that react when mixed (sometimes requiring a catalyst or heat) to form elastomers with relatively infrequent cross-links between molecules. These synthetic rubbers are known as ***reactive system elastomers.*** Certain polymers that cure by this means, such as urethanes and silicones, can be classified as either thermosets or elastomers, depending on the degree of cross-linking achieved during the reaction.

A relatively new class of elastomers, ***thermoplastic elastomers,*** possesses elastomeric properties that result from the mixture of two phases, both thermoplastic. One is above its T_g at room temperature while the other is below its T_g. Thus, we have a polymer that includes soft rubbery regions intermixed with hard particles that act as cross-links. The composite material is elastic in its mechanical behavior, although not as extensible as most other elastomers. Since both phases are thermoplastic, the aggregate

FIGURE 8.12 Model of long elastomer molecules, with low degree of cross-linking: (a) unstretched, and (b) under tensile stress.

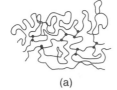

(a)

(b)

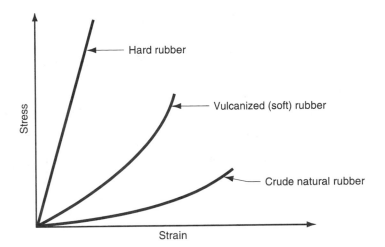

FIGURE 8.13 Increase in stiffness as a function of strain for three grades of rubber: natural rubber, vulcanized rubber, and hard rubber.

material can be heated above its T_m for forming, using processes that are generally more economical than those used for rubber.

We discuss the elastomers in two sections that follow. The first deals with natural rubber and how it is vulcanized to create a useful commercial material; the second examines the synthetic rubbers.

8.4.2 Natural Rubber

Natural rubber (NR) consists primarily of polyisoprene, a high-molecular-weight polymer of isoprene (C_5H_8). It is derived from latex, a milky substance produced by various plants, the most important of which is the rubber tree (Hevea brasiliensis) that grows in tropical climates (Historical Note 8.2). Latex is a water emulsion of polyisoprene (about one third by weight), plus various other ingredients. Rubber is extracted from the latex by various methods (e.g., coagulation, drying, and spraying) that remove the water.

Historical Note 8.2 *Natural rubber* [1], [3].

The first use of natural rubber seems to have been in the form of rubber balls used for sport by the Indians of Central and South America at least 500 hundred years ago. Columbus noted this during his second voyage to the New World in 1493–96. The balls were made from the dried gum of a rubber tree. The first white men in South America called the tree **caoutchouc,** which was their way of pronouncing the Indian name for it. The name **rubber** came from the English chemist J. Priestley, who discovered (around 1770) that gum rubber would "rub" away pencil marks.

Early rubber goods were less than satisfactory; they melted in summer heat and hardened in winter cold. One of those in the business of making and selling rubber goods was Charles Goodyear. Recognizing the deficiencies of the natural material, he experimented with ways to improve its properties and discovered that rubber could be cured by heating it with sulfur. This was in

1839, and the process, later called **vulcanization,** was patented by him in 1844.

Vulcanization and the emerging demand for rubber products led to tremendous growth in rubber production and the industry that supported it. In 1876, Henry Wickham collected thousands of rubber tree seeds from the Brazilian jungle and planted them in England; the sprouts were later transplanted to Ceylon and Malaya (then British colonies) to form rubber plantations. Soon, other countries in the region followed the British example. Southeast Asia became the base of the rubber industry.

In 1888, a British veterinary surgeon named John Dunlop patented pneumatic tires for bicycles. By the twentieth century, the motorcar industry was developing in the United States and Europe. Together, the automobile and rubber industries grew to occupy positions of unimagined importance.

Natural crude rubber (without vulcanization) is sticky in hot weather, stiff and brittle in cold weather. To form an elastomer with useful properties, natural rubber must be vulcanized. Traditionally, vulcanization has been accomplished by mixing small amounts of sulfur and other chemicals with the crude rubber and heating. The chemical effect of vulcanization is cross-linking; the mechanical result is increased strength and stiffness, yet maintenance of extensibility. The dramatic change in properties caused by vulcanization can be seen in the stress–strain curves of our previous Figure 8.13.

Sulfur alone can cause cross-linking, but the process is slow, taking hours to complete. Other chemicals are added to sulfur during vulcanization to accelerate the process and serve other beneficial functions. Also, rubber can be vulcanized using chemicals other than sulfur. Today, curing times have been reduced significantly compared to the original sulfur curing of years ago.

As an engineering material, vulcanized rubber is noted among elastomers for its high tensile strength, tear strength, resilience (capacity to recover shape after deformation), and resistance to wear and fatigue. Its weaknesses are that it degrades when subjected to heat, sunlight, oxygen, ozone, and oil. Some of these limitations can be reduced through the use of additives. Typical properties and other data for vulcanized natural rubber are listed in Table 8.5. Market share is relative to total annual rubber volume, natural plus synthetic. Rubber volume is about 15% of total polymer market.

The largest single market for natural rubber is automotive tires. In tires, carbon black is an important additive; it reinforces the rubber, serving to increase tensile strength and resistance to tearing and abrasion. Other products made of rubber include shoe soles, bushings, seals, and shock-absorbing components. In each case, the rubber is compounded to achieve the specific properties required in the application. Besides carbon black, other additives used in rubber and some of the synthetic elastomers include clay, kaolin, silica, talc, and calcium carbonate, as well as chemicals that accelerate and promote vulcanization.

8.4.3 Synthetic Rubbers

Today, the tonnage of synthetic rubbers is more than three times that of natural rubber. Development of these synthetic materials was motivated largely by the world wars when NR was difficult to obtain (Historical Note 8.3). The most important of the synthetics is styrene–butadiene rubber (SBR), a copolymer of butadiene (C_4H_6) and styrene (C_8H_8). As with most other polymers, the predominant raw material for the synthetic rubbers is petroleum. Only the synthetic rubbers of greatest commercial importance are discussed here. Technical data are presented in Table 8.6. Market share data are for total volume of natural and synthetic rubbers. About 10% of total volume of rubber production is reclaimed; thus, total tonnages in Tables 8.5 and 8.6 do not sum to 100%.

TABLE 8.5   Characteristics and typical properties of vulcanized rubber.

Polymer:	Polyisoprene (C_5H_8)$_n$			
Symbol:	NR		Specific gravity:	0.93
Modulus of elasticity:	18 MPa (2500 lb/in.^{2}) at 300% elongation		High temperature limit:	80°C (180°F)
			Low temperature limit:	−50°C (−60°F)
Tensile strength:	25 MPa (3500 lb/in.^{2})		Approx. market share:	22%
Elongation:	700% at failure			

Compiled from [2], [6], [9], and other sources.

Historical Note 8.3 *Synthetic rubbers [3] and other sources.*

In 1826, Faraday recognized the formula of natural rubber to be C_5H_8. Subsequent attempts at reproducing this molecule over many years were generally unsuccessful. Regrettably, it was the world wars that created the necessity which became the mother of invention for synthetic rubber. In World War I, the Germans, denied access to natural rubber, developed a methyl-based substitute. This material was not very successful, but it marks the first large-scale production of synthetic rubber.

After World War I, the price of natural rubber was so low that many attempts at fabricating synthetics were abandoned. The Germans, however, perhaps anticipating a future conflict, renewed their development efforts. The firm I. G. Farben developed two synthetic rubbers, starting in the early 1930s, called Buna-S and Buna-N. **Buna** is derived from ***bu**tadiene (C_4H_6), which has become the critical ingredient in many modern synthetic rubbers, and **Na,** the symbol for sodium, used to accelerate or catalyze the polymerization process (Natrium is the German word for sodium). The symbol **S**

in Buna-S stands for styrene. Buna-S is the copolymer we know today as ***styrene-butadiene rubber,*** or SBR. The **N** in Buna-N stands for acrylo **N**itrile, and the synthetic rubber is called ***nitrile rubber*** in current usage.

Other efforts included the work at the Du Pont Company in the United States, which led to the development of polychloroprene, first marketed in 1932 under the name Duprene, later changed to **Neoprene,** its current name. It was (and is) a synthetic rubber that is more resistant to oils than natural rubber.

During World War II, the Japanese cut off the supply of natural rubber from Southeast Asia to the United States. Production of Buna-S synthetic rubber was begun on a large scale in America. The federal government preferred to use the name **GR-S** (Government Rubber-Styrene) rather than Buna-S (the German name). By 1944, the United States was outproducing Germany in SBR ten-to-one. Since the early 1960s, worldwide production of synthetic rubbers has exceeded that of natural rubbers.

Butadiene Rubber *Polybutadiene* (BR) [Table 8.6(a)] is important mainly in combination with other rubbers. It is compounded with natural rubber and with styrene (styrene–butadiene rubber is discussed later) in the production of automotive tires. Alone, its tear resistance, tensile strength, and ease of processing are less than desirable.

TABLE 8.6 Characteristics and typical properties of synthetic rubbers: (a) butadiene rubber.

Polymer:	Polybutadiene $(C_4H_6)_n$		
Symbol:	BR	Specific gravity:	0.93
Tensile strength:	15 MPa (2000 lb/in.^{2})	High temperature limit:	100°C (210°F)
Elongation:	500% at failure	Low temperature limit:	−50°C (−60°F)
		Approx. market share:	12%

Compiled from [2], [6], [9], [11], and other sources.

Butyl Rubber Butyl rubber [Table 8.6(b)] is a copolymer of polyisobutylene (98–99%) and polyisoprene (1–2%). It can be vulcanized to provide a rubber with very low air permeability, which has led to applications in inflatable products such as inner tubes, liners in tubeless tires, and sporting goods.

TABLE 8.6 (continued): (b) butyl rubber.

Polymer:	Copolymer of isobutylene $(C_4H_8)_n$ and isoprene $(C_5H_8)_n$		
Symbol:	PIB	Specific gravity:	0.92
Modulus of elasticity:	7 MPa (1000 lb/in.^{2}) at 300% elongation	High temperature limit:	110°C (220°F)
		Low temperature limit:	−50°C (−60°F)
Tensile strength:	20 MPa (3000 lb/in.^{2})	Approx. market share:	about 3%
Elongation:	700%		

Chloroprene Rubber   Polychloroprene [Table 8.6(c)] was one of the first synthetic rubbers to be developed (early 1930s). Commonly known today as *Neoprene,* it is an important special-purpose rubber. It crystallizes when strained to provide good mechanical properties. Chloroprene rubber (CR) is more resistant to oils, weather, ozone, heat, and flame (chlorine makes this rubber self-extinguishing) than NR, but somewhat more expensive. Its applications include fuel hoses (and other automotive parts), conveyor belts, and gaskets, but not tires.

TABLE 8.6   (continued): (c) chloroprene rubber (neoprene).

Polymer:	Polychloroprene $(C_4H_5Cl)_n$		
Symbol:	CR	Specific gravity:	1.23
Modulus of elasticity:	7 MPa (1000 lb/in.^{2}) at	High temperature limit:	120°C (250°F)
	300% elongation.	Low temperature limit:	−20°C (−10°F)
Tensile strength:	25 MPa (3500 lb/in.^{2})	Approx. market share:	2%
Elongation:	500% at failure		

Ethylene–Propylene Rubber   Polymerization of ethylene and propylene with small proportions (3–8%) of a diene monomer produces the terpolymer ethylene–propylene–diene (EPDM), a useful synthetic rubber [Table 8.6(d).] Applications are in nontire parts, mostly in the automotive industry. Other uses are wire and cable insulation.

TABLE 8.6   (continued): (d) ethylene–propylene–diene rubber.

Representative polymer:	Terpolymer of ethylene (C_2H_4), propylene (C_3H_6), and a diene monomer (3–8%) for cross-linking		
Symbol:	EPDM	Specific gravity:	0.86
Tensile strength:	15 MPa (2000 lb/in.^{2})	High temperature limit:	150°C (300°F)
Elongation:	300% at failure	Low temperature limit:	−50°C (−60°F)
		Approx. market share:	5%

Isoprene Rubber   Isoprene can be polymerized to synthesize a chemical equivalent of natural rubber. Synthetic (unvulcanized) *polyisoprene* is softer and more easily molded than raw natural rubber [Table 8.6(e)]. Applications of the synthetic material are similar to those of its natural counterpart, car tires being the largest single market. It is also used for footwear, conveyor belts, and caulking compound. Cost per unit weight is about 35% higher than for NR.

TABLE 8.6   (continued): (e) isoprene rubber (synthetic).

Polymer:	Polyisoprene $(C_5H_8)_n$		
Symbol:	IR	Specific gravity:	0.93
Modulus of elasticity:	17 MPa (2500 lb/in.^{2}) at	High temperature limit:	80°C (180°F)
	300% elongation	Low temperature limit:	−50°C (−60°F)
Tensile strength:	25 MPa (3500 lb/in.^{2})	Approx. market share:	2%
Elongation:	500% at failure		

Nitrile Rubber   This is a vulcanizable copolymer of butadiene (50–75%) and acrylonitrile (25–50%) [see Table 8.6(f)]. Its more technical name is *butadiene-acryloni-*

trile rubber. It has good strength and resistance to abrasion, oil, gasoline, and water. These properties make it ideal for applications such as gasoline hoses and seals; and also for footwear.

TABLE 8.6   (continued): (f) nitrile rubber.

Polymer:	Copolymer of butadiene (C_4H_6) and acrylonitrile (C_3H_3N)		
Symbol:	NBR	Specific gravity:	1.00 (without fillers)
Modulus of elasticity:	10 MPa (1500 lb/in.^{2}) at 300% elongation.	High temperature limit:	120°C (250°F)
		Low temperature limit:	−50°F (−60°F)
Tensile strength:	30 MPa (4000 lb/in.^{2})	Approx. market share:	2%
Elongation:	500% at failure		

Polyurethanes   Thermosetting polyurethanes (Section 8.3.2) with minimum cross-linking are elastomers, most commonly produced as flexible foams [Table 8.6(g)]. In this form, they are widely used as cushion materials for furniture and automobile seats. Unfoamed polyurethane can be molded into products ranging from shoe soles to car bumpers, with cross-linking adjusted to achieve the desired properties for the application. With no cross-linking, the material is a thermoplastic elastomer that can be injection molded. As an elastomer or thermoset, reaction injection molding and other shaping methods are used.

TABLE 8.6   (continued): (g) polyurethane.

Polymer:	Polyurethane (chemistry varies)		
Symbol:	PUR	Specific gravity:	1.25
Modulus of elasticity:	10 MPa (1200 lb/in.^{2}) at 300% elongation.	High temperature limit:	100°C (210°F)
		Low temperature limit:	−50°C (−60°F)
Tensile strength:	60 MPa (8000 lb/in.^{2})	Approx. market share:	listed under thermosets, Table 8.4(e).
Elongation:	700% at failure		

Silicones   Like the polyurethanes, silicones can be elastomeric or thermosetting, depending on the degree of cross-linking. Silicone elastomers are noted for the wide temperature range over which they can be used. Their resistance to oils is poor. The silicones possess various chemistries, the most common being *polydimethylsiloxane,* Table 8.6(h). To obtain acceptable mechanical properties, silicone elastomers must be reinforced, usually with fine silica powders. Owing to their high cost, they are considered special-purpose rubbers for applications such as gaskets, seals, wire and cable insulation, prosthetic devices, and bases for caulking materials.

TABLE 8.6   (continued): (h) silicone rubber.

Representative polymer:	Polydimethylsiloxane ($SiO(CH_3)_2)_n$		
Symbol:	VMQ	Specific gravity:	0.98
Tensile strength:	10 MPa (1500 lb/in.^{2})	High temperature limit:	230°C (450°F)
Elongation:	700% at failure	Low temperature limit:	−50°C (−60°F)
		Approx. market share:	Less than 1%

Styrene–Butadiene Rubber   SBR [Table 8.6(i)] is a random copolymer of styrene (about 25%) and butadiene (about 75%). It was originally developed in Germany as Buna-S rubber before World War II. Today, it is the largest tonnage elastomer, totaling about 40% of all rubbers produced (natural rubber is second in tonnage). Its attractive features are low cost, resistance to abrasion, and better uniformity than NR. When reinforced with carbon black and vulcanized, its characteristics and applications are very similar to those of natural rubber. Cost is also similar. A close comparison of properties reveals that most of its mechanical properties except wear resistance are inferior to NR, but its resistance to heat aging, ozone, weather, and oils is superior. Applications include automotive tires, footwear, wire and cable insulation. A material chemically related to SBR is styrene–butadiene–styrene block copolymer, a thermoplastic elastomer discussed below.

TABLE 8.6   (continued): (i) styrene-butadiene rubber.

Polymer:	Copolymer of styrene (C_8H_8) and butadiene (C_4H_6)		
Symbol:	SBR	Elongation:	700% at failure
Modulus of elasticity:	17 MPa (2500 lb/in.^{2})	Specific gravity:	0.94
	at 300% elongation	High temperature limit:	110°C (230°F)
Tensile strength:	20 MPa (3000 lb/in.^{2})	Low temperature limit:	-50°C (-60°F)
	reinforced.	Approx. market share:	Slightly less than 30%

Thermoplastic Elastomers   As previously described, a thermoplastic elastomer (TPE) is a thermoplastic that behaves like an elastomer. It constitutes a family of polymers that is a fast-growing segment of the elastomer market. TPEs derive their elastomeric properties not from chemical cross-links, but from physical connections between soft and hard phases that make up the material. Thermoplastic elastomers include *styrene–butadiene–styrene* block copolymer (SBS), as opposed to styrene–butadiene rubber (SBR) which is a random copolymer (Section 8.1.2); *thermoplastic polyurethanes; thermoplastic polyester copolymers*; and other copolymers and polymer blends. Table 8.6(j) gives data on SBS. The chemistry and structure of these materials is generally complex, involving two materials that are incompatible so that they form distinct phases whose room temperature properties are different. Owing to their thermoplasticity, the TPEs cannot match conventional cross-linked elastomers in elevated temperature strength and creep resistance. Typical applications include footwear; rubber bands; extruded tubing, wire coating; molded parts for automotive and other uses in which elastomeric properties are required. TPEs are not suitable for tires.

TABLE 8.6   (continued): (j) thermoplastic elastomers (TPE).

Representative polymer:	Styrene–butadiene–styrene block copolymer		
Symbol:	SBS (also YSBR)	Specific gravity:	1.0
Tensile strength:	14 MPa (2000 lb/in.^{2})	High temperature limit:	65°C (150°F)
Elongation:	400%	Low temperature limit:	-50°C (-60°F)
		Approx. market share:	12%

8.5 GUIDE TO THE PROCESSING OF POLYMERS

Polymers are nearly always shaped in a heated, highly plastic consistency. Common operations are extrusion and molding. The molding of thermosets is generally more complicated because they require curing (cross-linking). Thermoplastics are easier to mold, and a greater variety of molding operations are available to process them (Chapter 13). Although plastics readily lend themselves to net shape processing, machining is sometimes required (Chapter 22). And plastic parts can be assembled into products by permanent joining techniques such as welding (Chapter 31), adhesive bonding (Section 32.3), or mechanical assembly (Chapter 33).

Rubber processing has a longer history than plastics, and the industries associated with these polymer materials have traditionally been separated, even though their processing is similar in many ways. We cover rubber processing technology in Chapter 14.

REFERENCES

[1] Alliger, G., and Sjothun, I. J. (eds). *Vulcanization of Elastomers.* Krieger Publishing Company, New York, 1978.

[2] Billmeyer, F., W., Jr. *Textbook of Polymer Science.* 3rd ed. Wiley, New York, 1984.

[3] Blow, C. M., and Hepburn, C. *Rubber Technology and Manufacture.* 2nd ed. Butterworth Scientific, London, 1982.

[4] Brandrup, J., and Immergut, E. E. (eds). *Polymer Handbook.* 4th ed. Wiley, New York, 1999.

[5] Brydson, J. A. *Plastics Materials.* 4th ed. Butterworths & Co., Ltd., London, 1999.

[6] Chanda, M., and Roy, S. K. *Plastics Technology Handbook.* Marcel Dekker, Inc., New York, 1998.

[7] Charrier, J-M. *Polymeric Materials and Processing.* Oxford University Press, New York, 1991.

[8] *Engineering Materials Handbook.* Vol. 2. *Engineering Plastics.* ASM International, Metals Park, Ohio, 1988.

[9] Flinn, R. A., and Trojan, P. K. *Engineering Materials and Their Applications.* 4th ed. Houghton Mifflin, Boston, 1990.

[10] Hall, C. *Polymer Materials.* 2nd ed. Wiley, New York, 1989.

[11] Hofmann, W. *Rubber Technology Handbook.* Hanser Publishers, Munich, Germany, 1988.

[12] Margolis, J. M. *Engineering Thermoplastics—Properties and Applications.* Marcel Dekker, Inc., New York, 1985.

[13] McCrum, N. G., Buckley, C. P., and Bucknall, C. B. *Principles of Polymer Engineering.* 2nd ed. Oxford University Press, Oxford, U.K., 1997.

[14] *Modern Plastics Encyclopedia.* Modern Plastics, McGraw-Hill, Hightstown, New Jersey, 1990.

[15] Rudin, A. *The Elements of Polymer Science and Engineering.* 2nd ed. Academic Press, Inc., Orlando, Fl. 1998.

[16] Seymour, R. B., and Carraher, C. E. *Seymour/Carraher's Polymer Chemistry.* 5th ed. Marcel Dekker, Inc., New York, 2000.

[17] Seymour, R. B. *Engineering Polymer Sourcebook.* McGraw-Hill, New York, 1990.

REVIEW QUESTIONS

8.1. What is a polymer?

8.2. What are the three basic categories of polymers?

8.3. How do the properties of polymers compare with those of metals?

8.4. What are the two methods by which polymerization occurs? Briefly describe the two methods.

8.5. What does the *degree of polymerization* indicate?

8.6. Define the term *tacticity* as it applies to polymers.

8.7. What is *cross-linking* in a polymer, and what is its significance?

8.8. What is a *copolymer*?

8.9. The arrangement of repeating units in a copolymer can vary. What are some of the possible arrangements?

8.10. What is a *terpolymer*?

8.11. How are a polyer's properties affected when it takes on a crystalline structure?

8.12. Does any polymer ver become 100% crystalline?

8.13. What are some of the factors that influence a polymer's tendency to crystallize?

8.14. Why are fillers added to a polymer?

8.15. What is a *plasticizer*?

8.16. In addition to fillers and plasticizers, what are some other additives used with polymers?

8.17. Describe the difference in mechanical properties as a function of temperature between a highly crystalline thermoplastic and an amorphous thermoplastic.

8.18. What is unique about the polymer *cellulose*?

8.19. The nylons are members of which polymer group?

8.20. What is the chemical formula of ethylene, the monomer for polyethylene?

8.21. What is the basic difference between low-density and high-density polyethylene?

8.22. How do the properties of thermosetting polymers differ from those of thermoplastics?

8.23. Cross-linking (curing) of thermosetting plastics is accomplished by one of three ways. Name the three ways.

8.24. Elastomers and thermosetting polymers are both cross-linked. Why are their properties so different?

8.25. What happens to an elastomer when it is below its glass transition temperature?

8.26. What is the primary polymer ingredient in natural rubber?

8.27. How are thermoplastic elastomers different from conventional rubbers?

MULTIPLE CHOICE QUIZ

There is a total of 25 correct answers in the following multiple choice questions (some questions have multiple answers that are correct). To attain a perfect score on the quiz, all correct answers must be given, since each correct answer is worth 1 point. For each question, each omitted answer or wrong answer reduces the score by 1 point, and each additional answer beyond the number of answers required reduces the score by 1 point. Percentage score on the quiz is based on the total number of correct answers.

8.1. Of the three polymer types, which one is the most important commercially? (a) thermoplastics, (b) thermosets, or (c) elastomers.

8.2. Which one of the three polymer types is not normally considered to be a plastic? (a) thermoplastics, (b) thermosets, or (c) elastomers.

8.3. Which one of the three polymer types does not involve cross-linking? (a) thermoplastics, (b) thermosets, or (c) elastomers.

8.4. As the degree of crystallinity in a given polymer increases, the polymer becomes denser and stiffer, and its melting temperature decreases: (a) true or (b) false.

8.5. Which of the following is the chemical formula for the repeating unit in polyethylene? (a) CH_2, (b) C_2H_4, (c) C_3H_6, (d) C_5H_8, or (e) C_8H_8.

8.6. Degree of polymerization is which one of the following? (a) average number of mers in the molecule chain; (b) proportion of the monomer that has been polymerized; (c) sum of the molecule weights of the mers in the molecule; or (d) none of the above.

8.7. A branched molecular structure is stronger in the solid state and more viscous in the molten state than a linear structure for the same polymer: (a) true or (b) false.

8.8. A copolymer is a mixture consisting of macromolecules of two different homopolymers: (a) true or (b) false.

8.9. As temperature of a polymer increases, its density (a) increases, (b) decreases, or (c) remains fairly constant.

8.10. Which answers complete the following sentence correctly (more than one)? As the temperature of an amorphous thermoplastic polymer is gradually reduced, the glass transition temperature T_g is indicated when (a) the polymer transforms to a crystalline structure, (b) the coefficient of thermal expansion increases markedly, (c) the slope of specific volume versus temperature changes markedly, (d) the polymer becomes stiff, strong, and elastic, or (e) the polymer solidifies from the molten state.

8.11. Which one of the following plastics has the highest market share? (a) phenolics, (b) polyethylene, (c) polypropylene, (d) polystyrene, or (e) polyvinylchloride.

8.12. Which of the following polymers are normally thermoplastic (more than one)? (a) acrylics, (b) cellulose acetate, (c) nylon, (d) polychloroprene, (e) polyethylene, or (f) polyurethane.

8.13. Polystyrene (without plasticizers) is amorphous, transparent, and brittle: (a) true or (b) false.

8.14. The fiber **rayon** used in textiles is based on which one of the following polymers? (a) cellulose, (b) nylon, (c) polyester, (d) polyethylene, or (e) polypropylene.

8.15. The basic difference between low-density polyethylene and high-density polyethylene is that the latter has a much higher degree of crystallinity: (a) true or (b) false.

8.16. Among the thermosetting polymers, the most widely used commercially is which one of the following? (a) epoxies, (b) phenolics, (c) silicones, or (d) urethanes.

8.17. Polyurethanes can be which of the following (more than one)? (a) thermoplastic, (b) thermosetting, or (c) elastomeric.

8.18. The chemical formula for polyisoprene in natural rubber is which one of the following? (a) CH_2, (b) C_2H_4, (c) C_3H_6, (d) C_5H_8, or (e) C_8H_8.

8.19. The leading commercial synthetic rubber is which one of the following? (a) butyl rubber, (b) isoprene rubber, (c) polybutadiene, (d) polyurethane, (e) styrene–butadiene rubber, or (f) thermoplastic elastomers.

COMPOSITE MATERIALS

CHAPTER CONTENTS

9.1 Technology and Classification of Composite Materials
    9.1.1 Components in a Composite Material
    9.1.2 The Reinforcing Phase
    9.1.3 Properties of Composite Materials
    9.1.4 Other Composite Structures
9.2 Metal Matrix Composites
    9.2.1 Cermets
    9.2.2 Fiber-Reinforced Metal Matrix Composites
9.3 Ceramic Matrix Composites
9.4 Polymer Matrix Composites
    9.4.1 Fiber-Reinforced Polymers
    9.4.2 Other Polymer Matrix Composites
9.5 Guide to Processing Composite Materials

In addition to metals, ceramics, and polymers, a fourth material category can be distinguished: composites. In some ways, these are the most interesting of the engineering materials because their structure is more complex than the other three types. Although general agreement on a definition of composite materials is elusive, let us use the following for now: a *composite material* is a materials system composed of two or more physically distinct phases whose combination produces aggregate properties that are different from those of its constituents.

The technological and commercial interest in composite materials derives from the fact that their properties are not just different from their components but are often far superior. Some of the possibilities include

➤ Composites can be designed that are very strong and stiff, yet very light in weight, giving them strength-to-weight and stiffness-to-weight ratios several times greater than steel or aluminum. These properties are highly desirable in applications ranging from commercial aircraft to sports equipment.

➤ Fatigue properties are generally better than for the common engineering metals. Toughness is often greater, too.

➤ Composites can be designed that do not corrode like steel; this is important in automotive and other applications.

➤ With composite materials, it is possible to achieve combinations of properties not attainable with metals, ceramics, or polymers alone.

➤ Better appearance and control of surface smoothness is possible with certain composite materials.

Along with the advantages, there are disadvantages and limitations associated with composite materials. These include: (1) properties of many important composites are anisotropic; the properties differ depending on the direction in which they are measured; (2) many of the polymer-based composites are subject to attack by chemicals or solvents, just as the polymers themselves are susceptible to attack; (3) composite materials are generally expensive, although prices may drop as volume increases; and (4) certain of the manufacturing methods for shaping composite materials are slow and costly.

We have already encountered several composite materials in our coverage of the three other material types. Examples include cemented carbides (tungsten carbide with cobalt binder), plastic molding compounds that contain fillers (e.g., cellulose fibers, wood flour), and rubber mixed with carbon black. We did not always identify these materials as composites; however, technically, they fit the above definition. It could even be argued that a two-phase metal alloy (e.g., $Fe + Fe_3C$) is a composite material, although it is not classified as such. Perhaps the most important composite material of all is wood.

In our presentation of composite materials, we first examine their technology and classification. There are many different materials and structures that can be used to form composites; we survey the various categories, devoting the most time to fiber-reinforced plastics—commercially the most important type. In the final section, we provide a guide to the manufacturing processes for composites.

9.1 TECHNOLOGY AND CLASSIFICATION OF COMPOSITE MATERIALS

As noted in our definition, a composite material consists of two or more distinct phases. The term *phase* indicates a homogeneous material, such as a metal or ceramic in which all of the grains have the same crystal structure, or a polymer with no fillers. By combining the phases, using methods yet to be described, a new material is created with aggregate performance exceeding that of its parts. The effect is synergistic.

Composite materials can be classified in various ways. One possible classification distinguishes between (1) traditional and (2) synthetic composites. *Traditional composites* are those that occur in nature or have been produced by civilizations for many years. Wood is a naturally occurring composite material, while concrete (Portland cement plus sand or gravel) and asphalt mixed with gravel are traditional composites used in construction. *Synthetic composites* are modern material systems normally associated with the manufacturing industries, in which the components are first produced separately and then combined in a controlled way to achieve the desired structure, properties, and part geometry. These synthetic materials are the composites normally thought of in the context of engineered products. Our attention in this chapter is focused on these materials.

9.1.1 Components in a Composite Material

In the simplest manifestation of our definition, a composite material consists of two phases: a primary phase and a secondary phase. The primary phase forms the *matrix* within which the secondary phase is imbedded. The imbedded phase is sometimes referred to as a *reinforcing agent* (or similar term), because it usually serves to strengthen the composite. The reinforcing phase may be in the form of fibers, particles, or various other geometries, as we shall see. The phases are generally insoluble in each other, but strong adhesion must exist at their interface(s).

The matrix phase can be any of three basic material types: polymers, metals, or ceramics. The secondary phase may also be one of the three basic materials, or it may be an element such as carbon or boron. Possible combinations in a two-component composite material can be organized as a 3×4 chart, as in Table 9.1. We see that certain combinations are not feasible, such as a polymer in a metallic matrix. We also see that the possibilities include two-phase structures consisting of components of the same material type, such as fibers of Kevlar (polymer) in a plastic (polymer) matrix. In other composites the imbedded material is an element such as carbon or boron.

The classification system for composite materials used in this book is based on the matrix phase. We list the classes here and discuss them in Sections 9.2 through 9.4:

1. *Metal Matrix Composites* (MMCs)—These composites include mixtures of ceramics and metals, such as cemented carbides and other cermets, as well as aluminum or magnesium reinforced by strong, high stiffness fibers.
2. *Ceramic Matrix Composites* (CMCs)—This is the least common composite matrix. Aluminum oxide and silicon carbide are materials that can be imbedded with fibers for improved properties, especially in high temperature applications.
3. *Polymer Matrix Composites* (PMCs)—Thermosetting resins are the most widely used polymers in PMCs. Epoxy and polyester are commonly mixed with fiber reinforcement, and phenolic is mixed with powders such as wood flour. Thermoplastics are also reinforced, usually with powders (Section 8.1.5). Virtually all elastomers are reinforced with carbon black.

The classification can be applied to traditional composites as well as synthetics. Concrete is a ceramic matrix composite, while asphalt and wood are polymer matrix composites.

The matrix material serves several functions in the composite. First, it provides the bulk form of the part or product made of the composite material. Second, it holds the imbedded phase in place, usually enclosing and often concealing it. Third, when a load is applied, the matrix shares the load with the secondary phase, in some cases deforming so that the stress is essentially born by the reinforcing agent.

TABLE 9.1   Possible combinations of two-component composite materials.

Secondary phase, reinforcement:	Primary Phase, Matrix		
	Metal	Ceramic	Polymer
Metal	Infiltrated powder metallurgy parts	Cermets	Plastic molding compounds Steel-belted radial tires
Ceramic	Cermets[a] Fiber-reinforced metals	SiC whisker-reinforced Al_2O_3	Plastic molding compounds Fiberglass-reinforced plastic
Polymer	Powder metal part infiltrated with polymer	NA	Plastic molding compounds Kevlar-reinforced epoxy
Elements (C, B)	Fiber-reinforced metals	NA	Rubber with carbon black B or C fiber-reinforced plastic

NA = not applicable currently.

[a] Cermets include cemented carbides.

FIGURE 9.1   Possible physical shapes of imbedded phases in composite materials: (a) fiber, (b) particle, and (c) flake.

9.1.2 The Reinforcing Phase

It is important to understand that the role played by the secondary phase is to reinforce the primary phase. The imbedded phase is most commonly one of the shapes illustrated in Figure 9.1—fibers, particles, or flakes. In addition, the secondary phase can take the form of an infiltrated phase in a skeletal or porous matrix.

Fibers   *Fibers* are filaments of reinforcing material, generally circular in cross-section, although alternative shapes are sometimes used (e.g., tubular, rectangular, hexagonal). Diameters range from less than 0.0025 mm (0.0001 in.) to about 0.13 mm (0.005 in.), depending on material.

Fiber reinforcement provides the greatest opportunity for strength enhancement of composite structures. In fiber reinforced composites, the fiber is often considered to be the principal constituent since it bears the major share of the load. Fibers are of interest as reinforcing agents because the filament form of most materials is significantly stronger than the bulk form. The effect of fiber diameter on tensile strength can be seen in Figure 9.2. As diameter is reduced, the material becomes oriented in the direction of the fiber axis and the probability of defects in the structure decreases significantly. As a result, tensile strength increases dramatically.

Fibers used in composites can be either continuous or discontinuous. *Continuous fibers* are very long; in theory, they offer a continuous path by which a load can be carried by the composite part. In reality, this is difficult to achieve due to variations in the fibrous material and processing. *Discontinuous fibers* (chopped sections of continuous fibers) are short lengths (L/D = roughly 100). An important type of discontinuous fiber are *whiskers*—hair-like single crystals with diameters down to about 0.001 mm (0.00004 in.) and very high strength.

FIGURE 9.2   Relationship between tensile strength and diameter for a carbon fiber (source: [1]). Other filament materials show similar relationships.

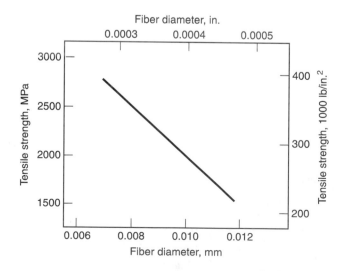

FIGURE 9.3 Fiber orientation in composite materials: (a) one-dimensional, continuous fibers; (b) planar, continuous fibers in the form of a woven fabric; and (c) random, discontinuous fibers.

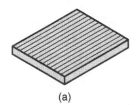
(a)

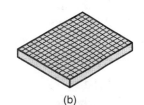
(b)

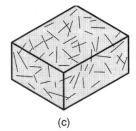
(c)

Fiber orientation is another factor in composite parts. We can distinguish three cases, illustrated in Figure 9.3: (a) one-dimensional reinforcement, in which maximum strength and stiffness are obtained in the direction of the fiber; (b) planar reinforcement, in some cases in the form of a two-dimensional woven fabric; and (c) random or three-dimensional in which the composite material tends to possess isotropic properties.

A variety of materials is used as fibers in fiber-reinforced composites: metals, ceramics, polymers, carbon, and boron. The most important commercial use of fibers is in polymer composites. However, use of fiber-reinforced metals and ceramics is growing. Following is a survey of the important types of fiber materials, with properties listed in Table 9.2:

> *Glass*—The most widely used fiber in polymers, the term *fiberglass* is applied to denote glass fiber-reinforced plastic (GFRP). The two common glass fibers are E-glass and S-glass (compositions listed in Table 7.5) E-glass is strong and low cost, but its modulus is less than other fibers. S-glass is stiffer and its tensile strength is one of the highest of all fiber materials; however, it is more expensive than E-glass.

> *Carbon*—Carbon (Section 7.5.1) can be made into high modulus fibers. Besides stiffness, other attractive properties include low density and low thermal expansion. C-fibers are generally a combination of graphite and amorphous carbon.

> *Boron*—Boron (Section 7.5.3) has a very high elastic modulus, but its high cost limits its applications to aerospace components in which this property (and others) are critical.

> *Kevlar 49*—This is the most important polymer fiber; it is a highly crystalline aramid, a member of the polyamide family (Section 8.2.2). Its specific gravity is low, giving it one of the highest strength-to-weight ratios of all fibers.

TABLE 9.2 Typical properties of fiber materials used as reinforcement in composites.

Fiber material	Diameter		Tensile Strength		Elastic Modulus	
	mm	(mils[a])	MPa	(lb/in.^{2})	GPa	(lb/in.^{2})
Metal: Steel	0.13	(5.0)	1000	(150,000)	206	(30×10^6)
Metal: Tungsten	0.013	(0.5)	4000	(580,000)	407	(59×10^6)
Ceramic: Al$_2$O$_3$	0.02	(0.8)	1900	(275,000)	380	(55×10^6)
Ceramic: SiC	0.13	(5.0)	3275	(475,000)	400	(58×10^6)
Ceramic: E-glass	0.01	(0.4)	3450	(500,000)	73	(10×10^6)
Ceramic: S-glass	0.01	(0.4)	4480	(650,000)	86	(12×10^6)
Polymer: Kevlar	0.013	(0.5)	3450	(500,000)	130	(19×10^6)
Element: Carbon	0.01	(0.4)	2750	(400,000)	240	(35×10^6)
Element: Boron	0.14	(5.5)	3100	(450,000)	393	(57×10^6)

Compiled from [3], [6], [10], and other sources. Note that strength depends on fiber diameter (Figure 9.2); the properties in this table must be interpreted accordingly. [a] 1 mil = 0.001 in.

> ➤ *Ceramics*—Silicon carbide (SiC) and aluminum oxide (Al_2O_3) are the main fiber materials among ceramics. Both have high elastic moduli and can be used to strengthen low-density, low-modulus metals such as aluminum and magnesium.
> ➤ *Metal*—Steel filaments, both continuous and discontinuous, are used as reinforcing fibers in plastics. Other metals are currently less common as reinforcing fibers.

Particles and Flakes   A second common shape of the imbedded phase is *particulate,* ranging in size from microscopic to macroscopic. Particles are an important material form for metals and ceramics; we discuss the characterization and production of engineering powders in Chapters 16 and 17.

The distribution of particles in the composite matrix is random, and therefore strength and other properties of the composite material are usually isotropic. The strengthening mechanism depends on particle size. The microscopic size is represented by very fine powders (less than 1 μm) distributed in the matrix in concentrations of 15% or less. The presence of these powders results in dispersion-hardening of the matrix, in which dislocation movement in the matrix material is restricted by the microscopic particles. In effect, the matrix itself is strengthened, and no significant portion of the applied load is carried by the particles.

As particle size increases to the macroscopic range (greater than 1 μm), and the proportion of imbedded material increases to 25% and more, the strengthening mechanism changes. In this case, the applied load is shared between the matrix and the imbedded phase. Strengthening occurs due to the load-carrying ability of the particles and the bonding of particles in the matrix. This form of composite strengthening occurs in cemented carbides, in which tungsten carbide is held in a cobalt binder. The proportion of WC in the Co matrix is typically 80% or more.

Flakes are basically two-dimensional particles—small flat platelets. Two examples of this shape are the minerals mica (silicate of K and Al) and talc ($Mg_3Si_4O_{10}(OH)_2$), used as reinforcing agents in plastics. They are generally lower cost materials than polymers, and they add strength and stiffness to the plastic molding compounds. Platelet sizes are usually in the range 0.01–1 mm (0.0004–0.040 in.) across the flake, with a thickness of 0.001–0.005 mm (0.04–0.20 mils).

Infiltrated Phase   The fourth form of imbedded phase occurs when the matrix has the form of a porous skeleton (like a sponge), and the second phase is simply a *filler.* In this case, the imbedded phase assumes the shape of the pores in the matrix. Metallic fillers are sometimes used to infiltrate the open porous structure of parts made by powder metallurgy techniques (Section 16.3.4), in effect creating a composite material. Oil-impregnated sintered PM components, such as bearings and gears, might be considered another example of this category.

The Interface   There is always an *interface* between constituent phases in a composite material. For the composite to operate effectively, the phases must bond where they join. In some cases, there is a direct bonding between the two ingredients, as suggested by Figure 9.4(a). In other cases, a third ingredient is added to promote bonding of the two primary phases. Called an *interphase,* this third ingredient can be thought of as an adhesive. An important example is the coating of glass fibers to achieve adhesion with the thermosetting resin in fiberglass-reinforced plastics. As illustrated in Figure 9.4(b), this case results in two interfaces, one on either boundary of the interphase. Finally, a third form of interface occurs when the two primary components are not completely insoluble in each other; in this case, the interphase consists of a solution of the phases, shown in Figure 9.4(c). An example occurs in cemented carbides (Section 9.2.1); at the

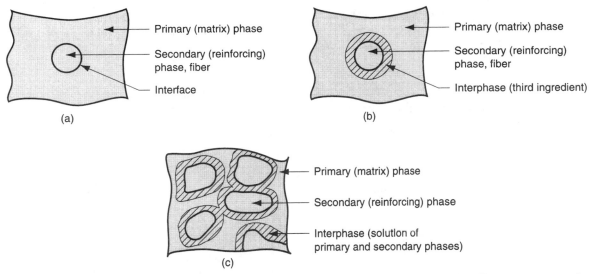

FIGURE 9.4   Interfaces and interphases between phases in a composite material: (a) direct bonding between primary and secondary phases; (b) addition of a third ingredient to bond the primary and secondary phases and form an interphase; and (c) formation of an interphase by solution of the primary and secondary phases at their boundary.

high sintering temperatures used on these materials, some solubility results at the boundaries to create an interphase.

9.1.3 Properties of Composite Materials

In the selection of a composite material, an optimum combination of properties is usually being sought, rather than one particular property. For example, the fuselage and wings of an aircraft must be lightweight as well as strong, stiff, and tough. Finding a monolithic material that satisfies these requirements is difficult. Several fiber-reinforced polymers possess this combination of properties.

Another example is rubber. Natural rubber is a relatively weak material. In the early 1900s, it was discovered that by adding significant amounts of carbon black (almost pure carbon) to natural rubber, its strength is increased dramatically. The two ingredients interact to provide a composite material that is significantly stronger than either one alone. Rubber, of course, must also be vulcanized to achieve full strength.

Rubber itself is a useful additive in polystyrene. One of the distinctive properties of polystyrene is its brittleness. Although most other polymers have considerable ductility, PS has virtually none. Rubber (natural or synthetic) can be added in modest amounts (5–15%) to produce high-impact polystyrene, which has much superior toughness and impact strength.

Properties of a composite material are determined by three factors: (1) the materials used as component phases in the composite, (2) the geometric shapes of the constituents and resulting structure of the composite system, and (3) the manner in which the phases interact with one another.

Rule of Mixtures   The properties of a composite material are a function of the starting materials. Certain properties of a composite material can be computed by means of a ***rule of mixtures,*** which involves calculating a weighted average of the constituent material properties. Density is an example of this averaging rule. The mass of a composite material is the sum of the masses of the matrix and reinforcing phases:

$$m_c = m_m + m_r \qquad (9.1)$$

where m = mass, kg (lb); and the subscripts c, m, and r indicate composite, matrix, and reinforcing phases, respectively. Similarly, the volume of the composite is the sum of its constituents:

$$V_c = V_m + V_r + V_v \tag{9.2}$$

where V = volume, cm^3 (in.^{3}). V_v is the volume of any voids in the composite (e.g., pores). The density of the composite is the mass divided by the volume.

$$\rho_c = \frac{m_c}{V_c} = \frac{m_m + m_r}{V_c} \tag{9.3}$$

Since the masses of the matrix and reinforcing phase are their respective densities multiplied by their volumes,

$$m_m = \rho_m V_m \text{ and } m_r = \rho_r V_r$$

We can substitute these terms into Eq. (9.3) and conclude that

$$\rho_c = f_m \rho_m + f_r \rho_r \tag{9.4}$$

where $f_m = V_m/V_c$ and $f_r = V_r/V_c$ are simply the volume fractions of the matrix and reinforcing phases.

Fiber-Reinforced Composites   Determining mechanical properties of composites from constituent properties is usually more involved. The rule of mixtures can sometimes be used to estimate the modulus of elasticity of a fiber-reinforced composite made of continuous fibers where E_c is measured in the longitudinal direction. The situation is depicted in Figure 9.5(a); we assume that the fiber material is much stiffer than the matrix and that the bonding between the two phases is secure. Under this model, the modulus of the composite can be predicted as follows:

$$E_c = f_m E_m + f_r E_r \tag{9.5}$$

where E_c, E_m, and E_r are the elastic moduli of the composite and its constituents, MPa (lb/in.^{2}); and f_m and f_r are again the volume fractions of the matrix and reinforcing phase. The effect of Eq. (9.5) is seen in Figure 9.5(b).

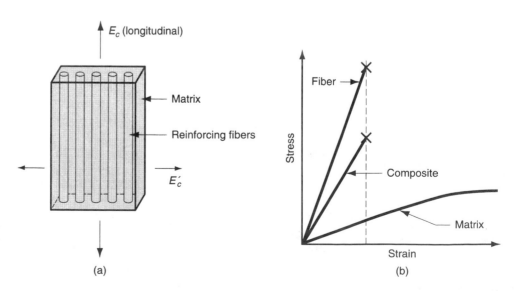

FIGURE 9.5   (a) Model of a fiber-reinforced composite material showing direction in which elastic modulus is being estimated by the rule of mixtures. (b) Stress–strain relationships for the composite material and its constituents. The fiber is stiff but brittle, while the matrix (commonly a polymer) is soft but ductile. The composite's modulus is a weighted average of its components' moduli. But when the reinforcing fibers fail, the composite does likewise.

(a)

(b)

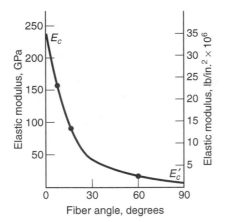
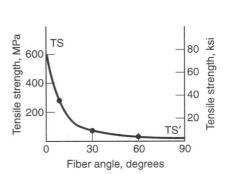

FIGURE 9.6   Variation in elastic modulus and tensile strength as a function of direction of measurement relative to longitudinal axis of carbon fiber-reinforced epoxy composite (source: [6]).

Perpendicular to the longitudinal direction, the fibers contribute little to the overall stiffness except for their filling effect. The composite modulus can be estimated in this direction using the following:

$$E_c' = \frac{E_m E_r}{f_m E_r + f_r E_m} \tag{9.6}$$

where E_c' = elastic modulus perpendicular to the fiber direction, MPa (lb/in.²). Our two equations for E_c demonstrate the significant anisotropy of fiber-reinforced composites. This directional effect can be seen in Figure 9.6 for a fiber-reinforced polymer composite, in which both elastic modulus and tensile strength are measured relative to fiber direction.

Fibers illustrate the importance of geometric shape. Most materials have tensile strengths several times greater in a fibrous form than in bulk. However, applications of fibers are limited by surface flaws, buckling when subjected to compression, and the inconvenience of the filament geometry when a solid component is needed. By imbedding the fibers in a polymer matrix, a composite material is obtained that avoids the problems of fibers but utilizes their strengths. The matrix provides the bulk shape to protect the fiber surfaces and resist buckling; and the fibers lend their high strength to the composite. When a load is applied, the low-strength matrix deforms and distributes the stress to the high-strength fibers, which then carry the load. If individual fibers break, the load is redistributed through the matrix to other fibers.

FIGURE 9.7   Laminar composite structures: (a) conventional laminar structure; (b) sandwich structure using a foam core, and (c) honeycomb sandwich structure.

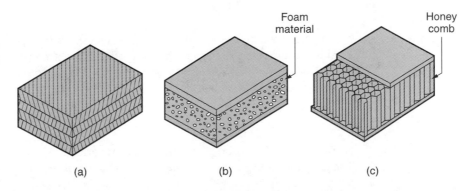

(a)   (b)   (c)

TABLE 9.3   Examples of laminar composite structures.

Laminar composite	Description (reference in text if applicable)
Automotive tires	A tire consists of multiple layers bonded together; the layers are composite materials (rubber reinforced with carbon black), and the plies consist of rubber-impregnated fabrics (Chapter 14).
Honeycomb sandwich	A lightweight honeycomb structures is bonded on either face to thin sheets, Figure 9.7(c).
FRPs	Multilayered fiber-reinforced plastic panels are used for aircraft, automobile body panels, boat hulls (Chapter 15).
Plywood	Alternating sheets of wood are bonded together at different orientations for improved strength.
Printed circuits	Layers of reinforced plastic and copper are used for electrical conductivity and insulation in alternating layers (Section 36.2).
Snow skis	Skis are laminar composite structures consisting of multiple layers of metals, particle board, and phenolic plastic.
Windshield glass	Two layers of glass on either side of a sheet of tough plastic (Section 12.3.1).

9.1.4 Other Composite Structures

Our model of a composite material is one in which a reinforcing phase is imbedded in a matrix phase, the combination having properties that are superior in certain respects to either of the constituents alone. However, composites can take alternative forms that do not fit this model, some of which are of considerable commercial and technological importance.

A *laminar composite structure* consists of two or more layers bonded together to form an integral piece, as in Figure 9.7(a). The layers are usually thick enough that this composite can be readily identified—not always the case with other composites. The layers are often of different materials, but not necessarily. Plywood is such an example; the layers are of the same wood, but the grains are oriented differently to increase overall strength of the laminated piece. A laminar composite often uses different materials in its layers to gain the advantage of combining the particular properties of each. In some cases, the layers themselves may be composite materials. We have mentioned that wood is a composite material; therefore, plywood is a laminar composite structure in which the layers themselves are composite materials. A list of examples of laminar composites is compiled in Table 9.3.

The *sandwich structure* is sometimes distinguished as a special case of the laminar composite structure. It consists of a relatively thick core of low-density material bonded on both faces to thin sheets of a different material. The low-density core may be a *foamed material,* as in Figure 9.7(b), or a *honeycomb,* as in (c). The reason for using a sandwich structure is to obtain a material with high strength-to-weight and stiffness-to-weight ratios.

9.2 METAL MATRIX COMPOSITES

Metal matrix composites (MMCs) consist of a metal matrix reinforced by a second phase. Common reinforcing phases include (1) particles of ceramic and (2) fibers of various materials, including other metals, ceramics, carbon, and boron. MMCs of the first type are commonly called *cermets.*

9.2.1 Cermets

A *cermet* is a composite material in which a ceramic is contained in a metallic matrix. The ceramic often dominates the mixture, sometimes ranging up to 96% by volume.

Bonding can be enhanced by slight solubility between phases at the elevated temperatures used in processing these composites. Cermets can be subdivided into (1) cemented carbides and (2) oxide-based cermets.

Cemented Carbides *Cemented carbides* are composed of one or more carbide compounds bonded in a metallic matrix. The term *cermet* is not used for all of these materials, even though it is technically correct. The common cemented carbides are based on tungsten carbide (WC), titanium carbide (TiC), and chromium carbide (Cr_3C_2). Tantalum carbide (TaC) and others are also used but less commonly. The principal metallic binders are cobalt and nickel. We have previously discussed the carbide ceramics (Section 7.3.2); they constitute the principal ingredient in cemented carbides, typically ranging in content from 80% to 95% of total weight.

Cemented carbide parts are produced by particulate processing techniques (Section 17.3). Cobalt is the binder used for WC (see Figure 9.8), and nickel is the common binder for TiC and Cr_3C_2. Even though the binder constitutes only about 5% to 15%, its effect on mechanical properties is significant in the composite material. Using WC–Co as an example, as the percentage of Co is increased, hardness is decreased and transverse rupture strength (TRS) is increased, as shown in Figure 9.9. TRS correlates with toughness of the WC–Co composite.

Cutting tools are the most common application of cemented carbides based on *tungsten carbide.* Other applications of WC–Co cemented carbides include wire drawing dies, rock-drilling bits and other mining tools, dies for powder metallurgy, indenters for hardness testers, cutting tools for sheetmetal operations, and other applications where hardness and wear resistance are critical factors.

Titanium carbide cermets are used principally for high temperature applications. Nickel is the preferred binder; its oxidation resistance at high temperatures is superior to that of cobalt. Applications include gas-turbine nozzle vanes, valve seats, thermocouple protection tubes, torch tips, and hot-working spinning tools [10]. TiC–Ni is also used as a cutting tool material for machining steels.

FIGURE 9.8
Photomicrograph (about 1500X) of cemented carbide with 85% WC and 15% Co (photo courtesy of Kennametal Inc.).

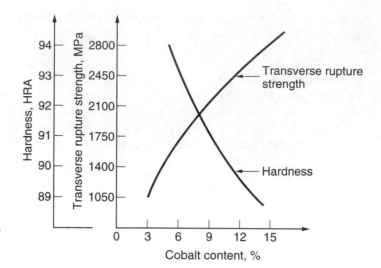

FIGURE 9.9 Typical plot of hardness and transverse rupture strength as a function of cobalt content.

Compared with WC–Co cemented carbides, nickel-bonded **chromium carbides** are more brittle, but have excellent chemical stability and corrosion resistance. This combination, together with good wear resistance, makes it suitable for applications such as gage blocks, valve liners, spray nozzles, and bearing seal rings [10].

Oxide-based Cermets Most of these composites utilize Al_2O_3 as the particulate phase; MgO is another oxide sometimes used. A common metal matrix is chromium, although other metals can also be used as binders. Relative proportions of the two phases vary significantly, with the possibility for the metal binder to be the major ingredient. Applications include cutting tools, mechanical seals, and thermocouple shields.

9.2.2 Fiber-Reinforced Metal Matrix Composites

These MMCs are of interest because they combine the high tensile strength and modulus of elasticity of a fiber with metals of low density, thus achieving good strength-to-weight and modulus-to-weight ratios in the resulting composite material. Typical metals used as the low-density matrix are aluminum, magnesium, and titanium. Some of the important fiber materials used in the composite include Al_2O_3, boron, carbon, and SiC.

Fiber-reinforced MMCs are anisotropic, as expected. Maximum tensile strength in the preferred direction is obtained by using continuous fibers bonded strongly to the matrix metal. Elastic modulus and tensile strength of the composite material increase with increasing fiber volume. MMCs with fiber reinforcement have good high-temperature strength properties, and they are good electrical and thermal conductors. Applications have largely been components in aircraft and turbine machinery, where these properties can be exploited.

9.3 CERAMIC MATRIX COMPOSITES

Ceramics have certain attractive properties: high stiffness, hardness, hot hardness, compressive strength, and relatively low density. Ceramics also have several faults—low tough-

FIGURE 9.10 Highly magnified electron microscopy photograph (about 3000X) showing fracture surface of SiC whisker reinforced ceramic (Al_2O_3) used as cutting tool material (courtesy Greenleaf Corporation).

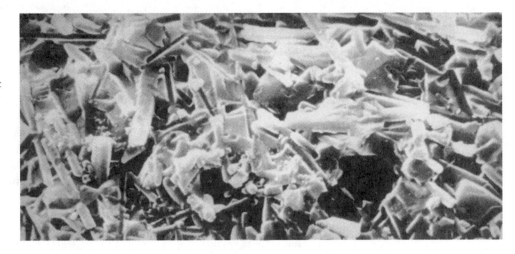

ness and bulk tensile strength, and susceptibility to thermal cracking. Ceramic matrix composites (CMCs) represent an attempt to retain the desirable properties of ceramics while compensating for their weaknesses. CMCs consist of a ceramic primary phase imbedded with a secondary phase. To date, most development work has focused on the use of fibers as the secondary phase. Success has been elusive. Technical difficulties include thermal and chemical compatibility of the constituents in CMCs during processing. Also, as with any ceramic material, limitations on part geometry must be considered.

Ceramic materials used as matrices include alumina (Al_2O_3), boron carbide (B_4C), boron nitride (BN), silicon carbide (SiC), silicon nitride (Si_3N_4), titanium carbide (TiC), and several types of glass [9]. Some of these materials are still in the development stage as CMC matrices. Fiber materials in CMCs include carbon, SiC, and Al_2O_3.

The reinforcing phase in current CMC technology consists of either short fibers, such as whiskers, or long fibers. Short fibers have been successfully fabricated using particulate processing methods (Chapter 17); they are treated as a form of powder in these materials. Although there are performance advantages in using long fibers as reinforcement in ceramic matrix composites, development of economical processing techniques for these materials has been difficult. One promising commercial application of CMCs is in metal-cutting tools as a competitor of cemented carbides, illustrated in Figure 9.10. The composite tool material has whiskers of SiC in a matrix of Al_2O_3. Other potential applications are in elevated temperatures and environments that are chemically corrosive to other materials.

9.4 POLYMER MATRIX COMPOSITES

A ***polymer matrix composite*** (PMC) consists of a polymer primary phase in which a secondary phase is imbedded in the form of fibers, particles, or flakes. Commercially, PMCs are the most important of the three classes of synthetic composites. They include most plastic molding compounds, rubber reinforced with carbon black, and fiber-reinforced polymers (FRPs). Of the three, FRPs are most closely identified with the term *composite*. If one mentions "composite material" to a design engineer, FRP is usually the composite that comes to mind.

9.4.1 Fiber-Reinforced Polymers

A *fiber-reinforced polymer* is a composite material consisting of a polymer matrix imbedded with high-strength fibers. The polymer is usually a thermosetting (TS) plastic such as unsaturated polyester or epoxy. The matrix can also be thermoplastic (TP) polymers, such as nylons (polyamides), polycarbonate, polystyrene, and polyvinylchloride. In addition, fiber reinforcement is widely used in rubber products such as tires and conveyor belts.

Fibers in PMCs can be in various forms: discontinuous (chopped), continuous, or woven as a fabric. Principal fiber materials in FRPs are glass, carbon, and Kevlar 49. Less common fibers include boron, SiC, and Al_2O_3, and steel. Glass (in particular E-glass) is the most common fiber material in today's FRPs; its use to reinforce plastics dates from around 1920.

The term *advanced composites* is sometimes used in connection with FRPs developed since the late 1960s that use boron, carbon, or Kevlar, as the reinforcing fibers [11]. Epoxy is the common matrix polymer. These composites generally have high fiber content (more than 50% by volume) and possess high strength and modulus of elasticity. When two or more fiber materials are combined in the FRP composite, it is called a *hybrid composite.* Advantages cited for hybrids over conventional or advanced FRPs include balanced strength and stiffness, improved toughness and impact resistance, and reduced weight [10]. Advanced and hybrid composites are used in aerospace applications.

The most widely used form of the FRP itself is a laminar structure, made by stacking and bonding thin layers of fiber and polymer until the desired thickness is obtained. By varying the fiber orientation among the layers, a specified level of anisotropy in properties can be achieved in the laminate. This method is used to form parts of thin cross-section, such as aircraft wing and fuselage sections, automobile and truck body panels, and boat hulls.

Properties   There are a number of attractive features that distinguish fiber-reinforced plastics as engineering materials. Most notable are (1) high strength-to-weight ratio, (2) high modulus-to-weight ratio, and (3) low specific gravity. A typical FRP weighs only about one fifth as much as steel; yet strength and modulus are comparable in the fiber direction. Table 9.4 compares these properties for several FRPs, steels, and an aluminum alloy. Properties listed in Table 9.4 depend on the proportion of fibers in the composite. Both tensile strength and elastic modulus increase as the fiber content is increased,

TABLE 9.4   Comparison of typical properties of fiber-reinforced plastics and representative metal alloys.

Material	Specific gravity (SG)	Tensile strength (TS)		Elastic modulus (E)		Index[a]	
		MPa	(lb/in.²)	GPa	(lb/in.²)	TS/SG	E/SG
Low-C steel	7.87	345	(50,000)	207	(30×10^6)	1.0	1.0
Alloy steel (heat treated)	7.87	3450	(500,000)	207	(30×10^6)	10.0	1.0
Aluminum alloy (heat treated)	2.70	415	(60,000)	69	(10×10^6)	3.5	1.0
FRP: fiberglass in polyester	1.50	205	(30,000)	69	(10×10^6)	3.1	1.7
FRP: Carbon in epoxy[b]	1.55	1500	(220,000)	140	(20×10^6)	22.3	3.4
FRP: Carbon in epoxy[c]	1.65	1200	(175,000)	214	(31×10^6)	16.7	4.9
FRP: Kevlar in epoxy matrix	1.40	1380	(200,000)	76	(11×10^6)	22.5	2.1

Compiled from [3], [6], and other sources. Properties are measured in the fiber direction.

[a]Indices are relative tensile strength-to-weight (TS/SG) and elastic modulus-to-weight (E/SG) ratios compared to low-C steel as the base (index = 1.0 for the base).

[b]High tensile strength carbon fibers used in FRP.

[c]High modulus carbon fibers used in FRP.

by Eq. (9.5). Other properties and characteristics of fiber-reinforced plastics include (4) good fatigue strength; (5) good corrosion resistance, although polymers are soluble in various chemicals; (6) low thermal expansion for many FRPs, leading to good dimensional stability; and (7) significant anisotropy in properties. With regard to this last feature, the mechanical properties of the FRPs given in Table 9.4 are in the direction of the fiber. As previously noted, their values are significantly less when measured in a different direction.

Applications   During the last three decades there has been a steady growth in the application of fiber-reinforced polymers in products requiring high strength and low weight, often as substitutions for metals. The aerospace industry is one of the biggest users of advanced composites. Designers are continually striving to reduce aircraft weight to increase fuel efficiency and payload capacity. Applications of advanced composites in both military and commercial aircraft have increased steadily. Much of the structural weight of today's airplanes and helicopters consists of FRPs. Figure 9.11 identifies the composites used in the Boeing 757.

The automotive industry is another important user of FRPs. The most obvious applications are FRP body panels for cars and truck cabs. A notable example is the Chevrolet Corvette that has been produced with FRP bodies for decades. Less apparent applications are in certain chassis and engine parts. Automotive applications differ from those in aerospace in two significant respects. First, the requirement for high strength-to-weight ratio is less demanding than for aircraft. Car and truck applications can use conventional fiberglass reinforced plastics rather than advanced composites. Second, production quantities are much higher in automotive applications, requiring more economical methods of fabrication. Continued use of low-carbon sheet steel in automobiles in the face of FRP's advantages is evidence of the low cost and processability of steel.

FIGURE 9.11   Composite materials in the Boeing 757 (courtesy of Boeing Commercial Airplane Group).

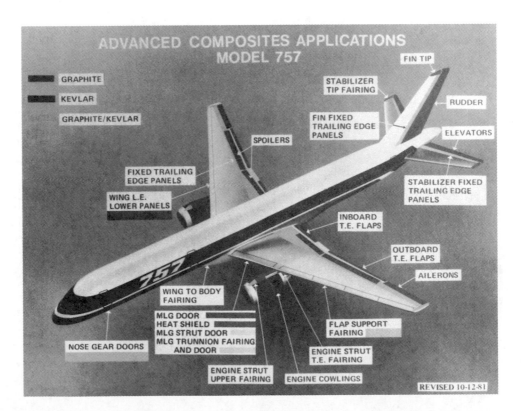

FRPs have been widely adopted for sports and recreational equipment. Fiberglass reinforced plastic has been used for boat hulls since the 1940s. Fishing rods were another early application. Today, FRPs are represented in a wide assortment of sports products, including tennis rackets, golf club shafts, football helmets, bows and arrows, skis, and bicycle wheels.

9.4.2 Other Polymer Matrix Composites

In addition to FRPs, other PMCs contain particles, flakes, and short fibers. Ingredients of the secondary phase are called *fillers* when used in polymer molding compounds (Section 8.1.5). Fillers divide into two categories: (1) reinforcements and (2) extenders. *Reinforcing fillers* serve to strengthen or otherwise improve mechanical properties of the polymer. Common examples include wood flour and powdered mica in phenolic and amino resins to increase strength, abrasion resistance, and dimensional stability; and carbon black in rubber to improve strength, wear, and tear resistance. *Extenders* simply increase the bulk and reduce the cost-per-unit weight of the polymer, but have little or no effect on mechanical properties. Extenders may be formulated to improve molding characteristics of the resin.

Foamed polymers (Section 13.11) are a form of composite in which gas bubbles are imbedded in a polymer matrix. Styrofoam and polyurethane foam are the most common examples. The combination of near-zero density of the gas and relatively low density of the matrix makes these materials extremely light weight. The gas mixture also lends very low thermal conductivity for applications in which heat insulation is required.

9.5 GUIDE TO PROCESSING COMPOSITE MATERIALS

Composite materials are formed into shapes by many different processing technologies. The two phases are typically produced separately before being combined into the composite part geometry. The matrix phases are generally processed by the technologies described in Chapters 6, 7, and 8 for metals, ceramics, and polymers.

Processing methods for the imbedded phase depend on geometry. Fiber production is described in Section 12.2.3 for glass and Section 13.4 for polymers. Fiber production methods for carbon, boron, and other materials are summarized in Table 15.1. Powder production for metals is described in Section 16.2 and for ceramics in Section 17.1.1. Processing techniques to fabricate MMC and CMC components, are similar to those used for powdered metals and ceramics (Chapters 16 and 17). We deal with the processing of cermets specifically in Section 17.3.

Molding processes are commonly performed on PMCs, both particle and chopped fiber types. Molding processes for these composites are the same as those used for polymers (Chapter 13). Other more specialized processes for polymer matrix composites, fiber-reinforced polymers in particular, are described in Chapter 15. Many laminated composite and honeycomb structures are assembled by adhesive bonding (Section 32.3).

REFERENCES

[1] Chawla, K. K. *Composite Materials: Science and Engineering.* 2nd ed. Springer-Verlag, New York, 1998.

[2] Delmonte, J. *Metal-Polymer Composites.* Van Nostrand Reinhold, New York, 1990.

[3] *Engineering Materials Handbook.* Vol. 1. *Composites.* ASM International, Metals Park, Oh., 1987.

[4] Flinn, R. A., and Trojan, P. K. *Engineering Materials*

and Their Applications. 4th ed. Houghton Mifflin, Boston, 1990.

[5] Greenleaf Corporation. *WG-300—Whisker Reinforced Ceramic/Ceramic Composites* (Marketing literature). Saegertown, Penn.

[6] Mallick, P. K. *Fiber-Reinforced Composites: Materials, Manufacturing, and Designs.* 2nd ed. Marcel Dekker, Inc., New York, 1993.

[7] McCrum, N. G., Buckley, C. P., and Bucknall, C. B. *Principles of Polymer Engineering.* 2nd ed. Oxford University Press, Inc., Oxford, UK, 1997.

[8] Morton-Jones, D. H. *Polymer Processing.* Chapman and Hall, London, UK, 1989.

[9] Naslain, R., and Harris, B. (eds.). *Ceramic Matrix Composites.* Elsevier Applied Science, London and New York, 1990.

[10] Schwartz, M. M. *Composite Materials Handbook.* 2nd ed. McGraw-Hill, New York, 1992.

[11] Wick, C., and Veilleux, R. F. *Tool and Manufacturing Engineers Handbook.* 4th ed. Vol. III—Materials, Finishing, and Coating, 1985, Chapter 8.

[12] Zweben, C., Hahn, H. T., and Chou, T-W. *Delaware Composites Design Encyclopedia.* Vol. 1. *Mechanical Behavior and Properties of Composite Materials.* Technomic Publishing Co., Inc., Lancaster, Penn., 1989.

REVIEW QUESTIONS

9.1. What is a *composite material*?

9.2. Identify some of the characteristic properties of composite materials.

9.3. What does the term *anisotropic* mean?

9.4. How are traditional composites distinguished from synthetic composites?

9.5. Name the three basic categories of composite materials.

9.6. What are the common forms of the reinforcing phase in composite materials?

9.7. What is a *whisker*?

9.8. What are the two forms of sandwich structure among laminar composite strutures? Briefly describe each.

9.9. Give some examples of commercial products that are laminar composite structures.

9.10. What are the three general factors that determine the properties of a composite material?

9.11. What is the *rule of mixtures*?

9.12. What is a *cermet*?

9.13. Cemented carbides are what class of composites?

9.14. What are some of the weaknesses of ceramics that might be corrected in fiber-reinforced ceramic matrix composites?

9.15. What is the most common fiber material in fiber-reinforced plastics?

9.16. What does the term *advanced composites* mean?

9.17. What is a *hybrid composite*?

9.18. Identify some of the important properties of fiber-reinforced plastic composite materials.

9.19. Name some of the important applications of FRPs.

9.20. What is meant by the term interface in the context of composite materials?

MULTIPLE CHOICE QUIZ

There is a total of 22 correct answers in the following multiple choice questions (some questions have multiple answers that are correct). To attain a perfect score on the quiz, all correct answers must be given, since each correct answer is worth 1 point. For each question, each omitted answer or wrong answer reduces the score by 1 point, and each additional answer beyond the number of answers required reduces the score by 1 point. Percentage score on the quiz is based on the total number of correct answers.

9.1. Anisotropic means which one of the following? (a) composite materials with composition consisting of more than two materials, (b) properties are the same in every direction, (c) properties vary depending on the direction in which they are measured, or (d) strength and other properties as a function of curing temperature.

9.2. The reinforcing phase is the matrix within which the secondary phase is imbedded: (a) true or (b) false.

9.3. Which one of the following reinforcing geometries offers the greatest potential for strength and stiffness improvement in the resulting composite material? (a) fibers, (b) flakes, or (c) particles.

9.4. Wood is which one of the following composite types? (a) CMC, (b) MMC, or (c) PMC.

9.5. Which of the following materials are used as fibers in fiber-reinforced plastics (more than one)? (a) aluminum oxide, (b) boron, (c) carbon/graphite, (d) epoxy, (e) Kevlar 49, (f) S-glass, and (g) unsaturated polyester.

9.6. Which of the following metals are most commonly used as the matrix material in fiber-reinforced MMCs (name three)? (a) aluminum, (b) copper, (c) iron, (d) magnesium, (e) titanium, or (f) zinc.

9.7. Which one of the following metals is used as the matrix metal in nearly all WC cemented carbides? (a) aluminum, (b) chromium, (c) cobalt, (d) lead, (e) nickel, (f) tungsten, or (g) tungsten carbide.

9.8. Ceramic matrix composites are designed to overcome which of the following weaknesses of ceramics (more than one)? (a) compressive strength, (b) hardness, (c) hot hardness, (d) modulus of elasticity, (e) tensile strength, or (f) toughness.

9.9. Which one of the following polymer types is most commonly used in polymer matrix composites? (a) elastomers, (b) thermoplastics, or (c) thermosets.

9.10. Which one of the following is the most common reinforcing material in FRPs? (a) Al_2O_3, (b) boron, (c) carbon, (d) cobalt, (e) graphite, (f) Kevlar 49, or (g) SiO_2.

9.11. Identify which of the following materials are composites (more than one). (a) cemented carbide, (b) phenolic molding compound, (c) plywood, (d) Portland cement, (e) rubber in automobile tires, (f) wood, or (g) 1020 steel.

Part III
Solidification Processes

10 FUNDAMENTALS OF METAL CASTING

CHAPTER CONTENTS

10.1 Overview of Casting Technology
    10.1.1 Casting Processes
    10.1.2 Sand Casting Molds
10.2 Heating and Pouring
    10.2.1 Heating the Metal
    10.2.2 Pouring the Molten Metal
    10.2.3 Engineering Analysis of Pouring
    10.2.4 Fluidity
10.3 Solidification and Cooling
    10.3.1 Solidification of Metals
    10.3.2 Solidification Time
    10.3.3 Shrinkage
    10.3.4 Directional Solidification
    10.3.5 Riser Design

In this part of the book, we consider those manufacturing processes in which the starting work material is either a liquid or is in a highly plastic condition, and a part is created through solidification of the material. Casting and molding processes dominate this category of shaping operations. With reference to Figure 10.1, the solidification processes can be classified according to the engineering material that is processed: (1) metals, (2) ceramics, specifically glasses,[1] and (3) polymers and polymer matrix composites (PMCs). Casting of metals is covered in this and the following chapter. Glassworking is covered in Chapter 12, and polymer and PMC processing is treated in Chapters 13, 14, and 15.

Casting is a process in which molten metal flows by gravity or other force into a mold where it solidifies in the shape of the mold cavity. The term *casting* is also applied to the part that is made by this process. It is one of the oldest shaping processes, dating back 6,000 years (Historical Note 10.1). The principle of casting seems simple: melt the metal, pour it into a mold, and let it freeze; yet there are many factors and variables that must be considered in order to accomplish a successful casting operation.

Casting includes both the casting of ingots and the casting of shapes. The term *ingot* is usually associated with the primary metals industries; it describes a large casting that

---

[1]Among the ceramics, only glass is processed by solidification; traditional and new ceramics are shaped using particulate processes, Chapter 17.

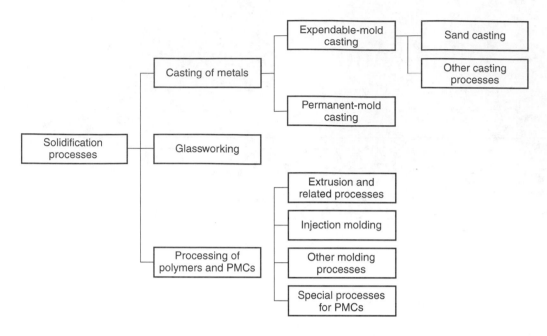

FIGURE 10.1   Classification of solidification processes.

Historical Note 10.1:   *Origins of casting* [11].

C asting of metals can be traced back to around 4000 B.C. Gold was the first metal to be discovered and used by the early civilizations; it was malleable and could be readily hammered into shape at room temperature. There seemed to be no need for other ways to shape gold. It was the subsequent discovery of copper that gave rise to the need for casting. Although copper could be forged to shape, the process was more difficult (due to strain hardening) and limited to relatively simple forms. Historians believe that hundreds of years elapsed before the process of casting copper was first performed, probably by accident during the reduction of copper ore in preparation for hammering the metal into some useful form. Thus, through serendipity, the art of casting was born. It is likely that the discovery occurred in Mesopotamia, and the "technology" quickly spread throughout the rest of the ancient world.

It was an innovation of significant importance in the history of mankind. Shapes much more intricate could be formed by casting than by hammering. More sophisticated tools and weapons could be fabricated. More detailed implements and ornaments could be fashioned. Fine gold jewelry could be made more beautiful and valuable than by previous methods. Alloys were

first used for casting when it was discovered that mixtures of copper and tin (the alloy thus formed was bronze) yielded much better castings than copper alone. Casting permitted the creation of wealth to those nations that could perform it best. Egypt ruled the Western civilized world during the Bronze Age (nearly 2,000 years) largely due to its ability to perform the casting process.

Religion provided an important influence during the Dark Ages (circa 400 to 1400 A.D.) for perpetuating the foundryman's skills. Construction of cathedrals and churches required the casting of bells that were used in these structures. Indeed, the time and effort needed to cast the large bronze bells of the period helped to move the casting process from the realm of art toward the regiment of technology. Advances in melting and mold-making techniques were made. Pit molding, in which the molds were formed in a deep pit located in front of the furnace to simplify the pouring process, was improved as a casting procedure. In addition, the bell-founder learned the relationships between the tone of the bell, which was the important measure of product quality, and its size, shape, thickness, and metal composition.

Another important product associated with the development of casting was the cannon. Chronologically, it followed the bell, and therefore many of the casting techniques developed for bellfounding were applied to cannon making. The first cast cannon was made in Ghent, Belgium, in the year 1313—by a religious monk, of all people. It was made of bronze, and the bore was formed by means of a core during casting. Because of the rough bore surface created by the casting process, these early guns were not accurate and had to be fired at relatively close range to be effective. It was soon realized that accuracy and range could be improved if the bore were made smooth by machining the surface. Quite appropriately, this machining process was called ***boring*** (Section 22.1.5).

is simple in shape and intended for subsequent reshaping by processes such as rolling or forging. Ingot casting was discussed in Chapter 6. ***Shape casting*** involves the production of more complex geometries that are much closer to the final desired shape of the part or product. It is with the casting of shapes rather than ingots that this chapter and the next are concerned.

A variety of shape casting methods are available, thus making it one of the most versatile of all manufacturing processes. Among its capabilities and advantages are the following:

➢ Casting can be used to create complex part geometries, including both external and internal shapes.
➢ Some casting processes are capable of producing parts to ***net shape.*** No further manufacturing operations are required to achieve the required geometry and dimensions of the parts. Other casting processes are ***near net shape,*** for which some additional shape processing is required (usually machining) in order to achieve accurate dimensions and details.
➢ Casting can be used to produce very large parts. Castings weighing more than a hundred tons have been made.
➢ The casting process can be performed on any metal that can be heated to the liquid state.
➢ Some casting methods are quite suited to mass production.

There are also disadvantages associated with casting—different disadvantages for different casting methods. These include limitations on mechanical properties, porosity, poor dimensional accuracy and surface finish for some casting processes, safety hazards to humans when processing hot molten metals, and environmental problems.

Parts made by casting processes range in size from small components weighing only a few ounces up to very large products weighing tons. The list of parts includes dental crowns, jewelry, wood-burning stoves, engine blocks and heads for automotive vehicles, machine frames, railway wheels, frying pans, pipes, and pump housings. All varieties of metals can be cast, ferrous and nonferrous.

Casting can also be used on other materials such as polymers and ceramics; however, the details are sufficiently different that we postpone discussion of the casting processes for these materials until later chapters. This chapter and the next deal exclusively with metal casting. In this chapter, we discuss the fundamentals that apply to virtually all casting operations. In the following chapter, the individual casting processes are described, along with some of the product design issues that must be considered when making parts out of castings.

10.1 OVERVIEW OF CASTING TECHNOLOGY

As a production process, casting is usually carried out in a foundry. A *foundry* is a factory equipped for making molds, melting and handling metal in molten form, performing the casting process, and cleaning the finished casting. The workers who perform the casting operations in these factories are called *foundrymen.*

10.1.1 Casting Processes

Discussion of casting logically begins with the mold. The *mold* contains a cavity whose geometry determines the shape of the cast part. The actual size and shape of the cavity must be slightly oversized to allow for shrinkage that occurs in the metal during solidification and cooling. Different metals undergo different amounts of shrinkage, so the mold cavity must be designed for the particular metal to be cast if dimensional accuracy is critical. Molds are made of a variety of materials, including sand, plaster, ceramic, and metal. The various casting processes are often classified according to these different types of molds.

To accomplish a casting operation, the metal is first heated to a temperature high enough to completely transform it into a liquid state. It is then poured, or otherwise directed, into the cavity of the mold. In an *open mold,* Figure 10.2(a), the liquid metal is simply poured until it fills the open cavity. In a *closed mold,* Figure 10.2(b), a passageway, called the gating system, is provided to permit the molten metal to flow from outside the mold into the cavity. The closed mold is by far the more important form in production casting operations.

As soon as the molten metal is in the mold, it begins to cool. When the temperature drops sufficiently (e.g., to the freezing point for a pure metal), solidification begins. Solidification involves a change of phase of the metal. Time is required to complete the phase change, and considerable heat is given up in the process. It is during this process that the metal assumes the solid shape of the mold cavity and many of the properties and characteristics of the casting are established.

Once the casting has cooled sufficiently, it is removed from the mold. Depending on the casting method and metal used, further processing may be required. This may in-

FIGURE 10.2   Two forms of mold: (a) open mold, simply a container in the shape of the desired part; and (b) closed mold, in which the mold geometry is more complex and requires a gating system (passageway) leading into the cavity.

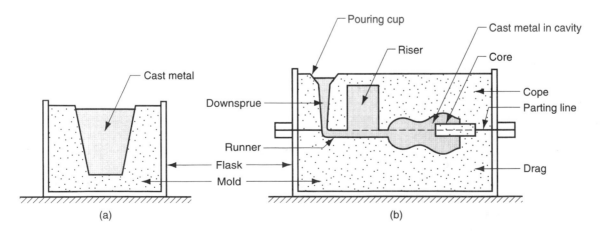

clude trimming the excess metal from the actual cast part, clear
ing the product, and heat treatment to enhance properties.
(Chapter 22) may be required to achieve closer tolerances on
to remove the cast surface and its associated metallurgical micr

Casting processes divide into two broad categories, accord
expendable mold casting processes, and permanent mold castir
able mold must be destroyed when the molten metal solidifie
These molds are made out of sand, plaster, and similar materi..., ...
tained by using binders of various kinds. Sand casting is the most prominent example of
the expendable mold processes. In sand casting, the liquid metal is poured into a mold
made of sand. After the metal hardens, the mold must be sacrificed in order to recover
the casting.

A *permanent mold* is one that can be used over and over to produce many castings. It is made of metal (or, less commonly, a ceramic refractory material) that can withstand the high temperatures of the casting operation. In permanent mold casting, the mold
consists of two (or more) sections that can be opened to permit removal of the finished
part. Die casting is the most familiar process in this group.

More intricate casting geometries are generally possible with the expendable mold
processes. Part shapes in the permanent mold processes are limited by the need to open
the mold. On the other hand, some of the permanent mold processes have certain economic advantages in high production operations. We discuss the expendable mold and
permanent mold casting processes in Chapter 11.

10.1.2 Sand Casting Molds

Sand casting is by far the most important casting process. A sand-casting mold will be
used to describe the basic features of a mold. Many of these features and terms are common to the molds used in other casting processes.

Figure 10.2(b) shows the cross-sectional view of a typical sand casting mold,
indicating some of the terminology. The mold consists of two halves: cope and drag.
The *cope* is the upper half of the mold, and the *drag* is the bottom half. These two mold
parts are contained in a box, called a *flask,* which is also divided into two halves, one
for the cope and the other for the drag. The two halves of the mold separate at the
parting line.

In sand casting (and other expendable mold processes) the mold cavity is formed
by means of a *pattern,* which is made of wood, metal, plastic, or other material and has
the shape of the part to be cast. The cavity is formed by packing sand around the pattern, about half each in the cope and drag, so that when the pattern is removed, the remaining void has the desired shape of the cast part. The pattern is usually made oversized to allow for shrinkage of the metal as it solidifies and cools. The sand for the mold
is moist and contains a binder to maintain its shape.

The cavity in the mold provides the external surfaces of the cast part. In addition,
a casting may have internal surfaces. These surfaces are determined by means of a *core,*
a form placed inside the mold cavity to define the interior geometry of the part. In sand
casting, cores are generally made of sand, although other materials can be used, such as
metals, plaster, and ceramics.

The *gating system* in a casting mold is the channel, or network of channels, by
which molten metal flows into the cavity from outside the mold. As shown in the figure,
the gating system typically consists of a *downsprue* (also called simply the *sprue*), through
which the metal enters a *runner* that leads into the main cavity. At the top of the down-

sprue, a ***pouring cup*** is often used to minimize splash and turbulence as the metal flows into the downsprue. It is shown in our diagram as a simple cone-shaped funnel. Some pouring cups are designed in the shape of a bowl, with an open channel leading to the downsprue.

In addition to the gating system, any casting in which shrinkage is significant requires a riser connected to the main cavity. The ***riser*** is a reservoir in the mold that serves as a source of liquid metal for the casting to compensate for shrinkage during solidification. The riser must be designed to freeze after the main casting in order to satisfy its function.

As the metal flows into the mold, the air that previously occupied the cavity, as well as hot gases formed by reactions of the molten metal, must be evacuated so that the metal will completely fill the empty space. In sand casting, for example, the natural porosity of the sand mold permits the air and gases to escape through the walls of the cavity. In permanent metal molds, small vent holes are drilled into the mold or machined into the parting line to permit removal of air and gases.

10.2 HEATING AND POURING

To perform a casting operation, the metal must be heated to a temperature somewhat above its melting point and then poured into the mold cavity to solidify. In this section, we consider several aspects of these two steps in casting.

10.2.1 Heating the Metal

Heating furnaces of various kinds (Section 11.4.1) are used to heat the metal to a molten temperature sufficient for casting. The heat energy required is the sum of (1) the heat to raise the temperature to the melting point, (2) the heat of fusion to convert it from solid to liquid, and (3) the heat to raise the molten metal to the desired temperature for pouring. This can be expressed:

$$H = \rho V\{C_s(T_m - T_o) + H_f + C_l(Tp - T_m)\} \qquad (10.1)$$

where H = total heat required to raise the temperature of the metal to the pouring temperature, J (Btu); ρ = density, g/cm^3 (lbm/in.^{3}); C_s = weight specific heat for the solid metal, J/g-°C (Btu/lbm-°F); T_m = melting temperature of the metal, °C (°F); T_o = starting temperature—usually ambient, °C (°F); H_f = heat of fusion, J/g (Btu/lbm); C_l = weight specific heat of the liquid metal, J/g-°C (Btu/lbm-°F); T_p = pouring temperature, °C (°F); and V = volume of metal being heated, cm^3 (in.^{3}).

**EXAMPLE 10.1
Heating metal for casting.**

One cubic meter of a certain eutectic alloy will be heated in a crucible from room temperature to 100°C above its melting point for casting. Density = 7.5 g/cm^3; melting point = 800°C; specific heat of the metal = 0.33 J/g-°C in the solid state and 0.29 J/g-°C in the liquid state; and heat of fusion = 160 J/g. How much heat energy must be added to accomplish the heating, assuming no losses?

Solution: Assume ambient temperature in the foundry = 25°C, and assume that the density of liquid and solid states of the metal are the same. Noting that one m^3 = 10^6 cm^3. and substituting the property values into Eq. (10.1), we have

$$H = (7.5)(10^6)\{0.33(800 - 25) + 160 + 0.29(100)\} = 3335(10^6) \text{ J}$$

The above equation is of conceptual value; however, its computational value is limited, notwithstanding our example calculation. Use of Eq. (10.1) is complicated by the following factors:

1. Specific heat and other thermal properties of a solid metal vary with temperature, especially if the metal undergoes a change of phase during heating.
2. A metal's specific heat may be different in the solid and liquid states.
3. Most casting metals are alloys, and most alloys melt over a temperature range between a solidus and liquidus rather than at a single melting point; thus, the heat of fusion cannot be applied so simply as indicated above.
4. The property values required in the equation for a particular alloy are not readily available in most cases.
5. There are significant heat losses to the environment during heating.

10.2.2 Pouring the Molten Metal

After heating, the metal is ready for pouring. Introduction of molten metal into the mold, including its flow through the gating system and into the cavity, is a critical step in the casting process. For this step to be successful, the metal must flow into all regions of the mold, including—most importantly—the main cavity, before solidifying. Factors affecting the pouring operation include: pouring temperature, pouring rate, and turbulence.

The *pouring temperature* is the temperature of the molten metal as it is introduced into the mold. What is important here is the difference between the temperature at pouring and the temperature at which freezing begins (the melting point for a pure metal or the liquidus temperature for an alloy). This temperature difference is sometimes referred to as the *superheat.* This term is also used for the amount of heat that must be removed from the molten metal between pouring and when solidification commences [6].

Pouring rate refers to the volumetric rate at which the molten metal is poured into the mold. If the rate is too slow, the metal will chill and freeze before filling the cavity. If the pouring rate is excessive, turbulence can become a serious problem.

Turbulence in fluid flow is characterized by erratic variations in the magnitude and direction of the velocity throughout the fluid. The flow is agitated and irregular rather than smooth and streamlined, as in laminar flow. Turbulent flow should be avoided during pouring for several reasons. It tends to accelerate the formation of metal oxides which can become entrapped during solidification, thus degrading the quality of the casting. Turbulence also aggravates *mold erosion,* the gradual wearing away of the mold surfaces due to impact of the flowing molten metal. The densities of most molten metals are much higher than water and other fluids we normally deal with. These molten metals are also much more chemically reactive than at room temperature. Consequently, the wear caused by the flow of these metals in the mold is significant, especially under turbulent conditions. Erosion is especially serious when it occurs in the main cavity because the geometry of the cast part is affected.

10.2.3 Engineering Analysis of Pouring

There are several relationships that govern the flow of liquid metal through the gating system and into the mold. An important relationship is *Bernoulli's theorem,* which states that the sum of the energies (head, pressure, kinetic, and friction) at any two points in a flowing liquid are equal. This can be written in the following form:

$$h_1 + \frac{p_1}{\rho} = \frac{v_1^2}{2g} + F_1 = h_2 + \frac{P_2}{\rho} + \frac{v_2^2}{2g} + F_2 \qquad (10.2)$$

where h = head, cm (in.); p = pressure on the liquid, N/cm^2 (lb/in.^{2}); ρ = density, g/cm^3 (lbm/in.^{3}); v = flow velocity, cm/s (in./sec); g = gravitational acceleration constant, 981 cm/s/s (32.2 × 12 = 386 in./sec/sec); and F = head losses due to friction, cm (in.). Subscripts 1 and 2 indicate any two locations in the liquid flow.

Bernoulli's equation can be simplified in several ways. If we ignore friction losses (to be sure, friction will affect the liquid flow through a sand mold), and assume that the system remains at atmospheric pressure throughout, then the equation can be reduced to

$$h_1 + \frac{v_1^2}{2g} = h_2 + \frac{v_2^2}{2g} \qquad (10.3)$$

This can be used to determine the velocity of the molten metal at the base of the sprue. Let us define point 1 at the top of the sprue and point 2 at its base. If point 2 is used as the reference plane, then the head at that point is zero ($h_2 = 0$) and h_1 is the height (length) of the sprue. When the metal is poured into the pouring cup and overflows down the sprue, its initial velocity at the top is zero ($v_1 = 0$). Hence, Eq. (10.3) further simplifies to

$$h_1 = \frac{v_2^2}{2g}$$

which can be solved for the flow velocity:

$$v = \sqrt{2gh} \qquad (10.4)$$

where v = the velocity of the liquid metal at the base of the sprue, cm/s (in./sec); g = 981 cm/s/s (386 in./sec); and h = the height of the sprue, cm (in.).

Another relationship of importance during pouring is the **continuity law,** which states that the volume rate of flow remains constant throughout the liquid. The volume flow rate is equal to the velocity multiplied by the cross-sectional area of the flowing liquid. The continuity law can be expressed:

$$Q = v_1 A_1 = v_2 A_2 \qquad (10.5)$$

where Q = volumetric flow rate, cm^3/s (in.^{3}/sec); v = velocity as before; A = cross-sectional area of the liquid, cm^2 (in.^{2}); and the subscripts refer to any two points in the flow system. Thus, an increase in area results in a decrease in velocity, and vice versa.

Eqs. (10.4) and (10.5) indicate that the sprue should be tapered. As the metal accelerates during its descent into the sprue opening, the cross-sectional area of the channel must be reduced; otherwise, as the velocity of the flowing metal increases toward the base of the sprue, air can be aspirated into the liquid and conducted into the mold cavity. To prevent this condition, the sprue is designed with a taper, so that the volume flow rate vA is the same at the top and bottom of the sprue.

Assuming that the runner from the sprue base to the mold cavity is horizontal (and therefore the head h is the same as at the sprue base), then the volume rate of flow through the gate and into the mold cavity remains equal to vA at the base. Accordingly, we can estimate the time required to fill a mold cavity of volume V as

$$MFT = \frac{V}{Q} \qquad (10.6)$$

where MFT = mold filling time, s (sec); V = volume of mold cavity, cm^3 (in.^{3}); and Q = volume flow rate, as before. The mold filling time computed by Eq. (10.6) must be considered a minimum time. This is because the analysis ignores friction losses and possible constriction of flow in the gating system; thus, the mold filling time will be longer than what is given by Eq. (10.6).

EXAMPLE 10.2
Pouring
calculations.

A certain mold has a sprue whose length is 20 cm and the cross-sectional area at the base of the sprue is 2.5 cm^2. The sprue feeds a horizontal runner leading into a mold cavity whose volume is 1560 cm^3. Determine: (a) velocity of the molten metal at the base of the sprue, (b) volume rate of flow, and (c) time to fill the mold.

Solution: (a) The velocity of the flowing metal at the base of the sprue is given by Eq. (10.4):

$$v = \sqrt{2(981)(20)} = 198.1 \text{ cm/s}$$

(b) The volumetric flow rate is

$$Q = (2.5 \text{ cm}^2)(198.1 \text{ cm/s}) = 495 \text{ cm}^3/\text{s}$$

(c) Time required to fill a mold cavity of 1560 in.^{3} at this flow rate is

$$MFT = 1560/495 = 3.2 \text{ s}$$

10.2.4 Fluidity

The molten metal flow characteristics are often described by the term ***fluidity,*** a measure of the capability of a metal to flow into and fill the mold before freezing. Fluidity is the inverse of viscosity (Section 3.4); as viscosity increases, fluidity decreases. Standard testing methods are available to assess fluidity, including the spiral mold test shown in Figure 10.3, in which fluidity is indicated by the length of the solidified metal in the spiral channel. A longer cast spiral means greater fluidity of the molten metal.

Factors affecting fluidity include pouring temperature, metal composition, viscosity of the liquid metal, and heat transfer to the surroundings. A higher pouring temperature relative to the freezing point of the metal increases the time it remains in the liquid state, allowing it to flow further before freezing. This tends to aggravate certain casting problems such as oxide formation, gas porosity, and penetration of liquid metal into the interstitial spaces between the grains of sand forming the mold. This last problem causes the surface of the casting to contain imbedded sand particles, thus making it rougher and more abrasive than normal.

FIGURE 10.3   Spiral mold test for fluidity, in which fluidity is measured as the length of the spiral channel that is filled by the molten metal prior to solidification.

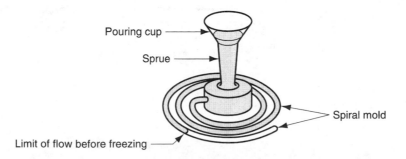

Composition also affects fluidity, particularly with respect to the metal's solidification mechanism. The best fluidity is obtained by metals that freeze at a constant temperature (e.g., pure metals and eutectic alloys). When solidification occurs over a temperature range (most alloys are in this category), the partially solidified portion interferes with the flow of the liquid portion, thereby reducing fluidity. In addition to the freezing mechanism, metal composition also determines **heat of fusion**—the amount of heat required to solidify the metal from the liquid state. A higher heat of fusion tends to increase the measured fluidity in casting.

10.3 SOLIDIFICATION AND COOLING

After pouring into the mold, the molten metal cools and solidifies. In this section we examine the physical mechanism of solidification that occurs during casting. Issues associated with solidification include the time for a metal to freeze, shrinkage, directional solidification, and riser design.

10.3.1 Solidification of Metals

Solidification involves the transformation of the molten metal back into the solid state. The solidification process differs depending on whether the metal is a pure element or an alloy.

Pure Metals A pure metal solidifies at a constant temperature equal to its freezing point, which is the same as its melting point. The melting points of pure metals are well known and documented (Table 4.1). The process occurs over time as shown in the plot of Figure 10.4, called a cooling curve. The actual freezing takes time, called the **local solidification time** in casting, during which the metal's latent heat of fusion is released into the surrounding mold. The **total solidification time** is the time taken between pouring and complete solidification. After the casting has completely solidified, cooling continues at a rate indicated by the downward slope of the cooling curve.

FIGURE 10.4 Cooling curve for a pure metal during casting.

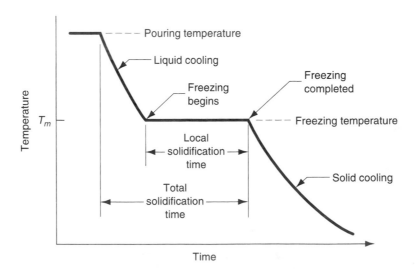

Because of the chilling action of the mold wall, a thin skin of solid metal is initially formed at the interface immediately after pouring. Thickness of the skin increases to form a shell around the molten metal as solidification progresses inward toward the center of the cavity. The rate at which freezing proceeds depends on heat transfer into the mold, as well as the thermal properties of the metal.

It is of interest to examine the metallic grain formation and growth during this solidification process. The metal which forms the initial skin has been rapidly cooled by the extraction of heat through the mold wall. This cooling action causes the grains in the skin to be fine, equiaxed, and randomly oriented. As cooling continues, further grain formation and growth occurs in a direction away from the heat transfer. Since the heat transfer is through the skin and mold wall, the grains grow inwardly as needles or spines of solid metal. As these spines enlarge, lateral branches form, and as these branches grow, further branches form at right angles to the first branches. This type of grain growth is referred to as *dendritic growth,* and it occurs not only in the freezing of pure metals but alloys as well. These treelike structures are gradually filled in during freezing, as additional metal is continually deposited onto the dendrites until complete solidification has occurred. The grains resulting from this dendritic growth take on a preferred orientation, tending to be coarse, columnar grains aligned toward the center of the casting. The resulting grain formation is illustrated in Figure 10.5.

Most Alloys   Most alloys freeze over a temperature range rather than at a single temperature. The exact range depends on the alloy system and the particular composition. Solidification of an alloy can be explained with reference to Figure 10.6, which shows the phase diagram for a particular alloy system (Section 6.1.2) and the cooling curve for a given composition. As temperature drops, freezing begins at the temperature indicated by the *liquidus* and is completed when the *solidus* is reached. The start of freezing is similar to that of the pure metal. A thin skin is formed at the mold wall due to the large temperature gradient at this surface. Freezing then progresses as before through the formation of dendrites that grow away from the walls. However, owing to the temperature spread between the liquidus and solidus, the nature of the dendritic growth is such that an advancing zone is formed in which both liquid and solid metal coexist. The solid portions are the dendrite structures that have formed sufficiently that small islands of liquid metal are trapped in the matrix. This solid–liquid region has a soft consistency leading to its name as the *mushy zone.* Depending on the conditions of freezing, the mushy zone can be relatively narrow, or it can exist throughout most of the casting. The latter condition is promoted by factors such as slow heat transfer out of the hot metal and a wide difference between liquidus and solidus temperatures. Gradually, the liquid islands in the dendrite matrix solidify as the temperature of the casting drops to the solidus for the given alloy composition.

Another factor complicating solidification of alloys is that the composition of forming dendrites favors the metal that has the higher melting point. As freezing continues

FIGURE 10.5   Characteristic grain structure in a casting of a pure metal, showing randomly oriented grains of small size near the mold wall, and large columnar grains oriented toward the center of the casting.

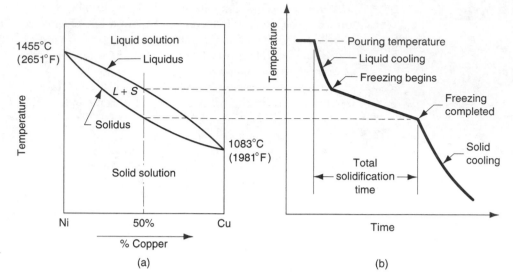

FIGURE 10.6 (a) Phase diagram for a copper–nickel alloy system and (b) associated cooling curve for a 50%Ni–50%Cu composition during casting.

and the dendrites grow, there develops an imbalance in composition between the metal that has solidified and the remaining molten metal. This composition imbalance is finally manifested in the completed casting in the form of segregation of the elements. The segregation is of two types, microscopic and macroscopic. At the microscopic level, the chemical composition varies throughout each individual grain. This is due to the fact that the beginning spine of each dendrite has a higher proportion of one of the elements in the alloy. As the dendrite grows in its local vicinity, it must expand using the remaining liquid metal that has been partially depleted of the first component. Finally, the last metal to freeze in each grain is that which has been trapped by the branches of the dendrite, and its composition is even further out of balance. Thus, we have a variation in chemical composition within single grains of the casting.

At the macroscopic level, the chemical composition varies throughout the casting itself. Since the regions of the casting that freeze first (at the outside near the mold walls) are richer in one component than the other, the remaining molten alloy is deprived of that component by the time freezing occurs at the interior. Thus, there is a general segregation through the cross-section of the casting, sometimes called ***ingot segregation,*** as illustrated in Figure 10.7.

Eutectic Alloys   Eutectic alloys constitute an exception to the general process by which alloys solidify. A ***eutectic alloy*** is a particular composition in an alloy system for which the solidus and liquidus are at the same temperature. Hence, solidification occurs at a constant temperature rather than over a temperature range, as described above. The effect can be seen in the phase diagram of the lead–tin system shown in Figure 6.3. Pure lead has a melting point of 327°C (621°F), while pure tin melts at 232°C (450°F).

FIGURE 10.7   Characteristic grain structure in an alloy casting, showing segregation of alloying components in center of casting.

Although most lead–tin alloys exhibit the typical solidus–liquidus temperature range, the particular composition of 61.9% tin and 38.1% lead has a melting (freezing) point of 183°C (362°F). This composition is the ***eutectic composition*** of the lead–tin alloy system, and 183°C is its ***eutectic temperature.*** Lead–tin alloys are not commonly used in casting; Pb–Sn compositions near the eutectic are used for electrical soldering, where the low melting point is an advantage. Examples of eutectic alloys encountered in casting include aluminum–silicon (11.6% Si) and cast iron (4.3% C).

10.3.2 Solidification Time

Whether the casting is pure metal or alloy, solidification takes time. The total solidification time is the time required for the casting to solidify after pouring. This time is dependent on the size and shape of the casting by an empirical relationship known as ***Chvorinov's Rule,*** which states the following:

$$TST = C_m \left(\frac{V}{A} \right)^n \tag{10.7}$$

where TST = total solidification time, min; V = volume of the casting, cm^3 (in.^{3}); A = surface area of the casting, cm^2 (in.^{2}); n is an exponent usually taken to have a value = 2; and C_m is the ***mold constant.*** Given that $n = 2$, the units of C_m are min/cm^2 (min/in.^{2}), and its value depends on the particular conditions of the casting operation, including: mold material (e.g., specific heat, thermal conductivity), thermal properties of the cast metal (e.g., heat of fusion, specific heat, thermal conductivity), and pouring temperature relative to the melting point of the metal. The value of C_m for a given casting operation can be based on experimental data from previous operations carried out using the same mold material, metal, and pouring temperature, even though the shape of the part may be quite different.

Chvorinov's Rule indicates that a casting with a higher volume-to-surface area ratio will cool and solidify more slowly than one with a lower ratio. This principle is put to good use in designing the riser in a mold. To perform its function of feeding molten metal to the main cavity, the metal in the riser must remain in the liquid phase longer than the casting. In other words, the TST for the riser must exceed the TST for the main casting. Since the mold conditions for both riser and casting are the same, their mold constants will be equal. By designing the riser to have a larger volume-to-area ratio, we can be fairly sure that the main casting solidifies first and that the effects of shrinkage are minimized. Before considering how the riser might be designed using Chvorinov's Rule, let us consider the topic of shrinkage, which is the reason why risers are needed.

10.3.3 Shrinkage

Our discussion of solidification has neglected the impact of shrinkage that occurs during cooling and freezing. Shrinkage occurs in three steps: (1) liquid contraction during cooling prior to solidification; (2) contraction during the phase change from liquid to solid, called ***solidification shrinkage;*** and (3) thermal contraction of the solidified casting during cooling to room temperature. The three steps can be explained with reference to a hypothetical cylindrical casting made in an open mold, as shown in Figure 10.8. The molten metal immediately after pouring is shown in part (0) of the series. Contraction of the liquid metal during cooling from pouring temperature to freezing temperature causes the height of the liquid to be reduced from its starting level as in (1) of the figure. The amount of this liquid contraction is usually around 0.5%. Solidification shrinkage, seen in part (2), has two effects. First, contraction causes a further reduction in the

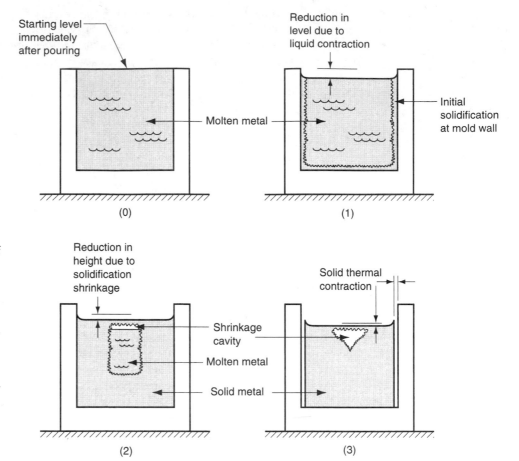

FIGURE 10.8 Shrinkage of a cylindrical casting during solidification and cooling: (0) starting level of molten metal immediately after pouring; (1) reduction in level caused by liquid contraction during cooling; (2) reduction in height and formation of shrinkage cavity caused by solidification shrinkage; and (3) further reduction in height and diameter due to thermal contraction during cooling of the solid metal. Dimensional reductions are exaggerated for clarity in our sketches.

height of the casting. Second, the amount of liquid metal available to feed the top center portion of the casting becomes restricted. This is usually the last region to freeze, and the absence of metal creates a void in the casting at this location. This shrinkage cavity is called a *pipe* by foundrymen. Once solidified, the casting experiences further contraction in height and diameter while cooling, as in (3). This shrinkage is determined by the solid metal's coefficient of thermal expansion, which in this case is applied in reverse to determine contraction.

Table 10.1 presents some typical values of volumetric contraction for different casting metals due to solidification shrinkage and solid contraction—steps (2) and (3). Solidification shrinkage occurs in nearly all metals because the solid phase has a higher density than the liquid phase. The phase transformation that accompanies solidification causes a reduction in the volume per unit weight of metal. The exception in Table 10.1 is cast iron containing high carbon content, whose solidification is complicated by a period of graphitization during the final stages of freezing, which causes expansion that tends to counteract the volumetric decrease associated with the phase change [6].

Pattern makers account for solidification shrinkage and thermal contraction by making the mold cavities oversized. The amount by which the mold must be made larger relative to the final casting size is called the **pattern shrinkage allowance.** Although the shrinkage is volumetric, the dimensions of the casting are almost always expressed linearly, so the allowances must be applied accordingly. Special "shrink rules" with slightly elongated scales are used to make the patterns and molds larger than the desired casting

TABLE 10.1 Volumetric contraction for different casting metals due to solidification shrinkage and solid contraction.

Metal	Volumetric contraction due to:	
	Solidification Shrinkage, %	Solid Thermal Contraction, %
Aluminum	7.0	5.6
Al alloy (typical)	7.0	5.0
Gray cast iron	1.8	3.0
Gray cast iron, high C	0	3.0
Low C cast steel	3.0	7.2
Copper	4.5	7.5
Bronze (Cu-Sn)	5.5	6.0

Compiled from [3].

by the appropriate amount. Depending on the metal to be cast, these shrink rules vary in elongation from less than 1% to more than 5% compared to a standard rule.

10.3.4 Directional Solidification

In order to minimize the damaging effects of shrinkage, it is desirable for the regions of the casting most distant from the liquid metal supply to freeze first and for solidification to progress from these remote regions toward the riser(s). In this way, molten metal will continually be available from the risers to prevent shrinkage voids during freezing. The term *directional solidification* is used to describe this aspect of the freezing process and the methods by which it is controlled. The desired directional solidification is achieved by observing Chvorinov's Rule in the design of the casting itself, its orientation within the mold, and the design of the riser system that feeds it. For example, by locating sections of the casting with lower V/A ratios away from the riser, freezing will occur first in these regions and the supply of liquid metal for the rest of the casting will remain open until these bulkier sections solidify.

Another way to encourage directional solidification is to use *chills*—internal or external heat sinks that cause rapid freezing in certain regions of the casting. *Internal chills* are small metal parts placed inside the cavity before pouring so that the molten metal will solidify first around these parts. The internal chill should have a chemical composition that is approximately the same as the metal being poured. This can be achieved by making the chill out of the same metal as the casting itself.

External chills are metal inserts in the walls of the mold cavity that can remove heat from the molten metal more rapidly than the surrounding sand in order to promote solidification. They are often used effectively in sections of the casting that are difficult to feed with liquid metal, thus encouraging rapid freezing in these sections while the connection to liquid metal is still open. Figure 10.9 illustrates a possible application of external chills and the likely result in the casting if the chill were not used.

As important as it is to initiate freezing in the appropriate regions of the cavity, it is also important to avoid premature solidification in sections of the mold nearest the riser. Of particular concern is the passageway between the riser and the main cavity. This connection must be designed in such a way that it does not freeze before the casting, thus isolating the molten metal in the riser. Although it is generally desirable to minimize the volume in the connection (to reduce wasted metal), the cross-sectional area must be sufficient to delay the onset of freezing. This goal is usually aided by making the passageway short in length, so that it absorbs heat from the molten metal in the riser and the casting.

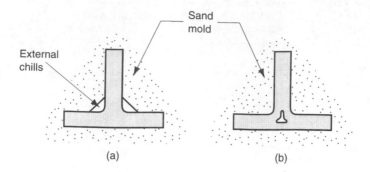

(a)   (b)

10.3.5 Riser Design

As described earlier, a riser, Figure 10.2(b), is used in a sand-casting mold to feed liquid metal to the casting during freezing in order to compensate for solidification shrinkage. To function, the riser must remain molten until after the casting solidifies. Chvorinov's Rule can be used to compute the size of a riser that will satisfy this requirement. The following example illustrates the calculation.

EXAMPLE 10.3
Riser design using Chvorinov's Rule.

A cylindrical riser must be designed for a sand-casting mold. The casting itself is a steel rectangular plate with dimensions 7.5 cm × 12.5 cm × 2.0 cm. Previous observations have indicated that the total solidification time (TST) for this casting = 1.6 min. The cylinder for the riser will have a diameter-to-height ratio = 1.0. Determine the dimensions of the riser so that its TST = 2.0 min.

Solution: First determine the V/A ratio for the plate. Its volume $V = 7.5 \times 12.5 \times 2.0 = 187.5 \, cm^3$, and its surface area A = $2(7.5 \times 12.5 + 7.5 \times 2.0 + 12.5 \times 2.0) = 267.5$ cm^2. Given that TST = 1.6 min, we can determine the mold constant C_m from Eq. (10.7), using a value of $n = 2$ in the equation.

$$C_m = \frac{TST}{(V/A)^2} = \frac{1.6}{(187.5/267.5)^2} = 3.26 \text{ min/cm}^2$$

Next we must design the riser so that its total solidification time is 2.0 min, using the same value of mold constant since both the casting and the riser are in the same mold. The volume of the riser is given by

$$V = \frac{\pi D^2 h}{4}$$

and the surface area is given by $A = \pi Dh + \dfrac{2\pi D^2}{4}$

Since we are using a D/H ratio = 1.0, then $D = H$. Substituting D for H in the volume and area formulas, we get

$$V = \frac{\pi D^3}{4}$$

and $A = \pi D^2 + \dfrac{2\pi D^2}{4} = 1.5\pi D^2$

Thus, the V/A ratio = $D/6$. Using this ratio in Chvorinov's equation, we have

$$TST = 2.0 = 3.26\left(\frac{D}{6}\right)^2 = 0.09056\ D^2$$

$$D^2 = 2.0/0.09056 = 22.086\ \text{cm}^2$$

$$D = 4.7\ \text{cm}$$

Since $H = D$, then $H = 4.7$ cm also.

The riser represents waste metal that will be separated from the cast part and remelted to make subsequent castings. It is desirable for the volume of metal in the riser to be a minimum. Since the geometry of the riser is normally selected to maximize the V/A ratio, this tends to reduce the riser volume as much as possible. Note that the volume of the riser in our example problem is $V = \pi(4.7)^3/4 = 81.5$ cm^3, only 44% of the volume of the plate (casting), even though its total solidification time is longer by 25%.

Risers can be designed in different forms. The design shown in Figure 10.2(b) is a *side riser*. It is attached to the side of the casting by means of a small channel. A *top riser* is one that is connected to the top surface of the casting. Risers can be open or blind. An *open riser* is exposed to the outside at the top surface of the cope. This has the disadvantage of allowing more heat to escape, promoting faster solidification. A *blind riser* is entirely enclosed within the mold, as in Figure 10.1(b).

REFERENCES

[1] Amstead, B. H., Ostwald, P. F., and Begeman, M. L. *Manufacturing Processes.* Wiley, New York, 1987.

[2] Beeley, P. R. *Foundry Technology.* Newnes-Butterworths, London, 1972.

[3] Datsko, J. *Material Properties and Manufacturing Processes.* Wiley, New York, 1966.

[4] Edwards, L., and Endean, M. *Manufacturing With Word Materials.* Open University, Milton Keynes, and Butterworth Scientific Ltd., London, 1990.

[5] Flinn, R. A. *Fundamentals of Metal Casting.* American Foundrymen's Society, Inc., Des Plaines, Ill., 1987.

[6] Heine, R. W., Loper, Jr., C. R., and Rosenthal, C. *Principles of Metal Casting.* 2nd ed. McGraw-Hill, New York, 1967.

[7] Kotzin, E. L. (ed.). *Metalcasting and Molding Processes.* American Foundrymen's Society, Inc., Des Plaines, Ill., 1981.

[8] *Metals Handbook.* 9th ed. Volume 15: *Casting.* American Society for Metals, Metals Park, Oh., 1988.

[9] Mikelonis, P. J. (ed.). *Foundry Technology.* American Society for Metals, Metals Park, Oh., 1982.

[10] Niebel, B. W., Draper, A. B., Wysk, R. A. *Modern Manufacturing Process Engineering.* McGraw-Hill, New York, 1989.

[11] Simpson, B. L. *History of the Metalcasting Industry.* 2nd ed. American Foundrymen's Society, Inc., Des Plaines, Ill., 1970.

[12] Taylor, H. F., Flemings, M. C., and Wulff, J. *Foundry Engineering.* 2nd ed. American Foundrymen's Society, Inc., Des Plaines, Ill., 1987.

[13] Wick, C., Benedict, J. T., and Veilleux, R. F. *Tool and Manufacturing Engineers Handbook.* 4th ed. Vol. II, *Forming.* Society of Manufacturing Engineers, Dearborn, Mich., 1984.

REVIEW QUESTIONS

10.1. Identify some of the important advantages of shape-casting processes.

10.2. What are some of the limitations and disadvantages of casting?

10.3. What is a factory that performs casting operations called?

10.4. What is the difference between an open mold and a closed mold?

10.5. Name the two basic mold types that distinguish casting processes.

10.6. Which casting process is the most important commercially?

10.7. What is the difference between a pattern and a core in sand molding?

10.8. What is meant by the term *superheat*?

10.9. Why should turbulent flow of molten metal into the mold be avoided?

10.10. What is the *continuity law* as it applies to the flow of molten metal in casting?

10.11. What are some of the factors affecting the fluidity of a molten metal during pouring into a mold cavity?

10.12. What does *heat of fusion* mean in casting?

10.13. How does solidification of alloys differ from solidification of pure metals?

10.14. What is a eutectic alloy?

10.15. What is the relationship known as *Chvorinov's Rule* in casting?

10.16. Identify the three sources of contraction in a metal casting after pouring.

10.17. What is a *chill* in casting?

MULTIPLE CHOICE QUIZ

There is a total of 13 correct answers in the following multiple choice questions (some questions have multiple answers that are correct). To attain a perfect score on the quiz, all correct answers must be given, since each correct answer is worth 1 point. For each question, each omitted answer or wrong answer reduces the score by 1 point, and each additional answer beyond the number of answers required reduces the score by 1 point. Percentage score on the quiz is based on the total number of correct answers.

10.1. Sand casting is which of the following types? (a) expendable mold or (b) permanent mold.

10.2. The upper half of a sand casting mold is called which of the following? (a) cope or (b) drag.

10.3. In casting, a flask is which one of the following? (a) beverage bottle for foundrymen, (b) box that holds the cope and drag, (c) container for holding liquid metal, or (d) metal that extrudes between the mold halves.

10.4. In foundry work, a runner is which one of the following? (a) channel in the mold leading from the downsprue to the main mold cavity, (b) foundryman who moves the molten metal to the mold, or (c) vertical channel into which molten metal is poured into the mold.

10.5. Total solidification time is defined as which one of the following? (a) time between pouring and complete solidification, (b) time between pouring and cooling to room temperature, (c) time between solidification and cooling to room temperature, or (d) time to give up the heat of fusion.

10.6. During solidification of an alloy when a mixture of solid and liquid metals are present, the solid–liquid mixture is referred to as which one of the following? (a) eutectic composition, (b) ingot segregation, (c) liq-

uidus, (d) mushy zone, or (e) solidus.

10.7. Chvorinov's Rule states that total solidification time is proportional to which one of the following quantities? (a) $(A/V)^n$, (b) H_f, (c) T_m, (d) V, (e) V/A, or (f) $(V/A)^2$; where A = surface area of casting, H_f = heat of fusion, T_m = melting temperature, and V = volume of casting.

10.8. A riser in casting is described by which of the following (more than one)? (a) an insert in the casting that inhibits buoyancy of the core, (b) gating system in which the sprue feeds directly into the cavity, (c) metal that is not part of the casting, (d) source of molten metal to feed the casting and compensate for shrinkage during solidification, and (e) waste metal that is usually recycled.

10.9. In a sand casting mold, the V/A ratio of the riser should be which one of the following relative to the V/A ratio of the casting itself? (a) equal, (b) greater, or (c) smaller.

10.10. A riser that is completely enclosed within the sand mold and connected to the main cavity by a channel to feed the molten metal is called which of the following (more than one)? (a) blind riser, (b) open riser, (c) side riser, and (d) top riser.

PROBLEMS

Heating and Pouring

10.1. A disk 40 cm in diameter and 5 cm thick is to be casted of pure aluminum in an open mold operation. The melting temperature of aluminum is 660°C, and the pouring temperature will be 800°C. Assume that the amount of aluminum heated will be 5% more than needed to fill the mold cavity. Compute the amount of heat that must be added to the metal to heat it to the pouring temperature, starting from a room temperature of 25°C. The heat of fusion of aluminum = 389.3 J/g. Other properties can be obtained from Tables 4.1 and 4.2. Assume that the specific heat has the same value for solid and molten aluminum.

10.2. A sufficient amount of pure copper is to be heated for casting a large plate in an open mold. The plate has dimensions: $L = 20$ in., $W = 10$ in., and $D = 3$ in. Compute the amount of heat that must be added to the metal to heat it to a temperature of 2150°F for pouring. Assume that the amount of metal heated will be 10% more than needed to fill the mold cavity. Properties of the metal are: density = 0.324 lbm/in.^{3}, melting point = 1981°F, specific heat of the metal = 0.093 Btu/lbm-°F in the solid state and 0.090 Btu/lbm-°F in the liquid state, and heat of fusion = 80 Btu/lbm.

10.3. The downsprue leading into the runner of a certain mold has a length of 175 mm. The cross-sectional area at the base of the sprue is 400 mm^2. The mold cavity has a volume of 0.001 m^3. Determine (a) the velocity of the molten metal flowing through the base of the downsprue, (b) the volume rate of flow, and (c) the time required to fill the mold cavity.

10.4. A mold has a downsprue of length = 6.0 in. The cross-sectional area at the bottom of the sprue is 0.5 in.^{2}. The sprue leads into a horizontal runner which feeds the mold cavity, whose volume = 75 in.^{3}. Determine (a) the velocity of the molten metal flowing through the base of the downsprue, (b) the volume rate of flow, and (c) the time required to fill the mold cavity.

10.5. The flow rate of liquid metal into the downsprue of a mold = 1 liter/sec. The cross-sectional area at the top of the sprue = 800 mm^2 and its length = 175 mm. What area should be used at the base of the sprue to avoid aspiration of the molten metal?

10.6. The volume rate of flow of molten metal into the downsprue from the pouring cup is 50 in.^{3}/sec. At the top where the pouring cup leads into the downsprue, the cross-sectional area is 1.0 in.^{2}. Determine what the area should be at the bottom of the sprue if its length is 8.0 in. It is desired to maintain a constant flow rate, top and bottom, in order to avoid aspiration of the liquid metal.

10.7. Molten metal can be poured into the pouring cup of a sand mold at a steady rate of 1000 cm^3/s. The molten metal overflows the pouring cup and flows into the downsprue. The cross section of the sprue is round, with a diameter at the top = 3.4 cm. If the sprue is 25 cm long, determine the proper diameter at its base so as to maintain the same volume flow rate.

10.8. During pouring into a sand mold, the molten metal can be poured into the downsprue at a constant flow rate during the time it takes to fill the mold. At the end of pouring the sprue is filled and there is negligible metal in the pouring cup. The downsprue is 6.0 in. long. Its cross-sectional area at the top is 0.8 in.^{2} and at the base = 0.6 in.^{2}. The cross-sectional area of the runner leading from the sprue also = 0.6 in.^{2}, and it is 8.0 in. long before leading into the mold cavity, whose volume = 65 in.^{3}. The volume of the riser located along the runner near the mold cavity = 25 in.^{3}. It takes a total of 3.0 sec to fill the entire mold (including cavity, riser, runner, and sprue). This is more than the theoretical time required, indicating a loss of velocity due to friction in the sprue and runner. Find (a) the theoretical velocity and flow rate at the base of the downsprue; (b) the total volume of the mold; (c) the actual velocity and flow rate at the base of the sprue; and (d) the loss of head in the gating system due to friction.

Shrinkage

10.9. A mold cavity has the shape of a cube, 100 mm on a side. Determine the dimensions and volume of the final cube after cooling to room temperature if the cast metal is copper. Assume that the mold is full at the start of solidification and that shrinkage occurs uniformly in all directions.

10.10. The cavity of a casting mold has dimensions: $L = 250$ mm, $W = 125$ mm, and $H = 20$ mm. Determine the dimensions of the final casting after cooling to room temperature if the cast metal is aluminum. Assume that the mold is full at the start of solidification and that shrinkage occurs uniformly in all directions.

10.11. Determine the scale of a "shrink rule" that is to be used by pattern makers for low carbon steel. Express your answer in terms of decimal fraction inches of elongation per foot of length compared to a standard rule.

10.12. Determine the scale of a "shrink rule" that is to be used by pattern makers for brass that is 70% copper and 30% zinc. Express your answer in terms of millimeters of elongation per meter of length compared to a standard rule.

10.13. Determine the scale of a "shrink rule" that is to be used by pattern makers for gray cast iron. The gray cast iron has a volumetric contraction of −2.5%, which means it expands during solidification. Express your answer in terms of millimeters of elongation per meter of length compared to a standard rule.

10.14. The final dimensions of a disk-shaped casting of 1.0% carbon steel are: diameter = 12.0 in. and thickness = 0.75 in. Determine the dimensions of the mold cavity to take shrinkage into account. Assume that shrinkage occurs uniformly in all directions.

Solidification Time and Riser Design

10.15. In the casting of steel under certain mold conditions, the mold constant in Chvorinov's Rule is known to be $C_m = 4.0$ min/cm^2, based on previous experience. The casting is a flat plate whose length = 30 cm, width = 10 cm, and thickness = 20 mm. Determine how long it will take for the casting to solidify.

10.16. Solve for total solidification time in the previous problem only using a value of $n = 1.9$ in Chvorinov's Rule. What adjustment must be made in the units of C_m?

10.17. A disk-shaped part is to be cast out of aluminum. The diameter of the disk = 500 mm and its thickness = 20 mm. If $C_m = 2.0$ sec/mm^2 in Chvorinov's Rule, how long will it take the casting to solidify?

10.18. In casting experiments performed using a certain alloy and type of sand mold, it took 155 sec for a cube-shaped casting to solidify. The cube was 50 mm on a side. (a) Determine the value of the mold constant C_m in Chvorinov's Rule. (b) If the same alloy and mold type were used, find the total solidification time for a cylindrical casting in which the diameter is 30 mm and length is 50 mm.

10.19. A steel casting has a cylindrical geometry with 4.0 in. diameter and weighs 20 lb. This casting takes 6.0 min to completely solidify. Another cylindrical-shaped casting with the same diameter-to-length ratio weighs 12 lb. This casting is made of the same steel, and the same conditions of mold and pouring were used. Determine (a) the mold constant in Chvorinov's Rule, (b) the dimensions, and (c) the total solidification time of the lighter casting. *Note*: The density of steel = 490 lb/ft^3.

10.20. The total solidification times of three casting shapes are to be compared: (1) a sphere with diameter = 10 cm, (2) a cylinder with diameter and length both = 10 cm, and (3) a cube with each side = 10 cm. The same casting alloy is used in the three cases. (a) Determine the relative solidification times for each geometry. (b) Based on the results of part (a), which geometric element would make the best riser? (c) If $C_m = 3.5$ min/cm^2 in Chvorinov's Rule, compute the total solidification time for each casting.

10.21. The total solidification times of three casting shapes are to be compared: (1) a sphere, (2) a cylinder, in which the L/D ratio = 1.0, and (3) a cube. For all three geometries, the volume $V = 1000$ cm^3. The same casting alloy is used in the three cases. (a) Determine the relative solidification times for each geometry. (b) Based on the results of part (a), which geometric element would make the best riser? (c) If $C_m = 3.5$ min/cm^2 in Chvorinov's Rule, compute the total solidification time for each casting.

10.22. A cylindrical riser is to be used for a sand casting mold. For a given cylinder volume, determine the diameter-to-length ratio that will maximize the time to solidify.

10.23. A riser in the shape of a sphere is to be designed for a sand casting mold. The casting is a rectangular plate, with length = 200 mm, width = 100 mm, and thickness = 18 mm. If the total solidification time of the casting itself is known to be 3.5 min, determine the diameter of the riser so that it will take 25% longer for the riser to solidify.

10.24. A cylindrical riser is to be designed for a sand casting mold. The length of the cylinder is to be 1.25 times its diameter. The casting is a square plate, each side = 10 in. and thickness = 0.75 in. If the metal is cast iron, and $C_m = 16.0$ min/in.^{2} in Chvorinov's Rule, determine the dimensions of the riser so that it will take 30% longer to solidify.

10.25. A cylindrical riser with diameter-to-length ratio = 1.0 is to be designed for a sand casting mold. The casting geometry is illustrated in Figure P10.25, in which units are inches. If $C_m = 19.5$ min/in.^{2} in Chvorinov's Rule, determine the dimensions of the riser so that the riser will take 0.5 minute longer to freeze than the casting itself.

FIGURE P10.25   Casting geometry for Problem 10.25 (units are inches).

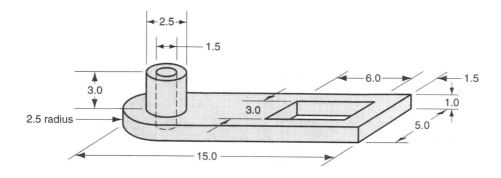

11

METAL CASTING PROCESSES

CHAPTER CONTENTS

11.1 Sand Casting
    11.1.1 Patterns and Cores
    11.1.2 Molds and Mold Making
    11.1.3 The Casting Operation
11.2 Other Expendable-Mold-Casting Processes
    11.2.1 Shell Molding
    11.2.2 Vacuum Molding
    11.2.3 Expanded Polystyrene Process
    11.2.4 Investment Casting
    11.2.5 Plaster-Mold and Ceramic-Mold Casting
11.3 Permanent-Mold-Casting Processes
    11.3.1 The Basic Permanent-Mold Process
    11.3.2 Variations of Permanent-Mold Casting
    11.3.3 Die Casting
    11.3.4 Centrifugal Casting
11.4 Foundry Practice
    11.4.1 Furnaces
    11.4.2 Pouring, Cleaning, and Heat Treatment
11.5 Casting Quality
11.6 Metals for Casting
11.7 Product Design Considerations

Metal casting processes divide into two categories, according to type of mold: (1) expendable mold and (2) permanent mold. In expendable mold casting operations, the mold must be sacrificed in order to remove the cast part. Since a new mold is required for each new casting, production rates in expendable mold processes are often limited by the time required to make the mold rather than the time to make the casting itself. However, for certain part geometries, sand molds can be produced and castings made at rates of 400 parts per hour and higher. In permanent mold casting processes, the mold is fabricated out of metal (or other durable material) and can be used many times to make many castings. Accordingly, these processes possess a natural advantage in terms of higher production rates.

    Our discussion of casting processes in this chapter is organized as follows: (1) sand casting, (2) other expendable-mold-casting processes, and (3) permanent-mold-casting processes. The chapter also includes casting equipment and practices used in foundries. Another section deals with inspection and quality issues. Product design guidelines are presented in the final section.

11.1 SAND CASTING

Sand casting is by far the most widely used casting process, accounting for a significant majority of the total tonnage cast. Nearly all casting alloys can be sand casted; indeed, it is one of the few processes that can be used for metals with high melting temperatures, such as steels, nickels, and titaniums. Its versatility permits the casting of parts ranging in size from small to very large (Figure 11.1) and in production quantities from one to millions.

Sand casting consists of pouring molten metal into a sand mold, allowing the metal to solidify, and then breaking up the mold to remove the casting. The casting must then be cleaned and inspected, and heat treatment is sometimes required to improve metallurgical properties. The cavity in the sand mold is formed by packing sand around a pattern (an approximate duplicate of the part to be cast), and then removing the pattern by separating the mold into two halves. The mold also contains the gating and riser system. In addition, if the casting is to have internal surfaces (e.g., hollow parts or parts with holes), a core must be included in the mold. Since the mold is sacrificed to remove the casting, a new sand mold must be made for each part that is produced. From this brief description, sand casting is seen to include not only the casting operation itself, but also

FIGURE 11.1 A large sand casting weighing more than 680 kg (1500 lb) for an air compressor frame (courtesy Elkhart Foundry, photo by Paragon Inc., Elkhart, Indiana).

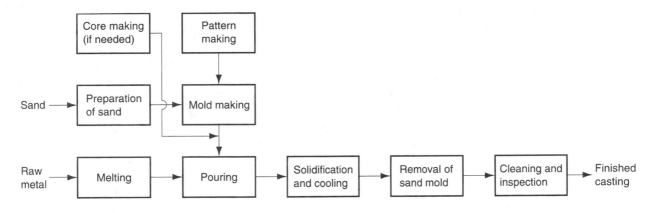

FIGURE 11.2   Steps in the production sequence in sand casting. The steps include not only the casting operation but also pattern making and mold making.

the fabrication of the pattern and the making of the mold. The production sequence is outlined in Figure 11.2.

Our discussion in the following sections deals with: patterns, cores, molds and mold making, the casting operation, and cleaning and inspection.

11.1.1 Patterns and Cores

Sand casting requires a ***pattern***—a full-sized model of the part, enlarged to account for shrinkage and machining allowances in the final casting. Materials used to make patterns include wood, plastics, and metals. Wood is a common pattern material because it is easily worked into shape. Its disadvantages are that it tends to warp, and it is abraded by the sand being compacted around it, thus limiting the number of times it can be reused. Metal patterns are more expensive to make, but they last much longer. Plastics represent a compromise between wood and metal. Selection of the appropriate pattern material depends to a large extent on the total quantity of castings to be made.

There are various types of patterns, as illustrated in Figure 11.3. The simplest is made of one piece, called a ***solid pattern***—same geometry as the casting, adjusted in size for shrinkage and machining. Although it is the easiest pattern to fabricate, it is not the

FIGURE 11.3   Types of patterns used in sand casting: (a) solid pattern, (b) split pattern, (c) match-plate pattern, and (d) cope-and-drag pattern.

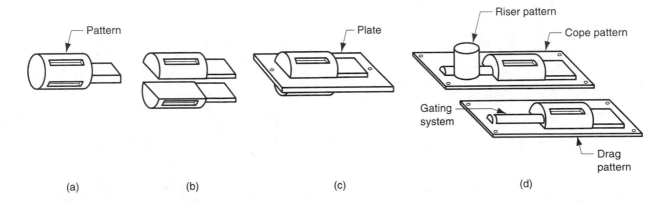

(a)   (b)   (c)   (d)

easiest to use in making the sand mold. Determining the location of the parting line between the two halves of the mold for a solid pattern can be a problem, and incorporating the gating system and sprue into the mold is left to the judgment and skill of the foundry worker. Consequently, solid patterns are generally limited to very low production quantities.

Split patterns consist of two pieces, dividing the part along a plane coinciding with the parting line of the mold. Split patterns are appropriate for complex part geometries and moderate production quantities. The parting line of the mold is predetermined by the two pattern halves, rather than by operator judgment.

For higher production quantities, match-plate patterns or cope-and-drag patterns are used. In *match-plate* patterns, the two pieces of the split pattern are attached to opposite sides of a wood or metal plate. Holes in the plate allow the top and bottom (cope and drag) sections of the mold to be aligned accurately. *Cope-and-drag patterns* are similar to match-plate patterns except that split pattern halves are attached to separate plates, so that the cope-and-drag sections of the mold can be fabricated independently, instead of using the same tooling for both. Part (d) of the figure includes the gating and riser system in the cope-and-drag patterns.

Patterns define the external shape of the cast part. If the casting is to have internal surfaces, a core is required. A *core* is a full-scale model of the interior surfaces of the part. It is inserted into the mold cavity prior to pouring, so that the molten metal will flow and solidify between the mold cavity and the core to form the casting's external and internal surfaces. The core is usually made of sand, compacted into the desired shape. As with the pattern, the actual size of the core must include allowances for shrinkage and machining. Depending on the geometry of the part, the core may or may not require supports to hold it in position in the mold cavity during pouring. These supports, called *chaplets,* are made of a metal with a higher melting temperature than the casting metal. For example, steel chaplets would be used for cast iron castings. On pouring and solidification, the chaplets become bonded into the casting. A possible arrangement of a core in a mold using chaplets is sketched in Figure 11.4. The portion of the chaplet protruding from the casting is subsequently cut off.

11.1.2 Molds and Mold Making

Foundry sands are silica (SiO$_2$) or silica mixed with other minerals. The sand should possess good refractory properties—capacity to stand up under high temperatures without

FIGURE 11.4   (a) Core held in place in the mold cavity by chaplets, (b) possible chaplet design, and (c) casting with internal cavity.

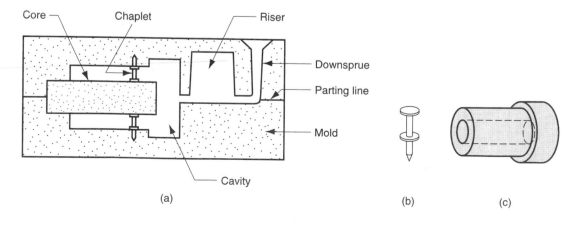

(a)   (b)   (c)

melting or otherwise degrading. Other important features of the sand include grain size, distribution of grain size in the mixture, and shape of the individual grains (Section 16.1). Small grain size provides a better surface finish on the cast part, but large grain size is more permeable (to allow escape of gases during pouring). Molds made from grains of irregular shape tend to be stronger than molds of round grains because of interlocking, yet interlocking tends to restrict permeability.

In making the mold, the grains of sand are held together by a mixture of water and bonding clay. A typical mixture (by volume) is 90% sand, 3% water, and 7% clay. Other bonding agents can be used in place of clay, including organic resins (e.g., phenolic resins) and inorganic binders (e.g., sodium silicate and phosphate). Besides sand and binder, additives are sometimes combined with the mixture to enhance properties such as strength and/or permeability of the mold.

To form the mold cavity, the traditional method is to compact the molding sand around the pattern for both cope and drag in a container called a ***flask.*** The packing process is performed by various methods. The simplest is hand ramming, accomplished manually by a foundry worker. In addition, various machines have been developed to mechanize the packing procedure. These machines operate by any of several mechanisms, including (1) squeezing the sand around the pattern by pneumatic pressure; (2) a jolting action in which the sand, contained in the flask with the pattern, is dropped repeatedly in order to pack it into place; and (3) a slinging action, in which the sand grains are impacted against the pattern at high speed.

An alternative to traditional flasks for each sand mold is ***flaskless molding,*** which refers to the use of one master flask in a mechanized system of mold production. Each sand mold is produced using the same master flask. Mold production rates up to 600 per hour are claimed for this more automated method [6].

Several indicators are used to determine the quality of the sand mold [5]: (1) ***strength***—the mold's ability to maintain its shape and resist erosion caused by the flow of molten metal; it depends on grain shape, adhesive qualities of the binder, and other factors; (2) ***permeability***—capacity of the mold to allow hot air and gases from the casting operation to pass through the voids in the sand; (3) ***thermal stability***—ability of the sand at the surface of the mold cavity to resist cracking and buckling upon contact with the molten metal; (4) ***collapsibility***—ability of the mold to give way and allow the casting to shrink without cracking the casting; it also refers to the ability to remove the sand from the casting during cleaning; and (5) ***reusability***—can the sand from the broken mold be reused to make other molds? These measures are sometimes incompatible; for example, a mold with greater strength is less collapsible.

Sand molds are often classified as green-sand, dry-sand, or skin-dried molds. ***Green-sand molds*** are made of a mixture of sand, clay, and water, the word *green* referring to the fact that the mold contains moisture at the time of pouring. Green-sand molds possess sufficient strength for most applications, good collapsibility, good permeability, good reusability, and are the least expensive of the molds. They are the most widely used mold type, but they are not without problems. Moisture in the sand can cause defects in some castings, depending on the metal and geometry of the part. A ***dry-sand mold*** is made using organic binders rather than clay, and the mold is baked in a large oven at temperatures ranging from 200°C to 320°C (400°F to 600°F) [6]. Oven baking strengthens the mold and hardens the cavity surface. A dry-sand mold provides better dimensional control in the cast product, compared to green-sand molding. However, dry-sand molding is more expensive, and production rate is reduced because of drying time. Applications are generally limited to medium and large castings in low to medium production rates. In a ***skin-dried mold,*** the advantages of a dry-sand mold is partially achieved by drying the surface of a green-sand mold to a depth of 10 to 25 mm (0.4 to 1 in.) at the mold cavity surface, using torches, heating lamps, or other means. Special bonding materials must be added to the sand mixture to strengthen the cavity surface.

The preceding mold classifications refer to the use of conventional binders consisting of either clay-and-water or ones that require heating to cure. In addition to these classifications, chemically bonded molds have been developed that are not based on either of these traditional binder ingredients. Some of the binder materials used in these so-called "no-bake" systems include furan resins (consisting of furfural alcohol, urea, and formaldehyde), phenolics, and alkyd oils. No-bake molds are growing in popularity due to their good dimensional control in high production applications.

11.1.3 The Casting Operation

After the core is positioned (if one is used) and the two halves of the mold are clamped together, then casting is performed. Casting consists of pouring, solidification, and cooling of the cast part (Sections 10.2 and 10.3). The gating and riser system in the mold must be designed to deliver liquid metal into the cavity and provide for a sufficient reservoir of molten metal during solidification shrinkage. Air and gases must be allowed to escape.

One of the hazards during pouring is that the buoyancy of the molten metal will displace the core. Buoyancy results from the weight of molten metal being displaced by the core, according to Archimedes' principle. The force tending to lift the core is equal to the weight of the displaced liquid less the weight of the core itself. Expressing the situation in equation form,

$$F_b = W_m - W_c \tag{11.1}$$

where F_b = buoyancy force, N (lb); W_m = weight of molten metal displaced, N (lb); and W_c = weight of the core, N (lb). Weights are determined as the volume of the core multiplied by the respective densities of the core material (typically sand) and the metal being cast. The density of a sand core is approximately 1.6 g/cm^3 (0.058 lb/in.^{3}). Densities of several common casting alloys are given in Table 11.1.

**EXAMPLE 11.1
Buoyancy in sand casting.**

A sand core has a volume = 1875 cm^3 and is located inside a sand mold cavity. Determine the buoyancy force tending to lift the core during pouring of molten lead into the mold.

Solution: The density of the sand core is 1.6 g/cm^3. Weight of the core is 1875(1.6) = 3000 g = 3.0 kg. The density of lead, based on Table 11.1, is 11.3 g/cm^3. The weight of

TABLE 11.1   Densities of selected casting alloys.

Material	Density g/cm^3	(lb/in.^{3})	Material	Density g/cm^3	(lb/in.^{3})
Aluminum (99% = pure)	2.70	(0.098)	Cast iron, gray[a]	7.16	(0.260)
Aluminum–silicon alloy	2.65	(0.096)	Copper (99% = pure)	8.73	(0.317)
Aluminum-copper (92% Al)	2.81	(0.102)	Lead (pure)	11.30	(0.410)
Brass[a]	8.62	(0.313)	Steel	7.82	(0.284)

Source: [5].

[a] Density depends on composition of alloy; value given is typical.

lead displaced by the core is $1875(11.3) = 21{,}188$ g $= 21.19$ kg. The difference $= 21.19 - 3.0 = 18.19$ kg. Given that 1 kg $\equiv 9.81$ N, the buoyancy force is therefore $F_b = 9.81(18.19) = $ **178.4 N.**

Following solidification and cooling, the sand mold is broken away from the casting to retrieve the part. The part is then cleaned—gating and riser system is separated and sand is removed. The casting is then inspected (Section 11.5).

11.2 OTHER EXPENDABLE-MOLD-CASTING PROCESSES

As versatile as sand casting is, there are other casting processes that have been developed to meet special needs. The differences between these methods are in the composition of the mold material, or the manner in which the mold is made, or in the way the pattern is made.

11.2.1 Shell Molding

Shell molding is a casting process in which the mold is a thin shell (typically 9 mm or 3/8 in.) made of sand held together by a thermosetting resin binder. This process, developed in Germany during the early 1940s, is described and illustrated in Figure 11.5.

FIGURE 11.5   Steps in shell-molding: (1) a match-plate or cope-and-drag metal pattern is heated and placed over a box containing sand mixed with thermosetting resin; (2) box is inverted so that sand and resin fall onto the hot pattern, causing a layer of the mixture to partially cure on the surface to form a hard shell; (3) box is repositioned so that loose, uncured particles drop away; (4) sand shell is heated in oven for several minutes to complete curing; (5) shell mold is stripped from the pattern; (6) two halves of the shell mold are assembled, supported by sand or metal shot in a box, and pouring is accomplished. The finished casting with sprue removed is shown in (7).

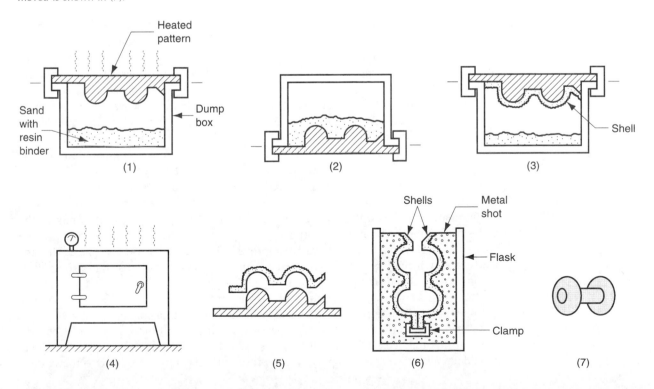

There are many advantages to the shell-molding process. The surface of the shell-mold cavity is smoother than a conventional green-sand mold, and this smoothness permits easier flow of molten metal during pouring and better surface finish on the final casting. Finishes of 2.5 μm (100 μ-in.) can be obtained. Good dimensional accuracy is also achieved, with tolerances of ± 0.25 mm (± 0.010 in.) possible on small- to medium-sized parts. The good finish and accuracy often precludes the need for further machining. Collapsibility of the mold is generally sufficient to avoid tearing and cracking of the casting.

Disadvantages of shell molding include a more expensive metal pattern than the corresponding pattern for green-sand molding. This makes shell molding difficult to justify for small quantities of parts. Shell molding can be mechanized for mass production and is very economical for large quantities. It seems particularly suited to steel castings of less than 20 lbs. Examples of parts made using shell molding include gears, valve bodies, bushings, and camshafts.

11.2.2 Vacuum Molding

Vacuum molding, also called the **V-process,** was developed in Japan around 1970. It uses a sand mold held together by vacuum pressure rather than by a chemical binder. Accordingly, the term *vacuum* in this process refers to the making of the mold rather than the casting operation itself. The steps of the process are explained in Figure 11.6.

Sand recovery is one of several advantages of vacuum molding, since no binders are used. Also, the sand does not require extensive mechanical reconditioning normally done when binders are used in the molding sand. Since no water is mixed with the sand, moisture-related defects are absent from the product. Disadvantages of the V-process are that it is relatively slow and not readily adaptable to mechanization.

11.2.3 Expanded Polystyrene Process

The **expanded polystyrene casting process** uses a mold of sand packed around a polystyrene foam pattern, which vaporizes when the molten metal is poured into the mold. The process and variations of it are known by other names, including **lost-foam process, lost pattern process, evaporative-foam process,** and **full-mold process** (the last being a trade name). The polystyrene pattern includes the sprue, risers, and gating system, and it may also contain internal cores (if needed), thus eliminating the need for a separate core in the mold. Also, since the foam pattern itself becomes the cavity in the mold, considerations of draft and parting lines can be ignored. The mold does not have to be opened into cope-and-drag sections. The sequence in this casting process is illustrated and described in Figure 11.7. Various methods for making the pattern can be used, depending on the quantities of castings to be produced. For one-of-a-kind castings, the foam is manually cut from large strips and assembled to form the pattern. For large production runs, an automated molding operation can be set up to mold the patterns prior to making the molds for casting. The pattern is normally coated with a refractory compound to provide a smoother surface on the pattern and to improve its high temperature resistance. Molding sands usually include bonding agents. However, dry sand is used in certain processes in this group, which aids recovery and reuse.

A significant advantage for this process is that the pattern need not be removed from the mold. This simplifies and expedites mold making. In a conventional green-sand mold, two halves are required with proper parting lines, draft allowances must be provided in the mold design, cores must be inserted, and the gating and riser system must be added. With the expanded polystyrene process, these steps are built into the pattern

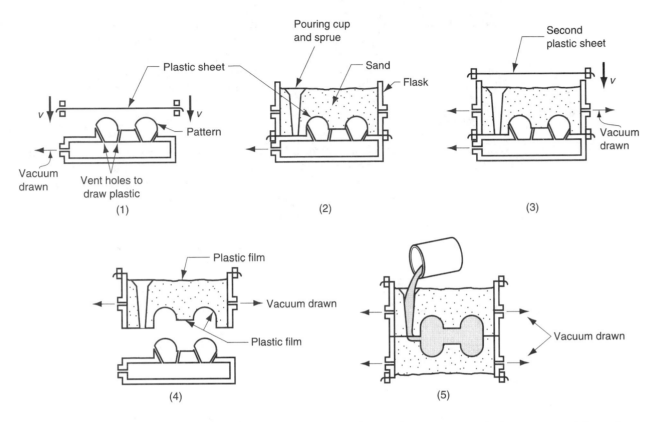

FIGURE 11.6   Steps in vacuum molding: (1) a thin sheet of preheated plastic is drawn over a match-plate or cope-and-drag pattern by vacuum—the pattern has small vent holes to facilitate vacuum forming; (2) a specially designed flask is placed over the pattern plate and filled with sand, and a sprue and pouring cup are formed in the sand; (3) another thin plastic sheet is placed over the flask, and a vacuum is drawn that causes the sand grains to be held together, forming a rigid mold; (4) the vacuum on the mold pattern is released to permit pattern to be stripped from the mold; (5) this mold is assembled with its matching half to form the cope and drag, and with vacuum maintained on both halves, pouring is accomplished. The plastic sheet quickly burns away on contacting the molten metal. After solidification, nearly all of the sand can be recovered for reuse.

FIGURE 11.7   Expanded polystyrene casting process: (1) pattern of polystyrene is coated with refractory compound; (2) foam pattern is placed in mold box, and sand is compacted around the pattern; and (3) molten metal is poured into the portion of the pattern that forms the pouring cup and sprue. As the metal enters the mold, the polystyrene foam is vaporized ahead of the advancing liquid, thus allowing the resulting mold cavity to be filled.

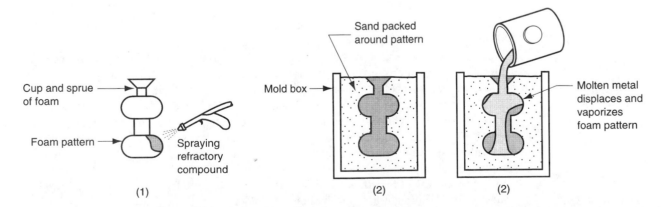

itself. The disadvantage of the process is that a new pattern is needed for every casting. The economic justification of the expanded polystyrene process is highly dependent on the cost of producing the patterns. The expanded polystyrene casting process has been applied to mass produce castings for automobiles engines. Automated production systems are installed to mold the polystyrene foam patterns for these applications.

11.2.4 Investment Casting

In *investment casting,* a pattern made of wax is coated with a refractory material to make the mold, after which the wax is melted away prior to pouring the molten metal. The term *investment* comes from one of the less familiar definitions of the word *invest,* which is "to cover completely," this referring to the coating of the refractory material around the wax pattern. It is a precision casting process, because it is capable of making castings of high accuracy and intricate detail. The process dates back to ancient Egypt (Historical Note 11.1) and is also known as the *lost-wax process,* because the wax pattern is lost from the mold prior to casting.

Historical Note 11.1   *Investment casting.*

The lost wax casting process was developed by the ancient Egyptians some 3,500 years ago. Although written records do not identify when the invention occurred or the artisan responsible, historians speculate that the process resulted from the close association between pottery and molding in early times. It was the potter who crafted the molds that were used for casting. The idea for the lost wax process must have originated with a potter who was familiar with the casting process. As he was working one day on a ceramic piece—perhaps an ornate vase or bowl—it occurred to him that the article might be more attractive and durable if made of metal. So he fashioned a core in the general shape of the piece, but smaller than the desired final dimensions, and coated it with wax to establish the size. The

wax proved to be an easy material to form, and intricate designs and shapes could be created by the craftsman. On the wax surface, he carefully plastered several layers of clay and devised a means of holding the resulting components together. He then baked the mold in a kiln, so that the clay hardened and the wax melted and drained out to form a cavity. At last, he poured molten bronze into the cavity and, after the casting had solidified and cooled, broke away the mold to recover the part. Considering the education and experience of this early pottery maker and the tools he had to work with, development of the lost wax-casting process demonstrated great innovation and insight. "No other process can be named by archeologists so crowded with deduction, engineering ability and ingenuity" [10].

Steps in investment casting are described in Figure 11.8. Since the wax pattern is melted off after the refractory mold is made, a separate pattern must be made for every casting. Pattern production is usually accomplished by a molding operation—pouring or injecting the hot wax into a *master die* that has been designed with proper allowances for shrinkage of both wax and subsequent metal casting. In cases where the part geometry is complicated, several separate wax pieces must be joined to make the pattern. In high-production operations, several patterns are attached to a sprue, also made of wax, to form a *pattern tree;* this is the geometry that will be cast out of metal.

Coating with refractory (step 3) is usually accomplished by dipping the pattern tree into a slurry of very fine-grained silica or other refractory (almost in powder form) mixed with plaster to bond the mold into shape. The small grain size of the refractory material provides a smooth surface and captures the intricate details of the wax pattern. The final mold (step 4) is accomplished by repeatedly dipping the tree into the refractory slurry or by gently packing the refractory around the tree in a container. The mold is allowed to air dry for about eight hours to harden the binder.

FIGURE 11.8   Steps in investment casting: (1) wax patterns are produced; (2) several patterns are attached to a sprue to form a pattern tree; (3) the pattern tree is coated with a thin layer of refractory material; (4) the full mold is formed by covering the coated tree with sufficient refractory material to make it rigid; (5) the mold is held in an inverted position and heated to melt the wax and permit it to drip out of the cavity; (6) the mold is preheated to a high temperature, which ensures that all contaminants are eliminated from the mold; it also permits the liquid metal to flow more easily into the detailed cavity; the molten metal is poured; it solidifies; and (7) the mold is broken away from the finished casting. Parts are separated from the sprue.

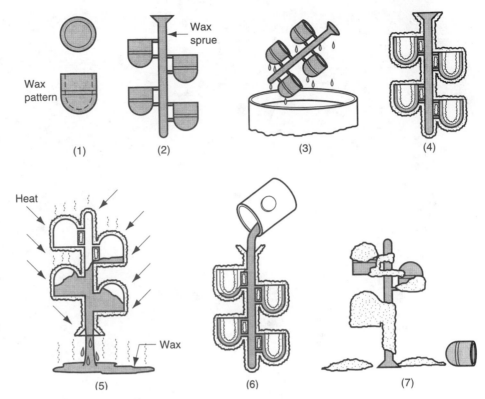

There are several advantages of investment casting: (1) parts of great complexity and intricacy can be cast; (2) close dimensional control—tolerances of ± 0.075 mm (± 0.003 in.) are possible; (3) good surface finish is possible; (4) the wax can usually be recovered for reuse; and (5) additional machining is not normally required—this is a net shape process. Because many steps are involved in this casting operation, it is a relatively expensive process. Parts made by investment casting are normally small in size, although parts with complex geometries weighing up to 75 lbs have been successfully cast. All types of metals, including steels, stainless steels, and other high temperature alloys, can be investment cast. Examples of parts include complex machinery parts, blades and other components for turbine engines, jewelry, and dental fixtures. Shown in Figure 11.9 is a part illustrating the intricate features possible with investment casting.

11.2.5 Plaster-Mold and Ceramic-Mold Casting

Plaster-mold casting is similar to sand casting except that the mold is made of plaster of Paris (gypsum—$CaSO_4$–$2H_2O$) instead of sand. Additives such as talc and silica flour are mixed with the plaster to control contraction and setting time, reduce cracking, and increase strength. To make the mold, the plaster mixture combined with water is poured over a plastic or metal pattern in a flask and allowed to set. Wood patterns are generally unsatisfactory due to the extended contact with water in the plaster. The fluid consistency permits the plaster mixture to readily flow around the pattern, capturing its details and surface finish. Thus, the cast product in plaster molding is noted for these attributes.

Curing of the plaster mold is one of the disadvantages of this process, at least in high production. The mold must set for about 20 minutes before the pattern is stripped.

FIGURE 11.9   A one-piece compressor stator with 108 separate airfoils made by investment casting (courtesy of Howmet Corp.).

The mold is then baked for several hours to remove moisture. Even with the baking, not all of the moisture content is removed from the plaster. The dilemma faced by foundrymen is that mold strength is lost when the plaster becomes too dehydrated, and yet moisture content can cause casting defects in the product. A balance must be achieved between these undesirable alternatives. Another disadvantage with the plaster mold is that it is not permeable, thus limiting escape of gases from the mold cavity. This problem can be solved in a number of ways: (1) evacuating air from the mold cavity before pouring; (2) aerating the plaster slurry prior to mold making so that the resulting hard plaster contains finely dispersed voids; and (3) using a special mold composition and treatment known as the *Antioch process.* This process involves using about 50% sand mixed with the plaster, heating the mold in an autoclave (an oven that uses superheated steam under pressure), and then drying. The resulting mold has considerably greater permeability than a conventional plaster mold.

Plaster molds cannot withstand the same high temperatures as sand molds. They are therefore limited to the casting of lower-melting-point alloys, such as aluminum, magnesium, and some copper-base alloys. Applications include metal molds for plastic and rubber molding, pump and turbine impellers, and other parts of relatively intricate geometry. Casting sizes range from about 20 g (less than one ounce) to more than 100 kg (more than 200 lb). Parts weighing less than about 10 kg (20 lb) are most common. Advantages of plaster molding for these applications are good surface finish and dimensional accuracy and the capability to make thin cross-sections in the casting.

Ceramic-mold casting is similar to plaster-mold casting except that the mold is made of refractory ceramic materials that can withstand higher temperatures than plaster. Thus, ceramic molding can be used to cast steels, cast irons, and other high-temperature alloys. Its applications (molds and relatively intricate parts) are similar to those of plaster-mold casting except for the metals cast. Its advantages (good accuracy and finish) are also similar.

11.3 PERMANENT-MOLD-CASTING PROCESSES

The economic disadvantage of any of the expendable mold processes is that a new mold is required for every casting. In permanent-mold casting, the mold is reused many times. In this section, we treat permanent-mold casting as the basic process in the group of casting processes that all use reusable metal molds. Other processes in the group include die casting and centrifugal casting.

11.3.1 The Basic Permanent-Mold Process

Permanent-mold casting uses a metal mold constructed of two sections that are designed for easy, precise opening and closing. These molds are commonly made of steel or cast iron. The cavity, with gating system included, is machined into the two halves to provide accurate dimensions and good surface finish. Metals commonly cast in permanent molds include aluminum, magnesium, copper-base alloys, and cast iron. However, cast iron requires a high pouring temperature, 1250°C to 1500°C (2300°F to 2700°F), which takes a heavy toll on mold life. The very high pouring temperatures of steel make permanent molds unsuitable for this metal, unless the mold is made of refractory material.

Cores can be used in permanent molds to form interior surfaces in the cast product. The cores can be made of metal, but either their shape must allow for removal from the casting, or they must be mechanically collapsible to permit removal. If withdrawal of a metal core would be difficult or impossible, sand cores can be used, in which case the casting process is often referred to as *semipermanent-mold casting.*

Steps in the basic permanent-mold-casting process are described in Figure 11.10. In preparation for casting, the mold is first preheated and one or more coatings are sprayed on the cavity. Preheating facilitates metal flow through the gating system and into the cavity. The coatings aid heat dissipation and lubricate the mold surfaces for easier separation of the cast product. After pouring, as soon as the metal solidifies, the mold is opened and the casting is removed. Unlike expendable molds, permanent molds do not collapse, so the mold must be opened before appreciable cooling contraction occurs in order to prevent cracks from developing in the casting.

Advantages of permanent-mold casting include good surface finish and close dimensional control, as previously indicated. In addition, more rapid solidification caused by the metal mold results in a finer grain structure, so stronger castings are produced. The process is generally limited to metals of lower melting points. Other limitations include simple part geometries compared to sand casting (because of the need to open the mold), and the expense of the mold. Because mold cost is substantial, the process is best suited to high-volume production and can be automated accordingly. Typical parts include automotive pistons, pump bodies, and certain castings for aircraft and missiles.

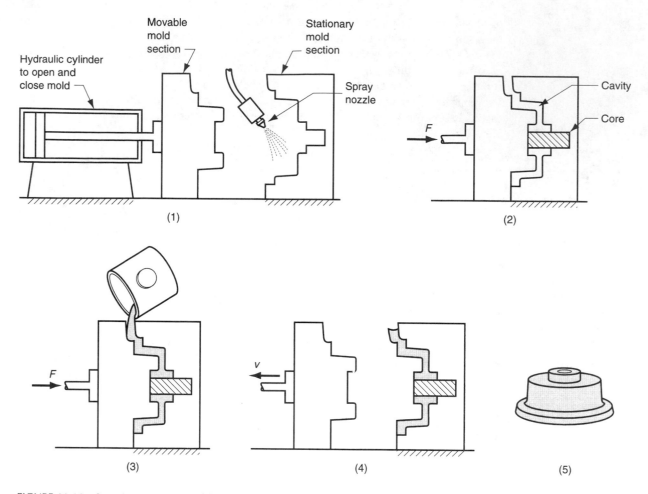

FIGURE 11.10 Steps in permanent-mold casting: (1) mold is preheated and coated; (2) cores (if used) are inserted and mold is closed; (3) molten metal is poured into the mold; and (4) mold is opened. Finished part is shown in (5).

11.3.2 Variations of Permanent-Mold Casting

Several casting processes are quite similar to the basic permanent-mold method. These include slush casting, low-pressure casting, and vacuum permanent-mold casting.

Slush Casting *Slush casting* is a permanent mold process in which a hollow casting is formed by inverting the mold after partial freezing at the surface to drain out the liquid metal in the center. Solidification begins at the mold walls because they are relatively cool, and it progresses over time toward the middle of the casting (Section 10.3.1). Thickness of the shell is controlled by the length of time allowed before draining. Slush casting is used to make statues, lamp pedestals, and toys out of low-melting-point metals such as lead, zinc, and tin. In these items, the exterior appearance is important, but the strength and interior geometry of the casting are minor considerations.

Low-Pressure Casting In the basic permanent-mold casting process and in slush casting, the flow of metal into the mold cavity is caused by gravity. In *low-pressure casting,* the liquid metal is forced into the cavity under low pressure—approximately 0.1 MPa

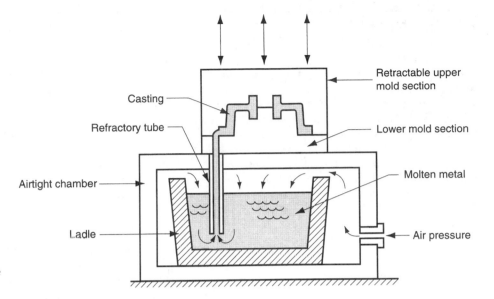

FIGURE 11.11 Low-pressure casting. The diagram shows how air pressure is used to force the molten metal in the ladle upward into the mold cavity. Pressure is maintained until the casting has solidified.

(15 lb/in.^{2})—from beneath so that the flow is upward, as illustrated in Figure 11.11. The advantage of this approach over traditional pouring is that clean molten metal from the center of the ladle is introduced into the mold, rather than metal that has been exposed to air. Gas porosity and oxidation defects are thereby minimized, and mechanical properties are improved.

Vacuum Permanent-Mold Casting   *Vacuum permanent-mold casting* (not to be confused with vacuum molding—Section 11.2.2) is a variation of low-pressure casting in which a vacuum is used to draw the molten metal into the mold cavity. The general configuration of the vacuum permanent-mold casting process is similar to the low-pressure casting operation. The difference is that reduced air pressure from the vacuum in the mold is used to draw the liquid metal into the cavity, rather than forcing it by positive air pressure from below. There are several benefits of the vacuum technique relative to low-pressure casting: air porosity and related defects are reduced, and greater strength is given to the cast product.

11.3.3 Die Casting

Die casting is a permanent-mold-casting process in which the molten metal is injected into the mold cavity under high pressure. Typical pressures are 7 to 350 MPa (1000 to 50,000 lb/in.^{2}). The pressure is maintained during solidification, after which the mold is opened and the part is removed. Molds in this casting operation are called dies; hence the name die casting. The use of high pressure to force the metal into the die cavity is the most notable feature that distinguishes this process from others in the permanent mold category.

Die-casting operations are carried out in special die-casting machines (Historical Note 11.2). Modern die-casting machines are designed to hold and accurately close the two halves of the mold, and keep them closed while the liquid metal is forced into the cavity. The general configuration is shown in Figure 11.12. There are two main types of die-casting machines: (1) hot-chamber and (2) cold-chamber, differentiated by how the molten metal is injected into the cavity.

Historical Note 11.2   *Die-casting machines.*

The modern die-casting machine has its origins in the printing industry and the need in the mid- to late 1800s to satisfy an increasingly literate population with a growing appetite for reading. The Linotype, invented and developed by O. Mergenthaler in the late 1800s, is a machine that produces printing type. It is a casting machine because it casts a line of type characters out of lead to be used in preparing printing plates. The name *linotype* derives from the fact that the machine produces a line of type characters during each cycle of operation.

The machine was first used successfully on a commercial basis in New York City by **The Tribune** in 1886.

The linotype proved the feasibility of mechanized casting machines. The first diecasting machine was patented by H. Doehler in 1905 (this machine is displayed in the Smithsonian Institute in Washington, D.C.). In 1907, E. Wagner developed the first die-casting machine to utilize the hot-chamber design. It was first used during World War I to cast parts for binoculars and gas masks.

In *hot-chamber machines,* the metal is melted in a container attached to the machine, and a piston is used to inject the liquid metal under high pressure into the die. Typical injection pressures are 7 to 35 MPa (1000 to 5000 lb/in.^{2}). The casting cycle is summarized in Figure 11.13. Production rates up to 500 parts per hour are not uncommon. Hot-chamber die casting imposes a special hardship on the injection system because much of it is submerged in the molten metal. The process is therefore limited in its applications to low-melting-point metals that do not chemically attack the plunger and other mechanical components. The metals include zinc, tin, lead, and sometimes magnesium.

In *cold-chamber die-casting machines,* molten metal is poured into an unheated chamber from an external melting container, and a piston is used to inject the metal under high pressure into the die cavity. Injection pressures used in these machines are typically 14 to 140 MPa (2000 to 20,000 lb/in.^{2}) The production cycle is explained in Figure 11.14. Compared to hot-chamber machines, cycle rates are not usually as fast because of the need to ladle the liquid metal into the chamber from an external source. Nevertheless, this casting process is a high-production operation. Cold-chamber machines are typically used for casting aluminum, brass, and magnesium alloys. Low-melting-point alloys (zinc, tin, lead) can also be cast on cold-chamber machines, but the advantages of the hot-chamber process usually favor its use on these metals.

Molds used in die-casting operations are usually made of tool steel, mold steel, or maraging steel. Tungsten and molybdenum with good refractory qualities are also being used, especially in attempts to die-cast steel and cast iron. Dies can be single-cavity or multiple-cavity. Single-cavity dies are shown in Figures 11.13 and 11.14. Ejector pins are required to remove the part from the die when it opens, as in our diagrams. These pins

FIGURE 11.12   General configuration of a (cold-chamber) die casting machine.

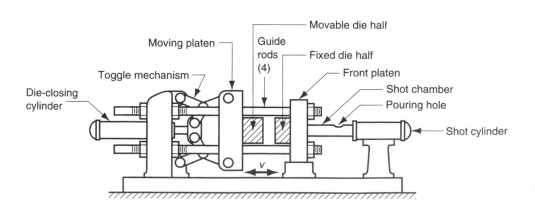

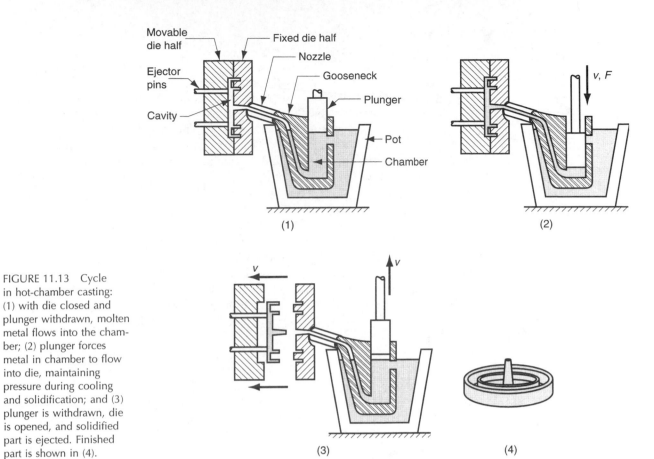

FIGURE 11.13   Cycle in hot-chamber casting: (1) with die closed and plunger withdrawn, molten metal flows into the chamber; (2) plunger forces metal in chamber to flow into die, maintaining pressure during cooling and solidification; and (3) plunger is withdrawn, die is opened, and solidified part is ejected. Finished part is shown in (4).

FIGURE 11.14   Cycle in cold-chamber casting: (1) with die closed and ram withdrawn, molten metal is poured into the chamber; (2) ram forces metal to flow into die, maintaining pressure during cooling and solidification; and (3) ram is withdrawn, die is opened, and part is ejected. (Gating system is simplified.)

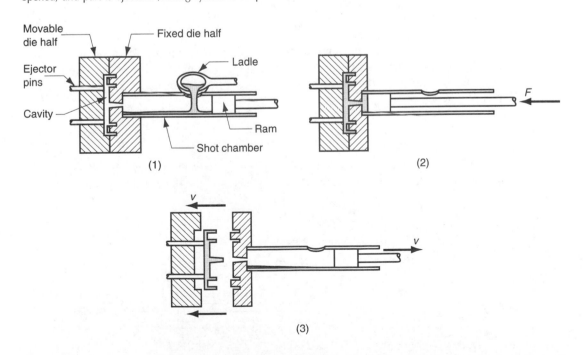

push the part away from the mold surface so that it can be removed. Lubricants must also be sprayed into the cavities to prevent sticking.

Since the die materials have no natural porosity and the molten metal rapidly flows into the die during injection, venting holes and passageways must be built into the dies at the parting line to evacuate the air and gases in the cavity. The vents are quite small; yet they fill with metal during injection. This metal must later be trimmed from the part. Also, formation of *flash* is common in die casting, in which the liquid metal under high pressure squeezes into the small space between the die halves at the parting line or into the clearances around the cores and ejector pins. This flash must be trimmed from the casting, along with the sprue and gating system.

Advantages of die casting include (1) high production rates possible; (2) economical for large production quantities; (3) close tolerances possible, on the order of ±0.076 mm (±0.003 in.) on small parts; (4) good surface finish; (5) thin sections are possible, down to about 0.5 mm (0.020 in.); and (6) rapid cooling provides small grain size and good strength to the casting. The limitation of this process, in addition to the metals cast, is the shape restriction. The part geometry must be such that it can be removed from the die cavity.

11.3.4 Centrifugal Casting

Centrifugal casting refers to several casting methods in which the mold is rotated at high speed so that centrifugal force distributes the molten metal to the outer regions of the die cavity. The group includes (1) true centrifugal casting, (2) semicentrifugal casting, and (3) centrifuge casting.

True Centrifugal Casting In true centrifugal casting, molten metal is poured into a rotating mold to produce a tubular part. Examples of parts made by this process include pipes, tubes, bushings, and rings. One possible setup is illustrated in Figure 11.15. Molten metal is poured into a horizontal rotating mold at one end. In some operations, mold rotation commences after pouring has occurred rather than beforehand. The high speed rotation results in centrifugal forces that cause the metal to take the shape of the mold cavity. Thus, the outside shape of the casting can be round, octagonal, hexagonal, and so on. However, the inside shape of the casting is (theoretically) perfectly round, due to the radially symmetric forces at work.

Orientation of the axis of mold rotation can be either horizontal or vertical, the former being more common. Let us consider how fast the mold must rotate in *horizontal centrifugal casting* for the process to work successfully. Centrifugal force is defined by this physics equation:

$$F = \frac{mv^2}{R} \tag{11.2}$$

where F = force, N (lb); m = mass, kg (lbm); v = velocity, m/s (ft/sec); and R = inside radius of the mold, m (ft). The force of gravity is its weight $W = mg$, where W is given in

FIGURE 11.15 Setup for true centrifugal casting.

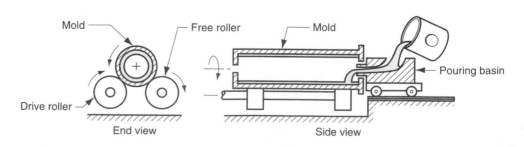

Mold — Free roller — Mold

Drive roller —

End view

Pouring basin

Side view

kg (lb), and g = acceleration of gravity, 9.8 m/s^2 (32.2 ft/sec^2). The so-called G-factor GF is the ratio of centrifugal force divided by weight:

$$GF = \frac{F}{W} = \frac{mv^2}{Rmg} = \frac{v^2}{Rg} \qquad (11.3)$$

Velocity v can be expressed as $2\pi RN/60 = \pi RN/30$, where N = rotational speed, rev/min. Substituting this expression into Eq. (11.3), we obtain

$$GF = \frac{R\left(\dfrac{\pi N}{30}\right)^2}{g} \qquad (11.4)$$

Rearranging this to solve for rotational speed N, and using diameter D rather than radius in the resulting equation, we have

$$N = \frac{30}{\pi}\sqrt{\frac{2gGF}{D}} \qquad (11.5)$$

where D = inside diameter of the mold, m (ft). If the G-factor is too low in centrifugal casting, the liquid metal will not remain forced against the mold wall during the upper half of the circular path but will "rain" inside the cavity. Slipping occurs between the molten metal and the mold wall, which means that the rotational speed of the metal is less than that of the mold. On an empirical basis, values of GF = 60 to 80 are found to be appropriate for horizontal centrifugal casting [2], although this depends to some extent on the metal being cast.

EXAMPLE 11.2 Rotation speed in true centrifugal casting.

A true centrifugal casting operation is to be performed horizontally to make copper tube sections with OD = 25 cm and ID = 22.5 cm. What rotational speed is required if a G-factor of 65 is used to cast the tubing?

Solution: The inside diameter of the mold D = OD of the casting = 25 cm = 0.25 m. We can compute the required rotational speed from Eq. (11.5) as follows:

$$N = \frac{30}{\pi}\sqrt{\frac{2(9.8)(65)}{0.25}} = \textbf{681.7 rev/min.}$$

In *vertical centrifugal casting,* the effect of gravity acting on the liquid metal causes the casting wall to be thicker at the base than at the top. The inside profile of the casting wall takes on a parabolic shape. The difference in inside radius between top and bottom is related to speed of rotation as follows:

$$N = \frac{30}{\pi}\sqrt{\frac{2gL}{R_t^2 - R_b^2}} \qquad (11.6)$$

where L = vertical length of the casting, m (ft); R_t = inside radius at the top of the casting, m (ft); and R_b = inside radius at the bottom of the casting, m (ft). Eq. (11.6) can be used to determine the required rotational speed for vertical centrifugal casting, given specifications on the inside radii at top and bottom. One can see from the formula that for R_t to equal R_b, the speed of rotation N would have to be infinite, which is impossible, of course. As a practical matter, part lengths made by vertical centrifugal casting are usually no more than about twice their diameters. This is quite satisfactory for bushings and other parts that have large diameters relatively to their lengths, especially if machining will be used to accurately size the inside diameter.

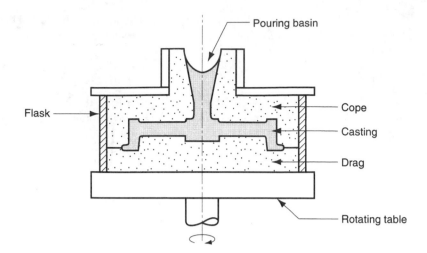

FIGURE 11.16
Semicentrifugal casting.

Castings made by true centrifugal casting are characterized by high density, especially in the outer regions of the part where F is greatest. Solidification shrinkage at the exterior of the cast tube is not a factor, because the centrifugal force continually reallocates molten metal toward the mold wall during freezing. Any impurities in the casting tend to be on the inner wall and can be removed by machining if necessary.

Semicentrifugal Casting   In this method, centrifugal force is used to produce solid castings, as in Figure 11.16, rather than tubular parts. The rotation speed in semicentrifugal casting is usually set so that G-factors of around 15 are obtained [2], and the molds are designed with risers at the center to supply feed metal. Density of metal in the final casting is greater in the outer sections than at the center of rotation. The process is often used on parts in which the center of the casting is machined away, thus eliminating the portion of the casting where the quality is lowest. Wheels and pulleys are examples of castings that can be made by this process. Expendable molds are often used in semicentrifugal casting, as suggested by our illustration of the process.

Centrifuge Casting   In centrifuge casting, Figure 11.17, the mold is designed with part cavities located away from the axis of rotation, so that the molten metal poured into

FIGURE 11.17   (a)
Centrifuge casting—centrifugal force causes metal to flow to the mold cavities away from the axis of rotation; and (b) the casting.

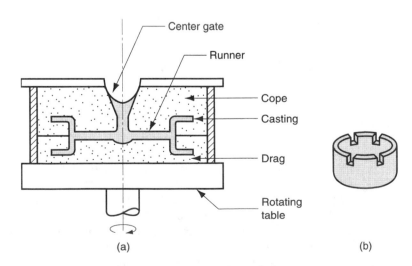

(a)   (b)

the mold is distributed to these cavities by centrifugal force. The process is used for smaller parts, and radial symmetry of the part is not a requirement as it is for the other two centrifugal casting methods.

11.4 FOUNDRY PRACTICE

In all casting processes, the metal must be heated to the molten state to be poured or otherwise forced into the mold. Heating and melting is accomplished in a furnace. This section covers various types of furnaces used in foundries and the pouring practices for delivering the molten metal from furnace to mold.

11.4.1 Furnaces

Several types of furnaces are most commonly used in foundries: (1) cupolas, (2) direct fuel-fired furnaces, (3) crucible furnaces, (4) electric-arc furnaces, and (5) induction furnaces. Selection of the most appropriate furnace type depends on factors such as the casting alloy; its melting and pouring temperatures; capacity requirements of the furnace; costs of investment, operation, and maintenance; and environmental pollution considerations.

Cupolas   A cupola is a vertical cylindrical furnace equipped with a tapping spout near its base. Cupolas are used only for melting cast irons, and although other furnaces are also used, the largest tonnage of cast iron is melted in cupolas. General construction and operating features of the cupola are illustrated in Figure 11.18. It consists of a large shell of steel plate lined with refractory. The "charge," consisting of iron, coke, flux, and possible alloying elements, is loaded through a charging door located less than halfway up the height of the cupola. The iron is usually a mixture of pig iron and scrap (including risers, runners, and sprues left over from previous castings). Coke is the fuel used to heat the furnace. Forced air is introduced through openings near the bottom of the shell for combustion of the coke. The flux is a basic compound such as limestone that reacts with coke ash and other impurities to form slag. The slag serves to cover the melt, protecting it from reaction with the environment inside the cupola and reducing heat loss. As the mixture is heated and melting of the iron occurs, the furnace is periodically tapped to provide liquid metal for the pour.

Direct Fuel-Fired Furnaces   A direct fuel-fired furnace contains a small open-hearth, in which the metal charge is heated by fuel burners located on the side of the furnace. The roof of the furnace assists the heating action by reflecting the flame down against the charge. Typical fuel is natural gas, and the combustion products exit the furnace through a stack. At the bottom of the hearth is a tap hole to release the molten metal. Direct fuel-fired furnaces are generally used in casting for melting nonferrous metals such as copper-base alloys and aluminum.

Crucible Furnaces   These furnaces melt the metal without direct contact with a burning fuel mixture. For this reason, they are sometimes called *indirect fuel-fired furnaces.* Three types of crucible furnaces are used in foundries: (a) lift-out type, (b) stationary, and (c) tilting, illustrated in Figure 11.19. They all utilize a container (the crucible) made out of a suitable refractory material (e.g., a clay-graphite mixture) or high-temperature steel alloy to hold the charge. In the *lift-out crucible furnace,* the crucible is placed in a furnace and heated sufficiently to melt the metal charge. Oil, gas, or powdered coal are

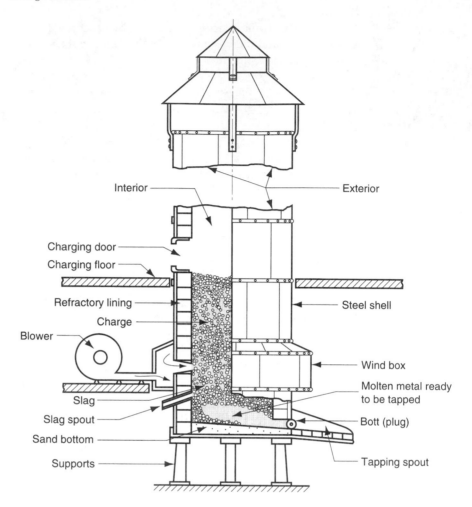

FIGURE 11.18 Cupola used for melting cast iron. Furnace shown is typical for a small foundry and omits details of emissions control system required in a modern cupola.

typical fuels for these furnaces. When the metal is melted, the crucible is lifted out of the furnace and used as a pouring ladle. The other two types, sometimes referred to as *pot furnaces,* have the heating furnace and container as one integral unit. In the *stationary pot furnace,* the furnace is stationary and the molten metal is ladled out of the container. In the *tilting-pot furnace,* the entire assembly can be tilted for pouring. Crucible furnaces

FIGURE 11.19 Three types of crucible furnaces: (a) lift-out crucible, (b) stationary pot, and (c) tilting-pot furnace.

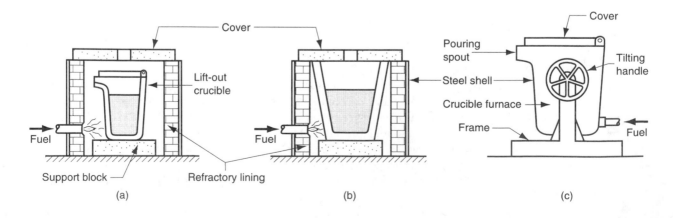

are used for nonferrous metals such as bronze, brass, and alloys of zinc and aluminum. Furnace capacities are generally limited to several hundred pounds.

Electric-Arc Furnaces   In this furnace type, the charge is melted by heat generated from an electric arc. Various configurations are available, with two or three electrodes (Figure 6.9). Power consumption is high, but electric-arc furnaces can be designed for high melting capacity (23,000 to 45,000 kg/hr or 25 to 50 tons/hr), and they are used primarily for casting steel.

Induction Furnaces   An induction furnace uses alternating current passing through a coil to develop a magnetic field in the metal, and the resulting induced current causes rapid heating and melting of the metal. Features of an induction furnace for foundry operations are illustrated in Figure 11.20. The electromagnetic force field causes a mixing action to occur in the liquid metal. Also, since the metal does not come in direct contact with the heating elements, the environment in which melting takes place can be closely controlled. All of this results in molten metals of high quality and purity, and induction furnaces are used for nearly any casting alloy when these requirements are important. Melting steel, cast iron, and aluminum alloys are common applications in foundry work.

11.4.2 Pouring, Cleaning, and Heat Treatment

Moving the molten metal from the melting furnace to the mold is sometimes done using crucibles. More often, the transfer is accomplished by *ladles* of various kinds. These ladles receive the metal from the furnace and allow for convenient pouring into the molds. Two common ladles are illustrated in Figure 11.21, one for handling large volumes of molten metal using an overhead crane, and the other a "two-man ladle" for manually moving and pouring smaller amounts.

One of the problems in pouring is that oxidized molten metal can be introduced into the mold. Metal oxides reduce product quality, perhaps rendering the casting defective, so measures are taken to minimize the entry of these oxides into the mold during pouring. Filters are sometimes used to catch the oxides and other impurities as the metal is poured from the spout, and fluxes are used to cover the molten metal to retard oxidation. In addition, ladles have been devised to pour the liquid metal from the bottom, since the top surface is where the oxides accumulate.

After the casting has solidified and been removed from the mold, a number of additional steps are usually required. These operations include (1) trimming, (2) removing

FIGURE 11.20   Induction furnace.

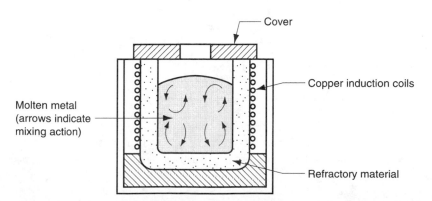

Cover

Copper induction coils

Molten metal (arrows indicate mixing action)

Refractory material

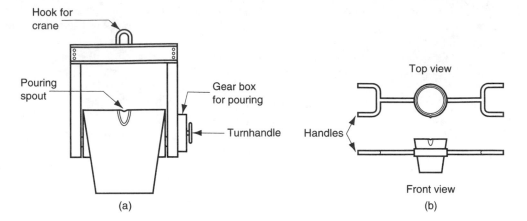

FIGURE 11.21 Two common types of ladles: (a) crane ladle and (b) two-man ladle.

the core, (3) surface cleaning, (4) inspection, (5) repair, if required, and (6) heat treatment. Steps (1) through (5) are collectively referred to in foundry work as "cleaning." The extent to which these additional operations are required varies with casting processes and metals. When required, they are usually labor-intensive and costly.

Trimming involves removal of sprues, runners, risers, parting-line flash, fins, chaplets, and any other excess metal from the cast part. In the case of brittle casting alloys and when the cross-sections are relatively small, these appendages on the casting can be broken off. Otherwise, hammering, shearing, hack-sawing, band-sawing, abrasive wheel cutting, or various torch cutting methods are used.

If cores have been used to cast the part, they must be removed. Most cores are chemically bonded or oil-bonded sand, and they often fall out of the casting as the binder deteriorates. In some cases, they are removed by shaking the casting, either manually or mechanically. In rare instances, cores are removed by chemically dissolving the bonding agent used in the sand core. Solid cores must be hammered or pressed out.

Surface cleaning is most important in the case of sand casting. In many of the other casting methods, especially the permanent mold processes, this step can be avoided. *Surface cleaning* involves removal of sand from the surface of the casting and otherwise enhancing the appearance of the surface. Methods used to clean the surface include tumbling, air-blasting with coarse sand grit or metal shot, wire brushing, buffing, and chemical pickling (Chapter 28).

Defects are possible in casting, and inspection is needed to detect their presence. We consider these quality issues in the following section.

Castings are often heat treated to enhance their properties, either for subsequent processing operations such as machining, or to bring out the desired properties for application of the part in service.

11.5 CASTING QUALITY

There are numerous opportunities for things to go wrong in a casting operation, resulting in quality defects in the cast product. In this section, we compile a list of the common defects that occur in casting, and we indicate the inspection procedures to detect them.

Casting Defects   Some defects are common to any and all casting processes. These defects are illustrated in Figure 11.22 and are briefly described in the following:

(a) *Misruns*—A misrun is a casting that has solidified before completely filling the mold cavity. Typical causes include (1) fluidity of the molten metal is insufficient, (2) pouring temperature is too low, (3) pouring is done too slowly, and/or (4) cross section of the mold cavity is too thin.

(b) *Cold shut*—A cold shut occurs when two portions of the metal flow together but there is a lack of fusion between them due to premature freezing. Its causes are similar to those of a misrun.

(c) *Cold shots*—When splattering occurs during pouring, solid globules of metal are formed that become entrapped in the casting. Pouring procedures and gating system designs that avoid splattering can prevent this defect.

(d) *Shrinkage cavity*—This defect is a depression in the surface or an internal void in the casting, caused by solidification shrinkage that restricts the amount of molten metal available in the last region to freeze. It often occurs near the top of the casting, in which case it is referred to as a "pipe" (Figure 10.8(3)). The problem can often be solved by proper riser design.

(e) *Microporosity*—This refers to a network of small voids distributed throughout the casting caused by localized solidification shrinkage of the final molten metal in the dendritic structure. The defect is usually associated with alloys, because of the protracted manner in which freezing occurs in these metals.

(f) *Hot tearing*—This defect, also called *hot cracking,* occurs when the casting is restrained from contraction by an unyielding mold during the final stages of solidification or early stages of cooling after solidification. The defect is manifested as a separation of the metal (hence, the terms *tearing* or *cracking*) at a point of high tensile stress caused by the metal's inability to shrink naturally. In sand casting and other expendable mold processes, it is prevented by compounding the mold to be collapsible. In permanent mold processes, hot tearing is reduced by removing the part from the mold immediately after freezing.

FIGURE 11.22   Some common defects in castings: (a) misrun, (b) cold shut, (c) cold shot, (d) shrinkage cavity, (e) microporosity, and (f) hot tearing.

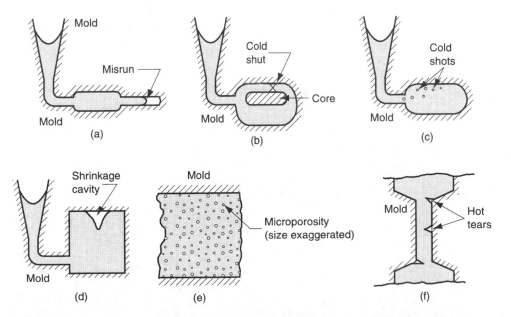

Some defects are related to the use of sand molds, and therefore they occur only in sand castings. To a lesser degree, other expendable mold processes are also susceptible to these problems. Defects found primarily in sand castings are shown in Figure 11.23 and are described here:

(a) *Sand blow*—This defect consists of a balloon-shaped gas cavity caused by release of mold gases during pouring. It occurs at or below the casting surface near the top of the casting. Low permeability, poor venting, and high moisture content of the sand mold are the usual causes.

(b) *Pinhole*—A defect similar to a sand blow involves the formation of many small gas cavities at or slightly below the surface of the casting.

(c) *Sand wash*—A wash is an irregularity in the surface of the casting that results from erosion of the sand mold during pouring. The contour of the erosion is imprinted into the surface of the final cast part.

(d) *Scab*—This is a rough area on the surface of the casting due to encrustations of sand and metal. It is caused by portions of the mold surface flaking off during solidification and becoming imbedded in the casting surface.

(e) *Penetration*—When the fluidity of the liquid metal is high, it may penetrate into the sand mold or sand core. After freezing, the surface of the casting consists of a mixture of sand grains and metal. Harder packing of the sand mold helps to alleviate this condition.

(f) *Mold shift*—This is manifested as a step in the cast product at the parting line caused by sidewise displacement of the cope with respect to the drag.

(g) *Core shift*—A similar movement can happen with the core, but the displacement is usually vertical. Core shift and mold shift are caused by buoyancy of the molten metal (Section 11.1.3).

(h) *Mold crack*—If mold strength is insufficient, a crack may develop, into which liquid metal can seep to form a "fin" on the final casting.

Inspection Methods Foundry inspection procedures include [5]: (1) visual inspection to detect obvious defects such as misruns, cold shuts, and severe surface flaws; (2) dimensional measurements to ensure that tolerances have been met; and (3) metallurgical, chemical, physical, and other tests concerned with the inherent quality of the cast metal.

FIGURE 11.23 Common defects in sand castings: (a) sand blow, (b) pin holes, (c) sand wash, (d) scabs, (e) penetration, (f) mold shift, (g) core shift, and (h) mold crack.

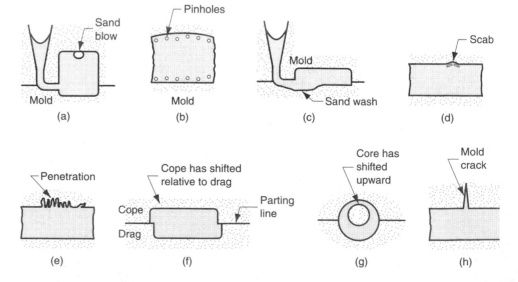

Tests in category (3) include: (a) pressure testing to locate leaks in the casting; (b) radiographic methods, magnetic particle tests, the use of fluorescent penetrants, and supersonic testing to detect either surface or internal defects in the casting; and (c) mechanical testing to determine properties such as tensile strength and hardness. If defects are discovered but are not too serious, it is often possible to save the casting by welding, grinding, or other salvage methods to which the customer has agreed.

11.6 METALS FOR CASTING

Most commercial castings are made of alloys rather than pure metals. Alloys are generally easier to cast, and properties of the resulting product are better. Casting alloys can be classified as: (1) ferrous or (2) nonferrous. The ferrous category is subdivided into cast iron and cast steel.

Ferrous Casting Alloys: Cast Iron   Cast iron is the most important of all casting alloys (Historical Note 11.3). The tonnage of cast iron castings is several times that of all other metals combined. There are several types of cast iron: (1) gray cast iron, (2) nodular iron, (3) white cast iron, (4) malleable iron, and (5) alloy cast irons (Section 6.2.4). Typical pouring temperatures for cast iron are around 1400°C (2500°F), depending on composition.

Historical Note 11.3   *Early cast iron products.*

In the early centuries of casting, bronze and brass were preferred over cast iron as foundry metals. Iron was more difficult to cast, due to its higher melting temperatures and lack of knowledge about its metallurgy. Also, there was little demand for cast iron products. This all changed starting in the sixteenth and seventeenth centuries.

The art of sand casting iron entered Europe from China, where iron was cast in sand molds more than 2,500 years ago. In 1550 the first cannons were cast from iron in Europe. Cannon balls for these guns were made of cast iron starting around 1568. Guns and their projectiles created a large demand for cast iron. But these items were for military rather than civilian use. Two cast iron products that became significant to the general public in the sixteenth and seventeenth centuries were the cast iron stove and cast iron water pipe.

As unspectacular a product as it may seem today, the cast iron stove brought comfort, health, and improved living conditions to many people in Europe and America. During the 1700s, the manufacture of cast iron stoves was one of the largest and most profitable industries on these two continents. The commercial success of stove making was due to the large demand for the product and the art and technology of casting iron that had been developed to produce it.

Cast iron water pipe was another product that spurred the growth of the iron casting industry. Until the advent of cast iron pipes, a variety of methods had been tried to supply water directly to homes and shops, including hollow wooden pipes (which quickly rotted), lead pipes (too expensive), and open trenches (susceptible to pollution). Development of the iron-casting process provided the capability to fabricate water pipe sections at relatively low cost. Cast iron water pipes were used in France starting in 1664, and later in other parts of Europe. By the early 1800s, cast iron pipe lines were being widely installed in England for water and gas delivery. The first significant water pipe installation in the United States was in Philadelphia in 1817, using pipe imported from England.

Ferrous Casting Alloys: Steel   The mechanical properties of steel make it an attractive engineering material (Section 6.2.3), and the capability to create complex geometries make casting an appealing process. However, great difficulties are faced by the foundry specializing in steel. First, the melting point of steel is considerably higher than for most other metals that are commonly cast. The solidification range for low carbon steels (Figure 6.4) begins at just under 1540°C (2800°F). This means that the pouring temperature required for steel is very high—about 1650°C (3000°F). At these high temperatures, steel is chemically very reactive. It readily oxidizes, so special procedures must be used during melting and pouring to isolate the molten metal from air. Also, molten steel has relatively poor fluidity, and this limits the design of thin sections in components cast out of steel.

Several characteristics of steel castings make it worth the effort to solve these problems. Tensile strength is higher than for most other casting metals, ranging upward from about 410 MPa (60,000 lb/in.^{2}) [7]. Steel castings have better toughness than most other casting alloys. The properties of steel castings are isotropic; strength is virtually the same in all directions. By contrast, mechanically formed parts (e.g., rolling, forging) exhibit directionality in their properties. Depending on the requirements of the product, isotropic behavior of the material may be desirable. Another advantage of steel castings is ease of welding. They can be readily welded without significant loss of strength, to repair the casting, or to fabricate structures with other steel components.

Nonferrous Casting Alloys   Nonferrous casting metals include alloys of aluminum, magnesium, copper, tin, zinc, nickel, and titanium (Section 6.3). *Aluminum alloys* are generally considered to be very castable. The melting point of pure aluminum is 660°C (1220°F), so pouring temperatures for aluminum casting alloys are low compared to cast iron and steel. Their properties make them attractive for castings: light weight, wide range of strength properties attainable through heat treatment, and ease of machining. *Magnesium alloys* are the lightest of all casting metals. Other properties include corrosion resistance, as well as high strength-to-weight and stiffness-to-weight ratios.

Copper alloys include bronze, brass, and aluminum bronze. Properties that make them attractive include corrosion resistance, attractive appearance, and good bearing qualities. The high cost of copper is a limitation on the use of its alloys. Applications include pipe fittings, marine propeller blades, pump components, and ornamental jewelry.

Tin has the lowest melting point of the casting metals. *Tin-based alloys* are generally easy to cast. They have good corrosion resistant but poor mechanical strength, which limits their applications to pewter mugs and similar products not requiring high strength. *Zinc alloys* are commonly used in die casting. Zinc has a low melting point and good fluidity, making it highly castable. Its major weakness is low creep strength, so its castings cannot be subjected to prolonged high stresses.

Nickel alloys have good hot strength and corrosion resistance, which make them suited to high-temperature applications such as jet engine and rocket components, heat shields, and similar components. Nickel alloys also have a high melting point and are not easy to cast. *Titanium alloys* for casting are corrosion resistant and possess high strength-to-weight ratios. However, titanium has a high melting point, low fluidity, and a propensity to oxidize at high temperatures. These properties make it and its alloys difficult to cast.

11.7 PRODUCT DESIGN CONSIDERATIONS

If casting is selected by the product designer as the primary manufacturing process for a particular component, then certain guidelines should be followed to facilitate production of the part and avoid many of the defects enumerated in Section 11.5. Some of the important guidelines and considerations for casting are presented next.

➤ *Geometric Simplicity*—Although casting is a process that can be used to produce complex part geometries, simplifying the part design will improve its castability.

FIGURE 11.24   (a) Thick section at intersection can result in a shrinkage cavity. Remedies include (b) redesign to reduce thickness, and (c) use of a core.

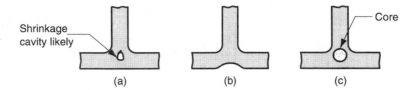

Avoiding unnecessary complexities simplifies mold making, reduces the need for cores, and improves the strength of the casting.

- ➤ *Corners*—Sharp corners and angles should be avoided, since they are sources of stress concentrations and may cause hot tearing and cracks in the casting. Generous fillets should be designed on inside corners and sharp edges should be blended.

- ➤ *Section Thicknesses*—Section thicknesses should be uniform in order to avoid shrinkage cavities. Thicker sections create *hot spots* in the casting, because greater volume requires more time for solidification and cooling. These are likely locations of shrinkage cavities. Figure 11.24 illustrates the problem and offers some possible solutions.

- ➤ *Draft*—Part sections that project into the mold should have a draft or taper, as defined in Figure 11.25. In expendable mold casting, the purpose of this draft is to facilitate removal of the pattern from the mold. In permanent mold casting, its purpose is to aid in removal of the part from the mold. Similar tapers should be allowed if solid cores are used in the casting process. The required draft need only be about $1°$ for sand casting and $2°$ to $3°$ for permanent mold processes.

- ➤ *Use of Cores*—Minor changes in part design can reduce the need for coring, as shown in Figure 11.25.

- ➤ *Dimensional Tolerances and Surface Finishes*—There are significant differences in the dimensional accuracies and finishes that can be achieved in castings, depending on which process is used. Table 11.2 provides a compilation of typical values for these parameters.

- ➤ *Machining Allowances*—Tolerances achievable in many casting processes are insufficient to meet functional needs in many applications. Sand casting is the most prominent example of this deficiency. In these cases, portions of the casting must be machined to the required dimensions. Almost all sand castings must be machined to some extent in order for the part to be made functional. Therefore, additional material, called the *machining allowance,* must be left on the casting for the machining operation in those surfaces where machining is necessary. Typical machining allowances for sand castings range between 1.5 and 3 mm (1/16 and 1/4 in.).

FIGURE 11.25   Design change to eliminate the need for using a core: (a) original design and (b) redesign.

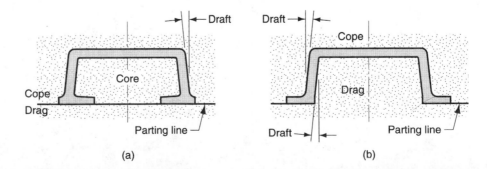

TABLE 11.2 Typical dimensional tolerances and surface finishes for different casting processes and metals.

Casting Process	Part Size	Tolerance (mm)	Tolerance (in.)	Surface Roughness (μm)	Surface Roughness (μ-in.)
Sand casting[a]					
Aluminum[b]	Small	±0.5	(±0.020)	>6	(>250)
Cast iron	Small	±1.0	(±0.040)		
	Large	±1.5	(±0.060)		
Copper alloys	Small	±0.4	(±0.015)		
Steel	Small	±1.3	(±0.050)		
	Large	±2.0	(±0.080)		
Shell molding[b]					
Aluminum[b]	Small	±0.25	(±0.010)	6.4	(250)
Cast iron	Small	±0.5	(±0.020)		
Copper alloys	Small	±0.4	(±0.015)		
Steel	Small	±0.8	(±0.030)		
Plaster mold	Small	±0.12	(±0.005)	0.75	(30)
	Large	±0.4	(±0.015)		

Casting Process	Part size	Tolerance (mm)	Tolerance (in.)	Surface Roughness (μm)	Surface Roughness (μ-in.)
Perm. mold					
Aluminum[b]	Small	±0.25	(±0.010)	3.2	(125)
Cast iron	Small	±0.8	(±0.030)		
Copper alloys	Small	±0.4	(±0.015)		
Steel	Small	±0.5	(±0.020)		
Die casting					
Aluminum[b]	Small	±0.12	(±0.005)	>1	(>40)
Copper alloys	Small	±0.12	(±0.005)		
Investment					
Aluminum[b]	Small	±0.12	(±0.005)	>0.75	(>30)
Cast iron	Small	±0.25	(±0.010)		
Copper alloys	Small	±0.12	(±0.005)		
Steel	Small	±0.25	(±0.010)		

Compiled from [5], [11], and other sources.

[a]Values of surface roughness are for green sand molding; for other sand mold processes, surface finish is better.

[b]Values for aluminum also apply to magnesium.

REFERENCES

[1] Amstead, B. H., Ostwald, P. F., and Begeman, M. L. *Manufacturing Processes.* Wiley, New York, 1987.

[2] Beeley, P. R. *Foundry Technology.* Newnes-Butterworths, London, 1972.

[3] Datsko, J. *Material Properties and Manufacturing Processes.* Wiley, New York, 1966.

[4] Flinn, R. A. *Fundamentals of Metal Casting.* American Foundrymen's Society, Inc., Des Plaines, Ill., 1987.

[5] Heine, R. W., Loper, Jr., C. R., and Rosenthal, C. *Principles of Metal Casting.* 2nd ed. McGraw-Hill, New York, 1967.

[6] Kotzin, E. L. *Metalcasting & Molding Processes.* American Foundrymen's Society, Inc., Des Plaines, Ill., 1981.

[7] *Metals Handbook.* 9th ed. Vol. 15: *Casting,* American Society for Metals, Metals Park, Ohio, 1988.

[8] Mikelonis, P. J. (ed.). *Foundry Technology.* American Society for Metals, Metals Park, Ohio, 1982.

[9] Niebel, B. W., Draper, A. B., Wysk, R. A. *Modern Manufacturing Process Engineering.* McGraw-Hill, New York, 1989.

[10] Simpson, B. L. *History of the Metalcasting Industry.* 2nd ed. American Foundrymen's Society, Inc., Des Plaines, Ill., 1970.

[11] Wick, C., Benedict, J. T., and Veilleux, R. F. *Tool and Manufacturing Engineers Handbook.* 4th ed. Volume II-*Forming.* Society of Manufacturing Engineers, Dearborn, Mich., 1984, Chapter 16.

REVIEW QUESTIONS

11.1. Name the two basic categories of casting processes.

11.2. There are various types of patterns used in sand casting. What is the difference between a split pattern and a match-plate pattern?

11.3. What is a *chaplet*?

11.4. What properties determine the quality of a sand mold for sand casting?

11.5. What is the *Antioch process*?

11.6. What is the difference between vacuum permanent-mold casting and vacuum molding?

11.7. What are the most common metals processed using die casting?

11.8. Which die casting machines usually have a higher production rate, cold-chamber or hot-chamber, and why?

11.9. What is *flash* in die casting?

11.10. What is the difference between true centrifugal casting and semicentrifugal casting?

11.11. What is a *cupola*?

11.12. What are some of the operations required of sand castings after removal from the mold?

11.13. What are some of the general defects encountered in casting processes?

MULTIPLE CHOICE QUIZ

There is a total of 28 correct answers in the following multiple choice questions (some questions have multiple answers that are correct). To attain a perfect score on the quiz, all correct answers must be given, since each correct answer is worth 1 point. For each question, each omitted answer or wrong answer reduces the score by 1 point, and each additional answer beyond the number of answers required reduces the score by 1 point. Percentage score on the quiz is based on the total number of correct answers.

11.1. Which one of the following casting processes is most widely used? (a) centrifugal casting, (b) die casting, (c) investment casting, (d) sand casting, or (e) shell casting.

11.2. In sand casting, the volumetric size of the pattern is which of the following relative to the cast part? (a) bigger, (b) same size, or (c) smaller.

11.3. Silica sand has which one of the following compositions? (a) Al_2O_3, (b) SiO, (c) SiO_2, or (d) $SiSO_4$.

11.4. For which of the following reasons is a *green mold* so named? (a) green is the color of the mold, (b) moisture is contained in the mold, (c) mold is cured, or (d) mold is dry.

11.5. Given that W_m = weight of the molten metal displaced by a core and W_c = weight of the core, the buoyancy force is which one of the following? (a) downward force = $W_m + W_c$, (b) downward force = $W_m - W_c$, (c) upward force = $W_m + W_c$, or (d) upward force = $W_m - W_c$.

11.6. Which of the following casting processes are expendable mold operations (more than one)? (a) investment casting, (b) low-pressure casting, (c) sand casting,

(d) shell molding, (e) slush casting, and (f) vacuum molding.

11.7. Shell molding is which one of the following? (a) casting operation in which the molten metal has been poured out after a thin shell has been solidified in the mold, (b) casting operation used to make artificial sea shells, (c) casting process in which the mold is a thin shell of sand binded by a thermosetting resin, or (d) sand-casting operation in which the pattern is a shell rather than a solid form.

11.8. Investment casting is also known by which one of the following names? (a) fast-payback molding, (b) full-mold process, (c) lost-foam process, (d) lost pattern process, or (e) lost-wax process.

11.9. In plaster mold casting, the mold is made of which one of the following materials? (a) Al_2O_3, (b) $CaSO_4$–H_2O, (c) SiC, or (d) SiO_2.

11.10. Which of the following qualifies as a precision casting process (more than one)? (a) ingot casting, (b) investment casting, (c) plaster-mold casting, (d) sand casting, and (e) shell molding.

11.11. Which of the following casting processes are permanent-mold operations (more than one)? (a) centrifu-gal casting, (b) die casting, (c) low-pressure casting, (d) shell molding, (e) slush casting, and (f) vacuum permanent-mold casting.

11.12. Which of the following metals would typically be die casted (more than one)? (a) aluminum, (b) cast iron, (c) steel, (d) tin, (e) tungsten, and (f) zinc.

11.13. Which of the following are advantages of die casting over sand casting (more than one)? (a) better surface finish, (b) higher melting temperature metals, (c) higher production rates, (d) larger parts can be casted, and (e) mold can be reused.

11.14. Cupolas are furnaces used to melt which of the following metals (one best answer)? (a) aluminum, (b) cast iron, (c) steel, or (d) zinc.

11.15. A misrun is which one of the following defects in casting? (a) globules of metal becoming entrapped in the casting, (b) metal is not properly poured into the down-sprue, (c) metal solidifies before filling the cavity, (d) microporosity, and (e) "pipe" formation.

11.16. Which one of the following casting metals is most important commercially? (a) aluminum and its alloys, (b) bronze, (c) cast iron, (d) cast steel, or (e) zinc alloys.

PROBLEMS

Buoyancy Force

11.1. An aluminum–copper alloy casting is made in a sand mold using a sand core that weighs 20 kg. Determine the buoyancy force in Newtons tending to lift the core during pouring.

11.2. A sand core located inside a mold cavity has a volume of 157.0 in.^{3} It is used in the casting of a cast iron pump housing. Determine the buoyancy force that will tend to lift the core during pouring.

11.3. Caplets are used to support a sand core inside a sand mold cavity. The design of the caplets and the manner in which they are placed in the mold cavity surface allow each caplet to sustain a force of 10 lb. Several caplets are located beneath the core to support it before pouring; and several other caplets are placed above the core to resist the buoyancy force during pouring. If the volume of the core = 325 cu in., and the metal poured is brass, determine the minimum number of caplets that should be placed: (a) beneath the core and (b) above the core.

11.4. A sand core used to form the internal surfaces of a steel casting experiences a buoyancy force of 23 kg. The volume of the mold cavity forming the outside surface of the casting is 5000 cm^3. What is the weight of the final casting? Ignore considerations of shrinkage.

Centrifugal Casting

11.5. A horizontal true centrifugal casting operation will be used to make copper tubing. The lengths will be 1.5 m with outside diameter = 15.0 cm, and inside diameter = 12.5 cm. If the rotational speed of the pipe = 1000 rev/min, determine the G-factor.

11.6. A true centrifugal casting operation is to be performed in a horizontal configuration to make cast iron pipe sections. The sections will have length = 42.0 in., outside diameter = 8.0 in., and wall thickness = 0.50 in. If the rotational speed of the pipe = 500 rev/min, determine the G-factor. Is the operation likely to be successful?

11.7. A horizontal true centrifugal casting process is used to make brass bushings with these dimensions: L = 10 cm, OD = 15 cm, and ID = 12 cm. (a) Determine the required rotational speed in order to obtain a G-factor of 70. (b) When operating at this speed, what is the centrifugal force per square meter (Pa) imposed by the molten metal on the inside wall of the mold?

11.8. True centrifugal casting operation is performed horizontally to make large-diameter copper tube sections. The tubes have a length = 1.0 m, diameter = 0.25 m, and wall thickness = 15 mm. If the rotational speed of the pipe = 700 rev/min, (a) determine the G-factor on the molten metal. (b) Is the rotational speed sufficient to avoid "rain"? (c) What volume of molten metal must be poured into the mold to make the casting if solidification shrinkage and contraction after solidification are considered?

11.9. If a true centrifugal casting operation were to be performed in a space station circling the Earth, how would weightlessness affect the process?

11.10. A horizontal true centrifugal casting process is used to make aluminum rings with dimensions: $L = 5$ cm, OD = 65 cm, and ID = 60 cm. (a) Determine the rotational speed that will provide a G-factor = 60. (b) Suppose that the ring were made out of steel instead of aluminum. If the rotational speed computed in that problem were used in the steel casting operation, determine the G-factor and (c) centrifugal force per square meter (Pa) on the mold wall. (d) Would this rotational speed result in a successful operation?

11.11. For the steel ring of preceding Problem 11.10(b), determine the volume of molten metal that must be poured into the mold, given that the liquid shrinkage is 0.5 percent, and the solidification shrinkage and solid contraction after freezing can be determined from Table 10.1.

11.12. A horizontal true centrifugal casting process is used to make lead pipe for chemical plants. The pipe has length = 0.5 m, outside diameter = 70 mm, and thickness = 6.0 mm. Determine the rotational speed that will provide a G-factor = 60.

11.13. A vertical true centrifugal casting process is used to make tube sections with length = 10.0 in. and outside diameter = 6.0 in. The inside diameter of the tube = 5.5 in. at the top and 5.0 in. at the bottom. At what speed must the tube be rotated during the operation in order to achieve these specifications?

11.14. A vertical true centrifugal casting process is used to produce bushings that are 200 mm long and 200 mm in outside diameter. If the rotational speed during solidification is 500 rpm, determine the inside diameter at the top of the bushing if the diameter at the bottom is 150 mm.

11.15. A vertical true centrifugal casting process is used to cast brass tubing that is 15.0 in. long and whose outside diameter = 8.0 in. If the speed of rotation during solidification is 1000 rpm, determine the inside diameters at the top and bottom of the tubing if the total weight of the final casting = 75.0 lb.

Defects and Design Considerations

11.16. The housing for a certain machinery product is made of two components, both aluminum castings. The larger component has the shape of a dish sink and the second component is a flat cover that is attached to the first component to create an enclosed space for the machine parts. Sand casting is used to produce the two castings, both of which are plagued by defects in the form of misruns and cold shuts. The foreman complains that the thickness of the parts are too thin, and that is the reason for the defects. However, it is known that the same components are cast successfully in other foundries. What other explanation can be given for the defects?

11.17. A large steel sand casting shows the characteristic signs of penetration defect—a surface consisting of a mixture of sand and metal. (a) What steps can be taken to correct the defect? (b) What other possible defects might result from taking each of these steps?

12 GLASSWORKING

CHAPTER CONTENTS

12.1   Raw Materials Preparation and Melting
12.2   Shaping Processes in Glassworking
          12.2.1   Shaping of Piece Ware
          12.2.2   Shaping of Flat and Tubular Glass
          12.2.3   Forming of Glass Fibers
12.3   Heat Treatment and Finishing
          12.3.1   Heat Treatment
          12.3.2   Finishing
12.4   Product Design Considerations

Glass products are commercially manufactured in an almost unlimited variety of shapes. Many are produced in very large quantities, such as light bulbs, beverage bottles, and window glass. Others, such as giant telescope lenses, are made individually.

Glass is one of three basic types of ceramics (Chapter 7); the others are traditional ceramics and new ceramics. Glass is distinguished by its noncrystalline (vitreous) structure, whereas the other ceramic materials have a crystalline structure. The methods by which glass is shaped into useful products are quite different from those used for the other types. In glassworking, the principal starting material is silica (SiO_2); this is usually combined with other oxide ceramics which form glasses. The starting material is heated to transform it from a hard solid into a viscous liquid; it is then shaped into the desired geometry while in this fluid condition. When cooled and hard, the material remains in the glassy state rather than crystallizing.

The typical process sequence in glassworking consists of the steps pictured in Figure 12.1. Shaping is accomplished by various processes, including casting, pressing-and-blowing (to produce bottles and other containers), and rolling (to make plate glass). A finishing step is required for certain products.

12.1 RAW MATERIALS PREPARATION AND MELTING

The principal component in nearly all glasses is silica (SiO_2), the primary source of which is natural quartz in sand. The sand must be washed and classified. Washing removes im-

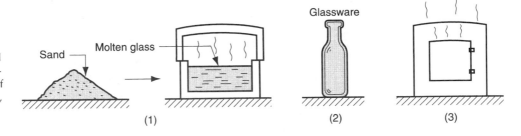

FIGURE 12.1   The typical process sequence in glassworking: (1) preparation of raw materials and melting, (2) shaping, and (3) heat treatment.

purities such as clay and certain minerals that would cause undesirable coloring of the glass. ***Classifying*** the sand means grouping the grains according to size. The most desirable particle size for glassmaking is in the range 0.1 to 0.6 mm (0.004 to 0.025 in.) [3]. The various other components, such as soda ash (source of Na_2O), limestone (source of CaO), aluminum oxide, potash (source of K_2O), and other minerals are added in the correct proportions to achieve the desired composition. The mixing is usually done in batches, in amounts that are compatible with the capacities of the available melting furnaces.

Recycled glass is usually added to the mixture in modern practice. In addition to preserving the environment, recycled glass facilitates melting. Depending on the amount of waste glass available and the specifications of the final composition, the proportion of recycled glass may be up to 100%.

The batch of starting materials to be melted is referred to as a ***charge,*** and the procedure of loading it into the melting furnace is called ***charging*** the furnace. Glass-melting furnaces can be divided into the following types [3]: (1) ***pot furnaces***—ceramic pots of limited capacity in which melting occurs by heating the walls of the pot; (2) ***day tanks***—larger-capacity vessels for batch production in which heating is done by burning fuels above the charge; (3) ***continuous tank furnaces***—long tank furnaces in which raw materials are fed in one end and melted as they move to the other end, where molten glass is drawn out for high production; and (4) ***electric furnaces*** of various designs for a wide range of production rates.

Glass melting is generally carried out at temperatures around 1500°C to 1600°C (2700°F to 2900°F). The melting cycle for a typical charge takes 24 to 48 hours. This is the time required for all of the sand grains to become a clear liquid and for the molten glass to be refined and cooled to the appropriate temperature for working. Molten glass is a viscous liquid, the viscosity being inversely related to temperature. Since the shaping operation immediately follows the melting cycle, the temperature at which the glass is tapped from the furnace depends on the viscosity required for the subsequent process.

12.2 SHAPING PROCESSES IN GLASSWORKING

The major categories of glass products were identified (Section 7.4.2) as window glass, containers, light bulbs, laboratory glassware, glass fibers, and optical glass. Despite the variety represented by this list, the shaping processes to fabricate these products can be grouped into three categories: (1) discrete processes for piece ware, which includes bottles, light bulbs, and other individual items; (2) continuous processes for making flat glass (sheet and plate glass for windows) and tubing (for laboratory ware and fluorescent lights); and (3) fiber-making processes to produce fibers for insulation, fiberglass composite materials, and fiber optics.

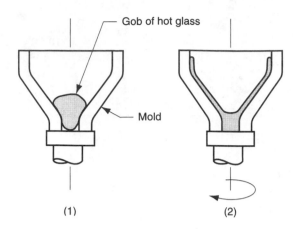

FIGURE 12.2 Spinning of funnel-shaped glass parts: (1) gob of glass dropped into mold; and (2) rotation of mold to cause spreading of molten glass on mold surface.

12.2.1 Shaping of Piece Ware

The ancient methods of hand-working glass, such as glass blowing, were briefly described in our Historical Note 7.3. Handicraft methods are still employed today for making glassware items of high value in small quantities. Most of the processes discussed in this section are highly mechanized technologies for producing discrete pieces such as jars, bottles, and light bulbs in high quantities.

Spinning Glass spinning is similar to *centrifugal casting* of metals, and is also known by that name in glassworking. It is used to produce funnel-shaped components such as the back sections of cathode ray tubes for televisions and computer monitors. The setup is pictured in Figure 12.2. A gob of molten glass is dropped into a conical mold made of steel. The mold is rotated so that centrifugal force causes the glass to flow upward and spread itself on the mold surface. The faceplate (i.e., the front viewing screen) is later assembled to the funnel using a sealing glass of low melting point.

Pressing This is a widely used process for mass production of glass pieces such as dishes, bake ware, headlight lenses, TV tube faceplates, and similar items that are relatively flat. The process is illustrated and described in Figure 12.3. The large quantities of most pressed products justify a high level of automation in this production sequence.

FIGURE 12.3 Pressing of a flat glass piece: (1) a gob of glass fed into mold from the furnace; (2) pressing into shape by plunger; and (3) plunger is retracted and the finished product is removed. Symbols v and F indicate motion (v = velocity) and applied force, respectively.

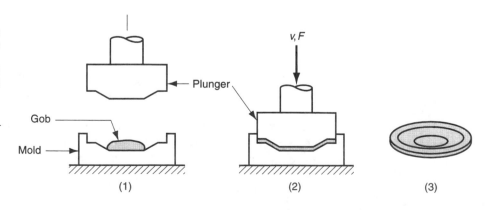

Blowing Several shaping sequences include blowing as one or more of the steps. Instead of a manual operation, blowing is performed on highly automated equipment. The two sequences we describe here are the press-and-blow and blow-and-blow methods.

As the name indicates, the ***press-and-blow*** method is a pressing operation followed by a blowing operation, as portrayed in Figure 12.4. The process is suited to the production of wide-mouth containers. A split mold is used in the blowing operation for part removal.

The ***blow-and-blow*** method is used to produce smaller-mouthed bottles. The sequence is similar to the preceding, except that two (or more) blowing operations are used rather than pressing and blowing. There are variations to the process, depending on the geometry of the product, with one possible sequence shown in Figure 12.5. Reheating is sometimes required between blowing steps. Duplicate and triplicate molds are sometimes used along with matching gob feeders to increase production rates. Press-and-blow and blow-and-blow methods are used to make jars, beverage bottles, incandescent light bulb enclosures, and similar geometries.

Casting If the molten glass is sufficiently fluid, it can be poured into a mold. Relatively massive objects, such as astronomical lenses and mirrors, are made by this method. These pieces must be cooled very slowly to avoid internal stresses and possible cracking due to temperature gradients that would otherwise be set up in the glass. After cooling and solidifying, the piece must be finished by lapping and polishing. Casting is not much used in glassworking except for these kinds of special jobs. Not only is cooling and cracking a problem, but also molten glass is relatively viscous at normal working temperatures, and does not flow through small orifices or into small sections as well as molten metals or heated thermoplastics. Smaller lenses are usually made by pressing, as already discussed.

FIGURE 12.4 Press-and-blow forming sequence: (1) molten gob is fed into mold cavity; (2) pressing to form a ***parison***; (3) the partially formed parison, held in a neck ring, is transferred to the blow mold, and (4) blown into final shape. Symbols *v* and *F* indicate motion (*v* = velocity) and applied force, respectively.

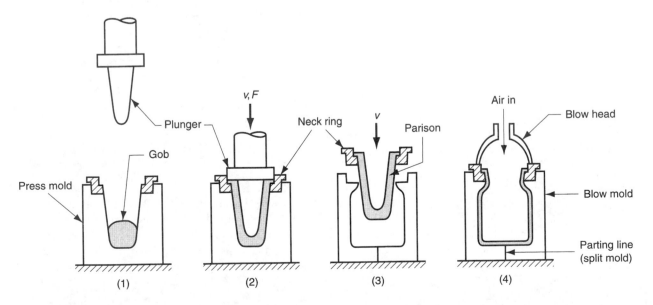

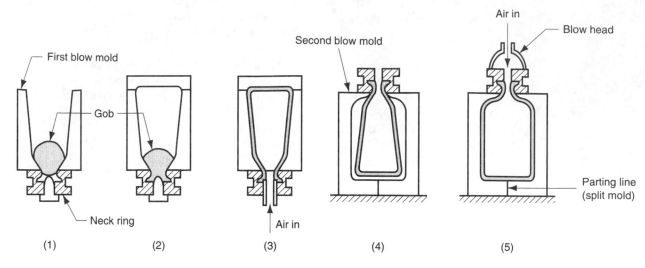

FIGURE 12.5 Blow-and-blow forming sequence: (1) gob is fed into inverted mold cavity; (2) mold is covered; (3) first blowing step; (4) partially formed piece is reoriented and transferred to second blow mold, and (5) blown to final shape.

12.2.2 Shaping of Flat and Tubular Glass

Here we describe two methods for making plate glass and one method for producing tube stock. They are continuous processes, in which long sections of flat window glass or glass tubing are made and later cut into appropriate sizes and lengths. They are modern technologies in contrast to the ancient method described in our historical note.

Historical Note 12.1   *Ancient methods of making flat glass* [7].

G lass windows have been used in buildings for many centuries. The oldest process for making flat window glass was by manual glassblowing. The procedure consisted of the following: (1) a glass globe was blown on a blowpipe; (2) a portion of the globe was made to stick to the end of a "punty," a metal rod used by glassblowers, and then detached from the blowpipe; and (3) after reheating the glass, the punty was rotated with sufficient speed for centrifugal force to shape the open globe into a flat disk. The disk, whose maximum possible size was only about 1 m (3 ft), was later cut into small panes for windows.

At the center of the disk, where the glass was attached to the punty during the third step in the process, a lump would tend to form which had the appearance of a crown. The name "crown glass" was derived from this resemblance. Lenses for spectacles were ground from glass made by this method. Today, we still use the name crown glass for certain types of optical and ophthalmic glass, even though the ancient method has been replaced by modern production technology.

Rolling of Flat Plate   Flat plate glass can be produced by rolling, as illustrated in Figure 12.6. The starting glass, in a suitably plastic condition from the furnace, is squeezed through opposing rolls whose separation determines the thickness of the sheet. The rolling operation is usually set up so that the flat glass is moved directly into an annealing furnace. The rolled glass sheet must later be ground and polished for parallelism and smoothness.

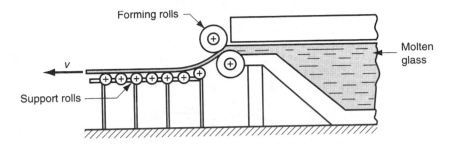

FIGURE 12.6　Rolling of flat glass.

Float Process　This process was developed in the late 1950s. Its advantage over other methods such as rolling is that it obtains smooth surfaces that need no subsequent finishing. In the *float process,* illustrated in Figure 12.7, the glass flows directly from its melting furnace onto the surface of a molten tin bath. The highly fluid glass spreads evenly across the molten tin surface, achieving a uniform thickness and smoothness. After moving into a cooler region of the bath, the glass hardens and travels through an annealing furnace, after which it is cut to size.

Drawing of Glass Tubes　Glass tubing is manufactured by a drawing process known as the *Danner process,* illustrated in Figure 12.8. Molten glass flows around a rotating hollow mandrel through which air is blown while the glass is being drawn. The air temperature and its volumetric flow rate, as well as the drawing velocity, determine the diameter and wall thickness of the tubular cross section. During hardening, the glass tube is supported by a series of rollers extending about 30 m (100 ft) beyond the mandrel. The continuous tubing is then cut into standard lengths. Tubular glass products include laboratory glassware, fluorescent light tubes, and thermometers.

12.2.3　Forming of Glass Fibers

Glass fibers are used in applications ranging from insulation wool to fiber optics communications lines (Section 9.4.2). Glass fiber products can be divided into two categories [6]: (1) fibrous glass for thermal insulation, acoustical insulation, and air filtration, in which the fibers are in a random, wool-like condition; and (2) long continuous filaments suitable for fiber-reinforced plastics, yarns and fabrics, and fiber optics. Different production methods are used for the two categories; we describe two methods below, representing each of the product categories, respectively.

FIGURE 12.7　The float process for producing sheet glass.

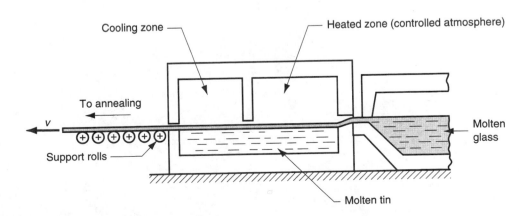

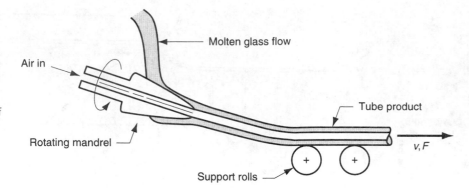

FIGURE 12.8   Drawing of glass tubes by the Danner process. Symbols *v* and *F* indicate motion (*v* = velocity) and applied force, respectively.

Centrifugal Spraying   In a typical process for making glass wool, molten glass flows into a rotating bowl with many small orifices around its periphery. Centrifugal force causes the glass to flow through the holes to become a fibrous mass suitable for thermal and acoustical insulation.

Drawing of Continuous Filaments   In this process, illustrated in Figure 12.9, continuous glass fibers of small diameter (lower size limit is around 0.0025 mm (0.0001 in.) are produced by drawing (pulling) strands of molten glass through small orifices in a heated plate made of a platinum alloy. The plate may have several hundred holes, each making one fiber. The individual fibers are collected into a strand by reeling them onto a spool. Prior to spooling, the fibers are coated with various chemicals to lubricate and protect them. Drawing speeds of around 50 m/s (10,000 ft/min) or more are not unusual.

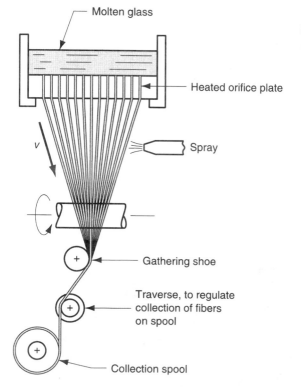

FIGURE 12.9   Drawing of continuous glass fibers.

Woodham,

Lectures

Harrison, M. (2013) The Ba

Harrison, M. (2013) Interna
2013.

Websites

http://www.artquid.com/

http://distinctbuild.ca/l

http://www.masters

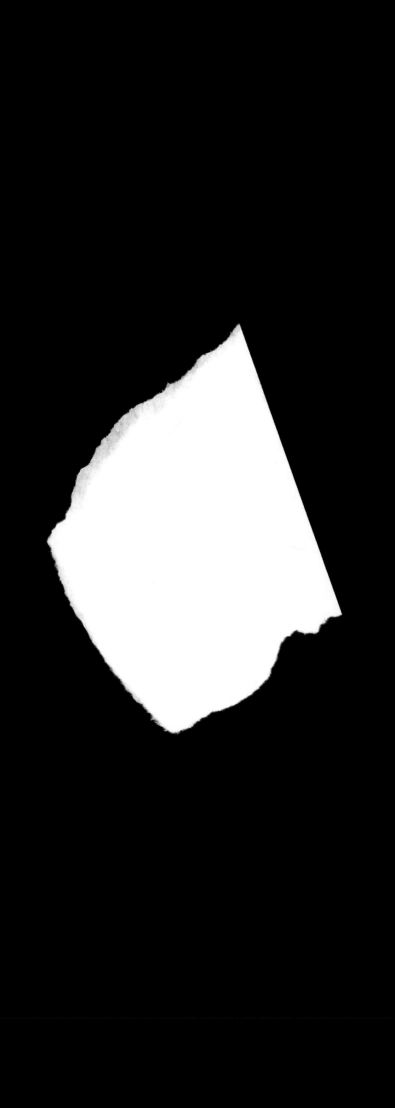

12.3 HEAT TREATMENT AND FINISHING

Heat treatment of the glass product is the third step in the glassworking sequence. For some products, additional finishing operations are performed.

12.3.1 Heat Treatment

We discussed glass-ceramics in Section 7.4.3. This unique material is made by a special heat treatment that transforms most of the vitreous state into a polycrystalline ceramic. Other heat treatments performed on glass cause changes that are less dramatic technologically but perhaps more important commercially; examples include annealing and tempering.

Annealing   Glass products usually have undesirable internal stresses after forming, which reduce their strength. Annealing is done to relieve these stresses; the treatment therefore has the same function in glassworking as it does in metalworking. *Annealing* involves heating the glass to an elevated temperature and holding it for a certain period to eliminate stresses and temperature gradients; then slowly cooling the glass to suppress stress formation, followed by more rapid cooling to room temperature. Common annealing temperatures are around 500°C (900°F). The length of time the product is held at the temperature, as well as the heating and cooling rates during the cycle, depend on thickness of the glass, the usual rule being that the required annealing time varies with the square of thickness.

Annealing in modern glass factories is performed in tunnel-like furnaces, called *lehrs,* in which the products flow slowly through the hot chamber on conveyors. Burners are located only at the front end of the chamber, so that the glass experiences the required heating and cooling cycle.

Tempered Glass and Related Products   A beneficial internal stress pattern can be developed in glass products by a heat treatment known as *tempering,* and the resulting material is called *tempered glass.* As in the treatment of hardened steel, tempering increases the toughness of glass. The process involves heating the glass to a temperature somewhat above its annealing temperature and into the plastic range, followed by quenching of the surfaces, usually with air jets. When the surfaces cool, they contract and harden while the interior is still plastic and compliant. As the internal glass slowly cools, it contracts, thus putting the hard surfaces in compression. Like other ceramics, glass is much stronger when subjected to compressive stresses than tensile stresses. Accordingly, tempered glass is much more resistant to scratching and breaking because of the compressive stresses on its surfaces. Applications include windows for tall buildings, all-glass doors, safety glasses, and other products requiring toughened glass.

When tempered glass fails, it does so by shattering into numerous small fragments which are less likely to cut someone than conventional (annealed) window glass. Interestingly, automobile windshields are not made of tempered glass, due to the danger posed to the driver by this fragmentation. Instead, conventional glass is used; however, it is fabricated by sandwiching two pieces of glass on either side of a tough polymer sheet. Should this *laminated glass* fracture, the glass splinters are retained by the polymer sheet and the windshield remains relatively transparent.

12.3.2 Finishing

Finishing operations are sometimes required for glassware products. These secondary operations include grinding, polishing, and cutting. When glass sheets are produced by drawing and rolling, the opposite sides are not necessarily parallel, and the surfaces contain defects and scratch marks caused by the use of hard tooling on soft glass. The glass sheets must be ground and polished for most commercial applications. In pressing and blowing operations when split dies are used, polishing is often required to remove the seam marks from the container product.

In continuous glassworking processes, such as plate and tube production, the continuous sections must be cut into smaller pieces. This is accomplished by first scoring the glass with a glass-cutting wheel or cutting diamond and then breaking the section along the score line. Cutting is generally done as the glass exits the annealing lehr.

Decorative and surface processes are performed on certain glassware products. These processes include mechanical cutting and polishing operations; sandblasting; chemical etching (with hydrofluoric acid, often in combination with other chemicals); and coating (e.g., coating of plate glass with aluminum or silver to produce mirrors).

12.4 PRODUCT DESIGN CONSIDERATIONS

Glass possesses special properties that make it desirable in certain applications. The following design recommendations are compiled from Bralla [1] and other sources.

- Glass is transparent and has certain optical properties that are unusual if not unique among engineering materials. For applications requiring transparency, light transmittance, magnification, and similar optical properties, glass is likely to be the material of choice. Certain polymers are transparent and may be competitive, depending on design requirements.
- Glass is several times stronger in compression than in tension; components should be designed so that they are subjected to compressive stresses, not tensile stresses.
- Ceramics, including glass, are brittle. Glass parts should not be used in applications that involve impact loading or high stresses that might cause fracture.
- Certain glass compositions have very low thermal expansion coefficients and are therefore tolerant of thermal shock. These glasses can be selected for applications where this characteristic is important.
- Outside edges and corners on glass parts should have large radii or chamfers; likewise, inside corners should have large radii. Both outside and inside corners are potential points of stress concentration.
- Unlike parts made of traditional and new ceramics, threads may be included in the design of glass parts; they are technically feasible with the press-and-blow shaping processes. However, the threads should be coarse.

REFERENCES

[1] Bralla, J. G. (Editor in Chief). *Design for Manufacturability Handbook.* 2nd ed. McGraw-Hill, New York, 1998.

[2] Flinn, R. A., and Trojan, P. K. *Engineering Materials and Their Applications.* 4th ed. Houghton Mifflin, Boston, 1990.

[3] Hlavac, J. *The Technology of Glass and Ceramics.* Elsevier Scientific Publishing Company, New York, 1983.

[4] McLellan, G., and Shand, E. B. *Glass Engineering Handbook.* 3rd ed. McGraw-Hill, New York, 1984.

[5] McColm, I. J. *Ceramic Science for Materials Technologists.* Chapman and Hall, New York, 1983.

[6] Mohr, J. G., and Rowe, W. P. *Fiber Glass.* Krieger Publishing Company, New York, 1990.

[7] Scholes, S. R., and Greene, C. H. *Modern Glass Practice.* 7th ed. TechBooks, Marietta, Georgia, 1993.

REVIEW QUESTIONS

12.1. We have classified glass as a ceramic material; yet glass is different from the traditional and new ceramics. What is the difference?

12.2. What is the predominant chemical compound in almost all glass products?

12.3. What are the three basic steps in the glassworking sequence?

12.4. Melting furnaces for glassworking can be divided into four types. Name three of the four types.

12.5. Describe the spinning process in glassworking.

12.6. What is the main difference between the press-and-blow and the blow-and-blow shaping processes in glassworking?

12.7. Two ways of producing plate or sheet glass are described in the text. Name and briefly describe one of them.

12.8. What is the Danner process?

12.9. Two processes for forming glass fibers are discussed in the text. Name and briefly describe one of them.

12.10. What is the purpose of annealing in glassworking?

12.11. Describe how a piece of glass is heat-treated to produce tempered glass.

12.12. Describe the type of material that is commonly used to make windshields for automobiles.

12.13. What are some of the design recommendations for glass parts?

MULTIPLE CHOICE QUIZ

There is a total of 10 correct answers in the following multiple choice questions (some questions have multiple answers that are correct). To attain a perfect score on the quiz, all correct answers must be given, since each correct answer is worth 1 point. For each question, each omitted answer or wrong answer reduces the score by 1 point, and each additional answer beyond the number of answers required reduces the score by 1 point. Percentage score on the quiz is based on the total number of correct answers.

12.1. Which one of the following terms refers to the glassy state of a material? (a) crystalline, (b) devitrified, (c) polycrystalline, (d) vitiated, or (e) vitreous.

12.2. Besides helping to preserve the environment, the use of recycled glass as an ingredient of the starting material in glassmaking serves what other useful purpose (one only)? (a) adds coloring variations to the glass for aesthetic value, (b) makes the glass easier to melt, (c) makes the glass stronger, or (d) reduces odors in the plant.

12.3. The charge in glassworking is which one of the following? (a) the duration of the melting cycle, (b) the electric energy required to melt the glass, (c) the name given to the melting furnace, or (d) the starting materials in melting.

12.4. Typical glass melting temperatures are in which of the following ranges? (a) 400°C to 500°C, (b) 900°C to 1000°C, (c) 1500°C to 1600°C, or (d) 2000°C to 2200°C.

12.5. Casting is a glassworking process used for high production: (a) true or (b) false.

12.6. Which one of the following processes or processing steps is not applicable in glassworking? (a) annealing, (b) pressing, (c) quenching, (d) sintering, and (e) spinning.

12.7. The press-and-blow process is best suited to the production of (narrow-necked) beverage bottles, while the blow-and-blow process is more appropriate for producing (wide-mouthed) jars: (a) true or (b) false.

12.8. Which one of the following processes is used to produce glass tubing? (a) Danner process, (b) pressing, (c) rolling, or (d) spinning.

12.9. If a glass part with a wall thickness of 5 mm (0.20 in.) takes 10 minutes to anneal, how much time would a glass part of similar geometry but with a wall thickness of 7.5 mm (0.30 in.) take to anneal (choose the closest answer)? (a) 10 minutes, (b) 15 minutes, (c) 20 minutes, or (c) 30 minutes.

12.10. A lehr is which of the following? (a) a lion's den, (b) a melting furnace, (c) a sintering furnace, (d) an annealing furnace, or (e) none of the above.

13 SHAPING PROCESSES FOR PLASTICS

CHAPTER CONTENTS

13.1 Properties of Polymer Melts
13.2 Extrusion
    13.2.1 Process and Equipment
    13.2.2 Analysis of Extrusion
    13.2.3 Die Configurations and Extruded Products
    13.2.4 Defects in Extrusion
13.3 Production of Sheet and Film
13.4 Fiber and Filament Production (Spinning)
13.5 Coating Processes
13.6 Injection Molding
    13.6.1 Process and Equipment
    13.6.2 The Mold
    13.6.3 Injection Molding Machines
    13.6.4 Shrinkage
    13.6.5 Defects in Injection Molding
    13.6.6 Other Injection Molding Processes
13.7 Compression and Transfer Molding
    13.7.1 Compression Molding
    13.7.2 Transfer Molding
13.8 Blow Molding and Rotational Molding
    13.8.1 Blow Molding
    13.8.2 Rotational Molding
13.9 Thermoforming
13.10 Casting
13.11 Polymer Foam Processing and Forming
    13.11.1 Foaming Processes
    13.11.2 Shaping Processes
13.12 Product Design Considerations
    13.12.1 General Considerations
    13.12.2 Extruded Plastics
    13.12.3 Molded Parts

Plastics can be shaped into a wide variety of products, such as molded parts, extruded sections, films and sheets, insulation coatings on electrical wires, and fibers for textiles. In addition, plastics are often the principal ingredient in other materials, such as paints and varnishes; adhesives; and various polymer matrix composites. In this chapter, we consider the technologies by which these products are shaped, postponing paints and varnishes, adhesives, and composites until later chapters. Many plastic shaping processes can be adapted to rubbers (Chapter 14) and polymer matrix composites (Chapter 15).

The commercial and technological importance of these shaping processes derives from the growing importance of the materials being processed. Applications of plastics have increased at a much faster rate than either metals or ceramics during the last 50 years. Indeed, many parts previously made of metals are today being made of plastics and plastic composites. The same is true of glass; plastic containers have been largely substituted for glass bottles and jars in product packaging. The total volume of polymers (plastics and rubbers) now exceeds that of metals (the tonnage is still much less, though, because the density of metals is considerably greater). We can identify several reasons why the plastic shaping processes are important:

> The variety of shaping processes, and the ease with which polymers can be processed, allows an almost unlimited variety of part geometries to be formed.

> Many plastic parts are formed by molding, which is a *net shape* process; further shaping is generally not needed.

> Although heating is usually required to form plastics, *less energy* is required than for metals because the processing temperatures are much lower for plastics.

> Because lower temperatures are used in processing, handling of the product is simplified during production. Because many plastic processing methods are one-step operations (e.g., molding), the amount of product handling required is substantially reduced compared to metals.

> Finishing by painting or plating is not required (except in unusual circumstances) for plastics.

As discussed in Chapter 8, the two types of plastics are *thermoplastics* and *thermosets.* The difference is that thermosets undergo a curing process during heating and shaping, which causes a permanent chemical change (*cross-linking*) in their molecular structure. Once they have been cured, they cannot be melted through reheating. By contrast, thermoplastics do not cure, and their chemical structure remains basically unchanged upon reheating even though they transform from solid to fluid. Of the two types, thermoplastics are by far the more important type commercially, comprising more than 80% of the total plastics tonnage.

Plastic shaping processes can be classified according to the resulting product geometry as follows: (1) continuous extruded products with constant cross section other than sheets, films, and filaments; (2) continuous sheets and films; (3) continuous filaments (fibers); (4) molded parts that are mostly solid; (5) hollow molded parts with relatively thin walls; (6) discrete parts made of formed sheets and films; (7) castings; and (8) foamed products. This chapter will examine each of these categories. The most important processes commercially are those associated with thermoplastics; the two processes of greatest significance are extrusion and injection molding. A brief history of the plastic shaping processes is presented in Historical Note 13.1.

Historical Note 13.1 *Plastic shaping processes.*

E quipment for shaping plastics evolved largely from rubber processing technology. Noteworthy among the early contributors was Edwin Chaffee, an American who developed a two-roll steam-heated mill for mixing additives into rubber around 1835 (Section 14.1.3). He was also responsible for a similar device called a calender, which consists of a series of heated rolls for coating rubber onto cloth (Section 13.3). Both machines are still used today for plastics as well as rubbers.

The first extruders, dating from around 1845 in England, were ram-driven machines for extruding rubber and coating rubber onto electrical wire. The trouble with ram-type extruders is that they operate in an intermittent fashion. An extruder that could operate continuously, especially for wire and cable coating, was highly desirable. Although several individuals worked with varying degrees of success on a screw-type extruder (Section 13.2.1), Mathew Gray in England is credited with the invention; his patent is dated 1879. As thermoplastics were subsequently developed, these screw extruders, originally designed for rubber, were adapted. An

extrusion machine specifically designed for thermoplastics was introduced in 1935.

Injection molding machines for plastics were adaptations of equipment designed for metal die casting (Historical Note 11.2). Around 1872, John Hyatt, an important figure in the development of plastics (Historical Note 8.1), patented a molding machine specifically for plastics. It was a plunger-type machine (Section 13.6.3). The injection molding machine in its modern form was introduced in 1921, with semiautomatic controls added in 1937. Ram-type machines were the standard in the plastic molding industry for many decades, until the superiority of the reciprocating screw machine, developed by William Willert in the United States in 1952, became obvious.

We begin our coverage of the plastic-shaping processes by examining the properties of polymer melts, since nearly all of the thermoplastic-shaping processes share the common step of heating the plastic so that it flows.

13.1 PROPERTIES OF POLYMER MELTS

To shape a thermoplastic polymer it must be heated so that it softens to the consistency of a liquid. In this form, it is called a *polymer melt*. Polymer melts exhibit several unique properties and characteristics, considered in this section.

Viscosity Because of its high molecular weight, a polymer melt is a thick fluid with high viscosity. As we defined the term in Section 3.4, *viscosity* is a fluid property which relates the shear stress experienced during flow of the fluid to the rate of shear. Viscosity is important in polymer processing because most of the shaping methods involve flow of the polymer melt through small channels or die openings. The flow rates are often large, thus leading to high rates of shear; and the shear stresses increase with shear rate, so that significant pressures are required to accomplish the processes.

Figure 13.1 shows viscosity as a function of shear rate for two types of fluids. For a *Newtonian fluid* (which includes most simple fluids such as water and oil), viscosity is a constant at a given temperature; it does not change with shear rate. The relationship

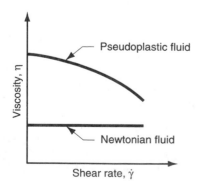

FIGURE 13.1 Viscosity relationships for Newtonian fluid and typical polymer melt.

between shear stress and shear strain is proportional, with viscosity as the constant of proportionality:

$$\tau = \eta \dot{\gamma} \quad \text{or} \quad \eta = \frac{\tau}{\dot{\gamma}} \tag{13.1}$$

where τ = shear stress, Pa (lb/in.^{2}); η = coefficient of shear viscosity, Ns/m^2, or Pas (lb-sec/in.^{2}); and $\dot{\gamma}$ = shear rate, 1/s (1/sec). However, for a polymer melt, viscosity decreases with shear rate, indicating that the fluid becomes thinner at higher rates of shear. This behavior is called ***pseudoplasticity*** and can be modeled to a reasonable approximation by the expression

$$\tau = k(\dot{\gamma})^n \tag{13.2}$$

where k = a constant corresponding to the viscosity coefficient and n = flow behavior index. For $n = 1$, the equation reduces to the previous Eq. (13.1) for a Newtonian fluid, and k becomes η. For a polymer melt, values of n are less than 1.

In addition to the effect of shear rate (fluid flow rate), viscosity of a polymer melt is also affected by temperature. Like most fluids, the value decreases with increasing temperature. This is shown in Figure 13.2 for several common polymers at the same shear rate of 10^3 s^{-1}. This shear rate approximates to those encountered in injection molding and high speed extrusion operations.

Thus we see that the viscosity of a polymer melt decreases with increasing values of shear rate and temperature. Our previous Eq. (13.2) can be applied, except that k depends on temperature as shown in Figure 13.2.

Viscoelasticity   Another property possessed by polymer melts is ***viscoelasticity.*** We discussed this property in the context of solids (solid polymers) in Section 3.5. However, liquid polymers manifest it also. A good example is ***die swell*** in extrusion, in which the hot plastic expands when exiting the die opening. The phenomenon, illustrated in Figure 13.3, can be explained by noting that the polymer was contained in a much larger cross section before entering the narrow die channel. In effect, the extruded material "remembers" its former shape and attempts to return to it after leaving the die orifice. More technically, the compressive stresses acting on the material as it enters the small die opening

FIGURE 13.2   Viscosity as a function of temperatures for selected polymers at a shear rate of 10^3 s^{-1} (data compiled from [12]).

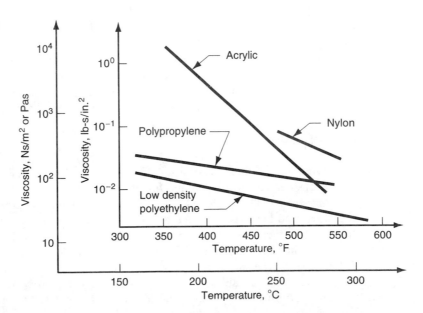

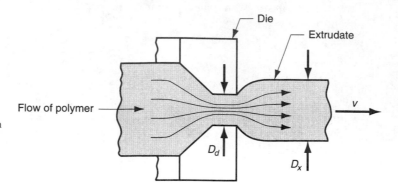

FIGURE 13.3 Die swell, a manifestation of viscoelasticity in polymer melts, as depicted here on exiting an extrusion die.

do not relax immediately. When the material subsequently exits the orifice and the restriction is removed, the unrelaxed stresses cause the cross section to expand.

Die swell can be most easily measured for a circular cross section by means of the **swell ratio,** defined as

$$r_s = \frac{D_x}{D_d} \tag{13.3}$$

where r_s = swell ratio; D_x = diameter of the extruded cross-section, mm (in.); and D_d = diameter of the die orifice, mm (in.). The amount of die swell depends on the time the polymer melt spends in the die channel. Increasing the time in the channel, by means of a longer channel, reduces die swell.

13.2 EXTRUSION

Extrusion is one of the fundamental shaping processes, for metals and ceramics as well as polymers. **Extrusion** is a compression process in which material is forced to flow through a die orifice to provide long continuous product whose cross-sectional shape is determined by the shape of the orifice. As a polymer-shaping process, it is widely used for thermoplastics and elastomers (but rarely for thermosets) to mass produce items such as tubing, pipes, hose, structural shapes (such as window and door molding), sheet and film, continuous filaments, and coated electrical wire and cable. For these types of products, extrusion is carried out as a continuous process; the **extrudate** (extruded product) is subsequently cut into desired lengths. In this section we cover the basic extrusion process, and in several subsequent sections we examine processes based on extrusion.

13.2.1 Process and Equipment

In polymer extrusion, feedstock in pellet or powder form is fed into an extrusion barrel where it is heated and melted and forced to flow through a die opening by means of a rotating screw, as illustrated in Figure 13.4. The two main components of the extruder are the barrel and the screw. The die is not a component of the extruder; it is a special tool that must be fabricated for the particular profile to be produced.

The internal diameter of the extruder barrel typically ranges from 25 to 150 mm (1.0 to 6.0 in.). The barrel is long relative to its diameter, with L/D ratios usually between 10 and 30. The L/D ratio is reduced in Figure 13.4 for clarity of drawing. The higher ratios are used for thermoplastic materials, while lower L/D values are for elastomers. A

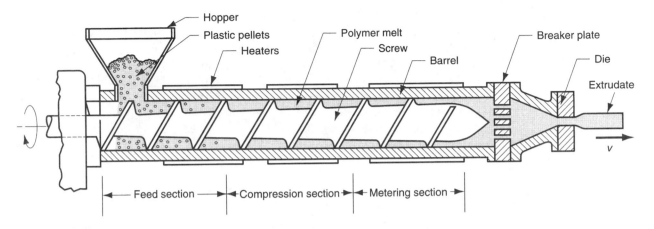

FIGURE 13.4   Components and features of a (single-screw) extruder for plastics and elastomers.

hopper containing the feedstock is located at the end of the barrel opposite the die. The pellets are fed by gravity onto the rotating screw whose turning moves the material along the barrel. Electric heaters are utilized to initially melt the solid pellets; subsequent mixing and mechanical working of the material generates additional heat, which maintains the melt. In some cases, enough heat is supplied through the mixing and shearing action that external heating is not required. Indeed, in some cases the barrel must be externally cooled in order to prevent overheating of the polymer.

The material is conveyed through the barrel toward the die opening by the action of the extruder screw, which rotates at about 60 rev/min. The screw serves several functions and is divided into sections that correspond to these functions. The sections and functions are (1) *feed section,* in which the stock is moved from the hopper port and preheated; (2) *compression section,* where the polymer is transformed into liquid consistency, air entrapped amongst the pellets is extracted from the melt, and the material is compressed; and (3) *metering section,* in which the melt is homogenized and sufficient pressure is developed to pump it through the die opening.

The operation of the screw is determined by its geometry and speed of rotation. A typical extruder screw geometry is depicted in Figure 13.5. The screw consists of spiraled "flights" (threads) with channels between them through which the polymer melt is moved.

FIGURE 13.5   Details of an extruder screw inside the barrel.

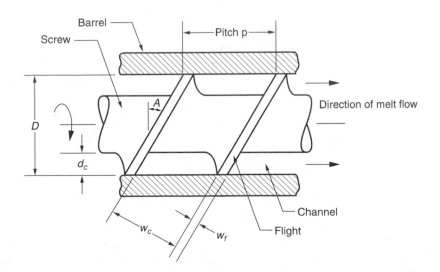

The channel has a width w_c and depth d_c. As the screw rotates, the flights push the material forward through the channel from the hopper end of the barrel toward the die. Although not discernible in our diagram, the flight diameter is smaller than the barrel diameter D by a very small clearance—around 0.05 mm (0.002 in.). The function of the clearance is to limit leakage of the melt backward to the trailing channel. The flight land has a width w_f and is made of hardened steel to resist wear as it turns and rubs against the inside of the barrel. The screw has a pitch whose value is usually close to the diameter D. The flight angle A is the helix angle of the screw and can be determined from the relation

$$\tan A = \frac{p}{\pi D} \tag{13.4}$$

where p = pitch of the screw.[1]

The increase in pressure applied to the polymer melt in the three sections of the barrel is determined largely by the channel depth d_c. In Figure 13.4, d_c is relatively large in the feed section to allow large amounts of granular polymer to be admitted into the barrel. In the compression section, d_c is gradually reduced, thus applying increased pressure on the polymer as it melts. In the metering section, d_c is small and pressure reaches a maximum as flow is restrained by the screen pack and backer plate. The three sections of the screw are shown as being about equal in length in Figure 13.4; this is appropriate for a polymer which melts gradually, such as low-density polyethylene. For other polymers, the optimal section lengths are different. For crystalline polymers such as nylon, melting occurs rather abruptly at a specific melting point, and therefore a short compression section is appropriate. Amorphous polymers such as polyvinylchloride melt more slowly than LDPE, and the compression zone for these materials must take almost the entire length of the screw. Although the optimal screw design for each material type is different, it is common practice to use general purpose screws. These designs represent a compromise among the different materials, and they avoid the need to make frequent screw changes with the associated equipment downtime.

Progress of the polymer along the barrel leads ultimately to the die zone. Before reaching the die, the melt passes through a screen pack—a series of wire meshes supported by a stiff plate (called a **breaker plate**) containing small axial holes. The screen pack assembly functions to (1) filter contaminants and hard lumps from the melt; (2) build pressure in the metering section; and (3) straighten the flow of the polymer melt and remove its "memory" of the circular motion imposed by the screw. This last function is concerned with the polymer's viscoelastic property; if the flow were left unstraightened, the polymer would play back its history of turning inside the extrusion chamber, tending to twist and distort the extrudate.

What we have described here is the conventional **single-screw extrusion machine**. We should also mention **twin-screw extruders** because they occupy an important place in the industry. In these machines, the screws are parallel and side-by-side inside the barrel. The twin-screw extruder seems especially suited for rigid PVC, normally a difficult polymer to extrude, and for materials that require greater mixing.

13.2.2 Analysis of Extrusion

In this section, we develop mathematical models to describe, in a simplified way, several aspects of polymer extrusion.

---

[1]Do not be confused. We use the same symbol, p, for pressure later in the chapter.

Melt Flow in the Extruder   As the screw rotates inside the barrel, the polymer melt is forced to move forward toward the die; the system operates much like an Archimedian screw. The principal transport mechanism is ***drag flow,*** resulting from friction between the viscous liquid and two opposing surfaces moving relative to each other: (1) the stationary barrel and (2) the channel of the turning screw. The arrangement can be likened to the fluid flow that occurs between a stationary plate and a moving plate separated by a viscous liquid, as illustrated in Figure 3.17. Given that the moving plate has a velocity v, it can be reasoned that the average velocity of the fluid is $v/2$, resulting in a volume flow rate of

$$Q_d = 0.5 \, vdw \tag{13.5}$$

where Q_d = volume drag flow rate, m³/s (in.³/sec.); v = velocity of the moving plate, m/s (in./sec.); d = distance separating the two plates, m (in.); and w = the width of the plates perpendicular to velocity direction, m (in.). These parameters can be compared to those in the channel defined by the rotating extrusion screw and the stationary barrel surface.

$$v = \pi DN \cos A \tag{13.6}$$

$$d = d_c \tag{13.7}$$

$$\text{and} \ \ w = w_c = (\pi D \tan A - w_f) \cos A \tag{13.8}$$

where D = screw flight diameter, m (in.); N = screw rotational speed, rev/s; d_c = screw channel depth, m (in.); w_c = screw channel width, m (in.); A = flight angle; and w_f = flight land width, m (in.). If we assume that the flight land width is negligibly small, then the last of these equations reduces to

$$w_c = \pi D \tan A \cos A = \pi D \sin A \tag{13.9}$$

Substituting Eqs. (13.6), (13.7), and (13.9) into Eq. (13.5), and using several trigonometric identities, we get

$$Q_d = 0.5 \, \pi^2 D^2 N d_c \sin A \cos A \tag{13.10}$$

If no forces were present to resist the forward motion of the fluid, this equation would provide a reasonable description of the melt flow rate inside the extruder. However, compressing the polymer melt through the downstream die creates a ***back pressure*** in the barrel that reduces the material moved by drag flow in Eq. (13.10). This flow reduction, which we shall call the ***back pressure flow,*** depends on the screw dimensions, viscosity of the polymer melt, and pressure gradient along the barrel. These dependencies can be summarized in the equation [12]:

$$Q_b = \frac{\pi D d_c^3 \sin^2 A}{12\eta} \left(\frac{dp}{dl} \right) \tag{13.11}$$

where Q_b = back pressure flow, m³/s (in.³/sec); η = viscosity, N-s/m² (lb-sec/in.²); dp/dl = the pressure gradient, MPa/m (lb/in.²/in.); and the other terms were previously defined. The actual pressure gradient in the barrel is a function of the shape of the screw over its length; a typical pressure profile is given in Figure 13.6. If we assume as an approximation that the profile is a straight line, indicated by the dashed line in the figure, then the pressure gradient becomes a constant p/L, and the previous equation reduces to

$$Q_b = \frac{p\pi D d_c^3 \sin^2 A}{12\eta L} \tag{13.12}$$

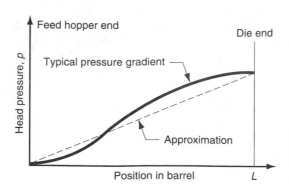

FIGURE 13.6   Typical pressure gradient in an extruder; dashed line indicates a straight-line approximation to facilitate computations.

where p = head pressure in the barrel, MPa (lb/in.^{2}); and L = the length of the barrel, m (in.). Recall that this back pressure flow is really not an actual flow by itself; it is a reduction in the drag flow. Thus, we can compute the magnitude of the melt flow in an extruder as the difference between the drag flow and the back pressure flow:

$$Q_x = Q_d - Q_b$$

$$Q_x = 0.5 \ \pi^2 D^2 N d_c \sin A \cos A - \frac{p\pi D d_c^3 \sin^2 A}{12\eta L} \tag{13.13}$$

where Q_x = the resulting flow rate of polymer melt in the extruder. Eq. (13.13) assumes that there is minimal **leak flow** through the clearance between flights and barrel. Leak flow of melt will be small compared to drag and back pressure flow except in badly worn extruders.

Equation (13.13) contains many parameters, which can be divided into two types: (1) design parameters, and (2) operating parameters. The design parameters are those that define the geometry of the screw and barrel: diameter D, channel depth d_c, and helix angle A. For a given extruder operation, these factors cannot be changed during the process. The operating parameters are those that can be changed during the process to affect output flow; they include rotational speed N, head pressure p, and melt viscosity η. Of course, melt viscosity is controllable only to the extent to which temperature and shear rate can be manipulated to affect this property. Let us see how the parameters play out their roles in the following example.

**EXAMPLE 13.1
Extrusion flow
rates.**

An extruder barrel has a diameter D = 75 mm. The screw rotates at N = 1 rev/s. Channel depth d_c = 6.0 mm and flight angle A = 20°. Head pressure at the end of the barrel p = 7.0×10^6 Pa, length of the barrel L = 1.9 m, and viscosity of the polymer melt is assumed to be η = 100 Pas. Determine the volume flow rate of the plastic in the barrel Q_x.

Solution: Using Eq. (13.13) we can compute the drag flow and opposing back pressure flow in the barrel.

$$Q_d = 0.5 \ \pi^2 (75 \times 10^{-3})^2 (1.0)(6 \times 10^{-3})(\sin 20)(\cos 20) = 53{,}525(10^{-9}) \ \text{m}^3/\text{s}$$

$$Q_b = \frac{\pi(7 \times 10^6)(75 \times 10^{-3})(6 \times 10^{-3})^3(\sin 20)^2}{12(100)(1.9)} = 18.276(10^{-6}) = 18{,}276(10^{-9}) \ \text{m}^3/\text{s}$$

$$Q_x = Q_d - Q_b = (53{,}525 - 18{,}276)(10^{-9}) = \mathbf{35{,}249(10^{-9}) \ m^3/s}$$

Extruder and Die Characteristics   If back pressure is zero, so that melt flow is unrestrained in the extruder, then the flow would equal drag flow Q_d given by Eq. (13.10). Given the design and operating parameters (D, A, N, etc.), this is the maximum possible flow capacity of the extruder. Let us denote it as Q_{max}:

$$Q_{max} = 0.5\ \pi^2 D^2 N d_c \sin A \cos A \tag{13.14}$$

On the other hand, if back pressure were so great as to cause zero flow, then back pressure flow would equal drag flow; that is,

$$Q_x = Q_d - Q_b = 0,\ \text{so}\ Q_d = Q_b$$

Using the expressions for Q_d and Q_b in Eq. (13.13), we can solve for p to determine what this maximum head pressure p_{max} would have to be to cause no flow in the extruder:

$$p_{max} = \frac{6\pi DNL\eta\ \cot A}{d_c^2} \tag{13.15}$$

The two values Q_{max} and p_{max} are points along the axes of a diagram known as the ***extruder characteristic*** (or ***screw characteristic***), as in Figure 13.7. It defines the relationship between head pressure and flow rate in an extrusion machine with given operating parameters.

With a die in the machine and the extrusion process underway, the actual values of Q_x and p will lie somewhere between the extreme values, the location determined by the characteristics of the die. Flow rate through the die depends on the size and shape of the opening and the pressure applied to force the melt through it. This can be expressed:

$$Q_x = K_s\ p \tag{13.16}$$

where Q_x = flow rate, m³/s (in.³/sec.); p = head pressure, Pa (lb/in.²); and K_s = shape factor for the die, m⁵/Ns (in.⁵/lb-sec.). For a circular die opening of a given channel length, the shape factor can be computed [12] as

$$K_s = \frac{\pi D_d^4}{128\eta L_d} \tag{13.17}$$

where D_d = die opening diameter, m (in.); η = melt viscosity, N-s/m² (lb-sec/in.²); and L_d = die opening length, m (in.). For shapes other than round, the die shape factor is less than for a round of the same cross-sectional area, meaning that greater pressure is required to achieve the same flow rate.

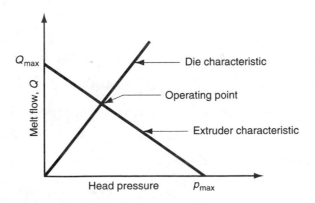

FIGURE 13.7 Extruder characteristic (also called the screw characteristic) and die characteristic. Extruder operating point is at the intersection of the two lines.

The relationship between Q_x and p in Eq. (13.16) is called the *die characteristic*. In Figure 13.7, this is drawn as a straight line, adding to the previous extruder characteristic. The two plots intersect; the corresponding values of Q_x and p are known as the *operating point* for the extrusion process.

EXAMPLE 13.2 Extruder and die characteristics.

Consider the extruder from Example 13.1, in which $D = 75$ mm, $L = 1.9$ m, $N = 1$ rev/s, $d_c = 6$ mm, and $A = 20°$. The plastic melt has a shear viscosity $\eta = 100$ Pas. Determine: (a) Q_{max} and p_{max}, (b) shape factor K_s for a circular die opening in which $D_d = 6.5$ mm and $L_d = 20$ mm, and (c) values of Q_x and p at the operating point.

Solution: (a) Q_{max} is given by Eq. (13.14).

$$Q_{max} = 0.5\pi^2 D^2 N d_c \sin A \cos A$$

$$= 0.5\,\pi^2 (75 \times 10^{-3})^2 (1.0)(6 \times 10^{-3})(\sin 20)(\cos 20)$$

$$= \mathbf{53{,}525(10^{-9})\ m^3/s}$$

p_{max} is given by Eq. (13.15).

$$p_{max} = \frac{6\pi DNL\eta \cot A}{d_c^2} = \frac{6\pi(75 \times 10^{-3})(1.9)(1.0)(100) \cot 20}{(6 \times 10^{-3})^2} = \mathbf{20{,}499{,}874\ Pa}$$

These two values define the intersection with the ordinate and abscissa for the extruder characteristic.

(b) The shape factor for a circular die opening with $D_d = 6.5$ mm and $L_d = 20$ mm can be determined from Eq. (13.17).

$$K_s = \frac{\pi(6.5 \times 10^{-3})^4}{128(100)(20 \times 10^{-3})} = \mathbf{21.9(10^{-12})\ m^5/Ns}$$

This shape factor defines the slope of the die characteristic.

(c) The operating point is defined by the values of Q_x and p at which the screw characteristic intersects with the die characteristic. The screw characteristic can be expressed as the equation of the straight line between Q_{max} and p_{max}, which is

$$Q_x = Q_{max} - (Q_{max}/p_{max})p \qquad (13.18)$$

$$= 53{,}525(10^{-9}) - (53{,}525(10^{-9})/20{,}499{,}874)p = 53{,}525(10^{-9}) - 2.611(10^{-12})p$$

The die characteristic is given by Eq. (13.16) using the value of K_S computed in part (b).

$$Q_x = 21.9(10^{-12})p$$

Setting the two equations equal, we have

$$53{,}525(10^{-9}) - 2.611(10^{-12})p = 21.9(10^{-12})p$$

$$p = \mathbf{2.184(10^6)\ Pa}$$

Solving for Q_x using one of the starting equations, we obtain

$$Q_x = 53.525(10^{-6}) - 2.611(10^{-12})(2.184)(10^6) = \mathbf{47.822(10^{-6})\ m^3/s}$$

Checking this with the other equation for verification,

$$Q_x = 21.9(10^{-12})(2.184)(10^6) = 47.82(10^{-6})\ m^3/s$$

13.2.3 Die Configurations and Extruded Products

The shape of the die orifice determines the cross-sectional shape of the extrudate. We can enumerate the common die profiles and corresponding extruded shapes as follows: (1) solid profiles; (2) hollow profiles, such as tubes; (3) wire and cable coating; (4) sheet and film; and (5) filaments. The first three categories are covered in the present section. Methods for producing sheet and film are examined in Section 13.3; and filament production is discussed in Section 13.4. These latter shapes sometimes involve forming processes other than extrusion.

Solid Profiles   Solid profiles include regular shapes such as rounds and squares and irregular cross sections such as structural shapes, door and window moldings, automobile trim, and house siding. The side view cross section of a die for these solid shapes is illustrated in Figure 13.8. Just beyond the end of the screw and before the die, the polymer melt passes through the screen pack and breaker plate to straighten the flow lines. Then it flows into a (usually) converging die entrance, the shape designed to maintain laminar flow and avoid dead spots in the corners that would otherwise be present near the orifice. The melt then flows through the die opening itself.

When the material exits the die, it is still soft. Polymers with high melt viscosities are the best candidates for extrusion, since they hold shape better during cooling. Cooling is accomplished by air blowing, water spray, or by passing the extrudate through a water trough. To compensate for die swell, the die opening is made long enough to remove some of the memory in the polymer melt. In addition, the extrudate is often drawn (stretched) to offset expansion from die swell.

For shapes other than round, the die opening is designed with a cross section that is slightly different from the desired profile, so that the effect of die swell is to provide shape correction. This correction is illustrated in Figure 13.9 for a square cross section. Since different polymers exhibit varying degrees of die swell, the shape of the die profile depends on the material to be extruded. Considerable skill and judgment are required by the die designer for complex cross sections.

FIGURE 13.8   (a) Side view cross section of an extrusion die for solid regular shapes, such as round stock; (b) front view of die, with profile of extrudate. Die swell is evident in both views. (Some die construction details are simplified or omitted for clarity.)

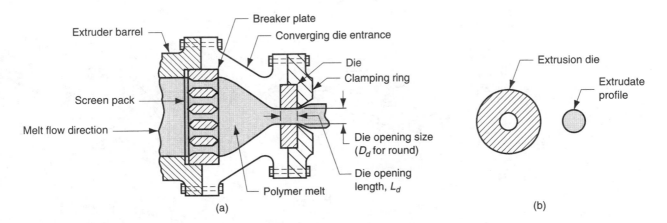

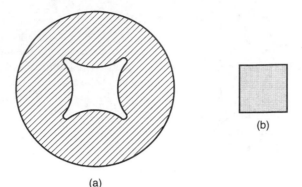

FIGURE 13.9 (a) Die cross section showing required orifice profile to obtain (b) a square extruded profile.

(a)

(b)

Hollow Profiles Extrusion of hollow profiles, such as tubes, pipes, hoses, and other cross-sections containing holes requires a mandrel to form the hollow shape. A typical die configuration is shown in Figure 13.10. The mandrel is held in place using a spider, seen in Section A-A of the figure. The polymer melt flows around the legs supporting the mandrel to reunite into a monolithic tube wall. The mandrel often includes an air channel through which air is blown to maintain the hollow form of the extrudate during hardening. Pipes and tubes are cooled using open water troughs or by pulling the soft extrudate through a water-filled tank with sizing sleeves that limit the OD of the tube while air pressure is maintained on the inside.

Wire and Cable Coating The coating of wire and cable for insulation is one of the most important polymer extrusion processes. As shown in Figure 13.11 for wire coating, the polymer melt is applied to the bare wire as it is pulled at high speed through a die. A slight vacuum is drawn between the wire and the polymer to promote adhesion of the coating. The taught wire provides rigidity during cooling, which is usually aided

FIGURE 13.10 Side view cross section of extrusion die for shaping hollow cross sections such as tubes and pipes; Section A-A is a front view cross section showing how the mandrel is held in place; Section B-B shows the tubular cross section just prior to exiting the die; die swell causes an enlargement of the diameter. (Some die construction details are simplified.)

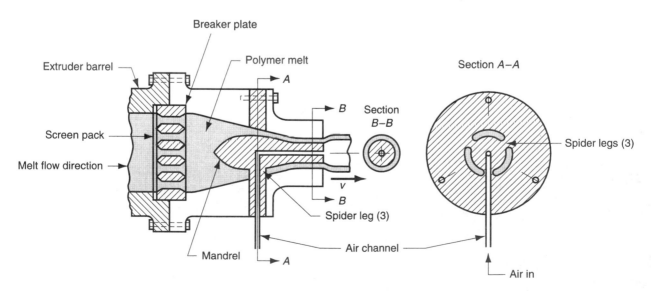

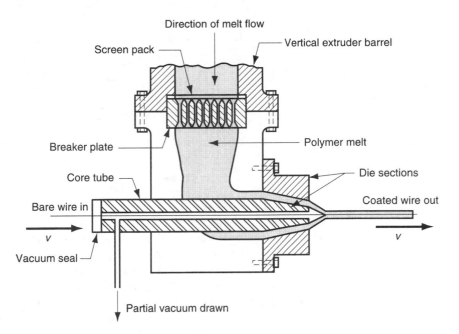

FIGURE 13.11   Side view cross section of die for coating of electrical wire by extrusion. (Some die construction details are simplified.)

by passing the coated wire through a water trough. The product is wound onto large spools at speeds up to 50 m/s (10,000 ft/min).

13.2.4 Defects in Extrusion

A number of defects can afflict extruded products. One of the worst is ***melt fracture,*** in which the stresses acting on the melt immediately before and during its flow through the die are so high as to cause failure, manifested in the form of a highly irregular surface on the extrudate. As suggested by Figure 13.12, melt fracture can be caused by a sharp reduction at the die entrance, causing turbulent flow that breaks up the melt. This contrasts with the streamlined, laminar flow in the gradually converging die in Figure 13.8.

A more common defect in extrusion is ***sharkskin,*** in which the surface of the product becomes roughened upon exiting the die. As the melt flows through the die opening, friction at the interface results in a velocity profile across the cross-section, Figure 13.13. Tensile stresses develop at the surface as this material is stretched to keep up with the faster moving center core. These stresses cause minor ruptures that roughen the surface. If the velocity gradient becomes extreme, prominent marks occur on the surface, giving it the appearance of a bamboo pole; hence, the name ***bambooing*** for this more severe defect.

FIGURE 13.12   Melt fracture, caused by turbulent flow of the melt through a sharply reduced die entrance.

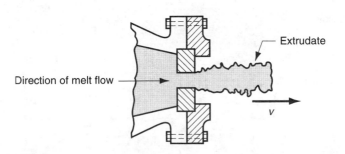

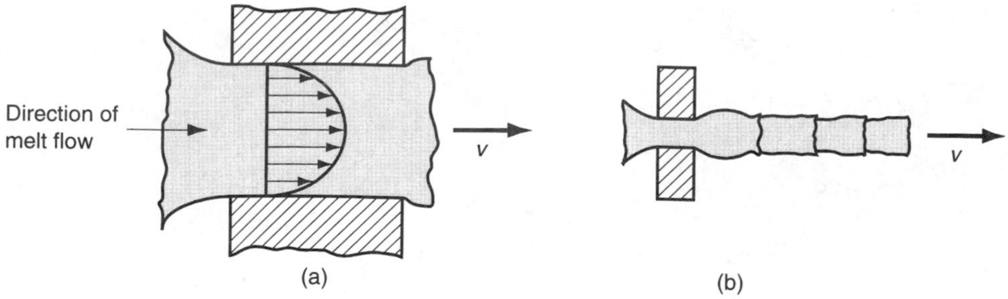

FIGURE 13.13 (a) Velocity profile of the melt as it flows through the die opening, which can lead to defects called sharkskin and (b) bambooing.

(a)

(b)

13.3 PRODUCTION OF SHEET AND FILM

Thermoplastic sheet and film are produced by a number of processes, most important of which are two methods based on extrusion. The term *sheet* refers to stock with a thickness ranging from 0.5 mm (0.020 in.) to about 12.5 mm (0.5 in.) and used for products such as flat window glazing and stock for thermoforming (Section 13.9). *Film* refers to thicknesses below 0.5 mm (0.020 in.). Thin films are used for packaging (product wrapping material, grocery bags, and garbage bags); thicker film applications include covers and liners (pool covers and liners for irrigation ditches).

All of the processes covered in this section are continuous, high-production operations. More than half of the films produced today are polyethylene, mostly low-density PE. The principal other materials are polypropylene, polyvinylchloride, and regenerated cellulose (cellophane). These are all thermoplastic polymers.

Slit-Die Extrusion of Sheet and Film Sheet and film of various thickness are produced by conventional extrusion, using a narrow slit as the die opening. The slit may be up to 3 m (10 ft) wide and as narrow as around 0.4 mm (0.015 in.). One possible die configuration is illustrated in Figure 13.14. The die includes a manifold which spreads the polymer melt laterally before it flows through the slit (die orifice). One of the difficulties in this extrusion method is uniformity of thickness throughout the width of the stock. This is due to the drastic shape change experienced by the polymer melt during its flow through the die and to temperature and pressure variations in the die. Usually, the edges of the film must be trimmed because of thickening at the edges. To help compensate for these variations, the dies include adjustable lips (not shown in our diagram) which allow the width of the slit to be altered.

FIGURE 13.14 One of several die configurations for extruding sheet and film.

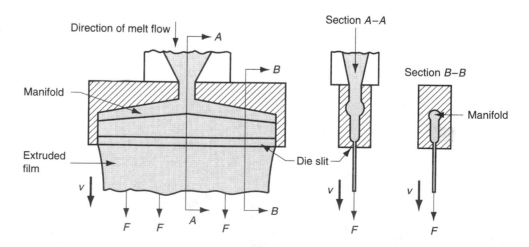

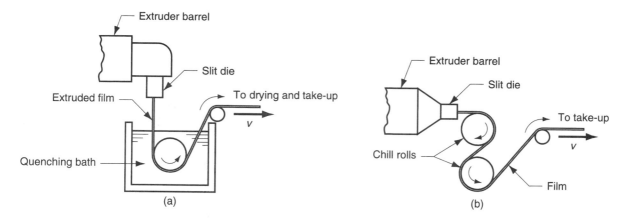

FIGURE 13.15   Use of (a) a water quenching bath or (b) chill rolls to achieve fast solidification of the molten film after extrusion.

To achieve high production rates, an efficient method of cooling and collecting the film must be integrated with the extrusion process. This is usually done by immediately directing the extrudate into a quenching bath of water or onto chill rolls, as shown in Figure 13.15. The chill roll method seems to be the more important commercially. Contact with the cold rolls quickly quenches and solidifies the extrudate; in effect, the extruder serves as a feeding device for the chill rolls which actually form the film. The process is noted for very high production speeds—5 m/s (1000 ft/min). In addition, close tolerances on film thickness can be achieved. Owing to the cooling method used in this process, it is known as ***chill-roll extrusion.***

Blown-Film Extrusion Process   This is the other widely used process for making thin polyethylene film for packaging. It is a complex process, combining extrusion and blowing to produce a tube of thin film; it is best explained with reference to the diagram in Figure 13.16. The process begins with the extrusion of a tube that is immediately drawn upward while still molten and simultaneously expanded in size by air inflated into it through the die mandrel. A "frost line" marks the position along the upward moving bubble where solidification of the polymer occurs. Air pressure in the bubble must be kept constant to maintain uniform film thickness and tube diameter. The air is contained in the tube by pinch rolls which squeeze the tube back together after it has cooled. Guide rolls and collapsing rolls are also used to restrain the blown tube and direct it into the pinch rolls. The flat tube is then collected onto a windup reel.

The effect of air inflation is to stretch the film in both directions as it cools from the molten state. This results in isotropic strength properties, which is an advantage over other processes in which the material is stretched primarily in one direction. Other advantages include the ease with which extrusion rate and air pressure can be changed to control stock width and gage. Comparing this process with slit-die extrusion, the blown-film method produces stronger film (so that a thinner film can be used to package a product), but thickness control and production rates are lower. The final blown film can be left in tubular form (e.g., for garbage bags), or it can be subsequently cut at the edges to provide two parallel thin films.

Calendering   Calendering is a process for producing sheet and film stock out of rubber (Section 14.1.4) or rubbery thermoplastics such as plasticized PVC. In the process, the initial feedstock is passed through a series of rolls to work the material and reduce its thickness to the desired gage. A typical setup is illustrated in Figure 13.17.

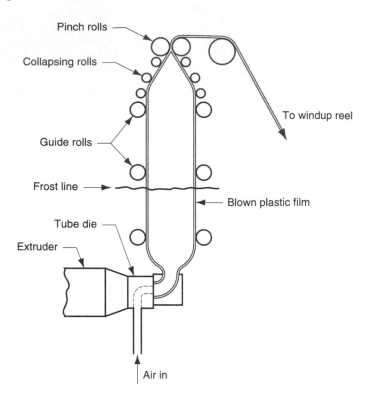

FIGURE 13.16   Blown-film process for high production of thin tubular film.

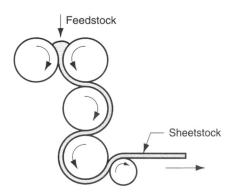

FIGURE 13.17   A typical roll configuration in calendering.

The equipment is expensive, but production rate is high; speeds approaching 2.5 m/s (500 ft/min) are possible. Close control is required over roll temperatures, pressures, and rotational speed. The process is noted for its good surface finish and high gage accuracy in the film. Plastic products made by the calendering process include PVC floor covering, shower curtains, vinyl table cloths, pool liners, and inflatable boats and toys.

13.4 FIBER AND FILAMENT PRODUCTION (SPINNING)

The most important application of fibers and filaments is in textiles. Their use as reinforcing materials in plastics (composites) is a growing application, but still small compared to textiles. A *fiber* can be defined as a long, thin strand of material whose length is at least 100 times its cross-sectional dimension. A *filament* is a fiber of continuous length.

Fibers can be natural or synthetic. Synthetic fibers constitute about 75% of the total fiber market today, polyester being the most important, followed by nylon, acrylics, and rayon. Natural fibers are about 25% of the total produced, with cotton by far the most important staple (wool production is significantly less than cotton).

The term *spinning* is a holdover from the methods used to draw and twist natural fibers into yarn or thread. In the production of synthetic fibers, the term refers to the process of extruding a polymer melt or solution through a *spinneret* (a die with multiple small holes) to make filaments which are then drawn and wound onto a *bobbin.* There are three principal variations in the spinning of synthetic fibers, depending on the polymer being processed: (1) melt spinning, (2) dry spinning, and (3) wet spinning.

Melt spinning is used when the starting polymer can best be processed by heating to the molten state and pumping through the spinneret, much in the manner of conventional extrusion. A typical spinneret is 6 mm (0.25 in.) thick and contains approximately 50 holes of diameter 0.25 mm (0.010 in.); the holes are countersunk, so that the resulting bore has a L/D ratio of only 5/1 or less. The filaments that emanate from the die are drawn and simultaneously air cooled before being collected together and spooled onto the bobbin, as shown in Figure 13.18. Significant extension and thinning of the filaments occur while the polymer is still molten, so that the final diameter wound onto the bobbin may be only 1/10 of the extruded size. Melt spinning is used for polyesters and nylons; since these are the most important synthetic fibers, melt spinning is the most important of the three processes for synthetic fibers.

In *dry spinning,* the starting polymer is in solution and the solvent can be separated by evaporation. The extrudate is pulled through a heated chamber that removes the solvent; otherwise, the sequence is similar to the previous. Fibers of cellulose acetate and acrylic are produced by this process. In *wet spinning,* the polymer is also in solution, only

FIGURE 13.18   Melt spinning of continuous filaments.

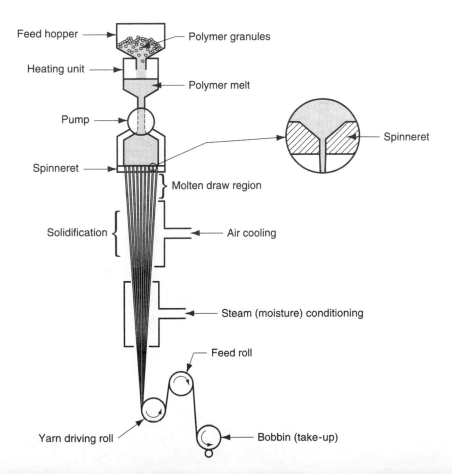

the solvent is nonvolatile. To separate the polymer, the extrudate must be passed through a liquid chemical that coagulates or precipitates the polymer into coherent strands that are then collected onto bobbins. This method is used to produce rayon (regenerated cellulose fibers).

Filaments produced by any of the three processes are usually subjected to further cold drawing to align the crystal structure along the direction of the filament axis. Extensions of 2 to 8 are typical [13]. This has the effect of significantly increasing the tensile strength of the fibers. Drawing is accomplished by pulling the thread between two spools, where the winding spool is driven at a faster speed than the unwinding spool.

13.5 COATING PROCESSES

Plastic (or rubber) coating involves application of a layer of the given polymer onto a substrate material. Three categories are distinguished [6]: (1) wire and cable coating; (2) planar coating, which involves the coating of a flat film; and (3) contour coating—the coating of a three-dimensional object. We have already examined wire and cable coating (Section 13.2.3); it is basically an extrusion process. The other two categories are surveyed in the following paragraphs. In addition, there is the technology of applying paints, varnishes, lacquers, and other similar coatings (Section 29.5).

Planar coating is used to coat fabrics, paper, cardboard, and metal foil; these items are major products for some plastics. The important polymers include polyethylene and polypropylene, with lesser applications for nylon, PVC, and polyester. In most cases, the coating is only 0.01 to 0.05 mm (0.0005 to 0.002 in.) thick. The two major planar coating techniques are illustrated in Figure 13.19. In the *roll method,* the polymer coating material is squeezed against the substrate by means of opposing rolls. In the *doctor blade method,* a sharp knife edge controls the amount of polymer melt that is coated onto the substrate. In both cases, the coating material is supplied either by a slit-die extrusion process or by calendering.

Contour coating of three-dimensional objects can be accomplished by dipping or spraying. *Dipping* involves submersion of the object into a suitable bath of polymer melt or solution, followed by cooling or drying. *Spraying* (such as spray-painting) is an alternative method for applying a polymer coating to a solid object.

FIGURE 13.19   Planar coating processes: (a) roll method, and (b) doctor-blade method.

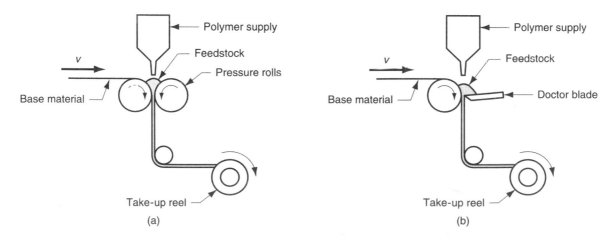

(a)   (b)

13.6 INJECTION MOLDING

Injection molding is a process in which a polymer is heated to a highly plastic state and forced to flow under high pressure into a mold cavity, where it solidifies. The molded part, called a *molding,* is then removed from the cavity. The process produces discrete components that are almost always net shape. The production cycle time is typically in the range 10 to 30 seconds, although cycles of one minute or longer are not uncommon. Also, the mold may contain more than one cavity, so that multiple moldings are produced each cycle.

Complex and intricate shapes are possible with injection molding. The challenge in these cases is to design and fabricate a mold whose cavity is the same geometry as the part and which also allows for part removal. Part size can range from about 50 g (2 oz) up to about 25 kg (more than 50 lb), the upper limit represented by components such as refrigerator doors and automobile bumpers. The mold determines the part shape and size and is the special tooling in injection molding. For large complex parts, the mold can cost hundreds of thousands of dollars. For small parts, the mold can be built to contain multiple cavities, also making the mold expensive. Thus, injection molding is economical only for large production quantities.

Injection molding is the most widely used molding process for thermoplastics. Some thermosets and elastomers are injection molded, with modifications in equipment and operating parameters to allow for cross-linking of these materials. We discuss these and other variations of injection molding in Section 13.6.6.

13.6.1 Process and Equipment

Equipment for injection molding evolved from metal die casting (Historical Note 13.1). A large injection molding machine is shown in Figure 13.20. As illustrated in our schematic in Figure 13.21, an injection molding machine consists of two principal components: (1) the plastic injection unit and (2) the mold clamping unit. The *injection unit* is much like an extruder. It consists of a barrel that is fed from one end by a hopper containing a supply of plastic pellets. Inside the barrel is a screw whose operation surpasses that of an extruder screw in the following respect: in addition to turning for mixing and heating the polymer, it also acts as a ram which rapidly moves forward to inject molten plastic into the mold. A nonreturn valve mounted near the tip of the screw prevents the melt from flowing backward along the screw threads. Later in the molding cycle the ram retracts to its former position. Because of its dual action, it is called a *reciprocating screw,* which name also identifies the machine type. Older injection molding machines used a simple ram (without screw flights), but the superiority of the reciprocating screw design has led to its widespread adoption in today's molding plants. To summarize, the functions of the injection unit are to melt and homogenize the polymer, and then inject it into the mold cavity.

The *clamping unit* is concerned with the operation of the mold. Its functions are to (1) hold the two halves of the mold in proper alignment with each other; (2) keep the mold closed during injection by applying a clamping force sufficient to resist the injection force; and (3) open and close the mold at the appropriate times in the molding cycle. The clamping unit consists of two platens, a fixed platen and a movable platen, and a mechanism for translating the latter. The mechanism is basically a power press that is operated by hydraulic piston or mechanical toggle devices of various types. Clamping forces of several thousand tons are available on large machines.

FIGURE 13.20 A large (3000 ton capacity) injection molding machine (courtesy Cincinnati Milacron).

The cycle for injection molding of a thermoplastic polymer proceeds in the following sequence, illustrated in Figure 13.22. Let us pick up the action with the mold open and the machine ready to start a new molding: (1) Mold is closed and clamped. (2) A *shot* of melt, which has been brought to the right temperature and viscosity by heating and by the mechanical working of the screw, is injected under high pressure into the mold cavity. The plastic cools and begins to solidify when it encounters the cold surface

FIGURE 13.21   Diagram of an injection molding machine, reciprocating screw type (some mechanical details are simplified).

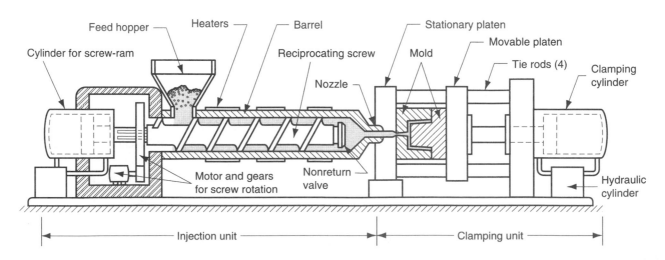

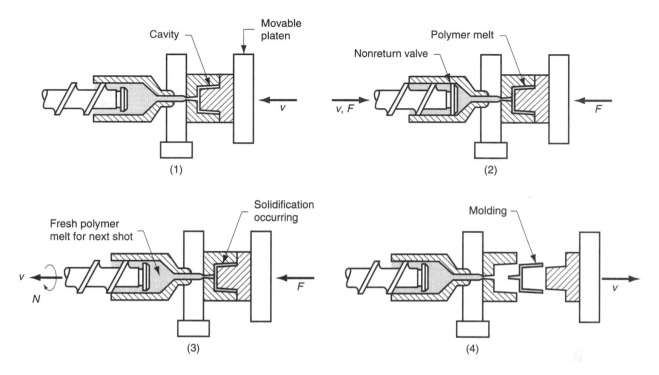

FIGURE 13.22   Typical molding cycle: (1) mold is closed, (2) melt is injected into cavity, (3) screw is retracted, and (4) mold opens and part is ejected.

of the mold. Ram pressure is maintained to pack additional melt into the cavity to compensate for contraction during cooling. (3) The screw is rotated and retracted with the nonreturn valve open to permit fresh polymer to flow into the forward portion of the barrel. Meanwhile, the polymer in the mold has completely solidified. (4) The mold is opened, and the part is ejected and removed.

13.6.2 The Mold

The mold is the special tool in injection molding; it is custom-designed and fabricated for the given part to be produced. When the production run for that part is finished, the mold is replaced with a new mold for the next part. In this section we examine several types of mold for injection molding.

Two-Plate Mold   The conventional *two-plate mold,* illustrated in Figure 13.23, consists of two halves fastened to the two platens of the molding machine's clamping unit. When the clamping unit is opened, the two mold halves open, as shown in (b). The most obvious feature of the mold is the *cavity,* which is usually formed by removing metal from the mating surfaces of the two halves. Molds can contain a single cavity or multiple cavities to produce more than one part in a single shot. The figure shows a mold with two cavities. The *parting surfaces* (or *parting line* in a cross-sectional view of the mold) is where the mold opens to remove the part(s).

In addition to the cavity, there are other features of the mold that serve indispensable functions during the molding cycle. A mold must have a distribution channel through which the polymer melt flows from the nozzle of the injection barrel into the mold cavity. The distribution channel consists of (1) a *sprue,* which leads from the nozzle into the mold;

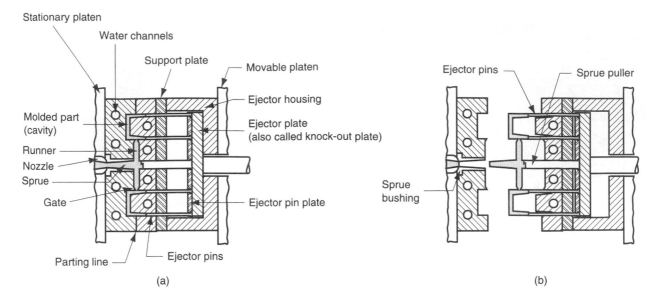

FIGURE 13.23 Details of a two-plate mold for thermoplastic injection molding: (a) closed and (b) open. Mold has two cavities to produce two cup-shaped parts (cross section shown) with each injection shot.

(2) *runners,* which lead from the sprue to the cavity (or cavities); and (3) *gates* that constrict the flow of plastic into the cavity. There are one or more gates for each cavity in the mold.

An *ejection system* is needed to eject the molded part from the cavity at the end of the molding cycle. *Ejector pins* built into the moving half of the mold usually accomplish this function. The cavity is divided between the two mold halves in such a way that the natural shrinkage of the molding causes the part to stick to the moving half. When the mold opens, the ejector pins push the part out of the mold cavity.

A *cooling system* is required for the mold. This consists of an external pump connected to passageways in the mold, through which water is circulated to remove heat from the hot plastic. Air must be evacuated from the mold cavity as the polymer rushes in. Much of the air passes through the small ejector pin clearances in the mold. In addition, narrow *air vents* are often machined into the parting surface; only about 0.03 mm (0.001 in.) deep and 12 to 25 mm (0.5 to 1.0 in.) wide, these channels permit air to escape to the outside but are too small for the viscous polymer melt to flow through.

To summarize, a mold consists of (1) one or more cavities that determine part geometry; (2) distribution channels through which the polymer melt flows to the cavities; (3) an ejection system for part removal; (4) a cooling system; and (5) vents to permit evacuation of air from the cavities.

Other Mold Types   The two-plate mold is the most common mold in injection molding. An alternative is a *three-plate mold,* shown in Figure 13.24, for the same part geometry as before. There are advantages to this mold design. First, the flow of molten plastic is through a gate located at the base of the cup-shaped part, rather than at the side. This allows more even distribution of melt into the sides of the cup. In the side gate design in the two-plate mold of Figure 13.23, the plastic must flow around the core and join on the opposite side, possibly creating a weakness at the weld line. Second, the three-plate mold allows more automatic operation of the mold-

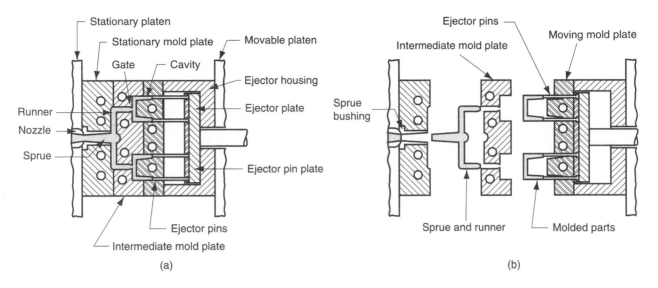

FIGURE 13.24   Three-plate mold: (a) closed and (b) open.

ing machine. As the mold opens, it divides into three plates with two openings between them. This forces disconnection of runner and parts, which drop by gravity (with possible assistance from blown air or a robotic arm) into different containers beneath the mold.

The sprue and runner in a conventional two-plate or three-plate mold represent waste material. In many instances they can be ground and reused; however, in some cases the product must be made of "virgin" plastic (that which has not been previously molded). The ***hot-runner mold*** eliminates the solidification of the sprue and runner by locating heaters around the corresponding runner channels. While the plastic in the mold cavity solidifies, the material in the sprue and runner channels remains molten, ready to be injected into the cavity in the next cycle.

13.6.3 Injection Molding Machines

Injection molding machines differ in both injection unit and clamping unit. This section discusses the important types of machines available today. The name of the injection molding machine is generally based on the type of injection unit used.

Injection Units   Two types of injection units are widely used today. The ***reciprocating-screw machine*** (Section 13.6.1, Figures 13.21 and 13.22) is the most common. This design uses the same barrel for melting and injection of plastic. The alternative unit involves the use of separate barrels for plasticizing and injecting the polymer, as shown in Figure 13.25(a). This type is called a ***screw-preplasticizer machine*** or ***two-stage machine.*** Plastic pellets are fed from a hopper into the first stage, which uses a screw to drive the polymer forward and melt it. This barrel feeds a second barrel, which uses a plunger to inject the melt into the mold. Older machines used one plunger-driven barrel to melt and inject the plastic. These machines are referred to as ***plunger-type injection molding machines,*** Figure 13.25(b).

Clamping Units   Clamping designs are of three types [11]: toggle, hydraulic, and hydromechanical. ***Toggle clamps*** include various designs, one of which is illustrated in Figure 13.26(a). An actuator moves the crosshead forward, extending the toggle links

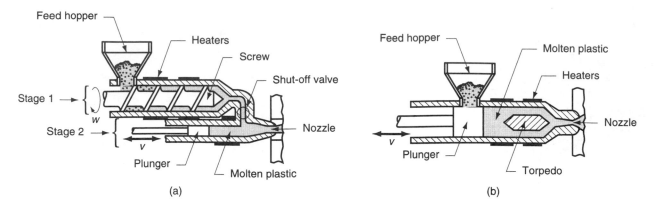

FIGURE 13.25   Two alternative injection systems to the reciprocating screw shown in Figure 13.21: (a) screw preplasticizer, and (b) plunger type.

to push the moving platen toward a closed position. At the beginning of the movement, mechanical advantage is low and speed is high; but near the end of the stroke, the reverse is true. Thus, toggle clamps provide both high speed and high force at different points in the cycle when they are desirable. They are actuated either by hydraulic cylinders or ball screws driven by electric motors. Toggle-clamp units seem most suited to relatively low-tonnage machines. **Hydraulic clamps,** shown in Figure 13.26(b), are used on higher-tonnage injection-molding machines, typically in the range 1300 to 8900 kN

FIGURE 13.26   Two clamping designs: (a) one possible toggle clamp design: (1) open and (2) closed; and (b) hydraulic clamping: (1) open and (2) closed. Tie rods used to guide moving platens not shown.

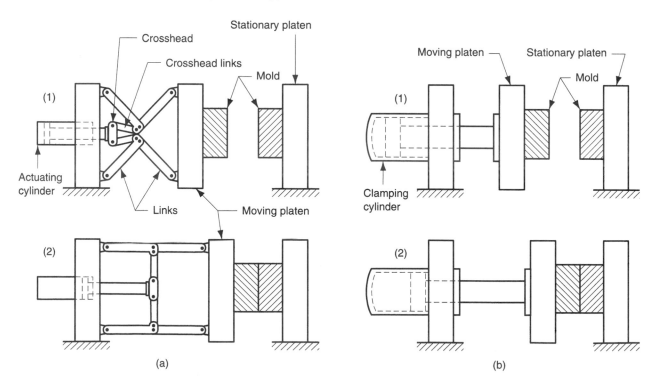

(150 to 1000 tons). These units are also more flexible than toggle clamps in terms of setting the tonnage at given positions during the stroke. *Hydromechanical clamps* are designed for large tonnages, usually above 8900 kN (1000 tons); they operate by (1) using hydraulic cylinders to rapidly move the mold toward closing position, (2) locking the position by mechanical means, and (3) using high pressure hydraulic cylinders to finally close the mold and build tonnage.

13.6.4 Shrinkage

Polymers have high thermal expansion coefficients, and significant shrinkage occurs during cooling of the plastic in the mold. Some thermoplastics undergo volumetric contractions of around 10% after injection into the mold. Contraction of crystalline plastics tends to be greater than for amorphous polymers. Shrinkage is usually expressed as the reduction in linear size that occurs during cooling to room temperature from the molding temperature for the given polymer. Appropriate units are therefore mm/mm (in./in.) of the dimension under consideration. Typical values for selected polymers are given in Table 13.1.

Fillers in the plastic tend to reduce shrinkage. In commercial molding practice, shrinkage values for the specific molding compound should be obtained from the producer prior to making the mold. To compensate for shrinkage, the dimensions of the mold cavity must be made larger than the specified part dimensions. The following formula can be used [14]:

$$D_c = D_p + D_pS + D_pS^2 \tag{13.19}$$

where D_c = dimension of cavity, mm (in.); D_p = molded part dimension, mm (in.), and S = shrinkage values obtained from Table 13.1. The third term on the right-hand side corrects for shrinkage that occurs in the shrinkage.

Example 13.3 Shrinkage in injection molding.

The nominal length of a part made of polyethylene is to be 80 mm. Determine the corresponding dimension of the mold cavity that will compensate for shrinkage.

Solution: From Table 13.1, the shrinkage for polyethylene is $S = 0.025$. Using Eq. (13.19), the mold cavity diameter should be:

$$D_c = 80.0 + 80.0(0.025) + 80.0(0.025)^2$$

$$= 80.0 + 2.0 + 0.05 = \textbf{82.05 mm}$$

TABLE 13.1 Typical values of shrinkage for moldings of selected thermoplastics.

Plastic	Shrinkage, mm/mm (in./in.)
ABS	0.006
Nylon-6,6	0.020
Polycarbonate	0.007
Polyethylene	0.025
Polystyrene	0.004
PVC	0.005

Compiled from [14].

It is clear that the mold dimensions must be determined for the particular polymer to be molded. The same mold will produce different part sizes for different polymer types.

Values in Table 13.1 represent a gross simplification of the shrinkage issue. In reality, shrinkage is affected by a number of factors, any of which can alter the amount of contraction experienced by a given polymer. The most important factors are injection pressure, compaction time, molding temperature, and part thickness. As injection pressure is increased, forcing more material into the mold cavity, shrinkage is reduced. Increasing compaction time has a similar effect, assuming the polymer in the gate does not solidify and seal off the cavity; maintaining pressure forces more material into the cavity while shrinkage is taking place. Net shrinkage is thereby reduced.

Molding temperature refers to the temperature of the polymer in the cylinder immediately prior to injection. One might expect that a higher polymer temperature would increase shrinkage, on the reasoning that the difference between molding and room temperatures is greater. However, shrinkage is actually lower at higher molding temperatures. The explanation is that higher temperatures significantly lower the viscosity of the polymer melt, allowing more material to be packed into the mold; the effect is the same as higher injection pressures. Thus, the effect on viscosity more than compensates for the larger temperature difference.

Finally, thicker parts show greater shrinkage. A molding solidifies from the outside; the polymer in contact with the mold surface forms a skin that grows toward the center of the part. At some point during solidification, the gate solidifies, isolating the material in the cavity from the runner system and from compaction pressure. When this happens, the molten polymer inside the skin accounts for most of the remaining shrinkage that occurs in the part. A thicker part section, since it contains a higher proportion of molten material, experiences greater shrinkage.

13.6.5 Defects in Injection Molding

Injection molding is a complicated process, and many things can go wrong. Here are some common defects in injection molded parts:

> *Short shots*—As in casting, a short shot is a molding that has solidified before completely filling the cavity. The defect can be corrected by increasing temperature and/or pressure. The defect may also result from use of a machine with insufficient shot capacity, in which case a larger machine is needed.

> *Flashing*—Flashing occurs when the polymer melt is squeezed into the parting surface between mold plates; it can also occur around ejection pins. The defect is usually caused by (1) vents and clearances in the mold that are too large; (2) injection pressure too high compared to clamping force; (3) melt temperature too high; or (4) excessive shot size.

> *Sink marks and voids*—These are defects usually related to thick molded sections. A *sink mark* occurs when the outer surface on the molding solidifies, but contraction of the internal material causes the skin to be depressed below its intended profile. A *void* is caused by the same basic phenomenon; however, the surface material retains its form and the shrinkage manifests itself as an internal void due to high tensile stresses on the still-molten polymer. These defects can be addressed by increasing the packing pressure following injection. A better solution is to design the part to have uniform section thicknesses and to use thinner sections.

> *Weld lines*—Weld lines occur when polymer melt flows around a core or other convex detail in the mold cavity and meets from opposite directions; the boundary thus formed is called a weld line, and it may have mechanical properties that are inferior to those in the rest of the part. Higher melt temperatures, higher injection pressures,

alternative gating locations on the part, and better venting are ways of dealing with this defect.

13.6.6 Other Injection Molding Processes

The vast majority of injection molding applications involve thermoplastics. Several variants of the process are described in this section.

Thermoplastic Foam Injection Molding   Plastic foams have a variety of applications, and we discuss these materials and their processing in Section 13.11. One of the processes, sometimes called *structural foam molding,* is appropriate to discuss here because it is injection molding. It involves the molding of thermoplastic parts that possess a dense outer skin surrounding a lightweight foam center. Such a part has a high stiffness-to-weight ratio suitable for structural applications.

A structural foam part can be produced either by introducing a gas into the molten plastic in the injection unit or by mixing a gas-producing ingredient with the starting pellets. During injection, an insufficient amount of melt is forced into the mold cavity, where it expands (foams) to fill the mold. The foam cells in contact with the cold mold surface collapse to form a dense skin, while the material in the core retains its cellular structure. Items made of structural foam include electronic cases, business machine housings, furniture components, and washing machine tanks. Advantages cited for structural foam molding include lower injection pressures and clamping forces, and thus the capability to produce large components, as suggested by our preceding list. A disadvantage of the process is that the resulting part surfaces tend to be rough, with occasional voids. If good surface finish is needed for the application, then additional processing is required, such as sanding, painting, and adhesion of a veneer.

Multi-Injection Molding Processes   Unusual effects can be achieved by multiple injection of different polymers to mold a part. The polymers are injected either simultaneously or sequentially, and there may be more than one mold cavity involved. Several processes fall under this heading, all characterized by two or more injection units—thus, the equipment for these processes is expensive.

Sandwich molding involves injection of two separate polymers—one is the outer skin of the part and the other is the inner core, which is typically a polymer foam. A specially designed nozzle controls the flow sequence of the two polymers into the mold. The sequence is designed so that the core polymer is completely surrounded by the skin material inside the mold cavity. The final structure is similar to that of a structural foam molding. However, the molding possesses a smooth surface, thus overcoming one of the major shortcomings of the previous process. In addition, it consists of two distinct plastics, each with its own characteristics suited to the application.

Another multi-injection molding process involves sequential injection of two polymers into a two-position mold. With the mold in the first position, the first polymer is injected into the cavity. Then the mold opens to the second position, and the second melt is injected into the enlarged cavity. The resulting part consists of two integrally connected plastics. *Bi-injection molding* is used to combine plastics of two different colors (e.g., automobile tail light covers) or to achieve different properties in different sections of the same part.

Injection Molding of Thermosets   Injection molding is used for thermosetting (TS) plastics, with certain modifications in equipment and operating procedure to allow for cross-linking. The machines for thermoset injection molding are similar to those used for thermoplastics. They utilize a reciprocating-screw injection unit, but the barrel length is shorter to avoid premature curing and solidification of the TS polymer. For the same reason, temperatures in the barrel are kept at relatively low levels—usually 50°C to

125°C (120°F to 260°F), depending on the polymer. The plastic, usually in the form of pellets or granules, is fed into the barrel through a hopper. Plasticizing occurs by the action of the rotating screw as the material is moved forward toward the nozzle. When sufficient melt has accumulated ahead of the screw, it is injected into a mold that is heated to 150°C to 230°C (300°F to 450°F), where cross-linking occurs to harden the plastic. The mold is then opened and the part is ejected and removed. Molding cycle times typically range from 20 sec to 2 min, depending on polymer type and part size.

Curing is the most time-consuming step in the cycle. In many cases, the part can be removed from the mold before curing is completed, so that final hardening occurs due to retained heat within a minute or two after removal. An alternative approach is to use a multiple-mold machine, in which two or more molds are attached to an indexing head served by a single injection unit.

The principal thermosets for injection molding are phenolics, unsaturated polyesters, melamines, epoxies, and urea-formaldehyde. Elastomers are also injected molded (Section 14.1.4). More than 50% of the phenolic moldings currently produced in the United States are made by this process [11], representing a shift away from compression and transfer molding, the traditional processes used for thermosets (Section 13.7). Most of the TS molding materials contain large proportions of fillers (up to 70% by weight), including glass fibers, clay, wood fibers, and carbon black. In effect, these are composite materials that are being injected molded.

Reaction Injection Molding *Reaction injection molding* (RIM) involves the mixing of two highly reactive liquid ingredients and immediately injecting the mixture into a mold cavity, where chemical reactions leading to solidification occur. The two ingredients form the components used in catalyst-activated or mixing-activated thermoset systems (Section 8.3.1). Urethanes, epoxies, and urea-formaldehyde are examples of these systems. RIM was developed with polyurethane to produce large automotive components such as bumpers, spoilers, and fenders; these kinds of parts still constitute the major application of the process. RIM-molded polyurethane parts typically possess a foam internal structure surrounded by a dense outer skin.

As shown in Figure 13.27, liquid ingredients are pumped in precisely measured amounts from separate holding tanks into a mixing head. The ingredients are rapidly

FIGURE 13.27   Reaction injection molding (RIM) system, shown immediately after ingredients A and B have been pumped into the mixing head prior to injection into the mold cavity (some details of processing equipment omitted).

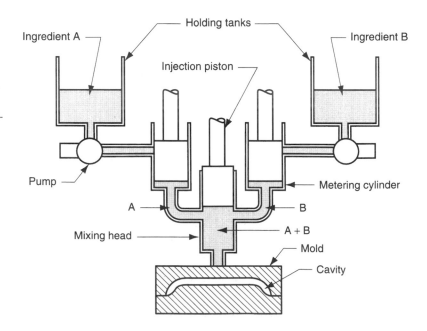

mixed and then injected into the mold cavity at relatively low pressure where polymerization and curing occur. A typical cycle time is around two minutes. For relatively large cavities the molds for RIM are much less costly than corresponding molds for conventional injection molding. This is due to the low clamping forces required in RIM and the opportunity to use light-weight components in the molds.

Advantages of RIM include [16]: (1) low energy is required in the process; (2) equipment and mold costs are less than injection molding; (3) a variety of chemical systems are available that enable specific properties to be obtained in the molded product; and (4) the production equipment is reliable and the chemical systems and machine relationships are well understood.

13.7 COMPRESSION AND TRANSFER MOLDING

Discussed in this section are two molding techniques widely used for thermosetting polymers and elastomers. For thermoplastics, these techniques cannot match the efficiency of injection molding, except for very special applications.

13.7.1 Compression Molding

Compression molding is an old and widely used molding process for thermosetting plastics. Its applications also include thermoplastic phonograph records, rubber tires, and various polymer matrix composite parts. The process, illustrated in Figure 13.28 for a TS plastic, consists of (1) loading a precise amount of molding compound, called the *charge,* into the bottom half of a heated mold; (2) bringing the mold halves together to compress the charge, forcing it to flow and conform to the shape of the cavity; (3) heating the charge by means of the hot mold to polymerize and cure the material into a solidified part; and (4) opening the mold halves and removing the part from the cavity.

FIGURE 13.28 Compression molding for thermosetting plastics: (1) charge is loaded, (2) and (3) charge is compressed and cured, and (4) part is ejected and removed (some details omitted).

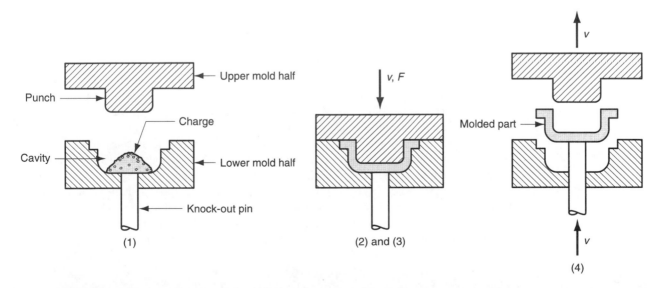

The initial charge of molding compound can be in any of several forms, including powders or pellets, liquid, or preform. The amount of polymer must be precisely controlled to obtain repeatable consistency in the molded product. It has become common practice to preheat the charge prior to its placement into the mold; this softens the polymer and shortens the production cycle time. Preheating methods include infrared heaters, convection heating in an oven, and use of a heated rotating screw in a barrel. The latter technique (borrowed from injection molding) is also used to meter the amount of the charge.

Compression molding presses are oriented vertically and contain two platens to which the mold halves are fastened. The presses involve either of two types of actuation: (1) upstroke of the bottom platen or (2) downstroke of the top platen, the former being the more common machine configuration. They are generally powered by a hydraulic cylinder that can be designed to provide clamping capacities up to several hundred tons.

Molds for compression molding are generally simpler than their injection mold counterparts. There is no sprue and runner system in a compression mold, and the process itself is generally limited to simpler part geometries due to the lower flow capabilities of the starting thermosetting materials. However, provision must be made for heating the mold, usually accomplished by electric resistance heating, steam, or hot oil circulation. Compression molds can be classified as *hand molds,* used for trial runs; *semiautomatic,* in which the press follows a programmed cycle but the operator manually loads and unloads the press; and *automatic,* which operate under a fully automatic press cycle (including automatic loading and unloading).

Materials for compression molding include phenolics, melamine, urea-formaldehyde, epoxies, urethanes, and elastomers. Typical TS plastic moldings include electric plugs, sockets, and housings; pot handles, and dinnerware plates. Advantages noted for compression molding in these applications include: molds that are simpler, less expensive, and require low maintenance; less scrap; and low residual stresses in the molded parts (thus favoring this process for flat thin parts such as phonograph records). A typical disadvantage is longer cycle times and therefore lower production rates than injection molding.

13.7.2 Transfer Molding

In this process, a thermosetting charge (preform) is loaded into a chamber immediately ahead of the mold cavity, where it is heated; pressure is then applied to force the softened polymer to flow into the heated mold where curing occurs. There are two variants of the process, illustrated in Figure 13.29: (a) *pot transfer molding,* in which the charge is injected from a "pot" through a vertical sprue channel into the cavity; and (b) *plunger transfer molding,* in which the charge is injected by means of a plunger from a heated well through lateral channels into the mold cavity. In both cases, scrap is produced each cycle in the form of the leftover material in the base of the well and lateral channels, called the *cull.* In addition, the sprue in pot transfer is scrap material. Because the polymers are thermosetting, the scrap cannot be recovered.

Transfer molding is closely related to compression molding, because it is utilized on the same polymer types (thermosets and elastomers). One can also see similarities to injection molding, in the way the charge is preheated in a separate chamber and then injected into the mold. Transfer molding is capable of molding part shapes that are more intricate than compression molding but not as intricate as injection molding. Transfer molding also lends itself to molding with inserts, in which a metal or ceramic insert is placed into the cavity prior to injection, and the heated plastic bonds to the insert during molding.

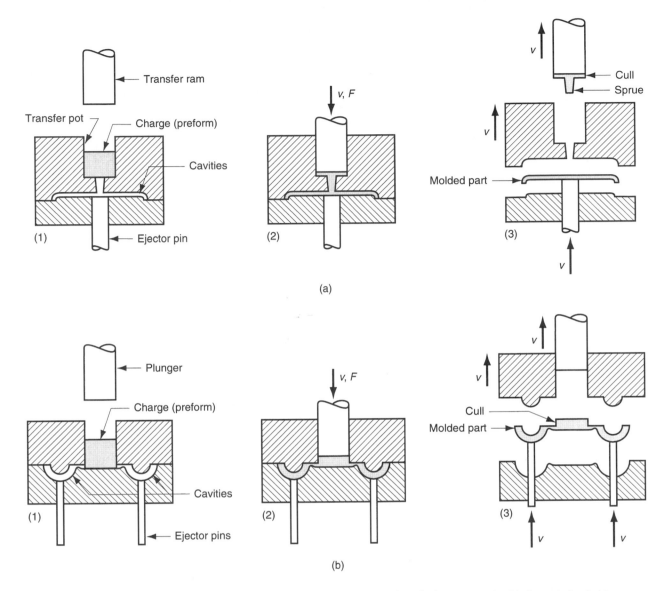

FIGURE 13.29   (a) Pot transfer molding, and (b) plunger transfer molding. Cycle in both processes is: (1) charge is loaded into pot; (2) softened polymer is pressed into mold cavity and cured; and (3) part is ejected.

13.8 BLOW MOLDING AND ROTATIONAL MOLDING

Both of these processes are used to make hollow, seamless parts out of thermoplastic polymers. Rotational molding can also be used for thermosets. Parts range in size from small plastic bottles of only 5 ml (0.15 oz) to large storage drums of 38,000 l (10,000 gal) capacity. Although the two processes compete in certain cases, generally they have found their own niches. Blow molding is more suited to the mass production of small disposable containers, while rotational molding favors large hollow shapes.

13.8.1 Blow Molding

Blow molding is a molding process in which air pressure is used to inflate soft plastic into a mold cavity. It is an important industrial process for making one-piece hollow plastic parts with thin walls, such as bottles and similar containers. Since many of these items are used for consumer beverages for mass markets, production is typically organized for very high quantities. The technology is borrowed from the glass industry (Section 12.2.1) with which plastics compete in the disposable or recyclable bottle market.

Blow molding is accomplished in two steps: (1) fabrication of a starting tube of molten plastic, called a **parison** (same as in glass-blowing); and (2) inflation of the tube to the desired final shape. Forming the parison is accomplished by either of two processes: extrusion or injection molding.

Extrusion Blow Molding   This form of blow molding consists of the cycle illustrated in Figure 13.30. In most cases, the process is organized as a very high production operation for making plastic bottles. The sequence is automated and usually integrated with downstream operations such as bottle filling and labeling.

It is usually a requirement that the blown container be rigid, and rigidity depends on wall thickness among other factors. We can relate wall thickness of the blown container to the starting extruded parison [12], assuming a cylindrical shape for the final product. The effect of die swell on the parison is shown in Figure 13.31. The mean diameter of the tube as it exits the die is determined by the mean die diameter D_d. Die swell causes expansion to a mean parison diameter D_p. At the same time, wall thickness swells from t_d to t_p. The swell ratio of the parison diameter is given by

$$r_{sd} = \frac{D_p}{D_d} \tag{13.20}$$

FIGURE 13.30   Extrusion blow molding: (1) extrusion of parison; (2) parison is pinched at the top and sealed at the bottom around a metal blow pin as the two halves of the mold come together; (3) the tube is inflated so that it takes the shape of the mold cavity; and (4) mold is opened to remove the solidified part.

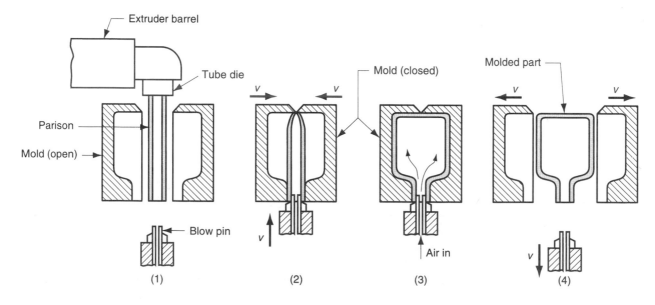

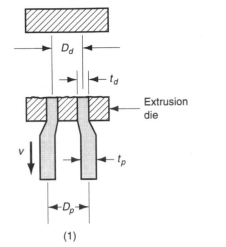 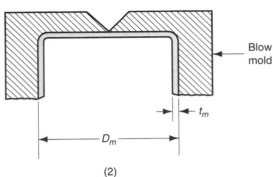

FIGURE 13.31 (1) Dimensions of extrusion die, showing parison after die swell; and (2) final blow-molded container in extrusion blow molding.

while the swell ratio for the wall thickness is

$$r_{st} = \frac{t_p}{t_d} \qquad (13.21)$$

The swelling of the wall thickness is proportional to the square of the diameter swelling; that is,

$$r_{st} = r_{sd}^2 \qquad (13.22)$$

and therefore,

$$t_p = r_{sd}^2 \, t_d \qquad (13.23)$$

When the parison is inflated to the blow mold diameter D_m with the corresponding reduction in wall thickness to t_m, and assuming constant volume of cross section, we have

$$\pi D_p t_p = \pi D_m t_m \qquad (13.24)$$

Solving for t_m, we obtain

$$t_m = \frac{D_p t_p}{D_m}$$

Substituting Eqs. (13.20) and (13.23) into this equation, we get

$$t_m = \frac{r_{sd}^3 t_d D_d}{D_m} \qquad (13.25)$$

The amount of die swell in the initial extrusion process can be measured by direct observation; and the dimensions of the die are known. Thus, we can determine the wall thickness on the blow-molded container.

Given the wall thickness of the molded container, an expression can be developed for the maximum air pressure that avoids bursting of the parison during inflation [12]. An equation borrowed from strength of materials relates stress to internal pressure p in a pipe, given its diameter D and wall thickness t:

$$\sigma = \frac{pD}{2t} \qquad (13.26)$$

Reasoning that the maximum stress will occur just before the parison is expanded to the size of the blow mold diameter (this is when D will be maximum and t will be minimum), and rearranging Eq. (13.26) to solve for p, we get

$$p = \frac{2\sigma t_m}{D_m} \qquad (13.27)$$

where p = air pressure used during blow molding, Pa (lb/in.^{2}); σ = maximum allowable tensile stress in the polymer during inflation, Pa (lb/in.^{2}); and t_m and D_m are wall thickness and diameter, respectively, of the molding, m (in.). The difficulty in using the formula is in determining the allowable stress since the polymer is in a heated and highly plastic condition. In an industrial operation, the process parameters are tuned by trial and error.

Injection Blow Molding   In this process, the starting parison is injection molded rather than extruded. A simplified sequence is outlined in Figure 13.32. Compared to its extrusion-based competitor, the injection blow-molding process has a lower production rate, which explains why it is less widely used.

In a variation of injection blow molding, called *stretch blow molding* (Figure 13.33), the blowing rod extends downward into the injection molded parison during step 2, thus stretching the soft plastic and creating a more favorable stressing of the polymer than conventional injection blow molding or extrusion blow molding. The resulting structure is more rigid, with higher transparency and better impact resistance. The most widely used material for stretch blow molding is polyethylene terephthalate (PET), a polyester that has very low permeability and is strengthened by the stretch-blow-molding process. The combination of properties makes it ideal as a container for carbonated beverages.

Materials and Products   Blow molding is limited to thermoplastics. Polyethylene is the polymer most commonly used for blow molding; in particular, high density and high molecular weight polyethylene (HDPE and HMWPE). In comparing their properties with those of low density PE given the requirement for stiffness in the final product, it is more economical to use these more expensive materials because the container walls can be made thinner. Other blow moldings are made of polypropylene (PP), polyvinylchloride (PVC), and polyethylene terephthalate.

FIGURE 13.32   Injection blow molding: (1) parison is injection molded around a blowing rod; (2) injection mold is opened and parison is transferred to a blow mold; (3) soft polymer is inflated to conform to the blow mold; and (4) blow mold is opened and blown product is removed.

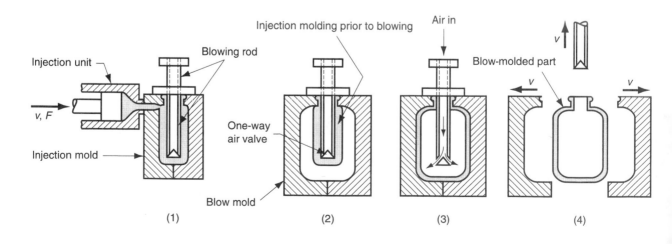

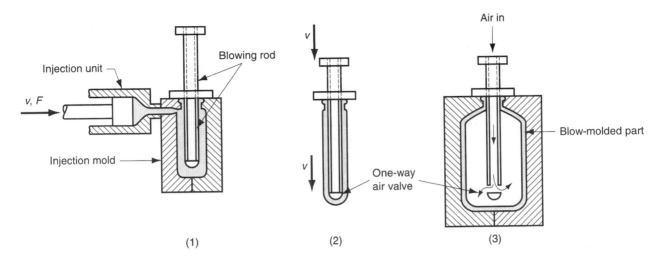

FIGURE 13.33   Stretch blow molding: (1) injection molding of parison; (2) stretching; and (3) blowing.

Disposable containers for packaging liquid consumer goods constitute the major share of products made by blow molding; but they are not the only products. Other items include large shipping drums (55 gallon) for liquids and powders, large storage tanks (2000 gallon), automotive gasoline tanks, toys, and hulls for sail boards and small boats. In the latter case, two boat hulls are made in a single blow molding and subsequently cut into two open hulls.

13.8.2 Rotational Molding

Rotational molding uses gravity inside a rotating mold to achieve a hollow form. Also called ***rotomolding,*** it is an alternative to blow molding for making large, hollow shapes. It is used principally for thermoplastic polymers, but applications for thermosets and elastomers are becoming more common. Rotomolding tends to favor more complex external geometries, larger parts, and lower production quantities than blow molding. The process consists of the following steps: (1) A predetermined amount of polymer powder is loaded into the cavity of a split mold. (2) The mold is then heated and simultaneously rotated on two perpendicular axes, so that the powder impinges on all internal surfaces of the mold, gradually forming a fused layer of uniform thickness. (3) While still rotating, the mold is cooled so that the plastic skin solidifies. (4) The mold is opened, and the part is unloaded. Rotational speeds used in the process are relatively slow. It is gravity, not centrifugal force, that causes uniform coating of the mold surfaces.

Molds in rotational molding are simple and inexpensive compared to injection molding or blow molding, but the production cycle is much longer, lasting perhaps ten minutes or more. To balance these advantages and disadvantages in production, rotational molding is often performed on a multicavity indexing machine, such as the three-station machine shown in Figure 13.34. The machine is designed so that three molds are indexed in sequence through three workstations. Thus, all three molds are working simultaneously. The first workstation is an unload–load station where the finished part is unloaded from the mold, and the powder for the next part is loaded into the cavity. The second station consists of a heating chamber where hot-air convection heats the mold while it is simultaneously rotated. Temperatures inside the chamber are around 375°C

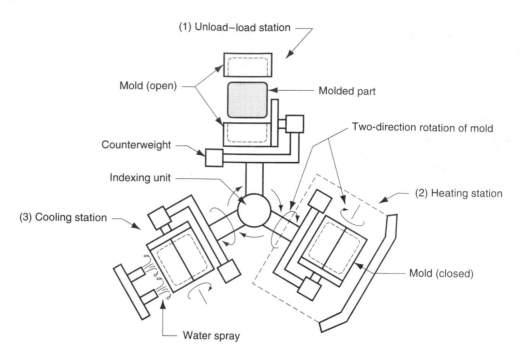

(1) Unload–load station

Mold (open)

Molded part

Two-direction rotation of mold

Counterweight

Indexing unit

(2) Heating station

(3) Cooling station

FIGURE 13.34 Rotational molding cycle performed on a three-station indexing machine: (1) unload–load station; (2) heat and rotate mold; (3) cool the mold.

Mold (closed)

Water spray

(700°F), depending on the polymer and the item being molded. The third station cools the mold, using forced cold air or water spray, to cool and solidify the plastic molding inside.

A fascinating variety of articles are made by rotational molding. The list includes hollow toys such as hobby horses and playing balls; boat and canoe hulls, sandboxes, small swimming pools; buoys and other flotation devices; truck body parts, automotive dashboards, fuel tanks; luggage pieces, furniture, garbage cans; fashion mannequins; large industrial barrels, containers, and storage tanks; portable outhouses, and septic tanks. The most popular molding material is polyethylene, especially HDPE. Other plastics include polypropylene, ABS, and high-impact polystyrene.

13.9 THERMOFORMING

Thermoforming is a process in which a flat thermoplastic sheet is heated and deformed into the desired shape. The process is widely used in packaging of consumer products and to fabricate large items such as bathtubs, contoured skylights, and internal door liners for refrigerators.

Thermoforming consists of two main steps: heating and forming. Heating is usually accomplished by radiant electric heaters, located on one or both sides of the starting plastic sheet at a distance of roughly 125 mm (5 in.). Duration of the heating cycle needed to sufficiently soften the sheet depends on the polymer, its thickness and color. The methods by which the forming step is accomplished can be classified into three basic categories: (1) vacuum thermoforming, (2) pressure thermoforming, and (3) mechanical thermoforming. In our discussion of these methods, we describe the forming of sheet stock; in the packaging industry, most thermoforming operations are performed on thin films.

Vacuum Thermoforming   The earliest method was *vacuum thermoforming* (called simply *vacuum forming* when it was developed in the 1950s), in which negative pressure is used to draw a preheated sheet into a mold cavity. The process is explained in Figure 13.35 in its most basic form. The holes for drawing the vacuum in the mold are on the order of 0.8 mm (0.031 in.) in diameter, so their effect on the plastic surface is minor.

Pressure Thermoforming   An alternative to vacuum forming involves positive pressure to force the heated plastic into the mold cavity. This is called *pressure thermoforming* or *blow forming*; its advantage over vacuum forming is that higher pressures can be developed because the latter is limited to a theoretical maximum of 1 atm. Blow-forming pressures of 3 to 4 atm are common. The process sequence is similar to the previous, the difference being that the sheet is pressurized from above into the mold cavity. Vent holes are provided in the mold to exhaust the trapped air. The forming portion of the sequence (steps 2 and 3) is illustrated in Figure 13.36.

At this point it is useful to distinguish between negative and positive molds. The molds shown in Figures 13.35 and 13.36 are *negative molds* because they have concave cavities. A *positive mold* has a convex shape. Both types are used in thermoforming. In the case of the positive mold, the heated sheet is draped over the convex form and

FIGURE 13.35   Vacuum thermoforming: (1) a flat plastic sheet is softened by heating; (2) the softened sheet is placed over a concave mold cavity; (3) a vacuum draws the sheet into the cavity; and (4) the plastic hardens on contact with the cold mold surface, and the part is removed and subsequently trimmed from the web.

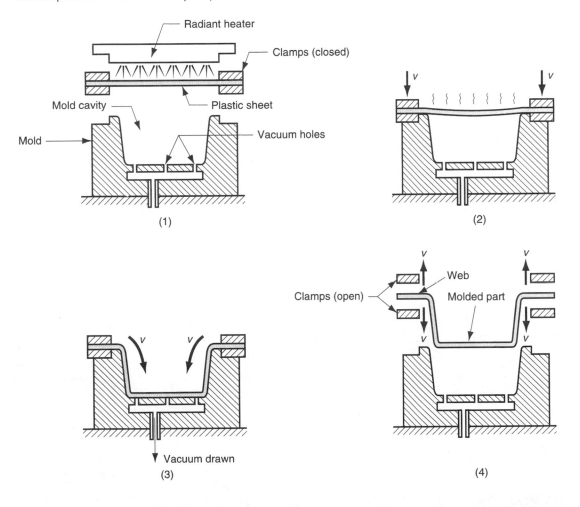

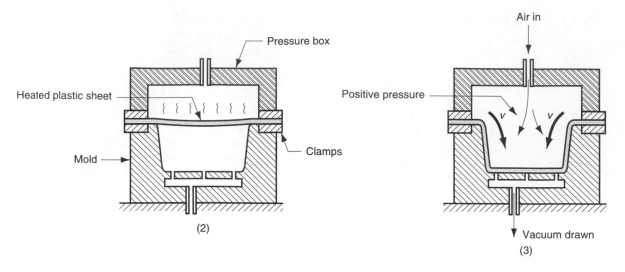

FIGURE 13.36   Pressure thermoforming. The sequence is similar to previous figure, the difference being: (2) sheet is placed over a mold cavity; and (3) positive pressure forces the sheet into the cavity.

negative or positive pressure is used to force the plastic against the mold surface. The positive mold is shown in Figure 13.37 for the case of vacuum forming.

The difference between positive and negative molds may seem unimportant, since the part shapes are virtually identical, as shown in our diagrams. However, if the part is drawn into the negative mold, then its exterior surface will have the exact surface contour of the mold cavity. The inside surface will be an approximation of the contour and will possess a finish corresponding to that of the starting sheet. By contrast, if the sheet is draped over a positive mold, then its interior surface will be identical to that of the convex mold; and its outside surface will follow approximately. Depending on the requirements of the product, this distinction might be important.

Another difference is in the thinning of the plastic sheet, one of the problems in thermoforming. Unless the contour of the mold is very shallow, there will be significant thinning of the sheet as it is stretched to conform to the mold contour. Positive and negative molds produce a different pattern of thinning in a given part. Consider our tub-shaped part as an example. In the positive mold, as the sheet is draped over the convex form, the portion making contact with the top surface (corresponding to the base of the tub) solidifies quickly and experiences virtually no stretching. This results in a thick base but with significant thinning in the walls of the tub. By contrast, a negative mold results

FIGURE 13.37   Use of a positive mold in vacuum thermoforming: (1) the heated plastic sheet is positioned above the convex mold and (2) the clamp is lowered into position, draping the sheet over the mold as a vacuum forces the sheet against the mold surface.

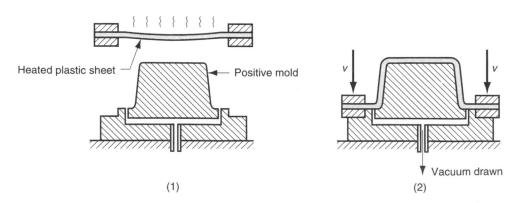

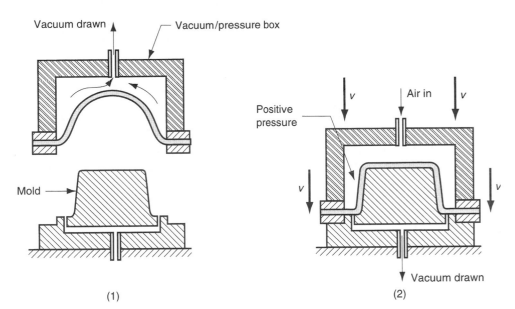

FIGURE 13.38   Pre-stretching the sheet in (1) prior to draping and vacuuming it over a positive mold in (2).

in a more even distribution of stretching and thinning in the sheet before contact is made with the cold surface.

A way to improve the thinning distribution with a positive mold is to prestretch the sheet before draping it over the convex form. As shown in Figure 13.38, the heated plastic sheet is stretched uniformly by vacuum pressure into a spherical shape prior to drawing it over the mold.

The first step depicted in frame (1) of Figure 13.38 can be utilized alone as a method to produce globe-shaped parts such as skylight windows and transparent domes. In the process, closely controlled air pressure is applied to inflate the soft sheet. The pressure is maintained until the blown shape has solidified.

Mechanical Thermoforming   The third method, called ***mechanical thermoforming,*** uses matching positive and negative molds that are brought together against the heated plastic sheet, forcing it to assume their shape. In the pure mechanical forming method, air pressure (positive or negative) is not used at all. The process is illustrated in Figure 13.39. Its advantages are better dimensional control and the opportunity for surface

FIGURE 13.39
Mechanical thermoforming: (1) heated sheet is placed above a negative mold, and (2) mold is closed to shape the sheet.

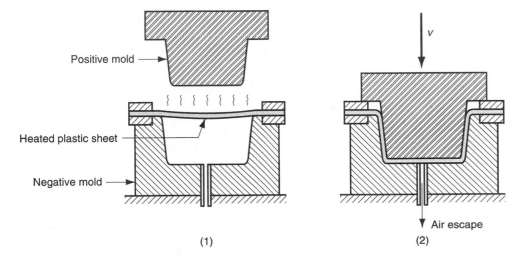

detailing on both sides of the part. The disadvantage is that two mold halves are required; the molds for the other two methods are therefore less costly.

Applications Thermoforming is a secondary shaping process, the primary process being that which produces the sheet or film (Section 13.3). Only thermoplastics can be thermoformed, since extruded sheets of thermosetting or elastomeric polymers have already been cross-linked and cannot be softened by reheating. Common thermoforming plastics are polystyrene, cellulose acetate and cellulose acetate butyrate, ABS, PVC, acrylic (polymethylmethacrylate), polyethylene, and polypropylene.

Mass production thermoforming operations are performed in the packaging industry. The starting sheet or film is rapidly fed through a heating chamber and then mechanically formed into the desired shape. The operations are often designed to produce multiple parts with each stroke of the press using molds with multiple punches and cavities. In some cases, the extrusion machine that produces the sheet or film is located directly upstream from the thermoforming process, thereby eliminating the need to reheat the plastic. And for best efficiency, the filling process to put the consumable food item into the container is placed immediately downstream from thermoforming.

Thin film packaging items that are mass produced by thermoforming include blister packs and skin packs. They offer an attractive way to display certain commodity products such as cosmetics, toiletries, small tools, and fasteners (nails, screws, etc.). Thermoforming applications include large parts that can be produced from thicker sheet stock. Examples include covers for business machines, boat hulls, shower stalls, diffusers for lights, advertising displays and signs, bathtubs, and certain toys. We had previously mentioned contoured skylights and internal door liners for refrigerators. These would be made, respectively, out of acrylic (because of its transparency) and ABS (because of its ease in forming and resistance to oils and fats found in refrigerators).

13.10 CASTING

In polymer shaping, *casting* involves pouring of a liquid resin into a mold, using gravity to fill the cavity, and allowing the polymer to harden. Both thermoplastics and thermosets are cast. Examples of the former include acrylics, polystyrene, polyamides (nylons) and vinyls (PVC). Conversion of the liquid resin into a hardened thermoplastic can be accomplished in several ways, which include (1) heating the thermoplastic resin to a highly fluid state so that it readily pours and fills the mold cavity, and then permitting it to cool and solidify in the mold; (2) using a low-molecular-weight prepolymer (or monomer) and polymerizing it in the mold to form a high-molecular-weight thermoplastic; and (3) pouring a plastisol (a liquid suspension of fine particles of a thermoplastic resin such as PVC in a plasticizer) into a heated mold so that it gels and solidifies.

Thermosetting polymers shaped by casting include polyurethane, unsaturated polyesters, phenolics, and epoxies. The process involves pouring the liquid ingredients that form the thermoset into a mold so that polymerization and cross-linking occur. Heat and/or catalysts may be required depending on the resin system. The reactions must be sufficiently slow to allow mold pouring to be completed. Fast reacting thermosetting systems, such as certain polyurethane systems, require alternative shaping processes like reaction injection molding (Section 13.6.6).

Advantages of casting over alternative processes such as injection molding include (1) the mold is simpler and less costly, (2) the cast item is relatively free of residual stresses and viscoelastic memory, and (3) the process is suited to low production quantities. Focusing on advantage (2), acrylic sheets (Plexiglas, Lucite) are generally cast between two pieces of highly polished plate glass. The casting process permits a high degree of flatness and desirable optical qualities to be achieved in the clear plastic sheets. Such flatness and clarity cannot be obtained by flat sheet extrusion. A disadvantage in some applications is significant shrinkage of the cast part during solidification. For example, acrylic sheets undergo a volumetric contraction of about 20% when cast. This is much more than in injection molding in which high pressures are used to pack the mold cavity in order to reduce shrinkage.

Slush casting is an alternative to conventional casting, borrowed from metal casting technology. In **slush casting,** a liquid plastisol is poured into the cavity of a heated split mold, so that a skin forms at the surface of the mold. After some duration that depends on the desired thickness of the skin, the excess liquid is poured out of the mold; the mold is then opened for part removal. The process is also referred to by the term **shell casting** [6].

An important application of casting in electronics is **encapsulation,** in which items such as transformers, coils, connectors, and other electrical components are encased in plastic by casting.

13.11 POLYMER FOAM PROCESSING AND FORMING

A **polymer foam** is a polymer-and-gas mixture, which gives the material a porous or cellular structure. Other terms used for polymer foams include **cellular polymer, blown polymer,** and **expanded polymer.** The most common polymer foams are polystyrene (Styrofoam, a trademark) and polyurethane. Other polymers used to make foams include natural rubber ("foamed rubber") and polyvinylchloride (PVC).

The characteristic properties of a foamed polymer include (1) low density, (2) high strength per unit weight, (3) good thermal insulation, and (4) good energy absorbing qualities. The elasticity of the base polymer determines the corresponding property of the foam. Polymer foams can be classified [6] as (1) **elastomeric,** in which the matrix polymer is a rubber, capable of large elastic deformation; (2) **flexible,** in which the matrix is a highly plasticized polymer such as soft PVC; and (3) **rigid,** in which the polymer is a stiff thermoplastic such as polystyrene or a thermosetting plastic such as a phenolic. Depending on chemical formulation and degree of cross-linking, polyurethanes can range over all three categories.

The characteristic properties of polymer foams, and the ability to control their elastic behavior through selection of the base polymer, make these materials highly suitable for certain types of applications, including hot beverage cups, heat insulating structural materials and cores for structural panels, packaging materials, cushion materials for furniture and bedding, padding for automobile dashboards, and products requiring buoyancy.

13.11.1 Foaming Processes

Common gases used in polymer foams are air, nitrogen, and carbon dioxide. The proportion of gas can range up to 90% or more. The gas is introduced into the polymer by several methods, called foaming processes. These include (1) mixing a liquid resin with air by **mechanical agitation,** then hardening the polymer by means of heat or chemical reaction; (2) mixing a **physical blowing agent** with the polymer—a gas such as nitrogen (N_2) or pentane (C_5H_{12}), which can be dissolved in the polymer melt under pressure, so that the gas comes out of solution and expands when the pressure is subsequently reduced; and (3) mixing the polymer with chemical compounds, called ***chemical blowing agents,*** that decompose at elevated temperatures to liberate gases such as CO_2 or N_2 within the melt.

The way the gas is distributed throughout the polymer matrix distinguishes two basic foam structures, illustrated in Figure 13.40: (a) ***closed cell,*** where the gas pores are roughly spherical and completely separated from each other by the polymer matrix; and (b) ***open cell,*** in which the pores are interconnected to some extent, allowing passage of a fluid through the foam. A closed cell structure makes a satisfactory life jacket; an open cell structure would become waterlogged. Other attributes that characterize the structure include the relative proportions of polymer and gas (already mentioned) and the cell density (number of cells per unit volume) which is inversely related to the size of the individual air cells in the foam.

13.11.2 Shaping Processes

There are many shaping processes for polymer foam products. Since the two most important foams are polystyrene and polyurethane, we limit our discussion to shaping processes for these two materials. Because polystyrene is a thermoplastic and polyurethane can be either a thermoset or an elastomer (it can also be a thermoplastic but is less important in this form), the processes covered here for these materials are representative of those used for other polymer foams.

Polystyrene Foams Polystyrene foams are shaped by extrusion and molding. In ***extrusion,*** a physical or chemical blowing agent is fed into the polymer melt near the die end of the extruder barrel; thus, the extrudate consists of the expanded polymer. Large sheets and boards are made in this way and are subsequently cut to size for heat insulation panels and sections.

Several molding processes are available for polystyrene foam. We have previously discussed ***structural foam molding*** and ***sandwich molding*** (Section 13.6.6). A more widely used process is ***expandable foam molding,*** in which the molding material usually consists of prefoamed polystyrene beads. The prefoamed beads are produced from pellets of solid polystyrene that have been impregnated with a physical blowing agent. Prefoaming is performed in a large tank by applying steam heat to partially expand the pellets, simultaneously agitating them to prevent fusion. Then, in the molding process, the prefoamed beads are fed into a mold cavity, where they are further expanded and fused together to form the molded product. Hot beverage cups of polystyrene foam are produced in this way. In some processes, the prefoaming step is omitted, and the impregnated beads are fed directly into the mold cavity where they are heated, expanded, and fused. In other operations, the expandable foam is first formed into a flat sheet by the ***blown-film extrusion***

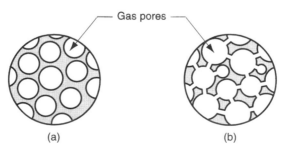

Gas pores

(a)   (b)

FIGURE 13.40 Two polymer foam structures: (a) closed cell and (b) open cell.

process (Section 13.3) and then shaped by *thermoforming* (Section 13.9) into packaging containers such as egg cartons.

Polyurethane Foams   Polyurethane foam products are made in a one-step process in which the two liquid ingredients (polyol and isocyanate) are mixed and immediately fed into a mold or other form, so that the polymer is synthesized and the part geometry is created at the same time. Shaping processes for polyurethane foam can be divided into two basic types [11]: spraying and pouring. *Spraying* involves use of a spray gun into which the two ingredients are continuously fed, mixed, and then sprayed onto a target surface. The reactions leading to polymerization and foaming occur after application on the surface. This method is used to apply rigid insulating foams onto construction panels, railway cars, and similar large items. *Pouring* involves dispensing the ingredients from a mixing head into an open or closed mold in which the reactions occur. An open mold can be a container with the required contour (e.g., for an automobile seat cushion) or a long channel that is slowly moved past the pouring spout to make long continuous sections of foam. The closed mold is a completely enclosed cavity into which a certain amount of the mixture is dispensed. Expansion of the reactants completely fills the cavity to shape the part. For fast-reacting polyurethanes, the mixture must be rapidly injected into the mold cavity using *reaction injection molding* (Section 13.6.6). The degree of cross-linking, controlled by the starting ingredients, determines the relative stiffness of the resulting foam.

13.12 PRODUCT DESIGN CONSIDERATIONS

Plastics are an important design material, but the designer must be aware of their limitations. In this section we list some design guidelines for plastic components, beginning with those that apply in general, and then ones applicable to extrusion and molding (injection molding, compression and transfer molding).

13.12.1 General Considerations

These general guidelines apply, irrespective of the shaping process. They are mostly limitations of plastic materials that must be considered by the designer.

> *Strength and stiffness*—Plastics are not as strong or stiff as metals. They should not be used in applications where high stresses will be encountered. Creep resistance is also a limitation. Strength properties vary significantly among plastics, and strength-to-weight ratios for some plastics are competitive with metals in certain applications.

> *Impact Resistance*—The capacity of plastics to absorb impact is generally good; plastics compare favorably with most metals.

> *Service temperatures* of plastics are limited relative to engineering metals and ceramics.

> *Thermal expansion* is greater for plastics than metals; so dimensional changes due to temperature variations are much more significant than for metals.

> Many types of plastics are subject to *degradation* from sunlight and certain other forms of radiation. Also, some plastics degrade in oxygen and ozone atmospheres. Finally, plastics are soluble in many common solvents. On the positive side, plastics are resistant to conventional corrosion mechanisms that afflict many metals. The weaknesses of specific plastics must be taken into account by the designer.

13.12.2 Extruded Plastics

Extrusion is one of the most widely used plastic shaping processes. Several design recommendations are presented here for the conventional process (compiled mostly from [3]).

> *Wall thickness*—Uniform wall thickness is desirable in an extruded cross section. Variations in wall thickness result in nonuniform plastic flow and uneven cooling that tend to warp the extrudate.
> *Hollow sections*—Hollow sections complicate die design and plastic flow. It is desirable to use extruded cross sections that are not hollow yet satisfy functional requirements.
> *Corners*—Sharp corners, inside and outside, should be avoided in the cross section, since they result in uneven flow during processing and stress concentrations in the final product.

13.12.3 Molded Parts

There are many plastic-molding processes. In this article, we list guidelines that apply to injection molding (the most popular molding process), compression molding, and transfer molding (compiled from Bralla [3], McCrum [10], and other sources).

> *Economic production quantities*—Each molded part requires a unique mold, and the mold for any of these processes can be costly, particularly for injection molding. Minimum production quantities for injection molding are usually around 10,000 pieces; for compression molding, minimum quantities are around 1,000 parts, due to the simpler mold designs involved. Transfer molding lies between the other two.
> *Part complexity*—Although more complex part geometries mean more costly molds, it may nevertheless be economical to design a complex molding if the alternative involves many individual components assembled together. An advantage of plastic molding is that it allows multiple functional features to be combined into one part.
> *Wall thickness*—Thick cross sections are generally undesirable; they are wasteful of material, more likely to cause warping due to shrinkage, and take longer to harden. *Reinforcing ribs* can be used in molded plastic parts to achieve increased stiffness without excessive wall thickness. The ribs should be made thinner than the walls they reinforce, to minimize sink marks on the outside wall.
> *Corner radii and fillets*—Sharp corners, both external and internal, are undesirable in molded parts; they interrupt smooth flow of the melt, tend to create surface defects, and cause stress concentrations in the finished part.
> *Holes*—Holes are quite feasible in plastic moldings, but they complicate mold design and part removal. They also cause interruptions in melt flow.
> *Draft*—A molded part should be designed with a draft on its sides to facilitate removal from the mold. This is especially important on the inside wall of a cup-shaped part because the molded plastic contracts against the positive mold shape. The recommended draft for thermosets is around $1/2°$ to $1°$; for thermoplastics it usually ranges between $1/8°$ and $1/2°$. Suppliers of plastic molding compounds provide recommended draft values for their products.
> *Tolerances*—Tolerances specify the allowable manufacturing variations for a part. Although shrinkage is predictable under closely controlled conditions, generous tolerances are desirable for injection moldings because of variations in process parameters that affect shrinkage and diversity of part geometries encountered. Table 13.2 lists typical tolerances for molded part dimensions of selected plastics.

TABLE 13.2 Typical tolerances on molded parts for selected plastics.

Plastic	Tolerances for:[a]	
	50 mm (2.0 in.) Dimension	10 mm (3/8 in.) Hole
Thermoplastic:		
ABS	±0.2 mm (±0.007 in.)	±0.08 mm (±0.003 in.)
Polyethylene	±0.3 mm (±0.010 in.)	±0.13 mm (±0.005 in.)
Polystyrene	±0.15 mm (±0.006 in.)	±0.1 mm (±0.004 in.)
Thermosetting:		
Epoxies	±0.15 mm (±0.006 in.)	±0.05 mm (±0.002 in.)
Phenolics	±0.2 mm (±0.008 in.)	±0.08 mm (±0.003 in.)

Compiled from [3], [7], [14], and [18].

Values represent typical commercial molding practice.

[a]For smaller sizes, tolerances can be reduced. For larger sizes, more generous tolerances are required.

REFERENCES

[1] Baird, D. G., and Collias, D. I. *Polymer Processing Principles and Design.* John Wiley & Sons, Inc., New York, 1998.

[2] Billmeyer, Fred, W., Jr. *Textbook of Polymer Science.* 3rd ed. John Wiley & Sons, New York, 1984.

[3] Bralla, J. G. (Editor in Chief). *Design for Manufacturability Handbook.* 2nd ed. McGraw-Hill Book Company, New York, 1998.

[4] Briston, J. H. *Plastic Films.* 3rd ed. Longman Group UK Ltd., Essex, England, 1989.

[5] Chanda, M., and Roy, S. K. *Plastics Technology Handbook.* Marcel Dekker, Inc., New York, 1998.

[6] Charrier, J-M. *Polymeric Materials and Processing.* Oxford University Press, New York, 1991.

[7] *Engineering Materials Handbook.* Vol. 2. *Engineering Plastics,* ASM International, Metals Park, Ohio, 1988.

[8] Hall, C. *Polymer Materials.* 2nd ed. John Wiley & Sons, New York, 1989.

[9] Hensen, F. (Editor). *Plastic Extrusion Technology.* Hanser Publishers, Munich, FRG, 1988 (Distributed in United States by Oxford University Press, New York).

[10] McCrum, N. G., Buckley, C. P., and Bucknall, C. B. *Principles of Polymer Engineering.* 2nd ed. Oxford University Press, Oxford, U.K., 1997.

[11] *Modern Plastics Encyclopedia.* Modern Plastics, McGraw-Hill, Inc. Hightstown, New Jersey, 1991.

[12] Morton-Jones, D. H. *Polymer Processing.* Chapman and Hall, London, U.K., 1989.

[13] Pearson, J. R. A. *Mechanics of Polymer Processing.* Elsevier Applied Science Publishers, London, 1985.

[14] Rubin, I. I. *Injection Molding: Theory and Practice.* John Wiley & Sons, New York, 1972.

[15] Rudin, A. *The Elements of Polymer Science and Engineering.* 2nd ed. Academic Press, Inc., Orlando, Florida, 1998.

[16] Sweeney, F. M. *Reaction Injection Molding Machinery and Processes.* Marcel Dekker, Inc., New York, 1987.

[17] Tadmor, Z., and Costas, G. G. *Principles of Polymer Processing.* John Wiley & Sons, New York, 1979.

[18] Wick, C., Benedict, J. T., and Veilleux, R. F. *Tool and Manufacturing Engineers Handbook.* 4th ed. Volume II. *Forming.* Society of Manufacturing Engineers, Dearborn, Michigan, 1984, Chapter 18.

REVIEW QUESTIONS

13.1. What are some of the reasons why the plastic shaping processes are important?

13.2. Identify the main categories of plastics shaping processes, as classified by the resulting product geometry.

13.3. Viscosity is an important property of a polymer melt in plastics shaping processes. Upon what two parameters does the viscosity of a given polymer depend?

13.4. How does the viscosity of a polymer melt differ from most fluids that are Newtonian.

13.5. Besides viscosity, what other property of a polymer melt is important in plastics processing? Briefly define the property.

13.6. Define *die swell* in extrusion.

13.7. Briefly describe the plastic extrusion process.

13.8. The barrel and screw of an extruder are generally divided into three sections; identify the sections.

13.9. What are the functions of the screen pack and breaker plate at the die end of the extruder barrel?

13.10. What are the various forms of extruded shapes and corresponding dies?

13.11. What is the distinction between plastic sheet and film?

13.12. What is the **blown-film process** for producing film stock?

13.13. Describe the **calendering process.**

13.14. Polymer fibers and filaments are used in several applications; what is the most important application?

13.15. Technically, what is the difference between a **fiber** and a **filament?**

13.16. Among the synthetic fiber materials, which are the most important?

13.17. Briefly describe the **injection molding** process.

13.18. An injection molding machine is divided into two principal components; identify them.

13.19. What are the two basic types of clamping units?

13.20. Gates in injection molds have several functions; name them.

13.21. What are the advantages of a three-plate mold over a two-plate mold in injection molding?

13.22. Discuss some of the defects that can occur in plastic injection molding.

13.23. Describe **structural foam molding.**

13.24. What are the significant differences in the equipment and operating procedures between injection molding of thermoplastics and injection molding of thermosets?

13.25. What is **reaction injection molding?**

13.26. What kinds of products are produced by blow molding?

13.27. What is the starting material form in thermoforming?

13.28. What is the difference between a positive mold and a negative mold in thermoforming?

13.29. Why are the molds generally more costly in mechanical thermoforming than in pressure or vacuum thermoforming?

13.30. What are the processes by which polymer foams are produced?

13.31. What are some of the general considerations that product designers must keep in mind when designing components out of plastics?

MULTIPLE CHOICE QUIZ

There is a total of 36 correct answers in the following multiple choice questions (some questions have multiple answers that are correct). To attain a perfect score on the quiz, all correct answers must be given, since each correct answer is worth 1 point. For each question, each omitted answer or wrong answer reduces the score by 1 point, and each additional answer beyond the number of answers required reduces the score by 1 point. Percentage score on the quiz is based on the total number of correct answers.

13.1. The shear viscosity of a polymer melt is affected by which of the following (more than one)? (a) degree of polymerization, (b) polymer type, (c) rate of flow, (d) temperature.

13.2. The forward movement of polymer melt in an extruder barrel is resisted by drag flow, which is caused by the resistance to flow through the die orifice: (a) true or (b) false.

13.3. Which three of the following are sections of a conventional extruder barrel for thermoplastics? (a) compression section, (b) die section, (c) feed section, (d) heating section, (e) metering section, and (f) shaping section.

13.4. Which of the following processes is not associated with the production of plastic sheet and film (more than one)? (a) blown-film extrusion process, (b) calendering, (c) chill-roll extrusion, (d) doctor blade method, and (e) slit-die extrusion.

13.5. **Spinning** in the production of synthetic fibers refers to which one of the following? (a) extrusion of polymer melt through small die openings, (b) drawing the strands to elongate and thin them, (c) both of the above, or (d) none of the above.

13.6. The principal components of an injection molding machine are which two of the following? (a) clamping unit, (b) hopper, (c) injection unit, (d) mold, and (e) part ejection unit.

13.7. The **parting line** in injection molding is which one of the following? (a) the lines formed where polymer melt meets after flowing around a core in the mold, (b) the narrow gate sections where the parts are separated from the runner, (c) where the clamping unit is joined to the injection unit in the molding machine, (d) where the two mold halves come together, or (e) none of the above.

13.8. The function of the ejection system is to (one best answer): (a) move polymer melt into the mold cavity, (b) open the mold halves after the cavity is filled, (c) remove the molded parts from the runner system after molding, (d) separate the part from the cavity after molding, or (e) none of the above.

13.9. A three-plate mold offers which of the following advantages when compared to a two-plate mold (may be more than one)? (a) automatic separation of parts from runners, (b) gating is usually at the base of the part to reduce weld lines, (c) sprue does not solidify, (d) stronger molded parts, (e) none of the above.

13.10. Which of the following defects or problems are associated with injection molding (more than one)? (a) bambooing, (b) die swell, (c) drag flow, (d) flash, (e) melt fracture, (f) short shots, (g) sink marks, or (h) unbalanced runner.

13.11. In rotational molding, centrifugal force is used to force the polymer melt against the surfaces of the mold cavity where solidification occurs: (a) true or (b) false.

13.12. Use of a parison is associated with which one of the following plastic shaping processes? (a) bi-injection molding, (b) blow molding, (c) compression molding, (d) pressure thermoforming, or (e) sandwich molding.

13.13. A thermoforming mold with a convex form is called which one of the following? (a) a die, (b) a negative mold, (c) a positive mold, or (d) a three-plate mold.

13.14. The term *encapsulation* refers to which one of the following plastics-shaping processes? (a) casting, (b) compression molding, (c) extrusion of hollow forms, (d) injection molding in which a metal insert is encased in the molded part, or (e) vacuum thermoforming using a positive mold.

13.15. Which of the following terms applies to the processing of foam plastics (more than one)? (a) chemical blowing agents, (b) open cell structure, (c) powder injection molding, (d) sandwich molding, or (e) structural foam molding.

13.16. The two most common polymer foams are which of the following? (a) polyacetal, (b) polyethylene, (c) polystyrene, (d) polyurethane, and (e) polyvinylchloride.

13.17. In which of the following property categories do plastic parts compare favorably with metals (more than one)? (a) impact resistance, (b) resistance to ultraviolet radiation, (c) stiffness, (d) strength, (e) strength-to-weight ratio, or (f) temperature resistance.

13.18. Which of the following processes are generally limited to thermoplastic polymers (more than one)? (a) blow molding, (b) compression molding, (c) reaction injection molding, (d) thermoforming, (e) transfer molding, (f) wire coating.

13.19. Which of the following processes would be applicable to produce hulls for small boats (more than one answer)? (a) blow molding, (b) compression molding, (c) injection molding, (d) rotational molding, or (e) vacuum thermoforming.

PROBLEMS

Extrusion

13.1. The diameter of an extruder barrel is 65 mm, and its length is 1.75 m. The screw rotates at 55 rev/min. The screw channel depth is 5.0 mm, and the flight angle is 18°. The head pressure at the die end of the barrel is 5.0×10^6 Pa. The viscosity of the polymer melt is given as 100 Pas. Find the volume flow rate of the plastic in the barrel.

13.2. An extruder barrel has a diameter of 120 mm, and a length of 3.0 m. The screw channel depth is 8.0 mm, and its pitch is 95 mm. The viscosity of the polymer melt is 75 Pas, and the head pressure in the barrel is 4.0 MPa. What rotational speed of the screw is required to achieve a volumetric flow rate of 90 cm³/s?

13.3. An extruder has a diameter of 80 mm and a length of 2.0 m. Its screw has a channel depth of 5 mm, flight angle is 18 degrees, and it rotates at 1 rev/sec. The plastic melt has a shear viscosity of 150 Pas. Determine the extruder characteristic by computing Q_{max} and p_{max} and then finding the equation of the straight line between them.

13.4. Determine the helix angle A such that the screw pitch p is equal to the screw diameter D. This is called the "square" angle in plastics extrusion—the angle that provides a flight advance equal to one diameter for each rotation of the screw.

13.5. An extruder barrel has a diameter of 2.5 in. The screw rotates at 60 rev/min; its channel depth is 0.20 in., and its flight angle is 17.5°. The head pressure at the die end of the barrel is 800 lb/in.², and the length of the barrel is 50 in. The viscosity of the polymer melt is 122×10^{-4} lb-sec/in.² Determine the volume flow rate of the plastic in the barrel.

13.6. An extruder barrel has a diameter of 4.0 in. and an L/D ratio of 28. The screw channel depth is 0.25 in., and its pitch is 4.8 in. It rotates at 60 rev/min. The viscosity of the polymer melt is 100×10^{-4} lb-sec/in.² What head pressure is required to obtain a volume flow rate of 150 in.³/min?

13.7. An extrusion operation produces continuous tubing with an outside diameter of 2.0 in. and an inside diameter of 1.7 in. The extruder barrel has a diameter of 4.0 in. and a length of 10 ft. The screw rotates at 50 rev/min; it has a channel depth of 0.25 in. and flight angle of 16°. The head pressure has a value of 350 lb/in.², and the viscosity of the polymer melt is 80×10^{-4} lb-sec/in.² Under these conditions, what is the production rate in length of tube/min, assuming the extrudate is pulled at a rate that eliminates the effect of die swell (i.e., the tubing has the same OD and ID as the die profile)?

13.8. An extruder barrel has a diameter and length of 100 mm and 2.8 m, respectively. The screw rotational speed is 50 rev/min, channel depth of 7.5 mm, and flight angle of 17°. The plastic melt has a shear viscosity of 175 Pas. Determine (a) the extruder characteristic, (b) the shape factor K_s for a circular die opening with a diameter of 3.0 mm and a length of 12.0 mm, and (c) the operating point (Q and p).

13.9. Consider an extruder in which the barrel diameter is 4.5 in. and length is 11 ft. The extruder screw rotates at 60 rev/min; it has channel depth of 0.35 in. and a flight angle of 20°. The plastic melt has a shear viscosity of 125×10^{-4} lb-sec/in.^{2} Determine (a) Q_{max} and p_{max}, (b) the shape factor K_s for a circular die opening in which D_d is 0.312 in. and L_d is 0.75 in., and (c) the values of Q and p at the operating point.

13.10. An extruder has a barrel diameter of 5.0 in. and a length of 12 ft. The extruder screw rotates at 50 rev/min;

it has a channel depth of 0.30 in. and a flight angle of 17.7°. The plastic melt has a shear viscosity of 100×10^{-4} lb-sec/in.^{2} Find (a) the extruder characteristic, (b) the values of Q and p at the operating point, given that the die characteristic is $Q_x = 0.00150\,p$.

13.11. An extruder has a barrel diameter of 4.0 in. and a length of 5.0 ft. The extruder screw rotates at 80 rev/min. It has a channel with a depth of 0.15 in. and a flight angle of 20°. The polymer melt has a shear viscosity of 60×10^{-4} lb-sec/in.^{2} at the operating temperature of the process. The specific gravity of the polymer is 1.2. (a) Find the equation for the extruder characteristic. If a T-shaped cross section is extruded at a rate of 0.13 lb/sec, determine (b) the operating point (Q and p), and (c) the die characteristic that is indicated by the operating point.

Injection Molding

13.12. Compute the percentage volumetric contraction of a polyethylene molded part, based on the value of shrinkage given in Table 13.1.

13.13. The specified dimension = 100.00 mm for a certain injection molded part made of nylon-6,6. Compute the corresponding dimension to which the mold cavity should be machined, using the value of shrinkage given in Table 13.1.

13.14. The part dimension for a certain injection molded part made of polycarbonate is specified as 3.75 in. Compute the corresponding dimension to which the mold cavity should be machined, using the value of shrinkage given in Table 13.1.

13.15. The foreman in the injection-molding department says that a polyethylene part produced in one of the operations has greater shrinkage than the calculations indicate it should have. The important dimension of the part is specified as 112.5 ± 0.25 mm. However, the actual molded part measures 112.02 mm. (a) As a first step, the corresponding mold cavity dimension should be checked. Compute the correct value of the mold dimension, given that the shrinkage value for polyethylene is 0.025 (from Table 13.1). (b) What adjustments in process parameters could be made to reduce the amount of shrinkage?

Other Molding Operations and Thermoforming

13.16. The extrusion die for a polyethylene parison used in blow molding has a mean diameter of 16.0 mm. The size of the ring opening in the die is 1.5 mm. The mean diameter of the parison is observed to swell to a size of 20.5 mm after exiting the die orifice. If the diameter of the blow molded container is to be 100 mm, determine (a) the corresponding wall thickness of the container and (b) the wall thickness of the parison.

13.17. A blow-molding operation produces a 6.25-in-diameter bottle from a parison that is extruded in a die whose outside diameter is 1.25 in. and inside diameter is 1.00 in. The observed diameter swell ratio is 1.24. What is the maximum air pressure that can be used if the maximum allowable tensile stress for the polymer is 1000 lb/in.^{2}?

13.18. A parison is extruded from a die with an outside diameter of 11.5 mm and an inside diameter of 7.5 mm. The observed die swell is 1.25. The parison is used to blow

mold a beverage container whose outside diameter is 112 mm (a standard size 2-liter soda bottle). (a) What is the corresponding wall thickness of the container? (b) Obtain an empty 2-liter plastic soda bottle and (carefully) cut it across the diameter. Using a micrometer, measure the wall thickness to compare with your answer in (a).

13.19. An extrusion operation is used to produce a parison whose mean diameter is 27 mm. The inside and outside diameters of the die that produced the parison are 18 mm and 22 mm, respectively. If the minimum wall thickness of the blow molded container is to be 0.40 mm, what is the maximum possible diameter of the blow mold?

13.20. A rotational-molding operation is to be used to mold a hollow playing ball out of polyethylene. The ball will be 1.5 ft in diameter, and its wall thickness should be 1/16 in. What weight of PE powder should be loaded into the mold in order to meet these specifications? The specific gravity of the PE grade is 0.95.

13.21. The problem in a certain thermoforming operation is that there is too much thinning in the walls of the large cup-shaped part. The operation is conventional pressure thermoforming using a positive mold, and the plastic is an ABS sheet with an initial thickness of 3.2 mm. (a) Why is thinning occurring in the walls of the cup? (b) What changes could be made in the operation to correct the problem?

14 RUBBER-PROCESSING TECHNOLOGY

CHAPTER CONTENTS

14.1 Rubber Processing and Shaping
    14.1.1 Production of Rubber
    14.1.2 Compounding
    14.1.3 Mixing
    14.1.4 Shaping and Related Processes
    14.1.5 Vulcanization
14.2 Manufacture of Tires and Other Rubber Products
    14.2.1 Tires
    14.2.2 Other Rubber Products
    14.2.3 Processing of Thermoplastic Elastomers
14.3 Product Design Considerations

Many of the production methods used for plastics (Chapter 13) are also applicable to rubbers. However, rubber-processing technology is different in certain respects, and the rubber industry is largely separate from the plastics industry. The rubber industry and goods made of rubber are dominated by one product: tires. Tires are used in large numbers for automobiles, trucks, aircraft, and bicycles. Although pneumatic tires date from the late 1880s, rubber technology can be traced to the discovery in 1839 of vulcanization (Historical Note 8.2), the process by which raw natural rubber is transformed into a usable material through cross-linking of the polymer molecules. During its first century, the rubber industry was concerned only with the processing of natural rubber. Around World War II, synthetic rubbers were developed (Historical Note 8.3); today they account for the majority of rubber production.

14.1 RUBBER PROCESSING AND SHAPING

Production of rubber goods can be divided into two basic steps: (1) production of the rubber itself, and (2) processing of the rubber into finished goods. Production of rubber differs, depending on whether it is natural or synthetic. The difference is due to the source of the raw materials. Natural rubber (NR) is produced as an agricultural crop, whereas most synthetic rubbers are made from petroleum.

Production of rubber is followed by processing into final products; this consists of (1) compounding, (2) mixing, (3) shaping, and (4) vulcanizing. Processing techniques for natural and synthetic rubbers are virtually the same, differences being in the chemicals used to effect vulcanization (cross-linking). This sequence does not apply to thermoplastic elastomers, whose shaping techniques are the same as for other thermoplastic polymers.

There are several distinct industries involved in the production and processing of rubber. The production of raw natural rubber might be classified as an agricultural industry because latex, the starting ingredient for natural rubber, is grown on large plantations located in tropical climates. By contrast, synthetic rubbers are produced by the petrochemical industry. Finally, the processing of these materials into tires, shoe soles, and other rubber products occurs at processor (fabricator) plants. The processors are commonly known as the rubber industry. Some of the great names in this industry include Goodyear, B. F. Goodrich, and Michelin. The importance of the tire is reflected in these names.

14.1.1 Production of Rubber

In this section we briefly survey the production of rubber before it goes to the processor. Our coverage distinguishes natural rubber and synthetic rubber.

Natural Rubber   Natural rubber is tapped from rubber trees (***Hevea brasiliensis***) as latex. The trees are grown on plantations in Southeast Asia and other parts of the world. Latex is a colloidal dispersion of solid particles of the polymer polyisoprene (Section 8.4.2) in water. Polyisoprene is the chemical substance that comprises rubber, and its content in the emulsion is about 30%. The latex is collected in large tanks, thus blending the yield of many trees together.

The preferred method of recovering rubber from the latex involves coagulation. The latex is first diluted with water to about half its natural concentration. An acid such as formic acid (HCOOH) or acetic acid (CH_3COOH) is added to cause the latex to coagulate after about 12 hours. The coagulum, now in the form of soft solid slabs, is then squeezed through a series of rolls which drive out most of the water and reduce the thickness to about 3 mm (1/8 in.). The final rolls have grooves that impart a criss-cross pattern to the resulting sheets. The sheets are then draped over wooden frames and dried in smokehouses. The hot smoke contains creosote that prevents mildew and oxidation of the rubber. Several days are normally required to complete the drying process. The resulting rubber, now in a form called ***ribbed smoked sheet,*** is folded into large bales for shipment to the processor. This raw rubber has a characteristic dark brown color. In some cases, the sheets are dried in hot air rather than smokehouses, and the term ***air-dried sheet*** is applied; this is considered to be a better grade of rubber. A still better grade, called ***pale crepe*** rubber, involves two coagulation steps, the first to remove undesirable components of the latex; then the resulting coagulum is subjected to a more involved washing and mechanical working procedure, followed by warm air drying. The color of pale crepe rubber approaches a light tan.

Synthetic Rubber   The various types of synthetic rubber were identified in Section 8.4.3. Most synthetics are produced from petroleum by the same polymerization techniques used to synthesize other polymers (Section 8.1.1). However, unlike thermoplastic and thermosetting polymers, which are normally supplied to the fabricator as pellets or liquid resins, synthetic rubbers are supplied to rubber processors in the form of large bales. The industry has developed a long tradition of handling natural rubber in these unit loads.

14.1.2 Compounding

Rubber is always compounded with additives. It is through compounding that the specific rubber is designed to satisfy the given application in terms of properties, cost, and processability. Compounding adds chemicals for vulcanization. Sulfur has traditionally been used for this purpose. The vulcanization process and the chemicals used to accomplish it are discussed in Section 14.1.5.

Additives include fillers that act either to enhance the rubber's mechanical properties (reinforcing fillers) or to extend the rubber to reduce cost (nonreinforcing fillers). The single most important reinforcing filler in rubber is *carbon black*, a colloidal form of carbon, black in color, obtained from the thermal decomposition of hydrocarbons (soot). Its effect is to increase tensile strength and resistance to abrasion and tearing of the final rubber product. Carbon black also provides protection from ultraviolet radiation. The importance of these enhancements in tires is obvious. Most rubber parts are black in color because of their carbon black content.

Although carbon black is the most important filler, others are also used. They include china clays—hydrous aluminum silicates $(Al_2Si_2O_5(OH)_4)$, which provide less reinforcing than carbon black but are used when the black color is not acceptable; calcium carbonates $(CaCO_3)$, which are classified as nonreinforcing fillers; and silica (SiO_2), which can serve reinforcing or nonreinforcing functions, depending on particle size; and other polymers, such as styrene, PVC, and phenolics. Reclaimed (recycled) rubber is also added as a filler in some rubber products, but usually not in proportions exceeding 10%.

Other additives compounded with the rubber include antioxidants, to retard aging by oxidation; fatigue- and ozone-protective chemicals; coloring pigments; plasticizers and softening oils; blowing agents in the production of foamed rubber; and mold-release compounds.

Many products require filament reinforcement to reduce extensibility but retain the other desirable properties of rubber. Tires and conveyor belts are notable examples. Filaments used for this purpose include cellulose, nylon, and polyester. Fiberglass and steel are also used as reinforcements (e.g., steel-belted radial tires). These continuous fiber materials must be added as part of the shaping process; they are not mixed with the other additives.

14.1.3 Mixing

The additives must be thoroughly mixed with the base rubber to achieve uniform dispersion of the ingredients. Uncured rubbers possess high viscosity. Mechanical working experienced by the rubber can increase its temperature up to 150°C (300°F). If vulcanizing agents were present from the start of mixing, premature vulcanization would result—the rubber processor's nightmare [7]. Accordingly, a two-stage mixing process is usually employed. In the first stage, carbon black and other non-vulcanizing additives are combined with the raw rubber. The term *masterbatch* is used for this first-stage mixture. After thorough mixing has been accomplished, and time for cooling has been allowed, the second stage is carried out in which the vulcanizing agents are added.

Equipment for mixing includes the two-roll mill and internal mixers such as the Banbury mixer, Figure 14.1. The *two-roll mill* consists of two parallel rolls, supported in a frame so they can be brought together to obtain a desired "nip" (gap size), and driven to rotate at the same or slightly different speeds. An *internal mixer* has two rotors encased in a jacket, as in Figure 14.1(b) for the Banbury-type internal mixer. The rotors have blades and rotate in opposite directions at different speeds, causing a complex flow pattern in the contained mixture.

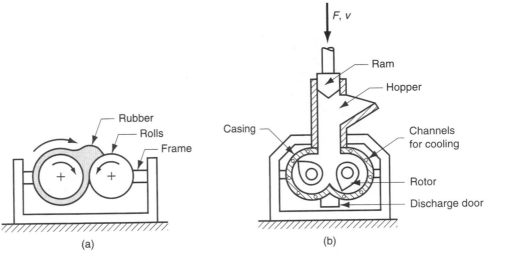

FIGURE 14.1   Mixers used in rubber processing: (a) two-roll mill and (b) Banbury-type internal mixer. These machines can also be used for mastication of natural rubber.

14.1.4 Shaping and Related Processes

Shaping processes for rubber products can be divided into four basic categories: (1) extrusion, (2) calendering, (3) coating, and (4) molding and casting. Most of these processes have been discussed in the previous chapter. We will examine here the special issues that arise when they are applied to rubber. Some products require several basic processes plus assembly work in their manufacture, tires being an example.

Extrusion   Extrusion of polymers was discussed in the preceding chapter. Screw extruders are generally used for extrusion of rubber. As with extrusion of thermosetting plastics, the L/D ratio of the extruder barrels is less than for thermoplastics, typically in the range 10 to 15, to reduce the risk of premature cross-linking. Die swell occurs in rubber extrudates, since the polymer is in a highly plastic condition and exhibits the memory property. It has not yet been vulcanized.

Calendering   This process involves passing rubber stock through a series of gaps of decreasing size made by a stand of rotating rolls (Section 13.3). The rubber process must be operated at lower temperatures than for thermoplastic polymers, to avoid premature vulcanization. Also, equipment used in the rubber industry is of heavier construction than that used for thermoplastics, since rubber is more viscous and harder to form. The output of the process is a rubber sheet of thickness determined by the final roll gap; again, swelling occurs in the sheet, causing its thickness to be slightly greater than the gap size. Calendering can also be used to coat or impregnate textile fabrics to produce rubberized fabrics.

There are problems in producing thick sheet by either extrusion or calendering. Thickness control is difficult in the former process, and air entrapment occurs in the latter. These problems are largely solved when extrusion and calendering are combined in the ***roller die*** process, Figure 14.2. The extruder die is a slit which feeds the calender rolls.

Coating   Coating or impregnating fabrics with rubber is an important process in the rubber industry. These composite materials are used in automobile tires, conveyor belts, inflatable rafts, and waterproof cloth for tarpaulins, tents, and rain coats. The ***coating*** of rubber onto substrate fabrics includes a variety of processes. We have previously

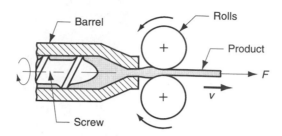

FIGURE 14.2 Roller die process rubber extrusion followed by rolling.

seen that calendering is one of the coating methods. Figure 14.3 illustrates one possible way in which the fabric is fed into the calendering rolls to obtain a reinforced rubber sheet.

Alternatives to calendering include skimming, dipping, and spraying. In the **skimming** process, a thick solution of rubber compound in an organic solvent is applied to the fabric as it is unreeled from a supply spool. The coated fabric passes under a doctor blade that skims the solvent to the proper thickness, and then moves into a steam chamber where the solvent is driven off by heat. As its name suggests, **dipping** involves temporary immersion of the fabric into a highly fluid solution of rubber, followed by drying. Likewise, in **spraying,** a spray gun is used to apply the rubber solution.

Molding and Casting   Molded articles include shoe soles and heals, gaskets and seals, suction cups, and bottle stops. Many foamed rubber parts are produced by molding. In addition, molding is an important process in tire production. Principal molding processes for rubber are (1) compression molding, (2) transfer molding, and (3) injection molding. Compression molding is the most important technique because of its use in tire manufacture. Curing (vulcanizing) is accomplished in the mold in all three processes, this representing a departure from the shaping methods already discussed, which require a separate vulcanizing step. With injection molding of rubber, there are risks of premature curing similar to those faced in the same process when applied to thermosetting plastics. Advantages of injection molding over traditional methods for producing rubber parts include better dimensional control, less scrap, and shorter cycle times. In addition to its use in the molding of conventional rubbers, injection molding is also applied for thermoplastic elastomers. Because of high mold costs, large production quantities are required to justify injection molding.

Dip casting is used for producing rubber gloves and overshoes. It involves submersion of a positive mold in a liquid polymer (or a heated form into plastisol) for a certain duration (the process may involve repeated dippings) to form the desired thickness. The coating is then stripped from the form and cured to cross-link the rubber.

FIGURE 14.3 Coating of fabric with rubber using a calendering process.

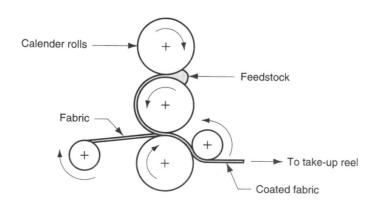

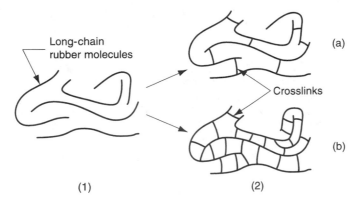

FIGURE 14.4   Effect of vulcanization on the rubber molecules: (1) raw rubber; (2) vulcanized (cross-linked) rubber—variations of (2) include (a) soft rubber, low degree of cross-linking; and (b) hard rubber, high degree of cross-linking.

14.1.5 Vulcanization

Vulcanization is the treatment that accomplishes cross-linking of elastomer molecules, so that the rubber becomes stiffer and stronger but retains extensibility. It is a critical step in the rubber-processing sequence. On a submicroscopic scale, the process can be pictured as in Figure 14.4, in which the long-chain molecules of the rubber become joined at certain tie points, the effect of which is to reduce the ability of the elastomer to flow. A typical soft rubber has one or two cross-links per thousand units (mers). As the number of cross-links increases, the polymer becomes stiffer and behaves more like a thermosetting plastic (hard rubber).

Vulcanization, as it was first invented by Goodyear, involved the use of sulfur (about 8 parts by weight of S mixed with 100 parts of natural rubber) at a temperature of 140°C (280°F) for about 5 hours. No other chemicals were included in the process. Vulcanization with sulfur alone is no longer used as a commercial treatment today, due to the long curing times. Various other chemicals, including zinc oxide (ZnO) and stearic acid ($C_{18}H_{36}O_2$), are combined with smaller doses of sulfur to accelerate and strengthen the treatment. The resulting cure time is 15 to 20 minutes. In addition, a variety of nonsulfur vulcanizing treatments have been developed.

In rubber-molding processes, vulcanization is accomplished in the mold; the mold temperature is maintained at the proper level for curing. In the other forming processes, vulcanization is performed after the part has been shaped. The treatments generally divide between batch processes and continuous processes. Batch methods include the use of an *autoclave,* a steam-heated pressure vessel; and *gas curing,* in which a heated inert gas such as nitrogen cures the rubber. Many of the basic processes make a continuous product; and if the output is not cut into discrete pieces, continuous vulcanization is appropriate. Continuous methods include *high-pressure steam,* suited to the curing of rubber coated wire and cable; *hot-air tunnel,* for cellular extrusions and carpet underlays [3]; and *continuous drum cure,* in which continuous rubber sheets (e.g., belts and flooring materials) pass through one or more heated rolls to effect vulcanization.

14.2 MANUFACTURE OF TIRES AND OTHER RUBBER PRODUCTS

Tires are the principal product of the rubber industry, accounting for about three-fourths of total tonnage. Other important products include footwear, hose, conveyor belts, seals, shock-absorbing components, foamed rubber products, and sports equipment.

14.2.1 Tires

Pneumatic tires are critical components of the vehicles on which they are used. They support the weight of the vehicle and the passengers and cargo on board; they transmit the motor torque to propel the vehicle; and they absorb road vibrations and shock to provide a comfortable ride. Tires are used on automobiles, trucks, buses, farm tractors, earth-moving equipment, military vehicles, bicycles, motorcycles, and aircraft.

Tire Construction and Production Sequence   A tire is an assembly of many parts, whose manufacture is unexpectedly complex. A passenger car tire consists of about 50 individual pieces; a large earthmover tire may have as many as 175. To begin with, there are three basic tire constructions: (a) diagonal ply, (b) belted bias, and (c) radial ply, pictured in Figure 14.5. In all three cases, the internal structure of the tire, known as the **carcass,** consists of multiple layers of rubber-coated cords, called **plies.** The cords are strands of various materials such as nylon, polyester, fiberglass, and steel, which provide inextensibility to reinforce the rubber in the carcass. The **diagonal ply tire** has the cords running diagonally, but in perpendicular directions in adjacent layers. A typical diagonal ply tire may have four plies. The **belted bias tire** is constructed of diagonal plies with opposite bias but adds several more layers around the outside periphery of the carcass. These **belts** increase the stiffness of the tire in the tread area and limit its diametric expansion during inflation. The cords in the belt also run diagonally, as indicated in our sketch.

A **radial tire** has plies running radially rather than diagonally; it also uses belts around the periphery for support. A **steel-belted radial** is a tire in which the circum-

FIGURE 14.5   Three principal tire constructions: (a) diagonal ply, (b) belted bias, and (c) radial ply.

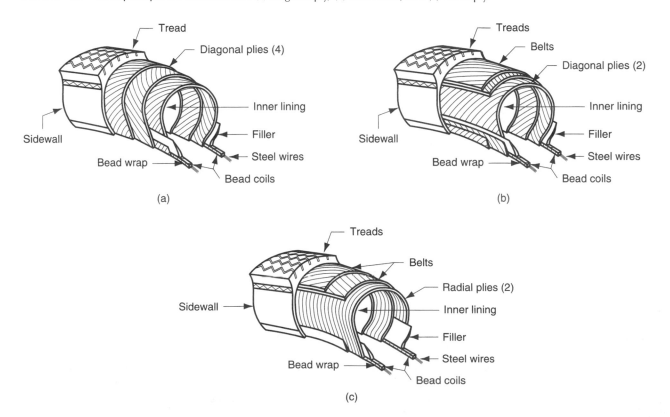

ferential belts have cords made of steel. The radial construction provides a more flexible sidewall which tends to reduce stress on the belts and treads as they continually deform on contact with the flat road surface during rotation. This effect is accompanied by greater tread life, improved cornering and driving stability, and a better ride at high speeds.

In each construction, the carcass is covered by solid rubber that reaches a maximum thickness in the tread area. The carcass is also lined on the inside with a rubber coating. For tires with inner tubes, the inner liner is a thin coating applied to the innermost ply during its fabrication. For tubeless tires, the inner liner must have low permeability since it holds the air pressure; it is generally a laminated rubber.

Tire production can be summarized in three steps: (1) preforming of components, (2) building the carcass and adding rubber strips to form the sidewalls and treads, and (3) molding and curing the components into one integral piece. The descriptions of these steps that follow are typical; there are variations in processing depending on construction, tire size, and type of vehicle on which the tire will be used.

Preforming of Components   As Figure 14.5 shows, the carcass consists of a number of separate components, most of which are rubber or reinforced rubber. These, as well as the sidewall and tread rubber, are produced by continuous processes and then pre-cut to size and shape for subsequent assembly. The components, labeled in Figure 14.5, and the preforming processes to fabricate them are:

> *Bead coil*—Continuous steel wire is rubber-coated, cut, coiled, and the ends joined.
> *Plies*—Continuous fabric (textile, nylon, fiber glass, steel) is rubber coated in a calendering process and pre-cut to size and shape.
> *Inner lining*—For tube tires, the inner liner is calendered onto innermost ply. For tubeless tires, liner is calendered as a two-layered laminate.
> *Belts*—Continuous fabric is rubber coated (similar to plies, above), but cut at different angles for better reinforcement; then made into multiply belt.
> *Tread*—Extruded as continuous strip; then cut and preassembled to belts.
> *Sidewall*—Extruded as continuous strip; then cut to size and shape.

Building the Carcass   The carcass is traditionally assembled using a machine known as a *building drum,* whose main element is a cylindrical arbor that rotates. Pre-cut strips that form the carcass are built up around this arbor in a step-by-step procedure. The layered plies that form the cross section of the tire are anchored on opposite sides of the rim by two bead coils. The *bead coils* consist of multiple strands of high strength steel wire. Their function is to provide a rigid support when the finished tire is mounted on the wheel rim. Other components are combined with the plies and bead coils. These include various wrappings and filler pieces to give the tire the proper strength, heat resistance, air retention, and fitting to the wheel rim. After these parts are placed around the arbor and the proper number of plies have been added, the belts are applied. This is followed by the outside rubber that will become the sidewall and tread. At this point in the process, the treads are rubber strips of uniform cross section—the tread design is added later in molding. The building drum is collapsible, so that the unfinished tire can be removed when finished. The form of the tire at this stage is roughly tubular, as portrayed in Figure 14.6.

Molding and Curing   Tire molds are usually two-piece construction (split molds) and contain the tread pattern to be impressed on the tire. The mold is bolted into a press, one half attached to the upper platen (the lid) and the bottom half fastened to the lower

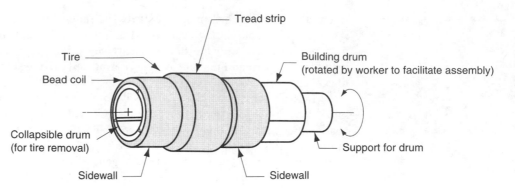

FIGURE 14.6 Tire just before removal from building drum, prior to molding and curing.

platen (the base). The uncured tire is placed over an expandable diaphragm and inserted between the mold halves, as in Figure 14.7. The press is then closed and the diaphragm expanded, so that the soft rubber is pressed against the cavity of the mold. This causes the tread pattern to be imparted to the rubber. At the same time, the rubber is heated, both from the outside by the mold and from the inside by the diaphragm. Circulating hot water or steam under pressure are used to heat the diaphragm. The duration of this curing step depends on the thickness of the tire wall. A typical passenger tire can be cured in about 15 minutes. Bicycle tires cure in about four minutes, whereas tires for large earth moving equipment take several hours to cure. After curing is completed, the tire is cooled and removed from the press.

14.2.2 Other Rubber Products

Most other rubber products are made by less complex processes. **Rubber belts** are widely used in conveyors and mechanical power transmission systems. Like tires, rubber is an ideal material for these products, but the belt must have little or no extensibility in order to function. Accordingly, it is reinforced with fibers, commonly polyester or nylon. Fabrics of these polymers are usually coated in calendering operations, assembled together to obtain the required number of plies and thickness, and subsequently vulcanized by continuous or batch heating processes.

Rubber hose can be either plain or reinforced. Plain hose is extruded tubing. Reinforced tube consists of an inner tube, a reinforcing layer (sometimes called the carcass), and a cover. The internal tubing is extruded of a rubber that has been compounded for the particular substance that will flow through it. The reinforcement layer is applied to the tube in the form of a fabric, or by spiraling, knitting, braiding, or other

FIGURE 14.7 Tire molding (tire is shown in cross-sectional view): (1) the uncured tire is placed over expandable diaphragm; (2) the mold is closed and the diaphragm is expanded to force uncured rubber against mold cavity, impressing tread pattern into rubber; mold and diaphragm are heated to cure rubber.

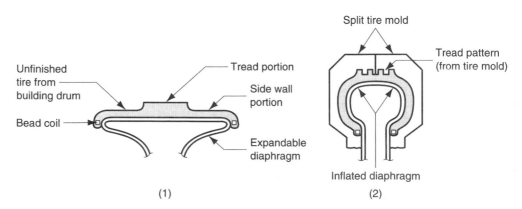

(1)   (2)

application method. The outer layer is compounded to resist environmental conditions. It is applied by extrusion, using rollers, or other techniques.

Footwear components include soles, heels, rubber overshoes, and certain upper parts. A variety of rubbers are used to make footwear components (Section 8.4). Molded parts are produced by injection molding, compression molding, and certain special molding techniques developed by the shoe industry; the rubbers include both solid and foamed varieties. In some cases, for low volume production, manual methods are used to cut rubber from flat stock.

Rubber is widely used in sports equipment and supplies, including ping pong paddle surfaces, golf club grips, football bladders, and sports balls of various kinds. Tennis balls, for example, are made in significant numbers. Production of these sports products relies on the various shaping processes discussed in Section 14.1.4, as well as special techniques that have been developed for particular items.

14.2.3 Processing of Thermoplastic Elastomers

A *thermoplastic elastomer* (TPE) is a thermoplastic polymer that possesses the properties of a rubber (Section 8.4.3); the term *thermoplastic rubber* is also used. TPEs can be processed like thermoplastics, but their applications are those of an elastomer. The most common shaping processes are injection molding and extrusion, which are generally more economical and faster than the traditional processes used for rubbers that must be vulcanized. Molded products include shoe soles, athletic footwear, and automotive components such as fender extensions and corner panels (but no tires—TPEs have been found to be unsatisfactory for that application). Extruded items include insulation coating for electrical wire, tubing for medical applications, conveyor belts, sheet and film stock. Other shaping techniques for TPEs include blow molding and thermoforming (Sections 13.8 and 13.9); these processes cannot be used for vulcanized rubbers.

14.3 PRODUCT DESIGN CONSIDERATIONS

Many of the same guidelines used for plastics apply to rubber products. There are differences, owing to the elastomeric properties of rubber. The following are compiled largely from Bralla [4]; they apply to conventional soft rubber, not hard rubber.

- > *Economic Production Quantities*—Rubber parts produced by compression molding (the traditional process) can often be produced in quantities of a thousand or less. The mold cost is relatively low compared to other molding methods. Injection molding, as with plastic parts, requires higher production quantities to justify the more expensive mold.
- > *Draft*—Draft is usually unnecessary for rubber molded parts. The flexibility of the material allows it to deform for removal from the mold. Shallow undercuts, although undesirable, are possible with rubber-molded parts, for the same reason. The low stiffness and high elasticity of the material permits removal from the mold.
- > *Holes*—Holes are difficult to cut into the rubber after initial forming, due the flexibility of the material. It is generally desirable to mold holes into the rubber during the primary shaping process.
- > *Screw threads*—Screw threads are generally not incorporated into molded rubber parts; the elastic deformability of rubber makes it difficult to assemble parts using the threads, and stripping is a problem once inserted.

REFERENCES

[1] Alliger, G., and Sjothun, I. J. (eds.). *Vulcanization of Elastomers.* Krieger Publishing Company, New York, 1978.

[2] Billmeyer, Fred, W., Jr. *Textbook of Polymer Science.* 3rd ed. John Wiley & Sons, New York, 1984.

[3] Blow, C. M., and Hepburn, C. *Rubber Technology and Manufacture.* 2nd ed. Butterworth-Heinemann, London, 1982.

[4] Bralla, J. G. (Editor in Chief). *Design for Manufacturability Handbook.* 2nd ed. McGraw-Hill Book Company, New York, 1998.

[5] Hofmann, W. *Rubber Technology Handbook.* Hanser-Gardner Publications, Cincinnati, Ohio, 1989.

[6] *Modern Plastics Encyclopedia.* (1990 Edition). Mid-October Issue, 1989, McGraw-Hill, Hightstown, New Jersey.

[7] Morton-Jones, D. H. *Polymer Processing.* Chapman and Hall, London, U.K., 1989.

REVIEW QUESTIONS

14.1. How is the rubber industry organized?

14.2. How is raw rubber recovered from the latex that is tapped from a rubber tree?

14.3. What is the sequence of processing steps required to produce finished rubber goods?

14.4. What are some of the functions of the additives that are combined with rubber during compounding?

14.5. Name the four basic categories of processes used to shape rubber.

14.6. What are some of the operations used to coat rubber onto a fabric to produce reinforced rubber?

14.7. What does vulcanization do to the rubber?

14.8. Name the three basic tire constructions and briefly identify the differences in their construction.

14.9. What are the three basic steps in the manufacture of a pneumatic tire?

14.10. Why are tire plies, conveyor belts, and most rubber hoses reinforced with fabric?

14.11. What is a TPE?

14.12. Many of the design guidelines that are applicable to plastics are also applicable to rubber. However, the extreme flexibility of rubber results in certain differences. What are some examples of these differences?

MULTIPLE CHOICE QUIZ

There is a total of 11 correct answers in the following multiple choice questions (some questions have multiple answers that are correct). To attain a perfect score on the quiz, all correct answers must be given, since each correct answer is worth 1 point. For each question, each omitted answer or wrong answer reduces the score by 1 point, and each additional answer beyond the number of answers required reduces the score by 1 point. Percentage score on the quiz is based on the total number of correct answers.

14.1. The most important rubber product is (a) conveyor belts, (b) footwear, (c) pneumatic tires, or (d) tennis balls.

14.2. The chemical name of the ingredient recovered from the latex of the rubber tree is which one of the following? (a) polybutadiene, (b) polyisobutylene, (c) polyisoprene, or (d) polystyrene.

14.3. Of the following rubber additives, which one would you rank as the single most important? (a) antioxidants, (b) carbon black, (c) clays and other hydrous aluminum silicates, (d) plasticizers and softening oils, or (e) reclaimed rubber.

14.4. Which of the following molding processes is the most important in the production of products made of conventional rubber? (a) compression molding, (b) injection molding, (c) thermoforming, or (d) transfer molding.

14.5. Which of the following ingredients do not contribute to the vulcanizing process (more than one answer)? (a) calcium carbonate, (b) carbon black, (c) stearic acid, (d) sulfur, and (e) zinc oxide.

14.6. How many minutes are required to cure (vulcanize) a modern passenger car tire? (a) 5, (b) 15, (c) 25, or (d) 45.

14.7. When is the tread pattern imprinted onto the circumference of the tire? (a) during preforming, (b) while building the carcass, (c) during molding, or (d) during curing.

14.8. Which of the following are not normally used in the processing of thermoplastic elastomers (more than one)? (a) blow molding, (b) compression molding, (c) extrusion, (d) injection molding, or (e) vulcanization.

14.9. Screw threads are not normally molded into rubber parts: (a) true or (b) false.

15

SHAPING PROCESSES FOR POLYMER MATRIX COMPOSITES

CHAPTER CONTENTS

15.1   Starting Materials for PMCs
    15.1.1   Polymer Matrix
    15.1.2   Reinforcing Agent
    15.1.3   Combining Matrix and Reinforcement
15.2   Open Mold Processes
    15.2.1   Hand Lay-up
    15.2.2   Spray-Up
    15.2.3   Automated Tape-Laying Machines
    15.2.4   Curing
15.3   Closed Mold Processes
    15.3.1   Compression Molding PMC Processes
    15.3.2   Transfer Molding PMC Processes
    15.3.3   Injection Molding PMC Processes
15.4   Filament Winding
15.5   Pultrusion Processes
    15.5.1   Pultrusion
    15.5.2   Pulforming
15.6   Other PMC Shaping Processes

In this chapter, we consider manufacturing processes by which polymer matrix composites are shaped into useful components and products. A *polymer matrix composite* (PMC) is a composite material consisting of a polymer imbedded with a reinforcing phase such as fibers or powders. The technological and commercial importance of the processes in this chapter derive from the growing use of this class of material, especially *fiber-reinforced polymers* (FRPs). In popular usage, PMC generally refers to fiber-reinforced polymers. FRP composites can be designed with very high strength-to-weight and modulus-to-weight ratios. These features make them attractive in aircraft, cars, trucks, boats, and sports equipment.

Some of the shaping processes described in this chapter are slow and labor intensive. In general, techniques for shaping composites are less efficient than manufacturing processes for other materials. There are two reasons for this: (1) composite materials are more complex than other materials, consisting as they do of two or more phases and the need to orient the reinforcing phase in the case of fiber-reinforced plastics; and (2) processing technologies for composites have not been the object of improvement and refinement over as many years as processes for other materials.

The variety of shaping methods for fiber-reinforced polymers is sometimes bewildering to students on first reading. Let us provide a road map for the reader entering this new territory. FRP composite shaping processes can be divided into five categories: (1) open mold processes, (2) closed mold processes, (3) filament winding, (4) pultrusion processes, and (5) other. ***Open mold processes*** include some of the original manual procedures for laying resins and fibers onto forms. ***Closed mold processes*** are much the same as those used in plastic molding; the reader will recognize the names—compression molding, transfer molding, and injection molding—although the names are sometimes changed and modifications are sometimes made for PMCs. In ***filament winding,*** continuous filaments that have been dipped in liquid resin are wrapped around a rotating mandrel; when the resin cures, a rigid, hollow, generally cylindrical shape is created. ***Pultrusion*** is a shaping process for producing long, straight sections of constant cross section; it is similar to extrusion, but it is adapted to include continuous fiber reinforcement. The "other" category includes several operations that do not fit into the previous categories.

Some of these processes are used to shape composites with continuous fibers, while others are used for short-fiber PMCs. Figure 15.1 provides an overview of the processes

FIGURE 15.1
Classification of manufacturing processes for fiber-reinforced polymer composites.

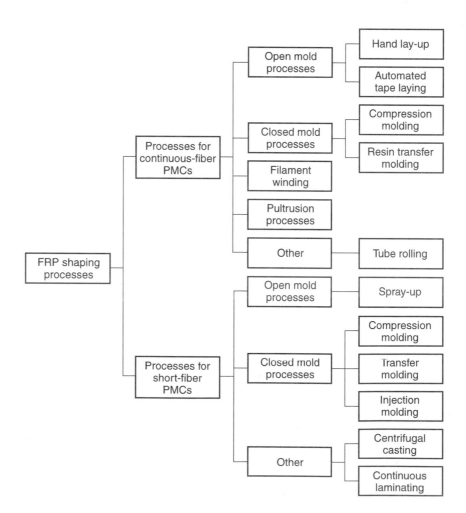

in each division. Let us begin our coverage by exploring how the individual phases in a PMC are produced and how these phases are combined into the starting materials for shaping.

15.1 STARTING MATERIALS FOR PMCs

In a PMC, the starting materials are a polymer and a reinforcing phase. They are processed separately before becoming phases in the composite. In this section, we consider how these materials are produced before being combined to make the composite part.

15.1.1 Polymer Matrix

All three basic polymer types—thermoplastics, thermosets, and elastomers—are used as matrices in PMCs. Thermosetting (TS) polymers are the most common matrix materials. The principal TS polymers are phenolics, unsaturated polyesters, and epoxies. Phenolics are associated with the use of particulate reinforcing phases, while polyesters and epoxies are more closely associated with FRPs. Thermoplastic (TP) polymers are also used in PMCs, and in fact, most molding compounds are composite materials that include fillers and/or reinforcing agents. Most elastomers are composite materials because nearly all rubbers are reinforced with carbon black. Shaping processes for rubbers are covered in Chapter 14. In this chapter, coverage is limited to the processing of PMCs that use TS and TP polymers as the matrix. Many of the polymer-shaping processes discussed in Chapter 13 are applicable to polymer matrix composites. However, combining the polymer with the reinforcing agent sometimes complicates the operations.

15.1.2 Reinforcing Agent

The reinforcing phase can be any of several geometries (fibers, particles, and flakes) and materials (ceramics, metals, other polymers, or elements such as carbon or boron). The role of the reinforcing phase and some of its technical features are discussed in Section 9.1.2.

Fibers   Common fiber materials in FRPs are glass, carbon, and the polymer Kevlar. Fibers of these materials are produced by various techniques, some of which we have covered in other chapters. Glass fibers are produced by drawing through small orifices (Section 12.2.3). For carbon, a series of heating treatments are performed to convert a precursor filament containing a carbon compound into a more pure carbon form. The precursor can be any of several substances, including polyacrylonitrile (PAN), pitch (a black carbon resin formed in the distillation of coal tar, wood tar, petroleum, etc.), or rayon (cellulose). Kevlar fibers are produced by extrusion combined with drawing through small orifices in a spinneret (Section 13.4).

Starting as continuous filaments, the fibers are combined with the polymer matrix in any of several forms, depending on the properties desired in the material and the processing method to be used to shape the composite. In some fabrication processes, the filaments are continuous, while in others, they are chopped into short lengths. In the continuous form, individual filaments are usually available as rovings. A *roving* is a collection of untwisted (parallel) continuous strands; this is a convenient form for handling and processing. Rovings typically contain from 12 to 120 individual strands. By contrast, a *yarn* is a twisted collection of filaments. Continuous rovings are used in several PMC processes, including filament winding and pultrusion.

The most familiar form of continuous fiber is a **cloth**—a fabric of woven yarns. Very similar to a cloth, but distinguished here, is a **woven roving,** a fabric consisting of untwisted filaments rather than yarns. Woven rovings can be produced with unequal numbers of strands in the two directions so that they possess greater strength in one direction than the other. Such unidirectional woven rovings are often preferred in laminated FRP composites.

Fibers can also be prepared in the form of a **mat**—a felt consisting of randomly oriented short fibers held loosely together with a binder, sometimes in a carrier fabric. Mats are commercially available as blankets of various weights, thicknesses, and widths. Mats can be cut and shaped for use as **preforms** in some of the closed mold processes. During molding, the resin impregnates the preform and then cures, thus yielding a fiber-reinforced molding.

Particles and Flakes   Particles and flakes are really in the same class. Flakes are particles whose length and width are large relative to thickness. We discuss these and other issues on characterization of engineering powders in Section 16.1. Production methods for metal powders are discussed in Section 16.2, and techniques for producing ceramic powders are discussed in Section 17.1.1.

15.1.3 Combining Matrix and Reinforcement

Incorporation of the reinforcing agent into the polymer matrix either occurs during the shaping process or beforehand. In the first case, the starting materials arrive at the fabrication operation as separate entities and are combined into the composite during shaping. Examples of this case are filament winding and pultrusion. The starting reinforcement in these processes consists of continuous fibers. In the second case, the two component materials are combined into some preliminary form that is convenient for use in the shaping process. Nearly all of the thermoplastics and thermosets used in the plastic-shaping processes are really polymers combined with fillers (Section 8.1.5). The fillers are either short fibers or particulate (including flakes).

Of greater interest in this chapter are the starting forms used in shaping processes specifically designed for FRP composites. We might think of the starting forms as pre-fabricated composites that arrive ready for use at the shaping process. These forms are molding compounds and prepregs.

Molding Compounds   Molding compounds are similar to those used in plastic molding. They are designed for use in molding operations, and so they must be capable of flowing. Most molding compounds for composite processing are thermosetting polymers. Accordingly, they have not been cured prior to shape processing. Curing is done during and/or after final shaping. FRP composite molding compounds consist of the resin matrix with short, randomly dispersed fibers. They come in several forms.

Sheet-molding compound (SMC) is a combination of TS polymer resin, fillers and other additives, and chopped glass fibers (randomly oriented), all rolled into a sheet of typical thickness = 6.5 mm (0.250 in.). The most common resin is unsaturated polyester; fillers are usually mineral powders such as talc, silica, limestone; and the glass fibers are typically 12–75 mm (0.5–3.0 in.) long and account for about 30% of the SMC by volume. SMCs are very convenient for handling and cutting to proper size as molding charges. Sheet-molding compounds are generally produced between thin layers of polyethylene to limit evaporation of volatiles from the thermosetting resin. The protective coating also improves surface finish on subsequent molded parts. The process for fabricating continuous SMC sheets is depicted in Figure 15.2.

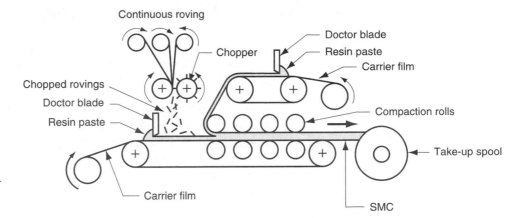

FIGURE 15.2   Process for producing sheet molding compound (SMC).

Bulk-molding compound (BMC) consists of similar ingredients as those in SMC, but the compounded polymer is in billet form rather than sheet. The fibers in BMC are shorter, typically 2–12 mm (0.1–0.5 in.), because greater fluidity is required in the molding operations for which these materials are designed. Billet diameter is usually 25–50 mm (1–2 in.). The process for producing BMC is similar to that for SMC, except extrusion is used to obtain the final billet form. BMC is also known as *dough-molding compound* (DMC), due to its dough-like consistency. Other FRP molding compounds include *thick molding compound* (TMC), similar to SMC but thicker—up to 50 mm (2 in.); and *pelletized molding compounds*—basically, conventional plastic-molding compounds containing short fibers.

Prepregs   Another prefabricated form for FRP shaping operations is *prepreg,* which consists of fibers impregnated with partially cured thermosetting resins to facilitate shape processing. Completion of curing must be accomplished during and/or after shaping. Prepregs are available in the form of tapes or cross-plied sheets or fabrics. The advantage of prepregs is that they are fabricated with continuous filaments rather than chopped random fibers, thus increasing strength and modulus. Prepreg tapes and sheets are associated with advanced composites (reinforced with boron, carbon/graphite, and Kevlar) as well as fiberglass.

15.2   OPEN MOLD PROCESSES

The distinguishing feature of this family of FRP shaping processes is its use of a single positive or negative mold surface (Figure 15.3) to produce laminated FRP structures. Other names for open mold processes include *contact lamination* and *contact molding.* The starting materials (resins, fibers, mats, and woven rovings) are applied to the mold in layers, building up to the desired thickness. This is followed by curing and part removal.

FIGURE 15.3   Types of open mold: (a) positive and (b) negative.

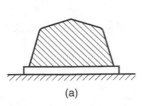
(a)

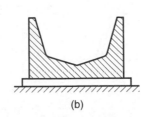
(b)

Common resins are unsaturated polyesters and epoxies, using fiberglass as the reinforcement. The moldings are usually large (e.g., for boat hulls). The advantage of using an open mold is that the mold costs much less than if two matching molds were used. The disadvantage is that only the part surface in contact with the mold surface is finished; the other side is rough. For the best possible part surface on the finished side, the mold itself must be very smooth.

There are several important open mold FRP processes. The differences are in the methods of applying the laminations to the mold, alternative curing techniques, and other differences. In this section we describe the family of open mold processes for shaping fiber-reinforced plastics: (1) hand lay-up, (2) spray-up, (3) automated tape-laying machines, and (4) bag molding. We treat hand lay-up as the base process and the others as modifications and refinements.

15.2.1 Hand Lay-Up

Hand lay-up is the oldest open mold method for FRP laminates, dating to the 1940s when it was first used to fabricate boat hulls. It is also the most labor-intensive method. As the name suggests, *hand lay-up* is a shaping method in which successive layers of resin and reinforcement are manually applied to an open mold to build the laminated FRP composite structure. The basic procedure consists of five steps, as illustrated in Figure 15.4. The finished molding must usually be trimmed with a power saw to size the outside edges. In general, these same five steps are required for all of the open mold processes; the differences between methods occur in steps (3) and (4), as we shall see.

In step (3), each layer of fiber reinforcement is dry when placed onto the mold. The liquid (uncured) resin is then applied by pouring, brushing, or spraying. Impregnation of resin into the fiber mat or fabric is accomplished by hand rolling. This approach is referred to as *wet lay-up.* An alternative approach is to use *prepregs,* in which the impregnated layers of fiber reinforcement are first prepared outside the mold and then laid onto the mold surface. Advantages cited for the prepregs include closer control over fiber-resin mixture and more efficient methods of adding the laminations [10].

FIGURE 15.4 Hand lay-up procedure: (1) mold is cleaned and treated with a mold release agent; (2) a thin gel coat (resin, possibly pigmented to color) is applied, which will become the outside surface of the molding; (3) when the gel coat has partially set, successive layers of resin and fiber are applied, the fiber being in the form of mat or cloth; each layer is rolled to fully impregnate the fiber with resin and remove air bubbles; (4) the part is cured; and (5) the fully hardened part is removed from the mold.

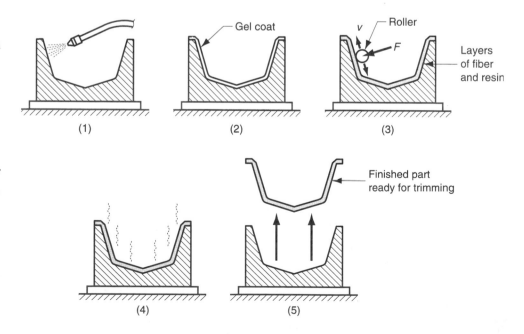

Molds for open mold contact laminating can be made of plaster, metal, glass fiber-reinforced plastic or other materials. Selection of material depends on economics, surface quality, and other technical factors. For prototype fabrication, in which only one part is produced, plaster molds are usually adequate. For medium quantities, the mold can be made of fiberglass reinforced plastic. High production generally requires metal molds. Aluminum, steel, and nickel are used, sometimes with surface hardening on the mold face to resist wear. An advantage of metal, in addition to durability, is its high thermal conductivity that can be used to implement a heat-curing system, or simply to dissipate heat from the laminate while it cures at room temperature.

Products suited to hand lay-up are generally large in size but low in production quantity. In addition to boat hulls, other applications include: swimming pools, large container tanks, stage props, radomes, and other formed sheets. Automotive parts have also been made, but the method is not economical for high production. The largest moldings ever made by this process were ship hulls for the British Royal Navy: 85 m (280 ft) long [1].

15.2.2 Spray-Up

This represents an attempt to mechanize the application of resin-fiber layers and to reduce the time for lay-up. It is an alternative for step (3) in the hand lay-up procedure. In the **spray-up method,** liquid resin and chopped fibers are sprayed onto an open mold to build successive FRP laminations, as in Figure 15.5. The spray gun is equipped with a chopper mechanism that feeds in continuous filament rovings and cuts them into fibers of length 25–75 mm (1–3 in.), which are added to the resin stream as it exits the nozzle. The mixing action results in random orientation of the fibers in the layer—unlike hand lay-up, in which the filaments can be oriented if desired. Another difference is that the fiber content in spray-up is limited to about 35% (compared to a maximum of around 65% in hand lay-up). This is a shortcoming of the spraying and mixing process.

Spraying can be accomplished manually using a portable spray gun or by an automated machine in which the path of the spray gun is preprogrammed and computer controlled. The automated procedure is advantageous for labor efficiency and environmental protection. Some of the volatile emissions from the liquid resins are hazardous, and the path-controlled machines can operate in sealed-off areas without humans present. However, rolling is generally required for each layer, as in hand lay-up.

Products made by the spray-up method include boat hulls, bathtubs, shower stalls, automobile and truck body parts, recreational vehicle components, furniture, large structural panels, and containers. Movie and stage props are sometimes made by this method.

FIGURE 15.5 Spray-up method.

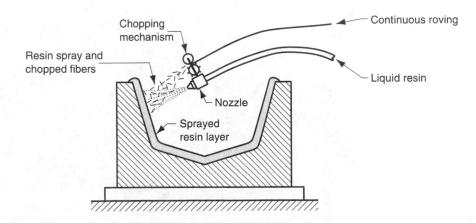

Since products made by spray-up have randomly oriented short fibers, they are not as strong as those made by lay-up, in which the fibers are continuous and directed.

15.2.3 Automated Tape-Laying Machines

This is another attempt to automate and accelerate step (3) in the lay-up procedure. Automated tape-laying machines operate by dispensing a prepreg tape onto an open mold following a programmed path. The typical machine consists of an overhead gantry, to which is attached the dispensing head, as shown in Figure 15.6. The gantry permits x-y-z travel of the head, for positioning and following a defined continuous path. The head itself has several rotational axes, plus a shearing device to cut the tape at the end of each path. Prepreg tape widths are commonly 75 mm (3 in.), although 300 mm (12 in.) widths have been reported [9]; thickness is around 0.13 mm (0.005 in.). The tape is stored on the machine in rolls, which are unwound and deposited along the defined path. Each lamination is placed by following a series of back-and-forth passes across the mold surface until the parallel rows of tape complete the layer.

Much of the work to develop automated tape-laying machines has been pioneered by the aircraft industry, which is eager to save labor costs and at the same time achieve the highest possible quality and uniformity in its manufactured components. The disadvantage of this and other computer numerically controlled machines is that it must be programmed, and programming takes time.

15.2.4 Curing

Curing (Step 4) is required of all thermosetting resins used in FRP laminated composites. Curing accomplishes cross-linking of the polymer, transforming it from its liquid or highly plastic condition into a hardened product. There are three principal process parameters in curing: time, temperature, and pressure.

Curing normally occurs at room temperature for the TS resins used in hand lay-up and spray-up procedures. Moldings made by these processes are often large (e.g., boat

FIGURE 15.6   Automated tape-laying machine (courtesy Cincinnati Milacron).

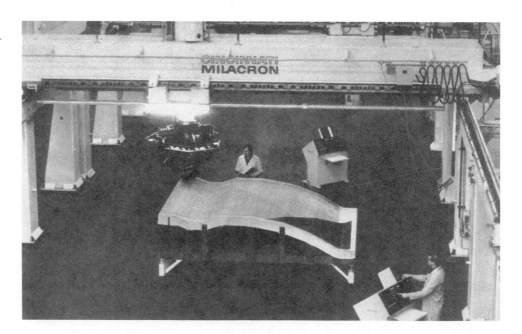

hulls), and heating would be difficult for such parts. In some cases, days are required before room temperature curing is sufficiently complete to remove the part. If feasible, heat is added to speed the curing reaction.

Heating is accomplished by several means. Oven curing provides heat at closely controlled temperatures; some curing ovens are equipped to draw a partial vacuum. Infrared heating can be used in applications where it is impractical or inconvenient to place the molding in an oven.

Curing in an autoclave provides control over both temperature and pressure. An *autoclave* is an enclosed chamber equipped to apply heat and/or pressure at controlled levels. In FRP composites processing, it is usually a large horizontal cylinder with doors at either end. The term *autoclave molding* is sometimes used to refer to the curing of a prepreg laminate in an autoclave. This procedure is used extensively in the aerospace industry to produce advanced composite components of very high quality.

15.3 CLOSED MOLD PROCESSES

These molding operations are performed in molds consisting of two sections that open and close during each molding cycle. The name *matched die molding* is used for some of these processes. One might think that a closed mold is about twice the cost of a comparable open mold. However, tooling cost is even greater due to the more complex equipment required in these processes. Despite their higher cost, advantages of a closed mold are (1) good finish on all part surfaces, (2) higher production rates, (3) closer control over tolerances, and (4) more complex three-dimensional shapes are possible.

We divide the closed mold processes into three classes based on their counterparts in conventional plastic molding, even though the terminology is often different when polymer matrix composites are molded: (1) compression molding, (2) transfer molding, and (3) injection molding.

15.3.1 Compression Molding PMC Processes

In compression molding of conventional molding compounds (Section 13.7.1), a charge is placed in the lower mold section, and the sections are brought together under pressure, causing the charge to take the shape of the cavity. The mold halves are heated to effect curing of the thermosetting polymer. When the molding is sufficiently cured, the mold is opened and the part is removed. There are several shaping processes for PMCs based on compression molding; the differences are mostly in the form of the starting materials. The flow of the resin, fibers, and other ingredients during the process is a critical factor in compression molding of FRP composites.

SMC, TMC, and BMC Molding Several of the FRP molding compounds, namely sheet-molding compound (SMC), bulk-molding compound (BMC), and thick-molding compound (TMC), can be cut to proper size and used as the starting charge in compression molding. Refrigeration is often required to store these materials prior to shape processing. The names of the molding processes are based on the starting molding compound (i.e., *SMC molding* refers to a molding operation in which the starting charge is precut sheet molding compound; and *BMC molding* uses BMC cut to size as the charge).

Preform Molding Another form of compression molding, called *preform molding* [10], involves placement of a precut mat into the lower mold section along with a polymer resin charge (e.g., pellets or sheet). The materials are then pressed between heated mold halves, causing the resin to flow and impregnate the fiber mat to produce a fiber reinforced molding. Variations of the process use either thermoplastic or thermosetting polymers.

Elastic Reservoir Molding The starting charge in elastic reservoir molding (ERM) is a sandwich consisting of a center of polymer foam between two dry fiber layers. The foam core is commonly open-cell polyurethane, impregnated with liquid resin such as epoxy or polyester; and the dry fiber layers can be cloth, woven roving, or other starting fibrous form. As depicted in Figure 15.7, the sandwich is placed in the lower mold section and pressed at moderate pressure—around 0.7 MPa (100 lb/in.²). As the core is compressed, it releases the resin to wet the dry surface layers. Curing produces a lightweight part consisting of a low-density core and thin FRP skins.

15.3.2 Transfer Molding PMC Processes

In conventional transfer molding (Section 13.7.2), a charge of thermosetting resin is placed in a pot or chamber, heated, and squeezed by ram action into one or more mold cavities. The mold is heated to cure the resin. The name of the process derives from the fact that the fluid polymer is transferred from the pot into the mold. It can be used to mold TS resins in which the fillers include short fibers to produce a FRP composite part. Another form of transfer molding for PMCs is called ***resin transfer molding***(RTM) [3], [10]; it refers to a closed mold process in which a preform mat is placed in the lower mold section, the mold is closed, and a thermosetting resin (e.g., polyester resin) is transferred into the cavity under moderate pressure to impregnate the preform. To confuse matters, RTM is sometimes called ***resin injection molding*** [3], [11] (the distinction between transfer molding and injection molding is blurry anyway, as the reader may have noted in Chapter 13). RTM has been utilized to manufacture such products as bathtubs, swimming pool shells, bench and chair seats, and hulls for small boats.

Several enhancements of the basic RTM process have been developed [4]. One enhancement, called ***advanced RTM,*** uses high-strength polymers such as epoxy resins and continuous fiber reinforcement instead of mats. Applications include aerospace components, missile fins, and snow skis. Two additional processes are thermal expansion resin transfer molding and ultimately reinforced thermoset resin injection. ***Thermal expansion resin transfer molding*** (TERTM) is a patented process of TERTM, Inc., which consists of the following steps [4]: (1) A rigid polymer foam (e.g., polyurethane) is shaped into a preform. (2) The preform is enclosed in a fabric reinforcement and placed in a closed mold. (3) A thermosetting resin (e.g., epoxy) is injected into the mold to impregnate the

FIGURE 15.7   Elastic reservoir molding: (1) foam is placed into mold between two fiber layers; (2) mold is closed, releasing resin from foam into fiber layers.

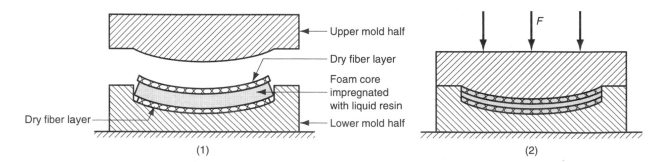

| (1) | (2) |

Upper mold half
Dry fiber layer
Foam core impregnated with liquid resin
Dry fiber layer
Lower mold half

fabric and surround the foam. (4) The mold is heated to expand the foam, fill the mold cavity, and cure the resin. ***Ultimately reinforced thermoset resin injection*** (URTRI) is similar to TERTM except that the starting foam core is cast epoxy imbedded with miniature hollow glass spheres.

15.3.3 Injection Molding PMC Processes

Injection molding is noted for low-cost production of plastic parts in large quantities. Although it is most closely associated with thermoplastics, the process can also be adapted to thermosets (Section 13.6.6).

Conventional Injection Molding   In PMC-shape processing, injection molding is used for both TP and TS type FRPs. In the TP category, virtually all thermoplastic polymers can be reinforced with fibers. Chopped fibers must be used; if continuous fibers were used, they would be reduced anyway by the action of the rotating screw in the barrel. During injection from the chamber into the mold cavity, the fibers tend to become aligned during their journey through the nozzle. Designers can sometimes exploit this feature to optimize directional properties through part design, location of gates, and cavity orientation relative to the gate [7].

Whereas TP molding compounds are heated and then injected into a cold mold, TS polymers are injected into a heated mold for curing. Control of the process with thermosets is trickier due to the risk of premature cross-linking in the injection chamber. Subject to the same risk, injection molding can be applied to fiber-reinforced TS plastics in the form of pelletized molding compound and dough-molding compound.

Reinforced Reaction Injection Molding   Some thermosets cure by chemical reaction rather than heat; these resins can be molded by reaction injection molding (Section 13.6.6). In RIM, two reactive ingredients are mixed and immediately injected into a mold cavity where curing and solidification of the chemicals occur rapidly. A closely related process includes reinforcing fibers, typically glass, in the mixture. In this case, the process is called ***reinforced reaction injection molding*** (RRIM). Its advantages are similar to those in RIM, with the added benefit of fiber reinforcement. RRIM is used extensively in auto body and truck cab applications for bumpers, fenders, and other body parts.

15.4 FILAMENT WINDING

Filament winding is a process in which resin-impregnated continuous fibers are wrapped around a rotating mandrel that has the internal shape of the desired FRP product. The resin is subsequently cured and the mandrel removed. Hollow axisymmetric components (usually circular in cross section) are produced, as well as some irregular shapes. The most common form of the process is depicted in Figure 15.8. A band of fiber rovings is pulled through a resin bath immediately before being wound in a helical pattern onto a cylindrical mandrel. Continuation of the winding pattern finally completes a surface layer of one filament thickness on the mandrel. The operation is repeated to form additional layers, each having a crisscross pattern with the previous, until the desired part thickness has been obtained.

There are several methods by which the fibers can be impregnated with resin: (1) ***wet winding,*** in which the filament is pulled through the liquid resin just prior to winding, as in our figure; (2) ***prepreg winding*** (also called ***dry winding***), in which filaments preimpregnated with partially cured resin are wrapped around a heated mandrel; and

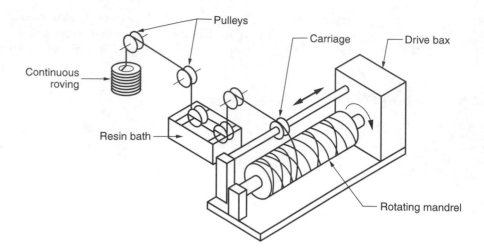

FIGURE 15.8 Filament winding.

(3) *postimpregnation,* in which filaments are wound onto a mandrel and then impregnated with resin by brushing or other technique.

Two basic winding patterns are used in filament winding: (a) helical and (b) polar, Figure 15.9. In *helical winding,* the filament band is applied in a spiral pattern around the mandrel, at a helix angle θ. If the band is wrapped with a helix angle approaching 90°, so that the winding advance is one bandwidth per revolution, this is referred to as a *hoop winding* since the filaments form nearly circular rings about the mandrel; it is a special case of helical winding. In *polar winding,* the filament is wrapped around the long axis of the mandrel, as in Figure 15.9(b); after each longitudinal revolution, the mandrel is indexed (partially rotated) by one bandwidth, so that a hollow enclosed shape is gradually created. Hoop and polar patterns can be combined in successive windings of the mandrel to produce adjacent layers with filament directions that are approximately perpendicular; this is called a *bi-axial winding* [1].

Filament winding machines have motion capabilities similar to those of an engine lathe (Section 22.1.3). The typical machine has a drive motor to rotate the mandrel and a powered feed mechanism to move the carriage. Relative motion between mandrel and carriage must be controlled in order to accomplish a given winding pattern. In helical winding, the relationship between helix angle and the machine parameters can be expressed as follows:

$$\tan \theta = \frac{v_c}{\pi D N} \qquad (15.1)$$

where θ = helix angle of the windings on the mandrel, as in Figure 15.9(a); v_c = speed at which the carriage traverses in the axial direction, m/s (in./sec); D = diameter of the mandrel, m (in.); and N = rotational speed, 1/s (rev/sec).

Various levels of control sophistication are available in filament winding machines. Two main types predominate: (1) *mechanical control,* which operates by means of direct linkages between mandrel and carriage drives, for the simplest and lowest-cost form of

FIGURE 15.9 Two basic winding patterns in filament winding: (a) helical and (b) polar.

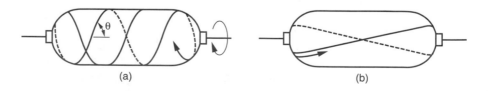

FIGURE 15.10   Filament winding machine (courtesy Cincinnati Milacron).

control; and (2) *computer numerical control* (CNC), in which mandrel rotation and carriage speed are independently controlled to permit greater adjustment and flexibility in the relative motions. CNC is especially useful in helical winding of contoured shapes, as in Figure 15.10. As indicated in Eq. (15.1), the ratio v_c/DN must remain fixed to maintain a constant helix angle θ. Thus, either v_c and/or N must be adjusted online to compensate for changes in D.

The *mandrel* is the special tooling that determines the geometry of the filament-wound part. Mandrels must be capable of collapsing after winding and curing, for part removal. Various mandrel designs are possible, including inflatable/deflatable mandrels, collapsible metal mandrels, and mandrels made of soluble salts or plasters.

Applications of filament winding are often classified as aerospace or commercial [9], the engineering requirements being more demanding in the first category. Aerospace applications include rocket-motor cases, missile bodies, radomes, helicopter blades, and airplane tail sections and stabilizers. These components are made of advanced composites and hybrid composites (Section 9.4.1), with epoxy resins being most common and reinforced with fibers of carbon, boron, Kevlar, and glass. Commercial applications include storage tanks, reinforced pipes and tubing, drive shafts, wind-turbine blades, and lightning rods. These are made of conventional FRPs. Polymers include polyester, epoxy, and phenolic resins; glass is the common reinforcing fiber.

15.5 PULTRUSION PROCESSES

The basic pultrusion process was developed around 1950 for making fishing rods of glass fiber reinforced polymer (GFRP). The process is similar to extrusion (hence the similarity in name), but it involves pulling of the workpiece (so the prefix "pul-" is used in place of "ex-"). Like extrusion, pultrusion produces continuous straight sections of constant cross section. A related process, called pulforming, can be used to make parts that are curved and that may have variations in cross section throughout their lengths.

15.5.1 Pultrusion

Pultrusion is a process in which continuous fiber rovings are dipped into a resin bath and pulled through a shaping die where the impregnated resin cures. The setup is sketched in Figure 15.11, which shows the cured product being cut into long, straight sections. The sections are reinforced throughout their length by continuous fibers. Like extrusion, the pieces have a constant cross section, whose profile is determined by the shape of the die opening.

The process consists of five steps (identified in our sketch) performed in a continuous sequence [1]: (1) *filament feeding,* in which the fibers are unreeled from a creel (shelves with skewers that hold filament bobbins); (2) *resin impregnation,* in which the fibers are dipped in the uncured liquid resin; (3) *pre-die forming*—the collection of filaments is gradually shaped into the approximate cross-section desired; (4) *shaping and curing,* in which the impregnated fibers are pulled through the heated die whose length is 1–1.5 m (3–5 ft) and whose inside surfaces are highly polished; and (5) *pulling and cutting*—pullers are used to draw the cured length through the die, after which it is cut by a cut-off wheel with SiC or diamond grits.

Common resins used in pultrusion are unsaturated polyesters, epoxies, and silicones, all thermosetting polymers. There are difficulties in processing with epoxy polymers due to sticking on the die surface. Thermoplastics have also been studied for possible applications [1]. E-glass is by far the most widely used reinforcing material; proportions range from 30% to 70%. Modulus of elasticity and tensile strength increase with reinforcement content. Products made by pultrusion include solid rods, tubing, long flat sheets, structural sections (such as channels, angled and flanged beams), tool handles for high voltage work, and third rail covers for subways.

15.5.2 Pulforming

The pultrusion process is limited to straight sections of constant cross section. There is also a need for long parts with continuous fiber reinforcement that are curved rather than straight and whose cross sections may vary throughout the length. The pulforming process is suited to these less regular shapes. *Pulforming* can be defined as pultrusion with addi-

FIGURE 15.11   Pultrusion process (see text for interpretation of sequence numbers).

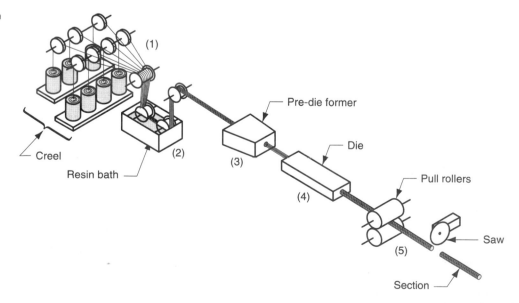

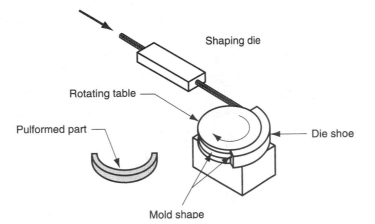

FIGURE 15.12
Pulforming process (not shown in the sketch is the cut-off of the pul-formed part).

tional steps to form the length into a semicircular contour and alter the cross section at one or more locations along the length. A sketch of the equipment is illustrated in Figure 15.12. After exiting the shaping die, the continuous workpiece is fed into a rotating table with negative molds positioned around its periphery. The work is forced into the mold cavities by a die shoe, which squeezes the cross section at various locations and forms the curvature in the length. The diameter of the table determines the radius of the part. As the work leaves the die table it is cut to length to provide discrete parts. Resins and fibers similar to those for pultrusion are used in pulforming. An important application of the process is production of automobile leaf springs.

15.6 OTHER PMC SHAPING PROCESSES

Additional PMC-shaping processes worth noting include centrifugal casting, tube rolling, continuous laminating, and cutting. In addition, many of the traditional thermoplastic-shaping processes are applicable to (short fiber) FRPs based on TP polymers; these include blow molding, thermoforming, and extrusion.

Centrifugal Casting   This process is ideal for cylindrical products such as pipes and tanks. The process is the same as its counterpart in metal casting (Section 11.3.4). Chopped fibers combined with liquid resin are poured into a fast-rotating cylindrical mold. Centrifugal force presses the ingredients against the mold wall, where curing takes place. The resulting inner surfaces are quite smooth. Part shrinkage or use of split molds permits part removal.

Tube rolling   FRP tubes can be fabricated from prepreg sheets by a rolling technique [6], shown in Figure 15.13. Such tubes are used in bicycle frames and space trusses. In the process, a precut prepreg sheet is wrapped around a cylindrical mandrel several times to obtain a tube wall of multiple sheet thicknesses. The rolled sheets are then encased in a heat-shrinking sleeve and oven cured. As the sleeve contracts, entrapped gases are squeezed out the ends of the tube. When curing is complete, the mandrel is removed to yield a rolled FRP tube. The operation is simple, and tooling cost is low. There are variations in the process, such as using different wrapping methods or using a steel mold to enclose the rolled prepreg tube for better dimensional control.

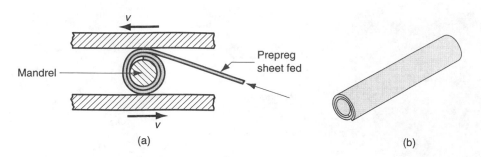

FIGURE 15.13 Tube rolling, showing: (a) one possible means of wrapping FRP prepregs around a mandrel and (b) the completed tube after curing and removal of mandrel.

Continuous Laminating Fiber-reinforced plastic panels, sometimes translucent and/or corrugated, are used in construction. The process to produce them consists of (1) impregnating layers of glass fiber mat or woven fabric by dipping in liquid resin or by passing beneath a doctor blade, (2) gathering between cover films (cellophane, polyester, or other polymer), and (3) compacting between squeeze rolls and curing. Corrugation (4) is added by formed rollers or mold shoes.

Cutting Methods FRP-laminated composites must be cut in both uncured and cured states. Uncured materials (prepregs, preforms, SMCs, and other starting forms) must be cut to size for lay-up, molding, and so on. Typical cutting tools include knives, scissors, power shears, and steel-rule blanking dies. Also used are nontraditional cutting methods, such as laser-beam cutting and water-jet cutting (Chapter 26).

Cured FRPs are hard, tough, abrasive, and difficult to cut. But cutting is necessary in many FRP-shaping processes to trim excess material, cut holes and outlines, and for other purposes. For fiberglass-reinforced plastics, cemented carbide cutting tools and high-speed steel saw blades must be used. For some advanced composites (e.g., boron-epoxy), diamond-cutting tools obtain best results. Water-jet cutting is also used with good success on cured FRPs; this process reduces the dust and noise problems associated with conventional sawing methods.

REFERENCES

[1] Bader, M. G., Smith, W., Isham, A. B., Rolston, J. A., and Metzner, A. B. *Delaware Composites Design Encyclopedia.* Vol. 3. *Processing and Fabrication Technology.* Technomic Publishing Co., Inc., Lancaster, Pennsylvania, 1990.

[2] Chawla, K. K. *Composite Materials: Science and Engineering.* 2nd ed. Springer-Verlag, New York, 1998.

[3] Charrier, J-M. *Polymeric Materials and Processing.* Oxford University Press, New York, 1991.

[4] Coulter, J. P. "Resin Impregnation During The Manufacture of Composite Materials." *PhD Dissertation.* University of Delaware, 1988.

[5] *Engineering Materials Handbook.* Vol. 1. *Composites,* ASM International, Metals Park, Ohio (USA), 1987.

[6] Mallick, P. K. *Fiber-Reinforced Composites: Materials, Manufacturing, and Design.* 2nd ed. Marcel Dekker, Inc., New York, 1993.

[7] McCrum, N. G., Buckley, C. P., and Bucknall, C. B. *Principles of Polymer Engineering.* Oxford University Press, Inc., Oxford, U.K., 1988.

[8] Morton-Jones, D. H. *Polymer Processing.* Chapman and Hall, London, U.K., 1989.

[9] Schwartz, M. M. *Composite Materials Handbook.* 2nd ed. McGraw-Hill Book Company, New York, 1992.

[10] Strong, A. B. *Fundamentals of Composites Manufacturing: Materials, Methods, and Applications.* Society of Manufacturing Engineers, Dearborn, Michigan, 1989.

[11] Wick, C., Benedict, J. T., and Veilleux, R. F. *Tool and Manufacturing Engineers Handbook.* 4th ed. Volume II—*Forming,* 1984.

[12] Wick, C., and Veilleux, R. F. *Tool and Manufacturing Engineers Handbook.* 4th ed. Volume III—*Materials, Finishing, and Coating,* 1985.

REVIEW QUESTIONS

15.1. What are the principal polymers used in fiber-reinforced polymers?

15.2. What is the difference between a roving and a yarn?

15.3. In the context of fiber reinforcement, what is a *mat*?

15.4. Why do we say that particles and flakes are members of the same basic class of reinforcing material?

15.5. What is *sheet molding compound* (SMC)?

15.6. How is a prepreg different from a molding compound?

15.7. Why are laminated FRP products made by the spray-up method not as strong as similar products made by hand lay-up?

15.8. What is the difference between the wet lay-up approach and the prepreg approach in hand lay-up?

15.9. What is an *autoclave*?

15.10. What are some of the distinguishing characteristics of the closed mold processes for PMCs?

15.11. Identify some of the different forms of PMC molding compounds.

15.12. What is *preform molding*?

15.13. Describe *reinforced reaction injection molding* (RRIM).

15.14. What is *filament winding*?

15.15. What is the advantage of computer numerical control over mechanical control in filament winding?

15.16. Describe the *pultrusion* process.

15.17. How does *pulforming* differ from pultrusion?

15.18. With what kinds of products is *tube rolling* associated?

15.19. How are FRPs cut?

MULTIPLE CHOICE QUIZ

There is a total of 12 correct answers in the following multiple choice questions (some questions have multiple answers that are correct). To attain a perfect score on the quiz, all correct answers must be given, since each correct answer is worth 1 point. For each question, each omitted answer or wrong answer reduces the score by 1 point, and each additional answer beyond the number of answers required reduces the score by 1 point. Percentage score on the quiz is based on the total number of correct answers.

15.1. Which one of the following is the most common polymer type in fiber-reinforced polymer composites? (a) elastomers, (b) thermoplastics, or (c) thermosetting plastics.

15.2. Most rubber products are properly classified into which of the following categories (more than one answer)? (a) elastomer reinforced with carbon black, (b) fiber-reinforced composite, (c) particle-reinforced composite, (d) polymer matrix composite, (e) pure elastomer, and (f) pure polymer.

15.3. Hand lay-up is classified in which of the following general categories of PMC shaping processes (more than one)? (a) closed mold process, (b) compression molding, (c) contact molding, (d) filament winding, or (e) open mold process.

15.4. A positive mold with a smooth surface will produce a good finish on which surface of the laminated product in the hand lay-up method? (a) inside surface or (b) outside surface.

15.5. SMC molding is a form of which one of the following? (a) compression molding, (b) contact molding, (c) injection molding, (d) open mold processing, (e) pultrusion, or (f) transfer molding.

15.6. Filament winding involves the use of which one of the following fiber reinforcements? (a) continuous filaments, (b) fabrics, (c) mats, (d) prepregs, (e) short fibers, or (f) woven rovings.

15.7. In filament winding, when the continuous filament is wound around the cylindrical mandrel at a helix angle close to 90°, it is called which of the following (one best answer)? (a) bi-axial winding, (b) helical winding, (c) hoop winding, (d) perpendicular winding, (e) polar winding, or (f) radial winding.

15.8. Pultrusion is most similar to which of the following plastic shaping processes? (a) blow-molding, (b) extrusion, (c) injection molding, or (d) thermoforming.

15.9. Water-jet cutting is one of several ways of cutting or trimming uncured or cured FRPs; in the case of cured FRPs, the process is noted for its reduction of dust and noise: (a) true or (b) false.

Part IV
Particulate Processing of Metals and Ceramics

16 POWDER METALLURGY

CHAPTER CONTENTS

16.1   Characterization of Engineering Powders
    16.1.1   Geometric Features
    16.1.2   Other Features

16.2   Production of Metallic Powders
    16.2.1   Atomization
    16.2.2   Other Production Methods

16.3   Conventional Pressing and Sintering
    16.3.1   Blending and Mixing of the Powders
    16.3.2   Compaction
    16.3.3   Sintering
    16.3.4   Secondary Operations

16.4   Alternative Pressing and Sintering Techniques
    16.4.1   Isostatic Pressing
    16.4.2   Powder Injection Molding
    16.4.3   Powder Rolling, Extrusion, and Forging
    16.4.4   Combined Pressing and Sintering
    16.4.5   Liquid Phase Sintering

16.5   Materials and Products for PM
    16.5.1   PM Materials
    16.5.2   PM Products

16.6   Design Considerations in Powder Metallurgy
    16.6.1   Parts Classification System
    16.6.2   Design Guidelines for PM Parts

This part of the book is concerned with the processing of metals and ceramics that are in the form of powders—very small particulate solids. In the case of traditional ceramics, the powders are produced by crushing and grinding common materials that are found in nature, such as silicate minerals (clay) and quartz. In the case of metals and the new ceramics (Section 7.3), the powders are produced by a variety of industrial processes. We cover these processes as well as the methods used to shape products out of these materials in two chapters: Chapter 16 is devoted to powder metallurgy, and Chapter 17 covers particulate processing of ceramics.

Powder metallurgy (PM) is a metal-processing technology in which parts are produced from metallic powders. In the usual PM production sequence, the powders are compressed into the desired shape and then heated to cause bonding of the particles into a hard, rigid mass. Compression, called *pressing,* is accomplished in a press-type machine using tools designed specifically for the part to be manufactured. The tooling, which typically consists of a die and one or more punches, can be expensive, and PM is therefore most appropriate for medium and high production. The heating treatment, called *sintering,* is performed at a temperature below the melting point of the metal. Considerations that make powder metallurgy an important commercial technology include:

➢ PM parts can be mass produced to *net shape* or *near net shape,* eliminating or reducing the need for subsequent processing.
➢ The PM process itself involves very little waste of material; about 97% of the starting powders are converted to product. This compares favorably to casting processes in which sprues, runners, and risers are wasted material in the production cycle.
➢ Owing to the nature of the starting material in PM, parts having a specified level of porosity can be made. This feature lends itself to the production of porous metal parts, such as filters, and oil-impregnated bearings and gears.
➢ Certain metals that are difficult to fabricate by other methods can be shaped by powder metallurgy. Tungsten is an example; tungsten filaments used in incandescent lamp bulbs are made using PM technology.
➢ Certain metal alloy combinations and cermets can be formed by PM that cannot be produced by other methods.
➢ PM compares favorably to most casting processes in terms of dimensional control of the product. Tolerances of ±0.13 mm (±0.005 in.) are held routinely.
➢ PM production methods can be automated for economical production.

There are limitations and disadvantages associated with PM processing. These include: (1) high tooling and equipment costs, (2) expensive metallic powders, and (3) difficulties with storing and handling metal powders (such as degradation of the metal over time, and fire hazards with particular metals). Also, (4) there are limitations on part geometry because metal powders do not readily flow laterally in the die during pressing, and allowances must be provided for ejection of the part from the die after pressing. In addition, (5) variations in material density throughout the part may be a problem in PM, especially for complex part geometries.

Although parts as large as 22 kg (50 lb) can be produced, most PM components are less than 2.2 kg (5 lbs). A collection of typical PM parts is shown in Figure 16.1. The largest tonnage of metals for PM are alloys of iron, steel, and aluminum. Other PM metals include copper, nickel, and refractory metals such as molybdenum and tungsten. Metallic carbides such as tungsten carbide are often included within the scope of powder metallurgy; however, since these materials are ceramics, we defer their consideration until the next chapter.

The development of the modern field of powder metallurgy dates back to the 1800s (Historical Note 16.1). The scope of the modern technology includes not only parts production but also preparation of the starting powders. Success in powder metallurgy depends to a large degree on the characteristics of the starting powders; we discuss this topic in Section 16.1. Subsequent sections describe powder production, pressing, and sintering. There is a close correlation between PM technology and aspects of ceramics processing (Chapter 17). In ceramics, the starting material is also powder, so the methods for characterizing the powders are closely related to those in PM. Several of the shape forming methods are similar, also.

FIGURE 16.1 A collection of powder metallurgy parts (courtesy of Dorst America, Inc.).

Historical Note 16.1   *Powder metallurgy [7].*

Powders of metals such as gold and copper, as well as some of the metallic oxides, have been used for decorative purposes since ancient times. The uses included decorations on pottery, bases for paints, and in cosmetics. It is believed that the Egyptians used PM to make tools as far back as 3000 B.C.

The modern field of powder metallurgy dates to the early nineteenth century, when there was a strong interest in the metal platinum. Around 1815, Englishman William Wollaston developed a technique for preparing platinum powders, compacting them under high pressure, and baking (sintering) them at red heat. The Wollaston process marks the beginning of powder metallurgy as it is practiced today.

U.S. patents were issued in 1870 to Gwynn that relate to PM self-lubricating bearings. He used a mixture of 99% powdered tin and 1% petroleum, mixing, heating, and finally subjecting the mixture to extreme pressures to form it into the desired shape inside a mold cavity.

By the early 1900s, the incandescent lamp had become an important commercial product. A variety of fil-ament materials had been tried, including carbon, zirconium, vanadium, and osmium; but it was concluded that tungsten was the best filament material. The problem was that tungsten was difficult to process due to its high melting point and unique properties. In 1908, William Coolidge developed a procedure that made production of tungsten incandescent lamp filaments feasible. In his process, fine powders of tungsten oxide (WO_3) were reduced to metallic powders, pressed into compacts, presintered, hot-forged into rounds, sintered, and finally drawn into filament wire. The Coolidge process is still used today to make filaments for incandescent light bulbs.

In the 1920s, cemented carbide tools (WC–Co) were being fabricated by PM techniques (Historical Note 7.2). Self-lubricating bearings were produced in large quantities starting in the 1930s. Powder metal gears and other components were mass produced in the 1960s and 1970s, especially in the automotive industry. And in the 1980s, PM parts for aircraft turbine engines were developed

---

16.1 CHARACTERIZATION OF ENGINEERING POWDERS

A ***powder*** can be defined as a finely divided particulate solid. In this section we characterize metallic powders. However, most of our discussion applies to ceramic powders as well.

16.1.1 Geometric Features

The geometry of the individual powders can be defined by the following attributes: (1) particle size and distribution, (2) particle shape and internal structure, and (3) surface area.

Particle Size and Distribution   Particle size refers to the dimensions of the individual powders. If the particle shape is spherical, a single dimension is adequate. For other shapes, two or more dimensions are needed. There are various methods available to obtain particle size data. The most common method uses screens of different mesh sizes. The term ***mesh count*** is used to refer to the number of openings per linear inch of screen. A mesh count of 200 means there are 200 openings per linear inch. Since the mesh is square, the count is the same in both directions, and the total number of openings per square inch is $200^2 = 40,000$. Thus, higher mesh count indicates smaller particle size.

Particles are sorted by passing them through a series of screens of progressively smaller mesh size. The powders are placed on a screen of a certain mesh count and vibrated so that particles small enough to fit through the openings pass through to the next screen below. The second screen empties into a third, and so forth, so that the particles are sorted according to size. A certain powder size might be called size 230 through 200, indicating that the powders have passed through the 200 mesh, but not 230. To make the specification easier, we simply say that the particle size is 200. The procedure of separating the powders by size is called ***classification.***

The openings in the screen are less than the reciprocal of the mesh count because of the thickness of the wire in the screen, as illustrated in Figure 16.2. Assuming that the limiting dimension of the particle is equal to the screen opening, we have

$$PS = \frac{1}{MC} - t_w \qquad (16.1)$$

where PS = particle size, in; MC = mesh count, openings per linear inch; and t_w = wire thickness of screen mesh, in. The figure shows how smaller particles would pass through the openings, while larger powders would not. Variations occur in the powder sizes sorted by screening due to differences in particle shapes, the range of sizes between mesh count steps, and variations in screen openings within a given mesh count. Also, the screening method has a practical upper limit of $MC = 400$ (approximately), due to the difficulty in making such fine screens and because of agglomeration of the small powders. Other methods to measure particle size include microscopy and X-ray techniques.

FIGURE 16.2   Screen mesh for sorting particle sizes.

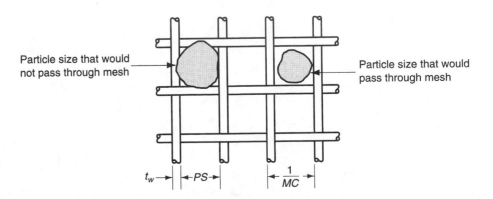

Particle Shape and Internal Structure Metal powder shapes can be cataloged into various types, several of which are illustrated in Figure 16.3. There will be a variation in the particle shapes in a collection of powders, just as the particle size will vary. A simple and useful measure of shape is the aspect ratio—the ratio of maximum dimension to minimum dimension for a given particle. The aspect ratio for a spherical particle is 1.0, but for an acicular grain the ratio might be 2 to 4. Microscopic techniques are required to determine shape characteristics.

Any volume of loose powders will contain pores between the particles. These are called *open pores* because they are external to the individual particles. Open pores are spaces into which a fluid such as water, oil, or a molten metal, can penetrate. In addition, there are *closed pores*—internal voids in the structure of an individual particle. The existence of these internal pores is usually minimal, and their effect when they do exist is minor, but they can influence density measurements, as we shall see later.

Surface Area Assuming that the particle shape is a perfect sphere, its area A and volume V are given by

$$A = \pi D^2 \tag{16.2}$$

$$V = \frac{\pi D^3}{6} \tag{16.3}$$

where D = diameter of the spherical particle, mm (in.). The area-to-volume ratio A/V for a sphere is then given by

$$\frac{A}{V} = \frac{6}{D} \tag{16.4}$$

In general, the area-to-volume ratio can be expressed for any particle shape—spherical or nonspherical—as follows:

$$\frac{A}{V} = \frac{K_s}{D} \text{ or } K_s = \frac{AD}{V} \tag{16.5}$$

where K_s = shape factor; D in the general case = the diameter of a sphere of equivalent volume as the nonspherical particle, mm (in.). Thus, K_s = 6.0 for a sphere. For particle shapes other than spherical, $K_s > 6$.

We can infer the following from these equations. Smaller particle size and higher shape factor (K_s) mean higher surface area for the same total weight of metal powders. This means greater area for surface oxidation to occur. Small powder size also leads to more agglomeration of the particles, which is a disadvantage in automatic feeding of the powders. The reason for using smaller particle sizes is that they provide more uniform shrinkage and better mechanical properties in the final PM product.

FIGURE 16.3 Several of the possible (ideal) particle shapes in powder metallurgy.

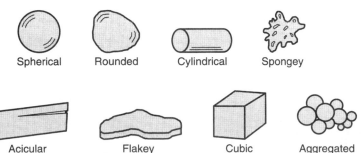

16.1.2 Other Features

Other features of engineering powders include interparticle friction, flow characteristics, packing, density, porosity, chemistry, and surface films.

Interparticle Friction and Flow Characteristics   Friction between particles affects the ability of a powder to flow readily and pack tightly. A common measure of interparticle friction is the *angle of repose,* which is the angle formed by a pile of powders as they are poured from a narrow funnel, as in Figure 16.4. Larger angles indicate greater friction between particles. Smaller particle sizes generally show greater friction and steeper angles. Spherical shapes result in the lowest interpartical friction; as shape deviates more from spherical, friction between particles tends to increase.

Flow characteristics are important in die filling and pressing. Automatic die filling depends on easy and consistent flow of the powders. In pressing, resistance to flow increases density variations in the compacted part; these density gradients are generally undesirable. A common measure of flow is the time required for a certain amount of powder (by weight) to flow through a standard-sized funnel. Smaller flow times indicate easier flow and lower interparticle friction. To reduce interparticle friction and facilitate flow during pressing, lubricants are often added to the powders in small amounts.

Packing, Density, and Porosity   Packing characteristics depend on two density measures. First, there is the *true density,* which is the density of the true volume of the material. This would be the density of the material if the powders were melted into a solid mass, values of which are given in Table 4.1. Second is the *bulk density,* the density of the powders in the loose state after pouring; this includes the effect of pores between particles. Because of the pores, bulk density is less than true density.

The *packing factor* is the bulk density divided by the true density. Typical values for loose powders range between 0.5 and 0.7. The packing factor depends on particle shape and the distribution of particle sizes. If powders of various sizes are present, the smaller powders will fit into the interstices of the larger ones that would otherwise be taken up by air, thus resulting in a higher packing factor. Packing can also be increased by vibrating the powders, causing them to settle more tightly. Finally, we should note that external pressure, as applied during compaction, greatly increases packing of powders through rearrangement and deformation of the particles.

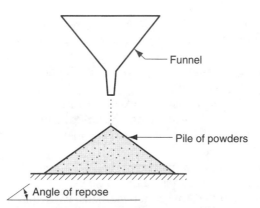

FIGURE 16.4   Interparticle friction as indicated by the angle of repose of a pile of powders poured from a narrow funnel. Larger angles indicate greater interparticle friction.

Porosity represents an alternative way of considering the packing characteristics of a powder. ***Porosity*** is defined as the ratio of the volume of the pores (empty spaces) in the powder to the bulk volume. In principle,

$$\text{Porosity} + \text{Packing factor} = 1.0 \tag{16.6}$$

The issue is complicated by the possible existence of closed pores in some of the particles. If these internal pore volumes are included in the above porosity, then the equation is exact.

Chemistry and Surface Films Characterization of the powder would not be complete without an identification of its chemistry. Metallic powders are classified as either elemental, consisting of a pure metal; or pre-alloyed, wherein each particle is an alloy. We discuss these classes and the metals commonly used in PM more thoroughly in Section 16.5.1.

Surface films are a problem in powder metallurgy because of the large area per unit weight of metal when dealing with powders. The possible films include oxides, silica, adsorbed organic materials, and moisture [5]. Generally, these films must be removed prior to shape processing.

16.2 PRODUCTION OF METALLIC POWDERS

In general, producers of metallic powders are not the same companies as those which make PM parts. The powder producers are the suppliers; the plants that manufacture components out of powder metals are the customers. It is therefore appropriate to separate the discussion of powder production (this section) from the processes used to make PM products (later sections).

Virtually any metal can be made into powder form. There are three principal methods by which metallic powders are commercially produced, each of which involves energy input to increase the surface area of the metal. The methods are [8]: (1) atomization, (2) chemical, and (3) electrolytic. In addition, mechanical methods are occasionally used to reduce powder sizes; however, these methods are much more commonly associated with ceramic powder production and we treat them in the next chapter.

16.2.1 Atomization

This method involves the conversion of molten metal into a spray of droplets that solidify into powders. It is the most versatile and popular method for producing metal powders today, applicable to almost all metals, alloys as well as pure metals. There are multiple ways of creating the molten metal spray, several of which are illustrated in Figure 16.5. Two of the methods shown are based on ***gas atomization,*** in which a high velocity gas stream (air or inert gas) is utilized to atomize the liquid metal. In Figure 16.5(a), the gas flows through an expansion nozzle, siphoning molten metal from the melt below and spraying it into a container. The droplets solidify into powder form. In a closely related method shown in Figure 16.5(b), molten metal flows by gravity through a nozzle and is immediately atomized by air jets. The resulting metal powders, which tend to be spherical, are collected in a chamber below.

The approach shown in part (c) of the figure is similar to (b), except that a high-velocity water stream is used instead of air. This is known as ***water atomization*** and is the most common of the atomization methods, particularly suited to metals that melt

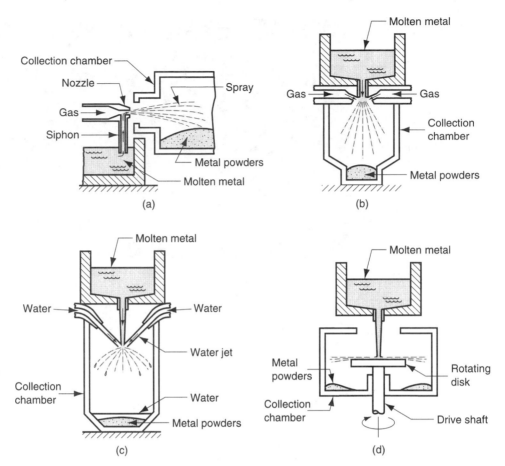

FIGURE 16.5 Several atomization methods for producing metallic powders: (a) and (b) two gas atomization methods; (c) water atomization; and (d) centrifugal atomization by the rotating disk method.

below 1600°C (2900°F). Cooling is more rapid, and the resulting powder shape is irregular rather than spherical. The disadvantage of using water is oxidation on the particle surface. A recent innovation involves the use of synthetic oil rather than water to reduce oxidation. In both air and water atomization processes, particle size is controlled largely by the velocity of the fluid stream; particle size is inversely related to velocity.

Several methods are based on *centrifugal atomization.* In one approach, the *rotating disk method* shown in Figure 16.5(d), the liquid metal stream pours onto a rapidly rotating disk that sprays the metal in all directions to produce powders.

16.2.2 Other Production Methods

Other metal powder production methods include various chemical reduction processes, precipitation methods, and electrolysis.

Chemical reduction includes a variety of chemical reactions by which metallic compounds are reduced to elemental metal powders. A common process involves liberation of metals from their oxides by use of reducing agents such as hydrogen or carbon monoxide. The reducing agent is made to combine with the oxygen in the compound to free the metallic element. This approach is used to produce powders of iron, tungsten, and copper. Another chemical process for iron powders involves the decomposition of iron pentacarbonyl to produce spherical particles of high purity. Powders produced by this method are illustrated in the photomicrograph of Figure 16.6. Other chemical processes include

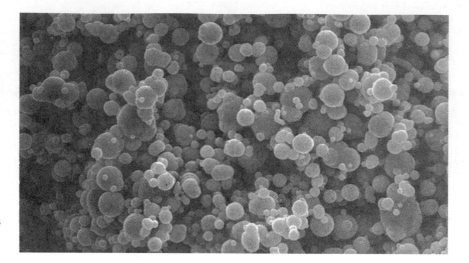

FIGURE 16.6   Iron powders produced by decomposition of iron pentacarbonyl; particle sizes range from about 0.25–3.0 μm (10–125 μ-in.) (photo courtesy of GAF Chemicals Corporation, Advanced Materials Division).

precipitation of metallic elements from salts dissolved in water. Powders of copper, nickel, and cobalt can be produced by this approach.

In *electrolysis,* an electrolytic cell is set up in which the source of the desired metal is the anode. The anode is slowly dissolved under an applied voltage, transported through the electrolyte, and deposited on the cathode. The deposit is removed, washed, and dried to yield a metallic powder of very high purity. The technique is used for producing powders of beryllium, copper, iron, silver, tantalum, and titanium.

16.3 CONVENTIONAL PRESSING AND SINTERING

After the metallic powders have been produced, the conventional PM sequence consists of three steps: (1) blending and mixing of the powders; (2) compaction, in which the powders are pressed into the desired part shape; and (3) sintering, which involves heating to a temperature below the melting point to cause solid-state bonding of the particles and strengthening of the part. The three steps, sometimes referred to as primary operations in PM, are portrayed in Figure 16.7. In addition, secondary operations are sometimes performed to improve dimensional accuracy, increase density, and for other reasons.

16.3.1 Blending and Mixing of the Powders

To achieve successful results in compaction and sintering, the metallic powders must be thoroughly homogenized beforehand. The terms *blending* and *mixing* are both used in this context. *Blending* refers to when powders of the same chemical composition but possibly different particle sizes are intermingled. Different particle sizes are often blended to reduce porosity. *Mixing* refers to powders of different chemistries being combined. An advantage of PM technology is the opportunity to mix various metals into alloys that would be difficult or impossible to produce by other means. The distinction between blending and mixing is not always precise in industrial practice.

Blending and mixing are accomplished by mechanical means. Figure 16.8 illustrates four alternatives: (a) rotation in a drum; (b) rotation in a double-cone container; (c) agitation in a screw mixer; and (d) stirring in a blade mixer. There is more science to these

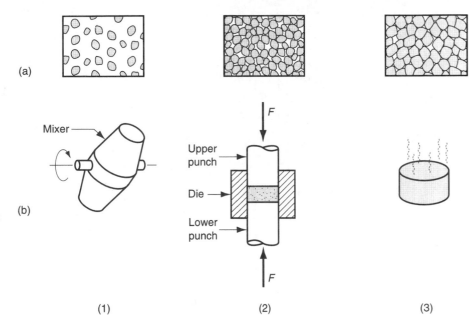

FIGURE 16.7   The conventional powder metallurgy production sequence: (1) blending, (2) compacting, and (3) sintering; (a) shows the condition of the particles while (b) shows the operation and/or workpart during the sequence.

devices than one would suspect. Best results seem to occur when the container is between 20% and 40% full. The containers are usually designed with internal baffles or other ways of preventing free-fall during blending of powders of different sizes, because variations in settling rates between sizes result in segregation—just the opposite of what is wanted in blending. Vibration of the powder is undesirable, since it also causes segregation.

Other ingredients are usually added to the metallic powders during the blending and/or mixing step. These additives include (1) *lubricants,* such as stearates of zinc and aluminum, in small amounts to reduce friction between particles and at the die wall during compaction; (2) *binders,* which are required in some cases to achieve adequate strength in the pressed but unsintered part; and (3) *deflocculants,* which inhibit agglomeration of powders for better flow characteristics during subsequent processing.

16.3.2 Compaction

In compaction, high pressure is applied to the powders to form them into the required shape. The conventional compaction method is *pressing,* in which opposing punches squeeze the powders contained in a die. The steps in the pressing cycle are shown in

FIGURE 16.8   Several blending and mixing devices: (a) rotating drum, (b) rotating double-cone, (c) screw mixer, and (d) blade mixer.

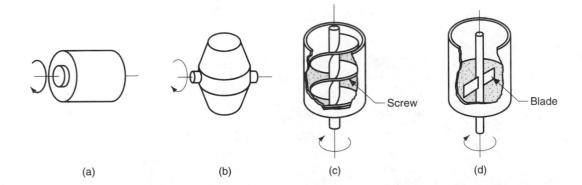

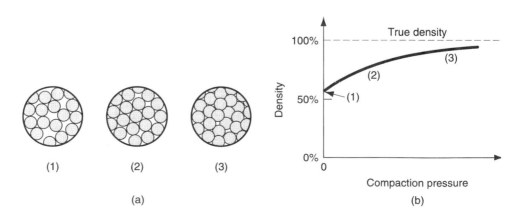

FIGURE 16.9 Pressing, the conventional method of compacting metal powders in PM: (1) filling the die cavity with powder, done by automatic feed in production, (2) initial, and (3) final positions of upper and lower punches during compaction, and (4) ejection of part.

Upper punch

Powders

Feeder

Die

Lower punch

(1)　(2)　(3)　(4)

Figure 16.9. The workpart after pressing is called a ***green compact,*** the word *green* meaning not yet fully processed. As a result of pressing, the density of the part, called the ***green density*** is much greater than the starting bulk density. The ***green strength*** of the part when pressed is adequate for handling but far less than that achieved after sintering.

The applied pressure in compaction results initially in repacking of the powders into a more efficient arrangement, eliminating "bridges" formed during filling, reducing pore space, and increasing the number of contacting points between particles. As pressure increases, the particles are plastically deformed, causing interparticle contact area to increase and additional particles to make contact. This is accompanied by a further reduction in pore volume. The progression is illustrated in three views in Figure 16.10 for starting particles of spherical shape. Also shown is the associated density represented by the three views as a function of applied pressure.

Presses used in conventional PM compaction are mechanical, hydraulic, or a combination of the two. A 450 kN (50 ton) hydraulic unit is shown in Figure 16.11. Because

FIGURE 16.10　(a) Effect of applied pressure during compaction: (1) initial loose powders after filling, (2) repacking, and (3) deformation of particles; and (b) density of the powders as a function of pressure. The sequence here corresponds to steps (1), (2), and (3) in Figure 16.9.

(1)　(2)　(3)

(a)

(b)

FIGURE 16.11 A 450-kN (50-ton) hydraulic press for compaction of powder metallurgy components. This press has the capability to actuate multiple levels to produce complex PM part geometries (photo courtesy Dorst America, Inc.).

of differences in part complexity and associated pressing requirements, presses can be distinguished as (1) pressing from one direction, referred to as single-action presses; or (2) pressing from two directions, any of several types including opposed ram, double-action, and multiple action. Current available press technology can provide up to 10 separate action controls to produce parts of significant geometric complexity [4]. We examine part complexity and other design issues in Section 16.6.

The capacity of a press for PM production is generally given in tons or kN or MN. The required force for pressing depends on the projected area of the PM part (area in the horizontal plane for a vertical press) multiplied by the pressure needed to compact the given metal powders. Reducing this to equation form,

$$F = A_p \, p_c \qquad\qquad (16.8)$$

where F = required force, N (lb); A_p = projected area of the part, mm^2 (in.^{2}); and p_c = compaction pressure required for the given powder material, MPa (lb/in.^{2}). Compaction pressures typically range from 70 MPa (10,000 lb/in.^{2}) for aluminum powders to 700 MPa (100,000 lb/in.^{2}) for iron and steel powders.

16.3.3 Sintering

After pressing, the green compact lacks strength and hardness; it is easily crumbled under low stresses. *Sintering* is a heat treatment operation performed on the compact to bond its metallic particles, thereby increasing strength and hardness. The treatment is usually carried out at temperatures between 0.7 and 0.9 of the metal's melting point (absolute scale). The terms *solid-state sintering,* or *solid-phase sintering,* are sometimes used for this conventional sintering because the metal remains unmelted at these treatment temperatures.

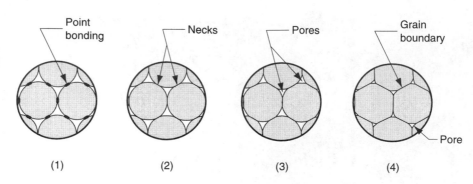

FIGURE 16.12   Sintering on a microscopic scale: (1) particle bonding is initiated at contact points; (2) contact points grow into "necks"; (3) the pores between particles are reduced in size; and (4) grain boundaries develop between particles in place of the necked regions.

It is generally agreed among researchers that the primary driving force for sintering is reduction of surface energy [5], [10]. The green compact consists of many distinct particles, each with its own individual surface, and so the total surface area contained in the compact is very high. Under the influence of heat, the surface area is reduced through the formation and growth of bonds between the particles, with associated reduction in surface energy. The finer the initial powder size, the higher the total surface area, and the greater the driving force behind the process.

The series of sketches in Figure 16.12 shows on a microscopic scale the changes that occur during sintering of metallic powders. Sintering involves mass transport to create the necks and transform them into grain boundaries. The principal mechanism by which this occurs is diffusion; other possible mechanisms include plastic flow. Shrinkage occurs during sintering as a result of pore size reduction. This depends to a large extent on the density of the green compact, which depends on the pressure during compaction. Shrinkage is generally predictable when processing conditions are closely controlled.

Since PM applications usually involve medium-to-high production, most sintering furnaces are designed with mechanized flow-through capability for the workparts. The heat treatment consists of three steps, accomplished in three chambers in these continuous furnaces: (1) preheat, in which lubricants and binders are burned off; (2) sinter; and (3) cool down. The treatment is illustrated in Figure 16.13. Typical sintering temperatures and times are given for selected metals in Table 16.1.

In modern sintering practice, the atmosphere in the furnace is controlled. The purposes of a controlled atmosphere include (1) protection from oxidation, (2) providing a reducing atmosphere to remove existing oxides, (3) providing a carburizing atmosphere, and (4) assisting in removing lubricants and binders used in pressing. Common sintering furnace atmospheres are [5] inert gas, nitrogen-based, dissociated ammonia, hydrogen, and natural gas based. Vacuum atmospheres are used for certain metals, such as stainless steel and tungsten.

16.3.4 Secondary Operations

The functions of secondary operations are varied; they include densification, sizing, impregnation, infiltration, heat treatment, and finishing.

Densification and Sizing   A number of secondary operations are performed to increase density, improve accuracy, or accomplish additional shaping of the sintered part. *Repressing* is a pressing operation in which the part is squeezed in a closed die to increase density and improve physical properties. *Sizing* is the pressing of a sintered part to improve dimensional accuracy. *Coining* is a pressworking operation on a sintered part to press details into its surface.

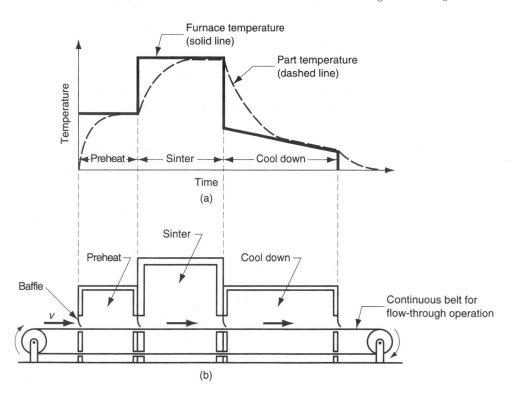

FIGURE 16.13   (a) Typical heat treatment cycle in sintering; and (b) schematic cross-section of a continuous sintering furnace.

Some PM parts require *machining* after sintering. Machining is rarely done to size the part, but rather to create geometric features that cannot be achieved by pressing, such as internal and external threads, side holes, and other details.

Impregnation and Infiltration   Porosity is a unique and inherent characteristic of powder metallurgy technology. It can be exploited to create special products by filling the available pore space with oils, polymers, or metals that have lower melting temperatures than the base powder metal.

Impregnation is the term used when oil or other fluid is permeated into the pores of a sintered PM part. The most common products of this process are oil-impregnated bearings, gears, and similar machinery components. Self-lubricating bearings, usually made of bronze or iron with 10% to 30% oil by volume, are widely used in the automotive industry. The treatment is accomplished by immersing the sintered parts in a bath of hot oil.

TABLE 16.1   Typical sintering temperatures and times for selected powder metals.

Metal	Sintering Temperature		Time (min)
	°C	(°F)	
Brass	850	(1600)	25
Bronze	820	(1500)	15
Copper	850	(1600)	25
Iron	1100	(2000)	30
Stainless steel	1200	(2200)	45
Tungsten	2300	(4200)	480

Compiled from [7] and [11].

An alternative application of impregnation involves PM parts that must be made pressure tight or impervious to fluids. In this case, the parts are impregnated with various types of polymer resins that seep into the pore spaces in liquid form and then solidify. In some cases, resin impregnation is used to facilitate subsequent processing; for example, to permit the use of processing solutions (such as plating chemicals) that would otherwise soak into the pores and degrade the product, or to improve machinability of the PM workpart.

Infiltration is an operation in which the pores of the PM part are filled with a molten metal. The melting point of the filler metal must be below that of the PM part. The process involves heating the filler metal in contact with the sintered component so that capillary action draws the filler into the pores. The resulting structure is relatively nonporous, and the infiltrated part has a more uniform density, as well as improved toughness and strength. An application of the process is copper infiltration of iron PM parts.

Heat Treatment and Finishing Powder metal components can be heat treated (Chapter 27) and finished (electroplated or painted, Chapter 29) by most of the operations used on parts fabricated by casting and other metalworking processes. Special care must be exercised in heat treatment because of porosity; for example, salt baths are not used for heating PM parts. Plating and coating operations are applied to sintered parts for appearance purposes and corrosion resistance. Again, precautions must be taken to avoid entrapment of chemical solutions in the pores; impregnation and infiltration are frequently used for this purpose. Common platings for PM parts include copper, nickel, chromium, zinc, and cadmium.

16.4 ALTERNATIVE PRESSING AND SINTERING TECHNIQUES

The conventional press and sinter sequence is the most widely used shaping technology in powder metallurgy. Additional methods for processing PM parts are discussed in this section. The methods fall into one of three categories: (1) alternative compaction methods, (2) combined compaction and sintering, and (3) alternative sintering methods.

16.4.1 Isostatic Pressing

A feature of conventional pressing is that pressure is applied uniaxially. This imposes limitations on part geometry, since metallic powders do not readily flow in directions perpendicular to the applied pressure. Uniaxial pressing also leads to density variations in the compact after pressing. In *isostatic pressing,* pressure is applied from all directions against the powders that are contained in a flexible mold; hydraulic pressure is used to achieve compaction. Isostatic pressing takes two alternative forms: (1) cold isostatic pressing and (2) hot isostatic pressing.

Cold isostatic pressing (CIP) involves compaction performed at room temperature. The mold, made of rubber or other elastomer material, is oversized to compensate for shrinkage. Water or oil is used to provide the hydrostatic pressure against the mold inside the chamber. Figure 16.14 illustrates the processing sequence in cold isostatic pressing. Advantages of CIP include more uniform density, less expensive tooling, and greater applicability to shorter production runs. Good dimensional accuracy is difficult to achieve in isostatic pressing, due to the flexible mold. Consequently, subsequent finish shaping operations are often required to obtain the required dimensions, either before or after sintering.

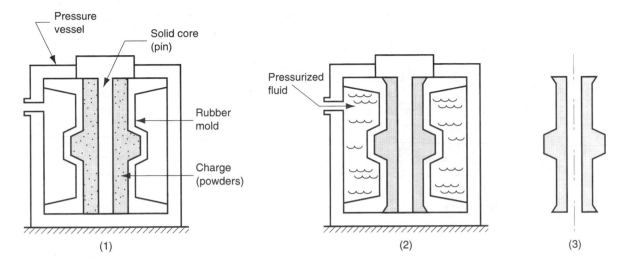

FIGURE 16.14   Cold isostatic pressing: (1) powders are placed in the flexible mold; (2) hydrostatic pressure is applied against the mold to compact the powders; and (3) pressure is reduced, and the part is removed.

Hot isostatic pressing (HIP) is carried out at high temperatures and pressures, using a gas such as argon or helium as the compression medium. The mold in which the powders are contained is made of sheetmetal to withstand the high temperatures. HIP accomplishes pressing and sintering in one step. Despite this apparent advantage, it is a relatively expensive process and its applications seem to be concentrated in the aerospace industry. PM parts made by HIP are characterized by high density (porosity near zero), thorough interparticle bonding, and good mechanical strength.

16.4.2 Powder Injection Molding

Injection molding is closely associated with the plastics industry (Section 13.6). The same basic process can be applied to form parts of metal or ceramic powders, the difference being that the starting polymer contains a high content of particulate matter, typically from 50% to 85% by volume. When used in powder metallurgy, the term *metal injection molding* (MIM) is used. The more general process is *powder injection molding* (PIM), which includes both metal and ceramic powders. The steps in MIM proceed as follows [6]: (1) Metallic powders are mixed with an appropriate binder. (2) Granular pellets are formed from the mixture. (3) The pellets are heated to molding temperature, injected into a mold cavity, and the part is cooled and removed from the mold. (4) The part is processed to remove the binder using any of several thermal or solvent techniques. (5) The part is sintered. (6) Secondary operations are performed as appropriate.

The binder in powder injection molding acts as a carrier for the particles. Its functions are to provide proper flow characteristics during molding and to hold the powders in the molded shape until sintering. The five basic types of binders in PIM are [6]: (1) thermosetting polymers, such as phenolics, (2) thermoplastic polymers, such as polyethylene, (3) water, (4) gels, and (5) inorganic materials. Polymers are the most frequently used types.

Powder injection molding is suited to part geometries similar to those in plastic injection molding. It is not cost competitive for simple axisymmetric parts, because the conventional press and sinter process is quite adequate for these cases. PIM seems most economical for small, complex parts of high value. Dimensional accuracy is limited by the shrinkage that accompanies densification during sintering.

16.4.3 Powder Rolling, Extrusion, and Forging

Rolling, extrusion, and forging are familiar bulk metal-forming processes (Chapter 19). We describe them here in the context of powder metallurgy.

Powder Rolling   Powders can be compressed in a rolling mill operation to form metal strip stock. The process is usually set up to run continuously or semicontinuously, as shown in Figure 16.15. The metallic powders are compacted between rolls into a green strip which is fed directly into a sintering furnace. It is then cold rolled and resintered.

Powder Extrusion   Extrusion is one of the basic manufacturing processes (Section 1.3.1). In PM extrusion, the starting powders can be in different forms. In the most popular method, powders are placed in a vacuum-tight sheet metal can, heated, and extruded with the container. In another variation, billets are preformed by a conventional press and sinter process, and then the billet is hot extruded. These methods achieve a high degree of densification in the PM product.

Powder Forging   Forging is an important metal-forming process (Section 1.3.1). In *powder forging,* the starting workpart is a powder metallurgy part preformed to proper size by pressing and sintering. Advantages of this approach are: (1) densification of the PM part; (2) lower tooling costs and fewer forging "hits" (and therefore higher production rate) because the starting workpart is preformed; and (3) reduced material waste.

16.4.4 Combined Pressing and Sintering

Hot isostatic pressing (Section 16.4.1) accomplishes compaction and sintering in one step. Other techniques which combine the two steps are hot pressing and spark sintering.

Hot Pressing   The setup in uniaxial hot pressing is very similar to conventional PM pressing, except that heat is applied during compaction. The resulting product is gen-

FIGURE 16.15   Powder rolling: (1) powders are fed through compaction rolls to form a green strip; (2) sintering; (3) cold rolling; and (4) resintering.

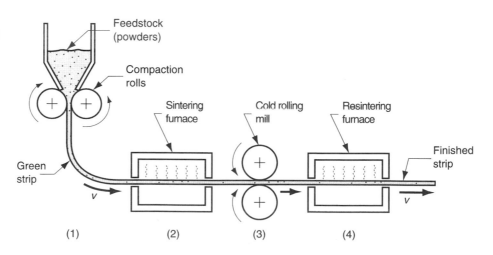

erally dense, strong, hard, and dimensionally accurate. Despite these advantages, the process presents certain technical problems which limit its adoption. Principal among these are [1]: (1) selecting a suitable mold material that can withstand the high sintering temperatures; (2) longer production cycle required to accomplish sintering; and (3) heating and maintaining atmospheric control in the process. Hot pressing has found some application in the production of sintered carbide products using graphite molds.

Spark Sintering An alternative approach that combines pressing and sintering but overcomes some of the problems in hot pressing is spark sintering. The process consists of two basic steps [1], [11]: (1) powder or green compacted preform is placed in a die; and (2) upper and lower punches, which also serve as electrodes, compress the part and simultaneously apply a high-energy electrical current that burns off surface contaminants and sinters the powders, forming a dense, solid part in about 15 seconds. The process has been applied to a variety of metals.

16.4.5 Liquid Phase Sintering

Conventional sintering (Section 16.3.3) is solid-state sintering; the metal is sintered at a temperature below its melting point. In systems involving a mixture of two powder metals, in which there is a difference in melting temperature between the metals, an alternative type of sintering is used, called liquid-phase sintering. In this process, the two powders are initially mixed, and then heated to a temperature that is high enough to melt the lower-melting-point metal but not the other. The melted metal thoroughly wets the solid particles, creating a dense structure with strong bonding between the metals upon solidification. Depending on the metals involved, prolonged heating may result in alloying of the metals by gradually dissolving the solid particles into the liquid melt and/or diffusion of the liquid metal into the solid. In either case, the resulting product is fully densified (no pores) and is strong. Examples of systems that involve liquid phase sintering include Fe–Cu, W–Cu, and Cu–Co [5].

16.5 MATERIALS AND PRODUCTS FOR PM

The raw materials for PM processing are more expensive than for other metalworking because of the additional energy required to reduce the metal to powder form. Accordingly, PM is competitive only in a certain range of applications. In this section we identify the materials and products that seem most suited to powder metallurgy.

16.5.1 PM Materials

From a chemistry standpoint, metal powders can be classified as either elemental or prealloyed. *Elemental* powders consist of a pure metal and are used in applications where high purity is important. For example, pure iron might be used where its magnetic properties are important. The most common elemental powders are those of iron, aluminum, and copper.

Elemental powders are also mixed with other metal powders to produce special alloys that are difficult to formulate using conventional processing methods. Tool steels are an example; PM permits blending of ingredients that is difficult or impossible by traditional alloying techniques. Using mixtures of elemental powders to form an alloy provides a processing benefit, even where special alloys are not involved. Since the powders are pure metals, they are not as strong as pre-alloyed metals. Therefore, they deform more readily during pressing, so that density and green strength are higher than with pre-alloyed compacts.

In *pre-alloyed* powders, each particle is an alloy composed of the desired chemical composition. Pre-alloyed powders are used for alloys that cannot be formulated by mixing elemental powders; stainless steel is an important example. The most common pre-alloyed powders are certain copper alloys, stainless steel, and high-speed steel.

The commonly used elemental and pre-alloyed powdered metals, in approximate order of tonnage usage, are: (1) iron, by far the most widely used PM metal, frequently mixed with graphite to make steel parts, (2) aluminum, (3) copper and its alloys, (4) nickel, (5) stainless steel, (6) high-speed steel, and (7) other PM materials such as tungsten, molybdenum, titanium, tin, and precious metals.

16.5.2 PM Products

The substantial advantage offered by PM technology is that parts can be made to near net shape or net shape; they require little or no additional shaping after PM processing. Some of the components commonly manufactured by powder metallurgy are gears, bearings, sprockets, fasteners, electrical contacts, cutting tools, and various machinery parts. When produced in large quantities, gears and bearings are particularly well suited to PM for two reasons: (1) the geometry is defined principally in two dimensions (the part has a top surface of a certain shape, and there are few or no features along the sides); and (2) there is a need for porosity in the material to serve as a reservoir for lubricant. More complex parts with true three dimensional geometries are also feasible in powder metallurgy, by adding secondary operations such as machining to complete the shape of the pressed and sintered part, and by observing certain design guidelines such as those outlined in the following section.

16.6 DESIGN CONSIDERATIONS IN POWDER METALLURGY

Use of PM techniques is generally suited to a certain class of production situations and part designs. In this section we attempt to define the characteristics of this class of applications for which powder metallurgy is most appropriate. We first present a classification system for PM parts, and then offer some guidelines on component design.

16.6.1 Parts Classification System

The Metal Powder Industries Federation (MPIF) defines four classes of powder metallurgy part designs, by level of difficulty in conventional pressing. The system is useful because it indicates some of the limitations on shape that can be achieved with conventional PM processing. The four part classes are illustrated in Figure 16.16.

16.6.2 Design Guidelines for PM Parts

The MPIF classification system provides some guidance concerning part geometries that are suited to conventional PM pressing techniques. Additional advice is offered in the following design guidelines, compiled from [2], [8], and [11].

> ➤ Economics of PM processing usually require large part quantities to justify the cost of equipment and special tooling required. Minimum quantities of 10,000 units are suggested [11], although exceptions exist.

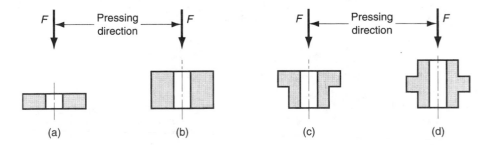

FIGURE 16.16 Four classes of PM parts (side view shown; cross section is circular): (a) Class I—simple thin shapes that can be pressed from one direction; (b) Class II—simple but thicker shapes that require pressing from two directions; (c) Class III—two levels of thickness, pressed from two directions; and (d) Class IV—multiple levels of thickness, pressed from two directions, with separate controls for each level to achieve proper densification throughout the compact.

➤ Powder metallurgy is unique in its capability to fabricate parts with a controlled level of porosity. Porosities up to 50% are possible.

➤ PM can be used to make parts out of unusual metals and alloys—materials that would be difficult if not impossible to fabricate by other means.

➤ The geometry of the part must permit ejection from the die after pressing; this generally means that the part must have vertical or near-vertical sides, although steps in the part are permissible as suggested by the MPIF classification system (Figure 16.16). Design features such as undercuts and holes on the part sides, as shown in Figure 16.17, must be avoided. Vertical undercuts and holes, as in Figure 16.18, are permissible because they do not interfere with ejection. Vertical holes can be of cross-sectional shapes other than round (e.g., squares, keyways) without significant increases in tooling or processing difficulty.

➤ Screw threads cannot be fabricated by PM pressing; if required, they must be machined into the PM component.

➤ Chamfers and corner radii are possible by PM pressing, as shown in Figure 16.19. Problems are encountered in punch rigidity when angles are too acute.

➤ Wall thickness should be a minimum of 1.5 mm (0.060 in.) between holes or a hole and the outside part wall, as indicated in Figure 16.20. Minimum recommended hole diameter is 1.5 mm (0.060 in.). Examples exist in which both of these guidelines have been violated [4].

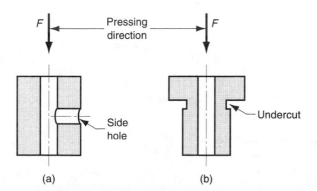

FIGURE 16.17 Part features to be avoided in PM: (a) side holes and (b) side undercuts. Part ejection is impossible.

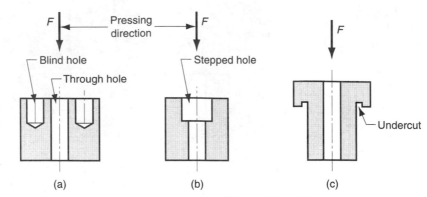

FIGURE 16.18
Permissible part features in PM: (a) vertical hole, blind and through; (b) vertical stepped hole; and (c) undercut in vertical direction. These features allow part ejection.

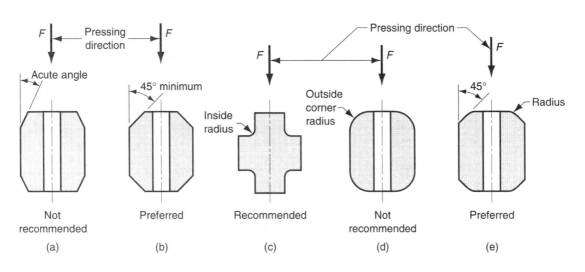

FIGURE 16.19   Chamfers and corner radii are accomplished but certain rules should be observed: (a) avoid acute chamfer angles; (b) larger angles are preferred for punch rigidity; (c) small inside radius is desirable; (d) full outside corner radius is difficult because punch is fragile at corner's edge; (e) outside corner problem can be solved by combining radius and chamfer.

FIGURE 16.20   Minimum recommended wall thickness (a) between holes or (b) between a hole and an outside wall should be 1.5 mm (0.060 in.).

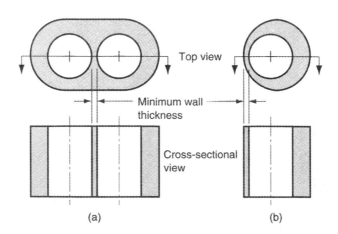

REFERENCES

[1] Amstead, B. H., Ostwald, P. F., and Begeman, M. L. *Manufacturing Processes.* 8th ed. John Wiley & Sons, New York, 1987.

[2] Bralla, J. G. (Editor in Chief). *Design for Manufacturability Handbook.* 2nd ed. McGraw-Hill Book Company, New York, 1998.

[3] Dixon, R. H. T., and Clayton, A. *Powder Metallurgy for Engineers.* The Machinery Publishing Co. Ltd., Brighton, U.K., 1971.

[4] Personal communications with technical personnel at Dorst America, Inc., 1991.

[5] German, R. M. *Powder Metallurgy Science.* 2nd ed. Metal Powder Industries Federation, Princeton, New Jersey, 1994.

[6] German, R. M. *Powder Injection Molding.* Metal Powder Industries Federation, Princeton, New Jersey, 1990.

[7] *Metals Handbook.* 9th ed. Vol 7: *Powder Metallurgy.* American Society for Metals, Metals Park, Ohio, 1984.

[8] *Powder Metallurgy Design Handbook.* Metal Powder Industries Federation, Princeton, New Jersey, 1989.

[9] Schey, J. A. *Introduction to Manufacturing Processes.* 3rd ed. McGraw-Hill Book Company, New York, 1999.

[10] Waldron, M. B., and Daniell, B. L. *Sintering.* Heyden, London, U.K., 1978.

[11] Wick, C., Benedict, J. T., and Veilleux, R. F. *Tool and Manufacturing Engineers Handbook.* 4th ed. Vol II: *Forming,* Society of Manufacturing Engineers, Dearborn, Michigan, 1984.

REVIEW QUESTIONS

16.1. Name some of the reasons for the commercial importance of powder metallurgy technology.

16.2. What are some of the disadvantages of PM methods?

16.3. In the screening of powders for sizing, what is meant by the term **mesh count?**

16.4. What is the difference between open pores and closed pores in metallic powders?

16.5. What is meant by the term **aspect ratio** for a metallic particle?

16.6. How would one measure the angle of repose for a given amount of metallic powder?

16.7. Define bulk density and true density for metallic powders.

16.8. What are the principal methods used to produce metallic powders?

16.9. What are the three basic steps in the conventional powder-metallurgy-shaping process?

16.10. What is the technical difference between mixing and blending in powder metallurgy?

16.11. What are some of the ingredients usually added to metallic powders during blending and/or mixing?

16.12. What is meant by the term **green compact?**

16.13. Describe what happens to the individual particles during compaction.

16.14. Which of the following most closely typifies the usual sintering temperatures in PM? (a) $0.5\ T_m$, (b) $0.8\ T_m$, (c) T_m.

16.15. What are the three steps in the sintering cycle in PM?

16.16. What are some of the reasons why a controlled furnace is desirable in sintering?

16.17. What are the advantages of infiltration in PM?

16.18. What is the difference between powder injection molding and metal injection molding?

16.19. How is isostatic pressing distinguished from conventional pressing and sintering in PM?

16.20. Describe liquid phase sintering.

16.21. What are the two basic classes of metal powders as far as chemistry is concerned?

16.22. Why is PM technology so well suited to the production of gears and bearings?

MULTIPLE CHOICE QUIZ

There is a total of 18 correct answers in the following multiple choice questions (some questions have multiple answers that are correct). To attain a perfect score on the quiz, all correct answers must be given, since each correct answer is worth 1 point. For each question, each omitted answer or wrong answer reduces the score by 1 point, and each additional answer beyond the number of answers required reduces the score by 1 point. Percentage score on the quiz is based on the total number of correct answers.

16.1. The particle size that can pass through a screen is obtained by taking the reciprocal of the mesh count of the screen: (a) true or (b) false.

16.2. Identify which of the phrases make the following statement correct: For a given weight of metallic powders, the total surface area of the powders is increased by (more than one): (a) larger particle size, (b) smaller particle size, (c) higher shape factor, and (d) smaller shape factor.

16.3. As particle size increases, interparticle friction (a) increases or (b) decreases.

16.4. Which one of the following powder shapes would tend to have the lowest interparticle friction? (a) acicular, (b) cubic, (c) flakey, (d) spherical, and (e) rounded.

16.5. Which of the following statements is correct in the context of metallic powders (more than one)? (a) porosity + packing factor = 1.0, (b) packing factor = 1/porosity, (c) packing factor = 1.0 − porosity, (d) packing factor = −porosity, (e) packing factor = bulk density/ true density.

16.6. Repressing refers to a pressworking operation used to compress an unsintered part in a closed die to achieve sizing and better surface finish. (a) true or (b) false.

16.7. Impregnation refers to which of the following (more than one)? (a) soaking oil by capillary action into the pores of a PM part, (b) putting polymers into the pores of a PM part, or (c) filling the pores of the PM part with a molten metal.

16.8. In cold isostatic pressing, the mold is most typically made of which of the following? (a) rubber, (b) sheetmetal, (c) tool steel, (d) textile, or (e) thermosetting polymer.

16.9. Which of the following processes combines pressing and sintering of the metal powders (may be more than one answer)? (a) metal injection molding, (b) hot pressing, (c) spark sintering, and (d) hot isostatic pressing.

16.10. Which of the following design features would be difficult or impossible to achieve by conventional pressing and sintering (more than one)? (a) side holes, (b) threaded holes, (c) outside rounded corners, (d) vertical stepped holes, or (e) vertical wall thickness of 1/8 in (3 mm).

PROBLEMS

Characterization of Engineering Powders

16.1. A screen with 325 mesh count has wires with a diameter of 0.001377 in. Using Eq. (16.1), determine (a) the maximum particle size that will pass through the wire mesh, and (b) the proportion of open space in the screen.

16.2. A screen with 10 mesh count has wires with a diameter of 0.0213 in. Using Eq. (16.1), determine: (a) the maximum particle size that will pass through the wire mesh, and (b) the proportion of open space in the screen.

16.3. What is the aspect ratio of a cubic particle shape?

16.4. Determine the shape factor for metallic particles of the following ideal shapes: (a) sphere, (b) cubic, (c) cylindrical with length-to-diameter ratio of 1 : 1, (d) cylindrical with length-to-diameter ratio of 2 : 1, and (e) a disk-shaped flake whose thickness-to-diameter ratio is 1 : 10.

16.5. A pile of iron powder weighs 2 lb. The particles are spherical in shape and all have the same diameter of 0.002 in. (a) Determine the total surface area of all the particles in the pile. (b) If the packing factor = 0.6, determine the volume taken by the pile. *Note*: the density of iron = 0.284 lb/in.^{3}

16.6. Solve Problem 16.5, except that the diameter of the particles is 0.004 in. Assume the same packing factor.

16.7. Suppose in Problem 16.5 that the average particle diameter = 0.002 in.; however, the sizes vary, forming a statistical distribution as follows: 25% of the particles by weight are 0.001 in, 50% are 0.002 in., and 25% are 0.003 in. Given this distribution, what is the total surface area of all the particles in the pile?

16.8. A solid cube of copper with each side = 1.0 ft is converted into metallic powders of spherical shape by gas atomization. What is the percentage increase in total surface area if the diameter of each particle is 0.004 in. (assume that all particles are the same size)?

16.9. A solid cube of aluminum with each side = 1.0 m is converted into metallic powders of spherical shape by gas atomization. How much total surface area is added by the process if the diameter of each particle is 100 microns (assume that all particles are the same size)?

16.10. Given a large volume of metallic powders, all of which are perfectly spherical and have the same exact diameter, what is the maximum possible packing factor that the powders can take?

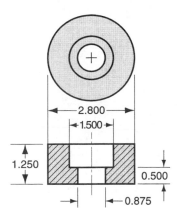

FIGURE P16.13   Part for Problem 16.13
(dimensions are inches).

Compaction and Design Considerations

16.11. In a certain pressing operation, the metallic powder fed into the open die has a packing factor of 0.5. The pressing operation reduces the powder to two-thirds of its starting volume. In the subsequent sintering operation, shrinkage amounts to 10% on a volume basis. Given that these are the only factors that affect the structure of the finished part, determine its final porosity.

16.12. A bearing of simple geometry is to be pressed out of bronze powders, using a compacting pressure of 207 MPa. The outside diameter = 44 mm, the inside diameter = 22 mm, and the length of the bearing = 25 mm. What is the required press tonnage to perform this operation?

16.13. The part shown in Figure P16.13 is to be pressed of iron powders using a compaction pressure of 75,000 lb/in.[2] Dimensions are inches. Determine: (a) the most appropriate pressing direction, (b) the required press tonnage to perform this operation, and (c) the final weight of the part if the porosity is 10%. Assume shrinkage during sintering can be neglected.

16.14. For each of the four part drawings in Figure P16.14, indicate which PM class the parts belong to, whether the part must be pressed from one or two directions, and how many levels of press control will be required. Dimensions are in mm.

FIGURE P16.14   Parts for Problem 16.14 (dimensions are mm).

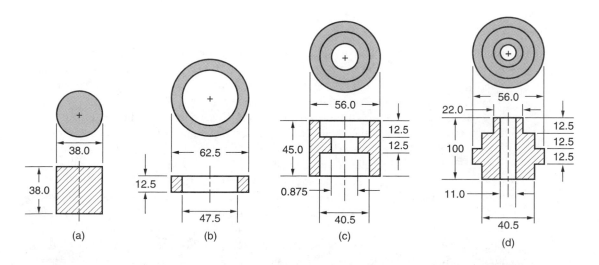

17 PROCESSING OF CERAMICS AND CERMETS

CHAPTER CONTENTS

17.1   Processing of Traditional Ceramics
    17.1.1   Preparation of the Raw Material
    17.1.2   Shaping Processes
    17.1.3   Drying
    17.1.4   Firing (Sintering)
17.2   Processing of New Ceramics
    17.2.1   Preparation of Starting Materials
    17.2.2   Shaping
    17.2.3   Sintering
    17.2.4   Finishing
17.3   Processing of Cermets
    17.3.1   Cemented Carbides
    17.3.2   Other Cermets and Ceramic Matrix Composites
17.4   Product Design Considerations

Ceramic materials divide into three categories (Chapter 7): (1) traditional ceramics, (2) new ceramics, and (3) glasses. The processing of glass involves solidification primarily and is covered in Chapter 12. In this chapter, we consider the particulate processing methods used for traditional and new ceramics. We also consider the processing of certain composite materials that are combinations of ceramics and/or metals.

Traditional ceramics are made from minerals occurring in nature and include pottery, porcelain, bricks, and cement. New ceramics are made from synthetically produced raw materials and cover a wide spectrum of products such as cutting tools, artificial bones, nuclear fuels, and substrates for electronic circuits. The starting material for all of these items is powder. In the case of the traditional ceramics, the powders are usually mixed with water to temporarily bind the particles together and achieve the proper consistency for shaping. For new ceramics, other substances are used as binders during shaping. After shaping, the green parts are sintered. This is often called *firing* in ceramics, but the function is the same as in powder metallurgy: to effect a solid-state reaction that bonds the material into a hard, solid mass.

The processing methods discussed in this chapter are commercially and technologically important because virtually all ceramic products are formed by these methods (except, of course, glass products). The manufacturing sequence is similar for traditional and new ceramics because the form of the starting material is the same: powder. However, the processing methods for the two categories are sufficiently different that we discuss them separately.

17.1 PROCESSING OF TRADITIONAL CERAMICS

In this section we discuss the production technology used to make traditional ceramic products such as pottery, stoneware and other dinnerware, bricks, tile, and ceramic refractories. Bonded grinding wheels are also produced by the same basic methods. What these products have in common is that their raw materials consist primarily of silicate ceramics—clays. The processing sequence for most of the traditional ceramics consists of the steps depicted in Figure 17.1.

17.1.1 Preparation of the Raw Material

The shaping processes for traditional ceramics require that the starting material be in the form of a plastic paste. This paste is made of fine ceramic powders mixed with water, and its consistency determines the ease of forming the material and the quality of the final product. The raw ceramic material usually occurs in nature as rocky lumps, and reduction to powder is the purpose of the preparation step in ceramics processing.

Techniques for reducing particle size in ceramics processing deliver mechanical energy in various forms, such as impact, compression, and attrition. The term *comminution* is used for these techniques, which are most effective on brittle materials, including cement, metallic ores, and brittle metals. Two general types of comminution operations are distinguished: crushing and grinding.

Crushing refers to the reduction of large lumps from the mine to smaller sizes for subsequent further reduction. Several stages may be required (e.g., primary crushing,

FIGURE 17.1   Usual steps in traditional ceramics processing: (1) preparation of raw materials, (2) shaping, (3) drying, and (4) firing. Part (a) shows the workpart during the sequence, while (b) shows the condition of the powders.

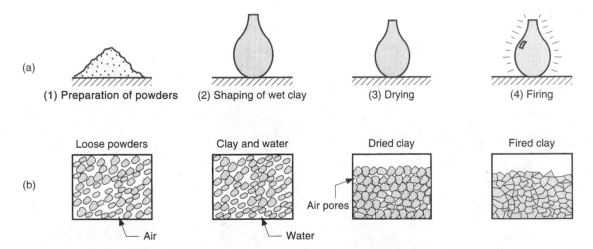

secondary crushing), the reduction ratio in each stage being in the range 3 to 6. Mineral crushing is accomplished by compression against rigid surfaces or by impact against surfaces in a rigid constrained motion [1]. Figure 17.2 presents equipment used to accomplish crushing: (a) jaw crushers, in which a large jaw toggles back and forth to crush lumps against a hard, rigid surface; (b) gyratory crushers, which use a gyrating cone to compress lumps against a rigid surface; (c) roll crushers, in which the ceramic lumps are squeezed between rotating rolls; and (d) hammer mills, which use rotating hammers impacting the material to break up the lumps.

Grinding, in our context here, refers to the operation of reducing the small pieces after crushing to a fine powder. Grinding is accomplished by abrasion and impact of the crushed mineral by the free motion of unconnected hard media such as balls, pebbles, or rods [1]. Examples of grinding include (a) ball mill, (b) roller mill, and (c) impact grinding, illustrated in Figure 17.3.

In a *ball mill,* hard spheres mixed with the stock to be comminuted are rotated inside a large cylindrical container. The motion causes the balls and stock to be carried up

FIGURE 17.2   Crushing operations: (a) jaw crusher, (b) gyratory crusher, (c) roll crusher, and (d) hammer mill.

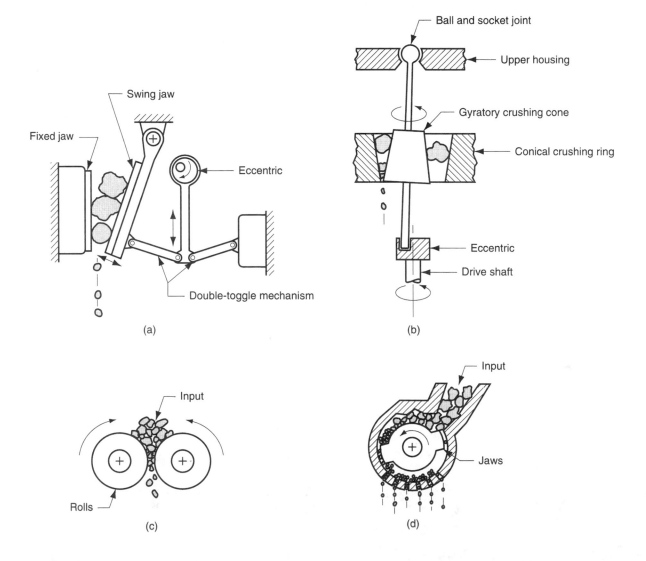

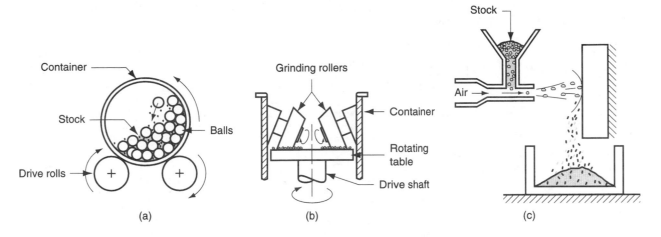

FIGURE 17.3   Mechanical methods of producing ceramic powders: (a) ball mill, (b) roller mill, and (c) impact grinding.

the container wall, and then pulled back down by gravity to accomplish a grinding action by a combination of impact and attrition. These operations are often carried out with water added to the mixture, so that the ceramic is in the form of a slurry. In a ***roller mill,*** stock is compressed against a flat horizontal grinding table by rollers riding over the table surface. Although not clearly shown in our sketch, the pressure of the grinding rollers against the table is regulated by mechanical springs or hydraulic-pneumatic means. In ***impact grinding,*** which seems to be less frequently used, particles of stock are thrown against a hard flat surface, either in a high velocity air stream or in a high-speed slurry. The impact fractures the pieces into smaller particles.

The plastic paste required for shaping consists of ceramic powders and water. Clay is usually the main ingredient in the paste because it has ideal forming characteristics. The more water there is in the mixture, the more plastic and easily formed is the clay paste. However, when the formed part is later dried and fired, shrinkage occurs that can lead to cracking in the product. To address this problem, other ceramic raw materials which do not shrink on drying and firing are usually added to the paste, often in significant amounts. Also, other components can be included to serve special functions. Thus, the ingredients of the ceramic paste can be divided into the following three categories [3]: (1) clay, which provides the consistency and plasticity required for shaping; (2) nonplastic raw materials, such as alumina and silica, which do not shrink in drying and firing but unfortunately reduce plasticity in the mixture during forming; and (3) other ingredients, such as fluxes that melt (vitrify) during firing and promote sintering of the ceramic material (feldspar is an example), and wetting agents that improve mixing of ingredients.

These ingredients must be thoroughly mixed, either wet or dry. The ball mill often serves this purpose in addition to its grinding function. Also, the proper amounts of powder and water in the paste must be attained, so water must be added or removed, depending on the prior condition of the paste and its desired final consistency.

17.1.2 Shaping Processes

The optimum proportions of powder and water depend on the shaping process used. Some shaping processes require high fluidity; others act on a composition which contains very low water content. At about 50% water, the mixture is a slurry which flows like a liquid.

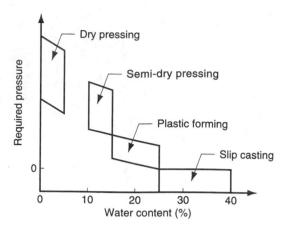

FIGURE 17.4 Four categories of shaping processes used for traditional ceramics, compared to water content and pressure required to form the clay.

As the water content is reduced, increased pressure is required on the paste to produce a similar flow. Thus, the shaping processes can be divided according to the consistency of the mixture: (1) slip casting, in which the mixture is a slurry; (2) plastic-forming methods that shape the clay in a plastic condition; (3) semi-dry pressing, in which the clay is moist but possesses low plasticity; and (4) dry pressing, where the clay is basically dry, containing less than 5% water. Dry clay has no plasticity. The four categories are represented in the chart of Figure 17.4, which compares the categories with the condition of the clay used as starting material. Each category includes several different shaping processes.

Slip Casting Slip casting is used in powder metallurgy, but its application in ceramics shaping is much more common. In slip casting, a suspension of ceramic powders in water, called a *slip,* is poured into a porous plaster of paris ($CaSO_4$–$2H_2O$) mold so that water from the mix is gradually absorbed into the plaster to form a firm layer of clay at the mold surface. The composition of the slip is typically 25% to 40% water, the remainder being clay often mixed with other ingredients. It must be sufficiently fluid to flow into the crevices of the mold cavity, yet lower water content is desirable for faster production rates. Slip casting has two principal variations: drain casting and solid casting. In *drain casting,* which is the traditional process, the mold is inverted to drain excess slip after the semi-solid layer has been formed, thus leaving a hollow part in the mold; the mold is then opened and the part removed. The sequence, which is very similar to slush casting of metals, is illustrated in Figure 17.5. It is used to make tea pots,

FIGURE 17.5 Sequence of steps in drain casting, a form of slip casting: (1) slip is poured into mold cavity; (2) water is absorbed into plaster mold to form a firm layer; (3) excess slip is poured out; and (4) part is removed from mold and trimmed.

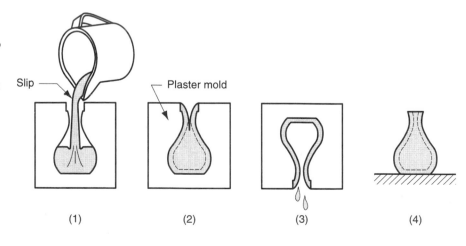

Slip

Plaster mold

(1)   (2)   (3)   (4)

vases, art objects, and other hollow-ware products. In ***solid casting,*** used to produce solid products, adequate time is allowed for the entire body to become firm. The mold must be periodically resupplied with additional slip to account for shrinkage due to absorbed water.

Plastic Forming   This category includes a variety of methods, both manual and mechanized. They all require the starting mixture to have a plastic consistency, which is generally achieved with 15% to 25% water. Manual methods generally make use of clay at the upper end of the range because it provides a material that is more easily formed; however, this is accompanied by greater shrinkage in drying. Mechanized methods generally employ a mixture with lower water content so that the starting clay is stiffer.

Although manual forming methods date back thousands of years, they are still used today by skilled artisans, either in production or for artworks. ***Hand modeling*** involves the creation of the ceramic product by manipulating the mass of plastic clay into the desired geometry. In addition to art pieces, patterns for plaster molds in slip casting are often made this way. ***Hand molding*** is a similar method, only a mold or form is used to define portions of the geometry. ***Hand throwing*** on a potter's wheel is another refinement of the handicraft methods. The ***potter's wheel*** is a round table that rotates on a vertical spindle, powered either by motor or foot-operated treadle. Ceramic products of circular cross-section can be formed on the rotating table by throwing and shaping the clay, sometimes using a mold to provide the internal shape.

Strictly speaking, use of a motor-driven potter's wheel is a mechanized method. However, most mechanized clay-forming methods are characterized by much less manual participation than the hand-throwing method just described. These more mechanized methods include jiggering, plastic pressing, and extrusion. ***Jiggering*** is an extension of the potter's wheel methods, in which hand throwing is replaced by mechanized techniques. It is used to produce large numbers of identical items such as houseware plates and bowls. Although there are variations in the tools and methods used, reflecting different levels of automation and refinements to the basic process, a typical sequence is as follows, depicted in Figure 17.6: (1) a wet clay slug is placed on a convex mold; (2) a forming tool is pressed into the slug to provide the initial rough shape—the operation is called ***batting*** and the workpiece thus created is called a ***bat;*** and (3) a heated jigger tool is used to impart the final contoured shape to the product by pressing the profile into the surface during rotation of the workpart. The reason for heating the tool is to produce steam from the wet clay which prevents sticking. Closely related to jiggering is ***jolleying,*** in which the basic mold shape is concave, rather than convex [7]. In both of these processes, a rolling tool is sometimes used in place of the nonrotating jigger (or jolley) tool; this rolls the clay into shape, avoiding the need to first bat the slug.

FIGURE 17.6 Sequence in jiggering: (1) wet clay slug is placed on a convex mold; (2) batting; and (3) a jigger tool imparts the final product shape. Symbols *v* and *F* indicate motion (*v* = velocity) and applied force respectively.

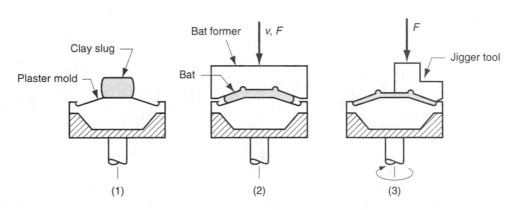

Plastic pressing is a forming process in which a plastic clay slug is pressed between upper and lower molds, contained in metal rings. The molds are made of a porous material such as gypsum, so that when a vacuum is drawn on the backs of the mold halves, moisture is removed from the clay. The mold sections are then opened, using positive air pressure to prevent sticking of the part in the mold. Plastic pressing achieves a higher production rate than jiggering and is not limited to radially symmetric parts.

Extrusion is used in ceramics processing to produce long sections of uniform cross-section, which are then cut to required piece length. The extrusion equipment utilizes a screw-type action to assist in mixing the clay and pushing the plastic material through the die opening. This production sequence is widely used to make hollow bricks, shaped tiles, drain pipes, tubes, and insulators. It is also used to make the starting clay slugs for other ceramics processing methods such as jiggering and plastic pressing.

Semi-dry Pressing   In semi-dry pressing, the proportion of water in the starting clay is typically in the range 10% to 15%. This results in low plasticity, precluding the use of plastic forming methods which require a very plastic clay. Semi-dry pressing uses high pressure to overcome the material's low plasticity and force it to flow into a die cavity, as depicted in Figure 17.7. Flash is often formed due to excess clay being squeezed between the die sections.

Dry Pressing   The main distinction between semi-dry and dry pressing is the moisture content of the starting mix. The moisture content of the starting clay in dry pressing is typically below 5%. Binders are usually added to the dry powder mix to provide sufficient strength in the pressed part for subsequent handling. Lubricants are also added to prevent die sticking during pressing and ejection. Because dry clay has no plasticity and is very abrasive, there are differences in die design and operating procedures, compared to semi-dry pressing. The dies must be made of hardened tool steel or cemented tungsten carbide to reduce wear. Since dry clay will not flow during pressing, the geometry of the part must be relatively simple, and the amount and distribution of starting powder in the die cavity must be right. No flash is formed in dry pressing, and no drying shrinkage occurs, so drying time is eliminated and good accuracy can be achieved

FIGURE 17.7   Semi-dry pressing: (1) depositing moist powder into die cavity, (2) pressing, and (3) opening the die sections and ejection. Symbols *v* and *F* indicate motion (*v* = velocity) and applied force respectively.

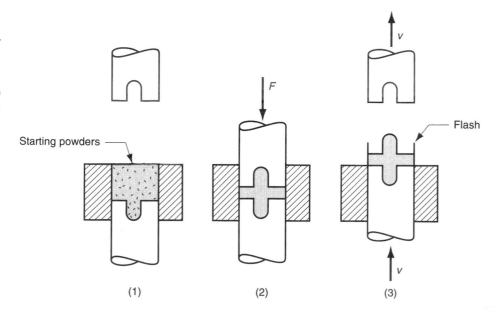

Starting powders

Flash

(1)   (2)   (3)

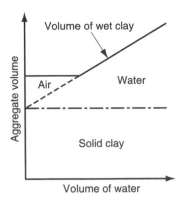

FIGURE 17.8 Volume of clay as a function of water content. Relationship shown here is typical; it varies for different clay compositions.

in the dimensions of the final product. The process sequence in dry pressing is similar to semi-dry pressing. Typical products include bathroom tile, electrical insulators, and refractory brick.

17.1.3 Drying

Water plays an important role in most of the traditional ceramics shaping processes. Thereafter, it serves no purpose and must be removed from the body of the clay piece before firing. Shrinkage is a problem during this step in the processing sequence because water contributes volume to the piece, and when it is removed, the volume is reduced. The effect can be seen in Figure 17.8. As water is initially added to dry clay, it simply replaces the air in the pores between ceramic grains, and there is no volumetric change. Increasing the water content above a certain point causes the grains to become separated and the volume to grow, resulting in a wet clay that has plasticity and formability. As more water is added, the mixture eventually becomes a liquid suspension of clay particles in water.

The reverse of this process occurs in drying. As water is removed from the wet clay, the volume of the piece shrinks. The drying process occurs in two stages, as depicted in Figure 17.9. In the first stage, the rate of drying is rapid and constant, as water is evaporated from the surface of the clay into the surrounding air and water from the interior migrates by capillary action toward the surface to replace it. It is during this stage that shrinkage occurs, with the associated risk of warping and cracking due to variations in drying in different sections of the piece. In the second stage of drying, the moisture con-

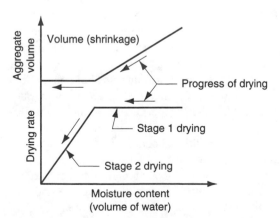

FIGURE 17.9 Typical drying rate curve and associated volume reduction (drying shrinkage) for a ceramic body in drying. Drying rate in the second stage of drying is depicted here as a straight line (constant rate decrease as a function of water content); the function is variously shown as concave or convex in the literature [3], [7].

tent has been reduced to where the ceramic grains are in contact, and little or no further shrinkage occurs. The drying process slows, and this is seen in the decreasing rate in the plot.

In production, drying is usually accomplished in drying chambers in which temperature and humidity are controlled to achieve the proper drying schedule. Care must be taken so that water is not removed too rapidly from the piece, lest large moisture gradients be set up in the piece, making it more prone to crack. Heating is usually by a combination of convection and radiation, using infrared sources. Typical drying times range between a quarter hour for thin sections to several days for very thick sections.

17.1.4 Firing (Sintering)

After shaping but prior to firing, the ceramic piece is said to be **green** (same term as in powder metallurgy), meaning not fully processed or treated. The green piece lacks hardness and strength; it must be fired to fix the part shape and achieve hardness and strength in the finished ware. **Firing** is the heat treatment process that sinters the ceramic material; it is performed in a furnace called a **kiln.** In **sintering,** bonds are developed between the ceramic grains, and this is accompanied by densification and reduction of porosity. Therefore, shrinkage occurs in the polycrystalline material in addition to that which has already occurred in drying. Sintering in ceramics is basically the same mechanism as in powder metallurgy. In the firing of traditional ceramics, certain chemical reactions between the components in the mixture may also take place, and a glassy phase also forms among the crystals which acts as a binder. Both of these phenomena depend on the chemical composition of the ceramic material and the firing temperatures used.

Unglazed ceramic ware is fired only once; glazed products are fired twice. **Glazing** refers to the application of a ceramic surface coating to make the piece more impervious to water and enhance its appearance (Section 7.2.2). The usual processing sequence with glazed ware is: (1) fire the ware once before glazing to harden the body of the piece, (2) apply the glaze, and (3) fire the piece a second time to harden the glaze.

17.2 PROCESSING OF NEW CERAMICS

Most of the traditional ceramics are based on clay, which possesses a unique capacity to be plastic when mixed with water but hard when dried and fired. Clay consists of various formulations of hydrous aluminum silicate, usually mixed with other ceramic materials, to form a rather complex chemistry. New ceramics (Section 7.3) are based on simpler chemical compounds, such as oxides, carbides, and nitrides. These materials do not possess the plasticity and formability of traditional clay when mixed with water. Accordingly, other ingredients must be combined with the ceramic powders to achieve plasticity and other desirable properties during forming, so that conventional shaping methods can be used. The new ceramics are generally designed for applications that require higher strength, hardness, and other properties not found in the traditional ceramic materials. These requirements have motivated the introduction of several new processing techniques not previously used for traditional ceramics.

The manufacturing sequence for the new ceramics can be summarized in the following steps: (1) preparation of starting materials, (2) shaping, (3) sintering, and (4) finishing. Although the sequence is nearly the same as for the traditional ceramics, the details are often quite different, as we shall see in the following sections.

17.2.1 Preparation of Starting Materials

Since the strength specified for these materials is usually much greater than for traditional ceramics, the starting powders must be more homogeneous in size and composition, and particle size must be smaller (strength of the resulting ceramic product is inversely related to grain size). All of this means that greater control of the starting powders is required. Powder preparation includes mechanical and chemical methods.

The mechanical methods consist of the same ball mill grinding operations used for traditional ceramics. The trouble with these methods is that the ceramic particles become contaminated from the materials used in the balls and walls of the mill. This compromises the purity of the ceramic powders and results in microscopic flaws that reduce the strength of the final product.

Two chemical methods are used to achieve greater homogeneity in the powders of new ceramics: freeze drying and precipitation from solution. In *freeze drying,* salts of the appropriate starting chemistry are dissolved in water and the solution is sprayed to form small droplets, which are rapidly frozen. The water is then removed from the droplets in a vacuum chamber, and the resulting freeze-dried salt is decomposed by heating to form the ceramic powders. Freeze drying is not applicable to all ceramics, because in some cases a suitable water-soluble salt cannot be identified as a starting material.

Precipitation from solution is another preparation method used for new ceramics. In the typical process, the desired ceramic compound is dissolved from the starting mineral, thus permitting impurities to be filtered out. An intermediate compound is then precipitated from solution, which is converted into the desired compound by heating. An example of the precipitation method is the *Bayer process* for producing high purity alumina (also used in the production of aluminum). In this process, aluminum oxide is dissolved from the mineral bauxite so that iron compounds and other impurities can be removed. Then, aluminum hydroxide ($Al(OH)_3$) is precipitated from solution and reduced to Al_2O_3 by heating.

Further preparation of the powders includes classification by size and mixing before shaping. Very fine powders are required for new ceramics applications, and so the grains must be separated and classified according to size. Thorough mixing of the particles, especially when different ceramic powders are combined, is required to avoid segregation.

Various additives are often combined with the starting powders, usually in small amounts. The additives include (1) *plasticizers* to improve plasticity and workability; (2) *binders* to bond the ceramic particles into a solid mass in the final product, (3) *wetting agents* for better mixing; (4) *deflocculants,* which help to prevent clumping and premature bonding of the powders; and (5) *lubricants,* to reduce friction between ceramic grains during forming and to reduce sticking during mold release.

17.2.2 Shaping

Many of the shaping processes for new ceramics are borrowed from powder metallurgy (PM) and traditional ceramics. The press and sinter methods discussed in Section 16.3 have been adapted to the new ceramic materials. And some of the traditional ceramics-forming techniques (Section 17.1.2) are used to shape the new ceramics, including: slip casting, extrusion, and dry pressing. The following processes are not normally associated with the forming of traditional ceramics, although several are associated with PM.

Hot Pressing   Hot pressing is similar to dry pressing (Section 17.1.2), except that the process is carried out at elevated temperatures, so that sintering of the product is accomplished simultaneously with pressing. This eliminates the need for a separate firing step in the sequence. Higher densities and finer grain size are obtained, but die life is reduced by the hot abrasive particles against the die surfaces.

Isostatic Pressing   Isostatic pressing of ceramics is the same process used in powder metallurgy (Section 16.4.1). It uses hydrostatic pressure to compact the ceramic powders from all directions, thus avoiding the problem of nonuniform density in the final product that is often observed in the traditional uniaxial pressing method.

Doctor-blade Process   This process is used for making thin sheets of ceramic. One common application of the sheets is in the electronics industry as a substrate material for integrated circuits. The process is diagrammed in Figure 17.10. A ceramic slurry is introduced onto a moving carrier film such as cellophane. Thickness of the ceramic on the carrier is determined by a wiper, called a *doctor-blade.* As the slurry moves down the line, it is dried into a flexible green ceramic tape. At the end of the line, a take-up spool reels in the tape for later processing. In its green condition, the tape can be cut or otherwise processed before firing.

Powder-Injection Molding (PIM)   This is the same as the PM process (Section 16.4.2), except that the powders are ceramic rather than metallic. Ceramic particles are mixed with a thermoplastic polymer, which acts as a carrier and provides the proper flow characteristics at molding temperatures. The mix is then heated and injected into a mold cavity. Upon cooling, which hardens the polymer, the mold is opened and the part is removed. Because the temperatures needed to plasticize the carrier are much lower than those required for sintering the ceramic, the piece is green after molding. Before sintering, the plastic binder must be removed. This is called *debinding,* which is usually accomplished by a combination of thermal and solvent treatments.

Applications of ceramic PIM are currently inhibited by difficulties in debinding and sintering. Burning off the polymer is relatively slow, and its removal significantly weakens the green strength of the molded part. Warping and cracking often occur during sintering. Further, ceramic products made by powder-injection molding are especially vulnerable to microstructural flaws which limit their strength.

17.2.3 Sintering

Since the plasticity needed to shape the new ceramics is not normally based on a water mixture, the drying step so commonly required to remove water from the traditional green ceramics can be omitted in the processing of most new ceramic products. The sintering step, however, is still very much required to obtain maximum possible strength and hard-

FIGURE 17.10 The doctor-blade process, used to fabricate thin ceramic sheets. Symbol v indicates motion (v = velocity).

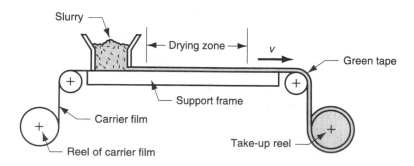

ness. The functions of sintering are the same as before: (1) to bond individual grains into a solid mass, (2) to increase density, and (3) to reduce or eliminate porosity.

Temperatures around 80% to 90% of the melting temperature of the material are commonly used in sintering ceramics. Sintering mechanisms differ somewhat between the new ceramics, which are based predominantly on a single chemical compound (e.g., Al_2O_3), and the clay-based ceramics, which are usually made of several compounds having different melting points. In the case of the new ceramics, the sintering mechanism is mass diffusion across the contacting particle surfaces, probably accompanied by some plastic flow. This mechanism causes the centers of the particles to move closer together, resulting in densification of the final material. In the sintering of traditional ceramics, this mechanism is complicated by the melting of some constituents and the formation of a glassy phase that acts as a binder between the grains.

17.2.4 Finishing

Parts made of new ceramics sometimes require finishing. In general, these operations have one or more of the following purposes: (1) to increase dimensional accuracy, (2) to improve surface finish, and (3) to make minor changes in part geometry. Finishing operations usually involve grinding and other abrasive processes (Chapter 25). Diamond abrasives must be used to cut the hardened ceramic materials.

17.3 PROCESSING OF CERMETS

Many metal matrix composites (MMCs) and ceramic matrix composites (CMCs) are processed by particulate processing methods. The most prominent examples are cemented carbides and other cermets.

17.3.1 Cemented Carbides

The cemented carbides are a family of composite materials consisting of carbide ceramic particles imbedded in a metallic binder. They are classified as metal matrix composites because the metallic binder is the matrix which holds the bulk material together; however, the carbide particles constitute the largest proportion of the composite material, normally ranging between 80% and 95% by volume. Cemented carbides are technically classified as cermets, although they are often distinguished from the other materials in this class.

The most important cemented carbide is tungsten carbide in a cobalt binder (WC–Co). Generally included within this category are certain mixtures of WC, TiC, and TaC in a Co matrix, in which tungsten carbide is the major component. Other cemented carbides include titanium carbide in nickel (TiC–Ni) and chromium carbide in nickel (Cr_3C_2–Ni). These composites were discussed in Section 9.2.1, and the carbide ingredients were described in Section 7.3.2. Here, we are concerned with the processing of cemented carbide, which is based on particulate technologies.

To provide a strong and pore-free part, the carbide powders must be sintered with a metal binder. Cobalt works best with WC, while nickel is better with TiC and Cr_3C_2. The usual proportion of binder metal is from around 4% up to 20%. Powders of carbide and binder metal are thoroughly mixed wet in a ball mill (or other suitable mixing machine) to form a homogeneous sludge. Milling also serves to refine particle size. The sludge is then dried in a vacuum or controlled atmosphere to prevent oxidation in preparation for compaction.

Compaction Various methods are used to shape the powder mix into a green compact of the desired geometry. The most common process is cold pressing, described earlier and used for high production of cemented carbide parts such as cutting tool inserts. The dies used in cold pressing must be made oversized to account for shrinkage during sintering. Linear shrinkage can be 20% or more. For high production, the dies are made with WC–Co liners to reduce wear, due to the abrasive nature of carbide particles. For smaller quantities, large flat sections are sometimes pressed and then cut into smaller pieces of the specified size.

Other compaction methods used for cemented carbide products include *isostatic pressing* and *hot pressing* for large pieces, such as draw dies and ball mill balls; and *extrusion,* for long sections of circular, rectangular, or other cross-section. Each of these processes has been described previously, either in this or the preceding chapter.

Sintering Although it is possible to sinter WC and TiC without a binder metal, the resulting material is somewhat less than 100% of true density. Use of a binder yields a structure that is virtually free of porosity.

Sintering of WC–Co involves liquid phase sintering (Section 16.4.5). The process can be explained with reference to the binary phase diagram for these constituents in Figure 17.11. The typical composition range for commercial cemented carbide products is identified in the diagram. The usual sintering temperatures for WC–Co are in the range 1370–1425°C (2500–2600°F), which is below cobalt's melting point of 1495°C (2716°F). Thus, the pure binder metal does not melt at the sintering temperature. However, as the phase diagram shows, WC dissolves in Co in the solid state. During the heat treatment, WC is gradually dissolved into the gamma phase, and its melting point is reduced so that melting finally occurs. As the liquid phase forms, it flows and wets the WC particles, further dissolving the solid. The presence of the molten metal also serves to remove gases from the internal regions of the compact. These mechanisms combine to effect a rearrangement of the remaining WC particles into a closer packing, which results in significant densification and shrinkage of the WC–Co mass. Later, during cooling in the sintering cycle, the dissolved carbide is precipitated and deposited onto the existing crystals to form a coherent WC skeleton, throughout which is imbedded the Co binder.

FIGURE 17.11 WC–Co phase diagram (source: [6]).

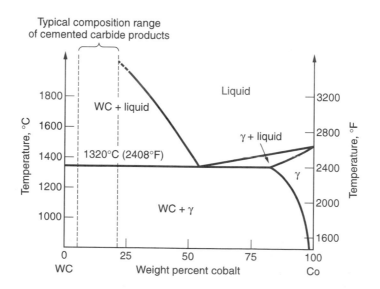

Secondary Operations Subsequent processing is usually required after sintering to achieve adequate dimensional control of cemented carbide parts. Grinding with a diamond or other very hard abrasive wheel is the most common secondary operation performed for this purpose. Other processes used to shape the hard cemented carbides include electric discharge machining and ultrasonic machining, two nontraditional material removal processes discussed in Chapter 26.

17.3.2 Other Cermets and Ceramic Matrix Composites

In addition to cemented carbides, other cermets are based on oxide ceramics such as Al_2O_3 and MgO. Chromium is a common metal binder used in these composite materials. The ceramic-to-metal proportions cover a wider range than those of the cemented carbides, in some cases, the metal being the major ingredient. These cermets are formed into useful products by the same basic shaping methods used for cemented carbides.

The current technology of ceramic matrix composites (Section 9.3) includes ceramic materials (e.g., Al_2O_3, BN, Si_3N_4, and glass) reinforced by fibers of carbon, SiC, or Al_2O_3. If the fibers are whiskers (fibers consisting of single crystals), these CMCs can be processed by particulate methods used for new ceramics (Section 17.2).

17.4 PRODUCT DESIGN CONSIDERATIONS

Ceramic materials have special properties that make them attractive to designers if the application is right. The following design recommendations, compiled from Bralla [2] and other sources, apply to both new and traditional ceramic materials, although designers are more likely to find opportunities for new ceramics in engineered products. In general, the same guidelines apply to cemented carbides.

- ➢ Ceramic materials are several times stronger in compression than in tension; components should be designed to be subjected to compressive stresses, not tensile stresses.
- ➢ Ceramics are brittle and possess almost no ductility. Ceramic parts should not be used in applications that involve impact loading or high stresses that might cause fracture.
- ➢ Although many of the ceramic-shaping processes allow complex geometries to be formed, it is desirable to keep shapes simple for both economic and technical reasons. Deep holes, channels, and undercuts should be avoided, as should large cantilevered projections.
- ➢ Outside edges and corners should have radii or chamfers; likewise, inside corners should have radii. This guideline is, of course, violated in cutting tool applications, in which the cutting edge must be sharp in order to function. The cutting edge is often fabricated with a very small radius to protect it from microscopic chipping, which could lead to failure.
- ➢ Part shrinkage in drying and firing (for traditional ceramics) and sintering (for new ceramics) may be significant and must be taken into account by the designer in dimensioning and tolerancing. This is mostly a problem for manufacturing engineers, who must determine appropriate size allowances so that the final dimensions will be within the tolerances specified.
- ➢ Screw threads in ceramic parts should be avoided. They are difficult to fabricate and do not have adequate strength in service after fabrication.

REFERENCES

[1] Bhowmick, A. K. Bradley Pulverizer Company. Allentown, Pennsylvania, personal communication, February, 1992.

[2] Bralla, J. G. (Editor in Chief). *Design for Manufacturability Handbook.* 2nd ed. McGraw-Hill Book Company, New York, 1998.

[3] Hlavac, J. *The Technology of Glass and Ceramics.* Elsevier Scientific Publishing Company, New York, 1983.

[4] Kingery, W. D., Bowen, H. K., and Uhlmann, D. R. *Introduction to Ceramics.* 2nd ed. John Wiley & Sons, Inc., New York, 1995.

[5] Richerson, D. W. *Modern Ceramic Engineering.* 2nd ed. Marcel Dekker, New York, 1992.

[6] Schwarzkopf, P. and Kieffer, R. *Cemented Carbides.* The Macmillan Company, New York, 1960.

[7] Singer, F., and Singer, S. S. *Industrial Ceramics.* Chemical Publishing Company, New York, 1963.

[8] Somiya, S., (ed.). *Advanced Technical Ceramics.* Academic Press, Inc., San Diego, Calif., 1989.

REVIEW QUESTIONS

17.1. What is the difference between the traditional ceramics and the new ceramics, as far as raw materials are concerned?

17.2. List the basic steps in the traditional ceramics processing sequence.

17.3. What is the technical difference between crushing and grinding in the preparation of traditional ceramic raw materials?

17.4. Describe the slip casting process in traditional ceramics processing.

17.5. List and briefly describe some of the plastic-forming methods used to shape traditional ceramics products.

17.6. What is the process of jiggering?

17.7. What is the difference between dry pressing and semi-dry pressing of traditional ceramics parts?

17.8. What happens to a ceramic material when it is sintered?

17.9. What is the name given to the furnace used to fire ceramic ware?

17.10. What is glazing in traditional ceramics processing?

17.11. Why is the drying step, so important in the processing of traditional ceramics, usually not required in processing of new ceramics?

17.12. Why is raw material preparation more important in the processing of new ceramics than for traditional ceramics?

17.13. What is the freeze drying process used to make certain new ceramic powders?

17.14. Describe the doctor-blade process.

17.15. Liquid phase sintering is used for WC–Co compacts, even though the sintering temperatures are below the melting points of either WC or Co. How is this possible?

17.16. What are some design recommendations for ceramic parts?

MULTIPLE CHOICE QUIZ

There is a total of 15 correct answers in the following multiple choice questions (some questions have multiple answers that are correct). To attain a perfect score on the quiz, all correct answers must be given, since each correct answer is worth 1 point. For each question, each omitted answer or wrong answer reduces the score by 1 point, and each additional answer beyond the number of answers required reduces the score by 1 point. Percentage score on the quiz is based on the total number of correct answers.

17.1. The following equipment is used for crushing and grinding of minerals in the preparation of traditional ceramics raw materials. Which one of the pieces listed is used for grinding? (a) ball mill, (b) hammer mill, (c) jaw crusher, or (d) roll crusher.

17.2. Which one of the following compounds becomes a plastic and formable material when mixed with suitable proportions of water? (a) aluminum oxide, (b) hydrogen oxide, (c) hydrous aluminum silicate, or (d) silicon dioxide.

17.3. At which one of the following water contents does clay become a suitably plastic material for the traditional ceramics plastic forming processes? (a) 5%, (b) 10%, (c) 20%, or (d) 40%.

17.4. Which of the following processes are not plastic-forming methods used in the shaping of traditional ceramics (more than one)? (a) extrusion, (b) jangling, (c) jiggering, (d) jolleying, or (e) spinning.

17.5. The term *green piece* in ceramics refers to a part that has been shaped but not yet fired: (a) true or (b) false.

17.6. In the final product made of a polycrystalline new ceramic material, strength increases with grain size: (a) true or (b) false.

17.7. Which one of the following processes for the new ceramic materials accomplishes shaping and sintering simultaneously? (a) doctor-blade process, (b) freeze drying, (c) hot pressing, (d) injection molding, or (e) isostatic pressing.

17.8. Which of the following are not the purposes of finishing operations used for parts made of the new ceramics (more than one answer)? (a) apply a surface coating, (b) improve surface finish, (c) increase dimensional accuracy, (d) remove material, or (e) work harden the surface.

17.9. Which one of the following terms best describes what a cemented carbide is? (a) ceramic, (b) cermet, (c) composite, or (d) metal.

17.10. Which of the following geometric features should be avoided if possible in the design of structural components made of new ceramics (more than one answer)? (a) complicated shapes, (b) rounded inside corners, (c) sharp edges, (d) thin sections, or (e) threads.

Part V
Metal Forming and Sheet Metalworking

18 FUNDAMENTALS OF METAL FORMING

CHAPTER CONTENTS

18.1 Overview of Metal Forming
18.2 Material Behavior in Metal Forming
18.3 Temperature in Metal Forming
18.4 Strain Rate Sensitivity
18.5 Friction and Lubrication in Metal Forming

Metal forming includes a large group of manufacturing processes in which plastic deformation is used to change the shape of metal workpieces. Deformation results from the use of a tool, usually called a *die* in metal forming, which applies stresses that exceed the yield strength of the metal. The metal therefore deforms to take a shape determined by the geometry of the die. Metal forming dominates the class of shaping operations identified in Chapter 1 as the *deformation processes* (Figure 1.4).

Stresses applied to plastically deform the metal are usually compressive. However, some forming processes stretch the metal, while others bend the metal, and still others apply shear stresses to the metal. To be successfully formed, a metal must possess certain properties. Desirable properties for forming include low yield strength and high ductility. These properties are affected by temperature. Ductility is increased and yield strength is reduced when work temperature is raised. The effect of temperature gives rise to distinctions between cold working, warm working, and hot working. Strain rate and friction are additional factors that affect performance in metal forming. We examine all of these issues in this chapter, but first let us provide an overview of the metal forming processes.

18.1 OVERVIEW OF METAL FORMING

Metal-forming processes can be classified as (1) bulk deformation processes or (2) sheet metalworking processes. These two categories are covered in detail in Chapters 21 and 22, respectively. Each category includes several major classes of shaping operations, as indicated in Figure 18.1.

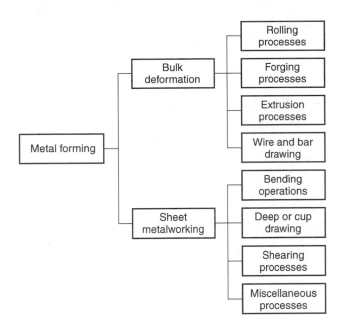

FIGURE 18.1
Classification of metal
forming operations.

Bulk Deformation Processes   Bulk deformation processes are generally characterized by significant deformations and massive shape changes; and the surface area-to-
volume of the work is relatively small. The term **bulk** describes the workparts that have
this low area-to-volume ratio. Starting work shapes for these processes include cylindrical billets and rectangular bars. Figure 18.2 illustrates the following basic operations
in bulk deformation:

FIGURE 18.2   Basic bulk
deformation processes: (a)
rolling, (b) forging, (c) extrusion, and (d) drawing.
Relative motion in the operations is indicated by *v*,
and forces are indicated
by *F*.

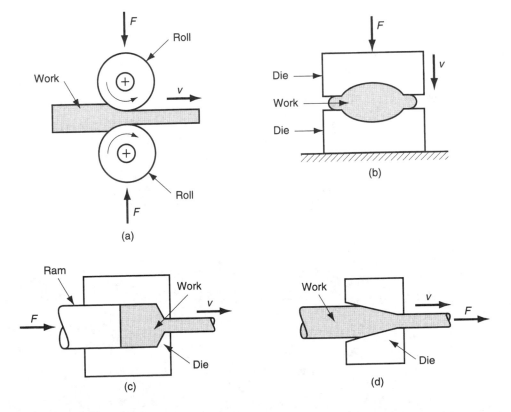

> *Rolling*—This is a compressive deformation process in which the thickness of a slab or plate is reduced by two opposing cylindrical tools called rolls. The rolls rotate so as to draw the work into the gap between them and squeeze it.

> *Forging*—In forging, a workpiece is compressed between two opposing dies, so that the die shapes are imparted to the work. Forging is traditionally a hot working process, but many types of forging are performed cold.

> *Extrusion*—This is a compression process in which the work metal is forced to flow through a die opening, thereby taking the shape of the opening as its own cross-section.

> *Drawing*—In this forming process, the diameter of a wire or bar is reduced by pulling it through a die opening.

Sheet Metalworking   Sheet metalworking processes are forming and related operations performed on metal sheets, strips, and coils. The surface area-to-volume ratio of the starting metal is high; thus, this ratio is a useful means to distinguish bulk deformation from sheet metal processes. *Pressworking* is the term often applied to sheet metal operations because the machines used to perform these operations are presses (presses of various types are also used in other manufacturing processes). A part produced in a sheet metal operation is often called a *stamping.*

Sheet metal operations are always performed as cold working processes and are accomplished using a set of tools called a *punch* and *die.* The punch is the positive portion and the die is the negative portion of the tool set. The basic sheet metal operations are sketched in Figure 18.3 and are defined as follows:

FIGURE 18.3   Basic sheet metalworking operations: (a) bending, (b) drawing, and (c) shearing: (1) as punch first contacts sheet, and (2) after cutting. Force and relative motion in these operations are indicated by *F* and *v*.

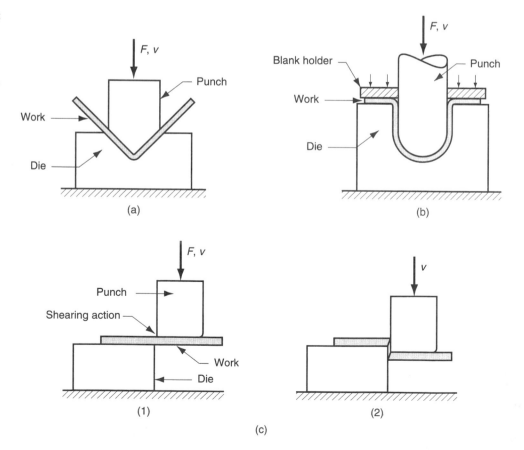

> *Bending*—Bending involves straining of a metal sheet or plate to take an angle along a (usually) straight axis.

> *Drawing*—In sheet metalworking, drawing refers to the forming of a flat metal sheet into a hollow or concave shape, such as a cup, by stretching the metal. A blankholder is used to hold down the blank while the punch pushes into the sheet metal, as shown in Figure 18.3(b). To distinguish this operation from bar and wire drawing, the terms *cup drawing* or *deep drawing* are often used.

> *Shearing*—This process is somewhat out-of-place in our list of deformation processes, because it involves cutting rather than forming of the metal. A shearing operation cuts the work using a punch and die, as in Figure 18.3(c). Although it is not a forming process, it is included here because it is a necessary and very common operation in sheet metalworking.

The miscellaneous processes within the sheet metalworking classification in Figure 18.1 include a variety of sheet and tube-shaping processes that do not use punch and die tooling. Examples of these processes are stretch forming, roll bending, spinning, and bending of tube stock.

18.2 MATERIAL BEHAVIOR IN METAL FORMING

Considerable insight about the behavior of metals during forming can be obtained from the stress–strain curve. The typical stress–strain curve for most metals is divided into an elastic region and a plastic region (Section 3.1.1). In metal forming, the plastic region is of primary interest because the material is plastically and permanently deformed in these processes.

The typical stress–strain relationship for a metal exhibits elasticity below the yield point and strain hardening above it. Figures 3.4 and 3.5 indicate this behavior in linear and logarithmic axes. In the plastic region, the metal's behavior is expressed by the flow curve:

$$\sigma = K\epsilon^n$$

where K = the strength coefficient, MPa (lb/in.^{2}); and n is the strain-hardening exponent. The stress and strain in the flow curve are true stress and true strain. The flow curve is generally valid as a relationship that defines a metal's plastic behavior in cold working. Typical values of K and n for different metals at room temperature are listed in Table 3.4.

Flow Stress   The flow curve describes the stress–strain relationship in the region in which metal forming takes place. It indicates the flow stress of the metal—the strength property that determines forces and power required to accomplish a particular forming operation. For most metals at room temperature, the stress–strain plot of Figure 3.5 indicates that as the metal is deformed, its strength increases due to strain hardening. The stress required to continue deformation must be increased to match this increase in strength. *Flow stress* is defined as the instantaneous value of stress required to continue deforming the material—to keep the metal "flowing." It is the yield strength of the metal as a function of strain, which can be expressed in this way:

$$Y_f = K\epsilon^n \tag{18.1}$$

where Y_f = flow stress, MPa (lb/in.^{2}).

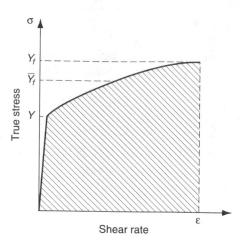

FIGURE 18.4 Stress–strain curve indicating location of average flow stress $\overline{Y}_f$ in relation to yield strength Y and final flow stress Y_f.

In the individual forming operations discussed in the following two chapters, the instantaneous flow stress can be used to analyze the process as it is occurring. For example, in certain forging operations, the instantaneous force during compression can be determined from the flow stress value. Maximum force can be calculated based on the flow stress that results from the final strain at the end of the forging stroke.

In other cases, the analysis is based on the average stresses and strains that occur during deformation rather than instantaneous values. Extrusion represents this case, Figure 18.2(c). As the billet is reduced in cross-section to pass through the extrusion die opening, the metal gradually strain hardens to reach a maximum value. Rather than determine a sequence of instantaneous stress–strain values during the reduction, which would be not only difficult but also of limited interest, it is more useful to analyze the process based on the average flow stress during deformation.

Average Flow Stress   The *average flow stress* (also called the *mean flow stress*) is the average value of stress over the stress–strain curve from the beginning of strain to the final (maximum) value that occurs during deformation. The value is illustrated in the stress–strain plot of Figure 18.4.

The average flow stress is determined by integrating the flow curve equation, Eq. (18.1), between zero and the final strain value defining the range of interest. This yields the equation:

$$\overline{Y}_f = \frac{K\epsilon^n}{1 + n} \tag{18.2}$$

where $\overline{Y}_f$ = average flow stress, MPa (lb/in.^{2}); and ϵ = maximum strain value during the deformation process.

We make extensive use of the average flow stress in our study of the bulk deformation processes in the following chapter. Given values of K and n for the work material, a method of computing final strain will be developed for each process. Based on this strain, Eq. (18.2) can be used to determine the average flow stress to which the metal is subjected during the operation.

18.3 TEMPERATURE IN METAL FORMING

The flow curve is a valid representation of stress–strain behavior of a metal during plastic deformation, particularly for cold working operations. For any metal, the values of K and n depend on temperature. Both strength and strain hardening are reduced at higher temperatures. In addition, ductility is increased at higher temperatures. These property changes are important because any deformation operation can be accomplished with lower forces and power at elevated temperature. There are three temperature ranges: cold, warm, and hot working.

Cold Working   *Cold working* (also known as *cold forming*) is metal forming performed at room temperature or slightly above. Significant advantages of cold forming compared to hot working are (1) better accuracy, meaning closer tolerances; (2) better surface finish; (3) strain hardening increases strength and hardness of the part; (4) grain flow during deformation provides the opportunity for desirable directional properties to be obtained in the resulting product; and (5) no heating of the work is required, which saves on furnace and fuel costs and permits higher production rates to be achieved. Owing to this combination of advantages, many cold forming processes have developed into important mass-production operations. They provide close tolerances and good surfaces, minimizing the amount of machining required and permitting these operations to be classified as net shape or near net shape processes (Section 1.3.1).

There are certain disadvantages or limitations associated with cold-forming operations: (1) higher forces and power are required to perform the operation; (2) care must be taken to ensure that the surfaces of the starting workpiece are free of scale and dirt; and (3) ductility and strain hardening of the work metal limit the amount of forming that can be done to the part. In some operations, the metal must be annealed (Section 27.1) in order to allow further deformation to be accomplished. In other cases, the metal is simply not ductile enough to be cold worked.

To overcome the strain hardening problem and reduce force and power requirements, many forming operations are performed at elevated temperatures. There are two elevated temperature ranges involved, giving rise to the terms warm working and hot working.

Warm Working   Because plastic deformation properties are normally enhanced by increasing workpiece temperature, forming operations are sometimes performed at temperatures somewhat above room temperature but below the recrystallization temperature. The term *warm working* is applied to this second temperature range. The dividing line between cold working and warm working is often expressed in terms of the melting point for the metal. The dividing line is usually taken to be $0.3T_m$, where T_m is the melting point (absolute temperature) for the particular metal.

The lower strength and strain hardening, as well as higher ductility of the metal at the intermediate temperatures, provide warm working with the following advantages over cold working: (1) lower forces and power, (2) more intricate work geometries possible, and (3) need for annealing may be reduced or eliminated.

Hot Working   *Hot working* (also called *hot forming*) involves deformation at temperatures above the recrystallization temperature. The recrystallization temperature for a given metal is about one-half of its melting point on the absolute scale. In practice, hot working is usually carried out at temperatures somewhat above $0.5T_m$. The work metal continues to soften as temperature is increased beyond $0.5T_m$, thus enhancing the advantage of hot working above this level. However, the deformation process itself generates heat which increases work temperatures in localized regions of the part. This can cause melting in these regions, which is highly undesirable. Also, scale on the work surface is accelerated at higher temperatures. Accordingly, hot working temperatures are usually maintained within the range $0.5T_m$ to $0.75T_m$.

The most significant advantage of hot working is the capability to produce substantial plastic deformation of the metal—far more than is possible with cold working or warm working. The principal reason for this is that the flow curve of the hot-worked metal has a strength coefficient that is substantially less than at room temperature, the strain hardening exponent is zero (at least theoretically), and the ductility of the metal is significantly increased. All of this results in the following advantages relative to cold working: (1) the shape of the workpart can be significantly altered; (2) lower forces and power are required to deform the metal; (3) metals that usually fracture in cold working can be hot formed; (4) strength properties are generally isotropic because of the absence of the oriented grain structure typically created in cold working; and (5) no strengthening of the part occurs from work hardening. This last advantage may seem inconsistent, since strengthening of the metal is often considered an advantage for cold working. However, there are applications in which it is undesirable for the metal to be work hardened because it reduces ductility; for example, if the part is to be subsequently processed by cold forming. Disadvantages of hot working include lower dimensional accuracy, higher total energy required (due to the thermal energy to heat the workpiece), work surface oxidation (scale), poorer surface finish, and shorter tool life.

Recrystallization of the metal in hot working involves atomic diffusion, which is a time-dependent process. Metal-forming operations are often performed at high speeds that do not allow sufficient time for complete recrystallization of the grain structure during the deformation cycle itself. However, because of the high temperatures, recrystallization eventually does occur. It may occur immediately following the forming process or later, as the workpiece cools. Even if recrystallization occurs after the actual deformation, its eventual occurrence—together with the substantial softening of the metal at high temperatures—distinguishes hot working from warm working or cold working.

Isothermal Forming Certain metals such as highly alloyed steels (e.g., high speed steel), many titanium alloys, and high-temperature nickel alloys possess good hot hardness, a property that makes them useful for high-temperature service. However, the very property that makes them attractive in these applications also makes them difficult to form with conventional methods. The problem is that when these metals are heated to their hot working temperatures and then come in contact with the relatively cold forming tools, heat is quickly transferred away from the part surfaces, thus raising the strength in these regions. The variations in temperature and strength in different regions of the workpiece cause irregular flow patterns in the metal during deformation, leading to high residual stresses and possible surface cracking.

Isothermal forming refers to forming operations that are carried out in such a way as to eliminate surface cooling and the resulting thermal gradients in the workpart. It is accomplished by preheating the tools that come in contact with the part to the same temperature as the work metal. This weakens the tools and reduces tool life, but it avoids the problems described above when these difficult metals are formed by conventional methods. In some cases, isothermal forming represents the only way in which these work materials can be formed. The procedure is most closely associated with forging, and we discuss isothermal forging in the following chapter.

18.4 STRAIN RATE SENSITIVITY

Theoretically, a metal in hot working behaves like a perfectly plastic material, with strain hardening exponent $n = 0$. This means that the metal should continue to flow under the same level of flow stress, once that stress level is reached. However, there is an additional phenomenon that characterizes the behavior of metals during deformation, especially at the elevated temperatures of hot working. That phenomenon is strain-rate sensitivity. Let us begin our discussion of this topic by defining strain rate.

The rate at which the metal is strained in a forming process is directly related to the speed of deformation v. In many forming operations, deformation speed is equal to the velocity of the ram or other moving element of the equipment. It is most easily visualized in a tensile test as the velocity of the testing machine head relative to its fixed base. Given the deformation speed, ***strain rate*** is defined:

$$\dot{\epsilon} = \frac{v}{h} \tag{18.3}$$

where $\dot{\epsilon}$ = true strain rate, m/s/m (in/sec/in.), or simply s^{-1}; and h = instantaneous height of the workpiece being deformed, m (in.). If deformation speed v is constant during the operation, strain rate will change as h changes. In most practical forming operations, valuation of strain rate is complicated by the geometry of the workpart and variations in strain rate in different regions of the part. Strain rate can reach $1000 \ s^{-1}$ or more for some metal forming processes such as high speed rolling and forging.

We have already observed that the flow stress of a metal is a function of temperature. At the temperatures of hot working, flow stress depends on strain rate. The effect of strain rate on strength properties is known as ***strain-rate sensitivity.*** The effect can be seen in Figure 18.5. As strain rate is increased, resistance to deformation increases. This usually plots approximately as a straight line on a log–log graph, thus leading to the relationship:

$$Y_f = C\dot{\epsilon}^{\,m} \tag{18.4}$$

where C is the strength constant (similar but not equal to the strength coefficient in the flow curve equation), and m is the strain-rate sensitivity exponent. The value of C is determined at a strain rate of 1.0 and m is the slope of the curve in Figure 18.5(b).

The effect of temperature on the parameters of Eq. (18.4) is pronounced. Increasing temperature decreases the value of C (consistent with its effect on K in the flow curve equation) and increases the value of m. The general result can be seen in Figure 18.6. At room temperature, the effect of strain rate is almost negligible, indicating that the flow curve is a good representation of the material behavior. As temperature is increased,

FIGURE 18.5   (a) Effect of strain rate on flow stress at an elevated work temperature. (b) Same relationship plotted on log–log coordinates.

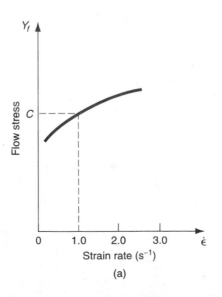

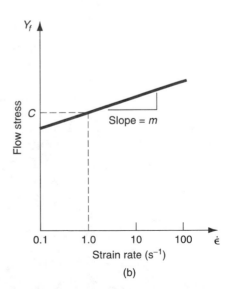

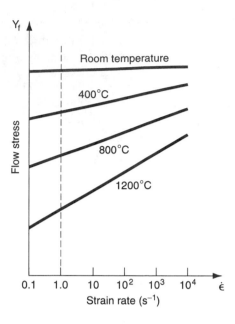

FIGURE 18.6 Effect of temperature on flow stress for a typical metal. The constant C in Eq. (18.4), indicated by the intersection of each plot with the vertical dashed line at strain rate = 1.0, decreases, and m (slope of each plot) increases with increasing temperature.

strain rate plays a more important role in determining flow stress, as indicated by the steeper slopes of the strain rate relationships. This is important in hot working because deformation resistance of the material increases so dramatically as strain rate is increased. To give a sense of the effect, typical values of m for the three temperature ranges of metal working are given in Table 18.1.

Thus, we see that even in cold working, strain rate can have an effect, if small, on flow stress. In hot working, the effect can be significant. A more complete expression for flow stress as a function of both strain and strain rate would be the following:

$$Y_f = A\epsilon^n \dot{\epsilon}^m \tag{18.5}$$

where A = a strength coefficient, combining the effects of the previous K and C values. Of course, A, n, and m would all be functions of temperature, and the enormous task of testing and compiling the values of these parameters for different metals and various temperatures would be forbidding.

In our coverage of the various bulk deformation processes in Chapter 19, many of which are performed hot, we neglect the effect of strain rate in analyzing forces and power. For cold working and warm working, and for hot working operations at relatively low deformation speeds, this neglect represents a reasonable assumption.

TABLE 18.1 Typical values of temperature, strain-rate sensitivity, and coefficient of friction in cold, warm, and hot working.

Category	Temperature Range	Strain-Rate Sensitivity Exponent	Coefficient Of Friction
Cold working	$\leq 0.3T_m$	$0 \leq m \leq 0.05$	0.1
Warm working	$0.3T_m - 0.5T_m$	$0.05 \leq m \leq 0.1$	0.2
Hot working	$0.5T_m - 0.75T_m$	$0.05 \leq m \leq 0.4$	$0.4 - 0.5$

18.5 FRICTION AND LUBRICATION IN METAL FORMING

Friction in metal forming arises because of the close contact between the tool and work surfaces and the high pressures that drive the surfaces together in these operations. In most metal-forming processes, friction is undesirable for the following reasons: (1) metal flow in the work is retarded, causing residual stresses and sometimes defects in the product; (2) forces and power to perform the operation are increased; and (3) tool wear can lead to loss of dimensional accuracy, resulting in defective parts and requiring replacement of the tooling. Since tools in metal forming are generally expensive, this is a major concern. Friction and tool wear are more severe in hot working because of the much harsher environment.

Friction in metal forming is different from that encountered in most mechanical systems, such as gear trains, shafts and bearings, and other components involving relative motion between surfaces. These other cases are generally characterized by low contact pressures, low to moderate temperatures, and ample lubrication to minimize metal-to-metal contact. By contrast, the metal-forming environment features high pressures between a hardened tool and a soft workpart, plastic deformation of the softer material, and high temperatures (at least in hot working). These conditions can result in relatively high coefficients of friction in metal working, even in the presence of lubricants. Typical values of coefficient of friction for the three categories of metal forming are listed in Table 18.1.

If the coefficient of friction becomes large enough, a condition known as sticking occurs. *Sticking* in metal working (also called *sticking friction*) is the tendency for the two surfaces in relative motion to adhere to each other rather than slide. It means that the friction stress between the surfaces exceeds the shear flow stress of the work metal, thus causing the metal to deform by a shear process beneath the surface rather than slip at the surface. Sticking occurs in metal forming operations and is a prominent problem in rolling; we discuss it in that context in the following chapter.

Metalworking lubricants are applied to the tool-work interface in many forming operations to reduce the harmful effects of friction. Benefits include reduced sticking, forces, power, and tool wear; and better surface finish on the product. Lubricants also serve other functions, such as removing heat from the tooling. Considerations in choosing an appropriate metalworking lubricant include: type of forming process (rolling, forging, sheet metal drawing, and so on); whether used in hot working or cold working; work material; chemical reactivity with the tool and work metals (it is generally desirable for the lubricant to adhere to the surfaces to be most effective in reducing friction); ease of application; toxicity; flammability; and cost.

Lubricants used for cold working operations include [3], [5]: mineral oils, fats and fatty oils, water-based emulsions, soaps, and other coatings. Hot working is sometimes performed dry for certain operations and materials (e.g., hot rolling of steel and extrusion of aluminum). When lubricants are used in hot working, they include mineral oils, graphite, and glass. Molten glass becomes an effective lubricant for hot extrusion of steel alloys. Graphite contained in water or mineral oil is a common lubricant for hot forging of various work materials. More detailed treatments of lubricants in metalworking are found in references [5] and [7].

REFERENCES

[1] Altan, T., Oh, S-I, and Gegel, H. L. *Metal Forming: Fundamentals and Applications.* ASM International, Materials Park, Ohio, 1983.

[2] Cook, N. H. *Manufacturing Analysis.* Addison-Wesley Publishing Company, Inc., Reading, Massachusetts, 1966.

[3] Lange, K., et al. (eds.). *Handbook of Metal Forming.* Society of Manufacturing Engineers, Dearborn, Michigan, 1995.

[4] Mielnik, E. M. *Metalworking Science and Engineering.* McGraw-Hill, Inc., New York, 1991.

[5] Nachtman, E. S., and Kalpakjian, S. *Lubricants and Lubrication in Metalworking Operations.* Marcel Dekker, Inc., New York, 1985.

[6] Wagoner, R. H., and Chenot, J-L. *Fundamentals of Metal Forming.* John Wiley & Sons, Inc., New York, 1997.

[7] Wick, C. et al. (eds.). *Tool and Manufacturing Engineers Handbook.* 4th ed. Volume II: *Forming.* Society of Manufacturing Engineers, Dearborn, Michigan, 1984.

REVIEW QUESTIONS

18.1. What are the characteristics that distinguish bulk deformation processes from sheet metal processes?

18.2. Extrusion is a fundamental shaping process. Describe it.

18.3. Why is the term *pressworking* often used for sheet metal processes?

18.4. What is the difference between deep drawing and bar drawing?

18.5. Indicate the mathematical equation for the flow curve.

18.6. How does increasing temperature affect the parameters in the flow curve equation?

18.7. Indicate some of the advantages of cold working relative to warm and hot working.

18.8. What is *isothermal forming*?

18.9. Describe the effect of strain rate in metal forming.

18.10. Why is friction generally undesirable in metal forming operations?

18.11. What is *sticking friction* in metalworking?

MULTIPLE CHOICE QUIZ

There is a total of 13 correct answers in the following multiple choice questions (some questions have multiple answers that are correct). To attain a perfect score on the quiz, all correct answers must be given, since each correct answer is worth 1 point. For each question, each omitted answer or wrong answer reduces the score by 1 point, and each additional answer beyond the number of answers required reduces the score by 1 point. Percentage score on the quiz is based on the total number of correct answers.

18.1. Which of the following are bulk deformation processes (more than one)? (a) bending, (b) deep drawing, (c) extrusion, (d) forging, and (e) rolling.

18.2. Which of the following is typical of the work geometry in sheet metal processes? (a) high volume-to-area ratio, or (b) low volume-to-area ratio.

18.3. The flow curve expresses the behavior of a metal in which of the following regions of the stress–strain curve? (a) elastic region or (b) plastic region.

18.4. The average flow stress is the flow stress multiplied by which of the following factors? (a) n, (b) $(1 + n)$, (c) $1/n$, or (d) $1/(1 + n)$, where n is the strain-hardening exponent.

18.5. Hot working of metals refers to which of the following temperature regions relative to the melting point of the given metal on an absolute temperature scale? (a) room temperature, (b) $0.2T_m$, (c) $0.4T_m$, or (d) $0.6T_m$.

18.6. Which of the following are advantages and characteristics of hot working relative to cold working (more than one answer)? (a) fracture of workpart less likely, (b) increased strength properties, (c) isotropic mechanical properties, (d) less overall energy required, (e) lower deformation forces required, and (f) more significant shape changes are possible.

18.7. Increasing strain rate tends to have which of the following effects on flow stress during hot forming of metal? (a) decreases flow stress, (b) has no effect, or (c) increases flow stress.

18.8. The coefficient of friction between the part and the tool in cold working tends to be which of the following relative to its value in hot working? (a) higher, (b) lower, or (c) no effect.

PROBLEMS

Flow Curve in Forming

18.1. $K = 600$ MPa and $n = 0.20$ for a certain metal. During a forming operation, the final true strain that the metal experiences = 0.73. Determine the flow stress at this strain and the average flow stress that the metal experienced during the operation.

18.2. A metal has a flow curve with parameters: $K = 850$ MPa and strain hardening exponent $n = 0.30$. A tensile specimen of the metal with gage length = 100 mm is stretched to a length = 157 mm. Determine the flow stress at the new length and the average flow stress that the metal has been subjected to during the deformation.

18.3. A particular metal has a flow curve with parameters: strength coefficient $K = 35{,}000$ lb/in.^{2} and strain-hardening exponent $n = 0.26$. A tensile specimen of the metal with gage length = 2.0 in. is stretched to a length of 3.3 in. Determine the flow stress at this new length and the average flow stress that the metal has been subjected to during deformation.

18.4. The strength coefficient and strain-hardening exponent of a certain test metal are $K = 40{,}000$ lb/in.^{2} and $n = 0.19$. A cylindrical specimen of the metal with starting diameter = 2.5 in. and length = 3.0 in. is compressed to a length of 1.5 in. Determine the flow stress at this compressed length and the average flow stress that the metal has experienced during deformation.

18.5. Derive the equation for average flow stress, Eq. (18.2) in the text.

18.6. For a certain metal, $K = 700$ MPa and $n = 0.27$. Determine the average flow stress that the metal experiences if it is subjected to a stress that is equal to its strength coefficient K.

18.7. Determine the value of the strain-hardening exponent for a metal that will cause the average flow stress to be three-fourths of the final flow stress after deformation.

18.8. $K = 35{,}000$ lb/in.^{2} and $n = 0.40$ for a metal used in a forming operation in which the workpart is reduced in cross-sectional area by stretching. If the average flow stress on the part is 20,000 lb/in.^{2}, determine the amount of reduction in cross-sectional area experienced by the part.

Strain Rate

18.9. The gage length of a tensile test specimen = 150 mm. It is subjected to a tensile test in which the grips holding the end of the test specimen are moved with a relative velocity = 0.1 m/s. Construct a plot of the strain rate as a function of length as the specimen is pulled to a length = 200 mm.

18.10. A specimen with 6.0 in. starting gage length is subjected to a tensile test in which the grips holding the end of the test specimen are moved with a relative velocity = 1.0 in./sec. Construct a plot of the strain rate as a function of length as the specimen is pulled to a length = 8.0 in.

18.11. A workpart with starting height $h = 100$ mm is compressed to a final height of 50 mm. During the deformation, the relative speed of the plattens compressing the part = 200 mm/s. Determine the strain rate at (a) $h = 100$ mm, (b) $h = 75$ mm, and (c) $h = 51$ mm.

18.12. A hot working operation is carried out at various speeds. The strength constant $C = 30{,}000$ lb/in.^{2} and the strain-rate sensitivity exponent $m = 0.15$. Determine the flow stress if the strain rate is (a) 0.01/sec (b) 1.0/sec, (c) 100/sec.

18.13. A tensile test is performed to determine the parameters C and m in Eq. (18.4) for a certain metal. The temperature at which the test is performed = 500°C. At a strain rate = 12/s, the stress is measured at 160 MPa; and at a strain rate = 250/sec, the stress = 300 MPa. (a) Determine C and m. (b) If the temperature were 600°C, what changes would you expect in the values of C and m?

18.14. A tensile test is carried out to determine the strength constant C and strain-rate sensitivity exponent m for a certain metal at 1000°F. At a strain rate = 10/sec, the stress is measured at 23,000 lb/in.^{2}; and at a strain rate = 300/sec, the stress = 45,000 lb/in.^{2} (a) Determine C and m. (b) If the temperature were 900°F, what changes would you expect in the values of C and m?

19 BULK DEFORMATION PROCESSES IN METALWORKING

CHAPTER CONTENTS

19.1   Rolling
    19.1.1   Flat Rolling and Its Analysis
    19.1.2   Shape Rolling
    19.1.3   Rolling Mills
19.2   Other Deformation Processes Related to Rolling
19.3   Forging
    19.3.1   Open-die Forging
    19.3.2   Impression-die Forging
    19.3.3   Closed-die Forging
    19.3.4   Forging Dies, Hammers, and Presses
19.4   Other Deformation Processes Related to Forging
19.5   Extrusion
    19.5.1   Types of Extrusion
    19.5.2   Analysis of Extrusion
    19.5.3   Dies and Presses for Extrusion
    19.5.4   Other Extrusion Processes
    19.5.5   Defects in Extruded Products
19.6   Wire and Bar Drawing
    19.6.1   Analysis of Drawing
    19.6.2   Drawing Practice
    19.6.3   Tube Drawing

The deformation processes described in this chapter accomplish significant shape change in metal parts whose initial form is bulk rather than sheet. The starting forms include cylindrical bars and billets, rectangular billets and slabs, and similar elementary shapes. The bulk deformation processes refine the raw shapes, often adding geometric features, sometimes improving mechanical properties, and always adding commercial value. Deformation processes work by stressing the metal sufficiently to cause it to plastically flow into the desired shape.

Bulk deformation processes are performed as cold, warm, and hot working operations. Cold and warm working is appropriate when the shape change is less severe, and there is a need to improve mechanical properties and achieve good finish on the part. Hot working is generally required when massive deformation of large workparts is involved.

The commercial and technological importance of bulk deformation processes derives from the following:

> ➤ When performed as hot working operations, they can achieve significant change in shape of the workpart.
> ➤ When performed as cold working operations, they can be used not only to shape the product, but also to increase its strength.
> ➤ These processes produce little or no waste as a by-product of the operation. Some bulk deformation operations are *near net shape* or *net shape* processes; they achieve final product geometry with little or no subsequent machining.

The bulk deformation processes covered in this chapter are (1) rolling, (2) forging, (3) extrusion, and (4) wire and bar drawing. The chapter also documents the variations and related operations of the four basic processes that have been developed over the years.

19.1 ROLLING

Rolling is a deformation process in which the thickness of the work is reduced by compressive forces exerted by two opposing rolls. The rolls rotate as illustrated in Figure 19.1 to pull and simultaneously squeeze the work between them. The basic process shown in our figure is flat rolling, used to reduce the thickness of a rectangular cross section. A closely related process is shape rolling, in which a square cross section is formed into a shape such as an I-beam.

Most rolling processes are very capital intensive, requiring massive pieces of equipment, called rolling mills, to perform them. The high investment cost requires the mills to be used for production in large quantities of standard items such as sheets and plates. Most rolling is carried out by hot working, called *hot rolling,* owing to the large amount of deformation required. Hot-rolled metal is generally free of residual stresses, and its properties are isotropic. Disadvantages of hot rolling are that the product cannot be held to close tolerances, and the surface has a characteristic oxide scale.

FIGURE 19.1   The rolling process (specifically, flat rolling).

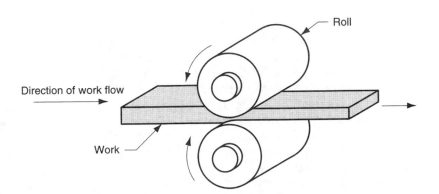

Steelmaking provides the most common application of rolling mill operations (Historical Note 19.1). Let us follow the sequence of steps in a steel rolling mill to illustrate the variety of products made. Similar steps occur in other basic metal industries. The work starts out as a cast steel ingot that has just solidified. While it is still hot, the ingot is placed in a furnace where it remains for many hours until it has reached a uniform temperature throughout, so that the metal will flow consistently during rolling. For steel, the desired temperature for rolling is around 1200°C (2200°F). The heating operation is called *soaking,* and the furnaces in which it is carried out are called *soaking pits.*

Historical Note 19.1   *Rolling*

Rolling of gold and silver by manual methods dates from the fourteenth century. Leonardo da Vinci designed one of the first rolling mills in 1480, but it is doubtful that his design was ever built. By around 1600, cold rolling of lead and tin was accomplished on manually operated rolling mills. By around 1700, hot rolling of iron was being done in Belgium, England, France, Germany, and Sweden. These mills were used to roll iron bars into sheets. Prior to this time, the only rolls in steelmaking were slitting mills—pairs of opposing rolls with collars (cutting disks) used to slit iron and steel into narrow strips for making nails and similar products. Slitting mills were not intended to reduce thickness.

Modern rolling practice dates from 1783 when a patent was issued in England for using grooved rolls to produce iron bars. The industrial revolution created a tremendous demand for iron and steel, stimulating developments in rolling. The first mill for rolling railway rails was started in 1820 in England. The first I-beams were rolled in France in 1849. In addition, the size and capacity of flat rolling mills increased dramatically during this period.

Rolling is a process that requires a very large power source. Water wheels were used to power rolling mills until the eighteenth century. Steam engines increased the capacity of these rolling mills until soon after 1900 when electric motors replaced steam.

From soaking, the ingot is moved to the rolling mill, where it is rolled into one of three intermediate shapes called blooms, billets, or slabs. A *bloom* has a square cross-section 150 mm × 150 mm (6 × 6 in.) or larger. A *slab* is rolled from an ingot or a bloom and has a rectangular cross-section of width 250 mm (10 in.) or more and thickness 40 mm (1.5 in.) or more. A *billet* is rolled from a bloom and is square with dimensions 40 mm (1.5 in.) on a side or larger. These intermediate shapes are subsequently rolled into final product shapes.

Blooms are rolled into structural shapes and rails for railroad tracks. Billets are rolled into bars and rods. These shapes are the raw materials for machining, wire drawing, forging, and other metalworking processes. Slabs are rolled into plates, sheets, and strips. Hot-rolled plates are used in shipbuilding, bridges, boilers, welded structures for various heavy machines, tubes and pipes, and many other products. Figure 19.2 shows some of these rolled steel products. Further flattening of hot-rolled plates and sheets is often accomplished by *cold rolling,* in order to prepare them for subsequent sheet metal operations (Chapter 20). Cold rolling strengthens the metal and permits a tighter tolerance on thickness. In addition, the surface of the cold-rolled sheet is absent of scale and generally superior to the corresponding hot-rolled product. These characteristics make cold-rolled sheets, strips, and coils ideal for stampings, exterior panels, and other parts of products ranging from automobiles to appliances and office furniture.

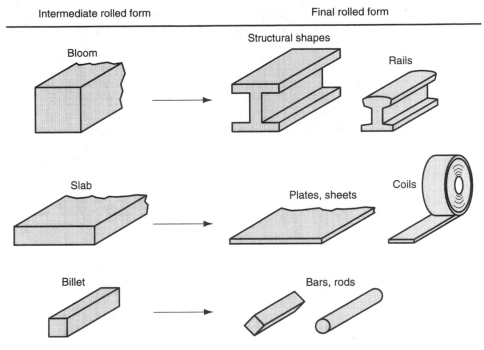

Intermediate rolled form   Final rolled form

FIGURE 19.2   Some of the steel products made in a rolling mill.

19.1.1 Flat Rolling and Its Analysis

Flat rolling is illustrated in Figures 19.1 and 19.3. It involves the rolling of slabs, strips, sheets, and plates—workparts of rectangular cross-section in which the width is greater than the thickness. In flat rolling, the work is squeezed between two rolls so that its thickness is reduced by an amount called the **draft**:

$$d = t_o - t_f \tag{19.1}$$

FIGURE 19.3   Side view of flat rolling, indicating before and after thicknesses, work velocities, angle of contact with rolls, and other features.

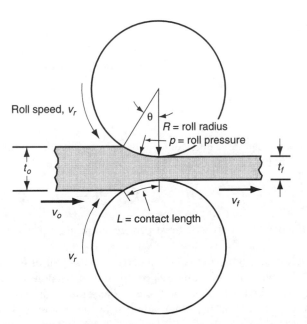

where d = draft, mm (in.); t_o = starting thickness, mm (in.); and t_f = final thickness, mm (in.). Draft is sometimes expressed as a fraction of the starting stock thickness, called the **reduction:**

$$r = \frac{d}{t_o} \tag{19.2}$$

where r = reduction. When a series of rolling operations are used, reduction is taken as the sum of the drafts divided by the original thickness.

In addition to thickness reduction, rolling usually increases work width. This is called **spreading,** and it tends to be most pronounced with low width-to-thickness ratios and low coefficients of friction. Conservation of material is preserved, so the volume of metal exiting the rolls equals the volume entering:

$$t_o w_o L_o = t_f w_f L_f \tag{19.3}$$

where w_o and w_f are the before and after work widths, mm (in.); and L_o and L_f are the before and after work lengths, mm (in.). Similarly, before and after volume rates of material flow must be the same, so the before and after velocities can be related:

$$t_o w_o v_o = t_f w_f v_f \tag{19.4}$$

where v_o and v_f are the entering and exiting velocities of the work.

The rolls contact the work along a contact arc defined by the angle θ. Each roll has radius R, and its rotational speed gives it a surface velocity v_r. This velocity is greater than the entering speed of the work v_o and less than its exiting speed v_f. Since the metal flow is continuous, there is a gradual change in velocity of the work between the rolls. However, there is one point along the arc where work velocity equals roll velocity. This is called the **no-slip point,** also known as the **neutral point.** On either side of this point, slipping and friction occur between roll and work. The amount of slip between the rolls and the work can be measured by means of the **forward slip,** a term used in rolling that is defined:

$$s = \frac{v_f - v_r}{v_r} \tag{19.5}$$

where s = forward slip; v_f = final (exiting) work velocity, m/s (ft/sec); and v_r = roll speed, m/s (ft/sec).

The true strain experienced by the work in rolling is based on before and after stock thicknesses. In equation form,

$$\epsilon = \ln \frac{t_o}{t_f} \tag{19.6}$$

The true strain can be used to determine the average flow stress $\overline{Y}_f$ applied to the work material in flat rolling. Recall from the previous chapter, Eq. (18.2), that

$$\overline{Y}_f = \frac{K \epsilon^n}{1 + n} \tag{19.7}$$

The average flow stress will be useful to compute estimates of force and power in rolling.

Friction in rolling occurs with a certain coefficient of friction, and the compression force of the rolls, multiplied by this coefficient of friction, results in a friction force between the rolls and the work. On the entrance side of the no-slip point, friction force is in one direction, and on the other side it is in the opposite direction. However, the two forces are not equal. The friction force on the entrance side is greater, so that the net force pulls the work through the rolls. If this were not the case, rolling would not be possible. There is a limit to the maximum possible draft that can be accomplished in flat rolling with a given coefficient of friction, given by:

$$d_{max} = \mu^2 R \tag{19.8}$$

where d_{max} = maximum draft, mm (in.); μ = coefficient of friction; and R = roll radius mm (in.). The equation indicates that if friction were zero, draft would be zero, and it would be impossible to accomplish the rolling operation.

Coefficient of friction in rolling depends on lubrication, work material, and working temperature. In cold rolling, the value is around 0.1; in warm working, a typical value is around 0.2; and in hot rolling, μ is around 0.4 [15]. Hot rolling is often characterized by a condition called *sticking,* in which the hot work surface adheres to the rolls over the contact arc. This condition often occurs in the rolling of steels and high-temperature alloys. When sticking occurs, the coefficient of friction can be as high as 0.7. The consequence of sticking is that the surface layers of the work are restricted to move at the same speed as the roll speed v_r; and below the surface, deformation is more severe in order to allow passage of the piece through the roll gap.

Given a coefficient of friction sufficient to perform rolling, roll force F required to maintain separation between the two rolls can be computed by integrating the unit roll pressure (shown as p in Figure 19.3) over the roll-work contact area. This can be expressed:

$$F = w \int_0^L p \, dL \tag{19.9}$$

where F = rolling force, N (lb); w = the width of the work being rolled, mm (in.); p = roll pressure, MPa (lb/in.^{2}); and L = length of contact between rolls and work, mm (in.). The integration requires two separate terms, one for either side of the neutral point. Variation in roll pressure along the contact length is significant. A sense of this variation can be obtained from the plot in Figure 19.4. Pressure reaches a maximum at the neutral point, and trails off on either side to the entrance and exit points. As friction increases, maximum pressure increases relative to entrance and exit values. As friction decreases,

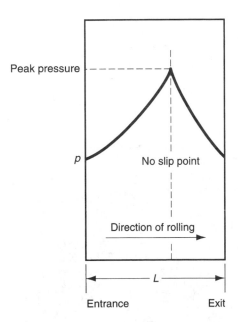

FIGURE 19.4 Typical variation in pressure along the contact length in flat rolling. The peak pressure is located at the neutral point. The area beneath the curve, representing the integration in Eq. (19.9), is the roll force F.

the neutral point shifts away from the entrance and toward the exit in order to maintain a net pull force in the direction of rolling. Otherwise, with low friction, the work would slip rather than passing between the rolls.

An approximation of the results obtained by Eq. (19.9) can be calculated based on the average flow stress experienced by the work material in the roll gap. That is,

$$F = \overline{Y}_f wL \qquad (19.10)$$

where $\overline{Y}_f$ = average flow stress from Eq. (19.7), MPa (lb/in.^{2}); and the product wL is the roll-work contact area, mm^2 (in.^{2}). Contact length can be approximated by

$$L = \sqrt{R(t_o - t_f)} \qquad (19.11)$$

The torque in rolling can be estimated by assuming that the roll force is centered on the work as it passes between the rolls, and that it acts with a moment arm of one-half the contact length L. Thus, torque for each roll is

$$T = 0.5FL \qquad (19.12)$$

The power required to drive each roll is the product of torque and angular velocity. Angular velocity is $2N$, where N = rotational speed of the roll. Thus, the power for each roll is $2NT$. Substituting Eq. (19.12) for torque in this expression for power, and doubling the value to account for the fact that a rolling mill consists of two powered rolls, we get the following expression:

$$P = 2\pi NFL \qquad (19.13)$$

where P = power, J/s or W (in.-lb/min); N = rotational speed, 1/s (rev/min); F = rolling force, N (lb); and L = contact length, m (in.).

EXAMPLE 19.1
Flat rolling.

A 300-mm-wide strip 25 mm thick is fed through a rolling mill with two powered rolls each of radius = 250 mm. The work thickness is to be reduced to 22 mm in one pass at a roll speed of 50 rev/min. The work material has a flow curve defined by $K = 275$ MPa and $n = 0.15$, and the coefficient of friction between the rolls and the work is assumed to be 0.12. Determine if the friction is sufficient to permit the rolling operation to be accomplished. If so, calculate the roll force, torque, and horsepower.

Solution: The draft attempted in this rolling operation is

$$d = 25 - 22 = 3 \text{ mm}$$

From Eq. (19.8), the maximum possible draft for the given coefficient of friction is

$$d_{\max} = (0.12)^2(250) = 3.6 \text{ mm}$$

Since the maximum allowable draft exceeds the attempted reduction, the rolling operation is feasible. To compute rolling force, we need the contact length L and the average flow stress $\overline{Y}_f$. The contact length is given by Eq. (19.11):

$$L = \sqrt{250(25 - 22)} = 27.4 \text{ mm}$$

$\overline{Y}_f$ is determined from the true strain:

$$\epsilon = \ln \frac{25}{22} = 0.128$$

$$\overline{Y}_f = \frac{275(0.128)^{0.15}}{1.15} = 175.7 \text{ MPa}$$

Rolling force is determined from Eq. (19.10):

$$F = 175.7(300)(27.4) = 1{,}444{,}786 \text{ N}$$

Torque required to drive each roll is given by Eq. (19.12):

$$T = 0.5(1{,}444{,}786)(27.4)(10^{-3}) = 19{,}786 \text{ N-m}$$

and the power is obtained from Eq. (19.13):

$$P = 2\pi(50)(1{,}444{,}786)(27.4)(10^{-3}) = 12{,}432{,}086 \text{ N-m/min} = 207{,}201 \text{ N-m/s } (W)$$

For comparison, let us convert this to horsepower (we note that one horsepower = 745.7 W):

$$HP = \frac{207{,}201}{745.7} = 278 \text{ hp}$$

It can be seen from this example that large forces and power are required in rolling. Inspection of Eqs. (19.10) and (19.13) indicates that force and/or power to roll a strip of a given width and work material can be reduced by any of the following: (1) using hot rolling rather than cold rolling to reduce strength and strain hardening (K and n) of the work material; (2) reducing the draft in each pass; (3) using a smaller roll radius R to reduce force; and (4) using a lower rolling speed N to reduce power.

19.1.2 Shape Rolling

In shape rolling, the work is deformed into a contoured cross section. Products made by shape rolling include construction shapes such as I-beams, L-beams, and U-channels; rails for railroad tracks; and round and square bars and rods (see Figure 19.2). The process is accomplished by passing the work through rolls that have the reverse of the desired shape.

Most of the principles that apply in flat rolling are also applicable to shape rolling. Shaping rolls are more complicated; and the work, usually starting as a square shape, requires a gradual transformation through several rolls in order to achieve the final cross section. Designing the sequence of intermediate shapes and corresponding rolls is called **roll-pass design.** Its goal is to achieve uniform deformation throughout the cross section in each reduction. Otherwise, certain portions of the work are reduced more than others, causing greater elongation in these sections. The consequence of nonuniform reduction can be warping and cracking of the rolled product. Both horizontal and vertical rolls are utilized to achieve consistent reduction of the work material.

19.1.3 Rolling Mills

Various rolling-mill configurations are available to deal with the variety of applications and technical problems in the rolling process. The basic rolling mill consists of two opposing rolls and is referred to as a **two-high** rolling mill, shown in Figures 19.5 and 19.6(a). The rolls in these mills have diameters in the range 0.6–1.4 m (2.0–4.5 ft). The two-high configuration can be either reversing or nonreversing. In the **nonreversing mill,** the rolls always rotate in the same direction, and the work always passes through from the same side. The **reversing mill** allows the direction of roll rotation to be reversed, so that the work can be passed through in either direction. This permits a series of reductions to be made through the same set of rolls, simply by passing through the work from opposite directions multiple times. The disadvantage of the reversing configuration is the significant angular momentum possessed by large rotating rolls and the associated technical problems involved in reversing the direction.

FIGURE 19.5   A rolling mill for hot flat rolling; the steel plate is seen as the glowing strip extending diagonally from the lower-left corner (photo courtesy of Bethlehem Steel Company).

Several alternative arrangements are illustrated in Figure 19.6. In the ***three-high*** configuration, Figure 19.6(b), there are three rolls in a vertical column, and the direction of rotation of each roll remains unchanged. To achieve a series of reductions, the work can be passed through from either side by raising or lowering the strip after each pass. The equipment in a three-high rolling mill becomes more complicated, because an elevator mechanism is needed to raise and lower the work.

As several of the previous equations indicate, advantages are gained in reducing roll diameter. Roll-work contact length is reduced with a lower roll radius, and this leads to lower forces, torque, and power. The ***four-high*** rolling mill uses two smaller-diameter rolls to contact the work and two backing rolls behind them, as in Figure 19.6(c). Owing to the high roll forces, these smaller rolls would deflect elastically between their end bearings as the work passes through unless the larger backing rolls were used to support them. Another roll configuration that allows smaller working rolls against the work is the ***cluster rolling mill,*** Figure 19.6(d).

To achieve higher throughput rates in standard products, a ***tandem rolling mill*** is often used. This configuration consists of a series of rolling stands, as represented in Figure 19.6(e). Although only three stands are shown in our sketch, a typical tandem rolling mill may have eight or ten stands, each making a reduction in thickness or a refinement in the

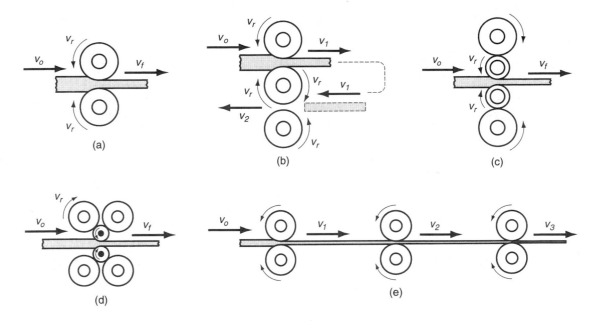

FIGURE 19.6   Various configurations of rolling mills: (a) two-high, (b) three-high, (c) four-high, (d) cluster mill, and (e) tandem rolling mill.

shape of the work passing through. With each rolling step, work velocity increases, and the problem of synchronizing the roll speeds at each stand is a significant one.

Modern tandem rolling mills are often supplied directly by continuous casting operations (Section 7.2.2). These setups achieve a high degree of integration among the processes required to transform starting raw materials into finished products. Advantages include elimination of soaking pits, reduction in floor space, and shorter manufacturing lead times. These technical advantages translate into economic benefits for a mill that can accomplish continuous casting and rolling.

19.2 OTHER DEFORMATION PROCESSES RELATED TO ROLLING

Several other bulk deformation processes use rolls to form the workpart. The operations include thread rolling, ring rolling, gear rolling, and roll piercing.

Thread Rolling   Thread rolling is used to form threads on cylindrical parts by rolling them between two dies. It is the most important commercial process for mass producing external threaded components (e.g., bolts and screws). The competing process is thread cutting (Section 22.1.2). Most thread-rolling operations are performed by cold working in thread-rolling machines. These machines are equipped with special dies that determine the size and form of the thread. The dies are of two types: (1) flat dies, which reciprocate relative to each other, as illustrated in Figure 19.7; and (2) round dies, which rotate relative to each other to accomplish the rolling action.

Production rates in thread rolling can be high, ranging up to eight parts per second for small bolts and screws. Not only are these rates significantly higher than thread cutting, but there are other advantages over machining as well: (1) better material utilization, (2) stronger threads due to work hardening, (3) smoother surface, and (4) better fatigue resistance due to compressive stresses introduced by rolling.

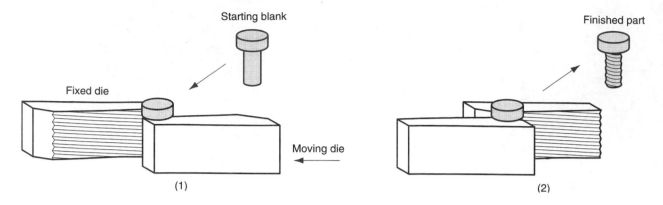

FIGURE 19.7   Thread rolling with flat dies: (1) start of cycle, and (2) end of cycle.

Ring Rolling   Ring rolling is a deformation process in which a thick-walled ring of smaller diameter is rolled into a thin-walled ring of larger diameter. The before and after views of the process are illustrated in Figure 19.8. As the thick-walled ring is compressed, the deformed material elongates, causing the diameter of the ring to be enlarged. Ring rolling is usually performed as a hot-working process for large rings and as a cold-working process for smaller rings.

Applications of ring rolling include ball and roller bearing races, steel tires for railroad wheels, and rings for pipes, pressure vessels, and rotating machinery. The ring walls are not limited to rectangular cross-sections; the process permits rolling of more complex shapes. There are several advantages of ring rolling over alternative methods of making the same parts: raw material savings, ideal grain orientation for the application, and strengthening through cold working.

Gear Rolling   Gear rolling is a cold working process to produce certain gears. The automotive industry is an important user of these products. The setup in gear rolling is similar to thread rolling, except that the deformed features of the cylindrical blank or disk are oriented parallel to its axis (or at an angle in the case of helical gears) rather than spiraled as in thread rolling. Advantages of gear rolling compared to machining are similar to the advantages in thread rolling: higher production rates, better strength and fatigue resistance, and less material waste.

Roll Piercing   Roll piercing is a specialized hot working process for making seamless thick-walled tubes. It utilizes two opposing rolls, and hence it is grouped with the rolling

FIGURE 19.8   Ring rolling used to reduce the wall thickness and increase the diameter of a ring: (1) start and (2) completion of process.

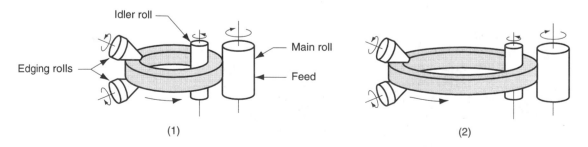

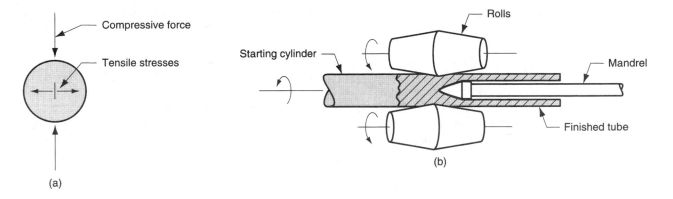

FIGURE 19.9   Roll piercing: (a) formation of internal stresses and cavity by compression of cylindrical part; and (b) setup of Mannesmann roll mill for producing seamless tubing.

processes. The process is based on the principle that when a solid cylindrical part is compressed on its circumference, as in Figure 19.9(a), high tensile stresses are developed at its center. If compression is high enough, an internal crack is formed. In roll piercing, this principle is exploited by the setup shown in Figure 19.9(b). Compressive stresses on a solid cylindrical billet are applied by two rolls, whose axes are oriented at slight angles (about 6°) from the axis of the billet, so that their rotation tends to pull the billet through the rolls. A mandrel is used to control the size and finish of the hole created by the action. The terms *rotary tube piercing* and *Mannesmann process* are also used for this tube-making operation.

19.3 FORGING

Forging is a deformation process in which the work is compressed between two dies, using either impact or gradual pressure to form the part. It is the oldest of the metal forming operations, dating back to perhaps 5000 B.C. (Historical Note 19.2). Today, forging is an important industrial process used to make a variety of high-strength components for automotive, aerospace, and other applications. These components include engine crankshafts and connecting rods, gears, aircraft structural components, and jet engine turbine parts. In addition, steel and other basic metals industries use forging to establish the basic form of large components that are subsequently machined to final shape and dimensions.

Historical Note 19.2   *Forging.*

The forging process dates from the earliest written records of man, around 7,000 years ago. There is evidence that forging was used in ancient Egypt, Greece, Persia, India, China, and Japan to make weapons, jewelry, and a variety of implements. Craftsmen in the art of forging during these times were held in high regard.

Engraved stone platens were used as impression dies in the hammering of gold and silver in ancient Crete around 1600 B.C. This evolved into the fabrication of coins by a similar process around 800 B.C. More complicated impression dies were used in Rome around 200 A.D. The blacksmith's trade remained relatively unchanged for many centuries until the drop hammer with guided ram was introduced near the end of the eighteenth century. This development brought forging practice into the Industrial Age.

Forging is carried out in many different ways. One of the ways to classify forging is by working temperature. Most forging operations are performed hot or warm, owing to the significant deformation demanded by the process and the need to reduce strength and increase ductility of the work metal. However, cold forging is also very common for certain products. The advantage of cold forging is the increased strength that results from strain hardening of the component.

Either impact or gradual pressure is used in forging. The distinction derives more from the type of equipment used than differences in process technology. A forging machine that applies an impact load is called a *forging hammer,* while one that applies gradual pressure is called a *forging press.*

Another difference among forging operations is the degree to which the flow of the work metal is constrained by the dies. By this classification, there are three types of forging operations, shown in Figure 19.10: (a) open-die forging, (b) impression-die forging, and (c) flashless forging. In *open-die forging,* the work is compressed between two flat (or almost flat) dies, thus allowing the metal to flow without constraint in a lateral direction relative to the die surfaces. In *impression-die forging,* the die surfaces contain a shape or impression that is imparted to the work during compression, thus constraining metal flow to a significant degree. In this type of operation, a portion of the work metal flows beyond the die impression to form *flash,* as shown in the figure. Flash is excess metal that must be trimmed off later. In *flashless forging,* the work is completely constrained within the die and no excess flash is produced. The volume of the starting workpiece must be controlled very closely so that it matches the volume of the die cavity.

FIGURE 19.10 Three types of forging operation illustrated by cross-sectional sketches: (a) open-die forging, (b) impression-die forging, and (c) flashless forging.

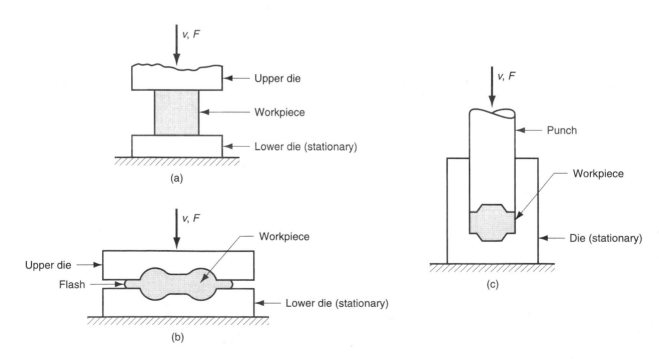

19.3.1 Open-Die Forging

The simplest case of open-die forging involves compression of a workpart of cylindrical cross section between two flat dies, much in the manner of a compression test (Section 3.1.2). This forging operation, known as **upsetting** or **upset forging,** reduces the height of the work and increases its diameter.

Analysis of Open-Die Forging   If open-die forging is carried out under ideal conditions of no friction between work and die surfaces, then homogeneous deformation occurs, and the radial flow of the material is uniform throughout its height, as pictured in Figure 19.11. Under these ideal conditions, the true strain experienced by the work during the process can be determined by

$$\epsilon = \ln \frac{h_o}{h} \tag{19.14}$$

where h_o = starting height of the work, mm (in.); and h = the height at some intermediate point in the process, mm (in.). At the end of the compression stroke, h = its final value h_f, and the true strain reaches its maximum value.

Estimates of force to perform upsetting can be calculated. The force required to continue the compression at any given height h during the process can be obtained by multiplying the corresponding cross-sectional area by the flow stress:

$$F = Y_f A \tag{19.15}$$

where F = force, lb (N); A = cross-sectional area of the part, mm^2 (in.^{2}); and Y_f = flow stress corresponding to the strain given by Eq. (19.14), MPa (lb/in.^{2}). Area A continuously increases during the operation as height is reduced. Flow stress Y_f also increases as a result of work hardening, except when the metal is perfectly plastic (e.g., in hot working). In this case, the strain-hardening exponent $n = 0$, and flow stress Y_f equals the metal's yield strength Y. Force reaches a maximum value at the end of the forging stroke, when both area and flow stress are at their highest values.

An actual upsetting operation does not occur quite as shown in Figure 19.11 because friction opposes the flow of work metal at the die surfaces. This creates the barreling effect shown in Figure 19.12. When performed on a hot workpart with cold dies, the barreling effect is even more pronounced. This results from a higher coefficient of friction typical in hot working and heat transfer at and near the die surfaces, which cools the metal and increases its resistance to deformation. The hotter metal in the middle of

FIGURE 19.11
Homogeneous deformation of a cylindrical workpart under ideal conditions in an open-die forging operation: (1) start of process with workpiece at its original length and diameter, (2) partial compression, and (3) final size.

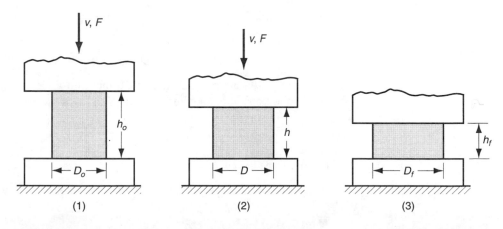

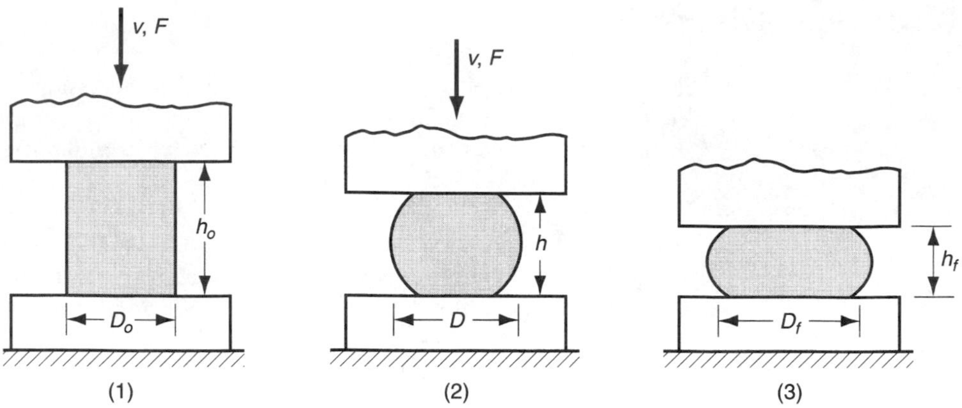

FIGURE 19.12   Actual deformation of a cylindrical workpart in open-die forging, showing pronounced barreling: (1) start of process, (2) partial deformation, and (3) final shape.

the part flows more readily than the cooler metal at the ends. These effects are more significant as the diameter-to-height ratio of the workpart increases, due to the greater contact area at the work-die interface.

All of these factors cause the actual upsetting force to be greater than what is predicted by Eq. (19.15). As an approximation, we can apply a shape factor to Eq. (19.15) to account for effects of the D/h ratio and friction:

$$F = K_f Y_f A \tag{19.16}$$

where F, Y_f, and A have the same definitions as in the previous equation; and K_f is the forging shape factor, defined as:

$$K_f = 1 + \frac{0.4\,\mu D}{h} \tag{19.17}$$

where μ = coefficient of friction; D = workpart diameter or other dimension representing contact length with die surface, mm (in.); and h = workpart height, mm (in.).

**EXAMPLE 19.2
Open-die forging.**

A cylindrical workpiece is subjected to a cold upset forging operation. The starting piece is 75 mm in height and 50 mm in diameter. It is reduced in the operation to a height of 36 mm. The work material has a flow curve defined by $K = 350$ MPa and $n = 0.17$. Assume a coefficient of friction of 0.1. Determine the force as the process begins, at intermediate heights of 62 mm, 49 mm, and at the final height of 36 mm.

Solution: Workpiece volume $V = 75\pi(50^2/4) = 147{,}262$ mm^3. At the moment contact is made by the upper die, $h = 75$ mm and the force $F = 0$. At the start of yielding, h is slightly less than 75 mm, and we assume that strain = 0.002, at which the flow stress is

$$Y_f = K\epsilon^n = 350(0.002)^{0.17} = 121.7 \text{ MPa}$$

The diameter is still approximately $D = 50$ mm and area $A = \pi(50^2/4) = 1963.5$ mm^2. For these conditions, the adjustment factor K_f is computed as

$$K_f = 1 + \frac{0.4(0.1)(50)}{75} = 1.027$$

The forging force is

$$F = 1.027(121.7)(1963.5) = 245{,}410 \text{ MPa}$$

At h = 62 mm,

$$\epsilon = \ln\frac{75}{62} = \ln(1.21) = 0.1904$$

$$Y_f = 350(0.1904)^{.17} = 264.0 \text{ MPa}$$

Assuming constant volume, and neglecting barreling,

$$A = 147{,}262/62 = 2375.2 \text{ mm}^2 \text{ and } D = 55.0 \text{ mm}$$

$$K_f = 1 + \frac{0.4(0.1)(55)}{62} = 1.035$$

$$F = 1.035(264)(2375.2) = 649{,}303 \text{ N}$$

Similarly, at h = 49 mm, F = 955,642 N. And at h = 36 mm, F = 1,467,422 N.
The load–stroke curve in Figure 19.13 was developed from the values in this example.

Open-Die Forging Practice   Open-die hot forging is an important industrial process. Shapes generated by open-die operations are simple; examples include shafts, disks, and rings. In some applications, the dies have slightly contoured surfaces that help to shape the work. In addition, the work must often be manipulated (e.g., rotating in steps) to effect the desired shape change. Skill of the human operator is a factor in the success of these operations. An example of open-die forging in the steel industry is the shaping of a large square cast ingot into a round cross section. Open-die forging operations produce rough forms, and subsequent operations are required to refine the parts to final geometry and dimensions. An important contribution of open-die hot-forging is that it creates a favorable grain flow and metallurgical structure in the metal.

Operations classified as open-die forging or related operations include fullering, edging, and cogging, illustrated in Figure 19.14. *Fullering* is a forging operation performed

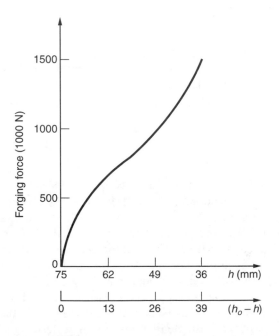

FIGURE 19.13   Upsetting force as a function of height h and height reduction $(h_o - h)$. This plot is sometimes called the load–stroke curve.

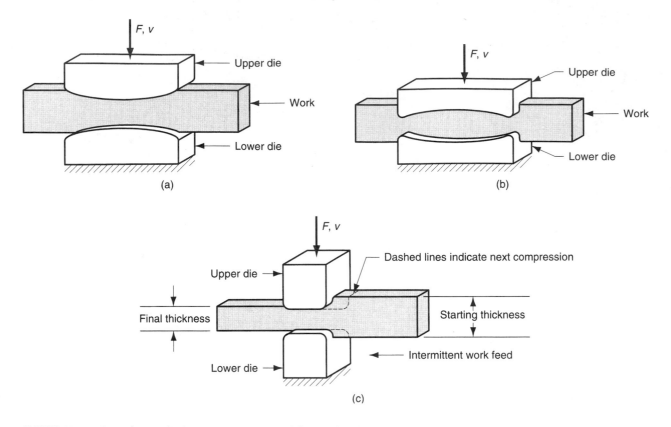

FIGURE 19.14 Several open-die forging operations: (a) fullering, (b) edging, and (c) cogging.

to reduce the cross section and redistribute the metal in a workpart in preparation for subsequent shape forging. It is accomplished by dies with convex surfaces. Fullering die cavities are often designed into multi-cavity impression dies, so that the starting bar can be rough formed before final shaping. *Edging* is similar to fullering, except that the dies have concave surfaces.

A *cogging* operation consists of a sequence of forging compressions along the length of a workpiece to reduce cross section and increase length. It is used in the steel industry to produce blooms and slabs from cast ingots. It is accomplished using open dies with flat or slightly contoured surfaces. The term *incremental forging* is sometimes used for this process.

19.3.2 Impression-Die Forging

Impression-die forging, sometimes called *closed-die forging,* is performed with dies that contain the inverse of the desired shape of the part. The process is illustrated in a three-step sequence in Figure 19.15. The raw workpiece is shown as a cylindrical part similar to that used in the previous open-die operation. As the die closes to its final position, flash is formed by metal that flows beyond the die cavity and into the small gap between the die plates. Although this flash must be cut away from the part in a subsequent trimming operation, it actually serves an important function in impression-die forging. As the flash begins to form in the die gap, friction resists continued flow of metal into the gap, thus constraining the bulk of the work material to remain in the die cavity. In hot forging, metal flow is further restricted because the thin flash cools quickly against the die

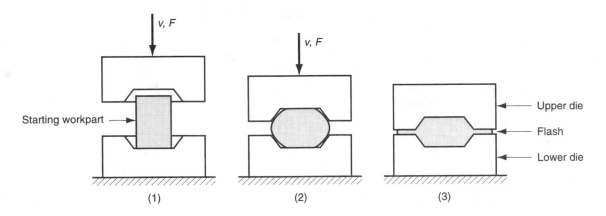

FIGURE 19.15   Sequence in impression-die forging: (1) just prior to initial contact with raw workpiece, (2) partial compression, and (3) final die closure, causing flash to form in gap between die plates.

plates, thereby increasing its resistance to deformation. Restricting metal flow in the gap causes the compression pressures on the part to increase significantly, thus forcing the material to fill the sometimes intricate details of the die cavity to ensure a high-quality product.

Several forming steps are often required in impression die forging to transform the starting blank into the desired final geometry. Separate cavities in the die are needed for each step. The beginning steps are designed to redistribute the metal in the workpart to achieve a uniform deformation and desired metallurgical structure in the subsequent steps. The final steps bring the part to its final geometry. In addition, when drop forging is used, several blows of the hammer may be required for each step. When impression-die drop forging is done manually, as it often is, considerable operator skill is required under adverse conditions to achieve consistent results.

Because of flash formation in impression-die forging and the more complex part shapes made with these dies, forces in this process are significantly greater and more difficult to analyze than in open-die forging. Relatively simple formulas and design factors are often used to estimate forces in impression-die forging. The force formula is the same as previous Eq. (19.16) for open-die forging, but its interpretation is slightly different:

$$F = K_f Y_f A \tag{19.18}$$

where F = maximum force in the operation, N (lb); A = projected area of the part including flash, mm^2 (in.^{2}); Y_f = flow stress of the material, MPa (lb/in.^{2}); and K_f = forging shape factor. In hot forging, the appropriate value of Y_f is the yield strength of the metal at the elevated temperature. In other cases, selecting the proper value of flow stress is difficult because the strain varies throughout the workpiece for complex shapes. K_f in Eq. (19.18) is a factor intended to account for increases in force required to forge part shapes of various complexities. Table 19.1 indicates the range of values of K_f for different part geometries. Obviously, the problem of specifying the proper K_f value for a given workpart limits the accuracy of the force estimate.

Eq. (19.18) applies to the maximum force during the operation, since this is the load that will determine the required capacity of the press or hammer used in the operation. The maximum force is reached at the end of the forging stroke, when the projected area is greatest and friction is maximum.

Impression-die forging is not capable of close tolerance work, and machining is often required to achieve the accuracies needed. The basic geometry of the part is

TABLE 19.1 Typical K_f values for various part shapes in impression-die and closed-die forging.

Part Shape	K_f	Part Shape	K_f
Impression-die forging:		Flashless forging:	
Simple shapes with flash	6.0	Coining (top and bottom surfaces)	6.0
Complex shapes with flash	8.0	Complex shapes	8.0
Very complex shapes with flash	10.0		

obtained from the forging process, with machining performed on those portions of the part that require precision finishing (e.g., holes, threads, and surfaces that mate with other components). The advantages of forging, compared to machining the part completely, are higher production rates, conservation of metal, greater strength, and favorable grain orientation of the metal that results from forging. A comparison of the grain flow in forging and machining is illustrated in Figure 19.16.

Improvements in the technology of impression-die forging have resulted in the capability to produce forgings with thinner sections, more complex geometries, drastic reductions in draft requirements on the dies (Section 19.2.4), closer tolerances, and the virtual elimination of machining allowances. Forging processes with these features are known as *precision forging.* Common work metals used for precision forging include aluminum and titanium. A comparison of precision and conventional impression-die forging is presented in Figure 19.17. Note that precision forging in this example does not eliminate flash, although it reduces it. Some precision forging operations are accomplished without producing flash. Depending on whether machining is required to finish the part geometry, precision forgings are properly classified as *near net shape* or *net shape* processes.

19.3.3 Flashless Forging

As already mentioned, impression-die forging is sometimes called closed-die forging in industry terminology. However, there is a technical distinction between impression-die forging and true closed-die forging. The distinction is that in closed-die forging, the raw workpiece is completely contained within the die cavity during compression, and no flash is formed. The process sequence is illustrated in Figure 19.18. The term *flashless forging* is appropriate to identify this process.

Flashless forging imposes requirements on process control that are more demanding than impression-die forging. Most important is that the work volume must equal the space in the die cavity within a very close tolerance. If the starting blank is too large, excessive pressures may cause damage to the die or press. If the blank is too small, the cavity will not be filled. Because of the special demands made by flashless forging, the process lends itself best to part geometries that are usually simple and symmetrical, and to work materials such as aluminum and magnesium and their alloys. Flashless forging is often classified as a *precision forging* process [3].

FIGURE 19.16
Comparison of metal grain flow in a part that is: (a) hot forged with finish machining, and (b) machined complete.

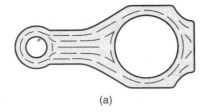

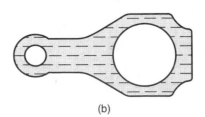

(a)   (b)

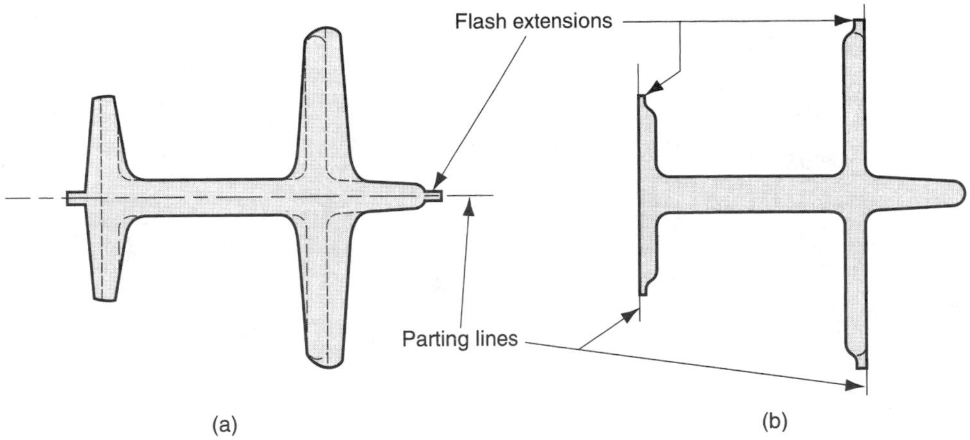

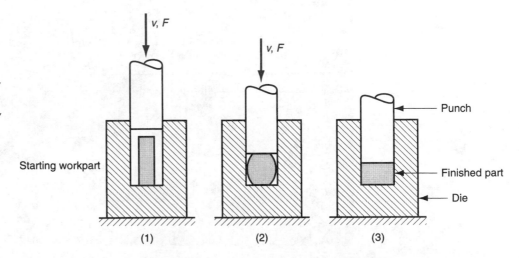

FIGURE 19.17   Cross sections of (a) conventional and (b) precision forgings. Dashed lines in (a) indicate subsequent machining required to make the conventional forging equivalent in geometry to the precision forging. In both cases, flash extensions must be trimmed.

Flash extensions

Parting lines

(a)   (b)

Forces in flashless forging reach values comparable to those in impression-die forging. Estimates of these forces can be computed using the same methods as for impression-die forging: Eq. (19.18) and Table 19.1.

Coining is a special application of closed-die forging in which fine details in the die are impressed into the top and bottom surfaces of the workpart. There is little flow of metal in coining, yet the pressures required to reproduce the surface details in the die cavity are high, as indicated by the value of K_f in Table 19.1. A common application of coining is, of course, in the minting of coins, shown in Figure 19.19. The process is also used to provide good surface finish and dimensional accuracy on workparts made by other operations.

19.3.4 Forging Dies, Hammers, and Presses

Equipment used in forging consists of forging machines, classified as hammers and presses; and forging dies, which are the special tooling used in these machines. In addition, auxiliary equipment is needed, such as furnaces to heat the work, mechanical devices to load and unload the work, and trimming stations to cut away the flash in impression-die forging.

FIGURE 19.18   Flashless forging: (1) just before initial contact with workpiece, (2) partial compression, and (3) final punch and die closure. Symbols v and F indicate motion (v = velocity) and applied force, respectively.

v, F

v, F

Punch

Starting workpart

Finished part

Die

(1)   (2)   (3)

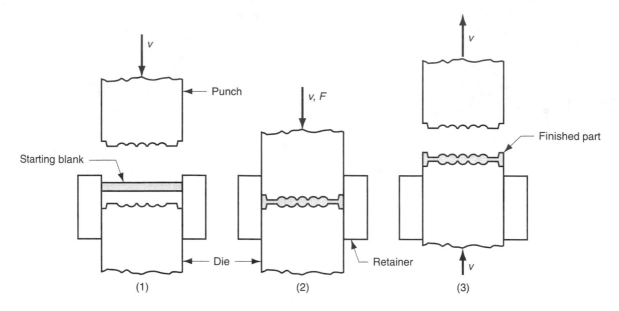

FIGURE 19.19 Coining operation: (1) start of cycle, (2) compression stroke, and (3) ejection of finished part.

Forging Hammers Forging hammers operate by applying an impact loading against the work. The term *drop hammer* is often used for these machines, owing to the means of delivering impact energy—see Figures 19.20 and 19.21. Drop hammers are most frequently used for impression-die forging. The upper portion of the forging die is attached to the ram, and the lower portion is attached to the anvil. In the operation, the work is placed on the lower die, and the ram is lifted and then dropped. When the upper die

FIGURE 19.20 Drop forging hammer, fed by conveyor and heating units at the right of the scene (photo courtesy of Chambersburg Engineering Company).

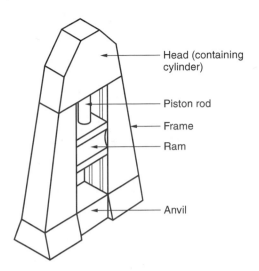

FIGURE 19.21   Diagram showing details of a drop hammer for impression-die forging.

Head (containing cylinder)

Piston rod

Frame

Ram

Anvil

strikes the work, the impact energy causes the part to assume the form of the die cavity. Several blows of the hammer are often required to achieve the desired change in shape. Drop hammers can be classified as gravity drop hammers and power drop hammers. ***Gravity drop hammers*** achieve their energy by the falling weight of a heavy ram. The force of the blow is determined by the height of the drop and the weight of the ram. ***Power drop hammers*** accelerate the ram by pressurized air or steam. One of the disadvantages of drop hammers is that a large amount of the impact energy is transmitted through the anvil and into the floor of the building.

Forging Presses   Presses apply gradual pressure, rather than sudden impact, to accomplish the forging operation. Forging presses include mechanical presses, hydraulic presses, and screw presses. ***Mechanical presses*** operate by means of eccentrics, cranks, or knuckle joints, which convert the rotating motion of a drive motor into the translation motion of the ram. These mechanisms are very similar to those used in stamping presses (Section 20.5.2). Mechanical presses typically achieve very high forces at the bottom of the forging stroke. ***Hydraulic presses*** use a hydraulically driven piston to actuate the ram. ***Screw presses*** apply force by a screw mechanism that drives the vertical ram. Both screw drive and hydraulic drive operate at relatively low ram speeds and can provide a constant force throughout the stroke. These machines are therefore suitable for forging (and other forming) operations that require a long stroke.

Forging Dies   Proper die design is important in the success of a forging operation. Parts to be forged must be designed based on knowledge of the principles and limitations of this process. Our purpose here is to describe some of the terminology and guidelines used in the design of forgings and forging dies. Design of open dies is generally straightforward because the dies are relatively simple in shape. Our comments apply to impression dies and closed dies. Figure 19.22 defines some of the terminology in an impression die.

We indicate some of the principles and limitations that must be considered in the part design or in the selection of forging as the manufacturing process to make the part in the following discussion of forging die terminology [3]:

> ***Parting line***—The parting line is the plane that divides the upper die from the lower die. Called the flash line in impression-die forging, it is the plane where the two die

FIGURE 19.22
Terminology for a conventional impression-die in forging.

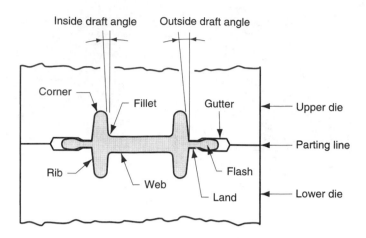

halves meet. Its selection by the designer affects grain flow in the part, required load, and flash formation.

> *Draft*—Draft is the amount of taper on the sides of the part required to remove it from the die. The term also applies to the taper on the sides of the die cavity. Typical draft angles are 3 degrees on aluminum and magnesium parts and 5 to 7 degrees on steel parts. Draft angles on precision forgings are near zero.

> *Webs and ribs*—A web is a thin portion of the forging that is parallel to the parting line, while a rib is a thin portion that is perpendicular to the parting line. These part features cause difficulty in metal flow as they become thinner.

> *Fillet and corner radii*—Fillet and corner radii are illustrated in Figure 19.22. Small radii tend to limit metal flow and increase stresses on die surfaces during forging.

> *Flash*—Flash formation plays a critical role in impression-die forging by causing pressure build-up inside the die to promote filling of the cavity. This pressure build-up is controlled by designing a flash land and gutter into the die, as pictured in Figure 19.22. The land determines the surface area along which lateral flow of metal occurs, thereby controlling the pressure increase inside the die. The gutter permits excess metal to escape without causing the forging load to reach extreme values.

19.4 OTHER DEFORMATION PROCESSES RELATED TO FORGING

In addition to the conventional forging operations discussed in the preceding sections, other metal-forming operations are closely associated with forging.

Upsetting and Heading Upsetting (also called *upset forging*) is a deformation operation in which a cylindrical workpart is increased in diameter and reduced in length. This operation was analyzed in our discussion of open-die forging (Section 19.3.1). However, as an industrial operation, it can also be performed as closed-die forging, as seen in Figure 19.23.

Upsetting is widely used in the fastener industry to form heads on nails, bolts, and similar hardware products. In these applications, the term *heading* is often used to denote the operation. Figure 19.24 illustrates a variety of heading applications, indicating various possible die configurations. Owing to these types of applications, more parts are produced by upsetting than by any other forging operation. It is performed as a mass production

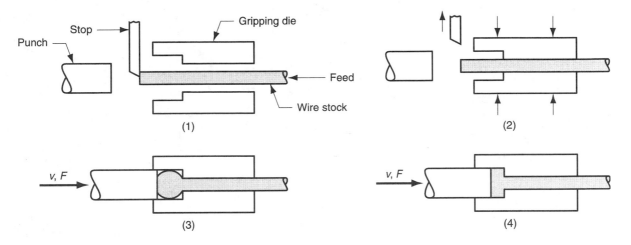

FIGURE 19.23   An upset forging operation to form a head on a bolt or similar hardware item. The cycle is as follows: (1) wire stock is fed to the stop; (2) gripping dies close on the stock and the stop is retracted; (3) punch moves forward and (4) bottoms to form the head.

operation—cold, warm, or hot—on special upset forging machines, called headers or formers. These machines are usually equipped with horizontal slides, rather than vertical slides as in conventional forging hammers and presses. Long wire or bar stock is fed into the machines, the end of the stock is upset forged, and then the piece is cut to length to make the desired hardware item. For bolts and screws, thread rolling (Section 19.2) is used to form the threads.

There are limits on the amount of deformation that can be achieved in upsetting, usually defined as the maximum length of stock to be forged. The maximum length that can be upset in one blow is three times the diameter of the starting stock. Otherwise, the metal bends or buckles instead of compressing properly to fill the cavity.

Swaging and Radial Forging   Swaging and radial forging are forging processes used to reduce the diameter of a tube or solid rod. Swaging is often performed on the end of a workpiece to create a tapered section. The *swaging* process, shown in Figure 19.25,

FIGURE 19.24   Examples of heading (upset forging) operations: (a) heading a nail using open dies, (b) round head formed by punch, (c) and (d) heads formed by die, and (e) carriage bolt head formed by punch and die.

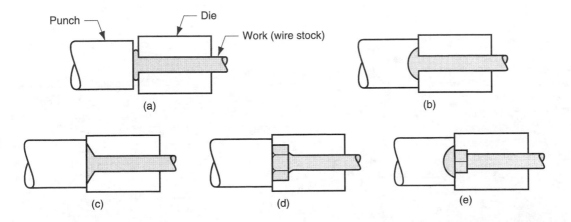

FIGURE 19.25 Swaging process to reduce solid rod stock; the dies rotate as they hammer the work. In radial forging, the workpiece rotates while the dies remain in a fixed orientation as they hammer the work.

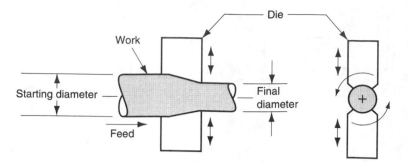

is accomplished by means of rotating dies that hammer a workpiece radially inward to taper it as the piece is fed into the dies. Figure 19.26 illustrates some of the shapes and products that are made by swaging. A mandrel is sometimes required to control the shape and size of the internal diameter of tubular parts that are swaged. *Radial forging* is similar to swaging in its action against the work and is used to create similar part shapes. The difference is that in radial forging the dies do not rotate around the workpiece; instead the work is rotated as it feeds into the hammering dies.

Roll Forging Roll forging is a deformation process used to reduce the cross-section of a cylindrical (or rectangular) workpiece by passing it through a set of opposing rolls that have grooves matching the desired shape of the final part. The typical operation is illustrated in Figure 19.27. Roll forging is generally classified as a forging process even though it utilizes rolls. The rolls do not turn continuously in roll forging, but rotate through only a portion of one revolution corresponding to the desired deformation to be accomplished on the part. Roll-forged parts are generally stronger and possess favorable grain structure compared to competing processes such as machining that might be used to produce the same part geometry.

Orbital Forging In this process, deformation occurs by means of a cone-shaped upper die that is simultaneously rolled and pressed into the workpart. As illustrated in Figure 19.28, the work is supported on a lower die, which has a cavity into which the work is compressed. Because the axis of the cone is inclined, only a small area of the work surface is compressed at any moment. As the upper die revolves, the area under compression also revolves. These operating characteristics of orbital forging result in a substantial reduction in press load required to accomplish deformation of the work.

FIGURE 19.26 Examples of parts made by swaging: (a) reduction of solid stock, (b) tapering a tube, (c) swaging to form a groove on a tube, (d) pointing of a tube, and (e) swaging of neck on a gas cylinder.

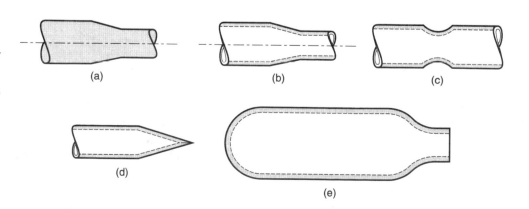

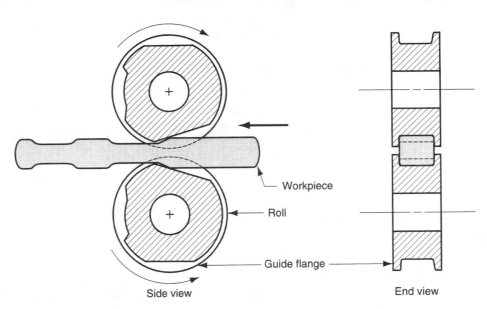

FIGURE 19.27   Roll forging.

Hubbing   As a forging operation, *hubbing* is a deformation process in which a hardened steel form is pressed into a soft steel (or other soft metal) block. The process is often used to make mold cavities for plastic molding and die casting, as sketched in Figure 19.29. The hardened steel form, called the *hub,* is machined to the geometry of the part to be molded. Substantial pressures are required to force the hub into the soft block, and this is usually accomplished by a hydraulic press. Complete formation of the

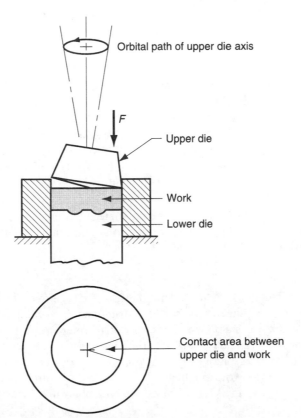

FIGURE 19.28   Orbital forging. At end of deformation cycle, lower die lifts to eject part.

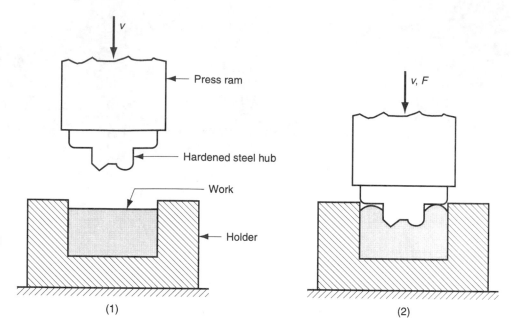

FIGURE 19.29 Hubbing: (1) before deformation, and (2) as the process is completed. Note that the excess material formed by the penetration of the hob must be machined away.

die cavity in the block often requires several steps—hubbing followed by annealing to recover the work metal from strain hardening. When significant amounts of material are deformed in the block, as shown in our figure, the excess must be machined away. The advantage of hubbing in this application is that it is generally easier to machine the positive form than the mating negative cavity. This advantage multiplies in cases where more than one cavity are to be made in the die block.

Isothermal and Hot Die Forging *Isothermal forging* is a term applied to a hot-forging operation in which the workpart is maintained at or near its starting elevated temperature during deformation, usually by heating the forging dies to the same elevated temperature. By avoiding chill of the workpiece on contact with the cold die surfaces as in conventional forging, the metal flows more readily and the force required to perform the process is reduced. Isothermal forging is more expensive than conventional forging and is usually reserved for difficult-to-forge metals, such as titanium and superalloys, and for complex part shapes. The process is sometimes carried out in a vacuum to avoid rapid oxidation of the die material. Similar to isothermal forging is *hot-die forging,* in which the dies are heated to a temperature that is somewhat below that of the work metal.

Trimming Trimming is an operation used to remove flash on the workpart in impression-die forging. In most cases, trimming is accomplished by shearing, as in Figure 19.30, in which a punch forces the work through a cutting die, the blades for which have the profile of the desired part. Trimming is usually done while the work is still hot, which means that a separate trimming press is included at each forging hammer or press. In cases where the work might be damaged by the cutting process, trimming may be done by alternative methods, such as grinding or sawing.

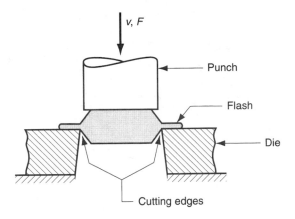

FIGURE 19.30   Trimming operation (shearing process) to remove the flash after impression-die forging.

19.5 EXTRUSION

Extrusion is a compression forming process in which the work metal is forced to flow through a die opening to produce a desired cross-sectional shape. The process can be likened to squeezing toothpaste out of a toothpaste tube. Extrusion dates from around 1800 (Historical Note 19.3). There are several advantages of the modern process: (1) a variety of shapes are possible, especially with hot extrusion; however, a limitation of the geometry is that the cross section of the part must be uniform throughout its extruded length; (2) grain structure and strength properties are enhanced in cold and warm extrusion; (3) fairly close tolerances are possible, especially in cold extrusion; and (4) in some extrusion operations, little or no wasted material is created.

Historical Note 19.3   *Extrusion.*

Extrusion as an industrial process was invented around 1800 in England during the Industrial Revolution when that country was leading the world in technological innovations. The invention consisted of the first hydraulic press for extruding lead pipes. An important step forward was made in Germany around 1890, when the first horizontal extrusion press was built for extruding metals with higher melting points than lead. The feature that made this possible was the use of a dummy block that separated the ram from the work billet.

19.5.1 Types of Extrusion

Extrusion is carried out in various ways. One way of classifying the operations is by physical configuration, in which the two principal types are direct extrusion and indirect extrusion. Another classification is by working temperature: cold, warm, or hot extrusion. Finally, extrusion is performed as either a continuous process or a discrete process.

Direct versus Indirect Extrusion   *Direct extrusion* (also called *forward extrusion*) is illustrated in Figure 19.31. A metal billet is loaded into a container, and a ram compresses the material, forcing it to flow through one or more openings in a die at the opposite end of the container. As the ram approaches the die, a small portion of the billet remains that cannot be forced through the die opening. This extra portion, called the *butt,* is separated from the product by cutting it just beyond the exit of the die.

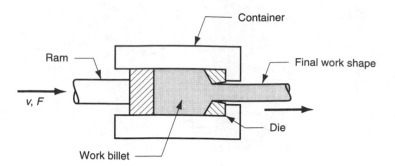

FIGURE 19.31   Direct extrusion.

One of the problems in direct extrusion is the significant friction that exists between the work surface and the walls of the container as the billet is forced to slide toward the die opening. This friction causes a substantial increase in the ram force required in direct extrusion. In hot extrusion, the friction problem is aggravated by the presence of an oxide layer on the surface of the billet. This oxide layer can cause defects in the extruded product. To address these problems, a dummy block is often used between the ram and the work billet. The diameter of the dummy block is slightly smaller than the billet diameter, so that a narrow ring of work metal (mostly the oxide layer) is left in the container, leaving the final product free of oxides.

Hollow sections (e.g., tubes) are possible in direct extrusion by the process setup in Figure 19.32. The starting billet is prepared with a hole parallel to its axis. This allows passage of a mandrel that is attached to the dummy block. As the billet is compressed, the material is forced to flow through the clearance between the mandrel and the die opening. The resulting cross section is tubular. Semi-hollow cross-sectional shapes are usually extruded in the same way.

The starting billet in direct extrusion is usually round in cross section, but the final shape is determined by the shape of the die opening. Obviously, the largest dimension of the die opening must be smaller than the diameter of the billet.

FIGURE 19.32   (a) Direct extrusion to produce a hollow or semi-hollow cross section; (b) hollow and (c) semi-hollow cross sections.

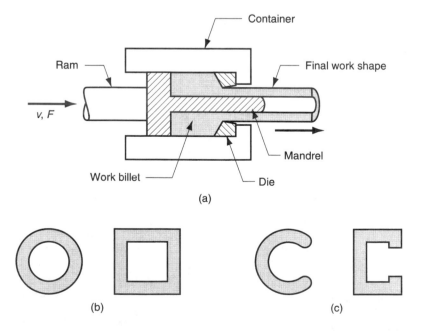

In *indirect extrusion,* also called *backward extrusion* and *reverse extrusion,* Figure 19.33(a), the die is mounted to the ram rather than at the opposite end of the container. As the ram penetrates into the work, the metal is forced to flow through the clearance in a direction opposite to the motion of the ram. Since the billet is not forced to move relative to the container, there is no friction at the container walls, and the ram force is therefore lower than in direct extrusion. Limitations of indirect extrusion are imposed by the lower rigidity of the hollow ram and the difficulty in supporting the extruded product as it exits the die.

Indirect extrusion can produce hollow (tubular) cross sections, as in Figure 19.33(b). In this method, the ram is pressed into the billet, forcing the material to flow around the ram and take a cup shape. There are practical limitations on the length of the extruded part that can be made by this method. Support of the ram becomes a problem as work length increases.

Hot versus Cold Extrusion   Extrusion can be performed either hot or cold, depending on work metal and amount of strain to which it is subjected during deformation. Metals that are typically extruded hot include aluminum, copper, magnesium, zinc, tin, and their alloys. These same metals are sometimes extruded cold. Steel alloys are usually extruded hot, although the softer, more ductile grades are sometimes cold extruded (e.g., low carbon steels and stainless steel). Aluminum is probably the most ideal metal for extrusion (hot and cold), and many commercial aluminum products are made by this process (e.g., structural shapes, door and window frames).

Hot extrusion involves prior heating of the billet to a temperature above its recrystallization temperature. This reduces strength and increases ductility of the metal, permitting more extreme size reductions and more complex shapes to be achieved in the process. Additional advantages include reduction of ram force, increased ram speed, and reduction of grain flow characteristics in the final product. Isothermal extrusion is sometimes used to overcome the problem of cooling the billet as it contacts the container walls. Lubrication is critical in hot extrusion for certain metals (e.g., steels), and special lubricants have been developed that are effective under the harsh conditions in hot extrusion. Glass is sometimes used as a lubricant in hot extrusion; in addition to reducing friction, it also provides effective thermal insulation between the billet and the extrusion container.

Cold extrusion and warm extrusion are generally used to produce discrete parts, often in finished (or near finished) form. The term *impact extrusion* is used to indicate high-speed cold extrusion, and this method is described in more detail in Section 19.5.4. Some important advantages of cold extrusion include increased strength due to strain hardening, close

FIGURE 19.33   Indirect extrusion to produce (a) a solid cross section and (b) a hollow cross section.

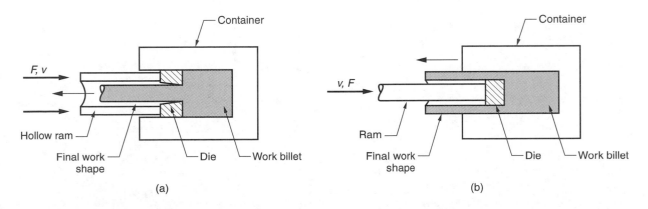

tolerances, improved surface finish, absence of oxide layers, and high production rates. Cold extrusion at room temperature also eliminates the need for heating the starting billet.

Continuous versus Discrete Processing A true continuous process operates in steady state mode for an indefinite period of time. Some extrusion operations approach this ideal by producing very long sections in one cycle, but these operations are ultimately limited by the size of the billet that can be loaded into the extrusion container. These processes are more accurately described as semi-continuous operations. In nearly all cases, the long section is cut into smaller lengths in a subsequent sawing or shearing operation.

In a discrete extrusion operation, a single part is produced in each extrusion cycle. Impact extrusion is an example of the discrete processing case.

19.5.2 Analysis of Extrusion

Let us use Figure 19.34 as a reference in discussing some of the parameters in extrusion. The diagram assumes that both billet and extrudate are round in cross section. One important parameter is the *extrusion ratio,* also called the *reduction ratio.* The ratio is defined:

$$r_x = \frac{A_o}{A_f} \qquad (19.19)$$

where r_x = extrusion ratio; A_o = cross-sectional area of the starting billet, mm^2 (in.^{2}); and A_f = final cross-sectional area of the extruded section, mm^2 (in.^{2}). The ratio applies for both direct and indirect extrusion. The value of r_x can be used to determine true strain in extrusion, given that ideal deformation occurs with no friction and no redundant work:

$$\epsilon = \ln r_x = \ln \frac{A_o}{A_f} \qquad (19.20)$$

Under the assumption of ideal deformation (no friction and no redundant work), the pressure applied by the ram to compress the billet through the die opening depicted in our figure can be computed as follows:

$$p = \overline{Y}_f \ln r_x \qquad (19.21)$$

FIGURE 19.34 Pressure and other variables in direct extrusion.

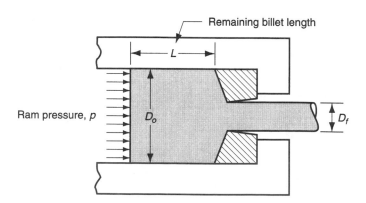

where $\overline{Y}_f$ = average flow stress during deformation, MPa (lb/in.^{2}). For convenience, we restate Eq. (18.2) from the previous chapter:

$$\overline{Y}_f = \frac{K\epsilon^n}{1+n}$$

In fact, extrusion is not a frictionless process, and the previous equations grossly underestimate the strain and pressure in an extrusion operation. Friction exists between the die and the work as the billet squeezes down and passes through the die opening. In direct extrusion, friction also exists between the container wall and the billet surface. The effect of friction is to increase the strain experienced by the metal. Thus, the actual pressure is greater than that given by Eq. (19.21), which assumes frictionless extrusion.

Various methods have been suggested to calculate the actual true strain and associated ram pressure in extrusion [1], [2], [4], [10], [11], and [18]. The following empirical equation proposed by Johnson [10] for estimating extrusion strain has gained considerable recognition:

$$\epsilon_x = a + b \ln r_x \tag{19.22}$$

where ϵ_x = extrusion strain; and a and b are empirical constants for a given die angle. Typical values of these constants are: $a = 0.8$ and $b = 1.2$ to 1.5. Values of a and b tend to increase with increasing die angle.

The ram pressure to perform **indirect extrusion** can be estimated based on Johnson's extrusion strain formula as follows:

$$p = \overline{Y}_f \epsilon_x \tag{19.23a}$$

where $\overline{Y}_f$ is calculated based on ideal strain from Eq. (19.20), rather than extrusion strain in Eq. (19.22).

In **direct extrusion,** the effect of friction between the container walls and the billet causes the ram pressure to be greater than for indirect extrusion. We can write the following expression which isolates the friction force in the direct extrusion container:

$$\frac{p_f \pi D_o^2}{4} = \mu p_c \pi D_o L$$

where p_f = additional pressure required to overcome friction, MPa (lb/in.^{2}); $(\pi D_o^2)/4$ = billet cross-sectional area, mm^2 (in.^{2}); μ = coefficient of friction at the container wall; p_c = pressure of the billet against the container wall, MPa (lb/in.^{2}); and $\pi D_o L$ = area of the interface between billet and container wall, mm^2 (in.^{2}). The right-hand side of this equation indicates the billet-container friction force, and the left-hand side gives the additional ram force to overcome that friction. In the worst case, sticking occurs at the container wall so that friction stress equals shear yield strength of the work metal:

$$\mu p_c \pi D_o L = Y_s \pi D_o L$$

where Y_s = shear yield strength, MPa (lb/in.^{2}). If we assume that $Y_s = \overline{Y}_f/2$, then p_f reduces to the following:

$$p_f = \overline{Y}_f \frac{2L}{D_o}$$

Based on this reasoning, the following formula can be used to compute ram pressure in direct extrusion:

$$p = \overline{Y}_f \left(\epsilon_x + \frac{2L}{D_o} \right) \tag{19.23b}$$

where the term $2L/D_o$ accounts for the additional pressure due to friction at the container-billet interface. L is the portion of the billet length remaining to be extruded, and D_o is the original diameter of the billet. Note that p is reduced as the remaining billet length decreases during the process. Typical plots of ram pressure as a function of ram stroke for direct and indirect extrusion are presented in Figure 19.35. Eq. (19.23b) probably overestimates ram pressure. With good lubrication, ram pressures would be lower than values calculated by this equation.

Ram force in indirect or direct extrusion is simply pressure p from Eqs. (19.23a) or (19.23b), respectively, multiplied by billet area A_o:

$$F = pA_o \qquad (19.24)$$

where F = ram force in extrusion, N (lb). Power required to carry out the extrusion operation is simply

$$P = Fv \qquad (19.25)$$

where P = power, J/s (in.-lb/min); F = ram force, N (lb); and v = ram velocity, m/s (in./min).

EXAMPLE 19.3
Extrusion pressures.

A billet 75 mm long and 25 mm in diameter is to be extruded in a direct extrusion operation with extrusion ratio $r_x = 4.0$. The extrudate has a round cross section. The die angle (half-angle) = 90°. The work metal has a strength coefficient = 415 MPa, and strain hardening exponent = 0.18. Use the Johnson formula with $a = 0.8$ and $b = 1.5$ to estimate extrusion strain. Determine the pressure applied to the end of the billet as the ram moves forward.

Solution: Let us examine the ram pressure at billet lengths of $L = 75$ mm (starting value), $L = 50$ mm, $L = 25$ mm, and $L = 0$. We compute the ideal true strain, extrusion strain using Johnson's formula, and average flow stress:

$$\epsilon = \ln r_x = \ln 4.0 = 1.3863$$

$$\epsilon_x = 0.8 + 1.5(1.3863) = 2.8795$$

$$\overline{Y}_f = \frac{415(1.3863)^{0.18}}{1.18} = 373 \text{ MPa}$$

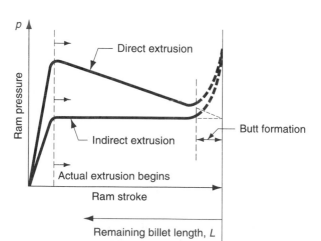

FIGURE 19.35 Typical plots of ram pressure versus ram stroke (and remaining billet length) for direct and indirect extrusion. The higher values in direct extrusion result from friction at the container wall. The shape of the initial pressure build-up at the beginning of the plot depends on die angle (higher die angles cause steeper pressure build-ups). The pressure increase at the end of the stroke is related to formation of the butt.

$L = 75$ mm: With a die angle of 90°, the billet metal is assumed to be forced through the die opening almost immediately; thus, our calculation assumes that maximum pressure is reached at the billet length of 75 mm. For die angles less than 90°, the pressure would build to a maximum as in Figure 19.35 as the starting billet is squeezed into the cone-shaped portion of the extrusion die. Using Eq. (19.23b),

$$p = 373\left(2.8795 + 2\frac{75}{25}\right) = 3312 \text{ MPa}$$

$$L = 50 \text{ mm: } p = 373\left(2.8795 + 2\frac{50}{25}\right) = 2566 \text{ MPa}$$

$$L = 25 \text{ mm: } p = 373\left(2.8795 + 2\frac{25}{25}\right) = 1820 \text{ MPa}$$

$L = 0$: Zero length is a hypothetical value in direct extrusion. In reality, it is impossible to squeeze all of the metal through the die opening. Instead, a portion of the billet (the "butt") remains unextruded and the pressure begins to increase rapidly as L approaches zero. This increase in pressure at the end of the stroke is seen in the plot of ram pressure versus ram stroke in Figure 19.35. Calculated below is the hypothetical minimum value of ram pressure that would result at L = 0.

$$p = 373\left(2.8795 + 2\frac{0}{25}\right) = 1074 \text{ MPa}$$

This is also the value of ram pressure that would be associated with indirect extrusion throughout the length of the billet.

19.5.3 Extrusion Dies and Presses

Important factors in an extrusion die are die angle and orifice shape. Die angle, more precisely die half-angle, is shown as α in Figure 19.36(a). For low angles, surface area of the die is large, leading to increased friction at the die-billet interface. Higher friction results in larger ram force. On the other hand, a large die angle causes more turbulence in the metal flow during reduction, increasing the ram force required. Thus, the effect of die angle on ram force is a U-shaped function, as in Figure 19.36(b). An optimum die angle exists, as suggested by our hypothetical plot. The optimum angle depends on various factors (e.g., work material, billet temperature, and lubrication) and is therefore difficult to

FIGURE 19.36 (a) Definition of die angle in direct extrusion; (b) effect of die angle on ram force.

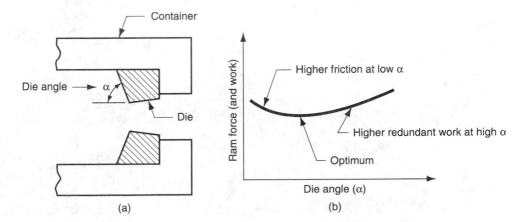

(a)

(b)

determine for a given extrusion job. Die designers rely on rules of thumb and judgment to decide the appropriate angle.

Our previous equations for ram pressure, Eqs. (19.23), apply to a circular die orifice. The shape of the die orifice affects the ram pressure required to perform an extrusion operation. A complex cross section, such as the one shown in Figure 19.37, requires a higher pressure and greater force than a circular shape. The effect of the die orifice shape can be assessed by the die *shape factor,* defined as the ratio of the pressure required to extrude a cross section of a given shape relative to the extrusion pressure for a round cross section of the same area. We can express the shape factor as follows:

$$K_x = 0.98 + 0.02\left(\frac{C_x}{C_c}\right)^{2.25} \tag{19.26}$$

where K_x = die shape factor in extrusion; C_x = perimeter of the extruded cross section, mm (in.); and C_c = perimeter of a circle of the same area as the extruded shape, mm (in.). Eq. (19.26) is based on empirical data in Altan et al. [1] over a range of C_x/C_c values from 1.0 to about 6.0. The equation may be invalid much beyond the upper limit of this range.

As indicated by Eq. (19.26), the shape factor is a function of the perimeter of the extruded cross section divided by the perimeter of a circular cross section of equal area. A circular shape is the simplest shape, with a value of $K_x = 1.0$. Hollow, thin-walled sections have higher shape factors and are more difficult to extrude. The increase in pressure is not included in our previous pressure equations, Eqs. (19.23), which apply only to round cross sections. For shapes other than round, the corresponding expression for indirect extrusion is

$$p = K_x \overline{Y}_f \tag{19.27a}$$

FIGURE 19.37 A complex extruded cross section for a heat sink (photo courtesy of Aluminum Company of America).

and for direct extrusion is

$$p = K_x \overline{Y}_f \left(\epsilon_x + \frac{2L}{D_o} \right) \tag{19.27b}$$

where p = extrusion pressure, MPa (lb/in.²); K_x = shape factor; and the other terms have the same interpretation as before. Values of pressure given by these equations can be used in Eq. (19.24) to determine ram force.

Die materials used for hot extrusion include tool and alloy steels. Important properties of these die materials include high wear resistance, high hot hardness, and high thermal conductivity to remove heat from the process. Die materials for cold extrusion include tool steels and cemented carbides. Wear resistance and ability to retain shape under high stress are desirable properties. Carbides are used when high production rates, long die life, and good dimensional control are required.

Extrusion presses are either horizontal or vertical, depending on orientation of the work axis. Horizontal types are more common. Extrusion presses are usually hydraulically driven. This drive is especially suited to semi-continuous production of long sections, as in direct extrusion. Mechanical drives are often used for cold extrusion of individual parts, such as in impact extrusion.

19.5.4 Other Extrusion Processes

Direct and indirect extrusion are the principal methods of extrusion. Various names are given to operations that are special cases of the direct and indirect methods described here. Other extrusion operations are unique. In this section we examine some of these special forms of extrusion and related processes.

Impact Extrusion   Impact extrusion is performed at higher speeds and shorter strokes than conventional extrusion. It is used to make individual components. As the name suggests, the punch impacts the workpart rather than simply applying pressure to it. Impacting can be carried out as forward extrusion, backward extrusion, or combinations of these. Some representative examples are shown in Figure 19.38.

Impact extrusion is usually done cold on a variety of metals. Backward impact extrusion is most common. Products made by this process include toothpaste tubes and battery cases. As indicated by these examples, very thin walls are possible on impact extruded parts. The high-speed characteristics of impacting permit large reductions and high production rates, making this an important commercial process.

Hydrostatic Extrusion   One of the problems in direct extrusion is friction along the billet-container interface. This problem can be addressed by surrounding the billet with fluid inside the container and pressurizing the fluid by the forward motion of the ram, as in Figure 19.39. This way, there is no friction inside the container, and friction at the die opening is reduced. Consequently, ram force is significantly lower than in direct extrusion. The fluid pressure acting on all surfaces of the billet gives the process its name. It can be carried out at room temperature or at elevated temperatures. Special fluids and procedures must be used at elevated temperatures. Hydrostatic extrusion is an adaptation of direct extrusion.

Hydrostatic pressure on the work increases the material's ductility. Accordingly, this process can be used on metals that would be too brittle for conventional extrusion operations. Ductile metals can also be hydrostatically extruded, and high reduction ratios are possible on these materials. One of the disadvantages of the process is the required preparation of the starting work billet. The billet must be formed with a taper at one end to

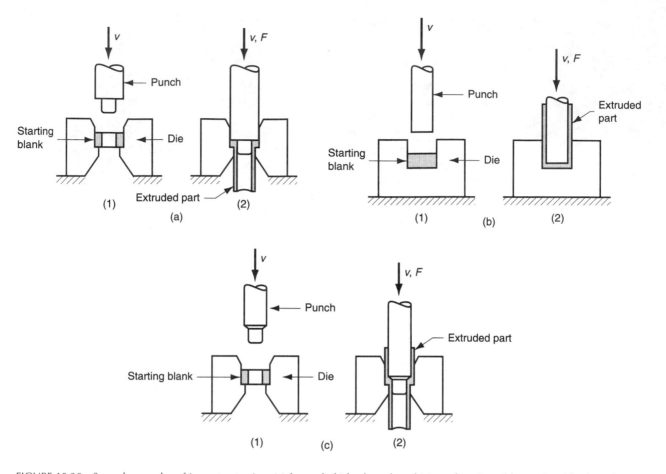

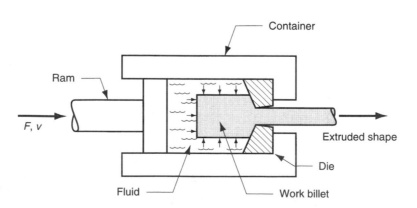

FIGURE 19.38 Several examples of impact extrusion: (a) forward, (b) backward, and (c) combination of forward and backward.

fit snugly into the die entry angle. This establishes a seal to prevent fluid from squirting out the die hole when the container is initially pressurized.

19.5.5 Defects in Extruded Products

Owing to the considerable deformation associated with extrusion operations, a number of defects can occur in extruded products. The defects can be classified into the following categories, illustrated in Figure 19.40:

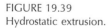

FIGURE 19.39
Hydrostatic extrusion.

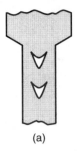
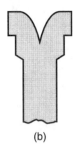
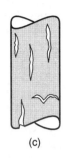

(a)   (b)   (c)

FIGURE 19.40 Some common defects in extrusion: (a) centerburst, (b) piping, and (c) surface cracking.

(a) ***Centerburst***—This defect is an internal crack that develops as a result of tensile stresses along the centerline of the workpart during extrusion. Although tensile stresses may seem unlikely in a compression process such as extrusion, they tend to occur under conditions that cause large deformation in the regions of the work away from the central axis. The significant material movement in these outer regions stretches the material along the center of the work. If stresses are great enough, bursting occurs. Conditions that promote centerburst are high die angles, low extrusion ratios, and impurities in the work metal that serve as starting points for crack defects. The difficult aspect of centerburst is its detection. It is an internal defect that is usually not noticeable by visual observation. Other names sometimes used for this defect include ***arrowhead fracture, center cracking,*** and ***chevron cracking.***

(b) ***Piping***—Piping is a defect associated with direct extrusion. As in Figure 19.40(b), it is the formation of a sink hole in the end of the billet. The use of a dummy block whose diameter is slightly less than that of the billet helps to avoid piping. Other names given to this defect include ***tailpipe*** and ***fishtailing.***

(c) ***Surface cracking***—This defect results from high workpart temperatures that cause cracks to develop at the surface. They often occur when extrusion speed is too high, leading to high strain rates and associated heat generation. Other factors contributing to surface cracking are high friction and surface chilling of high temperature billets in hot extrusion.

19.6 WIRE AND BAR DRAWING

In the context of bulk deformation, ***drawing*** is an operation in which the cross section of a bar, rod, or wire is reduced by pulling it through a die opening, as in Figure 19.41. The general features of the process are similar to those of extrusion. The difference is that the work is pulled through the die in drawing, whereas it is pushed through the die in extrusion. Although the presence of tensile stresses is obvious in drawing, compression also plays a significant role because the metal is squeezed down as it passes through the die

FIGURE 19.41 Drawing of bar, rod, or wire.

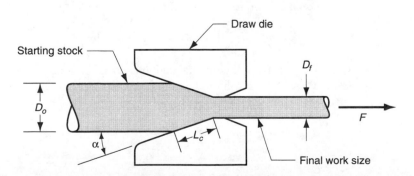

opening. For this reason, the deformation that occurs in drawing is sometimes referred to as indirect compression. Drawing is a term also used in sheet metalworking (Section 20.3). The term *wire and bar drawing* is used to distinguish the drawing process discussed here from the sheet metal process of the same name.

The basic difference between bar drawing and wire drawing is the stock size that is processed. *Bar drawing* is the term used for large diameter bar and rod stock, while *wire drawing* applies to small diameter stock. Wire sizes down to 0.03 mm (0.001 in.) are possible in wire drawing. Although the mechanics of the process are the same for the two cases, the methods, equipment, and even the terminology are somewhat different.

Bar drawing is generally accomplished as a *single-draft* operation—the stock is pulled through one die opening. Because the beginning stock has a large diameter, it is in the form of a straight cylindrical piece rather than coiled. This limits the length of the work that can be drawn, necessitating a batch type operation. By contrast, wire is drawn from coils consisting of several hundred (or even several thousand) feet of wire and is passed through a series of draw dies. The number of dies varies typically between 4 and 12. The term *continuous drawing* is used to describe this type of operation because of the long production runs that are achieved with the wire coils, which can be butt-welded each one to the next to make the operation truly continuous.

In a drawing operation, the change in size of the work is usually given by the area reduction, defined as follows:

$$r = \frac{A_o - A_f}{A_o} \tag{19.28}$$

where r = area reduction in drawing; A_o = original area of work, mm^2 (in.^{2}); and A_f = final area, mm^2 (in.^{2}). Area reduction is often expressed as a percentage.

In bar drawing, rod drawing, and in drawing of large diameter wire for upsetting and heading operations, the term draft is used to denote the before and after difference in size of the processed work. The *draft* is simply the difference between original and final stock diameters:

$$d = D_o - D_f \tag{19.29}$$

where d = draft, mm (in.); D_o = original diameter of work, mm (in.); and D_f = final work diameter, mm (in.).

19.6.1 Analysis of Drawing

In this section, we consider the mechanics of wire and bar drawing—how are stresses and forces computed in the process? We also consider how large a reduction is possible in a drawing operation.

Mechanics of Drawing   If no friction or redundant work occurred in drawing, true strain could be determined as follows:

$$\epsilon = \ln \frac{A_o}{A_f} = \ln \frac{1}{1 - r} \tag{19.30}$$

where A_o and A_f are the original and final cross-sectional areas of the work, as previously defined; and r = drawing reduction as given by Eq. (19.28). The stress that results from this ideal deformation is given by

$$\sigma = \overline{Y}_f \epsilon = \overline{Y}_f \ln \frac{A_o}{A_f} \tag{19.31}$$

where $\bar{Y}_f = \dfrac{K\epsilon^n}{1+n}$ = average flow stress based on the value of strain given by Eq. (19.30).

Because friction is present in drawing and the work metal experiences inhomogeneous deformation, the actual stress is larger than provided by Eq. (19.31). In addition to the ratio A_o/A_f, other variables that influence draw stress are die angle and coefficient of friction at the work–die interface. A number of methods have been proposed for predicting draw stress based on values of these parameters [1], [2], [12], and [18]. We present the equation suggested by Schey [18]:

$$\sigma_d = \bar{Y}_f \left(1 + \frac{\mu}{\tan \alpha}\right) \phi \ln \frac{A_o}{A_f} \qquad (19.32)$$

where σ_d = draw stress, MPa (lb/in.^{2}); μ = die-work coefficient of friction; α = die angle (half angle) as defined in Figure 19.41; and ϕ is a factor that accounts for inhomogeneous deformation, which is determined as follows for a round cross section:

$$\phi = 0.88 + 0.12 \frac{D}{L_c} \qquad (19.33)$$

where D = average diameter of work during drawing, mm (in.); and L_c = contact length of the work with the draw die in Figure 19.41, mm (in.). Values of D and L_c can be determined from the following:

$$D = \frac{D_o + D_f}{2} \qquad (19.34a)$$

$$L_c = \frac{D_o - D_f}{2 \sin \alpha} \qquad (19.34b)$$

The corresponding draw force is then the area of the drawn cross section multiplied by the draw stress:

$$F = A_f \sigma_d = A_f \bar{Y}_f \left(1 + \frac{\mu}{\tan \alpha}\right) \phi \ln \frac{A_o}{A_f} \qquad (19.35)$$

where F = draw force, N (lb); and the other terms are defined above. The power required in a drawing operation is the draw force multiplied by exit velocity of the work.

**EXAMPLE 19.4
Stress and force in
wire drawing.**

Wire is drawn through a draw die with entrance angle = 15°. Starting diameter is 2.5 mm, and final diameter = 2.0 mm. The coefficient of friction at the work-die interface = 0.07. The metal has a strength coefficient K = 205 MPa and a strain hardening exponent n = 0.20. Determine the draw stress and draw force in this operation.

Solution: The values of D and L_c for Eq. (19.33) can be determined using Eq. (19.34). D = 2.25 mm and L_c = 1.0 mm. Thus,

$$\phi = 0.88 + 0.12 \frac{2.25}{1.0} = 1.15$$

The areas before and after drawing are computed as A_o = 4.91 mm^2 and A_f = 3.14 mm^2. The resulting true strain ϵ = ln(4.91/3.14) = 0.446. And the average flow stress in the operation is computed:

$$\bar{Y}_f = \frac{205(0.446)^{0.20}}{1.20} = 145.4 \text{ MPa}$$

Draw stress is given by Eq. (19.32):

$$\sigma_d = (145.4)\left(1 + \frac{0.07}{\tan 15}\right)(1.15)(0.446) = 94.1 \text{ MPa}$$

Finally, the draw force is this stress multiplied by the cross-sectional area of the exiting wire:

$$F = 94.1(3.14) = 295.5 \text{ N}.$$

Maximum Reduction per Pass   A question that may occur to the reader is: Why is more than one step required to achieve the desired reduction in wire drawing? Why not take the entire reduction in a single pass through one die, as in extrusion? The answer can be explained as follows. From the preceding equations, it is clear that as the reduction increases, draw stress increases. If the reduction is large enough, draw stress will exceed the yield strength of the exiting metal. When that happens, the drawn wire will simply elongate instead of new material being squeezed through the die opening. For wire drawing to be successful, maximum draw stress must be less than the yield strength of the exiting metal.

It is a straightforward matter to determine this maximum draw stress and the resulting maximum possible reduction that can be made in one pass, under certain assumptions. Let us assume a perfectly plastic metal ($n = 0$), no friction, and no redundant work. In this ideal case, the maximum possible draw stress is equal to the yield strength of the work material. Expressing this using the equation for draw stress under conditions of ideal deformation, Eq. (19.31), and setting $\overline{Y}_f = Y$ (because $n = 0$),

$$\sigma_d = \overline{Y}_f \ln \frac{A_o}{A_f} = Y \ln \frac{A_o}{A_f} = Y \ln \frac{1}{1 - r} = Y$$

This means that $\ln (A_o/A_f) = \ln (1/(1 - r)) = 1$. Hence, $A_o/A_f = 1/(1 - r)$ must equal the natural logarithm base e. That is, the maximum possible strain is 1.0:

$$\epsilon_{max} = 1.0 \tag{19.36}$$

The maximum possible area ratio is

$$\frac{A_o}{A_f} = e = 2.7183 \tag{19.37}$$

and the maximum possible reduction is

$$r_{max} = \frac{e - 1}{e} = 0.632 \tag{19.38}$$

The value given by Eq. (19.38) is often used as the theoretical maximum reduction possible in a single draw, even though it ignores the effects of friction and redundant work, which would reduce the maximum possible value, and strain hardening, which would increase the maximum possible reduction because the exiting wire would be stronger than the starting metal. In practice, draw reductions per pass are quite below the theoretical limit. Reductions of 0.50 for single-draft bar drawing and 0.30 for multiple-draft wire drawing seem to be the upper limits in industrial operations.

19.6.2 Drawing Practice

Drawing is usually performed as a cold-working operation. It is most frequently used to produce round cross sections, but squares and other shapes are also drawn. Wire drawing is an important industrial process, providing commercial products such as electrical

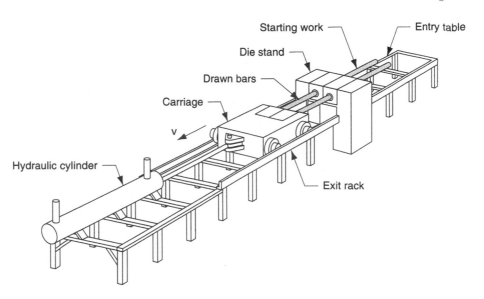

FIGURE 19.42
Hydraulically operated
draw bench for drawing
metal bars.

wire and cable; wire stock for fences, coat hangers, and shopping carts; and rod stock to produce nails, screws, rivets, springs, and other hardware items. Bar drawing is used to produce metal bars for machining, forging, and other processes.

Advantages of drawing in these applications include: (1) close dimensional control, (2) good surface finish, (3) improved mechanical properties such as strength and hardness, and (4) adaptability to economical batch or mass production. Drawing speeds are as high as 50 m/s (10,000 ft/min) for very fine wire. In the case of bar drawing to provide stock for machining, the operation improves the machinability of the bar (Section 24.1).

Drawing Equipment   Bar drawing is accomplished on a machine called a *draw bench,* consisting of an entry table, die stand (which contains the draw die), carriage, and exit rack. The arrangement is shown in Figure 19.42. The carriage is used to pull the stock through the draw die. It is powered by hydraulic cylinders or motor-driven chains. The die stand is often designed to hold more than one die, so that several bars can be pulled simultaneously through their respective dies.

Wire drawing is done on continuous drawing machines that consist of multiple draw dies, separated by accumulating drums between the dies, as in Figure 19.43. Each drum, called a *capstan,* is motor-driven to provide the proper pull force to draw the wire stock through the upstream die. It also maintains a modest tension on the wire as it proceeds to the next draw die in the series. Each die provides a certain amount of reduction in the wire, so that the desired total reduction is achieved by the series. Depending on the metal to be processed and the total reduction, annealing of the wire is sometimes required between groups of dies in the series.

Draw Dies   Figure 19.44 identifies the features of a typical draw die. Four regions of the die can be distinguished: (1) entry, (2) approach angle, (3) bearing surface (land), and (4) back relief. The *entry* region is usually a bell-shaped mouth that does not contact the work. Its purpose is to funnel the lubricant into the die and prevent scoring of work and die surfaces. The *approach* is where the drawing process occurs. It is cone-shaped with an angle (half-angle) normally ranging from about 6° to 20°. The proper angle varies according to work material. The *bearing surface,* or *land,* determines the size of the final drawn stock. Finally, the *back relief* is the exit zone. It

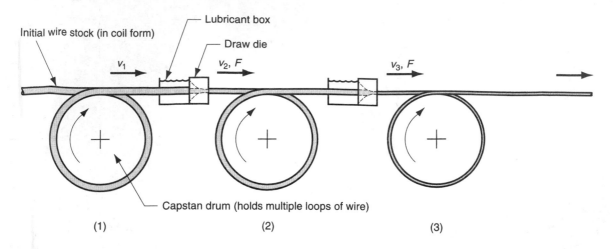

FIGURE 19.43 Continuous drawing of wire.

is provided with a back relief angle (half-angle) of about 30°. Draw dies are made of tool steels or cemented carbides. Dies for high-speed wire drawing operations frequently use inserts made of diamond (both synthetic and natural) for the wear surfaces.

Preparation of the Work Prior to drawing, the beginning stock must be properly prepared. This involves three steps: (1) annealing, (2) cleaning, and (3) pointing. The purpose of annealing is to increase the ductility of the stock to accept deformation during drawing. As previously mentioned, annealing is sometimes needed between steps in continuous drawing. Cleaning of the stock is required to prevent damage of the work surface and draw die. It involves removal of surface contaminants (e.g., scale and rust) by means of chemical pickling or shot blasting. In some cases, prelubrication of the work surface is accomplished subsequent to cleaning.

   Pointing involves the reduction in diameter of the starting end of the stock so that it can be inserted through the draw die to start the process. This is usually accomplished

FIGURE 19.44 Draw die for drawing of round rod or wire.

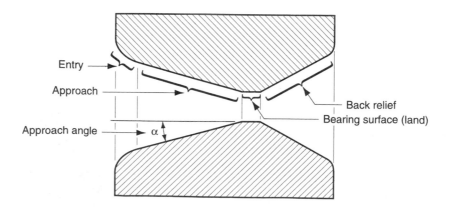

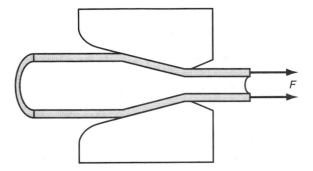

FIGURE 19.45   Tube drawing with no mandrel (tube sinking).

by swaging, rolling, or turning. The pointed end of the stock is then gripped by the carriage jaws or other device to initiate the drawing process.

19.6.3 Tube Drawing

Drawing can be used to reduce the diameter or wall thickness of seamless tubes and pipes, after the initial tubing has been produced by some other process such as extrusion. Tube drawing can be carried out either with or without a mandrel. The simplest method uses no mandrel and is used for diameter reduction, as in Figure 19.45. The term ***tube sinking*** is sometimes applied to this operation.

   The problem with tube drawing in which no mandrel is used, as shown in Figure 19.45, is that it lacks control over the inside diameter and wall thickness of the tube. This is why mandrels of various types are used, two of which are illustrated in Figure 19.46. The first, Figure 19.46(a), uses a ***fixed mandrel*** attached to a long support bar to establish inside diameter and wall thickness during the operation. Practical limitations on the length of the support bar in this method restrict the length of the tube that can be drawn. The second type, shown in (b), uses a ***floating plug*** whose shape is designed so that it finds a "natural" position in the reduction zone of the die. This method removes the limitations on work length present with the fixed mandrel.

FIGURE 19.46   Tube drawing with mandrels: (a) fixed mandrel, (b) floating plug.

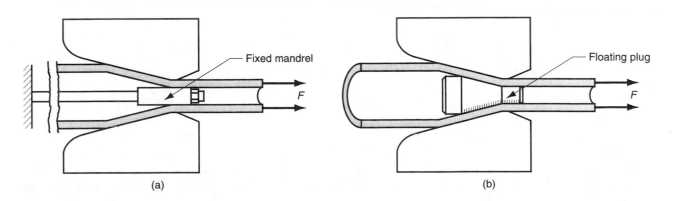

(a)   (b)

REFERENCES

[1] Altan, T., Oh, S-I, and Gegel, H. L. *Metal Forming: Fundamentals and Applications.* ASM International, Materials Park, Ohio, 1983.

[2] Avitzur, B. *Metal Forming: Processes and Analysis.* Robert E. Krieger Publishing Company, Huntington, New York, 1979.

[3] Byrer, T. G., et al. (eds.). *Forging Handbook.* Forging Industry Association, Cleveland, Ohio; and American Society for Metals, Metals Park, Ohio, 1985.

[4] Cook, N. H. *Manufacturing Analysis.* Addison-Wesley Publishing Company, Inc., Reading, Mass., 1966.

[5] DeGarmo, E. P., Black, J. T., and Kohser, R. A. *Materials and Processes in Manufacturing.* 8th ed. John Wiley & Sons, Inc., New York, 1997.

[6] Groover, M. P. "An Experimental Study of the Work Components and Extrusion Strain in the Cold Forward Extrusion of Steel." *Research Report.* Bethlehem Steel Corporation, 1966.

[7] Harris, J. N. *Mechanical Working of Metals.* Pergamon Press, Oxford, England, 1983.

[8] Hosford, W. F., and R. M. Cadell. *Metal Forming: Mechanics and Metallurgy.* 2nd ed. Prentice-Hall, Upper Saddle River, N.J., 1993.

[9] Jensen, J. E. (ed.). *Forging Industry Handbook.* Forging Industry Association, Cleveland, Ohio, 1970.

[10] Johnson, W. "The Pressure for the Cold Extrusion of Lubricated Rod Through Square Dies of Moderate Reduction at Slow Speeds." *Journal of the Institute of Metals.* Vol. 85, 1956–57.

[11] Kalpakjian, S. *Mechanical Processing of Materials.* D. Van Nostrand Company, Inc., Princeton, N.J., 1967, Chapter 5.

[12] Kalpakjian, S., and Schmid, S. R. *Manufacturing Processes for Engineering Materials.* 4th ed. Prentice Hall, Upper Saddle River, N.J., 2001.

[13] Lange, K., et al. (eds.). *Handbook of Metal Forming.* Society of Manufacturing Engineers, Dearborn, Mich., 1995.

[14] Laue, K., and Stenger, H. *Extrusion: Processes, Machinery, and Tooling.* American Society for Metals, Metals Park, Ohio, 1981.

[15] Mielnik, E. M. *Metalworking Science and Engineering.* McGraw-Hill, Inc., New York, 1991.

[16] Roberts, W. L. *Hot Rolling of Steel.* Marcel Dekker, Inc., New York, 1983.

[17] Roberts, W. L. *Cold Rolling of Steel.* Marcel Dekker, Inc., New York, 1978.

[18] Schey, J. A. *Introduction to Manufacturing Processes.* 3rd ed. McGraw-Hill Book Company, New York, 1999.

[19] Wick, C., et al. (eds.). *Tool and Manufacturing Engineers Handbook.* 4th ed. Volume II: *Forming.* Society of Manufacturing Engineers, Dearborn, Mich., 1984.

REVIEW QUESTIONS

19.1. Why are the bulk deformation processes important commercially and technologically?

19.2. List some of the products produced on a rolling mill.

19.3. Identify some of the ways in which force in flat rolling can be reduced.

19.4. What is a *two-high rolling mill*?

19.5. What is a *reversing mill* in rolling?

19.6. Besides flat rolling and shape rolling, identify some additional bulk-forming processes that use rolls to effect the deformation.

19.7. One way to classify forging operations is by the degree to which the work is constrained in the die. By this classification, name the three basic types.

19.8. Why is flash desirable in impression-die forging?

19.9. What are the two basic types of forging equipment?

19.10. What is *isothermal forging*?

19.11. Distinguish between direct and indirect extrusion.

19.12. Name some products that are produced by extrusion.

19.13. What does the *centerburst* defect in extrusion have in common with the *roll-piercing* process?

19.14. In a wire drawing operation, why must the drawing stress never exceed the yield strength of the work metal?

MULTIPLE CHOICE QUIZ

There is a total of 22 correct answers in the following multiple choice questions (some questions have multiple answers that are correct). To attain a perfect score on the quiz, all correct answers must be given, since each correct answer is worth 1 point. For each question, each omitted answer or wrong answer reduces the score by 1 point, and each additional answer beyond the number of answers required reduces the score by 1 point. Percentage score on the quiz is based on the total number of correct answers.

19.1. The maximum possible draft in a rolling operation depends on which of the following parameters (more than one)? (a) coefficient of friction between roll and work, (b) roll diameter, (c) roll velocity, (d) stock thickness, (e) strain, and (f) strength coefficient of the work metal.

19.2. Which of the following rolling mill types are associated with relatively small diameter rolls in contact with the work (more than one)? (a) cluster mill, (b) continuous rolling mill, (c) four-high mill, (d) reversing mill, or (e) three-high configuration.

19.3. Production of pipes and tubes can be accomplished by which of the following bulk deformation processes (more than one)? (a) extrusion, (b) hobbing, (c) ring rolling, (d) roll forging, (e) roll piercing, (f) tube sinking, or (g) upsetting.

19.4. Which of the four basic bulk deformation processes use compression to effect shape change (more than one answer)? (a) bar and wire drawing, (b) extrusion, (c) forging, and (d) rolling.

19.5. Flash in impression-die forging serves no useful purpose and is undesirable because it must be trimmed from the part after forming: (a) true or (b) false.

19.6. Which of the following are classified as forging operations (more than one)? (a) coining, (b) fullering, (c) impact extrusion, (d) roll forging, (e) thread rolling, and (f) upsetting.

19.7. The production of tubing is possible in indirect extrusion but not in direct extrusion: (a) true or (b) false.

19.8. Theoretically, the maximum reduction possible in a wire-drawing operation, under the assumptions of a perfectly plastic metal, no friction, and no redundant work, is which of the following (one answer)? (a) zero, (b) 0.632, (c) 1.0, or (d) 2.7183.

19.9. Which of the following bulk deformation processes are involved in the production of nails for lumber construction (more than one answer)? (a) bar and wire drawing, (b) extrusion, (c) forging, and (d) rolling.

19.10. Johnson's formula is associated with which of the four bulk deformation processes (one answer)? (a) bar and wire drawing, (b) extrusion, (c) forging, and (d) rolling.

PROBLEMS
Rolling

19.1. A 40-mm-thick plate is to be reduced to 30 mm in one pass in a rolling operation. Entrance speed = 16 m/min. Roll radius = 300 mm, and rotational speed = 18.5 rev/min. Determine (a) the minimum required coefficient of friction that would make this rolling operation possible, (b) exit velocity under the assumption that the plate widens by 2% during the operation, and (c) forward slip.

19.2. A 2.0-in.-thick slab is 10.0 in. wide and 12.0 ft long. Thickness is to be reduced in three steps in a hot rolling operation. Each step will reduce the slab to 75% of its previous thickness. It is expected that for this metal and reduction, the slab will widen by 3% in each step. If the entry speed of the slab in the first step is 40 ft/min, and roll speed is the same for the three steps, determine: (a) length and (b) exit velocity of the slab after the final reduction.

19.3. A series of cold-rolling operations are to be used to reduce the thickness of a plate from 50 mm down to 25 mm in a reversing two-high mill. Roll diameter = 700 mm and coefficient of friction between rolls and work = 0.15. The specification is that the draft is to be equal on each pass. Determine (a) minimum number of passes required, and (b) draft for each pass.

19.4. In the previous problem, suppose that the percent reduction were specified to be equal for each pass, rather than the draft. (a) What is the minimum number of passes required? (b) What is the draft for each pass?

19.5. A continuous hot-rolling mill has two stands. Thickness of the starting plate = 25 mm and width = 300 mm. Final thickness is to be 13 mm. Roll radius at each stand = 250 mm. Rotational speed at the first stand = 20 rev/min. Equal drafts of 6 mm are to be taken at each stand. The plate is wide enough relative to its thickness that no increase in width occurs. Under the assumption that the forward slip is equal at each stand, determine: (a) speed v_r at each stand, and (b) forward slip s. (c) Also, determine the exiting speeds at each rolling stand, if the entering speed at the first stand = 26 m/min.

19.6. A continuous hot rolling mill has eight stands. The dimensions of the starting slab are thickness = 3.0 in., width = 15.0 in., and length = 10 ft. The final thickness is to be 0.3 in. Roll diameter at each stand = 36 in., and rotational speed at stand number 1 = 30 rev/min. It is observed that the speed of the slab entering stand 1 = 240 ft/min. Assume that no widening of the slab occurs during the rolling sequence. Percent reduction in thickness is to be equal at all stands, and it is assumed that the forward slip will be equal at each stand. Determine (a) percent reduction at each stand, (b) rotational speed of the rolls at stands 2 through 8, and (c) forward slip. (d) What is the draft at stands 1 and 8? (e) What is the length and exit speed of the final strip exiting stand 8?

19.7. A plate that is 250 mm wide and 25 mm thick is to be reduced in a single pass in a two-high rolling mill to a thickness of 20 mm. The roll has a radius = 500 mm, and its speed = 30 m/min. The work material has a strength coefficient = 240 MPa and a strain hardening exponent = 0.2. Determine (a) roll force, (b) roll torque, and (c) power required to accomplish this operation.

19.8. Solve Problem 19.7 using a roll radius = 250 mm.

19.9. Solve Problem 19.7, only assume a cluster mill with working rolls of radius = 50 mm. Compare the results with the previous two problems, and note the important effect of roll radius on force, torque, and power.

19.10. A 3.0-in.-thick slab that is 9 in. wide is to be reduced in a single pass in a two-high rolling mill to a thickness of 2.50 in. The roll has a radius = 15 in., and its speed = 30 ft/min. The work material has a strength coeffi-

cient = 25,000 lb/in.^{2} and a strain hardening exponent = 0.16. Determine (a) roll force, (b) roll torque, and (c) power required to accomplish this operation.

19.11. A single-pass rolling operation reduces a 20-mm-thick plate to 18 mm. The starting plate is 200 mm wide. Roll radius = 250 mm and rotational speed = 12 rev/min. The work material has a strength coefficient = 600 MPa and a strength coefficient = 0.22. Determine (a) roll force, (b) roll torque, and (c) power required for this operation.

19.12. A hot-rolling mill has rolls of diameter = 24 in. It can exert a maximum force = 400,000 lb. The mill has a maximum horsepower = 100 hp. It is desired to reduce a 1.5 in. thick plate by the maximum possible draft in one pass. The starting plate is 10 in. wide. In the heated condition, the work material has a strength coefficient = 20,000 lb/in.^{2} and a strain-hardening exponent = 0. Determine (a) maximum possible draft, (b) associated true strain, and (c) maximum speed of the rolls for the operation.

19.13. Solve Problem 19.12 except that the operation is warm rolling and the strain-hardening exponent $n = 0.15$. Assume the strength coefficient remains $K = 20,000$ lb/in.^{2}.

Forging

19.14. A cylindrical part is warm upset forged in an open die. $D_o = 50$ mm and $h_o = 40$ mm. Final height = 20 mm. Coefficient of friction at the die–work interface = 0.20. The work material has a flow curve defined by: $K = 600$ MPa and $n = 0.12$. Determine the force in the operation (a) just as the yield point is reached (yield at strain = 0.002), (b) at $h = 30$ mm, and (c) at $h = 20$ mm.

19.15. A cylindrical workpart with $D = 2.5$ in. and $h = 2.5$ in. is upset forged in an open die to a height = 1.5 in. Coefficient of friction at the die–work interface = 0.10. The work material has a flow curve defined by: $K = 40,000$ lb/in.^{2} and $n = 0.15$. Determine the instantaneous force in the operation: (a) just as the yield point is reached (yield at strain = 0.002), (b) at height $h = 2.3$ in., (c) $h = 1.9$ in., and (d) $h = 1.5$ in.

19.16. A cylindrical workpart has a diameter = 2.0 in. and a height = 4.0 in. It is upset forged to a height = 2.5 in. Coefficient of friction at the die–work interface = 0.10. The work material has a flow curve with strength coefficient = 25,000 lb/in.^{2} and strain-hardening exponent = 0.22. Determine the plot of force versus work height.

19.17. A cold-heading operation is performed to produce the head on a steel nail. The strength coefficient for this steel is $K = 550$ MPa, and the strain-hardening exponent $n = 0.24$. Coefficient of friction at the die–work

interface = 0.10. The wire stock out of which the nail is made is 4.75 mm in diameter. The head is to have a diameter = 9.5 mm and a thickness = 1.5 mm. (a) What length of stock must project out of the die in order to provide sufficient volume of material for this upsetting operation? (b) Compute the maximum force that the punch must apply to form the head in this open-die operation.

19.18. Obtain a large common nail (flat head). Measure the head diameter and thickness, as well as the diameter of the nail shank. (a) What stock length must project out of the die in order to provide sufficient material to produce the nail? (b) Using appropriate values for strength coefficient and strain-hardening exponent for the metal out of which the nail is made (Table 3.5), compute the maximum force in the heading operation to form the head.

19.19. A hot upset forging operation is performed in an open die. The initial size of the workpart is $D_o = 25$ mm, and $h_o = 50$ mm. The part is upset to a diameter = 50 mm. The work metal at this elevated temperature yields at 85 MPa ($n = 0$). Coefficient of friction at the die–work interface = 0.40. Determine (a) final height of the part, and (b) maximum force in the operation.

19.20. A hydraulic forging press is capable of exerting a maximum force = 1,000,000 N. A cylindrical workpart is to be cold upset forged. The starting part has diameter = 30 mm and height = 30 mm. The flow curve of the metal is defined by $K = 400$ MPa and $n = 0.2$. Determine the maximum reduction in height to which the part can be compressed with this forging press, if the coefficient of friction = 0.1.

19.21. A part is designed to be hot forged in an impression die. The projected area of the part, including flash, is 15 in.^{2} After trimming, the part has a projected area = 10 in.^{2} Part geometry is relatively simple. As heated the work material yields at 9,000 lb/in.^{2}, and has no tendency to strain harden. Determine the maximum force required to perform the forging operation.

19.22. A connecting rod is designed to be hot forged in an impression die. The projected area of the part is 6,500 mm^2. The design of the die will cause flash to form during forging, so that the area, including flash, will be 9,000 mm^2. The part geometry is considered to be complex. As heated the work material yields at 75 MPa, and has no tendency to strain harden. Determine the maximum force required to perform the operation.

Extrusion

19.23. A cylindrical billet that is 100 mm long and 40 mm in diameter is reduced by indirect (backward) extrusion to a 15 mm diameter. Die angle = 90°. If the Johnson equation has $a = 0.8$ and $b = 1.5$, and the flow curve for the work metal has $K = 750$ MPa and $n = 0.15$, determine (a) extrusion ratio, (b) true strain (homogeneous deformation), (c) extrusion strain, (d) ram pressure, and (e) ram force.

19.24. A 3.0-in.-long cylindrical billet whose diameter = 1.5 in. is reduced by indirect extrusion to a diameter = 0.375 in. Die angle = 90°. In the Johnson equation, $a = 0.8$ and $b = 1.5$. In the flow curve for the work metal, $K = 75,000$ lb/in.^{2} and $n = 0.25$. Determine (a) extrusion ratio, (b) true strain (homogeneous deformation), (c) extrusion strain, (d) ram pressure, (e) ram force, and (f) power if the ram speed = 20 in./min.

19.25. A billet that is 75 mm long with diameter = 35 mm is direct extruded to a diameter of 20 mm. The extrusion die has a die angle = 75°. For the work metal, $K = 600$ MPa and $n = 0.25$. In the Johnson extrusion strain equation, $a = 0.8$ and $b = 1.4$. Determine (a) extrusion ratio, (b) true strain (homogeneous deformation), (c) extrusion strain, and (d) ram pressure at $L = 70$, 40, and 10 mm.

19.26. A 2.0-in.-long billet with diameter of 1.25 in. is direct extruded to a diameter of 0.50 in. The extrusion die angle is 90°. For the work metal, $K = 45,000$ lb/in.^{2} and $n = 0.20$. In the Johnson extrusion strain equation, $a = 0.8$ and $b = 1.5$. Determine (a) extrusion ratio, (b) true strain (homogeneous deformation), (c) extrusion strain, and (d) ram pressure at $L = 2.0$, 1.5, 1.0, 0.5, and 0 in.

19.27. A direct extrusion operation is performed on a cylindrical billet with $L_o = 3.0$ in. and $D_o = 2.0$ in. Die angle = 45° and orifice diameter = 0.50 in. In the Johnson extrusion strain equation, $a = 0.8$ and $b = 1.3$. The operation is carried out hot, and the hot metal yields at 15,000 lb/in.^{2} ($n = 0$). (a) What is the extrusion ratio? (b) Determine the ram position at the point when the metal has been compressed into the cone of the die and starts to extrude through the die opening.

(c) What is the ram pressure corresponding to this position? (d) Also determine the length of the final part if the ram stops its forward movement at the start of the die cone.

19.28. An indirect extrusion process starts with an aluminum billet with a diameter of 2.0 in. and a length of 3.0 in. Final cross section after extrusion is a square with 1.0 in. on a side. The die angle = 90°. The operation is performed cold and the strength coefficient of the metal $K = 26,000$ lb/in.^{2} and strain-hardening exponent $n = 0.20$. In the Johnson extrusion strain equation, $a = 0.8$ and $b = 1.2$. (a) Compute the extrusion ratio, true strain, and extrusion strain. (b) What is the shape factor of the product? (c) If the butt left in the container at the end of the stroke is 0.5 in. thick, what is the length of the extruded section? (d) Determine the ram pressure in the process.

19.29. An L-shaped structural section is direct extruded from an aluminum billet in which $L_o = 250$ mm and $D_o = 88$ mm. Dimensions of the cross section are given in Figure P19.29. Die angle = 90°. Determine (a) extrusion ratio, (b) shape factor, and (c) length of the extruded section if the butt remaining in the container at the end of the ram stroke is 25 mm.

19.30. The flow curve parameters for the aluminum alloy of Problem 19.29 are $K = 240$ MPa and $n = 0.16$. If the die angle in this operation is 90°, and the corresponding Johnson strain equation has constants $a = 0.8$ and $b = 1.5$, compute the maximum force required to drive the ram forward at the start of extrusion.

19.31. A cup-shaped part is backward extruded from an aluminum slug that is 50 mm in diameter. The final dimensions of the cup are OD = 50 mm, ID = 40 mm, height = 100 mm, and thickness of base = 5 mm. Determine (a) extrusion ratio, (b) shape factor, and (c) height of starting slug required to achieve the final dimensions. (d) If the metal has flow curve parameters $K = 400$ MPa and $n = 0.25$, and the constants in the Johnson extrusion strain equation are $a = 0.8$ and $b = 1.5$, determine the extrusion force.

19.32. Determine the shape factor for each of the extrusion die orifice shapes in Figure P19.32.

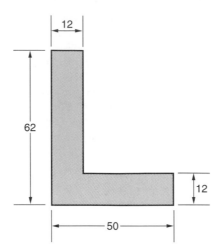

FIGURE P19.29   Part for Problem 19.29 (dimensions are in mm).

Drawing

19.33. Wire of starting diameter = 3.0 mm is drawn to 2.5 mm in a die with entrance angle = 15°. Coefficient of friction at the work–die interface = 0.07. For the work metal, $K = 500$ MPa and $n = 0.30$. Determine (a) area reduction, (b) draw stress, and (c) draw force required for the operation.

19.34. Rod stock is drawn through a draw die with an entrance angle of 12°. Starting diameter = 0.50 in., and final diameter = 0.35 in. Coefficient of friction at the work–die interface = 0.1. The metal has a strength coefficient = 45,000 lb/in.^{2} and a strain-hardening exponent = 0.22. Determine (a) area reduction, (b) draw force for the operation, and (c) horsepower to perform the operation if the exit velocity of the stock = 2 ft/sec.

19.35. Bar stock of initial diameter = 90 mm is drawn with a draft = 15 mm. The draw die has an entrance angle = 18°, and the coefficient of friction at the work–die interface = 0.08. The metal behaves as a perfectly plastic material with yield stress = 105 MPa. Determine (a) area reduction, (b) draw stress, (c) draw force required for the operation, and (d) power to perform the operation if exit velocity = 1.0 m/min.

19.36. Wire stock of initial diameter = 0.125 in. is drawn through two dies each providing a 0.20 area reduction. The starting metal has a strength coefficient = 40,000 lb/in.^{2} and a strain-hardening exponent = 0.15. Each die has an entrance angle of 12°, and the coefficient of friction at the work–die interface is estimated to be 0.10. The motors driving the capstans at the die exits can each deliver 1.50 hp at 90% efficiency. Determine the maximum possible speed of the wire as it exits the second die.

FIGURE P19.32   Cross-sectional shapes for Problem 19.32 (dimensions are mm): (a) rectangular bar, (b) tube, (c) channel, and (d) cooling fins.

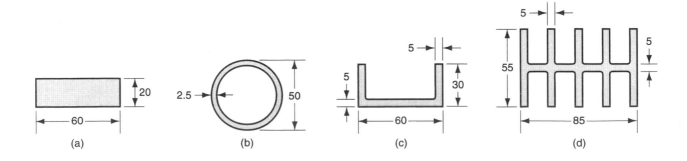

(a)   (b)   (c)   (d)

20 SHEET METALWORKING

CHAPTER CONTENTS

20.1 Cutting Operations
    20.1.1 Shearing, Blanking, and Punching
    20.1.2 Engineering Analysis of Sheet Metal Cutting
    20.1.3 Other Sheet-Metal-Cutting Operations
20.2 Bending Operations
    20.2.1 V-Bending and Edge Bending
    20.2.2 Engineering Analysis of Bending
    20.2.3 Other Bending and Related Forming Operations
20.3 Drawing
    20.3.1 Mechanics of Drawing
    20.3.2 Engineering Analysis of Drawing
    20.3.3 Other Drawing Operations
    20.3.4 Defects in Drawing
20.4 Other Sheet-Metal-Forming Operations
    20.4.1 Operations Performed with Conventional Dies
    20.4.2 Rubber Forming Processes
20.5 Dies and Presses for Sheet Metal Processes
    20.5.1 Dies
    20.5.2 Presses
20.6 Sheet-Metal Operations Not Performed on Presses
    20.6.1 Stretch Forming
    20.6.2 Roll Bending and Roll Forming
    20.6.3 Spinning
    20.6.4 High-Energy-Rate Forming
20.7 Bending of Tube Stock

Sheet metalworking includes cutting and forming operations performed on relatively thin sheets of metal. Typical sheet metal thicknesses are between 0.4 mm (1/64 in.) and 6 mm (1/4 in.). When thickness exceeds about 6 mm, the stock is usually referred to as plate rather than sheet. The sheet or plate stock used in sheet metalworking is produced by rolling (Section 19.1).

The commercial importance of sheet metalworking is significant. Consider the number of consumer and industrial products that include sheet or plate metal parts: automobile and truck bodies, airplanes, railway cars, farm and construction equipment, appliances, office furniture, computers, and more. Although these examples are conspicuous because they have sheet-metal exteriors, many of their internal components are also made of sheet or plate stock. Sheet metal parts are generally characterized by high strength, good dimensional accuracy, good surface finish, and relatively low cost. For components that must be made in large quantities, economical mass production operations can be designed to process the sheet metal.

Most sheet-metal processing is performed at room temperature (cold working). The exceptions are when the stock is thick, the metal is brittle, or the deformation is significant. These are usually cases of warm working rather than hot working.

The three major categories of sheet metal processes are cutting, bending, and drawing. Cutting is used to separate large sheets into smaller pieces, to cut out a part perimeter, or to make holes in a part. Bending and drawing are used to form sheet metal parts into their required shapes.

The tooling used to perform sheet metalwork is called a **punch-and-die.** Most sheet-metal operations are performed on machine tools called *presses.* The term **stamping press** is used to distinguish these presses from forging and extrusion presses. The sheet metal products are called **stampings.** To facilitate mass production, the sheet metal is often presented to the press as long strips or coils. Various types of punch-and-die tooling and stamping presses are described in Section 20.5. Final sections of the chapter cover various operations that do not utilize conventional punch-and-die tooling, and most of them are not performed on stamping presses.

20.1 CUTTING OPERATIONS

Cutting of sheet metal is accomplished by a shearing action between two sharp cutting edges. The shearing action is depicted in the four stop-action sketches of Figure 20.1, in which the upper cutting edge (the punch) sweeps down past a stationary lower cutting

FIGURE 20.1   Shearing of sheet metal between two cutting edges: (1) just before the punch contacts work; (2) punch begins to push into work, causing plastic deformation; (3) punch compresses and penetrates into work causing a smooth cut surface; and (4) fracture is initiated at the opposing cutting edges that separate the sheet. Symbols v and F indicate motion and applied force, respectively, t = stock thickness, c = clearance.

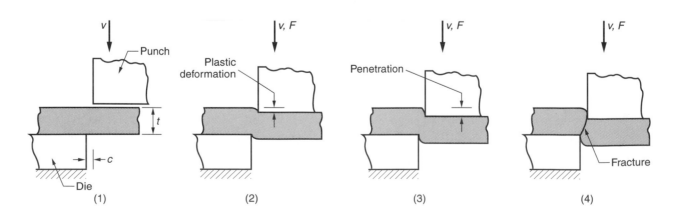

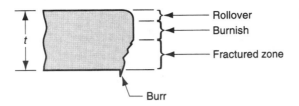

FIGURE 20.2 Characteristic sheared edges of the work.

edge (the die). As the punch begins to push into the work, ***plastic deformation*** occurs in the surfaces of the sheet. As the punch moves downward, ***penetration*** occurs in which the punch compresses the sheet and cuts into the metal. This penetration zone is generally about one-third the thickness of the sheet. As the punch continues to travel into the work, *fracture* is initiated in the work at the two cutting edges. If the clearance between the punch and die is correct, the two fracture lines meet, resulting in a clean separation of the work into two pieces.

The sheared edges of the sheet have characteristic features as in Figure 20.2. At the top of the cut surface is a region called the ***rollover.*** This corresponds to the depression made by the punch in the work prior to cutting. It is where initial plastic deformation occurred in the work. Just below the rollover is a relatively smooth region called the ***burnish.*** This results from penetration of the punch into the work before fracture began. Beneath the burnish is the ***fractured zone,*** a relatively rough surface of the cut edge where continued downward movement of the punch caused fracture of the metal. Finally, at the bottom of the edge is a ***burr,*** a sharp corner on the edge caused by elongation of the metal during final separation of the two pieces.

20.1.1 Shearing, Blanking, and Punching

There are three principal operations in pressworking that cut metal by the shearing mechanism just described: shearing, blanking, and punching.

Shearing is a sheet-metal cutting operation along a straight line between two cutting edges, as shown in Figure 20.3(a). Shearing is typically used to cut large sheets into smaller sections for subsequent pressworking operations. It is performed on a machine called a ***power shears,*** or ***squaring shears.*** The upper blade of the power shears is often inclined, as shown in Figure 20.3(b), to reduce the required cutting force.

Blanking involves cutting of the sheet metal along a closed outline in a single step to separate the piece from the surrounding stock, as in Figure 20.4(a). The part that is cut out is the desired product in the operation and is called the ***blank. Punching*** is similar to blanking

FIGURE 20.3 Shearing operation: (a) side view of the shearing operation; (b) front view of power shears equipped with inclined upper cutting blade. Symbol *v* indicates motion.

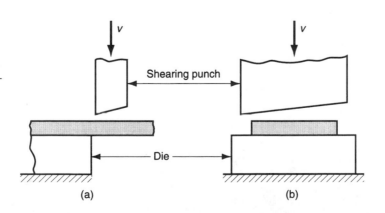

(a) (b)

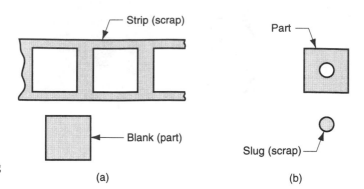

FIGURE 20.4 (a) Blanking and (b) punching.

(a)

(b)

except that the piece that is cut out is scrap, called the *slug.* The remaining stock is the desired part. The distinction is illustrated in Figure 20.4(b).

20.1.2 Engineering Analysis of Sheet-Metal Cutting

Important parameters in sheet-metal cutting are clearance between punch and die, stock thickness, type of metal and its strength, and length of the cut. Let us examine some of the aspects of the relationships involved.

Clearance The *clearance c* in a shearing operation is the distance between the punch and die, as shown in Figure 20.1(a). Typical clearances in conventional presswork range between 4% and 8% of the sheet metal thickness t. The effect of improper clearances is illustrated in Figure 20.5. If the clearance is too small, then the fracture lines tend to pass each other, causing a double burnishing and larger cutting forces. If the clearance is too large, the metal becomes pinched between the cutting edges and an excessive burr results. In special operations requiring very straight edges, such as shaving and fine blanking (Section 20.1.3), clearance is only about 1% of stock thickness.

The correct clearance depends on sheet metal type and thickness. The recommended clearance can be calculated by the following formula:

$$c = at \qquad (20.1)$$

FIGURE 20.5 Effect of clearance: (a) clearance too small causes less-than-optimal fracture and excessive forces; and (b) clearance too large causes oversized burr. Symbols v and F indicate motion and applied force, respectively.

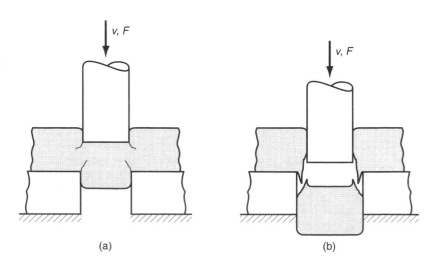

(a)

(b)

TABLE 20.1   Allowance value *a* for three sheet metal groups.

Metal group	*a*
1100S and 5052S aluminum alloys, all tempers.	0.045
2024ST and 6061ST aluminum alloys; brass, all tempers; soft cold rolled steel, soft stainless steel.	0.060
Cold rolled steel, half hard; stainless steel, half hard and full hard.	0.075

Compiled from [2].

where c = clearance, mm (in.); a = allowance; and t = stock thickness, mm (in.). Allowance is determined according to type of metal. For convenience, metals are classified into three groups given in Table 20.1, with an associated allowance value for each group.

These calculated clearance values can be applied to conventional blanking and hole-punching operations to determine the proper punch and die sizes. The die opening must always be larger than the punch size. Whether to add the clearance value to the die size or subtract it from the punch size depends on whether the part being cut out is a blank or a slug, as illustrated in Figure 20.6 for a circular part. Because of the geometry of the sheared edge, the outer dimension of the part cut out of the sheet will be larger than the hole size. Thus, punch and die sizes for a round blank of diameter D_b are determined as

$$\text{Blanking punch diameter} = D_b - 2c \qquad (20.2a)$$

$$\text{Blanking die diameter} = D_b \qquad (20.2b)$$

Punch and die sizes for a round hole of diameter D_h are determined as

$$\text{Hole punch diameter} = D_h \qquad (20.3a)$$

$$\text{Hole die diameter} = D_h + 2c \qquad (20.3b)$$

In order for the slug or blank to drop through the die, the die opening must have an *angular clearance* (see Figure 20.7) of 0.25° to 1.5° on each side.

Cutting Forces   Estimates of cutting force are important because this force determines the size (tonnage) of the press needed. Cutting force F in sheet metalworking can be determined by:

$$F = StL \qquad (20.4)$$

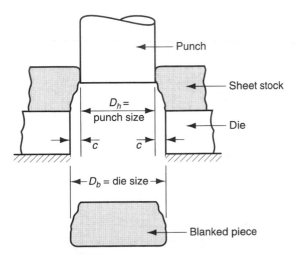

FIGURE 20.6   Die size determines blank size D_b; punch size determines hole size D_h.; c = clearance.

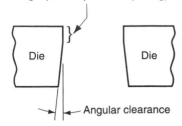

Straight portion (for resharpening)   FIGURE 20.7   Angular clearance.

Angular clearance

where S = shear strength of the sheet metal, MPa (lb/in.^{2}); t = stock thickness, mm (in.), and L = length of the cut edge, mm (in.). In blanking, punching, slotting, and similar operations, L is the perimeter length of the blank or hole being cut. The minor effect of clearance in determining the value of L can be neglected.

If shear strength is unknown, an alternative way of estimating the cutting force is to use the tensile strength, as follows:

$$F = 0.7 \; TStL \qquad (20.5)$$

where TS = ultimate tensile strength MPa (lb/in.^{2}).

These equations for estimating cutting force assume that the entire cut along the sheared edge length L is made at the same time. In this case the cutting force will be a maximum. It is possible to reduce the maximum force by using an angled cutting edge on the punch or die, as in Figure 20.3(b). The angle (called the **shear angle**), spreads the cut over time and reduces the force experienced at any one moment. However, the total energy required in the operation is the same, whether it is concentrated into a brief moment or distributed over a longer time period.

EXAMPLE 20.1 Blanking clearance and force.

A round disk of 150 mm diameter is to be blanked from a strip of 3.2 mm half hard cold-rolled steel whose shear strength = 310 MPa. Determine (a) the appropriate punch and die diameters, and (b) blanking force.

Solution: (a) The clearance allowance for half hard cold-rolled steel is a = 0.075. Accordingly,

$$c = 0.075(3.2 \text{ mm}) = \textbf{0.24 mm}$$

The blank is to have a diameter = 150 mm, and die size determines blank size. Therefore,

Die opening diameter = **150.00 mm**

Punch diameter = 150 − 2(0.24) = **149.52 mm**

(b) To determine the blanking force, we assume that the entire perimeter of the part is blanked at one time. The length of the cut edge is

$$L = \pi D_b = 150\pi = 471.2 \text{ mm}$$

and the force is

$$F = 310(471.2)(3.2) = \textbf{467,469 N} \text{ (This is about 53 tons.)}$$

20.1.3 Other Sheet-Metal-Cutting Operations

In addition to shearing, blanking, and punching, there are several other cutting operations in pressworking. The cutting mechanism in each case involves the same shearing action discussed above.

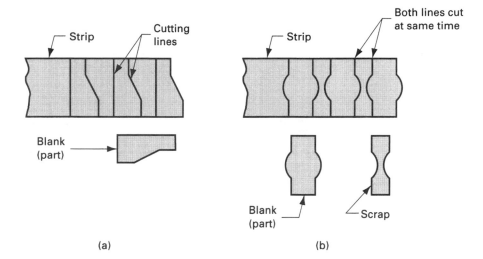

FIGURE 20.8   (a) Cutoff and (b) parting.

(a)   (b)

Cutoff and Parting   *Cutoff* is a shearing operation in which blanks are separated from a sheet-metal strip by cutting the opposite sides of the part in sequence, as shown in Figure 20.8(a). With each cut, a new part is produced. The features of a cutoff operation that distinguish it from a conventional shearing operation are: (1) the cut edges are not necessarily straight; and (2) the blanks can be nested on the strip in such a way that scrap is avoided.

   Parting involves cutting a sheet-metal strip by a punch with two cutting edges that match the opposite sides of the blank, as shown in Figure 20.8(b). This might be required because the part outline has an irregular shape that precludes perfect nesting of the blanks on the strip. Parting is less efficient than cutoff in the sense that it results in some wasted material.

Slotting, Perforating, and Notching   *Slotting* is a punching operation that cuts out an elongated or rectangular hole, as pictured in Figure 20.9(a). *Perforating* involves the simultaneous punching of a pattern of holes in sheet metal, as in Figure 20.9(b). The hole pattern is usually for decorative purposes, or to allow passage of light, gas, or fluid.

   To obtain the desired outline of a blank, portions of the sheet metal are often removed by notching and seminotching. *Notching* involves cutting out a portion of metal from the side of the sheet or strip. *Seminotching* removes a portion of metal from the interior of the sheet. These operations are depicted in Figure 20.9(c). Seminotching might seem to the reader to be the same as a punching or slotting operation. The difference is that the metal removed by seminotching creates part of the blank outline, while punching and slotting create holes in the blank.

FIGURE 20.9   (a) Slotting, (b) perforating, (c) notching and seminotching. Symbol *v* indicates motion of strip.

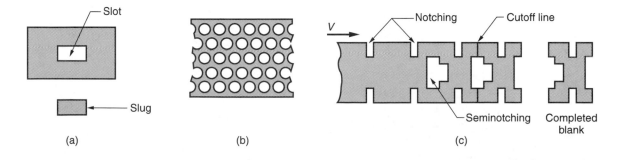

(a)   (b)   (c)

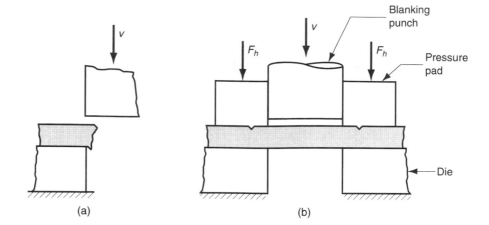

FIGURE 20.10 (a) Shaving, and (b) fine blanking. Symbols: v = motion of punch, F_h = blank holding force.

Trimming, Shaving, and Fine Blanking *Trimming* is a cutting operation performed on a formed part to remove excess metal and establish size. The term has the same basic meaning here as in forging (Section 19.4). A typical example in sheet metalwork is trimming the upper portion of a deep drawn cup to leave the desired dimensions on the cup.

Shaving is a shearing operation performed with very small clearance to obtain accurate dimensions and cut edges that are smooth and straight, as pictured in Figure 20.10(a). Shaving is typically performed as a secondary or finishing operation on parts that have been previously cut.

Fine blanking is a shearing operation used to blank sheet metal parts with close tolerances and smooth, straight edges in one step, illustrated in Figure 20.10(b). At the start of the cycle, a pressure pad with a V-shaped projection applies a holding force F_h against the work adjacent to the punch in order to compress the metal and prevent distortion. The punch then descends with a slower-than-normal velocity and smaller clearances to provide the desired dimensions and cut edges. The process is usually reserved for relatively small stock thicknesses.

20.2 BENDING OPERATIONS

Bending in sheet-metal work is defined as the straining of the metal around a straight axis, as in Figure 20.11. During the bending operation, the metal on the inside of the neutral plane is compressed, while the metal on the outside of the neutral plane is stretched. These strain conditions can be seen in Figure 20.11(b). The metal is plastically deformed

FIGURE 20.11 (a) Bending of sheet metal; (b) both compression and tensile elongation of the metal occur in bending.

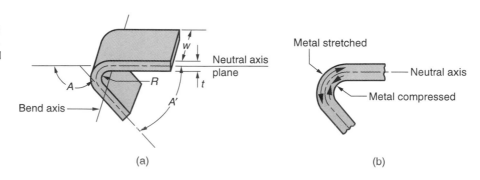

so that the bend takes a permanent set upon removal of the stresses that caused it. Bending produces little or no change in the thickness of the sheet metal.

20.2.1 V-Bending and Edge Bending

Bending operations are performed using punch and die tooling. The two common bending methods and associated tooling are V-bending, performed with a V-die; and edge bending, performed with a wiping die. These methods are illustrated in Figure 20.12.

In ***V-bending,*** the sheet metal is bent between a V-shaped punch and die. Included angles ranging from very obtuse to very acute can be made with V-dies. V-bending is generally used for low-production operations. It is often performed on a press brake (Section 20.5.2), and the associated V-dies are relatively simple and inexpensive.

Edge bending involves cantilever loading of the sheet metal. A pressure pad is used to apply a force F_h to hold the base of the part against the die, while the punch forces the part to yield and bend over the edge of the die. In the setup shown in Figure 20.12(b), edge bending is limited to bends of 90° or less. More complicated wiping dies can be designed for bend angles greater than 90°. Because of the pressure pad, wiping dies are more complicated and costly than V-dies and are generally used for high production work.

20.2.2 Engineering Analysis of Bending

Some of the important terms in sheet-metal bending are identified in Figure 20.11. The metal of thickness t is bent through an angle called the bend angle A. This results in a sheet-metal part with an included angle A', where $A + A' = 180°$. The bend radius R is normally specified on the inside of the part, rather than at the neutral axis, and is determined by the radius on the tooling used to perform the operation. The bend is made over the width of the workpiece w.

Bend Allowance   If the bend radius is small relative to stock thickness, the metal tends to stretch during bending. It is important to be able to estimate the amount of stretching that occurs, if any, so that the final part length will match the specified dimension.

FIGURE 20.12   Two common bending methods: (a) V-bending and (b) edge bending; (1) before and (2) after bending. Symbols: v = motion, F = applied bending force, F_h = blank holding force.

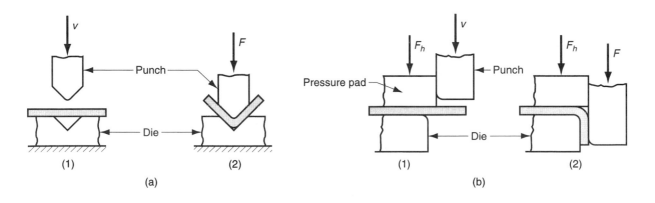

The problem is to determine the length of the neutral axis before bending to account for stretching of the final bent section. This length is called the **bend allowance,** and it can be estimated as follows:

$$BA = 2\pi\frac{A}{360}(R + K_{ba}t)$$
(20.6)

where BA = bend allowance, mm (in.); A = bend angle, degrees; R = bend radius, mm (in.); t = stock thickness, mm (in.); and K_{ba} is factor to estimate stretching. The following design values are recommended for K_{ba} [2]: if $R < 2t$, $K_{ba} = 0.33$; and if $R \geq 2t$, $K_{ba} = 0.50$. The values of K_{ba} predict that stretching occurs only if bend radius is small relative to sheet thickness.

Springback   When the bending pressure is removed at the end of the deformation operation, elastic energy remains in the bent part, causing it to recover partially toward its original shape. This elastic recovery is called **springback,** defined as the increase in included angle of the bent part relative to the included angle of the forming tool after the tool is removed. This is illustrated in Figure 20.13 and is expressed as

$$SB = \frac{A' - A'_b}{A'_b}$$
(20.7)

where SB = springback; A' = included angle of the sheet metal part, degrees; and A_b' = included angle of the bending tool, degrees. Although not as obvious, an increase in the bend radius also occurs due to elastic recovery. The amount of springback increases with modulus of elasticity E and yield strength Y of the work metal.

Compensation for springback can be accomplished by several methods. Two common methods are overbending and bottoming. In **overbending,** the punch angle and radius are fabricated slightly smaller than the specified angle on the final part so that the sheet metal springs back to the desired value. **Bottoming** involves squeezing the part at the end of the stroke, thus plastically deforming it in the bend region.

Bending Force   The force required to perform bending depends on the geometry of the punch-and-die and the strength, thickness, and length of the sheet metal. The max-

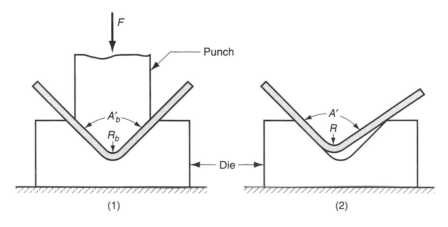

FIGURE 20.13   Springback in bending shows itself as a decrease in bend angle and an increase in bend radius: (1) during the operation, the work is forced to take the radius R_b and included angle A_b' determined by the bending tool (punch in V-bending); (2) after the punch is removed, the work springs back to radius R and included angle A'. Symbol: F = applied bending force.

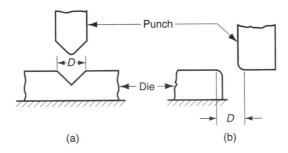

FIGURE 20.14 Die opening dimension D: (a) V-die, (b) wiping die.

(a)

(b)

imum bending force can be estimated by means of an equation based on bending of a simple beam in mechanics:

$$F = \frac{K_{bf}TSwt^2}{D} \qquad (20.8)$$

where F = bending force, N (lb); TS = tensile strength of the sheet metal, MPa (lb/in.^{2}); w = width of part in the direction of the bend axis, mm (in.); t = stock thickness, mm (in.); and D = die opening dimension as defined in Figure 20.14, mm (in.). Eq. (20.8) is based on bending of a simple beam in mechanics, and K_{bf} is a constant that accounts for differences encountered in an actual bending process. Its value depends on type of bending: for V-bending, K_{bf} = 1.33; and for edge bending, K_{bf} = 0.33.

EXAMPLE 20.2 Sheet-metal bending.

A sheet-metal blank is to be bent as shown in Figure 20.15. The metal has a modulus of elasticity E = 205 (10^3) MPa, yield strength Y = 275 MPa, and tensile strength TS = 450 MPa. Determine (a) the starting blank size and (b) the bending force if a V-die is used with a die opening width = 25 mm.

Solution: (a) The starting blank = 44.5 mm wide. Its length = 38 + BA + 25 (mm). For the included angle A' = 120°, the bend angle A = 60°. The value of K_{ba} in Eq. (20.6) = 0.33 since R/t = 4.75/3.2 = 1.48 (less than 2.0).

$$BA = 2\pi \frac{60}{360}(4.75 + 0.33 \times 3.2) = 6.08 \text{ mm.}$$

Length of the blank is therefore 38 + 6.08 + 25 = 69.08 mm.
(b) Force is obtained from Eq. (20.8) using K_{bf} = 1.33.

$$F = \frac{1.33(450)(44.5)(3.2)^2}{25} = 10,909 \text{ N}$$

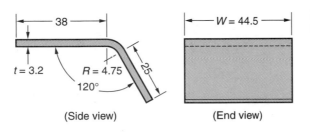

$t = 3.2$ $R = 4.75$

$120°$

(Side view)

$W = 44.5$

(End view)

FIGURE 20.15 Sheet-metal part of Example 20.2 (dimensions in mm).

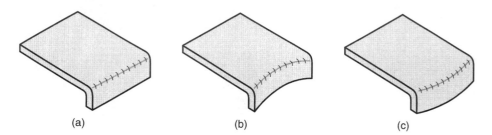

FIGURE 20.16   Flanging:
(a) straight flanging,
(b) stretch flanging, and
(c) shrink flanging.

(a)   (b)   (c)

20.2.3 Other Bending and Related Forming Operations

Some operations to bend sheet metal involve bending over a curved axis rather than a straight axis, or they have other features that differentiate them from the basic operations already described.

Flanging, Hemming, Seaming, and Curling   *Flanging* is a bending operation in which the edge of a sheet metal part is bent at a 90° angle (usually) to form a rim or flange. It is often used to strengthen or stiffen sheet metal. The flange can be formed over a straight bend axis, as illustrated in Figure 20.16(a), or it can involve some stretching or shrinking of the metal, as in (b) and (c).

Hemming involves bending the edge of the sheet over on itself, in more than one bending step. This is often done to eliminate the sharp edge on the piece, to increase stiffness, and to improve appearance. *Seaming* is a related operation in which two sheet-metal edges are assembled. Hemming and seaming are illustrated in Figure 20.17(a) and (b).

Curling, also called *beading,* forms the edges of the part into a roll or curl, as in Figure 20.17(c). As in hemming, it is done for purposes of safety, strength, and aesthetics. Examples of products in which curling is used include hinges, pots and pans, and pocket-watch cases. These examples show that curling can be performed over straight or curved bend axes.

Miscellaneous Bending Operations   Various other bending operations are depicted in Figure 20.18 to illustrate the variety of shapes that can be bent. Most of these operations are performed in relatively simple dies similar to V-dies.

FIGURE 20.17
(a) Hemming, (b) seaming,
and (c) curling.

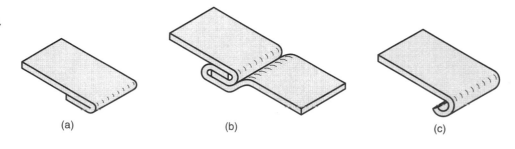

(a)   (b)   (c)

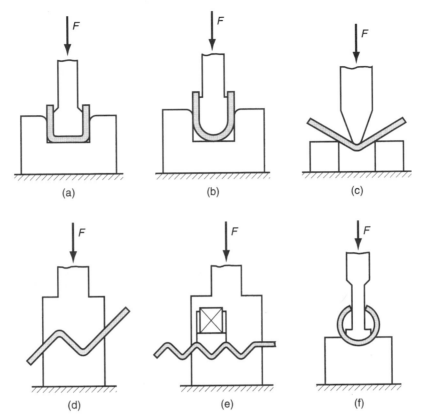

FIGURE 20.18
Miscellaneous bending operations: (a) channel bending, (b) U-bending, (c) air bending, (d) offset bending, (e) corrugating, and (f) tube forming. Symbol: F = applied force.

20.3 DRAWING

Drawing is a sheet-metal-forming operation used to make cup-shaped, box-shaped, or other complex-curved, hollow-shaped parts. It is performed by placing a piece of sheet metal over a die cavity and then pushing the metal into the opening with a punch, as in Figure 20.19. The blank must usually be held down flat against the die by a blankholder. Common parts made by drawing include beverage cans, ammunition shells, sinks, cooking pots, and automobile body panels.

20.3.1 Mechanics of Drawing

Drawing of a cup-shaped part is the basic drawing operation, with dimensions and parameters as pictured in Figure 20.19. A blank of diameter D_b is drawn into a die by means of a punch of diameter D_p. The punch and die must have corner radii, given by R_p and R_d. If the punch and die were to have sharp corners (R_p and $R_d = 0$), a hole-punching operation (and not a very good one) would be accomplished rather than a drawing operation. The sides of the punch and die are separated by a clearance c. This clearance in drawing is about 10% greater than the stock thickness:

$$c = 1.1\ t \tag{20.9}$$

The punch applies a downward force F to accomplish the deformation of the metal, and a downward holding force F_h is applied by the blankholder, as shown in the sketch.

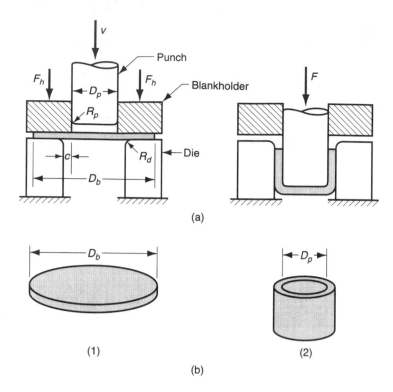

FIGURE 20.19 (a) Drawing of a cup-shaped part: (1) start of operation before punch contacts work, and (2) near end of stroke; and (b) corresponding workpart: (1) starting blank, and (2) drawn part. Symbols: c = clearance, D_b = blank diameter, D_p = punch diameter, R_d = die corner radius, R_p = punch corner radius, F = drawing force, F_h = holding force.

As the punch proceeds downward toward its final bottom position, the work experiences a complex sequence of stresses and strains as it is gradually formed into the shape defined by the punch and die cavity. The stages in the deformation process are illustrated in Figure 20.20. As the punch first begins to push into the work, the metal is subjected to a **bending** operation. The sheet is simply bent over the corner of the punch and the corner of the die, as in Figure 20.20(2). The outside perimeter of the blank moves in toward the center in this first stage, but only slightly.

As the punch moves further down, a **straightening** action occurs in the metal that was previously bent over the die radius, as in Figure 20.20(3). The metal at the bottom of the cup, as well as along the punch radius, has been moved downward with the punch, but the metal that was bent over the die radius must now be straightened in order to be pulled into the clearance to form the wall of the cylinder. At the same time, more metal must be added to replace that being used in the cylinder wall. This new metal comes from the outside edge of the blank. The metal in the outer portions of the blank is pulled or **drawn** toward the die opening to resupply the previously bent and straightened metal now forming the cylinder wall. This type of metal flow through a constricted space gives the drawing process its name.

During this stage of the process, friction and compression play important roles in the flange of the blank. In order for the material in the flange to move toward the die opening, **friction** between the sheet metal and the surfaces of the blankholder and the die must be overcome. Initially, static friction is involved until the metal starts to move; then, after metal flow begins, dynamic friction governs the process. The magnitude of the holding force applied by the blankholder, as well as the friction conditions at the two interfaces, are determining factors in the success of this aspect of the drawing operation. Lubricants or drawing compounds are generally used to reduce friction forces. In addition to friction, **compression** is also occurring in the outer edge of the

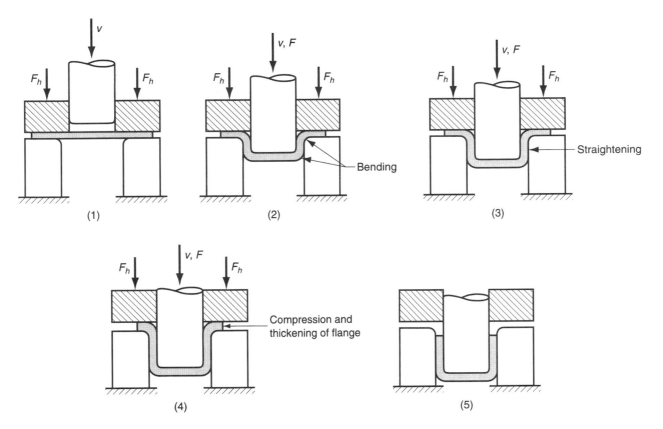

FIGURE 20.20   Stages in deformation of the work in deep drawing: (1) punch makes initial contact with work, (2) bending, (3) straightening, (4) friction and compression, and (5) final cup shape showing effects of thinning in the cup walls. Symbols: v = motion of punch, F = punch force, F_h = blankholder force.

blank. As the metal in this portion of the blank is drawn toward the center, the outer perimeter becomes smaller. Because the volume of metal remains constant, the metal is squeezed and becomes thicker as the perimeter is reduced. This often results in wrinkling of the remaining flange of the blank, especially when thin sheet metal is drawn, or when the blankholder force is too low. It is a condition which cannot be corrected once it has occurred. The friction and compression effects are illustrated in Figure 20.20(4).

The holding force applied by the blankholder is now seen to be a critical factor in deep drawing. If it is too small, wrinkling occurs. If it is too large, it prevents the metal from flowing properly toward the die cavity, resulting in stretching and possible tearing of the sheet metal. Determining the proper holding force involves a delicate balance between these opposing factors.

Progressive downward motion of the punch results in a continuation of the metal flow caused by drawing and compression. In addition, some *thinning* of the cylinder wall occurs, as in Figure 20.20(5). The force being applied by the punch is opposed by the metal in the form of deformation and friction in the operation. A portion of the deformation involves stretching and thinning of the metal as it is pulled over the edge of the die opening. Up to 25% thinning of the side wall may occur in a successful drawing operation, mostly near the base of the cup.

20.3.2 Engineering Analysis of Drawing

It is important to assess the limitations on the amount of drawing that can be accomplished. This is often guided by simple measures that can be readily calculated for a given operation. In addition, drawing force and holding force are important process variables. Finally, the starting blank size must be determined.

Measures of Drawing   One of the measures of the severity of a deep drawing operation is the *drawing ratio DR*. This is most easily defined for a cylindrical shape as the ratio of blank diameter D_b to punch diameter D_p. In equation form,

$$DR = \frac{D_b}{D_p} \tag{20.10}$$

The drawing ratio provides an indication, albeit a crude one, of the severity of a given drawing operation. The greater the ratio, the more severe is the operation. An approximate upper limit on the drawing ratio is a value of 2.0. The actual limiting value for a given operation depends on punch and die corner radii (R_p and R_d), friction conditions, depth of draw, and characteristics of the sheet metal (e.g., ductility, degree of directionality of strength properties in the metal).

Another way to characterize a given drawing operation is by the *reduction r*, where

$$r = \frac{D_b - D_p}{D_b} \tag{20.11}$$

It is very closely related to drawing ratio. Consistent with the previous limit on $DR (DR \le 2.0)$, the value of reduction r should be less than 0.50.

A third measure in deep drawing is the *thickness-to-diameter ratio t/D_b* (thickness of the starting blank t divided by the blank diameter D_b). Often expressed as a percent, it is desirable for the t/D_b ratio to be greater than 1%. As t/D_b decreases, tendency for wrinkling (Section 20.3.4) increases.

In cases where these limits on drawing ratio, reduction, and t/D_b ratio are exceeded by the design of the drawn part, the blank must be drawn in two or more steps, sometimes with annealing between the steps.

**EXAMPLE 20.3
Cup drawing.**

A drawing operation is used to form a cylindrical cup with inside diameter = 75 mm and height = 50 mm. The starting blank size = 138 mm and the stock thickness = 2.4 mm. Based on these data, is the operation feasible?

Solution: To assess feasibility, we determine the drawing ratio, reduction, and thickness-to-diameter ratio.

$$DR = 138/75 = 1.84$$

$$r = (138 - 75)/138 = 0.4565 = 45.65\%$$

$$t/D_b = 2.4/138 = 0.017 = 1.7\%$$

According to these measures, the drawing operation is feasible. The drawing ratio is less than 2.0, the reduction is less than 50%, and the t/D_b ratio is greater than 1%. These are general guidelines frequently used to indicate technical feasibility.   ▪

Forces The *drawing force* required to perform a given operation can be estimated roughly by this formula:

$$F = \pi D_p t \, (TS) \left(\frac{D_b}{D_p} - 0.7 \right) \qquad (20.12)$$

where F = drawing force, N (lb); t = original blank thickness, mm (in.); TS = tensile strength, MPa (lb/in.^{2}); and D_b and D_p are the starting blank diameter and punch diameter, respectively, mm (in.). The constant 0.7 is a correction factor to account for friction. Eq. (20.12) estimates the maximum force in the operation. The drawing force varies throughout the downward movement of the punch, usually reaching its maximum value at about one-third the length of the punch stroke.

The *holding force* is an important factor in a drawing operation. As a rough approximation, the holding pressure can be set at a value = 0.015 of the yield strength of the sheet metal [7]. This value is then multiplied by that portion of the starting area of the blank that is to be held by the blankholder. In equation form,

$$F_h = 0.015 \, Y \pi \, \{D_b{}^2 - (D_p + 2.2t + 2R_d)^2\} \qquad (20.13)$$

where F_h = holding force in drawing, N (lb); Y = yield strength of the sheet metal, MPa (lb/in.^{2}); t = starting stock thickness, mm (in.); R_d = die corner radius, mm (in.); and the other terms have been previously defined. The holding force is usually about one-third the drawing force [8].

EXAMPLE 20.4 **Forces in drawing.**	For the drawing operation of Example 20.3, determine (a) drawing force and (b) holding force, given that the tensile strength of the sheet metal (low-carbon steel) = 300 MPa and yield strength = 175 MPa. The die corner radius = 6 mm.

Solution: (a) Maximum drawing force is given by Eq. (20.12):

$$F = \pi(75)(2.4)(300)\left(\frac{138}{75} - 0.7 \right) = 193,396 \text{ N}$$

(b) Holding force is estimated by Eq. (20.13):

$$F_h = 0.015(175) \, \pi(138^2 - (75 + 2.2 \times 2.4 + 2 \times 6)^2) = 86,824 \text{ N}$$

Blank Size Determination For the final dimensions to be achieved on the cylindrical drawn shape, the correct starting blank diameter is needed. It must be large enough to supply sufficient metal to complete the cup. Yet if there is too much material, unnecessary waste will result. For drawn shapes other than cylindrical cups, the same problem of estimating the starting blank size exists, only the shape of the blank may be other than round.

The following is a reasonable method for estimating the starting blank diameter in a deep drawing operation that produces a round part (e.g., cylindrical cup and more complex shapes so long as they are axisymmetric). Because the volume of the final product is the same as that of the starting sheet metal blank, then the blank diameter can be calculated by setting the initial blank volume equal to the final volume of the product and solving for diameter D_b. To facilitate the calculation, it is often assumed that negligible thinning of the part wall occurs.

20.3.3 Other Drawing Operations

Our discussion has focused on a conventional cup-drawing operation that produces a simple cylindrical shape in a single step and uses a blankholder to facilitate the process. Let us consider some of the variations of this basic operation.

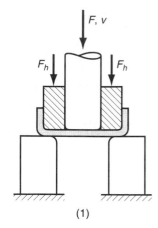

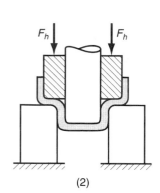

FIGURE 20.21 Redrawing of a cup: (1) start of redraw, and (2) end of stroke. Symbols: v = punch velocity, F = applied punch force, F_h = blankholder force.

(1)   (2)

Redrawing If the shape change required by the part design is too severe (drawing ratio is too high), complete forming of the part may require more than one drawing step. The second drawing step, and any further drawing steps if needed, are referred to as *redrawing.* A redrawing operation is illustrated in Figure 20.21.

When the part design indicates a drawing ratio that is too large to form the part in a single step, the following is a general guide to the amount of reduction that can be taken in each drawing operation [8]: For the first draw, the maximum reduction of the starting blank should be 40 to 45%; for the second draw (first redraw), the maximum reduction should be 30%; and for the third draw (second redraw), the maximum reduction should be 16%.

A related operation is *reverse drawing,* in which a drawn part is positioned face down on the die so that the second drawing operation produces a configuration such as that shown in Figure 20.22. Although it may seem that reverse drawing would produce a more severe deformation than redrawing, it is actually easier on the metal. The reason is that the sheet metal is bent in the same direction at the outside and inside corners of the die in reverse drawing; while in redrawing the metal is bent in the opposite directions at the two corners. Because of this difference, the metal experiences less strain hardening in reverse drawing and the drawing force is lower.

Drawing of Shapes Other than Cylindrical Cups Many products require drawing of shapes other than cylindrical cups. The variety of drawn shapes include square or rec-

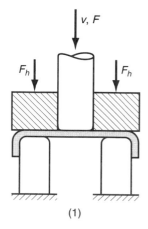

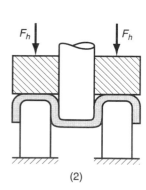

FIGURE 20.22 Reverse drawing: (1) start and (2) completion. Symbols: v = punch velocity, F = applied punch force, F_h = blankholder force.

(1)   (2)

tangular boxes (as in sinks), stepped cups, cones, cups with spherical rather than flat bases, and irregular curved forms (as in automobile body panels). Each of these shapes presents unique technical problems in drawing. Eary [1] provides a detailed discussion of the drawing of these kinds of shapes.

Drawing Without a Blankholder   One of the primary functions of the blankholder is to prevent wrinkling of the flange while the cup is being drawn. The tendency for wrinkling is reduced as the thickness-to-diameter ratio of the blank increases. If the t/D_b ratio is large enough, drawing can be accomplished without a blankholder, as in Figure 20.23. The limiting condition for drawing without a blankholder can be estimated from the following [4]:

$$D_b - D_p < 5t \qquad (20.14)$$

The draw die must have the shape of a funnel or cone to permit the material to be drawn properly into the die cavity. When drawing without a blankholder is feasible, it has the advantages of lower cost tooling and a simpler press, because the need to separately control the movement of the blankholder and punch can be avoided.

20.3.4 Defects in Drawing

Drawing is a more complex form of sheet metalworking than cutting or bending, and more things can go wrong. A number of defects can occur in a drawn product, some of which we have already alluded to. Following is a list of common defects, with sketches in Figure 20.24:

(a) *Wrinkling* in the flange—Wrinkling in a drawn part consists of a series of ridges that form radially in the undrawn flange of the workpart due to compressive buckling.

(b) *Wrinkling* in the wall—If and when the flange is drawn into the cup, these ridges appear in the vertical wall.

(c) *Tearing*—Tearing is an open crack in the vertical wall, usually near the base of the drawn cup, due to high tensile stresses that cause thinning and failure of the metal at this location. This type of failure can also occur as the metal is pulled over a sharp die corner.

(d) *Earing*—This is the formation of irregularities (called *ears*) in the upper edge of a deep drawn cup, caused by anisotropy in the sheet metal. If the material is perfectly isotropic, ears do not form.

FIGURE 20.23   Drawing without a blankholder: (1) start of process, (2) end of stroke. Symbols *v* and *F* indicate motion and applied force, respectively.

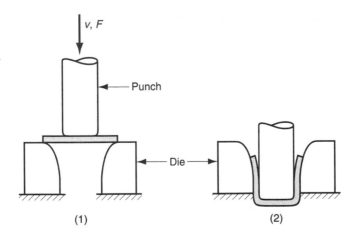

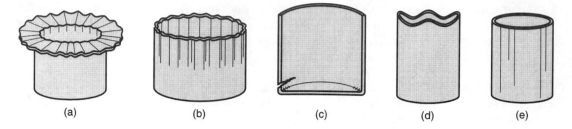

FIGURE 20.24   Common defects in drawn parts: (a) wrinkling can occur either in the flange or (b) in the wall, (c) tearing, (d) earing, and (e) surface scratches.

(e) **Surface scratches**—Surface scratches can occur on the drawn part if the punch and die are not smooth or if lubrication is insufficient.

20.4 OTHER SHEET-METAL-FORMING OPERATIONS

Operations performed with metal tooling and operations performed with flexible rubber tooling round out the sheet-metal-forming operations done on conventional presses.

20.4.1 Operations Performed with Metal Tooling

Operations performed with metal tooling include (1) ironing, (2) coining and embossing, (3) lancing, and (4) twisting.

Ironing   In deep drawing the flange is compressed by the squeezing action of the blank perimeter seeking a smaller circumference as it is drawn toward the die opening. Because of this compression, the sheet metal near the outer edge of the blank becomes thicker as it moves inward. If the thickness of this stock is greater than the clearance between the punch and die, it will be squeezed to the size of the clearance, a process known as *ironing.*

Sometimes ironing is performed as a separate step that follows drawing. This case is illustrated in Figure 20.25. Ironing makes the cylindrical cup more uniform in wall thickness. The drawn part is therefore longer and more efficient in terms of material usage. Beverage cans and artillery shells, two very high production items, include ironing among their processing steps to achieve economy in material usage.

FIGURE 20.25   Ironing to achieve a more uniform wall thickness in a drawn cup: (1) start of process; (2) during process. Note thinning and elongation of walls. Symbols v and F indicate motion and applied force, respectively.

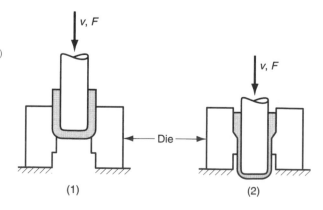

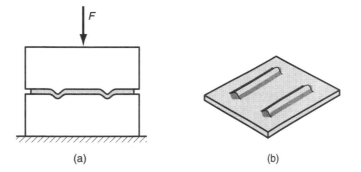

FIGURE 20.26 Embossing: (a) cross section of punch and die configuration during pressing; (b) finished part with embossed ribs.

(a)   (b)

Coining and Embossing   Coining is a bulk-deformation operation discussed in the previous chapter. It is frequently used in sheet-metal work to form indentations and raised sections in the part. The indentations result in thinning of the sheet metal, and the raised sections result in thickening of the metal.

Embossing is a forming operation used to create indentations in the sheet, such as raised (or indented) lettering or strengthening ribs, as depicted in Figure 20.26. Some stretching and thinning of the metal is involved. This operation may seem similar to coining. However, embossing dies possess matching cavity contours, the punch containing the positive contour and the die containing the negative; whereas coining dies may have quite different cavities in the two die halves, thus causing more significant metal deformation than embossing.

Lancing   Lancing is a combined cutting and bending or cutting and forming operation performed in one step to partially separate the metal from the sheet. Several examples are shown in Figure 20.27. Among other applications, lancing is used to make louvers in sheet metal for venting of heat from the interiors of electrical cabinets.

Twisting   *Twisting* subjects the sheet metal to a torsion loading rather than a bending load, thus causing a twist in the sheet over its length. This type of operation has limited applications. It is used to make such products as fan and propeller blades. It can be performed in a conventional punch and die, which has been designed to deform the part in the required twist shape.

20.4.2 Rubber Forming Processes

The two operations discussed in this article are performed on conventional presses, but the tooling is unusual in that it uses a flexible element (made of rubber or similar material) to effect the forming operation. The operations are: (1) the Guerin process, and (2) hydroforming.

Guerin Process   The *Guerin process* uses a thick rubber pad (or other flexible material) to form sheet metal over a positive form block, as in Figure 20.28. The rubber pad

FIGURE 20.27   Lancing in several forms: (a) cutting and bending; (b) and (c) two types of cutting and forming.

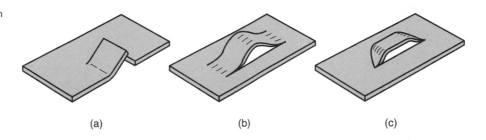

(a)   (b)   (c)

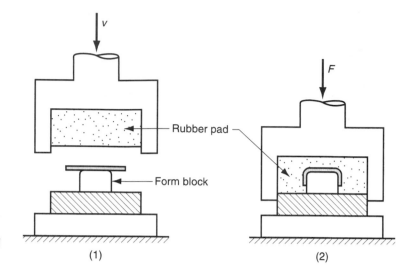

FIGURE 20.28 Guerin process: (1) before and (2) after. Symbols v and F indicate motion and applied force, respectively.

is confined in a steel container. As the ram descends, the rubber gradually surrounds the sheet, applying pressure to deform it to the shape of the form block. It is limited to relatively shallow forms, because the pressures developed by the rubber—up to about 10 MPa (1500 lb/in.^{2})—are not sufficient to prevent wrinkling in deeper formed parts.

The advantage of the Guerin process is the relatively low cost of the tooling. The form block can be made of wood, plastic, or other materials that are easy to shape, and the rubber pad can be used with different form blocks. These factors make rubber forming attractive in small-quantity production (e.g., the aircraft industry, where the process was developed).

Hydroforming *Hydroforming* is similar to the Guerin process; the difference is that it substitutes a rubber diaphragm filled with hydraulic fluid in place of the thick rubber pad, as illustrated in Figure 20.29. This allows the pressure that forms the workpart to

FIGURE 20.29 Hydroform process: (1) start-up, no fluid in cavity; (2) press closed, cavity pressurized with hydraulic fluid; (3) punch pressed into work to form part. Symbols: v = velocity, F = applied force, p = hydraulic pressure.

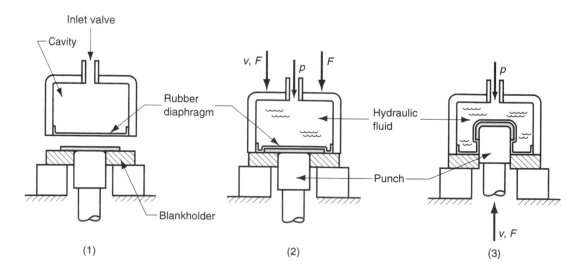

be increased—to around 100 MPa (15,000 lb/in.^{2})—thus preventing wrinkling in deep formed parts. In fact, deeper draws can be achieved with the Hydroform process than with conventional deep drawing. This is because the uniform pressure in hydroforming forces the work to contact the punch throughout its length, thus increasing friction and reducing the tensile stresses that cause tearing at the base of the drawn cup.

20.5 DIES AND PRESSES FOR SHEET METAL PROCESSES

In this section we examine the punch-and-die tooling and production equipment used in conventional sheet-metal processing.

20.5.1 Dies

Nearly all of the preceding pressworking operations are performed with conventional punch-and-die tooling. The tooling is referred to as a *die.* It is custom-designed for the particular part to be produced. The term *stamping die* is sometimes used for high-production dies.

Components of a Stamping Die   The components of a stamping die to perform a simple blanking operation are illustrated in Figure 20.30. The working components are the *punch* and *die,* which perform the cutting operation. They are attached to the upper and lower portions of the *die set,* respectively called the *punch holder* (or *upper shoe*) and *die holder* (*lower shoe*). The die set also includes guide pins and bushings to ensure proper alignment between the punch and die during the stamping operation. The die holder is attached to the base of the press, and the punch holder is attached to the ram. Actuation of the ram accomplishes the pressworking operation.

In addition to these components, a die used for blanking or hole-punching must include a means of preventing the sheet metal from sticking to the punch when it is retracted upward after the operation. The newly created hole in the stock is the same size as the punch, and it tends to cling to the punch on its withdrawal. The device in the die that strips the sheet metal from the punch is called a *stripper.* It is often a simple plate attached to the die as in Figure 20.30, with a hole slightly larger than the punch diameter.

FIGURE 20.30
Components of a punch and die for a blanking operation.

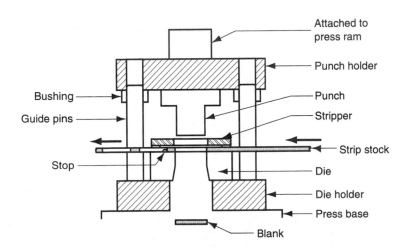

For dies that process strips or coils of sheet metal, a device is required to stop the sheet metal as it advances through the die between press cycles. That device is called (try to guess) a *stop.* Stops range from simple solid pins located in the path of the strip to block its forward motion, to more complex mechanisms synchronized to rise and retract with the actuation of the press. The simpler stop is shown in Figure 20.30.

There are other components in pressworking dies, but the preceding description provides an introduction to the terminology.

Types of Stamping Dies Aside from differences in stamping dies related to the operations they perform (e.g., cutting, bending, drawing), other differences deal with the number of separate operations to be performed in each press actuation and how they are accomplished.

The type of die considered above performs a single blanking operation with each stroke of the press and is called a *simple die.* Other dies that perform a single operation include V-dies (Section 20.2.1). More complicated pressworking dies include compound dies, combination dies, and progressive dies. A *compound die* performs two operations at a single station, such as blanking and punching, or blanking and drawing [1]. A *combination die* is less common; it performs two operations at two different stations in the die. Examples of applications include blanking two different parts, or blanking and then bending the same part [1].

A *progressive die* performs two or more operations on a sheet metal coil at two or more stations with each press stroke. The part is fabricated progressively. The coil is fed from one station to the next and different operations (e.g., punching, notching, bending, and blanking) are performed at each station. When the part exits the final station, it has been completed and separated (cut) from the remaining coil. Design of a progressive die begins with the layout of the part on the strip or coil and the determination of which operations are to be performed at each station. The result of this procedure is called the *strip development.* A progressive die and associated strip development are illustrated in Figure 20.31. Progressive dies can have a dozen or more stations. They are the most complicated and most costly stamping dies, economically justified only for complex parts requiring multiple operations at high production rates.

20.5.2 Presses

A *press* used for sheet metalworking is a machine tool with a stationary *bed* and a powered *ram* (or *slide*) that can be driven toward and away from the bed to perform various cutting and forming operations. A typical press, with principal components labeled, is diagrammed in Figure 20.32. The relative positions of the bed and ram are established by the *frame,* and the ram is driven by mechanical or hydraulic power. When a die is mounted in the press, the punch holder is attached to the ram, and the die holder is attached to a *bolster plate* of the press bed.

Presses are available in a variety of capacities, power systems, and frame types. The capacity of a press is its ability to deliver the required force and energy to accomplish the stamping operation. This is determined by the physical size of the press and by its power system. The power system refers to whether mechanical or hydraulic power is used and the type of drive used to transmit the power to the ram. Production rate is another important aspect of capacity. Type of press frame refers to the physical construction of the press. There are two types of frames in common use: gap frame and straight-sided frame.

Gap Frame Presses The *gap frame* has the general configuration of the letter C and is often referred to as a *C-frame.* Gap frame presses provide good access to the die,

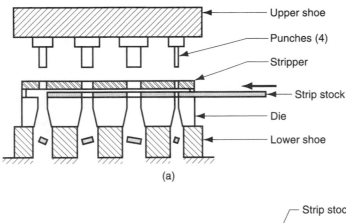

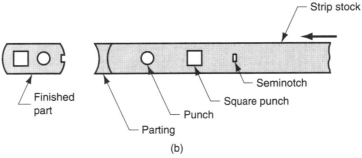

FIGURE 20.31   (a) Progressive die and (b) associated strip development.

and they are usually open in the back to permit convenient ejection of stampings or scrap. The principal types of gap frame press are (a) solid gap frame, (b) adjustable bed, (c) open back inclinable, (d) press brake, and (e) turret press.

The **solid gap frame** (sometimes called simply a **gap press**) has one-piece construction, as shown in Figure 20.32. Presses with this frame are rigid, yet the C-shape allows convenient access from the sides for feeding strip or coil stock. They are available in a range of sizes, with capacities up to around 9000 kN (1000 tons). The model shown in Figure 20.33 has a capacity of 1350 kN (150 tons). The **adjustable bed frame** press is a variation of the gap frame, in which an adjustable bed is added to accommodate various die sizes. The adjustment feature results in some sacrifice of tonnage capacity. The

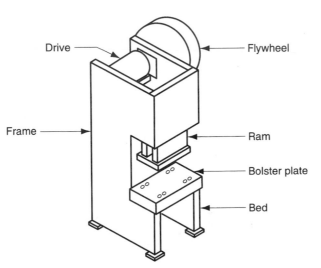

FIGURE 20.32   Components of a typical (mechanical drive) stamping press.

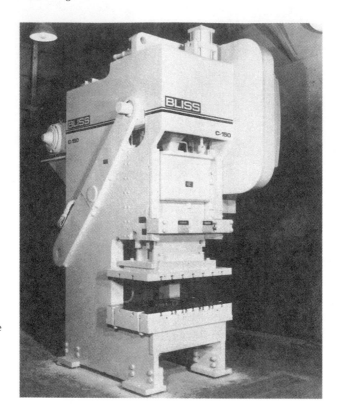

FIGURE 20.33   Gap frame press for sheet metalworking (photo courtesy of E. W. Bliss Company). Capacity = 1350 kN (150 tons).

open back inclinable press has a C-frame assembled to a base in such a way that the frame can be tilted back to various angles so that the stampings fall through the rear opening by gravity. Capacities of OBI presses range between one ton and around 2250 kN (250 tons). They can be operated at high speeds—up to around 1,000 strokes per minute.

The *press brake* is a gap frame press with a very wide bed. The model in Figure 20.34 has a bed width of 9.15 m (30 ft). This allows a number of separate dies (simple V-bending dies are typical) to be set up in the bed, so that small quantities of stampings can be made economically. These low quantities of parts, sometimes requiring multiple bends at different angles, necessitate a manual operation. For a part requiring a series of bends, the operator moves the starting piece of sheet metal through the desired sequence of bending dies, actuating the press at each die, to complete the work needed.

Whereas press brakes are well adapted to bending operations, *turret presses* are suited to situations in which a sequence of punching, notching, and related cutting operations must be accomplished on sheet metal parts, as in Figure 20.35. Turret presses have a C-frame, although this construction is not obvious in Figure 20.36. The conventional ram and punch is replaced by a turret containing many punches of different sizes and shapes. The turret works by indexing (rotating) to the position holding the punch to perform the required operation. Beneath the punch turret is a corresponding die turret that positions the die opening for each punch. Between the punch and die is the sheet metal blank, held by an *x*–*y* positioning system that operates by computer numerical control (Section 38.1). The blank is moved to the required coordinate position for each cutting operation.

Straight-sided Frame Presses   For jobs requiring high tonnage, press frames with greater structural rigidity are needed. Straight-sided presses have full sides, giving it a box-like appearance as in Figure 20.37. This construction increases the strength and

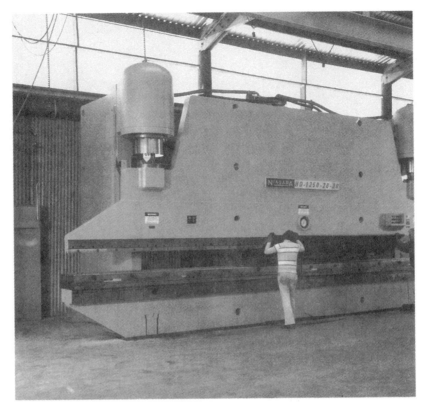

FIGURE 20.34   Press brake with bed width of 9.15 m (30 ft) and capacity of 11,200 kN (1250 tons); two workers are shown positioning plate stock for bending (photo courtesy of Niagara Machine & Tool Works).

FIGURE 20.35   Several sheet metal parts produced on a turret press, showing a variety of hole shapes (photo courtesy of Strippet, Inc.).

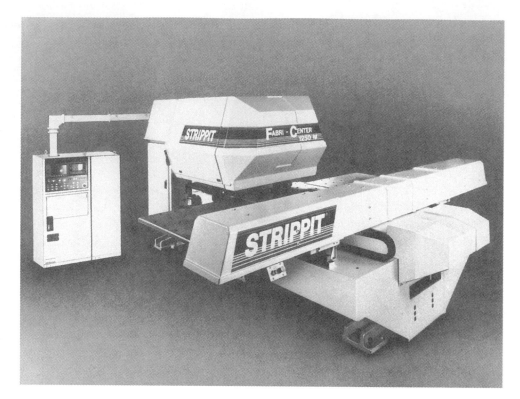

FIGURE 20.36 Computer numerical control turret press (photo courtesy of Strippet, Inc.).

FIGURE 20.37 Straight-sided frame press (photo courtesy Greenerd Press & Machine Company, Inc.).

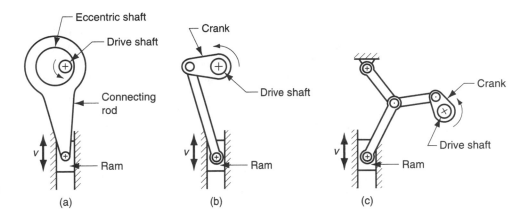

FIGURE 20.38 Types of drives for sheet metal presses: (a) eccentric, (b) crankshaft, and (c) knuckle joint.

stiffness of the frame. As a result, capacities up to 35,000 kN (4000 tons) are available in straight-sided presses for sheet metalwork. Large presses of this frame type are used for forging (Section 19.3).

In all of these presses, gap frame and straight-sided frame, the size is closely correlated to tonnage capacity. Larger presses are built to withstand higher forces in pressworking. Press size is also related to the speed at which it can operate. Smaller presses are generally capable of higher production rates than larger presses.

Power and Drive Systems Power systems on presses are either hydraulic or mechanical. *Hydraulic presses* use a large piston and cylinder to drive the ram. This power system typically provides longer ram strokes than mechanical drives and can develop the full tonnage force throughout the entire stroke. However, it is slower. Its application for sheet metal is normally limited to deep drawing and other forming operations where these characteristics are advantageous. These presses are available with one or more independently operated slides, called single action (single slide), double action (two slides), and so on. Double action presses are useful in deep drawing operations where it is required to separately control the punch force and the blankholder force.

There are several types of drive mechanisms used on *mechanical presses.* These include eccentric, crankshaft, and knuckle joint, illustrated in Figure 20.38. They convert the rotational motion of a drive motor into the linear motion of the ram. A *flywheel* is used to store the energy of the drive motor for use in the stamping operation. Mechanical presses using these drives achieve very high forces at the bottom of their strokes, and are therefore quite suited to blanking and punching operations. The knuckle joint delivers very high force when it bottoms, and is therefore often used in coining operations.

20.6 SHEET-METAL OPERATIONS NOT PERFORMED ON PRESSES

A number of sheet-metal operations are not performed on conventional stamping presses. In this section we examine several of these processes: (1) stretch forming, (2) roll bending and forming, (3) spinning, and (4) high-energy-rate forming processes.

20.6.1 Stretch Forming

Stretch forming is a sheet-metal deformation process in which the sheet metal is intentionally stretched and simultaneously bent in order to achieve shape change. The process is illustrated in Figure 20.39 for a relatively simple and gradual bend. The workpart is gripped by one or more jaws on each end and then stretched and bent over a positive die containing the desired form. The metal is stressed in tension to a level above its yield point. When the tension loading is released, the metal has been plastically deformed. The combination of stretching and bending results in relatively little springback in the part. An estimate of the force required in stretch forming can be obtained by multiplying the cross-sectional area of the sheet in the direction of pulling by the flow stress of the metal. In equation form,

$$F = LtY_f \qquad (20.15)$$

where F = stretching force, N (lb); L = length of the sheet in the direction perpendicular to stretching, mm (in.); t = instantaneous stock thickness, mm (in.); and Y_f = flow stress of the work metal, MPa (lb/in.^{2}). The die force F_{die} shown in the figure can be determined by balancing vertical force components.

More complex contours than that shown in our figure are possible by stretch forming, but there are limitations on how sharp the curves in the sheet can be. Stretch forming is widely used in the aircraft and aerospace industries to economically produce large sheet-metal parts in the low quantities characteristic of those industries.

20.6.2 Roll Bending and Roll Forming

The operations described in this section use rolls to form sheet metal. **Roll bending** is an operation in which (usually) large sheet-metal parts are formed into curved sections by means of rolls. One possible arrangement of the rolls is pictured in Figure 20.40. As the sheet passes between the rolls, the rolls are brought toward each other to a configuration that achieves the desired radius of curvature on the work. Components for large storage tanks and pressure vessels are fabricated by roll bending. The operation can also be used to bend plate metal parts, structural shapes, railroad rails, and tubes.

A related operation is **roll straightening** in which nonflat sheets (or other cross-sectional forms) are straightened by passing them between a series of rolls. The rolls subject the work to a sequence of decreasing small bends in opposite directions, thus causing it to be straight at the exit.

Roll forming (also called **contour roll forming**) is a continuous bending process in which opposing rolls are used to produce long sections of formed shapes from coil or strip stock. Several pairs of rolls are usually required to progressively accomplish the bending of the stock into the desired shape. The process is illustrated in Figure 20.41 for a U-shaped section. Products made by roll forming include channels, gutters, metal siding sections (for homes), pipes and tubing with seams, and various structural sections. Although roll forming has the general appearance of a rolling operation (and the tooling certainly looks similar), the difference is that roll forming involves bending rather than compressing the work.

FIGURE 20.39   Stretch forming: (1) start of process; (2) form die is pressed into the work with force F_{die}, causing it to be stretched and bent over the form. F = stretching force.

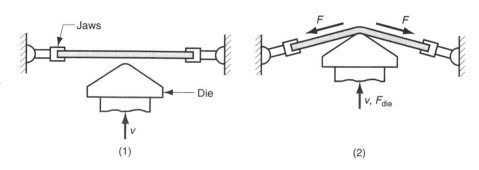

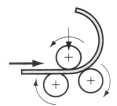

FIGURE 20.40   Roll bending.

20.6.3 Spinning

Spinning is a metal-forming process in which an axially symmetric part is gradually shaped over a mandrel or form by means of a rounded tool or roller. The tool or roller applies a very localized pressure (almost a point contact) to deform the work by axial and radial motions over the surface of the part. Basic geometric shapes typically produced by spinning include cups, cones, hemispheres, tubes, and cylinders. There are three types of spinning operations: (1) conventional spinning, (2) shear spinning, and (3) tube spinning.

Conventional Spinning   Conventional spinning is the basic spinning operation. As illustrated in Figure 20.42, a sheet metal disk is held against the end of a rotating mandrel of the desired inside shape of the final part, while the tool or roller deforms the metal against the mandrel. In some cases, the starting workpart is other than a flat disk. The process requires a series of steps, as indicated in the figure, to complete the shaping of the part. The tool position is controlled either by a human operator, using a fixed fulcrum to achieve the required leverage, or by an automatic method such as numerical control. These alternatives are *manual spinning* and *power spinning.* Power spinning has the capability to apply higher forces to the operation, resulting in faster cycle times and greater work size capacity. It also achieves better process control than manual spinning.

Conventional spinning bends the metal around a moving circular axis to conform to the outside surface of the axisymmetric mandrel. The thickness of the metal therefore remains unchanged (more or less) relative to the starting disk thickness. The diameter of the disk must therefore be somewhat larger than the diameter of the resulting part. The required starting diameter can be figured by assuming constant volume, before and after spinning.

FIGURE 20.41   Roll forming of a continuous channel section: (1) straight rolls, (2) partial form, and (3) final form.

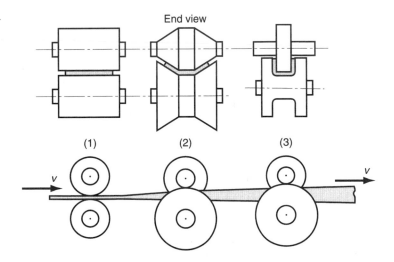

End view

(1)   (2)   (3)

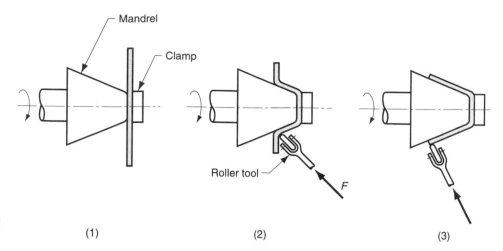

FIGURE 20.42
Conventional spinning: (1)
setup at start of process;
(2) during spinning; and (3)
completion of process.

(1)   (2)   (3)

Applications of conventional spinning include production of conical and curved shapes in low quantities. Very large diameter parts—up to 5 m (15 ft) or more—can be made by spinning. Alternative sheet-metal processes would require excessively high die costs. The form mandrel in spinning can be made of wood or other soft materials that are easy to shape. It is therefore a low-cost tool compared to the punch and die required for deep drawing, which might be a substitute process for some parts.

Shear Spinning   In **shear spinning,** the part is formed over the mandrel by a shear deformation process in which the outside diameter remains constant and the wall thickness is therefore reduced, as in Figure 20.43. This shear straining (and consequent thinning of the metal) distinguishes this process from the bending action in conventional spinning. Several other names have been used for shear spinning, including **flow turning, shear forming,** and **spin forging.** The process has been applied in the aerospace industry to form large parts such as rocket nose cones.

For the simple conical shape in our figure, the resulting thickness of the spun wall can readily be determined by the sine law relationship:

$$t_f = t \sin \alpha \qquad (20.16)$$

FIGURE 20.43   Shear
spinning: (1) setup and (2)
completion of process.

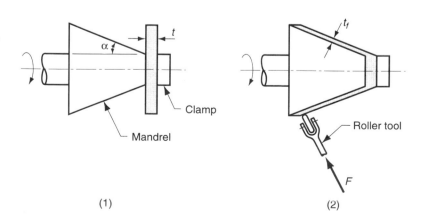

(1)   (2)

where t_f = the final thickness of the wall after spinning, t = the starting thickness of the disk, and α = the mandrel angle (actually the half angle). Thinning is sometimes quantified by the spinning reduction r:

$$r = \frac{t - t_f}{t} \qquad (20.17)$$

There are limits to the amount of thinning that the metal will endure in a spinning operation before fracture occurs. The maximum reduction correlates well with reduction of area in a tension test [7].

Tube Spinning *Tube spinning* is used to reduce the wall thickness and increase the length of a tube by means of a roller applied to the work over a cylindrical mandrel, as in Figure 20.44. Tube spinning is similar to shear spinning except that the starting workpiece is a tube rather than a flat disk. The operation can be performed by applying the roller against the work externally (using a cylindrical mandrel on the inside of the tube) or internally (using a die to surround the tube). It is also possible to form profiles in the walls of the cylinder, as in Figure 20.44(c), by controlling the path of the roller as it moves tangentially along the wall.

Spinning reduction for a tube-spinning operation that produces a wall of uniform thickness can be determined as in shear spinning by Eq. (20.17).

20.6.4 High-Energy-Rate Forming

Several processes have been developed to form metals using large amounts of energy applied in a very short time. Owing to this feature, these operations are called *high-energy-rate forming* (HERF) processes. They include explosive forming, electrohydraulic forming, and electromagnetic forming.

Explosive Forming *Explosive forming* involves the use of an explosive charge to form sheet (or plate) metal into a die cavity. One method of implementing the process is illustrated in Figure 20.45. The workpart is clamped and sealed over the die, and a vacuum is created in the cavity beneath. The apparatus is then placed in a large vessel of water. An explosive charge is placed in the water at a certain distance above the work.

FIGURE 20.44   Tube spinning: (a) external; (b) internal; and (c) profiling.

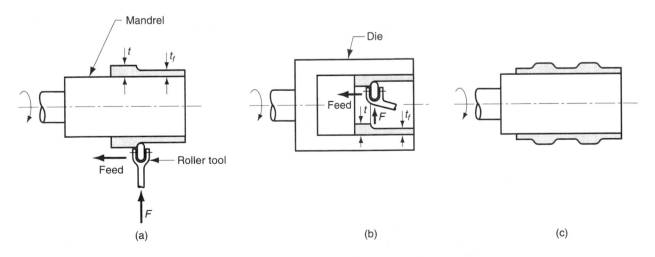

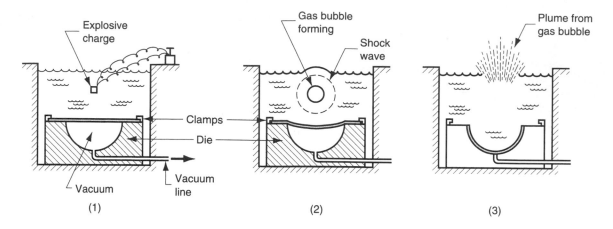

FIGURE 20.45 Explosive forming: (1) setup, (2) explosive is detonated, and (3) shock wave forms part and plume escapes water surface.

Detonation of the charge results in a shock wave whose energy is transmitted by the water to cause rapid forming of the part into the cavity. The size of the explosive charge and the distance at which it is placed above the part is largely a matter of art and experience. Explosive forming is reserved for large parts, typical of the aerospace industry.

Electrohydraulic Forming   *Electrohydraulic forming* is a HERF process in which a shock wave to deform the work into a die cavity is generated by the discharge of electrical energy between two electrodes submerged in a transmission fluid (water). Owing to its principle of operation, this process is also called *electric discharge forming.* The setup for the process is illustrated in Figure 20.46. Electrical energy is accumulated in large capacitors and then released to the electrodes. Electrohydraulic forming is similar to explosive forming. The difference is in the method of generating the energy and the smaller amounts of energy that are released. This limits electrohydraulic forming to much smaller part sizes.

Electromagnetic Forming   *Electromagnetic forming,* also called *magnetic pulse forming,* is a process in which sheet metal is deformed by the mechanical force of an electromagnetic field induced in the workpart by an energized coil. The coil, energized by a capacitor, produces a magnetic field. This generates eddy currents in the work that produce their own magnetic field. The induced field opposes the primary field, pro-

FIGURE 20.46
Electrohydraulic forming
setup.

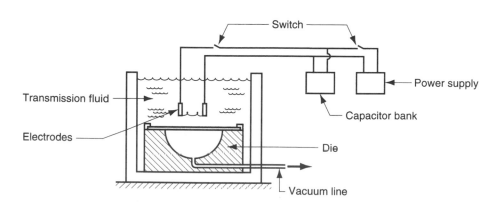

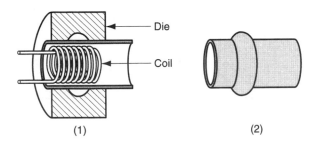

FIGURE 20.47   Electromagnetic forming: (1) setup in which coil is inserted into tubular workpart surrounded by die; (2) formed part.

ducing a mechanical force which deforms the part into the surrounding cavity. Developed in the 1960s, electromagnetic forming is the most widely used HERF process [8]. It is typically used to form tubular parts, as illustrated in Figure 20.47.

20.7 BENDING OF TUBE STOCK

Several methods of producing tubes and pipes are discussed in the previous chapter, and tube spinning is described in Section 20.6.3. In this section, we examine methods by which tubes are bent and otherwise formed. Bending of tube stock is more difficult than sheet stock because a tube tends to collapse and fold when attempts are made to bend it. Special flexible mandrels are usually inserted into the tube prior to bending to support the walls during the operation.

Some of the terms in tube bending are defined in Figure 20.48. The radius of the bend R is defined with respect to the centerline of the tube. When the tube is bent, the wall on the inside of the bend is in compression, and the wall at the outside is in tension. These stress conditions cause thinning and elongation of the outer wall and thickening and shortening of the inner wall. As a result, there is a tendency for the inner and outer walls to be forced toward each other to cause the cross section of the tube to flatten. Because of this flattening tendency, the minimum bend radius R that the tube can be bent is about 1.5 times the diameter D when a mandrel is used and 3.0 times D when no mandrel is used [8]. The exact value depends on the wall factor WF, which is the diameter D divided by wall thickness t. Higher values of WF increase the minimum bend radius; that is, tube bending is more difficult for thin walls. Ductility of the work material is also an important factor in the process.

A number of methods used to bend tubes (and similar sections) are illustrated in Figure 20.49. **Stretch bending** is accomplished by pulling and bending the tube around a fixed form block, as in Figure 20.49(a). **Draw bending** is performed by clamping the tube

FIGURE 20.48
Dimensions and terms for a bent tube: D = outside diameter of tube, R = bend radius, t = wall thickness.

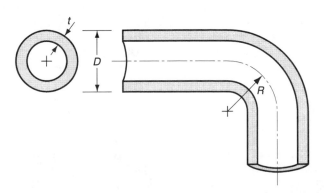

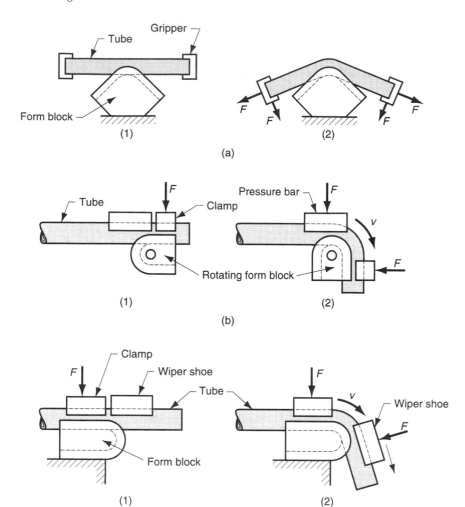

FIGURE 20.49 Tube bending methods: (a) stretch bending, (b) draw bending, and (c) compression bending. For each method: (1) start of process, and (2) during bending. Symbols *v* and *F* indicate motion and applied force.

against a form block, and then pulling the tube through the bend by rotating the block as in (b). A pressure bar is used to support the work as it is being bent. In **compression bending,** a wiper shoe is used to wrap the tube around the contour of a fixed form block, as seen in (c). **Roll bending** (Section 20.6.2), generally associated with the forming of sheet stock, is also used for bending tubes and other cross sections.

REFERENCES

[1] Eary, D. F., and E. A. Reed. *Techniques of Pressworking Sheet Metal.* 2nd ed. Prentice-Hall, Inc., Englewood Cliffs, N.J., 1974.

[2] Hoffman, E. G. *Fundamentals of Tool Design.* 2nd ed. Society of Manufacturing Engineers, Dearborn, Mich., 1984.

[3] Hosford, W. F. and Cadell, R. M. *Metal Forming: Mechanics and Metallurgy.* 2nd ed. Prentice-Hall, Upper Saddle River, N.J., 1993.

[4] Kalpakjian, S. *Manufacturing Processes for Engineering Materials.* 3rd ed. Addison-Wesley Publishing Company, Inc., Reading, Mass., 1996.

[5] Lange, K., et al. (eds.). *Handbook of Metal Forming.* Society of Manufacturing Engineers, Dearborn, Mich., 1995.

[6] Mielnik, E. M. *Metalworking Science and Engineering.* McGraw-Hill, Inc., New York, 1991.

[7] Schey, J. A. *Introduction to Manufacturing Processes.* 3rd ed. McGraw-Hill Book Company, New York, 1999.

[8] Wick, C., et al. (eds.), *Tool and Manufacturing Engineers Handbook.* 4th ed. Volume II, *Forming.* Society of Manufacturing Engineers, Dearborn, Mich., 1984.

REVIEW QUESTIONS

20.1. Identify the three basic types of sheet metalworking operations.

20.2. In blanking of a round sheet metal part, indicate how the clearance should be applied to the punch and die diameters.

20.3. What is the difference between a cutoff operation and a parting operation?

20.4. Describe V-bending and edge bending.

20.5. What is *springback* in sheet metal bending?

20.6. What are some of the simple measures used to assess the feasibility of a proposed cup drawing operation?

20.7. Distinguish between *redrawing* and *reverse drawing.*

20.8. What are some of the possible defects in drawn sheet-metal parts?

20.9. What is *stretch forming*?

20.10. Identify the principal components of a stamping die that performs blanking.

20.11. What are the two basic categories of structural frames used in stamping presses?

20.12. What are the relative advantages and disadvantages of mechanical versus hydraulic presses in sheet metalworking?

20.13. What is the *Guerin process*?

20.14. Identify a major technical problem in tube bending.

20.15. Distinguish between roll bending and roll forming.

MULTIPLE CHOICE QUIZ

There is a total of 17 correct answers in the following multiple choice questions (some questions have multiple answers that are correct). To attain a perfect score on the quiz, all correct answers must be given, since each correct answer is worth 1 point. For each question, each omitted answer or wrong answer reduces the score by 1 point, and each additional answer beyond the number of answers required reduces the score by 1 point. Percentage score on the quiz is based on the total number of correct answers.

20.1. As sheet-metal stock hardness increases, the clearance between punch and die in a blanking operation should: (a) be decreased, (b) be increased, or (c) be unaffected.

20.2. A round sheet-metal slug produced in a hole-punching operation will have the same diameter as which of the following? (a) die opening, or (b) punch.

20.3. The cutting force in a blanking operation depends on which mechanical property of the sheet metal (choose one best answer)? (a) compressive strength, (b) modulus of elasticity, (c) shear strength, (d) tensile strength, or (e) yield strength.

20.4. Sheet-metal bending involves which of the following stresses and strains (more than one)? (a) compressive, (b) shear, and (c) tensile.

20.5. Which one of the following is the best definition of *bend allowance*? (a) amount by which the die is larger than the punch, (b) amount of elastic recovery experienced by the metal after bending, (c) safety factor used in calculating bending force, or (d) length before bending of the straight sheet metal section to be bent.

20.6. Which of the following are variations of sheet metal bending operations (more than one)? (a) coining, (b) flanging, (c) hemming, (d) ironing, (e) notching, (f) shear spinning, (g) trimming, (h) tube bending, and (i) tube forming.

20.7. The following are measures of feasibility for several proposed cup drawing operations; which of the operations are likely to be feasible (more than one)? (a) $DR = 1.7$, (b) $DR = 2.7$, (c) $r = 0.35$, (d) $r = 65\%$, and (e) $t/D_b = 2\%$.

20.8. Holding force in drawing is most likely to be which of the following relative to maximum drawing force? (a) less than, (b) equal to, or (c) greater than.

20.9. Which one of the following stamping dies is the most complicated? (a) blanking die, (b) combination die, (c) compound die, (d) wiping die for edge bending, (e) progressive die, or (f) V-die.

20.10. Which one of the following press types is usually associated with the highest production rates in sheet metal stamping operations? (a) adjustable bed, (b) open back inclinable, (c) press brake, (d) solid gap, and (e) straight-sided.

20.11. Which of the following processes are classified as high-energy-rate forming processes (more than one)? (a) electrochemical machining, (b) electromagnetic forming, (c) electron beam cutting, (d) explosive forming, (e) Guerin process, (f) hydroforming, (g) redrawing, and (h) shear spinning.

PROBLEMS

Cutting Operations

20.1. A power shears is used to cut soft cold-rolled steel that is 4.75 mm thick. At what clearance should the shears be set to yield an optimum cut?

20.2. A blanking operation is to be performed on 2.0-mm-thick cold-rolled steel (half hard). The part is circular with diameter = 75.0 mm. Determine the appropriate punch and die sizes for this operation.

20.3. A compound die will be used to blank and punch a large washer out of aluminum alloy sheet stock 3.2 mm thick. The outside diameter of the washer = 65 mm and the inside diameter = 30 mm. Determine (a) the punch and die sizes for the blanking operation, and (b) the punch and die sizes for the punching operation.

20.4. A blanking die is to be designed to blank the part outline shown in Figure P20.4. The material is 5/32-inch-thick stainless steel (half hard). Determine the dimensions of the blanking punch and the die opening.

20.5. Determine the blanking force required in Problem 20.2, if the steel has a shear strength = 350 MPa.

20.6. Determine the minimum tonnage press to perform the blanking and punching operation in Problem 20.3, if the aluminum sheetmetal has a tensile strength = 290 MPa. Assume that blanking and punching occur simultaneously.

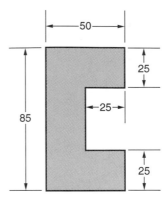

FIGURE P20.4   Blanked part for Problem 20.4 (dimensions in mm).

20.7. Determine the tonnage requirement for the blanking operation in Problem 20.4, given that the stainless steel has a shear strength = 62,000 lb/in.^{2}.

20.8. The foreman in the pressworking section comes to you with the problem of a blanking operation that is producing parts with excessive burrs. What are the possible reasons for the burrs, and what can be done to correct the condition?

Bending

20.9. A bending operation is to be performed on 4.75-mm-thick cold-rolled steel. The part drawing is given in Figure P20.9. Determine the blank size required.

20.10. Solve Problem 20.9 except that the bend radius R = 6.35 mm.

20.11. An L-shaped part is to be bent in a V-bending operation on a press brake from a flat blank 4.0 in. by 1.5 in. that is 5/32 in. thick. The bend of 90° is to be made in the middle of the 4-in. length. (a) Determine the dimensions of the two equal sides that will result after the bend, if the bend radius = 3/16 in. For convenience, these sides should be measured to the beginning of the bend radius. (b) Also, determine the length of the part's neutral axis after the bend. (c) Where should the machine operator set the stop on the press brake relative to the starting length of the part?

20.12. Determine the bending force required in Problem 20.9

Drawing Operations

20.17. Derive an expression for the reduction r in drawing as a function of drawing ratio DR.

20.18. A cup is to be drawn in a deep drawing operation. The

if the bend is to be performed in a V-die with a die opening width = 38 mm. The material has a tensile strength = 620 MPa.

20.13. Solve Problem 20.12 except that the operation is performed using a wiping die with die opening W = 25 mm.

20.14. Determine the bending force required in Problem 20.11 if the bend is to be performed in a V-die with a die opening width W = 1.25 inches. The material has a tensile strength = 70,000 lb/in.^{2}

20.15. Solve Problem 20.14 except that the operation is performed using a wiping die with die opening W = 0.75 inch.

20.16. A sheet-metal part 3.0 mm thick and 20.0 mm long is bent to an included angle = 60° and a bend radius = 7.5 mm in a V-die. The metal has a tensile strength = 340 MPa. Compute the required force to bend the part, given that the die opening = 15 mm.

height of the cup is 75 mm, and its inside diameter is 100 mm. The sheetmetal thickness is 2 mm. If the blank diameter is 225 mm, determine (a) drawing

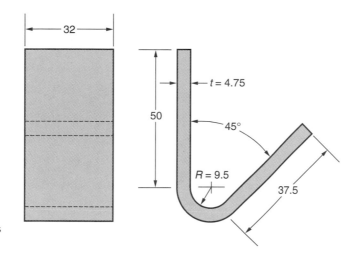

FIGURE P20.9 Part in bending operation of Problem 20.9 (dimensions in mm).

ratio, (b) reduction, and (c) thickness-to-diameter ratio. (d) Does the operation seem feasible?

20.19. Solve Problem 20.18, except that the starting blank size diameter is 175 mm.

20.20. A deep drawing operation is performed in which the inside of the cylindrical cup has a diameter = 4.0 in. and a height = 2.5 in. The stock thickness = 1/8 in., and the starting blank diameter = 7.5 in. Punch and die radii = 5/32 in. The metal has a tensile strength = 60,000 lb/in.^{2} and a yield strength = 30,000 lb/in.^{2} Determine (a) drawing ratio, (b) reduction, (c) drawing force, and (d) blankholder force.

20.21. Solve Problem 20.20, except that the stock thickness $t = 3/16$ in.

20.22. A cup-drawing operation is performed in which the inside diameter = 80 mm and the height = 50 mm. The stock thickness = 3.0 mm, and the starting blank diameter = 150 mm. Punch and die radii = 4 mm. Tensile strength = 400 MPa, and a yield strength = 180 MPa for this sheet metal. Determine (a) drawing ratio, (b) reduction, (c) drawing force, and (d) blankholder force.

20.23. A deep drawing operation is to be performed on a sheet-metal blank that is 1/8 in. thick. The height (inside dimension) of the cup = 3.8 in., and the diameter (inside dimension) = 5.0 in. Assuming the punch radius = 0, compute the starting diameter of the blank to complete the operation with no material left in the flange. Is the operation feasible (ignoring the fact that the punch radius is too small)?

20.24. Solve Problem 20.23, except use a punch radius = 0.375 in.

20.25. A drawing operation is performed on 3.0 mm stock. The part is a cylindrical cup with height = 50 mm and inside diameter = 70 mm. Assume the corner radius on the punch is zero. (a) Find the required starting blank size D_b. (b) Is the drawing operation feasible?

20.26. Solve Problem 20.25, except that the height = 60 mm.

20.27. Solve Problem 20.26, except that the corner radius on the punch = 10 mm.

20.28. The foreman in the drawing section of the shop brings to you several samples of parts that have been drawn in the shop. The samples have various defects. One has ears, another has wrinkles, and still a third has torn sections at its base. What are the causes of each of these defects, and what remedies would you propose?

20.29. A cup-shaped part is to be drawn without a blankholder from sheet metal whose thickness is 0.25 inches. The inside diameter of the cup is 2.5 inches, its height is 1.5 inches, and the corner radius at the base is 0.375 inch. (a) What is the minimum starting blank diameter that can be used, according to Eq. (20.14)? (b) Does this blank diameter provide sufficient material to complete the cup?

Other Operations

20.30. A 20-in.-long sheetmetal workpiece is stretched in a stretch forming operation to the dimensions shown in Figure P20.30. The thickness of the beginning stock $t = 0.125$ in. and the width = 10 in. The metal has a flow curve defined by $K = 70,000$ lb/in.^{2} and $n = 0.25$. (a) Find the stretching force F required near the beginning of the operation when yielding first occurs. Determine (b) true strain experienced by the metal, (c) stretching force F, and (d) die force F_{die} at the very end when the part is formed as indicated in Figure P20.30(b).

20.31. Determine the starting disk diameter required to spin the part in Figure P20.31 using a conventional spinning operation. The starting thickness = 2.4 mm.

20.32. If the part illustrated in Figure P20.31 were made by shear spinning, determine (a) the wall thickness along the cone-shaped portion, and (b) the spinning reduction r.

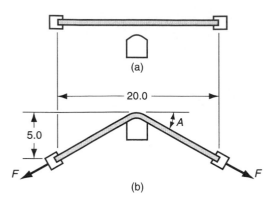

FIGURE P20.30 Stretch forming operation: (a) before, and (b) after (dimensions in inches).

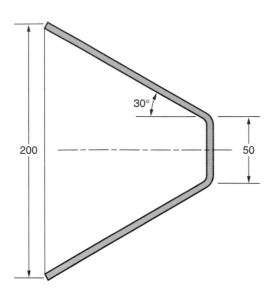

FIGURE P20.31 Part (cross section) in conventional spinning (dimensions in mm).

20.33. Determine the shear strain that is experienced by the material that is shear spun in Problem 20.32.

20.34. A 75-mm-diameter tube is bent into a rather complex shape with a series of simple tube bending operations. The wall thickness on the tube = 4.75 mm. The tubes will be used to deliver fluids in a chemical plant. In one of the bends where the bend radius is 125 mm, the walls of the tube are flattening badly. What can be done to correct the condition?

Part VI
Material Removal Processes

21 THEORY OF METAL MACHINING

CHAPTER CONTENTS

21.1 Overview of Machining Technology
21.2 Theory of Chip Formation in Metal Machining
    21.2.1 The Orthogonal Cutting Model
    21.2.2 Actual Chip Formation
21.3 Force Relationships and the Merchant Equation
    21.3.1 Forces in Metal Cutting
    21.3.2 The Merchant Equation
21.4 Power and Energy Relationships in Machining
21.5 Cutting Temperature
    21.5.1 Analytical Methods
    21.5.2 Measurement of Cutting Temperature

The *material removal processes* consist of a family of shaping operations (Figure 1.4), the common feature of which is that excess material is removed from a starting workpart so that the remaining part has the desired shape. The "family tree" is shown in Figure 21.1. The most important member of the family is *machining,* in which a sharp cutting tool is used to mechanically remove material to achieve the desired geometry. The three principal machining processes are turning, drilling, and milling. The "other machining operations" in Figure 21.1 include shaping, planing, broaching, and sawing. This chapter begins our coverage of machining, which runs through Chapter 24.

Another group of material removal processes is the *abrasive processes,* which mechanically remove material by the action of hard, abrasive particles. This process group, which includes grinding, is covered in Chapter 25. The "other abrasive processes" in Figure 21.1 includes honing, lapping, and superfinishing. Finally, there are the *nontraditional processes,* which use a variety of energy forms other than a sharp cutting tool or abrasive particles to remove material. The energy forms include mechanical, electrochemical, thermal, and chemical. The nontraditional processes are discussed in Chapter 26.

Machining is a manufacturing process in which a sharp cutting tool is used to cut away material to leave the desired part shape. The predominant cutting action in machining involves shear deformation of the work material to form a chip; as the chip is removed, a new surface is exposed. Machining is most frequently applied to shape metals. The process is illustrated in Color Plate 6 in Chapter 1 and in the diagram of Figure 21.2.

Machining is one of the most important manufacturing processes. The Industrial Revolution and the growth of the manufacturing-based economies of the world can be

475

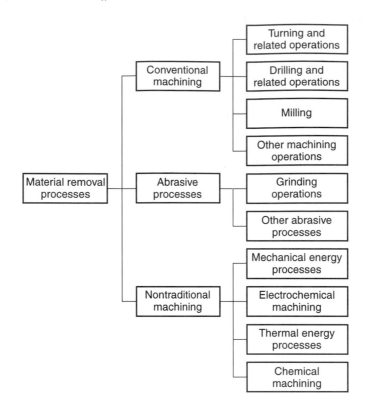

FIGURE 21.1 Classification of material removal processes.

traced largely to the development of the various machining operations (Historical Note 22.1). Machining is important commercially and technologically for several reasons:

> *Variety of work materials*—Machining can be applied to a wide variety of work materials. Virtually all solid metals can be machined. Plastics and plastic composites can also be cut by machining. Ceramics pose difficulties because of their high hardness

FIGURE 21.2 (a) A cross-sectional view of the machining process. (b) Tool with negative rake angle; compare with positive rake angle in (a).

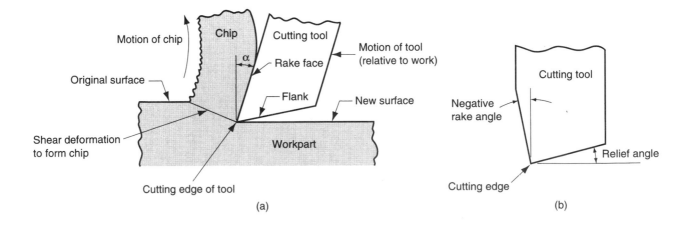

and brittleness; however, most ceramics can be successfully cut by the abrasive machining processes discussed in Chapter 26.

> *Variety of part shapes and geometric features*—Machining can be used to create any regular geometries, such as flat planes, round holes, and cylinders. By introducing variations in tool shapes and tool paths, irregular geometries can be created, such as screw threads and T-slots. By combining several machining operations in sequence, shapes of almost unlimited complexity and variety can be produced.

> *Dimensional accuracy*—Machining can produce dimensions to very close tolerances. Some machining processes can achieve tolerances of ±0.025 mm (±0.001 in.), much more accurate than most other processes.

> *Good surface finishes*—Machining is capable of creating very smooth surface finishes. Roughness values less than 0.4 microns (16 microinches) can be achieved in conventional machining operations. Some abrasive processes can achieve even better finishes.

On the other hand, certain disadvantages are associated with machining and other material removal processes:

> *Wasteful of material*—Machining is inherently wasteful of material. The chips generated in a machining operation are wasted material. Although these chips can usually be recycled, in terms of the unit operations, the material that is removed is waste.

> *Time consuming*—A machining operation generally takes more time to shape a given part than alternative shaping processes such as casting or forging.

Machining is generally performed after other manufacturing processes such as casting or bulk deformation (e.g., forging, bar drawing). The other processes create the general shape of the starting workpart, and machining provides the final geometry, dimensions, and finish.

21.1 OVERVIEW OF MACHINING TECHNOLOGY

Machining is not just one process; it is a group of processes. The common feature is the use of a cutting tool to form a chip that is removed from the workpart. To perform the operation, relative motion is required between the tool and work. This relative motion is achieved in most machining operations by means of a primary motion, called the *speed,* and a secondary motion, called the *feed.* The shape of the tool and its penetration into the work surface, combined with these motions, produces the desired shape of the resulting work surface.

Types of Machining Operations   There are many kinds of machining operations, each of which is capable of generating a certain part geometry and surface texture. We discuss these operations in considerable detail in Chapter 22, but for now it is appropriate to identify and define the three most common types: turning, drilling, and milling, illustrated in Figure 21.3.

In *turning,* a cutting tool with a single cutting edge is used to remove material from a rotating workpiece to generate a cylindrical shape, as in Figure 21.3(a). The speed motion in turning is provided by the rotating workpart, and the feed motion is achieved by the cutting tool moving slowly in a direction parallel to the axis of rotation of the workpiece.

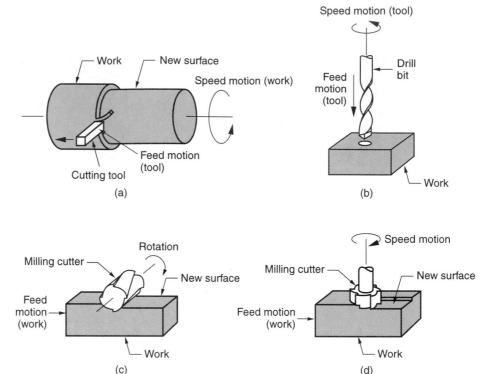

FIGURE 21.3 The three most common types of machining process: (a) turning, (b) drilling, and two forms of milling: (c) peripheral milling, and (d) face milling.

Drilling is used to create a round hole. It is usually accomplished by a rotating tool that has two cutting edges. The tool is fed in a direction parallel to its axis of rotation into the workpart to form the round hole, as in Figure 21.3(b).

In *milling,* a rotating tool with multiple cutting edges is moved slowly relative to the material to generate a plane or straight surface. The direction of the feed motion is perpendicular to the tool's axis of rotation. The speed motion is provided by the rotating milling cutter. The most basic forms of milling are peripheral milling and face milling, as in Figure 21.3(c) and (d).

Other conventional machining operations include shaping, planing, broaching, and sawing (Section 22.5). Also, grinding and similar abrasive operations are often included within the category of machining. These processes commonly follow the conventional machining operations and are used to achieve a superior surface finish on the workpart.

The Cutting Tool A cutting tool has one or more sharp cutting edges. The cutting edge serves to separate a chip from the parent work material, as in Figure 21.2. Connected to the cutting edge are two surfaces of the tool: the rake face and the flank. The rake face, which directs the flow of the newly formed chip, is oriented at a certain angle called the *rake angle* α. It is measured relative to a plane perpendicular to the work surface. The rake angle can be positive, as in Figure 21.2(a), or negative as in (b). The flank of the tool provides a clearance between the tool and the newly generated work surface, thus protecting the surface from abrasion, which would degrade the finish. This flank surface is oriented at an angle called the *relief angle.*

Because of the harsh environment in which the tool operates, its design is very important. It must be of the proper tool geometry to effectively cut the work material, and it must be made of a material that is harder than the work material.

Most cutting tools in practice have more complex geometries than those in Figure 21.2. There are two basic types, examples of which are illustrated in Figure 21.4: (a) single-point tools and (b) multiple cutting edge tools. A ***single-point tool*** has one cutting edge and is used for operations such as turning. In addition to the tool features shown in Figure 21.3, there is a tool point from which the name of this cutting tool is derived. During machining, the point of the tool penetrates below the original work surface of the workpart. The point is usually rounded to a certain radius, called the nose radius. We present a more detailed discussion of single-point cutting tools in Section 23.3.1.

Multiple cutting edge tools have more than one cutting-edge and usually achieve their motion relative to the workpart by rotating. Drilling and milling use rotating multiple cutting-edge tools. Section 23.3.1 explores the significant variety in these tools and their geometries, but for now let us illustrate the multiple cutting edge tooling with a typical milling cutter. Figure 21.4(b) shows a helical milling cutter used in peripheral milling. Although the shape is quite different from a single-point tool, many elements of tool geometry are similar.

Cutting Conditions   Relative motion is required between tool and work to perform a machining operation. The primary motion is accomplished at a certain ***cutting speed, v***. In addition, the tool must be moved laterally across the work. This is a much slower motion, called the ***feed, f***. The remaining dimension of the cut is the penetration of the cutting tool below the original work surface, called the ***depth of cut, d***. Collectively, speed, feed, and depth of cut are called the ***cutting conditions***. They form the three dimensions of the machining process, and for certain operations (e.g., most single point tool operations) their product can be used to obtain the material removal rate for the process:

$$MRR = vfd \qquad (21.1)$$

where MRR = material removal rate, mm^3/s (in.^{3}/min); v = cutting speed, m/s (ft/min), which must be converted to mm/s (in./min); f = feed, mm (in.); and d = depth of cut, mm (in.).

The cutting conditions for a turning operation are depicted in Figure 21.5. Typical units used for cutting speed are m/s (ft/min). Feed in turning is usually expressed in

FIGURE 21.4   (a) A single-point tool showing rake face, flank, and tool point; and (b) a helical milling cutter, representative of tools with multiple cutting edges.

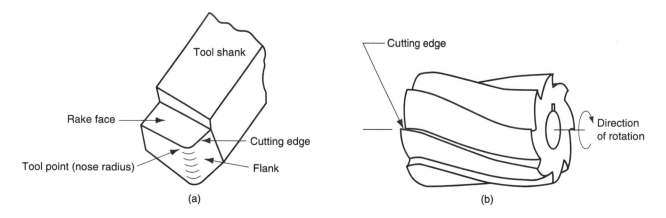

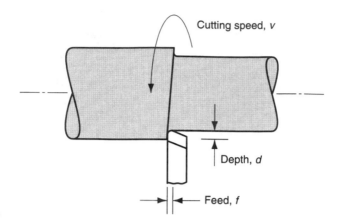

FIGURE 21.5 Cutting speed, feed, and depth of cut for a turning operation.

mm/rev (in./rev), and depth of cut is expressed in mm (in.). In other machining operations, these units may be different. For example, in a drilling operation the depth is normally interpreted as the depth of the drilled hole.

Machining operations usually divide into two categories, distinguished by purpose and cutting conditions: roughing cuts and finishing cuts. *Roughing* cuts are used to remove large amounts of material from the starting workpart as rapidly as possible, in order to produce a shape close to the desired form, but leaving some material on the piece for a subsequent finishing operation. *Finishing* cuts are used to complete the part and achieve the final dimensions, tolerances, and surface finish. In production machining jobs, one or more roughing cuts are usually performed on the work, followed by one or two finishing cuts. Roughing operations are performed at high feeds and depths—feeds of 0.4 to 1.25 mm/rev (0.015–0.050 in./rev) and depths of 2.5 to 20 mm (0.100–0.750 in.) are typical. Finishing operations are carried out at low feeds and depths—feeds of 0.125 to 0.4 mm (0.005–0.015 in./rev) and depths of 0.75 to 2.0 mm (0.030–0.075 in.) are typical. Cutting speeds are lower in roughing than in finishing.

A *cutting fluid* is often applied to the machining operation to cool and lubricate the cutting tool (cutting fluids are discussed in Section 23.4). Determining whether a cutting fluid should be used, and, if so, choosing the proper cutting fluid, is usually included within the scope of cutting conditions. Given the work material and tooling, the selection of these conditions is very influential in determining the success of a machining operation.

Machine Tools A machine tool is used to hold the workpart, position the tool relative to the work, and provide power for the machining process at the speed, feed, and depth that have been set. By controlling the tool, work, and cutting conditions, machine tools permit parts to be made with great accuracy and repeatability, to tolerances of 0.025 mm (0.001 in.) and better. The term *machine tool* applies to any power-driven machine that performs a machining operation, including grinding. The term is also frequently applied to machines that perform metal forming and pressworking operations (Chapters 19 and 20).

The machine tools traditionally used to perform the three common machining operations are identified in Table 21.1. The speed and feed motions accomplished on these machine tools are also indicated. Conventional machine tools are usually tended by a human operator, although modern machine tools are often designed to accomplish their processes with a high degree of automation. These automated machines operate under a form of control called numerical control (Section 38.1).

TABLE 21.1 Conventional machine tools used for the three common machining operations.

Operation	Machine tool	Speed	Feed	Depth of Cut
Turning	Lathe	Workpiece rotates. Speed = surface speed of workpiece.	Tool is fed parallel to work axis.	Tool penetration beneath original work surface.
Drilling	Drill press	Tool rotates. Drill bit diameter determines hole diameter.	Tool feeds in direction parallel to tool axis.	Depth of cut = depth of hole.
Milling	Milling machine	Tool rotates. Speed = surface speed of tool.	Work feeds in direction perpendicular to tool axis.	Tool penetration beneath original work surface.

21.2 THEORY OF CHIP FORMATION IN METAL MACHINING

The geometry of most practical machining operations is somewhat complex. A simplified model of machining is available which neglects many of the geometric complexities, yet describes the mechanics of the process fairly accurately. It is called the **orthogonal** cutting model, Figure 21.6. Although an actual machining process is three-dimensional, the orthogonal model has only two dimensions that play active roles in the analysis.

21.2.1 The Orthogonal Cutting Model

By definition, orthogonal cutting uses a wedge-shaped tool in which the cutting edge is perpendicular to the direction of cutting speed. As the tool is forced into the material, the chip is formed by shear deformation along a plane called the **shear plane,** which is oriented at an angle ϕ with the surface of the work. Only at the sharp cutting edge of the tool does failure of the material occur, resulting in separation of the chip from the parent material. Along the shear plane, where the bulk of the mechanical energy is consumed in machining, the material is plastically deformed.

FIGURE 21.6 Orthogonal cutting: (a) as a three-dimensional process, and (b) how it reduces to two dimensions in the side view.

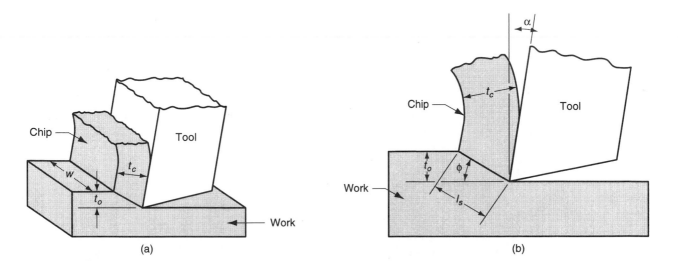

The tool in orthogonal cutting has only two elements of geometry, rake angle and clearance angle. As indicated previously, the rake angle α determines the direction that the chip flows as it is formed from the workpart; and the clearance angle provides a small clearance between the tool flank and the newly generated work surface.

During cutting, the cutting edge of the tool is positioned a certain distance below the original work surface. This corresponds to the thickness of the chip prior to chip formation, t_o. As the chip is formed along the shear plane, its thickness is increased to t_c. The ratio of t_o to t_c is called the **chip thickness ratio** (or simply the **chip ratio**) r:

$$r = \frac{t_o}{t_c} \tag{21.2}$$

Since the chip thickness after cutting is always greater than the corresponding thickness before cutting, the chip ratio will always be less than 1.0.

In addition to t_o, the orthogonal cut has a width dimension w, as shown in Figure 21.6(a), even though this dimension does not contribute much to the analysis in orthogonal cutting.

The geometry of the orthogonal cutting model allows us to establish an important relationship between the chip thickness ratio, the rake angle, and the shear plane angle. Let l_s be the length of the shear plane. We can make the substitutions: $t_o = l_s \sin\phi$, and $t_c = l_s \cos(\phi - \alpha)$. Thus,

$$r = \frac{l_s \sin\phi}{l_s \cos(\phi - \alpha)} = \frac{\sin\phi}{\cos(\phi - \alpha)}$$

This can be rearranged to determine ϕ as follows:

$$\tan\phi = \frac{r \cos\alpha}{1 - r \sin\alpha} \tag{21.3}$$

The shear strain that occurs along the shear plane can be estimated by examining Figure 21.7. Part (a) shows shear deformation approximated by a series of parallel plates sliding against one another to form the chip. Consistent with our definition of shear strain (Section 3.1.4), each plate experiences the shear strain shown in Figure 21.7(b). Referring to part (c), this can be expressed as

$$\gamma = \frac{AC}{BD} = \frac{AD + DC}{BD}$$

which can be reduced to the following definition of shear strain in metal cutting:

$$\gamma = \tan(\phi - \alpha) + \cot\phi \tag{21.4}$$

EXAMPLE 21.1 Orthogonal cutting.

In a machining operation that approximates orthogonal cutting, the cutting tool has a rake angle = 10°. The chip thickness before the cut $t_o = 0.50$ mm and the chip thickness after the cut $t_c = 1.125$ in. Calculate the shear plane angle and the shear strain in the operation.

Solution: The chip thickness ratio can be determined from Eq. (21.2):

$$r = \frac{0.50}{1.125} = 0.444$$

The shear plane angle is given by Eq. (21.3):

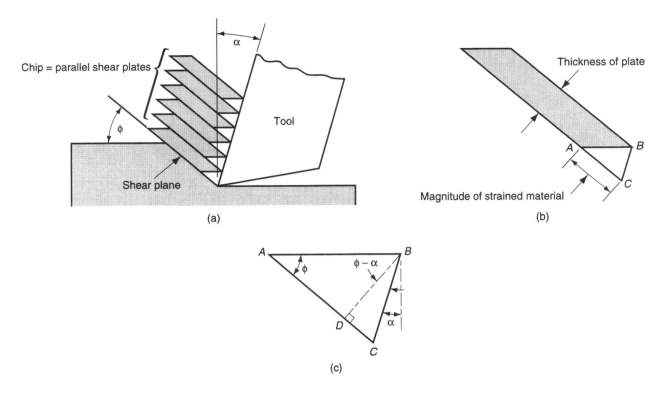

FIGURE 21.7   Shear strain during chip formation: (a) chip formation depicted as a series of parallel plates sliding relative to each other; (b) one of the plates isolated to illustrate the definition of shear strain based on this parallel plate model; and (c) shear strain triangle used to derive Eq. (21.4).

$$\tan \phi = \frac{0.444 \cos 10}{1 - 0.444 \sin 10} = 0.4738$$

$$\phi = 25.4°$$

Finally, the shear strain is calculated from Eq. (21.4):

$$\gamma = \tan(25.4 - 10) + \cot 25.4$$

$$\gamma = 0.275 + 2.111 = 2.386$$

Shear strains in metal cutting are very high.

21.2.2 Actual Chip Formation

We should note that there are differences between the orthogonal model and an actual machining process. First, the shear deformation process does not occur along a plane, but within a zone. If shearing were to take place across a plane of zero thickness, it would imply that the shearing action must occur instantaneously as it passes through the plane, rather than over some finite (although brief) time period. For the material to behave in a realistic way, the shear deformation must occur within a thin shear zone. This more realistic model of the shear deformation process in machining is illustrated in Figure 21.8. Metal-cutting experiments have indicated that the thickness of the shear zone is only a

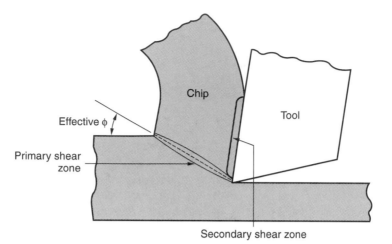

FIGURE 21.8   More realistic view of chip formation, showing shear zone rather than shear plane. Also shown is the secondary shear zone resulting from tool-chip friction.

few thousandths of an inch. Since the shear zone is so thin, there is not a great loss of accuracy in most cases by referring to it as a plane.

Second, another shearing action occurs in the chip after it has been formed. This additional shear is referred to as secondary shear to distinguish it from primary shear. Secondary shear results from friction between the chip and the tool as the chip slides along the rake face of the tool. Its effect increases with increased friction between the tool and chip. The primary and secondary shear zones can be seen in Figure 21.8.

Third, formation of the chip depends on the type of material being machined and the cutting conditions of the operation. Four basic types of chip can be distinguished, illustrated in Figure 21.9:

(a) *Discontinuous chip*—When relatively brittle materials (e.g., cast irons) are machined at low cutting speeds, the chips often form into separate segments (sometimes the segments are loosely attached). This tends to impart an irregular texture to the machined surface. High tool-chip friction and large feed and depth of cut promote the formation of this chip type.

(b) *Continuous chip*—When ductile work materials are cut at high speeds and relatively small feeds and depths, long continuous chips are formed. A good surface finish typically results when this chip type is formed. A sharp cutting edge on the tool and low tool-chip friction encourage the formation of continuous chips. Long continuous chips (as in turning) can cause problems with regard to chip disposal and/or tangling about the tool. To solve these problems, turning tools are often equipped with chip breakers (Section 23.3.1).

(c) *Continuous chip with built-up edge*—When machining ductile materials at low-to-medium cutting speeds, friction between tool and chip tends to cause portions of the work material to adhere to the rake face of the tool near the cutting edge. This formation is called a built-up edge (BUE). The formation of a BUE is cyclical; it forms and grows, then becomes unstable and breaks off. Much of the detached BUE is carried away with the chip, sometimes taking portions of the tool rake face with it, which reduces the life of the cutting tool. Portions of the detached BUE that are not carried off with the chip become imbedded in the newly created work surface, causing the surface to become rough.

The preceding chip types were first classified by Ernst in the late 1930s [12]. Since then, the available metals used in machining, cutting tool materials, and cutting speeds have all increased, and a fourth chip type has been identified:

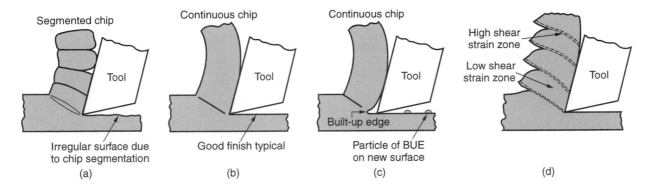

FIGURE 21.9   Four types of chip formation in metal cutting: (a) discontinuous, (b) continuous, (c) continuous with built-up edge, and (d) serrated.

(d) **Serrated chips** (the terms **shear-localized** and **segmented** are also used for this fourth chip type)—These chips are semicontinuous in the sense that they possess a sawtooth appearance that is produced by a cyclical chip formation of alternating high shear strain followed by low shear strain. This chip is most closely associated with certain difficult-to-machine metals such as titanium alloys, nickel-base superalloys, and austenitic stainless steels when they are machined at higher cutting speeds. However, the phenomenon is also found with more common work metals (e.g., steels), when they are cut at high speeds [12].[1]

21.3 FORCE RELATIONSHIPS AND THE MERCHANT EQUATION

Several forces can be defined relative to the orthogonal cutting model. Based on these forces, shear stress, coefficient of friction, and certain other relationships can be defined.

21.3.1 Forces in Metal Cutting

Consider the forces acting on the chip during orthogonal cutting in Figure 21.10(a). The forces applied against the chip by the tool can be separated into two mutually perpendicular components: The **friction force,** F, is the friction force between the tool and chip resisting the flow of the chip along the rake face of the tool. The **normal force to friction,** N, is the force that is normal to the friction force.

These two components can be used to define the coefficient of friction between the tool and the chip:

$$\mu = \frac{F}{N} \tag{21.5}$$

The friction force and its normal force can be added vectorially to form a resultant force, R. R is oriented at an angle, called the friction angle. The friction angle is related to the coefficient of friction as

---

[1] A more complete description of the serrated chip type can be found in Trent & Wright [12], pages 348–367.

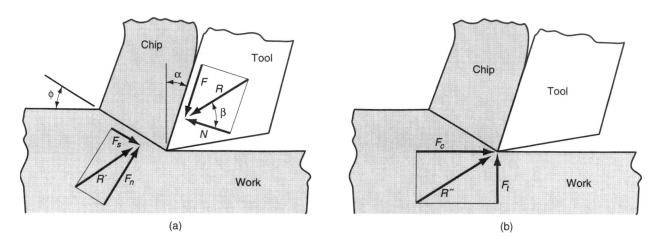

FIGURE 21.10 Forces in metal cutting: (a) forces acting on the chip in orthogonal cutting, and (b) forces acting on the tool that can be measured.

$$\mu = \tan \beta \tag{21.6}$$

In addition to the tool forces acting on the chip, there are two force components applied by the workpiece on the chip: The **shear force,** F_s is the force that causes shear deformation to occur in the shear plane. **Normal force to shear,** F_n is normal to the shear force.

Based on the shear force, we can define the shear stress which acts along the shear plane between the work and the chip:

$$S = \frac{F_s}{A_s} \tag{21.7}$$

where A_s = area of the shear plane. This shear plane area can be calculated as

$$A_s = \frac{t_o w}{\sin \phi} \tag{21.8}$$

The shear stress in Eq. (21.7) represents the level of stress required to perform the machining operation. Therefore, this stress is equal to the shear strength of the work material under the conditions at which cutting occurs.

Vector addition of the two force components F_s and F_n yields the resultant force R'. In order for the forces acting on the chip to be in balance, this resultant R' must be equal in magnitude, opposite in direction, and collinear with the resultant R.

None of the four force components F, N, F_s, and F_n can be directly measured in a machining operation, because the directions in which they are applied vary with different tool geometries and cutting conditions. However, it is possible for the cutting tool to be instrumented using a force measuring device called a dynamometer, so that two additional force components acting against the tool can be directly measured. **Cutting force,** F_c, is in the direction of cutting, the same direction as the cutting speed v. **Thrust force,** F_t, is associated with the chip thickness before the cut t_o. It is perpendicular to the cutting force.

The cutting force and thrust force are shown in Figure 21.10(b) together with their resultant force R''. The respective directions of these forces are known, so the force transducers in the dynamometer can be aligned accordingly.

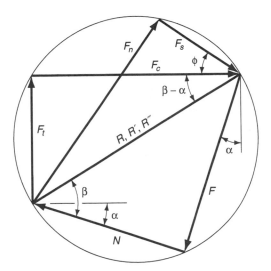

FIGURE 21.11   Force diagram showing geometric relationships between F, N, F_s, F_n, F_c, and F_t.

Equations can be derived to relate the four force components that cannot be measured to the two forces that can be measured. Using the force diagram in Figure 21.11, the following trigonometric relationships can be derived:

$$F = F_c \sin\alpha + F_t \cos\alpha \qquad (21.9)$$

$$N = F_c \cos\alpha - F_t \sin\alpha \qquad (21.10)$$

$$F_s = F_c \cos\phi - F_t \sin\phi \qquad (21.11)$$

$$F_n = F_c \sin\phi + F_t \cos\phi \qquad (21.12)$$

If cutting force and thrust force are known, these four equations can be used to calculate estimates of shear force, friction force, and normal force to friction. Based on these force estimates, shear stress and coefficient of friction can be determined.

Note that in the special case of orthogonal cutting when the rake angle $\alpha = 0$, Eqs (21.9) and (21.10) reduce to $F = F_t$ and $N = F_c$, respectively. Thus, in this special case, friction force and its normal force could be directly measured by the dynamometer.

**EXAMPLE 21.2
Shear stress in machining.**

Suppose in Example 21.1 that cutting force and thrust force are measured during an orthogonal cutting operation with values: $F_c = 1559$ N and $F_t = 1271$ N. The width of the orthogonal cutting operation $w = 3.0$ mm. Based on these data, determine the shear strength of the work material.

Solution: From Example 21.1, rake angle $\alpha = 10°$, and shear plane angle $\phi = 25.4°$. Shear force can be computed from Eq. (21.11):

$$F_s = 1559 \cos 25.4 - 1271 \sin 25.4 = 863 \text{ N}$$

The shear plane area is given by Eq. (21.8):

$$A_s = \frac{(0.5)(3.0)}{\sin 25.4} = 3.497 \text{ mm}^2$$

Thus the shear stress, which equals the shear strength of the work material, is

$$\tau = S = \frac{863}{3.497} = 247 \text{ N/mm}^2 = 247 \text{ MPa}$$

This example demonstrates that cutting force and thrust force are related to the shear strength of the work material. The relationships can be established in a more direct way. Recalling from Eq. (21.7) that the shear force $F_s = S A_s$, the force diagram of Figure 21.11 can be used to derive the following equations:

$$F_c = \frac{St_o \, w \, \cos(\beta - \alpha)}{\sin \phi \, \cos(\phi + \beta - \alpha)} = \frac{F_s \cos(\beta - \alpha)}{\cos(\phi + \beta - \alpha)} \qquad (21.13)$$

and

$$F_t = \frac{St_o \, w \, \sin(\beta - \alpha)}{\sin \phi \, \cos(\phi + \beta - \alpha)} = \frac{F_s \sin(\beta - \alpha)}{\cos(\phi + \beta - \alpha)} \qquad (21.14)$$

These equations allow one to estimate cutting force and thrust forces in an orthogonal cutting operation if the shear strength of the work material is known.

21.3.2 The Merchant Equation

One of the important relationships in metal cutting was derived by Eugene Merchant [9]. Its derivation was based on the assumption of orthogonal cutting, but its general validity extends to three-dimensional machining operations. Merchant started with the definition of shear stress expressed in the form of the following relationship derived by combining Eqs. (21.7), (21.8), and (21.11):

$$\tau = \frac{F_c \cos \phi - F_t \sin \phi}{(t_o w / \sin \phi)} \qquad (21.15)$$

He reasoned that, out of all the possible angles emanating from the cutting edge of the tool at which shear deformation could occur, there is one angle ϕ that predominates. This is the angle at which shear stress is just equal to the shear strength of the work material, and so shear deformation occurs at this angle. For all other possible shear angles, the shear stress is less than the shear strength, so chip formation cannot occur at these other angles. In effect, the work material will select a shear plane angle that minimizes energy. This angle can be determined by taking the derivative of the shear stress S in Eq. (21.15) with respect to ϕ and setting the derivative to zero. Solving for ϕ, we get the relationship named after Merchant:

$$\phi = 45 + \frac{\alpha}{2} - \frac{\beta}{2} \qquad (21.16)$$

Among the assumptions upon which the Merchant equation is based is that shear strength of the work material is a constant—unaffected by strain rate, temperature, and other factors. Because this assumption is violated in practical machining operations, Eq. (21.16) must be considered an approximate relationship rather than an accurate mathematical equation. Let us nevertheless consider its application in the following example.

**EXAMPLE 21.3
Estimating friction angle.**

Using the data and results from our previous examples, compute (a) the friction angle using the Merchant equation, and (b) the coefficient of friction.

Solution: (a) From Example 21.1, $\alpha = 10°$, and $\phi = 25.4°$. Rearranging Eq. (21.16), the friction angle can be estimated:

$$\beta = 2(45) + 10 - 2(25.4) = 49.2°$$

(b) The coefficient of friction is given by Eq. (21.6):

$$\mu = \tan 49.2 = 1.16$$

Lessons Based on the Merchant Equation   The real value of the Merchant equation is that it defines the general relationship between rake angle, tool-chip friction, and shear plane angle. Two conclusions follow from this relationship: (1) an increase in the rake angle causes the shear plane angle to increase; and (2) a decrease in the friction angle (or a decrease in the coefficient of friction) causes the shear plane angle to increase.

The importance of increasing the shear plane angle can be seen in Figure 21.12. If all other factors remain the same, a higher shear plane angle results in a smaller shear plane area. Since the shear strength is applied across this area, the shear force required to form the chip will decrease when the shear plane area is decreased. This tends to make machining easier to perform. Although it is not as obvious in the figure, a higher shear plane angle results in lower cutting energy and cutting temperature. These are good reasons to try to make the shear plane angle as large as possible during machining. How? By increasing rake angle and decreasing friction angle, perhaps by using a lubricant cutting fluid.

Approximation of Turning by Orthogonal Cutting   The orthogonal model can be used to approximate turning and certain other single point machining operations so long as the feed in these operations is small relative to depth of cut. Thus, most of the cutting will take place in the direction of the feed, and cutting on the nose of the tool will be negligible. Figure 21.13 indicates the conversion from one cutting situation to the other: (a) shows a turning operation while (b) depicts the corresponding orthogonal case.

The interpretation of cutting conditions is different in the two cases. The chip thickness before the cut t_o in orthogonal cutting corresponds to the feed f in turning, and the width of cut w in orthogonal cutting corresponds to the depth of cut d in turning. In addition, the thrust force in the orthogonal model corresponds to the feed force F_f (i.e., the force on the tool in the direction of feed) in turning. Cutting speed and cutting force have the same interpretations in the two cases. Table 21.2 summarizes the conversions.

FIGURE 21.12   Effect of shear plane angle ϕ: (a) higher ϕ with a resulting lower shear plane area; (b) smaller ϕ with a corresponding larger shear plane area. Note that the rake angle is larger in (a), which tends to increase shear angle according to the Merchant equation.

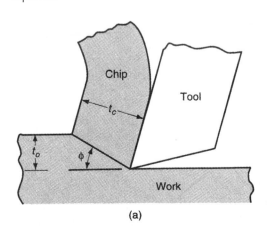

(a)

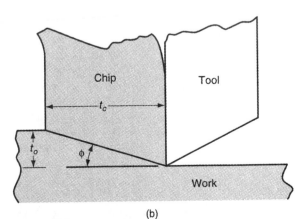

(b)

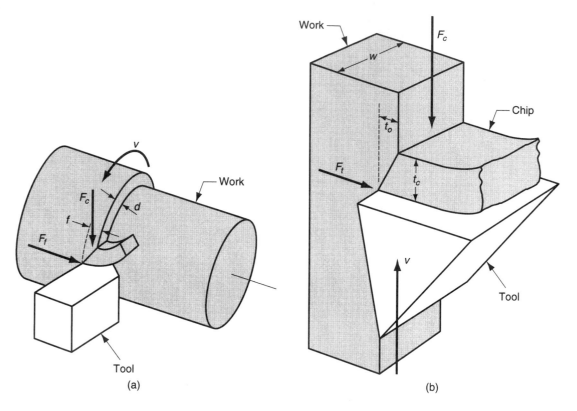

FIGURE 21.13 Approximation of turning by the orthogonal model: (a) turning; and (b) the corresponding orthogonal cutting.

TABLE 21.2 Conversion key: turning operation vs. orthogonal cutting.

Turning Operation	Orthogonal Cutting Model
Feed $f =$	Chip thickness before cut t_o
Depth $d =$	Width of cut w
Cutting speed $v =$	Cutting speed v
Cutting force $F_c =$	Cutting force F_c
Feed force $F_f =$	Thrust force F_t

21.4 POWER AND ENERGY RELATIONSHIPS IN MACHINING

A machining operation requires power. The cutting force in a production machining operation might exceed 1,000 N (several hundred lbs), as suggested by Example 21.2. Typical cutting speeds are several hundred m/min. The product of cutting force and speed gives the power (energy per unit time) required to perform a machining operation:

$$P_c = F_c v \qquad (21.17)$$

where P_c = cutting power, N-m/s or W (ft-lb/min); F_c = cutting force, N (lb); and v = cutting speed, m/s (ft/min). In U.S. customary units, power is traditional expressed as horsepower by dividing ft-lb/min by 33,000. Hence,

$$HP_c = \frac{F_c v}{33,000} \tag{21.18}$$

where HP_c = cutting horsepower, hp. The gross power required to operate the machine tool is greater than the power delivered to the cutting process because of mechanical losses in the motor and drive train in the machine. These losses can be accounted for by the mechanical efficiency of the machine tool.

$$P_g = \frac{P_c}{E} \text{ or } HP_g = \frac{HP_c}{E} \tag{21.19}$$

where P_g = gross power of the machine tool motor, W; HP_g = gross horsepower; and E = mechanical efficiency of the machine tool. Typical values of E for machine tools are around 90%.

It is often useful to convert power into power per unit volume rate of metal cut. This is called the **unit power**, P_u (or **unit horsepower**, HP_u), defined:

$$P_u = \frac{P_c}{MRR} \text{ or } HP_u = \frac{HP_c}{MRR} \tag{21.20}$$

where MRR = material removal rate, mm³/s (in.³/min). The material removal rate can be calculated as the product of $v t_o w$. This is Eq. (21.1) using the conversions from Table 21.2. Unit power is also known as the **specific energy**, U.

$$U = P_u = \frac{P_c}{MRR} = \frac{F_c v}{v t_o w} = \frac{F_c}{t_o w} \tag{21.21}$$

The units for specific energy are typically N-m/mm³ (in.-lb/in.³), although the last expression in Eq. (21.21) suggests units of N/mm² (lb/in.²). It is more meaningful to retain the units as N-m/mm³ or J/mm³ (in.-lb/in.³).

EXAMPLE 21.4 **Power relationships in machining.**	Continuing with our previous examples, let us determine cutting power and specific energy required to perform the machining process if the cutting speed = 100 m/min. Summarizing the data and results from previous examples, t_o = 0.50 mm., w = 3.0 mm., F_c = 1557 N.

Solution: From Eq. (21.18), power in the operation is

$$P_c = (1557 \text{ N})(100 \text{ m/min}) = 155,700 \text{ N-m/min} = 155,700 \text{ J/min} = 2595 \text{ J/s} = 2595 \text{ W}$$

Specific energy is calculated from Eq. (21.21):

$$U = \frac{155,700}{100(10^3)(3.0)(0.5)} = \frac{155,700}{150,000} = 1.038 \text{ N-m/mm}^3$$

Unit power and specific energy provide a useful measure of how much power (or energy) is required to remove one cubic inch of metal during machining. Using this measure, different work materials can be compared in terms of their power and energy requirements. Table 21.3 presents a listing of unit horsepower and specific energy values for selected work materials.

The values in Table 21.3 are based on two assumptions: (1) the cutting tool is sharp, and (2) the chip thickness before the cut t_o = 0.25 mm (0.010 in.). If these assumptions are not satisfied, some adjustments must be made. For worn tools, the power required to perform the cut is greater, and this is reflected in higher specific energy and unit horsepower values. As an approximate guide, the values in the table should be multiplied by

TABLE 21.3 Values of unit horsepower and specific energy for selected work materials using sharp cutting tools and chip thickness before the cut t_o = 0.25 mm (0.010 in.).

Material	Brinell Hardness	Specific Energy, U, or Unit Power P_u		Unit Horsepower, HP_u
		N-m/mm³	(in.-lb/in.³)	(hp/in.³/min)
Carbon steel	150–200	1.6	(240,000)	(0.6)
	201–250	2.2	(320,000)	(0.8)
	251–300	2.8	(400,000)	(1.0)
Alloy steels	200–250	2.2	(320,000)	(0.8)
	251–300	2.8	(400,000)	(1.0)
	301–350	3.6	(520,000)	(1.3)
	351–400	4.4	(640,000)	(1.6)
Cast irons	125–175	1.1	(160,000)	(0.4)
	175–250	1.6	(240,000)	(0.6)
Stainless steel	150–250	2.8	(400,000)	(1.0)
Aluminum	50–100	0.7	(100,000)	(0.25)
Aluminum alloys	100–150	0.8	(120,000)	(0.3)
Copper (pure)		1.9	(280,000)	(0.7)
Brass	100–150	2.2	(320,000)	(0.8)
Bronze	100–150	2.2	(320,000)	(0.8)
Magnesium alloys	50–100	0.4	(60,000)	(0.15)

Data compiled from [5], [7], [10], and other sources.

a factor between 1.00 and 1.25 depending on the degree of dullness of the tool. For sharp tools, the factor is 1.00. For tools in a finishing operation that are nearly worn out, the factor is around 1.10, and for tools in a roughing operation that are nearly worn out, the factor is 1.25.

Chip thickness before the cut t_o also affects the specific energy and unit horsepower values. As t_o is reduced, unit power requirements increase. This relationship is referred to as the *size effect.* For example, grinding, in which the chips are extremely small by comparison to most other machining operations, requires very high specific energy values. The U and HP_u values in Table 21.3 can still be used to estimate horsepower and energy for situations in which t_o is not equal to 0.25 mm (0.010 in.) by applying a correction factor to account for any difference in chip thickness before the cut. Figure 21.14 provides values of this correction factor as a function of t_o. The unit horsepower and specific energy values in Table 21.3 should be multiplied by the appropriate correction factor when t_o is different from 0.25 mm (0.010 in.).

It should be noted that in addition to tool sharpness and size effect, other factors influence the values of specific energy and unit horsepower for a given operation. These other factors include rake angle, cutting speed, and cutting fluid. As rake angle or cutting speed are increased, or when cutting fluid is added, the U and HP_u values are reduced slightly. For our purposes in the end-of-chapter exercises, the effects of these additional factors can be ignored.

The distribution of cutting energy between the tool, work, and chip varies with cutting speed, as indicated in Figure 21.15. At low speeds, a significant portion of the total energy is absorbed into the tool. But at higher speeds (and higher energy levels), the rapid motion of the chip across the rake face of the tool affords less opportunity for the heat generated in the primary shear zone to be conducted across the tool–chip interface into the tool. Hence, the proportion of total energy absorbed by the tool is reduced, and most of it is carried off with the chip. This helps to prolong the life of the cutting tool.

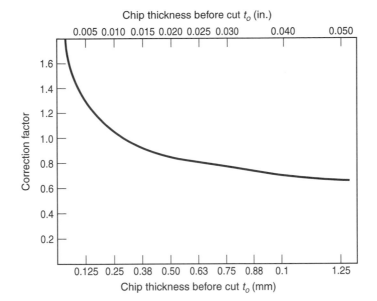

FIGURE 21.14   Correction factor for unit horsepower and specific energy when values of chip thickness before the cut t_o are different from 0.25 mm (0.010 in.).

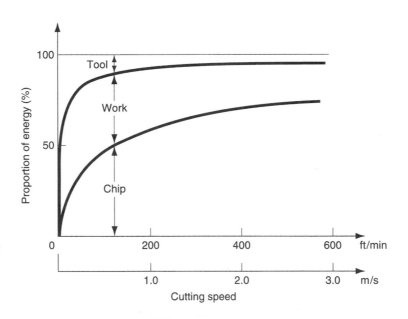

FIGURE 21.15   Typical distribution of total cutting energy among the tool, work, and chip as a function of cutting speed. (Based on data in [8].)

21.5 CUTTING TEMPERATURE

Of the total energy consumed in machining, nearly all of it (approximately 98%) is converted into heat. This heat can cause temperatures to be very high at the tool-chip interface—over 600°C (1100°F) is not unusual. The remaining energy (about 2%) is retained as elastic energy in the chip. In this section, methods of calculating and measuring machining temperatures are discussed.

21.5.1 Analytical Methods

There are several analytical methods to calculate estimates of cutting temperature. References [1], [3], [8], and [13] present some of these approaches. We describe the method by Cook [3]. This method was derived from dimensional analysis, using experimental data for a variety of work materials to establish parameter values for the resulting equation. The equation can be used to predict the rise in temperature at the tool–chip interface during machining.

$$T = \frac{0.4U}{\rho C}\left(\frac{vt_o}{K}\right)^{0.333} \tag{21.22}$$

where T = mean temperature rise at the tool-chip interface, C° (F°); U = specific energy in the operation, N-m/mm^3 or J/mm^3 (in.-lb/in.^{3}); v = cutting speed, m/s (in./sec); t_o = chip thickness before the cut, m (in.); ρC = volumetric specific heat of the work material, J/mm^3-C (in.-lb/in.^{3}-F); K = thermal diffusivity of the work material, m^2/s (in.^{2}/sec).

EXAMPLE 21.5 Cutting temperature.

For the specific energy obtained in Example 21.4, calculate the increase in temperature above ambient temperature of 20°C. Use the given data from the previous examples in this chapter: v = 100 m/min, t_o = 0.50 mm. In addition, the volumetric specific heat for the work material = 3.0 (10^{-3}) J/mm^3-C, and thermal diffusivity = 50 (10^{-6}) m^2/s (= 50 mm^2/s).

Solution: Cutting speed must be converted to mm/s: v = (100 m/min)(10^3 mm/m)/(60 s/min) = 1667 mm/s. Eq. (21.22) can now be used to compute the mean temperature rise:

$$T = \frac{0.4(1.038)}{3.0(10^{-3})}\,°C\left(\frac{1667(0.5)}{50}\right)^{0.333} = (138.4)(2.552) = 353\ C°$$

Adding this to the ambient temperature, the resulting cutting temperature is 20 + 353 = 373°C.

21.5.2 Measurement of Cutting Temperature

Experimental methods have been developed to measure temperatures in machining. The most frequently used measuring technique is the *tool–chip thermocouple.* This thermocouple consists of the tool and the chip as the two dissimilar metals forming the thermocouple junction. By properly connecting electrical leads to the tool and workpart (which is connected to the chip), the emf generated at the tool–chip interface during cutting can be monitored using a recording potentiometer or other appropriate data-collection device. The emf output of the tool–chip thermocouple can be converted into the corresponding temperature value by means of calibration equations for the particular tool-work combination.

The tool–chip thermocouple has been utilized by researchers to investigate the relationship between temperature and cutting conditions such as speed and feed. Trigger [13] determined the speed–temperature relationship to be of the following general form:

$$T = K v^m \tag{21.23}$$

where T = measured tool–chip interface temperature and v = cutting speed. The parameters K and m depend on cutting conditions (other than v) and work material. Figure 21.16 plots temperature versus cutting speed for several work materials, with equations of the form of Eq. (21.23) determined for each material. A similar relationship ex-

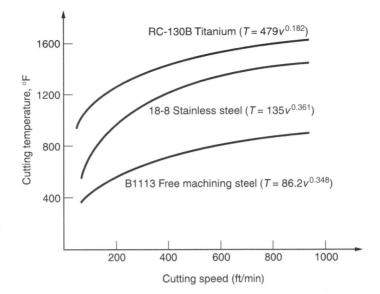

FIGURE 21.16
Experimentally measured cutting temperatures plotted against speed for three work materials, indicating general agreement with Eq. (21.22). (Based on data in [8].)[2]

ists between cutting temperature and feed; however, the effect of feed on temperature is not as strong as cutting speed. These empirical results tend to support the general validity of the Cook equation: Eq. (21.22).

REFERENCES

[1] Boothroyd, G., and Knight, W. A. *Fundamentals of Metal Machining and Machine Tools.* 2nd ed. Marcel Dekker, Inc., New York, 1989.

[2] Chao, B. T., and Trigger, K. J. "Temperature Distribution at the Tool–Chip Interface in Metal Cutting." *ASME Transactions.* Vol. 77, October 1955, pp. 1107–1121.

[3] Cook, N. "Tool Wear and Tool Life." *ASME Transactions, J. Engrg. for Industry.* Vol. 95, November 1973, pp. 931–938.

[4] DeGarmo, E. P., Black, J. T., and Kohser, R. A. *Materials and Processes in Manufacturing.* 7th ed. Macmillan Publishing Company, New York, 1988.

[5] Drozda, T. J., and Wick, C. (eds.). *Tool and Manufacturing Engineers Handbook.* 4th ed. Volume I: *Machining,* Society of Manufacturing Engineers, Dearborn, Michigan, 1983.

[6] Kalpakjian, S. *Manufacturing Processes for Engineering Materials.* 2nd ed. Addison-Wesley Publishing Company, Reading, Massachusetts, 1991.

[7] Lindberg, R. A. *Processes and Materials of Manufacture.* 4th ed. Allyn and Bacon, Inc., Boston, Massachusetts, 1990.

[8] Loewen, E. G., and Shaw, M. C. "On the Analysis of Cutting Tool Temperatures." *ASME Transactions.* Vol. 76, No. 2, February 1954, pp. 217–225.

[9] Merchant, M. E. "Mechanics of the Metal Cutting Process. II. Plasticity Conditions in Orthogonal Cutting." *Journal of Applied Physics.* Volume 16, June 1945, pp. 318–324.

[10] Schey, J. A. *Introduction to Manufacturing Processes.* 2nd ed. McGraw-Hill Book Company, New York, 1987.

[11] Shaw, M. C. *Metal Cutting Principles.* Clarendon Press, Oxford, England, 1984.

[12] Trent, E. M., and Wright, P. K. *Metal Cutting.* 4th ed. Butterworth Heinemann, Boston, Massachusetts, 2000.

[13] Trigger, K. J. "Progress Report No. 2 on Tool-Chip Interface Temperatures." *ASME Transactions.* Vol. 71, No. 2, February 1949, pp. 163–174.

[14] Trigger, K. J., and Chao, B. T. "An Analytical Evaluation of Metal Cutting Temperatures." *ASME Transactions.* Vol. 73, No. 1, January 1951, pp. 57–68.

[2] The units reported in the Loewen and Shaw ASME paper [Loewen54] were °F for cutting temperature and ft/min for cutting speed. We have retained those units in the plots and equations of our figure.

REVIEW QUESTIONS

21.1. What distinguishes machining from other manufacturing processes?

21.2. Identify some of the reasons why machining is commercially and technologically important.

21.3. Name the three most common machining processes.

21.4. What are the two basic categories of cutting tools in machining? Give an example of a machining operation that uses each of the tooling types.

21.5. Identify the parameters of a machining operation that are included within the scope of cutting conditions.

21.6. Define the difference between roughing and finishing operations in machining.

21.7. What is a machine tool?

21.8. What is an orthogonal cutting operation?

21.9. Name and briefly describe the four types of chips that occur in metal cutting.

21.10. Describe in words what the Merchant equation tells us.

21.11. What is the specific energy in metal machining?

21.12. What does the term size effect mean in metal cutting?

21.13. What is a tool-chip thermocouple?

MULTIPLE CHOICE QUIZ

There is a total of 11 correct answers in the following multiple choice questions (some questions have multiple answers that are correct). To attain a perfect score on the quiz, all correct answers must be given, since each correct answer is worth 1 point. For each question, each omitted answer or wrong answer reduces the score by 1 point, and each additional answer beyond the number of answers required reduces the score by 1 point. Percentage score on the quiz is based on the total number of correct answers.

21.1. A lathe is normally used to perform which of the following machining operations (one best answer)? (a) broaching, (b) drilling, (c) milling, or (d) turning.

21.2. With which one of the following geometric forms is the drilling operation most closely associated? (a) external cylinder, (b) flat plane, (c) round hole, (d) screw threads, or (e) sphere.

21.3. If the cutting conditions in a turning operation are $v = 300$ m/min, $f = 0.25$ mm/rev, and $d = 4.0$ mm, which one of the following is the material removal rate? (a) 3×10^{-3} m^3/min, (b) 30×10^3 mm^3/min, (c) 5000 mm^3/s, or (d) 5×10^4 mm^3/s.

21.4. A roughing operation generally involves which one of the following combinations of cutting conditions? (a) high v, f, and d; (b) high v, low f and d; (c) low v, high f and d; or (d) low v, f, and d.

21.5. The chip thickness ratio is which of the following? (a) t_c/t_o, (b) t_o/t_c. (c) f/d, or (d) t_o/w.

21.6. Which of the following types of chip would be expected in a turning operation conducted at low cutting speeds

on a brittle work material (one answer)? (a) continuous, (b) continuous with built-up edge, or (c) discontinuous.

21.7. According to the Merchant equation, an increase in rake angle would have which of the following results, all other factors remaining the same (more than one answer)? (a) decrease in friction angle, (b) decrease in power requirements, (c) decrease in shear plane angle, (d) increase in cutting temperature, or (e) increase in shear plane angle.

21.8. Which of the following metals would usually have the lowest unit horsepower (one answer)? (a) aluminum, (b) brass, (c) cast iron, or (d) steel.

21.9. For which one of the following values of chip thickness before the cut t_o would you expect the specific energy to be the greatest? (a) 0.010 in. (b) 0.025 in. (c) 0.12 mm, or (d) 2.0 mm.

21.10. Which of the following cutting conditions has the strongest effect on cutting temperature? (a) feed or (b) speed.

PROBLEMS

Chip Formation and Forces in Machining

21.1. In an orthogonal cutting operation, the tool has a rake angle = 15°. The chip thickness before the cut = 0.30 mm and the cut yields a deformed chip thickness = 0.65 mm. Calculate (a) the shear plane angle and (b) the shear strain for the operation.

21.2. In Problem 21.1, suppose the rake angle were changed to $\alpha = 0°$. Assuming that the friction angle remains the

same, determine (a) the shear plane angle, (b) the chip thickness, and (c) the shear strain for the operation.

21.3. In an orthogonal cutting operation, the tool has a rake angle = −5°. The chip thickness before the cut = 0.012 in. and the cut yields a deformed chip thickness = 0.028 in. Calculate (a) the shear plane angle and (b) the shear strain for the operation.

21.4. The cutting conditions in a turning operation are $v = 2$ m/s, $f = 0.25$ mm, and $d = 3.0$ mm. The tool rake angle = 10°, which produces a deformed chip thickness $t_c = 0.54$ mm. Determine (a) shear plane angle, (b) shear strain, and (c) material removal rate. Use the orthogonal cutting model as an approximation of the turning process.

21.5. The cutting force and thrust force in an orthogonal cutting operation are $F_c = 1470$ N and $F_t = 1589$ N. The rake angle = 5°, the width of the cut = 5.0 mm, the chip thickness before the cut = 0.6, and the chip thickness ratio = 0.38. Determine (a) the shear strength of the work material and (b) the coefficient of friction in the operation.

21.6. The cutting force and thrust force have been measured in an orthogonal cutting operation: $F_c = 300$ lb and $F_t = 291$ lb. The rake angle = 10°, the width of the cut = 0.200 in. the chip thickness before the cut = 0.015, and the chip thickness ratio = 0.4. Determine (a) the shear strength of the work material and (b) the coefficient of friction in the operation.

21.7. An orthogonal cutting operation is performed using a rake angle of 15°, $t_o = 0.012$ in. and $w = 0.100$ in. The chip thickness ratio is measured after the cut to be 0.55. Determine (a) the chip thickness after the cut, (b) the shear angle, (c) the friction angle, (d) the coefficient of friction, and (e) the shear strain.

21.8. The orthogonal cutting operation described in previous Problem 21.7 involves a work material whose shear strength is 40,000 lb/in.^{2} Based on your answers to the previous problem, compute (a) the shear force, (b) the cutting force, (c) the thrust force, and (d) the friction force.

21.9. In an orthogonal cutting operation, the rake angle = −5°, $t_o = 0.2$ mm and $w = 4.0$ mm. The chip ratio $r = 0.4$. Determine (a) the chip thickness after the cut, (b) the shear angle, (c) the friction angle, (d) the coefficient of friction, and (e) the shear strain.

21.10. The shear strength of a certain work material = 50,000 lb/in.^{2} An orthogonal cutting operation is performed using a tool with a rake angle = 20° at the following cutting conditions: Speed = 100 ft/min, chip thickness before the cut = 0.015 in. and width of cut = 0.150 in. The resulting chip thickness ratio = 0.50. Determine (a) the shear plane angle; (b) the shear force; (c) cutting force and thrust force, and (d) friction force.

21.11. Solve the previous problem, except that the rake angle has been changed to −5° and the resulting chip thickness ratio = 0.35.

21.12. A turning operation is performed using the following cutting conditions: $v = 300$ ft/min, $f = 0.010$ in./rev, and $d = 0.100$ in. The rake angle on the tool in the direction of chip flow = 10°, resulting in a chip ratio = 0.42. The shear strength of the work material = 40,000 lb/in.^{2} Using the orthogonal model as an approximation of turning, determine (a) the shear plane angle; (b) the shear force; (c) cutting force and feed force.

21.13. Turning is performed on a work material with shear strength of 250 MPa. The following conditions are used: $v = 3.0$ m/s, $f = 0.20$ mm/rev, $d = 3.0$ mm, and rake angle = 7° in the direction of chip flow. The resulting chip ratio = 0.5. Using the orthogonal model as an approximation of turning, determine (a) the shear plane angle; (b) the shear force; (c) cutting force and feed force.

21.14. A turning operation is made with a rake angle of 10°, a feed of 0.010 in./rev and a depth of cut = 0.100 in. The shear strength of the work material is known to be 50,000 lb/in.^{2}, and the chip thickness ratio is measured after the cut to be 0.40. Determine the cutting force and the feed force. Use the orthogonal cutting model as an approximation of the turning process.

21.15. Show how Eq. (21.3) is derived from the definition of chip ratio, Eq. (21.2) and Figure 21.6(b).

21.16. Show how Eq. (21.4) is derived from Figure 21.7.

21.17. Derive the force equations for F, N, F_s, and F_n (Eqs. (21.9) through (21.12) in the text) using the force diagram of Figure 21.11.

Power and Energy in Machining

21.18. In a turning operation on stainless steel with hardness = 200 HB, the cutting speed = 200 m/min, feed = 0.25 mm/rev, and depth of cut = 7.5 mm. How much power will the lathe draw in performing this operation if its mechanical efficiency = 90%? Use Table 21.3 to obtain the appropriate specific energy value.

21.19. In previous Problem 21.18, compute the lathe power requirements if feed = 0.50 mm/rev.

21.20. In a turning operation on aluminum, cutting conditions are as follows: $v = 900$ ft/min, $f = 0.020$ in./rev, and $d = 0.250$ in. What horsepower is required of the drive motor, if the lathe has a mechanical efficiency = 87%? Use Table 21.3 to obtain the appropriate unit horsepower value.

21.21. In a turning operation on plain carbon steel whose Brinell hardness = 275 HB, the cutting speed is set at 200 m/min and depth of cut = 6.0 mm. The lathe motor is rated at 25 kW, and its mechanical efficiency = 90%. Using the appropriate specific energy value from Table 21.3, determine the maximum feed that can be set for this operation.

21.22. A turning operation is to be performed on a 20 hp lathe with efficiency = 90%. The work material is an alloy steel whose hardness is in the range 360 to 380 HB. Cutting conditions are $v = 400$ ft/min., feed = 0.010 in./rev, and depth of cut = 0.150 in. Based on these values, can the job be performed on the 20 hp lathe? Use Table 21.3 to obtain the appropriate unit horsepower value.

21.23. Suppose the cutting speed in Problems 21.7 and 21.8 is $v = 200$ ft/min. From your answers to those problems, find (a) the horsepower consumed in the operation, (b) the metal removal rate in in.^{3}/min, (c) the unit horsepower (hp-min/(in.^{3})), and (d) the specific energy (in-lb/in.^{3}).

21.24. For Problem 21.12, the lathe has a mechanical efficiency = 0.80. Determine (a) the horsepower consumed by the turning operation; (b) the horsepower that must be generated by the lathe; (c) the unit horsepower and specific energy for the work material in this operation.

21.25. In a turning operation on low carbon steel (175 BHN), the cutting conditions are $v = 400$ ft/min, $f = 0.010$ in./rev, and $d = 0.075$ in. The lathe has a mechanical efficiency = 0.85. Based on the unit horsepower values in Table 21.3, determine (a) the horsepower consumed by the turning operation; (b) the horsepower that must be generated by the lathe.

21.26. Solve Problem 21.24 except that the feed $f = 0.005$ in./rev and the work material is stainless steel (Brinell Hardness = 225 HB).

21.27. A turning operation is carried out on aluminum (100 BHN), the cutting conditions are $v = 5.6$ m/s, $f = 0.25$ mm/rev, and $d = 2.0$ mm. The lathe has a mechanical efficiency = 0.85. Based on the specific energy values in Table 21.3, determine (a) the cutting power and (b) the gross power in the turning operation, in Watts.

21.28. Solve Problem 21.27 but with the following changes: $v = 1.3$ m/s, $f = 0.75$ mm/rev, and $d = 4.0$ mm. Note that although the power used in this operation is vir- tually the same as in the previous problem, the metal removal rate is about 40% greater.

21.29. A turning operation is performed on an engine lathe using a tool with zero rake angle in the direction of chip flow. The work material is an alloy steel with hardness = 325 Brinell hardness. The feed is .015 in./rev, the depth of cut is 0.125 in., and the cutting speed is 300 ft/min. After the cut, the chip thickness ratio is measured to be 0.45. (a) Using the appropriate value of specific energy from Table 21.3, compute the horsepower at the drive motor, if the lathe has an efficiency = 85%. (b) Based on horsepower, compute your best estimate of the cutting force for this turning operation. Use the orthogonal cutting model as an approximation of the turning process.

21.30. A lathe performs a turning operation on a workpiece of 6.0 in. diameter. The shear strength of the work = 40,000 lb/in.^{2} The rake angle of the tool = 10°. The machine settings are rotational speed = 500 rev/min, feed = 0.0075 in./rev, and depth = 0.075 in. The chip thickness after the cut is 0.015 in. Determine (a) the horsepower required in the operation, (b) the unit horsepower for this material under these conditions, and (c) the unit horsepower as it would be listed in Table 21.3 for a t_o of 0.010 in. Use the orthogonal cutting model as an approximation of the turning process.

Cutting Temperature

21.31. Orthogonal cutting is performed on a metal whose mass specific heat = 1.1 J/g-C, density = 2.7 g/cm^3, and thermal diffusivity = 0.9 cm^2/s. The following cutting conditions are used: $v = 4.0$ m/s, $t_o = 0.3$ mm, and $w = 2.0$ mm. The cutting force is measured at $F_c = 1100$ N. Using Cook's equation, determine the cutting temperature if the ambient temperature = 20°C.

21.32. Consider a turning operation performed on steel whose hardness = 225 HB at a speed = 3.0 m/s, feed = 0.25 mm, and depth = 4.0 mm. Using values of thermal properties found in the tables and definitions of Section 4.1 and the appropriate specific energy value from Table 21.3, compute an estimate of cutting temperature using the Cook equation. Assume ambient temperature = 20°C.

21.33. An orthogonal cutting operation is performed on a certain metal whose volumetric specific heat = 110 in.-lb/in.^{3}-F, and thermal diffusivity = 0.140 in.^{2}/sec. The following cutting conditions are used: $v = 350$ ft/min, $t_o = 0.008$ in., and $w = 0.100$ in. The cutting force is measured at $F_c = 200$ lb. Using Cook's equation, determine the cutting temperature if the ambient temperature = 70°F.

21.34. It is desired to estimate the cutting temperature for a certain alloy steel whose hardness = 275 Brinell. Use the appropriate value of specific energy from Table 21.3 and compute the cutting temperature by means of the Cook equation for a turning operation in which the following cutting conditions are used: speed $v = 300$ ft/min, feed $f = 0.0075$ in./rev, and depth $d = 0.100$ in. The thermal properties of the work material are volumetric specific heat = 200 in lb/in.^{3} F, and thermal diffusivity = 0.14 in.^{2}/sec. Assume ambient temperature = 70°F.

21.35. An orthogonal machining operation removes metal at 1.8 in.^{3}/min. The cutting force in the process = 300 lb. The work material has a thermal diffusivity = 0.18 in.^{2}/sec and a volumetric specific heat = 124 in.-lb/in.^{3} F. If the feed $f = t_o = 0.010$ in. and width of cut = 0.100 in., use the Cook formula to compute the cutting temperature in the operation given that ambient temperature = 70°F.

21.36. A turning operation uses a cutting speed = 200 m/min, feed = 0.25 mm/rev, and depth of cut = 4.00 mm. The thermal diffusivity of the work material = 20 mm^2/s, and the volumetric specific heat = 3.5 (10^{-3}) J/mm^3-C. If the temperature increase above ambient temperature (20°F) is measured by a tool–chip thermocouple to be 700°C, determine the specific energy for the work material in this operation.

21.37. During a turning operation, a tool–chip thermocouple was used to measure cutting temperature. The following temperature data were collected during the cuts at three different cutting speeds (feed and depth were held constant): (1) $v = 100$ m/min, $T = 505$°C, (2) $v = 130$ m/min, $T = 552$°C, (3) $v = 160$ m/min, $T = 592$°C. Determine an equation for temperature as a function of cutting speed that is in the form of the Trigger equation, Eq. (21.23).

MACHINING OPERATIONS AND MACHINE TOOLS

CHAPTER CONTENTS

22.1 Turning and Related Operations
   22.1.1 Cutting Conditions in Turning
   22.1.2 Operations Related to Turning
   22.1.3 The Engine Lathe
   22.1.4 Other Lathes and Turning Machines
   22.1.5 Boring Mills
22.2 Drilling and Related Operations
   22.2.1 Cutting Conditions in Drilling
   22.2.2 Operations Related to Drilling
   22.2.3 Drill Presses
22.3 Milling
   22.3.1 Types of Milling Operations
   22.3.2 Cutting Conditions in Milling
   22.3.3 Milling Machines
22.4 Machining Centers and Turning Centers
22.5 Other Machining Operations
   22.5.1 Shaping and Planing
   22.5.2 Broaching
   22.5.3 Sawing
22.6 High-Speed Machining

Machining is the most versatile and accurate of all manufacturing processes in its capability to produce a diversity of part geometries and geometric features (e.g., screw threads, gear teeth, flat surfaces). Casting can also produce a variety of shapes, but it lacks the precision and accuracy of machining. As an introduction to this chapter, let us discuss some of the issues related to the creation of workpart shapes by machining.

Machined parts can be classified as rotational or nonrotational, Figure 22.1. A *rotational* workpart has a cylindrical or disk-like shape. The characteristic operation that produces this geometry is one in which a cutting tool removes material from a rotating workpart. Examples include turning and boring. Drilling is closely related except that an internal cylindrical shape is created and the tool rotates (rather than the work) in most drilling operations. A *nonrotational* (also called *prismatic*) workpart is block-like or plate-like, as in Figure 22.1(b). This geometry is achieved by linear motions of the workpart, combined with either rotating or linear tool motions. Operations in this category include milling, shaping, planing, and sawing.

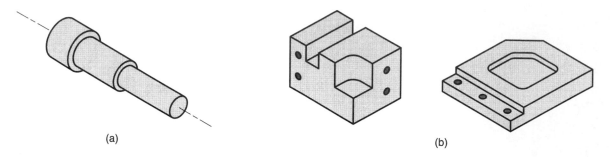

(a)

(b)

FIGURE 22.1   Machined parts are classified as (a) rotational, or (b) nonrotational, shown here by block and flat parts.

Each machining operation produces a characteristic geometry due to two factors: (1) the relative motions between the tool and the workpart and (2) the shape of the cutting tool. We classify these operations by which part shape is created as generating and forming. In **generating,** the geometry of the workpart is determined by the feed trajectory of the cutting tool. The path followed by the tool during its feed motion is imparted to the work surface in order to create shape. Examples of generating the work shape in machining include straight turning, taper turning, contour turning, peripheral milling, and profile milling, all illustrated in Figure 22.2. In each of these operations, material removal is accomplished

FIGURE 22.2   Generating shape in machining: (a) straight turning, (b) taper turning, (c) contour turning, (d) plain milling, and (e) profile milling.

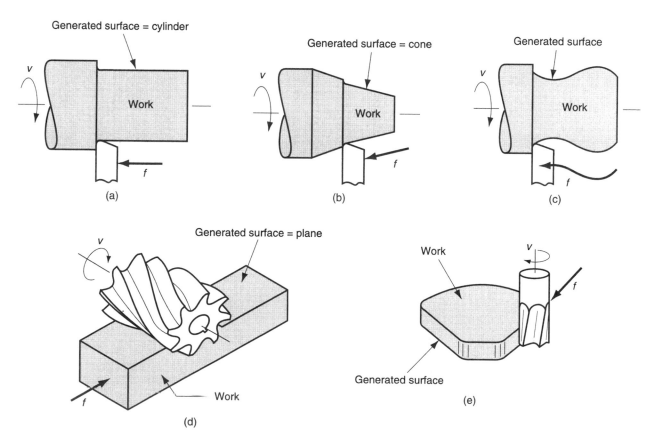

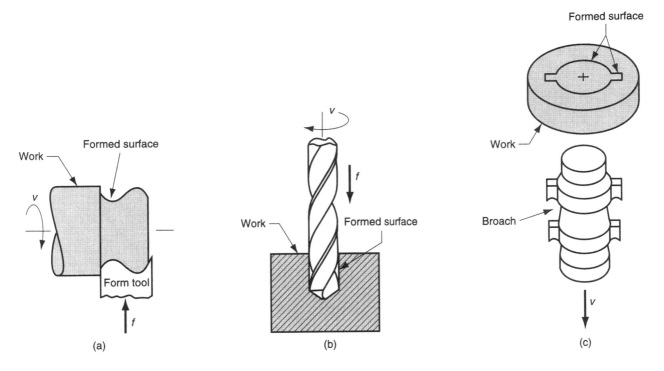

FIGURE 22.3   Forming to create shape in machining: (a) form turning, (b) drilling, and (c) broaching.

by the speed motion in the operation, but part shape is determined by the feed motion. The feed trajectory may involve variations in depth or width of cut during the operation. For example, in the contour turning and profile milling operations shown in our figure, the feed motion results in changes in depth and width, respectively, as cutting proceeds.

In *forming,* the shape of the part is created by the geometry of the cutting tool. In effect, the cutting edge of the tool has the reverse of the shape to be produced on the part surface. Form turning, drilling, and broaching are examples of this case. In these operations, illustrated in Figure 22.3, the shape of the cutting tool is imparted to the work

FIGURE 22.4
Combination of forming and generating to create shape: (a) thread cutting on a lathe, and (b) slot milling.

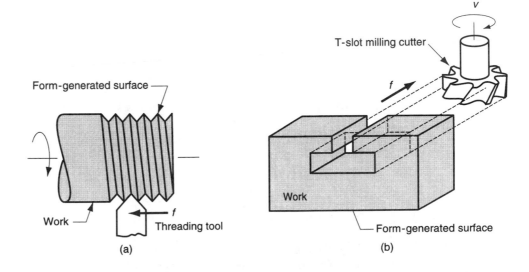

in order to create part geometry. The cutting conditions in forming usually include the primary speed motion combined with a feeding motion that is directed into the work. Depth of cut in this category of machining usually refers to the final penetration into the work after the feed motion has been completed.

Forming and generating are sometimes combined in one operation, such as in thread cutting on a lathe and slot milling. In thread cutting, the pointed shape of the cutting tool determines the form of the threads, but the large feed rate generates the threads. In slotting, the width of the cutter determines the width of the slot, but the feed motion creates the slot. These two cases are illustrated in Figure 22.4.

This chapter describes the important machining operations and the machine tools used to perform these operations. Historical Note 22.1 provides a brief narrative of the development of machine-tool technology.

Historical Note 22.1 *Machine tools* [8], [14], [15].

Material removal as a means of making things dates back to prehistoric times, when man learned to carve wood and chip stones to make hunting and farming implements. There is archaeological evidence that the ancient Egyptians used a rotating bowstring mechanism to drill holes.

Development of modern machine tools is closely related to the Industrial Revolution. When James Watt designed his steam engine in England around 1763, he had to make the bore of the cylinder sufficiently accurate to prevent steam from escaping around the piston. John Wilkinson built a water-wheel powered *boring machine* around 1775, which permitted Watt to build his steam engine. This boring machine is often recognized as the first machine tool.

Another Englishman, Henry Maudsley, developed the first *screw-cutting lathe* around 1800. Although the turning of wood had been accomplished for many centuries, Maudsley's machine added a mechanized tool carriage with which feeding and threading operations could be performed with much greater precision than any means before.

Eli Whitney is credited with developing the first *milling machine* in the United States around 1818. Development of the *planer* and *shaper* occurred in England between 1800 and 1835, in response to the need to make components for the steam engine, textile equipment, and other machines associated with the industrial revolution. The powered *drill press* was developed by James Nasmyth around 1846, which permitted drilling of accurate holes in metal.

Most of the conventional boring machines, lathes, milling machines, planers, shapers, and drill presses used today have the same basic designs as the early versions developed during the last two centuries. Modern machining centers—machine tools capable of performing more than one type of cutting operation—were introduced in the late 1950s, after numerical control had been developed (Historical Note 38.1).

22.1 TURNING AND RELATED OPERATIONS

Turning is a machining process in which a single point tool removes material from the surface of a rotating cylindrical workpiece; the tool is fed linearly in a direction parallel to the axis of rotation, as illustrated in Color Plate 6 (Chapter 1), and Figures 21.3(a), 21.5, and 22.5. Turning is traditionally carried out on a machine tool called a *lathe,* which provides power to turn the part at a given rotational speed and to feed the tool at a specified rate and depth of cut.

22.1.1 Cutting Conditions in Turning

The rotational speed in turning is related to the desired cutting speed at the surface of the cylindrical workpiece by the equation:

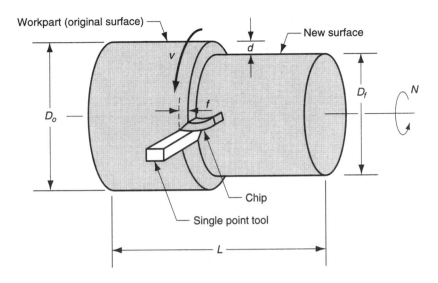

FIGURE 22.5 Turning operation.

$$N = \frac{v}{\pi D_o} \qquad (22.1)$$

where N = rotational speed, rev/min; v = cutting speed, m/min (ft/min); and D_o = original diameter of the part, m (ft).

The turning operation reduces the diameter of the work from D_o to final diameter D_f. The change in diameter is determined by the depth of cut d:

$$D_o - D_f = 2d \qquad (22.2)$$

The feed in turning is generally expressed in mm/rev (in./rev). This feed can be converted to a linear travel rate in mm/min (in./min) by the formula:

$$f_r = Nf \qquad (22.3)$$

where f_r = feed rate, mm/min (in./min); and f = feed mm/rev (in./rev).

The time to machine from one end of a cylindrical workpart to the other is given by

$$T_m = \frac{L}{f_r} \qquad (22.4)$$

where T_m = time of actual machining, minutes; and L = length of the cylindrical workpart, mm (in.). As a practical matter, a small distance is usually added to the length at the beginning and end of the workpiece to allow for approach and overtravel of the tool.

The volumetric rate of material removal can be most conveniently determined by the following equation:

$$MRR = vfd \qquad (22.5)$$

where MRR = material removal rate, mm^3/min (in.^{3}/min). In using this equation, the units for f are expressed simply as mm (in.), in effect neglecting the rotational character of turning. Also, care must be exercised to assure that the units for speed are consistent with those for f and d.

22.1.2 Operations Related to Turning

A variety of other machining operations can be performed on a lathe in addition to turning; these include the following, illustrated in Figure 22.6:

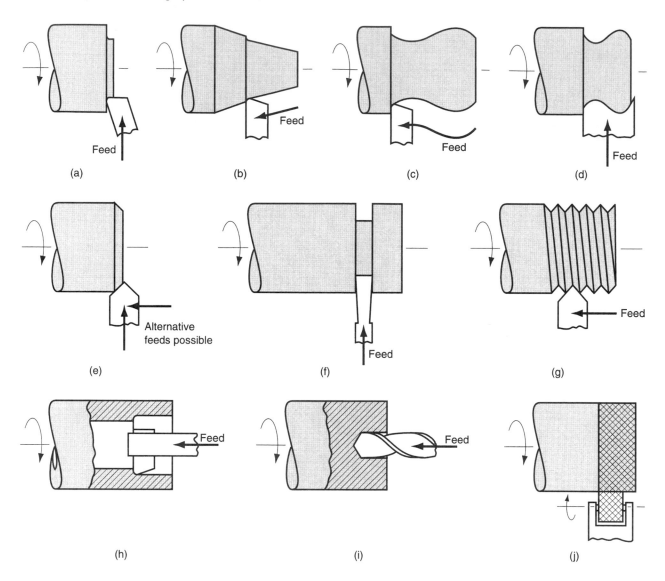

FIGURE 22.6   Machining operations other than turning that are performed on a lathe: (a) facing, (b) taper turning, (c) contour turning, (d) form turning, (e) chamfering, (f) cutoff, (g) threading, (h) boring, (i) drilling, and (j) knurling.

(a) ***Facing***—The tool is fed radially into the rotating work on one end to create a flat surface on the end.

(b) ***Taper turning***—Instead of feeding the tool parallel to the axis of rotation of the work, the tool is fed at an angle, thus creating a tapered cylinder or conical shape.

(c) ***Contour turning***—Instead of feeding the tool along a straight line parallel to the axis of rotation as in turning, the tool follows a contour that is other than straight, thus creating a contoured form in the turned part.

(d) ***Form turning***—In this operation, sometimes called ***forming,*** the tool has a shape that is imparted to the work by plunging the tool radially into the work.

(e) ***Chamfering***—The cutting edge of the tool is used to cut an angle on the corner of the cylinder, forming what is called a "chamfer."

(f) *Cutoff*—The tool is fed radially into the rotating work at some location along its length to cut off the end of the part. This operation is sometimes referred to as *parting*.

(g) *Threading*—A pointed tool is fed linearly across the outside surface of the rotating workpart in a direction parallel to the axis of rotation at a large effective feed rate, thus creating threads in the cylinder.

(h) *Boring*—A single-point tool is fed linearly, parallel to the axis of rotation, on the inside diameter of an existing hole in the part.

(i) *Drilling*—Drilling can be performed on a lathe by feeding the drill into the rotating work along its axis. *Reaming* can be performed in a similar way.

(j) *Knurling*—This is not a machining operation because it does not involve cutting of material. Instead, it is a metal forming operation used to produce a regular crosshatched pattern in the work surface.

Most lathe operations use single-point tools, which we discuss in Section 23.3.1. Turning, facing, taper turning, contour turning, chamfering, and boring are all performed with single-point tools. A threading operation is accomplished using a single-point tool designed with a geometry that shapes the thread. Certain operations require tools other than single-point. Form turning is performed with a specially designed tool called a form tool. The profile shape ground into the tool establishes the shape of the workpart. A cutoff tool is basically a form tool. Drilling is accomplished by a drill bit (Section 23.3.2). Knurling is performed by a knurling tool, consisting of two hardened forming rolls, each mounted between centers. The forming rolls have the desired knurling pattern on their surfaces. To perform knurling, the tool is pressed against the rotating workpart with sufficient pressure to impress the pattern onto the work surface.

22.1.3 The Engine Lathe

The basic lathe used for turning and related operations is an *engine lathe.* It is a versatile machine tool, manually operated, and widely used in low and medium production. The term *engine* dates from the time when these machines were driven by steam engines.

Engine Lathe Technology   Figure 22.7 is a sketch of an engine lathe showing its principal components. The *headstock* contains the drive unit to rotate the spindle, which rotates the work. Opposite the headstock is the *tailstock,* in which a center is mounted to support the other end of the workpiece.

The cutting tool is held in a *tool post* fastened to the *cross-slide,* which is assembled to the *carriage.* The carriage is designed to slide along the *ways* of the lathe in order to feed the tool parallel to the axis of rotation. The ways are like tracks along which the carriage rides, and they are made with great precision to achieve a high degree of parallelism relative to the spindle axis. The ways are built into the *bed* of the lathe, providing a rigid frame for the machine tool.

The carriage is driven by a leadscrew that rotates at the proper speed to obtain the desired feed rate. The cross-slide is designed to feed in a direction perpendicular to the carriage movement. Thus, by moving the carriage, the tool can be fed parallel to the work axis to perform straight turning; or by moving the cross-slide, the tool can be fed radially into the work to perform facing, form turning, or cutoff operations.

The conventional engine lathe and most other machines described in this section are *horizontal turning machines;* that is, the spindle axis is horizontal. This is appropriate for the majority of turning jobs, in which the length is greater than the diameter. For

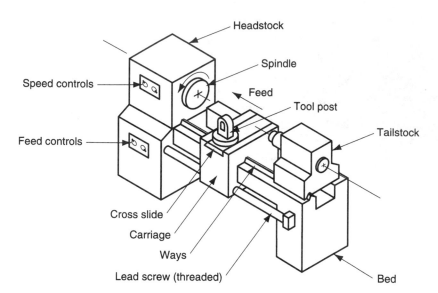

FIGURE 22.7   Diagram of an engine lathe, indicating its principal components.

jobs in which the diameter is large relative to length and the work is heavy, it is more convenient to orient the work so that it rotates about a vertical axis; these are *vertical turning machines.*

The size of a lathe is designated by swing and maximum distance between centers. The *swing* is the maximum workpart diameter that can be rotated in the spindle, determined as twice the distance between the centerline of the spindle and the ways of the machine. The actual maximum size of a cylindrical workpiece that can be accommodated on the lathe is smaller than the swing because the carriage and cross-slide assembly are in the way. The *maximum distance between centers* indicates the maximum length of a workpiece that can be mounted between headstock and tailstock centers. For example, a 350 mm × 1.2 m (14 in. × 48 in.) lathe designates that the swing is 350 mm (14 in.) and the maximum distance between centers is 1.2 m (48 in.).

Methods of Holding the Work in a Lathe   There are four common methods used to hold workparts in turning. These workholding methods consist of various mechanisms to grasp the work, center and support it in position along the spindle axis, and rotate it. The methods, illustrated in Figure 22.8, are (a) mounting the work between centers, (b) chuck, (c) collet, and (d) face plate.

Holding the work *between centers* refers to the use of two centers, one in the headstock and the other in the tailstock, as in Figure 22.8(a). This method is appropriate for parts with large length-to-diameter ratios. At the headstock center, a device called a *dog* is attached to the outside of the work and is used to drive the rotation from the spindle. The tailstock center has a cone-shaped point which is inserted into a tapered hole in the end of the work. The tailstock center is either a "live" center or a "dead" center. A *live center* rotates in a bearing in the tailstock, so that there is no relative rotation between the work and the live center, hence, no friction. In contrast, a *dead center* is fixed to the tailstock, so that it does not rotate; instead, the workpiece rotates about it. Because of friction, and the heat build-up that results, this setup is normally used at lower rotational speeds. The live center can be used at higher speeds.

The *chuck,* Figure 22.8(b), is available in several designs, with three or four jaws to grasp the cylindrical workpart on its outside diameter. The jaws are often designed so

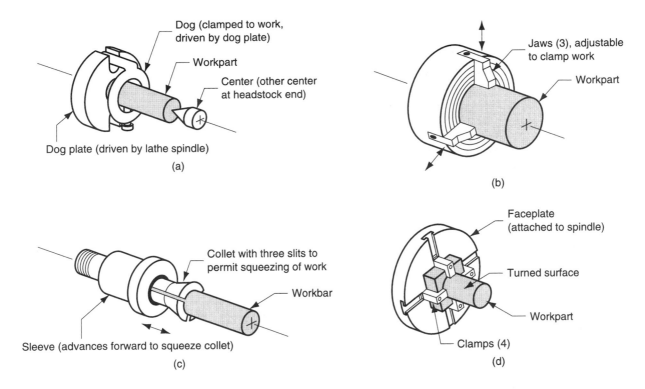

FIGURE 22.8 Four workholding methods used in lathes: (a) mounting the work between centers using a "dog," (b) three-jaw chuck, (c) collet, and (d) face plate for noncylindrical workparts.

they can also grasp the inside diameter of a tubular part. A *self-centering* chuck has a mechanism to move the jaws in or out simultaneously, thus centering the work at the spindle axis. Other chucks allow independent operation of each jaw. Chucks can be used with or without a tailstock center. For parts with low length-to-diameter ratios, holding the part in the chuck in a cantilever fashion is usually sufficient to withstand the cutting forces. For long workbars, the tailstock center is needed for support.

A *collet* consists of a tubular bushing with longitudinal slits running over half its length and equally spaced around its circumference, as in Figure 22.8(c). The inside diameter of the collet is used to hold cylindrical work such as barstock. Owing to the slits, one end of the collet can be squeezed to reduce its diameter and provide a secure grasping pressure against the work. Because there is a limit to the reduction obtainable in a collet of any given diameter, these workholding devices must be made in various sizes to match the particular workpart size in the operation.

A *face plate,* Figure 22.8(d), is a workholding device that fastens to the lathe spindle and is used to grasp parts with irregular shapes. Because of their irregular shape, these parts cannot be held by other workholding methods. The face plate is therefore equipped with the custom-designed clamps for the particular geometry of the part.

22.1.4 Other Lathes and Turning Machines

In addition to the engine lathe, other turning machines have been developed to satisfy particular functions or to automate the turning process. Among these machines are (1) toolroom lathe, (2) speed lathe, (3) turret lathe, (4) chucking machine, (5) automatic screw machine, and (6) numerically controlled lathe.

The toolroom lathe and speed lathe are closely related to the engine lathe. The *toolroom lathe* is smaller and has a wider available range of speeds and feeds. It is also built for higher accuracy, consistent with its purpose of fabricating components for tools, fixtures, and other high-precision devices.

The *speed lathe* is simpler in construction than the engine lathe. It has no carriage and cross-slide assembly, and therefore no leadscrew to drive the carriage. The cutting tool is held by the operator using a rest attached to the lathe for support. The speeds are higher on a speed lathe, but the number of speed settings is limited. Applications of this machine type include wood turning, metal spinning, and polishing operations.

A *turret lathe* is a manually operated lathe in which the tailstock is replaced by a turret that holds up to six cutting tools. These tools can be rapidly brought into action against the work one by one by indexing the turret. In addition, the conventional tool post used on an engine lathe is replaced by a four-sided turret that is capable of indexing up to four tools into position. Hence, because of the capacity to quickly change from one cutting tool to the next, the turret lathe is used for high-production work that requires a sequence of cuts to be made on the part.

As the name suggests, a *chucking machine* uses a chuck in its spindle to hold the workpart. The tailstock is absent on a chucker, so parts cannot be mounted between centers. This restricts the use of a chucking machine to short, light-weight parts. The setup and operation is similar to a turret lathe except that the feeding actions of the cutting tools are controlled automatically rather than by a human operator. The function of the operator is to load and unload the parts.

A *bar machine* is similar to a chucking machine except that a collet is used (instead of a chuck), which permits long bar stock to be fed through the headstock into position. At the end of each machining cycle, a cutoff operation separates the new part. The bar stock is then indexed forward to present stock for the next part. Feeding the stock as well as indexing and feeding the cutting tools is accomplished automatically. Owing to its high level of automatic operation, it is often called an *automatic bar machine.* One of its important applications is in the production of screws and similar small hardware items; the name *automatic screw machine* is frequently used for machines used in these applications.

Bar machines can be classified as single spindle or multiple spindle. A *single spindle bar machine* has one spindle that normally allows only one cutting tool to be used at a time on the single workpart being machined. Thus, while each tool is cutting the work, the other tools are idle. (Turret lathes and chucking machines are also limited by this sequential, rather than simultaneous, tool operation). To increase cutting tool utilization and production rate, *multiple spindle bar machines* are available. These machines have more than one spindle, so multiple parts are machined simultaneously by multiple tools. For example, a six-spindle automatic bar machine works on six parts at a time, as in Figure 22.9. At the end of each machining cycle, the spindles (including collets and workbars) are indexed (rotated) to the next position. In our illustration, each part is cut sequentially by five sets of cutting tools which takes six cycles (position 1 is for advancing the bar stock to a "stop"). With this arrangement, a part is completed at the end of each cycle. As a result, a six-spindle automatic screw machine has a very high production rate.

The sequencing and actuation of the motions on screw machines and chucking machines has traditionally been controlled by cams and other mechanical devices. The modern form of control is *computer numerical control* (CNC), in which the machine tool operations are controlled by a "program of instructions" (Section 38.1.4). CNC provides a more sophisticated and versatile means of control than mechanical devices. CNC has led to the development of machine tools capable of more complex machining cycles and part

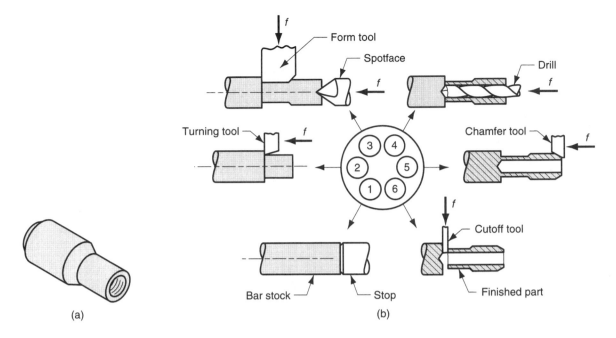

FIGURE 22.9   (a) Type of part produced on a six-spindle automatic bar machine; and (b) sequence of operations to produce the part: (1) feed stock to stop, (2) turn main diameter, (3) form second diameter and spotface, (4) drill, (5) chamfer, and (6) cutoff.

geometries, and a higher level of automated operation than conventional screw machines and chucking machines. The CNC lathe is an example of these machines in turning. It is especially useful for contour turning operations and close tolerance work. Today, automatic chuckers and bar machines are implemented by CNC.

22.1.5 Boring Mills

Boring is similar to turning. It uses a single-point tool against a rotating workpart. The difference is that boring is performed on the inside diameter of an existing hole rather than the outside diameter of an existing cylinder. In effect, boring is an internal turning operation. Machine tools used to perform boring operations are called **boring machines** (also **boring mills**). One might expect that boring machines would have features in common with turning machines; indeed, as previously indicated, lathes are sometimes used to accomplish boring.

Boring mills can be horizontal or vertical. The designation refers to the orientation of the axis of rotation of the machine spindle or workpart. In a **horizontal boring** operation, the setup can be arranged in either of two ways. The first setup is one in which the work is fixtured to a rotating spindle, and the tool is attached to a cantilevered boring bar that feeds into the work, as illustrated in Figure 22.10(a). The boring bar in this setup must be very stiff to avoid deflection and vibration during cutting. To achieve high stiffness, boring bars are often made of cemented carbide, whose modulus of elasticity approaches 620×10^3 MPa (90×10^6 lb/in.²). Figure 22.11 shows a carbide boring bar.

The second possible setup is one in which the tool is mounted to a boring bar, and the boring bar is supported and rotated between centers. The work is fastened to a feeding mechanism that feeds it past the tool. This setup, Figure 22.10(b), can be used to perform a boring operation on a conventional engine lathe.

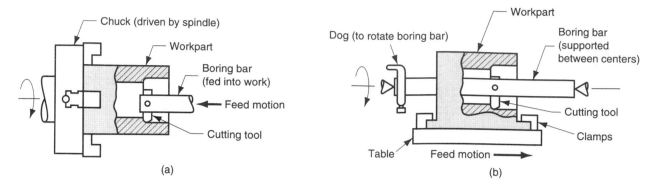

(a)

(b)

FIGURE 22.10 Two forms of horizontal boring: (a) boring bar is fed into a rotating workpart, and (b) work is fed past a rotating boring bar.

A ***vertical boring machine*** (VBM) is used for large, heavy workparts with large diameters; usually the workpart diameter is greater than its length. As in Figure 22.12, the part is fixed to a worktable that rotates relative to the machine base. Worktables up to 40 ft in diameter are available. The typical boring machine can position and feed several cutting tools simultaneously. The tools are mounted on tool heads that can be fed horizontally and vertically relative to the worktable. One or two heads are mounted on a horizontal cross-rail assembled to the machine tool housing above the worktable. The cutting tools mounted above the work can be used for facing and boring. In addition to the tools on the cross-rail, one or two additional tool heads can be mounted on the side columns of the housing to enable turning on the outside diameter of the work.

FIGURE 22.11 Boring bar made of cemented carbide (WC–Co) that uses indexable cemented carbide inserts (courtesy of Kennametal Inc.).

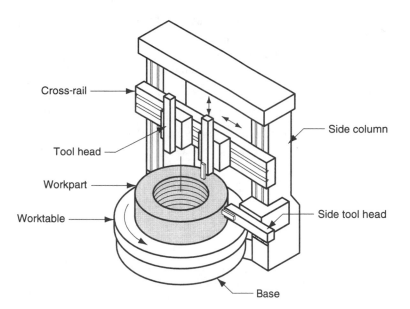

FIGURE 22.12   A vertical boring mill.

The tool heads used on a vertical boring machine often include turrets to accommodate several cutting tools. This results in a loss of distinction between this machine and a *vertical turret lathe* (VTL). Some machine tool builders make the distinction that the VTL is used for work diameters up to 2.5 m (100 in.), while the VBM is used for larger diameters [4]. Also, vertical boring mills are often applied to one-of-a-kind jobs, while vertical turret lathes are used for batch production.

22.2   DRILLING AND RELATED OPERATIONS

Drilling, Figure 21.3(b), is a machining operation used to create a round hole in a workpart. This contrasts with boring, which can only be used to enlarge an existing hole. Drilling is usually performed with a rotating cylindrical tool which has two cutting edges on its working end. The tool is called a *drill* or *drill bit* (described in Section 23.3.2). The rotating drill feeds into the stationary workpart to form a hole whose diameter is equal to the drill diameter. Drilling is customarily performed on a *drill press,* although other machine tools can also perform this operation.

22.2.1   Cutting Conditions in Drilling

The cutting speed in a drilling operation is the surface speed at the outside diameter of the drill. It is specified in this way for convenience, even though nearly all of the cutting is actually performed at lower speeds closer to the axis of rotation. To set the desired cutting speed in drilling, it is necessary to determine the rotational speed of the drill for its diameter. Letting N represent the spindle rev/min,

$$N = \frac{v}{\pi D} \qquad (22.6)$$

where v = cutting speed, mm/min (in./min); and D = the drill diameter, mm (in.). In some drilling operations, the workpiece is rotated about a stationary tool, but the same formula applies.

Feed f in drilling is specified in mm/rev (in./rev). Recommended feeds are roughly proportional to drill diameter; higher feeds are used with larger diameter drills. Since there are (usually) two cutting edges at the drill point, the uncut chip thickness (chip load) taken by each cutting edge is half the feed. Feed can be converted to feed rate using the same equation as for turning:

$$f_r = Nf \tag{22.7}$$

where f_r = feed rate, mm/min (in./min).

Drilled holes are either through holes or blind holes, Figure 22.13. In ***through holes,*** the drill exits the opposite side of the work; in ***blind holes,*** it does not. The machining time required to drill a through hole can be determined by the following formula:

$$T_m = \frac{t + A}{f_r} \tag{22.8}$$

where T_m = machining (drilling) time, min; t = work thickness, mm (in.); f_r = feed rate, mm/min (in./min); and A = an approach allowance that accounts for the drill point angle, representing the distance the drill must feed into the work before reaching full diameter, Figure 22.10(a). This allowance is given by

$$A = 0.5\, D \tan\left(90 - \frac{\theta}{2}\right) \tag{22.9}$$

where A = approach allowance, mm (in.); and θ = drill point angle.

In a blind hole, hole depth d is defined as the distance from the work surface to the "point" of the hole, Figure 22.13(b). By this definition, the drill point angle allowance does not affect the time to drill the hole. Thus, for a blind hole, machining time is given by

$$T_m = \frac{d}{f_r} \tag{22.10}$$

The rate of metal removal in drilling is determined as the product of the drill cross-sectional area and the feed rate:

$$MRR = \frac{\pi D^2 f_r}{4} \tag{22.11}$$

FIGURE 22.13 Two hole types: (a) through hole and (b) blind hole.

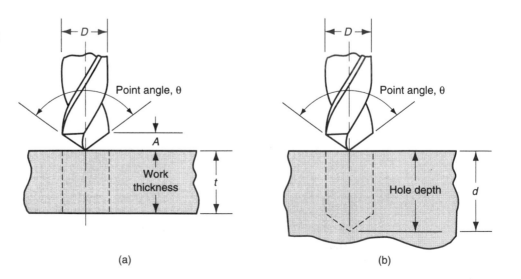

(a)   (b)

This equation is valid only after the drill reaches full diameter and excludes the initial approach of the drill into the work.

22.2.2 Operations Related to Drilling

Several operations are related to drilling. These are illustrated in Figure 22.14 and described in this section. Most of the operations follow drilling; a hole must be made first by drilling, and then the hole is modified by one of the other operations. Centering and spotfacing are exceptions to this rule. All of the operations use rotating tools.

(a) *Reaming*—Reaming is used to slightly enlarge a hole, to provide a better tolerance on its diameter, and to improve its surface finish. The tool is called a *reamer,* and it usually has straight flutes.

(b) *Tapping*—This operation is performed by a *tap* and is used to provide internal screw threads on an existing hole.

(c) *Counterboring*—Counterboring provides a stepped hole, in which a larger diameter follows a smaller diameter partially into the hole. A counterbored hole is used to seat bolt heads into a hole so the heads do not protrude above the surface.

(d) *Countersinking*—This is similar to counterboring, except that the step in the hole is cone-shaped for flat head screws and bolts.

(e) *Centering*—Also called centerdrilling, this operation drills a starting hole to accurately establish its location for subsequent drilling. The tool is called a *centerdrill.*

(f) *Spotfacing*—Spotfacing is similar to milling. It is used to provide a flat machined surface on the workpart in a localized area.

FIGURE 22.14 Machining operations related to drilling: (a) reaming, (b) tapping, (c) counterboring, (d) countersinking, (e) center drilling, and (f) spot facing.

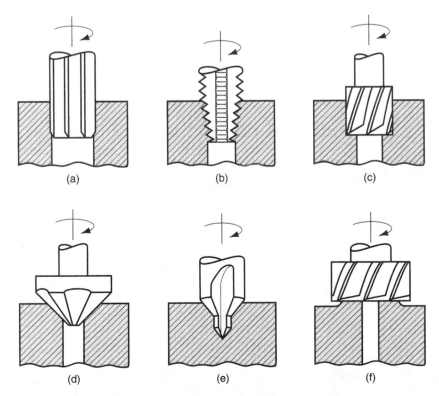

22.2.3 Drill Presses

The drill press is the standard machine tool for drilling. There are various types of drill press, the most basic of which is the upright drill, Figure 22.15. The **upright drill** stands on the floor and consists of a table for holding the workpart, a drilling head with powered spindle for the drill bit, and a base and column for support. A similar drill press, but smaller, is the **bench drill,** which is mounted on a table or bench rather than the floor.

The **radial drill,** Figure 22.16, is a large drill press designed to cut holes in large parts. It has a radial arm along which the drilling head can be moved and clamped. The head therefore can be positioned along the arm at locations that are a significant distance from the column to accommodate large work.

The **gang drill** is a drill press consisting basically of a series of two to six upright drills connected together in an in-line arrangement. Each spindle is powered and operated independently, and they share a common worktable, so that a series of drilling and related operations can be accomplished in sequence (e.g., centering, drilling, reaming, tapping) simply by sliding the workpart along the worktable from one spindle to the next. A related machine is the **multiple-spindle drill,** in which several drill spindles are connected together to drill multiple holes simultaneously into the workpart.

In addition, **numerical control drill presses** are available to control the positioning of the holes in the workparts. These drill presses are often equipped with turrets to hold multiple tools that can be indexed under control of the NC program. The term **CNC turret drill** is used for these machine tools.

Workholding on a drill press is accomplished by clamping the part in a vise, fixture, or jig. A **vise** is a general-purpose workholding device possessing two jaws that grasp the work in position. A **fixture** is a workholding device that is usually custom-designed for the particular workpart. The fixture can be designed to achieve higher accuracy in positioning the part relative to the machining operation, faster production rates, and greater operator convenience in use. A **jig** is a workholding device that is also specially designed for the workpart. The distinguishing feature between a jig and a fixture is that the jig provides a means of guiding the tool during the drilling operation. A fixture does not provide this tool guidance feature. A jig used for drilling is called a **drill jig.**

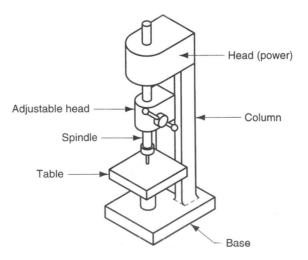

FIGURE 22.15   Upright drill press.

Head (power)

Adjustable head

Spindle

Table

Column

Base

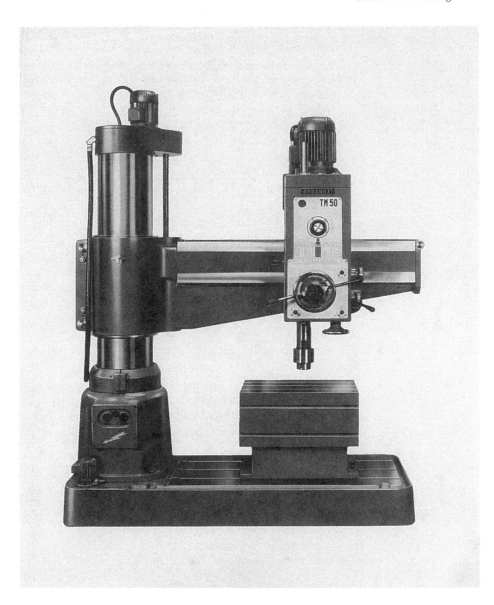

FIGURE 22.16 Radial drill press (courtesy Willis Machinery and Tools).

22.3 MILLING

Milling is a machining operation in which a workpart is fed past a rotating cylindrical tool with multiple cutting edges (in rare cases, a tool with one cutting edge, called a *fly-cutter,* is used). The axis of rotation of the cutting tool is perpendicular to the direction of feed. This orientation between the tool axis and the feed direction is one of the features that distinguishes milling from drilling. In drilling, the cutting tool is fed in a direction parallel to its axis of rotation. The cutting tool in milling is called a *milling cutter* and the cutting edges are called teeth. The machine tool that traditionally performs this operation is a *milling machine.*

The geometric form created by milling is a plane surface. Other work geometries can be created either by means of the cutter path or the cutter shape. Owing to the variety of shapes possible and its high production rates, milling is one of the most versatile and widely used machining operations.

Milling is an **interrupted cutting** operation; the teeth of the milling cutter enter and exit the work during each revolution. This interrupted cutting action subjects the teeth to a cycle of impact force and thermal shock on every rotation. The tool material and cutter geometry must be designed to withstand these conditions.

22.3.1 Types of Milling Operations

There are two basic types of milling operations, shown in Figure 22.17: (a) peripheral milling and (b) face milling.

Peripheral Milling   In *peripheral milling,* also called *plain milling,* the axis of the tool is parallel to the surface being machined, and the operation is performed by cutting edges on the outside periphery of the cutter. Several types of peripheral milling are shown in Figure 22.18: (a) *Slab milling*—The basic form of peripheral milling; the cutter width extends beyond the workpiece on both sides. (b) *Slotting,* also called *slot milling*—The width of the cutter is less than the workpiece width, creating a slot in the work. When the cutter is very thin, this operation can be used to mill narrow slots or cut a workpart in two, called *saw milling.* (c) *Side milling*—The cutter machines the side of the workpiece. (d) *Straddle milling*—The same as side milling, only cutting takes place on both sides of the work.

In peripheral milling, the rotation direction of the cutter distinguishes two forms of milling: up milling and down milling, illustrated in Figure 22.19. In *up milling,* also called *conventional milling,* the direction of motion of the cutter teeth is opposite the feed direction when the teeth cut into the work. It is milling "against the feed." In *down milling,* also called *climb milling,* the direction of cutter motion is the same as the feed direction when the teeth cut the work. It is milling "with the feed."

The relative geometries of these two forms of milling result in differences in their cutting actions. In up milling, the chip formed by each cutter tooth starts out very thin and increases in thickness during the sweep of the cutter. In down milling, each chip starts out thick and reduces in thickness throughout the cut. The length of a chip in down milling is less than in up milling (the difference is exaggerated in our figure). This means that the cutter is engaged in the work for less time per volume of material cut, and this tends to increase tool life in down milling.

FIGURE 22.17   Two basic types of milling operation: (a) peripheral or plain milling and (b) face milling.

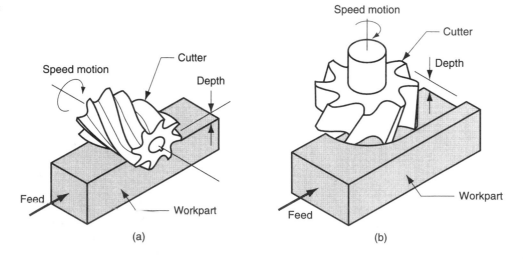

(a)   (b)

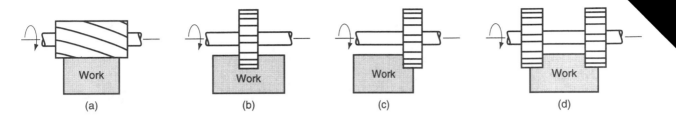

FIGURE 22.18   Peripheral milling: (a) slab milling, (b) slotting, (c) side milling, and (d) straddle milling.

The cutting force direction is tangential to the periphery of the cutter for the teeth that are engaged in the work. In up milling, this has a tendency to lift the workpart as the cutter teeth exit the material. In down milling this cutter force direction is downward, tending to hold the work against the milling machine table.

Face Milling   In *face milling,* the axis of the cutter is perpendicular to the surface being milled, and machining is performed by cutting edges on both the end and outside periphery of the cutter. As in peripheral milling, various forms of face milling exist, several of which are shown in Figure 22.20: (a) ***Conventional face milling***—The diameter of the cutter is greater than the workpart width, so that the cutter overhangs the work on both sides. (b) ***Partial face milling***—the cutter overhangs the work on only one side. (c) ***End milling***—the cutter diameter is less than the work width, so a slot is cut into the part. (d) ***Profile milling***—this is a form of end milling in which the outside periphery of a flat part is cut. (e) ***Pocket milling***—another form of end milling, this is used to mill shallow pockets into flat parts. (f) ***Surface contouring***—a ball-nose cutter (rather than square-end cutter) is fed back and forth across the work along a curvilinear path at close intervals to create a three-dimensional surface form. The same basic cutter control is required to machine the contours of molds and dies, in which case the operation is called ***die sinking***.

22.3.2 Cutting Conditions in Milling

The cutting speed is determined at the outside diameter of a milling cutter. This can be converted to spindle rotation speed using a formula that should now be familiar:

FIGURE 22.19   Two forms of milling with a 20-tooth cutter: (a) up milling, and (b) down milling.

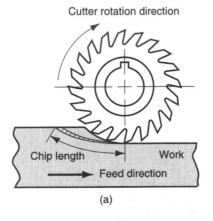

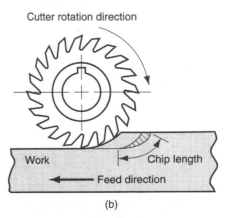

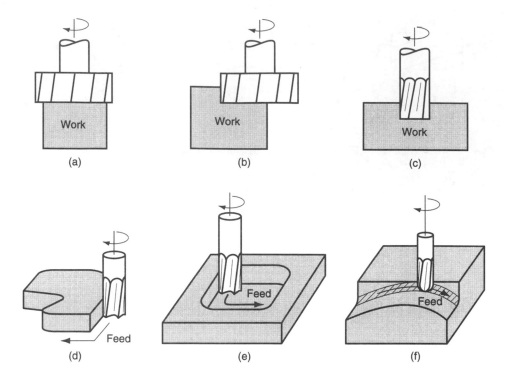

FIGURE 22.20 Face milling: (a) conventional face milling, (b) partial face milling, (c) end milling, (d) profile milling, (e) pocket milling, and (f) surface contouring.

$$N = \frac{v}{\pi D} \tag{22.12}$$

The feed f in milling is usually given as a feed per cutter tooth; called the **chip load**, it represents the size of the chip formed by each cutting edge. This can be converted to feed rate by taking into account the spindle speed and the number of teeth on the cutter as follows:

$$f_r = Nn_t f \tag{22.13}$$

where f_r = feed rate, mm/min (in./min); N = spindle speed, rev/min; n_t = number of teeth on the cutter; and f = chip load in mm/tooth (in./tooth).

Material removal rate in milling is determined using the product of the cross-sectional area of the cut and the feed rate. Accordingly, if a slab-milling operation is cutting a workpiece with width w at a depth d, the material removal rate is

$$MRR = wdf_r \tag{22.14}$$

This neglects the initial entry of the cutter before full engagement. Eq. (22.14) can be applied to end milling, side milling, face milling, and other milling operations, making the proper adjustments in the computation of cross-sectional area of cut.

The time required to mill a workpiece of length L must account for the approach distance required to fully engage the cutter. First, consider the case of slab milling, Figure 22.21. To determine the time to perform a slab milling operation, the approach distance A to reach full cutter depth is given by

$$A = \sqrt{d(D - d)} \tag{22.15}$$

where d = depth of cut, mm (in.); and D − diameter of the milling cutter, mm (in.). The time to mill the workpiece T_m is therefore

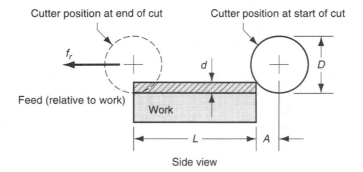

FIGURE 22.21 Slab (peripheral) milling showing entry of cutter into the workpiece.

$$T_m = \frac{L + A}{f_r} \qquad (22.16)$$

For face milling, it is customary to allow for the approach distance A plus an over-travel distance O. There are two possible cases as pictured in Figure 22.22. In both cases, $A = O$. The first case is when the cutter is centered over the rectangular workpiece. It is clear from Figure 22.22(a) that A and O are each equal to half the cutter diameter. That is,

$$A = O = \frac{D}{2} \qquad (22.17)$$

where D = cutter diameter, mm (in.).

The second case is when the cutter is offset to one side of the work, as shown in Figure 22.22(b). In this case, the approach and overtravel distances are given by

$$A = O = \sqrt{w(D - w)} \qquad (22.18)$$

where w = width of the cut, mm (in.). Machining time in either case is therefore given by

$$T_m = \frac{L + 2A}{f_r} \qquad (22.19)$$

FIGURE 22.22   Face milling showing approach and overtravel distances for two cases: (a) when cutter is centered over the workpiece, and (b) when cutter is offset to one side over the work.

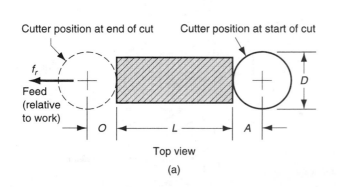

Top view

(a)

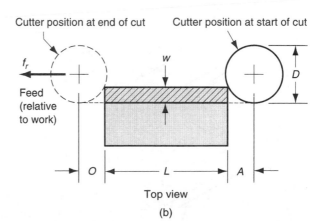

Top view

(b)

22.3.3 Milling Machines

Milling machines must provide a rotating spindle for the cutter and a table for fastening, positioning, and feeding the workpart. Various machine tool designs satisfy these requirements. To begin with, milling machines can be classified as horizontal or vertical. A *horizontal milling machine* has a horizontal spindle, and this design is well-suited for performing peripheral milling (e.g., slab milling, slotting, side and straddle milling) on workparts that are roughly cube-shaped. A *vertical milling machine* has a vertical spindle, and this orientation is appropriate for face milling, end milling, surface contouring, and diesinking on relatively flat workparts.

Other than spindle orientation, milling machines can be classified into the following types: (1) knee-and-column, (2) bed type, (3) planer type, (4) tracer mills, and (5) CNC milling machines.

The *knee-and-column milling machine* is the basic machine tool for milling. It derives its name from the fact that its two main components are a *column* that supports the spindle, and a *knee* (roughly resembling a human's knee) that supports the work table. It is available as either a horizontal or a vertical machine, as illustrated in Figure 22.23. In the horizontal version, an arbor usually supports the cutter. The *arbor* is basically a shaft that holds the milling cutter and is driven by the spindle. An overarm is provided on horizontal machines to support the arbor. On vertical knee-and-column machines, milling cutters can be mounted directly in the spindle without an arbor.

One of the features of the knee-and-column milling machine that makes it so versatile is its capability for worktable feed movement in any of the $x–y–z$ axes. The worktable can be moved in the x-direction, the saddle can be moved in the y-direction, and the knee can be moved vertically to achieve the z-movement.

Two special knee-and-column machines should be identified. One is the *universal* milling machine, Figure 22.24(a), which has a table that can be swiveled in a horizontal plane (about a vertical axis) to any specified angle. This facilitates the cutting of angular shapes and helixes on workparts. Another special machine is the *ram mill,* Figure 22.24(b), in which the toolhead containing the spindle is located on the end of a horizontal ram; the ram can be adjusted in and out over the worktable to locate the cutter relative to the work. The toolhead can also be swiveled to achieve an angular orientation of the cutter with respect to the work. These features provide considerable versatility in machining a variety of work shapes.

FIGURE 22.23   Two basic types of knee-and-column milling machine: (a) horizontal, and (b) vertical.

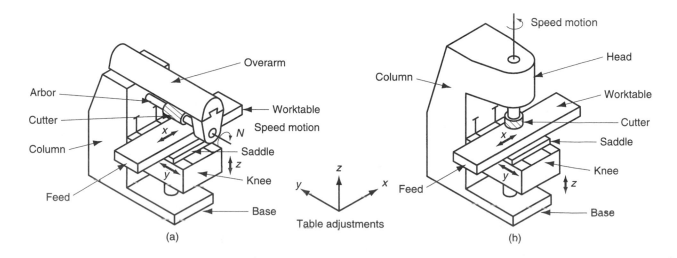

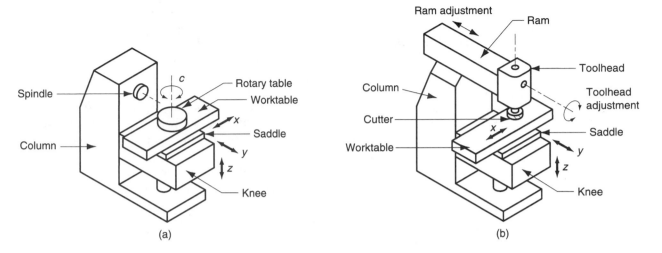

FIGURE 22.24   Special types of knee-and-column milling machine: (a) universal—overarm, arbor, and cutter omitted for clarity, and (b) ram type.

Bed type milling machines are designed for high production. They are constructed with greater rigidity than knee-and-column machines, thus permitting them to achieve heavier feed rates and depths of cut needed for high material removal rates. The characteristic construction of the bed-type milling machine is shown in Figure 22.25. The worktable is mounted directly to the bed of the machine tool, rather than using the less rigid knee type design. This construction limits the possible motion of the table to longitudinal feeding of the work past the milling cutter. The cutter is mounted in a spindle head that can be adjusted vertically along the machine column. Single spindle bed machines are called *simplex* mills, and are available in either horizontal or vertical models. *Duplex* mills use two spindle heads. The heads are usually positioned horizontally on opposite sides of the bed to perform simultaneous operations during one feeding pass of the work. *Triplex* mills add a third spindle mounted vertically over the bed to further increase machining capability.

Planer type mills are the largest milling machines. Their general appearance and construction are those of a large planer (Figure 22.31); the difference is that milling is performed instead of planing. Accordingly, one or more milling heads are substituted for the single-point cutting tools used on planers, and the motion of the work past the tool is a feed rate motion rather than a cutting speed motion. Planer mills are built to machine very large parts. The worktable and bed of the machine are heavy and relatively

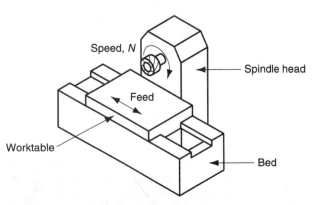

FIGURE 22.25   Simplex bed-type milling machine horizontal spindle.

low to the ground, and the milling heads are supported by a bridge structure that spans across the table.

A *tracer mill,* also called a *profiling mill,* is designed to reproduce an irregular part geometry that has been created on a template. Using either manual feed by a human operator or automatic feed by the machine tool, a tracing probe is controlled to follow the template while a milling head duplicates the path taken by the probe to machine the desired shape. Tracer mills can be divided into the following types: (1) *x–y tracing,* in which the template is a flat shape with an outline to be profile-milled using two-axis control; and (2) *x–y–z tracing,* in which the probe follows a three dimensional pattern using three-axis control.

Tracer mills have been used for creating shapes that cannot easily be generated by a simple feeding action of the workpart against the milling cutter. Their applications include the machining of molds and dies. In recent years, many of the applications previously accomplished on tracer mills have been taken over by computer numerical control (CNC) milling machines.

CNC milling machines are machines in which the cutter path is controlled by data rather than a physical template. They are especially suited to profile milling, pocket milling, surface contouring, and die sinking operations, in which two or three axes of the worktable must be simultaneously controlled to achieve the required cutter path. An operator is normally required to change cutters as well as load and unload workparts.

22.4 MACHINING CENTERS AND TURNING CENTERS

A *machining center* is a highly automated machine tool capable of performing multiple machining operations under CNC control in one setup with minimal human attention. Typical operations are those that use a rotating cutting tool, such as milling and drilling. Several features make a machining center a productive machine:

> *Automatic tool-changing*—To change from one machining operation to the next, the cutting tools must be changed. This is done on a machining center under NC program control by an automatic tool-changer designed to exchange cutters between the machine tool spindle and a *tool storage drum.* Capacities of these drums commonly range from 16 to 80 cutting tools.

> *Pallet shuttles*—Some machining centers are equipped with two or more pallet shuttles, which can be automatically transferred to the spindle to machine the workpart. With two shuttles, the operator can be unloading the previous part and loading the next part while the machine tool is engaged in machining the current part. This reduces nonproductive time on the machine.

> *Automatic workpart positioning*—Many machining centers have more than three axes. One of the additional axes is often designed as a rotary table to position the part at some specified angle relative to the spindle. The rotary table permits the cutter to perform machining on four sides of the part in a single setup.

Machining centers are classified as horizontal, vertical, or universal. The designation refers to spindle orientation. Horizontal machining centers (HMCs) normally machine cube-shaped parts, in which the four vertical sides of the cube can be accessed by the cutter. Vertical machining centers (VMCs) are suited to flat parts on which the tool can machine the top surface. Universal machining centers have workheads that swivel their spindle axes to any angle between horizontal and vertical, as illustrated in Figure 22.26.

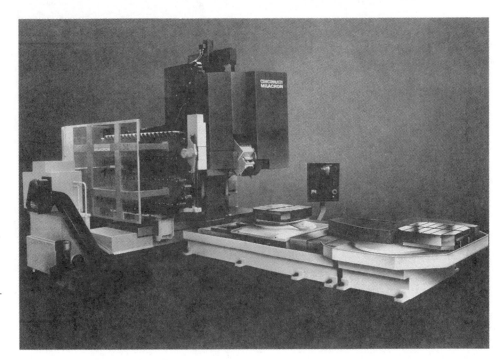

FIGURE 22.26   A universal machining center (courtesy Cincinnati Milacron). Capability to orient the workhead makes this a five-axis machine.

FIGURE 22.27   CNC 4-axis turning center (courtesy Cincinnati Milacron).

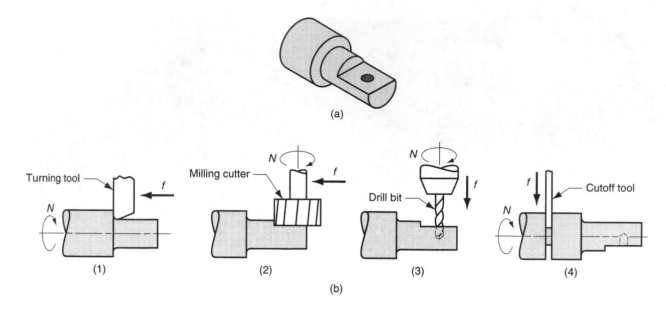

FIGURE 22.28 Operation of a mill-turn center: (a) example part with turned, milled, and drilled surfaces; and (b) sequence of operations on a mill-turn center: (1) turn second diameter, (2) mill flat with part in programmed angular position, (3) drill hole with part in same programmed position, and (4) cutoff.

Success of CNC machining centers has led to development of CNC turning centers. A modern **CNC turning center,** Figure 22.27, is capable of performing various turning and related operations, contour turning, and automatic tool indexing, all under computer control. In addition, the most sophisticated turning centers can accomplish (1) workpart gaging (checking key dimensions after machining), (2) tool monitoring (sensors to indicate when the tools are worn), (3) automatic tool changing when tools become worn, and even (4) automatic workpart changing at the completion of the work cycle [12].

A recent development in CNC machine-tool technology is the **CNC mill-turn center.** This machine has the general configuration of a turning center; in addition, it can position a cylindrical workpart at a specified angle so that a rotating cutting tool (e.g., milling cutter) can machine features into the outside surface of the part, as illustrated in Figure 22.28. An ordinary turning center does not have the capability to stop the workpart at a defined angular position, and it does not possess rotating tool spindles.

22.5 OTHER MACHINING OPERATIONS

In addition to turning, drilling, and milling, several other machining operations should be included in our survey: (1) shaping and planing, (2) broaching, and (3) sawing.

22.5.1 Shaping and Planing

Shaping and planing are similar operations, both involving the use of a single-point cutting tool moved linearly relative to the workpart. In conventional shaping and planing, a straight, flat surface is created by this action. The difference between the two operations

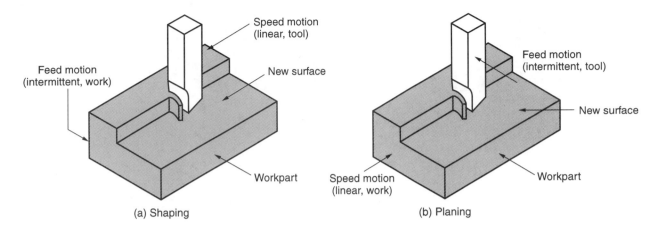

FIGURE 22.29   (a) Shaping, and (b) planing.

is illustrated in Figure 22.29. In shaping, the speed motion is accomplished by moving the cutting tool. In planing, the speed motion is accomplished by moving the workpart. Cutting tools used in shaping and planing are single point tools. Unlike turning, interrupted cutting occurs in shaping and planing, subjecting the tool to an impact loading upon entry into the work. Also, the tools are limited to low speeds due to their start-and-stop motion. The conditions normally dictate use of high-speed steel-cutting tools.

Shaping   Shaping is performed on a machine tool called a *shaper,* Figure 22.30. The components of the shaper include a *ram,* which moves relative to a *column* to provide the cutting motion, and a worktable that holds the part and accomplishes the feed motion. The motion of the ram consists of a forward stroke to achieve the cut, and a return stroke during which the tool is lifted slightly to clear the work and then reset for the next pass. On completion of each return stroke, the worktable is advanced laterally relative to the ram motion in order to feed the part. Feed is specified in mm/stroke (in./stroke). The drive mechanism for the ram can be either hydraulic or mechanical. Hydraulic drive has greater flexibility in adjusting the stroke length and a more uniform speed during the forward stroke, but it is more expensive than a mechanical drive unit. Both mechanical and hydraulic drives are designed to achieve higher speeds on

FIGURE 22.30
Components of a shaper.

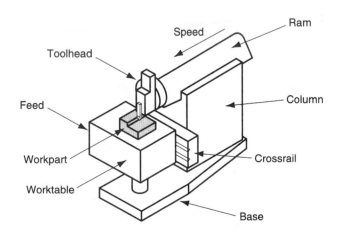

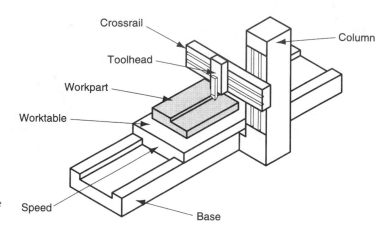

FIGURE 22.31   Open side planer.

the return (noncutting) stroke than on the forward (cutting) stroke, thereby increasing the proportion of time spent cutting.

Planing   The machine tool for planing is a *planer.* Cutting speed is achieved by a reciprocating worktable that moves the part past the single point cutting tool. Construction and motion capability of a planer permit much larger parts to be machined than on a shaper. Planers can be classified as either open side planers or double-column planers. The *open side planer,* also known as a *single column planer,* Figure 22.31, has a single column supporting the cross-rail on which a toolhead is mounted. Another toolhead can also be mounted and fed along the vertical column. Multiple toolheads permit more than one cut to be taken on each pass. At the completion of each stroke, each toolhead is moved relative to the cross-rail (or column) to achieve the intermittent feed motion. The configuration of the open side planer permits very wide workparts to be machined.

A *double-column planer* has two columns, one on either side of the base and worktable. The columns support the cross-rail, on which one or more toolheads are mounted. The two columns provide a more rigid structure for the operation; however, the two columns limit the width of the work that can be handled on this machine.

Shaping and planing can be used to machine shapes other than flat surfaces. The restriction is that the cut surface must be straight. This allows the cutting of grooves, slots, gear teeth, and other shapes as illustrated in Figure 22.32. Special tool geometries, other than the standard single point geometry, must be specified to cut some of these shapes. In fact, special machine tools are sometimes used for some of the shapes; an important example is the *gear shaper,* a vertical shaper with a specially designed rotary feed table and synchronized toolhead used to generate teeth on spur gears.

22.5.2   Broaching

Broaching is performed using a multiple tooth cutting tool by moving the tool linearly relative to the work in the direction of the tool axis, as in Figure 22.33. The cutting tool

FIGURE 22.32   Types of shapes that can cut by shaping and planing: (a) V-groove, (b) square groove, (c) T-slot, (d) dovetail slot, and (e) gear teeth.

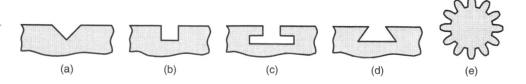

(a)   (b)   (c)   (d)   (e)

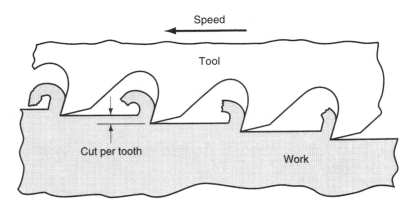

FIGURE 22.33   The broaching operation.

is called a ***broach,*** and the machine tool is called a ***broaching machine.*** In certain jobs for which broaching can be used, it is a highly productive method of machining. Advantages include good surface finish, close tolerances, and variety of work shapes possible. Owing to the complicated and often custom-shaped geometry of the broach, tooling is expensive.

There are two principal types of broaching: external (also called surface broaching) and internal. ***External broaching*** is performed on the outside surface of the work to create a certain cross-sectional shape on the surface. Figure 22.34(a) shows some possible cross sections that can be formed by external broaching. ***Internal broaching*** is accomplished on the internal surface of a hole in the part. Accordingly, a starting hole must be present in the part so as to insert the broach at the beginning of the broaching stroke. Figure 22.34(b) indicates some of the shapes that can be produced by internal broaching.

The basic function of a ***broaching machine*** is to provide a precise linear motion of the tool past a stationary work position, but there are various ways in which this can be done. Most broaching machines can be classified as either vertical or horizontal machines. The ***vertical broaching machine*** is designed to move the broach along a vertical path, while the ***horizontal broaching machine*** has a horizontal tool trajectory. Most broaching machines pull the broach past the work. However, there are exceptions to this pull action. One exception is a relatively simple type called a ***broaching press,*** used only for internal broaching, that pushes the tool through the workpart. Another exception is the ***continuous***

FIGURE 22.34   Work shapes that can be cut by: (a) external broaching, and (b) internal broaching. Cross-hatching indicates the surfaces broached.

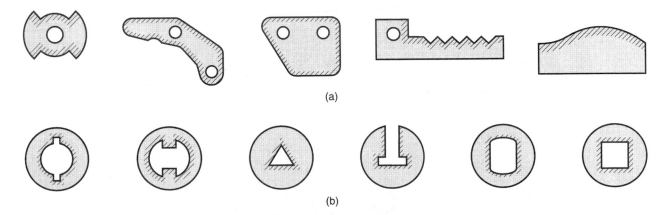

(a)

(b)

broaching machine, in which the workparts are fixtured to an endless belt loop and moved past a stationary broach. Because of its continuous operation, this machine can be used only for surface broaching.

22.5.3 Sawing

Sawing is a process in which a narrow slit is cut into the work by a tool consisting of a series of narrowly spaced teeth. Sawing is normally used to separate a workpart into two pieces, or to cut off an unwanted portion of a part. These operations are often referred to as *cutoff* operations. Since many factories require cutoff operations at some point in the production sequence, sawing is an important manufacturing process.

In most sawing operations, the work is held stationary and the *saw blade* is moved relative to it. There are three basic types of sawing, as in Figure 22.35, according to the type of blade motion involved: (a) hacksawing, (b) bandsawing, and (c) circular sawing.

Hacksawing, Figure 22.35(a), involves a linear reciprocating motion of the saw against the work. This method of sawing is often used in cutoff operations. Cutting is accomplished only on the forward stroke of the saw blade. Because of this intermittent cutting action, hacksawing is inherently less efficient than the other sawing methods, both of which are continuous. The *hacksaw* blade is a thin straight tool with cutting teeth on one edge. Hacksawing can be done either manually or with a power hacksaw. A *power hacksaw* provides a drive mechanism to operate the saw blade at a desired speed; it also applies a given feed rate or sawing pressure.

Bandsawing involves a linear continuous motion, using a *bandsaw blade* made in the form of an endless flexible loop with teeth on one edge. The sawing machine is a *bandsaw,* which provides a pulley-like drive mechanism to continuously move and guide the bandsaw blade past the work. Bandsaws are classified as vertical or horizontal. The designation refers to the direction of saw blade motion during cutting. Vertical bandsaws are used for cutoff as well as other operations such as contouring and slotting. *Contouring* on a bandsaw involves cutting a part profile from flat stock. *Slotting* is the cutting of a thin slot into a part, an operation for which bandsawing is well suited. Contour sawing and slotting are operations in which the work is fed into the saw blade.

FIGURE 22.35   Three types of sawing operations: (a) power hacksaw, (b) bandsaw (vertical), and (c) circular saw.

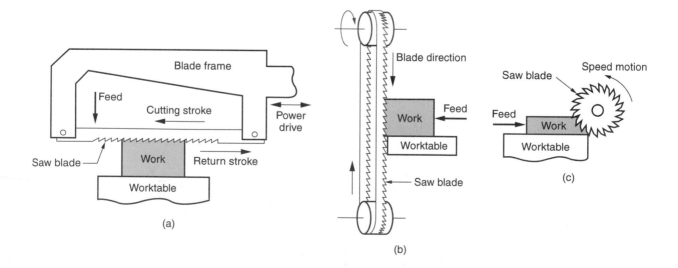

Vertical bandsaw machines can be operated either manually, in which the operator guides and feeds the work past the bandsaw blade, or automatically, in which the work is power fed past the blade. Recent innovations in bandsaw design have permitted the use of CNC to perform contouring of complex outlines. Some of the details of the vertical bandsawing operation are illustrated in Figure 22.35(b). Horizontal bandsaws are normally used for cutoff operations as alternatives to power hacksaws.

Circular sawing, Figure 22.35(c), uses a rotating saw blade to provide a continuous motion of the tool past the work. Circular sawing is often used to cut long bars, tubes, and similar shapes to specified length. The cutting action is similar to a slot milling operation, except that the saw blade is thinner and contains many more cutting teeth than a slot milling cutter. Circular sawing machines have powered spindles to rotate the saw blade and a feeding mechanism to drive the rotating blade into the work.

Two operations related to circular sawing are abrasive cutoff and friction sawing. In *abrasive cutoff,* an abrasive disk is used to perform cutoff operations on hard materials that would be difficult to saw with a conventional saw blade. In *friction sawing,* a steel disk is rotated against the work at very high speeds, resulting in friction heat that causes the material to soften sufficiently to permit penetration of the disk through the work. The cutting speeds in both of these operations are much faster than in circular sawing.

22.6 HIGH-SPEED MACHINING

One persistent trend throughout the history of metal machining has been the use of higher and higher cutting speeds. In recent years, there has been renewed interest in this area due to its potential for faster production rates, shorter lead times, reduced costs, and improved part quality. In its simplest definition, *high-speed machining* (HSM) means using cutting speeds that are significantly higher than those used in conventional machining operations. Some examples of cutting speed values for conventional and HSM are presented in Table 22.1, according to data compiled by Kennametal Inc.[1]

Other definitions of HSM have been developed to deal with the wide variety of work materials and tool materials used in machining. One popular HSM definition is by

TABLE 22.1   Comparison of cutting speeds used in conventional versus high-speed machining for selected work materials.

Work Material	Solid Tools (end mills, drills)[a]				Indexable Tools (face mills)[a]			
	Conventional Speed		High Cutting Speed		Conventional Speed		High Cutting Speed	
	m/min	ft/min	m/min	ft/min	m/min	ft/min	m/min	ft/min
Aluminum	300+	1000+	3000+	10,000+	600+	2000+	3600+	12,000+
Cast iron, soft	150	500	360	1200	360	1200	1200	4000
Cast iron, ductile	105	350	250	800	250	800	900	3000
Steel, free machining	105	350	360	1200	360	1200	600	2000
Steel, alloy	75	250	250	800	210	700	360	1200
Titanium	40	125	60	200	45	150	90	300

Source: Kennametal Inc. [1].

[a]Solid tools are made of one solid piece, indexable tools use indexable inserts. Appropriate tool materials include cemented carbide and coated carbide of various grades for all materials, ceramics for all materials, polycrystalline diamond tools for aluminum, and cubic boron nitride for steels (see Section 23.2 for discussion of these tool materials).

[1]Kennametal Inc. is a leading cutting tool producer.

the ***DN ratio***—the bearing bore diameter (mm) multiplied by the maximum spindle speed (rev/min). For high-speed machining, the typical DN ratio is between 500,000 and 1,000,000. This definition allows larger diameter bearings to fall within the HSM range, even though they operate at lower rotational speeds than smaller bearings. Typical HSM spindle velocities range between 8,000 and 35,000 rpm, although some spindles today are designed to rotate at 100,000 rpm.

Another HSM definition is based on the ratio of horsepower to maximum spindle speed, or ***hp/rpm ratio.*** Conventional machine tools usually have a higher hp/rpm ratio than machines equipped for high-speed machining. By this metric, the dividing line between conventional machining and HSM is around 0.005 hp/rpm. Thus, high-speed machining includes 50 hp spindles capable of 10,000 rpm (0.005 hp/rpm) and 15 hp spindles that can rotate at 30,000 rpm (0.0005 hp/rpm).

Other definitions emphasize higher production rates and shorter lead times. In this case, important noncutting factors come into play, such as high rapid traverse speeds and quick automatic tool changes ("chip-to-chip" times of 7 s and less).

Requirements for high-speed machining include the following: (1) high-speed spindles using special bearings designed for high rpm operation; (2) high feed rate capability, typically around 50 m/min (2000 in/min); (3) CNC motion controls with "look-ahead" features that allow the controller to see upcoming directional changes and to make adjustments to avoid "undershooting" or "overshooting" the desired tool path; (4) balanced cutting tools, toolholders, and spindles to minimize vibration effects; (5) coolant delivery systems that provide pressures an order of magnitude greater than in conventional machining; and (6) chip control and removal systems to cope with the much larger metal removal rates in HSM. Also important are the cutting tool materials. As listed in Table 22.1, various tool materials are used for high-speed machining, and these materials are discussed in the following chapter.

Applications of HSM seem to divide into three categories [1]. One is in the aircraft industry, by companies such as Boeing, in which long airframe structural components are machined from large aluminum blocks. Much metal removal is required, mostly by milling. The resulting pieces are characterized by thin walls and large surface-to-volume ratios, but they can be produced more quickly and are more reliable than assemblies involving multiple components and riveted joints. A second category involves the machining of aluminum by multiple operations to produce a variety of components for industries such as automotive, computer, and medical. Multiple cutting operations mean many tool changes as well as many accelerations and decelerations of the tooling. Thus, quick tool changes and tool path control are important in these applications. The third application category for HSM is in the die and mold industry, which fabricates complex geometries from hard materials. In this case, high-speed machining involves much metal removal to create the mold or die cavity and finishing operations to achieve fine surface finishes.

REFERENCES

[1] Ashley, S. "High-speed Machining Goes Mainstream." *Mechanical Engineering.* Vol. 117, No. 5, May 1995, pp. 56–61.

[2] Boston, O. W. *Metal Processing.* 2nd ed. John Wiley & Sons, Inc., New York, 1951.

[3] DeGarmo, E. P., Black, J. T., and Kohser, R. A. *Materials and Processes in Manufacturing.* 8th ed. Prentice Hall, Upper Saddle River, N.J., 1997.

[4] Drozda, T. J., and Wick, C., (eds.). ***Tool and Manufacturing Engineers Handbook.*** 4th ed. Volume I: ***Machining,*** Society of Manufacturing Engineers, Dearborn, Mich., 1983.

[5] Eary, D. F., and Johnson, G. E. ***Process Engineering: for Manufacturing.*** Prentice-Hall, Inc., Englewood Cliffs, N.J., 1962.

[6] Hogan, B. J. "No Speed Limits." ***Manufacturing Engineering.*** Vol. 122, No. 3, March 1999, pp. 66–79.

[7] Kalpakjian, S., and Schmid, S. R. *Manufacturing Engineering and Technology.* 4th ed. Prentice Hall, Upper Saddle River, N.J., 2001.

[8] Krar, S. F., and Ratterman, E. *Superabrasives: Grinding and Machining with CBN and Diamond.* McGraw-Hill, Inc., New York, 1990.

[9] Lindberg, R. A. *Processes and Materials of Manufacture.* 4th ed. Allyn and Bacon, Inc., Boston, Mass., 1990.

[10] Luer, K. "High Speed Machining of Aluminum for Use in Aerospace Applications." *MSE 427 Term Paper.* Lehigh University, Bethlehem, Penn., 1998.

[11] Marinac, D. "Smart Toolpaths for HSM." *Manufacturing Engineering.* Vol. 125, No. 5, November 2000, pp. 44–50.

[12] Mason, F., and Freeman, N. B. "Turning Centers Come of Age." Special Report 773, *American Machinist.* February 1985, pp. 97–116.

[13] *Metals Handbook.* 8th Edition. Volume 3: *Machining.* American Society for Metals, Metals Park, Ohio, 1967.

[14] Rolt, L. T. C. *A Short History of Machine Tools.* The M.I.T. Press, Cambridge, Mass., 1965.

[15] Steeds, W. *A History of Machine Tools—1700–1910.* Oxford University Press, London, 1969.

[16] Trent, E. M. and P. K. Wright. *Metal Cutting.* 4th ed. Butterworth Heinemann, Boston, Mass., 2000.

REVIEW QUESTIONS

22.1. Discuss the differences between rotational parts and prismatic parts in machining.

22.2. Distinguish between generating and forming when machining workpart geometries.

22.3. Give two examples of machining operations in which generating and forming are combined to create workpart geometry.

22.4. Describe the turning process.

22.5. What is the difference between threading and tapping?

22.6. How does a boring operation differ from a turning operation?

22.7. What is meant by the designation 12 × 36 in. lathe?

22.8. Name the various ways in which a workpart can be held in a lathe.

22.9. What is the difference between a live center and a dead center, when these terms are used in the context of workholding in a lathe?

22.10. How does a turret lathe differ from an engine lathe?

22.11. What is a blind hole?

22.12. What is the distinguishing feature of a radial drill press?

22.13. What is the difference between peripheral milling and face milling?

22.14. Describe profile milling.

22.15. What is pocket milling?

22.16. Describe the difference between up milling and down milling.

22.17. How does a universal milling machine differ from a conventional knee-and-column machine?

22.18. What is a machining center?

22.19. What is the difference between a machining center and a turning center?

22.20. What can a mill-turn center do that a conventional turning center cannot do?

22.21. How do shaping and planing differ?

22.22. What is the difference between internal broaching and external broaching?

22.23. Identify the three basic forms of sawing operation.

MULTIPLE CHOICE QUIZ

There is a total of 20 correct answers in the following multiple choice questions (some questions have multiple answers that are correct). To attain a perfect score on the quiz, all correct answers must be given, since each correct answer is worth 1 point. For each question, each omitted answer or wrong answer reduces the score by 1 point, and each additional answer beyond the number of answers required reduces the score by 1 point. Percentage score on the quiz is based on the total number of correct answers.

22.1. Which of the following are examples of generating the workpart geometry in machining, as compared to forming the geometry (more than one answer)? (a) boring, (b) contour turning, (c) drilling, (d) profile milling, and (e) tapping.

22.2. In a turning operation, the change in diameter of the workpart is equal to which one of the following? (a) 1 × depth of cut, (b) 2 × depth of cut, (c) 1 × feed, or (d) 2 × feed.

22.3. A lathe can be used to perform which of the following machining operations (more than one answer)? (a) boring, (b) broaching, (c) drilling, (d) milling, (e) planing, or (f) turning.

22.4. A facing operation is normally performed on which one of the following machine tools (one best answer)? (a) drill press, (b) lathe, (c) milling machine, (d) planer, or (e) shaper.

22.5. Knurling is performed on a lathe, but it is a metal forming operation rather than a metal removal operation: (a) true or (b) false.

22.6. Which of the following cutting tools can be used on a turret lathe (more than one answer)? (a) broach, (b) cut-off tool, (c) single point turning tool, (d) tap, or (e) threading tool.

22.7. Which of the following turning machines permits very long bar stock to be used (one best answer)? (a) chucking machine, (b) engine lathe, (c) screw machine, (d) speed lathe, or (e) turret lathe.

22.8. Reaming is used for which of the following functions (more than one answer)? (a) accurately locate a hole position, (b) enlarge a drilled hole, (c) improve surface finish on a hole, (d) improve tolerance on hole diameter, and (e) provide an internal thread.

22.9. End milling is most similar to which one of the following? (a) face milling, (b) peripheral milling, (c) plain milling, or (d) slab milling.

22.10. The basic milling machine is which one of the following? (a) bed type, (b) knee-and-column, (c) profiling mill, (d) ram mill, and (e) universal milling machine.

22.11. A broaching operation is best described by which one of the following? (a) a rotating tool moves past a stationary workpart, (b) a tool with multiple teeth moves linearly past a stationary workpart, (c) a workpart is fed past a rotating cutting tool, or (d) a workpart moves linearly past a stationary single-point tool.

22.12. A planing operation is best described by which one of the following? (a) a single-point tool moves linearly past a stationary workpart, (b) a tool with multiple teeth moves linearly past a stationary workpart, (c) a workpart is fed linearly past a rotating cutting tool, or (d) a workpart moves linearly past a single-point tool.

PROBLEMS

Turning and Related Operations

22.1. A cylindrical workpart 125 mm in diameter and 900 mm long is to be turned in an engine lathe. Cutting conditions are: $v = 2.5$ m/s, $f = 0.3$ mm/rev, and $d = 2.0$ mm. Determine (a) cutting time, and (b) metal removal rate.

22.2. In a production turning operation, the foreman has decreed that the single pass must be completed on the cylindrical workpiece in 5.0 min. The piece is 400 mm long and 150 mm in diameter. Using a feed = 0.30 mm/rev and a depth of cut = 4.0 mm, what cutting speed must be used to meet this machining time requirement?

22.3. A tapered surface is to be turned on an automatic lathe. The workpiece is 750 mm long with minimum and maximum diameters of 100 mm and 200 mm at opposite ends. The automatic controls on the lathe permit the surface speed to be maintained at a constant value of 200 m/min by adjusting the rotational speed as a function of workpiece diameter. Feed = 0.25 mm/rev, and depth of cut = 3.0 mm. The rough geometry of the piece has already been formed, and this operation will be the final cut. Determine (a) the time required to turn the taper and (b) the rotational speeds at the beginning and end of the cut.

22.4. In the taper turning job of previous Problem 22.3, suppose that the automatic lathe with surface speed control is not available and a conventional lathe must be used. Determine the rotational speed that would be required to complete the job in exactly the same time as your answer to part (a) of that problem.

22.5. A workbar with a 5.0 in. diameter and 48 in. length is chucked in an engine lathe and supported at the opposite end using a live center. A 40.0 in. portion of the length is to be turned to a diameter of 4.75 in. in one pass at a speed = 400 ft/min and a feed = 0.012 in./rev.

Determine: (a) the required depth of cut, (b) cutting time, and (c) metal removal rate.

22.6. A 4-in.-diameter workbar that is 25 in. long is to be turned down to a 3.5 in. diameter in two passes on an engine lathe using the following cutting conditions: $v = 300$ ft/min, $f = 0.015$ in./rev, and $d = 0.125$ in. The bar will be held in a chuck and supported on the opposite end in a live center. With this workholding setup, one end must be turned to diameter; then the bar must be reversed to turn the other end. Using an overhead crane available at the lathe, the time required to load and unload the bar is 5 minutes, and the time to reverse the bar is 3 minutes. For each turning cut an allowance must be added to the cut length for approach and overtravel. The total allowance (approach plus overtravel) is 0.50 in. Determine the total cycle time to complete this turning operation.

22.7. The end of a large tubular workpart is to be faced on a NC vertical boring mill. The part has an outside diameter = 45 in. and inside diameter = 25 in. If the facing operation is performed at a rotational speed = 30 rev/min, feed = 0.020 in./rev, and depth = 0.150 in., determine (a) the cutting time to complete the facing operation and (b) the cutting speeds and metal removal rates at the beginning and end of the cut.

22.8. Solve previous Problem 22.7 except that the machine tool controls operate at a constant cutting speed by continuously adjusting rotational speed for the position of the tool relative to the axis of rotation. The rotational speed at the beginning of the cut = 30 rev/min, and is continuously increased thereafter to maintain a constant cutting speed.

Drilling

22.9. A drilling operation is to be performed with a 25.4 mm diameter twist drill in a steel workpart. The hole is a blind hole at a depth = 50 mm, and the point angle = 118°. Cutting conditions are: speed = 25 m/min, feed = 0.25 mm/rev. Determine (a) the cutting time to complete the drilling operation, and (b) metal removal rate during the operation, after the drill bit reaches full diameter.

22.10. A NC drill press is to perform a series of through-hole drilling operations on a 1.75 in. thick aluminum plate that is a component in a heat exchanger. Each hole is 3/4 in. diameter. There are 100 holes in all, arranged in a 10 × 10 matrix pattern, and the distance between adjacent hole centers (along the square) = 1.5 in. The cutting speed = 300 ft/min, the penetration feed (z-direction) = 0.015 in./rev, and the feed rate between holes (x–y plane) = 15.0 in./min. Assume that x–y moves are made at a distance of 0.5 in above the work surface, and that this distance must be included in the penetration feed rate for each hole. Also, the rate at which the drill is retracted from each hole is twice the penetration feed rate. The drill has a point angle = 100 degrees. Determine the time required from the beginning of the first hole to the completion of the last hole, assuming the most efficient drilling sequence will be used to accomplish the job.

22.11. A gun-drilling operation is used to drill a 7/16-in. diameter hole to a certain depth. It takes 4.5 minutes to perform the drilling operation using high pressure fluid delivery of coolant to the drill point. The cutting conditions are: N = 3000 rev/min at a feed = 0.002 in./rev. In order to improve the surface finish in the hole, it has been decided to increase the speed by 20% and decrease the feed by 25%. How long will it take to perform the operation at the new cutting conditions?

Milling

22.12. A peripheral milling operation is performed on the top surface of a rectangular workpart which is 300 mm long by 100 mm wide. The milling cutter, which is 75 mm in diameter and has four teeth, overhangs the width of the part on both sides. Cutting conditions are: v = 80 m/min, f = 0.2 mm/tooth, and d = 7.0 mm. Determine (a) the time to make one pass across the surface, and (b) the material removal rate during the cut.

22.13. A face-milling operation is used to machine 5 mm from the top surface of a rectangular piece of aluminum 400 mm long by 100 mm wide. The cutter has four teeth (cemented carbide inserts) and is 150 mm in diameter. Cutting conditions are v = 3 m/s, f = 0.27 mm/tooth, and d = 5.0 mm. Determine (a) time to make one pass across the surface, and (b) metal removal rate during cutting.

22.14. A slab-milling operation is performed to finish the top surface of a steel rectangular workpiece 10.0 in. long by 3.0 in. wide. The helical milling cutter, which has a 2.5 in. diameter and eight teeth, is set up to overhang the width of the part on both sides. Cutting conditions are v = 100 ft/min, f = 0.009 in./tooth, and d = 0.250 in. Determine (a) the time to make one pass across the surface, and (b) the metal removal rate during the cut.

22.15. A face-milling operation is performed to finish the top surface of a steel rectangular workpiece 12.0 in. long by 2.0 in. wide. The milling cutter has four teeth (cemented carbide inserts) and a 3.0 in diameter. Cutting conditions are v = 500 ft/min, f = 0.010 in./tooth, and d = 0.150 in. Determine (a) the time to make one pass across the surface, and (b) the metal removal rate during the cut.

22.16. Solve Problem 22.15 except that the workpiece is 5.0 in. wide and the cutter is offset to one side so that the swath cut by the cutter = 1.0 in wide.

Other Operations

22.17. An open side planer is to be used to plane the top surface of a rectangular workpart, 25 in. by 40 in. Cutting conditions are: v = 25 ft/min, f = 0.020 in./pass, and d = 0.200 in. The length of the stroke across the work must be set up so that 10 in. are allowed at both the beginning and end of the stroke for approach and over-travel. The return stroke, including an allowance for acceleration and deceleration, takes 75% of the time for the forward stroke. How long will it take to complete the job, assuming that the part is oriented in such a way as to minimize the time?

23 CUTTING-TOOL TECHNOLOGY

CHAPTER CONTENTS

23.1 Tool Life
    23.1.2 Tool Wear
    23.1.2 Tool Life and the Taylor Tool Life Equation
23.2 Tool Materials
    23.2.1 High-Speed Steel and Its Predecessors
    23.2.2 Cast Cobalt Alloys
    23.2.3 Cemented Carbides, Cermets, and Coated Carbides
    23.2.4 Ceramics
    23.2.5 Synthetic Diamonds and Cubic Boron Nitride
23.3 Tool Geometry
    23.3.1 Single-Point Tool Geometry
    23.3.2 Multiple-Cutting-Edge Tools
23.4 Cutting Fluids
    23.4.1 Types of Cutting Fluids
    23.4.2 Application of Cutting Fluids

Machining operations are accomplished using cutting tools. The high forces and temperatures during machining create a very harsh environment for the tool. If cutting force becomes too large, the tool fractures. If cutting temperature becomes too high, the tool material softens and fails. And if neither of these conditions cause tool failure, continual wearing action on the cutting tool ultimately leads to failure.

Cutting-tool technology has two principal aspects: ***tool material*** and ***tool geometry.*** The first is concerned with developing materials that can withstand the forces, temperatures, and wearing action in the machining process. The second deals with optimizing the geometry of the cutting tool for the tool material and for a given operation. These are the issues we address in the present chapter. It is appropriate to begin by considering tool life, since this is a prerequisite for much of our subsequent discussion on tool materials. It also seems appropriate to include a section on cutting fluids at the end of this chapter; cutting fluids are often used in machining operations to prolong the life of a cutting tool.

23.1 TOOL LIFE

As suggested by our opening paragraph, there are three possible modes by which a cutting tool can fail in machining:

1. **Fracture failure**—This mode of failure occurs when the cutting force at the tool point becomes excessive, causing it to fail suddenly by brittle fracture.

2. **Temperature failure**—This failure occurs when the cutting temperature is too high for the tool material, causing the material at the tool point to soften, which leads to plastic deformation and loss of the sharp edge.

3. **Gradual wear**—Gradual wearing of the cutting edge causes loss of tool shape, reduction in cutting efficiency, an acceleration of wearing as the tool becomes heavily worn, and finally tool failure in a manner similar to a temperature failure.

Fracture and temperature failures result in premature loss of the cutting tool. These two modes of failure are therefore undesirable. Of the three possible tool failures, gradual wear is preferred because it leads to the longest possible use of the tool, with the associated economic advantage of that longer use.

Product quality must also be considered when attempting to control the mode of tool failure. When the tool point fails suddenly during a cut, it often causes damage to the work surface. This damage requires either rework of the surface or possible scrapping of the part. The damage can be avoided by selecting cutting conditions that favor gradual wearing of the tool rather than fracture or temperature failure, and by changing the tool before the final catastrophic loss of the cutting edge occurs.

23.1.1 Tool Wear

Gradual wear occurs at two principal locations on a cutting tool: the top rake face and the flank. Accordingly, two main types of tool wear can be distinguished: crater wear and flank wear, illustrated in Figures 23.1 and 23.2. We will use a single point tool to explain tool wear and the mechanisms which cause it. **Crater wear,** Figure 23.2(a), consists of a concave section on the rake face of the tool, formed by the action of the chip sliding against the surface. High stresses and temperatures characterize the tool–chip contact interface, contributing to the wearing action. The crater can be measured either by its depth or its area. **Flank wear,** Figure 23.2(b), occurs on the flank, or relief face, of the tool. It results from rubbing between the newly generated work surface and the flank face adjacent to the cutting edge. Flank wear is measured by the width of the wear band, FW. This wear band is sometimes called the flank wear *land*.

FIGURE 23.1   Diagram of worn cutting tool, showing the principal locations and types of wear that occur.

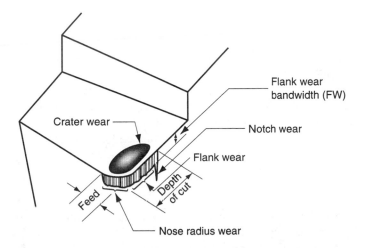

FIGURE 23.2 (a) Crater wear and (b) flank wear on a cemented carbide tool, as seen through a tool-maker's microscope. (Courtesy Manufacturing Technology Laboratory, Lehigh University, photos by J. C. Keefe.)

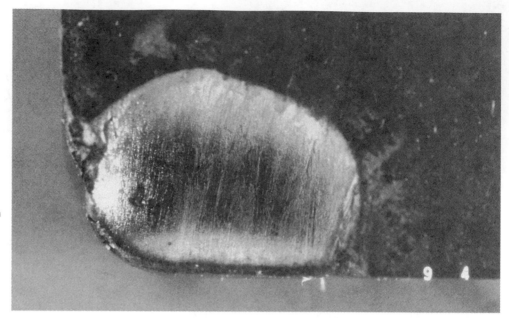

(a)

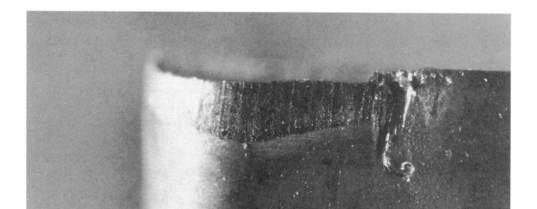

(b)

Certain features of flank wear can be identified. First, an extreme condition of flank wear often appears on the cutting edge at the location corresponding to the original surface of the workpart. This is called *notch wear.* It occurs because the original work surface is harder and/or more abrasive than the internal material, due to work hardening from cold drawing or previous machining, sand particles in the surface from casting, or other reasons. As a consequence of the harder surface, wear is accelerated at this location. A second region of flank wear that can be identified is *nose radius wear*; this occurs on the nose radius leading into the end cutting edge.

The mechanisms that cause wear at the tool–chip and tool-work interfaces in machining can be summarized as follows:

> *Abrasion*—This is a mechanical wearing action due to hard particles in the work material gouging and removing small portions of the tool. This abrasive action occurs in both flank wear and crater wear; it is a significant cause of flank wear.

> *Adhesion*—When two metals are forced into contact under high pressure and temperature, adhesion or welding occur between them. These conditions are present between the chip and the rake face of the tool. As the chip flows across the tool, small particles of the tool are broken away from the surface, resulting in attrition of the surface.

> *Diffusion*—This is a process in which an exchange of atoms takes place across a close contact boundary between two materials (Section 4.3). In the case of tool wear, diffusion occurs at the tool–chip boundary, causing the tool surface to become depleted of the atoms responsible for its hardness. As this process continues, the tool surface becomes more susceptible to abrasion and adhesion. Diffusion is believed to be a principal mechanism of crater wear.

> *Chemical reactions*—The high temperatures and clean surfaces at the tool–chip interface in machining at high speeds can result in chemical reactions, in particular, oxidation, on the rake face of the tool. The oxidized layer, being softer than the parent tool material, is sheared away, exposing new material to sustain the reaction process.

> *Plastic deformation*—Another mechanism that contributes to tool wear is plastic deformation of the cutting edge. The cutting forces acting on the cutting edge at high temperature cause the edge to deform plastically, making it more vulnerable to abrasion of the tool surface. Plastic deformation contributes mainly to flank wear.

Most of these tool-wear mechanisms are accelerated at higher cutting speeds and temperatures. Diffusion and chemical reaction are especially sensitive to elevated temperature.

23.1.2 Tool Life and the Taylor Tool Life Equation

As cutting proceeds, various wear mechanisms result in increasing levels of wear on the cutting tool. The general relationship of tool wear versus cutting time is shown in Figure 23.3. Although the relationship shown is for flank wear, a similar relationship occurs for crater wear. Three regions can usually be identified in the typical wear growth curve. The first is the **break-in period,** in which the sharp cutting edge wears rapidly at the beginning of its use. This first region occurs within the first few minutes of cutting. The break-in period is followed by wear that occurs at a fairly uniform rate. This is called the **steady state wear** region. In our figure, this region is pictured as a linear function of time, although there are deviations from the straight line in actual machining. Finally, wear reaches a level at which the wear rate begins to accelerate. This marks the beginning of the **failure region,** in which

FIGURE 23.3 Tool wear as a function of cutting time. Flank wear (FW) is used here as the measure of tool wear. Crater wear follows a similar growth curve.

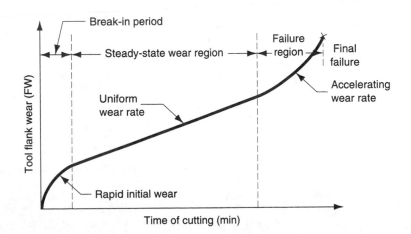

cutting temperatures are higher, and the general efficiency of the machining process is reduced. If allowed to continue, the tool finally fails by temperature failure.

The slope of the tool wear curve in the steady state region is affected by work material and cutting conditions. Harder work materials cause the wear rate (slope of the tool wear curve) to increase. Increased speed, feed, and depth of cut have a similar effect, with speed being the most important of the three. If the tool wear curves are plotted for several different cutting speeds, the results appear as in Figure 23.4. As cutting speed is increased, wear rate increases so the same level of wear is reached in less time.

Tool life is defined as the length of cutting time that the tool can be used. Operating the tool until final catastrophic failure is one way of defining tool life. This is indicated in Figure 23.4 by the end of each tool wear curve. However, in production, it is often a disadvantage to use the tool until this failure occurs because of difficulties in resharpening the tool and problems with workpart quality. As an alternative, a level of tool wear can be selected as a criterion of tool life, and the tool is replaced when wear reaches that level. A convenient tool life criterion is a certain flank wear value, such as 0.5 mm (0.020 in.), illustrated as the horizontal line on the graph. When each of the three wear curves intersects that line, the life of the corresponding tool is defined as ended. If the intersection points are projected down to the time axis, the values of tool life can be identified, as we have done.

Taylor Tool Life Equation   If the tool life values for the three wear curves in Figure 23.4 are plotted on a natural log-log graph of cutting speed versus tool life, the resulting relationship is a straight line as shown in Figure 23.5.[1]

The discovery of this relationship around 1900 is credited to F. W. Taylor. It can be expressed in equation form and is called the Taylor tool life equation:

$$vT^n = C \qquad (23.1)$$

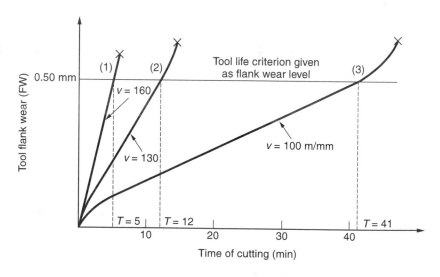

FIGURE 23.4   Effect of cutting speed on tool flank wear (FW) for three cutting speeds. Hypothetical values of speed and tool life are shown for a tool life criterion of 0.50 mm flank wear.

---

[1]The reader may have noted in Figure 23.5 that we have plotted the dependent variable (tool life) on the horizontal axis and the independent variable (cutting speed) on the vertical axis. Although this is a reversal of the normal plotting convention, it nevertheless is the way the Taylor tool life relationship is usually presented.

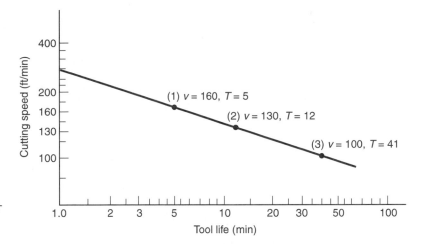

FIGURE 23.5   Natural log-log plot of cutting speed vs. tool life.

where v = cutting speed, m/min (ft/min); T = tool life, min; and n and C are parameters whose values depend on feed, depth of cut, work material, tooling (material in particular), and the tool life criterion used. The value of n is relative constant for a given tool material, while the value of C depends on tool material, work material, and cutting conditions. We will elaborate on these relationships when we discuss the various tool materials in Section 23.2.

Basically, Eq. (23.1) states that higher cutting speeds result in shorter tool lives. Relating the parameters n and C to Figure 23.5, n is the slope of the plot (expressed in linear terms rather than in the scale of the axes), and C is the intercept on the speed axis. C represents the cutting speed that results in a one-min tool life.

The problem with Eq. (23.1) is that the units on the right-hand side of the equation are not consistent with the units on the left-hand side. To make the units consistent, the equation should be expressed in the form:

$$vT^n = C(T_{ref}{}^n) \qquad (23.2)$$

where T_{ref} = a reference value for C. T_{ref} is simply 1 min when m/min (ft/min) and minutes are used for v and T respectively. The advantage of Eq. (23.2) is seen when it is desired to use the Taylor equation with units other than m/min (ft/min) and minutes; for example, if cutting speed were expressed as m/sec and tool life as sec. In this case, T_{ref} would be 60 sec and C would therefore be the same speed value as in Eq. (23.1), although converted to units of m/sec. The slope n would have the same numerical value as in Eq. (23.1).

EXAMPLE 23.1 Taylor tool life equation.

To determine the values of C and n in the plot of Figure 23.5, select two of the three points on the curve and solve simultaneous equations of the form of Eq. 23.1.

Solution: Choosing the two extreme points: v = 160 m/min, T = 5 min; and v = 100 m/min, T = 41 min, we have

$$160(5)^n = C$$

$$100(41)^n = C$$

Setting the left-hand sides of each equation equals,

$$160(5)^n = 100 (41)^n$$

Taking the natural logarithms of each term,

$$\ln(160) + n\ln(5) = \ln(100) + n\ln(41)$$

$$5.0752 + 1.6094\,n = 4.6052 + 3.7136\,n$$

$$0.4700 = 2.1042\,n$$

$$n = \frac{0.4700}{2.1042} = 0.223$$

Substituting this value of n into either starting equation, we obtain the value of C:

$$C = 160(5)^{0.223} = 229$$

or
$$C = 100(41)^{0.223} = 229$$

The Taylor tool life equation for the data of Figure 23.5 is therefore

$$vT^{0.223} = 229$$

An expanded version of Eq. (23.2) can be formulated to include the effects of feed, depth of cut, and even work material hardness:

$$vT^n f^m d^p H^q = KT_{\mathrm{ref}}{}^n f_{\mathrm{ref}}{}^m d_{\mathrm{ref}}{}^p H_{\mathrm{ref}}{}^q \tag{23.3}$$

where f = feed, mm (in.); d = depth of cut, mm (in.); H = hardness, expressed in an appropriate hardness scale; m, p, and q are exponents whose values are experimentally determined for the conditions of the operation; K = a constant analogous to C in Eq. (23.2); and f_{ref}, d_{ref}, and H_{ref} are reference values for feed, depth of cut, and hardness. The values of m and p, the exponents for feed and depth, are less than 1.0. This indicates the greater effect of cutting speed on tool life, since the exponent of v is 1.0. After speed, feed is next in importance, so m has a value greater than p. The exponent for work hardness, q, is also less than 1.0.

Perhaps the greatest difficulty in applying Eq. (23.3) in a practical machining operation is the tremendous amount of machining data required to determine the parameters of the equation. Variations in work materials and testing conditions also cause difficulties by introducing statistical variations in the data. Eq. (23.3) is valid in indicating general trends among its variables, but not in its ability to accurately predict tool life performance. To reduce these problems and to make the scope of the equation more manageable, some terms are usually eliminated:

$$vT^n f^m = KT_{\mathrm{ref}}{}^n f_{\mathrm{ref}}{}^m \tag{23.4}$$

where the terms have the same meaning as before, except that the constant K will have a slightly different interpretation.

Tool Life Criteria in Production   Although flank wear is the tool life criterion in our previous discussion of the Taylor equation, this criterion is not very practical in a factory environment because of the difficulties and time required to measure flank wear. Following are nine alternative tool life criteria that are more convenient to use in a production machining operation, some of which are admittedly subjective:

1. The cutting edge completely fails (fracture failure, temperature failure, or wearing until complete breakdown of the tool has occurred).
2. Machine operator gives a visual inspection of flank wear (or crater wear without a toolmaker's microscope). This criterion is limited by the operator's judgment and ability to observe tool wear with the naked eye.
3. The operator runs a fingernail across the cutting edge to test for irregularities.
4. The sound emitting from the operation changes, as judged by the operator.
5. Chips become ribbony, stringy, and difficult to dispose of.
6. The surface finish on the work degrades.
7. Power consumption in the operation increases, as measured by a wattmeter connected to the machine tool.
8. Workpiece count. The operator is instructed to change the tool after a certain specified number of parts have been machined.
9. Cumulative cutting time—this is similar to the previous workpiece count, except that the length of time the tool has been cutting is monitored. This is possible on machine tools controlled by computer; the computer is programmed to keep data on the total cutting time for each tool.

23.2 TOOL MATERIALS

The three modes of tool failure can be used to identify some of the important properties required in a tool material:

> *Toughness*—To avoid fracture failure, the tool material must possess high toughness. Toughness is the capacity of a material to absorb energy without failing. It is usually characterized by a combination of strength and ductility in the material.

> *Hot hardness*—Hot hardness is the ability of a material to retain its hardness at high temperatures. This is required because of the high temperature environment in which the tool operates.

> *Wear resistance*—Hardness is the single most important property needed to resist abrasive wear. All cutting tool materials must be hard. However, wear resistance in metal cutting depends on more than just tool hardness, because of the other tool wear mechanisms. Other characteristics affecting wear resistance include surface finish on the tool (a smoother surface means a lower coefficient of friction), chemistry of tool and work materials, and whether a cutting fluid is used.

Cutting tool materials achieve this combination of properties in varying degrees. In this section, the following cutting tool materials are discussed: (1) high-speed steel and its predecessors, plain carbon and low alloy steels, (2) cast cobalt alloys, (3) cemented carbides, cermets, and coated carbides, (4) ceramics, (5) synthetic diamond and cubic boron nitride. Before examining these individual materials, a brief overview and technical comparison will be helpful. The historical development of these materials is described in Historical Note 23.1. Commercially, the most important tool materials are high-speed steel and cemented carbides, cermets, and coated carbides. These two categories account for more than 90% of the cutting tools used in machining operations.

Historical Note 23.1 *Cutting tool materials* [10], [11], [13].

In 1800, England was leading the Industrial Revolution, and iron was the leading metal in the revolution. The best tools for cutting iron were made of cast steel by the crucible process, invented in 1742 by B. Huntsman. Cast steel, whose carbon content lies between wrought iron and cast iron, could be hardened by heat treatment to machine the other metals. In 1868, R. Mushet discovered that by alloying about 7% tungsten in crucible steel, a hardened tool steel was obtained by air quenching after heat treatment. Mushet's tool steel was far superior to its predecessor in machining.

F. W. Taylor stands as an important figure in the history of cutting tools. Starting around 1880 at Midvale Steel in Philadelphia and later at Bethlehem Steel in Bethlehem, Pennsylvania, he began a series of experiments that lasted a quarter century, yielding a much improved understanding of metal cutting. Among the developments resulting from the work of Taylor and colleague M. White at Bethlehem was *high-speed steel* (HSS), a class of highly alloyed tool steels that permitted substantially higher cutting speeds than previous cutting tools. The superiority of HSS resulted not only from greater alloying, but also from refinements in heat treatment. Tools of the new steel allowed cutting speeds that were more than twice those of Mushet's steel and almost four times those of plain carbon cast steels.

Tungsten carbide (WC) was first synthesized in the late 1890s. It took nearly three decades before a useful cutting tool material was developed by sintering the WC with a metallic binder to form *cemented carbides.* These were first used in metal cutting in the mid-1920s in Germany, and in the late 1920s in the United States (see Historical Note 7.2). *Cermet* cutting tools based on titanium carbide were first introduced in the 1950s, but their commercial importance dates from the 1970s. The first *coated carbides,* consisting of one coating on a WC–Co substrate, were first used around 1970. Coating materials included TiC, TiN, and Al_2O_3. Modern coated carbides have three or more coatings of these and other hard materials.

Attempts to use *alumina ceramics* in machining date from the early 1900s in Europe. Their brittleness inhibited success in these early applications. Processing refinements over many decades have resulted in property improvements in these materials. U.S. commercial use of ceramic cutting tools dates from the mid-1950s.

The first industrial diamonds were produced by the General Electric Company in 1954. They were single crystal diamonds that were applied with some success in grinding operations starting around 1957. Greater acceptance of diamond cutting tools has resulted from the use of *sintered polycrystalline diamond* (SPD), dating from the early 1970s. A similar tool material, sintered *cubic boron nitride,* was first introduced in 1969 by GE under the trade name BORAZON.

TABLE 23.1 Typical hardness values (at room temperature) and transverse rupture strengths for various tool materials.[a]

Material	Hardness	Transverse Rupture Strength	
		MPa	(lb/in.^{2})
Plain carbon steel	60 HRC	5200	(750,000)
High-speed steel	65 HRC	4100	(600,000)
Cast cobalt alloy	65 HRC	2250	(325,000)
Cemented carbide (WC)			
Low Co content	93 HRA, 1800 HK	1400	(200,000)
High Co content	90 HRA, 1700 HK	2400	(350,000)
Cermet (TiC)	2400 HK	1700	(250,000)
Alumina (Al_2O_3)	2100 HK	400	(60,000)
Cubic boron nitride	5000 HK	700	(100,000)
Polycrystalline diamond	6000 HK	1000	(150,000)
Natural diamond	8000 HK	1500	(215,000)

Compiled from [1], [4], [16], and other sources.

[a]*Note*: The values of hardness and TRS are intended to be comparative and typical. Variations in properties result from differences in composition and processing.

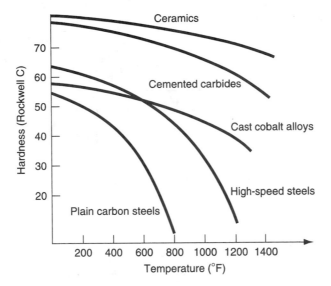

FIGURE 23.6 Typical hot hardness relationships for selected tool materials. Plain carbon steel shows a rapid loss of hardness as temperature increases. High-speed steel is substantially better, while cemented carbides and ceramics are significantly harder at elevated temperatures.

Table 23.1 and Figure 23.6 present data on properties of various tool materials. The properties are those related to the requirements of a cutting tool: hardness, toughness, and hot hardness. Table 23.1 lists room temperature hardness and transverse rupture strength for selected materials. Transverse rupture strength (Section 3.1.3) is a property used to indicate toughness for hard materials. Figure 23.6 shows hardness as a function of temperature for several of the tool materials discussed in this section.

In addition to these property comparisons, it is useful to compare the materials in terms of the parameters n and C in the Taylor tool life equation. In general, the development of new cutting tool materials has resulted in increases in the values of these two parameters. Table 23.2 provides a listing of representative values of n and C in the Taylor tool life equation for selected cutting tool materials.

The chronological development of tool materials has generally followed a path in which new materials have permitted higher and higher cutting speeds to be achieved. Table 23.3 identifies the cutting tool materials, together with their approximate year of introduction and typical maximum allowable cutting speeds at which they can be used.

TABLE 23.2   Representative values of n and C in the Taylor tool life equation, Eq. (23.1), for selected tool materials.

| | | C | | | |
| | | Nonsteel Cutting | | Steel Cutting | |
Tool material	n	m/min	(ft/min)	m/min	(ft/min)
Plain carbon tool steel	0.1	70	(200)	20	(60)
High-speed steel	0.125	120	(350)	70	(200)
Cemented carbide	0.25	900	(2700)	500	(1500)
Cermet	0.25			600	(2000)
Coated carbide	0.25			700	(2200)
Ceramic	0.6			3000	(10,000)

Compiled from [1], [4], and other sources.

The parameter values are approximated for turning at feed = 0.25 mm/rev (0.010 in./rev) and depth = 2.5 mm (0.100 in.). Nonsteel cutting refers to easy-to-machine metals such as aluminum, brass, and cast iron. Steel cutting refers to the machining of mild (unhardened) steel. It should be noted that significant variations in these values can be expected in practice.

TABLE 23.3 Cutting tool materials with their approximate dates of initial use and allowable cutting speeds.

Tool Material	Year of Initial Use	Allowable cutting speed[a]			
		Nonsteel Cutting		Steel Cutting	
		m/min	(ft/min)	m/min	(ft/min)
Plain carbon tool steel	1800s	Below 10	(Below 30)	Below 5	(Below 15)
High-speed steel	1900	25–65	(75–200)	17–33	(50–100)
Cast cobalt alloys	1915	50–200	(150–600)	33–100	(100–300)
Cemented carbides (WC)	1930	330–650	(1000–2000)	100–300	(300–900)
Cermets (TiC)	1950s			165–400	(500–1200)
Ceramics (Al_2O_3)	1955			330–650	(1000–2000)
Synthetic diamonds	1954, 1973	390–1300	(1200–4000)		
Cubic boron nitride	1969			500–800	(1500–2500)
Coated carbides	1970			165–400	(500–1200)

[a]Compiled from [4], [10], [15], [18], and other sources.

Dramatic increases in machining productivity have been made possible due to advances in tool material technology, as indicated in our table. Machine tool practice has not always kept pace with cutting tool technology. Limitations on horsepower, machine tool rigidity, spindle bearings, and the widespread use of older equipment in industry have acted to underutilize the possible upper speeds permitted by available cutting tools.

23.2.1 High-Speed Steel and Its Predecessors

Prior to the development of high-speed steel, plain carbon steel and Mushet's steel were the principal tool materials for metal cutting. Today, these steels are rarely used in industrial metal machining applications. The plain carbon steels used as cutting tools could be heat-treated to achieve relatively high hardness (Rockwell C 60), due to their fairly high carbon content. However, because of low alloying levels, they possess poor hot hardness (Figure 23.6), which renders them unusable in metal cutting except at speeds too low to be practical by today's standards. Mushet's steel has been displaced by advances in tool steel metallurgy.

High-speed steel (HSS) is a highly alloyed tool steel capable of maintaining hardness at elevated temperatures better than high carbon and low alloy steels. Its good hot hardness permits tools made of HSS to be used at higher cutting speeds. Compared to the other tool materials at the time of its development, it was truly deserving of its name high speed. A wide variety of high-speed steels is available, but they can be divided into two basic types: (1) tungsten-type, designated T-grades by the American Iron and Steel Institute (AISI), and (2) molybdenum-type, designated M-grades by AISI.

Tungsten-type HSS contains tungsten (W) as its principal alloying ingredient. Additional alloying elements are chromium (Cr), and vanadium (V). One of the original and best known HSS grades is T1, or 18-4-1 high speed steel, containing 18% W, 4% Cr, and 1% V. *Molybdenum HSS* grades contain combinations of tungsten and molybdenum (Mo), plus the same additional alloying elements as in the T-grades. Cobalt (Co) is sometimes added to HSS to enhance hot hardness. Of course, high-speed steel contains carbon, the element common to all steels. Typical alloying contents and functions of each alloying elements in HSS are listed in Table 23.4.

Commercially, high-speed steel is one of the most important cutting tool materials in use today, despite the fact that it was introduced a century ago. HSS is especially suited to applications involving complicated tool geometries, such as drills, taps, milling cutters, and broaches. These complex tools are generally easier and less expensive to produce

TABLE 23.4   Typical contents and functions of alloying elements in high-speed steel.

Alloying Element	Typical Content in HSS, % by Weight	Functions in High-Speed Steel
Tungsten	T-type HSS: 12–20	Increases hot hardness
	M-type HSS: 1.5–6	Improves abrasion resistance through formation of hard carbides in HSS
Molybdenum	T-type HSS: none	Increases hot hardness
	M-type HSS: 5–10	Improves abrasion resistance through formation of hard carbides in HSS
Chromium	3.75–4.5	Depth hardenability during heat treatment
		Improves abrasion resistance through formation of hard carbides in HSS
		Corrosion resistance (minor effect)
Vanadium	1–5	Combines with carbon for wear resistance
		Retards grain growth for better toughness
Cobalt	0–12	Increases hot hardness
Carbon	0.75–1.5	Principal hardening element in steel
		Provides available carbon to form carbides with other alloying elements for wear resistance

from HSS than other tool materials. They can be heat-treated so that cutting edge hardness is very good (Rockwell C 65) while toughness of the internal portions of the tool is also good. HSS cutters possess better toughness than any of the harder nonsteel tool materials used for machining, such as cemented carbides and ceramics. Even for single point tools, HSS is popular among machinists because of the ease with which a desired tool geometry can be ground into the tool point. Over the years, improvements have been made in the metallurgical formulation and processing of HSS so that this class of tool material remains competitive in many applications. Also, HSS tools, drills in particular, are often coated with a thin film of titanium nitride (TiN) to provide significant increases in cutting performance. Sputtering and ion plating, both physical vapor deposition processes (Section 29.3), are commonly used to coat these HSS tools.

23.2.2 Cast Cobalt Alloys

Cast cobalt alloy cutting tools consist of cobalt, around 40% to 50%; chromium, about 25% to 35%; and tungsten, usually 15% to 20%; with trace amounts of other elements. These tools are made into the desired shape by casting in graphite molds and then grinding to final size and cutting edge sharpness. High hardness is achieved as cast, an advantage over HSS, which requires heat treatment to achieve its hardness. Wear resistance of the cast cobalts is better than high-speed steel, but not as good as cemented carbide. Toughness of cast cobalt tools is better than carbides but not as good as HSS. Hot hardness also lies between these two materials.

As might be expected from their properties, applications of cast cobalt tools are generally between those of high-speed steel and cemented carbides. They are capable of heavy roughing cuts at speeds greater than HSS and feeds greater than carbides. Work materials include both steels and nonsteels, as well as nonmetallic materials such as plastics and graphite. Today, cast cobalt alloy tools are not nearly as important commercially as either high-speed steel or cemented carbides. They were introduced around 1915 as a tool material that would allow higher cutting speeds than HSS. The carbides were subsequently developed and proved to be superior to the cast Co alloys in most cutting situations.

23.2.3 Cemented Carbides, Cermets, and Coated Carbides

Cermets[2] are defined as composites of *cer*amic and *met*allic materials (Section 9.2.1). Technically speaking, cemented carbides are included within this definition; however, cermets based on WC–Co, including WC–TiC–TaC–Co, are known as carbides (cemented carbides) in common usage. In cutting-tool terminology, the term cermet is applied to ceramic-metal composites containing TiC, TiN, and certain other ceramics not including WC. One of the advances in cutting tool materials involves the application of a very thin coating to a WC–Co substrate. These tools are called coated carbides. Thus, we have three important and closely related tool materials to discuss: (1) cemented carbides, (2) cermets, and (3) coated carbides.

Cemented Carbides *Cemented carbides* (also called *sintered carbides*) are a class of hard tool material formulated from tungsten carbide (WC) using powder metallurgy techniques (Chapter 16) with cobalt (Co) as the binder (Sections 7.3.2, 9.2.1, and 17.3.1). There may be other carbide compounds in the mixture, such as titanium carbide (TiC) and/or tantalum carbide (TaC), in addition to WC.

The first cemented carbide cutting tools were made of WC–Co (Historical Note 7.2) and could be used to machine cast irons and nonsteel materials at cutting speeds faster than those possible with high-speed steel and cast cobalt alloys. However, when the straight WC–Co tools were used to machine steel, crater wear occurred rapidly, leading to early failure of the tools. A strong chemical affinity exists between steel and the carbon in WC, resulting in accelerated wear by diffusion and chemical reaction at the tool–chip interface for this work–tool combination. Consequently, straight WC–Co tools cannot be used effectively to machine steel. It was subsequently discovered that additions of titanium carbide and tantalum carbide to the WC–Co mix significantly retarded the rate of crater wear when cutting steel. These new WC–TiC–TaC–Co tools could be used for steel machining. The result is that cemented carbides are divided into two basic types: (1) nonsteel cutting grades, consisting of only WC–Co, and (2) steel cutting grades, with combinations of TiC and TaC added to the WC–Co.

The general properties of the two types of cemented carbides are similar: (1) high compressive strength but low-to-moderate tensile strength; (2) high hardness (90 to 95 HRA); (3) good hot hardness; (4) good wear resistance; (5) high thermal conductivity; (6) high modulus of elasticity—E values up to around 600×10^3 MPa (90×10^6 lb/in.^{2}); and (7) toughness lower than high-speed steel.

Nonsteel-cutting grades refer to those cemented carbides that are suitable for machining aluminum, brass, copper, magnesium, titanium, and other nonferrous metals; anomalously, grey cast iron is included in this group of work materials. In the nonsteel cutting grades, grain size and cobalt content are the factors that influence properties of the cemented carbide material. The typical grain size found in conventional cemented carbides ranges between 0.5 and 5 μm (20 and 200 μ-in.). As grain size is increased, hardness and hot hardness decrease, but transverse rupture strength increases.[3] The typical cobalt content in cemented carbides used for cutting tools is 3 to 12%. The effect of cobalt content on hardness and transverse rupture strength is shown in Figure 9.9. As

---

[2] The word "cermet" was first used around 1948.

[3] The effect of grain size (GS) on transverse rupture strength (TRS) is more complicated than we are reporting. Published data indicate that the effect of GS on TRS is influenced by cobalt content. At lower Co contents (< 10%), TRS does indeed increase as GS increases, but at higher Co contents (> 10%) TRS decreases as GS increases [3], [14].

cobalt content increases, TRS improves at the expense of hardness and wear resistance. Cemented carbides with low percentages of cobalt content (3%–6%) have high hardness and low TRS; while carbides with high Co (6%–12%) have high TRS but lower hardness (Table 23.1). Accordingly, cemented carbides with higher cobalt are used for roughing operations and interrupted cuts (such as milling), while carbides with lower cobalt (therefore, higher hardness and wear resistance) are used in finishing cuts.

Steel-cutting grades are used for low carbon, stainless, and other alloy steels. For these carbide grades, titanium carbide and/or tantalum carbide is substituted for some of the tungsten carbide. TiC is the more popular additive in most applications. Typically, from 10 to 25% of the WC might be replaced by combinations of TiC and TaC. This composition increases the crater wear resistance for steel cutting, but tends to adversely affect flank wear resistance for nonsteel-cutting applications. That is why two basic categories of cemented carbide are needed.

One of the important developments in cemented carbide technology in recent years is the use of very fine grain sizes (submicron sizes) of the various carbide ingredients (WC, TiC, and TaC). Although small grain size is usually associated with higher hardness but lower transverse rupture strength, the decrease in TRS is reduced or reversed at the submicron particle sizes. Therefore, these ultrafine grain carbides possess high hardness combined with good toughness.

Since the two basic types of cemented carbide were introduced in the 1920s and 1930s, the increasing number and variety of engineering materials have complicated the selection of the most appropriate cemented carbide. To address the problem of grade selection, two classification systems have been developed: (1) the ANSI[4] C-grade system, developed in the United States starting around 1942, and (2) the ISO R513-1975(E) system, introduced by the International Organization of Standardization (ISO) around 1964. In the C-grade system, summarized in Table 23.5, machining grades of cemented carbide are divided into two basic groups, corresponding to nonsteel-cutting and steel-cutting categories. Within each group there are four levels, corresponding to roughing, general purpose, finishing, and precision finishing.

The ISO R513-1975(E) system, titled "Application of Carbides for Machining by Chip Removal," classifies all machining grades of cemented carbides into three basic groups, each with its own letter and color code, as summarized in Table 23.6. Within each group, the grades are numbered on a scale that ranges from maximum hardness to maximum toughness. Harder grades are used for finishing operations (high speeds, low feeds and depths), while tougher grades are used for roughing operations. The ISO classification system can also be used to recommended applications for cermets and coated carbides.

TABLE 23.5 The ANSI C-grade classification system for cemented carbides.

Machining Application	Nonsteel-cutting Grades	Steel-cutting Grades	Cobalt and Properties
Roughing	C1	C5	High Co for max. toughness
General purpose	C2	C6	Medium to high Co
Finishing	C3	C7	Medium to low Co
Precision finishing	C4	C8	Low Co for max. hardness
Work materials	Al, brass, Ti, cast iron	Carbon & alloy steels	
Typical ingredients	WC–Co	WC–TiC–TaC–Co	

---

[4]ANSI = American National Standards Institute.

TABLE 23.6   ISO R513-1975(E) "Application of Carbides for Machining by Chip Removal."

Group	Carbide Type	Work Materials	Number Scheme (Cobalt and Properties)
P (blue)	Highly alloyed WC–TiC–TaC–Co	Steel, steel castings, ductile cast iron (ferrous metals with long chips)	P01 (low Co for maximum hardness) to P50 (high Co for maximum toughness)
M (yellow)	Alloyed WC–TiC–TaC–Co	Free-cutting steel, gray cast iron, austenitic stainless steel, superalloys	M10 (low Co for maximum hardness) to M40 (high Co for maximum toughness)
K (red)	Straight WC–Co	Nonferrous metals and alloys, gray cast iron (ferrous metals with short chips), nonmetallics	K01 (low Co for maximum hardness) to K40 (high Co for maximum toughness)

The two systems map into each other, as follows: The ANSI C1 through C4-grades map into the ISO K-grades, but in reverse numerical order, and the ANSI C5 through C8 grades translate into the ISO P-grades, but again in reverse numerical order.

Cermets   Although cemented carbides are technically classified as cermet composites, the term *cermet* in cutting tool technology is generally reserved for combinations of TiC, TiN, and titanium carbonitride (TiCN), with nickel and/or molybdenum as binders. Some of the cermet chemistries are more complex (e.g., ceramics such as Ta_xNb_yC and binders such as Mo_2C). However, cermets exclude metallic composites that are primarily based on WC–Co. Applications of cermets include high-speed finishing and semi-finishing of steels, stainless steels, and cast irons. Higher speeds are generally allowed with these tools compared to steel-cutting carbide grades. Lower feeds are typically used so that better surface finish is achieved, often eliminating the need for grinding.

Coated Carbides   The development of coated carbides around 1970 represented a significant advance in cutting tool technology. *Coated carbides* are a cemented carbide insert coated with one or more thin layers of wear resistant material, such as titanium carbide, titanium nitride, and/or aluminum oxide (Al_2O_3). The coating is applied to the substrate by chemical vapor deposition (Section 29.4) or physical vapor deposition (Section 29.3). The coating thickness is only 2.5 to 13 μm (0.0001–0.0005 in.). It has been found that thicker coatings tend to be brittle, resulting in cracking, chipping, and separation from the substrate.

The first generation of coated carbides had only a single layer coating (TiC, TiN, or Al_2O_3), and these tools are still in use. More recently, coated inserts have been developed that consist of multiple layers, as in Color Plate 7. The first layer applied to the WC–Co base is usually TiN or TiCN because of good adhesion and similar coefficient of thermal expansion. Additional layers of various combinations of TiN, TiCN, Al_2O_3 and TiAlN are subsequently applied.

Coated carbides are used to machine cast irons and steels in turning and milling operations. They are best applied at high cutting speeds in situations where dynamic force and thermal shock are minimal. If these conditions become too severe, as in some interrupted cut operations, chipping of the coating can occur, resulting in premature tool failure. In this situation, uncoated carbides formulated for toughness are preferred. When properly applied, coated carbide tools usually permit increases in allowable cutting speeds compared to uncoated cemented carbides.

Use of coated carbide tools is expanding to nonferrous metal and nonmetal applications for improved tool life and higher cutting speeds. Different coating materials are required, such as chromium carbide (CrC), zirconium nitride (ZrN), and diamond [9].

23.2.4 Ceramics

Cutting tools made from ceramics were first used commercially in the United States in the mid-1950s, although their development and use in Europe dates back to the early 1900s. Today's ceramic cutting tools are composed primarily of fine-grained *aluminum oxide* (Al_2O_3), pressed and sintered at high pressures and temperatures with no binder into insert form (Section 17.2). The aluminum oxide is usually very pure (99% is typical), although some manufacturers add other oxides (such as zirconium oxide) in small amounts. In producing ceramic tools, it is important to use a very fine grain size in the alumina powder, and to maximize density of the mix through high-pressure compaction in order to improve the material's low toughness.

Aluminum oxide cutting tools are most successful in high-speed turning of cast iron and steel. They can be used for finish turning operations on hardened steels if high cutting speeds, low feed and depth of cut, and rigid work setup are employed. Many premature fracture failures of ceramic tools are due to nonrigid machine tool setups, which subject the tools to mechanical shock. When properly applied, ceramic cutting tools can create very good surface finish. Ceramics are not recommended for heavy interrupted cut operations (e.g., rough milling) due to their low toughness. In addition to its use as inserts in conventional machining operations, Al_2O_3 is widely used as an abrasive in grinding and other abrasive processes (Chapter 25).

Other commercially available ceramic cutting tool materials include silicon nitride (SiN), *sialon* (silicon nitride and aluminum oxide, $SiN–Al_2O_3$), aluminum oxide and titanium carbide ($Al_2O_3–TiC$), and aluminum oxide reinforced with single-crystal whiskers of silicon carbide. These tools are usually intended for special applications, a discussion of which is beyond our scope.

23.2.5 Synthetic Diamonds and Cubic Boron Nitride

Diamond is the hardest material known (Section 7.6.1). By some measures of hardness, diamond is three to four times as hard as tungsten carbide or aluminum oxide. Since high hardness is one of the desirable properties of a cutting tool, it is natural to think of diamonds for machining and grinding applications. Synthetic diamond-cutting tools are made of sintered polycrystalline diamond (SPD), which dates from the early 1970s. *Sintered polycrystalline diamond* is fabricated by sintering fine-grained diamond crystals under high temperatures and pressures into the desired shape. Little or no binder is used. The crystals have a random orientation and this adds considerable toughness to the SPD tools compared to single crystal diamonds. Tool inserts are typically made by depositing a layer of SPD about 0.5 mm (0.020 in.) thick on the surface of a cemented carbide base. Very small inserts have also been made of 100% SPD.

Applications of diamond cutting tools include high speed machining of nonferrous metals and abrasive nonmetals such as fiberglass, graphite, and wood. Machining of steel, other ferrous metals, and nickel-based alloys with SPD tools is not practical because of the chemical affinity which exists between these metals and carbon (a diamond, after all, is carbon).

Next to diamond, *cubic boron nitride* (Section 7.3.3) is the hardest material known, and its fabrication into cutting tool inserts is basically the same as SPD; that is, coatings on WC–Co inserts. Cubic boron nitride (symbolized cBN) does not react chemically with iron and nickel as SPD does; therefore, the applications of cBN-coated tools are for machining steel and nickel-based alloys. Both SPD and cBN tools are expensive, as one might expect, and the applications must justify the additional tooling cost.

23.3 TOOL GEOMETRY

A cutting tool must possess a shape that is suited to the machining operation. One important way to classify cutting tools is according to the machining process. Thus, we have turning tools, cut-off tools, milling cutters, drill bits, reamers, taps, and many other cutting tools that are named for the operation in which they are used, each with its own tool geometry, in some cases quite unique.

As indicated in Section 21.1, cutting tools can be divided into two categories: single point and multiple cutting edge. Single-point tools are used in turning, boring, shaping, and planing. Multiple-cutting-edge tools are used in drilling, reaming, tapping, milling, broaching, and sawing. Most of these operations in the second category use rotating tools. Many of the principles that apply to single-point tools also apply to other tool types, simply because the mechanism of chip formation is basically the same for all machining operations.

23.3.1 Single-Point Tool Geometry

The general shape of a single-point tool is illustrated in Figure 21.4(a); a more detailed diagram is shown in Figure 23.7. We have previously treated the rake angle of a cutting tool as one parameter. In a single point tool, the orientation of the rake face is defined by two angles, **back rake angle** (α_b) and **side rake angle** (α_s). Together, these angles are influential in determining the direction of chip flow across the rake face. The flank surface of the tool is defined by the **end relief angle** (ERA) and **side relief angle** (SRA). These angles determine the amount of clearance between the tool and the freshly cut work surface. The cutting edge of a single-point tool is divided into two sections, side cutting edge and end cutting edge. These two sections are separated by the tool point, which has a certain radius, called the nose radius. The **side cutting edge angle** (SCEA) determines the entry of the tool into the work and can be used to reduce the sudden force the

FIGURE 23.7 (a) Seven elements of single-point tool geometry; and (b) the tool signature convention that defines the seven elements.

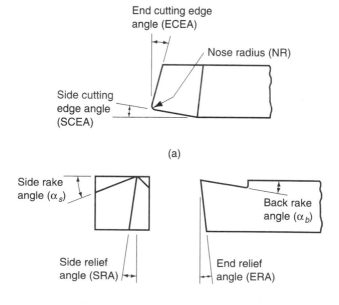

(a)

(b) Tool signature: α_b, α_s, ERA, SRA, ECEA, SCEA, NR

tool experiences as it enters a workpart. *Nose radius* (NR) determines to a large degree the texture of the surface generated in the operation. A very pointed tool (small nose radius) results in very pronounced feed marks on the surface. We return to this issue of surface roughness in machining in Section 24.2.2. *End cutting edge angle* (ECEA) provides a clearance between the trailing edge of the tool and the newly generated work surface, thus reducing rubbing and friction against the surface.

In all, there are seven elements of tool geometry for a single-point tool. When specified in the following order, they are collectively called the *tool geometry signature*: back rake angle, side rake angle, end relief angle, side relief angle, end cutting edge angle, side cutting edge angle, and nose radius. For example, a single-point tool used in turning might have the following signature: 5, 5, 7, 7, 20, 15, 2/64 in.

Chip Breakers   Chip disposal is a problem that is often encountered in turning and other continuous operations. Long stringy chips are often generated, especially when turning ductile materials at high speeds. These chips cause a hazard to the machine operator and to the workpart finish, and they interfere with automatic operation of the turning process. *Chip breakers* are frequently used with single-point tools to force the chips to curl more tightly than they would naturally be inclined to do, thus causing them to fracture. There are two principal forms of chip breaker design commonly used on single-point turning tools, illustrated in Figure 23.8: (a) groove-type chipbreaker designed into the cutting tool itself, and (b) obstruction-type chip breaker designed as an additional device on the rake face of the tool. The chip breaker distance can be adjusted in the obstruction-type device for different cutting conditions.

Effect of Tool Material on Tool Geometry   Our discussion of the Merchant equation (Section 21.3.2) noted that a positive rake angle is generally desirable because it reduces cutting forces, temperature, and power consumption. HSS cutting tools are almost always ground with positive rake angles, typically ranging from +5° to +20°. HSS has good strength and toughness, so that the thinner cross section of the tool created by high positive rake angles does not usually cause a problem with tool breakage. HSS tools are predominantly made of one piece. The heat treatment of high-speed steel can be controlled to provide a hard cutting edge while maintaining a tough inner core. With the development of the very hard tool materials (e.g., cemented carbides and ceramics), changes in tool geometry were required. As a group, these materials have higher hardness and lower toughness than HSS. Also, their shear and tensile strengths are low relative to their compressive strengths, and their properties cannot be manipulated through

FIGURE 23.8   Two methods of chip breaking in single-point tools: (a) groove-type and (b) obstruction-type chipbreakers.

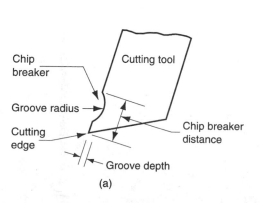

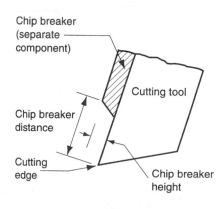

(a)

(b)

heat treatment like those of HSS. Finally, cost per unit weight for these very hard materials is higher than the cost of HSS. These factors have affected cutting tool design for the very hard tool materials in several ways.

First, the very hard materials must be designed with either negative rake or small positive angles. This change tends to load the tool more in compression and less in shear, thus favoring the high compressive strength of these harder materials. Cemented carbides, for example, are used with rake angles typically in the range from $-5°$ to $+10°$. Ceramics have rake angles between $-5°$ and $-15°$. Relief angles are made as small as possible ($5°$ is typical) to provide as much support for the cutting edge as possible.

Another difference is the way in which the cutting edge of the tool is held in position. The alternative ways of holding and presenting the cutting edge for a single point tool are illustrated in Figure 23.9. The geometry of an HSS tool is ground from a solid shank, as shown in part (a) of the figure. The higher cost and differences in properties and processing of the harder tool materials have given rise to the use of inserts that are either brazed or mechanically clamped to a toolholder. Part (b) shows a brazed insert, in which a cemented carbide insert is brazed to a tool shank. The shank is made of tool steel for strength and toughness. Part (c) illustrates one possible design for mechanically clamping an insert in a toolholder. Mechanical clamping is used for cemented carbides, ceramics, and the other hard materials. The significant advantage of the mechanically clamped insert is that each insert contains multiple cutting edges. When an edge wears out, the insert is unclamped, indexed (rotated in the toolholder) to the next edge, and reclamped in the toolholder. When all of the cutting edges are worn, the insert is discarded and replaced.

Inserts   Cutting tool inserts are widely used in machining because they are economical and adaptable to many different types of machining operations: turning, boring, threading, milling, and even drilling. They are available in a variety of shapes and sizes for the variety of cutting situations encountered in practice. A square insert is shown in Figure 23.9(c). Other common shapes used in turning operations are displayed in Figure 23.10. In general, the largest point angle should be selected for strength and economy. Round inserts possess large point angles (and large nose radii) just because of their shape. Inserts with large point angles are inherently stronger and less likely to chip or break during cutting, but they require more power, and there is a greater likelihood of vibration. The economic advantage of round inserts is that they can be indexed multiple times for more cuts per insert. Square inserts present four cutting edges, triangular shapes have three edges, whereas rhombus shapes have only two. Fewer edges are a

FIGURE 23.9   Three ways of holding and presenting the cutting edge for a single-point tool: (a) solid tool, typical of HSS; (b) brazed insert, one way of holding a cemented carbide insert; and (c) mechanically clamped insert, used for cemented carbides, ceramics, and other very hard tool materials.

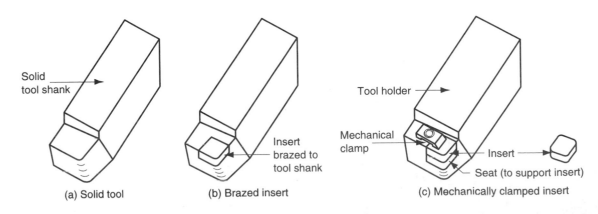

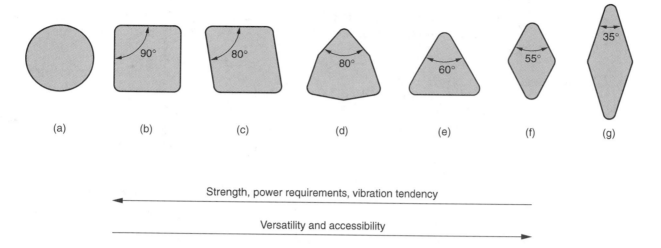

FIGURE 23.10   Common insert shapes: (a) round, (b) square, (c) rhombus with two 80° point angles, (d) hexagon with three 80° point angles, (e) triangle (equilateral), (f) rhombus with two 55° point angles, (g) rhombus with two 35° point angles. Also shown are typical features of the geometry. Strength, power requirements, and tendency for vibration increase as we move to the left; while versatility and accessibility tend to be better with the geometries at the right.

cost disadvantage. If both sides of the insert can be used (e.g., in most negative rake angle applications), then the number of cutting edges is doubled. Rhombus shapes are used (especially with acute point angles) because of their versatility and accessibility when a variety of operations are to be performed. These shapes can be more readily positioned in tight spaces and can be used not only for turning but also for facing, Figure 22.6(a), and contour turning, Figure 22.6(c).

Inserts are usually not made with perfectly sharp cutting edges, because a sharp edge is weaker and fractures more easily, especially for the very hard and brittle tool materials from which inserts are made (cemented carbides, coated carbides, cermets, ceramics, cBN, and diamond). Some kind of shape alteration is commonly performed on the cutting edge at an almost microscopic level. The effect of this ***edge preparation*** is to increase the strength of the cutting edge by providing a more gradual transition between the clearance edge and the rake face of the tool. Three common edge preparations are shown in Figure 23.11: (a) radius or edge rounding, also referred to as *honed edge,* (b) chamfer, and (c) land. For

FIGURE 23.11   Three types of edge preparation that are applied to the cutting edge of an insert: (a) radius, (b) chamfer, (c) land, and (d) perfectly sharp edge (no edge preparation).

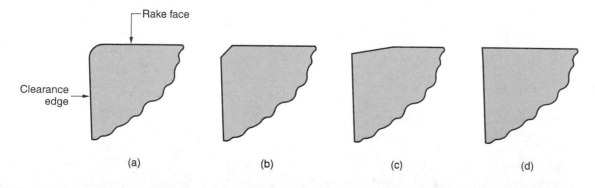

comparison a perfectly sharp cutting edge is shown in (d). The radius in (a) is typically only about 0.025 mm (0.001 in.), and the land in (c) is 15° or 20°. Combinations of these edge preparations are often applied to a single cutting edge to maximize the strengthening effect.

23.3.2 Multiple-Cutting-Edge Tools

Most multiple-cutting-edge tools are used in machining operations in which the tool is rotated. Primary examples are drilling and milling. On the other hand, broaching and some sawing operations (hacksawing and bandsawing) use multiple-cutting-edge tools that operate with a linear motion. Other sawing operations (circular sawing) use rotating saw blades.

Drilling with Twist Drills   Various cutting tools are available for hole making, but the *twist drill* is by far the most common. It comes in diameters ranging from about 0.15 mm (0.006 in.) to as large as 75 mm (3.0 in.). Twist drills are widely used in industry to produce holes rapidly and economically. The standard twist drill geometry is illustrated in Figure 23.12. The body of the drill has two spiral *flutes* (the spiral gives the twist drill its name). The angle of the spiral flutes is called the *helix angle,* a typical value of which is around 30°. While drilling, the flutes act as passageways for extraction of chips from the hole. Although it is desirable for the flute openings to be large to provide maximum clearance for the chips, the body of the drill must be supported over its length. This support is provided by the *web,* which is the thickness of the drill between the flutes.

The point of the twist drill has a conical shape. A typical value for the *point angle* is 118°. The point can be designed in various ways, but the most common design is a *chisel edge,* as in Figure 23.12. Connected to the chisel edge are two cutting edges (sometimes called lips) that lead into the flutes. The portion of each flute adjacent to the cutting edge acts as the rake face of the tool.

The cutting action of the twist drill is complex. The rotation and feeding of the drill bit result in relative motion between the cutting edges and the workpiece to form the chips. The cutting speed along each cutting edge varies as a function of the distance from the axis of rotation. Accordingly, the efficiency of the cutting action varies, being most efficient at the outer diameter of the drill and least efficient at the center. In fact, the relative velocity at the drill point is zero, so no cutting takes place. Instead, the chisel edge of the drill point pushes aside the material at the center as it penetrates into the hole; a large thrust force is required to drive the twist drill forward into the hole. Also, at the beginning of the operation, the rotating chisel edge tends to wander on the surface of the

FIGURE 23.12   Standard geometry of a twist drill.

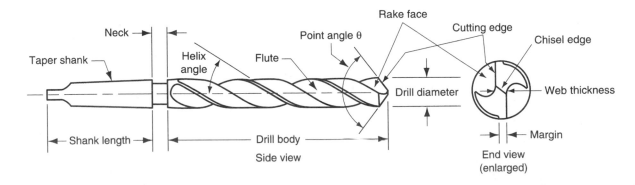

workpart, causing loss of positional accuracy. Various alternative drill point designs have been developed to address this problem.

Chip removal can be a problem in drilling. The cutting action takes place inside the hole, and the flutes must provide sufficient clearance throughout the length of the drill to allow the chips to be extracted from the hole. As the chip is formed it is forced through the flutes to the work surface. Friction makes matters worse in two ways. In addition to the usual friction in metal cutting between the chip and the rake face of the cutting edge, friction also results from rubbing between the outside diameter of the drill bit and the newly formed hole. This increases the temperature of the drill and work. Delivery of cutting fluid to the drill point to reduce the friction and heat is difficult because the chips are flowing in the opposite direction. Because of chip removal and heat, a twist drill is normally limited to a hole depth of about four times its diameter. Some twist drills are designed with internal holes running their lengths, through which cutting fluid can be pumped to the hole near the drill point, thus delivering the fluid directly to the cutting operation. An alternative approach with twist drills that do not have fluid holes is to use a "pecking" procedure during the drilling operation. In this procedure, the drill is periodically withdrawn from the hole to clear the chips before proceeding deeper.

Twist drills are normally made of high-speed steel. The geometry of the drill is fabricated before heat treatment, and then the outer shell of the drill (cutting edges and friction surfaces) is hardened while retaining an inner core that is relatively tough. Grinding is used to sharpen the cutting edges and shape the drill point.

Milling Cutters   Classification of milling cutters is closely associated with the milling operations described in Section 22.3.1. The types of milling cutters include the following:

> *Plain milling cutters*—These are used for peripheral or slab milling. As Figures 22.17(a) and 22.18(a) indicate, they are cylinder-shaped with several rows of teeth. The cutting edges are usually oriented at a helix angle (as in the figure) to reduce impact on entry into the work, and these cutters are called *helical milling cutters.* Tool geometry elements of a plain milling cutter are shown in Figure 23.13.

> *Form milling cutters*—These are peripheral milling cutters in which the cutting edges have a special profile that is to be imparted to the work. An important application is in gear making, in which the form milling cutter is shaped to cut the slots between adjacent gear teeth, thereby leaving the geometry of the gear teeth.

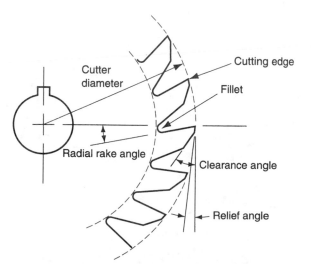

FIGURE 23.13   Tool geometry elements of an 18-tooth plain milling cutter.

> *Face milling cutters*—These are designed with teeth that cut on both the side as well as the periphery of the cutter. Face milling cutters can be made of HSS, as in Figure 22.17(b), or they can be designed to use cemented carbide inserts. Figure 23.14 shows a four-tooth face milling cutter that uses inserts.

> *End milling cutters*—As shown in Figure 22.20(c), an end milling cutter looks like a drill bit, but close inspection indicates that it is designed for primary cutting with its peripheral teeth rather than its end. (A drill bit cuts only on its end as it penetrates into the work.) End mills are designed with square ends, ends with radii, and ball ends. End mills can be used for face milling, profile milling and pocketing, cutting slots, engraving, surface contouring, and die sinking.

Broaches   The terminology and geometry of the broach is illustrated in Figure 23.15. The broach consists of a series of distinct cutting teeth along its length. Feed is accomplished by the increased step between successive teeth on the broach. This feeding action is unique among machining operations, since most operations accomplish feeding by a relative feed motion that is carried out by either the tool or the work. The total material removed in a single pass of the broach is the cumulative result of all the steps in the tool. The speed motion is accomplished by the linear travel of the tool past the work surface. The shape of the cut surface is determined by the contour of the cutting edges on the broach, particularly the final cutting edge. Owing to its complex geometry and the low speeds used in broaching, most broaches are made of HSS. In broaching of certain cast irons, the cutting edges are cemented carbide inserts either brazed or mechanically held in place on the broaching tool.

Saw Blades   For each of the three sawing operations (Section 22.5.3), the saw blades possess certain common features, including tooth form, tooth spacing, and tooth set, as seen in Figure 23.16. *Tooth form* is concerned with the geometry of each cutting tooth. Rake angle, clearance angle, tooth spacing, and other features of geometry are shown in Figure 23.16(a). *Tooth spacing* is the distance between adjacent teeth on the saw blade. This parameter determines the size of the teeth and the size of the gullet between teeth. The gullet allows space for the formation of the chip by the adjacent cutting tooth. Different tooth forms are appropriate for different work materials and cut-

FIGURE 23.14   Tool geometry elements of a four-tooth face milling cutter: (a) side view and (b) bottom view.

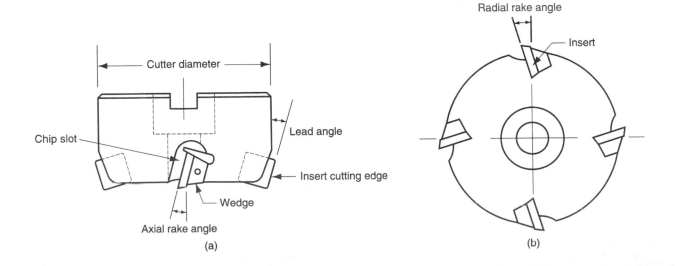

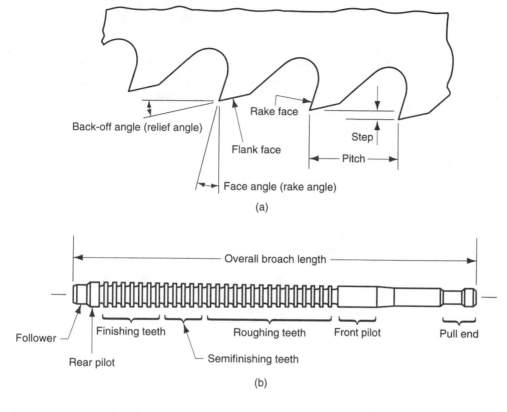

(a)

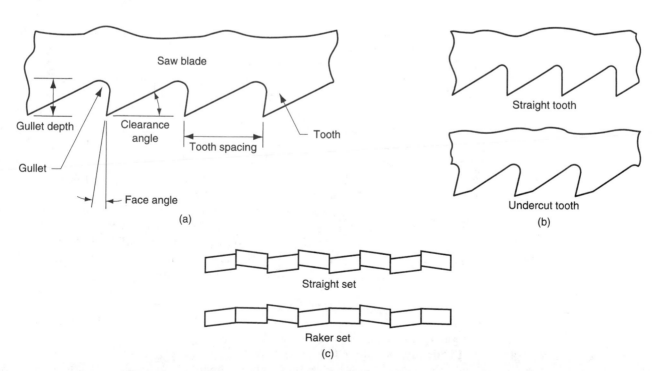

(b)

FIGURE 23.15 The broach: (a) terminology of the tooth geometry, and (b) a typical broach used for internal broaching.

FIGURE 23.16 Features of saw blades: (a) nomenclature for saw blade geometries, (b) two common tooth forms, and (c) two types of tooth set.

557

ting situations. Two forms commonly used in hacksaw and bandsaw blades are shown in Figure 23.16(b). The ***tooth set*** permits the kerf cut by the saw blade to be wider than the width of the blade itself; otherwise the blade would bind against the walls of the slit made by the saw. Two common tooth sets are illustrated in Figure 23.16(c).

23.4 CUTTING FLUIDS

A ***cutting fluid*** is any liquid or gas that is applied directly to the machining operation to improve cutting performance. Cutting fluids address two main problems: (1) heat generation at the shear zone and friction zone, and (2) friction at the tool–chip and tool–work interfaces. In addition to removing heat and reducing friction, cutting fluids wash away chips (especially in grinding and milling), reduce the temperature of the workpart for easier handling, reduce cutting forces and power requirements, improve dimensional stability of the workpart, and improve surface finish.

23.4.1 Types of Cutting Fluids

A variety of cutting fluids are commercially available. It is appropriate to discuss them first according to function and then to classify them according to chemical formulation.

Cutting Fluid Functions There are two general categories of cutting fluids, corresponding to the two main problems they are designed to address: coolants and lubricants. ***Coolants*** are cutting fluids designed to reduce the effects of heat in the machining operation. They have a limited effect on the amount of heat energy generated in cutting; instead, they carry away the heat that is generated, thereby reducing the temperature of tool and workpiece. This helps to prolong the life of the cutting tool. The capacity of a cutting fluid to reduce temperatures in machining depends on its thermal properties. Specific heat and thermal conductivity are the most important properties (Section 4.2.1). Water has high specific heat and thermal conductivity relative to other liquids, which is why water is used as the base in coolant-type cutting fluids. These properties allow the coolant to draw heat away from the operation, thereby reducing the temperature of the cutting tool.

Coolant-type cutting fluids seem to be most effective at relatively high cutting speeds where heat generation and high temperatures are problems. They are most effective on tool materials that are most susceptible to temperature failures, such as HSS, and are used frequently in turning and milling operations where large amounts of heat are generated. Coolants are usually water-based solutions or water emulsions, since water has thermal properties that are ideally suited for these cutting fluids.

Lubricants are usually oil-based fluids (since oils possess good lubricating qualities) formulated to reduce friction at the tool–chip and tool–work interfaces. Lubricant cutting fluids operate by ***extreme pressure lubrication,*** a special form of lubrication that involves formation of thin solid salt layers on the hot, clean metal surfaces through chemical reaction with the lubricant. Compounds of sulfur, chlorine, and phosphorous in the lubricant cause the formation of these surface layers, which act to separate the two metal surfaces (i.e., chip and tool). These extreme pressure films are significantly more effective in reducing friction in metal cutting than conventional lubrication, which is based on the presence of liquid films between the two surfaces.

Lubricant-type cutting fluids are most effective at lower cutting speeds. They tend to lose their effectiveness at high speeds—above about 120 m/min (400 ft/min)— because the motion of the chip at these speeds prevents the cutting fluid from reaching the tool–chip interface. In addition, high cutting temperatures at these speeds cause the oils to vaporize before they can lubricate. Machining operations such as drilling and tapping usually benefit from lubricants. In these operations, built-up edge formation is retarded, and torque on the tool is reduced.

Although the principal purpose of a lubricant is to reduce friction, it also reduces the temperature in the operation through several mechanisms. First, the specific heat and thermal conductivity of the lubricant help to remove heat from the operation, thereby reducing temperatures. Second, because friction is reduced, the heat generated from friction is also reduced. Third, a lower coefficient of friction means a lower friction angle. According to Merchant's equation, Eq. (21.16), a lower friction angle causes the shear plane angle to increase, hence reducing the amount of heat energy generated in the shear zone.

There is typically an overlapping effect between the two types of cutting fluids. Coolants are formulated with ingredients that help reduce friction. And lubricants have thermal properties that, although not as good as those of water, act to remove heat from the cutting operation. Cutting fluids (both coolants and lubricants) manifest their effect on the Taylor tool life equation through higher C values. Increases of 10 to 40% are typical. The slope n is not significantly affected.

Chemical Formulation of Cutting Fluids There are four categories of cutting fluids according to chemical formulation: (1) cutting oils, (2) emulsified oils, (3) semi-chemical fluids, and (4) chemical fluids. All of these cutting fluids provide both coolant and lubricating functions. The cutting oils are most effective as lubricants, while the other three categories are more effective as coolants because they are primarily water.

Cutting oils are based on oil derived from petroleum, animal, marine, or vegetable origin. Mineral oils (petroleum based) are the principal type because of their abundance and generally desirable lubricating characteristics. To achieve maximum lubricity, several types of oils are often combined in the same fluid. Chemical additives are also mixed with the oils to increase lubricating qualities. These additives contain compounds of sulfur, chlorine, and phosphorous, and are designed to react chemically with the chip and tool surfaces to form solid films (extreme pressure lubrication) that help to avoid metal-to-metal contact between the two.

Emulsified oils consist of oil droplets suspended in water. The fluid is made by blending oil (usually mineral oil) in water using an emulsifying agent to promote blending and stability of the emulsion. A typical ratio of water to oil is 30 : 1. Chemical additives based on sulfur, chlorine, and phosphorous are often used to promote extreme pressure lubrication. Because they contain both oil and water, the emulsified oils combine cooling and lubricating qualities in one cutting fluid.

Chemical fluids are chemicals in a water solution rather than oils in emulsion. The dissolved chemicals include compounds of sulfur, chlorine, and phosphorous, plus wetting agents. The chemicals are intended to provide some degree of lubrication to the solution. Chemical fluids provide good coolant qualities, but their lubricating qualities are less than the other cutting fluid types. *Semi-chemical fluids* have small amounts of emulsified oil added to increase the lubricating characteristics of the cutting fluid. In effect, they are a hybrid class between chemical fluids and emulsified oils.

23.4.2 Application of Cutting Fluids

Cutting fluids are applied to machining operations in various ways. In this section we consider these application techniques. We also consider the problem of cutting-fluid contamination and what steps can be taken to address this problem.

Application Methods   The most common method is *flooding,* sometimes called flood-cooling because it is generally used with coolant-type cutting fluids. In flooding, a steady stream of fluid is directed at the tool–work or tool–chip interface of the machining operation. A second method of delivery is *mist application,* primarily used for water-based cutting fluids. In this method the fluid is directed at the operation in the form of a high-speed mist carried by a pressurized air stream. Mist application is generally not as effective as flooding in cooling the tool. However, because of the high-velocity air stream, mist application may be more effective in delivering the cutting fluid to areas that are difficult to access by conventional flooding.

    Manual application by means of a squirt can or paint brush is sometimes used for applying lubricants in tapping and other operations where cutting speeds are low and friction is a problem. It is generally not preferred by most production machine shops because of its variability in application.

Cutting Fluid Filtration and Dry Machining   Cutting fluids become contaminated over time with a variety of foreign substances such as tramp oil (machine oil, hydraulic fluid, etc.), garbage (cigarette butts, food, etc.), small chips, molds, fungi, and bacteria. In addition to causing odors and health hazards, contaminated cutting fluids do not perform their lubricating function as well. Alternative ways of dealing with this problem are to (1) replace the cutting fluid at regular and frequent intervals (perhaps twice per month); (2) use a filtration system to continuously or periodically clean the fluid; or (3) machine without cutting fluids (dry machining). Because of growing concern about pollution and associated legislation, fluid disposal has become both costly and contrary to the general public welfare.

    Filtration systems are being installed in numerous machine shops today to solve the contamination problem. Advantages of these systems include: (1) prolonged cutting fluid life between changes—instead of replacing the fluid once or twice per month, coolant lives of one year have been reported; (2) reduced fluid disposal cost, since disposal is much less frequent when a filter is used; (3) cleaner cutting fluid for better working environment and reduced health hazards; (4) lower machine tool maintenance; and (5) longer tool life. There are various types of filtration systems for filtering cutting fluids. For the interested reader, these filtration systems are discussed in Reference [4].

    The third alternative is called *dry machining,* meaning that no cutting fluid is used. Dry machining avoids the problems of cutting fluid contamination, disposal, and filtration, but can lead to problems of its own: (1) overheating the tool; (2) operating at lower cutting speeds and production rates to prolong tool life; and (3) absence of chip removal benefits in grinding and milling. Cutting tool producers have developed certain grades of carbides and coated carbides for use in dry machining.

REFERENCES

[1] Brierley, R. G., and Siekman, H. J. *Machining Principles and Cost Control.* McGraw-Hill Book Company, New York, 1964.

[2] Cook, N. H. "Tool Wear and Tool Life." *ASME Transactions, J. Engrg. for Industry.* Vol. 95, November 1973, pp. 931–938.

[3] Davis, J. R. (ed.). *ASM Specialty Handbook ® Tool Materials.* ASM International, Materials Park, Ohio, 1995.

[4] Drozda, T. J., and Wick, C. (eds.). *Tool and Manu-facturing Engineers Handbook.* 4th ed. Volume I: *Machining.* Society of Manufacturing Engineers, Dearborn, Michigan, 1983.

[5] Esford, D. "Ceramics Take a Turn." *Cutting Tool Engineering.* Vol. 52, No. 7, July 2000, pp. 40–46.

[6] Graham, D. "Dry Out." *Cutting Tool Engineering.* Vol. 52, No. 3, March 2000, pp. 56–65.

[7] H & W Systems, marketing and technical literature, Gastonia, North Carolina.

[8] Hoffman, E. G. (ed.). *Fundamentals of Tool Design.*

2nd ed. Society of Manufacturing Engineers, Dearborn, Michigan, 1984.

[9] Koelsch, J. R. "Beyond TiN." *Manufacturing Engineering.* October 1992, pp. 27–32.

[10] Krar, S. F., and Ratterman, E. *Superabrasives: Grinding and Machining with CBN and Diamond.* McGraw-Hill, Inc., New York, 1990.

[11] Liebhold, P. "The History of Tools." *Cutting Tool Engineer.* June 1989, pp. 137–38.

[12] *Machining Data Handbook.* 3rd ed. Vols. I and II, Metcut Research Associates, Inc., Cincinnati, Ohio, 1980.

[13] *Metals Handbook.* 9th ed. Vol. 16: *Machining,* ASM International, Metals Park, Ohio, 1989.

[14] *Modern Metal Cutting.* AB Sandvik Coromant, Sandvik, Sweden, 1994.

[15] Owen, J. V. "Are Cermets for Real?" *Manufacturing Engineering.* October 1991, pp. 28–31.

[16] Pfouts, W. R. "Cutting Edge Coatings." *Manufacturing Engineering.* Vol. 125, No. 1, July 2000, pp. 98–107.

[17] Schey, J. A. *Introduction to Manufacturing Processes.* 3rd ed. McGraw-Hill Book Company, New York, 1999.

[18] Shaw, M. C. *Metal Cutting Principles.* Oxford University Press, Inc., Oxford, England, 1997.

[19] Tlusty, J. *Manufacturing Processes and Equipment.* Prentice Hall, Upper Saddle River, New Jersey, 2000.

REVIEW QUESTIONS

23.1. What are the two principal aspects of cutting tool technology?

23.2. Name the three modes of tool failure in machining.

23.3. What are the two principal locations on a cutting tool where tool wear occurs?

23.4. Identify the mechanisms by which cutting tools wear during machining.

23.5. What is the physical interpretation of the parameter C in the Taylor tool life equation?

23.6. In addition to cutting speed, what other cutting variables are included in the expanded version of the Taylor tool life equation?

23.7. What are some of the tool life criteria used in production machining operations?

23.8. Identify three desirable properties of a cutting tool material.

23.9. What are the principal alloying ingredients in high-speed steel?

23.10. What is the difference in ingredients between steel-cutting grades and nonsteel-cutting grades of cemented carbides?

23.11. Identify some of the common compounds that form the thin coatings on the surfaceof coated carbide inserts.

23.12. Name the seven elements of tool geometry for a single-point cutting tool.

23.13. Why are ceramic cutting tools generally designed with negative rake angles?

23.14. Identify the three ways of holding and presenting the cutting edge for single-point tools.

23.15. Name the two main categories of cutting fluid according to function.

23.16. Name the four types of cutting fluid according to chemistry.

23.17. What is the principal lubricating mechanism by which cutting fluids work?

23.18. What are the methods by which cutting fluids are applied in a machining operation?

23.19. Why are cutting-fluid filter systems becoming more common, and what are their advantages?

23.20. Dry machining is being considered by machine shops because of certain problems inherent in the use of cutting fluids. (a) What are those problems associated with the use of cutting fluids? (b) What are the new problems introduced by machining dry?

MULTIPLE CHOICE QUIZ

There is a total of 18 correct answers in the following multiple choice questions (some questions have multiple answers that are correct). To attain a perfect score on the quiz, all correct answers must be given, since each correct answer is worth 1 point. For each question, each omitted answer or wrong answer reduces the score by 1 point, and each additional answer beyond the number of answers required reduces the score by 1 point. Percentage score on the quiz is based on the total number of correct answers.

23.1. Of the following cutting conditions, which one has the greatest effect on tool wear? (a) cutting speed, (b) depth of cut, or (c) feed.

23.2. As an alloying ingredient in high-speed steel, tungsten serves which of the following functions (more than one answer)? (a) forms hard carbides to resist abrasion, (b) improves strength and hardness, (c) increases corrosion resistance, and (d) increases hot hardness.

23.3. Cast cobalt alloys typically contain which of the following main ingredients (more than one answer)? (a) aluminum, (b) cobalt, (c) chromium, (d) nickel, and (e) tungsten.

23.4. Which of the following is not a common ingredient in cemented carbide cutting tools (more than one answer)? (a) Al_2O_3, (b) Co, (c) CrC, (d) TiC, and (e) WC.

23.5. An increase in cobalt content has which of the following effects on WC–Co cemented carbides (one best answer)? (a) decreases transverse rupture strength, (b) increases hardness, (c) increases toughness.

23.6. Steel cutting grades of cemented carbide are typically characterized by which of the following ingredients (more than one answer)? (a) Co, (b) Ni, (c) TiC, (d) TaC, and (e) WC.

23.7. If you had to select a cemented carbide for an application involving the finish turning of steel, which of the following grades would you select (one best answer)? (a) K01, (b) K40, (c) P01, (d) P50.

23.8. Which of the following processes are used to provide the thin coatings on the surface of coated carbide inserts (more than one answer)? (a) chemical vapor deposition, (b) electroplating, (c) physical vapor deposition, or (d) pressing and sintering.

23.9. Which of the following materials has the highest hardness? (a) aluminum oxide, (b) cubic boron nitride, (c) high speed steel, (d) titanium carbide, or (e) tungsten carbide.

23.10. Which of the following are the two main functions of a cutting fluid in machining (two answers only)? (a) improve surface finish on the workpiece, (b) reduce forces and power, (c) reduce friction at the tool-chip interface, (d) remove heat from the process, and (e) wash away chips.

PROBLEMS

Tool Life and the Taylor Equation

23.1. The following flank wear data were collected in a series of turning tests using a coated carbide tool on hardened alloy steel. The feed rate was 0.30 mm/rev, and the depth was 4.0 mm. The last wear data value in each column is when final tool failure occurred. (a) On a single piece of linear graph paper, plot flank wear as a function of time. Using 0.75 mm of flank wear as the criterion of tool failure, determine the tool lives for the two cutting speeds. (b) On a piece of natural log-log paper, plot your results determined in the previous part. From the plot, determine the values of n and C in the Taylor tool life equation. (c) As a comparison, calculate the values of n and C in the Taylor equation solving simultaneous equations. Are the resulting n and C values the same?

Cutting time, min.	1	3	5	7	9	11	13	15	20	25
Flank wear, mm. at v = 125 m/min	0.12	0.20	0.27	0.33	0.40	0.45	0.50	0.58	0.73	0.97
Flank wear, mm. at v = 165 m/min	0.22	0.35	0.47	0.57	0.70	0.80	0.99			

23.2. Solve Problem 23.1 except that the tool life criterion is 0.50 mm of flank land wear.

23.3. Tool life tests on a lathe have resulted in the following data: (1) v = 350 ft/min, T = 7 min; (2) v = 250 ft/min, T = 50 min. (a) Determine the parameters n and C in the Taylor tool life equation. (b) Based on the n and C values, what is the likely tool material used in this operation? (c) Using your equation, compute the tool life that corresponds to a cutting speed v = 300 ft/min. (d) Compute the cutting speed that corresponds to a tool life T = 10 min.

23.4. Tool life tests in turning yield the following data: (1) v = 100 m/min, T = 10 min; (2) v = 75 m/min, T = 30 min. (a) Determine the n and C values in the Taylor tool life equation. Based on your equation, compute (b) the tool life for a speed of 90 m/min, and (c) the speed corresponding to a tool life of 20 min.

23.5. Turning tests have resulted in 1-min tool life for a cutting speed v = 4.0 m/s and a 20-min tool life at a speed v = 2.0 m/s. (a) Find the n and C values in the Taylor tool life equation. (b) Project how long the tool would last at a speed v = 1.0 m/s.

23.6. In a production turning operation, the workpart is 125 mm in diameter and 300 mm long. A feed rate of 0.225 mm/rev is used in the operation. If cutting speed = 3.0 m/s, the tool must be changed every five workparts; but if cutting speed = 2.0 m/s, the tool can be used to produce 25 pieces between tool changes. Determine the Taylor tool life equation for this job.

23.7. For the tool life plot of Figure 23.5, show that the middle data point (v = 130 m/min, T = 12 min) is consistent with the Taylor equation determined in Example Problem 23.1.

23.8. In the tool wear plots of Figure 23.4, complete failure of the cutting tool is indicated by the end of each wear curve. Using complete failure as the criterion of tool life instead of 0.50 mm flank wear, the resulting data would be (1) $v = 160$ m/min, $T = 5.75$ min; (2) $v = 130$ m/min, $T = 14.25$ min; (3) $v = 100$ m/min, $T = 47$ min. Determine the parameters n and C in the Taylor tool life equation for this data.

23.9. The Taylor equation for a certain set of test conditions is $vT^{.25} = 1000$, where the U.S. customary units are used: ft/min for v and min for T. Convert this equation to the equivalent Taylor equation in the International System of units (metric), where v is in m/sec and T is in sec. Validate the metric equation using a tool life = 16 min. That is, compute the corresponding cutting speeds in ft/min and m/sec using the two equations.

23.10. A series of turning tests are performed to determine the parameters n, m, and K in the expanded version of the Taylor equation, Eq. (23.4). The following data were obtained during the tests: (1) $v = 2.0$ m/s, $f = 0.20$ mm/rev, $T = 12$ min; (2) $v = 1.5$ m/s, $f = 0.20$ mm/rev, $T = 40$ min; and (3) $v = 2.0$ m/s, $f = 0.3$ mm/rev, $T = 10$ min. (a) Determine n, m, and K. (b) Using your equation, compute the tool life when $v = 1.5$ m/s and $f = 0.3$ mm/rev.

23.11. Eq. (23.4) in the text relates tool life to speed and feed. In a series of turning tests conducted to determine the parameters n, m, and K, the following data were collected: (1) $v = 400$ ft/min, $f = 0.010$ in./rev, $T = 10$ min; (2) $v = 300$ ft/min, $f = 0.010$ in./rev, $T = 35$ min; and (3) $v = 400$ ft/min, $f = 0.015$ in./rev, $T = 8$ min. Determine n, m, and K. What is the physical interpretation of the constant K?

23.12. The n and C values in Table 23.2 are based on a feed rate of 0.25 mm/rev and a depth of cut = 4.0 mm. Determine how many cubic mm of steel would be removed for each of the following tool materials, if a 10-min tool life were required in each case: (a) plain carbon steel, (b) high-speed steel, (c) cemented carbide, (d) ceramic, and (e) coated carbide.

23.13. A drilling operation is performed in which 0.5 in. diameter holes are drilled through cast iron plates that are 1.0 in. thick. Sample holes have been drilled to determine the tool life at two cutting speeds. At 80 surface ft/min, the tool lasted for exactly 50 holes. At 120 surface ft/min, the tool lasted for exactly five holes. The feed rate of the drill was 0.003 in./rev. (Ignore effects of drill entrance and exit from the hole. Consider the depth of cut to be exactly 1.00 in., corresponding to the plate thickness.) Determine the values of n and C in the Taylor tool life equation for the above sample data, where cutting speed v is expressed in ft/min, and tool life T is expressed in min.

23.14. The outside diameter of a cylinder made of titanium alloy is to be turned. The starting diameter = 500 mm, and the length = 1000 mm. Cutting conditions are: $f = 0.4$ mm/rev, and $d = 3.0$ mm. The cut will be made with a cemented carbide cutting tool whose Taylor tool life parameters are: $n = 0.23$ and $C = 400$. Units for the Taylor equation are min for tool life and m/min for cutting speed. Compute the cutting speed that will allow the tool life to be just equal to the cutting time for this part.

23.15. The outside diameter of a roll for a steel rolling mill is to be turned. In the final pass, the starting diameter = 26.25 in. and the length = 48.0 in. The cutting conditions will be feed = 0.0125 in./rev, and depth of cut = 0.125 in. A cemented carbide cutting tool is to be used and the parameters of the Taylor tool life equation for this setup are: $n = 0.25$ and $C = 1300$. Units for the Taylor equation are min for tool life and ft/min for cutting speed. It is desirable to operate at a cutting speed so that the tool will not need to be changed during the cut. Determine the cutting speed that will make the tool life equal to the time required to complete the turning operation.

Tooling Applications

23.16. A certain machine shop uses a limited number of cemented carbide grades in its operations. These grades are listed below by chemical composition. (a) Which grade should be used for finish turning of unhardened steel? (b) Which grade should be used for rough milling of aluminum? (c) Which grade should be used for finish turning of brass? (d) Which of the grades listed would be suitable for machining cast iron? For each case, explain your recommendation.

Grade	%WC	%Co	%TiC
1	95	5	0
2	82	4	14
3	80	10	10
4	89	11	0

23.17. A turning operation is performed on a steel shaft with diameter = 5.0 in. and length = 32 in. A slot or keyway has been milled along its entire length. The turning operation reduces the shaft diameter. For each of the following tool materials, indicate whether or not it is a reasonable candidate to use in the operation: (a) plain carbon steel, (b) high-speed steel, (c) cemented carbide, (d) ceramic, and (e) sintered polycrystalline diamond. For each material that is not a good candidate, give the reason why it is not.

Cutting Fluids

23.18. In a turning operation using high-speed steel tooling, a cutting speed $v = 90$ m/min is used. The Taylor tool life equation has parameters $n = 0.120$ and $C = 130$ (m/min) when the operation is conducted dry. When a coolant is used in the operation, the value of C is increased by 10%. Determine the percent increase in tool life that results if the cutting speed is maintained at $v = 90$ m/min.

23.19. A production turning operation on a steel workbar normally operates at a cutting speed of 125 ft/min using high-speed steel tooling with no cutting fluid. The appropriate n and C values in the Taylor equation are given in Table 23.2. It has been found that the use of a coolant type cutting fluid will allow an increase of 25 ft/min in the speed without any effect on tool life. If it can be assumed that the effect of the cutting fluid is simply to increase the constant C by 25, what would be the increase in tool life if the original cutting speed of 125 ft/min were used in the operation?

23.20. A high-speed steel 6.0 mm twist drill is being used in a production drilling operation on mild steel. A cutting oil is applied by the operator by brushing the lubricant onto the drill point and flutes prior to each hole. The cutting conditions are: speed = 25 m/min, and feed = 0.10 mm/rev, and hole depth = 40 mm. The foreman says that the "speed and feed are right out of the handbook" for this work material. Nevertheless, he says, "The chips are clogging in the flutes, resulting in friction heat, and the drill bit is failing prematurely due to overheating." What's the problem? What do you recommend to solve it?

24 ECONOMIC AND PRODUCT DESIGN CONSIDERATIONS IN MACHINING

CHAPTER CONTENTS

24.1   Machinability
24.2   Tolerances and Surface Finish
        24.2.1   Tolerances in Machining
        24.2.2   Surface Finish in Machining
24.3   Selection of Cutting Conditions
        24.3.1   Selecting Feed and Depth of Cut
        24.3.2   Optimizing Cutting Speed
24.4   Product Design Considerations in Machining

In this chapter, we conclude our coverage of traditional machining technology by discussing several remaining topics. The first topic is machinability, which is concerned with the properties of work materials used in machining and how these properties affect machining performance. The second topic is concerned with the tolerances and surface finishes (Chapter 5) that can be expected in machining processes. Third, we consider how to select cutting conditions (speed, feed, and depth of cut) in a machining operation. This selection determines to a large extent the economic success of a given operation. Finally, we provide some guidelines for product designers to consider when they design parts that are to be produced by machining.

24.1 MACHINABILITY

Properties of the work material have a significant influence on the success of the machining operation. These properties and other characteristics of the work are often summarized in the term machinability. *Machinability* denotes the relative ease with which a material (usually a metal) can be machined using appropriate tooling and cutting conditions.

There are various criteria used to evaluate machinability, the most important of which are (1) tool life, (2) forces and power, (3) surface finish, and (4) ease of chip disposal. Although machinability generally refers to the work material, it should be recognized that machining performance depends on more than just material. The type of machining operation, tooling, and cutting conditions, are also important factors as well as material properties. In addition, the machinability criterion is also a source of variation. One material may yield a longer tool life while another material provides a better surface finish. All of these factors make evaluation of machinability difficult.

Machinability testing usually involves a comparison of work materials. The machining performance of a test material is measured relative to that of a base (standard) material. Possible measures of performance in machinability testing include (1) tool life, (2) tool wear, (3) cutting force, (4) power in the operation, (5) cutting temperature, and (6) material removal rate under standard test conditions. The relative performance is expressed as an index number, called the machinability rating (MR). The base material used as the standard is given a machinability rating of 1.00. B1112 steel is often used as the base material in machinability comparisons. Materials that are easier to machine than the base have ratings greater than 1.00, and materials that are more difficult to machine have ratings less than 1.00. Machinability ratings are often expressed as percentages rather than index numbers. Let us illustrate how a machinability rating might be determined using a tool life test as the basis of comparison.

EXAMPLE 24.1 Machinability testing.

A series of tool life tests are conducted on two work materials under identical cutting conditions, varying only speed in the test procedure. The first material, defined as the base material, yields a Taylor tool life equation $vT^{0.28} = 350$, and the other material (test material) yields a Taylor equation $vT^{0.27} = 440$, where speed is in m/min and tool life is in min. Determine the machinability rating of the test material using the cutting speed that provides a 60-min tool life as the basis of comparison. This speed is denoted by v_{60}.

Solution: The base material has a machinability rating = 1.0. Its v_{60} value can be determined from the Taylor tool life equation as follows:

$$v_{60} = (350/60^{0.28}) = 111 \text{ m/min}$$

The cutting speed at a 60-min tool life for the test material is determined similarly:

$$v_{60} = (440/60^{0.27}) = 146 \text{ m/min}$$

Accordingly, the machinability rating can be calculated as

$$MR \text{ (for the test material)} = \frac{146}{111} = 1.31 \text{ (or } 131\%)$$

Many work material factors affect machining performance. Mechanical properties of a work material affecting machinability include hardness and strength. As hardness increases, abrasive wear of the tool increases so that tool life is reduced. Strength is usually indicated as tensile strength, even though machining involves shear stresses. Of course, shear strength and tensile strength are correlated. As work material strength increases, cutting forces, specific energy, and cutting temperature increase, making the material more difficult to machine. On the other hand, very low hardness can be detrimental to machining performance. For example, low carbon steel, which has relatively low hardness, is often too ductile to machine well. High ductility causes tearing of the metal as the chip is formed, resulting in poor finish, and problems with chip disposal. Cold drawing is often used on low carbon bars to increase surface hardness and promote chip-breaking during cutting.

A metal's chemistry has an important effect on properties; and in some cases, chemistry affects the wear mechanisms that act on the tool material. Through these relationships, chemistry affects machinability. Carbon content has a significant effect on the properties of steel. As carbon is increased, strength and hardness of the steel increases; this reduces machining performance. Many alloying elements added to steel to enhance properties are detrimental to machinability. Chromium, molybdenum, and tungsten form carbides in steel, which increase tool wear and reduce machinability. Manganese and nickel add strength and toughness to steel, which reduce machinability. Certain elements can be added to steel to improve machining performance, such as lead, sulfur, and phosphorous. The additives have the effect of reducing the coefficient of friction between the tool and chip, thereby reducing forces, temperature, and built-up edge formation. Better tool life and surface finish result from these effects. Steel alloys formulated to improve machinability are referred to as *free machining steels.*

Similar relationships exist for other work materials. Table 24.1 lists selected metals together with their approximate machinability ratings. These ratings are intended to summarize the machining performance of the materials.

TABLE 24.1   Approximate values of Brinell hardness and typical machinability ratings for selected work materials.

Work Material	Brinell Hardness	Machinability Rating[a]	Work Material	Brinell Hardness	Machinability Rating[a]
Base steel: B1112	180–220	1.00	Tool steel (unhardened)	200–250	0.30
Low carbon steel:	130–170	0.50	Cast iron		
C1008, C1010, C1015			Soft	60	0.70
Medium carbon steel:	140–210	0.65	Medium hardness	200	0.55
C1020, C1025, C1030			Hard	230	0.40
High carbon steel:	180–230	0.55	Super alloys		
C1040, C1045, C1050			Inconel	240–260	0.30
Alloy steels[b]			Inconel X	350–370	0.15
1320, 1330, 3130, 3140	170–230	0.55	Waspalloy	250–280	0.12
4130	180–200	0.65	Titanium		
4140	190–210	0.55	Plain	160	0.30
4340	200–230	0.45	Alloys	220–280	0.20
4340 (casting)	250–300	0.25	Aluminum		
6120, 6130, 6140	180–230	0.50	2-S, 11-S, 17-S	soft	5.00[c]
8620, 8630	190–200	0.60	Aluminum alloys (soft)	soft	2.00[d]
B1113	170–220	1.35	Aluminum alloys (hard)	hard	1.25[d]
Free machining steels	160–220	1.50	Copper	soft	0.60
Stainless steel			Brass	soft	2.00[d]
301, 302	170–190	0.50	Bronze	soft	0.65[d]
304	160–170	0.40			
316, 317	190–200	0.35			
403	190–210	0.55			
416	190–210	0.90			

Values are estimated average values based on [1], [3], [4], [7], and other sources. Ratings represent relative cutting speeds for a given tool life (see Example 24.1).

[a]Machinability ratings are often expressed as percents (index number $\times$ 100%).

[b]Our list of alloy steels is by no means complete. We have attempted to include some of the more common alloys and to indicate the range of machinability ratings among these steels.

[c]The machinability of aluminum varies widely. It is expressed here as $MR = 5.00$, but the range is probably from 3.00 to 10.00 or more.

[d]Aluminum alloys, brasses, and bronzes also vary significantly in machining performance. Different grades have different machinability ratings. For each case, we have attempted to reduce the variation to a single average value to indicate relative performance with other work materials.

24.2 TOLERANCES AND SURFACE FINISH

Machining operations are used to produce parts with defined geometries to tolerances and surface finishes specified by the product designer. In this section we examine these issues of tolerance and surface finish in machining. Table 24.2 lists typical tolerances and surface finishes that can be obtained in the various machining operations.

24.2.1 Tolerances in Machining

There is variability in any manufacturing process, and tolerances are used to set permissible limits on this variability (Section 5.1.1). Machining is often selected when tolerances are close, because machining operations provide high accuracy relative to most other shape making processes. Table 24.2 indicates typical tolerances that can be achieved for most machining operations examined in Chapter 22. It should be mentioned that the values in this tabulation represent ideal conditions, yet conditions that are readily achievable in a modern factory. If the machine tool is old and worn, process variability will likely be greater than the ideal, and these tolerances would be difficult to maintain. On the other hand, newer machine tools can achieve closer tolerances than those listed.

Closer tolerances usually mean higher costs. For example, if the product designer specifies a tolerance of ±0.10 mm on a hole diameter of 6.0 mm, this tolerance could be achieved by a drilling operation, according to Table 24.2. However, if the designer specifies a tolerance of ±0.025 mm, then an additional reaming operation is needed to satisfy this tighter requirement. The general relationship between tolerance and manufacturing costs is depicted in Figure 43.1.

This is not to suggest that looser tolerances are always good. It often happens that closer tolerances and lower variability in the machining of the individual components will lead to fewer problems in assembly, final product testing, field service, and customer acceptance. Although these costs are not always as easy to quantify as direct manufacturing costs, they can nevertheless be significant. Tighter tolerances that push the factory to achieve better control over its manufacturing processes may lead to lower total operating costs for the company over the long run.

TABLE 24.2 Typical tolerances and surface finishes (arithmetic average) achievable in machining operations.

Machining Operation	Tolerance Capability —Typical mm	(in.)	Surface Finish (AA)—Typical Best μm	μ-in.	Machining Operation	Tolerance Capability —Typical mm	(in.)	Surface Finish (AA)—Typical Best μm	μ-in.
Turning, boring			0.8	(32)	Reaming			0.4	(16)
Diameter D < 25 mm	±0.025	(±0.001)			Diameter D < 12 mm	±0.025	(±0.001)		
25 mm < D < 50 mm	±0.05	(±0.002)			12 mm < D < 25 mm	±0.05	(±0.002)		
Diameter D > 50 mm	±0.075	(±0.003)			Diameter D > 25 mm	±0.075	(±0.003)		
Drilling*			0.8	(32)	Milling			0.4	(16)
Diameter D < 2.5 mm	±0.05	(±0.002)			Peripheral	±0.025	(±0.001)		
2.5 mm < D < 6 mm	±0.075	(±0.003)			Face	±0.025	(±0.001)		
6 mm < D < 12 mm	±0.10	(±0.004)			End	±0.05	(±0.002)		
12 mm < D < 25 mm	±0.125	(±0.005)			Shaping, slotting	±0.025	(±0.001)	1.6	(63)
Diameter D > 25 mm	±0.20	(±0.008)			Planing	±0.075	(±0.003)	1.6	(63)
Broaching	±0.025	(±0.001)	0.2	(8)	Sawing	±0.50	(±0.02)	6.0	(250)

*Drilling tolerances are typically expressed as biased bilateral tolerances (e.g., +0.010 / −0.002). Values in this table are expressed as closest bilateral tolerance (e.g., ±0.006).

Compiled from various sources including [4], [6], [7], [8], [12], and [16].

24.2.2 Surface Finish in Machining

Since machining is often the manufacturing process that determines the final geometry and dimensions of the part, it is also the process that determines the part's surface texture (Section 5.2.2). Table 24.2 lists typical surface finishes that can be achieved in machining operations. These finishes should be readily achievable by modern, well-maintained machine tools.

Let us examine how surface finish is determined in a machining operation. The roughness of a machined surface depends on many factors that can be grouped as follows: (1) geometric factors, (2) work material factors, and (3) vibration and machine tool factors. Our discussion of surface finish in this section exams these factors and their effects.

Geometric Factors   These are the machining parameters that determine the surface geometry of a machined part. They include (1) type of machining operation; (2) cutting tool geometry, most importantly nose radius; and (3) feed. The surface geometry that would result from these factors is referred to as the "ideal" or "theoretical" surface roughness, which is the finish that would be obtained in the absence of work material, vibration, and machine tool factors.

Type of operation refers to the machining process used to generate the surface. For example, peripheral milling, facing milling, and shaping all produce a flat surface; however, the surface geometry is different for each operation because of differences in tool shape and the way the tool interacts with the surface.

Tool geometry and feed combine to form the surface geometry. In tool geometry, the shape of the tool point is most important. The effects can be seen for a single-point tool in Figure 24.1. With the same feed, a larger nose radius causes the feed marks to be less pronounced, thus leading to a better finish. If two feeds are compared with the same nose radius, the larger feed increases the separation between feed marks, leading to an

FIGURE 24.1   Effect of geometric factors in determining the theoretical finish on a work surface for single-point tools: (a) effect of nose radius, (b) effect of feed, and (c) effect of end cutting-edge angle.

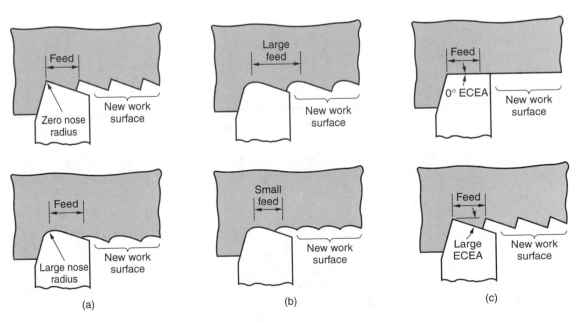

increase in the value of ideal surface roughness. If feed rate is large enough and the nose radius is small enough so that the end cutting edge participates in creating the new surface, then the end cutting-edge angle will affect surface geometry. In this case, a higher ECEA will result in a higher surface roughness value. In theory, a zero ECEA would yield a perfectly smooth surface; however, imperfections in the tool, work material, and machining process preclude achieving such an ideal finish.

The effects of nose radius and feed can be combined in an equation to predict the ideal average roughness for a surface produced by a single-point tool. The equation applies to operations such as turning, shaping, and planing:

$$R_i = \frac{f^2}{32\ NR} \tag{24.1}$$

where R_i = theoretical arithmetic average surface roughness, mm (in.); f = feed, mm (in.); and NR = nose radius on the tool point, mm (in.). The equation assumes that the nose radius is not zero and that feed and nose radius will be the principal factors that determine the geometry of the surface. The values for R_i will be in units of mm (in.), which can be converted to μm (μ-in.).

Eq. (24.1) can be used to estimate the ideal surface roughness in face milling with insert tooling, using f to represent the chip load (feed per tooth). However, both the leading and trailing edges of the rotating face milling cutter produce feed marks on the work surface, which complicates the surface geometry.

In slab milling, where the straight cutting edges of the cutter are utilized to generate surface geometry, the following relationship can be used to estimate the ideal surface roughness value, based on an analysis by Martellotti [13]:

$$R_i = \frac{0.125 f^2}{(D/2) \pm (f n_t / \pi)} \tag{24.2}$$

where f = chip load, mm/tooth (in./tooth); D = diameter of the milling cutter, mm (in.); and n_t = number of teeth on the cutter. The positive sign in the denominator is for up milling and the negative sign is for down milling. Eq. (24.2) assumes that each tooth is equally spaced around the cutter, that all of the cutting edges are equidistant from the axis of rotation, and that the arbor supporting the cutter is perfectly straight during rotation (zero run-out). These assumptions are seldom realized in practice. Consequently, there is often a waviness pattern superimposed on the surface, where the waviness corresponds to the cutter rotational speed.

The preceding relationships for ideal surface finish assume a sharp cutting tool. As the tool wears, the shape of the cutting point changes, which is reflected in the geometry of the work surface. For slight amounts of tool wear, the effect is not noticeable. However, when tool wear becomes significant, especially nose radius wear, surface roughness deteriorates compared to the ideal values given by the above equations.

Work Material Factors   Achieving the ideal surface finish is not possible in most machining operations because of factors related to the work material and its interaction with the tool. Work material factors that affect finish include (1) built-up edge effects—as the BUE cyclically forms and breaks away, particles are deposited on the newly created work surface, causing it to have a rough "sandpaper" texture; (2) damage to the surface caused by the chip curling back into the work; (3) tearing of the work surface during chip formation when machining ductile materials; (4) cracks in the surface caused by discontinuous chip formation when machining brittle materials; and (5) friction between the tool flank and the newly generated work surface. These work material fac-

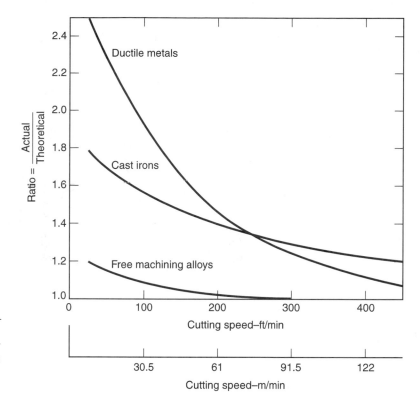

FIGURE 24.2 Ratio of actual surface roughness to ideal surface roughness for several classes of materials (source: General Electric Co. data [15]).

tors are influenced by cutting speed and rake angle, such that an increase in cutting speed or rake angle generally improves surface finish.

The work material factors usually cause the actual surface finish to be worse than the ideal. An empirical ratio can be developed to convert the ideal roughness value into an estimate of the actual surface roughness value. This ratio takes into account BUE formation, tearing, and other factors. The value of the ratio depends on cutting speed as well as work material. Figure 24.2 shows the ratio of actual to ideal surface roughness as a function of speed for several classes of work material.

The procedure for predicting the actual surface roughness in a machining operation is to compute the ideal surface roughness value and then multiply this value by the ratio of actual to ideal roughness for the appropriate class of work material. This can be summarized as

$$R_a = r_{ai}R_i \tag{24.3}$$

where R_a = the estimated value of actual roughness; r_{ai} = ratio of actual to ideal surface finish from Figure 24.2; and R_i = ideal roughness value from previous Eqs. (24.1) or (24.2).

EXAMPLE 24.2
Surface roughness.

A turning operation is performed on C1008 steel (a relatively ductile material) using a tool with a nose radius = 1.2 mm. The cutting conditions are speed = 100 m/min, and feed = 0.25 mm/rev. Compute an estimate of the surface roughness in this operation.

Solution: The ideal surface roughness can be calculated from Eq. (24.1):

$$R_i = (0.25)^2/(32 \times 1.2) = 0.0016 \text{ mm} = 1.6 \ \mu\text{m}.$$

From the chart in Figure 24.2, the ratio of actual to ideal roughness for ductile metals at 100 m/min is approximately 1.25. Accordingly, the actual surface roughness for the operation would be (approximately)

$$R_a = 1.25 \times 1.6 = 2.0 \ \mu m.$$

Vibration and Machine Tool Factors   These factors are related to the machine tool, tooling, and setup in the operation. They include chatter or vibration in the machine or cutting tool; deflections in the fixturing, often resulting in vibration; and backlash in the feed mechanism, particularly on older machine tools. If these machine tool factors can be minimized or eliminated, the surface roughness in machining will be determined primarily by geometric factors and work material factors already described.

Chatter or vibration in a machining operation can result in pronounced waviness in the work surface. When chatter occurs, a distinctive noise occurs that can be recognized by any experienced machinist. Possible steps to reduce or eliminate vibration include: (1) adding stiffness and/or damping to the setup; (2) operating at speeds that do not cause cyclical forces whose frequency approaches the natural frequency of the machine tool system; (3) reducing feeds and depths to reduce forces in cutting; and (4) changing the cutter design to reduce forces. Workpiece geometry can sometimes play a role in chatter. Thin cross sections tend to increase the likelihood of chatter, requiring additional supports to alleviate the condition.

24.3 SELECTION OF CUTTING CONDITIONS

One practical problem in machining is selecting the proper cutting conditions for a given operation. This is one of the tasks in process planning (Section 41.1). For each operation, decisions must be made about machine tool, cutting tool(s), and cutting conditions based on workpart machinability, part geometry, surface finish, and so forth.

24.3.1 Selecting Feed and Depth of Cut

Cutting conditions in a machining operation consist of speed, feed, depth of cut, and cutting fluid (whether a cutting fluid is to be used and, if so, type of cutting fluid). Tooling considerations are usually the dominant factor in decisions about cutting fluids (Section 23.4). Depth of cut is often predetermined by workpiece geometry and operation sequence. Many jobs require a series of roughing operations followed by a final finishing operation. In the roughing operations, depth is made as large as possible within the limitations of available horsepower, machine tool and setup rigidity, strength of the cutting tool, and so on. In the finishing cut, depth is set to achieve the final dimensions for the part.

The problem then reduces to selection of feed and speed. In general, parameter values should be decided in the order: *feed first, speed second.* Determining the appropriate feed rate for a given machining operation depends on the following factors:

> *Tooling*—What type of tooling will be used? Harder tool materials (cemented carbides, ceramics, etc.) tend to fracture more readily than high speed steel. These tools are normally used at lower feed rates. HSS can tolerate higher feeds due to its greater toughness.

> ➤ *Roughing or finishing*—Roughing operations involve high feeds, typically 0.5 to 1.25 mm/rev (0.020 to 0.050 in./rev) for turning; finishing operations involve low feeds, typically 0.125 to 0.4 mm/rev (0.005 to 0.015 in./rev) for turning.

> ➤ *Constraints on feed in roughing*—If the operation is roughing, how high can the feed rate be set? To maximize metal removal rate, feed should be set as high as possible. Upper limits on feed are imposed by cutting forces, setup rigidity, and sometimes horsepower.

> ➤ *Surface finish requirements in finishing*—If the operation is finishing, what is the desired surface finish? Feed is an important factor in surface finish. Computations like those in Example 24.2 can be used to estimate the feed that will produce a desired surface finish.

24.3.2 Optimizing Cutting Speed

Selection of cutting speed is based on making the best use of the particular cutting tool, which normally means choosing a speed that provides a high metal-removal rate yet suitably long tool life. Mathematical formulas have been derived to determine optimal cutting speed for a machining operation, given that the various time and cost components of the operation are known. The original derivation of these *machining economics* equations is credited to W. Gilbert [10]. The formulas allow the optimal cutting speed to be calculated for two objectives: (1) maximum production rate, or (2) minimum unit cost. Both objectives seek to achieve a balance between material removal rate and tool life. The formulas are based on a known Taylor tool life equation for the tool used in the operation. Accordingly, feed, depth of cut, and work material have already been set. The derivation will be illustrated for a turning operation. Similar derivations can be developed for other types of machining operations [2].

Maximizing Production Rate   For maximum production rate, the speed that minimizes machining time per production unit is determined. Minimizing cutting time per unit is equivalent to maximizing production rate. This objective is important in cases when the production order must be completed as quickly as possible.

In turning, three time elements contribute to the total production cycle time for one part:

1. *Part handling time,* T_h—This is the time the operator spends loading the part into the machine tool at the beginning of the production cycle and unloading the part after machining is completed.
2. *Machining time,* T_m—This is the time the tool is actually engaged in machining during the cycle.
3. *Tool change time,* T_t—At the end of the tool life, the tool must be changed, which takes time. This time must be apportioned over the number of parts cut during the tool life. Let n_p = the number of pieces cut in one tool life (the number of pieces cut with one cutting edge until the tool is changed); thus, the tool change time per part = T_t/n_p.

The sum of these three time elements gives the total time per unit product for the operation cycle:

$$T_c = T_h + T_m + \frac{T_t}{n_p} \tag{24.4}$$

where T_c = production cycle time per piece, min; and the other terms are defined above.

The cycle time T_c is a function of cutting speed. As cutting speed is increased, T_m decreases and T_t/n_p increases; T_h is unaffected by speed. These relationships are shown in Figure 24.3.

The total time per part is minimized at a certain value of cutting speed. This optimal speed can be identified by recasting Eq. (24.4) as a function of speed. It can be shown that the machining time in a straight turning operation is given by

$$T_m = \frac{\pi DL}{vf} \tag{24.5}$$

where T_m = machining time, min; D = workpart diameter, mm (in.); L = workpart length, mm (in.); f = feed, mm/rev (in./rev); and v = cutting speed, mm/min for consistency of units (in./min for consistency of units).

The number of pieces per tool n_p is also a function of speed. It can be shown that

$$n_p = \frac{T}{T_m} \tag{24.6}$$

where T = tool life, min/tool; and T_m = machining time per part, min/pc. Both T and T_m are functions of speed; hence, the ratio is a function of speed:

$$n_p = \frac{fC^{1/n}}{\pi DLv^{1/n-1}} \tag{24.7}$$

The effect of this relation is to cause T_t/n_p in Eq. (24.4) to increase as cutting speed increases. Substituting Eqs. (24.5) and (24.7) into Eq. (24.4) for T_c, we have

$$T_c = T_h + \frac{\pi DL}{fv} + \frac{T_t(\pi DLv^{1/n-1})}{fC^{1/n}} \tag{24.8}$$

The cycle time per piece is a minimum at the cutting speed at which the derivative of Eq. (24.8) is zero.

$$\frac{dT_c}{dv} = 0$$

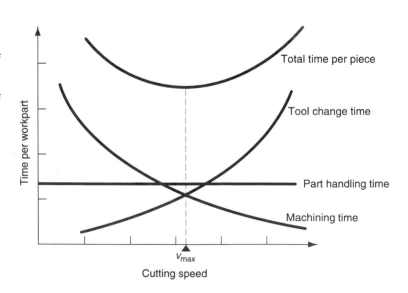

FIGURE 24.3 Time elements in a machining cycle plotted as a function of cutting speed. Total cycle time per piece is minimized at a certain value of cutting speed. This is the speed for maximum production rate.

Time per workpart

Total time per piece

Tool change time

Part handling time

Machining time

v_{max}

Cutting speed

Solving this equation yields the cutting speed for maximum production rate in the operation:

$$v_{\max} = \frac{C}{\left[\left(\frac{1}{n} - 1\right) T_t\right]^n}$$

(24.9)

where $v_{\max}$ is expressed in m/min (ft/min). The corresponding tool life for maximum production rate is

$$T_{\max} = \left(\frac{1}{n} - 1\right) T_t$$

(24.10)

Minimizing Cost per Unit   For minimum cost per unit, the speed that minimizes production cost per unit for the operation is determined. To derive the equations for this case, we begin with the four cost components that determine total cost of producing one part during a turning operation:

1. ***Cost of part handling time***—This is the cost of the time the operator spends loading and unloading the part. Let C_o = the cost rate (e.g., \$/min) for the operator and machine. Thus the cost of part handling time = $C_o T_h$.
2. ***Cost of machining time***—This is the cost of the time the tool is engaged in machining. Using C_o again to represent the cost per minute of the operator and machine tool, the cutting time cost = $C_o T_m$.
3. ***Cost of tool change time***—The cost of tool change time = $C_o T_t / n_p$.
4. ***Tooling cost***—In addition to the tool change time, the tool itself has a cost which must be added to the total operation cost. This cost is the cost per cutting edge C_t, divided by the number of pieces machined with that cutting edge n_p. Thus, tool cost per unit of product is given by C_t / n_p.

Tooling cost requires explanation, since it is affected by different tooling situations. For disposable inserts (e.g., cemented carbide inserts), tool cost is determined as

$$C_t = \frac{P_t}{n_e}$$

(24.11)

where C_t = cost per cutting edge, \$/tool life; P_t = price of the insert, \$/insert; and n_e = number of cutting edges per insert. This depends on the insert type; for example, triangular inserts that can be used only one side (positive rake tooling) yield three edges/insert; if both sides of the insert can be used (negative rake tooling), there are six edges/insert; and so forth.

For regrindable tooling (e.g., high speed steel solid shank tools, brazed carbide tools), the tool cost includes purchase price plus cost to regrind.

$$C_t = \frac{P_t}{n_g} + T_g C_g$$

(24.12)

where C_t = cost per tool life, \$/tool life; P_t = purchase price of the solid shank tool or brazed insert, \$/tool; n_g = number of tool lives per tool, which is the number of times the tool can be ground before it can no longer be used (5–10 times for roughing tools and 10–20 times for finishing tools); T_g = time to grind or regrind the tool, min/tool life; and C_g = grinder's rate, \$/min.

The sum of the four cost components gives the total cost per unit product C_c for the machining cycle.

$$C_c = C_oT_h + C_oT_m + \frac{C_oT_t}{n_p} + \frac{C_t}{n_p} \tag{24.13}$$

C_c is a function of cutting speed, just as T_c is a function of v. The relationships for the individual terms and total cost as a function of cutting speed are shown in Figure 24.4. Eq. (24.13) can be rewritten in terms of v to yield:

$$C_c = C_oT_h + \frac{C_o\pi DL}{fv} + \frac{(C_oT_t + C_t)(\pi DLv^{1/n-1})}{fC^{1/n}} \tag{24.14}$$

The cutting speed that obtains minimum cost per piece for the operation can be determined by taking the derivative of Eq. (24.14) with respect to v, setting it to zero, and solving for v_{min}.

$$v_{min} = C\left(\frac{n}{1-n} \cdot \frac{C_o}{C_oT_t + C_t}\right)^n \tag{24.15}$$

The corresponding tool life is given by

$$T_{min} = \left(\frac{1}{n} - 1\right)\left(\frac{C_oT_t + C_t}{C_o}\right) \tag{24.16}$$

**EXAMPLE 24.3
Determining
cutting speeds in
machining
economics.**

Suppose a turning operation is to be performed with HSS tooling on mild steel, with Taylor tool life parameters $n = 0.125$, $C = 70$ m/min (Table 23.2). The workpart has length = 500 mm and diameter = 100 mm. Feed = 0.25 mm/rev. Handling time per piece = 5.0 min and tool change time = 2.0 min. Cost of machine and operator = \$30/hr, and tooling cost = \$3 per cutting edge. Find: (a) cutting speed for maximum production rate, and (b) cutting speed for minimum cost.

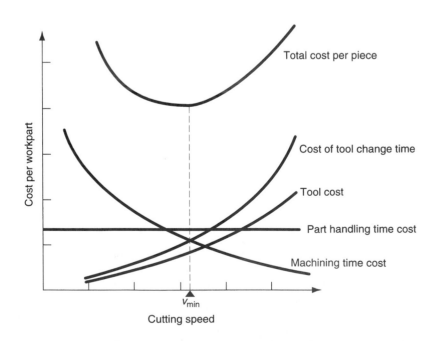

FIGURE 24.4 Cost components in a machining operation plotted as a function of cutting speed. Total cost per piece is minimized at a certain value of cutting speed. This is the speed for minimum cost per piece.

Solution: (a) Cutting speed for maximum production rate is given by Eq. (24.9).

$$v_{max} = 70 \left(\frac{0.125}{0.875} \cdot \frac{1}{2} \right)^{0.125} = 50 \text{ m/min}$$

(b) Converting C_o = \$30/hr to \$0.5/min, the cutting speed for minimum cost is given by Eq. (24.15).

$$v_{min} = 70 \left(\frac{0.125}{0.875} \cdot \frac{0.5}{0.5(2) + 3.00} \right)^{0.125} = 42 \text{ m/min}$$

**EXAMPLE 24.4
Production rate
and cost in
machining
economics.**

Determine the hourly production rate and cost per piece for the two cutting speeds computed in Example 24.3. We have the following additional data regarding tool cost: price of original HSS tool shank = \$25, number of regrinds = 16, time to grind/regrind = 7.0 min, and regrinder's rate = \$30/hr (\$0.50/min).

Solution: Let us compute the tooling cost first, which applies to both cutting speeds:

$$C_t = \frac{25}{16} + (7.0)(0.50) = \$5.06/\text{cutting edge}$$

(a) For the cutting speed for maximum production, v_{max} = 50 m/min, let us calculate machining time per piece and tool life.

$$\text{Machining time } T_m = \frac{\pi(0.5)(0.1)}{(0.25)(10^{-3})(50)} = 12.57 \text{ min/pc}$$

$$\text{Tool life } T = \left(\frac{70}{50} \right)^8 = 14.76 \text{ min/cutting edge}$$

From this we see that the number of pieces per tool n_p = 14.76/12.57 = 1.17. Use n_p = 1.0. From Eq. (24.4), average production cycle time for the operation is

$$T_c = 5.0 + 12.57 + 2.0/1 = 19.57 \text{ min/pc.}$$

Corresponding hourly production rate R_p = 60/19.57 = 3.1 pc/hr
From Eq. (24.13), average cost per piece for the operation is

$$C_c = 0.5(5.0) + 0.5(12.57) + 0.5(2.0)/1 + 5.06/1 = \$14.85/\text{pc}$$

(b) For the cutting speed for minimum production cost per piece, v_{min} = 42 m/min, the machining time per piece and tool life are calculated as follows:

$$\text{Machining time } T_m = \frac{\pi(0.5)(0.1)}{(0.25)(10^{-3})(42)} = 14.96 \text{ min/pc}$$

$$\text{Tool life } T = \left(\frac{70}{42} \right)^8 = 59.54 \text{ min/cutting edge}$$

The number of pieces per tool n_p = 59.54/14.96 = 3.98 → Use n_p = 3 to avoid failure during the operation. The average production cycle time for the operation is

$$T_c = 5.0 + 14.96 + 2.0/3 = 20.63 \text{ min/pc}$$

Corresponding hourly production rate R_p = 60/20.63 = 2.9 pc/hr

Average cost per piece for the operation is

$$C_c = 0.5(5.0) + 0.5(14.96) + 0.5(2.0)/3 + 5.06/3 = \$12.00/\text{pc}$$

Note that production rate is greater for v_{max} and cost per piece is minimum for v_{min}. ■

Some Comments on Machining Economics Some practical observations can be made relative to these optimum cutting speed equations. First, as the values of C and n increase in the Taylor tool life equation, the optimum cutting speed increases by either Eq. (24.9) or Eq. (24.15). Cemented carbides and ceramic cutting tools should be used at speeds that are significantly higher than for HSS tools.

Second, as the tool change time and/or tooling cost (T_{tc} and C_t) increase, the cutting speed equations yield lower values. Lower speeds allow the tools to last longer, and it is wasteful to change tools too frequently if either the cost of tools or the time to change them is high. An important effect of this tool cost factor is that disposable inserts usually possess a substantial economic advantage over regrindable tooling. Even though the cost per insert is significant, the number of edges per insert is large enough and the time required to change the cutting edge is low enough that disposable tooling generally achieves higher production rates and lower costs per unit product.

Third, v_{max} is always greater than v_{min}. The C_t/n_p term in Eq. (24.13) has the effect of pushing the optimum speed value to the left in Figure 24.4, resulting in a lower value than in Figure 24.3. Rather than taking the risk of cutting at a speed above v_{max} or below v_{min}, some machine shops strive to operate in the interval between v_{min} and v_{max}—an interval sometimes referred to as the "high-efficiency range."

The procedures outlined for selecting feeds and speeds in machining are often difficult to apply in practice. The best feed rate is difficult to determine because the relationships between feed and surface finish, force, horsepower, and other constraints are not readily available for each machine tool. Experience, judgment, and experimentation are required to select the proper feed. The optimum cutting speed is difficult to calculate because the Taylor equation parameters C and n are not usually known without prior testing. Testing of this kind in a production environment is expensive.

24.4 PRODUCT DESIGN CONSIDERATIONS IN MACHINING

Several important aspects of product design have already been considered in our discussion of tolerance and surface finish (Section 24.2). In this section, we present design guidelines for machining, compiled from sources [1, 4, and 16]:

➢ If possible, parts should be designed that do not need machining. If this is not possible, then minimize the amount of machining required on the parts. In general, a lower-cost product is achieved through the use of net shape processes such as precision casting, closed die forging, or (plastic) molding; or near net shape processes such as impression die forging. Reasons why machining may be required include close tolerances; good surface finish; and special geometric features such as threads, precision holes, cylindrical sections with high degree of roundness, and similar shapes that cannot be achieved except by machining.

➢ Tolerances should be specified to satisfy functional requirements, but capabilities of processes should also be considered. See Table 24.2 for tolerance capabilities in machining. Excessively close tolerances add cost but may not add value to the part. As tolerances become tighter (smaller), product costs generally increase due to additional processing, fixturing, inspection, sortation, rework, and scrap (see Figure 42.1).

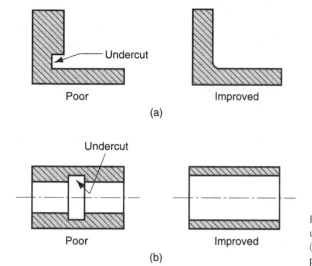

FIGURE 24.5   Two machined parts with undercuts: cross sections of (a) bracket and (b) rotational part. Also shown is how the part design might be improved.

> Surface finish should be specified to meet functional and/or aesthetic requirements, but better finishes generally increase processing costs by requiring additional operations such as grinding or lapping.

> Machined features such as sharp corners, edges, and points should be avoided; they are often difficult to accomplish by machining. Sharp internal corners require pointed cutting tools that tend to break during machining. Sharp corners and edges tend to create burrs and are dangerous to handle.

> Deep holes that must be bored should be avoided. Deep hole boring requires a long boring bar. Boring bars must be stiff, and this often requires use of high modulus materials such as cemented carbide, which is expensive.

> Machined parts should be designed so they can be produced from standard available stock. Choose exterior dimensions equal to or close to the standard stock size to minimize machining; for example, rotational parts with outside diameters that are equal to standard bar stock diameter.

> Parts should be designed to be rigid enough to withstand forces of cutting and workholder clamping. Machining of long narrow parts, large flat parts, parts with thin walls, and similar shapes should be avoided if possible.

> Undercuts as in Figure 24.5 should be avoided since they often require additional setups and operations and/or special tooling; they can also lead to stress concentrations in service.

> Materials with good machinability should be selected by the designer (Section 24.1). As a rough guide, the machinability rating of a material correlates with the allowable cutting speed and production rate that can be used. Thus, parts made of materials with low machinability cost more to produce. Parts that are hardened by heat treatment must usually be finish ground or machined with higher cost tools after hardening to achieve final size and tolerance.

> Machined parts should be designed with features that can be produced in a minimum number of setups—one setup if possible. This usually means geometric features that can be accessed from one side of the part—see Figure 24.6.

> Machined parts should be designed with features that can be achieved with standard cutting tools. This means avoiding unusual hole sizes, threads, and features with unusual shapes requiring special form tools. In addition, it is helpful to design parts such

FIGURE 24.6 Two parts with similar hole features: (a) holes that must be machined from two sides, requiring two setups, and (b) holes that can all be machined from one side.

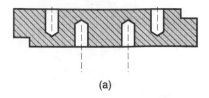

(a)

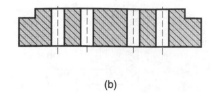

(b)

that the number of individual cutting tools needed in machining is minimized; this often allows the part to be completed in one setup on a machine such as a machining center with limited tool storage capacity.

REFERENCES

[1] Bakerjian, R. (ed.). *Tool and Manufacturing Engineers Handbook.* 4th ed. Volume VI: *Design for Manufacturability,* Society of Manufacturing Engineers, Dearborn, Mich., 1992.

[2] Boothroyd, G. and Knight, W. A., *Fundamentals of Machining and Machine Tools,* 2nd ed., Marcel Dekker, Inc., New York, 1989.

[3] Boston, O. W. *Metal Processing.* 2nd ed. Wiley. New York, 1951.

[4] Bralla, J. G. (Editor in Chief). *Handbook of Product Design for Manufacturing.* McGraw-Hill, New York, 1986.

[5] Brierley, R. G., and Siekman, H. J. *Machining Principles and Cost Control.* McGraw-Hill, New York, 1964.

[6] DeGarmo, E. P., Black, J. T., and Kohser, R. A. *Materials and Processes in Manufacturing.* 7th ed. Macmillan, New York, 1988.

[7] Drozda, T. J., and Wick, C. (eds.). *Tool and Manufacturing Engineers Handbook.* 4th ed. Vol. I: *Machining,* Society of Manufacturing Engineers, Dearborn, Mich., 1983.

[8] Eary, D. F., and Johnson, G. E. *Process Engineering: for Manufacturing.* Prentice-Hall, Inc., Englewood Cliffs, N. J., 1962.

[9] Ewell, J. R. "Thermal Coefficients—A Proposed Machinability Index." *Technical Paper MR67-200,* Society of Manufacturing Engineers, Dearborn, Mich., 1967.

[10] Gilbert, W. W. "Economics of Machining." *Machining—Theory and Practice.* American Society for Metals, Metals Park, Ohio, 1950, pp 465–485.

[11] Groover, M. P. "A Survey on the Machinability of Metals." *Technical Paper MR76-269.* Society of Manufacturing Engineers, Dearborn, Mich., 1976.

[12] *Machining Data Handbook.* 3rd ed. Vols. I and II, Metcut Research Associates, Inc., Cincinnati, Ohio, 1980.

[13] Martellotti, M. E. "An Analysis of the Milling Process." *ASME Transactions.* Volume 63, November 1941, pp 677–700.

[14] Schaffer, G. H. "The Many Faces of Surface Texture." Special Report 801, *American Machinist & Automated Manufacturing.* June 1988, pp 61–68.

[15] *Surface Finish.* Machining Development Service, Publication A-5, General Electric Company, Schenectady, New York (no date).

[16] Trucks, H. E. *Designing for Economical Production.* Society of Manufacturing Engineers, Dearborn, Mich., 1974.

[17] Van J., Voast. *United States Air Force Machinability Report.* Vol. 3, Curtiss-Wright Corporation, 1954.

REVIEW QUESTIONS

24.1. Define machinability.

24.2. What are the criteria by which machinability is commonly assessed in a production machining operation?

24.3. Name some of the important mechanical and physical properties that affect the machinability of a work material.

24.4. Why do costs tend to increase when better surface finish is required on a machined part?

24.5. What are the basic factors that affect surface finish in machining?

24.6. What are the parameters that have the greatest influence in determining the ideal surface roughness R_i?

24.7. Name some of the steps that can be taken to reduce or eliminate vibrations in machining.

24.8. What are the factors on which the selection of feed in a machining operation should be based?

24.9. The unit cost in a machining operation is the sum of four cost terms. The first three terms are (1) part load/unload cost, (2) cost of time the tool is actually cutting the work, and (3) the cost of the time to change the tool. What is the fourth term?

24.10. Which cutting speed is always lower for a given machining operation, cutting speed for minimum cost or cutting speed for maximum production rate? Why?

MULTIPLE CHOICE QUIZ

There is a total of 14 correct answers in the following multiple choice questions (some questions have multiple answers that are correct). To attain a perfect score on the quiz, all correct answers must be given, since each correct answer is worth 1 point. For each question, each omitted answer or wrong answer reduces the score by 1 point, and each additional answer beyond the number of answers required reduces the score by 1 point. Percentage score on the quiz is based on the total number of correct answers.

24.1. Which of the following criteria are generally recognized to indicate good machinability (more than one answer)? (a) all of the following, (b) ease of chip disposal, (c) high value of R_a, (d) long tool life, (e) low cutting forces, (f) low value of R_a, (g) zero shear plane angle.

24.2. Of the various methods for testing machinability, which of the following seems to be the most important (one answer)? (a) cutting forces, (b) cutting temperature, (c) horsepower consumed in the operation, (d) surface roughness, (e) tool life, or (f) tool wear.

24.3. A machinability rating of greater than 1.0 indicates that the work material is which of the following relative to the defined base material, whose rating = 1.0? (a) easier to machine than the base or (b) more difficult to machine than the base.

24.4. In general, which of the following materials has the highest machinability (one best answer)? (a) aluminum, (b) cast iron, (c) copper, (d) low carbon steel, (e) stainless steel, (f) titanium alloys, or (g) unhardened tool steel.

24.5. Which one of the following operations is generally capable of the closest tolerances (one best answer)? (a) broaching, (b) drilling, (c) end milling, (d) planing, or (e) sawing.

24.6. When cutting a ductile work material, an increase in cutting speed will generally have which effect on surface finish? (a) degrade surface finish, which means higher value of R_a or (b) improve surface finish, which means lower value of R_a.

24.7. Which one of the following operations is generally capable of the best surface finishes (lowest value of R_a (one best answer)? (a) broaching, (b) drilling, (c) end milling, (d) planing, or (e) turning.

24.8. Which of the following time components in the average production machining cycle is affected by cutting speed (more than one answer)? (a) part loading and unloading time, and (b) setup time for the machine tool, (c) time the tool is engaged in cutting, and (d) tool change time.

24.9. Which cutting speed is always lower for a given machining operation? (a) cutting speed for maximum production rate, or (b) cutting speed for minimum cost.

24.10. A high tooling cost and/or tool change time will tend to have which of the following effects on v_{max} or v_{min}? (a) decrease or (b) increase.

PROBLEMS

Machinability

24.1. A machinability rating is to be determined for a new work material using the cutting speed for a 60 min tool life as the basis of comparison. For the base material (B1112 steel), test data resulted in Taylor equation parameter values of $n = 0.29$ and $C = 500$, where speed is in m/min and tool life is min. For the new material, the parameter values were $n = 0.21$ and $C = 400$. These results were obtained using cemented carbide tooling. (a) Compute a machinability rating for the new material. (b) Suppose the machinability criterion was the cutting speed for a 10 min tool life. Compute the machinability rating for this case. (c) What do the results of the two calculations show about the difficulties in machinability measurement?

24.2. A machinability rating is to be determined for a new work material. For the base material (B1112), test data resulted in a Taylor equation with parameters $n = 0.27$ and $C = 450$. For the new material, the Taylor parameters were $n = 0.22$ and $C = 420$. Units in both cases are speed in m/min and tool life in min. These results were obtained using cemented carbide tooling. (a) Compute a machinability rating for the new material using cutting speed for a 30 min tool life as the basis of comparison. (b) If the machinability criterion were tool life for a cutting speed of 150 m/min, what is the machinability rating for the new material?

24.3. Tool life turning tests have been conducted on B1112 steel with high-speed steel tooling, and the resulting parameters of the Taylor equation are $n = 0.13$ and $C = 225$. The feed and depth during these tests were: $f =$ 0.010 in./rev and $d = 0.100$ in. Based on this information, and machinability data given in Table 24.1, determine the cutting speed you would recommend for the following work materials, if the tool life desired in operation is 30 min: (a) C1008 low carbon steel with 150 Brinell hardness, (b) 4130 alloy steel with 190 Brinell hardness, and (c) B1113 steel with 170 Brinell hardness. Assume that the same feed and depth of cut are to be used.

Surface Roughness

24.4. In a turning operation on cast iron, the nose radius on the tool = 1.0 mm, feed rate = 0.2 mm/rev, and speed = 2 m/s. Compute an estimate of the surface roughness for this cut.

24.5. A turning operation uses a 2/64 in nose radius cutting tool on a free machining steel with a feed rate = 0.010 in/rev and a cutting speed = 300 ft/min. Determine the surface roughness for this cut.

24.6. A single-point HSS tool with a 3/64 in nose radius is used in a shaping operation on a ductile steel workpart. Cutting speed = 100 ft/min, feed = 0.015 in/pass, and depth of cut = 0.125 in. Determine the surface roughness for this operation.

24.7. A part to be turned in an engine lathe must have a surface finish of 1.6 μm. The part is made of a free-machining aluminum alloy. Cutting speed = 150 m/min, and depth of cut = 4.0 mm. The nose radius on the tool = 0.75 mm. Determine the feed that will achieve the specified surface finish.

24.8. Solve previous Problem 24.7 except that the part is made of cast iron instead of aluminum and the cutting speed is reduced to 100 m/min.

24.9. A part to be turned in an engine lathe must have a surface finish of 1.6 μm. The part is made of a free-machining steel. Cutting conditions: $v = 1.5$ m/s and $d = 3.0$ mm. The nose radius on the tool = 1.2 mm. Determine the feed that will achieve the specified surface finish.

24.10. The surface finish specification in a turning job is 0.8 μm. The work material is cast iron. The cutting conditions have been selected as follows: $v = 75$ m/min, $f = 0.3$ mm/rev, and $d = 4.0$ mm. The nose radius of the cutting tool must be selected. Determine the minimum nose radius that will obtain the specified finish in this operation.

24.11. A face milling operation is to be performed on a cast iron part at 400 ft/min to finish the surface to 32 μ-in.

AA. The cutter uses four inserts and its diameter is 3.0 in. To obtain the best possible finish, a type of carbide insert with 4/64 in nose radius is to be used. Determine the required feed rate (in./min) that will achieve the 32 μ-in. finish.

24.12. A face milling operation is not yielding the required surface finish on the work. The cutter is a four-tooth insert type face milling cutter. The machine shop foreman thinks the problem is that the work material is too ductile for the job, but this property tests well within the ductility range for the material specified by the designer. Without knowing any more about the job, what changes in cutting conditions and tooling would you suggest to improve the surface finish?

24.13. A turning operation is to be performed on C1010 steel, which is a ductile grade. It is desired to achieve a surface finish of 64 μ-in. (AA), while at the same time maximizing the metal removal rate. It has been decided that the speed should be in the range 200 ft/min to 400 ft/min, and that the depth of cut will be 0.080 in. The tool nose radius = 3/64 in. Determine the speed and feed combination that meets these criteria.

24.14. Plain milling is performed to finish a cast iron workpart prior to plating. The milling cutter has four equally spaced teeth and the diameter = 60 mm. The chip load $f = 0.35$ mm/tooth, and cutting speed $v = 1.0$ m/s. Estimate the surface roughness for (a) up milling, and (b) down milling.

24.15. A peripheral milling operation is performed using a slab milling cutter with four teeth and a 2.50 in. diameter. Feed = 0.015 in./tooth, and cutting speed = 150 ft/min. Assuming first that the teeth are equally spaced around the cutter, and that each tooth projects an equal distance from the axis of rotation, determine the theoretical surface roughness for (a) up milling and (b) down-milling.

Machining Economics

24.16. A HSS tool is used to turn a steel workpart that is 300 mm long and 80 mm in diameter. The parameters in the Taylor equation are $n = 0.13$ and $C = 75$ (m/min) for a feed of 0.4 mm/rev. The operator and machine tool rate = $30/hr, and the tooling cost per cutting

edge = $4. It takes 2.0 min to load and unload the workpart and 3.5 min to change tools. Determine (a) cutting speed for maximum production rate, (b) tool life in min of cutting, and (c) cycle time and cost per unit of product.

24.17. Solve Problem 24.16 except that in part (a) determine cutting speed for minimum cost.

24.18. A cemented carbide tool is used to turn a part with length = 18 in. and diameter = 3 in. The parameters in the Taylor equation are $n = 0.27$ and $C = 1200$. The rate for the operator and machine tool = \$33/hr, and the tooling cost per cutting edge = \$2. It takes 3.0 min to load and unload the workpart and 1.5 min to change tools. The feed = 0.013 in./rev. Determine (a) cutting speed for maximum production rate, (b) tool life in min of cutting, and (c) cycle time and cost per unit of product.

24.19. Solve Problem 24.18 except that in part (a) determine cutting speed for minimum cost.

24.20. Compare disposable and regrindable tooling. The same grade of cemented carbide tooling is available in two forms for turning operations in a certain machine shop: disposable inserts and brazed inserts. The parameters in the Taylor equation for this grade are $n = 0.25$ and $C = 300$ (m/min) under the cutting conditions considered here. For the disposable inserts, each insert costs \$6, there are four cutting edges per insert, and the tool change time = 1 min (this is an average of the time to index the insert and the time to replace it when all edges have been used). For the brazed insert, the tool costs \$30 and it is estimated that it can be used a total of 15 times before it must be scrapped. The tool change time for the regrindable tooling = 3 min. The standard time to grind or regrind the cutting edge is 5 min, and the grinder is paid at a rate = \$20/hr. Machine time on the lathe costs \$24/hr. The workpart to be used in the comparison is 375 mm long and 62.5 mm in diameter, and it takes 2 min to load and unload the work. The feed = 0.3 mm/rev. For the two tooling cases, compare (a) cutting speeds for minimum cost, (b) tool lives, (c) cycle time and cost per unit of production. Which tool would you recommend?

24.21. Solve Problem 24.20 except that in part (a) determine the cutting speeds for maximum production rate.

24.22. Three tool materials are to be compared for the same finish turning operation on a batch of 100 steel parts: high-speed steel, cemented carbide, and ceramic. For the HSS tool, the Taylor equation parameters are $n = 0.125$ and $C = 70$. The price of the HSS tool is \$15, and it is estimated that it can be ground and reground 15 times at a cost of \$1.50. Tool change time = 3 min. Both carbide and ceramic tools are in insert form and can be held in the same mechanical toolholder. The Taylor equation parameters for the cemented carbide are $n = 0.25$ and $C = 500$; and for the ceramic, $n = 0.6$ and $C = 3,000$. The cost per insert for the carbide = \$6 and for the ceramic = \$8. Number of cutting edges per insert in both cases = 6. Tool change time = 1 min for both tools. Time to change parts = 2 min. Feed = 0.25 mm/rev, and depth = 3 mm. The cost of machine time = \$30/hr. The part dimensions are diameter = 56 mm and length = 290 mm. Setup time for the batch is 2 hr. For the three tooling cases, compare (a) cutting speeds for minimum cost, (b) tool lives, (c) cycle time, (d) cost per production unit, (e) total time to complete the batch and production rate. (f) What is the proportion of time spent actually cutting metal for each tooling?

24.23. Solve Problem 24.22 except that in parts (a) and (b) determine the cutting speeds and tool lives for maximum production rate.

24.24. A vertical boring mill is used to bore the inside diameter of a large batch of tube-shaped parts. The diameter = 28 in., and the length of the bore = 14 in. Current cutting conditions are speed = 200 ft/min, feed = 0.015 in./rev, and depth = 0.125 in. The parameters of the Taylor equation for the cutting tool in the operation are $n = 0.23$ and $C = 850$ (ft/min). Tool change time = 3 min, and tooling cost = \$3.50 per cutting edge. The time required to load and unload the parts = 12.0 min, and the cost of machine time on this boring mill = \$42/hr. Management has decreed that the production rate must be increased by 25%. Is that possible? Assume that feed must remain unchanged in order to achieve the required surface finish. What is the current production rate and the maximum possible production rate for this job?

24.25. An NC lathe cuts two passes across a cylindrical workpiece under automatic cycle. The operator loads and unloads the machine. The starting diameter of the work is 3 in., and its length = 10 in. The work cycle consists of the following steps (with element times given in parentheses where applicable): 1—Operator loads part into machine, starts cycle (1.00 min); 2—NC lathe positions tool for first pass (0.10 min); 3—NC lathe turns first pass (time depends on cutting speed); 4—NC lathe repositions tool for second pass (0.4 min); 5—NC lathe turns second pass (time depends on cutting speed); and 6—Operator unloads part and places in tote pan (1 min). In addition, the cutting tool must be periodically changed. This tool change time takes 1 min. The feed rate = 0.007 in./rev and the depth of cut for each pass = 0.100 in. The cost of the operator and machine = \$39/hr and the tool cost = \$2/cutting edge. The applicable Taylor tool life equation has parameters: $n = 0.26$ and $C = 900$ (ft/min). Determine (a) the cutting speed for minimum cost per piece, (b) the average time required to complete one production cycle, (c) cost of the production cycle. (d) If the setup time for this job is 3 hours and the batch size = 300 parts, how long will it take to complete the batch?

24.26. As indicated in Section 24.4, the effect of a cutting fluid is to increase the value of C in the Taylor tool life equation. In a certain machining situation using HSS tooling, the C value is increased from $C = 200$ to $C = 225$ due to the use of the cutting fluid. The n value is the same with or without fluid at $n = 0.125$. Cutting speed used in the operation is $v = 125$ ft/min. Feed = 0.010 in./rev, and depth = 0.100 in. The effect of the cutting fluid can be to either increase cutting speed (at the same tool life) or increase tool life (at the same cutting speed). (a) What is the cutting speed that would result from using the cutting fluid if tool life remains the same as with no fluid? (b) What is the tool life that would result if the cutting speed remains at 125 ft/min? (c) Economically, which effect is better, given that tooling cost = $2 per cutting edge, tool change time = 2.5 min., and operator and machine rate = $30/hr? Justify your answer with calculations, using cost per cubic in. of metal machined as the criterion of comparison. Ignore effects of workpart handling time.

24.27. In a turning operation on ductile steel, it is desired to obtain an actual surface roughness of 63 μ-in. with a 2/64 in. nose radius tool. The ideal roughness is given by Eq. (24.1) and an adjustment will have to be made using Figure 24.2 to convert the 63 μ-in. actual roughness to an ideal roughness, taking into account the material and cutting speed. Disposable inserts are used at a cost of $1.75 per cutting edge (each insert costs $7.00 and there are four edges per insert). Time to index each insert = 25 sec and to replace an insert every fourth index takes 45 sec. The workpiece length = 30.0 in. and its diameter = 3.5 in. The machine and operator's rate = $39 per hour including applicable overheads. The Taylor tool life equation for this tool and work combination is given by: $vT^{0.23} f^{0.55} = 40.75$, where T = tool life, min; v = cutting speed, ft/min; and f = feed, in./rev. Solve for (a) the feed in in/rev that will achieve the desired actual finish, (b) cutting speed for minimum cost per piece at the feed determined in (a). **Hint:** To solve (a) and (b) requires an iterative computational procedure.

24.28. Verify that the derivative of Eq. (24.8) results in Eq. (24.9).

24.29. Verify that the derivative of Eq. (24.14) results in Eq. (24.15).

GRINDING AND OTHER ABRASIVE PROCESSES

CHAPTER CONTENTS

25.1   Grinding
    25.1.1   The Grinding Wheel
    25.1.2   Analysis of the Grinding Process
    25.1.3   Application Considerations in Grinding
    25.1.4   Grinding Operations and Grinding Machines
25.2   Related Abrasive Process
    25.2.1   Honing
    25.2.2   Lapping
    25.2.3   Superfinishing
    25.2.4   Polishing and Buffing

Abrasive machining involves material removal by the action of hard, abrasive particles that are usually in the form of a bonded wheel. Grinding is the most important of the abrasive processes. In terms of number of machine tools in use, grinding is the most common of all metalworking operations [10]. Other abrasive processes include honing, lapping, superfinishing, polishing, and buffing. The abrasive machining processes are generally used as finishing operations, although some abrasive processes are capable of high material removal rates rivaling those of conventional machining operations.

The use of abrasives to shape parts is probably the oldest material removal process (Historical Note 25.1). Abrasive processes are important commercially and technologically for these reasons:

➤ They can be used on all types of materials ranging from soft metals to hardened steels and hard nonmetallic materials such as ceramics and silicon.

➤ Some of these processes can be used to produce extremely fine surface finishes, to 0.025 μm (1 μ-in.).

➤ For certain abrasive processes, dimensions can be held to extremely close tolerances.

Historical Note 25.1 *Development of abrasive processes* [14]

Use of abrasives predates any of the other machining operations. There is archaeological evidence that ancient man used abrasive stones such as sandstone found in nature to sharpen tools and weapons and to scrape away unwanted portions of softer materials to make domestic implements.

Grinding became an important technical trade in ancient Egypt. The large stones used to build the Egyptian pyramids were cut to size by a rudimentary grinding process. The grinding of metals dates to around 2000 BC and was a highly valued skill at that time.

Early abrasive materials were those found in nature, such as sandstone, which consists primarily of quartz (SiO_2); emery, consisting of corundum (Al_2O_3) plus equal or lesser amounts of the iron minerals hematite (Fe_2O_3) and magnetite (Fe_3O_4); and diamond. The first grinding wheels were likely cut out of sandstone and were no doubt rotated under manual power. However, grinding wheels made in this way were not consistent in quality.

In the early 1800s, the first solid bonded grinding wheels were produced in India. They were used to grind gems, an important trade in India at the time. The abrasives were corundum, emery, or diamond. The bonding material was natural gum-resin shellac. The technology was exported to Europe and the United States, and other bonding materials were subsequently introduced: rubber bond in the mid-1800s, vitrified bond around 1870, shellac bond around 1880, and resinoid bond in the 1920s with the development of the first thermosetting plastics (phenol-formaldehyde).

In the late 1800s, synthetic abrasives were first produced: silicon carbide (SiC) and aluminum oxide (Al_2O_3). By manufacturing the abrasives, chemistry and size of the individual abrasive grains could be controlled more closely, resulting in higher quality grinding wheels.

The first real grinding machines were made by the U.S. firm Brown & Sharpe in the 1860s for grinding parts for sewing machines, an important industry during the period. Grinding machines also contributed to the development of the bicycle industry in the 1890s and later the U.S. automobile industry. The grinding process was used to size and finish heat-treated (hardened) parts in these products.

The superabrasives diamond and cubic boron nitride are products of the twentieth century. Synthetic diamonds were first produced by the General Electric Company in 1955. These abrasives were used to grind cemented carbide cutting tools, and today this remains one of the important applications of diamond abrasives. Cubic boron nitride (cBN), second only to diamond in hardness, was first synthesized in 1957 by GE using a similar process to that for making artificial diamonds. Cubic BN has become an important abrasive for grinding hardened steels.

Abrasive water jet cutting and ultrasonic machining are sometimes classified as abrasive machining because they accomplish cutting by means of abrasives. However, these processes are commonly known as nontraditional material removal processes and are covered in the following chapter.

25.1 GRINDING

Grinding is a material-removal process in which abrasive particles are contained in a bonded grinding wheel that operates at very high surface speeds. The grinding wheel is usually disk-shaped, and is precisely balanced for high rotational speeds.

Grinding can be likened to the milling process. Cutting occurs on either the periphery or the face of the grinding wheel, similar to peripheral milling and face milling. Peripheral grinding is much more common than face grinding. The rotating grinding wheel consists of many cutting teeth (the abrasive particles), and the work is fed relative to the wheel to accomplish material removal. Despite these similarities, there are significant differences between grinding and milling: (1) the abrasive grains in the wheel are much smaller and more numerous than the teeth on a milling cutter; (2) cutting speeds in grinding are much higher than in milling; (3) the abrasive grits in a grinding wheel are randomly oriented and possess on average a very high negative rake angle; and (4) a grinding wheel is self-sharpening—as the wheel wears, the abrasive particles become dull and either fracture to create fresh cutting edges or are pulled out of the surface of the wheel to expose new grains.

25.1.1 The Grinding Wheel

A grinding wheel consists of abrasive particles and bonding material. The bonding material holds the particles in place and establishes the shape and structure of the wheel. These two ingredients, and the way they are fabricated, determine the parameters of the grinding wheel, which are: (1) abrasive material, (2) grain size, (3) bonding material, (4) wheel grade, and (5) wheel structure. These parameters are analogous to the material and geometry of a conventional cutting tool. To achieve the desired performance in a given application, each of the parameters must be carefully selected.

Abrasive Material Different abrasive materials are appropriate for grinding different work materials. General properties of an abrasive material used in grinding wheels include high hardness, wear resistance, toughness, and friability. Hardness, wear resistance, and toughness are desirable properties of any cutting tool material. *Friability* refers to the capacity of the abrasive material to fracture when the cutting edge of the grain becomes dull, thereby exposing a new sharp edge.

The development of grinding abrasives is described in our historical note. Today, the abrasive materials of greatest commercial importance are described below, with relative hardness values listed in Table 25.1:

> *Aluminum oxide* (Al_2O_3)—This is the most common abrasive material (Section 7.3.1). It is used to grind steel and other ferrous, high-strength alloys.
> *Silicon carbide* (SiC)—SiC is harder than Al_2O_3, but not as tough (Section 7.2.2). Its grinding applications include ductile metals such as aluminum, brass, and stainless steel, as well as brittle materials such as some cast irons and certain ceramics. It cannot be used effectively for grinding steel because of the strong chemical affinity between the carbon in SiC and the iron in steel.
> *Cubic boron nitride* (cBN)—When used as an abrasive, cBN (Section 7.3.3) is produced under the trade name Borazon by the General Electric Company. Borazon grinding wheels are used for hard materials such as hardened tool steels and aerospace alloys.
> *Diamond*—Diamond abrasives occur naturally and are also made synthetically (Section 7.5.1). Diamond wheels are generally used in grinding applications on hard, abrasive materials such as ceramics, cemented carbides, and glass.

Grain Size The grain size of the abrasive particle is an important parameter in determining surface finish and material removal rate. Small grit sizes produce better

TABLE 25.1 Hardness values of abrasive materials used in grinding wheels.

Abrasive Material	Knoop Hardness
Aluminum oxide	2100
Silicon carbide	2500
Cubic boron nitride	5000
Diamond (artificial)	7000

finishes, while larger grain sizes permit larger material removal rates. Thus, a choice must be made between these two objectives when selecting abrasive grain size. The selection of grit size also depends to some extent on the type of work material. Harder work materials require smaller grain sizes to cut effectively, while softer materials require larger grit sizes.

The grit size is measured using a screen mesh procedure, as explained in Section 16.1.1. In this procedure, smaller grit sizes have larger numbers and vice versa. Grain sizes used in grinding wheels typically range between 8 and 250. Grit size 8 is very coarse and size 250 is very fine. Finer grit sizes are used for lapping and superfinishing (Section 25.2).

Bonding Materials   The bonding material holds the abrasive grains and establishes the shape and structural integrity of the grinding wheel. Desirable properties of the bond material include strength, toughness, hardness, and temperature resistance. The bonding material must be able to withstand the centrifugal forces and high temperatures experienced by the grinding wheel, resist shattering in shock loading of the wheel, and hold the abrasive grains rigidly in place to accomplish the cutting action while allowing those grains which are worn to be dislodged so that new grains can be exposed. Bonding materials commonly used in grinding wheels include the following:

> *Vitrified bond*—Vitrified bonding material consists chiefly of baked clay and ceramic materials. Most grinding wheels in common use are vitrified bonded wheels. They are strong and rigid, resistant to elevated temperatures, and relatively unaffected by water and oil that might be used in grinding fluids.
> *Silicate bond*—This bond material consists of sodium silicate (Na_2SO_3). Its applications are generally limited to situations in which heat generation must be minimized, such as the grinding of cutting tools.
> *Rubber bond*—Rubber is the most flexible of the bonding materials. It is used as a bonding material in cut-off wheels.
> *Resinoid bond*—This bond is made of various thermosetting resin materials, such as phenol-formaldehyde. It has very high strength and is used for rough grinding and cut-off operations.
> *Shellac bond*—Shellac-bonded grinding wheels are relatively strong but not rigid. They are often used in applications requiring a good finish.
> *Metallic bond*—Metal bonds, usually bronze, are the common bond material for diamond and cubic boron nitride grinding wheels. Particulate processing techniques (Chapters 16 and 17) are used to bond the matrix of abrasive grains and bonding material to only the outside periphery of the wheel, thus conserving the costly abrasive materials.

Wheel Structure and Wheel Grade   *Wheel structure* refers to the relative spacing of the abrasive grains in the wheel. In addition to the abrasive grains and bond material, grinding wheels contain air gaps or pores, as illustrated in Figure 25.1. The volumetric proportions of grains, bond material, and pores can be expressed as

$$P_g + P_b + P_p = 1.0 \qquad (25.1)$$

where P_g = proportion of abrasive grains in the total wheel volume, P_b = proportion of bond material, and P_p = proportion of pores (air gaps).

Wheel structure is measured on a scale that ranges between "open" and "dense." An open structure is one in which P_p is relatively large, and P_g is relatively small. That

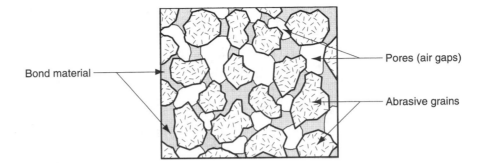

FIGURE 25.1 Typical structure of a grinding wheel.

is, there are more pores and fewer grains per unit volume in a wheel of open structure. By contrast, a dense structure is one in which P_p is relatively small, and P_g is larger. Generally, open structures are recommended in situations where clearance for chips must be provided. Dense structures are used to obtain better surface finish and dimensional control.

Wheel grade indicates the grinding wheel's bond strength in retaining the abrasive grits during cutting. This is largely dependent on the amount of bonding material present in the wheel structure—P_b in Eq. (25.1). Grade is measured on a scale that ranges between soft and hard. "Soft" wheels lose grains readily, while "hard" wheels retain their abrasive grains. Soft wheels are generally used for applications requiring low material removal rates and grinding of hard work materials. Hard wheels are typically used to achieve high stock removal rates and for grinding of relative soft work materials.

Grinding Wheel Specification The preceding parameters can be concisely designated in a standard grinding wheel marking system defined by the American National Standards Institute (ANSI)[2]. This marking system uses numbers and letters to specify abrasive type, grit size, grade, structure, and bond material. Table 25.2 presents an abbreviated version of the ANSI Standard, indicating how the numbers and letters are interpreted. The standard also provides for additional identifications that might be used by the grinding wheel manufacturers.

The ANSI Standard for diamond and cubic boron nitride grinding wheels is slightly different than for conventional wheels. The marking system for these newer grinding wheels is presented in Table 25.3.

TABLE 25.2   Marking system for conventional grinding wheels as defined by ANSI Standard B74.13–1977 [ANSI77].

30	A	46	H	6	V	XX

Manufacturer's private marking for wheel (optional).

Bond type: B = Resinoid, BF = resinoid reinforced, E = Shellac, R = Rubber, RF = rubber reinforced, S = Silicate, V = Vitrified.

Structure: Scale ranges from 1 to 15: 1 = very dense structure, 15 = very open structure.

Grade: Scale ranges from A to Z: A = soft, M = medium, A = hard.

Grain size: Coarse = grit sizes 8 to 24, Medium = grit sizes 30 to 60, Fine = grit sizes 70 to 180, Very fine = grit sizes 220 to 600.

Abrasive type: A = aluminum oxide, C = silicon carbide.

Prefix: Manufacturer's symbol for abrasive (optional).

TABLE 25.3 Marking system for diamond and cubic boron nitride grinding wheels as defined by ANSI Standard B74.13-1977 [2].

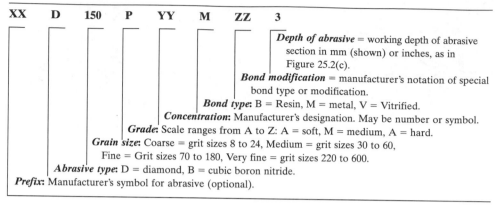

XX   D   150   P   YY   M   ZZ   3

Depth of abrasive = working depth of abrasive section in mm (shown) or inches, as in Figure 25.2(c).

Bond modification = manufacturer's notation of special bond type or modification.

Bond type: B = Resin, M = metal, V = Vitrified.

Concentration: Manufacturer's designation. May be number or symbol.

Grade: Scale ranges from A to Z: A = soft, M = medium, A = hard.

Grain size: Coarse = grit sizes 8 to 24, Medium = grit sizes 30 to 60, Fine = Grit sizes 70 to 180, Very fine = grit sizes 220 to 600.

Abrasive type: D = diamond, B = cubic boron nitride.

Prefix: Manufacturer's symbol for abrasive (optional).

FIGURE 25.2 Some of the standard grinding wheel shapes: (a) straight, (b) recessed two sides, (c) metal wheel frame with abrasive bonded to outside circumference, (d) abrasive cut-off wheel, (e) cylinder wheel, (f) straight cup wheel, and (g) flaring cup wheel.

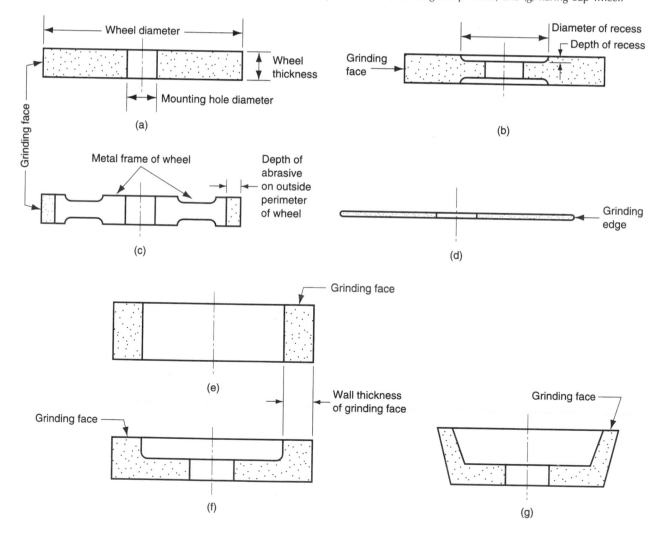

Grinding wheels come in a variety of shapes and sizes, as in Figure 25.2. Configurations (a), (b), and (c) are peripheral grinding wheels, in which material removal is accomplished by the outside circumference of the wheel. A typical abrasive cut-off wheel is shown in (d), which also involves peripheral cutting. Wheels (e), (f), and (g) are face grinding wheels, in which the flat face of the wheel removes material from the work surface.

25.1.2 Analysis of the Grinding Process

The cutting conditions in grinding are characterized by very high speeds and very small cut size, compared to milling and other traditional machining operations. Using surface grinding to illustrate, Figure 25.3(a) shows the principal features of the process. The peripheral speed of the grinding wheel is determined by the rotational speed of the wheel:

$$v = \pi DN \tag{25.2}$$

where v = surface speed of wheel, m/min (ft/min); N = spindle speed, rev/min; and D = wheel diameter, m (ft).

Depth of cut d, called the **infeed,** is the penetration of the wheel below the original work surface. As the operation proceeds, the grinding wheel is fed laterally across the surface on each pass by the work. This is called the **crossfeed,** and it determines the width of the grinding path w in Figure 25.3(a). This width, multiplied by depth d determines the cross-sectional area of the cut. In most grinding operations, the work moves past the wheel at a certain speed v_w, so that the material removal rate MRR is

$$MRR = v_w wd \tag{25.3}$$

Each grain in the grinding wheel cuts an individual chip whose longitudinal shape before cutting is shown in Figure 25.3(b) and whose assumed cross-sectional shape is tri-

FIGURE 25.3   (a) The geometry of surface grinding, showing the cutting conditions; (b) assumed longitudinal shape; and (c) the cross section of a single chip.

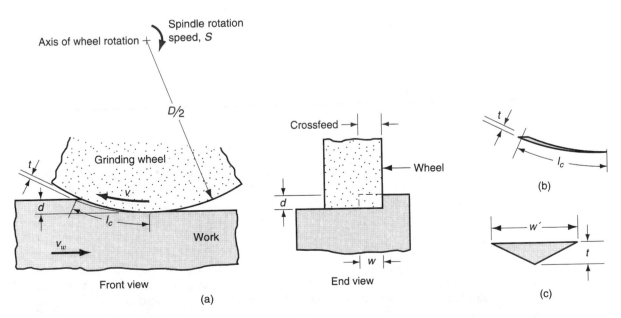

angular, as in Figure 25.3(c). At the exit point of the grit from the work, where the chip cross section is largest, this triangle has height t and width w'.

In a grinding operation, we are interested in how the cutting conditions combine with the grinding wheel parameters to affect the following: (1) surface finish, (2) forces and energy, (3) temperature of the work surface, and (4) wheel wear.

Surface Finish   Most commercial grinding is performed to achieve a surface finish that is superior to that which can be accomplished with conventional machining. The surface finish of the workpart is affected by the size of the individual chips formed during grinding. One obvious factor in determining chip size is grit size—smaller grit sizes yield better finishes.

Let us examine the dimensions of an individual chip. From the geometry of the grinding process in Figure 25.3, it can be shown that the average length of a chip is given by

$$l_c = \sqrt{Dd} \qquad (25.4)$$

where l_c is the length of the chip, mm (in.); D = wheel diameter, mm (in.); and d = depth of cut, or infeed, mm (in.). This assumes the chip is formed by a grit that acts throughout the entire sweep arc shown in the diagram.

Figure 25.3(c) shows the assumed cross section of a chip in grinding. The cross-sectional shape is triangular with width w' being greater than the thickness t by a factor called the grain aspect ratio r_g, defined by

$$r_g = \frac{w'}{t} \qquad (25.5)$$

Typical values of grain aspect ratio are between 10 and 20.

The number of active grits (cutting teeth) per square inch on the outside periphery of the grinding wheel is denoted by C, whose value is normally inversely proportional to grit size. C is also related to the wheel structure. A denser structure means more grits per area. Based on the value of C, the number of chips formed per time n_c is given by

$$n_c = vwC \qquad (25.6)$$

where v = wheel speed, mm/min (in./min); w = cross-feed, mm (in.); and C = grits per area on the grinding wheel surface, grits/mm^2 (grits/in.^{2}). It stands to reason that surface finish will be improved by increasing the number of chips formed per unit time on the work surface for a given width w. Therefore, according to Eq. (25.6), increasing v and/or C will improve finish. Recall that smaller grain sizes give larger C values.

Forces and Energy   If the force required to drive the work past the grinding wheel were known, the specific energy in grinding could be determined as

$$U = \frac{F_c v}{v_w w d} \qquad (25.7)$$

where U = specific energy, J/mm^3 (in.-lb/in.^{3}); F_c = cutting force, which is the force to drive the work past the wheel, N (lb); v = wheel speed, m/min (ft/min); v_w = work speed, mm/min (in./min); w = width of cut, mm (in.); and d = depth of cut, mm (in.).

In grinding, the specific energy is much greater than in conventional machining. There are several reasons for this. First is the *size effect* in machining. As discussed previously, the chip thickness in grinding is much smaller than for other machining operations, such as milling. According to the size effect (Section 21.4), the small chip sizes in grinding cause the energy required to remove each unit volume of material to be significantly higher than in conventional machining—roughly 10 times higher.

Second, the individual grains in a grinding wheel have extremely negative rake angles. The average rake angle is about $-30°$, with values on some individual grains believed to be as low as $-60°$. These very low rake angles result in low values of shear plane angle and high shear strains, both of which mean higher energy levels in grinding.

Third, specific energy is higher in grinding because not all of the individual grits are engaged in actual cutting. Because of the random positions and orientations of the grains in the wheel, some grains do not project far enough into the work surface to accomplish cutting. Three types of grain actions can be recognized, as illustrated in Figure 25.4: (a) *cutting,* in which the grit projects far enough into the work surface to form a chip and remove material; (b) *plowing,* in which the grit projects into the work, but not far enough to cause cutting; instead, the work surface is deformed and energy is consumed without any material removal; and (c) *rubbing,* where the grit contacts the surface during its sweep, but only rubbing friction occurs, thus consuming energy without removing any material.

The size effect, negative rake angles, and ineffective grain actions combine to make the grinding process inefficient in terms of energy consumption per volume of material removed.

Using the specific energy relationship in Eq. (25.7), and assuming that the cutting force acting on a single grain in the grinding wheel is proportional to $r_g t$, it can be shown [8] that

$$F_c' = K_1 \left(\frac{r_g v_w}{vC} \right)^{0.5} \left(\frac{d}{D} \right)^{0.25} \tag{25.8}$$

where F_c' is the cutting force acting on an individual grain, K_1 is a constant of proportionality that depends on the strength of the material being cut and the sharpness of the individual grain, and the other terms have been previously defined. The practical significance of this relationship is that F_c' affects whether or not an individual grain will be pulled out of the grinding wheel, an important factor in the wheel's capacity to "resharpen" itself. Referring back to our discussion on wheel grade, a hard wheel can be made to appear softer by increasing the cutting force acting on an individual grain through appropriate adjustments in v_w, v, and d, according to Eq. (25.8).

Temperatures at the Work Surface   Because of the size effect, high negative rake angles, and plowing and rubbing of the abrasive grits against the work surface, the grinding process is characterized by high temperatures and high friction. Unlike conventional machining operations in which most of the heat energy generated in the process is carried off in the chip, most of the energy in grinding remains in the ground surface [10], resulting in high work surface temperatures. The high surface temperatures have several possible damaging effects, primarily surface burns and cracks. The burn marks show

FIGURE 25.4   Three types of grain action in grinding: (a) cutting, (b) plowing, and (c) rubbing.

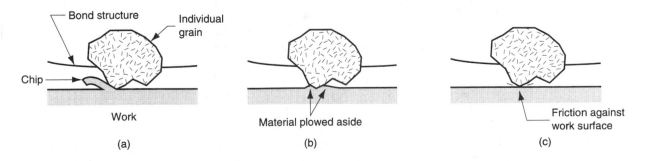

themselves as discolorations on the surface due to oxidation. Grinding burns are often a sign of metallurgical damage immediately beneath the surface. The surface cracks are perpendicular to the wheel speed direction. They indicate an extreme case of thermal damage to the work surface.

A second harmful thermal effect is softening of the work surface. Many grinding operations are carried out on parts that have been heat-treated to obtain high hardness. High grinding temperatures can cause the surface to lose some of its hardness. Third, thermal effects in grinding can cause residual stresses in the work surface, possibly decreasing the fatigue strength of the part.

It is important to understand which factors influence work surface temperatures in grinding. Experimentally, it has been observed that surface temperature is dependent on energy per surface area ground (closely related to specific energy U). Since this varies inversely with chip thickness, it can be shown that surface temperature T_s is related to grinding parameters as follows [8]:

$$T_s = K_2 d^{0.75} \left(\frac{r_g C v}{v_w} \right)^{0.5} D^{0.25} \tag{25.9}$$

where K_2 = a constant of proportionality. The practical implication of this relationship is that surface damage due to high work temperatures can be mitigated by decreasing depth of cut d, wheel speed v, and number of active grits per square inch on the grinding wheel C, or by increasing work speed v_w. In addition, dull grinding wheels and wheels that have a hard grade and dense structure tend to cause thermal problems. Of course, grinding temperatures can also be reduced by using a cutting fluid.

Wheel Wear Grinding wheels wear, just as conventional cutting tools wear. Three mechanisms are recognized as the principal causes of wear in grinding wheels: (1) grain fracture, (2) attritious wear, and (3) bond fracture. *Grain fracture* occurs when a portion of the grain breaks off, but the rest of the grain remains bonded in the wheel. The edges of the fractured area become new cutting edges on the grinding wheel. The tendency of the grain to fracture is called *friability.* High friability means the grains fracture more readily due to the cutting forces on the grains F_c'.

Attritious wear involves dulling of the individual grains, resulting in flat spots and rounded edges. Attritious wear is analogous to tool wear in a conventional cutting tool. It is caused by similar physical mechanisms including friction and diffusion, as well as chemical reactions between the abrasive material and the work material in the presence of very high temperatures.

Bond fracture occurs when the individual grains are pulled out of the bonding material. The tendency toward this mechanism depends on wheel grade, among other factors. Bond fracture usually occurs because the grain has become dull due to attritious wear, and the resulting cutting force is excessive. Sharp grains cut more efficiently with lower cutting forces; hence, they remain attached in the bond structure.

The three mechanisms combine to cause the grinding wheel to wear as depicted in Figure 25.5. Three wear regions can be identified. In the first region, the grains are initially sharp, and wear is accelerated due to grain fracture. This corresponds to the "break-in" period in conventional tool wear. In the second region, the wear rate is fairly constant, resulting in a linear relationship between wheel wear and volume of metal removed. This region is characterized by attritious wear, with some grain and bond fracture. In the third region of the wheel wear curve, the grains become dull, and the amount of plowing and rubbing increases relative to cutting. In addition, some of the chips become clogged in the pores of the wheel. This is called *wheel loading,* and it impairs the cutting action and leads to higher heat and work surface temperatures. As a consequence, grinding ef-

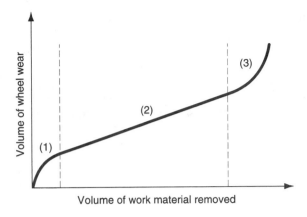

FIGURE 25.5   Typical wear curve of a grinding wheel. Wear is conveniently plotted as a function of volume of material removed, rather than as a function of time (based on [12].)

ficiency decreases, and the volume of wheel removed increases relative to the volume of metal removed.

The ***grinding ratio*** is a term used to indicate the slope of the wheel wear curve. Specifically,

$$GR = \frac{V_w}{V_g} \qquad (25.10)$$

where GR = the grinding ratio; V_w = the volume of work material removed; and V_g = the corresponding volume of the grinding wheel that is worn in the process. The grinding ratio has the most significance in the linear wear region of Figure 25.5. Typical values of GR range between 95 and 125 [4], which is about five orders of magnitude less than the analogous ratio in conventional machining. Grinding ratio is generally increased by increasing wheel speed v. The reason for this is that the size of the chip formed by each grit is smaller with higher speeds, so the amount of grain fracture is reduced. Since higher wheel speeds also improve surface finish, there is a general advantage in operating at high grinding speeds. However, when speeds become too high, attritious wear and surface temperatures increase. As a result, the grinding ratio is reduced and the surface finish is impaired. This effect was originally reported by Krabacher [12], as in Figure 25.6.

When the wheel is in the third region of the wear curve, it must be resharpened by a procedure called ***dressing,*** which consists of (1) breaking off the dulled grits on the out-

FIGURE 25.6   Grinding ratio and surface finish as a function of wheel speed. (Based on data in Krabacher [12].)

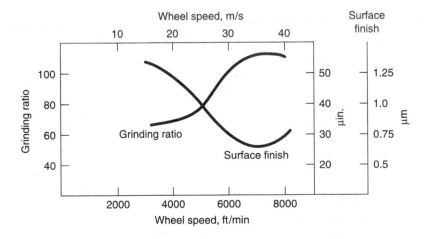

side periphery of the grinding wheel in order to expose fresh sharp grains and (2) removing chips that have become clogged in the wheel. It is accomplished by a rotating disk, an abrasive stick, or another grinding wheel operating at high speed, held against the wheel being dressed as it rotates. Although dressing sharpens the wheel, it does not guarantee the shape of the wheel. *Truing* is an alternative procedure which not only sharpens the wheel, but also restores its cylindrical shape and insures that it is straight across its outside perimeter. The procedure involves the use of a diamond-pointed tool (other types of truing tools can also be used) that is fed slowly and precisely across the wheel as it rotates. A very light depth is taken against the wheel.

25.1.3 Application Considerations in Grinding

In this section, we attempt to bring together the previous discussion of wheel parameters and theoretical analysis of grinding and consider their practical application. We also consider grinding fluids, which are commonly used in the grinding process.

Application Guidelines   There are many variables in grinding that affect the performance and success of the operation. The following guidelines are helpful in sorting out the many complexities and selecting the proper wheel parameters and grinding conditions (compiled from [6, 10, and 13]):

1. To optimize surface finish, select a small grit size and dense wheel structure. Also, use higher wheel speeds (v), and lower work speeds (v_w). Smaller depths of cut (d) and larger wheel diameters (D) will also help somewhat.
2. To maximize material removal rate, select a large grit size, more open wheel structure, and vitrified bond.
3. For grinding steel and most cast irons, select aluminum oxide as the abrasive.
4. For grinding most nonferrous metals, select silicon carbide as the abrasive.
5. For grinding hardened tool steels and certain aerospace alloys, choose cubic boron nitride as the abrasive.
6. For grinding hard abrasive materials such as ceramics, cemented carbides, and glass, choose diamond as the abrasive.
7. For soft metals, choose a large grit size and harder grade wheel. For hard metals, choose a small grit size and softer grade wheel.
8. To minimize heat damage, cracking, and warping of the work surface, maintain the sharpness of the wheel. Dress the wheel frequently. Also, use lighter depths of cut (d), lower wheel speeds (v), and faster work speeds (v_w).
9. If the grinding wheel glazes and burns, select a wheel with a softer grade and more open structure.
10. If the grinding wheel breaks down too rapidly, select a wheel with a harder grade and denser structure.

Grinding Fluids   The proper application of cutting fluids has been found to be effective in reducing the thermal effects and high work surface temperatures described previously. When used in grinding operations, cutting fluids are called grinding fluids. The functions performed by grinding fluids are similar to those performed by cutting fluids (Section 23.4). Reducing friction and removing heat from the process are the two common functions. In addition, washing away chips and reducing temperature of the work surface are very important in grinding.

Types of grinding fluids by chemistry include grinding oils and emulsified oils. The grinding oils are derived from petroleum and other sources. These products would seem to be attractive because friction is such an important factor in grinding. However, they pose hazards in terms of fire and operator health, and their cost is high relative to emulsified oils. In addition, their capacity to carry away heat is less than fluids based on water. Accordingly, mixtures of oil in water are most commonly recommended as grinding fluids. These are usually mixed with higher concentrations than emulsified oils used as conventional cutting fluids. In this way, the friction reduction mechanism is emphasized.

25.1.4 Grinding Operations and Grinding Machines

Grinding is traditionally used to finish parts whose geometries have already been created by other operations. Accordingly, grinding machines have been developed to grind plain flat surfaces, external and internal cylinders, and contour shapes such as threads. The contour shapes are often created by special formed wheels that have the opposite of the desired contour to be imparted to the work. Grinding is also used in tool rooms to form the geometries on cutting tools. In addition to these traditional uses, applications of grinding are expanding to include more high-speed, high-material removal operations. Our discussion of operations and machines in this section includes the following types: (1) surface grinding, (2) cylindrical grinding, (3) centerless grinding, (4) creep feed grinding, and (5) other grinding operations.

Surface Grinding   Surface grinding is normally used to grind plain flat surfaces. It is performed using either the periphery of the grinding wheel or the flat face of the wheel. Since the work is normally held in a horizontal orientation, peripheral grinding is performed by rotating the wheel about a horizontal axis, and face grinding is performed by rotating the wheel about a vertical axis. In either case, the relative motion of the workpart is achieved by reciprocating the work past the wheel or by rotating it. These possible combinations of wheel orientations and workpart motions yield the four types of surface grinding machines illustrated in Figure 25.7.

Of the four types, the horizontal spindle machine with reciprocating worktable is the most common, shown in Figure 25.8. Grinding is accomplished by reciprocating the work longitudinally under the wheel at a very small depth (infeed) and by feeding the wheel transversely into the work a certain distance between strokes. In these operations, the width of the wheel is usually less than that of the workpiece.

In addition to its conventional application, a grinding machine with horizontal spindle and reciprocating table can be used to form special contoured surfaces by employing a formed grinding wheel. Instead of feeding the wheel transversely across the work as it reciprocates, the wheel is *plunge-fed* vertically into the work. The shape of the formed wheel is therefore imparted to the work surface.

Grinding machines with vertical spindles and reciprocating tables are set up so that the wheel diameter is greater than the work width. Accordingly, these operations can be performed without using a transverse feed motion. Instead, grinding is accomplished by reciprocating the work past the wheel, and feeding the wheel vertically into the work to the desired dimension. This configuration is capable of achieving a very flat surface on the work.

Of the two types of rotary table grinding in Figure 25.7(b) and (d), the vertical spindle machines are more common. Owing to the relatively large surface contact area between wheel and workpart, vertical spindle-rotary table grinding machines are capable of high metal removal rates when equipped with appropriate grinding wheels.

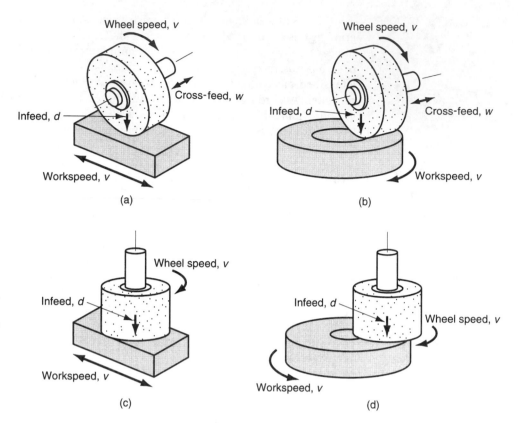

FIGURE 25.7   Four types of surface grinding: (a) horizontal spindle with reciprocating worktable, (b) horizontal spindle with rotating worktable, (c) vertical spindle with reciprocating worktable, and (d) vertical spindle with rotating worktable.

Cylindrical Grinding   As its name suggests, cylindrical grinding is used for rotational parts. These grinding operations divide into two basic types, Figure 25.9: (a) external cylindrical grinding and (b) internal cylindrical grinding.

   External cylindrical grinding (also called *center-type grinding* to distinguish it from centerless grinding) is performed much like a turning operation. The grinding machines

FIGURE 25.8   Surface grinder with horizontal spindle and reciprocating worktable.

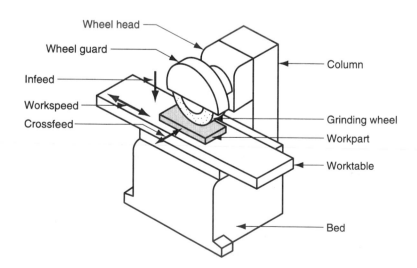

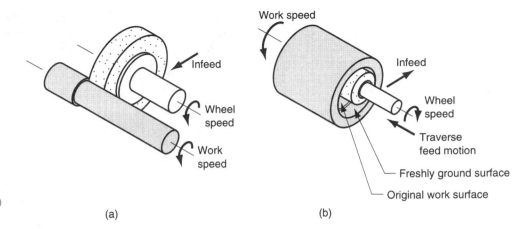

FIGURE 25.9 Two types of cylindrical grinding: (a) external, and (b) internal.

(a)   (b)

used for these operations closely resemble a lathe in which the tool post has been replaced by a high speed motor to rotate the grinding wheel. The cylindrical workpiece is rotated between centers to provide a surface speed of 18 to 30 m/min (60 to 100 ft/min) [13], and the grinding wheel, rotating at 1200 to 2000 m/min (4000 to 6500 ft/min), is engaged to perform the cut. There are two types of feed motion possible, traverse feed and plunge-cut, shown in Figure 25.10. In traverse feed, the grinding wheel is fed in a direction parallel to the axis of rotation of the workpart. The infeed is set within a range typically from 0.0075 to 0.075 mm (0.0003–0.003 in.). A longitudinal reciprocating motion is sometimes given to either the work or the wheel to improve surface finish. In plunge-cut, the grinding wheel is fed radially into the work. Formed grinding wheels use this type of feed motion.

External cylindrical grinding is used to finish parts that have been machined to approximate size and heat treated to desired hardness. Parts include axles, crankshafts, spindles, bearings and bushings, and rolls for rolling mills. The grinding operation produces the final size and required surface finish on these hardened parts.

Internal cylindrical grinding operates somewhat like a boring operation. The workpiece is usually held in a chuck and rotated to provide surface speeds of 20 to 60 m/min (75–200 ft/min) [13]. Wheel surface speeds similar to external cylindrical grinding are used. The wheel is fed in either of two ways: traverse feed, Figure 25.9(b), or

FIGURE 25.10 Two types of feed motion in external cylindrical grinding: (a) traverse feed, and (b) plunge-cut.

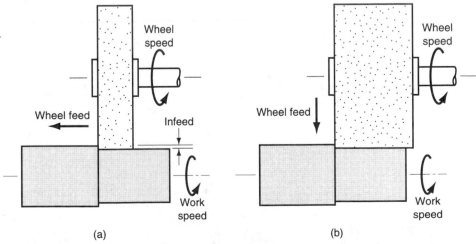

(a)   (b)

plunge feed. Obviously, the wheel diameter in internal cylindrical grinding must be smaller than the original bore hole. This often means that the wheel diameter is quite small, necessitating very high rotational speeds in order to achieve the desired surface speed. Internal cylindrical grinding is used to finish the hardened inside surfaces of bearing races and bushing surfaces.

Centerless Grinding   Centerless grinding is an alternative process for grinding external and internal cylindrical surfaces. As its name suggests, the workpiece is not held between centers. This results in a reduction in work handling time; hence, centerless grinding is often used for high production work. The setup for *external centerless grinding,* Figure 25.11, consists of two wheels: the grinding wheel and a regulating wheel. The workparts, which may be many individual short pieces or long rods (e.g., 3–4 m long), are supported by a rest blade and fed through between the two wheels. The grinding wheel does the cutting, rotating at surface speeds of 1200 to 1800 m/min (4000–6000 ft/min). The regulating wheel rotates at much lower speeds and is inclined at a slight angle I to control throughfeed of the work. The following equation can be used to predict throughfeed rate, based on inclination angle and other parameters of the process [13]:

$$f_r = \pi D_r N_r \sin I \qquad (25.11)$$

where f_r = throughfeed rate, mm/min (in./min); D_r = diameter of the regulating wheel, mm (in.); N_r = rotational speed of the regulating wheel, rev/min; and I = inclination angle of the regulating wheel.

The typical setup in *internal centerless grinding* is shown in Figure 25.12. In place of the rest blade, two support rolls are used to maintain the position of the work. The regulating wheel is tilted at a small inclination angle to control the feed of the work past the grinding wheel. Because of the need to support the grinding wheel, throughfeed of the work as in external centerless grinding is not possible. Therefore this grinding operation cannot achieve the same high production rates as in the external centerless process. Its advantage is that it is capable of providing very close concentricity between internal and external diameters on a tubular part such as a roller bearing race.

Creep Feed Grinding   A relatively new form of grinding is creep feed grinding, developed around 1958. Creep feed grinding is performed at very high depths of cut and very low feed rates; hence, the name creep feed. The comparison with conventional surface grinding is illustrated in Figure 25.13.

FIGURE 25.11 External centerless grinding.

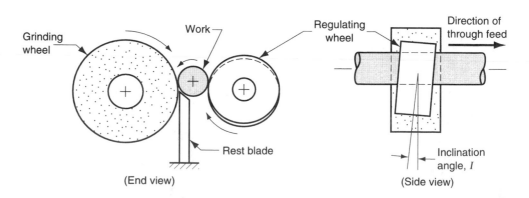

(End view)   (Side view)

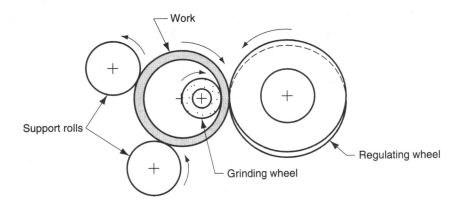

FIGURE 25.12 Internal centerless grinding.

Depths of cut in creep feed grinding are 1,000 to 10,000 times greater than in conventional surface grinding, and the feed rates are reduced by about the same proportion. However, material removal rate and productivity are increased in creep feed grinding because the wheel is continuously cutting. This contrasts with conventional surface grinding in which the reciprocating motion of the work results in significant lost time during each stroke.

Creep feed grinding can be applied in both surface grinding and external cylindrical grinding. Surface grinding applications include grinding of slots and profiles. The process seems especially suited to those cases in which depth-to-width ratios are relatively large. The cylindrical applications include threads, formed gear shapes, and other cylindrical components. The term *deep grinding* is used in Europe to describe these external cylindrical creep feed grinding applications.

The introduction of grinding machines designed with special features for creep feed grinding has spurred interest in the process. The features include [10] high static and dynamic stability, highly accurate slides with reduced tendency to stick-slip, increased spindle power (two to three times the power of conventional grinding machines), consistent table speeds for low feeds, high-pressure grinding fluid delivery systems, and dressing systems capable of dressing the grinding wheels during the process. Typical advantages of creep feed grinding include (1) high material removal rates, (2) improved accuracy for formed surfaces, and (3) reduced temperatures at the work surface.

FIGURE 25.13 Comparison of (a) conventional surface grinding and (b) creep feed grinding.

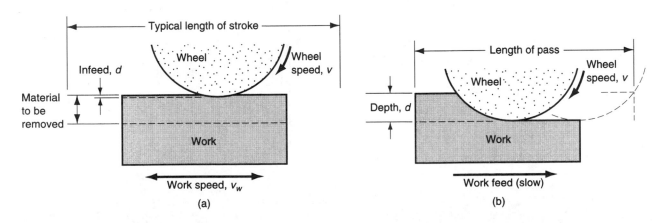

Other Grinding Operations   Several other grinding operations should be briefly mentioned to complete our review. These include tool grinding, jig grinding, disc grinding, snag grinding, and abrasive belt grinding.

Cutting tools are made of hardened tool steel and other hard materials. *Tool grinders* are special grinding machines of various designs to sharpen and recondition cutting tools. They have devices for positioning and orienting the tools to grind the desired surfaces at specified angles and radii. Some tool grinders are general purpose, while others cut unique geometries of specific tool types. General-purpose tool and cutter grinders use special attachments and adjustments to accommodate a variety of tool geometries. Single-purpose tool grinders include gear cutter sharpeners, milling cutter grinders of various types, broach sharpeners, and drill point grinders.

Jig grinders are grinding machines traditionally used to grind holes in hardened steel parts to high accuracies. The original applications included pressworking dies and tools. Although these applications are still important, jig grinders are used today in a broader range of applications where high accuracy and good finish are required on hardened components. Numerical control is available on modern jig grinding machines to achieve automated operation.

Disc grinders are grinding machines with large abrasive discs mounted on either end of a horizontal spindle as in Figure 25.14. The work is held (usually manually) against the flat surface of the wheel to accomplish the grinding operation. Some disc grinding machines have double opposing spindles. By setting the discs at the desired separation, the workpart can be fed automatically between the two discs and ground simultaneously on opposite sides. Advantages of the disc grinder are good flatness and parallelism at high production rates.

The *snag grinder* is similar in configuration to a disc grinder. The difference is that the grinding is done on the outside periphery of the wheel rather than on the side flat surface. The grinding wheels are therefore different in design than those in disc grinding. Snag grinding is generally a manual operation, used for rough grinding operations such as removing the flash from castings and forgings, and smoothing weld joints.

Abrasive belt grinding uses abrasive particles bonded to a flexible (cloth) belt. A typical setup is illustrated in Figure 25.15. Support of the belt is required when the work is pressed against it, and this support is provided by a roll or platten located behind the belt. A flat platten is used for work that will have a flat surface. A soft platten can be used if it is desirable for the abrasive belt to conform to the general contour of the part during grinding. Belt speed depends on the material being ground; a range of 750 to 1700 m/min (2500–5500 ft/min) is typical [13]. Owing to improvements in abrasives and bonding materials, abrasive belt grinding is being used increasingly for heavy stock

FIGURE 25.14 Typical configuration of a disc grinder.

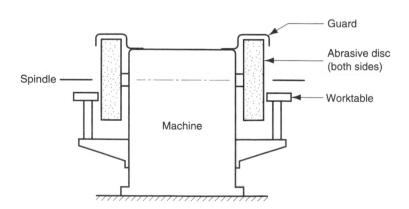

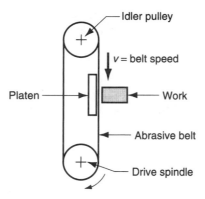

FIGURE 25.15   Abrasive belt grinding.

removal rates, rather than light grinding which was its traditional application. The term **belt sanding** refers to the light grinding applications in which the workpart is pressed against the belt to remove burrs and high spots, and to produce an improved finish quickly by hand.

25.2 RELATED ABRASIVE PROCESSES

Other abrasive processes include honing, lapping, superfinishing, polishing, and buffing. They are used exclusively as finishing operations. The initial part shape is created by some other process; then the part is finished by one of these operations to achieve superior surface finish. The usual part geometries and typical surface roughness values for these processes are indicated in Table 25.4. For comparison, we also present corresponding data for grinding.

Another class of finishing operations, called mass finishing (Section 28.2.2), is used to finish parts in bulk rather than individually. These mass finishing methods are also used for cleaning and deburring.

25.2.1 Honing

Honing is an abrasive process performed by a set of bonded abrasive sticks. A common application is to finish the bores of internal combustion engines. Other applications include bearings, hydraulic cylinders, and gun barrels. Surface finishes of around 0.12 μm (5 μ-in.) or slightly better are typically achieved in these applications. In addition, hon-

TABLE 25.4   Usual part geometries for honing, lapping, superfinishing, polishing and buffing.

Process	Usual Part Geometry	Surface Roughness, μm (μ-in.)
Grinding, medium grit size	Flat, external cylinders, round holes	0.4 to 1.6 (16 to 63)
Grinding, fine grit size	Flat, external cylinders, round holes	0.2 to 0.4 (8 to 16)
Honing	Round hole (e.g., engine bore)	0.1 to 0.8 (4 to 32)
Lapping	Flat or slightly spherical (e.g., lens)	0.025 to 0.4 (1 to 16)
Superfinishing	Flat surface, external cylinder	0.013 to 0.2 (0.5 to 8)
Polishing	Miscellaneous shapes	0.025 to 0.8 (1 to 32)
Buffing	Miscellaneous shapes	0.013 to 0.4 (0.5 to 16)

ing produces a characteristic cross-hatched surface that tends to retain lubrication during operation of the component, thus contributing to its function and service life.

The honing process for an internal cylindrical surface is illustrated in Figure 25.16. The honing tool consists of a set of bonded abrasive sticks. Four sticks are used on the tool shown in the figure, but the number depends on hole size. Two to four sticks would be used for small holes (e.g., gun barrels), and a dozen or more would be used for larger diameter holes. The motion of the honing tool is a combination of rotation and linear reciprocation, regulated in such a way that a given point on the abrasive stick does not trace the same path repeatedly. This rather complex motion accounts for the cross-hatched pattern on the bore surface. Honing speeds are 15 to 150 m/min (50–500 ft/min) [3]. During the process, the sticks are pressed outward against the hole surface to produce the desired abrasive cutting action. Hone pressures of 1 to 3 MPa (150–450 lb/in.^{2}) are typical, although pressures outside this range have also been reported [3]. The honing tool is supported in the hole by two universal joints, thus causing the tool to follow the previously defined hole axis. Honing enlarges and finishes the hole but cannot change its location.

Grit sizes in honing range between 30 and 600. The same trade-off between better finish and faster material removal rates exists in honing as in grinding. The amount of material removed from the work surface during a honing operation may be as much as 0.5 mm (0.020 in.), but is usually much less than this. A cutting fluid must be used in honing to cool and lubricate the tool and to help remove the chips.

25.2.2 Lapping

Lapping is an abrasive process used to produce surface finishes of extreme accuracy and smoothness. It is used in the production of optical lenses, metallic bearing surfaces, gages, and other parts requiring very good finishes. Metal parts that are subject to fatigue loading or surfaces that must be used to establish a seal with a mating part are often lapped.

Instead of a bonded abrasive tool, lapping uses a fluid suspension of very small abrasive particles between the workpiece and the lapping tool. The process is illustrated in Figure 25.17 as applied in lens-making. The fluid with abrasives is referred to as the

FIGURE 25.16 The honing process: (a) the honing tool used for internal bore surface, and (b) cross-hatched surface pattern created by the action of the honing tool.

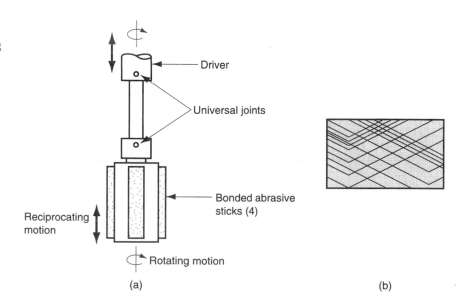

Driver

Universal joints

Bonded abrasive sticks (4)

Reciprocating motion

Rotating motion

(a)

(b)

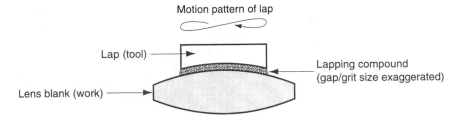

FIGURE 25.17 The lapping process in lens-making.

lapping compound and has the general appearance of a chalky paste. The fluids used to make the compound include oils and kerosene. Common abrasives are aluminum oxide and silicon carbide with typical grit sizes between 300 to 600. The lapping tool is called a *lap,* and it has the reverse of the desired shape of the workpart. To accomplish the process, the lap is pressed against the work and moved back and forth over the surface in a "figure-eight" or other motion pattern, subjecting all portions of the surface to the same action. Lapping is sometimes performed by hand, but lapping machines accomplish the process with greater consistency and efficiency.

Materials used to make the lap range from steel and cast iron to copper and lead. Wood laps have also been made. Because a lapping compound is used rather than a bonded abrasive tool, the mechanism by which this process works is somewhat different than grinding and honing. It is hypothesized that two alternative cutting mechanisms are at work in lapping [3]. The first mechanism is that the abrasive particles roll and slide between the lap and the work, with very small cuts occurring in both surfaces. The second mechanism is that the abrasives become imbedded in the lap surface and the cutting action is very similar to grinding. It is likely that lapping is a combination of these two mechanisms, depending on the relative hardnesses of the work and the lap. For laps made of soft materials, the embedded grit mechanism is emphasized; and for hard laps, the rolling and sliding mechanism dominates.

25.2.3 Superfinishing

Superfinishing is an abrasive process similar to honing. Both processes use a bonded abrasive stick moved with a reciprocating motion and pressed against the surface to be finished. Superfinishing differs from honing in the following respects [3]: (1) the strokes are shorter, 5 mm (3/16 in.); (2) higher frequencies are used, up to 1500 strokes per minute; (3) lower pressures are applied between the tool and the surface, below 0.28 MPa (40 lb/in.^{2}); (4) workpiece speeds are lower, 15 m/min (50 ft/min) or less; and (5) grit sizes are generally smaller. The relative motion between the abrasive stick and the work surface is varied so that individual grains do not retrace the same path. A cutting fluid is used to cool the work surface and wash away chips. In addition, the fluid tends to separate the abrasive stick from the work surface after a certain level of smoothness is achieved, thus preventing further cutting action. The result of these operating conditions is mirror-like finishes with surface roughness values around 0.025 μm (1 μ-in.). Superfinishing can be used to finish flat and external cylindrical surfaces. The process is illustrated in Figure 25.18 for the latter geometry.

25.2.4 Polishing and Buffing

Polishing is used to remove scratches and burrs and to smooth rough surfaces by means of abrasive grains attached to a polishing wheel rotating at high speed—around 2300 m/min (7500 ft/min). The wheels are made of canvas, leather, felt, and even paper; thus, the wheels

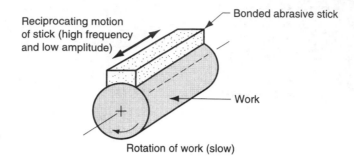

FIGURE 25.18
Superfinishing on an external cylindrical surface.

are somewhat flexible. The abrasive grains are glued to the outside periphery of the wheel. After the abrasives have been worn down and used up, the wheel is replenished with new grits. Grit sizes of 20 to 80 are used for rough polishing, 90 to 120 for finish polishing, and above 120 for fine finishing. Polishing operations are often accomplished manually.

Buffing is similar to polishing in appearance, but its function is different. Buffing is used to provide attractive surfaces with high luster. Buffing wheels are made of materials similar to those used for polishing wheels—leather, felt, cotton, etc.—but buffing wheels are generally softer. The abrasives are very fine and are contained in a buffing compound that is pressed into the outside surface of the wheel while it rotates. This contrasts with polishing in which the abrasive grits are glued to the wheel surface. As in polishing, the abrasive particles must be periodically replenished, and buffing is usually done manually, although machines have been designed to perform the process automatically. Buffing is performed at speeds generally ranging from 2400 to 5200 m/min (8000–17000 ft/min).

REFERENCES

[1] Andrew, C., Howes, T. D., and Pearce, T. R. A. *Creep Feed Grinding.* Holt, Rinehart and Winston Ltd., London, 1985.

[2] *ANSI Standard B74.13-1977.* "Markings for Identifying Grinding Wheels and Other Bonded Abrasives." American National Standards Institute, New York, 1977.

[3] Armarego, E. J. A., and Brown, R. H. *The Machining of Metals.* Prentice-Hall, Inc., Englewood Cliffs, N. J., 1969, Chapter 11.

[4] Bacher, W. R., and Merchant, M. E. "On the Basic Mechanics of the Grinding Process." *Transactions ASME.* Series B, Vol. 80, No. 1, 1958, pp 141.

[5] Black, P. H. *Theory of Metal Cutting.* McGraw-Hill, New York, 1961, Chapter 9.

[6] Boothroyd, G., and Knight, W. A. *Fundamentals of Metal Machining and Machine Tools.* 2nd ed. Marcel Dekker, Inc., New York, 1989, Chapter 10.

[7] Boston, O. W. *Metal Processing.* 2nd ed. Wiley, New York, 1951, Chapter 15.

[8] Cook, N. H. *Manufacturing Analysis.* Addison-Wesley, Reading, Mass., 1966. Chapter 3.

[9] DeGarmo, E. P., J. T. Black, and R. A. Kohser. *Materials and Processes in Manufacturing.* 8th ed. Macmillan, New York, 1997, Chapter 27.

[10] Drozda, T. J., and C. Wick (eds.). *Tool and Manufacturing Engineers Handbook.* 4th Ed. Volume I: *Machining.* Society of Manufacturing Engineers, Dearborn, Mich., 1983, Chapter 11.

[11] Eary, D. F., and Johnson, G. E. *Process Engineering: for Manufacturing.* Prentice-Hall, Inc., Englewood Cliffs, N. J., 1962, Chapter 12.

[12] Krabacher, E. J. "Factors Influencing the Performance of Grinding Wheels." *Transactions ASME.* Series B, Vol 81, No. 3, 1959, pp 187–199.

[13] *Machining Data Handbook.* 3rd ed. Vols. I and II, Metcut Research Associates, Inc., Cincinnati, Ohio, 1980.

[14] Malkin, S. *Grinding Technology.* Ellis Horwood Ltd., Wiley, New York, 1989.

REVIEW QUESTIONS

25.1. Why are abrasive processes technologically and commercially important?

25.2. What are the five principal parameters of a grinding wheel?

25.3. What are some of the principal abrasive materials used in grinding wheels?

25.4. Name some of the principal bonding materials used in grinding wheels.

25.5. What is *wheel structure*?

25.6. What is *wheel grade*?

25.7. Why are specific energy values so much higher in grinding than in traditional metal cutting processes?

25.8. Grinding creates high temperatures. How is temperature harmful in grinding?

25.9. What are the three mechanisms of grinding wheel wear?

25.10. What is *dressing*, in reference to grinding wheels?

25.11. What is *truing*, in reference to grinding wheels?

25.12. What abrasive material would one select for grinding a cemented carbide cutting tool?

25.13. What are the functions of a grinding fluid?

25.14. What is *centerless grinding*?

25.15. How does creep feed grinding differ from conventional grinding?

25.16. How does abrasive belt grinding differ from a conventional surface grinding operation?

25.17. Name some of the abrasive operations available to achieve very good surface finishes.

MULTIPLE CHOICE QUIZ

There is a total of 17 correct answers in the following multiple choice questions (some questions have multiple answers that are correct). To attain a perfect score on the quiz, all correct answers must be given, since each correct answer is worth 1 point. For each question, each omitted answer or wrong answer reduces the score by 1 point, and each additional answer beyond the number of answers required reduces the score by 1 point. Percentage score on the quiz is based on the total number of correct answers.

25.1. Grinding is geometrically closest to which of the following conventional machining processes (one answer)? (a) drilling, (b) milling, (c) shaping, or (d) turning.

25.2. Of the following abrasive materials, which has the highest hardness? (a) aluminum oxide, (b) cubic boron nitride, (c) silicon carbide.

25.3. Smaller abrasive grain size in a grinding wheel causes which one of the following? (a) improve surface finish, (b) have no effect on surface finish, or (c) degrade surface finish.

25.4. Which of the following results in higher material removal rates? (a) larger grain size, or (b) smaller grain size.

25.5. Which of the following improves surface finish in grinding (more than one answer)? (a) higher wheel speed, (b) larger infeed, (c) lower wheel speed, (d) lower work speed.

25.6. Which of the following abrasive materials is most appropriate for grinding steel and cast iron (one best answer)? (a) aluminum oxide, (b) cubic boron nitride, (c) diamond, or (d) silicon carbide.

25.7. Which of the following abrasive materials is most appropriate for grinding hardened tool steel (one best answer)? (a) aluminum oxide, (b) cubic boron nitride, (c) diamond, or (d) silicon carbide.

25.8. Which of the following abrasive materials is most appropriate for grinding nonferrous metals (one best answer)? (a) aluminum oxide, (b) cubic boron nitride, (c) diamond, or (d) silicon carbide.

25.9. Which of the following will help to reduce the incidence of heat damage to the work surface in grinding (more than one answer)? (a) frequent dressing or truing of the wheel, (b) higher infeeds, (c) higher work speeds, or (d) lower wheel speeds.

25.10. Which of the following abrasive processes achieves the best surface finish (one best answer)? (a) centerless grinding, (b) honing, (c) lapping, or (d) superfinishing.

25.11. Which of the following abrasive processes could be used to finish a hole or internal bore (more than one answer)? (a) centerless grinding, (b) honing, (c) cylindrical grinding, (d) lapping, or (e) superfinishing.

25.12. The term *deep grinding* refers to which of the following (one best answer)? (a) alternative name for any creep feed grinding operation, (b) external cylindrical creep feed grinding, (c) grinding operation performed at the bottom of a hole, (d) surface grinding which uses a large cross-feed, or (e) surface grinding which uses a large infeed.

PROBLEMS

25.1. In a surface grinding operation the wheel diameter = 150 mm and the infeed = 0.07 mm. The wheel speed = 1450 m/min, work speed = 0.25 m/s, and the cross-feed = 5 mm. The number of active grits per area of wheel surface C = 0.75 grits/mm². Determine: (a) average length per chip, (b) metal removal rate, and (c) number of chips formed per unit time for the portion of the operation when the wheel is engaged in the work.

25.2. The following conditions and settings are used in a certain surface grinding operation: wheel diameter = 6.0 in. infeed = 0.003 in. wheel speed = 4750 ft/min, work speed = 50 ft/min, and cross-feed = 0.20 in. The number of active grits per square inch of wheel surface C = 500. Determine (a) the average length per chip, (b) the metal removal rate, and (c) the number of chips formed per unit time for the portion of the operation when the wheel is engaged in the work.

25.3. An internal cylindrical grinding operation is used to finish an internal bore from an initial diameter of 250.00 mm to a final diameter of 252.5 mm. The bore is 125 mm long. A grinding wheel with an initial diameter of 150.00 mm and a width of 20.00 mm is used. After the operation, the diameter of the grinding wheel has been reduced to 149.75 mm. Determine the grinding ratio in this operation.

25.4. In a surface grinding operation performed on hardened plain carbon steel, the grinding wheel has a diameter = 200 mm and width = 25 mm. The wheel rotates at 2400 rev/min, with a depth of cut (infeed) = 0.05 mm/pass and a cross-feed = 3.50 mm. The reciprocating speed of the work is 6 m/min, and the operation is performed dry. Determine (a) the length of contact between the wheel and the work, (b) the volume rate of metal removed. (c) If C = 0.64 active grits/mm², estimate the number of chips formed per unit time. (d) What is the average volume per chip? (e) If the tangential cutting force on the work = 30 N, compute the specific energy in this operation?

25.5. An 8-in.-diameter grinding wheel, 1.0 in. wide, is used in a certain surface grinding job performed on a flat piece of heat-treated 4340 steel. The wheel is rotating to achieve a surface speed of 5000 ft/min, with a depth of cut (infeed) = 0.002 in. per pass and a cross-feed = 0.15 in. The reciprocating speed of the work is 20 ft/min, and the operation is performed dry. (a) What is the length of contact between the wheel and the work? (b) What is the volume rate of metal removed? (c) If C = 300 active grits/in.², estimate the number of chips formed per unit time. (d) What is the average volume per chip? (e) If the tangential cutting force on the workpiece = 10 lb, what is the specific energy calculated for this job?

25.6. A surface grinding operation is being performed on a 6150 steel workpart (annealed, approximately 200 BHN). The designation on the grinding wheel is 51-C-24-D-5-V-23. The wheel diameter = 7.0 in., and its width = 1.00 in. Rotational speed = 3000 rev/min. The depth (infeed) = 0.002 in. per pass, and the cross-feed = 0.5 in. Workpiece speed = 20 ft/min. This operation has been a source of trouble right from the beginning. The surface finish is not as good as the 16 μ-in. specified on the part print, and there are signs of metallurgical damage on the surface. In addition, the wheel seems to become clogged almost as soon as the operation begins. In short, nearly everything that can go wrong with the job has gone wrong. (a) Determine the rate of metal removal when the wheel is engaged in the work. (b) If the number of active grits per square inch = 200, determine the average chip length and the number of chips formed per time. (c) What changes would you recommend in the grinding wheel to help solve the problems encountered? Explain why you made each recommendation.

25.7. The grinding wheel in a centerless grinding operation has a diameter = 200 mm, and the regulating wheel diameter = 125 mm. The grinding wheel rotates at 3000 rev/min and the regulating wheel rotates at 200 rev/min. The inclination angle of the regulating wheel = 2.5°. Determine the throughfeed rate of cylindrical workparts that are 25.0 mm in diameter and 175 mm long.

25.8. A centerless grinding operation uses a regulating wheel that is 150 mm in diameter and rotates at 500 rev/min. At what inclination angle should the regulating wheel be set, if it is desired to feed a workpiece with length = 3.5 m and diameter = 18 mm through the operation in exactly 45 sec.

25.9. In a certain centerless grinding operation, the grinding wheel diameter = 8.5 in. and the regulating wheel diameter = 5.0 in. The grinding wheel rotates at 3500 rev/min and the regulating wheel rotates at 150 rev/min. The inclination angle of the regulating wheel = 3 degrees. Determine the throughfeed rate of cylindrical workparts that have the following dimensions: diameter = 1.25 in. and length = 8.0 in.

25.10. It is desired to compare the cycle times required to grind a particular workpiece using traditional surface grinding and using creep feed grinding. The workpiece is 200 mm long, 30 mm wide, and 75 mm thick. To make a fair comparison, the grinding wheel in both cases is 250 mm in diameter, 35 mm in width, and rotates at 1500 rev/min. It is desired to remove 25 mm of material from the surface. When traditional grinding is used, the infeed is set at 0.025 mm, and the wheel traverses twice (forward and back) across the work surface during each pass before resetting the infeed. There is no cross-feed since the wheel width is greater than the work width. Each pass is made at a work speed of 12 m/min, but the wheel overshoots the part on both sides. With acceleration and deceleration, the wheel is engaged in the work for 50% of the time on each pass. When creep feed grinding is used, the depth is increased by 1000 and the forward feed is decreased by 1000. How long will it take to complete the grinding operation (a) with traditional grinding and (b) with creep feed grinding?

25.11. In a certain grinding operation, the grade of the grinding wheel should be "M" (medium), but the only avail-

able wheel is grade "T" (hard). It is desired to make the wheel appear softer by making changes in cutting conditions. What changes would you recommend?

25.12. An aluminum alloy is to be ground in an external cylindrical grinding operation to obtain a good surface finish. Specify the appropriate grinding wheel parameters and the grinding conditions for this job.

25.13. A HSS broach (hardened) is to be resharpened to achieve a good finish. Specify the appropriate parameters of the grinding wheel for this job.

25.14. Based on equations in the text, derive an equation to compute the average volume per chip formed in the grinding process.

NONTRADITIONAL MACHINING AND THERMAL CUTTING PROCESSES

CHAPTER CONTENTS

26.1 Mechanical Energy Processes
    26.1.1 Ultrasonic Machining
    26.1.2 Processes Using Water and Abrasive Jets
26.2 Electrochemical Machining Processes
    26.2.1 Electrochemical Machining
    26.2.2 Electrochemical Deburring and Grinding
26.3 Thermal Energy Processes
    26.3.1 Electric Discharge Processes
    26.3.2 Electron Beam Machining
    26.3.3 Laser Beam Machining
    26.3.4 Arc Cutting Processes
    26.3.5 Oxyfuel Cutting Processes
26.4 Chemical Machining
    26.4.1 Mechanics and Chemistry of Chemical Machining
    26.4.2 CHM Processes
26.5 Application Considerations

Conventional machining processes (i.e., turning, drilling, milling, etc.) use a sharp cutting tool to form a chip from the work by shear deformation. In addition to these conventional methods, there is a group of processes that use other mechanisms to remove material. The term ***nontraditional machining*** refers to this group of processes, which remove excess material by various techniques involving mechanical, thermal, electrical, or chemical energy (or combinations of these energies). They do not use a sharp cutting tool in the conventional sense.

The nontraditional processes have been developed since World War II largely in response to new and unusual machining requirements that could not be satisfied by conventional methods. These requirements, and the resulting commercial and technological importance of the nontraditional processes, include:

➢ The need to machine newly developed metals and nonmetals. These new materials often have special properties (e.g., high strength, high hardness, high toughness) that make them difficult or impossible to machine by conventional methods.

> ➤ The need for unusual and/or complex part geometries that cannot easily be accomplished and in some cases are impossible to achieve by conventional machining.

> ➤ The need to avoid surface damage that often accompanies the stresses raised by conventional machining.

Many of these requirements are associated with the aerospace and electronics industries, which have grown significantly in recent decades.

There are literally dozens of nontraditional machining processes, most of which are unique in their range of applications. In the present chapter, we discuss those that are most important commercially. More detailed discussions of these nontraditional methods are presented in several of the references, in particular [4], [5], and [13].

The nontraditional processes are often classified according to principal form of energy used to effect material removal. By this classification, there are four types:

(1) *Mechanical*—Mechanical energy in some form different from the action of a conventional cutting tool is used in these nontraditional processes. Erosion of the work material by a high velocity stream of abrasives or fluid (or both) is the typical form of mechanical action in these processes.

(2) *Electrical*—These nontraditional processes use electrochemical energy to remove material; the mechanism is the reverse of electroplating.

(3) *Thermal*—These processes use thermal energy to cut or shape the workpart. The thermal energy is generally applied to a very small portion of the work surface, causing that portion to be removed by fusion and/or vaporization of the material. The thermal energy is generated by the conversion of electrical energy.

(4) *Chemical*—Most materials (in particular, metals) are susceptible to chemical attack by certain acids or other etchants. In chemical machining, chemicals selectively remove material from portions of the workpart, while other portions of the surface are protected by a mask.

26.1 MECHANICAL ENERGY PROCESSES

In this section we examine several of the mechanical energy nontraditional processes: (1) ultrasonic machining, (2) water jet cutting, (3) abrasive water jet cutting, and (4) abrasive jet machining.

26.1.1 Ultrasonic Machining

Ultrasonic machining (USM) is a nontraditional machining process in which abrasives contained in a slurry are driven at high velocity against the work by a tool vibrating at low amplitude—around 0.075 mm (0.003 in.) and high frequency—approximately 20,000 Hz. The tool oscillates in a direction perpendicular to the work surface, and is fed slowly into the work, so that the shape of the tool is formed in the part. However, it is the action of the abrasives, impinging against the work surface, that performs the cutting. The general arrangement of the USM process is depicted in Figure 26.1.

Common tool materials used in USM include soft steel and stainless steel. Abrasive materials in USM include boron nitride, boron carbide, aluminum oxide, silicon carbide,

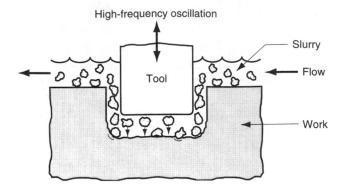

FIGURE 26.1   Ultrasonic machining.

and diamond. Grit size (Section 16.1.1) ranges between 100 and 2000. The vibration amplitude should be set approximately equal to the grit size, and the gap size should be maintained at about two times grit size. To a significant degree, grit size determines the surface finish on the new work surface.

In addition to surface finish, material removal rate is an important performance variable in ultrasonic machining. For a given work material, the removal rate in USM increases with increasing frequency and amplitude of vibration, as shown in Figure 26.2.

The cutting action in USM operates on the tool as well as the work. As the abrasive particles erode the work surface, they also erode the tool, thus affecting its shape. It is therefore important to know the relative volumes of work material and tool material removed during the process—similar to the grinding ratio (Section 25.1.2). This ratio of stock removed to tool wear varies for different work materials, ranging from around 100 : 1 for cutting glass down to about 1 : 1 for cutting tool steel.

The slurry in USM consists of a mixture of water and abrasive particles. Concentration of abrasives in water ranges from 20% to 60% [5]. The slurry must be continuously circulated in order to bring fresh grains into action at the tool–work gap. It also washes away chips and worn grits created by the cutting process.

The development of ultrasonic machining was motivated by the need to machine hard, brittle work materials, such as ceramics, glass, and carbides. It is also successfully used on certain metals such as stainless steel and titanium. Shapes obtained by USM include nonround holes, holes along a curved axis, and coining operations, in which an image pattern on the tool is imparted to a flat work surface.

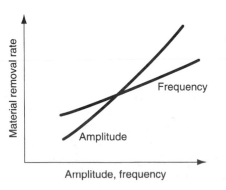

FIGURE 26.2   Effect of oscillation frequency and amplitude on material removal rate in USM.

26.1.2 Processes Using Water and Abrasive Jets

The processes described in this section remove material by means of high-velocity streams of water, abrasives, or combinations of the two.

Water Jet Cutting *Water jet cutting* (WJC) uses a fine, high-pressure, high-velocity stream of water directed at the work surface to cause cutting of the work, as illustrated in Figure 26.3. The name ***hydrodynamic machining*** is also used for this process, but water jet cutting seems to be the commonly used term in industry.

To obtain the fine stream of water a small nozzle opening of diameter 0.1 to 0.4 mm (0.004–0.016 in.) is used. To provide the stream with sufficient energy for cutting, pressures up to 400 MPa (60,000 lb/in.^{2}) are used, and the jet reaches velocities up to 900 m/s (3,000 ft/sec). The fluid is pressurized to the desired level by a hydraulic pump. The nozzle unit consists of a holder made of stainless steel and a jewel nozzle made of sapphire, ruby, or diamond. Diamond lasts the longest but costs the most. Filtration systems must be used in WJC to separate the swarf produced during cutting.

Cutting fluids in WJC are polymer solutions, preferred because of their tendency to produce a coherent stream. We have discussed cutting fluids before in the context of conventional machining (Section 23.4), but never has the term been more appropriately applied than in WJC.

Important process parameters include standoff distance, nozzle opening diameter, water pressure, and cutting feed rate. As in Figure 26.3, the ***standoff distance*** is the separation between the nozzle opening and the work surface. It is generally desirable for this distance to be small to minimize dispersion of the fluid stream before it strikes the surface. A typical standoff distance is 3.2 mm (0.125 in.). Size of the nozzle orifice affects the precision of the cut; smaller openings are used for finer cuts on thinner materials. To cut thicker stock, thicker jet streams and higher pressures are required. The cutting feed rate refers to the velocity at which the WJC nozzle is traversed along the cutting path. Typical feed rates range between 5 mm/s (12 in./min) to more than 500 mm/s (1200 in./min), depending on work material and its thickness [4]. The WJC process is usually automated using computer numerical control or industrial robots to manipulate the nozzle unit along the desired trajectory.

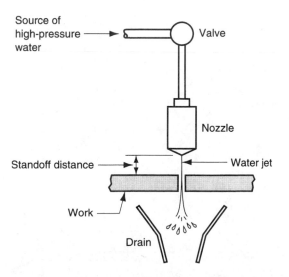

FIGURE 26.3   Water jet cutting.

WJC can be used effectively to cut narrow slits in flat stock such as plastic, textiles, composites, tile, carpet, leather, and cardboard. Robotic cells have been installed with WJC nozzles mounted as the robot's tool to follow cutting patterns that are irregular in three dimensions, such as cutting and trimming of automobile dashboards prior to assembly [6]. In these applications, the advantages of WJC include: no crushing or burning of the work surface typical in other mechanical or thermal processes, minimum material loss because of the narrow cut slit, no environmental pollution, and ease of automating the process using numerical control or industrial robots. A limitation of WJC is that the process is not suitable for cutting brittle materials (e.g., glass) because of their tendency to crack during cutting.

Abrasive Water Jet Cutting   When WJC is used on metallic workparts, abrasive particles must usually be added to the jet stream to facilitate cutting. This process is therefore called *abrasive water jet cutting* (AWJC). Introduction of the abrasive particles into the stream complicates the process by adding to the number of parameters that must be controlled (e.g., abrasive type, grit size, and flow rate). Aluminum oxide, silicon dioxide, and garnet (a silicate mineral) are typical abrasive materials used, at grit sizes ranging between 60 and 120. The abrasive particles are added to the water stream at approximately 0.25 kg/min (0.5 lb/min) after it has exited the WJC nozzle.

The remaining process parameters include those that are common to WJC: nozzle opening diameter, water pressure, and standoff distance. Nozzle orifice diameters are in the range 0.25 to 0.63 mm (0.010–0.025 in.)—somewhat larger than in WJC to permit higher flow rates and more energy to be contained in the stream prior to injection of abrasives. Water pressures are about the same as in WJC. Standoff distances are somewhat less to minimize the effect of dispersion of the cutting fluid which now contains abrasive particles. Typical standoff distances are between 1/4 and 1/2 of those in WJC.

Abrasive Jet Machining   Not to be confused with AWJC is the process called abrasive jet machining. *Abrasive jet machining* (AJM) is a material removal process that results from the action of a high-velocity stream of gas containing small abrasive particles, as shown in Figure 26.4. The gas is dry, and pressures of 0.2 to 1.4 MPa (25–200 lb/in.^{2}) are used to propel the gas through nozzle orifices of diameter 0.075 to 1.0 mm (0.003–0.040 in.) at velocities of 2.5 to 5.0 m/s (500–1000 ft/min). Gases include dry air, nitrogen, carbon dioxide, and helium.

Usually in AJM, an operator manually directs the nozzle at the work. Typical distances between nozzle tip and work surface range between 3 mm and 75 mm (0.125 in. and 3 in.). The workstation must be set up to provide proper ventilation for the operator.

FIGURE 26.4   Abrasive jet machining (AJM).

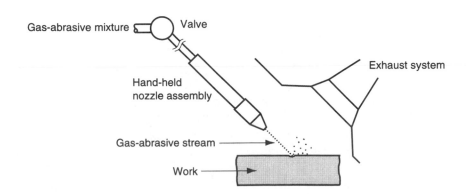

AJM is normally used as a finishing process rather than a production cutting process. Applications include deburring, trimming and deflashing, cleaning, and polishing. Cutting is accomplished successfully on hard, brittle materials (e.g., glass, silicon, mica, and ceramics) that are in the form of thin flat stock. Typical abrasives used in AJM include aluminum oxide (for aluminum and brass), silicon carbide (for stainless steel and ceramics), and glass beads (for polishing). Grit sizes are small, 15 to 40 μm (0.0006–0.0016 in.) in diameter, and must be uniform in size for a given application. It is important not to recycle the abrasives because used grains become fractured (and therefore smaller in size), worn, and contaminated.

26.2 ELECTROCHEMICAL MACHINING PROCESSES

An important group of nontraditional processes use electrical energy to remove material. This group is identified by the term *electrochemical processes,* because electrical energy is used in combination with chemical reactions to accomplish material removal. In effect, these processes are the reverse of electroplating (Section 29.1.1). The work material must be a conductor in electrochemical machining.

26.2.1 Electrochemical Machining

The basic process in this group is electrochemical machining (ECM). *Electrochemical machining* removes metal from an electrically conductive workpiece by anodic dissolution, in which the shape of the workpiece is obtained by a formed electrode tool in close proximity to, but separated from, the work by a rapidly flowing electrolyte. ECM is basically a deplating operation. As illustrated in Figure 26.5, the workpiece is the anode, and the tool is the cathode. The principle underlying the process is that material is deplated from the anode (the positive pole) and deposited onto the cathode (the negative pole) in the presence of an electrolyte bath (Section 4.5). The difference in ECM is that the electrolyte bath flows rapidly between the two poles to carry off the deplated material, so that it does not become plated onto the tool.

FIGURE 26.5
Electrochemical machining (ECM).

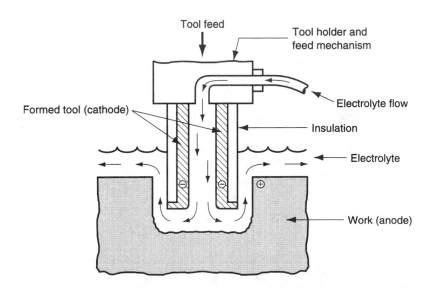

The electrode tool, usually made of copper, brass, or stainless steel, is designed to possess approximately the inverse of the desired final shape of the part. An allowance in the tool size must be provided for the gap that exists between the tool and the work. To accomplish metal removal, the electrode is fed into the work at a rate equal to the rate of metal removal from the work. Metal removal rate is determined by Faraday's First Law, which states that the amount of chemical change produced by an electric current (i.e., the amount of metal dissolved) is proportional to the quantity of electricity passed (current × time):

$$V = CIt \qquad (26.1)$$

where V = volume of metal removed, mm^3 ($in.^3$); C = a constant called the specific removal rate which depends on atomic weight, valence, and density of the work material, mm^3/amp-s ($in.^3$/amp-min); I = current, amps; and t = time, s (min).

Based on Ohm's Law, current $I = E/R$, where E = voltage and R = resistance. Under the conditions of the ECM operation, resistance is given by

$$R = \frac{gr}{A} \qquad (26.2)$$

where g = gap between electrode and work, mm (in.); r = resistivity of electrolyte, ohm-mm (ohm-in.); and A = surface area between work and tool in the working frontal gap, mm^2 ($in.^2$). Substituting this expression for R into Ohm's Law, we have

$$I = \frac{EA}{gr} \qquad (26.3)$$

And substituting this equation back into the equation defining Faraday's Law,

$$V = \frac{C(EAt)}{gr} \qquad (26.4)$$

It is convenient to convert this equation into an expression for feed rate, the rate at which the electrode (tool) can be advanced into the work. This conversion can be accomplished in two steps. First, let us divide Eq. (26.4) by At (area × time) to convert volume of metal removed into a linear travel rate:

$$\frac{V}{At} = f_r = \frac{CE}{gr} \qquad (26.5)$$

where f_r = feed rate, mm/s (in./min). Second, let us substitute I/A in place of $E/(gr)$, as provided by Eq. (26.3). Thus, feed rate in ECM is

$$f_r = \frac{CI}{A} \qquad (26.6)$$

where A = the frontal area of the electrode, mm^2 ($in.^2$). This is the projected area of the tool in the direction of the feed into the work. Values of specific removal rate C are presented in Table 26.1 for various work materials. We should note that this equation assumes 100% efficiency of metal removal. The actual efficiency is in the range 90% to 100% and depends on tool shape, voltage and current density, and other factors.

EXAMPLE 26.1
Electrochemical machining.

An ECM operation is to be used to cut a hole into a plate of aluminum that is 12 mm thick. The hole has a rectangular cross section, 10 mm by 30 mm. The ECM operation will be accomplished at a current = 1200 amps. Efficiency is expected to be 95%. Determine feed rate and time required to cut through the plate.

Solution: From Table 26.1, specific removal rate C for aluminum $= 3.44 \times 10^{-2}$ mm^3/A-s. The frontal area of the electrode $A = 10$ mm $\times$ 30 mm $= 300$ mm.^{2} At a current level of 1200 amps, feed rate is

$$f_r = 0.0344 \text{ mm}^3/\text{A-s}\left(\frac{1200}{300} \text{ A/mm}^2\right) = 0.1376 \text{ mm/s}$$

At an efficiency of 95%, the actual feed rate is

$$f_r = 0.1376 \text{ mm/s}(0.95) = 0.1307 \text{ mm/s}$$

Time to machine through the 12 mm plate is

$$T_m = \frac{12.0}{0.1307} = 91.8 \text{ s} = 1.53 \text{ min}$$

The preceding equations indicate the important process parameters for determining metal removal rate and feed rate in electrochemical machining: gap distance g, electrolyte resistivity r, current I, and electrode frontal area A. Gap distance needs to be controlled closely. If g becomes too large, the electrochemical process slows down. However, if the electrode touches the work, a short circuit occurs, which stops the process altogether. As a practical matter, gap distance is usually maintained within a range 0.075 to 0.75 mm (0.003–0.030 in.).

Water is used as the base for the electrolyte in ECM. To reduce electrolyte resistivity, salts such as NaCl or $NaNO_3$ are added in solution. In addition to carrying off the material that has been removed from the workpiece, the flowing electrolyte also serves the function of removing heat and hydrogen bubbles created in the chemical reactions of the process. The removed work material is in the form of microscopic particles, which must be separated from the electrolyte through centrifuge, sedimentation, or other means. The separated particles form a thick sludge whose disposal is an environmental problem associated with ECM.

Large amounts of electrical power are required to perform ECM. As the equations indicate, rate of metal removal is determined by electrical power, specifically the current density that can be supplied to the operation. The voltage in ECM is kept relatively low to minimize arcing across the gap.

Electrochemical machining is generally used in applications where the work metal is very hard or difficult-to-machine, or where the workpart geometry is difficult (or impossible) to accomplish by conventional machining methods. Work hardness makes no difference in ECM, because the metal removal is not mechanical. Typical ECM

TABLE 26.1   Typical values of specific removal rate C for selected work materials in electrochemical machining.

Work Material[a]	Specific Removal Rate C		Work Material[a]	Specific Removal Rate C	
	mm^3/amp-sec	(in.^{3}/amp-min)		mm^3/amp-sec	(in.^{3}/amp-min)
Aluminum (3)	3.44×10^{-2}	(1.26×10^{-4})	Steels:		
Copper (1)	7.35×10^{-2}	(2.69×10^{-4})	Low alloy	3.0×10^{-2}	(1.1×10^{-4})
Iron (2)	3.69×10^{-2}	(1.35×10^{-4})	High alloy	2.73×10^{-2}	(1.0×10^{-4})
Nickel (2)	3.42×10^{-2}	(1.25×10^{-4})	Stainless	2.46×10^{-2}	(0.9×10^{-4})
			Titanium (4)	2.73×10^{-2}	(1.0×10^{-4})

Compiled from data in [5].

[a]Most common valence given in parentheses ( )—assumed in determining specific removal rate C. For different valence, compute C by multiplying C by most common valence and dividing by actual valence.

applications include: (1) die sinking, which involves the machining of irregular shapes and contours into forging dies, plastic molds, and other shaping tools; (2) multiple hole drilling, where many holes can be drilled simultaneously with ECM and conventional drilling would probably require the holes to be made sequentially; (3) holes that are not round, since ECM does not use a rotating drill; and (4) deburring (Section 26.2.2).

Advantages of ECM include (1) little surface damage to the workpart, (2) no burrs as in conventional machining, (3) low tool wear (the only tool wear results from the flowing electrolyte), and (4) relatively high metal removal rates for hard and difficult-to-machine metals. Disadvantages of ECM are: (1) significant cost of electrical power to drive the operation, and (2) problems of disposing of the electrolyte sludge.

26.2.2 Electrochemical Deburring and Grinding

Electrochemical deburring (ECD) is an adaptation of ECM designed to remove burrs or to round sharp corners on metal workparts by anodic dissolution. One possible setup for ECD is shown in Figure 26.6. The hole in the workpart has a sharp burr of the type that is produced in a conventional through-hole drilling operation. The electrode tool is designed to focus the metal removal action on the burr. Portions of the tool not being used for machining are insulated. The electrolyte flows through the hole to carry away the burr particles. The same ECM principles of operation also apply to ECD. However, since much less material is removed in electrochemical deburring, cycle times are much shorter. A typical cycle time in ECD is less than a minute. The time can be increased if it is desired to round the corner in addition to removing the burr.

Electrochemical grinding (ECG) is a special form of ECM in which a rotating grinding wheel with a conductive bond material is used to augment the anodic dissolution of the metal workpart surface, as illustrated in Figure 26.7. Abrasives used in ECG include aluminum oxide and diamond. The bond material is either metallic (for diamond abrasives) or resin bond impregnated with metal particles to make it electrically conductive (for aluminum oxide). The abrasive grits protruding from the grinding wheel at the contact with the workpart establish the gap distance in ECG. The electrolyte flows through the gap between the grains to play its role in electrolysis.

Deplating is responsible for 95% or more of the metal removal in ECG, and the abrasive action of the grinding wheel removes the remaining 5% or less, mostly in the form of salt films that have been formed during the electrochemical reactions at the work surface. Because most of the machining is accomplished by electrochemical action, the grinding wheel in ECG lasts much longer than a wheel in conventional grinding. The result is a much higher grinding ratio. In addition, dressing of the grinding wheel is required much less frequently. These are the significant advantages of the process. Applications of ECG include sharpening of cemented carbide tools and grinding of surgical needles, other thin wall tubes, and fragile parts.

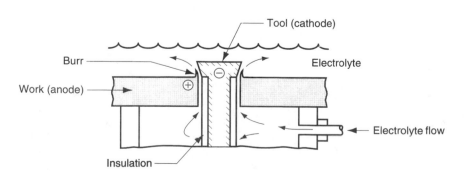

FIGURE 26.6
Electrochemical deburring (ECD).

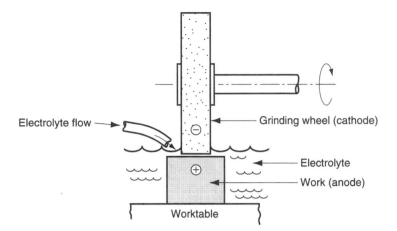

FIGURE 26.7
Electrochemical grinding
(ECG).

Electrolyte flow

Grinding wheel (cathode)

Electrolyte

Work (anode)

Worktable

26.3 THERMAL ENERGY PROCESSES

Material removal processes based on thermal energy are characterized by very high local temperatures—hot enough to remove material by fusion or vaporization. Because of the high temperatures, these processes cause physical and metallurgical damage to the new work surface. In some cases, the resulting finish is so poor that subsequent processing is required to smooth the surface. In this section we examine several thermal energy processes that have commercial importance: (1) electric discharge machining and electric discharge wire cutting, (2) electron beam machining, (3) laser beam machining, (4) plasma arc machining, and (5) conventional thermal cutting processes.

26.3.1 Electric Discharge Processes

Electric discharge material removal processes remove metal by a series of discrete electrical discharges (sparks) that cause localized temperatures high enough to melt or vaporize the metal in the immediate vicinity of the discharge. The two main processes in this category are (1) electric discharge machining and (2) wire electric discharge machining. These processes can be used only on electrically conducting work materials.

Electric Discharge Machining   Electric discharge machining (EDM) is one of the most widely used nontraditional processes. An EDM setup is illustrated in Figure 26.8. The shape of the finished work surface is produced by a formed electrode tool. The sparks occur across a small gap between tool and work surface. The EDM process must take place in the presence of a dielectric fluid, which creates a path for each discharge as the fluid becomes ionized in the gap. The discharges are generated by a pulsating direct current power supply connected to the work and the tool.

Figure 26.8(b) shows a close-up view of the gap between the tool and the work. The discharge occurs at the location where the two surfaces are closest. The dielectric fluid ionizes at this location to create a path for the discharge. The region in which discharge occurs is heated to extremely high temperatures, so that a small portion of the work surface is suddenly melted and removed. The flowing dielectric then flushes away the small particle (call it a "chip"). Since the surface of the work at the location of the previous discharge is now separated from the tool by a greater distance, this location is less likely to be the site of another spark until the surrounding regions have been reduced to the

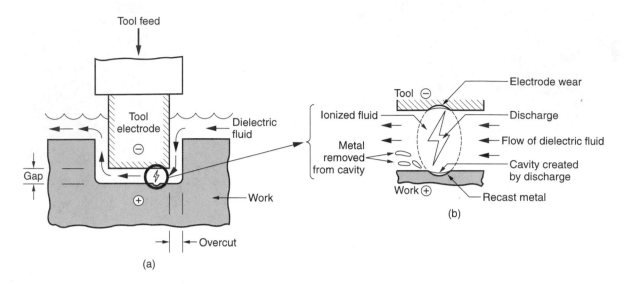

FIGURE 26.8   Electric discharge machining (EDM): (a) overall setup, and (b) close-up view of gap, showing discharge and metal removal.

same level or below. Although the individual discharges remove metal at very localized points, they occur hundreds or thousands of times per second so that a gradual erosion of the entire surface occurs in the area of the gap.

Two important process parameters in EDM are discharge current and frequency of discharges. As either of these parameters is increased, metal removal rate increases. Surface roughness is also affected by current and frequency, as shown in Figure 26.9(a). The best surface finish is obtained in EDM by operating at high frequencies and low discharge currents. As the electrode tool penetrates into the work, overcutting occurs. *Overcut* in EDM is the distance by which the machined hole in the workpart exceeds the size of the tool. It is produced because the electrical discharges occur at the sides of the tool as well as its end. Overcut is a function of current and frequency, as illustrated in Figure 26.9(b), and can amount to several hundredths of a mm.

Note that the high spark temperatures that melt the work also melt the tool, creating a small cavity in the surface opposite the cavity produced in the work. Tool wear is usually measured as the ratio of work material removed to tool material removed (similar to the grinding ratio). This wear ratio ranges between 1.0 and 100 or slightly above,

FIGURE 26.9   (a) Surface finish in EDM as a function of discharge current and frequency of discharges. (b) Overcut in EDM as a function of discharge current and frequency of discharges.

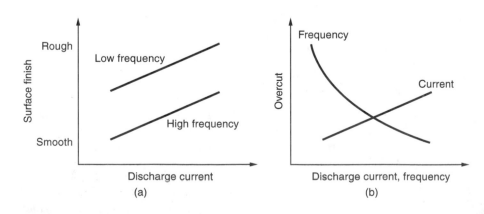

depending on the combination of work and electrode materials. Electrodes are made of graphite, copper, brass, copper tungsten, silver tungsten, and other materials. The selection depends on the type of power supply circuit available on the EDM machine, the type of work material that is to be machined, and whether roughing or finishing is to be done. Graphite is preferred for many applications because of its melting characteristics. In fact, graphite does not melt. It vaporizes at very high temperatures, and the cavity created by the spark is generally smaller than for most other EDM electrode materials. Consequently, a high ratio of work material removed to tool wear is usually obtained with graphite tools.

The hardness and strength of the work material are not factors in EDM, since the process is not a contest of hardness between tool and work. The melting point of the work material is an important property, and metal removal rate can be related to melting point approximately by the following empirical formula, based on an equation described in Weller [13]:

$$MRR = \frac{KI}{T_m^{1.23}} \tag{26.7}$$

where MRR = metal removal rate, mm^3/s (in.^{3}/min); K = constant of proportionality whose value = 664 in SI units (5.08 in U.S. customary units); I = discharge current, amps; and T_m = melting temperature of work metal, °C (°F). Melting points of selected metals are listed in Table 4.1.

EXAMPLE 26.2 Electric discharge machining.

A certain alloy whose melting point = 1100°C is to be machined in an EDM operation. If discharge current = 25 amps, what is the expected metal removal rate?

Solution: Using Eq. (26.7), the anticipated metal removal rate is

$$MRR = \frac{664(25)}{1100^{1.23}} = 3.01 \text{ mm}^3/\text{s}$$

Dielectric fluids used in EDM include hydrocarbon oils, kerosene, and distilled or deionized water. The dielectric fluid serves as an insulator in the gap except when ionization occurs in the presence of a spark. Its other functions are to flush debris out of the gap and to remove heat from tool and workpart.

Applications of electric discharge machining include both tool fabrication and parts production. The tooling for many of the mechanical processes discussed in this book are often made by EDM, including molds for plastic injection molding, extrusion dies, wire drawing dies, forging and heading dies, and sheetmetal stamping dies. For many of these applications, the materials used to fabricate the tooling are difficult (or impossible) to machine by conventional methods. Certain production parts also call for application of EDM. Examples include delicate parts that are not rigid enough to withstand conventional cutting forces, hole drilling where the axis of the hole is at an acute angle to the surface so that a conventional drill would be unable to start the hole, and production machining of hard and exotic metals.

Electric Discharge Wire Cutting   *Electric discharge wire cutting* (EDWC), commonly called *wire EDM,* is a special form of electric discharge machining that uses a small diameter wire as the electrode to cut a narrow kerf in the work. The cutting action in wire EDM is achieved by thermal energy from electric discharges between the electrode wire and the workpiece. Wire-EDM is illustrated in Figure 26.10. The workpiece is fed continuously and slowly past the wire in order to achieve the desired cutting path, somewhat in the manner of a bandsaw operation. Numerical control is used to control the

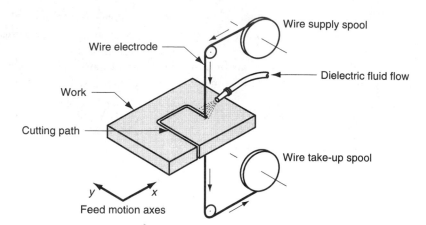

FIGURE 26.10 Electric discharge wire cutting (EDWC), also called wire EDM.

workpart motions during cutting. As it cuts, the wire is continuously advanced between a supply spool and a take-up spool to present a fresh electrode of constant diameter to the work. This helps to maintain a constant kerf width during cutting. As in EDM, wire EDM must be carried out in the presence of a dielectric. This is applied by nozzles directed at the tool–work interface as in our figure, or the workpart is submerged in a dielectric bath.

Wire diameters range from 0.076 to 0.30 mm (0.003–0.012 in.), depending on required kerf width. Materials used for the wire include brass, copper, tungsten, and molybdenum. Dielectric fluids include deionized water or oil. As in EDM, an overcut exists in wire EDM that makes the kerf larger than the wire diameter, as shown in Figure 26.11. This overcut is in the range 0.020 to 0.051 mm (0.0008–0.002 in.). Once cutting conditions have been established for a given cut, the overcut remains fairly constant and predictable.

Although EDWC seems similar to a bandsaw operation, its precision far exceeds that of a bandsaw. The kerf is much narrower, corners can be made much sharper, and the cutting forces against the work are nil. In addition, hardness and toughness of the work material do not affect cutting performance. The only requirement is that the work material must be electrically conductive.

The special features of wire EDM make it ideal for making components for stamping dies. Since the kerf is so narrow, it is often possible to fabricate punch and die in a single cut, as suggested by Figure 26.12. Other tools and parts with intricate outline shapes,

FIGURE 26.11 Definition of kerf and overcut in electric discharge wire cutting.

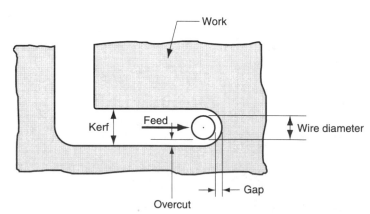

FIGURE 26.12   Irregular outline cut from a solid metal slab by wire-EDM (photo courtesy of LeBlond Makino Machine Tool Company).

such as lathe form tools, extrusion dies, and flat templates, are made with electric discharge wire cutting.

26.3.2 Electron Beam Machining

Electron beam machining (EBM) is one of several industrial processes that use electron beams. Besides machining, other applications of the technology include heat treating (Section 27.5.2) and welding (Section 31.4.1). **_Electron beam machining_** uses a high velocity stream of electrons focused on the workpiece surface to remove material by melting and vaporization. A schematic of the EBM process is illustrated in Figure 26.13. An

FIGURE 26.13   Electron beam machining (EBM).

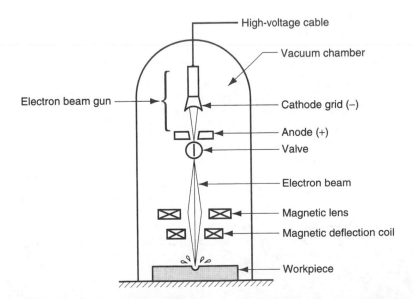

electron beam gun generates a continuous stream of electrons that is accelerated to approximately 75% of the speed of light and focused through an electromagnetic lens on the work surface. The lens is capable of reducing the area of the beam to a diameter as small as 0.025 mm (0.001 in.). On impinging the surface, the kinetic energy of the electrons is converted into thermal energy of extremely high density which melts or vaporizes the material in a very localized area.

Electron beam machining is used for a variety of high precision cutting applications on any known material. Applications include drilling of extremely small diameter holes—down to 0.05 mm (0.002 in.) diameter, drilling of holes with very high depth-to-diameter ratios—more than 100 : 1, and cutting of slots that are only about 0.025 mm (0.001 in.) wide. These cuts can be made to very close tolerances with no cutting forces or tool wear. The process is ideal for micromachining and is generally limited to cutting operations in thin parts—in the range 0.25 to 6.3 mm (0.010 to 0.250 in.) thick. EBM must be carried out in a vacuum chamber to eliminate collision of the electrons with gas molecules. Other limitations include the high energy required and the expensive equipment.

26.3.3 Laser Beam Machining

Lasers are being used for a variety of industrial applications, including heat treatment (Section 27.5.2), welding (Section 31.4.2), measurement (Section 44.5.2), as well as scribing, cutting, and drilling (described here). The term *laser* stands for "light amplification by stimulated emission of radiation." A laser is an optical transducer that converts electrical energy into a highly coherent light beam. A laser light beam has several properties that distinguish it from other forms of light. It is monochromatic (theoretically, the light has a single wave length) and highly collimated (the light rays in the beam are almost perfectly parallel). These properties allow the light generated by a laser to be focused, using conventional optical lenses, onto a very small spot with resulting high power densities. Depending on the amount of energy contained in the light beam, and its degree of concentration at the spot, the various laser processes identified above can be accomplished.

Laser beam machining (LBM) uses the light energy from a laser to remove material by vaporization and ablation. The setup for LBM is illustrated in Figure 26.14. The types of lasers used in LBM are carbon dioxide gas lasers and solid state lasers (of which there are several types). In laser beam machining, the energy of the coherent light beam is concentrated not only optically but also in terms of time. The light beam is pulsed so that the released energy results in an impulse against the work surface that produces a combination of evaporation and melting, with the melted material evacuating the surface at high velocity.

LBM is used to perform various types of drilling, slitting, slotting, scribing, and marking operations. Drilling small diameter holes is possible—down to 0.025 mm (0.001 in.). For larger holes, above 0.50 mm (0.020 in.) diameter, the laser beam is controlled to cut the outline of the hole. LBM is not considered a mass production process, and it is generally used on thin stock. The range of work materials that can be machined by LBM is virtually unlimited. Ideal properties of a material for LBM include high light energy absorption, poor reflectivity, good thermal conductivity, low specific heat, low fusion heat, and low vaporization heat. Of course, no material has this ideal combination of properties. The actual list of work materials processed by LBM includes metals with high hardness and strength, soft metals, ceramics, glass and glass epoxy, plastics, rubber, cloth, and wood.

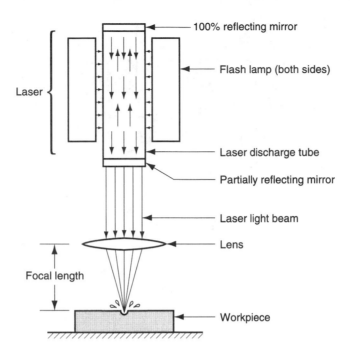

FIGURE 26.14   Laser beam machining (LBM).

26.3.4 Arc Cutting Processes

The intense heat from an electric arc can be used to melt virtually any metal for the purpose of welding or cutting. Most arc-cutting processes use the heat generated by an arc between an electrode and a metallic workpart (usually a flat plate or sheet) to melt a kerf that separates the part. The most common arc-cutting processes are [8] plasma arc cutting and air carbon arc cutting.

Plasma Arc Cutting   A *plasma* is defined as a superheated, electrically ionized gas. *Plasma arc cutting* (PAC) uses a plasma stream operating at temperatures in the range 10,000°C to 14,000°C (18,000°F–25,000°F) to cut metal by melting (Figure 26.15). The cutting action operates by directing the high-velocity plasma stream at the work, thus melting it and blowing the molten metal through the kerf. The plasma arc is generated between an electrode inside the torch and the anode workpiece. The plasma flows through a water-cooled nozzle, which constricts and directs the stream to the desired location on the work. The resulting plasma jet is a high-velocity, well-collimated stream with extremely high temperatures at its center, hot enough to cut through metal in some cases 150 mm (6 in.) thick.

Gases used to create the plasma in PAC include nitrogen, argon, hydrogen, or mixtures of these gases. These are referred to as the primary gases in the process. Secondary gases or water are often directed to surround the plasma jet to help confine the arc and clean the kerf of molten metal as it forms.

Most PAC applications involve cutting of flat metal sheets and plates. Operations include hole piercing and cutting along a defined path. The desired path can be cut either by use of a hand-held torch manipulated by a human operator, or by directing the cutting path of the torch under numerical control (NC). For faster production and higher accuracy, NC is preferred because of better control over the important process variables such as standoff distance and feed rate. Plasma arc cutting can be used to cut

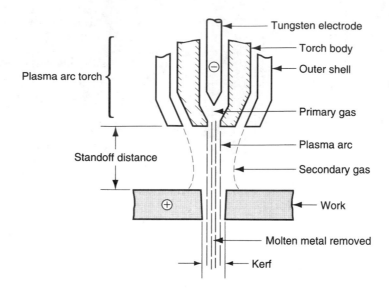

FIGURE 26.15   Plasma arc cutting (PAC).

nearly any electrically conductive metal. The metals frequently cut by PAC include plain carbon steel, stainless steel, and aluminum. The advantage of NC PAC in these applications is high productivity. Feed rates along the cutting path can be as high as 200 mm/s (450 in./min) for 6 mm (0.25 in.) aluminum plate and 85 mm/s (200 in./min) for 6 mm (0.25 in.) steel plate [5]. Feed rates must be reduced for thicker stock. For example, the maximum feed rate for cutting 100 mm (4 in.) thick aluminum stock is around 8 mm/s (20 in./min) [5]. Disadvantages of PAC are (1) the cut surface is rough, and (2) metallurgical damage at the surface is the most severe among the nontraditional metalworking processes.

Air Carbon Arc Cutting   In this process, the arc is generated between a carbon electrode and the metallic work, and a high-velocity air jet is used to blow away the melted portion of the metal. This procedure can be used to form a kerf for severing the piece, or to gouge a cavity in the part. Gouging is used to prepare the edges of plates for welding, for example to create a U-groove in a butt joint (Section 30.2). Air carbon arc cutting is used on a variety of metals, including cast iron, carbon steel, low alloy, and stainless steels, and various nonferrous alloys. Spattering of the molten metal is a hazard and a disadvantage of the process.

Other Arc Cutting Processes   Various other electric arc processes are used for cutting applications, although not as widely as plasma arc and air carbon arc cutting. These other processes include (1) gas metal arc cutting, (2) shielded metal arc cutting, (3) gas tungsten arc cutting, and (4) carbon arc cutting. The technologies are the same as those used in arc welding (Section 31.1), except that the heat of the electric arc is used for cutting.

26.3.5 Oxyfuel Cutting Processes

A widely used family of thermal cutting processes, popularly known as *flame cutting,* use the heat of combustion of certain fuel gases combined with the exothermic reaction of the metal with oxygen. The cutting torch used in these processes is designed to deliver a mixture of fuel gas and oxygen in the proper amounts, and to direct a stream of oxygen to the cutting region. The primary mechanism of material removal in oxyfuel cutting (OFC) is the chemical reaction of oxygen with the base metal. The purpose of the oxyfuel combustion is to raise the temperature in the region of cutting to support the reaction. These processes are commonly used to cut ferrous metal plates, in which the rapid oxidation of iron occurs according to the following reactions [8]:

$$Fe + O \rightarrow FeO + heat \tag{26.8a}$$

$$3Fe + 2O_2 \rightarrow Fe_3O_4 + heat \tag{26.8b}$$

$$2Fe + 1.5O_2 \rightarrow Fe_2O_3 + heat \tag{26.8c}$$

The second of these reactions, Eq. (26.8b), is the most significant in terms of heat generation.

The cutting mechanism for nonferrous metals is somewhat different. These metals are generally characterized by lower melting temperatures than the ferrous metals, and they are more oxidation resistant. In these cases, the heat of combustion of the oxyfuel mixture plays a more important role in creating the kerf. Also, to promote the metal oxidation reaction, chemical fluxes or metallic powders are often added to the oxygen stream.

Fuels used in OFC include acetylene (C_2H_2), MAPP (methylacetylene-propadiene—C_3H_4), propylene (C_3H_6), and propane (C_3H_8). Flame temperatures and heats of combustion for these fuels are listed in Table 31.2 in Chapter 31. Acetylene burns at the highest flame temperature and is the most widely used fuel for welding and cutting. However, there are certain hazards with the storage and handling of acetylene that must be considered (Section 31.3.1).

OFC processes are performed either manually or by machine. Manually operated torches are used for repair work, cutting of scrap metal, trimming of risers from sand castings, and similar operations which generally require minimal accuracy. For production work, machine flame cutting allows faster speeds and greater accuracies. This equipment is often numerically controlled to allow profiled shapes to be cut.

26.4 CHEMICAL MACHINING

Chemical machining (CHM) is a nontraditional process in which material removal occurs through contact with a strong chemical etchant. Applications as an industrial process began shortly after World War II in the aircraft industry. The use of chemicals to remove unwanted material from a workpart can be applied in several ways, and several different terms have been developed to distinguish the applications. These terms include chemical milling, chemical blanking, chemical engraving, and photochemical machining (PCM). They all utilize the same mechanism of material removal, and it is appropriate to discuss the general characteristics of chemical machining before defining the individual processes.

26.4.1 Mechanics and Chemistry of Chemical Machining

The chemical machining process consists of several steps. Differences in applications and the ways in which the steps are implemented account for the different forms of CHM:

(1) *Cleaning*—The first step is a cleaning operation to ensure that material will be removed uniformly from the surfaces to be etched.

(2) *Masking*—A protective coating called a maskant is applied to certain portions of the part surface. This maskant is made of a material that is chemically resistant to

the etchant (the term **resist** is used for this masking material). It is therefore applied to those portions of the work surface that are not to be etched.

(3) **Etching**—This is the material removal step. The part is immersed in an etchant which chemically attacks those portions of the part surface that are not masked. The usual method of attack is to convert the work material (e.g., a metal) into a salt that dissolves in the etchant and is thereby removed from the surface. When the desired amount of material has been removed, the part is withdrawn from the etchant and washed to stop the process.

(4) **Demasking**—The maskant is removed from the part.

The two steps in chemical machining that involve significant variations in methods, materials, and process parameters are masking and etching—steps (2) and (3).

Maskant materials include neoprene, polyvinylchloride, polyethylene, and other polymers. Masking can be accomplished by any of three methods: (1) cut and peel, (2) photographic resist, and (3) screen resist. The **cut and peel** method involves application of the maskant over the entire part by dipping, painting, or spraying. The resulting thickness of the maskant is 0.025 to 0.125 mm (0.001–0.005 in.) thick. After the maskant has hardened, it is cut using a scribing knife and peeled away in the areas of the work surface that are to be etched. The maskant cutting operation is performed by hand, usually guiding the knife with a template. The cut and peel method is generally used for large workparts, low production quantities, and where accuracy is not a critical factor. This method cannot hold tolerances tighter than ±0.125 mm (±0.005 in.) except with extreme care.

As the name suggests, the **photographic resist** method (called the **photoresist** method for short) uses photographic techniques to perform the masking step. The masking materials contain photosensitive chemicals. They are applied to the work surface and exposed to light through a negative image of the desired areas to be etched. These areas of the maskant can then be removed from the surface using photographic developing techniques. This procedure leaves the desired surfaces of the part protected by the maskant and the remaining areas unprotected, vulnerable to chemical etching. Photoresist masking techniques are normally applied where small parts are produced in high quantities, and close tolerances are required. Tolerances closer than ±0.0125 mm (±0.0005 in.) can be held [13].

The **screen resist** method applies the maskant by means of silk screening methods. In these methods, the maskant is painted onto the workpart surface through a silk or stainless steel mesh. Embedded in the mesh is a stencil that protects those areas to be etched from the painting application. The maskant is thus painted onto the work areas that are not to be etched. The screen resist method is generally used in applications that are between the other two masking methods in terms of accuracy, part size, and production quantities. Tolerances of ±0.075 mm (±0.003 in.) can be achieved with this masking method.

Selection of the **etchant** depends on work material to be etched, desired depth and rate of material removal, and surface finish requirements. The etchant must also be matched with the type of maskant that is used to ensure that the maskant material is not chemically attacked by the etchant. Table 26.2 lists some of the work materials machined by CHM together with the etchants that are generally used on these materials. Also included in the table are penetration rates and etch factors. These parameters are explained next.

Material removal rates in CHM are generally indicated as penetration rates, mm/min (in./min), since rate of chemical attack of the work material by the etchant is directed into the surface. The penetration rate is unaffected by surface area. Penetration rates listed in Table 26.2 are typical values for the given material and etchant.

TABLE 26.2   Common work materials and etchants in CHM, with typical penetration rates and etch factors.

Work Material	Etchant	Penetration Rates		Etch Factor
		mm/min	(in./min)	
Aluminum	$FeCl_3$	0.020	(0.0008)	1.75
and alloys	NaOH	0.025	(0.001)	1.75
Copper and alloys	$FeCl_3$	0.050	(0.002)	2.75
Magnesium	H_2SO_4	0.038	(0.0015)	1.0
and alloys				
Silicon	$HNO_3 : HF : H_2O$	very slow		NA
Mild steel	$HCl : HNO_3$	0.025	(0.001)	2.0
	$FeCl_3$	0.025	(0.001)	2.0
Titanium	HF	0.025	(0.001)	1.0
and alloys	$HF : HNO_3$	0.025	(0.001)	1.0

Compiled from [4], [5], and [13].

NA = data not available.

Depths of cut in chemical machining are as much as 12.5 mm (0.5 in.) for aircraft panels made out of metal plates. However, many applications require depths that are only several hundredths of a mm. Along with the penetration into the work, etching also occurs sideways under the maskant, as illustrated in Figure 26.16. The effect is referred to as the **undercut,** and it must be accounted for in the design of the mask in order for the resulting cut to have the specified dimensions. For a given work material, the undercut is directly related to the depth of cut. The constant of proportionality for the material is called the etch factor, defined as

$$F_e = \frac{d}{u} \qquad (26.9)$$

where F_e = etch factor; d = depth of cut, mm (in.); and u = undercut, mm (in.). The dimensions u and d are defined in Figure 26.16. Different work materials have different etch factors in chemical machining. Some typical values are presented in Table 26.2. The etch factor can be used to determine the dimensions of the cutaway areas in the maskant, so that the specified dimensions of the etched areas on the part can be achieved.

26.4.2 CHM Processes

In this section, we describe the principle chemical-machining processes: (1) chemical milling, (2) chemical blanking, (3) chemical engraving, and (4) photochemical machining.

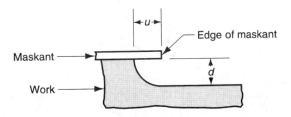

FIGURE 26.16   Undercut in chemical machining.

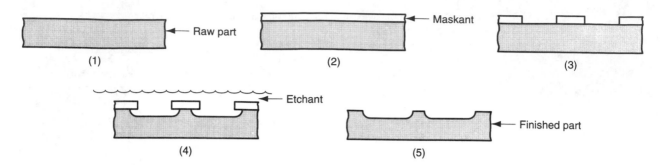

FIGURE 26.17   Sequence of processing steps in chemical milling: (1) clean raw part; (2) apply maskant; (3) scribe, cut, and peel the maskant from areas to be etched; (4) etch; and (5) remove maskant and clean to yield finished part.

Chemical Milling   Chemical milling was the first CHM process to be commercialized. During World War II, a U.S. aircraft company began to use chemical milling to remove metal from aircraft components. They referred to their process as the "Chem-Mill" process. Today, chemical milling is still used largely in the aircraft industry, to remove material from aircraft wing and fuselage panels for weight reduction. It is applicable to large parts where substantial amounts of metal are removed during the process. The cut and peel maskant method is employed. A template is generally used that takes into account the undercut that will result during etching. The sequence of processing steps is illustrated in Figure 26.17.

Chemical milling produces a surface finish that varies with different work materials. Table 26.3 provides a sampling of the values. Surface finish depends on depth of penetration. As depth increases, finish becomes worse, approaching the upper side of the ranges given in the table. Metallurgical damage from chemical milling is very small, perhaps around 0.005 mm (0.0002 in.) into the work surface.

Chemical Blanking   Chemical blanking uses chemical erosion to cut very thin sheet-metal parts—down to 0.025 mm (0.001 in.) thick and/or for intricate cutting patterns. In both instances, conventional punch-and-die methods do not work because the stamping forces damage the sheet metal, or the tooling cost would be prohibitive, or both. Chemical blanking produces parts that are burr free, an advantage over conventional shearing operations.

Methods used for applying the maskant in chemical blanking are either the photoresist method or the screen resist method. For small and/or intricate cutting patterns and close tolerances, the photoresist method is used; otherwise, the screen resist method is used. The small size of the work in chemical blanking excludes the cut and peel maskant method.

TABLE 26.3   Surface finishes expected in chemical milling.

Work Material	Surface Finish Range	
	μm	μ-in.
Aluminum and alloys	1.8–4.1	(70–160)
Magnesium	0.8–1.8	(30–70)
Mild steel	0.8–6.4	(30–250)
Titanium and alloys	0.4–2.5	(15–100)

Compiled from [5] and [13].

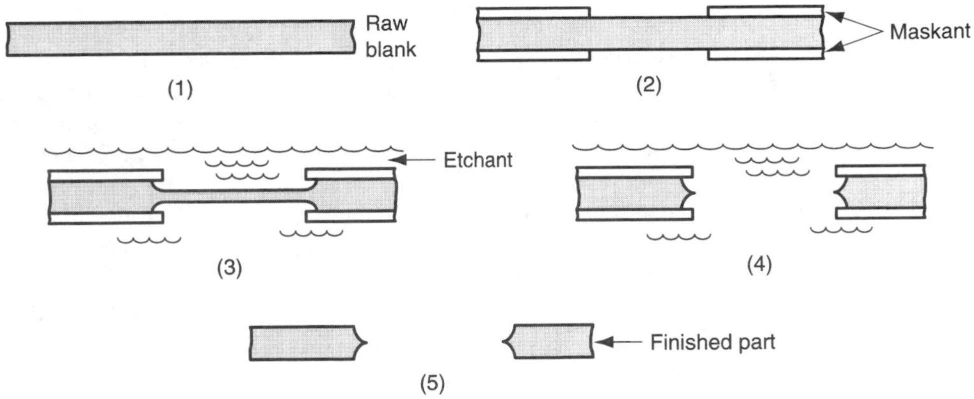

Raw
blank

(1)

Maskant

(2)

Etchant

(3)

(4)

Finished part

(5)

FIGURE 26.18   Sequence of processing steps in chemical blanking: (1) clean raw part, (2) apply resist (maskant) by painting through screen, (3) etch (partially completed), (4) etch (completed), and (5) remove resist and clean to yield finished part.

The steps in chemical blanking are shown in Figure 26.18, using the screen resist method. Since chemical etching takes place on both sides of the part in chemical blanking, it is important that the masking procedure provides accurate registration between the two sides. Otherwise, the erosion into the part from opposite directions will not line up. This is especially critical with small part sizes and intricate patterns.

Application of chemical blanking is generally limited to thin materials and/or intricate patterns for reasons given above. Maximum stock thickness is around 0.75 mm (0.030 in.). Also, hardened and brittle materials can be processed by chemical blanking where mechanical methods would surely fracture the work. Figure 26.19 presents a sampling of parts that were produced by the chemical blanking process.

FIGURE 26.19   Parts made by chemical blanking (courtesy Buckbee-Mears St. Paul).

Tolerances as close as ±0.0025 mm (±0.0001 in.) can be held on 0.025 mm (0.001 in.) thick stock when the photoresist method of masking is used. As stock thickness increases, more generous tolerances must be allowed. Screen resist masking methods are not nearly so accurate as photoresist. Accordingly, when close tolerances on the part are required, the photoresist method should be used to perform the masking step.

Chemical Engraving   Chemical engraving is a chemical machining process for making name plates and other flat panels that have lettering and/or artwork on one side. These plates and panels would otherwise be made using a conventional engraving machine or similar process. Chemical engraving can be used to make panels with either recessed lettering or raised lettering, simply by reversing the portions of the panel to be etched. Masking is done by either the photoresist or screen resist methods. The sequence in chemical engraving is similar to the other CHM processes, except that a filling operation follows etching. The purpose of filling is to apply paint or other coating into the recessed areas that have been created by etching. Then, the panel is immersed in a solution which dissolves the resist but does not attack the coating material. Thus, when the resist is removed, the coating remains in the etched areas but not in the areas that were masked. The effect is to highlight the pattern.

Photochemical Machining   Photochemical machining (PCM) is chemical machining in which the photoresist method of masking is used. The term can therefore be applied correctly to chemical blanking and chemical engraving when these methods use the photographic resist method. PCM is employed in metalworking when close tolerances

FIGURE 26.20   Sequence of processing steps in photochemical machining: (1) clean raw part, (2) apply resist (maskant) by dipping, spraying, or painting, (3) place negative on resist, (4) expose to ultraviolet light, (5) develop to remove resist from areas to be etched, (6) etch (shown partially etched), (7) etch (completed), (8) remove resist and clean to yield finished part.

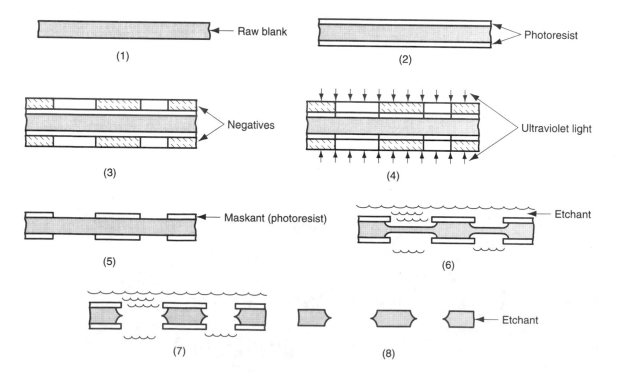

and/or intricate patterns are required on flat parts. Photochemical processes are also used extensively in the electronics industry to produce intricate circuit designs on semiconductor wafers (Section 35.3.1).

Figure 26.20 shows the sequence of steps in photochemical machining as it is applied to chemical blanking. There are various ways to photographically expose the desired image onto the resist. The figure shows the negative in contact with the surface of the resist during exposure. This is contact printing, but other photographic printing methods are available that expose the negative through a lens system to enlarge or reduce the size of the pattern printed on the resist surface. Photoresist materials in current use are sensitive to ultraviolet light but not to light of other wavelengths. Therefore, with proper lighting in the factory, there is no need to carry out the processing steps in a dark room environment. Once the masking operation is accomplished, the remaining steps in the procedure are similar to the other chemical machining methods.

In photochemical machining, the term corresponding to etch factor is *anisotropy,* which is defined as the depth of cut d divided by the undercut u (see Figure 26.18). It is defined the same way as in Eq. (26.9):

$$A = F_e = \frac{d}{u} \tag{26.10}$$

where A = degree of anisotropy; F_e = etch factor; d = depth of cut; and u = undercut.

26.5 APPLICATION CONSIDERATIONS

Typical applications of nontraditional processes include special part feature geometries and work materials that cannot be readily processed by conventional techniques. In this section, we examine these issues. We also summarize the general performance characteristics of nontraditional processes.

Workpart Geometry Features Some of the special workpart shapes for which nontraditional processes are well suited include the following, illustrated in Figure 26.21:

> *Very small holes*—diameters less than 0.125 mm (0.005 in.). This is generally smaller than the possible diameter range of conventional drill bits. LBM can be used to make holes down to diameters of 0.025 mm (0.001 in.).
> *Holes with large depth-to-diameter ratios* (e.g., $d/D > 20$). Except for gun drilling, these holes cannot be machined in conventional drilling operations. ECM and EDM are used successfully in these applications.
> *Holes that are not round* and therefore cannot be drilled with a rotating drill bit. EDM and ECM can be used for these applications because the tools do not rotate.
> Cutting *narrow slots* in slabs and plates of various materials, where the slots are not necessarily straight. EBM, LBM, wire EDM, WJC, and AJC can be used for various applications in this category. Some of these processes can be used to cut extremely intricate shapes.
> *Micromachining*—In addition to cutting small holes and narrow slits, there are other material removal applications where the workpart and/or areas to be cut are very small. Certain nontraditional processes such as PCM, LBM, and EBM, can be used for these micromachining applications.

FIGURE 26.21 Special shapes for which the non-traditional machining processes are appropriate: (a) very small diameter holes; (b) holes with large depth-to-diameter ratios; (c) non-round holes; (d) narrow, non-straight slots; (e) pockets; and (f) die-sinking.

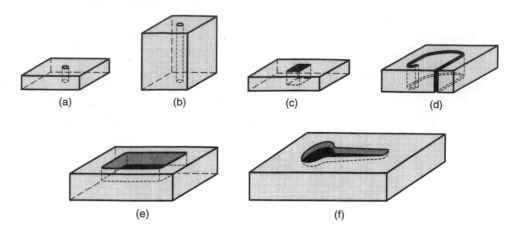

(a)   (b)   (c)   (d)

(e)   (f)

> *Shallow pockets and surface details in flat parts*—There is a significant range in the sizes of the parts in this category, from microscopic integrated circuit chips to large air-craft panels. Chemical machining and its variations are used to accomplish this work.

> Creation of *special contoured shapes for mold and die applications.* These applications are sometimes referred to as die-sinking. EDM and ECM are often favored for these situations.

Work Materials As a group the nontraditional processes can be applied to nearly all work materials, metals and nonmetals. However, certain processes are not suited to cer-tain work materials. Table 26.4 relates applicability of the nontraditional processes to various types of materials.

Several of the processes can be used on metals but not nonmetals. For example, ECM, EDM, and PAM require work materials that are electrical conductors. This gen-erally limits their applicability to metal parts. Chemical machining depends on the avail-

TABLE 26.4 Applicability of selected nontraditional machining processes to various work materials. For comparison, conventional milling and grinding are included in the compilation.

| Work Material | Nontraditional Processes | | | | | | | | Conventional Processes | |
| | Mech | | Elec | | Thermal | | | Chem | | |
	USM	WJC	ECM	EDM	EBM	LBM	PAC	CHM	Milling	Grinding
Aluminum	C	C	B	B	B	B	A	A	A	A
Steel	B	D	A	A	B	B	A	A	A	A
Super alloys	C	D	A	A	B	B	A	B	B	B
Ceramic	A	D	D	D	A	A	D	C	D	C
Glass	A	D	D	D	B	B	D	B	D	C
Silicon[a]			D	D	B	B	D	B	D	B
Plastics	B	B	D	D	B	B	D	C	B	C
Cardboard[b]	D	A	D	D				D	D	D
Textiles[c]	D	A	D	D				D	D	D

Compiled from [13] and other sources.

Key: A = good application, B = fair application, C = poor application, D = not applicable, and blank entries indicate no data available during compilation.

[a] Refers to silicon used in fabricating integrated circuit chips.

[b] Includes other paper products.

[c] Includes felt, leather, and similar materials.

TABLE 26.5 Machining characteristics of the nontraditional machining processes.

| Characteristic | Nontraditional Processes | | | | | | | | Conventional Processes | |
| | Mech | | Elec | | Thermal | | | Chem | | |
	USM	WJC	ECM	EDM	EBM	LBM	PAC	CHM	Milling	Grinding
Material removal rates	C	C	B	C	D	D	A	B–D[a]	A	B
Dimensional control	A	B	B	A–D[b]	A	A	D	A–B[b]	B	A
Surface finish	A	A	B	B–D[b]	B	B	D	B	B–C[b]	A
Surface damage[c]	B	B	A	D	D	D	D	A	B	B–C[b]

Compiled from [13].

Key: A = excellent, B = good, C = fair, and D = poor.

[a] Rating depends on size of work and masking method.

[b] Rating depends on cutting conditions.

[c] In surface damage a good rating means low surface damage and poor rating means deep penetration of surface damage; thermal processes can cause damage up to 0.020 in. (0.50 mm) below the new work surface.

ability of an appropriate etchant for the given work material. Since metals are more susceptible to chemical attack by various etchants, CHM is commonly used to process metals.

With some exceptions, USM, AJM, EBM, and LBM can be used on both metals and nonmetals. WJC is generally limited to the cutting of plastics, cardboards, textiles, and other materials that do not possess the strength of metals.

Performance of Nontraditional Processes The nontraditional processes are generally characterized by low material removal rates and high specific energies relative to conventional machining operations. The capabilities for dimensional control and surface finish of the nontraditional processes vary widely, with some of the processes providing high accuracies and good finishes, and others yielding poor accuracies and finishes. Surface damage is also a consideration. Some of these processes produce very little metallurgical damage at and immediately below the work surface, while others (mostly the thermal-based processes) do considerable damage to the surface. Table 26.5 compares these features of the prominent nontraditional methods, using conventional milling and surface grinding for comparison. Inspection of the data reveals wide differences in machining characteristics. In comparing the characteristics of nontraditional and conventional machining, it must be remembered that nontraditional processes are generally used where conventional methods are not practical or economical.

REFERENCES

[1] Bellows, G., and Kohls, J. B. "Drilling without Drills." Special Report 743, *American Machinist,* March 1982, pp 173–188.

[2] Benedict, G. F. *Nontraditional Manufacturing Processes.* Marcel Dekker, Inc., New York, 1987.

[3] Dini, J. W. "Fundamentals of Chemical Milling." Special Report 768, *American Machinist,* July 1984, pp 99–114.

[4] Drozda, T. J., and Wick, C. *Tool and Manufacturing Engineers Handbook.* 4th ed. Vol. I, *Machining.* Soc. of Manufacturing Engineers, Dearborn, Michigan, 1983, Chapter 14.

[5] *Machining Data Handbook.* 3rd ed. Vol. 2, Machinability Data Center, Metcut Research Associates Inc., Cincinnati, Ohio, 1980.

[6] Mason, F. "Water Jet Cuts Instrument Panels." *American Machinist & Automated Manufacturing,* July 1988, pp 126–127.

[7] McGeough, J. A. *Advanced Methods of Machining.* Chapman and Hall, London, England, 1988.

[8] O'Brien, R. L. *Welding Handbook.* 8th ed. Vol. 2, *Welding Processes.* American Welding Society, Miami, Florida, 1991.

[9] Pandey, P. C., and Shan, H. S. *Modern Machining*

Processes. Tata McGraw-Hill Publishing Company, New Delhi, India, 1980.

[10]   Vaccari, J. A. "The Laser's Edge in Metalworking." Special Report 768, *American Machinist,* August 1984, pp 99–114.

[11]   Vaccari, J. A. "Thermal Cutting." Special Report 778, *American Machinist.* July 1985, pp 111–126.

[12]   Vaccari, J. A. "Advances in Laser Cutting." *American Machinist & Automated Manufacturing.* March 1988, pp 59–61.

[13]   Weller, E. J.(ed.). *Nontraditional Machining Processes.* 2nd ed. Soc. of Manufacturing Engineers, Dearborn, Mich., 1984.

REVIEW QUESTIONS

26.1.   Why are the nontraditional material removal processes important?

26.2.   There are four categories of nontraditional machining processes, based on principal energy form. Name the four categories.

26.3.   How does the ultrasonic machining process work?

26.4.   Describe the water jet cutting process.

26.5.   What is the difference between water jet cutting, abrasive water jet cutting, and abrasive jet cutting?

26.6.   Name the three main types of electrochemical machining.

26.7.   Identify the significant disadvantages of electrochemical machining.

26.8.   How does increasing discharge current affect metal removal rate and surface finish in electric discharge machining?

26.9.   What is meant by the term *overcut* in electric discharge machining?

26.10.   Identify two major disadvantages of plasma arc cutting.

26.11.   What are some of the fuels used in oxyfuel cutting?

26.12.   Name the four principal steps in chemical machining.

26.13.   What are the three methods of performing the masking step in chemical machining?

26.14.   What is a *photoresist* in chemical machining?

MULTIPLE CHOICE QUIZ

There is a total of 18 correct answers in the following multiple choice questions (some questions have multiple answers that are correct). To attain a perfect score on the quiz, all correct answers must be given, since each correct answer is worth 1 point. For each question, each omitted answer or wrong answer reduces the score by 1 point, and each additional answer beyond the number of answers required reduces the score by 1 point. Percentage score on the quiz is based on the total number of correct answers.

26.1.   Which of the following processes use mechanical energy as the principal energy source (may be more than one)? (a) grinding, (b) laser beam machining, (c) milling, (d) ultrasonic machining, (e) water jet cutting, and (f) wire EDM.

26.2.   Ultrasonic machining can be used to machine both metallic and nonmetallic materials: (a) true or (b) false.

26.3.   Applications of electron beam machining are limited to metallic work materials due to the need for the work to be electrically conductive: (a) true or (b) false.

26.4.   Which one of the following is closest to the temperatures used in plasma arc cutting? (a) 2750°C (5000°F), (b) 5,500°C (10,000°F), (c) 8,300°C (15,000°F), (d) 11,000°C (20,000°F), (e) 16,500°C (30,000°F).

26.5.   Chemical milling is used in which of the following (more than one)? (a) drilling holes with high depth-to-diameter ratio, (b) making intricate patterns in thin sheet metal, (c) removing material to make shallow pockets in metal, (d) removing metal from aircraft wing panels, and (e) cutting of plastic sheets.

26.6.   Etch factor is which of the following in chemical machining (more than one)? (a) A, (b) $1/A$, (c) CIt, (d) d/u, and (e) u/d; where A = degree of anisotropy, C = specific removal rate, d = depth of cut, I = current, t = time, and u = undercut.

26.7.   Of the following processes, which one is noted for the highest material removal rates? (a) electric discharge machining, (b) electrochemical machining, (c) laser beam machining, (d) oxyfuel cutting, (e) plasma arc cutting, (f) ultrasonic machining, and (g) water jet cutting.

26.8.   Which of the following processes would be appropriate to drill a hole with a square cross section, 6.25 mm (0.25 in.) on a side and 25 mm (1.0 in.) deep (one answer)? (a) abrasive jet machining, (b) chemical milling, (c) EDM, (d) laser beam machining, (e) oxyfuel cutting, (f) water jet cutting, and (g) wire EDM.

26.9. Which of the following processes would be appropriate for cutting a narrow slot, less than 0.5 mm (0.020 in.) wide, in a 9 mm (3/8 in.) thick sheet of fiber-reinforced plastic (more than one answer)? (a) abrasive jet machining, (b) chemical milling, (c) EDM, (d) laser beam machining, (e) oxyfuel cutting, (f) water jet cutting, and (g) wire EDM.

26.10. Which of the following processes would be appropriate for cutting a hole of 0.075 mm (0.003 in.) diameter through a plate of aluminum that is 1.6 mm (0.063 in.) thick (one answer)? (a) abrasive jet machining, (b) chemical milling, (c) EDM, (d) laser beam machining, (e) oxyfuel cutting, (f) water jet cutting, and (g) wire EDM.

26.11. Which of the following processes could be used to cut a large piece of 12.5 mm (0.5 in.) plate steel into two sections (more than one answer)? (a) abrasive jet machining, (b) chemical milling, (c) EDM, (d) laser beam machining, (e) oxyfuel cutting, (f) water jet cutting, and (g) wire EDM.

PROBLEMS

General

26.1. For each of the following applications, identify one or more nontraditional machining processes that might be used, and present arguments to support your selection. Assume that either the part geometry or the work material (or both) preclude the use of conventional machining. (a) A matrix of 0.1 mm (0.004 in.) diameter holes in a plate of 3.2 mm (0.125 in.) thick hardened tool steel. The matrix is rectangular, 75 by 125 mm (3.0 by 5.0 in.) with the separation between holes in each direction = 1.6 mm (0.0625 in.). (b) An engraved aluminum printing plate to be used in an offset printing press to make 275 by 350 mm (11 by 14 in.) posters of Lincoln's Gettysburg address. (c) A through hole in the shape of the letter L in a 12.5-mm-(0.5-in.) thick plate of glass. The size of the L is 25 by 15 mm (1.0 by 0.6 in.) and the width of the hole is 3 mm (1/8 in.). (d) A blind hole in the shape of the letter G in a 50 mm (2.0 in.) cube of steel. The overall size of the "G" is 25 by 19 mm (1.0 by 0.75 in.), the depth of the hole is 3.8 mm (0.15 in.), and its width is 3 mm (1/8 in.).

26.2. Much of the work at the Cut-Anything Company involves cutting and forming flat sheets of fiberglass for the pleasure-boat industry. Manual methods based on portable saws are currently used to perform the cutting operation, but production is slow and scrap rates are high. The foreman says the company should invest in a plasma arc-cutting machine, but the plant manager thinks it would be too expensive. What do you think? Justify your answer by indicating the characteristics of the process that make PAC attractive or unattractive in this application.

26.3. A furniture company that makes upholstered chairs and sofas must cut large quantities of fabrics. Many of these fabrics are strong and wear-resistant, which make them difficult to cut. What nontraditional process(es) would you recommend to the company for this application? Justify your answer by indicating the characteristics of the process that make it attractive.

Electrochemical Machining

26.4. The frontal working area of the electrode is 2000 mm^2 in a certain ECM operation in which the applied current = 1800 amps and the voltage = 12 volts. The material being cut is nickel (valence = 2), whose specific removal rate C is given in Table 26.1. (a) If the process is 90% efficient, determine the rate of metal removal in mm^3/min. (b) If the resistivity of the electrolyte = 140 ohm-mm, determine the working gap.

26.5. In an electrochemical machining operation, the frontal working area of the electrode is 2.5 in.^{2} The applied current = 1500 amps, and the voltage = 12 volts. The material being cut is pure aluminum, whose specific removal rate C is indicated in Table 26.1. (a) If the ECM process is 90% efficient, determine the rate of metal removal in in.^{3}/hr. (b) If the resistivity of the electrolyte = 6.2 ohm-in., determine the working gap.

26.6. A square hole is to be cut using ECM through a plate of pure copper (valence = 1) that is 20 mm thick. The hole is 25 mm on each side, but the electrode that is used to cut the hole is slightly less that 25 mm on its sides to allow for overcut, and its shape includes a hole in its center to permit the flow of electrolyte and to reduce the area of the cut. This tool design results in a frontal area of 200 mm^2. The applied current = 1000 amps. Using an efficiency of 95%, determine how long it will take to cut the hole.

26.7. A 3.5-in.-diameter through hole is to be cut in a block of pure iron (valence = 2) by electrochemical machining. The block is 2.0 in. thick. To speed the cutting process, the electrode tool will have a center hole of 3.0 in., which will produce a center core that can be removed after the tool breaks through. The outside diameter of the electrode is undersized to allow for over-

cut. The overcut is expected to be 0.005 in. on a side. If the efficiency of the ECM operation is 90%, what current will be required to complete the cutting operation in 20 min?

Electric Discharge Machining

26.8. An electric discharge machining operation is being performed on tungsten. (a) Determine the amount of metal removed in the operation after one hour at a discharge amperage = 20 amps. (b) If the work material were tin, determine the amount of material removed in the same time. Use metric units and express the answer in mm^3.

26.9. Same as Problem 26.8, except the new material to be compared with tungsten is zinc. Use U.S. customary units and express the answer in $in.^3$

26.10. Suppose the hole in Problem 26.7 were to be cut using EDM rather than ECM. Using a discharge current = 20 amps (which would be typical for EDM), how long would it take to cut the hole?

26.11. A metal removal rate of 0.01 $in.^3$/min is achieved in a certain EDM operation on a pure iron workpart. What metal removal rate would be achieved on nickel in this EDM operation, if the same discharge current were used?

26.12. In a wire EDM operation performed on 7 mm thick C1080 steel using a tungsten wire electrode whose di-

ameter = 0.125 mm, past experience suggests that the overcut will be 0.02 mm, so that the kerf width will be 0.165 mm. Using a discharge current = 10 amps, what is the allowable feed rate that can be used in the operation? Estimate the melting temperature of 0.80% carbon steel from the phase diagram of Figure 7.4.

26.13. A wire EDM operation is to be performed on a slab of 3/4 in. thick aluminum using a brass wire electrode whose diameter = 0.005 in. It is anticipated that the overcut will be 0.001 in., so that the kerf width will be 0.007 in. Using a discharge current = 7 amps, what is the expected allowable feed rate that can be used in the operation?

26.14. A wire EDM operation is used to cut out punch and die components from 25 mm thick tool steel plates. However, in preliminary cuts, the surface finish on the cut edge is poor. What changes in discharge current and frequency of discharges should be made to improve the finish?

Chemical Machining

26.15. Chemical milling is used in an aircraft plant to create pockets in wing sections made of an aluminum alloy. The starting thickness of one workpart of interest is 20 mm. A series of rectangular-shaped pockets 12 mm deep are to be formed with dimensions 200 mm by 400 mm. The corners of each rectangle are radiused to 15 mm. The part is an aluminum alloy and the etchant is NaOH. The penetration rate for this combination is 0.024 mm/min and the etch factor is 1.75. Determine (a) metal removal rate in mm^3/min, (b) time required to machine to the specified depth, and (c) required dimensions of the opening in the cut and peel maskant to achieve the desired pocket size on the part.

26.16. In a chemical milling operation on a flat mild steel plate, it is desired to cut an ellipse-shaped pocket to a depth of 0.4 in. The semiaxes of the ellipse are $a = 9.0$ in. and $b = 6.0$ in. A solution of hydrochloric and nitric acids will be used as the etchant. Determine (a) metal removal rate in $in.^3$/hr, (b) time required to machine to depth, (c) required dimensions of the opening in cut and peel maskant required to achieve the desired pocket size on the part.

26.17. In a certain chemical blanking operation, a sulfuric acid etchant is used to remove material from a sheet of magnesium alloy. The sheet is 0.25 mm thick. The screen resist method of masking permits high production rates to be achieved. As it turns out, the process is producing a large proportion of scrap. Specified tolerances of ±0.025 mm are not being achieved. The foreman in the CHM department complains that there must be something wrong with the sulfuric acid. "Perhaps the concentration is incorrect," he suggests. Analyze the problem and recommend a solution.

26.18. In a chemical blanking operation, stock thickness of the aluminum sheet is 0.015 in. The pattern to be cut out of the sheet is a hole pattern, consisting of a matrix of 0.100. in. diameter holes. If photochemical machining is used to cut these holes, and contact printing is used to make the resist (maskant) pattern, determine the diameter of the holes that should be used in the pattern.

Part VII

Property Enhancing and Surface Processing Operations

HEAT TREATMENT OF METALS

CHAPTER CONTENTS

27.1 Annealing

27.2 Martensite Formation in Steel
    27.2.1 The Time-Temperature-Transformation Curve
    27.2.2 The Heat Treatment Process
    27.2.3 Hardenability

27.3 Precipitation Hardening

27.4 Surface Hardening

27.5 Heat Treatment Methods and Facilities
    27.5.1 Furnaces for Heat Treatment
    27.5.2 Selective Surface Hardening Methods

The manufacturing processes covered in the preceding chapters involved the creation of part geometry. We now consider processes that either enhance the properties of the workpart (Chapter 27) or apply some surface treatment to it, such as cleaning it (Chapter 28) or coating it (Chapter 29). Property-enhancing operations are performed to improve mechanical or physical properties of the work material. They do not alter part geometry, at least not intentionally. The most important property-enhancing operations are heat treatments. *Heat treatment* involves various heating and cooling procedures performed to effect structural changes in a material, which in turn affect its mechanical properties. Its most common applications are on metals, discussed in this chapter. Similar treatments are performed on glass ceramics (Section 7.4.3), tempered glass (Section 12.3.1), and powder metals and ceramics (Sections 16.3.3 and 17.2.3).

Heat treatment can be performed on a metallic workpart at various times during its manufacturing sequence. In some cases, the treatment is applied prior to shaping (e.g., to soften the metal so that it can be more easily formed while hot). In other cases, heat treatment is used to relieve the effects of strain hardening that occur during forming, so that the material can be subjected to further deformation. Heat treatment can also be accomplished at or near the end of the sequence to achieve the strength and hardness required in the finished product. The principal heat treatments are annealing, martensite formation in steel, precipitation hardening, and surface hardening.

27.1 ANNEALING

Annealing consists of heating the metal to a suitable temperature, holding at that temperature for a certain time (called *soaking*), and slowly cooling. It is performed on a metal for any of the following reasons: (1) to reduce hardness and brittleness, (2) to alter microstructure so that desirable mechanical properties can be obtained, (3) to soften metals for improved machinability or formability, (4) to recrystallize cold worked (strain-hardened) metals, and (5) to relieve residual stresses induced by prior shaping processes. Different terms are used in annealing, depending on the details of the process and the temperature used relative to the recrystallization temperature of the metal being treated.

Full annealing is associated with ferrous metals (usually low and medium carbon steels); it involves heating the alloy into the austenite region, followed by slow cooling in the furnace to produce coarse pearlite. *Normalizing* involves similar heating and soaking cycles, but the cooling rates are faster. The steel is allowed to cool in air to room temperature. This results in fine pearlite, higher strength and hardness, but lower ductility than the full anneal treatment.

Cold-worked parts are often annealed to reduce effects of strain hardening and to increase ductility. The treatment allows the strain-hardened metal to recrystallize partially or completely, depending on temperatures, soaking periods, and cooling rates. When annealing is performed to allow for further cold working of the part, it is called a *process anneal.* When performed on the completed (cold-worked) part to remove the effects of strain hardening and where no subsequent deformation will be accomplished, it is simply called an *anneal.* The process itself is pretty much the same, but different terms are used to indicate the purpose of the treatment.

If annealing conditions permit full recovery of the cold-worked metal to its original grain structure, then *recrystallization* has occurred. After this type of anneal, the metal has the new geometry created by the forming operation, but its grain structure and associated properties are essentially the same as before cold working. The conditions that tend to favor recrystallization are higher temperature, longer holding time, and slower cooling rate. If the annealing process only permits partial return of the grain structure toward its original state, it is termed a *recovery anneal.* Recovery allows the metal to retain most of the strain hardening obtained in cold working, but the toughness of the part is improved.

These annealing operations are performed primarily to accomplish functions other than stress relief. However, *stress-relief annealing* is performed solely to relieve residual stresses in the workpiece caused by prior shape processing. It helps to reduce distortion and dimensional variations that might otherwise result in the stressed parts.

27.2 MARTENSITE FORMATION IN STEEL

The iron–carbon phase diagram in Figure 6.4 indicates the phases of iron and iron carbide (cementite) present under equilibrium conditions. It assumes that cooling from high temperature has been slow enough to permit austenite to decompose into a mixture of ferrite and cementite (Fe_3C) at room temperature. This decomposition reaction requires diffusion and other processes that depend on time and temperature in order to transform the metal into its preferred final form. However, under conditions of rapid cooling, so that the equilibrium reaction is prevented, austenite transforms into a nonequilibrium phase called martensite. *Martensite* is a hard, brittle phase that gives steel its unique ability to be strengthened to very high values.

27.2.1 The Time-Temperature-Transformation Curve

The nature of the martensite transformation can best be understood using the time-temperature-transformation curve (TTT curve) for eutectoid steel, illustrated in Figure 27.1. The TTT curve shows how cooling rate affects the transformation of austenite into various possible phases. The phases can be divided between (1) alternative forms of ferrite and cementite and (2) martensite. Time is displayed (logarithmically for convenience) along the horizontal axis, and temperature is scaled on the vertical axis. The curve is interpreted by starting at time zero in the austenite region (somewhere above the A_1 temperature line for the given composition) and proceeding downward and to the right along a trajectory representing how the metal is cooled as a function of time. The TTT curve shown in the figure is for a specific composition of steel (0.80% carbon). The shape of the curve is different for other compositions.

At slow cooling rates, the trajectory proceeds through the region indicating transformation into pearlite or bainite, which are alternative forms of ferrite–carbide mixtures. Since these transformations take time, the TTT diagram shows two lines—the start and finish of the transformation as time passes, indicated for the different phase regions by the subscripts s and f, respectively. *Pearlite* is a mixture of ferrite and carbide phases in the form of thin parallel plates. It is obtained by slow cooling from austenite, so that the cooling trajectory passes through P_s above the "nose" of the TTT curve. *Bainite* is an alternative mixture of the same phases that can be produced by initial rapid cooling to a temperature somewhat above M_s, so that the nose of the TTT curve is avoided; this is followed by much slower cooling to pass through B_s and into the ferrite–carbide region. Bainite has a needle-like or feather-like structure consisting of fine carbide regions.

If cooling occurs at a sufficiently rapid rate (indicated by the dashed line in Figure 27.1), austenite is transformed into martensite. *Martensite* is a unique phase consisting of an iron–carbon solution whose composition is the same as the austenite from which it was derived. The face-centered cubic structure of austenite is transformed into

FIGURE 27.1   The TTT curve, showing the transformation of austenite into other phases as a function of time and temperature for a composition of about 0.80% C steel. The cooling trajectory shown here yields martensite.

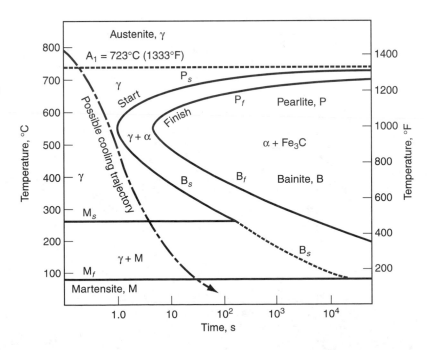

the body-centered tetragonal (BCT) structure of martensite almost instantly—without the time-dependent diffusion process needed to separate ferrite and iron carbide in the preceding transformations.

During cooling, the martensite transformation begins at a certain temperature M_s, and finishes at a lower temperature M_f, as shown in our TTT diagram. At points between these two levels, the steel is a mixture of austenite and martensite. If cooling is stopped at a temperature between the M_s and M_f lines, the austenite will transform to bainite as the time-temperature trajectory crosses the B_s threshold. The level of the M_s line is influenced by alloying elements, including carbon. In some cases, the M_s line is depressed below room temperature, making it impossible for these steels to form martensite by traditional heat treating methods.

The extreme hardness of martensite results from the lattice strain created by carbon atoms trapped in the BCT structure, thus providing a barrier to slip. Figure 27.2 shows the significant effect that the martensite transformation has on the hardness of steel for increasing carbon contents.

27.2.2 The Heat Treatment Process

The heat treatment to form martensite consists of two steps: austenitizing and quenching. These steps are often followed by tempering to produce tempered martensite. **Austenitizing** involves heating the steel to a sufficiently high temperature that it is converted entirely or partially to austenite. This temperature can be determined from the phase diagram for the particular alloy composition. The transformation to austenite involves a phase change, which requires time as well as heating. Accordingly, the steel must be held at the elevated temperature for a sufficient period of time to allow the new phase to form and the required homogeneity of composition to be achieved.

The **quenching** step involves cooling the austenite rapidly enough to avoid passing through the nose of the TTT curve, as indicated in the cooling trajectory shown in Figure 27.1. The cooling rate depends on the quenching medium and the rate of heat

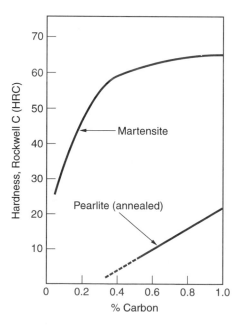

FIGURE 27.2 Hardness of plain carbon steel as a function of carbon content in (hardened) martensite and pearlite (annealed).

transfer within the steel workpiece. Various quenching media are used in commercial heat treatment operations. They include (1) brine—salt water, usually agitated; (2) still fresh water; (3) still oil; and (4) air. Quenching in agitated brine provides the fastest cooling of the heated part surface, while air quench is the slowest. Trouble is, the more effective the quenching media is at cooling, the more likely it is to cause internal stresses, distortion, and cracks in the product.

The rate of heat transfer within the part depends largely on its mass and geometry. A large cubic shape will cool much more slowly than a small, thin sheet. The coefficient of thermal conductivity k of the particular composition is also a factor in the flow of heat in the metal. There is considerable variation in k for different grades of steel; for example, plain low-carbon steel has a typical k value equal to 0.046 J/sec-mm-°C (2.2 Btu/hr-in.-°F), whereas a highly alloyed steel might have one-third that value.

Martensite is hard and brittle. *Tempering* is a heat treatment applied to hardened steels to reduce brittleness, increase ductility and toughness, and relieve stresses in the martensite structure. It involves heating and soaking at a temperature below the eutectoid for about an hour, followed by slow cooling. This results in precipitation of very fine carbide particles from the martensitic iron–carbon solution, and gradually transforms the crystal structure from BCT to BCC. This new structure is called *tempered martensite.* A slight reduction in strength and hardness accompanies the improvement in ductility and toughness. The temperature and tempering time control the degree of softening in the hardened steel, since the change from untempered to tempered martensite involves diffusion.

Taken together, the three steps in the heat treatment of steel to form tempered martensite can be pictured as in Figure 27.3. There are two heating and cooling cycles, the first to produce martensite and the second to temper the martensite.

27.2.3 Hardenability

Hardenability refers to the relative capacity of a steel to be hardened by transformation to martensite. It determines the depth below the quenched surface to which the steel is hardened, or the severity of the quench required to achieve a certain hardness penetration. Steels with good hardenability can be hardened more deeply below the surface and do not require high cooling rates. Hardenability does not refer to the maximum hardness that can be attained in the steel; that depends on the carbon content.

The hardenability of a steel is increased through alloying. Alloying elements having the greatest effect are chromium, manganese, molybdenum (and nickel, to a lesser extent).

FIGURE 27.3 Typical heat treatment of steel: austenitizing, quenching, and tempering.

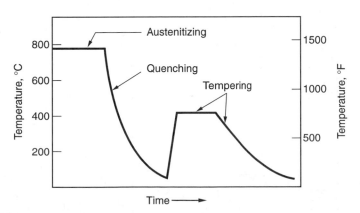

FIGURE 27.4 The Jominy end-quench test: (a) setup of the test, showing end quench of the test specimen; and (b) typical pattern of hardness readings as a function of distance from quenched end.

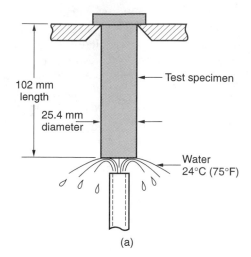

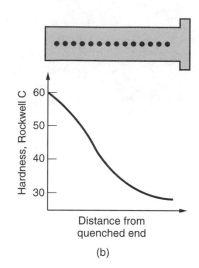

(a)

(b)

The mechanism by which these alloying ingredients operate is to extend the time before the start of the austenite-to-pearlite transformation in the TTT diagram. In effect, the TTT curve is moved to the right, thus permitting slower quenching rates during quenching. Thus the cooling trajectory is able to follow a less hastened path to the M_s line, more easily avoiding the obstacle imposed by the nose of the TTT curve.

The most common method for measuring hardenability is the **Jominy end-quench test.** The test involves heating a standard specimen of diameter = 25.4 mm (1.0 in.) and length = 102 mm (4.0 in.) into the austenite range, and then quenching one end with a stream of cold water while the specimen is supported vertically as shown in Figure 27.4(a). The cooling rate in the test specimen decreases with increased distance from the quenched end. Hardenability is indicated by the hardness of the specimen as a function of distance from quenched end, as in Figure 27.4(b).

27.3 PRECIPITATION HARDENING

Precipitation hardening involves the formation of fine particles (precipitates) that act to block the movement of dislocations and thus strengthen and harden the metal. It is the

FIGURE 27.5 Precipitation hardening: (a) phase diagram of an alloy system consisting of metals A and B that can be precipitation hardened; and (b) heat treatment: (1) solution treatment, (2) quenching, and (3) precipitation treatment.

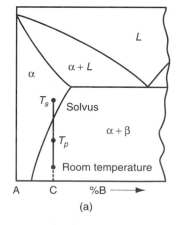

(a)

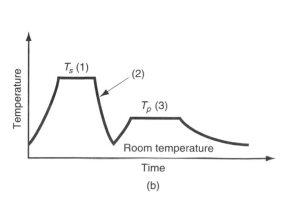

(b)

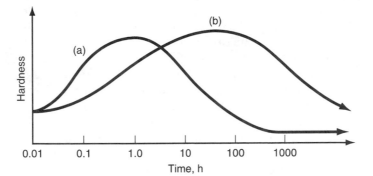

FIGURE 27.6   Effect of temperature and time during precipitation treatment (aging): (a) high precipitation temperature; and (b) lower precipitation temperature.

principal heat treatment for strengthening alloys of aluminum, copper, magnesium, nickel, and other nonferrous metals. Precipitation hardening can also be used to strengthen steel alloys that cannot form martensite by the usual method.

The necessary condition that determines whether an alloy system can be strengthened by precipitation hardening is the presence of a sloping solvus line, as shown in the phase diagram of Figure 27.5(a). A composition that can be precipitation hardened contains two phases at room temperature, but which can be heated to a temperature that dissolves the second phase. Composition C satisfies this requirement. The heat treatment process consists of three steps, illustrated in Figure 27.5(b): (1) *solution treatment,* in which the alloy is heated to a temperature T_s above the solvus line into the alpha phase region and held for a period sufficient to dissolve the beta phase; (2) *quenching* to room temperature to create a supersaturated solid solution; and (3) *precipitation treatment,* in which the alloy is heated to a temperature T_p, below T_s, to cause precipitation of fine particles of the beta phase. This third step is called *aging,* and for this reason the whole heat treatment is sometimes called *age hardening.* However, aging can occur in some alloys at room temperature, and so the term *precipitation hardening* seems more precise for the three-step heat treatment process under discussion here. When the aging step is performed at room temperature, it is called *natural aging;* when it is accomplished at an elevated temperature, as in our figure, the term *artificial aging* is often used.

During the aging step high strength and hardness are achieved in the alloy. The combination of temperature and time during the precipitation treatment (aging) is critical in bringing out the desired properties in the alloy. At higher precipitation treatment temperatures, as in Figure 27.6(a), the hardness peaks in a relatively short time; whereas at lower temperatures, as in (b), more time is required to harden the alloy but its maximum hardness is likely to be greater than in the first case. As seen in the plot, continuation of the aging process results in a reduction in hardness and strength properties, called *overaging.* Its overall effect is similar to annealing.

27.4 SURFACE HARDENING

Surface hardening refers to any of several thermochemical treatments applied to steels in which the composition of the part surface is altered by addition of carbon, nitrogen, or other elements. The most common treatments are carburizing, nitriding, and carbonitriding. These processes are commonly applied to low-carbon steel parts to achieve a hard, wear-resistant outer shell while retaining a tough inner core. The term *case hardening* is often used for these treatments.

Carburizing   Carburizing is the most common surface hardening treatment. It involves heating a part of low-carbon steel in the presence of a carbon-rich environment so that C is diffused into the surface. In effect the surface is converted to a high-carbon steel, capable of higher hardness than the low-C core. The carbon-rich environment can be created in several ways. One method involves the use of carbonaceous materials such as charcoal or coke packed in a closed container with the parts. This process, called *pack carburizing,* produces a relatively thick layer on the part surface, ranging from around 0.6 to 4 mm (0.025–0.150 in.). Another method, called *gas carburizing,* uses hydrocarbon fuels such as propane (C_3H_8) inside a sealed furnace to diffuse carbon into the parts. The case thickness in this treatment is thin, 0.13 to 0.75 mm (0.005–0.030 in.). Another process is *liquid carburizing,* which employs a molten salt bath containing sodium cyanide (NaCN), barium chloride ($BaCl_2$), and other compounds to diffuse carbon into the steel. This process produces surface layer thicknesses generally between those of the other two treatments. Typical carburizing temperatures are 875° to 925°C (1600°–1700°F), well into the austenite range.

Carburizing followed by quenching produces a case hardness of around HRC = 60. However, because the internal regions of the part consist of low-carbon steel, and its hardenability is low, it is unaffected by the quench and remains relatively tough and ductile to withstand impact and fatigue stresses.

Nitriding   Nitriding is a treatment in which nitrogen is diffused into the surfaces of special alloy steels to produce a thin hard casing without quenching. To be most effective, the steel must contain certain alloying ingredients such as aluminum (0.85–1.5%) or chromium (5% or more). These elements form nitride compounds that precipitate as very fine particles in the casing to harden the steel. Nitriding methods include: *gas nitriding,* in which the steel parts are heated in an atmosphere of ammonia (or other nitrogen rich gas mixture); and *liquid nitriding,* in which the parts are dipped in molten cyanide salt baths. Both processes are carried out at around 500°C (950°F). Case thicknesses range as low as 0.025 mm (0.001 in.) and up to around 0.5 mm (0.020 in.), with hardnesses up to HRC 70.

Carbonitriding   As its name suggests, carbonitriding is a treatment in which both carbon and nitrogen are absorbed into the steel surface, usually by heating in a furnace containing carbon and ammonia (NH_3). Case thicknesses are usually 0.07 to 0.5 mm (0.003–0.020 in.), with hardnesses comparable to those of the other two treatments.

Chromizing and Boronizing   Two additional surface-hardening treatments diffuse chromium and boron, respectively, into the steel to produce casings that are typically only 0.025 to 0.05 mm (0.001–0.002 in.) thick. *Chromizing* requires higher temperatures and longer treatment times than the preceding surface hardening treatments, but the resulting casing is not only hard and wear resistant, it is also heat and corrosion resistant. The process is usually applied to low carbon steels. Techniques for diffusing chromium into the surface include: packing the steel parts in chromium-rich powders or granules, dipping in a molten salt bath containing Cr and Cr salts, and chemical vapor deposition (Section 29.4).

Boronizing is performed on tool steels, nickel- and cobalt-based alloys, and cast irons, in addition to plain carbon steels, using powders, salts, or gas atmospheres containing boron. The process results in a thin casing with high abrasion resistance and low coefficient of friction. Casing hardnesses reach 70 HRC. When boronizing is used on low-carbon and low alloy steels, corrosion resistance is also improved.

27.5 HEAT TREATMENT METHODS AND FACILITIES

Most heat treatment operations are performed in furnaces. But other techniques selectively heat only the work surface or a portion of the work surface. Thus, we divide this section into two categories of methods and facilities for heat treatment [9]: furnaces, and selective surface-hardening methods.

It should be mentioned that some of the equipment described here is utilized for other processes in addition to heat treatment; these include melting metals for casting (Section 11.4.1); heating prior to warm and hot working (Section 18.3); brazing, soldering, and adhesive curing (Chapter 32); and semiconductor processing (Chapter 35).

27.5.1 Furnaces for Heat Treatment

Furnaces vary greatly in heating technology, size and capacity, construction, and atmosphere control. They usually heat the workparts by a combination of radiation, convection, and conduction. Heating technologies divide between fuel-fired and electric heating. *Fuel-fired furnaces* are normally *direct-fired,* which means that the work is exposed directly to the combustion products. Fuels include gases (such as natural gas or propane) and oils that can be atomized (such as diesel fuel and fuel oil). The chemistry of the combustion products can be controlled by adjusting the fuel–air or fuel–oxygen mixture to minimize scaling (oxide formation) on the work surface. *Electric furnaces* use electric resistance for heating; they are cleaner, quieter, and provide more uniform heating, but they are more expensive to purchase and operate.

A conventional furnace is an enclosure designed to resist heat loss and to accommodate the size of the work to be processed. Furnaces are classified as batch or continuous. *Batch furnaces* are simpler, basically consisting of a heating system in an insulated chamber, with a door for loading and unloading the work. Examples of this general type include *box furnaces,* which are constructed as a rectangular box, available in various sizes; *car-bottom furnaces,* which are much larger and use a railway-type flatcar to move large parts into the heating chamber; and *bell-type furnaces,* in which the cover, or bell, of the furnace can be lifted by a gantry crane to gain access for loading and unloading the hearth.

Continuous furnaces are generally used for higher production rates and provide a means of moving the work through the interior of the heating chamber. Alternative mechanisms for transporting the work include circular configurations that use rotating hearths, and straight-through types in which the work is moved by conveyor through one or more heating chambers in an in-line arrangement.

Special atmospheres are required in certain heat-treatment operations, such as some of the surface-hardening treatments we have discussed. These atmospheres include carbon- and nitrogen-rich environments for diffusion of these elements into the surface of the work. Atmosphere control is desirable in conventional heat treatment operations to avoid excessive oxidation or decarburization.

Vacuum furnaces are capable of creating a vacuum in the heating chamber, and radiant energy is used to heat the workparts. One of the advantages often cited for vacuum furnaces is that work surface oxidation is prevented; thus, this furnace type represents an attractive alternative to atmosphere control. A disadvantage is the time required in each cycle to draw the vacuum; this reduces production rate.

Other furnace types include salt bath and fluidized bed. *Salt bath furnaces* consist of vessels containing molten salts of chlorides and/or nitrates. Parts to be treated are immersed in the molten media. *Fluidized bed furnaces* have a container in which small inert particles are suspended by a high velocity stream of hot gas. Under proper conditions, the aggregate behavior of the particles is fluid-like; thus, rapid heating of parts immersed in the particle bed occurs.

27.5.2 Selective Surface Hardening Methods

These methods heat only the surface of the work, or local areas of the work surface. They differ from surface hardening methods (Section 27.4) in that no chemical changes occur. Here the treatments are only thermal. The selective surface hardening methods include flame hardening, induction hardening, high-frequency resistance heating, electron beam heating, and laser beam heating.

Flame Hardening As the name indicates, this method involves heating of the work surface by means of one or more torches followed by rapid quenching. As a hardening process, it is applied to carbon and alloy steels, tool steels, and cast irons. Fuels include acetylene (C_2H_2), propane (C_3H_8), and other gases. Flame hardening invokes images of a highly manual operation with general lack of control over the results; however, the process can be set up to include temperature control, fixtures for positioning the work relative to the flame, and indexing devices that operate on a precise cycle time, all of which provide close control over the resulting heat treatment. It is fast and versatile, lending itself to high production as well as big components such as large gears that exceed the capacity of furnaces. With proper controls, only the exterior surfaces are hardened, the part interior remaining unaffected. Typical hardness depth is about 2.5 mm (0.10 in.).

Induction Heating This method involves application of electromagnetically induced energy supplied by an induction coil to an electrically conductive workpart. Induction heating is widely used in industry for processes such as brazing, soldering, adhesive curing, and various heat treatments. When used for steel hardening, quenching follows heating. A typical setup is illustrated in Figure 27.7. The induction heating coil carries a high frequency alternating current that induces a current in the encircled workpart to effect heating. The surface, or a portion of the surface, or the entire mass of the part can be heated by the process. Induction heating provides a fast and efficient method of heating any electrically conductive material. Heating cycle times are short, so the process lends itself to high production as well as midrange production.

High-frequency (HF) Resistance Heating This method is used to harden specific areas of steel work surfaces by application of localized resistance heating at high frequency (400 kHz typical). A typical setup is shown in Figure 27.8. The apparatus consists of a water-cooled proximity conductor located over the area to be heated. Contacts are attached to the workpart at the outer edges of the area. When the HF current is applied, the region beneath the proximity conductor is heated rapidly to high temperature—heating to the austenite range typically requires less than a second. When the power is turned off, the area, usually a narrow line, is quenched by heat transfer to the surrounding metal. Depth of the treated area is around 0.63 mm (0.025 in.); hardness depends on carbon content of the steel and can range up to 60 HRC [9].

Electron Beam (EB) Heating EB technology in manufacturing is relatively new. The applications include cutting (Section 26.3.2), welding (Section 31.4.1), and heat treatment (discussed here). The attractive feature of EB processing is the concentration of high-energy densities in a small region of the part. EB heat treatment involves localized surface hardening of steel. The electron beam is generated by an EB gun and

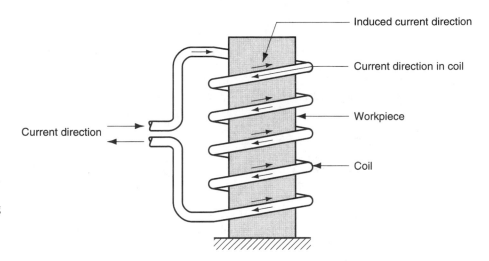

FIGURE 27.7   Typical induction heating setup. High frequency alternating current in a coil induces current in the workpart to effect heating.

focused onto a small area, resulting in rapid heat buildup. Austenitizing temperatures can often be achieved in less than a second. When the directed beam is removed, the heated area is immediately quenched and hardened by heat transfer to the surrounding cold metal.

A disadvantage of EB heating (the same disadvantage applies to other applications) is that best results are achieved when the process is performed in a vacuum. A special vacuum chamber is needed, and time is required to draw the vacuum, thus slowing production rates. When performed in this way, EB hardening eliminates oxidation scale on the work surface.

Laser Beam (LB) Heating   Lasers are another new technology, whose applications include cutting (Section 26.3.3), welding (Section 31.4.2), measurement and inspection (Section 44.5.2), and heat treatment. *Laser* is an acronym for *l*ight *a*mplification by *s*timulated *e*mission of *r*adiation. In LB hardening of steel, a high-density beam of coherent light is focused on a small area—the beam is usually moved along a defined path on the work surface. This causes heating of the steel into the austenite region. When the beam is moved, the area is immediately quenched by heat conduction to the surrounding metal. The advantage of LB over EB heating is that laser beams do not require a vacuum to achieve best results. Energy density levels in EB and LB heating are lower than in cutting or welding.

FIGURE 27.8   Typical setup for high-frequency resistance heating.

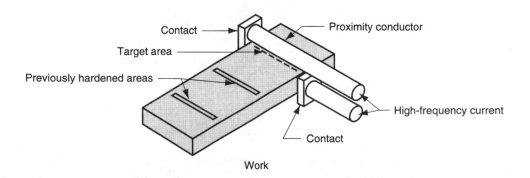

REFERENCES

[1]   Ostwald, P. F., and Munoz, J. *Manufacturing Processes and Systems.* 9th ed. Wiley, New York, 1997.

[2]   Brick, R. M., Pense, A. W., and Gordon, R. B. *Structure and Properties of Engineering Materials.* 4th ed. McGraw-Hill, New York, 1977.

[3]   Chandler, H. (ed.). *Heat Treater's Guide: Practices and Procedures for Irons and Steels.* ASM International, Materials Park, Oh., 1995.

[4]   Chandler, H. (ed.). *Heat Treater's Guide: Practices and Procedures for Nonferrous Alloys.* ASM International, Materials Park, Oh., 1996.

[5]   Flinn, R. A., and Trojan, P. K. *Engineering Materials and Their Applications.* 4th ed. Houghton Mifflin, Boston, 1990.

[6]   Guy, A. G., and Hren, J. J. *Elements of Physical Metallurgy.* 3rd ed. Addison-Wesley, Reading, Mass., 1974.

[7]   *Metals Handbook.* 9th ed. Vol. 4, *Heat Treating.* ASM International, Materials Park, Oh., 1981.

[8]   Vaccari, J. A. "Fundamentals of Heat Treating." Special Report 737, *American Machinist.* September 1981, p. 185–200.

[9]   Wick, C. and Veilleux, R. F. (eds.). *Tool and Manufacturing Engineers Handbook.* 4th ed. Vol. 3—*Materials, Finishing, and Coating*; Section 2—Heat Treatment; Society of Manufacturing Engineers, Dearborn, Mich., 1985.

REVIEW QUESTIONS

27.1. Why are metals heat treated?

27.2. Identify the important reasons why metals are annealed.

27.3. What is the most important heat treatment for hardening steels?

27.4. What is the mechanism by which carbon strengthens steel during heat treatment?

27.5. What information is conveyed by the TTT curve?

27.6. What function is served by *tempering*?

27.7. Define *hardenability.*

27.8. Name some of the elements that have the greatest effect on the hardenability of steel.

27.9. Indicate how the hardenability alloying elements in steel affect the TTT curve.

27.10. Define *precipitation hardening.*

27.11. How does *carburizing* work?

27.12. Identify the selective surface hardening methods.

MULTIPLE CHOICE QUIZ

There is a total of 14 correct answers in the following multiple choice questions (some questions have multiple answers that are correct). To attain a perfect score on the quiz, all correct answers must be given, since each correct answer is worth 1 point. For each question, each omitted answer or wrong answer reduces the score by 1 point, and each additional answer beyond the number of answers required reduces the score by 1 point. Percentage score on the quiz is based on the total number of correct answers.

27.1. Which of the following are the usual objectives of heat treatment (more than one)? (a) increase hardness, (b) increase toughness, (c) recrystallization of the metal, (d) reduce brittleness, (e) reduce density, or (f) relieve stresses.

27.2. Of the following quenching media, which one produces the most rapid cooling rate? (a) air, (b) brine, (c) oil, or (d) pure water.

27.3. On which one of the following metals can the treatment called austenitizing be performed? (a) aluminum alloys, (b) brass, (c) copper alloys, or (d) steel.

27.4. The treatment in which the brittleness of martensite is reduced is called which one of the following? (a) aging, (b) annealing, (c) austenitizing, (d) normalizing, (e) quenching, or (f) tempering.

27.5. The Jominy end-quench test is designed to indicate which one of the following? (a) cooling rate, (b) ductility, (c) hardenability, (d) hardness, or (e) strength.

27.6. In precipitation hardening, the hardening and strengthening of the metal occurs in which one of the following steps? (a) aging, (b) quenching, or (c) solution treatment.

27.7. Which one of the following surface hardening treatments is the most common? (a) boronizing, (b) carbonitriding, (c) carburizing, (d) chromizing, or (e) nitriding.

27.8. Which of the following are selective surface hardening methods (more than one)? (a) electron beam heating, (b) fluidized bed furnaces, (c) induction heating, (d) laser beam heating, or (e) vacuum furnaces.

28 CLEANING AND SURFACE TREATMENTS

CHAPTER CONTENTS

28.1 Chemical Cleaning
    28.1.1 General Considerations in Cleaning
    28.1.2 Chemical Cleaning Processes
28.2 Mechanical Cleaning and Surface Preparation
    28.2.1 Blast Finishing and Shot Peening
    28.2.2 Tumbling and Other Mass Finishing
28.3 Diffusion and Ion Implantation
    28.3.1 Diffusion
    28.3.2 Ion Implantation

Most surface-processing methods have little or no effect on part geometry, except on a microscopic level. In this chapter, we survey the following collection of industrial cleaning and surface treatments: (1) chemical cleaning, (2) mechanical cleaning and related surface treatments, and (3) diffusion and ion implantation. Chapter 29 covers coating and deposition processes.

Workparts must be cleaned one or more times during their manufacturing sequence. Chemical and/or mechanical processes are used to accomplish this cleaning. Chemical cleaning methods use chemicals to remove unwanted substances, such as oils and dirts, from the workpart surface. Mechanical cleaning involves removal of substances from a surface by mechanical operations of various kinds. These operations often serve other functions such as removing burrs, improving smoothness, adding luster, and enhancing surface properties. Other processes that enhance surface properties are diffusion and ion implantation. These processes impregnate the work surface with atoms of a foreign material to alter the surface chemistry and change its physical properties. Thus, the principal functions of the processes discussed in this chapter are to clean the work surface and/or to enhance its properties in some way.

28.1 CHEMICAL CLEANING

A typical surface is covered with various films, oils, dirt, and other contaminants (Section 5.2.1). Although some of these substances may operate in a beneficial way (such as the oxide film on aluminum), it is usually desirable to remove contaminants from the surface. In this section, we discuss some general considerations related to cleaning, and we survey the principal chemical cleaning processes used in industry.

Some of the important reasons why manufactured parts (and products) must be cleaned are: (1) to prepare the surface for subsequent industrial processing, such as a coating application or adhesive bonding; (2) to improve hygiene conditions for workers and customers; (3) to remove contaminants which might chemically react with the surface; and (4) to enhance appearance and performance of the product.

28.1.1 General Considerations in Cleaning

There is no single cleaning method that can be used for all cleaning tasks. Just as various soaps and detergents are required for different household jobs (laundry, dishwashing, pot scrubbing, bathtub cleaning, and so forth), various cleaning methods are also needed to solve different cleaning problems in industry. Important factors in selecting a cleaning method are: (1) the contaminant to be removed, (2) degree of cleanliness required, (3) substrate material to be cleaned, (4) purpose of the cleaning, (5) environmental and safety factors, (6) size and geometry of the part, and (7) production and cost requirements.

Various kinds of contaminants build up on part surfaces, either due to previous processing or the factory environment. To select the best cleaning method, one must first identify what must be cleaned. Surface contaminants found in the factory usually divide into one of the following categories: (1) oil and grease, which includes lubricants used in metalworking; (2) solid particles such as metal chips, abrasive grits, shop dirt, dust, and similar materials; (3) buffing and polishing compounds; and (4) oxide films, rust, and scale.

Degree of cleanliness refers to the amount of contaminant remaining after a given cleaning operation. Parts being prepared to accept a coating (e.g., paint, metallic film) or adhesive must be very clean; otherwise, adhesion of the coated material is jeopardized. In other cases, it may be desirable for the cleaning operation to leave a residue on the part surface for corrosion protection during storage, in effect replacing one contaminant on the surface by another that is beneficial. Degree of cleanliness is often difficult to measure in a quantifiable way. A simple test is a *wiping method,* in which the surface is wiped with a clean white cloth, and the amount of soil absorbed by the cloth is observed. It is a nonquantitative but easy test to use.

The substrate material must be considered in selecting a cleaning method, so that damaging reactions are not caused by the cleaning chemicals. To cite several examples: aluminum is dissolved by most acids and alkalis; magnesium is attacked by many acids; copper is attacked by oxidizing acids (e.g., nitric acid); steels are resistant to alkalis but react with virtually all acids.

Some cleaning methods are appropriate to prepare the surface for painting, while others are better for plating. Environmental protection and worker safety are becoming increasingly important in industrial processes. Cleaning methods and the associated chemicals should be selected to avoid pollution and health hazards.

28.1.2 Chemical Cleaning Processes

Chemical cleaning uses various types of chemicals to effect contaminant removal from the surface. The major chemical cleaning methods are (1) alkaline cleaning, (2) emulsion cleaning, (3) solvent cleaning, (4) acid cleaning, and (5) ultrasonic cleaning. In some cases, chemical action is augmented by other energy forms (e.g., ultrasonic cleaning uses high-frequency mechanical vibrations combined with chemical cleaning). In the following paragraphs, we review these chemical methods.

Alkaline Cleaning This is the most widely used industrial cleaning method. As its name indicates, *alkaline cleaning* employs an alkali to remove oils, grease, wax, and various types of particles (metal chips, silica, carbon, and light scale) from a metallic surface. Alkaline cleaning solutions consist of low-cost, water-soluble salts such as sodium and potassium hydroxide ($NaOH$, KOH), sodium carbonate (Na_2CO_3), borax ($Na_2B_4O_7$), phosphates and silicates of sodium and potassium, combined with dispersants and surfactants in water. The cleaning method is commonly by immersion or spraying, usually at temperatures of 50 to 95°C (120–200°F). Following application of the alkaline solution, a water rinse is used to remove the alkali residue. Metal surfaces cleaned by alkaline solutions are typically electroplated or conversion coated.

   Electrolytic cleaning, also called *electrocleaning,* is a related process in which a 3 to 12 V direct current is applied to an alkaline cleaning solution. The electrolytic action results in the generation of gas bubbles at the part surface, causing a scrubbing action that aids in removal of tenacious dirt films.

Emulsion Cleaning This cleaning method uses organic solvents (oils) dispersed in an aqueous solution. The use of suitable emulsifiers (soaps) results in a two-phase cleaning fluid (oil-in-water), which functions by dissolving or emulsifying the soils on the part surface. The process can be used on either metal or nonmetallic parts. Emulsion cleaning must be followed by alkaline cleaning to eliminate all residues of the organic solvent prior to plating.

Solvent Cleaning In *solvent cleaning,* organic soils such as oil and grease are removed from a metallic surface by means of chemicals that dissolve the soils. Common application techniques include hand-wiping, immersion, spraying, and vapor degreasing. *Vapor degreasing* uses hot vapors of chlorinated or fluorinated solvents to remove oils, greases, and other soils from parts. Because the chemicals used in this cleaning method are hazardous to humans and the environment, vapor degreasing has been largely discontinued, at least in the United States.

Acid Cleaning and Pickling *Acid cleaning* removes oils and light oxides from metal surfaces by soaking, spraying, or manual brushing or wiping. The process is carried out at ambient or elevated temperatures. Common cleaning fluids are acid solutions combined with water-miscible solvents, wetting and emulsifying agents. Cleaning acids include hydrochloric (HCl), nitric (HNO_3), phosphoric (H_3PO_4), and sulfuric (H_2SO_4), the selection depending on the base metal and purpose of the cleaning. For example, phosphoric acid produces a light phosphate film on the metallic surface, which can be a useful preparation for painting.

   The distinction between acid cleaning and acid pickling is a matter of degree. *Acid pickling* involves a more severe treatment to remove thicker oxides, rusts, and scales; it generally results in some etching of the metallic surface, which serves to improve organic paint adhesion.

Ultrasonic Cleaning Ultrasonic cleaning combines chemical cleaning and mechanical agitation of the cleaning fluid to provide a highly effective method for removing surface contaminants. The cleaning fluid is generally an aqueous solution containing alkaline detergents. The mechanical agitation is produced by high-frequency vibrations of sufficient amplitude to cause cavitation-formation of low pressure vapor bubbles or cavities. As the vibration wave passes a given point in the liquid, the low pressure region is followed by a high-pressure front that implodes the cavity, thereby producing a shock wave capable of penetrating contaminant particles adhering to the work surface. This rapid cycle of cavitation and implosion occurs throughout the liquid medium, thus making ultrasonic cleaning effective even on complex and intricate internal shapes. The cleaning process is performed at frequencies between 20 and 45 kHz, and the cleaning solution is usually at an elevated temperature, typically 65 to 85°C (150–190°F).

28.2 MECHANICAL CLEANING AND SURFACE PREPARATION

Mechanical cleaning involves the physical removal of soils, scales, or films from the work surface of the workpart by means of abrasives or similar mechanical action. The processes used for mechanical cleaning often serve other functions in addition to cleaning, such as deburring and improving surface finish.

28.2.1 Blast Finishing and Shot Peening

Blast finishing uses the high-velocity impact of particulate media to clean and finish a surface. The most well known of these methods is *sand blasting,* which uses grits of sand (SiO_2) as the blasting media; however, various other media are also used, including hard abrasives such as aluminum oxide (Al_2O_3) and silicon carbide (SiC), and soft media such as nylon beads and crushed nut shells. The media is propelled at the target surface by pressurized air or centrifugal force. In some applications, the process is performed wet, in which fine particles in a water slurry are directed under hydraulic pressure at the surface.

In *shot peening,* a high-velocity stream of small cast-steel pellets (called *shot*) is directed at a metallic surface with the effect of cold working and inducing compressive stresses into the surface layers. Shot peening is used primarily to improve fatigue strength of metal parts. Its principal purpose is therefore different from blast finishing, although surface cleaning is accomplished as a byproduct of the operation.

28.2.2 Tumbling and Other Mass Finishing

Tumbling, vibratory finishing, and similar operations form a group of finishing processes that have come to be known as mass finishing methods. *Mass finishing* involves the finishing of parts in bulk by a mixing action inside a container, usually in the presence of an abrasive media. The mixing causes the parts to rub against the media and each other to achieve the desired finishing action. Mass finishing methods are used for deburring, descaling, deflashing, polishing, radiusing, burnishing, and cleaning. The parts include stampings, castings, forgings, extrusions, and machined parts. Even plastic and ceramic parts are sometimes subjected to these mass finishing operations to achieve desired finishing results. The parts processed by these methods are usually small and are therefore uneconomical to finish individually.

Processes and Equipment   Mass finishing methods include tumbling, vibratory finishing, and several techniques that utilize centrifugal force. *Tumbling* (also called *barrel finishing* and *tumbling barrel finishing*) involves the use of a horizontally oriented barrel of hexagonal or octagonal cross section in which parts are mixed by rotating the barrel at speeds of 10 to 50 rev/min. Finishing is performed by a "landslide" action of the media and parts as the barrel revolves. As pictured in Figure 28.1, the contents rise in the barrel due to rotation, followed by a tumbling down of the top layer due to gravity. This cycle of rising and tumbling occurs continuously and, over time, subjects all of the parts to the same desired finishing operation. However,

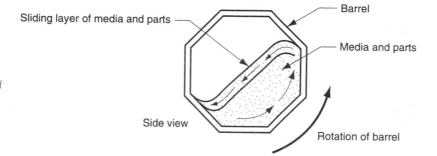

Barrel

Sliding layer of media and parts

Media and parts

Side view

Rotation of barrel

FIGURE 28.1 Diagram of tumbling (barrel finishing) operation showing "landslide" action of parts and abrasive media to finish the parts.

because only the top layer of parts is being finished at any moment, barrel finishing is a relatively slow process compared to other mass finishing methods. It often takes several hours of tumbling to complete the processing. Other drawbacks of barrel finishing include high noise levels and large floorspace requirements.

Vibratory finishing was introduced in the late 1950s as an alternative to tumbling. The vibrating vessel subjects all parts to agitation with the abrasive media, as opposed to only the top layer as in barrel finishing. Consequently, processing times for vibratory finishing are significantly reduced. The open tubs used in this method permit inspection of the parts during processing, and noise is reduced.

Media Most of the *media* in these operations are abrasive; however, some media perform nonabrasive finishing operations such as burnishing and surface hardening. The media may be natural or synthetic materials. Natural media include corundum, granite, limestone, and even hardwood. The problem with these materials is that they are generally softer (and therefore wear more rapidly) and nonuniform in size (and sometimes clog in the workparts). Synthetic media can be made with greater consistency, both in size and hardness. These materials include Al_2O_3 and SiC, compacted into a desired shape and size using a bonding material such as a polyester resin. Their shapes include spheres, cones, angle-cut cylinders, and other regular geometric forms, as in Figure 28.2(a). Steel is also used as a mass finishing media in shapes such as those shown in Figure 28.2(b) for burnishing, surface-hardening, and light deburring operations. The shapes shown in Figure 28.2 come in various sizes. Selection of media is based on part size and shape, as well as finishing requirements.

FIGURE 28.2 Typical preformed media shapes used in mass finishing operations: (a) abrasive media for finishing, and (b) steel media for burnishing.

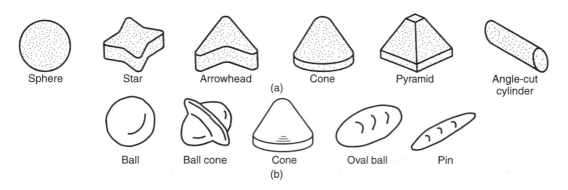

| Sphere | Star | Arrowhead | Cone | Pyramid | Angle-cut cylinder |

(a)

| Ball | Ball cone | Cone | Oval ball | Pin |

(b)

In most mass finishing processes, a compound is used with the media. The mass finishing **compound** is a combination of chemicals for specific functions such as cleaning, cooling, rust inhibiting (of steel parts and steel media), and enhancing brightness and color of the parts (especially in burnishing).

28.3 DIFFUSION AND ION IMPLANTATION

In this section, we discuss two processes in which the surface of a substrate is impregnated with foreign atoms that alter its properties.

28.3.1 Diffusion

Diffusion involves the alteration of surface layers of a material by diffusing atoms of a different material (usually an element) into the surface (Section 4.3). The process has important applications in metallurgy and semiconductor manufacture. The diffusion process impregnates the surface layers of the substrate with the foreign element, but the surface still contains a high proportion of substrate material. A typical profile of composition as a function of depth below the surface for a diffusion coated metal part is illustrated in Figure 28.3. The characteristic of a diffusion impregnated surface is that the diffused element has a maximum percentage at the surface and rapidly declines with distance below the surface.

Metallurgical Applications of Diffusion Coating Diffusion is used to alter the surface chemistry of metals in a number of processes and treatments. One important application is surface hardening, typified by **carburizing, nitriding, carbonitriding, chromizing,** and **boronizing** (Section 27.4). In these treatments, one or more elements (C and/or Ni, Cr, or Bo) are diffused into the surface of iron or steel. The principal purpose of the altered surface chemistry is to increase hardness and wear resistance.

There are other diffusion processes in which corrosion resistance and/or high-temperature oxidation resistance are main objectives. Chromizing (Section 27.4), aluminizing, and siliconizing are important examples. **Aluminizing,** also known as **calorizing,** involves diffusion of aluminum into carbon steel, alloy steels, and alloys of nickel and cobalt.

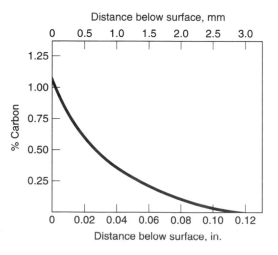

FIGURE 28.3 Characteristic profile of diffused element as a function of distance below surface in diffusion. The plot given here is for carbon diffused into iron (source: [2]).

The treatment is accomplished by either (1) ***pack diffusion,*** in which workparts are packed with Al powders and baked at high temperature to create the diffusion layer; or (2) a ***slurry method,*** in which the workparts are dipped or sprayed with a mixture of Al powders and binders, then dried and baked.

Siliconizing is a treatment of steel in which silicon is diffused into the part surface to create a layer with good corrosion and wear resistance and moderate heat resistance. The treatment is carried out by heating the work in powders of silicon carbide (SiC) in an atmosphere containing vapors of silicon tetrachloride ($SiCl_4$). Siliconizing is less common than aluminizing.

Semiconductor Applications   In semiconductor processing, diffusion of an impurity element into the surface of a silicon chip is used to change the electrical properties at the surface to create devices such as transistors and diodes. We examine how diffusion is used to accomplish this ***doping,*** as it is called, and other semiconductor processes in Chapter 35.

28.3.2 Ion Implantation

Ion implantation is an alternative when diffusion is not feasible because of the high temperatures required. The ion implantation process involves embedding atoms of one (or more) foreign element(s) into a substrate surface using a high-energy beam of ionized particles. The result is an alteration of the chemical and physical properties of the layers near the substrate surface. Penetration of atoms produces a much thinner altered layer than diffusion, as indicated by a comparison of Figures 28.3 and 28.4. Also, the concentration profile of the impregnated element is different than the characteristic diffusion layer.

Advantages of ion implantation include (1) low temperature processing, (2) good control and reproducibility of penetration depth of impurities, and (3) solubility limits can be exceeded without precipitation of excess atoms. Ion implantation finds some of its applications as a substitute for certain coating processes, where its advantages include (4) no problems with waste disposal as in electroplating and many coating processes, and

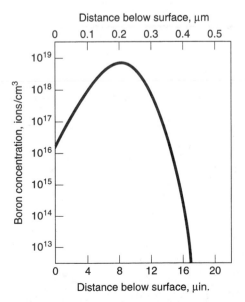

FIGURE 28.4   Profile of surface chemistry as treated by ion implantation (source: [5]). Shown here is a typical plot for boron implanted in silicon. Note the difference in profile shape and depth of altered layer compared to diffusion coating in Figure 28.3.

(5) no discontinuity between coating and substrate. Principal applications of ion implantation are in modifying metal surfaces to improve properties and fabrication of semiconductor devices.

REFERENCES

[1] Freeman, N. B. "A New Look at Mass Finishing." Special Report 757, *American Machinist.* August 1983, pp. 93–104.

[2] Hocking, M. G., Vasantasree, V., and Sidky, P. S. *Metallic and Ceramic Coatings.* Addison-Wesley, Reading, Mass., 1989.

[3] *Metal Finishing.* Guidebook and Directory Issue, Metals and Plastics Publications, Inc., Hackensack, New Jersey, 1991.

[4] *Metals Handbook.* 9th ed. Volume 5, *Surface Cleaning, Finishing, and Coating.* American Society for Metals, Metals Park, Ohio, 1982.

[5] Wick, C. and Veilleux, R. (eds.). *Tool and Manufacturing Engineers Handbook.* 4th ed. Volume III, *Materials, Finishes. and Coating.* Society of Manufacturing Engineers, Dearborn, Mich., 1985.

REVIEW QUESTIONS

28.1. What are some of the important reasons why manufactured parts must be cleaned?

28.2. Mechanical surface treatments are often performed for reasons other than or in addition to cleaning. What are the reasons?

28.3. What are the basic types of contaminants that must be cleaned from metallic surfaces in manufacturing?

28.4. Identify some of the mechanical cleaning methods.

28.5. In addition to surface cleaning, what is the main function performed by shot peening?

28.6. Name some of the important chemical cleaning methods.

28.7. What is meant by the term *mass finishing*?

28.8. What is the difference between diffusion and ion implantation?

28.9. What is *calorizing*?

MULTIPLE CHOICE QUIZ

There is a total of 16 correct answers in the following multiple choice questions (some questions have multiple answers that are correct). To attain a perfect score on the quiz, all correct answers must be given, since each correct answer is worth 1 point. For each question, each omitted answer or wrong answer reduces the score by 1 point, and each additional answer beyond the number of answers required reduces the score by 1 point. Percentage score on the quiz is based on the total number of correct answers.

28.1. Reasons why workparts must be cleaned include which of the following (more than one)? (a) for better appearance, (b) to enhance mechanical properties of the surface, (c) to improve hygiene conditions for worker, (d) to prepare the surface for subsequent processing, or (e) to remove contaminants that might chemically attack the surface.

28.2. Which of the following chemicals is associated with alkaline cleaning (more than one)? (a) borax, (b) sodium hydroxide, (c) sulfuric acid, or (d) trichlorethylene.

28.3. Shot peening is a mechanical cleaning method used primarily to remove surface scale from metallic parts: (a) true or (b) false.

28.4. In sand blasting, which one of the following abrasives is used? (a) Al_2O_3, (b) crushed nut shells, (c) nylon beads, (d) SiC, or (e) SiO_2.

28.5. The abrasive media used in mass finishing, such as barrel tumbling, include which of the following (more than one)? (a) Al_2O_3, (b) corundum, (c) emery, (d) limestone, and (e) SiC.

28.6. Which of the following processes generally produces a deeper penetration of atoms in the impregnated surface? (a) diffusion or (b) ion implantation.

28.7. Calorizing is the same as which one of the following? (a) aluminizing, (b) doping, (c) hot sand blasting, or (d) siliconizing.

28.8. Carburizing involves which one of the following? (a) acid pickling, (b) blast finishing, (c) diffusion, (d) tumbling, or (e) vapor degreasing.

29

COATING AND DEPOSITION PROCESSES

CHAPTER CONTENTS

29.1 Plating and Related Processes
    29.1.1 Electroplating
    29.1.2 Electroforming
    29.1.3 Electroless Plating
    29.1.4 Hot Dipping
29.2 Conversion Coatings
    29.2.1 Chemical Conversion Coatings
    29.2.2 Anodizing
29.3 Physical Vapor Deposition
    29.3.1 Vacuum Evaporation
    29.3.2 Sputtering
    29.3.3 Ion Plating
29.4 Chemical Vapor Deposition
29.5 Organic Coatings
    29.5.1 Application Methods
    29.5.2 Powder Coatings
29.6 Porcelain Enameling and Other Ceramic Coatings
29.7 Thermal and Mechanical Coating Processes
    29.7.1 Thermal Surfacing Processes
    29.7.2 Mechanical Plating

Products made of metal are almost always coated—by painting, plating, or other process. Principal reasons for coating a metal are: (1) to provide corrosion protection of the substrate; (2) to enhance product appearance (e.g., providing a specified color or texture); (3) to increase wear resistance and/or reduce friction of the surface; (4) to increase electrical conductivity; (5) to increase electrical resistance; (6) to prepare a metallic surface for subsequent processing; and (7) to rebuild surfaces worn or eroded during service.

Nonmetallic materials are also sometimes coated. Several examples are: (1) plastic parts coated to give them a metallic appearance; (2) antireflection coatings are commonly applied to optical glass lenses; and (3) certain coating and deposition processes are used in the fabrication of semiconductor chips (Chapter 35) and printed circuit boards (Chapter 36).

The most important industrial coating processes are covered (excuse the pun) in this chapter. The common feature of these processes is that they all produce a discrete coating on the surface of a substrate material. Good adhesion must be achieved between coating and substrate, and for this to occur the substrate surface must be very clean.

29.1 PLATING AND RELATED PROCESSES

Plating involves the coating of a thin metallic layer onto the surface of a substrate material. The substrate is usually metallic, although methods are available to plate plastic and ceramic parts. Reasons for plating a part include (1) corrosion protection, (2) attractive appearance, (3) wear resistance, (4) increased electrical conductivity, (5) improved solderability, and (6) enhanced lubricity of the surface. The most familiar and widely used plating technology is electroplating.

29.1.1 Electroplating

Electroplating, also known as *electrochemical plating,* is an electrolytic process (Section 4.5) in which metal ions in an electrolyte solution are deposited onto a cathode workpart. The setup is shown in Figure 29.1. The anode is generally made of the metal being plated and thus serves as the source of the plate metal. Direct current from an external power supply is passed between the anode and the cathode. The electrolyte is an aqueous solution of acids, bases, or salts; it conducts electric current by the movement of plate metal ions in solution. For optimum results, parts must be chemically cleaned just prior to electroplating.

Principles of Electroplating Electrochemical plating is based on Faraday's two physical laws. Briefly for our purposes, the laws state: (1) the mass of a substance liberated in electrolysis is proportional to the quantity of electricity passed through the cell; and (2) the mass of the material liberated is proportional to its electrochemical equivalent (ratio of atomic weight to valence). The effects can be summarized in the equation

$$V = CIt \qquad (29.1)$$

where V = volume of metal plated, $mm^3 (in.^3)$; C = plating constant which depends on electrochemical equivalent and density, $mm^3/amp\text{-}s$ ($in.^3/amp\text{-}min$); I = current, amps; and t = time during which current is applied, s (min). The product It (current $\times$ time) is the electrical charge passed in the cell, and the value of C indicates the amount of plating material deposited onto the cathodic workpart per electrical charge.

FIGURE 29.1 Setup for electroplating.

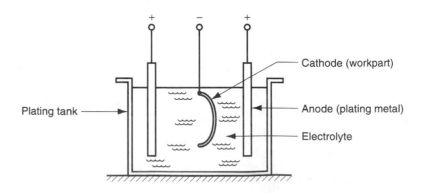

- Cathode (workpart)
- Plating tank
- Anode (plating metal)
- Electrolyte

For most plating metals, not all of the electrical energy in the process is used for deposition; some energy may be consumed in other reactions, such as the liberation of hydrogen at the cathode. This reduces the amount of metal plated. The actual amount of metal deposited on the cathode (workpart) divided by the theoretical amount given by Eq. (29.1) is called the *cathode efficiency.* Taking the cathode efficiency into account, a more accurate equation for determining the volume of metal plated is

$$V = ECIt \tag{29.2}$$

where E = cathode efficiency, and the other terms are defined as before. Typical values of cathode efficiency E and plating constant C for different metals are presented in Table 29.1. The average plating thickness can be determined from the following:

$$d = \frac{V}{A} \tag{29.3}$$

where d = plating thickness, mm (in.); V = volume of plate metal from Eq. (29.2); and A = surface area of plated part, mm^2 (in.^{2}).

EXAMPLE 29.1
Electroplating

A steel part with surface area $A = 125$ cm^2 is to be nickel plated. What average plating thickness will result if 12 amps are applied for 15 min in an acid sulfate electrolyte bath?

Solution: From Table 29.1, the cathode efficiency for nickel is $E = 0.95$ and the plating constant $C = 3.42(10^{-2})$ mm^3/amp-s. Using Eq. (29.2), the total amount of plating metal deposited onto the part surface in 15 min is given by

$$V = 0.95(3.42 \times 10^{-2})(12)(15)(60) = 350.9 \text{ mm}^3$$

This is spread across an area $A = 125$ cm^2 = 12,500 mm^2, so the average plate thickness is

$$d = \frac{350.9}{12500} = \textbf{0.028 mm}$$

Methods and Applications A variety of equipment is available for electroplating, the choice depending on part size and geometry, throughput requirements, and plating

TABLE 29.1 Typical cathode efficiencies in electroplating and values of plating constant C.

Plate Metal[a]	Electrolyte	Cathode Efficiency %	Plating Constant C[a] mm^3/amp-s	(in.^{3}/amp-min)
Cadmium (2)	Cyanide	90	6.73×10^{-2}	(2.47×10^{-4})
Chromium (3)	Chromium-acid-sulfate	15	2.50×10^{-2}	(0.92×10^{-4})
Copper (1)	Cyanide	98	7.35×10^{-2}	(2.69×10^{-4})
Gold (1)	Cyanide	80	10.6×10^{-2}	(3.87×10^{-4})
Nickel (2)	Acid sulfate	95	3.42×10^{-2}	(1.25×10^{-4})
Silver (1)	Cyanide	100	10.7×10^{-2}	(3.90×10^{-4})
Tin (4)	Acid sulfate	90	4.21×10^{-2}	(1.54×10^{-4})
Zinc (2)	Chloride	95	4.75×10^{-2}	(1.74×10^{-4})

Compiled from [12].
[a]Most common valence given in parenthesis ( ); this is the value assumed in determining the plating constant C. For a different valence, compute the new C by multiplying C value in the table by the most common valence and then dividing by the new valence.

metal. The principal methods are (1) barrel plating, (2) rack plating, and (3) strip plating. *Barrel plating* is performed in rotating barrels that are oriented either horizontally or at an oblique angle (35°). The method is suited to the plating of many small parts in a batch. Electrical contact is maintained through the tumbling action of the parts themselves and by means of an externally connected conductor that projects into the barrel. There are limitations to barrel plating; the tumbling action inherent in the process may damage soft metal parts, threaded components, parts requiring good finishes, and heavy parts with sharp edges.

Rack plating is used for parts that are too large, heavy, or complex for barrel plating. Racks are made of heavy-gauge copper wire, formed into suitable shapes for holding the parts and conducting current to them. The racks are fabricated so that workparts can be hung on hooks, or held by clips, or loaded into baskets. To avoid plating of the copper itself, the racks are covered with insulation except in locations where part contact occurs. *Strip plating* is a high-production method in which the work consists of a continuous strip which is pulled through the plating solution by means of a take-up reel. Plated wire is an example of a suitable application. Small sheet metal parts held in a long strip can also be plated by this method. The process can be set up so that only specific regions of the parts are plated; for example, contact points plated with gold on electrical connectors.

Common coating metals in electroplating include zinc, nickel, tin, copper, and chromium. Steel is the most common substrate metal. Precious metals (gold, silver, platinum) are plated on jewelry. Gold is also used for electrical contacts.

Zinc-plated steel products include fasteners, wire goods, electric switch boxes, and various sheet-metal parts. The zinc coating serves as a sacrificial barrier to the corrosion of the steel beneath. An alternative process for coating zinc onto steel is galvanizing (Section 29.1.4). *Nickel plating* is used for corrosion resistance and decorative purposes over steel, brass, zinc die castings, and other metals. Applications include automotive trim and other consumer goods. Nickel is also used as a base coat under a much thinner chrome plate. *Tin plate* is still widely used for corrosion protection in "tin cans" and other food containers. Tin plate is also used to improve solderability of electrical components.

Copper has several important applications as a plating metal. It is widely used as a decorative coating on steel and zinc, either alone or alloyed with zinc as brass plate. It also has important plating applications in printed circuit boards (Section 36.2). Finally, copper is often plated on steel as a base beneath nickel and/or chrome plate. *Chromium plate* (popularly known as *chrome plate*) is valued for its decorative appearance and is widely used in automotive, office furniture, and kitchen appliances. It also produces one of the hardest of all electroplated coatings, so it is widely used for parts requiring wear resistance (e.g., hydraulic pistons and cylinders, piston rings, aircraft engine components, and thread guides in textile machinery).

29.1.2 Electroforming

This process is virtually the same as electroplating, but its purpose is quite different. *Electroforming* involves electrolytic deposition of metal onto a pattern until the required thickness is achieved; the pattern is then removed to leave the formed part. Whereas typical plating thickness is only about 0.05 mm (0.002 in.) or less, electroformed parts are often substantially thicker, so the production cycle is proportionally longer.

Patterns used in electroforming are either solid or expendable. Solid patterns have a taper or other geometry that permits removal of the electroplated part. Expendable patterns are destroyed during part removal; they are used when part shape precludes a solid pattern. Expendable patterns are either fusible or soluble. Fusible patterns are made of low-melting alloys, plastic, wax, or other material that can be removed by melting. When nonconductive materials are used, the patterns must be metallized to accept the electrodeposited coating. Soluble patterns are made of a material that can be readily dissolved by chemicals; for example, aluminum can be dissolved in sodium hydroxide (NaOH).

Electroformed parts are commonly fabricated of copper, nickel, and nickel–cobalt alloys. Applications include fine molds and dies; examples include molds for lenses and phonograph records, and plates for embossing and printing. A recent and rather demanding application involves the production of molds for laser-read compact discs and video discs. The surface details that must be imprinted onto a compact disc are measured in μm or μ-in. (1 μm = 39.4 μ-in.). These details are readily obtained in the mold by electroforming.

29.1.3 Electroless Plating

Electroless plating is the name given to a plating process driven entirely by chemical reactions—no external source of electric current is required. Deposition of metal onto a part surface occurs in an aqueous solution containing ions of the desired plating metal. The process uses a reducing agent, and the workpart surface acts as a catalyst for the reaction.

The metals that can be electroless plated are limited; and for those that can be processed by this technique, the cost is generally greater than electrochemical plating. The most common electroless plating metal is nickel and certain of its alloys (Ni–Co, Ni–P, and Ni–B). Copper and, to a lesser degree, gold are also used as plating metals. Nickel plating by this process is used for applications requiring high resistance to corrosion and wear. Electroless copper plating is used to plate through holes of printed circuit boards (Section 36.2.4). Cu can also be plated onto plastic parts for decorative purposes. Advantages sometimes cited for electroless plating include: (1) uniform plate thickness on complex part geometries (a problem with electroplating); (2) the process can be used on both metallic and nonmetallic substrates; and (3) no need for a DC power supply to drive the process.

29.1.4 Hot Dipping

Hot dipping is a process in which a metal substrate is immersed in a molten bath of a second metal; upon removal, the second metal is coated onto the first. Of course, the first metal must possess a higher melting temperature than the second. The most common substrate metals are steel and iron. Zinc, aluminum, tin, and lead are the common coating metals. Hot dipping works by forming transition layers of varying alloy compositions. Next to the substrate are normally intermetallic compounds of the two metals; at the exterior are solid solution alloys consisting predominantly of the coating metal. The transition layers provide excellent adhesion of the coating.

The primary purpose of hot dipping is corrosion protection. Two mechanisms normally operate to provide this protection: (1) barrier protection—the coating simply serves as a shield for the metal beneath; and (2) sacrificial protection—the coating corrodes by an electrochemical process to preserve the substrate.

Hot dipping goes by different names, depending on coating metal: *galvanizing* is when zinc (Zn) is coated onto steel or iron; *aluminizing* refers to coating of aluminum (Al) onto a substrate; *tinning* is coating of tin (Sn); and *terneplate* describes the plating of lead–tin alloy onto steel. Galvanizing is by far the most important hot dipping process, dating back about 200 years. It is applied to finished steel and iron parts in a batch process, and to sheet, strip, piping, tubing, and wire in a continuous automated process. Coating thickness typically ranges between 0.04 and 0.09 mm (0.0016 and 0.0035 in.). Thickness is controlled largely by immersion time. Bath temperature is maintained at around 450°C (850°F). Commercial use of aluminizing is on the rise, gradually increasing in market share relative to galvanizing. Hot-dipped aluminum coatings provide excellent corrosion protection, in some cases five times more effective than galvanizing [12].

Tin plating by hot dipping provides a nontoxic corrosion protection for steel in applications for food containers, dairy equipment, and soldering applications. Hot dipping has gradually been overtaken by electroplating as the preferred commercial method for plating of tin onto steel. Terneplating involves hot dipping of a lead–tin alloy onto steel. The alloy is predominantly lead (only 2–15% Sn); however, tin is required to obtain satisfactory adhesion of the coating. Terneplate is the lowest cost of the coating methods for steel, but its corrosion protection is limited.

29.2 CONVERSION COATING

Conversion coating refers to a family of processes in which a thin film of oxide, phosphate, or chromate is formed on a metallic surface by chemical or electrochemical reaction. Immersion and spraying are the two common methods of exposing the metal surface to the reacting chemicals. The common metals treated by conversion coating are steel (including galvanized steel), zinc, and aluminum. However, nearly any metal product can benefit from the treatment. The important reasons for using a conversion coating process are [12]: (1) corrosion protection, (2) preparation for painting, (3) wear reduction, (4) permit the surface to better hold lubricants for metal forming processes, (5) increase electrical resistance of surface, (6) decorative finish, and (7) part identification.

The conversion coating processes divide into two categories: (1) chemical treatments, and (2) anodizing. The first category includes those processes which involve a chemical reaction only; phosphate and chromate conversion coatings are the common treatments. The second category is *anodizing,* in which an oxide coating is produced by electrochemical reaction (anodize is a contraction of *anodic oxidize*). This coating process is most commonly associated with aluminum and its alloys.

29.2.1 Chemical Conversion Coatings

Chemical conversion coating exposes the base metal to certain chemicals that form thin, nonmetallic surface films. Similar reactions occur in nature; the oxidation of iron and aluminum are examples. Whereas rusting is progressively destructive of iron, formation of a thin Al_2O_3 coating on aluminum protects the base metal. It is the purpose of these chemical conversion treatments to accomplish the latter effect. The two main processes are phosphate and chromate coating.

Phosphate coating involves transformation of the base metal surface into a protective phosphate film by exposure to solutions of certain phosphate salts (e.g., Zn, Mg, and Ca) together with dilute phosphoric acid (H_3PO_4). The coatings range in thickness from 0.0025 to 0.05 mm (0.0001–0.002 in.). The most common base metals are zinc and steel, including galvanized steel. The phosphate coating serves as a useful preparation for painting in the automotive and heavy appliance industries.

Chromate coating converts the base metal into various forms of chromate films using aqueous solutions of chromic acid, chromate salts, and other chemicals. Metals treated by this method include aluminum, cadmium, copper, magnesium, and zinc (and their alloys). Immersion of the base part is the common method of application. Chromate conversion coatings are somewhat thinner than phosphate, typically less than 0.0025 mm (0.0001 in.). Usual reasons for chromate coating are (1) corrosion protection, (2) base for painting, and (3) decorative purposes. Chromate coatings can be clear or colorful; available colors include olive drab, bronze, yellow, or bright blue.

29.2.2 Anodizing

Although the previous processes are normally performed without electrolysis, *anodizing* is an electrolytic treatment that produces a stable oxide layer on a metallic surface. Its most common applications are with aluminum and magnesium, but it is also applied to zinc, titanium, and other less common metals. Anodized coatings are used primarily for decorative purposes; they also provide corrosion protection.

It is instructive to compare anodizing to electroplating, since they are both electrolytic processes. Two differences stand out. (1) In electrochemical plating, the workpart to be coated is the cathode in the reaction. By contrast, in anodizing, the work is the anode, whereas the processing tank is cathodic. (2) In electroplating, the coating is grown by adhesion of ions of a second metal to the base metal surface. In anodizing, the surface coating is formed through chemical reaction of the substrate metal into an oxide layer.

Anodized coatings usually range in thickness between 0.0025 and 0.075 mm (0.0001 and 0.003 in.). Dyes can be incorporated into the anodizing process to create a wide variety of colors; this is especially common in aluminum anodizing. Very thick coatings up to 0.25 mm (0.010 in.) can also be formed on aluminum by a special process called *hard anodizing*; these coatings are noted for high resistance to wear and corrosion.

29.3 PHYSICAL VAPOR DEPOSITION

Physical vapor deposition (PVD) refers to a family of processes in which a material is converted to its vapor phase in a vacuum chamber and condensed onto a substrate surface as a very thin film. PVD can be used to apply a wide variety of coating materials: metals, alloys, ceramics and other inorganic compounds, and even certain polymers. Possible substrates include metals, glass, and plastics. Thus, PVD represents a versatile coating technology, applicable to an almost unlimited combination of coating substances and substrate materials.

Applications of PVD include thin decorative coatings on plastic and metal parts such as trophies, toys, pens and pencils, watchcases, and interior trim in automobiles. The coatings are thin films of aluminum (around 150 nm) coated with clear lacquer to give a high gloss silver or chrome appearance. Another use of PVD is to apply antireflection coatings of magnesium fluoride (MgF_2) onto optical lenses. PVD is applied in the fabrication of electronic devices, principally for depositing metal to form electrical connections in integrated circuits (Chapter 35). Finally, PVD is widely used to coat titanium nitride (TiN) onto cutting tools and plastic injection molds for wear resistance.

All physical vapor deposition processes consist of the following steps: (1) synthesis of the coating vapor, (2) vapor transport to the substrate, and (3) condensation of vapors onto the substrate surface. These steps are generally carried out inside a vacuum chamber, so evacuation of the chamber must preceed the actual PVD process.

Synthesis of the coating vapor can be accomplished by any of several methods, such as electric resistance heating or ion bombardment to vaporize an existing solid (or liquid). These and other variations result in several PVD processes. They are grouped into three

TABLE 29.2   Summary of physical vapor deposition (PVD) processes.

PVD Process	Features and Coating Materials
Vacuum Evaporation	*Features*: equipment is relatively low-cost and simple; deposition of compounds is difficult; coating adhesion not as good as other PVD processes. *Typical coating materials*: Ag, Al, Au, Cr, Cu, Mo, W.
Sputtering	*Features*: better throwing power and coating adhesion than vacuum evaporation, can coat compounds, slower deposition rates and more difficult process control than vacuum evaporation. *Typical coating materials*: Al_2O_3, Au, Cr, Mo, SiO_2, Si_3N_4, TiC, TiN.
Ion Plating	*Features*: best coverage and coating adhesion of PVD processes, most complex process control, higher deposition rates than sputtering. *Typical coating materials*: Ag, Au, Cr, Mo, Si_3N_4, TiC, TiN.

Compiled from [1].

principal types: (1) vacuum evaporation, (2) sputtering, and (3) ion plating. Table 29.2 presents a summary of these processes.

29.3.1 Vacuum Evaporation

Certain materials (mostly pure metals) can be deposited onto a substrate by first transforming them from solid to vapor state in a vacuum and then letting them condense on the substrate surface. The setup for the vacuum evaporation process is shown in Figure 29.2. The material to be deposited, called the source, is heated to a sufficiently high temperature that it evaporates (or sublimes). Since heating is accomplished in a vacuum, the temperature required for vaporization is significantly below the corresponding temperature required at atmospheric pressure. Also, the absence of air in the chamber prevents oxidation of the source material at the heating temperatures.

Various methods can be used to heat and vaporize the material. A container must be provided to hold the coating material (called the source material) before vaporization. Among the important vaporization methods are resistance heating and electron beam bombardment. **Resistance heating** is the simplest technology. A refractory metal (e.g., W, Mo) is formed into a suitable container to hold the source material. Current is applied to heat the container, which then heats the material in contact with it. One problem with

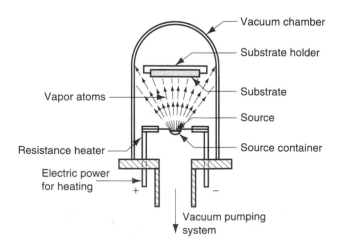

FIGURE 29.2   Setup for vacuum evaporation PVD.

this heating method is possible alloying between the holder and its contents, so that the deposited film becomes contaminated with the metal of the resistance heating container. In *electron beam evaporation,* a stream of electrons at high velocity is directed to bombard the surface of the source material to cause vaporization. By contrast with resistance heating, very little energy acts to heat the container, thus minimizing contamination of the container material with the coating.

Whatever the vaporization technique, evaporated atoms leave the source and follow straight-line paths until they collide with other gas molecules or strike a solid surface. The vacuum inside the chamber virtually eliminates other gas molecules, thus reducing the probability of collisions with source vapor atoms. The substrate surface to be coated is usually positioned relative to the source so that it is the likely solid surface on which the vapor atoms will be deposited. A mechanical manipulator is sometimes used to rotate the substrate so that all surfaces are coated. Upon contact with the relative cool substrate surface, the energy level of the impinging atoms is suddenly reduced to the point where they cannot remain in a vapor state; they condense and become attached to the solid surface, forming a deposited thin film.

29.3.2 Sputtering

If the surface of a solid (or liquid) is bombarded by atomic particles of sufficiently high energy, individual atoms of the surface may acquire enough energy due to the collision that they are ejected from the surface by transfer of momentum. This is the process known as sputtering. The most convenient form of high energy particle is an ionized gas, such as argon, energized by means of an electric field to form a plasma. As a PVD process, *sputtering* involves bombardment of the cathodic coating material with argon ions (Ar^+), causing surface atoms to escape and then be deposited onto a substrate, forming a thin film on the substrate surface. The substrate must be placed close to the cathode and is usually heated to improve bonding of the coating atoms. A typical arrangement is shown in Figure 29.3.

Whereas vacuum evaporation is generally limited to metals, sputtering can be applied to nearly any material—metallic and nonmetallic elements; alloys, ceramics, and polymers. Films of alloys and compounds can be sputtered without changing their chemical compositions. Films of chemical compounds can also be deposited by employing reactive gases that form oxides, carbides, or nitrides with the sputtered metal.

FIGURE 29.3   One possible setup for sputtering, a form of physical vapor deposition.

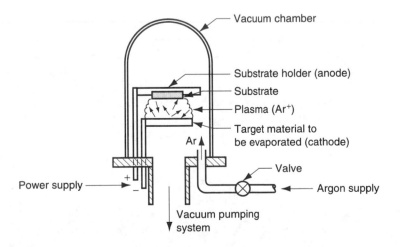

Drawbacks of sputtering PVD include (1) slow deposition rates and (2) since the ions bombarding the surface are a gas, traces of the gas can usually be found in the coated films; the entrapped gases sometimes affect mechanical properties adversely.

29.3.3 Ion Plating

Ion plating uses a combination of sputtering and vacuum evaporation to deposit a thin film onto a substrate. The process works as follows. The substrate is set up to be the cathode in the upper part of the chamber, and the source material is placed below it. A vacuum is then established in the chamber. Argon gas is admitted and an electric field is applied to ionize the gas (Ar^+) and establish a plasma. This results in ion bombardment (sputtering) of the substrate so that its surface is scrubbed to a condition of atomic cleanliness (interpret this as "very clean"). Next, the source material is heated sufficiently to generate coating vapors. The heating methods used here are similar to those used in vacuum evaporation—resistance heating, electron beam bombardment, etc. The vapor molecules pass through the plasma and coat the substrate. Sputtering is continued during deposition, so that the ion bombardment consists not only of the original argon ions but also source material ions that have been energized while being subjected to the same energy field as the argon. The effect of these processing conditions is to produce films of uniform thickness and excellent adherence to the substrate.

Ion plating is applicable to parts having irregular geometries, due to the scattering effects that exist in the plasma field. An example of interest here is TiN coating of high-speed steel cutting tools (e.g., drill bits). In addition to coating uniformity and good adherence, other advantages of the process include: high deposition rates, high film densities, and the capability to coat the inside walls of holes and other hollow shapes.

29.4 CHEMICAL VAPOR DEPOSITION

PVD involves deposition of a coating by condensation onto a substrate from the vapor phase; it is strictly a physical process. By comparison, *chemical vapor deposition* (CVD) involves the interaction between a mixture of gases and the surface of a heated substrate, causing chemical decomposition of some of the gas constituents and formation of a solid film on the substrate. The reactions take place in an enclosed reaction chamber. The reaction product (either a metal or a compound) nucleates and grows on the substrate surface to form the coating. Most CVD reactions require heat. However, depending on the chemicals involved, the reactions can be driven by other possible energy sources, such as ultraviolet light or plasma. CVD includes a wide range of pressures and temperatures; and it can be applied to a great variety of coating and substrate materials.

Industrial metallurgical processes based on chemical vapor deposition date back to the 1800s (e.g., the Mond process in Table 29.3). Modern interest in CVD is focused on its coating applications such as coated cemented carbide tools, solar cells, depositing refractory metals on jet engine turbine blades, and other applications where resistance to wear, corrosion, erosion, and thermal shock are important. In addition, CVD is used in integrated circuit fabrication.

Advantages typically cited for CVD include [6]: (1) capability to deposit refractory materials at temperatures below their melting or sintering temperatures; (2) control of grain size is possible; (3) the process is carried out at atmospheric pressure—it does not require vacuum equipment; and (4) good bonding of coating to substrate surface. Disadvantages include: (1) corrosive and/or toxic nature of chemicals generally necessi-

TABLE 29.3   Some examples of reactions in chemical vapor deposition.

1. The ***Mond Process*** includes a CVD process for decomposition of nickel from nickel carbonyl ($Ni(CO)_4$), which is an intermediate compound formed in reducing nickel ore:

$$Ni(CO)_4 \xrightarrow{\quad 200°C\ (400°F)\quad} Ni + 4CO \qquad (29.6)$$

2. Coating of titanium carbide (TiC) onto a substrate of cemented tungsten carbide (WC–Co) to produce a high-performance cutting tool:

$$TiCl_4 + CH_4 \xrightarrow[\text{excess } H_2]{1000°C\ (1800°F)} TiC + 4HCl \qquad (29.7)$$

3. Coating of titanium nitride (TiN) onto a substrate of cemented tungsten carbide (WC–Co) to produce a high-performance cutting tool:

$$TiCl_4 + 0.5N_2 + 2H_2 \xrightarrow{\quad 900°C\ (1650°F)\quad} TiN + 4HCl \qquad (29.8)$$

4. Coating of aluminum oxide (Al_2O_3) onto a substrate of cemented tungsten carbide (WC–Co) to produce a high-performance cutting tool:

$$2AlCl_3 + 3CO_2 + 3H_2 \xrightarrow{\quad 500°C\ (900°F)\quad} Al_2O_3 + 3CO + 6HCl \qquad (29.9)$$

5. Coating of silicon nitride (Si_3N_4) onto silicon (Si), a process in semiconductor manufacturing:

$$3SiF_4 + 4NH_3 \xrightarrow{\quad 1000°C\ (1800°F)\quad} Si_3N_4 + 12\ HF \qquad (29.10)$$

6. Coating of silicon dioxide (SiO_2) onto silicon (Si), a process in semiconductor manufacturing:

$$2SiCl_3 + 3H_2O + 0.5O_2 \xrightarrow{\quad 900°C\ (1600°F)\quad} 2SiO_2 + 6HCl \qquad (29.11)$$

7. Coating of the refractory metal tungsten (W) onto a substrate, such as a jet engine turbine blade:

$$WF_6 + 3H_2 \xrightarrow{\quad 600°C\ (1100°F)\quad} W + 6HF \qquad (29.12)$$

Compiled from [4], [11], and [12].

tates a closed chamber as well as special pumping and disposal equipment; (2) certain reaction ingredients are relatively expensive; and (3) low material utilization.

CVD Materials and Reactions   In general, metals that are readily electroplated are not good candidates for CVD, owing to the hazardous chemicals that must be used and the costs of safeguarding against them. Metals suitable for coating by CVD include tungsten, molybdenum, titanium, vanadium, and tantalum. Chemical vapor deposition is especially suited to the deposition of compounds, such as aluminum oxide (Al_2O_3), silicon dioxide (SiO_2), silicon nitride (Si_3N_4), titanium carbide (TiC), and titanium nitride (TiN). Figure 29.4 illustrates the application of both CVD and PVD to provide multiple wear-resistant coatings on a cemented carbide cutting tool. Also see Color Plate 7.

The commonly used reacting gases or vapors are metallic hydrides (MH_x), chlorides (MCl_x), fluorides (MF_x), and carbonyls ($M(CO)_x$), where M = the metal to be deposited and x is used to balance the valences in the compound. Other gases, such as hydrogen (H_2), nitrogen (N_2), methane (CH_4), carbon dioxide (CO_2), and ammonia (NH_3) are used in some of the reactions. Table 29.3 presents some examples of CVD reactions that result in deposition of a metal or ceramic coating onto a suitable substrate. Typical temperatures at which these reactions are carried out are also given.

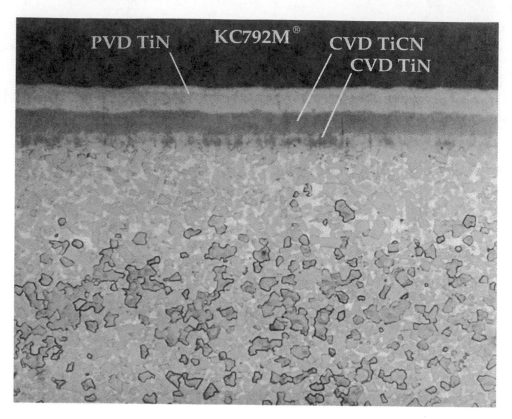

FIGURE 29.4 Photomicrograph of the cross section of a coated carbide cutting tool (Kennametal Grade KC792M); CVD was used to coat TiN and TiCN onto the surface of a WC–Co substrate, followed by a TiN coating applied by PVD (photo courtesy of Kennametal Inc.).

Processing Equipment Chemical vapor deposition processes are carried out in a reactor, which consists of [4] (1) a reactant supply system, (2) a deposition chamber, and (3) a recycle/disposal system. Although reactor configurations differ depending on the application, one possible CVD reactor is illustrated in Figure 29.5. The purpose of the reactant supply system is to deliver reactants to the deposition chamber in the proper proportions. Different types of supply system are required, depending on whether the reactants are delivered as gas, liquid, or solid (e.g., pellets, powders).

The deposition chamber contains the substrates and the chemical reactions that lead to the deposition of reaction products onto the substrate surfaces. Deposition occurs at elevated temperatures, and the substrate must be heated by induction heating, radiant heat, or other means. Deposition temperatures for different CVD reactions are 250°C to 1950°C (500°–3500°F), so the chamber must be designed to meet these temperature demands.

The third component of the reactor is the recycle/disposal system, whose function is to render harmless the byproducts of the CVD reaction. This includes a collection of materials that are toxic, corrosive, and/or flammable, followed by proper processing and disposition.

Alternative Forms of CVD What we have described is *atmospheric pressure chemical vapor deposition* (APCVD), in which the reactions are carried out at or near atmospheric pressure. For many reactions, there are advantages in performing the process at pressures well below atmospheric. This is called *low-pressure chemical vapor deposition* (LPCVD), in which the reactions occur in a partial vacuum. Advantages of LPCVD include [11] (1) uniform thickness, (2) good control over composition and structure, (3) low-temperature processing, (4) fast deposition rates, (5) high throughput and

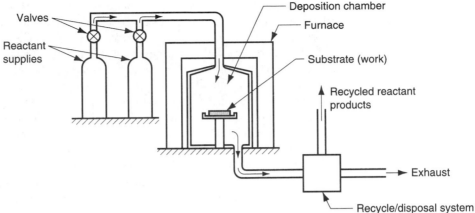

FIGURE 29.5 A typical reactor used in chemical vapor deposition.

lower processing costs. The technical problem in LPCVD is designing the vacuum pumps to create the partial vacuum when the reaction products are not only hot but may also be corrosive. These pumps must often include systems to cool and trap the corrosive gases before they reach the actual pumping unit.

Another variation of CVD is ***plasma assisted chemical vapor deposition*** (PACVD), in which deposition onto a substrate is accomplished by reacting the ingredients in a gas that has been ionized by means of electric dicharge (i.e., a plasma). In effect, the energy contained in the plasma rather than thermal energy is used to activate the chemical reactions. There are several advantages of PACVD [4]: (1) lower substrate temperatures, (2) better covering power, (3) better adhesion, and (4) faster deposition rates. Applications include deposition of silicon nitride (Si_3N_4) in semiconductor processing, TiN and TiC coatings for tools, and polymer coatings. The process is also known as plasma-enhanced chemical vapor deposition (PECVD), plasma chemical vapor deposition (PCVD), or just simply plasma deposition.

29.5 ORGANIC COATINGS

Organic coatings are polymers and resins, produced either naturally or synthetically, usually formulated to be applied as liquids which dry or harden as thin surface films on substrate materials. These coatings are valued for the variety of colors and textures possible, their capacity to protect the substrate surface, low cost, and ease with which they can be applied. In this section, we consider the compositions of organic coatings and the methods to apply them. Although most organic coatings are applied in liquid form, some are applied as powders; we consider this alternative in Section 29.5.2.

Organic coatings are formulated to contain the following: (1) binders, which give the coating its properties; (2) dyes or pigments, which lend color to the coating; (3) solvents, to dissolve the polymers and resins and add proper fluidity to the liquid; and (4) additives.

Binders in organic coatings are polymers and resins that determine the solid-state properties of the coating, such as strength, physical properties, and adhesion to the substrate surface. The binder holds the pigments and other ingredients in the coating during and after application to the surface. The most common binders in organic coatings are natural oils (used to produce oil-based paints), and resins of polyesters, polyurethanes, epoxies, acrylics, and cellulosics.

Dyes and pigments provide color to the coating. ***Dyes*** are soluble chemicals that color the coating liquid but do not conceal the surface beneath when applied. Thus, dye-colored coatings are generally transparent or translucent. ***Pigments*** are solid particles of uniform, microscopic size that are dispersed in the coating liquid but insoluble in it. They not only color the coating; they also hide the surface below. Since pigments are particulate matter, they also tend to strengthen the coating.

Solvents are used to dissolve the binder and certain other ingredients in the liquid coating composition. Common solvents used in organic coatings are aliphatic and aromatic hydrocarbons, alcohols, esters, ketones, and chlorinated solvents. Different solvents are required for different binders. ***Additives*** in organic coatings include surfactants (to facilitate spreading on the surface), biocides and fungicides, thickeners, freeze/thaw stabilizers, heat and light stabilizers, coalescing agents, plasticizers, defoamers, and catalysts to promote cross-linking. These ingredients are formulated to obtain a wide variety of coatings, such as paints, lacquers, and varnishes.

29.5.1 Application Methods

The method of applying an organic coating to a surface depends on factors such as composition of the coating liquid, required thickness of the coating, production rate and cost considerations, part size, and environmental requirements. For any of the application methods, it is of utmost importance that the surface be properly prepared. This includes cleaning and possible treatment of the surface such as phosphate coating. In some cases, metallic surfaces are plated prior to organic coating for maximum corrosion protection.

With any coating method, transfer efficiency is a critical measure. ***Transfer efficiency*** is the proportion of paint supplied to the process that is actually deposited onto the work surface. Some methods yield as low as a 30% transfer efficiency (meaning that 70% of the paint is wasted and cannot be recovered).

Available methods of applying liquid organic coatings include brushing and rolling, spraying, dip coating, and flow coating. In some cases, several successive coatings are applied to the substrate surface to achieve the desired result. An automobile car body is an important example; the following is a typical sequence applied to the sheet-metal car body in a mass-production automobile: (1) phosphate coat applied by dipping, (2) primer coat applied by dipping, (3) color paint coat applied by spray coating, and (4) clear coat (for high gloss and added protection) applied by spraying.

Brushing and Rolling   These are the two most familiar application methods to most people. They have a high transfer efficiency—approaching 100%. Manual brushing and rolling methods are suited to low production but not mass production. Although brushing is quite versatile, rolling is limited to flat surfaces.

Spraying   Spray coating is a widely used production method for applying organic coatings. The process forces the coating liquid to atomize into a fine mist immediately prior to deposition onto the part surface. When the droplets hit the surface, they spread and flow together to form a uniform coating within the localized region of the spray. If done properly, spray coating provides a uniform coating over the entire work surface.

Spray coating can be performed manually in spray painting booths, or it can be set up as an automated process. Transfer efficiency is relatively low (as low as 30%) with these methods. Efficiency can be improved by ***electrostatic spraying,*** in which the workpart is grounded electrically and the atomized droplets are electrostatically charged. This causes the droplets to be drawn to the part surfaces, increasing transfer efficiencies to values up to 90% [12]. Spraying is utilized extensively in the automotive industry for applying external paint coats to car bodies. It is also used for coating appliances and other consumer products.

Immersion and Flow Coating These methods apply large amounts of liquid coating to the workpart and allow the excess to drain off and be recycled. The simplest method is *dip coating,* in which a part is immersed in an open tank of liquid coating material; when the part is withdrawn, the excess liquid drains back into the tank. A variation of dip coating is *electrocoating,* in which the part is electrically charged and then dipped into a paint bath that has been given an opposite charge. This improves adhesion and permits use of water-based paints (which reduce fire and pollution hazards).

In *flow coating,* workparts are moved through an enclosed paint booth, where a series of nozzles shower the coating liquid onto the part surfaces. Excess liquid drains back into a sump, which allows it to be reused.

Drying and Curing Once applied, the organic coating must convert from liquid to solid. The term *drying* is often used to describe this conversion process. Many organic coatings dry by evaporation of their solvents. However, in order to form a durable film on the substrate surface, a further conversion is necessary, called curing. *Curing* involves a chemical change in the organic resin in which polymerization or cross-linking occurs to harden the coating.

The type of resin determines the type of chemical reaction that takes place in curing. The principal methods by which curing is effected in organic coatings are [12]: (1) *ambient temperature curing,* which involves evaporation of the solvent and oxidation of the resin (most lacquers cure by this method); (2) *elevated temperature curing,* in which elevated temperatures are used to accelerate solvent evaporation, as well as polymerization and cross-linking of the resin; (3) *catalytic curing,* in which the starting resins require reactive agents mixed immediately prior to application to bring about polymerization and cross-linking (epoxy and polyurethane paints are examples); and (4) *radiation curing,* in which various forms of radiation, such as microwaves, ultraviolet light, and electron beams, are required to cure the resin.

29.5.2 Powder Coating

The organic coatings discussed so far are liquid systems consisting of resins that are soluble (or at least miscible) in a suitable solvent. Powder coatings are different. They are applied as dry, finely pulverized, solid particles that are melted on the surface to form a uniform liquid film, after which they resolidify into a dry coating. Powder coating systems have grown significantly in commercial importance among organic coatings since the mid 1970s.

Powder coating systems include several resins that are not used in liquid organic coatings because the powder coating material is solid at room temperature. Powder coatings are classified as thermoplastic or thermosetting. Common thermoplastic powders include polyvinylchloride, nylon, polyester, polyethylene, and polyproplylene. They are generally applied as relatively thick coatings, 0.08 to 0.30 mm (0.003–0.012 in.). Common thermosetting coating powders are epoxy, polyester, and acrylic. They are applied as uncured resins that polymerize and cross-link on heating or reaction with other ingredients. Coating thicknesses are typically 0.025 to 0.075 mm (0.001–0.003 in.).

There are two principal application methods for powder coatings: spraying and fluidized bed. In the *spraying* method, an electrostatic charge is given to each particle in order to attract it to an electrically grounded part surface. Several spray gun designs are available to impart the charge to the powders. The spray guns can be operated manually or by industrial robots. Compressed air is used to propel the powders to the nozzle. The powders are dry when sprayed, and any excess particles that do not attach to the surface can be recycled (unless multiple paint colors are mixed in the same spray booth). Powders can be sprayed onto a part at room temperature, followed by heating of the part to melt the powders; or they can be sprayed onto a part that has been heated to above the melting point of the powder, which usually provides a thicker coating.

The *fluidized bed* is a less commonly used alternative to electrostatic spraying. In this method, the workpart to be coated is preheated and passed through a fluidized bed, in which powders are suspended (fluidized) by an airstream. The powders attach themselves to the part surface to form the coating. In some implementations of this coating method, the powders are electrostatically charged to increase attraction to the grounded part surface.

29.6 PORCELAIN ENAMELING AND OTHER CERAMIC COATINGS

Porcelain is a ceramic made from kaolin, feldspar, and quartz (Chapter 7). It can be applied to substrate metals such as steel, cast iron, and aluminum as a vitreous porcelain enamel. Porcelain coatings are valued for their beauty, color, smoothness, ease of cleaning, chemical inertness, and general durability. Porcelain enameling is the name given to the technology of these ceramic coating materials and the processes by which they are applied.

Porcelain enameling is used in a wide variety of products, including bathroom fixtures, household appliances, kitchen ware, hospital utensils, jet engine components, automotive mufflers, and electronic circuit boards. Compositions of the porcelains vary, depending on product requirements. Some porcelains are formulated for color and beauty, while others are designed for functions such as resistance to chemicals and weather, ability to withstand high service temperatures, hardness and abrasion resistance, and electrical resistance.

As a process, porcelain enameling consists of (1) preparation of the coating material, (2) application onto the surface, (3) drying, if needed, and (4) firing. Preparation involves converting the glassy porcelain into fine particles, called *frit,* that are milled to proper and consistent size. The methods for applying the frit are similar to methods used for applying organic coatings, even though the starting material is quite different. Some application methods involve mixing frit with water as a carrier (the mixture is called a *slip*), while other methods apply the porcelain as dry powder. The techniques include spraying, electrostatic spraying, flow coating, dipping, and electrodeposition. Firing is accomplished at temperatures around 800°C (1500°F). Firing is a *sintering* process (Section 17.1.4) in which the frit is transformed into nonporous vitreous porcelain. Coating thickness ranges from around 0.075 mm (0.003 in.) to about 2 mm (0.08 in.). The processing sequence may be repeated several times to obtain the desired thickness.

Other ceramics may also be used as coatings for special purposes. These coatings usually contain a high content of alumina, which makes them more suited to refractory applications. Techniques for applying the coatings are similar to the preceding, except firing temperatures are higher.

29.7 THERMAL AND MECHANICAL COATING PROCESSES

These processes apply discrete coatings that are generally thicker than coatings deposited by other processes considered in this chapter. They are based on either thermal or mechanical energy.

29.7.1 Thermal Surfacing Processes

These methods use thermal energy in various forms to apply a coating whose function is to provide resistance to corrosion, erosion, wear, and high temperature oxidation.

Thermal Spraying   In *thermal spraying,* molten and semimolten coating materials are sprayed onto a substrate, where they solidify and adhere to the surface. A wide variety of coating materials can be applied; the categories are pure metals and metal alloys; ceramics (oxides, carbides, and certain glasses); other metallic compounds (sulfides, silicides); cermet composites; and certain plastics (epoxy, nylon, teflon, and others). The substrates include metals, ceramics, glass, some plastics, wood, and paper. Not all coatings can be applied to all substrates. When the process is used to apply a metallic coating, the terms *metallizing* or *metal spraying* are used.

Technologies used to heat the coating material are oxyfuel flame, electric arc, and plasma arc. The starting coating material is in the form of wire or rod, or powders. When wire (or rod) is used, the heating source melts the leading end of the wire, thereby separating it from the solid stock. The molten material is then atomized by a high-velocity gas stream (compressed air or other source), and the droplets are spattered against the work surface. When powder stock is used, a powder feeder dispenses the fine particles into a gas stream, which transports them into the flame, where they are melted. The expanding gases in the flame propel the molten (or semimolten) powders against the workpiece. Coating thickness in thermal spraying is generally greater than in other deposition processes; the typical range is 0.05 mm to 2.5 mm (0.002–0.100 in.).

The first applications of thermal spray coating were to rebuild worn areas on used machinery components and to salvage workparts that had been machined undersize. Success of the technique has led to its use in manufacturing as a coating process for corrosion resistance, high temperature protection, wear resistance, electrical conductivity, electrical resistance, electromagnetic interference shielding, and other functions.

Hard Facing   *Hard facing* is a surfacing technique in which alloys are applied as welded deposits to substrate metals. What distinguishes hard facing is that fusion occurs between the coating and the substrate, whereas the bond in thermal spraying is typically mechanical interlocking that does not stand up as well to abrasive wear. Thus, hard facing is especially suited to applications requiring good wear resistance. Applications include coating of new parts and repair of used part surfaces that are heavily worn, eroded, or corroded. An advantage of hard facing that should be mentioned is that it is readily accomplished outside of the relatively controlled factory environment by many of the common welding processes, such as oxyacetylene gas welding and arc welding. Some of the common surfacing materials include steel and iron alloys, cobalt-based alloys, and nickel-based alloys. Coating thickness is usually 0.75 to 2.5 mm (0.030–0.125 in.), although thicknesses as great as 9 mm (3/8 in.) are possible.

Flexible Overlay Process   The flexible overlay process is capable of depositing a very hard coating material, such as tungsten carbide (WC), onto a substrate surface. This is an important advantage of the process compared to other methods, permitting coating hardness up to about 70 Rockwell C. The process can also be used to apply coatings only to selected regions of a workpart. In the *flexible overlay process,* a cloth impregnated with hard ceramic or metal powders and another cloth impregnated with brazing alloy are laid onto a substrate and heated to fuse the powders to the surface. Thickness of overlay coatings is usually 0.25 to 2.5 mm (0.010–0.100 in.). In addition to coatings of WC and WC–Co, cobalt-based and nickel-based alloys are also applied. Applications include chain saw teeth, rock drill bits, oil drill collars, extrusion dies, and similar parts requiring good wear resistance.

29.7.2 Mechanical Plating

In this coating process, mechanical energy is used to build a metallic coating onto the surface. In *mechanical plating,* the parts to be coated, together with plating metal powders, glass beads, and special chemicals to promote the plating action, are tumbled in a barrel. The metallic powders are microscopic in size—5 μm (0.0002 in.) in diameter; while the glass beads are much larger—2.5 mm (0.10 in.) in diameter. As the mixture is tumbled, the mechanical energy from the rotating barrel is transmitted through the glass beads to pound the metal powders against the part surface, causing a mechanical or metallurgical bond to result. The deposited metals must be malleable in order to achieve a satisfactory bond with the substrate. Plating metals include zinc, cadmium, tin, and lead. The term *mechanical galvanizing* is used for parts that are zinc coated. Ferrous metals are most commonly coated; other metals include brass and bronze. Typical applications include fasteners such as screws, bolts, nuts, and nails. Plating thickness in mechanical plating is usually 0.005 to 0.025 mm (0.0002–0.001 in.). Zinc is mechanically plated to a thickness of around 0.075 mm (0.003 in.).

REFERENCES

[1] Budinski, K. G. *Surface Engineering for Wear Resistance.* Prentice Hall, Inc., Englewood Cliffs, N. J. 1988.

[2] Durney, L. J. (ed.). *The Graham's Electroplating Engineering Handbook.* 4th ed. Chapman & Hall, London, 1996.

[3] George, J. *Preparation of Thin Films.* Marcel Dekker, Inc., New York, 1992.

[4] Hocking, M. G., Vasantasree, V., and Sidky, P. S. *Metallic and Ceramic Coatings.* Addison-Wesley, Reading, Mass., 1989.

[5] *Metal Finishing.* Guidebook and Directory Issue, Metals and Plastics Publications, Inc., Hackensack, N. J., 2000.

[6] *Metals Handbook.* 9th ed. Vol. 5, *Surface Cleaning, Finishing, and Coating.* American Society for Metals, Metals Park, Oh., 1982.

[7] Morosanu, C. E. *Thin Films by Chemical Vapour Deposition.* Elsevier, Amsterdam, Holland, 1990.

[8] Murphy, J. A. (ed.). *Surface Preparation and Finishes for Metals.* McGraw-Hill, New York, 1971.

[9] Satas, D. (ed.). *Coatings Technology Handbook.* 2nd ed. Marcel Dekker, Inc., New York, 2000.

[10] Stuart, R. V. *Vacuum Technology, Thin Films, and Sputtering.* Academic Press, New York, 1983.

[11] Sze, S. M. *VLSI Technology.* 2nd ed. McGraw-Hill, New York, 1988.

[12] Wick, C., and Veilleux, R. (eds.). *Tool and Manufacturing Engineers Handbook.* 4th ed. Vol. III, *Materials, Finishes, and Coating.* Society of Manufacturing Engineers, Dearborn, Mich., 1985.

REVIEW QUESTIONS

29.1. Why are metals coated?

29.2. Identify the most common types of coating processes.

29.3. What are the many reasons why a metallic surface is plated?

29.4. What is meant by the term *cathode efficiency* in electroplating?

29.5. What are the two basic mechanisms of corrosion protection?

29.6. What is the most commonly plated substrate metal?

29.7. One of the mandrel types in electroforming is a solid mandrel. How is the part removed from a solid mandrel?

29.8. How does electroless plating differ from electrochemical plating?

29.9. What is a *conversion coating?*

29.10. How does anodizing differ from other conversion coatings?

29.11. What is *physical vapor deposition?*

29.12. What is the difference between physical vapor deposition (PVD) and chemical vapor deposition (CVD)?

29.13. What are some of the applications of PVD?

29.14. Name the three basic types of PVD.

29.15. What is a commonly used coating material deposited by PVD onto cutting tools?

29.16. Define *sputtering yield.*

29.17. What are some of the advantages of chemical vapor deposition?

29.18. What are the two most common titanium compounds that are coated onto cutting tools by chemical vapor deposition?

29.19. Identify the four major ingredients in organic coatings.

29.20. What is meant by the term *transfer efficiency* in organic coating technology?

29.21. Describe the principal methods by which organic coatings are applied to a surface.

29.22. The terms *drying* and *curing* have different meanings; indicate the distinction.

29.23. In porcelain enameling, what is *frit?*

29.24. What does the term *mechanical galvanizing* refer to?

MULTIPLE CHOICE QUIZ

There is a total of 17 correct answers in the following multiple choice questions (some questions have multiple answers that are correct). To attain a perfect score on the quiz, all correct answers must be given, since each correct answer is worth 1 point. For each question, each omitted answer or wrong answer reduces the score by 1 point, and each additional answer beyond the number of answers required reduces the score by 1 point. Percentage score on the quiz is based on the total number of correct answers.

29.1. Which one of the following plate metals produces the hardest surface on a metallic substrate? (a) cadmium, (b) chromium, (c) copper, (d) nickel, or (e) tin.

29.2. Which one of the following terms is used in connection with dip coating of lead onto a substrate such as sheet steel? (a) aluminizing, (b) anodizing, (c) conversion coating, (d) galvanizing, or (e) terneplating.

29.3. Which one of the following plating metals is associated with the term *galvanizing?* (a) iron, (b) lead, (c) steel, (d) tin, or (e) zinc.

29.4. Which of the following is most typical of the thickness of an electroplated coating (choose either of two acceptable answers)? (a) 0.0025 mm or 0.0001 in., (b) 0.025 mm or 0.001 in., (c) 0.25 mm or 0.010 in., or (d) 2.5 mm or 0.100 in.

29.5. Which of the following processes involves electrochemical reactions (more than one)? (a) anodizing, (b) chromate coatings, (c) electroless plating, (d) electroplating, or (e) phosphate coatings.

29.6. With which one of the following metals is anodizing most commonly associated? (a) aluminum, (b) magnesium, (c) steel, (d) titanium, or (e) zinc.

29.7. Sputtering is a form of which one of the following? (a) chemical vapor deposition, (b) defect in arc welding, (c) diffusion, (d) ion implantation, or (e) physical vapor deposition.

29.8. Which of the following gases is the most commonly used in sputtering and ion plating? (a) argon, (b) chlorine, (c) neon, (d) nitrogen, or (e) oxygen.

29.9. The Mond process is used for which one of the following? (a) chemical vapor deposition of silicon nitride onto silicon, (b) an electroplating process, (c) physical vapor deposition for coating TiN onto cutting tools, or (d) reducing nickel carbonyl to metallic Ni.

29.10. Which of the following thin film processes is most common in semiconductor processing? (a) chemical vapor deposition or (b) physical vapor deposition.

29.11. The principal methods of applying powder coatings are which of the following (select two best answers)? (a) brushing, (b) electrostatic spraying, (c) fluidized bed, (d) immersion, and (e) roller coating.

29.12. Porcelain enamel is applied to a surface in which one of the following forms? (a) liquid emulsion, (b) liquid solution, (c) molten liquid, or (d) powders.

29.13. Which of the following are alternative names for thermal spraying (more than one aswer)? (a) flexible overlay process, (b) hard facing, (c) metallizing, or (d) metal spraying.

29.14. Hard facing utilizes which one of the following basic processes? (a) arc welding, (b) brazing, (c) dip coating, (d) electroplating, (e) mechanical deformation to work harden the surface.

PROBLEMS

Electroplating

29.1. What volume (cm^3) and weight (g) of zinc will be deposited onto a cathodic workpart if 10 amps of current are applied for one hour?

29.2. A sheet-metal steel part with surface area $A = 100 \, cm^2$ is to be zinc plated. What average plating thickness will result if 15 amps are applied for 12 min in a chloride electrolyte solution?

29.3. A sheet-metal steel part with surface area $A = 15.0 \, in^2$ is to be chrome plated. What average plating thickness will result if 15 amps are applied for 10 min in a chromic acid-sulfate bath?

29.4. Twenty-five jewelry pieces, each with a surface area = 0.5 in.^{2} are to be gold plated in a batch plating operation. (a) What average plating thickness will result if

8 amps are applied for 10 min in a cyanide bath? (b) What is the value of the gold that will be plated if one ounce of gold is valued at $300? The density of gold = 0.698 lb/in.^{3}

29.5. A part made of sheet steel is to be nickel plated. The part is a rectangular flat plate that is 0.075 cm thick and whose face dimensions are 14 cm by 19 cm. The plating operation is carried out in an acid sulfate electrolyte, using a current $I = 20$ amps for a duration $t = 30$ min. Determine the average thickness of the plated metal resulting from this operation.

29.6. A steel sheet-metal part has total surface area $A = 36$ in.^{2} How long will it take to deposit a copper plating (assume valence = +1) of thickness = 0.001 in. onto the surface if 15 amps of current are applied?

29.7. Increasing current is applied to a workpart surface in an electroplating process according to the relation $I =$ 12.0 + 0.2t, where I = current, amps; and t = time, min. The plating metal is chromium, and the part is submersed in the plating solution for a duration of 20 min. What volume of coating will be applied in the process?

29.8. A batch of 100 parts are to be nickel plated in a barrel-plating operation. The parts are identical, each with a surface area $A = 7.8$ in.^{2} The plating process applies a current $I = 120$ amps, and the batch takes 40 min to complete. Determine the average plating thickness on the parts.

29.9. A batch of 40 identical parts are to be chrome plated using racks. Each part has a surface area = 22.7 cm^2. If it is desired to plate an average thickness = 0.010 mm on the surface of each part, how long should the plating operation be allowed to run at a current = 80 amps?

30 FUNDAMENTALS OF WELDING

CHAPTER CONTENTS

30.1   Overview of Welding Technology
        30.1.1   Types of Welding Processes
        30.1.2   Welding as a Commercial Operation
30.2   The Weld Joint
        30.2.1   Types of Joints
        30.2.2   Types of Welds
30.3   Physics of Welding
30.4   Features of a Fusion Welded Joint

In this part of the book we consider the various processes that are used to join two or more parts into an assembled entity. These processes are labeled in the lower stem of Figure 1.4. The term *joining* is generally used to refer to welding, brazing, soldering, and adhesive bonding, which form a permanent joint between the parts—a joint that cannot easily be separated. The term *assembly* usually refers to mechanical methods of fastening the parts together. Some of these methods allow for easy disassembly, while others do not. Mechanical assembly is covered in Chapter 33. Brazing, soldering, and adhesive bonding are discussed in Chapter 32. We begin our coverage of the joining and assembly processes with welding, covered in this chapter and the following.

   Welding is a materials joining process in which two or more parts are coalesced at their contacting surfaces by a suitable application of heat and/or pressure. Many welding processes are accomplished by heat alone with no pressure applied; others by a combination of heat and pressure; and still others by pressure alone, with no external heat supplied. In some welding processes a *filler* material is added to facilitate coalescence. The assemblage of parts that are joined by welding is called a *weldment.* Welding is most commonly associated with metal parts, but the process is also used for joining plastics. Our discussion of welding will focus on the joining of metals.

   Welding is a relatively new process (Historical Note 30.1). Its commercial and technological importance derives from the following:

➤ Welding provides a permanent joint. The welded parts become a single entity.

➤ The welded joint can be stronger than the parent materials if a filler metal is used that has strength properties superior to those of the parents, and proper welding techniques are used.

➤ Welding is usually the most economical way to join components in terms of material usage and fabrication costs. Alternative mechanical methods of assembly require more complex shape alterations (e.g., drilling of holes) and addition of fasteners (e.g., rivets or bolts). The resulting mechanical assembly is usually heavier than a corresponding weldment.

➤ Welding is not restricted to the factory environment. It can be accomplished "in the field."

Although welding has the advantages just indicated, it also has certain limitations and drawbacks (or potential drawbacks):

➤ Most welding operations are performed manually and are expensive in terms of labor cost. Many welding operations are considered "skilled trades," and the labor to perform these operations may be scarce.

➤ Most welding processes, involving the use of high energy, are inherently dangerous.

➤ Since welding accomplishes a permanent bond between the components, it does not allow for convenient disassembly. If there is a need for occasional disassembly of the product (e.g., for repair or maintenance), then welding should not be used as the assembly method.

➤ The welded joint can suffer from certain quality defects that are difficult to detect. The defects can reduce the strength of the joint.

Historical Note 30.1   *Origins of welding.*

Although welding is considered a relatively new process as practiced today, its origins can be traced to ancient times. Around 1000 B.C., the Egyptians and others in the eastern Mediterranean area learned to accomplish forge welding (Section 31.5.2). It was a natural extension of hot forging, which they used to make weapons, tools, and other implements. Forge-welded articles of bronze have been recovered by archeologists from the pyramids of Egypt. From these early beginnings through the Middle Ages, the blacksmith trade developed the art of welding by hammering to a high level of maturity. Welded objects of iron and other metals dating from these times have been found in India and Europe.

It was not until the 1800s that the technological foundations of modern welding were established. Two important discoveries were made, both attributed to English scientist Sir Humphrey Davy: (1) the electric arc, and (2) acetylene gas.

Around 1801, Davy observed that an electric arc could be struck between two carbon electrodes. However, not until the mid-1800s, when the electric generator was invented, did electrical power become available in amounts sufficient to sustain **arc welding.** It was a Russian, Nikolai Benardos, working out of a laboratory in France, who was granted a series of patents for the carbon arc welding process (one in England in 1885, and another in the United States in

1887). By 1900, carbon arc welding had become a popular commercial process for joining metals.

Benardos' inventions seem to have been limited to carbon arc welding. In 1892, an American named Charles Coffin was awarded a U.S. patent for developing an arc welding process utilizing a metal electrode. The unique feature was that the electrode added filler metal to the weld joint (the carbon arc process does not deposit filler). The idea of coating the metal electrode (to shield the welding process from the atmosphere) was developed later, with enhancements to the metal arc welding process being made in England and Sweden starting around 1900.

Between 1885 and 1900, several forms of **resistance welding** were developed by E. Thompson. These included spot welding and seam welding, two joining methods widely used today in sheet metalworking.

Although Davy discovered acetylene gas early in the 1800s, **oxyfuel gas welding** required the subsequent development of torches for combining acetylene and oxygen around 1900. During the 1890s, hydrogen and natural gas were mixed with oxygen for welding, but the oxyacetylene flame achieved significantly higher temperatures.

These three welding processes—arc welding, resistance welding, and oxyfuel gas welding—constitute by far the majority of welding operations performed today.

30.1 OVERVIEW OF WELDING TECHNOLOGY

Welding involves localized coalescence or joining together of two metallic parts at their faying surfaces. The *faying surfaces* are the part surfaces in contact or close proximity that are to be joined. Welding is usually performed on parts made of the same metal, but some welding operations can be used to join dissimilar metals.

30.1.1 Types of Welding Processes

Some 50 different types of welding operations have been cataloged by the American Welding Society. They use various types or combinations of energy to provide the required power. We can divide the welding processes into two major groups: (1) fusion welding and (2) solid state welding.

Fusion Welding *Fusion welding* uses heat to melt the base metals. In many fusion welding operations, a filler metal is added to the molten pool to facilitate the process and provide bulk and strength to the welded joint. A fusion-welding operation in which no filler metal is added is referred to as an *autogenous* weld. The fusion category comprises the most widely used welding processes and includes the following general groups (initials in parentheses are designations of the American Welding Society):

> *Arc welding* (AW)—Arc welding refers to a group of welding processes in which heating of the metals is accomplished by an electric arc, as shown in Figure 30.1. Some arc welding operations also apply pressure during the process and most utilize a filler metal.

> *Resistance welding* (RW)—Resistance welding achieves coalescence using heat from electrical resistance to the flow of a current passing between the faying surfaces of two parts held together under pressure.

> *Oxyfuel gas welding* (OFW)—These joining processes use an oxyfuel gas, such as a mixture of oxygen and acetylene, to produce a hot flame for melting the base metal and filler metal, if one is used.

> Other fusion-welding processes—There are other welding processes that produce fusion of the metals joined. Examples include *electron beam welding* and *laser beam welding.*

FIGURE 30.1   Basics of arc welding: (1) before the weld; (2) during the weld, the base metal is melted and filler metal is added to the molten pool; and (3) the completed weldment. There are many variations of arc welding.

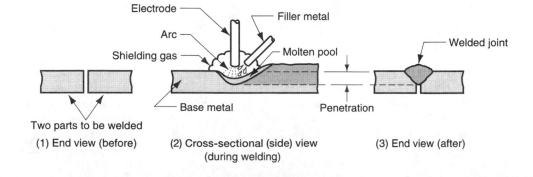

Certain arc and oxyfuel processes are also used for cutting metals (Sections 26.3.4 and 26.3.5).

Solid-State Welding   *Solid-state welding* refers to joining processes in which coalescence results from application of pressure alone or a combination of heat and pressure. If heat is used, the temperature in the process is below the melting point of the metals being welded. No filler metal is utilized. This group includes the following:

> *Diffusion welding* (DFW)—Two surfaces are held together under pressure at an elevated temperature and the parts coalesce by solid state fusion.

> *Friction welding* (FRW)—Coalescence is achieved by the heat of friction between two surfaces.

> *Ultrasonic welding* (USW)—Moderate pressure is applied between the two parts and an oscillating motion is used at ultrasonic frequencies in a direction parallel to the contacting surfaces. The combination of normal and vibratory forces results in shear stresses that remove surface films and achieve atomic bonding of the surfaces.

In Chapter 31, we describe the various welding processes in greater detail. The preceding survey should provide a sufficient framework for our current discussion of welding terminology and principles.

30.1.2 Welding as a Commercial Operation

The principal applications of welding are [4]: (1) construction (i.e., buildings and bridges); (2) piping, pressure vessels, boilers, and storage tanks; (3) shipbuilding; (4) aircraft and aerospace; and (5) automotive and railroad. Welding is performed in a variety of locations and in a variety of industries. Owing to its versatility as an assembly technique for commercial products, many welding operations are performed in factories. But several of the traditional welding processes, such as arc welding and oxyfuel gas welding, use equipment that can be readily moved, so these operations are not limited to the factory. They are performed at construction sites, in shipyards, at customers' plants, and in automotive repair shops.

Most welding operations are labor intensive. For example, arc welding is usually performed by a skilled worker, called a *welder,* who manually controls the path or placement of the weld to join individual parts into a larger unit. (See Color Plate 10.) In factory operations in which arc welding is manually performed, the welder often works with a second worker, called a *fitter.* It is the fitter's job to arrange the individual components for the welder prior to making the weld, using welding fixtures and positioners. A *welding fixture* is a device for clamping and holding the components in fixed position for welding. It is custom-fabricated for the particular geometry of the weldment and therefore must be economically justified on the basis of the quantities of assemblies to be produced. A *welding positioner* holds the parts and also moves the assemblage to the desired position for welding. This differs from a welding fixture that only holds the parts in a single fixed position. The desired position is usually one in which the weld path is flat and horizontal.

The Safety Issue   Welders must follow strict safety precautions. The high temperatures of the molten metals in welding are an obvious danger. In gas welding, the fuels (e.g., acetylene) are a fire hazard. Most of the processes use high energy to cause melting of the part surfaces to be joined. In many welding processes, electrical power is the source of thermal energy, so there is the hazard of electrical shock to the worker. Certain welding processes have their own particular perils. In arc welding, for example, ultraviolet radiation is emitted that is injurious to human vision. A special helmet that includes a dark viewing window must be worn by the welder, as in Color Plate 10. This window filters out the dangerous radiation but is so dark that it renders the welder virtually blind, except when the arc is struck. Sparks, spatters of molten metal, smoke, and fumes add to the risks associated with welding operations. Ventilation facilities must be used to exhaust the dangerous fumes generated by some of the fluxes and molten metals used in welding. If the operation is performed in an enclosed area, special ventilation suits or hoods are required.

Automation in Welding   Because of the hazards of manual welding, and to increase productivity and improve product quality, various forms of mechanization and automation have been developed. The categories include machine welding, automatic welding, and robotic welding.

Machine welding can be defined as mechanized welding with equipment that performs the operation under the continuous supervision of an operator. It is normally accomplished by a welding head that is moved by mechanical means relative to a stationary work, or by moving the work relative to a stationary welding head. The human worker must continually observe and interact with the equipment to control the operation.

If the equipment is capable of performing the operation without adjustment of the controls by a human operator, it is referred to as ***automatic welding.*** A human worker is usually present to oversee the process and detect variations from normal conditions. What distinguishes automatic welding from machine welding is a weld cycle controller to regulate the arc movement and workpiece positioning without continuous human attention. Automatic welding requires a welding fixture and/or positioner to position the work relative to the welding head. It also requires a higher degree of consistency and accuracy in the component parts used in the weldment. For these reasons, automatic welding can be justified only for large quantity production.

In ***robotic welding,*** an industrial robot or programmable manipulator is used to automatically control the movement of the welding head relative to the work (Section 38.2.3). The versatile reach of the robot arm permits the use of relatively simple fixtures, and the robot's capacity to be reprogrammed for new part configurations allows this form of automation to be justified for relatively low production quantities. A typical robotic arc welding cell consists of two welding fixtures and a human fitter to load and unload parts while the robot welds. In addition to arc welding, industrial robots are also used in automobile final assembly plants to perform resistance welding on car bodies (see Color Plate 11 and Figure 38.11).

30.2 THE WELD JOINT

Welding produces a solid connection between two pieces, called a weld joint. A ***weld joint*** is the junction of the edges or surfaces of parts that have been joined by welding. This section covers weld joints: types of welded joints and the various types of welds used to join the pieces that form the joints.

30.2.1 Types of Joints

There are five basic types of joints for bringing two parts together for joining. With reference to Figure 30.2, the five joint types can be defined as follows:

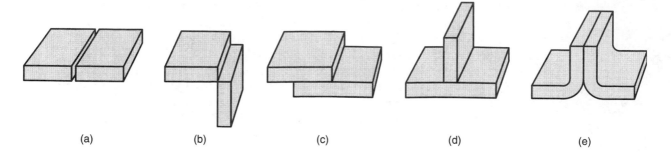

FIGURE 30.2   Five basic types of joints: (a) butt, (b) corner, (c) lap, (d) tee, and (e) edge.

(a) ***Butt joint***—In this joint type, the parts lie in the same plane and are joined at their edges.

(b) ***Corner joint***—The parts in a corner joint form a right angle and are joined at the corner of the angle.

(c) ***Lap joint***—This joint type consists of two overlapping parts.

(d) ***Tee joint***—In the tee joint, one part is perpendicular to the other in the approximate shape of the letter T.

(e) ***Edge joint***—The parts in an edge joint are parallel with at least one of their edges in common, and the joint is made at the common edge(s).

30.2.2   Types of Welds

Each of the preceding joints can be made by welding. Other joining processes can also be used for some of the joint types, but welding is the most universally applicable joining method. It is appropriate to distinguish between the joint type and the way in which it is welded—the weld type. Differences among weld types are in geometry (joint type) and welding process.

A ***fillet weld*** is used to fill in the edges of plates created by corner, lap, and tee joints, as in Figure 30.3. Filler metal is used to provide a cross section approximately the shape of a right triangle. It is the most common weld type in arc and oxyfuel welding because it requires minimum edge preparation—the basic square edges of the parts are used. Fillet welds can be single or double (i.e., welded on one side or both) and can be continuous or intermittent (i.e., welded along the entire length of the joint or with unwelded spaces along the length).

Groove welds usually require that the edges of the parts be shaped into a groove to facilitate weld penetration. The grooved shapes include square, bevel, V, U, and J, in single or double sides, as shown in Figure 30.4. Filler metal is used to fill in the joint, usually by arc or oxyfuel welding. Preparation of the part edges beyond the basic square edge, although requiring additional processing, is often done to increase the strength of the welded joint or where thicker parts are to be welded. Although most closely associated with a butt joint, groove welds are used on all joint types except lap.

Plug welds and ***slot welds*** are used for attaching flat plates, as shown in Figure 30.5, using one or more holes or slots in the top part and then filling with filler metal to fuse the two parts together.

Spot welds and seam welds, used for lap joints, are diagrammed in Figure 30.6. A ***spot weld*** is a small fused section between the surfaces of two sheets or plates. Multiple

FIGURE 30.3   Various forms of fillet welds: (a) inside single fillet corner joint; (b) outside single fillet corner joint; (c) double fillet lap joint; and (d) double fillet tee joint. Dashed lines show the original part edges.

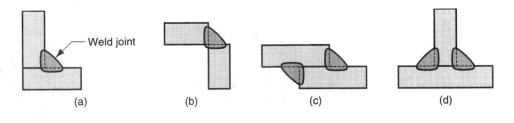

(a)   (b)   (c)   (d)

FIGURE 30.4   Some typical groove welds: (a) square groove weld, one side; (b) single bevel groove weld; (c) single V-groove weld; (d) single U-groove weld; (e) single J-groove weld; (f) double V-groove weld for thicker sections. Dashed lines show the original part edges.

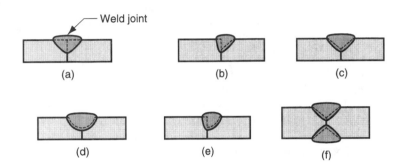

(a)   (b)   (c)

(d)   (e)   (f)

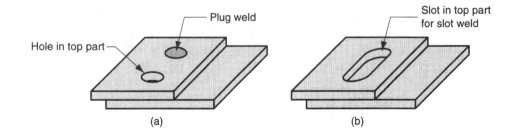

FIGURE 30.5   (a) Plug weld, and (b) slot weld.

(a)   (b)

FIGURE 30.6   (a) Spot weld, and (b) seam weld.

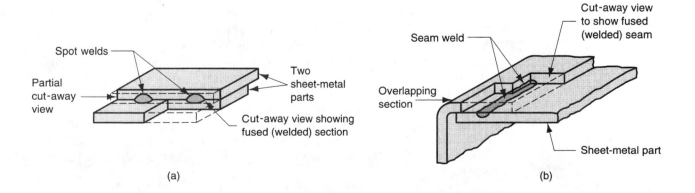

(a)   (b)

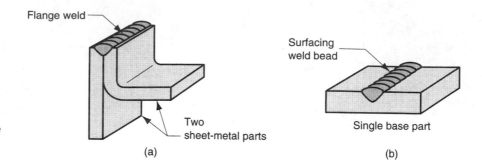

FIGURE 30.7 (a) Flange weld, and (b) surfacing weld.

spot welds are typically required to join the parts. It is most closely associated with resistance welding. A *seam weld* is similar to a spot weld except it consists of a more or less continuously fused section between the two sheets or plates.

Flange welds and surfacing welds are shown in Figure 30.7. A *flange weld* is made on the edges of two (or more) parts, usually sheet metal or thin plate, at least one of the parts being flanged as in Figure 30.7(a). A *surfacing weld* is not used to join parts, but rather to deposit filler metal onto the surface of a base part in one or more weld beads. The weld beads can be made in a series of overlapping parallel passes, thereby covering large areas of the base part. The purpose is to increase the thickness of the plate or to provide a protective coating on the surface.

30.3 PHYSICS OF WELDING

Although several coalescing mechanisms are available for welding, fusion is by far the most common means. To accomplish fusion, a source of high-density heat energy is supplied to the faying surfaces, and the resulting temperatures are sufficient to cause localized melting of the base metals. If a filler metal is added, the heat density must be high enough to melt it also. Heat density can be defined as the power transferred to the work per unit surface area, W/mm^2 ($Btu/sec-in.^2$). The time to melt the metal is inversely proportional to the power density. At low power densities, a significant amount of time is required to cause melting. If power density is too low, the heat is conducted into the work as rapidly as it is added at the surface, and melting never occurs. It has been found that the minimum power density required to melt most metals in welding is about 10 W/mm^2 (6 $Btu/sec-in.^2$). As heat density increases, melting time is reduced. If power density is too high—above around 10^5 W/mm^2 (60,000 $Btu/sec-in.^2$)—the localized temperatures vaporize the metal in the affected region. Thus, there is a practical range of values for power density within which welding can be performed. Differences among welding processes in this range are (1) the rate at which welding can be performed and/or (2) the size of the region that can be welded. Table 30.1 provides a comparison of power densities for the major groups of fusion welding processes (plus two very high-power density operations). Oxyfuel gas welding is capable of developing large amounts of heat, but the heat density is relatively low because it is spread over a large area. Oxyacetylene gas, the hottest of the OFW fuels, burns at a top temperature of around 3500°C (6300°F). By comparison, arc welding produces high energy over a smaller area, resulting in local temperatures of 5500° to 6600°C (10,000°–12,000°F). For metallurgical reasons, it is desirable to melt the metal with minimum energy, and high heat densities are generally preferable.

TABLE 30.1   Comparison of several fusion-welding
processes on the basis of their power densities.

Welding Process	Approximate Power Density W/mm²	(Btu/sec-in.²)
Oxyfuel welding	10	(6)
Arc welding	50	(30)
Resistance welding	1,000	(600)
Laser beam welding	9,000	(5,000)
Electron beam welding	10,000	(6,000)

Power density can be computed as the power entering the surface divided by the corresponding surface area:

$$PD = \frac{P}{A} \tag{30.1}$$

where PD = power density, W/mm² (Btu/sec-in.²); P = power entering the surface, W (Btu/sec); and A = surface area over which the energy is entering, mm² (in.²). The issue is more complicated than indicated by Eq. (30.1). One complication is that the power source (e.g., the arc) is moving in many welding processes, which results in preheating ahead of the operation and postheating behind it. Another complication is that power density is not uniform throughout the affected surface; it is distributed as a function of area, as demonstrated by the following example.

**EXAMPLE 30.1
Power Density in
Welding**

A heat source is capable of transferring 3000 W to the surface of a metal part. The heat impinges the surface in a circular area, with intensities varying inside the circle. The distribution is as follows: 70% of the power is transferred within a circle of diameter = 5 mm, and 90% is transferred within a concentric circle of diameter = 12 mm. What are the power densities in (a) the 5 mm diameter inner circle and (b) the 12 mm diameter ring that lies around the inner circle?

Solution: (a) The inner circle has an area $A = \dfrac{\pi(5)^2}{4} = 19.63$ mm².

The power inside this area $P = 0.70 \times 3000 = 2100$ W

Thus the power density $PD = \dfrac{2100}{19.63} = 107$ W/mm².

(b) The area of the ring outside the inner circle is $A = \dfrac{\pi(12^2 - 5^2)}{4} = 93.4$ mm²

The power in this region $P = 0.9 \times 3000 - 2100 = 810$ W

The power density is therefore $PD = \dfrac{810}{93.4} = 8.7$ W/mm².

Observation: The power density seems high enough for melting in the inner circle, but probably not sufficient in the ring that lies outside this inner circle.   ▪

The quantity of heat required to melt a given volume of metal is the sum of (1) the heat to raise the temperature of the solid metal to its melting point, which depends on the metal's volumetric specific heat, and (2) the heat to transform the metal from solid to liquid phase at the melting point, which depends on the metal's heat of fusion. To a reasonable approximation, this quantity of heat can be estimated by [5]:

$$U_m = KT_m^2 \qquad (30.2)$$

where U_m is the unit energy for melting – the quantity of heat required to melt a unit volume of metal starting from room temperature, J/mm^3 ($Btu/in.^3$); T_m = melting point of the metal on an absolute temperature scale, °K (°R); and K = constant whose value is 3.33×10^{-6} when the Kelvin scale is used (and $K = 1.467 \times 10^{-5}$ for the Rankine temperature scale). Absolute melting temperatures for selected metals are presented in Table 30.2.

Not all of the input energy is used to melt the weld metal. There are two heat transfer mechanisms at work, both of which reduce the amount of heat available to the welding process. The first mechanism is the transfer of heat between the heat source and the surface of the work. This process has a certain **heat transfer efficiency,** f_1, defined as the ratio of the actual heat received by the workpiece divided by the total heat generated at the source. The second mechanism involves the conduction of heat away from the weld area to be dissipated throughout the work metal, so that only a portion of the heat transferred to the surface is available for melting. This **melting efficiency,** f_2, is the proportion of heat received at the work surface that can be used for melting. The combined effect of these two efficiencies is to reduce the heat energy available for welding as follows:

$$H_w = f_1 f_2 H \qquad (30.3)$$

where H_w = net heat available for welding, J (Btu), f_1 = heat transfer efficiency, f_2 = the melting efficiency, and H = the total heat generated by the welding process, J (Btu).

It is appropriate to separate f_1 and f_2 in concept, even though they act in concert during the welding process. Heat transfer efficiency f_1 is determined largely by the welding process and the capacity to convert the power source (e.g., electrical energy) into usable heat at the work surface. Oxyfuel gas welding processes are relatively inefficient in this regard, while arc welding processes are relatively efficient.

Melting efficiency f_2 depends on the welding process, but it is also influenced by the thermal properties of the metal, joint configuration, and work thickness. Metals with

TABLE 30.2 Melting temperatures on the absolute temperature scale for selected metals.

Metal	Melting Temperature		Metal	Melting Temperature	
	°K[a]	°R[b]		°K[a]	°R[b]
Aluminum alloys	930	(1680)	Steels		
Cast iron	1530	(2760)	Low carbon	1760	(3160)
Copper and alloys			Medium carbon	1700	(3060)
Pure	1350	(2440)	High carbon	1650	(2960)
Brass, navy	1160	(2090)	Low alloy	1700	(3060)
Bronze (90 Cu–10 Sn)	1120	(2010)	Stainless steels		
Inconel	1660	(3000)	Austenitic	1670	(3010)
Magnesium	940	(1700)	Martensitic	1700	(3060)
Nickel	1720	(3110)	Titanium	2070	(3730)

Based on values in [1].

[a] Kelvin scale = Centigrade (Celsius) temperature + 273.

[b] Rankine scale = Fahrenheit temperature + 460.

high thermal conductivity, such as aluminum and copper, present a problem in welding because of the rapid dissipation of heat away from the heat contact area. The problem is exacerbated by welding heat sources with low energy densities (e.g., oxyfuel welding) because the heat input is spread over a larger area, thus facilitating conduction into the work. In general, a high-intensity welding heat source, combined with a low conductivity work material, results in a high melting efficiency.

We can now write a balance equation between the energy input and the energy needed for welding:

$$H_w = U_m V \qquad (30.4)$$

where H_w = net heat energy delivered to the operation, J (Btu); U_m = unit energy required to melt the metal, J/mm^3 (Btu/in.^{3}); and V = the volume of metal melted, mm^3 (in.^{3}). Most welding operations are rate processes; that is, the net heat energy H_w is delivered at a given rate, and the weld bead is made at a certain travel velocity. This is characteristic for example of most arc welding and many oxyfuel gas welding operations. It is therefore appropriate to express Eq. (30.4) in the form of a rate balance equation:

$$HR_w = U_m WVR \qquad (30.5)$$

where HR_w = rate of heat energy delivered to the operation, J/s = W (Btu/min); and WVR = volume rate of metal welded, mm^3/s (in.^{3}/min). In the welding of a continuous bead, the volume rate of metal welded is the product of weld area A_w and travel velocity v. Substituting these terms into the above equation, the rate balance equation can now be expressed as

$$HR_w = f_1 f_2 HR = U_m A_w v \qquad (30.6)$$

where f_1 and f_2 are the heat transfer and melting efficiencies; HR = rate of input energy generated by the welding power source, W (Btu/min); A_w = weld cross-sectional area, mm^2 (in.^{2}); and v = the travel velocity of the welding operation, mm/s (in./min).

EXAMPLE 30.2
Welding Travel
Speed

The power source in a particular welding setup is capable of generating 3500 W that can be transferred to the work surface with an efficiency $f_1 = 0.7$. The metal to be welded is low carbon steel, whose melting temperature, from Table 30.2, is $T_m = 1760°$K. Melting efficiency in the operation is $f_2 = 0.5$. A continuous fillet weld is to be made with a cross-sectional area $A_w = 20$ mm^2. Determine the travel speed at which the welding operation can be accomplished.

Solution: Let us first find the unit energy required to melt the metal U_m from Eq. (30.2).

$$U_m = 3.33(10^{-6}) \times 1760^2 = 10.3 \text{ J/mm}^3$$

Rearranging Eq. (30.6) to solve for travel velocity, we have $v = \dfrac{f_1 f_2 HR}{U_m A_w}$

and solving for the conditions of the problem, $v = \dfrac{0.7(0.5)(3500)}{10.3(20)} = 5.95$ mm/s

In Chapter 31, we examine how the power density in Eq. (30.1) and the input energy rate for Eq. (30.6) are generated for some of the individual welding processes.

30.4 FEATURES OF A FUSION-WELDED JOINT

Most of the weld joints considered above are fusion welded. As illustrated in the cross-sectional view of Figure 30.8(a), a typical fusion weld joint in which filler metal has been added consists of several zones: (1) fusion zone, (2) weld interface, (3) heat affected zone, and (4) unaffected base metal zone.

The *fusion zone* consists of a mixture of filler metal and base metal that have completely melted. This zone is characterized by a high degree of homogeneity among the component metals that have been melted during welding. The mixing of these components is motivated largely by convection in the molten weld pool. Solidification in the fusion zone has similarities to a casting process. In welding, the mold is formed by the unmelted edges or surfaces of the components being welded. The significant difference between solidification in casting and in welding is that epitaxial grain growth occurs in welding. The reader may recall that in casting, the metallic grains are formed from the melt by nucleation of solid particles at the mold wall, followed by grain growth. In welding, by contrast, the nucleation stage of solidification is avoided by the mechanism of *epitaxial grain growth,* in which atoms from the molten pool solidify on preexisting lattice sites of the adjacent solid base metal. Consequently, the grain structure in the fusion zone near the heat affected zone tends to mimic the crystallographic orientation of the surrounding heat affected zone. Further into the fusion zone, a preferred orientation develops in which the grains are roughly perpendicular to the boundaries of the weld interface. The resulting structure in the solidified fusion zone tends to feature coarse columnar grains, as depicted in Figure 30.8(b). The grain structure depends on various factors, including welding process, metals being welded (e.g., identical metals vs. dissimilar metals welded), whether a filler metal is used, and the feed rate at which welding is accomplished. A detailed discussion of welding metallurgy is beyond the scope of this text, and interested readers can consult any of several references [4], [5].

The second zone in the weld joint is the *weld interface,* a narrow boundary that separates the fusion zone from the heat-affected zone. The interface consists of a thin band of base metal that was melted or partially melted (localized melting within the grains) during the welding process but then immediately solidified before any mixing with the metal in the fusion zone. Its chemical composition is therefore identical to that of the base metal.

The third zone in the typical fusion weld is the *heat-affected zone* (HAZ). The metal in this zone has experienced temperatures that are below its melting point, yet high enough to cause microstructural changes in the solid metal. The chemical composition in the HAZ is the same as the base metal, but this region has been heat-treated due to the welding

FIGURE 30.8   Cross section of a typical fusion-welded joint: (a) principal zones in the joint, and (b) typical grain structure.

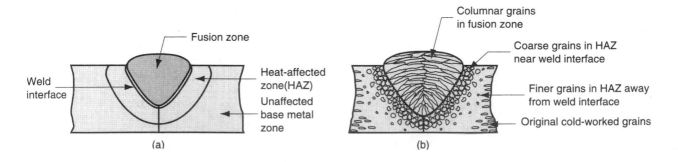

temperatures so that its properties and structure have been altered. The amount of metallurgical damage in the HAZ depends on factors such as the amount of heat input and peak temperatures reached, distance from the fusion zone, length of time the metal has been subjected to the high temperatures, cooling rate, and the metal's thermal properties. The effect on mechanical properties in the heat affected zone is usually negative, and it is in this region of the weld joint that welding failures often occur.

As the distance from the fusion zone increases, the **unaffected base metal zone** is finally reached, in which no metallurgical change has occurred. Nevertheless, the base metal surrounding the HAZ is likely to be in a state of high residual stress, the result of shrinkage in the fusion zone.

REFERENCES

[1] Cary, H. B. *Modern Welding Technology.* 4th ed. Prentice-Hall, Inc., Upper Saddle River, N.J., 1997.

[2] Datsko, J. *Material Properties and Manufacturing Processes.* Wiley, New York, 1966, Chap. 4.

[3] Messler, R. W., Jr. *Principles of Welding: Processes, Physics, Chemistry, and Metallurgy.* Wiley, New York, 1999.

[4] *Metals Handbook.* 9th. ed. Vol. 6: *Welding, Brazing,* *and Soldering.* ASM International, Materials Park, Ohio, 1993.

[5] *Welding Handbook.* 8th. ed. Volume 1: *Welding Technology.* American Welding Society, Miami, Fla., 1987.

[6] Wick, C., and Veilleux, R. F. *Tool and Manufacturing Engineers Handbook.* 4th. ed. Vol. IV, Quality Control and Assembly.

REVIEW QUESTIONS

30.1. What are the advantages and disadvantages of welding compared to other types of assembly operations?

30.2. What two discoveries of Sir Humphrey Davy led to the development of modern welding technology?

30.3. What is meant by the term *faying surface?*

30.4. Define the term *fusion weld.*

30.5. What is the fundamental difference between a fusion weld and a solid state weld?

30.6. What is an *autogenous weld?*

30.7. Discuss the reasons why most welding operations are inherently dangerous.

30.8. What is the difference between machine welding and automatic welding?

30.9. Name and sketch the five joint types.

30.10. Define and sketch a *fillet weld.*

30.11. Define and sketch a *groove weld.*

30.12. Why is a surfacing weld different from the other weld types?

30.13. What is the difference between a continuous weld and an intermittent weld as the terms apply to a fillet weld of a lap joint?

30.14. Why is it desirable to use energy sources for welding that have high heat densities?

30.15. What is the *unit melting energy* in welding, and what are the factors on which it depends?

30.16. Define and distinguish the two terms *heat transfer efficiency* and *melting efficiency* in welding.

30.17. What is *epitaxial grain growth,* and how is this form of solidification different from that which occurs in casting?

30.18. What is the *heat affected zone* (HAZ) in a fusion weld?

MULTIPLE CHOICE QUIZ

There is a total of 11 correct answers in the following multiple choice questions (some questions have multiple answers that are correct). To attain a perfect score on the quiz, all correct answers must be given, since each correct answer is worth 1 point. For each question, each omitted answer or wrong answer reduces the score by 1 point, and each additional answer beyond the number of answers required reduces the score by 1 point. Percentage score on the quiz is based on the total number of correct answers.

30.1. Welding can only be performed on metals that have the same melting point; otherwise, the metal with the lower melting temperature always melts while the other metal remains solid. (a) true or (b) false.

30.2. A fillet weld can be used to join which of the following joint types (more than one)? (a) butt, (b) corner, (c) lap, or (d) tee.

30.3. A fillet weld has a cross-sectional shape that is approximately which one of the following? (a) rectangular, (b) round, (c) square, or (d) triangular.

30.4. Groove welds are most closely associated with which one of the following joint types? (a) butt, (b) corner, (c) edge, (d) lap, or (e) tee.

30.5. A flange weld is most closely associated with which one of the following joint types? (a) butt, (b) corner, (c) edge, (d) lap, or (e) tee.

30.6. For metallurgical reasons, it is desirable to melt the weld metal with minimum energy input. Which one of the following heat sources is most consistent with this objective? (a) high power, (b) high-power density, (c) low power, or (d) low-power density.

30.7. The amount of heat required to melt a given volume of metal depends strongly on which of the following properties (more than one)? (a) coefficient of thermal expansion, (b) heat of fusion, (c) melting temperature, (d) modulus of elasticity, and (e) thermal conductivity.

30.8. Weld failures always occur in the fusion zone of the weld joint, since this is the part of the joint that has been melted. (a) true, or (b) false.

PROBLEMS

Joint Design

30.1. Prepare sketches showing how the part edges would be prepared and aligned with each other and also showing the weld cross section for the following welds: (a) square groove weld, both sides, for a butt weld; (b) single fillet weld for a lap joint; (c) single fillet weld for tee joint; and (d) double U-groove weld for a butt weld.

Power Density

30.2. A heat source can transfer 3000 J/sec to a metal part surface. The heated area is circular, and the heat intensity decreases as the radius increases, as follows: 60% of the heat is concentrated in a circular area that is 3 mm in diameter. Is the resulting power density enough to melt metal?

30.3. A welding heat source is capable of transferring 150 Btu/min to the surface of a metal part. The heated area is approximately circular, and the heat intensity decreases with increasing radius as follows: 50% of the power is transferred within a circle of diameter = 0.1 in., and 75% is transferred within a concentric circle of diameter = 0.25 in. What are the power densities in (a) the 0.1-in.-diameter inner circle and (b) the 0.25-in.-diameter ring that lies around the inner circle? (c) Are these power densities sufficient for melting metal?

Unit Melting Energy

30.4. Compute the unit energy for melting for the following metals: (a) aluminum and (b) plain low carbon steel.

30.5. Compute the unit energy for melting for the following metals: (a) copper and (b) titanium.

30.6. Make the calculations and plot on linearly scaled axes the relationship for unit melting energy as a function of temperature. Use temperatures as follows to construct the plot: 250°C, 500°C, 750°C, 1000°C, 1500°C, and 2000°C. On the plot, mark the positions of some of the welding metals in Table 30.2.

30.7. Make the calculations and plot on linearly scaled axes the relationship for unit melting energy as a function of temperature. Use temperatures as follows to construct the plot: 500°F, 1000°F, 1500°F, 2000°F, 2500°F, 3000°F, and 3500°F. On the plot, mark the positions of some of the welding metals in Table 30.2.

30.8. A fillet weld has a cross-sectional area $A_w = 20.0$ mm^2 and is 200 mm long. (a) What quantity of heat (in joules) is required to accomplish the weld, if the metal to be welded is austenitic stainless steel? (b) How much heat must be generated at the welding source, if the heat transfer efficiency = 0.8 and the melting efficiency = 0.6?

30.9. A certain groove weld has a cross-sectional area $A_w = 0.045$ in.^{2} and is 10 in. long. (a) What quantity of heat (in Btu) is required to accomplish the weld, if the metal to be welded is medium carbon steel? (b) How much heat must be generated at the welding source, if the heat transfer efficiency = 0.9 and the melting efficiency = 0.7?

30.10. Solve the previous problem, except that the metal to be welded is aluminum, and the corresponding melting efficiency is half the value for steel.

30.11. Compute the unit melting energy for (a) aluminum and (b) steel as the sum of (1) the heat required to raise the temperature of the metal from room temperature to its melting point, which is the product of the volumetric specific heat and the temperature rise; and (2) the heat of fusion, so that this value can be compared to the unit melting energy calculated by Eq. (30.2). Use either the U.S. Customary units or the International System. Find the values of the properties needed in these calculations either in this text or in other references. Are the values close enough to validate Eq. (30.2)?

Energy Balance in Welding

30.12. The welding power generated in a particular arc welding operation = 3000 W. This is transferred to the work surface with a heat transfer efficiency $f_1 = 0.9$. The metal to be welded is copper whose melting point is given in Table 30.2. Assume that the melting efficiency $f_2 = 0.25$. A continuous fillet weld is to be made with a cross-sectional area $A_w = 15.0$ mm^2. Determine the travel speed at which the welding operation can be accomplished.

30.13. Solve the previous problem except that the metal to be welded is high carbon steel, the cross-sectional area of the weld = 25.0 mm^2, and the melting efficiency $f_2 = 0.6$.

30.14. In a certain welding operation to make a groove weld, $A_w = 22.0$ mm^2 and $v = 5$ mm/sec. If $f_1 = 0.95$, $f_2 = 0.5$, and $T_m = 1000°C$ for the metal to be welded, determine the rate of heat generation required at the welding source to accomplish this weld.

30.15. The power source in a particular welding operation generates 125 Btu/min, which is transferred to the work surface with an efficiency $f_1 = 0.8$. The melting point for the metal to be welded $T_m = 1800°F$, and its melting efficiency $f_2 = 0.5$. A continuous fillet weld is to be made with a cross-sectional area $A_w = 0.04$ in.^{2} Determine the travel speed at which the welding operation can be accomplished.

30.16. In a certain welding operation to make a fillet weld, $A_w = 0.025$ in.^{2} and $v = 15$ in./min. If $f_1 = 0.95$ and $f_2 = 0.5$, and $T_m = 2000°F$ for the metal to be welded, determine the rate of heat generation required at the welding source to accomplish this weld.

30.17. A spot weld is to be made using an arc welding operation. The total volume of (melted) metal forming the weld = 0.005 in.^{3}, and the operation required the arc to be on for 4 sec. If $f_1 = 0.85$, $f_2 = 0.5$, and the metal to be welded was aluminum, determine the rate of heat generation that was required at the source to accomplish this weld.

30.18. A surfacing weld is to be applied to a rectangular low-carbon steel plate, which is 200 mm by 350 mm. The metal to be applied is a harder (alloy) grade of steel, whose melting point is assumed to be the same. A thickness of 2.0 mm will be added to the plate, but with penetration into the base metal, the total thickness melted during welding = 6.0 mm, on average. The surface will be applied by making a series of parallel, overlapped welding beads running lengthwise on the plate. The operation will be carried out automatically with the beads laid down in one long continuous operation at a travel speed $v = 7.0$ mm/s, using welding passes separated by 5 mm. Ignore the minor complications of the turnarounds at the ends of the plate. Assuming the heat transfer efficiency = 0.8 and the melting efficiency = 0.6, determine (a) the rate of heat that must be generated at the welding source, and (b) how long it will take to complete the surfacing operation.

WELDING PROCESSES

CHAPTER CONTENTS

31.1   Arc Welding
        31.1.1   General Technology of Arc Welding
        31.1.2   AW Processes that use Consumable Electrodes
        31.1.3   AW Processes that use Nonconsumable Electrodes
31.2   Resistance Welding
        31.2.1   Power Source in Resistance Welding
        31.2.2   Resistance Welding Processes
31.3   Oxyfuel Gas Welding
        31.3.1   Oxyacetylene Welding
        31.3.2   Alternative Gases for Oxyfuel Welding
31.4   Other Fusion-Welding Processes
        31.4.1   Electron-Beam Welding
        31.4.2   Laser-Beam Welding
        31.4.3   Electroslag Welding
        31.4.4   Thermit Welding
31.5   Solid-State Welding
        31.5.1   General Considerations in Solid-State Welding
        31.5.2   Solid-State Welding Processes
31.6   Weld Quality
        31.6.1   Residual Stresses and Distortion
        31.6.2   Welding Defects
        31.6.3   Inspection and Testing Methods
31.7   Weldability
31.8   Design Considerations in Welding

Welding processes are divided into two major categories: (1) *fusion welding,* in which co-alescence is accomplished by melting the two parts to be joined, in some cases adding filler metal to the joint; and (2) *solid-state welding,* in which heat and/or pressure are used to achieve coalescence, but no melting of the base metals occurs and no filler metal is added.

Fusion welding is by far the more important category. It includes (1) arc welding, (2) resistance welding, (3) oxyfuel gas welding, and (4) other fusion welding processes—ones that cannot be classified as any of the first three types. Fusion welding is discussed in the first four sections of this chapter. Section 31.5 covers solid-state welding. And in the final three sections of the chapter, we examine issues common to all welding operations: weld quality, weldability, and design for welding.

31.1 ARC WELDING

Arc welding (AW) is a fusion-welding process in which coalescence of the metals is achieved by the heat from an electric arc between an electrode and the work. The same basic process is also used in arc cutting (Section 26.3.4). A generic AW process is shown in Figure 31.1. An electric arc is a discharge of electric current across a gap in a circuit. It is sustained by the presence of a thermally ionized column of gas (called a plasma) through which current flows. To initiate the arc in an AW process, the electrode is brought into contact with the work and is then quickly separated from it by a short distance. The electric energy from the arc thus formed produces temperatures of 5500°C (10,000°F) or higher, sufficiently hot to melt any metal. A pool of molten metal, consisting of base metal(s) and filler metal (if one is used) is formed near the tip of the electrode. In most arc-welding processes, filler metal is added during the operation to increase the volume and strength of the weld joint. As the electrode is moved along the joint, the molten weld pool solidifies in its wake.

Movement of the electrode relative to the work is accomplished by either a human welder (manual welding) or by mechanical means (i.e., machine welding, automatic welding, or robotic welding). One of the troublesome aspects of manual arc welding is that the quality of the weld joint depends on the skill and work ethic of the human welder. Productivity is also an issue. Productivity is often measured as *arc time* (also called *arc-on time*)—the proportion of the hours worked that arc welding is being accomplished:

$$\text{Arc time} = (\text{time arc is on})/(\text{hours worked}) \tag{31.1}$$

This definition can be applied to an individual welder or to a mechanized workstation. For manual welding, arc time is usually around 20%. Frequent rest periods are needed by the welder to overcome fatigue in manual arc welding, which is demanding in hand-eye coordination under stressful conditions. Arc time increases to about 50% (more or less, depending on the operation) for machine, automatic, and robotic welding.

31.1.1 General Technology of Arc Welding

Before describing the individual AW processes, it is instructional to examine some of the general technical issues that apply to these processes.

FIGURE 31.1 The basic configuration and electrical circuit of an arc welding process.

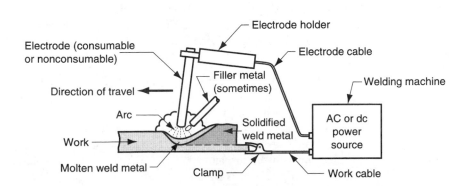

Electrodes   Electrodes used in AW processes are classified as consumable or nonconsumable. *Consumable electrodes* provide the source of the filler metal in arc welding. These electrodes are available in two principal forms: rods (also called sticks) and wire. Welding rods are typically 225 to 450 mm (9–18 in.) long and 9.5 mm (3/8 in.) or less in diameter. The problem with consumable welding rods, at least in production welding operations, is that they must be changed periodically, reducing arc time of the welder. Consumable weld wire has the advantage that it can be continuously fed into the weld pool from spools containing long lengths of wire, thus avoiding the frequent interruptions that occur when using welding sticks. In both rod and wire forms, the electrode is consumed by the arc during the welding process and added to the weld joint as filler metal.

Nonconsumable electrodes are made of tungsten (or carbon, rarely), which resists melting by the arc. Despite its name, a nonconsumable electrode is gradually depleted during the welding process (vaporization is the principal mechanism), analogous to the gradual wearing of a cutting tool in a machining operation. For AW processes that utilize nonconsumable electrodes, any filler metal used in the operation must be supplied by means of a separate wire that is fed into the weld pool.

Arc Shielding   At the high temperatures in arc welding, the metals being joined are very chemically reactive to oxygen, nitrogen, and hydrogen in the air. The mechanical properties of the weld joint can be seriously degraded by these reactions. Thus, some means is provided in nearly all AW processes to shield the arc from the surrounding air. Arc shielding is accomplished by covering the electrode tip, arc, and molten weld pool with a blanket of gas or flux, or both, which inhibit exposure of the weld metal to air.

Common shielding gases include argon and helium, both of which are inert. In the welding of ferrous metals with certain AW processes, oxygen and carbon dioxide are used, usually in combination with Ar and/or He, to produce an oxidizing atmosphere or to control weld shape.

A *flux* is used to prevent the formation of oxides and other unwanted contaminants, or to dissolve them and facilitate removal. During welding, the flux melts and becomes a liquid slag, covering the operation and protecting the molten weld metal. The slag hardens upon cooling and must be removed later by chipping or brushing. Flux is usually formulated to serve several additional functions, including (1) providing a protective atmosphere for welding, (2) stabilizing the arc, and (3) reducing spattering.

The method of flux application differs for each process. The delivery techniques include (1) pouring granular flux onto the welding operation, (2) using a stick electrode coated with flux material in which the coating melts during welding to cover the operation, and (3) using tubular electrodes in which flux is contained in the core and released as the electrode is consumed. These techniques are discussed further in our descriptions of the individual AW processes.

Power Source in Arc Welding   Both direct current (DC) and alternating current (AC) are used in arc welding. AC machines are less expensive to purchase and operate, but are generally restricted to welding of ferrous metals. DC equipment can be used on all metals with good results and is generally noted for better arc control.

In all AW processes, power to drive the operation is the product of the current I passing through the arc and the voltage E across it. This power is converted into heat, but not all of the heat is transferred to the surface of the work. Convection, conduction, radiation, and spatter account for losses that reduce the amount of usable heat. The effect of the losses is expressed by the heat transfer efficiency f_1 (Section 30.3). Some representative values of f_1 for several AW processes are given in Table 31.1. Heat transfer efficiency is greater for AW processes that use consumable electrodes because most of the heat consumed in melting the electrode is subsequently transferred to the work as molten metal. The process with the lowest f_1 value in Table 31.1 is gas tungsten arc welding, which uses a nonconsumable electrode. Melting efficiency f_2 (Section 30.3) further reduces the available heat for welding. The resulting power balance in arc welding is defined by

$$HR_w = f_1 f_2 IE = U_m A_w v \qquad (31.2)$$

where E = voltage, V; I = current, A; and the other terms were defined in Section 30.3. The units of HR_w that result from product of amps $\times$ voltage are watts, which equal joule/sec. This can be converted to Btu/sec by recalling that 1 Btu = 1055 joule.

EXAMPLE 31.1
Power in Arc
Welding

A gas tungsten AW operation is performed at a current of 300 A and voltage of 20 V. The melting efficiency $f_2 = 0.5$, and the unit melting energy for the metal $U_m = 10$ J/mm^3. Determine (a) power in the operation, (b) rate of heat generation at the weld, and (c) volume rate of metal welded.

Solution: (a) The power in this arc-welding operation is

$$P = IE = (300 \text{ A})(20 \text{ V}) = 6000 \text{ W}$$

(b) From Table 31.1, the heat transfer efficiency $f_1 = 0.7$. The rate of heat used for welding is given by

$$HR_w = f_1 f_2 IE = (0.7)(0.5)(6000) = 2100 \text{ W} = 2100 \text{ J/s}$$

(c) The volume rate of metal welded is

$$WVR = (2100 \text{ J/s})/(10 \text{ J/mm}^3) = 210 \text{ mm}^3/\text{s}$$

31.1.2 AW Processes—Consumable Electrodes

A number of important AW processes use consumable electrodes. These are discussed in this section. Symbols for the welding processes are those used by the American Welding Society.

TABLE 31.1   Heat transfer efficiencies for several arc-welding processes.

Arc Welding Process[a]	Typical Heat Transfer Efficiency f_1
Shielded metal arc welding	0.9
Gas metal arc welding	0.9
Flux-cored arc welding	0.9
Submerged arc welding	0.95
Gas tungsten arc welding	0.7

Compiled from [5].

[a] The AW processes are described in Section 31.1.2.

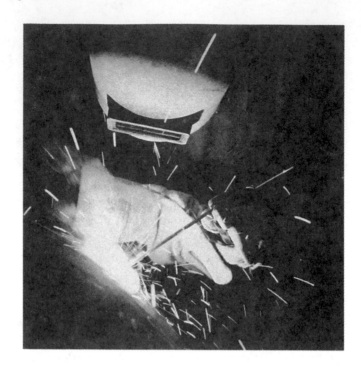

FIGURE 31.2 Shielded metal arc welding (stick welding) performed by a (human) welder (photo courtesy of Hobart Brothers Co.).

Shielded Metal Arc Welding *Shielded metal arc welding* (SMAW) is an AW process that uses a consumable electrode consisting of a filler metal rod coated with chemicals that provide flux and shielding. The process is illustrated in Figures 31.2 and 31.3. The welding stick (SMAW is sometimes called *stick welding*) is typically 225 to 450 mm (9–18 in.) long and 2.5 to 9.5 mm (3/32–3/8 in.) in diameter. The filler metal used in the rod must be compatible with the metal to be welded, the composition usually being very close to that of the base metal. The coating consists of powdered cellulose (i.e., cotton and wood powders) mixed with oxides, carbonates, and other ingredients, held together by a silicate binder. Metal powders are also sometimes included in the coating to increase the amount of filler metal and to add alloying elements. The heat of the welding process melts the coating to provide a protective atmosphere and slag for the welding operation. It also helps to stabilize the arc and regulate the rate at which the electrode melts.

During operation the bare metal end of the welding stick (opposite the welding tip) is clamped in an electrode holder that is connected to the power source. The holder has an insulated handle so that it can be held and manipulated by a human welder. Currents typically used in SMAW range between 30 and 300 A at voltages from 15 to 45 V. Selection

FIGURE 31.3 Shielded metal arc welding (SMAW).

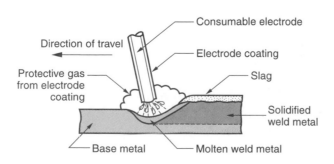

of the proper power parameters depends on the metals being welded, electrode type and length, and depth of weld penetration required. Power supply, connecting cables, and electrode holder can be bought for a few thousand dollars.

Shielded metal arc welding is usually performed manually. Common applications include construction, pipelines, machinery structures, shipbuilding, fabrication job shops, and repair work. It is preferred over oxyfuel welding for thicker sections—above 5 mm (3/16 in.)—because of its higher power density. The equipment is portable and low cost, making SMAW highly versatile and probably the most widely used of the AW processes. Base metals include steels, stainless steels, cast irons, and certain nonferrous alloys. It is not used or seldomly used for aluminum and its alloys, copper alloys, and titanium.

The disadvantage of shielded-metal AW as a production operation results from the use of the consumable electrode stick. As the sticks are used up, they must periodically be changed. This reduces the arc time with this welding process. Another limitation is the current level that can be used. Because the electrode length varies during the operation and this length affects the resistance heating of the electrode, current levels must be maintained within a safe range or the coating will overheat and melt prematurely when starting a new welding stick. Some of the other AW processes overcome the limitations of welding stick length in SMAW by using a continuously fed wire electrode.

Gas Metal Arc Welding   *Gas metal arc welding* (GMAW) is an AW process in which the electrode is a consumable bare metal wire, and shielding is accomplished by flooding the arc with a gas. The bare wire is fed continuously and automatically from a spool through the welding gun, as illustrated in Color Plate 10 and Figure 31.4. A welding gun is shown in Figure 31.5. Wire diameters ranging from 0.8 to 6.5 mm (1/32–1/4 in.) are used in GMAW, the size depending on the thickness of the parts being joined and the desired deposition rate. Gases used for shielding include inert gases such as argon and helium, and active gases such as carbon dioxide. Selection of gases (and mixtures of gases) depends on the metal being welded, as well as other factors. Inert gases are used for welding aluminum alloys and stainless steels, while CO_2 is commonly used for welding low and medium carbon steels. The combination of bare electrode wire and shielding gases eliminates the slag covering on the weld bead and thus precludes the need for manual grinding and cleaning of the slag. The GMAW process is therefore ideal for making multiple welding passes on the same joint.

The various metals on which GMAW is used and the variations of the process itself have given rise to a variety of names for GMAW. When the process was first introduced

FIGURE 31.4   Gas metal arc welding (GMAW).

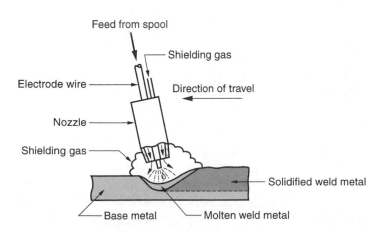

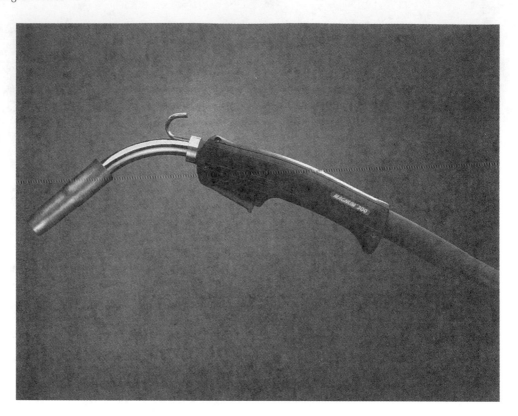

FIGURE 31.5 Welding gun for GMAW (courtesy Lincoln Electric Company).

in the late 1940s, it was applied to the welding of aluminum using inert gas (argon) for arc shielding. The name applied to this process was ***MIG welding*** (for Metal Inert Gas welding). When the same welding process was applied to steel, it was found that inert gases were expensive and CO_2 was used as a substitute. Hence the term ***CO_2 welding*** was applied. Refinements in GMAW for steel welding have led to the use of gas mixtures, including CO_2 and argon, and even oxygen and argon.

GMAW is widely used in fabrication operations in factories for welding a variety of ferrous and nonferrous metals. Because it uses continuous weld wire rather than welding sticks, it has a significant advantage over SMAW in terms of arc time when performed manually. For the same reason, it also lends itself to automation of arc welding. The electrode stubs remaining after stick welding also wastes filler metal, so the utilization of electrode material is higher with GMAW. Other features of GMAW include elimination of slag removal (since no flux is used), higher deposition rates than SMAW, and good versatility.

Flux-Cored Arc Welding   This AW process was developed in the early 1950s as an adaptation of SMAW to overcome the limitations imposed by the use of stick electrodes. ***Flux-cored arc welding*** (FCAW) is an AW process in which the electrode is a continuous consumable tubing that contains flux and other ingredients in its core. Other ingredients may include deoxidizers and alloying elements. The tubular flux-cored "wire" is flexible and can therefore be supplied in the form of coils to be continuously fed through the AW gun. There are two versions of FCAW: (1) self-shielded and (2) gas-shielded. In the first version of FCAW to be developed, arc shielding was provided by a flux core, thus leading to the name ***self-shielded flux-cored arc welding.*** The core in this form of FCAW includes not only fluxes but also ingredients that generate shield-

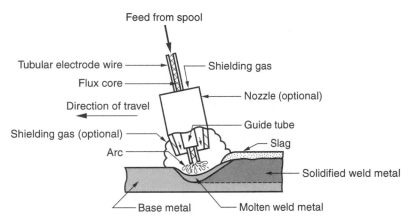

FIGURE 31.6   Flux-cored arc welding. The presence or absence of externally supplied shielding gas distinguishes the two types: (1) self-shielded, in which the core provides the ingredients for shielding; and (2) gas-shielded, in which external shielding gases are supplied.

ing gases for protecting the arc. The second version of FCAW, developed primarily for welding steels, obtains arc shielding from externally supplied gases, similar to GMAW. This version is called *gas-shielded flux-cored arc welding.* Because it utilizes an electrode containing its own flux together with separate shielding gases, it might be considered a hybrid of SMAW and GMAW. Shielding gases typically employed are carbon dioxide for mild steels or mixtures of argon and carbon dioxide for stainless steels. Figure 31.6 illustrates the FCAW process, with the gas (optional) distinguishing between the two types.

FCAW has advantages similar to GMAW, due to continuous feeding of the electrode. It is used primarily for welding steels and stainless steels over a wide stock thickness range. It is noted for its capability to produce very-high-quality weld joints that are smooth and uniform.

Electrogas Welding   *Electrogas welding* (EGW) is an AW process that uses a continuous consumable electrode (either flux-cored wire or bare wire with externally supplied shielding gases) and molding shoes to contain the molten metal. The process is primarily applied to vertical butt welding, as pictured in Figure 31.7. When the flux-cored electrode wire is employed, no external gases are supplied, and the process can be considered a special application of self-shielded FCAW. When a bare electrode wire is used with shielding gases from an external source, it is considered a special case of GMAW. The molding shoes are water cooled to prevent their being added to the weld pool. Together with the edges of the parts being welded, the shoes form a container, almost like a mold cavity, into which the molten metal from the electrode and base parts is gradually added. The process is performed automatically, with a moving weld head to travel vertically upward to fill the cavity in a single pass.

Principal applications of electrogas welding are steels (low- and medium-carbon, low-alloy, and certain stainless steels) in the construction of large storage tanks and in shipbuilding. Stock thicknesses from 12 to 75 mm (0.5–3.0 in.) are within the capacity of EGW. In addition to butt welding, it can also be used for fillet and groove welds, always in a vertical orientation. Specially designed molding shoes must sometimes be fabricated for the joint shapes involved.

Submerged Arc Welding   This process, developed during the 1930s, was one of the first AW processes to be automated. *Submerged arc welding* (SAW) is an AW process that uses a continuous, consumable bare wire electrode, and arc shielding is provided by a

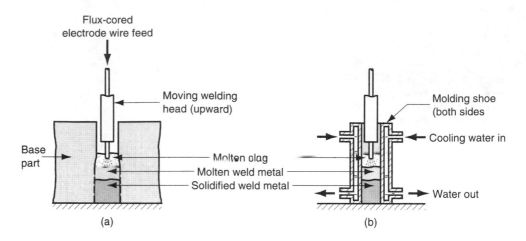

FIGURE 31.7   Electrogas welding using flux-cored electrode wire: (a) front view with molding shoes removed for clarity, and (b) side view showing molding shoes on both sides.

cover of granular flux. The electrode wire is fed automatically from a coil into the arc. The flux is introduced into the joint slightly ahead of the weld arc by gravity from a hopper, as shown in Figure 31.8. The blanket of granular flux completely submerges the AW operation, preventing sparks, spatter, and radiation that are so hazardous in other AW processes. Thus, the welding operator in SAW need not wear the somewhat cumbersome face shield required in the other operations (safety glasses and protective gloves, of course, are required). The portion of the flux closest to the arc is melted, mixing with the molten weld metal to remove impurities and then solidifying on top of the weld joint to form a glasslike slag. The slag and unfused flux granules on top provide good protection from the atmosphere and good thermal insulation for the weld area, resulting in relatively slow cooling and a high-quality weld joint, noted for toughness and ductility. As depicted in our sketch, the unfused flux remaining after welding can be recovered and reused. The solid slag covering the weld must be chipped away, usually by manual means.

Submerged arc welding is widely used in steel fabrication for structural shapes (e.g., welded I-beams); longitudinal and circumferential seams for large diameter pipes, tanks, and pressure vessels; and welded components for heavy machinery. In these kinds of applications, steel plates of 25 mm (1.0 in.) thickness and heavier are routinely welded by this process. Low-carbon, low alloy, and stainless steels can be readily welded by SAW;

FIGURE 31.8   Submerged arc welding.

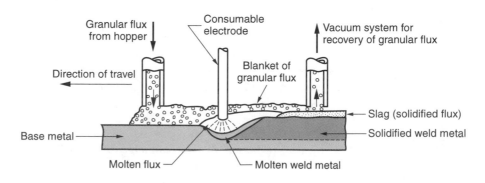

but not high-carbon steels, tool steels, and most nonferrous metals. Because of the gravity feed of the granular flux, the parts must always be in a horizontal orientation, and a backup plate is often required beneath the joint during the welding operation.

31.1.3 AW Processes—Nonconsumable Electrodes

The AW processes just discussed use consumable electrodes. Gas tungsten arc welding, plasma arc welding, and several other processes use nonconsumable electrodes.

Gas Tungsten Arc Welding   *Gas tungsten arc welding* (GTAW) is an AW process that uses a nonconsumable tungsten electrode and an inert gas for arc shielding. The term *TIG welding* (Tungsten Inert Gas welding) is often applied to this process (in Europe, *WIG welding* is the term—the chemical symbol for tungsten is W, for Wolfram). The GTAW process can be implemented with or without a filler metal. Figure 31.9 illustrates the latter case. When a filler metal is used, it is added to the weld pool from a separate rod or wire, being melted by the heat of the arc rather than transferred across the arc as in the consumable electrode AW processes. Tungsten is a good electrode material due to its high melting point of 3410°C (6170°F). Typical shielding gases include argon, helium, or a mixture of these gas elements.

GTAW is applicable to nearly all metals in a wide range of stock thicknesses. It can also be used for joining various combinations of dissimilar metals. Its most common applications are for aluminum and stainless steel. Cast irons, wrought irons, lead, and of course tungsten are difficult to weld by GTAW. In steel-welding applications, GTAW is generally slower and more costly than the consumable electrode AW processes, except when thin sections are involved and very-high-quality welds are required. When thin sheets are TIG welded to close tolerances, filler metal is usually not added. The process can be performed manually or by machine and automated methods for all joint types. Advantages of GTAW in the applications to which it is suited include high-quality welds, no weld spatter because no filler metal is transferred across the arc, and little or no postweld cleaning because no flux is used.

Plasma Arc Welding   *Plasma arc welding* (PAW) is a special form of gas tungsten arc welding in which a constricted plasma arc is directed at the weld area. In PAW, a tungsten electrode is contained in a specially designed nozzle that focuses a high-velocity stream of inert gas (e.g., argon or argon-hydrogen mixtures) into the region of the arc to form a high-velocity, intensely hot plasma arc stream, as in Figure 31.10. Argon, argon–hydrogen, and helium are also used as the arc shielding gases.

FIGURE 31.9   Gas tungsten arc welding.

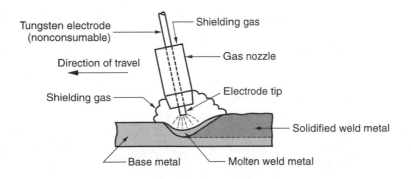

Tungsten electrode (nonconsumable)

Shielding gas

Direction of travel

Gas nozzle

Shielding gas

Electrode tip

Solidified weld metal

Base metal

Molten weld metal

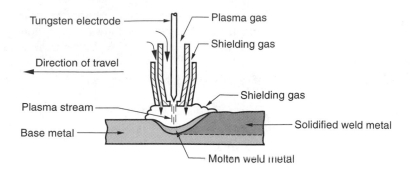

FIGURE 31.10 Plasma arc welding (PAW).

Temperatures in plasma arc welding reach 28,000°C (50,000°F) or greater, hot enough to melt any known metal. The reason the temperatures are so high in PAW (significantly higher than those in GTAW) derives from the constriction of the arc. Although the typical power levels used in PAW are below those used in GTAW, the power is highly concentrated to produce a plasma jet of small diameter and very high energy density.

Plasma arc welding was introduced around 1960 but was slow to catch on. In recent years its use is increasing as a substitute for GTAW in applications such as automobile subassemblies, metal cabinets, door and window frames, and home appliances. Owing to the special features of PAW, its advantages in these applications include good arc stability, better penetration control than most other AW processes, high travel speeds, and excellent weld quality. The process can be used to weld almost any metal, including tungsten. Difficult-to-weld metals with PAW include bronze, cast irons, lead, and magnesium. Other limitations include high equipment cost and larger torch size than other AW operations, which tend to restrict access in some joint configurations.

Other Arc Welding and Related Processes   The preceding AW processes are the most important commercially. There are several others that should be mentioned, which are special cases or variations of the principal AW processes.

Carbon arc welding (CAW) is an arc welding process in which a nonconsumable carbon (graphite) electrode is used. It has historical importance because it was the first AW process to be developed, but its commercial importance today is practically nil. The carbon arc process is used as a heat source for brazing and for repairing iron castings. It can also be used in some applications for depositing wear-resistant materials on surfaces. Graphite electrodes for welding have been largely superseded by tungsten (in GTAW and PAW).

Stud welding (SW) is a specialized AW process for joining studs or similar components to base parts. A typical SW operation is illustrated in Figure 31.11, in which shielding is obtained by the use of a ceramic ferrule. To begin with, the stud is chucked in a special weld gun that automatically controls the timing and power parameters of the steps shown in the sequence. The worker must only position the gun at the proper location against the base workpart to which the stud will be attached and pull the trigger. SW applications include threaded fasteners for attaching handles to cookware, heat radiation fins on machinery, and similar assembly situations. In high-production operations, stud welding usually has advantages over rivets, manually arc welded attachments, and drilled and tapped holes.

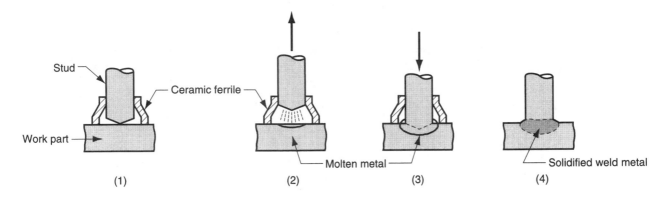

FIGURE 31.11   Stud arc welding (SW): (1) stud is positioned; (2) current flows from the gun and stud is pulled from base to establish arc and create a molten pool; (3) stud is plunged into molten pool; and (4) ceramic ferrule is removed after solidification.

31.2 RESISTANCE WELDING

Resistance welding (RW) is a group of fusion-welding processes that use a combination of heat and pressure to accomplish coalescence, the heat being generated by electrical resistance to current flow at the junction to be welded. The principal components in resistance welding are shown in Figure 31.12 for a resistance spot-welding operation, the most widely used process in the group. The components include workparts to be welded (usually sheet metal parts), two opposing electrodes, a means of applying pressure to squeeze the parts between the electrodes, and an AC power supply from which a controlled current can be applied. The operation results in a fused zone between the two parts, called a *weld nugget* in spot welding.

By comparison to arc welding, resistance welding uses no shielding gases, flux, or filler metal; and the electrodes that conduct electrical power to the process are nonconsumable. RW is classified as fusion welding because the applied heat almost always causes melting of the faying surfaces. However, there are exceptions. Some welding operations based on resistance heating use temperatures below the melting points of the base metals, so fusion does not occur.

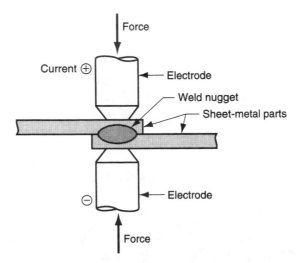

FIGURE 31.12   Resistance welding, showing the components in spot welding, the predominant process in the RW group.

31.2.1 Power Source in Resistance Welding

The heat energy supplied to the welding operation depends on current flow, resistance of the circuit, and length of time the current is applied. This can be expressed by the equation

$$H = I^2Rt \tag{31.3}$$

where H = heat generated, J (to convert to Btu divide by 1055); I = current, A; R = electrical resistance, Ω; and t = time, s.

The current used in resistance welding operations is very high (5,000 to 20,000 A, typically), although voltage is relatively low (usually below 10 V). The duration t of the current is short in most processes, perhaps lasting 0.1 to 0.4 s in a typical spot welding operation.

Current is so high in RW because the squared term in Eq. (31.3) amplifies the effect of current, and resistance is very low (around 0.0001 Ω). Resistance in the welding circuit is the sum of (1) resistance of the electrodes, (2) resistances of the workparts, (3) contact resistances between electrodes and workparts, and (4) contact resistance of the faying surfaces. The ideal situation is for the faying surfaces to be the largest resistance in the sum, since this is the desired location of the weld. The resistance of the electrodes is minimized by using metals with very low resistivities, such as copper. The resistances of the workparts is a function of the resistivities of the base metals involved and the thicknesses of the parts. The contact resistances between the electrodes and the parts is determined by the contact areas (i.e., size and shape of the electrode) and the condition of the surfaces (e.g., cleanliness of the work surfaces and scale on the electrode). Finally, the resistance at the faying surfaces depends on surface finish, cleanliness, contact area, and pressure. No paint, oil, dirt, or other contaminants should be present to separate the contacting surfaces.

EXAMPLE 31.2
Resistance Welding

A resistance spot welding operation is performed on two pieces of 1.5 mm thick sheet steel using 12,000 amps for a 0.20 second duration. The electrodes are 6 mm in diameter at the contacting surfaces. Resistance is assumed to be 0.0001 ohms, and the resulting weld nugget is 6 mm in diameter and 2.5 mm thick. The unit melting energy for the metal U_m = 12.0 J/mm^3. What portion of the heat generated was used to form the weld, and what portion was dissipated into the surrounding metal?

Solution: The heat generated in the operation is given by Eq. (31.3) as

$$H = (12,000)^2(0.0001)(0.2) = 2880 \text{ J}$$

The volume of the weld nugget (assumed disc-shaped) is $V = 2.5 \dfrac{\pi(6)^2}{4} = 70.7$ mm^3.

The heat required to melt this volume of metal is H_m = 70.7(12.0) = 848 J.

The remaining heat, 2880 − 848 = 2032 J (70.6% of the total), is absorbed into the surrounding metal.

Success in resistance welding depends on pressure, as well as heat. The principal functions of pressure in RW are to (1) force contact between the electrodes and the workparts and between the two work surfaces prior to applying current; and (2) press the faying surfaces together to accomplish coalescence when the proper welding temperature has been reached.

There are some general advantages of resistance welding: (1) no filler metal is required, (2) high production rates are possible, (3) it lends itself to mechanization and automation, (4) operator skill level is lower than that required for arc welding, and (5) good repeatability and reliability. Drawbacks are that initial equipment cost is high—usually much higher than most AW operations, and the types of joints that can be welded are limited to lap joints for most RW processes.

31.2.2 Resistance-Welding Processes

The resistance-welding processes of most commercial importance are spot, seam, and projection welding.

Resistance Spot Welding   Resistance spot welding is by far the predominant process in this group. It is widely used in mass production of automobiles, appliances, metal furniture, and other products made of sheet metal. If one considers that a typical car body has approximately 10,000 individual spot welds, and that the annual production of automobiles throughout the world is measured in tens of millions of units, the economic importance of resistance spot welding can be appreciated.

Resistance spot welding (RSW) is an RW process in which fusion of the faying surfaces of a lap joint is achieved at one location by opposing electrodes. The process is used to join sheet metal parts of thickness 3 mm (0.125 in.) or less, using a series of spot welds, in situations where an airtight assembly is not required. The size and shape of the weld spot is determined by the electrode tip, the most common electrode shape being round, but hexagonal, square, and other shapes are also used. The resulting weld nugget is typically 5 to 10 mm (0.2–0.4 in.) in diameter, with an HAZ extending slightly beyond the nugget into the base metals. If the weld is made properly, its strength will be comparable to that of the surrounding metal. The steps in a spot welding cycle are shown in Figure 31.13.

FIGURE 31.13   (a) Steps in a spot welding cycle, and (b) plot of squeezing force and current during cycle. The sequence is: (1) parts inserted between open electrodes, (2) electrodes close and force is applied, (3) weld time—current is switched on, (4) current is turned off but force is maintained or increased (a reduced current is sometimes applied near the end of this step for stress relief in the weld region), and (5) electrodes are opened, and the welded assembly is removed.

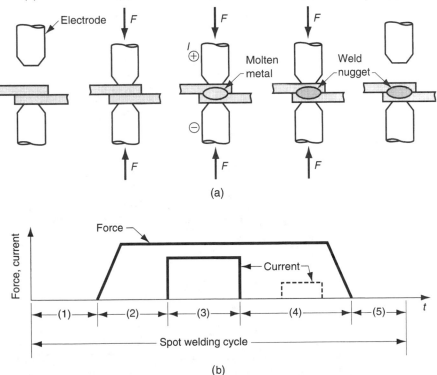

Materials used for RSW electrodes consist of two main groups: (1) copper-based alloys and (2) refractory metal compositions such as copper and tungsten combinations. The second group is noted for superior wear resistance. As in most manufacturing processes, the tooling in spot welding gradually wears out as it is used. Whenever practical, the electrodes are designed with internal passageways for water cooling.

Because of its widespread industrial use, various machines and methods are available to perform spot welding operations. The equipment includes rocker-arm and press-type spot-welding machines, and portable spot-welding guns. ***Rocker-arm spot welders,*** shown in Figure 31.14, have a stationary lower electrode and a movable upper electrode that can be raised and lowered for loading and unloading the work. The upper electrode is mounted on a rocker arm (hence the name) whose movement is controlled by a foot pedal operated by the worker. Modern machines can be programmed to control force and current during the weld cycle.

Press-type spot welders are intended for larger work. The upper electrode has a straight-line motion provided by a vertical press, which is pneumatically or hydraulically powered. The press action permits larger forces to be applied, and the controls usually permit programming of complex weld cycles.

The previous two machine types are both stationary spot welders, in which the work is brought to the machine. For large, heavy work it is difficult to move (and orient) the work to stationary machines. For these cases, ***portable spot welding guns*** are available in various sizes and configurations. These devices consist of two opposing electrodes contained in a pincer mechanism. Each unit is light weight so that it can be held and manipulated by a human worker or an industrial robot. The gun is connected to its own power and control source by means of flexible electrical cables and air hoses. Water cooling for the electrodes, if needed, can also be provided through a water hose. Portable spot-welding guns are widely used in automobile final assembly plants to spot weld car bodies. Some of these guns are operated by people, but industrial robots have become the preferred technology, illustrated in Color Plate 11 and Figure 38.11).

Resistance Seam Welding   In *resistance seam welding* (RSEW), the stick-shaped electrodes in spot welding are replaced by rotating wheels, as shown in Figure 31.15, and a series of overlapping spot welds are made along the lap joint. The process is capable of producing air-tight joints, and its industrial applications include the production of gasoline tanks, automobile mufflers, and various other fabricated sheet metal containers. Technically, RSEW is the same as spot welding, except that the wheel electrodes

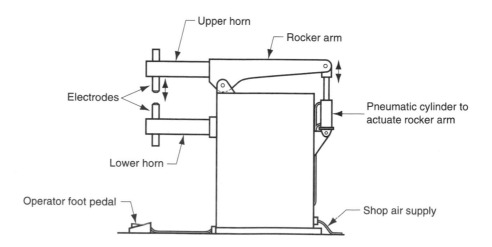

FIGURE 31.14   Rocker-arm spot welding machine.

Upper horn

Rocker arm

Electrodes

Pneumatic cylinder to actuate rocker arm

Lower horn

Operator foot pedal

Shop air supply

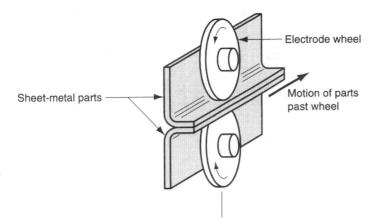

FIGURE 31.15   Resistance seam welding (RSEW).

introduce certain complexities. Since the operation is usually carried out continuously, rather than discretely, the seams should be along a straight or uniformly curved line. Sharp corners and similar discontinuities are difficult to deal with. Also, warpage of the parts becomes more of a factor in resistance seam welding, and well-designed fixtures are required to hold the work in position and minimize distortion.

The spacing between the weld nuggets in resistance seam welding depends on the motion of the electrode wheels relative to the application of the weld current. In the usual method of operation, called *continuous motion welding,* the wheel is rotated continuously at a constant velocity, and current is turned on at timing intervals consistent with the desired spacing between spot welds along the seam. Frequency of the current discharges is normally set so that overlapping weld spots are produced. But if the frequency is reduced sufficiently, then there will be spaces between the weld spots, and this method is termed *roll spot welding.* In another variation, the welding current remains on at a constant level (rather than being pulsed) so that a truly continuous welding seam is produced. These variations are depicted in Figure 31.16.

An alternative to continuous motion welding is *intermittent motion welding,* in which the electrode wheel is periodically stopped to make the spot weld. The amount of wheel rotation between stops determines the distance between weld spots along the seam, yielding patterns similar to (a) and (b) in Figure 31.16.

Seam-welding machines are similar to press-type spot welders except that electrode wheels are used rather than the usual stick-shaped electrodes. Cooling of the work and wheels is often necessary in RSEW, and this is accomplished by directing water at the top and underside of the workpart surfaces near the electrode wheels.

FIGURE 31.16   Different types of seams produced by electrode wheels: (a) conventional resistance seam welding, in which overlapping spots are produced; (b) roll spot welding; and (c) continuous resistance seam.

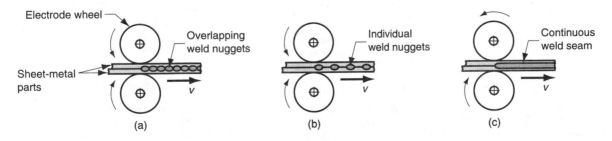

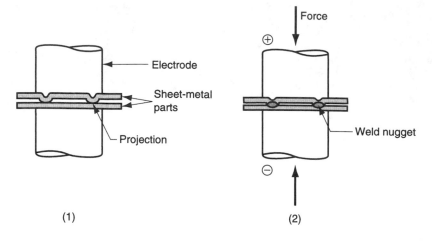

FIGURE 31.17 Resistance projection welding (RPW): (1) at start of operation, contact between parts is at projections; and (2) when current is applied, weld nuggets similar to those in spot welding are formed at the projections.

Resistance Projection Welding *Resistance projection welding* (RPW) is an RW process in which coalescence occurs at one or more relatively small contact points on the parts. These contact points are determined by the design of the parts to be joined, and may consist of projections, embossments, or localized intersections of the parts. A typical case in which two sheet-metal parts are welded together is described in Figure 31.17. The part on top has been fabricated with two embossed points to contact the other part at the start of the process. It might be argued that the embossing operation increases the cost of the part, but this increase may be more than offset by savings in welding cost.

There are variations of resistance projection welding, two of which are shown in Figure 31.18. In one variation, fasteners with machined or formed projections can be permanently joined to sheet or plate by RPW, facilitating subsequent assembly operations. Another variation, called *cross-wire welding,* is used to fabricate welded wire products such as wire fence, shopping carts, and stove grills. In this process, the contacting surfaces of the round wires serve as the projections to localize the resistance heat for welding.

Other Resistance-Welding Operations In addition to the principal resistance-welding processes, flash, upset, percussion, and high-frequency resistance welding are also types of RW.

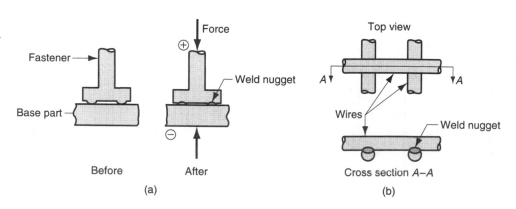

FIGURE 31.18 Two variations of resistance projection welding: (a) welding of a machined or formed fastener onto a sheet-metal part; and (b) cross-wire welding.

In *flash welding* (FW), normally used for butt joints, the two surfaces to be joined are brought into contact or near contact and electric current is applied to heat the surfaces to the melting point, after which the surfaces are forced together to form the weld. The two steps are outlined in Figure 31.19. In addition to resistance heating, some arcing occurs (called *flashing,* hence the name), depending on the extent of contact between the faying surfaces, so flash welding is sometimes classified in the AW group. Current is usually stopped during upsetting. Some metal, as well as contaminants on the surfaces, is squeezed out of the joint and must be subsequently machined to provide a joint of uniform size.

Applications of flash welding include butt welding of steel strips in rolling-mill operations, joining ends of wire in wire drawing, and welding of tubular parts. The ends to be joined must have the same cross sections. For these kinds of high-production applications, flash welding is fast and economical, but the equipment is expensive.

Upset welding (UW) is similar to FW except that in UW the faying surfaces are pressed together during heating and upsetting. In FW, the heating and pressing steps are separated during the cycle. Heating in UW is accomplished entirely by electrical resistance at the contacting surfaces; no arcing occurs. When the faying surfaces have been heated to a suitable temperature below the melting point, the force pressing the parts together is increased to cause upsetting and coalescence in the contact region. Thus, upset welding is not a fusion welding process in the same sense as the other welding processes we have discussed. Applications of UW are similar to those of flash welding: joining ends of wire, pipes, tubes, and so on.

Percussion welding (PEW) is also similar to flash welding, except that the duration of the weld cycle is extremely short, typically lasting only 1 to 10 ms. Fast heating is accomplished by rapid discharge of electrical energy between the two surfaces to be joined, followed immediately by percussion of one part against the other to form the weld. The heating is very localized, making this process attractive for electronic applications in which the dimensions are very small and nearby components may be sensitive to heat.

High-frequency resistance welding (HFRW) is a resistance-welding process in which a high-frequency alternating current is used for heating, followed by the rapid application of an upsetting force to cause coalescence, as in Figure 31.20(a). The frequencies are 10 to 500 kHz, and the electrodes make contact with the work in the immediate vicinity of the weld joint. In a variation of the process, called *high-frequency induction welding* (HFIW), the heating current is induced in the parts by a high-frequency induction coil, as in Figure 31.20(b). The coil does not make physical contact with the work. The principal applications of both HFRW and HFIW are continuous butt welding of the longitudinal seams of metal pipes and tubes.

FIGURE 31.19 Flash welding (FW): (1) heating by electrical resistance; and (2) upsetting—parts are forced together.

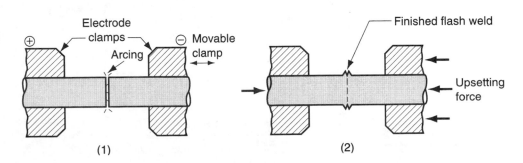

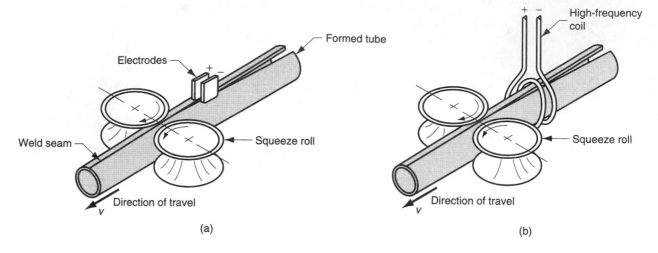

FIGURE 31.20   Welding of tube seams by (a) high-frequency resistance welding, and (b) high-frequency induction welding.

31.3 OXYFUEL GAS WELDING

Oxyfuel gas welding (OFW) is the term used to describe the group of fusion operations that burn various fuels mixed with oxygen to perform welding. The OFW processes employ several types of gases, which is the primary distinction among the members of this group. Oxyfuel gas is also commonly used in cutting torches to cut and separate metal plates and other parts (Section 27.3.5). The most important OFW process is oxyacetylene welding.

31.3.1 Oxyacetylene Welding

Oxyacetylene welding (OAW) is a fusion-welding process performed by a high-temperature flame from combustion of acetylene and oxygen. The flame is directed by a welding torch. A filler metal is sometimes added, and pressure is occasionally applied in OAW between the contacting part surfaces. A typical OAW operation is sketched in Figure 31.21. When filler metal is used, it is typically in the form of rod with diameters ranging from 1.6 mm (1/16 in.) to 9.5 mm (3/8 in.). Composition of the filler must be similar to that of the base metals. The filler is often coated with a ***flux*** that helps to clean the surfaces and prevent oxidation, thus creating a better weld joint.

FIGURE 31.21   A typical oxyacetylene welding operation (OAW).

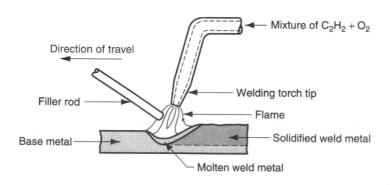

Acetylene (C_2H_2) is the most popular fuel among the OFW group because it is capable of higher temperatures than any of the others—up to 3480° C (6300° F). The flame in OAW is produced by the chemical reaction of acetylene and oxygen in two stages. The first stage is defined by the reaction

$$C_2H_2 + O_2 \rightarrow 2CO + H_2 + heat \qquad (31.4a)$$

the products of which are both combustible, which leads to the second-stage reaction:

$$2CO + H_2 + 1.5O_2 \rightarrow 2CO_2 + H_2O + heat \qquad (31.4b)$$

The two stages of combustion are visible in the oxyacetylene flame emitted from the torch. When the mixture of acetylene and oxygen is in the ratio 1 : 1, as described in Eqs. (31.4 a and b), the resulting **neutral flame** is as shown in Figure 31.22. The first-stage reaction is seen as the inner cone of the flame (which is bright white), while the second-stage reaction is exhibited by the outer envelope (which is nearly colorless but with tinges ranging from blue to orange). The maximum temperature of the flame is reached at the tip of the inner cone; the second-stage temperatures are somewhat below those of the inner cone. During welding, the outer envelope spreads out and covers the work surfaces being joined, thus shielding them from the surrounding atmosphere.

Total heat liberated during the two stages of combustion is 55×10^6 J/m^3 (1470 Btu/ft^3) of acetylene. However, because of the temperature distribution in the flame, the way in which the flame spreads over the work surface, and losses to the air, power densities and heat transfer efficiencies in oxyacetylene welding are relatively low; $f_1 = 0.10$ to 0.30.

**EXAMPLE 31.3
Heat Generation
in Oxyacetylene
Welding**

An oxyacetylene torch supplies 0.3 m^3 of acetylene per hour and an equal volume rate of oxygen for an OAW operation on 4.5 mm steel. Heat generated by combustion is transferred to the work surface with an efficiency $f_1 = 0.25$. If 75% of the heat from the flame is concentrated in a circular area on the work surface that is 9.0 mm in diameter, find (a) rate of heat liberated during combustion, (b) rate of heat transferred to the work surface, and (c) average power density in the circular area.

Solution: (a) The rate of heat generated by the torch is the product of the volume rate of acetylene times the heat of combustion:

$$HR = (0.3 \text{ m}^3/\text{hr})(55 \times 10^6 \text{ J/m}^3) = 16.5 \times 10^6 \text{ J/hr or } 4583 \text{ J/s}$$

(b) With a heat transfer efficiency $f_1 = 0.25$, the rate of heat received at the work surface is

$$f_1 HR = 0.25(4583) = 1146 \text{ J/s}$$

(c) The area of the circle in which 75% of the heat of the flame is concentrated is

$$A = \frac{\pi(9)^2}{4} = 63.6 \text{ mm}^2$$

FIGURE 31.22 The neutral flame from an oxyacetylene torch, indicating temperatures achieved.

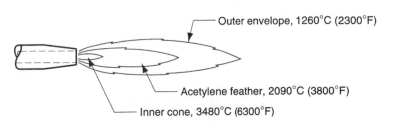

Outer envelope, 1260°C (2300°F)

Acetylene feather, 2090°C (3800°F)

Inner cone, 3480°C (6300°F)

The power density in the circle is found by dividing the available heat by the area of the circle:

$$PD = \frac{0.75(1146)}{63.6} = 13.5 \text{ W/mm}^2$$

The combination of acetylene and oxygen is highly flammable, and the environment in which OAW is performed is therefore hazardous. Some of the dangers relate specifically to the acetylene. Pure C_2H_2 is a colorless, odorless gas. For safety reasons, commercial acetylene is processed to have a characteristic garlic odor. One of the physical limitations of the gas is that it is unstable at pressures much above 1 atm (0.1 MPa or 15 lb/in.^{2}). Accordingly, acetylene storage cylinders are packed with a porous filler material (such as asbestos, balsa wood, and other materials) saturated with acetone (CH_3COCH_3). Acetylene dissolves in liquid acetone; in fact, acetone dissolves about 25 times its own volume of acetylene, thus providing a relatively safe means of storing this welding gas. The welder wears eye and skin protection (goggles, gloves, and protective clothing) as an additional safety precaution, and different screw threads are standard on the acetylene and oxygen cylinders and hoses to avoid accidental connection of the wrong gases. Proper maintenance of the equipment is imperative.

OAW equipment is relatively inexpensive and portable. It is therefore an economical, versatile process that is well suited to low-quantity production and repair jobs. It is rarely used to weld sheet and plate stock thicker than 6.4 mm (1/4 in.) because of the advantages of arc welding in such applications. Although OAW can be mechanized, it is usually performed manually and is hence dependent on the skill of the welder to produce a high-quality weld joint.

31.3.2 Alternative Gases for Oxyfuel Welding

Several members of the OFW group are based on gases other than acetylene. Most of the alternative fuels are listed in Table 31.2, together with their burning temperatures and

TABLE 31.2   Gases used in oxyfuel welding and/or cutting, with flame temperatures and heats of combustion.

Fuel	Temperature[a]		Heat of Combustion	
	°C	(°F)	MJ/m^3	(Btu/ft^3)
Acetylene (C_2H_2)	3087	(5589)	54.8	(1470)
MAPP[b] (C_3H_4)	2927	(5301)	91.7	(2460)
Hydrogen (H_2)	2660	(4820)	12.1	(325)
Propylene[c] (C_3H_6)	2900	(5250)	89.4	(2400)
Propane (C_3H_8)	2526	(4579)	93.1	(2498)
Natural gas[d]	2538	(4600)	37.3	(1000)

Compiled from [9], p. 354.

[a]Neutral flame temperatures are compared since this is the flame that would most commonly be used for welding.

[b]MAPP is the commercial abbreviation for methylacetylene-propadiene.

[c]Propylene is used primarily in flame cutting.

[d]Data are based on methane gas (CH_4); natural gas consists of ethane (C_2H_6) as well as methane; flame temperature and heat of combustion vary with composition.

combustion heats. For comparison, acetylene is included in the list. Although oxyacetylene is the most common OFW fuel, each of the other gases can be used in certain applications—typically limited to welding of sheet metal and metals with low melting temperatures, and brazing (Section 32.1). In addition, some users prefer these alternative gases for safety reasons.

The fuel that competes most closely with acetylene in burning temperature and heating value is methylacetylene-propadiene. It is a fuel developed by the Dow Chemical Company sold under the trade name ***MAPP*** (we are grateful to Dow for the abbreviation). MAPP (C_3H_4) has heating characteristics similar to acetylene and can be stored under pressure as a liquid, thus avoiding the special storage problems associated with C_2H_2.

When hydrogen is burned with oxygen as the fuel, the process is called ***oxyhydrogen welding*** (OHW). As shown in Table 31.2, the welding temperature in OHW is below that possible in oxyacetylene welding. In addition, the color of the flame is not affected by differences in the mixture of hydrogen and oxygen, and therefore it is more difficult for the welder to adjust the torch.

Other fuels used in OFW include propane and natural gas. Propane (C_3H_8) is more closely associated with brazing, soldering, and cutting operations than with welding. Natural gas consists mostly of ethane (C_2H_6) and methane (CH_4). When mixed with oxygen it achieves a high temperature flame and is becoming more common in small welding shops.

Pressure Gas Welding   This is a special OFW process, distinguished by type of application rather than fuel gas. ***Pressure gas welding*** (PGW) is a fusion-welding process in which coalescence is obtained over the entire contact surfaces of the two parts by heating them with an appropriate fuel mixture (usually oxyacetylene gas) and then applying pressure to bond the surfaces. A typical application is illustrated in Figure 31.23. Parts are heated until melting begins on the surfaces. The heating torch is then withdrawn, and the parts are pressed together and held at high pressure while solidification occurs. No filler metal is used in PGW.

FIGURE 31.23   An application of pressure gas welding: (a) heating of the two parts, and (b) applying pressure to form the weld.

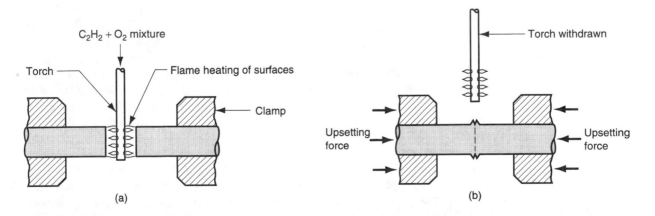

31.4 OTHER FUSION-WELDING PROCESSES

Some fusion-welding processes cannot be classified as arc, resistance, or oxyfuel welding. Each of these other processes uses a unique technology to develop heat for melting; and typically, the applications are unique.

31.4.1 Electron-Beam Welding

Electron beam welding (EBW) is a fusion-welding process in which the heat for welding is provided by a highly focused, high-intensity stream of electrons impinging against the work surface. The equipment is similar to that used for electron-beam machining (Section 26.3.2). The electron beam gun operates at high voltage to accelerate the electrons (e.g., 10 to 150 kV typical), and beam currents are low (measured in milliamps). The power in EBW is not exceptional, but power density is. High-power density is achieved by focusing the electron beam on a very small area of the work surface, so that the power density *PD* is based on

$$PD = \frac{f_1 EI}{A} \tag{31.5}$$

where PD = power density, W/mm^2 (W/in.^{2}, which can be converted to Btu/sec-in.^{2} by dividing by 1055.); f_1 = heat transfer efficiency (typical values for EBW range from 0.8 to 0.95 [8]); E = accelerating voltage, V; I = beam current, A; and A = the work surface area on which the electron beam is focused, mm^2 (in.^{2}). Typical weld areas for EBW range from 13×10^{-3} to 2000×10^{-3} mm^2 (20×10^{-6} to 3000×10^{-6} in.^{2}).

The process had its beginnings in the 1950s in the atomic power field. When first developed, welding had to be carried out in a vacuum chamber to minimize the disruption of the electron beam by air molecules. This requirement was, and still is, a serious inconvenience in production, due to the time required to evacuate the chamber prior to welding. The pump-down time, as it is called, can take as long as an hour, depending on the size of the chamber and the level of vacuum required. Today, EBW technology has progressed to where some operations are performed without a vacuum. Three categories can be distinguished: (1) *high-vacuum welding* (EBW-HV), in which welding is carried out in the same vacuum as beam generation; (2) *medium-vacuum welding* (EBW-MV), in which the operation is performed in a separate chamber where only a partial vacuum is achieved; and (3) *nonvacuum welding* (EBW-NV), in which welding is accomplished at or near atmospheric pressure. The pump-down time during workpart loading and unloading is reduced in medium-vacuum EBW and minimized in nonvacuum EBW, but there is a price paid for this advantage. In the latter two operations, the equipment must include one or more vacuum dividers (very small orifices that impede air flow but permit passage of the electron beam) to separate the beam generator (which requires a high vacuum) from the work chamber. Also, in nonvacuum EBW, the work must be located close to the orifice of the electron beam gun, approximately 13 mm (0.5 in.) or less. Finally, the lower vacuum processes cannot achieve the high weld qualities and depth-to-width ratios accomplished by EBW-HV.

Any metals that can be arc welded can be welded by EBW, as well as certain refractory and difficult-to-weld metals that are not suited to AW. Work sizes range from thin foil to thick plate. EBW is applied mostly in the automotive, aerospace, and nuclear industries. In the automotive industry, EBW assembly includes aluminum manifolds, steel torque converters, catalytic converters, and transmission components. In these and other applications, electron beam welding is noted for the following advantages: high-quality welds with deep and/or narrow profiles, limited heat affected zone, and low thermal distortion. Welding speeds are high compared to other continuous welding operations. No filler metal is used, and no flux or shielding gases are needed. Disadvantages of EBW include high equipment cost, need for precise joint preparation and alignment, and the limitations associated with performing the process in a vacuum, as we have already discussed. In addition, there are safety concerns because EBW generates x-rays from which humans must be shielded.

31.4.2 Laser-Beam Welding

Laser-beam welding (LBW) is a fusion-welding process in which coalescence is achieved by the energy of a highly concentrated, coherent light beam focused on the joint to be welded. The term *laser* is an acronym for "light amplification by stimulated emission of radiation." This same technology is used for laser-beam machining (Section 26.3.3). LBW is normally performed with shielding gases (e.g., helium, argon, nitrogen, and carbon dioxide) to prevent oxidation. Filler metal is not usually added.

LBW produces welds of high quality, deep penetration, and narrow HAZ. These features are similar to those achieved in electron beam welding (EBW), and the two processes are often compared. There are several advantages of LBW over EBW: no vacuum chamber is required, no x-rays are emitted, and laser beams can be focused and directed by optical lenses and mirrors. On the other hand, LBW does not possess the capability for the deep welds and high depth-to-width ratios of EBW. Maximum depth in laser welding is about 19 mm (0.75 in.), whereas EBW can be used for weld depths of 50 mm (2 in.) or more; and the depth-to-width ratios in LBW are typically limited to around 5:1. Because of the highly concentrated energy in the small area of the laser beam, the process is often used to join small parts.

31.4.3 Electroslag Welding

Electroslag welding (ESW) uses the same basic equipment as some of the AW processes, and it utilizes an arc to initiate the welding operation. However, it is not an AW process because an arc is not used during welding. *Electroslag welding* (ESW) is a fusion-welding process in which coalescence is achieved by hot, electrically conductive molten slag acting on the base parts and filler metal. As shown in Figure 31.24, the general configuration of ESW is similar to electrogas welding. It is performed in a vertical orientation (shown here for butt welding), using water-cooled molding shoes to contain the molten slag and weld metal. At the start of the process, granulated conductive flux is put into

FIGURE 31.24
Electroslag welding (ESW): (a) front view with molding shoe removed for clarity; (b) side view showing schematic of molding shoe. Setup is similar to electrogas welding (Figure 31.7) except that resistance heating of molten slag is used to melt the base and filler metals.

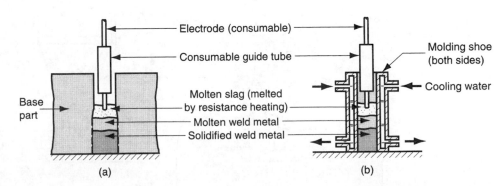

the cavity. The consumable electrode tip is positioned near the bottom of the cavity, and an arc is generated for a short while to start melting the flux. Once a pool of slag has been created, the arc is extinguished and the current passes from the electrode to the base metal through the conductive slag, so that its electrical resistance generates heat to maintain the welding process. Since the density of the slag is less than that of the molten metal, it remains on top to protect the weld pool. Solidification occurs from the bottom, while additional molten metal is supplied from above by the electrode and the edges of the base parts. The process gradually continues until it reaches the top of the joint.

31.4.4 Thermit Welding

Thermit is a trademark name for ***thermite,*** a mixture of aluminum powder and iron oxide that produces an exothermic reaction when ignited. It is used in incendiary bombs and for welding. As a welding process, the use of Thermit dates from around 1900. ***Thermit welding*** (TW) is a fusion-welding process in which the heat for coalescence is produced by superheated molten metal from the chemical reaction of Thermit. Filler metal is obtained from the liquid metal; and although the process is used for joining, it has more in common with casting than it does with welding.

Finely mixed powders of aluminum and iron oxide (in a 1:3 mixture), when ignited to a temperature of around 1300°C (2300°F), produce the following chemical reaction:

$$8Al + 3Fe_3O_4 \rightarrow 9Fe + 4Al_2O_3 + heat \qquad (31.6)$$

The temperature from the reaction is around 2500°C (4500°F), resulting in superheated molten iron plus aluminum oxide that floats to the top as a slag and protects the iron from the atmosphere. In Thermit welding, the superheated iron (or steel if the mixture of powders is formulated accordingly) is contained in a crucible located above the joint to be welded, as indicated by our diagram of the TW process in Figure 31.25. After the reaction is complete (about 30 s, irrespective of the amount of Thermit involved), the crucible is tapped and the liquid metal flows into a mold built specially to surround the weld joint. Because the entering metal is so hot, it melts the edges of the base parts, causing coalescence upon solidification. After cooling, the mold is broken away, and the gates and risers are removed by oxyacetylene torch or other method.

Thermit welding has applications in joining of railroad rails (as pictured in our figure), and repair of cracks in large steel castings and forgings such as ingot molds, large diameter shafts, and frames for machinery and, ship rudders. The surface of the weld in these applications is often sufficiently smooth so that no subsequent finishing is required.

FIGURE 31.25   Thermit welding: (1) Thermit ignited; (2) crucible tapped, superheated metal flows into mold; (3) metal solidifies to produce weld joint.

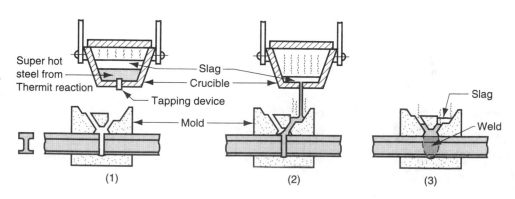

31.5 SOLID-STATE WELDING

In solid-state welding, coalescence of the part surfaces is achieved by pressure alone, or heat and pressure. For some solid-state processes, time is also a factor. If both heat and pressure are used, the amount of heat by itself is not sufficient to cause melting of the work surfaces. In other words, fusion of the parts would not occur using only the heat that is externally applied in these processes. In some cases, the combination of heat and pressure, or the particular manner in which pressure alone is applied, generates sufficient energy to cause localized melting of the faying surfaces. Filler metal is not added in solid-state welding.

31.5.1 General Considerations in Solid-State Welding

In most of the solid-state processes, a metallurgical bond is created with little or no melting of the base metals. To metallurgically bond two similar or dissimilar metals, the two metals must be brought into intimate contact so that their cohesive atomic forces attract each other. In normal physical contact between two surfaces, such intimate contact is prohibited by the presence of chemical films, gases, oils, etc. In order for atomic bonding to succeed, these films and other substances must be removed. In fusion welding (as well as other joining processes such as brazing and soldering), the films are dissolved or burned away by high temperatures, and atomic bonding is established by the melting and solidification of the metals in these processes. But in solid-state welding, the films and other contaminants must be removed by other means to allow metallurgical bonding to take place. In some cases, a thorough cleaning of the surfaces is done just before the welding process; while in other cases, the cleaning action is accomplished as an integral part of bringing the part surfaces together. To summarize, the essential ingredients for a successful solid-state weld are that the two surfaces must be very clean, and they must be brought into very close physical contact with each other to permit atomic bonding.

Welding processes that do not involve melting have several advantages over fusion-welding processes. If no melting occurs, then there is no heat-affected zone, and so the metal surrounding the joint retains its original properties. Many of these processes produce welded joints that make up the entire contact interface between the two parts, rather than at distinct spots or seams, as in most fusion welding operations. Also, some of these processes are quite applicable to bonding dissimilar metals, without concerns about relative thermal expansions, conductivities, and other problems that usually arise when dissimilar metals are melted and then solidified during joining.

31.5.2 Solid-State Welding Processes

The solid-state welding group includes the oldest joining process as well as some of the most modern. Each process in this group has its own unique way of creating the bond at the faying surfaces. We begin our coverage with forge welding, the first welding process.

Forge Welding Forge welding is of historic significance in the development of manufacturing technology. The process dates from about 1000 B.C., when blacksmiths of the ancient world learned to join two pieces of metal (Historical Note 30.1). *Forge welding* is a welding process in which the components to be joined are heated to hot working temperatures and then forged together by hammer or other means. Considerable skill was required by the craftsmen who practiced it in order to achieve a good weld by present-day standards. The process may be of historic interest; however, it is of minor commercial importance today except for its variants, which are discussed next.

Cold Welding   *Cold welding* (CW) is a solid-state welding process accomplished by applying high pressure between clean contacting surfaces at room temperature. The faying surfaces must be exceptionally clean for CW to work, and cleaning is usually done by degreasing and wire brushing immediately before joining. Also, at least one of the metals to be welded, and preferably both, must be very ductile and free of work hardening. Metals such as soft aluminum and copper can be readily cold welded. The applied compression forces in the process result in cold working of the metal parts, reducing thickness by as much as 50%; but they also cause localized plastic deformation at the contacting surfaces, resulting in coalescence. For small parts, the forces may be applied by simple hand-operated tools. For heavier work, powered presses are required to exert the necessary force. No heat is applied from external sources in CW, but the deformation process raises the temperature of the work somewhat. Applications of CW include making electrical connections.

Roll Welding   Roll welding is a variation of either forge welding or cold welding, depending on whether external heating of the workparts is accomplished prior to the process. *Roll welding* (ROW) is a solid-state welding process in which pressure sufficient to cause coalescence is applied by means of rolls, either with or without external application of heat. The process is illustrated in Figure 31.26. If no external heat is supplied, the process is called *cold roll welding*; if heat is supplied, the term *hot roll welding* is used. Applications of roll welding include cladding stainless steel to mild or low alloy steel for corrosion resistance, making bimetallic strips for measuring temperature, and producing "sandwich" coins for the U.S. mint.

Hot Pressure Welding   *Hot pressure welding* (HPW) is another variation of forge welding in which coalescence occurs from the application of heat and pressure sufficient to cause considerable deformation of the base metals. The deformation disrupts the surface oxide film, thus leaving clean metal to establish a good bond between the two parts. Time must be allowed for diffusion to occur across the faying surfaces. The operation is usually carried out in a vacuum chamber or in the presence of a shielding medium. Principal applications of HPW are in the aerospace industry.

Diffusion Welding   *Diffusion welding* (DFW) is a solid-state welding process that results from the application of heat and pressure, usually in a controlled atmosphere, with sufficient time allowed for diffusion and coalescence to occur. Temperatures are well below the melting points of the metals (about 0.5 T_m is the maximum), and plastic deformation at the surfaces is minimal. The primary mechanism of coalescence is solid-state diffusion, which involves migration of atoms across the interface between contacting surfaces. Applications of DFW include the joining of high-strength and refractory metals in the aerospace and nuclear industries. The process is used to join both similar

FIGURE 31.26   Roll welding (ROW).

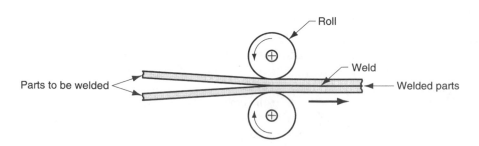

and dissimilar metals, and in the latter case a filler layer of a different metal is often sandwiched between the two base metals to promote diffusion. The time for diffusion to occur between the faying surfaces can be significant, requiring more than an hour in some applications [9].

Explosion Welding *Explosion welding* (EXW) is a solid-state welding process in which rapid coalescence of two metallic surfaces is caused by the energy of a detonated explosive. It is commonly used to bond two dissimilar metals, in particular to clad one metal on top of a base metal over large areas. Applications include production of corrosion-resistant sheet and plate stock for making processing equipment in the chemical and petroleum industries. The term *explosion cladding* is used in this context. No filler metal is used in EXW, and no external heat is applied. Also, no diffusion occurs during the process (the time is too short). The nature of the bond is metallurgical, in many cases combined with a mechanical interlocking that results from a rippled or wavy interface between the metals.

The process for cladding one metal plate on another can be described with reference to Figure 31.27. In this setup, the two plates are in a parallel configuration, separated by a certain gap distance, with the explosive charge above the upper plate, called the *flyer plate*. A buffer layer (e.g., rubber, plastic) is often used between the explosive and the flyer plate to protect its surface. The lower plate, called the *backer* metal, rests on an anvil for support. When detonation is initiated, the explosive charge propagates from one end of the flyer plate to the other, caught in the stop-action view shown in Figure 31.27 (2). One of the difficulties in comprehending what happens in EXW is the common misconception that an explosion occurs instantaneously; it is actually a progressive reaction, although admittedly very rapid—propagating at rates as high as 8,500 m/s (28,000 ft/s). The resulting high-pressure zone propels the flyer plate to collide with the backer metal progressively at high velocity, so that it takes on an angular shape as the explosion advances, as illustrated in our sketch. The upper plate remains in position in the region where the explosive has not yet detonated. The high-speed collision, occurring in a progressive and angular fashion as it does, causes the surfaces at the point of contact to become fluid, and any surface films are expelled forward from the apex of the angle. The colliding surfaces are thus chemically clean, and the fluid behavior of the metal, which involves some interfacial melting, provides intimate contact between the surfaces, leading to metallurgical bonding. Variations in collision velocity and impact angle during the process can result in a wavy or rippled interface between the two metals. This kind of

FIGURE 31.27 Explosive welding (EXW): (1) setup in the parallel configuration, and (2) during detonation of the explosive charge.

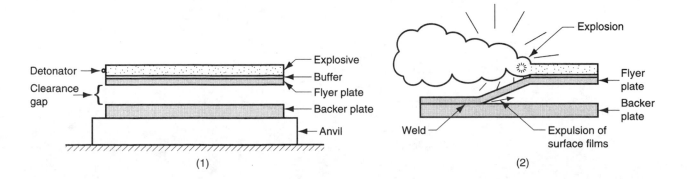

interface strengthens the bond because it increases the contact area and tends to mechanically interlock the two surfaces.

Friction Welding   Friction welding is a widely used commercial process, amenable to automated production methods. The process was developed in the (former) Soviet Union and introduced into the United States around 1960. *Friction welding* (FRW) is a solid-state welding process in which coalescence is achieved by frictional heat combined with pressure. The friction is induced by mechanical rubbing between the two surfaces, usually by rotation of one part relative to the other, to raise the temperature at the joint interface to the hot working range for the metals involved. Then the parts are driven toward each other with sufficient force to form a metallurgical bond. The sequence is portrayed in Figure 31.28 for welding two cylindrical parts, the typical application. The axial compression force upsets the parts and a flash is produced by the material displaced. Any surface films that may have been on the contacting surfaces are expunged during the process. The flash must be subsequently trimmed (e.g., by turning) to provide a smooth surface in the weld region. When properly carried out, no melting occurs at the faying surfaces. No filler metal, flux, or shielding gases are normally used.

Nearly all FRW operations use rotation to develop the frictional heat for welding. There are two principal drive systems, distinguishing two types of FRW: (1) continuous-drive friction welding, and (2) inertia friction welding. In *continuous-drive friction welding,* one part is driven at a constant rotational speed and forced into contact with the stationary part at a certain force level so that friction heat is generated at the interface. When the proper hot working temperature has been reached, braking is applied to stop the rotation abruptly, and simultaneously the pieces are forced together at forging pressures. In *inertia friction welding,* the rotating part is connected to a flywheel, which is brought up to a predetermined speed. Then the flywheel is disengaged from the drive motor, and

FIGURE 31.28   Friction welding (FRW): (1) rotating part, no contact; (2) parts brought into contact to generate friction heat; (3) rotation stopped and axial pressure applied; and (4) weld created.

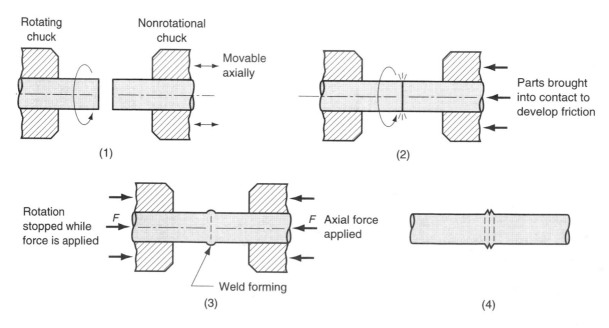

the parts are forced together. The kinetic energy stored in the flywheel is dissipated in the form of friction heat to cause coalescence at the abutting surfaces. The total cycle for these operations is about 20 seconds.

Machines used for friction welding have the appearance of an engine lathe. They require a powered spindle to turn one part at high speed, and a means of applying an axial force between the rotating part and the nonrotating part. With its short cycle times, the process lends itself to mass production. It is applied in the welding of various shafts and tubular parts in industries such as automotive, aircraft, farm equipment, petroleum, and natural gas. The process yields a narrow heat-affected zone and can be used to join dissimilar metals. However, at least one of the parts must be rotational, flash must usually be removed, and upsetting reduces the part lengths (which must be taken into consideration in product design).

Ultrasonic Welding   *Ultrasonic welding* (USW) is a solid-state welding process in which two components are held together under modest clamping force, and oscillatory shear stresses of ultrasonic frequency are applied to the interface to cause coalescence. The operation is illustrated in Figure 31.29 for lap welding, the typical application. The oscillatory motion between the two parts breaks down any surface films to allow intimate contact and strong metallurgical bonding between the surfaces. Although heating of the contacting surfaces occurs due to interfacial rubbing and plastic deformation, the resulting temperatures are well below the melting point. No filler metals, fluxes, or shielding gases are required in USW.

The oscillatory motion is transmitted to the upper workpart by means of a *sonotrode,* which is coupled to an ultrasonic transducer. This device converts electrical power into high-frequency vibratory motion. Typical frequencies used in USW are 15 to 75 kHz, with amplitudes of 0.018 to 0.13 mm (0.0007–0.005 in.). Clamping pressures are well below those used in cold welding and produce no significant plastic deformation between the surfaces. Welding times under these conditions are less than 1 sec.

USW operations are generally limited to lap joints on soft materials such as aluminum and copper. Welding harder materials causes rapid wear of the sonotrode contacting the upper workpart. Workparts should be relatively small, and welding of thicknesses less than 3 mm (1/8 in.) is the typical case. Applications include wire terminations and splicing in electrical and electronics industries (eliminates the need for soldering), assembly of aluminum sheet-metal panels, welding of tubes to sheets in solar panels, and other tasks in small-parts assembly.

FIGURE 31.29   Ultrasonic welding (USW): (a) general setup for a lap joint; and (b) close-up of weld area.

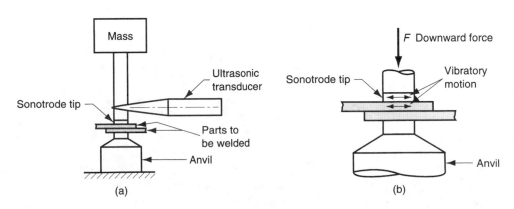

31.6 WELD QUALITY

The purpose of any welding process is to join two or more components into a single structure. The physical integrity of the structure thus formed depends on the quality of the weld. Our discussion of weld quality deals primarily with arc welding, the most widely used welding process and the one for which the quality issue is the most critical and complex.

31.6.1 Residual Stresses and Distortion

The rapid heating and cooling in localized regions of the work during fusion welding, especially arc welding, result in thermal expansion and contraction that cause residual stresses in the weldment. These stresses, in turn, cause distortion and warpage of the welded assembly.

The situation in welding is complicated because (1) the heating is very localized, (2) melting of the base metals occurs in these local regions, and (3) the location of heating and melting is in motion (at least in arc welding). Consider, for example, the butt welding of two plates by an AW operation as shown in Figure 31.30(a). The operation begins at one end and travels to the opposite end. As it proceeds, a molten pool is formed from the base metal (and filler metal, if used) that quickly solidifies behind the moving arc. The portions of the work immediately adjacent to the weld bead become extremely hot and expand, while portions removed from the weld remain relatively cool. The weld pool quickly solidifies in the cavity between the two parts, and as it and the surrounding metal cool and contract, shrinkage occurs across the width of the weldment, as seen in Figure 31.30(b). The weld seam is left in residual tension, and reactionary compressive stresses are set up in regions of the parts away from the weld. Residual stresses and shrinkage also occurs along the length of the weld bead. Since the outer regions of the

FIGURE 31.30 (a) Butt welding two plates; (b) shrinkage across the width of the welded assembly; (c) transverse and longitudinal residual stress pattern; and (d) likely warpage in the welded assembly.

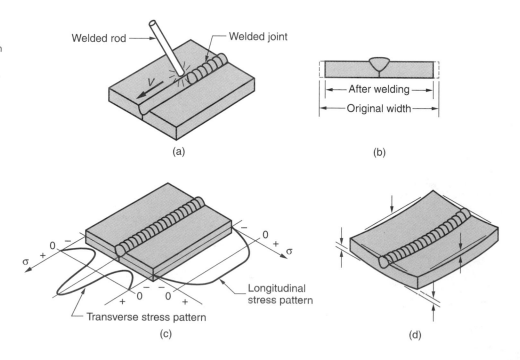

base parts have remained relatively cool and dimensionally unchanged, while the weld bead has solidified at very high temperatures and then contracted, residual tensile stresses remain longitudinally in the weld bead. These transverse and longitudinal stress patterns are depicted in Figure 31.30(c). The net result of these residual stresses, transversely and longitudinally, is likely to cause warpage in the welded assembly as shown in Figure 31.30(d).

The arc-welded butt joint in our example is only one of a variety of joint types and welding operations. Thermally induced residual stresses and the accompanying distortion are a potential problem in nearly all fusion-welding processes and in certain solid-state welding operations in which significant heating takes place.

Various techniques can be employed to minimize warpage in a weldment:

> *Welding fixtures* can be used to physically restrain movement of the parts during welding.
> *Heat sinks* can rapidly remove heat from sections of the welded parts to reduce distortion.
> *Tack welding* at multiple points along the joint can create a rigid structure prior to continuous seam welding.
> *Welding conditions* (speed, amount of filler metal used, etc.) can be chosen to reduce warpage.
> The base parts can be *preheated* to reduce the level of thermal stresses experienced by the parts.
> *Stress relief* heat treatment can be performed on the welded assembly, either in a furnace for small weldments, or other methods can be used for large structures in the field.
> *Proper design* can be used for the weldment itself (see Section 31.8).

31.6.2 Welding Defects

In addition to residual stresses and distortion in the final assembly, other defects can occur in welding. Following is a brief description of each of the major categories, based on a classification in Cary [2]:

Cracks   Cracks are fracture-type interruptions either in the weld itself or in the base metal adjacent to the weld. This type is perhaps the most serious welding defect because it constitutes a discontinuity in the metal that causes significant reduction in the strength of the weldment. Several forms are defined in Figure 31.31. Welding cracks

FIGURE 31.31   Various forms of welding cracks.

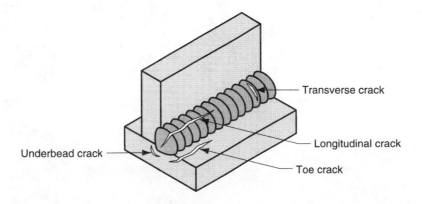

Transverse crack

Longitudinal crack

Toe crack

Underbead crack

are caused by embrittlement or low ductility of the weld and/or base metal combined with high restraint during contraction. Generally, this defect must be repaired.

Cavities   These include various porosity and shrinkage voids. ***Porosity*** consists of small voids in the weld metal formed by gases entrapped during solidification. The shapes of the voids vary between spherical (blow holes) to elongated (worm holes). Porosity usually results from inclusion of atmospheric gases, sulfur in the weld metal, or contaminants on the surfaces. ***Shrinkage voids*** are cavities formed by shrinkage during solidification. Both of these cavity-type defects are similar to defects found in castings and emphasize the close kinship between casting and welding.

Solid Inclusions   Solid inclusions are any nonmetallic solid material entrapped in the weld metal. The most common form is slag inclusions generated during the various arc-welding processes that use flux. Instead of floating to the top of the weld pool, globules of slag become encased during solidification of the metal. Another form of inclusion is metallic oxides that form during the welding of certain metals such as aluminum, which normally has a surface coating of Al_2O_3.

Incomplete Fusion   Several forms of this defect are illustrated in Figure 31.32. Also known as ***lack of fusion,*** it is simply a weld bead in which fusion has not occurred throughout the entire cross section of the joint. A related but different defect is lack of penetration. ***Penetration*** refers to the depth that the weld extends into the base metal of the joint. A ***lack of penetration*** means that fusion has not penetrated deeply enough into the root of the joint, relative to specified standards.

Imperfect Shape or Unacceptable Contour   The weld should have a certain desired profile for maximum strength, as indicated in Figure 31.33(a) for a single V-groove weld. This weld profile maximizes the strength of the welded joint and avoids incomplete fusion and lack of penetration. Some of the common defects in weld shape and contour are illustrated in Figure 31.33.

Miscellaneous Defects   The miscellaneous category includes: ***arc strikes,*** in which the welder accidentally allows the electrode to touch the base metal next to the joint, leaving a scar on the part; ***excessive spatter,*** in which drops of molten weld metal splash onto the surface of the base parts; and other defects not covered by the previous categories.

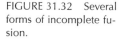

FIGURE 31.32   Several forms of incomplete fusion.

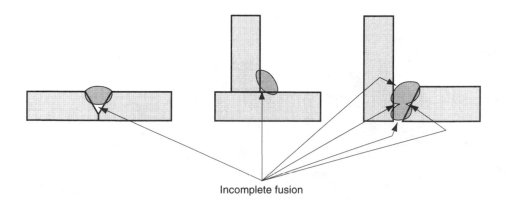

Incomplete fusion

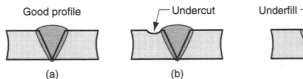
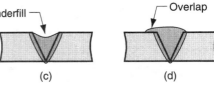

FIGURE 31.33   (a) Desired weld profile for single V-groove weld joint. Same joint but with several weld defects; (b) **undercut,** in which a portion of the base metal part is melted away; (c) **underfill,** a depression in the weld below the level of the adjacent base metal surface; and (d) **overlap,** in which the weld metal spills beyond the joint onto the surface of the base part but no fusion occurs.

31.6.3 Inspection and Testing Methods

A variety of inspection and testing methods are available to check the quality of the welded joint. Standardized procedures have been developed and specified over the years by engineering and trade societies such as the American Welding Society (AWS). For purposes of discussion, these inspection and testing procedures can be divided into three categories: (1) visual, (2) nondestructive, and (3) destructive.

Visual Inspection   Visual inspection is no doubt the most widely used welding inspection method. An inspector visually examines the weldment for (1) conformance to dimensional specifications on the part drawing, (2) warpage, and (3) cracks, cavities, incomplete fusion, and other defects described in the previous section. The welding inspector also determines if additional tests are warranted, usually in the nondestructive category. The limitation of visual inspection is that only surface defects are detectable; internal defects cannot be discovered by visual methods.

Nondestructive Evaluation   The nondestructive inspection group includes a variety of inspection methods that do not damage the specimen being evaluated. **Dye-penetrant** and **fluorescent-penetrant tests** are methods for detecting small defects such as cracks and cavities that are open to the surface. Fluorescent penetrants are highly visible when exposed to ultraviolet light. Their use is therefore a more sensitive technique than dyes.

   Magnetic particle testing is limited to ferromagnetic materials. A magnetic field is established in the subject part, and magnetic particles (e.g., iron filings) are sprinkled on the surface. Subsurface defects such as cracks and inclusions reveal themselves by distorting the magnetic field, causing the particles to be concentrated in certain regions on the surface. **Ultrasonic testing** involves the use of high-frequency sound waves (over 20 kHz) directed through the specimen. Discontinuities (e.g., cracks, inclusions, porosity) are detected by losses in sound transmission. **Radiographic testing** uses X-rays or gamma radiation to detect flaws internal to the weld metal. It provides a photographic film record of any defects.

Destructive Testing   These are methods in which the weld is destroyed either during the test or to prepare the test specimen. They include mechanical and metallurgical tests. **Mechanical tests** are similar in purpose to conventional testing methods such as tensile tests and shear tests (Chapter 3). The difference is that the test specimen is a weld joint. Figure 31.34 presents a sampling of the mechanical tests used in

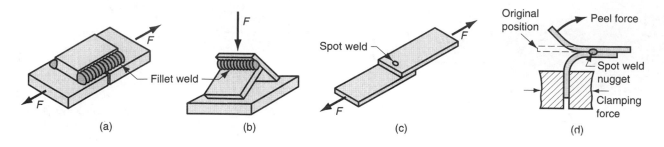

FIGURE 31.34   Mechanical tests used in welding: (a) tension–shear test of arc weldment, (b) fillet break test, (c) tension–shear test of spot weld, (d) peel test for spot weld.

welding. ***Metallurgical tests*** involve the preparation of metallurgical specimens of the weldment to examine such features as metallic structure, defects, extent and condition of heat affected zone, presence of other elements, and similar phenomena.

31.7 WELDABILITY

Weldability is the capacity of a metal or combination of metals to be welded into a suitably designed structure, and for the resulting weld joint(s) to possess the required metallurgical properties to perform satisfactorily in the intended service. Good weldability is characterized by the ease with which the welding process is accomplished, absence of weld defects, and acceptable strength, ductility, and toughness in the welded joint.

Factors that affect weldability include (1) welding process, (2) base metal properties, (3) filler metal, and (4) surface conditions. The welding process is significant. Some metals or metal combinations that can be readily welded by one process are difficult to weld by others. For example, stainless steel can be readily welded by most AW processes, but is considered a difficult metal for oxyfuel welding.

Properties of the base metal affect welding performance. Important properties include melting point, thermal conductivity, and coefficient of thermal expansion. One might think that a lower melting point would mean easier welding. However, some metals melt too easily for good welding (e.g., aluminum). Metals with high thermal conductivity tend to transfer heat away from the weld zone, which can make them hard to weld (e.g., copper). High thermal expansion and contraction in the metal causes distortion problems in the welded assembly.

Dissimilar metals pose special problems in welding when their physical and/or mechanical properties are substantially different. Differences in melting temperature are an obvious problem. Differences in strength or coefficient of thermal expansion may result in high residual stresses that can lead to cracking. If a filler metal is used, it must be compatible with the base metal(s). In general, elements mixed in the liquid state that form a solid solution upon solidification will not cause a problem. Embrittlement in the weld joint may occur if the solubility limits are exceeded.

Surface conditions of the base metals can adversely affect the operation. For example, moisture can result in porosity in the fusion zone. Oxides and other solid films on the metal surfaces can prevent adequate contact and fusion from occurring.

31.8 DESIGN CONSIDERATIONS IN WELDING

If an assembly is to be permanently welded, the designer must follow certain guidelines (compiled from Bralla [1], Cary [2], and other sources):

➤ **Design for welding**—The most basic guideline is that the product should be designed from the start as a welded assembly, and not as a casting or forging or other formed shape.

➤ **Minimum parts**—Welded assemblies should consist of the fewest number of parts possible. For example, it is usually more cost efficient to perform simple bending operations on a part than to weld an assembly from flat plates and sheets.

The following guidelines apply to arc welding:

➤ **Good fit-up of parts** to be welded is important to maintain dimensional control and minimize distortion. Machining is sometimes required to achieve satisfactory fit-up.

➤ The assembly must provide access room to allow the welding gun to reach the welding area.

➤ Whenever possible, design of the assembly should allow **flat welding** to be performed, since this is the fastest and most convenient welding position. The possible welding positions are defined in Figure 31.35. The overhead position is the most difficult.

The following design guidelines apply to resistance spot welding:

➤ Low-carbon sheet steel up to 3.2 mm (0.125 in.) is the ideal metal for resistance spot welding.

➤ Additional strength and stiffness can be obtained in large flat sheet metal components by (1) spot welding reinforcing parts into them, or (2) forming flanges and embossments into them.

➤ The spot welded assembly must provide access for the electrodes to reach the welding area.

➤ Sufficient overlap of the sheet metal parts is required for the electrode tip to make proper contact in spot welding. For example, for low-carbon sheet steel, the overlap distance should range from about six times stock thickness for thick sheets of 3.2 mm (0.125 in.) to about 20 times thickness for thin sheets, such as 0.5 mm (0.020 in.).

FIGURE 31.35 Welding positions (defined here for groove welds): (a) flat, (b) horizontal, (c) vertical, and (d) overhead.

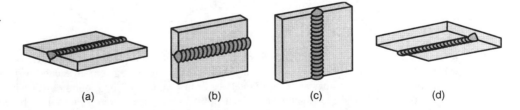

(a)   (b)   (c)   (d)

REFERENCES

[1] Bralla, J. G. (editor in chief). *Design for Manufacturability Handbook.* 2nd. ed. McGraw-Hill, New York, 1998.

[2] Cary, H. B. *Modern Welding Technology.* 4th ed. Prentice-Hall, Inc., Upper Saddle River, N.J., 1997.

[3] Galyen, J., Sear, G., and Tuttle, C. A. *Welding, Fundamentals and Procedures.* Prentice-Hall, Inc., Upper Saddle River, N.J., 1985.

[4] Messler, R. W., Jr. *Principles of Welding: Processes, Physics, Chemistry, and Metallurgy.* Wiley, New York, 1999.

[5] *Metals Handbook.* 9th ed. Volume 6: *Welding, Brazing, and Soldering.* American Society for Metals, Metals Park, Ohio, 1983.

[6] Rich, T., and Roberts, R. "The Forge Phase of Friction Welding." *Welding Journal.* March 1971.

[7] Stout, R. D., and Doty, W. D. *Weldability of Steels.* 3rd ed. Welding Research Council, New York, 1978.

[8] *Welding Handbook.* 8th. ed. Volume 1: *Welding Technology.* American Welding Society, Miami, Fla., 1987.

[9] *Welding Handbook.* 8th. ed. Vol. 2: *Welding Processes.* American Welding Society, Miami, Fla., 1991.

[10] Wick, C., and Veilleux, R. F. *Tool and Manufacturing Engineers Handbook.* 4th. ed. Vol. IV: *Quality Control and Assembly.*

REVIEW QUESTIONS

31.1. Name the principal groups of processes included in fusion welding.

31.2. What is the fundamental feature that distinguishes fusion welding from solid-state welding?

31.3. Define what an *electrical arc* is.

31.4. What does *arc time* mean?

31.5. Electrodes in arc welding are divided into two categories. Name and define the two types.

31.6. What are the two basic methods of arc shielding?

31.7. Why is the heat transfer efficiency greater in arc-welding processes that utilize consumable electrodes?

31.8. Describe the *shielded-metal arc-welding* (SMAW) process.

31.9. Why is the SMAW process difficult to automate?

31.10. Describe *submerged arc welding* (SAW).

31.11. Describe the *electrogas-welding* (EGW) process and identify its major application.

31.12. Why are the temperatures much higher in plasma arc welding than in other AW processes?

31.13. Define *resistance welding.*

31.14. What are the desirable properties of a metal that would provide good weldability for resistance welding?

31.15. Describe the sequence of steps in the cycle of a resistance spot-welding operation.

31.16. What is *resistance projection welding?*

31.17. Describe *cross-wire welding.*

31.18. Why is oxyacetylene welding favored over other oxyfuel-welding processes?

31.19. Define *pressure gas welding.*

31.20. Electron beam welding has a significant disadvantage in high-production applications. What is that disadvantage?

31.21. Laser beam welding and electron beam welding are often compared because they both produce very-high-power densities. LBM has certain advantages over EBM. What are they?

31.22. There are several modern-day variations of forge welding, the original welding process. Name the variations.

31.23. There are two basic types of friction welding. Describe and distinguish the two types.

31.24. What is a *sonotrode* in ultrasonic welding?

31.25. Distortion (warpage) is a serious problem in fusion welding, particularly arc welding. What measures can be taken to reduce the incidence and extent of distortion?

31.26. What are some of the important welding defects?

31.27. What are the three basic categories of inspection and testing techniques used for weldments? Name some typical inspections and/or tests in each category.

31.28. Identify the factors that affect weldability.

31.29. What are some of the design guidelines for weldments that are fabricated by arc welding?

MULTIPLE CHOICE QUIZ

There is a total of 27 correct answers in the following multiple choice questions (some questions have multiple answers that are correct). To attain a perfect score on the quiz, all correct answers must be given, since each correct answer is worth 1 point. For each question, each omitted answer or wrong answer reduces the score by 1 point, and each additional answer beyond the number of answers required reduces the score by 1 point. Percentage score on the quiz is based on the total number of correct answers.

31.1. The feature that distinguishes fusion welding processes from solid-state welding is that melting of the faying surfaces occurs during fusion welding: (a) true or (b) false.

31.2. Which of the following processes (more than one) are classified as fusion welding (more than one)? (a) electrogas welding, (b) electron beam welding, (c) explosive welding, and (d) percussion welding.

31.3. Which of the following processes (more than one) are classified as fusion welding (more than one)? (a) diffusion welding, (b) friction welding, (c) pressure gas welding, and (d) RSW.

31.4. Which of the following processes (more than one) are classified as solid-state welding (more than one)? (a) friction welding, (b) resistance spot welding, (c) roll welding, (d) Thermit welding, and (e) upset welding.

31.5. Which of the following processes (more than one) are classified as solid-state welding (more than one)? (a) CW, (b) HPW, (c) LBW, and (d) OAW.

31.6. An electric arc is a discharge of current across a gap in an electrical circuit. The arc is sustained in AW processes by the transfer of molten metal across the gap between the electrode and the work: (a) true or (b) false.

31.7. Which one of the following arc welding processes uses a nonconsumable electrode? (a) FCAW, (b) GMAW, (c) GTAW, or (d) SMAW.

31.8. MIG welding is a term sometimes applied when referring to which one of the following processes? (a) FCAW, (b) GMAW, (c) GTAW, or (d) SMAW.

31.9. "Stick" welding is a term sometimes applied when referring to which one of the following processes? (a) FCAW, (b) GMAW, (c) GTAW, or (d) SMAW.

31.10. Which one of the following AW processes uses an electrode consisting of continuous consumable tubing

containing flux and other ingredients in its core? (a) FCAW, (b) GMAW, (c) GTAW, or (d) SMAW.

31.11. Which one of the following AW processes produces the highest temperatures? (a) CAW, (b) PAW, (c) SAW, or (d) TIG.

31.12. Shielding gases used for welding do not include which of the following (more than one)? (a) argon, (b) carbon monoxide, (c) helium, (d) hydrogen, and (e) nitrogen.

31.13. Resistance-welding processes make use of the heat generated by electrical resistance to achieve fusion of the two parts to be joined; no pressure is used in these processes, and no filler metal is added: (a) true or (b) false.

31.14. Metals that are easiest to weld in resistance welding are ones that have low resistivities since low resistivity assists in the flow of electrical current: (a) true or (b) false.

31.15. Oxyacetylene welding is the most widely used oxyfuel welding process because acetylene mixed with an equal volume of air burns hotter than any other commercially available fuel: (a) true or (b) false.

31.16. The term *laser* stands for "light actuated system for effective reflection": (a) true or (b) false.

31.17. Which of the following solid-state welding processes applies heat from an external source (more than one)? (a) diffusion welding, and (b) forge welding, (c) friction welding, and (d) ultrasonic welding.

31.18. The term ***weldability*** takes into account not only the ease with which a welding operation can be performed, but also the quality of the resulting weld: (a) true or (b) false.

31.19. Copper is a relatively easy metal to weld because its thermal conductivity is high: (a) true or (b) false.

PROBLEMS
Arc Welding

31.1. A SMAW operation is accomplished in a work cell using a fitter and a welder. The fitter takes 5.5 min to place the unwelded components into the welding fixture at the beginning of the work cycle, and 2.5 min to unload the completed weldment at the end of the cycle. The total length of the several weld seams to be made is 2000 mm, and the travel speed used by the welder averages 400 mm/min. Every 750 mm of weld length, the welding stick must be changed, which takes 0.8 min. While the fitter is working, the welder is idle (resting); and while the welder is working, the fitter is idle. (a) Determine the average arc time in this welding cycle. (b) How much improvement in arc time would result if the welder used FCAW (manually operated), given that the spool of flux-cored weld wire must be changed every

five weldments, and it takes the welder 5.0 min to accomplish the change. (c) What are the production rates for these two cases (weldments completed per hour)?

31.2. In the previous problem, suppose an industrial robot cell were installed to replace the welder. The cell consists of the robot (using GMAW instead of SMAW or FCAW), two welding fixtures, and the fitter who loads and unloads the parts. With two fixtures, fitter and robot work simultaneously, the robot welding at one fixture while the fitter unloads and loads at the other. At the end of each work cycle, they switch places. The electrode wire spool must be changed every five workparts, which task requires 5.0 minutes and is accomplished by the fitter. Determine (a) arc time and (b) production rate for this work cell.

31.3. A shielded-metal arc-welding operation is performed on steel at $E = 30$ volts and $I = 225$ amps. The heat transfer efficiency $f_1 = 0.85$ and melting efficiency $f_2 = 0.75$. The unit melting energy for steel = 10.2 J/mm^3. Solve for (a) the rate of heat generation at the weld and (b) the volume rate of metal welded.

31.4. A GTAW operation is performed on stainless steel, whose unit melting energy $U_m = 9.3$ J/mm^3. The conditions are $E = 25$ volts, $I = 125$ amps, $f_1 = 0.65$, and $f_2 = 0.70$. If filler metal wire of 3.0 mm diameter is added to the operation, and the final weld bead is composed of equal volumes of filler and base metal. If the travel speed in the operation $v = 5$ mm/sec, determine (a) cross-sectional area of the weld bead, and (b) the feed rate (in mm/sec) at which the filler wire must be supplied.

31.5. A flux-cored arc welding operation is performed to butt weld two aluminum plates together, using the following conditions: $E = 20$ volts and $I = 250$ amps. The cross-sectional area of the weld seam = 80 mm^2 and the melting efficiency of the aluminum is assumed to be $f_2 = 0.5$. Using tabular data and equations given in this and the preceding chapter, determine the likely value for travel speed v in the operation.

31.6. A GMAW test is performed to determine the value of melting efficiency f_2 for a certain metal and operation. The welding conditions are $E = 25$ volts, $I = 125$ amps, and heat transfer efficiency is assumed to be $f_1 = 0.90$, a typical value for GMAW. The rate at which the filler metal is added to the weld is 0.50 in.^{3} per minute, and measurements indicate that the final weld bead consists of 57% filler metal and 43% base metal. The unit melting energy for the metal is known to be 75 Btu/in.^{3} (a) Find f_2. (b) What is the travel speed if the cross-sectional area of the weld bead is 0.05 in.^{2}?

31.7. A continuous weld is to be made around the circumference of a round steel tube of diameter = 6.0 ft, using a submerged arc-welding operation under automatic control at a voltage of 25 and current of 300 A. The tube is slowly rotated under a stationary welding head. The heat-transfer efficiency for SAW is $f_1 = 0.95$ and the assumed melting efficiency $f_2 = 0.7$. The cross-sectional area of the weld bead is 0.12 in.^{2} If the unit melting energy for the steel = 150 Btu/in.^{3}, determine (a) the rotational speed of tube and (b) the time required to complete the weld.

Resistance Welding

31.8. A RSW operation is used to make a series of spot welds between two pieces of aluminum, each 2.0 mm thick. The unit melting energy for aluminum $U_m = 2.90$ J/mm^3. Welding current $I = 6,000$ amps, time duration = 0.15 sec. Assume that the resistance = 75 micro-ohms. The resulting weld nugget measures 5.0 mm in diameter by 2.5 mm thick. How much of the total energy generated is used to form the weld nugget?

31.9. The unit melting energy for a certain sheet metal to be spot welded is $U_m = 10.0$ J/mm^3. The thickness of each of the two sheets to be welded is 3.0 mm. To achieve required strength, it is desired to form a weld nugget that is 6.0 mm in diameter and 4.5 mm thick. The weld duration will be set at 0.2 sec. If it is assumed that the electrical resistance between the surfaces is 125 micro-ohms, and that only one-third of the electrical energy generated will be used to form the weld nugget (the rest being dissipated into the work), determine the minimum current level required in this operation.

31.10. A resistance spot welding operation is performed on two pieces of 0.040 in. thick sheet steel (low carbon). The unit melting energy for steel = 150 Btu/in.^{3} Process parameters are current = 9500 A and time duration = 0.17 sec. This results in a weld nugget of diameter = 0.19 in. and thickness = 0.060 in. Assume the resistance = 100 micro-ohms. Determine (a) the average power density in the interface area defined by the weld nugget, and (b) the proportion of energy generated that went into formation of the weld nugget.

31.11. A resistance seam welding operation is performed on two pieces of 2.5-mm-thick austenitic stainless steel to fabricate a container. The weld current in the operation is 10,000 amps, the weld duration $t = 0.3$ sec, and the resistance at the interface is 75 micro-ohms. Continuous motion welding is used, with 200 mm diameter electrode wheels. The individual weld nuggets formed in this RSEW operation have dimensions: diameter = 6 mm and thickness = 3 mm (assume the weld nuggets are disc-shaped). These weld nuggets must be contiguous to form a sealed seam. The power unit driving the process requires an off-time between spot welds of s. Given these conditions, determine (a) the unit melting energy of stainless steel using the methods of the previous chapter, (b) the proportion of energy generated that goes into the formation of each weld nugget, and (c) the rotational speed of the electrode wheels.

31.12. Suppose in the previous problem that a roll spot-welding operation is performed instead of seam welding. The interface resistance increases to 100 micro-ohms, and the center-to-center separation between weld nuggets is 25 mm. Given the conditions from the previous problem, with the changes noted here, determine (a) the proportion of energy generated that goes into the formation of each weld nugget, and (b) the rotational speed of the electrode wheels. (c) At this higher rotational speed, how much does the wheel move during the current on-time, and might this have the effect of elongating the weld nugget (making it elliptical rather than round)?

31.13. An experimental power source for spot welding is designed to deliver current as a ramp function of time: $I = 100,000 \, t$, where I = amp and t = s. At the end of the power-on time, the current is stopped abruptly. The sheet metal being spot welded is low carbon steel whose unit melting energy = 10 J/mm^3. The resistance R = 85 micro-ohms. The desired weld nugget size is diameter = 4 mm and thickness = 2 mm (assume a disc-shaped nugget). It is assumed that one-fourth of the energy generated from the power source will be used to form the weld nugget. Determine the power-on time the current must be applied in order to perform this spot-welding operation.

Oxyfuel Welding

31.14. Suppose in Example 31.3 in the text that the fuel used in the welding operation is MAPP instead of acetylene, and the proportion of heat concentrated in the 9 mm circle is 60% instead of 75%. Compute (a) rate of heat liberated during combustion, (b) rate of heat transferred to the work surface, and (c) average power density in the circular area.

31.15. An oxyacetylene torch supplies 10 ft^3 of acetylene per hour and an equal volume rate of oxygen for an OAW operation on 3/16 in. steel. Heat generated by combustion is transferred to the work surface with an efficiency $f_1 = 0.25$. If 75% of the heat from the flame is concentrated in a circular area on the work surface whose diameter = 0.375 in., find (a) rate of heat liberated during combustion, (b) rate of heat transferred to the work surface, and (c) average power density in the circular area.

Electron Beam Welding

31.16. The voltage in an EBW operation = 50 kV and the beam current = 65 milliamp. The electron beam is focused on a circular area that is 0.3 mm in diameter. The heat transfer efficiency $f_1 = 0.85$. Calculate the average power density in the area in watt/mm^2.

31.17. An electron beam welding operation is to be accomplished to butt weld two sheet metal parts that are 3.0 mm thick. The unit melting energy = 5.0 J/mm^3. The weld joint is to be 0.35 mm wide, so that the cross section of the fused metal is 0.35 mm by 3.0 mm. If accelerating voltage = 25 kV, beam current = 30 milliamp, heat transfer efficiency $f_1 = 0.85$, and melting efficiency $f_2 = 0.75$, determine the travel speed at which this weld can be made along the seam.

31.18. An electron beam welding operation uses the following process parameters: accelerating voltage = 25 kV, beam current = 100 milliamp, and the circular area on which the beam is focused has a diameter = 0.020 in. If the heat transfer efficiency $f_1 = 90\%$, determine the average power density in the area in Btu/sec in.^{2}.

32

BRAZING, SOLDERING, AND ADHESIVE BONDING

CHAPTER CONTENTS

32.1   Brazing
        32.1.1   Brazed Joints
        32.1.2   Fluxes and Filler Metals
        32.1.3   Brazing Methods
32.2   Soldering
        32.2.1   Joint Designs in Soldering
        32.2.2   Solders and Fluxes
        32.2.3   Soldering Methods
32.3   Adhesive Bonding
        32.3.1   Joint Design
        32.3.2   Adhesive Types
        32.3.3   Application Technology

In this chapter, we consider three joining processes that are similar to welding in certain respects: brazing, soldering, and adhesive bonding.

Brazing and soldering both use filler metals to join and bond two (or more) metal parts to provide a permanent joint. It is difficult, although not impossible, to disassemble the parts after a brazed or soldered joint has been made. In the spectrum of joining processes, brazing and soldering lie between fusion welding and solid-state welding. A filler metal is added in brazing and soldering as in most fusion welding operations; however, no melting of the base metals occurs, which is similar to solid-state welding. Despite these incongruities, brazing and soldering are generally considered to be distinct from welding. Brazing and soldering are attractive compared to welding under circumstances where (1) the metals have poor weldability, (2) dissimilar metals are to be joined, (3) the intense heat of welding may damage the components being joined, (4) the geometry of the joint does not lend itself to any of the welding methods, and/or (5) high strength is not a requirement.

Adhesive bonding shares certain features in common with brazing and soldering. It utilizes the forces of attachment between a filler material and two closely spaced surfaces to bond the parts. The differences are that the filler material in adhesive bonding is not metallic, and the joining process is carried out at room temperature or only modestly above.

32.1 BRAZING

Brazing is a joining process in which a filler metal is melted and distributed by capillary action between the faying surfaces of the metal parts being joined. No melting of the base metals occurs in brazing; only the filler melts. In brazing, the filler metal (also called the *brazing metal*), has a melting temperature (liquidus) that is above 450°C (840°F) but below the melting point (solidus) of the base metal(s) to be joined. If the joint is properly designed and the brazing operation has been properly performed, the brazed joint will be stronger than the filler metal out of which it has been formed upon solidification. This rather remarkable result is due to the small part clearances used in brazing, the metallurgical bonding that occurs between base and filler metals, and the geometric constrictions that are imposed on the joint by the base parts.

Brazing has several advantages compared to welding: (1) any metals can be joined, including dissimilar metals; (2) certain brazing methods can be performed quickly and consistently, thus permitting high cycle rates and automated production; (3) some methods allow multiple joints to be brazed simultaneously; (4) brazing can be applied to join thin-walled parts that cannot be welded; (5) in general, less heat and power are required than in fusion welding; (6) problems with the heat-affected zone (HAZ) in the base metal near the joint are reduced; and (7) joint areas that are inaccessible by many welding processes can be brazed, since capillary action draws the molten filler metal into the joint.

There are some limitations to brazing: (1) joint strength is generally less than that of a welded joint; (2) although strength of a good brazed joint is greater than that of the filler metal, it is likely to be less than that of the base metals; (3) high service temperatures may weaken a brazed joint; and (4) the color of the metal in the brazed joint may not match the color of the base metal parts, a possible aesthetic disadvantage.

Brazing as a production process is widely used in a variety of industries, including automotive (e.g., joining tubes and pipes), electrical equipment (e.g., joining wires and cables), cutting tools (e.g., brazing cemented carbide inserts to shanks), and jewelry making. In addition, the chemical processing industry and plumbing and heating contractors join metal pipes and tubes by brazing. The process is used extensively for repair and maintenance work in nearly all industries.

32.1.1 Brazed Joints

Brazed joints are of two types: butt and lap. However, the two types have been adapted for the brazing process in several ways. The conventional butt joint provides a limited area for brazing, thus jeopardizing the strength of the joint. To increase the faying areas in brazed joints, the mating parts are often scarfed or stepped or otherwise altered, as shown in Figure 32.1. Of course, additional processing is usually required in the making of the parts for these special joints. One of the particular difficulties associated with a scarfed joint is the problem of maintaining the alignment of the parts before and during brazing.

Lap joints are more widely used in brazing, since they can provide a relatively large interface area between the parts. An overlap of at least three times the thickness of the thinner part is generally considered good design practice. Some adaptations of the lap joint for brazing are illustrated in Figure 32.2. An advantage of brazing over welding in lap joints is that the filler metal is bonded to the base parts throughout the entire interface area between the parts, rather than only at the edges (as in fillet welds made by arc welding) or at discrete spots (as in resistance spot welding).

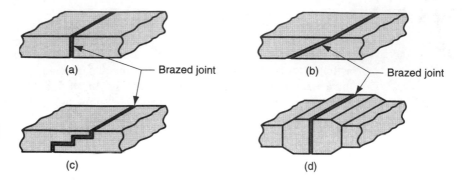

FIGURE 32.1 (a) Conventional butt joint, and adaptations of the butt joint for brazing: (b) scarf joint, (c) stepped butt joint, (d) increased cross section of the part at the joint.

Clearance between mating surfaces of the base parts is important in brazing. The clearance must be large enough so as not to restrict molten filler metal from flowing throughout the entire interface. Yet if the joint clearance is too great, capillary action will be reduced and there will be areas between the parts where no filler metal is present. Joint strength is affected by clearance, as depicted in Figure 32.3. There is an optimum clearance value at which joint strength is maximized. The issue is complicated by the fact that the optimum depends on base and filler metals, joint configuration, and processing conditions. Typical brazing clearances in practice are 0.025 to 0.25 mm (0.001–0.010 in.). These values represent the joint clearance at the brazing temperature, which may be different from room temperature clearance, depending on thermal expansion of the base metal(s).

Cleanliness of the joint surfaces prior to brazing is also important. Surfaces must be free of oxides, oils, and other contaminants in order to promote wetting and capillary attraction during the process, as well as bonding across the entire interface. Chemical treatments such as solvent cleaning (Section 28.1) and mechanical treatments such as wire brushing and sand blasting (Section 28.2) are used to clean the surfaces. After cleaning and during the brazing operation, fluxes are used to maintain surface cleanliness and promote wetting for capillary action in the clearance between faying surfaces.

FIGURE 32.2 (a) Conventional lap joint, and adaptations of the lap joint for brazing: (b) cylindrical parts, (c) sandwiched parts, and (d) use of sleeve to convert butt joint into lap joint.

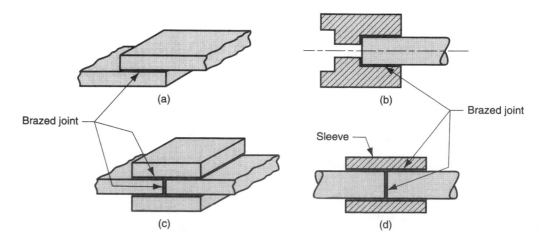

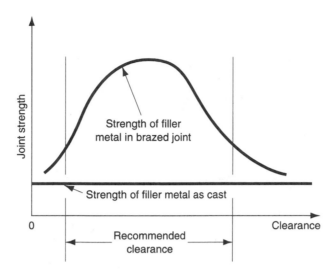

FIGURE 32.3   Joint strength as a function of joint clearance.

32.1.2 Filler Metals and Fluxes

Common ***filler metals*** used in brazing are listed in Table 32.1 along with the principal base metals on which they are typically used. To qualify as a brazing metal, the following characteristics are needed: (1) melting temperature must be compatible with base mctal, (2) surface tension in liquid phase must be low for good wettability, (3) fluidity for penetration into the interface must be high, (4) the metal must be capable of being brazed into a joint of adequate strength for the application, and (5) avoidance of chemical and physical interactions with base metal (e.g., galvanic reaction). Filler metals are applied to the brazing operation in various ways, including wire, rod, sheets and strips, powders, pastes, preformed parts made of braze metal designed to fit a particular joint configuration, and cladding on one of the surfaces to be brazed. Several of these techniques are illustrated in Figures 32.4 and 32.5. Braze metal pastes, shown in Figure 32.5, consist of filler metal powders mixed with fluid fluxes and binders.

Brazing fluxes serve a similar purpose as in welding; they dissolve, combine with, and otherwise inhibit the formation of oxides and other unwanted byproducts in the brazing process. Use of a flux does not substitute for the cleaning steps just described. Characteristics of a good flux include (1) low melting temperature, (2) low viscosity so

TABLE 32.1   Common filler metals used in brazing and the base metals on which they are used.

Filler Metal	Typical Composition	Approx. Brazing Temperature		Base Metals
		°C	(°F)	
Aluminum and silicon	90 Al, 10 Si	600	(1100)	Aluminum
Copper	99.9 Cu	1120	(2050)	Nickel copper
Copper and phosphorous	95 Cu, 5 P	850	(1550)	Copper
Copper and zinc	60 Cu, 40 Zn	925	(1700)	Steels, cast irons, nickel
Gold and silver	80 Au, 20 Ag	950	(1750)	Stainless steel, nickel alloys
Nickel alloys	Ni, Cr, others	1120	(2050)	Stainless steel, nickel alloys
Silver alloys	Ag, Cu, Zn, Cd	730	(1350)	Titanium, monel, inconel, tool steel, nickel

Compiled from [3], [4], and [8].

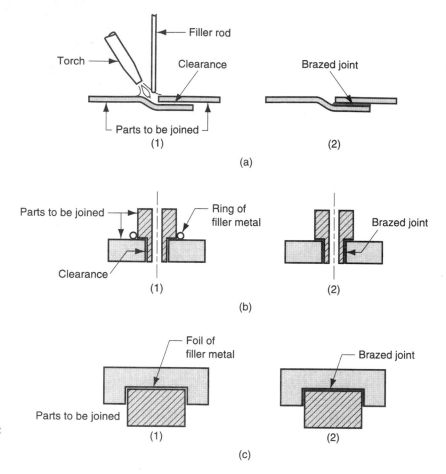

FIGURE 32.4 Several techniques for applying filler metal in brazing: (a) torch and filler rod; (b) ring of filler metal at entrance of gap; and (c) foil of filler metal between flat part surfaces. Sequence: (1) before and (2) after.

that it can be displaced by the filler metal, (3) facilitates wetting, and (4) protects the joint until solidification of the filler metal. The flux should also be easy to remove after brazing. Common ingredients for brazing fluxes include borax, borates, fluorides, and chlorides. Wetting agents are also included in the mix to reduce surface tension of the molten filler metal and to improve wettability. Forms of flux include powders, pastes, and slurries. Alternatives to using a flux are to perform the operation in a vacuum or a reducing atmosphere that inhibits oxide formation.

32.1.3 Brazing Methods

There are various methods used in brazing. Referred to as brazing processes, they are differentiated by their heating sources.

Torch Brazing In torch brazing, flux is applied to the part surfaces and a torch is used to direct a flame against the work in the vicinity of the joint. A reducing flame is typically used to inhibit oxidation. After the workpart joint areas have been heated to a suitable temperature, filler wire is added to the joint, usually in wire or rod form. Fuels used in torch brazing include acetylene, propane, and other gases, with air or oxygen. The selection of the mixture depends on heating requirements of the job. Torch brazing is often performed manually, and skilled workers must be employed to control the flame, manipulate the hand-held torches, and properly judge the temperatures; repair

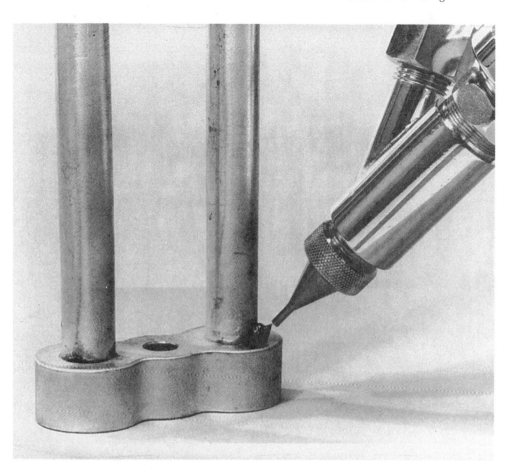

FIGURE 32.5   Application of brazing paste to joint by dispenser (courtesy of Fusion, Inc.).

work is a common application. The method can also be used in mechanized production operations, in which parts and brazing metal are loaded onto a conveyor or indexing table and passed under one or more torches.

Furnace Brazing   Furnace brazing uses a furnace to supply heat for brazing and is best suited to medium and high production. In medium production, usually in batches, the component parts and brazing metal are loaded into the furnace, heated to brazing temperature, and then cooled and removed. High-production operations use flow-through furnaces, in which parts are placed on a conveyor and are transported through the various heating and cooling sections. Temperature and atmosphere control are important in furnace brazing; the atmosphere must be neutral or reducing. Vacuum furnaces are sometimes used. Depending on the atmosphere and metals being brazed, the need for a flux may be eliminated.

Induction Brazing   Induction brazing utilizes heat from electrical resistance to a high-frequency current induced in the work. The parts are preloaded with filler metal and placed in a high-frequency AC field—the parts do not directly contact the induction coil. Frequencies range from 5 kHz to 5 MHz. High-frequency power sources tend to provide surface heating, while lower frequencies cause deeper heat penetration into the work and are appropriate for heavier sections. The process can be used to meet low- to high-production requirements.

FIGURE 32.6   Braze welding. The joint consists of braze (filler) metal; no base metal is fused in the joint.

Resistance Brazing   Heat to melt the filler metal in this process is obtained by resistance to flow of electrical current through the parts. As distinguished from induction brazing, the parts are directly connected to the electrical circuit in resistance brazing. The equipment is similar to that used in resistance welding, except that a lower power level is required for brazing. The parts with filler metal preplaced are held between electrodes while pressure and current are applied. Both induction and resistance brazing achieve rapid heating cycles and are used for relatively small parts. Induction brazing seems to be the more widely used of the two processes.

Dip Brazing   In dip brazing, either a molten salt bath or a molten metal bath accomplishes heating. In both methods, assembled parts are immersed in the baths contained in a heating pot. Solidification occurs when the parts are removed from the bath. In the *salt bath method,* the molten mixture contains fluxing ingredients and the filler metal is preloaded onto the assembly. In the *metal bath method,* the molten filler metal is the heating medium; it is drawn by capillary action into the joint during submersion. A flux cover is maintained on the surface of the molten metal bath. Dip brazing achieves fast heating cycles and can be used to braze many joints on a single part or on multiple parts simultaneously.

Infrared Brazing   This method uses heat from a high-intensity infrared lamp. Some IR lamps are capable of generating up to 5,000 W of radiant heat energy, which can be directed at the workparts for brazing. The process is slower than most of the other processes reviewed above, and is generally limited to thin sections.

Braze Welding   This process differs from the other brazing processes in the type of joint to which it is applied. As pictured in Figure 32.6, braze welding is used for filling a more conventional weld joint, such as the V-joint shown. A greater quantity of filler metal is deposited than in brazing, and no capillary action occurs. In braze welding, the joint consists entirely of filler metal; the base metal does not melt, and is therefore not fused into the joint as in a conventional fusion welding process. The principal application of braze welding is repair work.

32.2 SOLDERING

Soldering is similar to brazing and can be defined as a joining process in which a filler metal with melting point (liquidus) not exceeding 450°C (840°F) is melted and distributed by capillary action between the faying surfaces of the metal parts being joined. As in brazing, no melting of the base metals occurs, but the filler metal wets and combines with the base metal to form a metallurgical bond. Details of soldering are similar to those of brazing, and many of the heating methods are the same. Surfaces to be soldered must be precleaned so that they are free of oxides, oils, and so on. An appropriate flux must be applied to the faying surfaces, and the surfaces are heated. Filler metal, called ***solder,*** is added to the joint, which distributes itself between the closely fitting parts.

In some applications, the solder is precoated onto one or both of the surfaces—a process called ***tinning,*** irrespective of whether the solder contains any tin. Typical clearances in soldering range from 0.075 to 0.125 mm (0.003–0.005 in.), except when the surfaces are tinned, in which case a clearance of about 0.025 mm (0.001 in.) is used. After solidification, the flux residue must be removed.

As an industrial process, soldering is most closely associated with electronics assembly (Chapter 36). It is also used for mechanical joints, but not for joints subjected to elevated stresses or temperatures. Advantages attributed to soldering include: (1) low energy input relative to brazing and fusion welding; (2) variety of heating methods available; (3) good electrical and thermal conductivity in the joint; (4) capability to make airtight and liquid-tight seams for containers; and (5) easy to repair and rework.

The biggest disadvantages of soldering are: (1) low joint strength unless reinforced by mechanically means; and (2) possible weakening or melting of the joint in elevated temperature service.

32.2.1 Joint Designs in Soldering

As in brazing, soldered joints are limited to lap and butt types, although butt joints should not be used in load-bearing applications. Some of the brazing adaptations of these joints also apply to soldering, and soldering technology has added a few more variations of its own to deal with the special part geometries that occur in electrical connections. In soldered mechanical joints of sheet-metal parts, the edges of the sheets are often bent over and interlocked before soldering, as shown in Figure 32.7, to increase joint strength.

For electronics applications, the principal function of the soldered joint is to provide an electrically conductive path between two parts being joined. Other design considerations in these types of soldered joints include heat generation (from the electrical resistance of the joint) and vibration. Mechanical strength in a soldered

FIGURE 32.7
Mechanical interlocking in soldered joints for increased strength: (a) flat lock seam; (b) bolted or riveted joint; (c) copper pipe fittings—lap cylindrical joint; and (d) crimping (forming) of cylindrical lap joint.

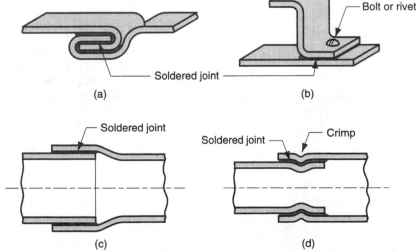

electrical connection is often achieved by deforming one or both of the metal parts to accomplish a mechanical joint between them, or by making the surface area larger to provide maximum support by the solder. Several possibilities are sketched in Figure 32.8.

32.2.2 Solders and Fluxes

Solders and fluxes are the materials used in soldering. Both are critically important in the joining process.

Solders   Most solders are alloys of tin and lead, since both metals have low melting points (see Figure 6.3). Their alloys possess a range of liquidus and solidus temperatures to achieve good control of the soldering process for a variety of applications. Lead is poisonous and its percentage is minimized in most solder compositions. Tin is chemically active at soldering temperatures and promotes the wetting action required for successful joining. In soldering copper, common in electrical connections, intermetallic compounds of copper and tin are formed that strengthen the bond. Silver and antimony are also sometimes used in soldering alloys. Table 32.2 lists various solder alloy compositions, indicating their approximate soldering temperatures and principal applications.

Soldering Fluxes   *Soldering fluxes* should do the following: (1) be molten at soldering temperatures, (2) remove oxide films and tarnish from the base part surfaces, (3) prevent oxidation during heating, (4) promote wetting of the faying surfaces, (5) be readily displaced by the molten solder during the process, and (6) leave a residue that is noncorrosive and nonconductive. Unfortunately, there is no single flux that serves all of these functions perfectly for all combinations of solder and base metals. The flux formulation must be selected for a given application.

Soldering fluxes can be classified as organic or inorganic. *Organic fluxes* are made of either rosin (i.e., natural rosin such as gum wood, which is not water-soluble) or water-soluble ingredients (e.g., alcohols, organic acids, and halogenated salts). The water-soluble type facilitates cleanup after soldering. Organic fluxes are most commonly used for electrical and electronics connections. They tend to be chemically reactive at elevated soldering temperatures but relatively noncorrosive at room temperatures. *Inorganic fluxes*

FIGURE 32.8   Techniques for securing the joint by mechanical means prior to soldering in electrical connections: (a) crimped lead wire on PC board; (b) plated through hole on PC board to maximize solder contact surface; (c) hooked wire on flat terminal; and (d) twisted wires.

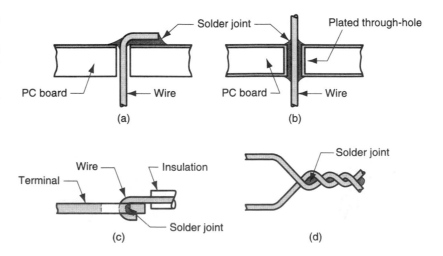

TABLE 32.2   Some common solder alloy compositions with their melting temperatures and applications.

Filler Metal	Approximate Composition	Approx. Melting Temperature		Principal Applications
		°C	(°F)	
Lead-silver	96 Pb, 4 Ag	305	(580)	Elevated temperature joints
Tin-antimony	95 Sn, 5 Sb	238	(460)	Plumbing & heating
Tin-lead	63 Sn, 37 Pb	183	(361)	Electronics[a]
	60 Sn, 40 Pb	188	(370)	Electronics
	50 Sn, 50 Pb	199	(390)	General purpose
	40 Sn, 60 Pb	207	(405)	Automobile radiators
Tin-silver	96 Sn, 4 Ag	221	(430)	Food containers
Tin-zinc	91 Sn, 9 Zn	199	(390)	Aluminum joining

Compiled from [1], [4], [7], [8].

[a]Eutectic composition—lowest melting point of tin-lead compositions.

consist of inorganic acids (e.g., muriatic acid) and salts (e.g., combinations of zinc and ammonium chlorides) and are used to achieve rapid and active fluxing where oxide films are a problem. The salts become active when melted, but are less corrosive than the acids. When solder wire is purchased with an *acid core* it is in this category.

Both organic and inorganic fluxes should be removed after soldering, but it is especially important in the case of inorganic acids to prevent continued corrosion of the metal surfaces. Flux removal is usually accomplished using water solutions except in the case of rosins, which require chemical solvents. Recent trends in industry favor water-soluble fluxes over rosins because chemical solvents used with rosins are harmful to the environment and to humans.

32.2.3 Soldering Methods

Many of the methods used in soldering are the same as those used in brazing, except that less heat and lower temperatures are required for soldering. These methods include torch soldering, furnace soldering, induction soldering, resistance soldering, dip soldering, and infrared soldering. There are other soldering methods, not used in brazing, that should be described here. These methods are hand soldering, wave soldering, and reflow soldering.

Hand Soldering   Hand soldering is performed manually using a hot soldering iron. A *bit,* made of copper, is the working end of a soldering iron. The bit (1) delivers heat to the parts being soldered, (2) melts the solder, (3) conveys molten solder to the joint, and (4) withdraws excess solder. Most modern soldering irons are heated by electrical resistance. Some are designed as fast-heating *soldering guns,* which are popular in electronics assembly for intermittent (on-off) operation. They are capable of making a solder joint in about a second.

Wave Soldering   Wave soldering is a mechanized technique that allows multiple lead wires to be soldered to a printed circuit board (PCB) as it passes over a wave of molten solder. The typical setup is one in which a PCB, on which electronic components have been placed with their lead wires extending through the holes in the board, is loaded onto a conveyor for transport through the wave-soldering equipment. The conveyor supports the PCB on its sides, so that its underside is exposed to the processing steps,

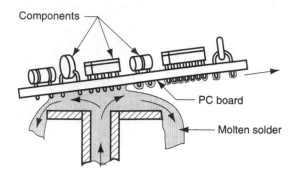

Components

PC board

Molten solder

FIGURE 32.9   Wave soldering, in which molten solder is delivered up through a narrow slot onto the underside of a printed circuit board to connect the component lead wires.

which consist of the following: (1) flux is applied using any of several methods, including foaming, spraying, or brushing; (2) preheating (using light bulbs, heating coils, and infrared devices) to evaporate solvents, activate the flux, and raise the temperature of the assembly; and (3) wave soldering, in which the liquid solder is pumped from a molten bath through a slit onto the bottom of the board to make the soldering connections between the lead wires and the metal circuit on the board. This third step is illustrated in Figure 32.9. The board is often inclined slightly, as depicted in the sketch, and a special tinning oil is mixed with the molten solder to lower its surface tension. Both of these measures help to inhibit build-up of excess solder and formation of "icicles" on the bottom of the board. Wave soldering is widely applied in electronics to produce printed circuit board assemblies (Section 36.3.2).

Reflow Soldering   This process is also widely used in electronics to assemble surface mount components to printed circuit boards (Section 36.4.2). In the process, a solder paste consisting of solder powders in a flux binder is applied to spots on the board where electrical contacts are to be made between surface mount components and the copper circuit. The components are then placed on the paste spots, and the board is heated to melt the solder, forming mechanical and electrical bonds between the component leads and the copper on the circuit board.

Heating methods for reflow soldering include vapor phase reflow and infrared reflow. In ***vapor phase reflow soldering,*** an inert fluorinated hydrocarbon liquid is vaporized by heating in an oven; it subsequently condenses on the board surface where it transfers its heat of vaporization to melt the solder paste and form solder joints on the printed circuit boards. In ***infrared reflow soldering,*** heat from an infrared lamp is used to melt the solder paste and form joints between component leads and circuit areas on the board. Additional heating methods to reflow the solder paste include the use of hot plates, hot air, and lasers.

32.3 ADHESIVE BONDING

Use of adhesives dates back to ancient times (Historical Note 32.1) and adhesive bonding was probably the first of the permanent joining methods. Today, adhesives are used in a wide range of bonding and sealing applications for joining similar and dissimilar materials such as metals, plastics, ceramics, wood, paper, and cardboard. Although well-established as a joining technique, adhesive bonding is considered a growth area among assembly technologies because of the tremendous opportunities for increased applications.

Historical Note 32.1 *Adhesive bonding.*

A dhesives date from ancient times. Carvings 3,300 years old show a glue pot and brush for gluing veneer to wood planks. The ancient Egyptians used gum from the Acacia tree for various assembly and sealing purposes. Bitumen, an asphalt adhesive, was used in ancient times as a cement and mortar for construction in Asia Minor. The Romans used pine wood tar and beeswax to caulk their ships. Glues derived from fish, stag horns, and cheese were used in the early centuries after Christ for assembling components of wood.

In more modern times, adhesives have become an important joining process. Plywood, which relies on the use of adhesives to bond multiple layers of wood, was developed around 1900. Phenol formaldehyde was the first synthetic adhesive developed, around 1910, and its primary use was in bonding of wood products such as plywood. During World War II, phenolic resins were developed for adhesive bonding of certain aircraft components. In the 1950s, epoxies were first formulated. And since the 1950s a variety of additional adhesives have been developed, including anaerobics, various new polymers, and second generation acrylics.

Adhesive bonding is a joining process in which a filler material is used to hold two (or more) closely spaced parts together by surface attachment. The filler material that binds the parts together is the *adhesive.* It is a nonmetallic substance—usually a polymer. The parts being joined are called *adherends.* Adhesives of greatest interest in engineering are *structural adhesives,* which are capable of forming strong, permanent joints between strong, rigid adherends. A large number of commercially available adhesives are cured by various mechanisms and suited to the bonding of various materials. *Curing* refers to the process by which the adhesive's physical properties are changed from a liquid to a solid, usually by chemical reaction, to accomplish the surface attachment of the parts. The chemical reaction may involve polymerization, condensation, or vulcanization. Curing is often motivated by heat and/or a catalyst, and pressure is sometimes applied between the two parts to activate the bonding process. If heat is required, the curing temperatures are relatively low, and so the materials being joined are usually unaffected—an advantage for adhesive bonding. The curing or hardening of the adhesive takes time, called *curing time* or *setting time.* In some cases this time is significant—generally a disadvantage in manufacturing.

Joint strength in adhesive bonding is determined by the strength of the adhesive itself and the strength of attachment between adhesive and each of the adherends. One of the criteria often used to define a satisfactory adhesive joint is that if a failure should occur due to excessive stresses, it occurs in one of the adherends rather than at an interface or within the adhesive itself. The strength of the attachment results from several mechanisms, all depending on the particular adhesive and adherends [3]: (1) chemical bonding, in which the adhesive unites with the adherends and forms a primary chemical bond upon hardening; (2) physical interactions, in which secondary bonding forces result between the atoms of the opposing surfaces; and (3) mechanical interlocking, in which the surface roughness of the adherend causes the hardened adhesive to become entangled or trapped in its microscopic surface asperities.

For these adhesion mechanisms to operate with best results, the following conditions must prevail: (1) surfaces of the adherend must be clean—free of dirt, oil, and oxide films that would interfere with achieving intimate contact between adhesive and adherend; special preparation of the surfaces is often required; (2) the adhesive in its initial liquid form must achieve thorough wetting of the adherend surface; and (3) it is usually helpful for the surfaces to be other than perfectly smooth—a slightly roughened surface increases the effective contact area and promotes mechanical interlocking. In addition, the joint must be designed to exploit the particular strengths of adhesive bonding and avoid its limitations.

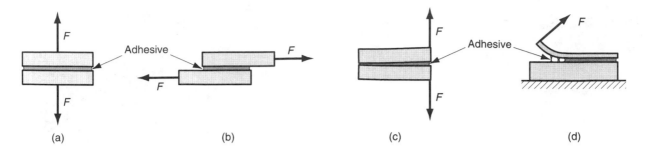

FIGURE 32.10 Types of stresses that must be considered in adhesive bonded joints: (a) tension, (b) shear, (c) cleavage, and (d) peeling.

32.3.1 Joint Design

Adhesive joints are not generally as strong as those by welding, brazing, or soldering. Accordingly, consideration must be given to the design of joints that are adhesively bonded. The following design principles are applicable [3]: (1) Joint contact area should be maximized; (2) Adhesive joints are strongest in shear and tension as in Figure 32.10(a) and (b), and joints should be designed so that the applied stresses are of these types; and (3) Adhesive bonded joints are weakest in cleavage or peeling as in Figure 32.10(c) and (d), and adhesive bonded joints should be designed to avoid these types of stresses.

Typical joint designs for adhesive bonding that illustrate these design principles are presented in Figure 32.11. Some joint designs combine adhesive bonding with other joining methods to increase strength or provide sealing between the two components. Some of the possibilities are shown in Figure 32.12. For example, the combination of adhesive bonding and spot welding is called **weldbonding.**

In addition to the mechanical configuration of the joint, the application must be selected so that the physical and chemical properties of adhesive and adherends are compatible under the service conditions to which the assembly will be subjected. Adherend

FIGURE 32.11 Some joint designs for adhesive bonding: (a) through (d) butt joints; (e) and (f) T-joints; and (g) through (j) corner joints.

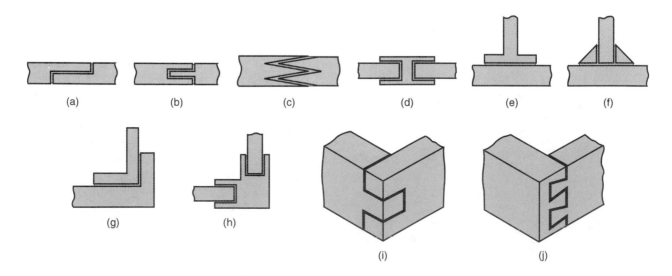

FIGURE 32.12 Adhesive bonding combined with other joining methods: (a) weldbonding—spot welded and adhesive bonded; (b) riveted (or bolted) and adhesive bonded; and (c) formed plus adhesive bonded.

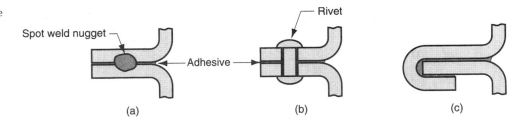

materials include metals, ceramics, glass, plastics, wood, rubber, leather, cloth, paper, and cardboard. Note that the list includes materials that are rigid and flexible, porous and nonporous, metallic and nonmetallic, and note that similar or dissimilar substances can be bonded together.

32.3.2 Adhesive Types

A large number of commercial adhesives are available. They can be classified into three categories: (1) natural, (2) inorganic, and (3) synthetic.

Natural adhesives are derived from natural sources (e.g., plants and animals), including gums, starch, dextrin, soya flour, and collagen. This category of adhesive is generally limited to low-stress applications, such as cardboard cartons, furniture, and bookbinding; or where large surface areas are involved (e.g., plywood). *Inorganic adhesives* are based principally on sodium silicate and magnesium oxychloride. Although relatively low in cost, they are also low in strength—a serious limitation in a structural adhesive.

Synthetic adhesives constitute the most important category in manufacturing. They include a variety of thermoplastic and thermosetting polymers, many of which are listed and briefly described in Table 32.3. They are cured by various mechanisms, including

TABLE 32.3   Important synthetic adhesives.

Adhesive	Description and Applications
Anaerobic	Single-component, thermosetting, acrylic-based. Cures by free radical mechanism at room temperature. Applications: sealant, structural assembly.
Modified acrylics	Two-component thermoset, consisting of acrylic-based resin and initiator/hardener. Cures at room temperature after mixing. Applications: fiberglass in boats, sheet metal in cars and aircraft.
Cyanoacrylate	Single-component, thermosetting, acrylic-based that cures at room temperature on alkaline surfaces. Applications: rubber to plastic, electronic components on circuit boards, plastic and metal cosmetic cases.
Epoxy	Includes a variety of widely used adhesives formulated from epoxy resins, curing agents, and filler/modifiers that harden upon mixing. Some are cured when heated. Applications: aluminum bonding applications and honeycomb panels for aircraft, sheet-metal reinforcements for cars, lamination of wooden beams, seals in electronics.
Hot melt	Single-component, thermoplastic adhesive hardens from molten state after cooling from elevated temperatures. Formulated from thermoplastic polymers including ethylene vinyl acetate, polyethylene, styrene block copolymer, butyl rubber, polyamide, polyurethane, and polyester. Applications: packaging (e.g., cartons, labels), furniture, footwear, bookbinding, carpeting, and assemblies in appliances and cars.
Pressure-sensitive tapes and films	Usually one component in solid form that possesses high tackiness resulting in bonding when pressure is applied. Formed from various polymers of high molecular weight. Can be single-sided or double-sided. Applications: solar panels, electronic assemblies, plastics to wood and metals.
Silicone	One or two components, thermosetting liquid, based on silicon polymers. Curing by room-temperature vulcanization to rubbery solid. Applications: seals in cars (e.g., windshields), electronic seals and insulation, gaskets, bonding of plastics.
Urethane	One or two components, thermosetting, based on urethane polymers. Applications: bonding of fiberglass and plastics.

Compiled from [5], [6], [8], and [9].

(1) mixing a catalyst or reactive ingredient with the polymer immediately prior to applying; (2) heating to initiate the chemical reaction; (3) radiation curing, such as ultraviolet light; and (4) curing by evaporation of water from the liquid or paste adhesive. In addition, some synthetic adhesives are applied as films or as pressure-sensitive coatings on the surface of one of the adherends.

32.3.3 Application Technology

Industrial applications of adhesive bonding are widespread and growing. Major users are automotive, aircraft, building products, and packaging industries; other industries include footwear, furniture, bookbinding, electrical, and shipbuilding [9]. Table 32.3 indicates some of the specific applications for which synthetic adhesives are used. In this section we consider several issues relating to adhesives application technology.

Surface Preparation   In order for adhesive bonding to succeed, part surfaces must be extremely clean. The strength of the bond depends on the degree of adhesion between adhesive and adherend, and this depends on the cleanliness of the surface. In most cases, additional processing steps are required for cleaning and surface preparation, the methods varying with different adherend materials. For metals, solvent wiping is often used for cleaning, and abrading the surface by sandblasting or other process usually improves adhesion. For nonmetallic parts, some type of solvent cleaning is generally used, and the surfaces are sometimes mechanically abraded or chemically etched to increase roughness. It is desirable to accomplish the adhesive bonding process as soon as possible after these treatments, since surface oxidation and dirt accumulation increase with time.

Application Methods   The actual application of the adhesive to one or both part surfaces is accomplished in a number of ways. The following list, though incomplete, provides a sampling of the techniques used in industry [7], [9]:

> *Brushing,* performed manually, uses a stiff-bristled brush. Coatings are often uneven.
> *Flowing,* using manually operated pressure-fed flow guns, has more consistent control than brushing.
> *Manual rollers,* similar to paint rollers, are used to apply adhesive from a flat container.
> *Silk screening* involves brushing the adhesive through the open areas of the screen onto the part surface, so that only selected areas are coated.
> *Spraying* uses an air-driven (or airless) spray gun for fast application over large or difficult-to-reach areas.
> *Automatic applicators* include various automatic dispensers and nozzles for use on medium- and high-speed production applications. Figure 32.13 illustrates the use of a dispenser for assembly.
> *Roll coating* is a mechanized technique in which a rotating roller is partially submersed in a pan of liquid adhesive and picks up a coating of the adhesive, which is then transferred to the work surface. Figure 32.14 shows one possible application, in which the work is a thin, flexible material (e.g., paper, cloth, leather, plastic). Variations of the method are used for coating adhesive onto wood, wood composite, cardboard, and similar materials with large surface areas.

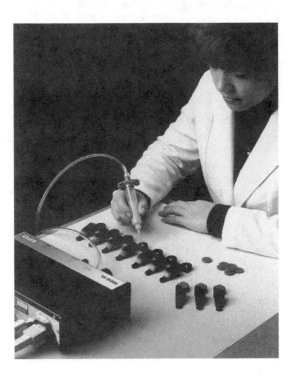

FIGURE 32.13   Adhesive is dispensed by a manually controlled dispenser to bond parts during assembly (courtesy EFD, Inc.).

Advantages and Limitations   Advantages of adhesive bonding are: (1) the process is applicable to a wide variety of materials; (2) parts of different sizes and cross sections can be joined—fragile parts can be joined by adhesive bonding; (3) bonding occurs over the entire surface area of the joint, rather than in discrete spots or along seams as in fusion welding, thereby distributing stresses over the entire area; (4) some adhesives are flexible after bonding and are thus tolerant of cyclical loading and differences in thermal expansion of adherends; (5) low temperature curing avoids damage to parts being joined; (6) sealing as well as bonding can be achieved; and (7) joint design is often simplified (e.g., two flat surfaces can be joined without providing special part features such as screw holes).

Here are the principal limitations of this technology: (1) joints are generally not as strong as other joining methods; (2) adhesive must be compatible with materials being

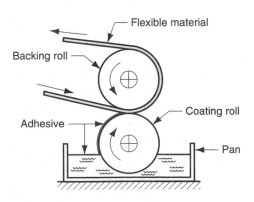

FIGURE 32.14   Roll coating of adhesive onto thin flexible material such as paper, cloth, or flexible polymer.

joined; (3) service temperatures are limited; (4) cleanliness and surface preparation prior to application of adhesive are important; (5) curing times can impose a limit on production rates; and (6) inspection of the bonded joint is difficult.

REFERENCES

[1] Bilotta, A. J. *Connections in Electronic Assemblies.* Marcel Dekker, Inc., New York, 1985.

[2] Bralla, J. G. (Editor in Chief). *Design for Manufacturability Handbook.* 2nd ed. McGraw-Hill, New York, 1998.

[3] *Brazing Manual.* American Welding Society, New York, 1963.

[4] Cary, H. B. *Modern Welding Technology.* 4th ed. Prentice-Hall, Inc., Upper Saddle River, N.J., 1997.

[5] Doyle, D. J. "The Sticky Six-Steps for Selecting Adhesives." *Manufacturing Engineering.* June 1991, pp. 39–43.

[6] Hartshorn, S. R. (ed.). *Structural Adhesives, Chemistry and Technology.* Plenum Press, New York, 1986.

[7] Lambert, L. P. *Soldering for Electronic Assemblies.* Marcel Dekker, Inc., New York, 1988.

[8] Lincoln, B., Gomes, K. J., and Braden, J. F. *Mechanical Fastening of Plastics.* Marcel Dekker, Inc., New York, 1984.

[9] Minniti, A. "Adhesives in Manufacturing." *Project Report.* Manufacturing Systems Engineering Program, Lehigh University, April 1990.

[10] O'brien, R. L. *Welding Handbook.* 8th ed. Vol. 2: *Welding Processes.* American Welding Society, Miami, Fla., 1991.

[11] Petrie, E. M. *Handbook of Adhesives and Sealants.* McGraw-Hill, Monterey, Calif., 1999.

[12] Schneberger, G. L. (ed.). *Adhesives in Manufacturing.* Marcel Dekker, Inc., New York, 1983.

[13] Shields, J. *Adhesives Handbook.* 3rd ed. Butterworths Heinemann., Woburn, England, 1984.

[14] Skeist, I. (ed.). *Handbook of Adhesives.* 3rd ed. Chapman & Hall, New York, 1989.

[15] *Soldering Manual.* 2nd ed. American Welding Society, Miami, Fla., 1978.

[16] Wick, C., and Veilleux, R. F. *Tool and Manufacturing Engineers Handbook.* 4th ed. Vol. 4: *Quality Control and Assembly.* Society of Manufacturing Engineers, Dearborn, Mich., 1987.

REVIEW QUESTIONS

32.1. How do brazing and soldering differ from fusion welding?

32.2. How do brazing and soldering differ from solid-state welding?

32.3. What is the technical difference between brazing and soldering?

32.4. Under what circumstances would brazing or soldering be preferred over welding?

32.5. What are the two joint types most commonly used in brazing?

32.6. Certain changes in joint configuration are usually made to improve the strength of brazed joints. What are some of these changes?

32.7. The molten filler metal in brazing is distributed throughout the joint by capillary action. What is capillary action?

32.8. What are the desirable characteristics of a brazing flux?

32.9. What is *dip brazing*?

32.10. Define *braze welding.*

32.11. What are some of the disadvantages and limitations of brazing?

32.12. What are the two most common alloying metals used in solders?

32.13. Why are rosins as soldering fluxes losing favor in industry?

32.14. What are the functions served by the bit of a soldering iron in hand soldering?

32.15. What is *wave soldering*?

32.16. List the advantages often attributed to soldering as an industrial joining process.

32.17. What are the disadvantages and drawbacks of soldering?

32.18. What is meant by the term *structural adhesive*?

32.19. An adhesive must cure in order to bond. What is meant by the term *curing*?

32.20. What are some of the methods used to cure adhesives?

32.21. Name the three basic categories of commercial adhesives.

32.22. What is an important precondition for the success of an adhesive bonding operation?

32.23. What are some of the methods used to apply adhesives in industrial production operations?

32.24. Identify some of the advantages of adhesive bonding compared to alternative joining methods.

32.25. What are some of the limitations of adhesive bonding?

MULTIPLE CHOICE QUIZ

There is a total of 24 correct answers in the following multiple choice questions (some questions have multiple answers that are correct). To attain a perfect score on the quiz, all correct answers must be given, since each correct answer is worth 1 point. For each question, each omitted answer or wrong answer reduces the score by 1 point, and each additional answer beyond the number of answers required reduces the score by 1 point. Percentage score on the quiz is based on the total number of correct answers.

32.1. In brazing, the base metals melt at temperatures above 450°C (840°F) while in soldering they melt at 450°C (840°F) or below: (a) true or (b) false.

32.2. The strength of a brazed joint is typically which one of the following relative to the filler metal out of which it is made: (a) equal to, (b) stronger than, or (c) weaker than.

32.3. Scarfing in the brazing of a butt joint involves the wrapping of a sheath around the two parts to be joined to contain the molten filler metal during the heating process: (a) true or (b) false.

32.4. Clearances between surfaces in brazing are which one of the following: (a) 0.0025 to 0.025 mm (0.0001–0.001 in.), (b) 0.025 to 0.250 mm (0.001–0.010 in.), (c) 0.250 to 2.50 mm (0.010–0.100 in.), or (d) 2.5 to 5.0 mm (0.10–0.20 in.).

32.5. Which of the following is an advantage of brazing (more than one)? (a) dissimilar metals can be joined, (b) less heat and energy required than fusion welding, (c) multiple joints can be brazed simultaneously, (d) stronger joint than welding.

32.6. Which of the following soldering methods are not used for brazing (more than one)? (a) dip soldering, (b) infrared soldering, (c) soldering iron, (d) torch soldering, and (e) wave soldering.

32.7. Which one of the following is not a function of a flux in brazing or soldering? (a) chemically etch the surfaces to increase roughness for better adhesion of the filler metal, (b) promote wetting of the surfaces, (c) protect the faying surfaces during the process, or (d) remove or inhibit formation of oxide films.

32.8. Which type of soldering flux is preferred for electrical and electronics connections? (a) inorganic fluxes such as zinc chloride, (b) natural rosin fluxes, or (c) water-soluble organic fluxes.

32.9. Which of the following metals are used in solder alloys (more than one)? (a) antimony, (b) gold, (c) lead, (d) silver, or (e) tin.

32.10. A soldering gun is capable of injecting molten solder metal into the joint area: (a) true or (b) false.

32.11. In adhesive bonding, which one of the following is the term used for the parts that are joined? (a) adherend, (b) adherent, (c) adhesive, (d) adhibit, or (e) ad infinitum.

32.12. Weldbonding is an adhesive joining method in which heat is used to melt the adhesive: (a) true, or (b) false.

32.13. Adhesively bonded joints are strongest under which type of stresses (pick two best answers)? (a) cleavage, (b) peeling, (c) shear, and (d) tension.

32.14. Which of the following are the mechanisms that operate in adhesive bonding (more than one)? (a) chemical bonding, in which a primary chemical bond is formed between the adhesive and the parts being joined, (b) mechanical interlocking, (c) secondary bonding forces between atoms of opposing surfaces, and (d) surface tension of the fluid adhesive.

32.15. Roughening of the faying surfaces tends to (a) increase or (b) reduce the strength of an adhesively bonded joint because it inhibits the adhesive from spreading itself across the entire area of the joint.

33 MECHANICAL ASSEMBLY

CHAPTER CONTENTS

33.1   Threaded Fasteners
    33.1.1   Screws, Bolts, and Nuts
    33.1.2   Other Threaded Fasteners and Related Hardware
    33.1.3   Stresses and Strengths in Bolted Joints
    33.1.4   Tools and Methods for Threaded Fasteners
33.2   Rivets and Eyelets
33.3   Assembly Methods Based on Interference Fits
33.4   Other Mechanical Fastening Methods
33.5   Molding Inserts and Integral Fasteners
33.6   Design for Assembly
    33.6.1   General Principles of DFA
    33.6.2   Design for Automated Assembly

Mechanical assembly involves the use of various fastening methods to mechanically attach two (or more) parts together. In most cases, the fastening method involves the use of discrete hardware components, called *fasteners,* that are added to the parts during the assembly operation. In other cases, the fastening mechanism involves the shaping or reshaping of one of the components being assembled, and no separate fasteners are required. Many consumer products are assembled largely by mechanical fastening methods: automobiles, large and small appliances, telephones, furniture, utensils — even wearing apparel is "assembled" by mechanical means. In addition, industrial products ranging from electronics to construction equipment almost always involve some mechanical assembly.

Mechanical fastening methods can be divided into two major classes: (1) those that allow for disassembly, and (2) those that create a permanent joint. Threaded fasteners (e.g., screws, bolts, and nuts) are examples of the first class, and rivets illustrate the second. There are good reasons why mechanical assembly is often preferred over other joining processes discussed in previous chapters. The main reasons are:

> *Ease of assembly*

> *Ease of disassembly* (for the fastening methods that permit disassembly)

Mechanical assembly is usually accomplished with relative ease by unskilled workers using a minimum of special tooling and in a relatively short time. The technology is simple (although there is more to it than one might think), and the results are easily inspected. These factors are advantageous not only in the factory, but also during field installation. Large products that are too big and heavy to be transported completely assembled can be shipped in smaller subassemblies and then put together at the customer's site.

Ease of disassembly applies, of course, only to the mechanical fastening methods that permit disassembly. Periodic disassembly is required for most products so that maintenance and repair can be performed; for example, to replace worn-out components, to make adjustments, and so forth. Permanent joining techniques such as welding do not allow for disassembly.

For purposes of organization, we divide mechanical assembly methods into the following categories: (1) threaded fasteners, (2) rivets, (3) interference fits, (4) other mechanical fastening methods, and (5) molded-in inserts and integral fasteners. These categories are described in Sections 33.1 through 33.5. In Section 33.6, we discuss an important issue in assembly: design for assembly. Assembly of electronic products includes mechanical techniques. However, electronics assembly represents a unique and specialized field, discussed in Chapter 36.

33.1 THREADED FASTENERS

Threaded fasteners are discrete hardware components that have external or internal threads for assembly of parts. In nearly all cases, they permit disassembly. Threaded fasteners are the most important category of mechanical assembly; the common threaded fastener types are screws, bolts, and nuts.

33.1.1 Screws, Bolts, and Nuts

Screws and bolts are threaded fasteners that have external threads. There is a technical distinction between a screw and a bolt that is often blurred in popular usage. A *screw* is an externally threaded fastener that is generally assembled into a blind threaded hole. Some types, called *self-tapping screws,* possess geometries that permit them to form or cut the matching threads in the hole. A *bolt* is an externally threaded fastener that is inserted through holes in the parts and "screwed" into a nut on the opposite side. A *nut* is an internally threaded fastener having standard threads that match those on bolts of the same diameter, pitch, and thread form. The typical assemblies that result from the use of screws and bolts are illustrated in Figure 33.1.

Screws and bolts come in a variety of sizes, threads, and shapes. An explanation and enumeration of these specifications goes beyond our purpose here and can be found

FIGURE 33.1 Typical assemblies using: (a) bolt and nut, and (b) screw.

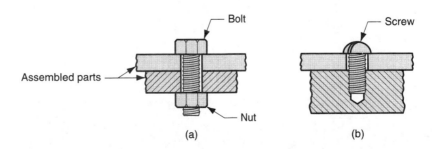

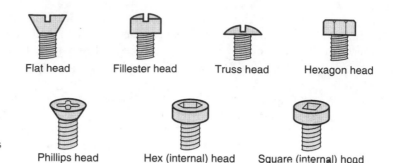

FIGURE 33.2 Various head styles available on screws and bolts. There are additional head styles not shown.

Flat head   Fillester head   Truss head   Hexagon head

Phillips head   Hex (internal) head   Square (internal) head

in design texts and standard handbooks. However, several comments are appropriate. First, despite the many variations, there is a good deal of standardization in the threaded fastener industry, which promotes interchangeability—an important issue in assembly. The United States has been gradually converting to metric fastener sizes, which will further reduce the variations among specifications. Second, the differences between threaded fasteners have tooling implications. To use a particular type of screw or bolt, the assembly operator must have tools that are designed for that fastener type. For example, there are numerous head styles available on bolts and screws, the most common of which are shown in Figure 33.2. The geometries of these heads, as well as the variety of sizes available, require different hand tools (e.g., screwdrivers) for the operator. One cannot easily turn a hex-head bolt with a conventional flat-blade screwdriver.

Screws come in a greater variety of configurations than bolts, since their functions vary more. The types include machine screws, capscrews, setscrews, and self-tapping screws. **Machine screws** are the generic type, designed for assembly into tapped holes. They are sometimes assembled to nuts, and in this usage they overlap with bolts. **Capscrews** have the same geometry as machine screws but are made of higher strength metals and to closer tolerances. **Setscrews** are hardened and designed for assembly functions such as fastening collars, gears, and pulleys to shafts as shown in Figure 33.3(a). They come in various geometries, some of which are illustrated in Figure 33.3(b). A **self-tapping screw** (also called a **tapping screw**) is designed to form or cut threads in a preexisting hole into which it is being turned. Figure 33.4 shows two of the typical thread geometries for self-tapping screws.

Most threaded fasteners are produced by cold forming (Section 19.2). Some are machined (Section 22.1.1), but this is usually a more expensive thread-making process. A variety of materials are used to make threaded fasteners, steels being the most common because of their good strength and low cost. These include low and medium carbon as

FIGURE 33.3   (a) Assembly of collar to shaft using a setscrew; (b) various setscrew geometries (head types and points).

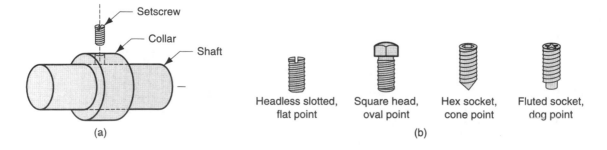

Setscrew
Collar
Shaft

(a)

Headless slotted, flat point

Square head, oval point

Hex socket, cone point

Fluted socket, dog point

(b)

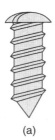 

(a)   (b)

FIGURE 33.4   Self-tapping screws: (a) thread-forming and (b) thread-cutting.

well as alloy steels. Fasteners made of steel are usually plated or coated for superficial resistance to corrosion. Nickel, chromium, zinc, black oxide, and similar coatings are used for this purpose. When corrosion or other factors deny the use of steel fasteners, other materials must be used, including stainless steels, aluminum alloys, nickel alloys, and plastics (however, plastics are suited to low stress applications only).

33.1.2 Other Threaded Fasteners and Related Hardware

Additional threaded fasteners and related hardware include studs, screw thread inserts, captive threaded fasteners, and washers. A *stud* (in the context of fasteners) is an externally threaded fastener, but without the usual head possessed by a bolt. Studs can be used to assemble two parts using two nuts as shown in Figure 33.5(a). They are available with threads on one end or both as in Figure 33.5(b) and (c).

Screw thread inserts are internally threaded plugs or wire coils made to be inserted into an unthreaded hole and to accept an externally threaded fastener. They are assembled into weaker materials (e.g., plastic, wood, and light-weight metals such as magnesium) to provide strong threads. There are many designs of screw thread inserts, one example of which is illustrated in Figure 33.6. Upon subsequent assembly of the screw into the insert, the insert barrel expands into the sides of the hole, securing the assembly.

Captive threaded fasteners are threaded fasteners that have been permanently preassembled to one of the parts to be joined. Possible preassembly processes include welding, brazing, press fitting, or cold forming. Two types of captive threaded fasteners are illustrated in Figure 33.7.

A *washer* is a hardware component often used with threaded fasteners to ensure tightness of the mechanical joint; in its simplest form, it is a flat, thin ring of sheet metal. Washers serve various functions [13]. They (1) distribute stresses that might otherwise be concentrated at the bolt or screw head and nut, (2) provide support for large clearance holes in the assembled parts, (3) increase spring tension, (4) protect part surfaces, (5) seal the joint, and (6) resist inadvertent unfastening. Three washer types are illustrated in Figure 33.8.

FIGURE 33.5   (a) Stud and nuts used for assembly. Other stud types: (b) threads on one end only and (c) double-end stud.

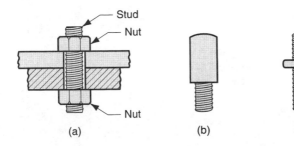

(a)   (b)   (c)

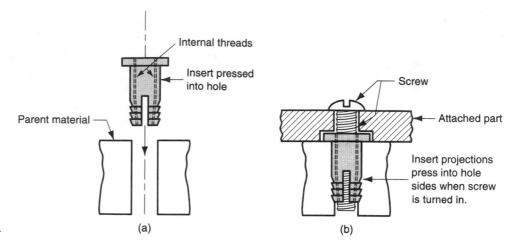

FIGURE 33.6   Screw thread inserts: (a) before insertion, and (b) after insertion into hole and screw is turned into insert.

33.1.3 Stresses and Strengths in Bolted Joints

Typical stresses acting on a bolted or screwed joint include both tensile and shear, as depicted in Figure 33.9. Shown in the figure is a bolt-and-nut assembly. Once tightened, the bolt is loaded in tension, and the parts are loaded in compression. In addition, forces may be acting in opposite directions on the parts, which results in a shear stress on the bolt cross section. Finally, there are stresses applied on the threads throughout their engagement length with the nut in a direction parallel to the axis of the bolt. These shear stresses can cause *stripping* of the threads (this failure can also occur on the internal threads of the nut).

The strength of a threaded fastener is generally specified by two measures: (1) tensile strength, which has the traditional definition, and (2) proof strength. *Proof strength* is roughly equivalent to yield strength; specifically, it is the maximum tensile

FIGURE 33.7   Captive threaded fasteners: (a) weld nut and (b) riveted nut.

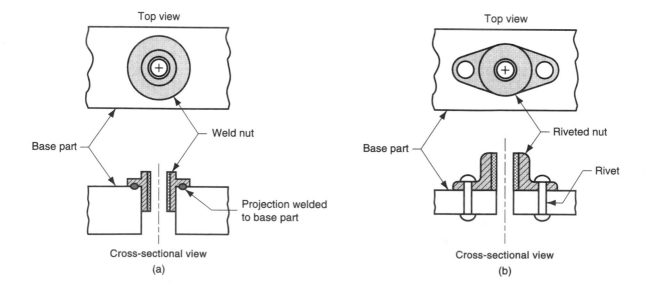

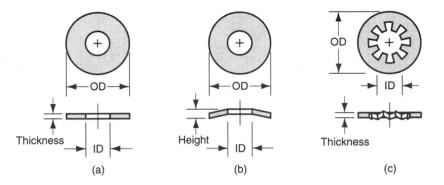

FIGURE 33.8 Types of washers: (a) plain (flat) washers; (b) spring washers, used to dampen vibration or compensate for wear; and (c) lockwasher designed to resist loosening of the bolt or screw.

stress to which an externally threaded fastener can be subjected without permanent deformation. Typical values of tensile and proof strength for steel bolts are given in Table 33.1.

The problem that can arise during assembly is that the threaded fasteners are overtightened, causing stresses that exceed the strength of the fastener material. Assuming a bolt and nut assembly as shown in the figure, failure can occur in one of the following ways: (1) external threads (e.g., bolt or screw) can strip, (2) internal threads (e.g., nut) can strip, or (3) the bolt can break due to excessive tensile stresses on its cross-sectional area. Thread stripping, failures (1) and (2), is a shear failure and occurs when the length of engagement is too short (less than about 60% of the nominal bolt diameter). This can be avoided by providing adequate thread engagement in the fastener design. Tensile failure (3) is the most common problem. The bolt breaks at about 85% of its rated tensile strength because of combined tensile and torsion stresses during tightening [2].

The tensile stress to which a bolt is subjected can be calculated as the tensile load applied to the joint divided by the applicable area:

$$\sigma = \frac{F}{A_s} \tag{33.1}$$

where σ = stress, MPa (lb/in.^{2}); F = load, N (lb); and A_s = tensile stress area, mm^2 (in.^{2}). This stress is compared to the bolt strength values listed in Table 33.1. The tensile stress area for a threaded fastener is the cross-sectional area of the threaded section. This area

FIGURE 33.9 Typical stresses acting on a bolted joint.

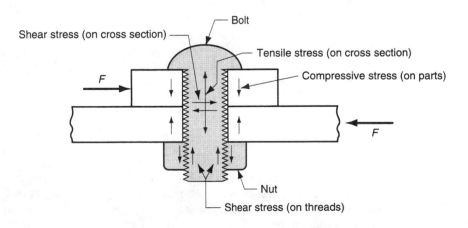

TABLE 33.1 Typical values of tensile and proof strengths for steel bolts and screws, diameters range from 6.4 mm (0.25 in.) to 38 mm (1.50 in.).

Material	Proof Stress		Tensile Stress	
	MPa	(lb/in.^{2})	MPa	(lb/in.^{2})
Low/medium C steel	228	(33,000)	414	(60,000)
Alloy steel	830	(120,000)	1030	(150,000)

Source: [13] pp. 8–11.

can be calculated directly from one of the following equations [2], depending on whether the bolt is metric standard or American standard. For the metric standard, the formula is

$$A_s = 0.25(D - 0.9382p)^2 \qquad (33.2)$$

where D = nominal size (basic major diameter) of the bolt or screw, mm; and p = thread pitch, mm. For the American (inch) standard, the formula is

$$A_s = 0.25\pi \left(D - \frac{0.9743}{n} \right)^2 \qquad (33.3)$$

where D = nominal size (basic major diameter) of the bolt or screw, in; and n = the number of threads per unit length, threads/in.

33.1.4 Tools and Methods for Threaded Fasteners

The basic function of the tools and methods for assembling threaded fasteners is to provide relative rotation between the external and internal threads, and to apply sufficient torque to secure the assembly. Available tools range from simple hand-held screwdrivers or wrenches to powered tools with sophisticated electronic sensors to ensure proper tightening. It is important that the tool match the screw or bolt and/or the nut in style and size, since there are so many bolt heads available. Hand tools are usually made with a single point or blade, but powered tools are generally designed to use interchangeable points. The powered tools operate by pneumatic, hydraulic, or electric power.

Whether a threaded fastener serves its intended purpose depends to a large degree on the amount of torque applied to tighten it. Once the bolt or screw (or nut) has been rotated until it is seated against the part surface, additional tightening will increase the tension in the fastener (and simultaneously the amount of compression in the parts being held together); and the tightening will be resisted by an increasing torque. Thus, there is a correlation between the torque required to tighten the fastener and the tensile stress experienced by it. To achieve the desired function in the assembled joint (e.g., to improve fatigue resistance) and to lock the threaded fasteners, the product designer will often specify the tension force that should be applied. This force is called the **preload**. The following relationship can be used to determine the required torque to obtain a specified preload [13]:

$$T = C_t D F \qquad (33.4)$$

where T = torque, N-mm (lb-in.); C_t = the torque coefficient whose value typically ranges between 0.15 and 0.25 depending on the thread surface conditions; D = nominal bolt or screw diameter, mm (in.); and F = specified tension force, N (lb).

Various methods are employed to apply the required torque, including (1) operator feel—not very accurate, but adequate for most assemblies; (2) torque wrenches; (3) stall-motors—motorized wrenches designed to stall when the required torque is reached, and

(4) torque-turn tightening, in which the fastener is initially tightened to a low torque level and then rotated a specified additional amount.

33.2 RIVETS AND EYELETS

Rivets are widely used for achieving a permanent mechanically fastened joint. Riveting is a fastening method that offers high production rates, simplicity, dependability, and low cost. Despite these apparent advantages, its applications have declined in recent decades in favor of threaded fasteners, welding, and adhesive bonding. Riveting is one of the primary fastening processes in the aircraft and aerospace industries for joining skins to channels and other structural members.

A *rivet* is an unthreaded, headed pin used to join two (or more) parts by passing the pin through holes in the parts and then forming (upsetting) a second head in the pin on the opposite side. The deforming operation can be performed hot or cold (hot working or cold working), and by hammering or steady pressing. Once the rivet has been deformed, it cannot be removed except by breaking one of the heads. Rivets are specified by their length, diameter, head, and type. Rivet type refers to five basic geometries that affect how the rivet will be upset to form the second head. The five basic types, illustrated in Figure 33.10, are (a) solid, (b) tubular, (c) semitubular, (d) bifurcated, and (e) compression. In addition, there are special rivets for special applications.

Rivets are used primarily for lap joints. The clearance hole into which the rivet is inserted must be close to the diameter of the rivet. If the hole is too small, rivet insertion will be difficult, thus reducing production rate. If the hole is too large, the rivet will not fill the hole and may bend during formation of the opposite head. Rivet design tables are available to specify the optimum hole sizes.

The tooling and methods used in riveting can be divided into the following categories: (1) impact, in which a pneumatic hammer delivers a succession of blows to upset the rivet; (2) steady compression, in which the riveting tool applies a continuous squeezing pressure to upset the rivet; and (3) a combination of impact and compression. Much of the equipment used in riveting is portable and manually operated. Automatic drilling and riveting machines are available for drilling the holes, and then inserting and upsetting the rivets.

FIGURE 33.10 Five basic rivet types, also shown in assembled configuration: (a) solid, (b) tubular, (c) semitubular, (d) bifurcated, and (e) compression.

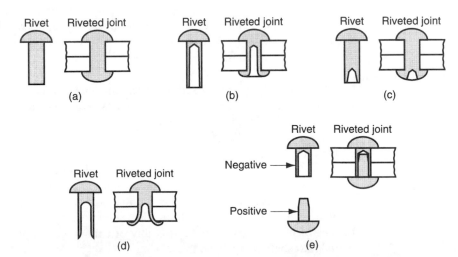

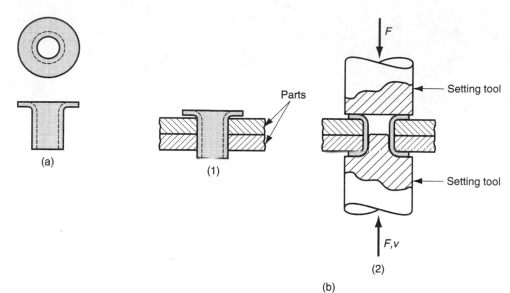

FIGURE 33.11 Fastening with an eyelet: (a) the eyelet, and (b) assembly sequence; (1) inserting the eyelet through the hole and (2) setting operation.

Eyelets are thin-walled tubular fasteners with a flange on one end, usually made from sheet metal, as in Figure 33.11(a). They are used to produce a permanent lap joint between two (or more) flat parts. Eyelets are substituted for rivets in low-stress applications to save material, weight, and cost. During fastening, the eyelet is inserted through the part holes, and the straight end is formed over to secure the assembly. The forming operation is called *setting* and is performed by opposing tools that hold the eyelet in position and curl the extended portion of its barrel. Figure 33.11(b) illustrates the sequence for a typical eyelet design. Applications of this fastening method include automotive subassemblies, electrical components, toys, and apparel.

33.3 ASSEMBLY METHODS BASED ON INTERFERENCE FITS

Several assembly methods are based on mechanical interference between the two mating parts being joined. This interference, either during assembly or after the parts are joined, holds the parts together. The methods include press fitting, shrink and expansion fits, snap fits, and retaining rings.

Press Fitting   A press-fit assembly is one in which the two components have an interference fit between them. The typical case is where a pin (e.g., a straight cylindrical pin) of a certain diameter is pressed into a hole of a slightly smaller diameter. Standard pin sizes are commercially available to accomplish a variety of functions, such as: (1) locating and locking the components—used to augment threaded fasteners by holding two (or more) parts in fixed alignment with each other; (2) pivot points, to permit rotation of one component about the other; and (3) shear pins. Except for (3), the pins are normally hardened. Shear pins are made of softer metals so as to break under a sudden or severe shearing load in order to save the rest of the assembly. Other applications of press fitting include assembly of collars, gears, pulleys, and similar components onto shafts.

The pressures and stresses in an interference fit can be estimated using several applicable formulas. If the fit consists of a round solid pin or shaft inside a collar (or similar component), as depicted in Figure 33.12, and the components are made of the same material, the radial pressure between the pin and the collar can be determined by [13]

$$p_f = \frac{Ei(D_c^2 - D_p^2)}{D_p D_c^2} \qquad (33.5)$$

where p_f = radial or interference fit pressure, MPa (lb/in.^{2}); E = modulus of elasticity for the material; i = interference between the pin (or shaft) and the collar; that is, the starting difference between the inside diameter of the collar hole and the outside diameter of the pin, mm (in.); D_c = outside diameter of the collar, mm (in.); and D_p = pin or shaft diameter, mm (in.).

The maximum effective stress occurs in the collar at its inside diameter and can be calculated as

$$\text{Max } \sigma_e = \frac{2p_f D_c^2}{D_c^2 - D_p^2} \qquad (33.6)$$

where Max σ_e = the maximum effective stress, MPa (lb/in.^{2}), and p_f is the interference fit pressure computed from Eq. (33.5).

In situations where a straight pin or shaft is pressed into the hole of a large part with geometry other than that of a collar, we can alter the previous equations by taking the outside diameter D_c to be infinite, thus reducing the equation for interference pressure to

$$p_f = \frac{Ei}{D_p} \qquad (33.7)$$

and the corresponding maximum effective stress becomes

$$\text{Max } \sigma_e = 2p_f \qquad (33.8)$$

In most cases, particularly for ductile metals, the maximum effective stress should be compared with the yield strength of the material, applying an appropriate safety factor, as in the following:

$$\text{Max } \sigma_e \leq \frac{Y}{SF} \qquad (33.9)$$

where Y = yield strength of the material, and SF is the applicable safety factor.

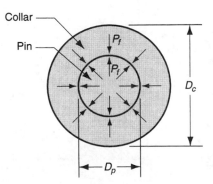

FIGURE 33.12 Cross section of a solid pin or shaft assembled to a collar by interference fit.

Various pin geometries are available for interference fits. The basic type is a ***straight pin***, usually made from cold-drawn carbon steel wire or bar stock, ranging in diameter from 1.6 to 25.0 mm (1/16–1 in.). They are unground, with chamfered or square ends (chamfered ends facilitate press fitting). ***Dowel pins*** are manufactured to more precise specifications than straight pins, and can be ground and hardened. They are used to fix the alignment of assembled components in dies, fixtures, and machinery. ***Taper pins*** possess a taper of 6.4 mm (0.25 in.) per foot and are driven into the hole to establish a fixed relative position between the parts. Their advantage is that they can readily be driven back out of the hole.

Additional pin geometries are commercially available, including ***grooved pins***—solid, straight pins with three longitudinal grooves in which the metal is raised on either side of each groove to cause interference when the pin is pressed into a hole; ***knurled pins***—pins with a knurled pattern that causes interference in the mating hole; and ***coiled pins***, or ***spiral pins***—pins made by rolling strip stock into a coiled spring.

Shrink and Expansion Fits   These terms refer to the assembly of two parts that have an interference fit at room temperature. The typical case is a cylindrical pin or shaft assembled into a collar. To assemble by ***shrink fitting***, the external part is heated to enlarge it by thermal expansion, and the internal part either remains at room temperature or is cooled to contract its size. The parts are then assembled and brought back to room temperature, so that the external part shrinks, and if previously cooled the internal part expands, to form a strong interference fit. An ***expansion fit*** is when only the internal part is cooled to contract it for assembly; once inserted into the mating component, it warms to room temperature, expanding to create the interference assembly. These assembly methods are used to fit gears, pulleys, sleeves, and other components onto solid and hollow shafts.

Various methods are used to heat and/or cool the workparts to change their diameters. Heating equipment includes torches, furnaces, electric resistance heaters, and electric induction heaters. Cooling methods include conventional refrigeration, packing in dry ice, and immersion in cold liquids, including liquid nitrogen. The resulting diameter change depends on the coefficient of thermal expansion and the temperature difference that is applied to the part. If we assume that the heating or cooling has produced a uniform temperature throughout the work, then the change in diameter is given by

$$D_2 - D_1 = \alpha D_1(T_2 - T_1) \tag{33.10}$$

FIGURE 33.13   Snap fit assembly, showing cross sections of two mating parts: (1) before assembly and (2) parts snapped together.

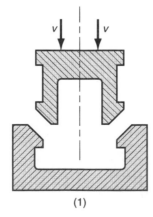

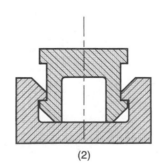

(1)   (2)

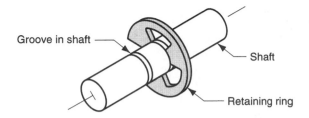

Groove in shaft

Shaft

Retaining ring

FIGURE 33.14   Retaining ring assembled into a groove on a shaft.

where α = the coefficient of linear thermal expansion, mm/mm-°C (in./in.-°F) for the material (see Table 4.1); T_2 = the temperature to which the parts have been heated or cooled, °C (°F); T_1 = starting ambient temperature; D_2 = diameter of the part at T_2, mm (in.); and D_1 = diameter of the part at T_1.

Eqs. (33.5) through (33.9) for computing interference pressures and effective stresses can be used to determine the corresponding values for shrink and expansion fits.

Snap Fits and Retaining Rings   Snap fits are a modification of interference fits. A *snap fit* involves joining two parts in which the mating elements possess a temporary interference while being pressed together, but once assembled they interlock to maintain the assembly. A typical example is shown in Figure 33.13: as the parts are pressed together, the mating elements elastically deform to accommodate the interference, subsequently allowing the parts to snap together; once in position, the elements become connected mechanically so that they cannot easily be disassembled. The parts are usually designed so that a slight interference exists after assembly.

Advantages of snap fit assembly include: (1) the parts can be designed with self-aligning features, (2) no special tooling is required, and (3) assembly can be accomplished very quickly. Snap fitting was originally conceived as a method that would be ideally suited to industrial robotics applications; however, it is no surprise that assembly techniques that are easier for robots are also easier for human assembly workers.

A *retaining ring,* also known as a *snap ring,* is a fastener that snaps into a circumferential groove on a shaft or tube to form a shoulder, as in Figure 33.14. The assembly can be used to locate or restrict the movement of parts mounted on the shaft. Retaining rings are available for both external (shaft) and internal (bore) applications. They are made from either sheet metal or wire stock, heat treated for hardness and stiffness. To assemble a retaining ring, a special pliers tool is used to elastically deform the ring so that it fits over the shaft (or into the bore) and then is released into the groove.

33.4 OTHER MECHANICAL FASTENING METHODS

In addition to the mechanical assembly techniques discussed above, there are several additional methods that involve the use of fasteners. These include stitching, stapling, sewing, and cotter pins.

Stitching, Stapling, and Sewing   Industrial stitching and stapling are similar operations involving the use of U-shaped metal fasteners. *Stitching* is a fastening operation in which a stitching machine is used to form the U-shaped stitches one-at-a-time from steel wire and immediately drive them through the two parts to be joined. Figure 33.15 illustrates

FIGURE 33.15 Common types of wire stitches: (a) unclinched, (b) standard loop, (c) bypass loop, and (d) flat clinch.

(a) (b) (c) (d)

several types of wire stitches. The parts to be joined must be relatively thin, consistent with the stitch size, and the assembly can involve various combinations of metal and nonmetal materials. Applications of industrial stitching include light sheet-metal assembly, metal hinges, electrical connections, magazine binding, corrugated boxes, and final product packaging. Conditions that favor stitching in these applications are high-speed operation, elimination of prefabricated holes in the parts, and fasteners that encircle the parts are desired.

In *stapling,* preformed U-shaped staples are punched through the two parts to be attached. The staples are supplied in convenient strips. The individual staples are lightly stuck together to form the strip, but they can be separated by the stapling tool for driving. The staples come with various point styles to facilitate their entry into the work. Staples are usually applied by means of portable pneumatic guns, into which strips containing several hundred staples can be loaded. Applications of industrial stapling include furniture and upholstery, assembly of car seats, and various light-gage sheetmetal and plastic assembly jobs.

Sewing is a common joining method for soft, flexible parts such as cloth and leather. The method involves the use of a long thread or cord interwoven with the parts so as to produce a continuous seam between them. The process is widely used in the needle trades industry for assembling garments.

Cotter Pins *Cotter pins* are fasteners formed from half-round wire into a single two-stem pin, as in Figure 33.16. They vary in diameter, ranging between 0.8 mm (0.031 in.) and 19 mm (3/4 in.), and in point style, several of which are shown in the figure. Cotter pins are inserted into holes in the mating parts and their legs are split to lock the assembly. They are used to secure parts onto shafts and similar applications.

FIGURE 33.16 Cotter pins: (a) offset head, standard point; (b) symmetric head, hammerlock point; (c) square point; (d) mitered point; and (e) chisel point.

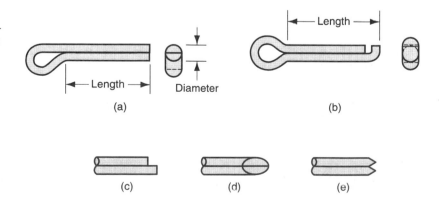

(a) (b)

(c) (d) (e)

33.5 MOLDING INSERTS AND INTEGRAL FASTENERS

These assembly methods form a permanent joint between parts by shaping or reshaping one of the components through a manufacturing process such as casting, molding, or sheet-metal forming.

Inserts in Moldings and Castings   This method involves the placement of a component into a mold prior to plastic molding or metal casting, so that it becomes a permanent and integral part of the molding or casting. Inserting a separate component is preferable to molding its shape if the superior properties of the insert (e.g., strength) are required or the geometry achieved through the use of the insert is too complex or intricate to incorporate into the mold. Examples of inserts in molded or cast parts include internally threaded bushings and nuts, externally threaded studs, bearings, and electrical contacts. Some of these are illustrated in Figure 33.17. Internally threaded inserts must be placed into the mold with threaded pins to prevent the molding material from flowing into the threaded hole.

Placing inserts into a mold has certain disadvantages in production [8]: (1) design of the mold becomes more complicated; (2) handling and placing the insert into the cavity takes time that reduces production rate; and (3) inserts introduce a foreign material into the casting or molding, and in the event of a defect, the cast metal or plastic cannot be easily reclaimed and recycled. Despite these disadvantages, use of inserts is often the most functional design and least-cost production method.

Integral Fasteners   *Integral fasteners* involve deformation of component parts so they interlock and create a mechanically fastened joint. This assembly method is most common for sheet-metal parts. The possibilities, Figure 33.18, include: (a) **lanced tabs** to attach wires or shafts to sheet-metal parts; (b) **embossed protrusions,** in which bosses are formed in one part and flattened over the mating assembled part; (c) **seaming,** where the edges of two separate sheet-metal parts or the opposite edges of the same part are bent over to form the fastening seam—the metal must be ductile in order for the bending to be feasible; (d) **beading,** in which a tube-shaped part is attached to a smaller shaft (or other round part) by deforming the outer diameter inward to cause an interference around the entire circumference; and (e) **dimpling**—forming of simple round indentations in an outer part to retain an inner part.

Crimping, in which the edges of one part are deformed over a mating component, is another example of integral assembly. A common example involves squeezing the barrel of an electrical terminal onto a wire (Section 36.5.1).

FIGURE 33.17   Examples of molded-in inserts: (a) threaded bushing and (b) threaded stud.

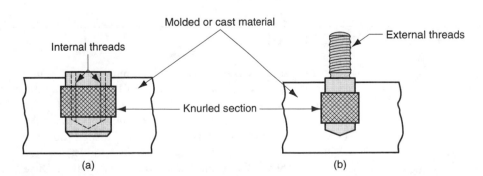

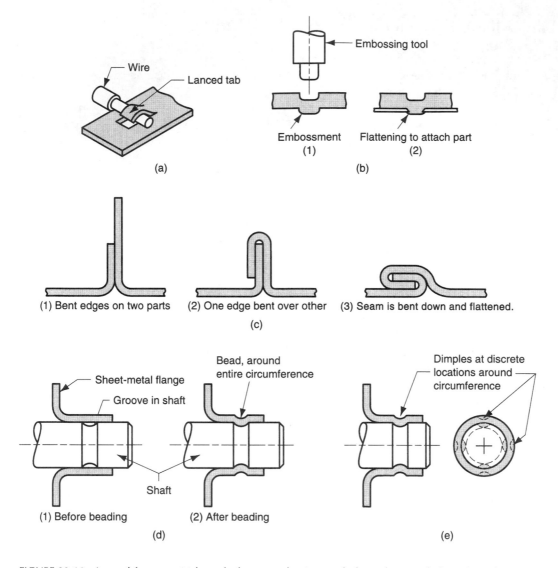

FIGURE 33.18   Integral fasteners: (a) lanced tabs to attach wires or shafts to sheetmetal, (b) embossed protrusions, similar to riveting, (c) single-lock seaming, (d) beading, and (e) dimpling. Numbers in parentheses indicate sequence in (b), (c), and (d).

33.6 DESIGN FOR ASSEMBLY

Design for assembly (DFA) has received much attention in recent years because assembly operations constitute a high labor cost for many manufacturing companies. The key to successful design assembly can be simply stated [3]: (1) design the product with as few parts as possible, and (2) design the remaining parts so that they are easy to assemble. The cost of assembly is determined largely during product design, because that is when the number of separate components in the product are determined, and decisions are made about how these components will be assembled. Once these decisions have been made, there is little that can be done in manufacturing to influence assembly costs (except, of course, to manage the operations well).

In this section, we consider some of the principles that can be applied during product design to facilitate assembly. Most have been developed in the context of mechanical assembly, although some principles apply to the other assembly and joining processes. Much of the research in design for assembly has been motivated by the increasing use of automated assembly systems in industry. Accordingly, our discussion is divided into two sections, the first dealing with general principles of DFA, and the second concerned specifically with design for automated assembly.

33.6.1 General Principles of DFA

Most of the general principles apply to both manual and automated assembly. Their goal is to achieve the required design function by the simplest and lowest cost means. The following recommendations have been compiled from [1], [3], [4], and [6]:

> ➤ *Use the fewest number of parts possible to reduce the amount of assembly required.* This principle is implemented by combining functions within the same part that might otherwise be accomplished by separate components (e.g., using a plastic molded part instead of an assembly of sheet metal parts).

> ➤ *Reduce the number of threaded fasteners required.* Instead of using separate threaded fasteners, design the component to utilize snap fits, retaining rings, integral fasteners, and similar fastening mechanisms that can be accomplished more rapidly. Use threaded fasteners only where justified (e.g., where disassembly or adjustment is required).

> ➤ *Standardize fasteners.* This is intended to reduce the number of sizes and styles of fasteners required in the product. Ordering and inventory problems are reduced, the assembly worker does not have to distinguish between so many separate fasteners, the workstation is simplified, and the variety of separate fastening tools is reduced.

> ➤ *Reduce parts orientation difficulties.* Orientation problems are generally reduced by designing a part to be symmetrical and minimizing the number of asymmetric features. This allows easier handling and insertion during assembly. This principle is illustrated in Figure 33.19.

> ➤ *Avoid parts that tangle.* Certain part configurations are more likely to become entangled in parts bins, frustrating assembly workers or jamming automatic feeders. Parts with hooks, holes, slots, and curls exhibit more of this tendency than parts without these features. See Figure 33.20.

FIGURE 33.19 Symmetrical parts are generally easier to insert and assemble: (a) only one rotational orientation possible for insertion; (b) two possible orientations; (c) four possible orientations; and (d) infinite rotational orientations.

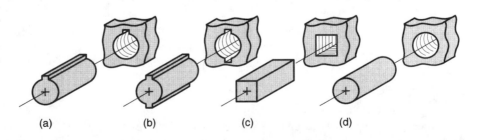

(a)   (b)   (c)   (d)

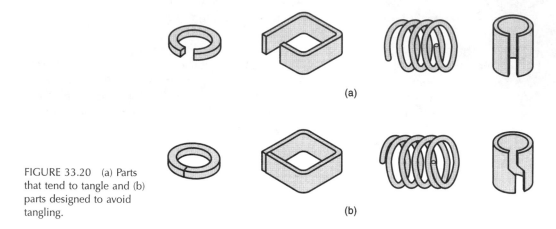

FIGURE 33.20 (a) Parts that tend to tangle and (b) parts designed to avoid tangling.

33.6.2 Design for Automated Assembly

Methods suitable for manual assembly are not necessarily the best methods for automated assembly. Some assembly operations readily performed by a human worker are quite difficult to automate (e.g., assembly using bolts and nuts). To automate the assembly process, parts fastening methods must be specified during product design that lend themselves to machine insertion and joining techniques and do not require the senses, dexterity, and intelligence of human assembly workers. Following are some recommendations and principles that can be applied in product design to facilitate automated assembly [6], [11]:

➤ *Use modularity in product design.* Increasing the number of separate tasks that are accomplished by an automated assembly system will reduce the reliability of the system. To alleviate the reliability problem, Riley [11] suggests that the design of the product be modular in which each module or subassembly has a maximum of 12 or 13 parts to be produced on a single assembly system. Also, the subassembly should be designed around a base part to which other components are added.

➤ *Reduce the need for multiple components to be handled at once.* The preferred practice for automated assembly is to separate the operations at different stations rather than to simultaneously handle and fasten multiple components at the same workstation.

➤ *Limit the required directions of access.* This means that the number of directions in which new components are added to the existing subassembly should be minimized. Ideally, all components should be added vertically from above, if possible.

➤ *High-quality components.* High performance of an automated assembly system requires that consistently good-quality components are added at each workstation. Poor quality components cause jams in feeding and assembly mechanisms that result in downtime.

➤ *Use of snap fit assembly.* This eliminates the need for threaded fasteners; assembly is by simple insertion, usually from above. It requires that the parts be designed with special positive and negative features to facilitate insertion and fastening.

REFERENCES

[1] Andreasen, M., Kahler, S., and Lund, T. *Design for Assembly.* Springer-Verlag, New York, 1988.

[2] Blake, A. *What Every Engineer Should Know About Threaded Fasteners.* Marcel Dekker, New York, 1986.

[3] Boothroyd, G., Dewhurst, P., and Knight, W. *Product Design for Manufacture and Assembly.* Marcel Dekker, New York, 1994.

[4] Bralla, J. G. (Editor in Chief). *Design for Manufacturability Handbook.* 2nd ed. McGraw-Hill, New York, 1998.

[5] Dewhurst, P., and Boothroyd, G. "Design for Assembly in Action." *Assembly Engineering.* January 1987, pp 64–68.

[6] Groover, M. P. *Automation, Production Systems, and Computer Integrated Manufacturing.* 2nd ed. Prentice-Hall, Upper Saddle River, N.J., 2001.

[7] Groover, M. P., Weiss, M., Nagel, R. N., and Odrey, N. G. *Industrial Robotics: Technology, Programming, and Applications.* McGraw-Hill, New York, 1986.

[8] Laughner, V. H., and Hargan, A. D. *Handbook of Fastening and Joining of Metal Parts.* McGraw-Hill, New York, 1956.

[9] Nof, S. Y., Wilhelm, W. E., and Warnecke, H-J. *Industrial Assembly.* Chapman & Hall, New York, 1997.

[10] Parmley, R. O. (ed.). *Standard Handbook of Fastening and Joining.* 2nd ed. McGraw-Hill, New York, 1989.

[11] Riley, F. J. *Assembly Automation, A Management Handbook.* 2nd ed. Industrial Press, New York, 1996.

[12] Speck, J. A. *Mechanical Fastening, Joining, and Assembly.* Marcel Dekker, New York, 1997.

[13] Wick, C., and Veilleux, R. F. *Tool and Manufacturing Engineers Handbook.* 4th ed. Vol. IV: *Quality Control and Assembly.* Society of Manufacturing Engineers, Dearborn, Mich., 1987.

REVIEW QUESTIONS

33.1. How does mechanical assembly differ from the other methods of assembly discussed in previous chapters (e.g., welding, brazing)?

33.2. What are some of the reasons why assemblies must be sometimes disassembled?

33.3. What is the technical difference between a screw and a bolt?

33.4. What is a stud (in the context of threaded fasteners)?

33.5. What is *torque-turn tightening*?

33.6. Define *proof strength* as the term applies in threaded fasteners.

33.7. What are the three ways in which a threaded fastener can fail during tightening?

33.8. What is a *rivet*?

33.9. What is the difference between a shrink fit and expansion fit in assembly?

33.10. What are the advantages of snap fitting?

33.11. What is the difference between industrial stitching and stapling?

33.12. What are *integral fasteners*?

33.13. Identify some of the general principles and guidelines for design for assembly.

33.14. Identify some of the principles and guidelines that apply specifically to automated assembly.

MULTIPLE CHOICE QUIZ

There is a total of 18 correct answers in the following multiple choice questions (some questions have multiple answers that are correct). To attain a perfect score on the quiz, all correct answers must be given, since each correct answer is worth 1 point. For each question, each omitted answer or wrong answer reduces the score by 1 point, and each additional answer beyond the number of answers required reduces the score by 1 point. Percentage score on the quiz is based on the total number of correct answers.

33.1. Most externally threaded fasteners are produced by: (a) cold forming or (b) machining.

33.2. Which of the following methods is used for applying the required torque to achieve a desired preload of a threaded fastener (more than one)? (a) sense of feel by a human operator, (b) snap fit, (c) stall-motor wrenches, or (d) torque wrench.

33.3. Which of the following are reasons for using mechanical assembly (more than one)? (a) ease of assembly, (b) ease of disassembly, (c) in some cases involves a melting of the base parts, and (d) no heat affected zone in the base parts.

33.4. Which of the following are the common ways in which threaded fasteners fail during tightening (more than one)? (a) excessive pressure applied to the bolt or screw head by the tightening tool (e.g., screwdriver) resulting in failure of the head, (b) excessive shearing stresses on the threads due to inadequate length of engagement, (c) excessive tensile stresses, or (d) stripping of the internal or external threads.

33.5. In a shrink fit the internal part is cooled to a sufficiently low temperature to reduce its size for assembly, whereas in an expansion fit, the external part is heated sufficiently to increase its size for assembly; when brought back to room temperature in either case, an interference fit is formed: (a) true or (b) false.

33.6. The advantages of snap fit assembly include which of the following (more than one)? (a) assembly can be accomplished quickly, (b) no special tools are required, (c) the components can be designed with features that facilitate parts mating, and (d) the resulting joint is stronger than with most other assembly methods.

33.7. The difference between industrial stitching and stapling is that the U-shaped fasteners are formed during the stitching process while in stapling the fasteners are preformed: (a) true or (b) false.

33.8. From the standpoint of assembly cost, it is more desirable to use many small threaded fasteners rather than few large ones in order to distribute the stresses more uniformly: (a) true or (b) false.

33.9. Which of the following are considered good product design rules for automated assembly (more than one)? (a) design the assembly with the fewest number of components possible; (b) design the product using bolts and nuts wherever possible to allow for disassembly; (c) design with as many different fastener types as possible to achieve maximum flexibility in design; (d) design parts with asymmetric features to mate with other parts having corresponding (but reverse) features, so as to minimize the number of ways the parts will go together; and (e) limit the required directions of access when adding components to a base part.

PROBLEMS

Threaded Fasteners

33.1. A 5-mm diameter bolt is to be tightened to produce a preload = 25 N. If the torque coefficient $C = 0.23$, determine the torque that should be applied.

33.2. A metric 10×1.5 screw (10 mm diameter, pitch $p = 1.5$ mm) is to be turned into a threaded hole and tightened to one/half of its proof strength, which is 300 MPa. Determine the maximum torque that should be used if the torque coefficient $C = 0.18$.

33.3. A metric 16×2 bolt (16 mm diameter, pitch $p = 2$ mm) is subjected to a torque of 12 N-m during tightening. If the torque coefficient $C = 0.20$, determine the tensile stress on the bolt.

33.4. A 1/2-in. diameter screw is to be preloaded to a tension force $F = 1000$ lb. Torque coefficient $C = 0.22$. Determine the torque that should be used to tighten the bolt.

33.5. A torque wrench is used on a 3/4-10 UNC screw (3/4 in. nominal diameter, 10 threads/in.) in an automobile final assembly plant. A torque of 125 in.-lb is generated by the wrench. If the torque coefficient $C = 0.20$, determine the tension in the bolt.

33.6. The designer has specified that a 3/8-16 UNC low-carbon bolt (3/8 in. nominal diameter, 16 threads/in.) in a certain application should be stressed to its proof stress of 33,000 lb/in.^{2} (see Table 33.1). Determine the maximum torque that should be used if $C = 0.25$.

33.7. A 1-8 UNC low carbon steel bolt (diameter = 1.0 in. 8 threads/in.) is currently planned for a certain application. It is to be preloaded to 75% of its proof strength, which is 33,000 lb/in.^{2} (Table 33.1). However, this bolt is too large for the size of the components involved, and a higher strength but smaller bolt would be preferable. Determine (a) the smallest nominal size of an alloy steel bolt (proof strength = 120,000 lb/in.^{2}) that could be used to achieve the same preload from the following standard UNC sizes used by the company: 1/4-20, 5/16-18, 3/8-16, 1/2-13, 5/8-11, or 3/4-10; and (b) compare the torque required to obtain the preload for the original 1-in. bolt and the alloy steel bolt selected in part (a) if the torque coefficient in both cases $C = 0.20$.

Interference Fits

33.8. A dowel pin made of steel ($E = 209,000$ MPa) is to be press fitted into a steel collar. The pin has a nominal diameter of 13.0 mm, and the collar has an outside diameter = 25.0 mm. (a) Compute the radial pressure and the maximum effective stress if the interference between the shaft OD and the collar ID is 0.02 mm. (b) Determine the effect of increasing the outside diameter of the collar to 35.0 mm on the radial pressure and the maximum effective stress.

33.9. A gear made of aluminum (modulus of elasticity $E = 69,000$ MPa) is press-fitted onto an aluminum shaft. The gear has a diameter of 55 mm at the base of its teeth. The nominal internal diameter of the gear = 30 mm and the interference = 0.10 mm. Compute (a) the radial pressure between the shaft and the gear, and (b) the maximum effective stress in the gear at its inside diameter.

33.10. A steel collar is press fitted onto a steel shaft. The modulus of elasticity of steel $E = 30 \times 10^6$ lb/in.^{2} The collar has an internal diameter = 0.998 in., and the shaft has an outside diameter = 1.000 in. The outside diameter of the collar is 1.750 in. Determine the radial (interference) pressure on the assembly, and (b) the maximum effective stress in the collar at its inside diameter.

33.11. The yield strength of a certain metal $Y = 50,000$ lb/in.^{2}, and its modulus of elasticity $E = 22 \times 10^6$ lb/in.^{2} It is to be used for the outer ring of a press-fit assembly with a mating shaft made of the same metal. The nominal inside diameter of the ring is 1.000 in., and its outside diameter = 2.500 in. Using a safety factor $SF = 2.0$, determine the maximum interference that should be used with this assembly.

33.12. A shaft made of aluminum is 40.0 mm in diameter at room temperature (21°C). Its coefficient of thermal expansion $\alpha = 24.8 \times 10^{-6}$ mm/mm per °C. If it must be reduced in size by 0.20 mm in order to be expansion fitted into a hole, determine the temperature to which the shaft must be cooled.

33.13. A steel ring has an inside diameter = 30 mm and an outside diameter = 50 mm at room temperature (21°C). If the coefficient of thermal expansion of steel $\alpha = 12.1 \times 10^{-6}$ mm/mm per °C, determine the inside diameter of the ring when heated to 500°C.

33.14. A 1-in. diameter steel pin is to be heated from room temperature (70°F) to 700°F. If the coefficient of thermal expansion of the pin is $\alpha = 6.7 \times 10^{-6}$ in./in. per °F, determine the increase in diameter of the pin.

33.15. A steel collar whose outside diameter = 3.000 in. at room temperature is to be shrink fitted onto a steel shaft by heating it to an elevated temperature while the shaft remains at room temperature. The shaft diameter = 1.500 in. For ease of assembly when the collar is heated to an elevated temperature of 1000°F, the clearance between the shaft and the collar is to be 0.007 in. Determine: (a) the initial inside diameter of the collar at room temperature so that this clearance is satisfied, (b) the radial pressure and (c) maximum effective stress on the resulting interference fit at room temperature (70°F). For steel, $E = 30,000,000$ lb/in.^{2} and $\alpha = 6.7 \times 10^{-6}$ in./in. per °F.

33.16. A pin is to be inserted into a collar using an expansion fit. Properties of the pin and collar metal are: coefficient of thermal expansion = 12×10^{-6} m/m/°C, yield strength = 450 MPa, and modulus of elasticity = 209 GPa. At room temperature (20°C), the outer and inner diameters of the collar = 75.00 mm and 40.00 mm, respectively, and the pin has a diameter = 40.02 mm. The pin is to be reduced in size for assembly into the collar by cooling to a sufficiently low temperature that there is a clearance of 0.04 mm. (a) What is the temperature to which the pin must be cooled for assembly? (b) What is the radial pressure at room temperature after assembly? (c) What is the safety factor in the resulting assembly?

34

RAPID PROTOTYPING

CHAPTER CONTENTS

34.1 Fundamentals of Rapid Prototyping
34.2 Rapid Prototyping Technologies
    34.2.1 Liquid-Based Rapid Prototyping Systems
    34.2.2 Solid-Based Rapid Prototyping Systems
    34.2.3 Powder-Based Rapid Prototyping Systems
34.3 Applications Issues in Rapid Prototyping

In this part of the book, we discuss a collection of processing and assembly technologies that do not fit neatly into our classification scheme in Figure 1.4. They are technologies that have been adapted from the more traditional manufacturing processes and assembly operations or developed from scratch to serve the special functions or needs of designers and manufacturers. Rapid prototyping, covered in this chapter, is a collection of processes used to fabricate a model, part, or tool quickly. Chapters 35 and 36 cover technologies used in electronics manufacturing, an activity of significant and growing economic importance. Chapter 35 covers integrated circuit processing, and Chapter 36 examines electronics assembly and packaging. Chapter 37 surveys some of the technologies used to produce very small parts and products. The processing and assembly topics covered in these four chapters are relatively new. Rapid prototyping dates from about 1988. Modern electronics production techniques date from around 1960 (Historical Note 35.1), although dramatic advances have been made in electronics processing since that time. The microfabrication technologies discussed in Chapter 37 followed soon after electronics processing.

Rapid prototyping (RP) consists of a family of unique fabrication processes developed to make engineering prototypes in minimum possible lead times based on a computer-aided design (CAD) model of the item. The traditional method of fabricating a prototype part is machining, which can require significant lead times — up to several weeks, sometimes longer, depending on part complexity and difficulty in ordering-materials. A number of rapid prototyping techniques are available that allow a part to be produced in hours or days rather than weeks, given that a computer model of the part has been generated on a CAD system.

34.1 FUNDAMENTALS OF RAPID PROTOTYPING

The special need that motivates the variety of rapid prototyping technologies arises because product designers would like to have a physical model of a new part or product design rather than a computer model or line drawing. The creation of a prototype is an integral step in the design procedure. A *virtual prototype,* which is a computer model of the part design on a CAD system, may not be sufficient for the designer to visualize the part adequately. It certainly is not adequate to conduct real physical tests on the part, although it is possible to perform simulated tests by finite element analysis or other methods. Using one of the available RP technologies, a solid physical part can be created in a relatively short time (hours if the company possesses the RP equipment or days if the part fabrication must be contracted to an outside firm specializing in RP). The designer can therefore visually examine and physically feel the part and begin to perform tests and experiments to assess its merits and shortcomings.

Available rapid prototyping technologies can be divided into two basic categories: (1) material removal processes and (2) material addition processes. The *material removal RP* alternative involves machining (Chapter 22), primarily milling and drilling, using a dedicated CNC machine that is available to the design department on short notice. Of course, the problem of preparing the NC part program from the CAD model must be solved (Section 38.1.4). If the part geometry can be analyzed by an automatic NC part programming algorithm, then this is one way to solve the problem. The alternative approach often taken for rapid prototyping is to apply the same kind of input format used in other RP technologies, which involves slicing the solid model into thin layers that approximate the part's solid geometry. The CNC milling machine then contours the part layer by layer from a solid block of starting material. The starting material is often wax, which can be melted and resolidified for reuse when the current prototype part is no longer needed. Wax is also very easy to machine. Other starting materials can also be used, such as wood, plastics, or metals (e.g., a machinable grade of aluminum or brass). The CNC machines used for rapid prototyping are often small, and the terms *desktop milling* or *desktop machining* are sometimes used for this technology. Maximum starting block sizes in desktop machining are typically 180 mm (7 in.) in the x-direction, 150 mm (6 in.) in the y-direction, and 150 mm (6 in.) in the z-direction [2].

The principal emphasis in this chapter is on *material-addition RP* technologies, all of which work by adding layers of material one at a time to build the solid part from bottom to top. Starting materials include: (1) liquid monomers that are cured layer by layer into solid polymers; (2) powders that are aggregated and bonded layer by layer; and (3) solid sheets that are laminated to create the solid part. In addition to starting material, what distinguishes the various material addition RP technologies is the method of building and adding the layers to create the solid part. Some techniques use lasers to solidify the starting material, another deposits a soft plastic filament in the outline of each layer, while others bond solid layers together. There is a correlation between the starting material and the part-building techniques, as we shall see in our discussion of the RP technologies.

The common approach to prepare the control instructions (part program) in all of the current material-addition RP techniques involves the following steps [4]:

1. *Geometric modeling.* This consists of modeling the component on a CAD system to define its enclosed volume. Solid modeling is the preferred technique because it provides a complete and unambiguous mathematical representation of the geometry. For rapid prototyping, the important issue is to distinguish the interior (mass) of the part from its exterior, and solid modeling provides for this distinction.

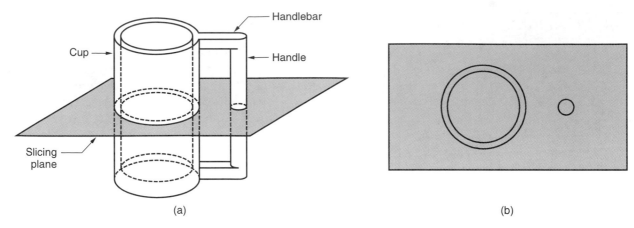

FIGURE 34.1   Conversion of a solid model of an object into layers (only one layer is shown).

2. **Tessellation of the geometric model.** In this step, the CAD model is converted into a format that approximates its surfaces by facets (triangles or polygons). More generally, tessellation involves the laying out or creation of a mosaic (e.g., one consisting of small colored tiles affixed to a surface for decoration). In the case of RP, the tiles (facets) are used to define the surface, at least approximately. The triangles or polygons have their vertices arranged to distinguish the object's interior from its exterior. The common tessellation format used in rapid prototyping is STL,[1] which has become the de facto standard input format for nearly all RP systems.

3. **Slicing of the model into layers.** In this step, the model in STL file format is sliced into closely spaced parallel horizontal layers. Conversion of a solid model into layers is illustrated in Figure 34.1. These layers are subsequently used by the RP system to construct the physical model. By convention, the layers are formed in the x-y plane orientation, and the layering procedure occurs in the z-axis direction. For each layer, a curing path is generated, called the STI file, which is the path that will be followed by the RP system to cure (or otherwise solidify) the layer.

As our brief overview indicates, there are several different technologies used for material addition rapid prototyping. This heterogeneity has spawned several alternative names for rapid prototyping, including **layer manufacturing, direct CAD manufacturing,** and **solid freeform fabrication.** The term **rapid prototyping and manufacturing** (RPM) is also being used more frequently to indicate that the RP technologies can be applied to make production parts and production tooling, not just prototypes.

34.2 RAPID PROTOTYPING TECHNOLOGIES

The 25 or so RP techniques currently developed can be classified in various ways. We adopt a classification system recommended in [4] and consistent with that used in this book for the part-shaping processes (after all, rapid prototyping is a part-shaping process). The classification method is based on the form of the starting material in the RP process:

---

[1]STL stands for STereoLithography, one of the primary technologies used for rapid prototyping, developed by 3D Systems Inc.

(1) liquid-based, (2) solid-based, and (3) powder-based. We discuss examples of each class in the following three sections.

34.2.1 Liquid-Based Rapid Prototyping Systems

The starting material in these technologies is a liquid. About a dozen RP technologies are in this category, of which we have selected the following to describe here: (1) stereolithography, (2) solid ground curing, and (3) droplet deposition manufacturing.

Stereolithography This was the first material addition RP technology, dating from about 1988 and introduced by 3D Systems Inc. based on the work of inventor Charles Hull. At time of writing, there are more installations of stereolithography than any other RP technology. *Stereolithography* (STL, also abbreviated SLA for stereolithography apparatus) is a process for fabricating a solid plastic part out of a photosensitive liquid polymer using a directed laser beam to solidify the polymer. The general setup for the process is illustrated in Figure 34.2. Part fabrication is accomplished as a series of layers, in which one layer is added onto the previous layer to gradually build the desired three-dimensional geometry. A part fabricated by STL is illustrated in Figure 34.3.

The stereolithography apparatus consists of (1) a platform that can be moved vertically inside a vessel containing the photosensitive polymer, and (2) a laser whose beam can be controlled in the *x-y* direction. At the start of the process, the platform is positioned vertically near the surface of the liquid photopolymer, and a laser beam is directed through a curing path that forms an area corresponding to the base (bottom layer) of the part. This and subsequent curing paths are defined by the STI file (step 3 in data preparation described above). The laser hardens (cures) the photosensitive polymer where the beam strikes the liquid, forming a solid layer of plastic that adheres to the platform. When the initial layer is completed, the platform is lowered by a distance equal to the layer thickness, and a second layer is formed on top of the first by the laser, and so on. Before each new layer is cured, a wiper blade is passed over the viscous liquid resin to ensure that its level is the same throughout the surface. Each layer consists of its own area shape, so that the succession of layers, one on top of the previous, creates the solid part shape. Each layer is 0.076 mm to 0.50 mm (0.003–0.020 in.) thick. Thinner layers provide better resolution and allow more intricate part shapes; but processing time is greater.

FIGURE 34.2
Stereolithography: (1) at the start of the process, in which the initial layer is added to the platform; and (2) after several layers have been added so that the part geometry gradually takes form.

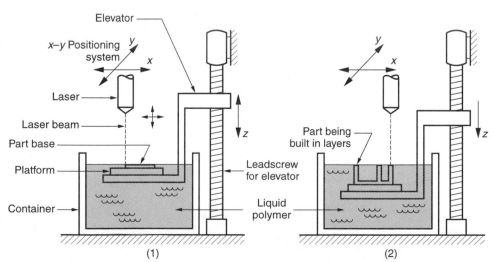

FIGURE 34.3   A part produced by
stereolithography (photo courtesy of
3D Systems, Inc.).

Photopolymers are typically acrylic [10], although use of epoxy for STL has also been reported [8]. The starting materials are liquid monomers. Polymerization occurs upon exposure to ultraviolet light produced by helium-cadmium or argon ion lasers. Scan speeds
of STL lasers typically range between 500 and 2500 mm/s.

The time required to build the part by this layering process ranges from one hour
for small parts of simple geometry up to several dozen hours for complex parts. Other
factors that affect cycle time are scan speed and layer thickness. The part build time in
stereolithography can be estimated by determining the time to complete each layer and
then summing the times for all layers. First, the time to complete a single layer is given
by the following equation:

$$T_i = \frac{A_i}{vD} + T_d \tag{34.1}$$

where T_i = time to complete layer i, seconds, where the subscript i is used to identify the
layer; A_i = area of layer i, mm^2 (in.^{2}); v = average scanning speed of the laser beam at
the surface, mm/s (in./sec); D = diameter of the laser beam at the surface (called the "spot
size," assumed circular), mm (in.); and T_d = delay time between layers to reposition the
worktable, s. The average scanning speed v must include any effects of interruptions in
the scanning path (e.g., due to gaps between areas of the part in a given layer). Once the
T_i values have been determined for all layers, then the build cycle time can be determined:

$$T_c = \sum_{i=1}^{n_l} T_i \tag{34.2}$$

where T_c = the STL build cycle time, s; and n_l = the number of layers used to approximate the part.[2]

---

[2]Although these equations have been developed here for stereolithography, similar formulas can be
developed for the other RP material addition technologies discussed in this chapter, since they all use
the same layer-by-layer fabrication method.

After all of the layers have been formed, the photopolymer is about 95% cured. The piece is therefore "baked" in a fluorescent oven to completely solidify the polymer. Excess polymer is removed with alcohol, and light sanding is sometimes used to improve smoothness and appearance.

Depending on its design and orientation, a part may contain overhanging features that have no means of support during the bottom-up approach used in stereolithography. For example, in the part of Figure 34.1, if the lower half of the handle and the lower handlebar were eliminated, the upper portion of the handle would be unsupported during fabrication. In these cases, extra pillars or webs may need to be added to the part simply for support purposes. Otherwise, the overhangs may float away or otherwise distort the desired part geometry. These extra features must be trimmed away after the process is completed.

Solid Ground Curing   Like stereolithography, solid ground curing (SGC) works by curing a photosensitive polymer layer by layer to create a solid model based on CAD geometric data. Instead of using a scanning laser beam to accomplish the curing of a given layer, SGC exposes the entire layer to an ultraviolet light source through a mask that is positioned above the surface of the liquid polymer. The hardening process takes 2 to 3 s for each layer. SGC systems are sold under the name ***Solider system*** by Cubital Ltd.

The starting data in SGC are similar to those used in stereolithography: a CAD geometric model of the part that has been sliced into layers. For each layer, the step-by-step procedure in SGC is illustrated in Figure 34.4 and described here: (1) A mask is created on a glass plate by electrostatically charging a negative image of the layer onto the surface. The imaging technology is basically the same as that used in photocopiers. (2) A thin flat layer of liquid photopolymer is distributed over the surface of the work platform. (3) The mask is positioned above the liquid polymer surface and exposed by a high powered (e.g., 2000 W) ultraviolet lamp. The portions of the liquid polymer layer that are unprotected by the mask are solidified in about 2 s. The shaded areas of the layer remain in the liquid state. (4) The mask is removed, the glass plate is cleaned and made ready for a subsequent layer in step (1). Meanwhile, the liquid polymer remaining on the surface is removed in a wiping and vacuuming procedure. (5) The now-open areas of the layer are filled in with hot wax. When hardened, the wax acts to support overhanging sections of the part. (6) When the wax has cooled and solidified, the polymer-wax surface is milled to form a flat layer of specified thickness, ready to receive the next application of liquid photopolymer in step (2). Although we have described SGC as a sequential process, certain steps are accomplished in parallel. Specifically, the mask preparation step (1) for the next layer is performed simultaneously with the layer fabrication steps (2) through (6), using two glass plates during alternating layers.

The sequence for each layer takes about 90 s. Throughput time to produce a part by SGC is claimed to be about eight times faster than competing RP systems [4]. The solid cubic form created in SGC consists of solid polymer and wax. The wax provides support for fragile and overhanging features of the part during fabrication, but can be melted away later to leave the free-standing part. No post-curing of the completed prototype model is required, as in stereolithography.

Droplet Deposition Manufacturing   These systems operate by melting the starting material and shooting small droplets onto a previously formed layer. The liquid droplets cold weld to the surface to form a new layer. The deposition of droplets for each new layer is controlled by a moving *x-y* spray nozzle workhead whose path is based on a cross section of a CAD geometric model that has been sliced into layers (similar to the

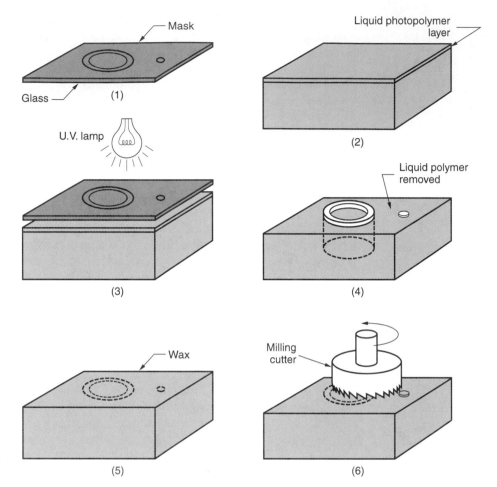

FIGURE 34.4 Solid ground curing process for each layer: (1) mask preparation, (2) applying liquid photopolymer layer, (3) mask positioning and exposure of layer, (4) uncured polymer removed from surface, (5) wax filling, (6) milling for flatness and thickness.

other RP systems). After each layer has been applied, the platform supporting the part is lowered a certain distance corresponding to the layer thickness, in preparation for the next layer. The term droplet deposition manufacturing (DDM) refers to the fact that small particles of work material are deposited as projectile droplets from the work-head nozzle.

Several commercial RP systems are based on this general operating principle, the differences being in the type of material that is deposited and the corresponding technique by which the workhead operates to melt and apply the material. An important criterion that must be satisfied by the starting material is that it be readily melted and solidified. Work materials used in DDM include wax, thermoplastics, and metals with low melting points, such as tin, zinc, lead, and aluminum. For example, the droplet deposition technique can be used to apply solder droplets for integrated circuit packaging (Section 35.6) and fine-line printed circuit boards (Section 36.2) [9].

One of the more popular DDM systems is the Personal Modeler®, available from BPM Technology, Inc. for about $40,000 (at time of writing), one of the lowest priced RP systems. Wax is commonly used as the work material. The ejector head operates using a piezoelectric oscillator that shoots droplets of wax at a rate of 10,000 to 15,000 per second. The droplets are of uniform size at about 0.076 mm (0.003 in.) diameter, which flatten to about 0.05 mm (0.002 in.) solidified thickness on impact against the existing part surface. After each layer has been deposited, the surface is milled or thermally smoothed to achieve accuracy in the z-direction. Layer thickness is about 0.09 mm (0.0035 in.).

34.2.2 Solid-Based Rapid Prototyping Systems

The common feature in these RP systems is that the starting material is solid. In this section, we discuss two solid-based RP systems: (1) laminated-object manufacturing and (2) fused-deposition modeling.

Laminated-Object Manufacturing   The principal company offering laminated object manufacturing (LOM) systems is Helisys, Inc. Of interest is that much of the early research and development work on LOM was funded by National Science Foundation. The first commercial LOM unit was shipped in 1991.

Laminated-object manufacturing produces a solid physical model by stacking layers of sheet stock that are each cut to an outline corresponding to the cross-sectional shape of a CAD model that has been sliced into layers. The layers are bonded one on top of the previous one prior to cutting. After cutting, the excess material in the layer remains in place to support the part during building. Starting material in LOM can be virtually any material in sheet stock form, such as paper, plastic, cellulose, metals, or fiber-reinforced materials. Stock thickness is 0.05 to 0.50 mm (0.002–0.020 in.). In LOM, the sheet material is usually supplied with adhesive backing as rolls that are spooled between two reels, as in Figure 34.5. Otherwise, the LOM process must include an adhesive coating step for each layer.

The data preparation phase in LOM consists of slicing the geometric model using the STL file for the given part. The slicing function is accomplished by LOMSlice™, the special software used in laminated object manufacturing. Slicing of the STL model in LOM is performed after each layer has been physically completed and the vertical height

FIGURE 34.5   Laminated object manufacturing.

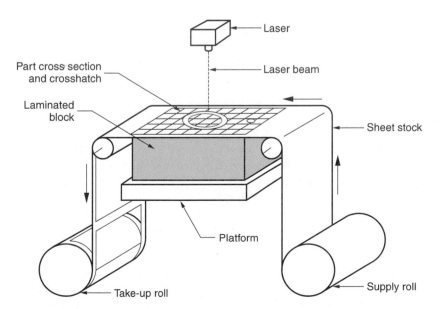

of the part has been measured. This provides a feedback correction to account for the actual thickness of the sheet stock being used, a feature unavailable on most other RP systems. With reference to Figure 34.5, the LOM process for each layer can be described as follows, picking up the action with a sheet of stock in place and bonded to the previous stack: (1) LOMSlice™ computes the cross-sectional perimeter of the STL model based on the measured height of the physical part at the current layer of completion. (2) laser beam is used to cut along the perimeter, as well as to crosshatch the exterior portions of the sheet for subsequent removal. The laser is typically a 25 or 50 W CO_2 laser. The cutting trajectory is controlled by means of an x-y positioning table. The cutting depth is controlled so that only the top layer is cut. (3) The platform holding the stack is lowered, and the sheet stock is advanced between supply roll and take-up spool for the next layer. The platform is then raised to a height consistent with the stock thickness and a heated roller moves across the new layer to bond it to the previous layer. The height of the physical stack is measured in preparation for the next slicing computation by LOMSlice™.

When all of the layers are completed, the new part is separated from the excess external material using a hammer, putty knife, and wood carving tools. The part can then be sanded to smooth and blend the layer edges. A sealing application is recommended, using a urethane, epoxy, or other polymer spray to prevent moisture absorption and damage. LOM part sizes can be relatively large among RP processes, with work volumes ranging up to 800 mm × 500 mm × 550 mm (32 in. × 20 in. × 22 in.). More common work volumes are 380 mm × 250 mm × 350 mm (15 in. × 10 in. × 14 in.).

Several low-cost systems based on the LOM build method are available. For example, the JP System 5, available from Schroff Development Corporation, uses a mechanical knife rather than a laser to cut the sheet stock for each layer. This system is intended as a teaching tool and requires manual assembly of the layers.

Fused-Deposition Modeling   Fused-deposition modeling (FDM) is an RP process in which a filament of wax or polymer is extruded onto the existing part surface from a workhead to complete each new layer. The workhead is controlled in the x-y plane during each layer and then moves up by a distance equal to one layer in the z-direction. The starting material is a solid filament with typical diameter = 1.25 mm (0.050 in.) fed from a spool into the workhead that heats the material to about 0.5°C (1°F) above its melting point before extruding it onto the part surface. The extrudate is solidified and cold welded to the cooler part surface in about 0.1 s. The part is fabricated from the base up, using a layer-by-layer procedure similar to other RP systems.

FDM was developed by Stratasys Inc., which sold its first machine in 1990. The starting data is a CAD geometric model that is processed by Stratasys' software modules QuickSlice® and SupportWork™. QuickSlice® is used to slice the model into layers, and SupportWork™ is used to generate any support structures that are required during the build process. If supports are needed, a dual extrusion head and a different material is used to create the supports. The second material is designed to readily be separated from the primary modeling material. The slice (layer) thickness can be set anywhere from 0.05 to 0.75 mm (0.002–0.030 in.). About 400 mm of filament material can be deposited per second by the extrusion workhead in widths (called the *road width*) that can be set between 0.25 and 2.5 mm (0.010 and 0.100 in.). Starting materials include investment casting wax and several polymers, including ABS, polyamide, polyethylene, and polypropylene. These materials are nontoxic, allowing the FDM machine to be set up in an office environment.

34.2.3 Powder-Based Rapid Prototyping Systems

The common feature of the RP technologies described in this section is that the starting material is powder.[3] We discuss two RP systems in this category: (1) selective laser sintering and (2) three-dimensional printing.

Selective Laser Sintering   Selective laser sintering (SLS) uses a moving laser beam to sinter heat-fusible powders in areas corresponding to the CAD geometric model one layer at a time to build the solid part. After each layer is completed, a new layer of loose powders is spread across the surface using a counter-rotating roller. The powders are preheated to just below their melting point in order to facilitate bonding and reduce distortion. Layer by layer, the powders are gradually bonded into a solid mass that forms the three-dimensional part geometry. In areas not sintered by the laser beam, the powders remain loose so they can be poured out of the completed part. In the meantime, they serve to support the solid regions of the part as fabrication proceeds. Layer thickness are 0.075 to 0.50 mm (0.003–0.020 in.).

SLS was developed at the University of Texas (Austin) as an alternative to stereolithography, and SLS machines are currently marketed by the DTM Corporation. It is a more versatile process than stereolithography in terms of possible work materials. Current materials used in selective laser sintering include polyvinylchloride, polycarbonate, polyester, polyurethane, ABS, nylon, and investment casting wax. These materials are less expensive than the photosensitive resins used in stereolithography. They are also nontoxic and can be sintered using low power (25 to 50 W) CO_2 lasers. Metal and ceramic powders are also being used in SLS.

Three-Dimensional Printing   This RP technology was developed at Massachusetts Institute of Technology. Three-dimensional printing (3DP) builds the part in the usual layer-by-layer fashion using an ink-jet printer to eject an adhesive bonding material onto successive layers of powders. The binder is deposited in areas corresponding to the cross sections of the solid part, as determined by slicing the CAD geometric model into layers. The binder holds the powders together to form the solid part, while the unbonded powders remain loose to be removed later. While the loose powders are in place during the build process, they provide support for overhanging and fragile features of the part. When the build process is completed, the part is heat treated to strengthen the bonding, followed by removal of the loose powders. To further strengthen the part, a sintering step can be applied to bond the individual powders.

The part is built on a platform whose level is controlled by a piston. Let us describe the process for one cross section with reference to Figure 34.6: (1) A layer of powder is spread on the existing part-in-process. (2) An ink-jet printing head moves across the surface, ejecting droplets of binder on those regions that are to become the solid part. (3) When the printing of the current layer is completed, the piston lowers the platform for the next layer.

Starting materials in 3DP are powders of ceramic, metal, or cermet, and binders that are polymeric or colloidal silica or silicon carbide [8], [10]. Typical layer thickness ranges between 0.10 and 0.18 mm (0.004 and 0.007 in.). The ink-jet printing head moves across the layer at a speed of about 1.5 m/sec (59 in./sec), with ejection of liquid binder determined during the sweep by raster scanning. The sweep time, together with the spreading of the powders, permits a cycle time per layer of about 2 sec [10].

---

[3]The definition, characteristics, and production of powders are described in Chapters 16 and 17.

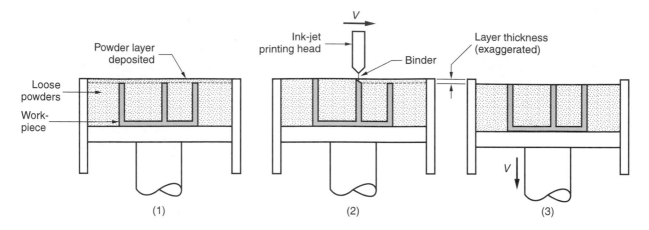

FIGURE 34.6   Three-dimensional printing: (1) powder layer is deposited, (2) ink-jet printing of areas that will become the part, and (3) piston is lowered for next layer (key: *v* = motion).

34.3 APPLICATIONS ISSUES IN RAPID PROTOTYPING

Applications of rapid prototyping can be classified into three categories: (1) design, (2) engineering analysis and planning, and (3) tooling and manufacturing.

Design   Design applications were the initial application emphasis for RP systems, and most of the early applications were in design. Designers are able to confirm their design by building a real physical model in minimum time using rapid prototyping. The features and functions of the part can be communicated to others more easily using a physical model than by a paper drawing or displaying it on a CAD system monitor. Benefits to design attributed to rapid prototyping include [2]: (1) reduced lead times to produce prototype components; (2) improved ability to visualize the part geometry due to its physical existence; (3) earlier detection and reduction of design errors; and (4) increased capability to compute mass properties of components and assemblies.

Engineering Analysis and Planning   The existence of an RP-fabricated part allows for certain types of engineering analysis and planning activities to be accomplished that would be more difficult without the physical entity, including: (1) comparison of different shapes and styles to optimize aesthetic appeal of the part; (2) analysis of fluid flow through different orifice designs in valves fabricated by RP; (3) wind tunnel testing of different streamline shapes using physical models created by RP; (4) stress analysis of a physical model; (5) fabrication of preproduction parts by RP as an aid in process planning and tool design; and (6) combining medical imaging technologies, such as MRI,[4] with RP to create models for doctors in planning surgical procedures or fabricating prostheses or implants.

Tooling and Manufacturing   The trend in RP applications is toward its greater use in the fabrication of production tooling and in the actual manufacture of parts. When RP is adopted to fabricate production tooling, the term ***rapid tool making*** (RTM) is often

---

[4]MRI stands for magnetic resonance imaging.

used. RTM applications divide into two approaches [3]: *indirect* RTM method, in which a pattern is created by RP and the pattern is used to fabricate the tool, and *direct* RTM method, in which RP is used to make the tool itself. Examples of indirect RTM include [4], [8]: (1) use of an RP-fabricated part as the master in making a silicon rubber mold that is subsequently used as a production mold; (2) RP patterns to make the sand molds in sand casting (Section 11.1); (3) fabrication of patterns of low-melting point materials (e.g., wax) in limited quantities for investment casting (Section 11.2.4); and (4) making electrodes for EDM (Section 26.3.1). Examples of direct RTM include [3], [4], [8]: (1) RP-fabricated mold cavity inserts that can be sprayed with metal to produce injection molds for a limited quantity of production plastic parts (Section 13.6); and (2) 3D Printing to create a die geometry in metallic powders followed by sintering and infiltration to complete the fabrication of the die.

Examples of actual part production include [8]: (1) small batch sizes of plastic parts that could not be economically injection molded because of the high cost of the mold; (2) parts with intricate internal geometries that could not be made using conventional technologies without assembly; and (3) one-of-a-kind parts such as bone replacements that must be made to correct size for each user.

Not all RP technologies can be used for all of these tooling and manufacturing examples. Interested readers should consult more complete treatments of the RP technologies for specific details on these and other examples.

Problems with Rapid Prototyping The principal problems with current RP technologies include: (1) part accuracy, (2) limited variety of material, and (3) mechanical performance of the fabricated parts.

Several sources of error limit part accuracy in RP systems [10]: (1) mathematical, (2) process-related, or (3) material-related. Mathematical errors include approximations of part surfaces used in RP data preparation and differences between the slicing thicknesses and actual layer thicknesses in the physical part. The latter differences result in z-axis dimensional errors. An inherent limitation in the physical part is the steps between layers, especially as layer thickness is increased, resulting in a staircase appearance for a sloping part surface. Process-related errors are those that result from the particular part building technology used in the RP system. These errors degrade the shape of each layer as well as the registration between adjacent layers. Process errors can also affect the z-dimension. Finally, material-related errors include shrinkage and distortion. An allowance for shrinkage can be made by enlarging the CAD model of the part based on previous experience with the process and materials.

Current rapid prototyping systems are limited in the variety of materials they can process. For example, the most common RP technology, stereolithography, is limited to photosensitive polymers. In general, the materials used in RP systems are not as strong as the production part materials that will be used in the actual product. This limits the mechanical performance of the prototypes and the amount of realistic testing that can be done to verify the design during product development.

REFERENCES

[1] Ashley, S. "Rapid Prototyping is Coming of Age." *Mechanical Engineering.* Vol. 117, No. 7, July 1995, pp. 62–68.

[2] Bakerjian, R., and Mitchell, P. *Tool and Manufacturing Engineers Handbook.* 4th ed. Vol. VI: *Design for Manufacturability.* Society of Manufacturing Engineers, Dearborn, Mich. 1992, Chapter 7.

[3]   Hilton, P. "Making the Leap to Rapid Tool Making." *Mechanical Engineering.* Vol. 117, No. 7, July 1995, pp. 75–76.

[4]   Kai, C. C., and Fai, L. K. *Rapid Prototyping: Principles and Applications in Manufacturing.* Wiley (Asia) Pte Ltd, Singapore, 1997.

[5]   Kai, C. C., and Fai, L. K. "Rapid Prototyping and Manufacturing: The Essential Link between Design and Manufacturing." Chapter 6 in *Integrated Product and Process Development: Methods, Tools, and Technologies.* Ed. J. M. Usher, U. Roy, and H. R. Parsaei. Wiley, New York, 1998, pp. 151–183.

[6]   Kochan, D., C. C. Kai, and Zhaohui, D. "Rapid Prototyping Issues in the 21st Century. *Computers in Industry.* Vol. 39, pp. 3–10, 1999.

[7]   Pacheco, J. M. *Rapid Prototyping.* Report MTIAC SOAR-93-01, Manufacturing Technology Information Analysis Center, IIT Research Institute, Chicago, Ill., 1993.

[8]   Pham, D. T, and Gault, R. S. "A Comparison of Rapid Prototyping Technologies," *International Journal of Machine Tools and Manufacture.* Vol. 38, pp. 1257–1287, 1998.

[9]   Tseng, A. A., Lee, M. H., and Zhao, B., "Design and Operation of a Droplet Deposition System for Freedom Fabrication of Metal Parts; *ASME J. Eng. Mat. Tech.,* Vol. 123, No. 1, 2001.

[10]   Yan, X., and Gu, P. "A Review of Rapid Prototyping Technologies and Systems." *Computer-Aided Design.* Vol. 28, No. 4, pp. 307–318, 1996.

REVIEW QUESTIONS

34.1. What is *rapid prototyping?* Provide a definition of the term.

34.2. What are the three types of starting materials in rapid prototyping?

34.3. Besides the starting material, what other feature distinguishes the rapid prototyping technologies?

34.4. What is the common approach used in all of the material addition technologies to prepare the control instructions for the RP system?

34.5. Of all of the current rapid prototyping technologies, which one is the most widely used?

34.6. Describe the RP technology called solid ground curing.

34.7. Describe the RP technology called laminated-object manufacturing.

34.8. What is the starting material in fused-deposition modeling?

MULTIPLE CHOICE QUIZ

There is a total of 14 correct answers in the following multiple choice questions (some questions have multiple answers that are correct). To attain a perfect score on the quiz, all correct answers must be given, since each correct answer is worth 1 point. For each question, each omitted answer or wrong answer reduces the score by 1 point, and each additional answer beyond the number of answers required reduces the score by 1 point. Percentage score on the quiz is based on the total number of correct answers.

34.1. Machining is never used for rapid prototyping because it takes too long: (a) true or (b) false.

34.2. Which of the following rapid prototyping processes starts with a photosensitive liquid polymer to fabricate a component (more than one)? (a) ballistic particle manufacturing, (b) fused deposition modeling, (c) selective laser sintering, (d) solid ground curing, and (e) stereolithography.

34.3. Of all of the current material addition RP technologies, which one is the most widely used? (a) ballistic particle manufacturing, (b) fused deposition modeling, (c) selective laser sintering, (d) solid ground curing, and (e) stereolithography.

34.4. Which of the following RP technologies use a liquid as the starting material (more than one)? (a) ballistic par-

ticle manufacturing, (b) fused deposition modeling, (c) laminated object manufacturing, (d) selective laser sintering, (e) solid ground curing, and (f) stereolithography.

34.5. Which one of the following RP technologies uses solid sheet stock as the starting material? (a) ballistic particle manufacturing, (b) fused deposition modeling, (c) laminated object manufacturing, (d) solid ground curing, and (e) stereolithography.

34.6. Which of the following RP technologies uses powders as the starting material (more than one)? (a) ballistic particle manufacturing, (b) fused deposition modeling, (c) selective laser sintering, (d) solid ground curing, and (e) three dimensional printing.

34.7. Rapid prototyping technologies are never used to make production parts: (a) true or (b) false.

34.8. Which of the following are problems with the current material addition rapid prototyping technologies (more than one)? (a) inability to convert a solid part into layers, (b) limited material variety, (c) part accuracy, and (d) part shrinkage.

PROBLEMS

34.1. A prototype of a tube with a square cross section is to be fabricated using stereolithography. The outside dimension of the square = 100 mm and the inside dimension = 90 mm (wall thickness = 5 mm except at corners). The height of the tube (z-direction) = 80 mm. Layer thickness = 0.10 mm. The diameter of the laser beam ("spot size") = 0.25 mm, and the beam is moved across the surface of the photopolymer at a velocity of 500 mm/s. Compute an estimate for the time required to build the part, if 10 s are lost from each layer to lower the height of the platform that holds the part. Neglect the time for postcuring.

34.2. Solve Problem 34.1, except that the layer thickness = 0.40 mm.

34.3. The part in Problem 34.1 is to be fabricated using fused deposition modeling instead of stereolithography. Layer thickness is to be 0.20 mm, and the width of the extrudate deposited on the surface of the part = 1.25 mm. The extruder workhead moves in the x-y plane at a speed of 150 mm/s. A delay of 10 s is experienced between each layer to reposition the workhead. Compute an estimate for the time required to build the part.

34.4. Solve Problem 34.3, except using the following additional information. It is known that the diameter of the filament fed into the extruder workhead is 1.25 mm, and the filament is fed into the workhead from its spool at a rate of 30.6 mm of length per second while the workhead is depositing material. Between layers, the feed rate from the spool is zero.

34.5. A cone-shaped part is to be fabricated using stereolithography. The radius of the cone at its base = 35 mm and its height = 40 mm. The layer thickness = 0.20 mm. The diameter of the laser beam = 0.22 mm, and the beam is moved across the surface of the photopolymer at a velocity of 500 mm/s. Compute an estimate for the time required to build the part, if 10 s are lost from each layer to lower the height of the platform that holds the part. Neglect postcuring time.

34.6. The cone-shaped part in Problem 34.5 is to be built using laminated object manufacturing. Layer thickness = 0.20 mm. The laser beam can cut the sheet stock at a velocity of 500 mm/s. Compute an estimate for the time required to build the part, if 10 s are lost each layer to lower the height of the platform that holds the part and advance the sheet stock in preparation for the next layer. Ignore cutting of the crosshatched areas outside of the part since the cone should readily drop out of the stack owing to its geometry.

34.7. Stereolithography is to be used to build the part in Figure 34.1 (in text). Dimensions of the part are: height = 125 mm, outside diameter = 75 mm, inside diameter = 65 mm, handle diameter = 12 mm, and handle distance from cup = 70 mm measured from center (axis) of cup to center of handle. The handle bars connecting the cup and handle at the top and bottom of the part have a rectangular cross section and are 10 mm thick and 12 mm wide. The thickness at the base of the cup is 10 mm. The laser beam diameter is 0.25 mm, and the beam can be moved across the surface of the photopolymer at is 500 mm/s. Layer thickness = 0.20 mm. Compute an estimate of the time required to build the part, if 10 s are lost each layer to lower the height of the platform that holds the part. Neglect postcuring time.

35 PROCESSING OF INTEGRATED CIRCUITS

CHAPTER CONTENTS

35.1 Overview of IC Processing
    35.1.1 Processing Sequence
    35.1.2 Clean Rooms
35.2 Silicon Processing
    35.2.1 Production of Electronic Grade Silicon
    35.2.2 Crystal Growing
    35.2.3 Shaping of Silicon into Wafers
35.3 Lithography
    35.3.1 Photolithography
    35.3.2 Other Lithography Techniques
35.4 Layer Processes Use in IC Fabrication
    35.4.1 Thermal Oxidation
    35.4.2 Chemical Vapor Deposition
    35.4.3 Introduction of Impurities into Silicon
    35.4.4 Metallization
    35.4.5 Etching
35.5 Integrating the Fabrication Steps
35.6 IC Packaging
    35.6.1 IC Package Design
    35.6.2 Processing Steps in IC Packaging
35.7 Yields in IC Processing

An *integrated circuit* (IC) is a collection of electronic devices such as transistors, diodes, and resistors that have been fabricated and electrically intraconnected onto a small flat chip of semiconductor material (see Color Plate 2). The IC was invented in 1959 and has been the subject of continual development ever since (Historical Note 35.1). Silicon (Si) is the most widely used semiconductor material for ICs, due to its combination of properties and low cost. Less common semiconductor chips are made of germanium (Ge) and gallium arsenide (GaAs). Since the circuits are fabricated into one solid piece of material, the term *solid-state* electronics is used to denote these devices.

Integrated circuits can be divided into two major types: analog and digital. Analog integrated circuits operate with voltages that are continuous and variable. Typical devices include amplifiers, oscillators, and voltage regulators. Digital ICs operate on signals that

---

Historical Note 35.1   *Integrated circuit technology* [11].

The history of integrated circuits includes inventions of electronic devices and the processes for making these devices. The development of radar immediately before World War II (1939–1945) identified germanium and silicon as important semiconductor elements for the diodes used in radar circuitry. Owing to the importance of radar technology in the war, commercial sources of germanium and silicon were developed.

In 1947, the transistor was developed at the Bell Telephone Laboratories by J. Bardeen and W. Brattain. An improved version was subsequently invented by W. Shockley of Bell Labs in 1952. These three inventors shared the 1956 Nobel Prize in Physics for their research on semiconductors and the discovery of the transistor. The interest of the Bell Labs was to develop electronic switching systems that were more reliable than the electromechanical relays and vacuum tubes used at that time.

In February 1959, J. Kilby of Texas Instruments Inc. filed a patent application for the fabrication of multiple electronic devices and their intraconnection to form a circuit on a single piece of semiconductor material. Kilby was describing an integrated circuit (IC). In May 1959, J. Hoerni of Fairchild Semiconductor Corp. applied for a patent describing the planar process for fabricating transistors. In July of the same year, R. Noyce also of Fairchild filed a patent application similar to the Kilby invention but specifying the use of planar technology and adherent leads.

Although filed later than Kilby's, Noyce's patent was issued first, in 1961 (the Kilby patent was awarded in 1964). This discrepancy in dates and similarity in inventions have resulted in considerable controversy over who was really the inventor of the IC. The issue was argued in legal suits stretching all the way to the U.S. Supreme Court. The high court refused to hear the case, leaving stand a lower court ruling that favored several of Noyce's claims. The result (at the risk of oversimplifying) is that Kilby is generally credited with the concept of the monolithic integrated circuit, while Noyce is credited with the method for fabricating it.

The first commercial ICs were introduced by Texas Instruments in March 1960. Early integrated circuits contained about 10 devices on a small silicon chip— about 3 mm (0.12 in.) square. By 1966, silicon had overtaken germanium as the preferred semiconductor material. Since that year, Si has been the predominant material in IC fabrication. Since the 1960s, a continual trend toward miniaturization and increased integration of multiple devices in a single chip has occurred in the electronics industry (the progress can be seen in Table 35.1), leading to the components described in this chapter.

---

have only two voltage levels, usually indicating the bit values 0 or 1. This second category is represented by microprocessors and memory devices for data storage.

The most fascinating aspect of microelectronics technology is the huge number of devices that can be packed onto a single small chip. Various terms have been developed to define the level of integration and density of packing, such as large-scale integration (LSI) and very large scale integration (VLSI). Table 35.1 lists these terms, their definitions (although there is not complete agreement over the dividing lines between levels), and the period during which the technology was or is being introduced.

In this chapter, we consider how integrated circuits are manufactured. Our presentation will focus almost exclusively on silicon processing, although similar processing techniques are used for ICs fabricated from other semiconductor materials. Some of the terminology used in this chapter may be unfamiliar to readers. In Table 35.2, we present a glossary of terms relating to semiconductor and IC technology.

35.1 OVERVIEW OF IC PROCESSING

Structurally, an integrated circuit consists of hundreds, thousands, or millions of microscopic electronic devices that have been fabricated and electrically intraconnected within the surface of a silicon chip. A ***chip,*** also called a ***die,*** is a square or rectangular flat plate that is about 0.5 mm (0.020 in.) thick and typically 5 to 25 mm (0.2–1.0 in.) on a side

TABLE 35.1 Levels of integration in microelectronics.

Integration Level	Number of Devices on a Chip	Approx. Year Introduced
Small scale integration (SSI)	10–50	1959
Medium scale integration (MSI)	$50–10^3$	1960s
Large scale integration (LSI)	$10^3–10^4$	1970s
Very large scale integration (VLSI)	$10^4–10^6$	1980s
Ultra large scale integration (ULSI)	$10^6–10^8$	1990s
Giga scale integration	$10^9–10^{10}$	2000s

(Color Plate 2). Each electronic device (i.e., transistor, diode, etc.) on the chip surface consists of separate layers and regions with different electrical properties combined to perform the particular electronic function of the device. A typical cross section of such a MOSFET device is illustrated in Figure 35.1. The devices are electrically connected to one another by very fine lines of conducting material, usually aluminum, so that the intraconnected devices (i.e., the integrated circuit) function in the specified way. Conducting lines and pads are also provided to electrically connect the IC to leads, which in turn permit the IC to be connected to external circuits.

TABLE 35.2 Glossary of basic terms in (silicon) semiconductor technology.

Bipolar Junction Transistor (BJT)—A transistor in which current flows between an emitter and a collector through a thin base region, the flow being controlled by voltage applied to the base. It is called bipolar because both positive (p-type) and negative (n-type) charge carriers are used to carry current.

Diode—Electronic device with two electrodes (anode and cathode) used as a rectifier; also used as a detector in radio and television receivers.

Doping—The process of introducing impurities into a semiconductor material to alter its electrical properties, transforming the material into either a n-type or a p-type semiconductor.

Field-Effect Transistor (FET)—A semiconductor transistor in which current flows between source and drain regions through a channel, the flow depending on the application of voltage to the channel gate. Field-effect transistors are available in several types, the most common of which is the metal-oxide-semiconductor field-effect transistor (MOSFET).

Metal-Oxide-Semiconductor Field-Effect Transistor (MOSFET)—A field effect transistor in which silicon dioxide (an insulator) is used to separate the channel and gate metallization in silicon semiconductors. There are three types of MOS devices: (1) NMOS, (2) PMOS, and (3) CMOS. A *NMOS* device is one in which the channel is n-type. A *PMOS* device is one in which the channel is p-type. CMOS stands for complementary metal-oxide-semiconductor. A CMOS device contains both NMOS and PMOS devices, which has the advantage of reduced power consumption and increased speed during operation of the circuit.

n-Type—A semiconductor material containing an excess of donor impurities that give rise to electrons in its atomic structure; these electrons possess a negative charge and serve as charge carriers to conduct electrical current.

p-Type—A semiconductor material containing an excess of acceptor impurities that give rise to holes (missing electrons) in its atomic structure; these holes possess a positive charge and serve as charge carriers to conduct electrical current.

Rectifier—Device that permits electrical current to flow in only one direction; it is thus capable of converting alternating current to direct current.

Semiconductor—Crystalline solid material whose electrical conductivity is between that of conductors (metals) and insulators. Common semiconductor materials are silicon (Si), germanium (Ge), and gallium arsenide (GaAs).

Transistor—Semiconductor device capable of performing various functions such as amplifying, controlling, or generating electrical signals. There are two types: (1) bipolar junction transistor, and (2) field effect transistor—see definitions.

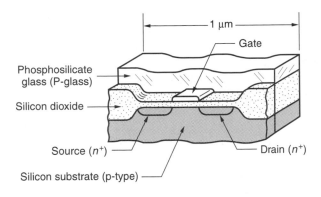

FIGURE 35.1   Cross section of a transistor (specifically, a MOSFET) in an integrated circuit. Approximate size of the device is shown; feature sizes within the device can be as small as 0.1 μm with current technology.

In order to allow the IC to be connected to the outside world, and to protect it from damage, the chip is attached to a lead frame and encapsulated inside a suitable package, as in Figure 35.2. The package is an enclosure, usually made of plastic or ceramic, that provides mechanical and environmental protection for the chip and includes leads by which the IC can be electrically connected to external circuits. The leads are attached to conducting pads on the chip that access the IC.

35.1.1 Processing Sequence

The sequence to fabricate a silicon-based IC chip begins with the processing of silicon (Section 7.5.2). Briefly, silicon of very high purity is reduced in several steps from sand (silicon dioxide, SiO_2). The silicon is grown from a melt into a large solid single crystal log, with typical length of 1 to 3 m (3 to 10 ft) and diameter up to 300 mm (12 in.). The log, called a *boule,* is then sliced into thin wafers, which are disks of thickness equal to about 0.5 mm (0.020 in.).

After suitable finishing and cleaning, the wafers are ready for the sequence of processes by which microscopic features of various chemistries will be created in their surface to form the electronic devices and their intraconnections. The sequence consists of several types of processes, most of them repeated many times. Basically, the objective in each step is to add, alter, or remove a layer of material in selected regions of the wafer surface. The layering steps in IC fabrication are sometimes referred to as the *planar process,* because the processing relies on the geometric form of the silicon wafer being a plane. The processes by which the layers are added include thin film deposition techniques such as physical vapor deposition (Section 29.3) and chemical vapor deposition (Section 29.4), and existing layers are altered by diffusion and ion implantation (Section 28.3). Additional layer-forming techniques, such as thermal oxidation, are also employed.

FIGURE 35.2   Packaging of an integrated circuit chip: (a) cutaway view showing the chip attached to a lead frame and encapsulated in a plastic enclosure, and (b) the package as it would appear to a user. This type of package is called a dual in-line package (DIP).

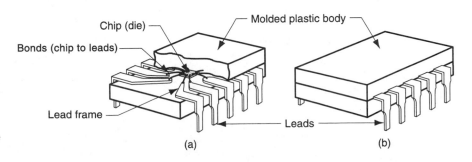

Layers are removed in selected regions by etching, using chemical etchants (usually acid solutions) and other more advanced technologies such as plasma etching.

The addition, alteration, and removal of layers must be done selectively; that is, only in certain extremely small regions of the wafer surface to create the device details such as in Figure 35.1. To distinguish which regions will be affected in each processing step, a procedure involving *lithography* is used. In this technique, masks are formed on the surface to protect certain areas and allow other areas to be exposed to the particular process (e.g., film deposition, etching). By repeating the steps many times, exposing different areas in each step, the starting silicon wafer is gradually transformed into many integrated circuits.

Processing of the wafer is organized in such a way that many individual chip surfaces are formed on a single wafer. Since the wafer is round with diameter ranging from 150 to 300 mm (6–12 in.), while the final chip may only be 12 mm (.5 in.) square, it is possible to produce hundreds of chips on a single wafer. At the conclusion of planar processing, each IC on the wafer is visually and functionally tested, the wafer is diced into individual chips, and each chip that passes the quality testing is packaged as in Figure 35.2.

Summarizing the preceding discussion, the production of silicon-based integrated circuits consists of the following stages, portrayed in Figure 35.3: (1) *Silicon processing,* in which sand is reduced to very pure silicon and then shaped into wafers; (2) *IC fabrication,* consisting of multiple processing steps that add, alter, and remove thin layers in selected regions to form the electronic devices; lithography is used to define the regions to be processed on the surface of the wafer; and (3) *IC packaging,* in which the wafer is tested, cut into individual dies (IC chips), and the dies are encapsulated in an appropriate package.

The presentation in subsequent sections of our chapter are concerned with the details of these processing stages. Section 35.2 deals with silicon processing. Section 35.3 discusses lithography, and Section 35.4 examines the processes used in conjunction with lithography to add, alter, or remove layers. We consider an example of IC fabrication in Section 35.5. Section 35.6 describes die cutting and packaging of the chips. And finally, Section 35.7 covers yield analysis in IC fabrication.

Before beginning our coverage of processing details, it is important to note that the microscopic dimensions of the devices in integrated circuits impose special requirements on the environment in which IC fabrication is carried out.

FIGURE 35.3   Sequence of processing steps in the production of integrated circuits: (1) pure silicon is formed from the molten state into an ingot and then sliced into wafers; (2) fabrication of integrated circuits on the wafer surface; and (3) wafer is cut into chips and packaged.

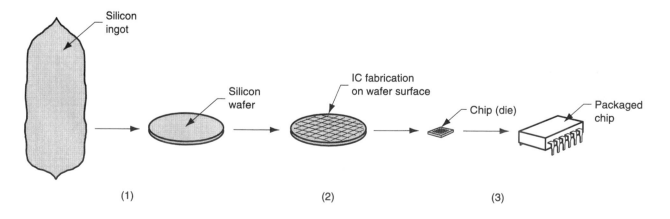

35.1.2 Clean Rooms

Much of the processing sequence for integrated circuits must be carried out in a clean room, the ambiance of which is more like a hospital operating room than a production factory. Cleanliness is dictated by the microscopic feature sizes in an IC, the scale of which continues to decrease with each passing year. Figure 35.4 shows the trend in IC device feature size; also displayed in the same figure are common airborne particles that are potential contaminants in IC processing. These particles can cause defects in the integrated circuits, reducing yields and increasing costs.

A clean room provides protection from these contaminants. The air is purified to remove most of the particles from the processing environment; temperature and humidity are also controlled. A standard classification system is used to specify the cleanliness of a clean room. In the system, a number (in increments of ten) is used to indicate the quantity of particles of size 0.5 μm or greater in one cubic foot of air (only in America would we mix metric units with the U.S. customary system). Thus, a **_class 100_** clean room must maintain a count of particles of size 0.5 μm or greater at less than 100/ft^3. Modern VLSI processing requires **_class 10_** clean rooms, which means that the number of particles of size equal to or greater than 0.5 μm is less than 10/ft^3. The air in the clean room is air conditioned to a temperature of 21°C (70°F) and 45% relative humidity. The air is passed through a high efficiency particulate air (HEPA) filter to capture particle contaminants.

Humans are perhaps the biggest source of contaminants in IC processing; emanating from humans are bacteria, tobacco smoke, viruses, hair, and other particles. Human workers in IC processing areas are required to wear special clothing, generally consisting of white cloaks, gloves, and hair nets. Where extreme cleanliness is required, workers are completely encased in bunny suits. Processing equipment is a second major source of contaminants; machinery produces wear particles, oil, dirt, and similar contaminants. IC processing is usually accomplished in laminar-flow hooded work areas, which can be purified to greater levels of cleanliness than the general environment of the clean room.

In addition to the very pure atmosphere provided by the clean room, the chemicals and water used in IC processing must be very clean and free of particles. Modern practice requires that chemicals and water be filtered prior to use.

FIGURE 35.4   Trend in device feature size in IC fabrication; also shown is the size of common airborne particles that can contaminate the processing environment.

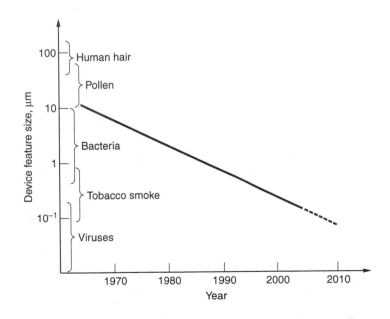

35.2 SILICON PROCESSING

Microelectronic chips are fabricated on a substrate of semiconductor material. Silicon is the leading semiconductor material today, constituting more than 95% of all semiconductor devices produced in the world. Our discussion in this introductory treatment will be limited to Si. The preparation of the silicon substrate can be divided into three steps: (1) production of electronic grade silicon, (2) crystal growing, and (3) shaping of Si into wafers.

35.2.1 Production of Electronic Grade Silicon

Silicon is one of the most abundant materials in the Earth's crust (Table 7.1), occurring naturally as silica (e.g., sand) and silicates (e.g., clay). Electronic grade silicon (EGS) is polycrystalline silicon of ultra high purity—so pure that the impurities are in the range of parts per billion (ppb). They cannot be measured by conventional chemical laboratory techniques but must be inferred from measurements of resistivity on test ingots. The reduction of the naturally occurring Si compound to EGS involves the following processing steps.

The first step is carried out in a submerged-electrode arc furnace. The principal raw material for silicon is **quartzite,** which is very pure SiO_2. The charge includes coal, coke, and wood chips as sources of carbon for the various chemical reactions that occur in the furnace. The net product consists of metallurgical grade silicon (MGS), and the gases SiO and CO. MGS is only about 98% Si, which is adequate for metallurgical alloying but not for electronics components. The major impurities (making up the remaining 2% of MGS) include aluminum, calcium, carbon, iron, and titanium.

The second step involves grinding the brittle MGS and reacting the Si powders with anhydrous HCl to form trichlorsilane:

$$Si + 3HCl(gas) \rightarrow SiHCl_3(gas) + H_2(gas) \qquad (35.1)$$

The reaction is performed in a fluidized-bed reactor at temperatures around 300°C (550°F). Trichlorsilane ($SiHCl_3$), although shown as a gas in Eq. (35.1), is a liquid at room temperature. Its low boiling point of 32°C (90°F) permits it to be separated from the leftover impurities of MGS by fractional distillation.

The final step in the process is reduction of the purified trichlorsilane by means of hydrogen gas. The process is carried out at temperatures up to 1000°C (1800°F), and a simplified equation of the reaction can be written as follows:

$$SiHCl_3(gas) + H_2(gas) \rightarrow Si + 3HCl(gas) \qquad (35.2)$$

The product of this reaction is electronic grade silicon—Si of nearly 100% purity. Two processes to accomplish this reaction will be mentioned here. The first is the **Siemens process,** in which Si is deposited onto a thin silicon rod by chemical vapor deposition. It is capable of producing polycrystalline silicon cylinders up to 200 mm (8 in.) in diameter and 3 m (10 ft) in length. Siemens has been the dominant process in the industry since around 1970, but it suffers from several disadvantages [6]: (1) high capital equipment cost; (2) high power consumption; (3) relatively low efficiency in yielding Si because of intermediate reaction products such as $SiCl_4$; and (4) high labor cost because it is a batch process.

The alternative process employs a fluidized-bed reactor rather than CVD, but similar reactions take place during the process. Use of the fluidized-bed process is expected to increase due to the following advantages over the Siemens method [6]: (1) higher Si yield in the product; (2) lower power consumption; and (3) continuous operation rather than batch.

35.2.2 Crystal Growing

The silicon substrate for microelectronic chips must be made of a single crystal whose unit cell is oriented in a certain direction. The properties of the substrate and the way it is processed are both influenced by these factors. Accordingly, silicon used as the raw material in semiconductor device fabrication must not only be of ultra high purity, as in electronic grade silicon; it must also be prepared in the form of a single crystal and then cut in a direction that achieves the desired planar orientation. The crystal-growing process is covered here, while the next section details the cutting operation.

The most widely used crystal-growing method in the semiconductor industry is the **_Czochralski process,_** illustrated in Figure 35.5, in which a single crystal ingot, called a **_boule,_** is pulled upward from a pool of molten silicon. The setup includes a furnace, a mechanical apparatus for pulling the boule, a vacuum system, and supporting controls. The furnace consists of a crucible and heating system contained in a vacuum chamber. The crucible is supported by a mechanism that permits rotation during the crystal-pulling procedure. Chunks of EGS are placed in the crucible and heated to a temperature slightly above the melting point of silicon: 1410°C (2570°F). Heating is by induction or resistance, the latter being used for large melt sizes. The molten silicon is doped (Table 35.2) prior to boule pulling to make the crystal either p-type or n-type.

To initiate crystal growing, a seed crystal of silicon is dipped into the molten pool and then withdrawn upward under carefully controlled conditions. At first the pulling rate

FIGURE 35.5 The Czochralski process for growing single-crystal ingots of silicon: (a) initial setup prior to start of crystal pulling, and (b) during crystal pulling to form the boule.

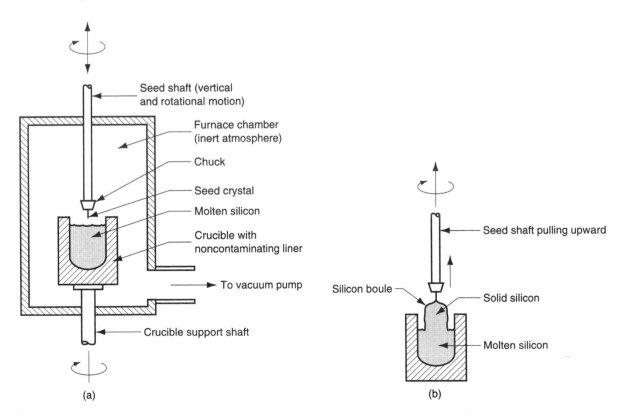

(a)

(b)

(vertical velocity of the pulling apparatus) is relatively rapid, which causes a single crystal of silicon to solidify against the seed, forming a thin neck. The velocity is then reduced, causing the neck to grow into the desired larger diameter of the boule while maintaining its single crystal structure. In addition to pulling rate, rotation of the crucible and other process parameters are used to control boule size. Single-crystal ingots of diameter = 200 mm (8 in.) or greater and up to 3 m (10 ft) long are commonly produced for subsequent fabrication of microelectronic chips.

It is important to avoid contamination of the silicon during crystal growing, since contaminants, even in small amounts, can dramatically alter the electrical properties of Si. To minimize unwanted reactions with silicon and the introduction of contaminants at the elevated temperatures of crystal growing, the procedure is carried out either in an inert gas (argon or helium) or a vacuum. Choice of crucible material is also important; fused silica (SiO_2), although not perfect for the application, represents the best available material and is used almost exclusively. Gradual dissolution of the crucible introduces oxygen as an unintentional impurity in the silicon boule. Unfortunately, the level of oxygen in the melt increases during the process, leading to a variation in concentration of the impurity throughout the length and diameter of the ingot.

35.2.3 Shaping of Silicon into Wafers

A series of processing steps are used to reduce the boule into thin, disc-shaped wafers. The steps can be grouped as follows: (1) ingot preparation, (2) wafer slicing, and (3) wafer preparation.

In ingot preparation, the seed and tang ends of the ingot are first cut off, as well as portions of the ingot that do not meet the strict resistivity and crystallographic requirements for subsequent IC processing. Next, a form of cylindrical grinding, as shown in Figure 35.6(a), is used to shape the ingot into a more perfect cylinder, since the crystal-growing process cannot achieve sufficient control over diameter and roundness. One or more flats are then ground along the length of the ingot, as in Figure 35.6(b). After the wafers have been cut from the ingot, these flats serve several functions: (1) identification, (2) orientation of the ICs relative to crystal structure, and (3) mechanical location during processing.

The ingot is now ready to be sliced into wafers, using the abrasive cut-off process illustrated in Figure 35.7. A very thin saw blade with diamond grit bonded to the internal diameter serves as the cutting edge. Use of the ID for slicing rather than the OD of the saw blade provides better control over flatness, thickness, parallelism, and surface

FIGURE 35.6 Grinding operations used in shaping the silicon ingot: (a) a form of cylindrical grinding provides diameter and roundness control, and (b) a flat ground on the cylinder.

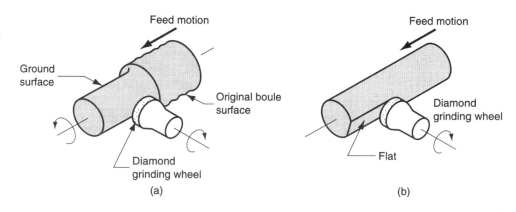

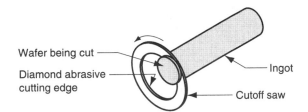

FIGURE 35.7   Wafer slicing using a diamond abrasive cut-off saw.

characteristics of the wafer. The wafers are cut to a thickness of around 0.5 to 0.7 mm (0.020–0.028 in.), depending on diameter (greater thicknesses for larger wafer diameters). For every wafer cut, a certain amount of silicon is wasted due to the kerf width of the saw blade. To minimize kerf loss, the blades are made as thin as possible—around 0.33 mm (0.013 in.).

Next the wafer must be prepared for the subsequent processes and handling in IC fabrication. After slicing, the rims of the wafers are rounded using a contour-grinding operation, such as in Figure 35.8(a). This reduces chipping of the wafer edges during handling and minimizes accumulation of photoresist solutions at the wafer rims. The wafers are then chemically etched to remove surface damage produced during slicing. This is followed by a flat polishing operation to provide surfaces of high smoothness to accept the photolithography processes to follow. The polishing step, seen in Figure 35.8(b), uses a slurry of very fine silica (SiO_2) particles in an aqueous solution of sodium hydroxide (NaOH). The NaOH oxidizes the Si wafer surface, and the abrasive particles

FIGURE 35.8 Two of the steps in wafer preparation: (a) contour grinding to round the wafer rim, and (b) surface polishing.

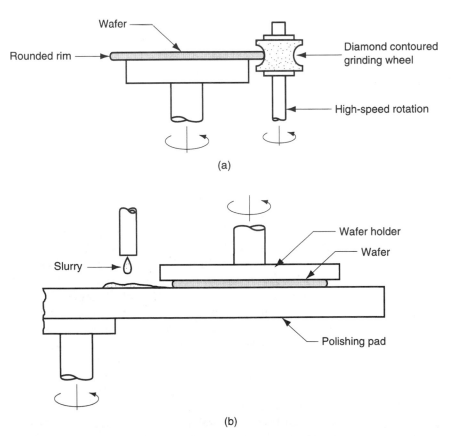

remove the oxidized surface layers—about 0.025 mm (0.001 in.) is removed from each side during polishing. Finally, the wafer is chemically cleaned to remove residues and organic films.

35.3 LITHOGRAPHY

An IC consists of many microscopic regions on the wafer surface that make up the transistors, other devices, and intraconnections as specified in the circuit design. In the planar process, the regions are fabricated by a sequence of steps, each step adding another layer to selected areas of the surface. The form of each layer is determined by a geometric pattern representing circuit design information that is transferred to the wafer surface by a procedure known as lithography—basically, the same procedure used by artists and printers for centuries.

Several lithographic technologies are used in semiconductor processing: (1) photolithography, (2) electron lithography, (3) x-ray lithography, and (4) ion lithography. As their names indicate, the differences are in the type of radiation used to transfer the mask pattern to the surface by exposing the photoresist. The traditional technique is photolithography, and most of our discussion will be directed at this topic. The reader may recall that photolithography is used in some chemical machining processes (Section 26.4).

35.3.1 Photolithography

Photolithography, also known as ***optical lithography,*** uses light radiation to expose a coating of photoresist on the surface of the silicon wafer; a mask containing the required geometric pattern for each layer separates the light source from the wafer, so that only the portions of the photoresist not blocked by the mask are exposed. The ***mask*** consists of a flat plate of transparent glass onto which a thin film of an opaque substance has been deposited in certain areas to form the desired pattern. Thickness of the glass plate is around 2 mm (0.080 in.), while the deposited film is only a few μm thick—for some film materials, less than a μm. The mask itself is fabricated by lithography, the pattern being based on circuit design data, usually in the form of digital output from the CAD system used by the circuit designer.

Photoresists   A ***photoresist*** is an organic polymer that is sensitive to light radiation in a certain wavelength range; the sensitivity causes either an increase or decrease in solubility of the polymer to certain chemicals. Typical practice in semiconductor processing is to use photoresists that are sensitive to ultraviolet light. UV light has a short wavelength compared to visible light, permitting sharper imaging of microscopic circuit details on the wafer surface. It also permits the fabrication and photoresist areas in the plant to be illuminated at low light levels outside the UV band.

The performance of a photoresist is characterized by the following measures [3]: (1) adhesion to the wafer surface, (2) etch resistance—how much the resist itself stands up to the etchant, (3) resolution—a term used to describe the minimum feature width and spacing that can be transferred from the mask to the wafer surface, and (4) photosensitivity—a measure of the response to increasing light intensities.

Two types of photoresists are available: positive and negative. A ***positive resist*** becomes more soluble in developing solutions after exposure to light. A ***negative resist*** becomes less soluble (the polymer cross-links and hardens) when exposed to light. Figure 35.9 illustrates the operation of both resist types. The principal advantage of the positive resist is better resolution. The negative resists have better adhesion to SiO_2 and metal surfaces, good etch resistance, high sensitivity, and low cost.

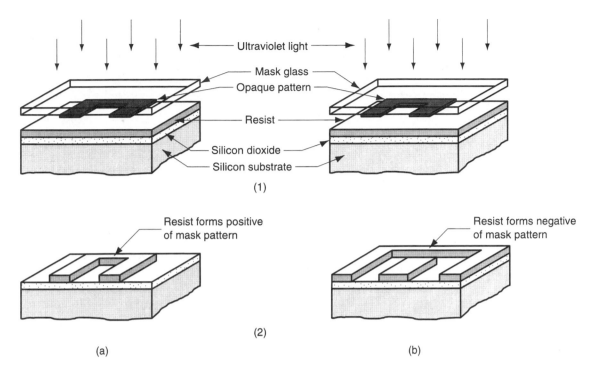

FIGURE 35.9 Application of (a) positive resist and (b) negative resist in photolithography; for both types, the sequence shows: (1) exposure through the mask and (2) remaining resist after developing.

Exposure Techniques The resists are exposed through the mask by one of three exposure techniques: (a) contact printing, (b) proximity printing, and (c) projection printing, illustrated in Figure 35.10. In ***contact printing,*** the mask is pressed against the resist coating during exposure. This results in high resolution of the pattern onto the wafer surface; an important disadvantage is that physical contact with the wafers gradually wears out the mask. In ***proximity printing,*** the mask is separated from the resist

FIGURE 35.10
Photolithography exposure techniques: (a) contact printing, (b) proximity printing, and (c) projection printing.

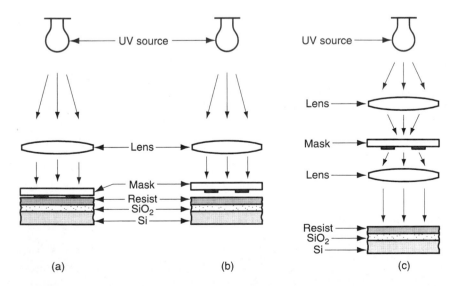

coating by a distance of 10 to 25 μm (0.0004–0.001 in.). This eliminates mask wear, but resolution of the image is slightly reduced. ***Projection printing*** involves the use of a high-quality lens (or mirror) system to project the image through the mask onto the wafer. This has become the preferred technique because it is noncontact (thus, no mask wear), and the mask pattern can be reduced through optical projection to obtain high resolution.

Processing Sequence in Photolithography   Let us examine a typical processing sequence for a silicon wafer in which photolithography is employed. The surface of the silicon has been oxidized to form a thin film of SiO_2 on the wafer. It is desired to remove the SiO_2 film in certain regions as defined by the mask pattern. The sequence for a negative resist proceeds as follows, illustrated in Figure 35.11. (1) ***Prepare surface.*** The wafer is properly cleaned to promote wetting and adhesion of the resist. (2) ***Apply photoresist.***   In semiconductor processing, photoresists are applied by feeding a metered amount of liquid resist onto the center of the wafer and then spinning the wafer to spread the liquid and achieve a uniform coating thickness. Desired thickness is around 1 μm (0.04 mils), which gives good resolution yet minimizes pinhole defects. (3) ***Soft-bake.***   Purpose of this pre-exposure bake is to remove solvents, promote adhesion, and harden the resist. Typical soft-bake temperatures are around 90°C (190°F) for 10 to 20 min. (4) ***Align mask and expose.***   The pattern mask is aligned relative to the wafer and the resist is exposed through the mask by one of the methods just described. Alignment must be accomplished with very high precision, using optical-mechanical equipment designed specifically for the purpose. If the wafer has been previously processed by lithography so that a pattern has already been formed in the wafer, then subsequent masks must be accurately registered relative to the existing pattern. Exposure of the resist depends on the same basic rule as in photography—the exposure is a function of light intensity $\times$ time. A mercury arc lamp or other source of UV light is used. (5) ***Develop resist.***   The exposed wafer is next immersed in a developing solution, or the solution is sprayed onto the wafer surface. For the negative resist in our example, the unexposed areas are dissolved in the developer, thus leaving the SiO_2 surface uncovered in these areas. Development is usually followed by a rinse to stop de-

FIGURE 35.11
Photolithography process applied to a silicon wafer: (1) prepare surface; (2) apply photoresist; (3) soft-bake; (4) align mask and expose; (5) develop resist; (6) hard-bake; (7) etch; (8) strip resist.

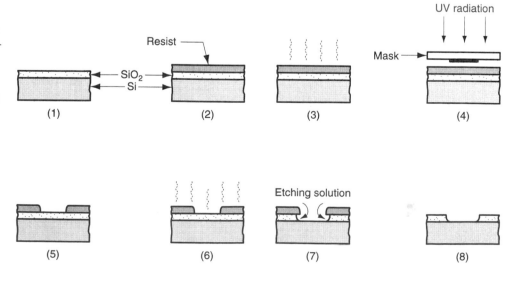

velopment and remove residual chemicals. (6) **Hard-bake.**   This baking step expels volatiles remaining from the developing solution and increases adhesion of the resist especially at the newly created edges of the resist film. (7) **Etch.**   Etching removes the SiO$_2$ layer at selected regions where the resist has been removed. (8) **Strip resist.**   After etching, the resist coating that remains on the surface must be removed. Stripping is accomplished using either wet or dry techniques. Wet stripping uses liquid chemicals; a mixture of sulfuric acid and hydrogen peroxide (H$_2$SO$_4$–H$_2$O$_2$) is common. Dry stripping uses plasma etching with oxygen as the reactive gas.

Although our example describes the use of photolithography to remove a thin film of SiO$_2$ from a silicon substrate, the same basic procedure is followed for other processing steps. The purpose of photolithography in all of these steps is to expose specific regions beneath the photoresist layer so that the process can be performed on these exposed regions. In the processing of a given wafer, photolithography is repeated as many times as needed to produce the desired integrated circuit, each time using a different mask to define the appropriate pattern.

35.3.2 Other Lithography Techniques

As feature size in integrated circuits continues to decrease and conventional UV photolithography becomes increasingly inadequate, other lithography techniques that offer higher resolution are growing in importance. These techniques are extreme ultraviolet lithography, electron beam lithography, x-ray lithography, and ion lithography. In the following paragraphs, we provide brief descriptions of these alternatives. For each technique, special resist materials are required that react to the particular type of radiation.

Extreme ultraviolet lithography (EUV) represents a refinement of current UV lithography through the use of shorter wavelengths during exposure. The ultraviolet wavelength spectrum ranges from about 10 nm to 380 nm (nm = nanometer = 10^{-9} m), the upper end of which is near the visible light range (approximately 400 to 700 nm wavelengths). EUV technology permits the feature size of an integrated circuit to be reduced to at least 0.03 μm (μm = micron or micrometer = 10^{-6} m), compared to about 0.1 μm with conventional UV exposure.

Electron-beam (E-beam) lithography has the advantage of a shorter wavelength compared to UV photolithography, thus virtually eliminating diffraction during exposure of the resist and permitting higher resolution of the image. Another potential advantage is that a scanning E-beam can be directed to expose only certain regions of the wafer surface, thus eliminating the need for a mask. Unfortunately, high-quality electron-beam systems are expensive. Also, due to the time-consuming sequential nature of the exposure method, production rates are low compared to the mask techniques of optical lithography. Accordingly, use of E-beam lithography tends to be limited to small production quantities. E-beam techniques are widely used for making the masks in UV lithography.

X-ray lithography has been under development since around 1972. As in E-beam lithography, the wavelengths of x-rays are much shorter than UV light (the x-ray wavelength ranges from 0.005 nm to several dozen nm, overlapping the lower end of the uv range). Thus, they hold the promise of sharper imaging during exposure of the resist. X-rays are difficult to focus during lithography. Consequently, contact or proximity printing must be used, and a small x-ray source must be used at a relatively large distance from the wafer surface in order to achieve good image resolution through the mask.

Ion lithography systems divide into two categories: (1) focused ion beam systems, whose operation is similar to a scanning E-beam system and avoids the need for a mask; and (2) masked ion beam systems, which expose the resist through a mask by proximity printing. As with E-beam and x-ray systems, ion lithography produces higher image resolution than conventional UV photolithography.

35.4 LAYER PROCESSES USED IN IC FABRICATION

The steps required to produce an integrated circuit consist of chemical and physical processes that add, alter, or remove regions on the silicon wafer that have been defined by photolithography. These regions constitute the insulating, semiconducting, and conducting areas that form the devices and their intraconnections in the integrated circuits. The layers are fabricated one at a time, step by step, each layer having a different configuration and each requiring a separate photolithography mask, until all of the microscopic details of the electronic devices and conducting paths have been constructed on the wafer surface.

In this section we consider the wafer processes used to add, alter, and subtract layers. Processes that add or alter layers to the surface include (1) thermal oxidation, used to grow a layer of silicon dioxide onto the silicon substrate; (2) chemical vapor deposition, a versatile process used to apply various types of layers in IC fabrication; (3) diffusion and ion implantation, used to alter the chemistry of an existing layer or substrate; and (4) various metallization processes that add metal layers to provide regions of electrical conduction on the wafer. Finally, (5) several etching processes are used to remove portions of the layers that have been added to achieve the desired details of the integrated circuit.

35.4.1 Thermal Oxidation

Oxidation of the silicon wafer may be performed multiple times during fabrication of an integrated circuit. Silicon dioxide (SiO_2) is an insulator, contrasted with the semiconducting properties of Si. The ease with which a thin film of SiO_2 can be produced on the surface of a silicon wafer is one of the attractive features of silicon as a semiconductor material.

Silicon dioxide serves a number of important functions in IC fabrication [13]: (1) it is used as a mask to prevent diffusion or ion implantation of dopants into silicon; (2) it can be used to isolate devices in the circuit; (3) it is a critical component in certain types of MOS devices; and (4) it provides electrical insulation between levels in multilevel metallization systems.

Several processes are used to form SiO_2 in semiconductor manufacturing, depending on when during chip fabrication the oxide must be added. The most common process is thermal oxidation, appropriate for growing SiO_2 films on silicon substrates. In ***thermal oxidation,*** the wafer is exposed to an oxidizing atmosphere at elevated temperature; either oxygen or steam atmospheres are used, with the following reactions, respectively:

$$Si + O_2 \rightarrow SiO_2 \tag{35.3}$$

or

$$Si + 2H_2O \rightarrow SiO_2 + 2H_2 \tag{35.4}$$

Typical temperatures used in thermal oxidation of silicon range from 900° to 1300°C (1650°–2350°F). By controlling temperature and time, oxide films of predictable thickness can be obtained. Films produced by thermal oxidation possess an amorphous structure, good uniformity, and low incidence of pinholes and similar defects. The equations

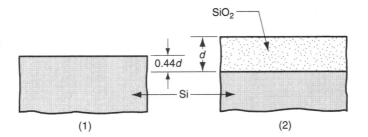

FIGURE 35.12 Growth of SiO_2 film on a silicon substrate by thermal oxidation, showing changes in thickness that occur: (1) before oxidation and (2) after thermal oxidation.

show that silicon at the surface of the wafer is consumed in the reaction, as seen in Figure 35.12. To grow a SiO_2 film of thickness d requires a layer of silicon that is $0.44d$ thick.

When a silicon dioxide film must be applied to surfaces other than silicon, then direct thermal oxidation is not appropriate. An alternative process must be used, such as chemical vapor deposition.

35.4.2 Chemical Vapor Deposition

Chemical vapor deposition (CVD) involves growth of a thin film on the surface of a heated substrate by chemical reactions or decomposition of gases (Section 29.4). CVD is widely used in the processing of integrated circuit wafers to add layers of silicon dioxide, silicon nitride (Si_3N_4), and silicon. Plasma-enhanced CVD is often used because it permits the reactions to take place at lower temperatures.

Typical CVD Reactions in IC Fabrication In the case of silicon dioxide, if the surface of the wafer is only silicon (e.g., at the start of IC fabrication), then thermal oxidation is the appropriate process by which to form a layer of SiO_2. If the oxide layer must be grown over materials other than silicon, such as aluminum or silicon nitride, then some alternative technique must be used, such as CVD. Chemical vapor deposition of SiO_2 is accomplished by reacting a silicon compound such as silane (SiH_4) with oxygen onto a heated substrate. The reaction is carried out at around 425°C (800°F) and can be summarized by

$$SiH_4 + O_2 \rightarrow SiO_2 + 2H_2 \tag{35.5}$$

The density of the silicon dioxide film and its bonding to the substrate is generally poorer than that achieved by thermal oxidation. Consequently, CVD is used only when the preferred process is not feasible—when the substrate surface is not silicon, or when the high temperatures used in thermal oxidation cannot be tolerated. CVD can be used to deposit layers of doped SiO_2, such as phosphorous-doped silicon dioxide (called P-glass).

Silicon nitride is used as a masking layer during oxidation of silicon. Si_3N_4 has a low oxidation rate compared to Si, so a nitride mask can be used to prevent oxidation in coated areas on the silicon surface. Silicon nitride is also used as a passivation layer (protecting against sodium diffusion and moisture). A conventional CVD process for coating Si_3N_4 onto a silicon wafer involves reaction of silane and ammonia (NH_3) at around 800°C (1700°F) as follows:

$$3SiH_4 + 4NH_3 \rightarrow Si_3N_4 + 12H_2 \tag{35.6}$$

Plasma-enhanced CVD is also used for basically the same coating reaction, the advantage being that it can be performed at much lower temperatures—around 300°C (600°F).

Polycrystalline silicon (called **polysilicon** to distinguish it from silicon having a single crystal structure such as the boule and wafer) has a number of uses in IC fabrication, including [13]: conducting material for leads, gate electrodes in MOS devices, and contact material in shallow junction devices. Chemical vapor deposition to coat polysilicon onto a wafer involves reduction of silane at temperatures around 600°C (1100°F), as expressed by the following:

$$SiH_4 \rightarrow Si + 2H_2 \tag{35.7}$$

Epitaxial Deposition   A related process for growing a film onto a substrate is called **epitaxial deposition,** distinguished by the feature that the film has a crystalline structure that is an extension of the substrate's structure. If the film material is the same as the substrate (e.g., silicon on silicon), then its crystal lattice will be identical to and a continuation of the wafer crystal. Several techniques are available to perform epitaxial deposition: (1) vapor-phase epitaxy, (2), liquid-phase epitaxy, and (3) molecular-beam epitaxy.

Vapor-phase epitaxy is the most important in semiconductor processing and is based on chemical vapor deposition. In growing silicon on silicon, the process is accomplished under closely controlled conditions at higher temperatures than conventional CVD of Si, using diluted reacting gases to slow the process so that an epitaxial layer can be successfully formed. Various reactions are possible, including Eq. (35.7), but the most widely used industrial process involves hydrogen reduction of silicon tetrachloride gas ($SiCl_4$) at around 1100°C (2000°F) as follows:

$$SiCl_4 + 2H_2 \rightarrow Si + 4HCl \tag{35.8}$$

The melting point of silicon is 1410°C (2570°F), so the preceding reaction is carried out at temperatures below T_m for Si, considered an advantage for vapor-phase epitaxy. If the epitaxial film is grown from the melt rather than from the vapor phase, the technique is called **liquid-phase epitaxy.** It is not a common technique in silicon processing, but it is used in gallium arsenide IC fabrication.

Molecular-beam epitaxy uses a vacuum evaporation process (Section 29.3.1), in which silicon together with one or more dopants are vaporized and transported to the substrate in a vacuum chamber. Its advantage is that it can be carried out at lower temperatures than CVD; processing temperatures are 400° to 800°C (750°–1450°F). However, throughput is relatively low and equipment is expensive.

35.4.3 Introduction of Impurities into Silicon

IC technology relies on the ability to alter the electrical properties of silicon by introducing impurities into selected regions at the surface. Adding impurities into the silicon surface is called **doping.** The doped regions are used to create p–n junctions that form the transistors, diodes, and other devices in the circuit. A silicon-dioxide mask produced by thermal oxidation and photolithography is used to define the silicon regions that are to be doped. Common elements used as impurities are boron (B) which forms electron acceptor regions in the silicon substrate (p-type regions); and phosphorous (P), arsenic (As), and antimony (Sb), which form electron donor regions (n-type regions). The techniques by which silicon is doped with these elements are diffusion and ion implantation.

Thermal Diffusion   Diffusion is a physical process in which atoms migrate from regions of high concentration into regions of lower concentration (Section 28.3.1). High temperatures accelerate the process. In semiconductor processing, diffusion is carried out to dope the silicon substrate with controlled amounts of a desired impurity. This is usually accomplished in two steps: (1) predeposition and (2) drive-in. In *predeposition,* the dopant source is deposited onto the surface of the wafer at a temperature of around 1000°C (1800°F). The dopant enters the crystal structure of the substrate, substituting for silicon atoms until a maximum concentration limit for the processing temperature is reached.

The *drive-in* step is basically a heat treatment in which the dopant introduced into the surface during predeposition is redistributed to obtain the desired depth and concentration profile. It can be performed in an oxidizing atmosphere to grow a protective SiO_2 film on top of the doped region.

Ion Implantation In *ion implantation,* vaporized ions of the impurity element are accelerated by an electric field and directed at the silicon substrate surface (Section 28.3.2). The atoms penetrate into the surface, losing energy and finally stopping at some depth in the crystal structure, average depth being determined by the mass of the ion and the acceleration voltage. Higher voltages produce greater depths of penetration, typically several hundred Angstroms (1 Angstrom = 10^{-8} cm). Advantages of ion implantation are that it can be accomplished at room temperature and it provides exact doping density.

The problem with ion implantation is that the ion collisions disrupt and damage the crystal lattice structure. Very-high-energy collisions can transform the starting crystalline material into an amorphous structure. This problem is solved by annealing at temperatures between 500° and 900°C (1000° and 1800°F), which allows the lattice structure to repair itself and return to its crystal state. Ion implantation results in penetrations that are less than those by diffusion—suited to very large-scale integration in which devices have shallow impurity depths. The controllability and reproducibility of ion implantation is superior to diffusion. These advantages have made ion implantation the preferred doping process in semiconductor technology since its introduction in the 1970s.

35.4.4 Metallization

Conductive materials must be deposited onto the wafer during processing to serve several functions: (1) form certain components (e.g., gates) of devices in the IC; (2) provide intraconnecting conduction paths between devices on the chip, and (3) connect the chip to external circuits. To satisfy these functions the conducting materials must be formed into very fine patterns. The process of fabricating these patterns is known as *metallization,* and it combines various thin film-deposition technologies with photolithography. In this section we consider the materials and processes used in metallization. Connecting the chip to external circuitry also involves IC packaging, which is explored in Section 35.6.

Metallization Materials Materials used in the metallization of silicon-based integrated circuits must have certain desirable properties, some of which relate to electrical function while others relate to manufacturing processing. The desirable properties of a metallization material are [3], [13]: (1) low resistivity, (2) low contact resistance with silicon, (3) good adherence to the underlying material, usually Si or SiO_2, (4) ease of deposition, compatible with photolithography, (5) chemical stability—noncorroding, nonreactive, and noncontaminating, (6) physical stability during temperatures encountered in processing, and (7) good lifetime stability.

Although no material meets all of these requirements perfectly, aluminum satisfies most of them either well or adequately, and it is the most widely used metallization material. Aluminum is usually alloyed with small amounts of (1) silicon to reduce reactivity with silicon in the substrate; and (2) copper to inhibit electromigration of Al atoms caused by current flow when the IC is in service. Other materials used for metallization in integrated circuits include polysilicon (Si); gold (Au); refractory metals (e.g., W, Mo); silicides (e.g., WSi_2, $MoSi_2$, $TaSi_2$); and nitrides (e.g., TaN, TiN, and ZrN). These other materials are generally used in applications such as gates and contacts. Aluminum is generally favored for device intraconnections and top level connections to external circuitry.

Metallization Processes A number of processes are available to accomplish metallization in IC fabrication: physical vapor deposition, chemical vapor deposition, and electroplating. Among PVD processes, vacuum evaporation and sputtering are applicable. *Vacuum evaporation* (Section 29.3.1) can be applied for aluminum metallization. Vaporization is usually accomplished by resistance heating or electron beam evaporation. Evaporation is difficult or impossible for depositing refractory metals and compounds. *Sputtering* (Section 29.3.2) can be used for depositing aluminum as well as refractory metals and certain metallizing compounds. It achieves better step coverage than evaporation, often important after many processing cycles when the surface contour has become irregular. However, deposition rates are lower and equipment is more expensive.

    Chemical vapor deposition is also applicable as a metallization technique. Its processing advantages include excellent step coverage and good deposition rates. Materials suited to CVD include tungsten, molybdenum, and most of the silicides used in semiconductor metallization. CVD for metallization in semiconductor processing is less common than PVD. Finally, *electroplating* (Section 29.1.1) is occasionally used in IC fabrication to increase the thickness of thin films.

35.4.5 Etching

All of the preceding processes in this section involve addition of material to the wafer surface, either in the form of a thin film or the doping of the surface with an impurity element. Certain steps in IC manufacturing require material removal from the surface; this is accomplished by etching away the unwanted material. Etching is usually done selectively, by coating surface areas that are to be protected and leaving other areas exposed for etching. The coating may be an etch-resistant photoresist, or it may be a previously applied layer of material such as silicon dioxide. We briefly encountered etching in our description of photolithography. This section gives some of the technical details of this step in IC fabrication.

    There are two main categories of etching process in semiconductor processing: wet chemical etching and dry plasma etching. Wet chemical etching is the older of the two processes and is easier to use. However, there are certain disadvantages that have resulted in growing use of dry plasma etching.

Wet Chemical Etching *Wet chemical etching* involves the use of an aqueous solution, usually an acid, to etch away a target material. The etching solution is selected because it chemically attacks the specific material to be removed and not the protective layer used as a mask. Some of the common etchants used to remove materials in wafer processing are listed in Table 35.3.

    In its simplest form, the process can be accomplished by immersing the masked wafers in an appropriate etchant for a specified time and then immediately transferring

TABLE 35.3   Some common chemical etchants used in semiconductor processing.

Material to Be Removed	Etchant (usually in aqueous solution)
Aluminum (Al)	Mixture of phosphoric acid (H_3PO_4), nitric acid (HNO_3), and acetic acid (CH_3COOH).
Silicon (Si)	Mixture of nitric acid (HNO_3) and hydrofluoric acid (HF)
Silicon dioxide (SiO_2)	Hydrofluoric acid (HF)
Silicon nitride (Si_3N_4)	Hot phosphoric acid (H_3PO_4)

them to a thorough rinsing procedure to stop the etching. Process variables such as immersion time, etchant concentration, and temperature are important in determining the amount of material removed. A properly etched layer will have a profile as shown in Figure 35.13. Note that the etching reaction is *isotropic* (it proceeds equally in all directions), resulting in an undercut below the protective mask. In general, wet chemical etching is isotropic, and so the mask pattern must be sized to compensate for this effect.

Note also that the etchant does not attack the layer below the target material in our illustration. In the ideal case, an etching solution can be formulated that will react only with the target material and not with other materials in contact with it. In practical cases, the other materials exposed to the etchant may be attacked but to a lesser degree than the target material. The *etch selectivity* of the etchant is the ratio of etching rates between the target material and some other material used as a mask or as the substrate material. For example, etch selectivity of hydrofluoric acid for SiO_2 over Si is infinite.

If process control is inadequate, either underetching or overetching can occur, as in Figure 35.14. Underetching, in which the target layer is not completely removed, results when the etching time is too short and/or the etching solution is too weak. Overetching involves too much of the target material being removed, resulting in loss of pattern definition and possible damage to the layer beneath the target layer. Overetching is caused by overexposure to the etchant.

Dry Plasma Etching   This etching process uses an ionized gas to etch a target material. The ionized gas is created by introducing an appropriate gas mixture into a vacuum chamber and using radio frequency (RF) electrical energy to ionize a portion of the gas, thus creating a plasma. The high-energy plasma reacts with the target surface, vaporizing the material to remove it. There are several ways in which a plasma can be used to etch a material; the two principal processes in IC fabrication are plasma etching and reactive ion etching.

In *plasma etching,* the function of the ionized gas is to generate atoms or molecules that are chemically very reactive, so that the target surface is chemically etched upon exposure. The plasma etchants are usually based on fluorine or chlorine gases. Etch selectivity is generally more of a problem in plasma etching than in wet chemical etching.

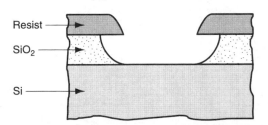

FIGURE 35.13   Profile of a properly etched layer.

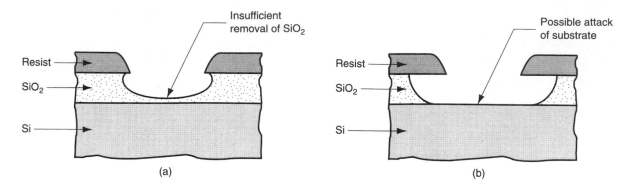

FIGURE 35.14 Two problems in etching: (a) underetching and (b) overetching.

For example, etch selectivity for SiO_2 over Si in a typical plasma etching process is 15 at best [5], compared to infinity with HF chemical etching.

An alternative function of the ionized gas can be to physically bombard the target material, causing atoms to be ejected from the surface. This is the process of sputtering, one of the techniques in physical vapor deposition. When utilized for etching, the process is called *sputter etching.* Although this form of etching has been applied in semiconductor processing, it is much more common to combine sputtering with plasma etching as described above, which results in the process known as *reactive ion etching.* This produces both chemical and physical etching of the target surface.

The advantage of the plasma-etching processes over wet chemical etching is that they are much more *anisotropic.* This property can be readily defined with reference to Figure 35.15. In (a), a fully anisotropic etch is shown; the undercut is zero. The degree to which an etching process is anisotropic is defined as the ratio:

$$A = \frac{d}{u} \tag{35.9}$$

where A = degree of anisotropy; d = depth of etch, which in most cases will be the thickness of the etched layer; and u = the undercut dimension, as illustrated in Figure 35.15(b). Wet chemical etching usually yields A values around 1.0, indicating isotropic etching. In sputter etching, ion bombardment of the surface is nearly perpendicular, resulting in A values approaching infinity—almost fully anisotropic. Plasma etching and reactive ion etching have high degrees of anisotropy, but below those achieved in sputter etching. As IC feature sizes continue to shrink, anisotropy becomes increasingly important for achieving the required dimensional tolerances.

FIGURE 35.15 (a) A fully anisotropic etch, with A = ∞; and (b) a partially anisotropic etch, with A = approximately 1.3.

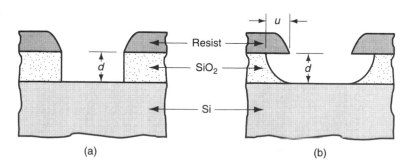

TABLE 35.4   Layer materials added or altered in IC fabrication and associated processes.

Layer Material (function)	Typical Fabrication Processes
Si, polysilicon (semiconductor)	CVD
Si, epitaxial (semiconductor)	Vapor phase epitaxy
Si doping (n-type or p-type)	Ion implantation, diffusion
SiO_2 (insulator, mask)	Thermal oxidation, CVD
Si_3N_4 (mask)	CVD
Al (conductor)	PVD, CVD
P-glass (protection)	CVD

35.5 INTEGRATING THE FABRICATION STEPS

In Sections 35.3 and 35.4, we examined the individual processing technologies used in IC fabrication. In this section, we show how these technologies are combined into the sequence of steps to produce an integrated circuit.

The planar processing sequence consists of fabricating a series of layers of various materials in selected areas on a silicon substrate. The layers form insulating, semiconducting, or conducting regions on the substrate to create the particular electronic devices required in the integrated circuit. The layers might also serve the temporary function of masking certain areas so that a particular process is only applied to desired portions of the surface. The masks are subsequently removed.

The layers are formed by thermal oxidation, epitaxial growth, deposition techniques (CVD and PVD), diffusion, and ion implantation. In Table 35.4, we summarize the processes typically used to add or alter a layer of a given material type. The use of lithography to apply a particular process only to selected regions of the surface is illustrated in Figure 35.16.

An example will be useful here to show the process integration in IC fabrication. We will use an n-channel metal oxide semiconductor (NMOS) logic device to illustrate the processing sequence. The sequence for NMOS integrated circuits is less complex than for CMOS or bipolar technologies, although the processes for these IC categories are basically similar. The device to be fabricated is illustrated in Figure 35.1.

The starting substrate is a lightly doped p-type silicon wafer, which will form the base of the n-channel transistor. The processing steps are illustrated in Figure 35.17 and described here (some details have been simplified and the metallization process for interconnecting devices has been omitted): (1) A layer of Si_3N_4 is deposited by CVD onto the Si substrate using photolithography to define the regions. This layer of Si_3N_4 will serve

FIGURE 35.16   Formation of layers selectively through the use of masks: (a) thermal oxidation of silicon, (b) selective doping, and (c) deposition of a material onto a substrate.

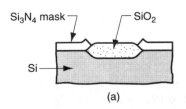
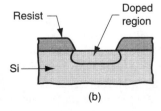
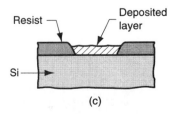

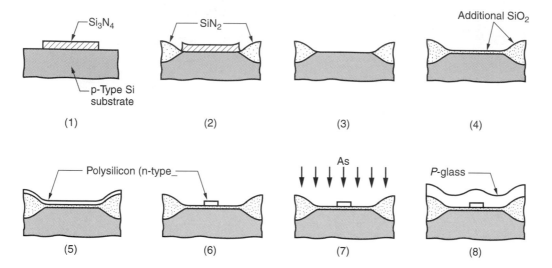

FIGURE 35.17   IC fabrication sequence: (1) Si_3N_4 mask is deposited by CVD on Si substrate; (2) SiO_2 is grown by thermal oxidation in unmasked regions; (3) the Si_3N_4 mask is stripped; (4) a thin layer of SiO_2 is grown by thermal oxidation; (5) polysilicon is deposited by CVD and doped n+ using ion implantation; (6) the poly-Si is selectively etched using photolithography to define the gate electrode; (7) source and drain regions are formed by doping n+ in the substrate; (8) P-glass is deposited onto the surface for protection.

as a mask for the thermal oxidation process in the next step. (2) SiO_2 is grown in the exposed regions of the surface by thermal oxidation. The SiO_2 regions are insulating and will become the means by which this device is isolated from other devices in the circuit. (3) The Si_3N_4 mask is stripped by etching. (4) Another thermal oxidation is done to add a thin gate oxide layer to previously uncoated surfaces and to increase the thickness of the previous SiO_2 layer. (5) Polysilicon is deposited by CVD onto the surface and then doped n-type using ion implantation. (6) The polysilicon is selectively etched using photolithography to leave the gate electrode of the transistor. (7) The source and drain regions (n+) are formed by ion implantation of arsenic (As) into the substrate. An implantation energy level is selected that will penetrate the thin SiO_2 layer but not the polysilicon gate or the thicker SiO_2 isolation layer. (8) Phosphosilicate glass (P-glass) is deposited onto the surface by CVD to protect the circuitry beneath.

35.6 IC PACKAGING

After all of the processing steps on the wafer have been completed, a final series of operations must be accomplished to transform the wafer into individual chips, ready to connect to external circuits and prepared to withstand the harsh environment of the world outside the clean room. These final steps are referred to as IC packaging. (As we shall see in the following chapter, packaging extends beyond the preparation of individual IC chips).

Packaging of integrated circuits is concerned with design issues such as (1) electrical connections to external circuits; (2) materials to encase the chip and protect it from the environment (humidity, corrosion, temperature, vibration, mechanical shock); (3) heat dissipation; (4) performance, reliability, and service life; and (5) cost.

There are also manufacturing issues in packaging, including: (1) chip separation—cutting the wafer into individual chips, (2) connecting it to the package, (3) encapsulating the chip, and (4) circuit testing. The manufacturing issues are the ones of greatest interest in this section. Although most of the design issues are properly left to other texts [7], [9], and [12], let us examine some of the engineering aspects of IC packages and the types of IC packages available, before describing the package processing steps to make them.

35.6.1 IC Package Design

In this section we consider three topics related to the design of an integrated circuit package: (1) the number of input/output terminals required for an IC of a given size, (2) the materials used in IC packages, and (3) package styles.

Determining the Number of Input/Output Terminals The basic engineering problem in IC packaging is to connect the many internal circuits to input/output (I/O) terminals so that the appropriate electrical signals can be communicated between the IC and the outside world. As the number of devices in an IC increases, the required number of I/O terminals (leads) also increases. The problem is of course aggravated by trends in semiconductor technology that have led to decreases in device size and increases in the number of devices that can be packed into an IC. Fortunately, the number of I/O terminals does not have to equal the number of devices in the IC. The dependency between the two values is given by Rent's Rule, named after the IBM engineer who defined the following relationship around 1960:

$$n_{io} = C n_c{}^m \tag{35.10}$$

where n_{io} = the number of input/output terminals required; n_c = the number of circuits in the IC, usually taken to be the number of logic gates; and C and m are parameters in the equation.

Commonly accepted C and m values are 4.5 and 0.5 for a modern VLSI microprocessor circuit [7], [13]. However, the parameters in Rent's Rule depend on the type of circuit. Memory devices require far fewer I/O terminals than microprocessors due to the column and row structure of memory units. Values for a static memory device published in [1] are $C = 6.0$ and $m = 0.12$. An alternative computation of the number of input/output pins in a static memory assumes that address decoding is used to design the device [7]. This allows the memory cells in the device to be configured in a two-dimensional array and a binary truth table to be used to access each cell. Based on this assumption, the value of n_{io} can be determined by:

$$n_{io} = 1.4427 \ln(n_c) \tag{35.11}$$

where n_c = the number of memory cells; and the constant 1.4427 is $1/\ln(2)$. It can be shown that the most efficient configuration of memory cells in the device is a square array (two dimensions equal) and that the total number of cells should be an integer power of 2, since the number of I/O pins n_{io} must be an integer.

IC Package Materials Package sealing involves encapsulating the IC chip in an appropriate packaging material. Two material types dominate current packaging technology: ceramic and plastic. Metal was used in early packaging designs but is today no longer important, except for lead frames.

The common ceramic packaging material is alumina (Al_2O_3). Advantages of ceramic packaging include hermetic sealing of the IC chip and the fact that highly complex packages can be produced. Disadvantages include poor dimensional control due to shrinkage during firing and the high dielectric constant of alumina.

Plastic IC packages are not hermetically sealed, but their cost is lower than ceramic. They are generally used for mass produced ICs, where very high reliability is not required. Plastics used in IC packaging include epoxies, polyimides, and silicones.

IC Package Styles   A wide variety of IC package styles is available to meet the input/output requirements indicated above. In nearly all applications, the IC is a component in a larger electronic system and must be attached to a printed circuit board (PCB). There are two broad categories of component mounting to a PCB, shown in Figure 35.18: through-hole and surface mount. In ***through-hole mounting,*** also known as ***pin-in-hole*** (PIH) technology, the IC package and other electronic components (e.g., discrete resistors, capacitors) have leads that are inserted through holes in the board and are soldered on the underside. In ***surface-mount technology*** (SMT), the components are attached to the surface of the board (or in some cases, both top and bottom surfaces). Several lead configurations are available in SMT, as illustrated in (b), (c), and (d) of the figure.

The major styles of IC packages include: (1) dual in-line package, (2) square package, and (3) pin grid array. Some of these are available in both through-hole and surface-mount styles, while others are designed for only one mounting method.

The ***dual in-line package*** (DIP) is currently the most common form of IC package, available in both through-hole and surface-mount configurations. It has two rows of leads (terminals) on either side of a rectangular body, as in Figure 35.19. Spacing between leads (center-to-center distance) in the conventional through-hole DIP is 2.54 mm (0.1 in.), and the number of leads ranges between 8 and 64. Hole spacing in the through-hole DIP style is limited by the ability to drill holes closely together in a printed circuit board. This limitation can be relaxed with surface mount technology because the leads are not inserted into the board; standard lead spacing on surface mount DIPs is 1.27 mm (0.05 in.).

The number of terminals in a DIP is limited by its rectangular shape in which leads project only from two sides; that means that the number of leads on either side is $n_{io}/2$. For high values of n_{io} (between 48 and 64), differences in conducting lengths between leads in the middle of the DIP and those on the ends cause problems in high-speed electrical characteristics. Some of these problems are addressed with a square package, in which the leads are arranged around the periphery so that the number of terminals on a side is $n_{io}/4$. A common example of the square package is the chip carrier. ***Chip carriers*** are used to reduce the space requirements of the package compared to a DIP and are often considered when the terminal count exceeds 48. A lead spacing of 1.27 mm (0.05 in.) is standard, and the number of leads can be as high as 124. Chip carriers come in several forms, the two main categories being ***leaded chip carrier*** (LCC), designed for either through-hole or surface mount; and ***leadless chip carrier*** (LLCC), which has no terminals and mounts on a mating base component. The surface mount LCC is illustrated in Figure 35.20. ***Quad flat packages*** ("quad packs") are a reduced version of the chip carrier, intended only for surface mount technology.

FIGURE 35.18   Types of component lead attachment on a printed circuit board: (a) through-hole, and several styles of surface-mount technology: (b) butt lead, (c) "J" lead, and (d) gull-wing.

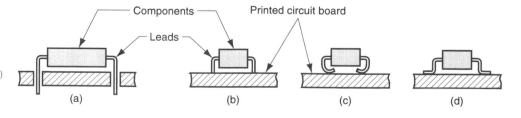

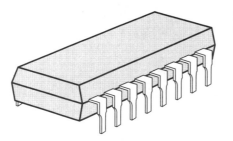

FIGURE 35.19   Dual-in-line package with 16 terminals, shown here in through-hole configuration.

They are thinner in profile; and their leads, which project outward rather than downward, have a smaller center-to-center distance than the chip carrier—as low as 0.5 mm (0.020 in.).

Even with a square chip package, there is still a practical upper limit on terminal count dictated by the manner in which the leads in the package are linearly allocated. The number of leads on a package can be maximized by using a square matrix of pins. A *pin grid array* (PGA) consists of a two-dimensional array of pin terminals on the underside of a square chip enclosure. The PGA is a through-hole package, with pin spacing of 2.54 mm (0.1 in.). In the ideal, the entire bottom surface of the package is fully occupied by pins, so that the pin count in each direction is square root of n_{io}. However, as a practical matter, the center area of the package has no pins because this region contains the IC chip.

35.6.2 Processing Steps in IC Packaging

The packaging of an IC chip in manufacturing can be divided into the following steps: (1) wafer testing, (2) chip separation, (3) die bonding, (4) wire bonding, and (5) package sealing. After packaging, a final functional test is performed on each packaged IC.

Wafer Testing   Current semiconductor processing techniques provide several hundred individual ICs per wafer. It is convenient to perform certain functional tests on the ICs while they are still together on the wafer—before chip separation. Testing is accomplished by computer-controlled test equipment that uses a set of needle probes configured to match the connecting pads on the surface of the chip; *multiprobe* is the term used for this testing procedure. When the probes contact the pads, a series of dc tests are carried out to indicate short circuits and other faults; this is followed by a functional test of the IC. Chips that fail the test are marked with an ink dot; these defects are not packaged. Each IC is positioned in its turn beneath the probes for testing, using a high precision *x-y* table to index the wafer from one chip site to the next.

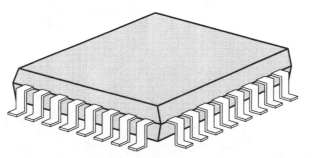

FIGURE 35.20   Square leaded chip carrier (LCC) for surface mounting with gull wing leads.

Chip Separation   The next step after testing is to cut the wafer into individual chips (dice). A thin, diamond-impregnated saw blade is used to perform the cutting operation. The sawing machine is highly automatic and its alignment with the "streets" between circuits is very accurate. The wafer is attached to a piece of adhesive tape, which itself is mounted in a frame. The adhesive tape holds the individual chips in place during and after sawing; the frame is a convenience in subsequent handling of the chips. Chips with ink dots are now discarded.

Die Bonding   The individual chips must next be attached to their individual packages, a procedure called die bonding. Owing to the miniature size of the chips, automated handling systems are used to pick the separated chips from the tape frame and place them for die bonding. Various techniques have been developed to bond the chip to the packaging substrate; we describe two methods that seem to be the most important today: eutectic die bonding and epoxy die bonding. *Eutectic die bonding,* used for ceramic packages, consists of the following steps: (1) a thin film of gold is deposited on the bottom surface of the chip; (2) the base of the ceramic package is heated to a temperature above 370°C (698°F), the eutectic temperature of the Au–Si system; and (3) the chip is bonded to the metallization pattern on the heated base. In *epoxy die bonding,* used for plastic VLSI packaging, a small amount of epoxy is dispensed on the package base (the lead frame), and the chip is positioned on the epoxy; the epoxy is then cured, bonding the chip to the surface.

Wire Bonding   After the die is bonded to the package, electrical connections are made between the contact pads on the chip surface and the package leads. The connections are generally made using small diameter wires of aluminum or gold, as illustrated in Figure 35.21. Typical wire diameters for aluminum are 0.05 mm (0.002 in.), and gold wire diameters are about half that diameter (Au has higher electrical conductivity than Al, but is more expensive). Aluminum wires are bonded by ultrasonic bonding, while gold wires are bonded by thermocompression, thermosonic, or ultrasonic means. *Ultrasonic bonding* uses ultrasonic energy to weld the wire to the pad surface. *Thermocompression bonding* involves heating the end of the wire to form a molten ball, and then the ball is pressed into the pad to form the bond. *Thermosonic bonding* combines ultrasonic and thermal energies to form the bond. Automatic wire bonding machines are used to perform these operations at rates up to 200 bonds per minute.

Package Sealing   As mentioned above, the two common packaging materials are ceramic and plastic. The processing methods are different for the two materials. *Ceramic packages* are made from a dispersion of ceramic powder (Al_2O_3 is most common) in a liquid binder (e.g., polymer and solvent). The mix is first formed into thin sheets and

FIGURE 35.21   Typical wire connection between chip contact pad and lead.

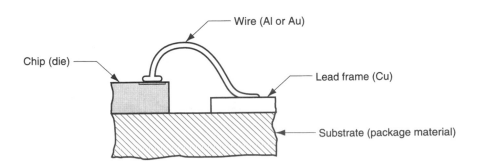

dried, and then cut to size. Holes are punched for interconnections. The required wiring paths are then fabricated onto each sheet, and metal is filled into the holes. The sheets are then laminated by pressing and sintering to form a monolithic (single-stone) body.

An alternative and lower cost ceramic package involves sealing of the IC chip between two ceramic plates using refractory glass—typically $PbO–ZnO–B_2O_3$ glass whose melting point is around 400°C (750°F). This provides hermetic sealing but is not capable of the complexity of the more conventional ceramic package. The technique goes by the name CERDIP (for glass-sealed ceramic DIPs) and CERQUADs for the corresponding quad packs.

Two types of *plastic package* are available, postmolded and premolded. In *postmolded packages,* an epoxy thermosetting plastic is transfer molded around the assembled chip and lead frame (after wire bonding), in effect transforming the pieces into one solid body. However, the molding process can be harsh on the delicate bond wires, and premolded packages are an alternative. In *premolded packaging,* an enclosure base is molded prior to encapsulation and then the chip and lead frame are connected to it, adding a solid lid or other material to provide protection. The extra assembly steps cause this production method to be more costly than postmolding.

Final Testing   Upon completion of the packaging sequence, each IC must undergo a final test, the purpose of which is: (1) to determine which units, if any, have been damaged during packaging; and (2) to measure performance characteristics of each device.

Burn-in test procedures sometimes include elevated temperature testing, in which the packaged IC is placed in an oven at temperatures around 125°C (250°F) for 24 hours and then tested. A device that fails such a test would have been likely to have failed early during service. If the device is intended for environments where wide temperature variations occur, a temperature cycle test is appropriate. This test subjects each device to a series of temperature reversals, between values around −50°C (−60°F) on the lower side and 125°C (250°F) on the upper side. Additional tests for devices requiring high reliability might include mechanical vibration tests and hermeticity (leak) tests.

35.7 YIELDS IN IC PROCESSING

The fabrication of integrated circuits consists of many processing steps performed in sequence. In wafer processing in particular, there may be hundreds of distinct operations through which the wafer passes. At each step, there is a chance that something may go wrong, resulting in the loss of the wafer or portions of it corresponding to individual chips. A simple probability model to predict the final yield of good product is

$$Y = Y_1 Y_2 \ldots Y_n \tag{35.12}$$

where Y = final yield; Y_1, Y_2, Y_n are the yields of each processing step, and n = total number of steps in the processing sequence.

As a practical matter, this model, although perfectly valid, is difficult to use due to the large number of steps involved and the variability of yields for each step. It is more convenient to divide the processing sequence into major phases, as we have organized our discussion of the sequence in this chapter, and to define the yields for each phase [11]. The first phase involves growth of the single-crystal boule. The term *crystal yield* Y_c refers to the amount of single-crystal material in the boule compared to the starting amount of electronic grade silicon. The typical crystal yield is about 50%;

with recycling this increases to about 65%. After crystal growing, the boule is sliced into wafers, the yield for which is described as the **crystal-to-slice yield** Y_s. This depends on the amount of material lost during grinding of the boule, the width of the saw blade relative to the wafer thickness during slicing, and other losses. A typical value might be 50%, although much of the lost silicon during grinding and slicing is recyclable.

The next phase is wafer processing to fabricate the individual ICs. From a yield viewpoint, this can be divided into wafer yield and multiprobe yield. **Wafer yield** Y_w refers to the number of wafers that survive processing compared to the starting quantity. Certain wafers are designated as test pieces or similar uses and therefore result in losses and a reduction in yield; in other cases, wafers are broken or processing conditions go awry. Typical values of wafer yield are around 70% if testing losses are included, and 90% or greater if excluded. For wafers that come through processing and are multiprobe tested, only a certain proportion pass the test, called the **multiprobe yield** Y_m. Multiprobe yield is highly variable and can range from very low values (less than 10%) to relatively high values (over 90%), depending on IC complexity and worker skill in the processing areas.

Following packaging, final testing of the IC is performed. This will invariably produce additional losses, resulting in a **final test yield** Y_t in the range 90% to 95%. If the five phase yields are combined as in Eq. (35.12), the final yield can be estimated by

$$Y = Y_c Y_s Y_w Y_m Y_t \qquad (35.13)$$

Given the typical values at each step, the final yield compared to the starting amount of silicon is quite low.

The heart of IC fabrication is wafer processing, the yield from which is measured in multiprobe testing Y_m. Yields in the other areas are fairly predictable, but not in wafer-fab. Two types of processing defects can be distinguished in wafer processing: (1) area defects and (2) point defects. **Area defects** are those that affect major areas of the wafer, possibly the entire surface. These are caused by variations or incorrect settings in process parameters. Examples include added layers that are too thin or too thick, insufficient diffusion depths in doping, and over- or under-etching. In general these defects are correctable by improved process control or development of alternative processes that are superior. For example, doping by ion implantation has largely replaced diffusion, and dry plasma etching has been substituted for wet chemical etching to better control feature dimensions.

Point defects occur at very localized areas on the wafer surface, affecting only one or a limited number of ICs in a particular area. They are commonly caused by dust particles either on the wafer surface or the lithographic masks. Point defects also include dislocations in the crystal lattice structure (Section 2.3.2). These point defects are distributed in some way over the surface of the wafer, resulting in a yield that is a function of the density of the defects, their distribution over the surface, and the processed area of the wafer. If the area defects are assumed negligible, and the point defects are assumed uniform over the surface area of the wafer, the resulting yield can be modeled by the equation

$$Y_m = e^{-AD} \qquad (35.14)$$

where Y_m = the yield of good chips as determined in multiprobe; A = the area processed, cm^2 (in.^{2}); and D = density of point defects, defects/cm^2 (defects/in.^{2}). This is sometimes referred to as the **Poisson yield** estimate because it is based on a Poisson distribution of defects over the surface area. Also referred to in the literature as the ***Boltzmann yield*** [11], it has been criticized as providing yield estimates that are overly pessimistic. As IC fabrication technologies improved over the years and wafer areas increased, actual yields turned out to be significantly better than those predicted by the Boltzmann yield equation.

An alternative prediction equation is based on ***Bose–Einstein*** statistics [10], in which point defects are indistinguishable on the surface of the wafer:

$$Y_m = \frac{1}{1 + AD} \tag{35.15}$$

where the symbols represent the same quantities as before. This equation has been found to be a better predictor of wafer processing performance than the Boltzmann formula, especially for large VLSI chips.

Wafer processing is the key to successful fabrication of integrated circuits. For an IC producer to be profitable, high yields must be achieved during this phase of manufacturing. This is accomplished using the purest possible starting materials, the latest equipment technologies, good process control over the individual processing steps, maintenance of clean room conditions, and efficient and effective inspection and testing procedures.

REFERENCES

[1] Bakoglu, H. B. ***Circuits, Interconnections, and Packaging for VLSI.*** Addison-Wesley Longman, Inc., Reading, Mass., 1990.

[2] Edwards, P. R. ***Manufacturing Technology in the Electronics Industry.*** Chapman & Hall, London, U.K., 1991.

[3] Gise, P., and Blanchard, R. ***Modern Semiconductor Fabrication Technology.*** Prentice-Hall, Upper Saddle River, N.J., 1986.

[4] Jackson, K. A., and Schroter, W. (eds.). ***Handbook of Semiconductor Technology.*** Vol. 2: ***Processing of Semiconductors.*** Wiley, New York, 2000.

[5] ***Encyclopedia of Chemical Technology.*** 4th ed., Wiley, New York, 2000.

[6] Lee, H. H. ***Fundamentals of Microelectronics Processing.*** McGraw-Hill, New York, 1990.

[7] Manzione, L. T. ***Plastic Packaging of Microelectronic Devices.*** AT&T Bell Laboratories, published by Van Nostrand Reinhold, New York, 1990.

[8] Moreau, W. M. ***Semiconductor Lithography Principles, Practices, and Materials.*** Plenum Press, New York, 1988.

[9] Pecht, M. (ed.). ***Handbook of Electronic Package Design.*** Marcel Dekker, Inc., New York, 1991.

[10] Price, J. E. "A New Look at Yield of Integrated Circuits." ***Proceedings, IEEE.*** Vol. 58, 1970, pp. 1290–91.

[11] Runyan, W. R., and Bean, K. E. ***Semiconductor Integrated Circuit Processing Technology.*** Addison-Wesley Longman, Inc., Reading, Mass., 1990.

[12] Seraphim, D. P., Lasky, R., and Li, C-Y., (eds.), ***Principles of Electronic Packaging.*** McGraw-Hill, New York, 1989.

[13] Sze, S. M. (ed.), ***VLSI Technology.*** McGraw-Hill, New York, 1988.

[14] Van Zant, P. ***Microchip Fabrication.*** 4th ed. McGraw-Hill, New York, 2000.

REVIEW QUESTIONS

35.1. What is an ***integrated circuit?***

35.2. Name some of the important semiconductor materials.

35.3. Describe the ***planar process.***

35.4. What are the three major stages in the production of silicon-based integrated circuits?

35.5. What is a ***clean room?*** Explain the classification system by which clean rooms are rated.

35.6. What are some of the significant sources of contaminants in IC processing?

35.7. What is the name of the process most commonly used to grow single-crystal ingots of silicon for semiconductor processing?

35.8. What are the alternatives to photolithography in IC processing?

35.9. What is a *photoresist?*
35.10. Why is ultraviolet light favored over visible light in photolithography?
35.11. Name the three exposure techniques in photolithography.
35.12. What layer material is produced by thermal oxidation in IC fabrication?
35.13. Define *epitaxial deposition.*

35.14. What are some of the important design functions of IC packaging?
35.15. What is *Rent's Rule?*
35.16. Name the two categories of component mounting to a printed circuit board.
35.17. What is a DIP?
35.18. What is the difference between postmolding and premolding in plastic IC chip packaging?

MULTIPLE CHOICE QUIZ

There is a total of 17 correct answers in the following multiple choice questions (some questions have multiple answers that are correct). To attain a perfect score on the quiz, all correct answers must be given, since each correct answer is worth 1 point. For each question, each omitted answer or wrong answer reduces the score by 1 point, and each additional answer beyond the number of answers required reduces the score by 1 point. Percentage score on the quiz is based on the total number of correct answers.

35.1. How many electronic devices would be contained in an IC chip in order for it to be classified in the VLSI category? (a) 1000, (b) 10,000, (c) 1 million, or (d) 100 million.
35.2. An alternative name for chip in semiconductor processing is which one of the following? (a) component, (b) device, (c) die, (d) package, or (e) wafer.
35.3. Which one of the following is the source of silicon for semiconductor processing? (a) pure Si in nature, (b) SiC, (c) Si_3N_4, or (d) SiO_2.
35.4. Which one of the following is the most common form of radiation used in photolithography? (a) electronic-beam radiation, (b) incandescent light, (c) infrared light, (d) ultraviolet light, or (e) X-ray.
35.5. After exposure to light, a positive resist becomes which of the following? (a) less soluble or (b) more soluble to the chemical developing fluid.
35.6. Which of the following processes are used to add layers of various materials in IC fabrication (more than one)? (a) chemical vapor deposition, (b) diffusion, (c) ion implantation, (d) physical vapor deposition, (e) plasma etching, (f) thermal oxidation, and (g) wet etching.
35.7. Which of the following are doping processes in IC fab-

rication (more than one)? (a) chemical vapor deposition, (b) diffusion, (c) ion implantation, (d) physical vapor deposition, (e) plasma etching, (f) thermal oxidation, and (g) wet etching.
35.8. Which one of the following impurity elements form electron acceptor (p-type) regions in silicon wafers? (a) antimony, (b) arsenic, (c) boron, (d) nitrogen, (e) phosphorous, or (f) potassium.
35.9. Which one of the following is the most common metal for intraconnection of devices in a silicon integrated circuit? (a) aluminum, (b) copper, (c) gold, (d) nickel, (e) silicon, or (f) silver.
35.10. Which etching process produces the more anisotropic etch in IC fabrication? (a) plasma etching or (b) wet chemical etching.
35.11. Which of the following are the two principal packaging materials used in IC packaging? (a) aluminum, (b) aluminum oxide, (c) copper, (d) epoxies, and (e) silicon dioxide.
35.12. Which of the following metals are commonly used for wire bonding of chip pads to the lead frame (two best answers)? (a) aluminum, (b) copper, (c) gold, (d) nickel, (e) silicon, and (f) silver.

PROBLEMS

Silicon Processing and IC Fabrication

35.1. A single-crystal boule of silicon is grown by the Czochralski process to an average diameter of 110 mm with length of 1200 mm. The seed and tang ends are removed, which reduces the length to 950 mm. The diameter is ground to 100 mm. A 30-mm-wide flat is ground on the surface that extends from one end to

the other. The ingot is then sliced into wafers 0.50 mm thick, using an abrasive saw blade whose thickness is 0.33 mm. Assuming that the seed and tang portions cut off the ends of the starting boule were conical in shape, determine (a) the original volume of the boule, mm^3; (b) how many wafers are cut from it, assuming the entire 950 mm length can be sliced; and (c) the volumetric proportion of silicon in the starting boule that is wasted during processing.

35.2. A silicon boule is grown by the Czochralski process to a diameter of 5.25 in. and a length of 5 ft. The seed and tang ends are cut off, reducing the effective length to 48.00 in. Assume that the seed and tang portions are conical in shape. The diameter is ground to 4.921 inch (125 mm). A primary flat of width 1.625 in. is ground on the surface the entire length of the ingot. The ingot is then sliced into wafers 0.025 in. thick, using an abrasive saw blade whose thickness = 0.0128 in. Determine (a) the original volume of the boule, in.^{3}; (b) how many wafers are cut from it, assuming the entire 4 ft length can be sliced; and (c) the volumetric proportion of silicon in the starting boule that is wasted during processing.

35.3. The processable area on a 125-mm-diameter wafer is a 110-mm-diameter circle. How many square IC chips can be processed within this area, if each chip is 5 mm on a side? All chips must lie completely within the processable area. Assume the cut lines (streets) between chips are of negligible width.

35.4. Solve the previous problem, only use a wafer size of 200 mm whose processable area is 175 in. in diameter. What is the percent increase in (a) number of chips; (b) wafer diameter, and (c) processable wafer

area, compared to the values in the previous problem?

35.5. A 4.00 in. wafer has a processable area that is only 3.65 inches in diameter. How many square IC chips can be fabricated within this area, if each chip is 0.25 in. on a side? All chips must lie completely within the processable area. Assume the cut lines (streets) between chips are of negligible width.

35.6. Solve the previous problem, only use a wafer size of 6.0 in. whose processable area is 5.50 in. in diameter. What is the percent increase in number of chips compared to the 50% increase in wafer diameter?

35.7. The surface of a silicon wafer is thermally oxidized, resulting in a SiO_2 film that is 3 μm thick. If the starting thickness of the wafer was exactly 0.400 mm thick, what is the final wafer thickness?

35.8. It is desired to etch out a region of a silicon dioxide film on the surface of a silicon wafer. The SiO_2 film is 3 μm thick. The width of the etched-out area is specified to be 10 μm. If the degree of anisotropy for the etchant in the process is known to be 0.8, what should be the size of the opening in the mask through which the etchant will operate?

35.9. In the previous problem, if plasma etching is used instead of wet etching, and the degree of anisotropy for plasma etching is infinity, what should be the size of the mask opening?

IC Packaging

35.10. An integrated circuit used in a microprocessor will contain 1000 logic gates. Use Rent's Rule ($C = 4.5$ and $m = 0.5$) to determine the approximate number of input/output pins required in the package.

35.11. A dual-in-line package has a total of 48 leads. Use Rent's Rule ($C = 4.5$ and $m = 0.5$) to determine the approximate number of logic gates that could be fabricated in the IC chip for this package.

35.12. It is desired to determine the effect of package style on the number of circuits (logic gates) that can be fabricated onto an IC chip to which the package is assembled. Using Rent's Rule ($C = 4.5$ and $m = 0.5$), compute the estimated number of devices (logic gates) that could be placed on the chip in the following cases: (a) a DIP with 16 I/O pins on a side—a total of 32 pins; (b) a square chip carrier with 16 pins on a side—a total of 64 I/O pins; and (c) a pin grid array with 16 by 16 pins—a total of 256 pins.

35.13. In the equation for Rent's Rule with $C = 4.5$ and $m =$

0.5, determine the value of n_{io} and n_c at which the number of logic gates equals the number of I/O terminals in the package.

35.14. A static memory device will have a two-dimensional array with 64 by 64 cells. Compare the number of input/output pins required using (a) Rent's Rule ($C = 6.0$ and $m = 0.12$), and (b) the alternative computation given in Eq. (35.11).

35.15. To produce a 1-megabit memory chip, how many I/O pins are predicted by (a) Rent's Rule ($C = 6.0$ and $m = 0.12$), and (b) the alternative computation method given in Eq. (35.11)?

35.16. Suppose it is desired to produce a memory device that will be contained in a dual-in-line package with 32 I/O leads. How many memory cells can be contained in the device, as estimated by: (a) Rent's Rule ($C = 6.0$ and $m = 0.12$); and (b) the alternative computation method given in Eq. (35.11)?

Yields in IC Processing

35.17. Given the following: crystal yield $Y_c = 50\%$, crystal-to-slice yield $Y_s = 50\%$, wafer yield $Y_w = 70\%$, multiprobe yield $Y_m = 60\%$, and final test yield $Y_t = 90\%$.

If a starting boule weighs 75 kg, what is the final weight of silicon that results after final test?

35.18. A silicon wafer with a nominal diameter of 100 mm

is processed to fabricate square chips of 5 mm on a side. The area of the processed chips occupies 65.25% of the total wafer area on one side. The density of point defects in the surface area is 0.027 defects/cm^2. Determine the number of good chips using (a) the Boltzmann yield estimate, Eq. (35.14), and (b) the Bose–Einstein yield estimate, Eq. (35.15).

35.19. A 5-in.-diameter wafer is processed over a circular area that is 4.75 in. in diameter. The density of point defects in the surface area is 0.32 defects/in.^{2} Determine the multiprobe yield using: (a) the Boltzmann yield estimate, Eq. (35.14), and (b) the Bose–Einstein yield estimate, Eq. (35.15).

35.20. The yield of good chips in multiprobe for a certain batch of wafers is 83%. The wafers have a nominal diameter of 150 mm with a processable area that is 135 mm in diameter. If the defects are all assumed to be point defects that are uniformly distributed over the surface (Poisson distribution), what is their density D?

35.21. In the previous problem, determine the density of point defects using Bose–Einstein statistics, Eq. (35.15), as the method of estimating yield.

35.22. A silicon wafer has a processable area of 20.0 in.^{2} The yield of good chips on this wafer is $Y_m = 75\%$. If the defects are all assumed to be point defects that are uniformly distributed over the surface (Poisson distribution), what is the density of point defects D?

ELECTRONICS ASSEMBLY
AND PACKAGING

CHAPTER CONTENTS

36.1 Electronics Packaging
36.2 Printed Circuit Boards
    36.2.1 Structures, Types, and Materials for PCBs
    36.2.2 Production of the Starting Boards
    36.2.3 Processes Used in PCB Fabrication
    36.2.4 PCB Fabrication Sequence
36.3 Printed Circuit Board Assembly
    36.3.1 Component Insertion
    36.3.2 Soldering
    36.3.3 Cleaning, Testing, and Rework
36.4 Surface Mount Technology
    36.4.1 Adhesive Bonding and Wave Soldering
    36.4.2 Solder Paste and Reflow Soldering
    36.4.3 Combined SMT-PIH Assembly
    36.4.4 Cleaning, Inspection, Testing, and Rework
36.5 Electrical Connector Technology
    36.5.1 Permanent Connections
    36.5.2 Separable Connectors

Integrated circuits constitute the heart of an electronic system, but the complete system consists of much more than packaged ICs. The ICs and other components are mounted and interconnected on printed circuit boards, which in turn are interconnected and contained in a chassis or cabinet. Chip packaging (Section 35.6) is only part of the total electronic package. In this chapter we consider the remaining levels of the package and how they are manufactured and assembled.

36.1 ELECTRONICS PACKAGING

The electronics package is the physical means by which the components in a system are electrically interconnected and interfaced to external devices; it includes the mechanical structure that holds and protects the circuitry. A well-designed electronics package serves the following functions: (1) power distribution and signal interconnection, (2) structural support, (3)

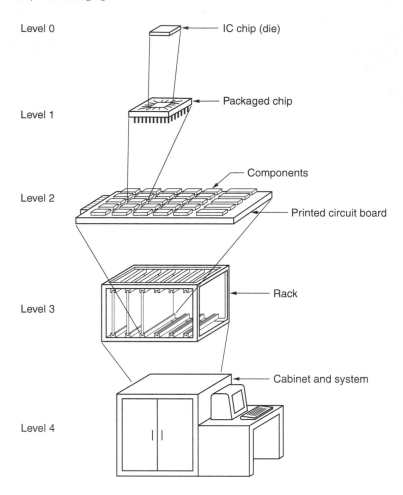

Level 0 — IC chip (die)

Level 1 — Packaged chip

Level 2 — Components
— Printed circuit board

Level 3 — Rack

Level 4 — Cabinet and system

FIGURE 36.1 Packaging hierarchy in a large electronic system.

circuit protection from physical and chemical hazards in the environment, (4) dissipation of heat generated by the circuits, and (5) minimum delays in signal transmission within the system.

For complex systems containing many components and interconnections, the electronics package is organized into levels which comprise a ***packaging hierarchy,*** illustrated in Figure 36.1 and summarized in Table 36.1. The lowest level is the ***zero level,*** which refers to the intraconnections on the semiconductor chip. The packaged chip, consisting of the IC in a plastic or ceramic enclosure and connected to the package leads, constitutes the ***first level of packaging.***

TABLE 36.1 Packaging hierarchy.

Level	Description of Interconnection
0	Intraconnections on the chip
1	Chip to package interconnections to form IC package
2	IC package to circuit board interconnections
3	Circuit board to rack; card-on-board packaging
4	Wiring and cabling connections in cabinet.

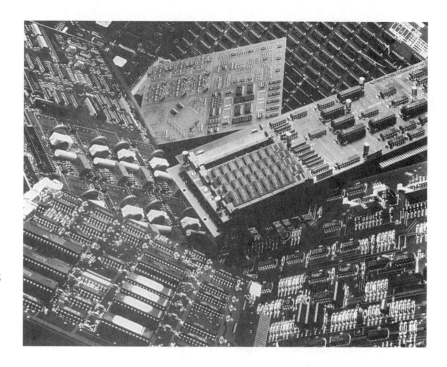

FIGURE 36.2   A collection of printed circuit board assemblies showing both pin-in-hole and surface-mount technologies (photo courtesy of Phoenix Technologies, Inc.).

Packaged chips and other components are assembled to a printed circuit board (PCB) using either of two technologies (Section 35.6.1): (1) *pin-in-hole* (PIH) technology, or (2) *surface-mount technology* (SMT). The chip package styles and assembly techniques are different for PIH and SMT. In many cases, both assembly technologies are employed in the same board. Printed circuit board assembly represents the *second level of packaging.* Figure 36.2 shows a variety of PCB assemblies of both PIH and SMT types.

The assembled PCBs are, in turn, connected to a chassis or other framework; this is the *third level of packaging.* This third level may consist of a *rack* that holds the boards, using wiring cables to make the interconnections. In major electronic systems, such as large computers, the PCBs are typically mounted onto a larger printed circuit board called a *back plane,* which has conduction paths to permit interconnection between the smaller boards attached to it. This latter configuration is known as *card-on-board* (COB) packaging, where the smaller PCBs are called cards and the back plane is the board.

The *fourth level of packaging* consists of wiring and cabling inside the cabinet that contains the electronic system. For systems of relatively low complexity, the packaging may not include all of the possible levels in the hierarchy.

36.2   PRINTED CIRCUIT BOARDS

A printed circuit board consists of one or more thin sheets of insulating material, with thin copper lines on one or both surfaces that interconnect the components attached to the board. In boards consisting of more than one layer, copper conducting paths are interleaved between the layers. PCBs are used in packaged electronic systems to hold components, provide electrical interconnections among them, and make connections to external circuits. They have become standard building blocks in virtually all electronic systems that contain packaged ICs and other components (Historical Note 36.1). PCBs are so important and widely used because: (1) they provide a convenient structural platform for the components; (2) a board with correctly routed interconnections can be mass produced consistently, without the variability usually associated with hand wiring; (3) all of the soldering connections between components and the PCB can be accomplished in a one-step mechanized operation; (4) an assembled PCB gives reliable performance; and (5) in complex electronic systems, each assembled PCB can be detached from the system for service and repair.

Historical Note 36.1   *Printed circuit boards* [1].

Before printed circuit boards, electrical and electronic components were manually fastened to a sheet-metal chassis and then hand wired and soldered to form the desired circuit. The usual sheet metal was aluminum. In the late 1950s, various plastic boards became commercially available. These boards, which provided electrical insulation, gradually replaced the aluminum chassis. The first plastics were phenolics, followed by glass-fiber reinforced epoxies. The boards came with predrilled holes spaced at standard intervals in both directions. This inspired the use of electronic components that matched these hole spacings. The dual-in-line package evolved during this period.

The components in these circuit boards were hand-wired, which proved increasingly difficult and prone to human error as component densities increased and circuits became more complex. The printed circuit board, with etched copper foil on its surface to form the wiring interconnections, was developed to solve these problems with manual wiring.

Initial techniques to design the circuit masks involved a manual inking procedure, in which the designer attempted to route the conducting tracks to provide the required connections and avoid short circuits on a large sheet of paper or vellum. This became more difficult as the number of components on the board increased and the conducting lines interconnecting the components became finer. Computer programs were developed to aid the designer in solving the routing problem. However, in many cases, it was impossible to find a solution with no intersecting tracks (short circuits). To solve the problem, jumper wires were hand-soldered to the board to make these connections. As the number of jumper wires increased, the problem of human error again appeared. Multilayer boards were introduced to deal with this routing problem.

The initial technique for "printing" the circuit pattern onto the copper-clad board was screen printing. As track widths became finer and finer, photolithography was introduced.

36.2.1 Structures, Types, and Materials for PCBs

A *printed circuit board* (PCB), also called a *printed wiring board* (PWB), is a laminated flat panel of insulating material designed to provide electrical interconnections between electronic components attached to it. Interconnections are made by thin conducting paths on the surface of the board or in alternating layers sandwiched between layers of insulating material. The conducting paths are made of copper and are called *tracks.* Other copper areas, called *lands,* are also available on the board surface for attaching and electrically connecting components.

Insulation materials in PCBs are usually polymer composites reinforced with glass fabrics or paper. Polymers include epoxy (most widely used), phenolic, and polyimide. E-glass is the usual fiber in glass-reinforcing fabrics, especially in epoxy PCBs; paper is a common reinforcing layer for phenolic boards. The usual thickness of the substrate layer is 0.8 to 3.2 mm (0.031–0.125 in.), and copper foil thickness is around 0.04 mm (0.0015 in.). The materials forming the PCB structure must be electrically insulating,

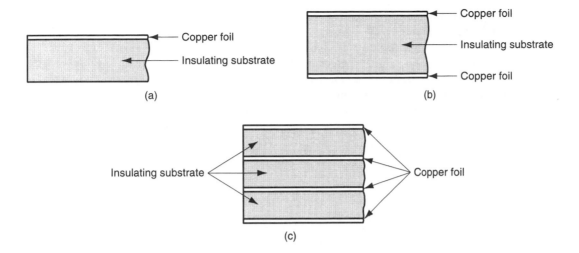

FIGURE 36.3   Three types of printed circuit board structure: (a) single-sided, (b) double-sided, and (c) multilayer.

strong and rigid, resistant to warpage, dimensionally stable, heat resistant, and flame retardent. Chemicals are often added to the polymer composite to obtain the last two characteristics.

There are three principal types of printed circuit board, shown in Figure 36.3: (a) *single-sided* board, in which copper foil is only on one side of the insulation substrate; (b) *double-sided* board, in which the copper foil is on both sides of the substrate; and (c) *multilayer* board, consisting of alternating layers of conducting foil and insulation. In all three structures, the insulation layers are constructed of multiple laminates of epoxy-glass sheets (or other composite) bonded together to form a strong and rigid structure. Multilayer boards are used for complex circuit assemblies in which a large number of components must be interconnected with many track routings, thus requiring more conducting paths than can be accommodated in one or two copper layers. Four layers is the most common multilayer configuration, but boards with up to 24 conducting layers are produced.

36.2.2 Production of the Starting Boards

Single-sided and double-sided boards can be purchased from suppliers that specialize in mass producing them in standard sizes. The boards are then custom-processed by a circuit fabricator to create the specified circuit pattern and board size for a given application. Multilayer boards are fabricated from standard single- and double-sided boards. The circuit fabricator processes the boards separately to form the required circuit pattern for each layer in the final structure, and then the individual boards are bonded together with additional layers of epoxy-fabric. Processing of multilayer boards is more involved and more expensive than the other types; the reason for using them is that they provide better performance for large systems than using a much greater number of lower-density boards of simpler construction.

The copper foil used to clad the starting boards is produced by a continuous electroforming process (Section 29.1.2), in which a rotating smooth metal drum is partially submersed in an electrolytic bath containing copper ions. The drum is the cathode in the circuit, causing the copper to plate onto its surface. As the drum rotates out of the bath, the thin copper foil is peeled from its surface. The process is ideal for producing the very thin copper foil needed for PCBs.

Production of the starting boards consists of pressing multiple sheets of woven glass fiber that have been impregnated with partially cured epoxy (or other thermosetting polymer). The number of sheets used in the starting sandwich determines the thickness of the final board. Copper foil is placed on one or both sides of the epoxy-glass laminated stack, depending on whether single- or double-sided boards are to be made. For single-sided boards, a thin release film is used on one side in place of the copper foil to prevent sticking of the epoxy in the press. Pressing is accomplished between two steam-heated platens of a hydraulic press. The combination of heat and pressure compacts and cures the epoxy-glass layers to bond and harden the laminates into a single-piece board. The board is then cooled and trimmed to remove excess epoxy that has been squeezed out around the edges.

The completed board consists of a glass-fabric-reinforced epoxy panel, clad with copper over its surface area on one or both sides. It is now ready for the circuit fabricator. Panels are usually produced in large standard widths designed to match the board handling systems on wave-soldering equipment, automatic insertion machines, and other PCB processing and assembly facilities. If the electronic design calls for a smaller size, several units can be processed together on the same larger board and then separated later.

36.2.3 Processes Used in PCB Fabrication

The circuit fabricator employs a variety of processing operations to produce a finished PCB, ready for assembly of components. The operations include cleaning, shearing, hole drilling or punching, pattern imaging, etching, and electroless and electrolytic plating. Most of these processes have been discussed previously. In this section we focus on the details that are relevant to PCB fabrication. Our discussion follows approximately the order in which the processes are performed on a board. However, there are differences in processing sequence between different board types, and we examine these differences in Section 36.2.4. Some of the operations in PCB fabrication must be performed under clean room conditions in order to avoid defects in the printed circuits, especially for boards with fine tracks and details.

Board Preparation   Initial preparation of the board consists of shearing, hole-making, and other shaping operations to create tabs, slots, and similar features in the board. If necessary, the starting panel may have to be sheared to size for compatibility with the circuit fabricator's equipment. The holes, called tooling holes, are made by drilling or punching and are used for positioning the board during subsequent processing. The sequence of fabrication steps requires close alignment from one process to the next, and these holes are used with locating pins at each operation to achieve accurate registration. Three tooling holes per board are usually sufficient for this purpose; hole size is about 3.2 mm (0.125 in.), larger than the circuit holes to be drilled later. The remaining machined features are for handling or other processing reasons that often relate to the particular equipment types used by the fabricator.

The board is typically bar coded for identification purposes in this preparation phase. Finally, a cleaning process removes dirt and grease from the board surface. Although cleanliness requirements are not as stringent as in IC fabrication, small particles of dirt and dust can cause defects in the circuit pattern of a printed circuit board; and surface films of grease can inhibit etching and other chemical processes. Cleanliness is essential for consistent manufacture of reliable PCBs.

Hole Drilling   In addition to tooling holes, functional circuit holes are required in PCBs as (1) *insertion holes* for insertion of component leads in through-hole boards; (2) *via holes,* which are copper-plated and used as conducting paths from one side of the board to the other; and (3) holes to fasten certain components such as heat sinks and connectors to the board. These holes are either drilled or punched, using the tooling holes for location. Cleaner holes can be produced by drilling, but faster production rates can be achieved by punching. The quality requirement seems to dominate the choice, and most holes in PCB fabrication are drilled. A stack of three or four panels may be drilled in the same operation, using a computer numerically controlled (CNC) drill press that receives its programming instructions from the design data base. For high-production jobs, multiple-spindle drills are sometimes used, permitting all of the holes in the board to be drilled in one feed motion.

Standard twist drills (Section 23.3.2) are used to drill the holes, but the process makes a number of unusual demands on the drill and drilling equipment. Perhaps the biggest single problem is the small hole size in printed circuit boards; drill diameter is generally less than 1.27 mm (0.050 in.), but some high-density boards require hole sizes of 0.15 mm (0.006 in.) or even less [7]. Such small drill bits lack strength, and their capacity to dissipate heat is low.

Another difficulty is the unique work material. The drill bit must first pass through a thin metallic foil and then proceed through an abrasive epoxy-glass composite. Different drills would normally be specified for these materials, but in the case of PCB drilling, a single drill must suffice. The small hole size, combined with the stacking of several boards or the drilling of multilayer boards, results in a high depth-to-diameter ratio, aggravating the problem of chip extraction from the hole. Other requirements placed on the operation include high accuracy in hole location, smooth hole walls, and absence of burrs on the holes. Burrs are usually formed when the drill enters or exits a hole; thin sheets of material are often placed on top of and beneath the stack of boards to inhibit burr formation on the boards themselves.

Finally, any cutting tool must be used at a certain cutting speed in order to operate at best efficiency. For a drill bit, cutting speed is measured at the diameter. For very small drill sizes, this means extremely high rotational speeds—up to 100,000 rev/min in some cases. Special spindle bearings and motors are required to achieve these speeds.

Circuit Pattern Imaging and Etching   There are two basic methods by which the circuit pattern is transferred to the copper surface of the board, screen printing and photolithography. Both methods involve the use of a resist coating on the board surface that determines where etching of the copper will occur to create the tracks and lands of the circuit.

Screen printing was the first method used for PCBs. It is indeed a printing technique, and the term printed circuit board can be traced to this method. In *screen printing* (also called *screening*), a stencil screen containing the circuit pattern is placed on the board, and liquid resist is squeezed through the screen mesh to the surface beneath. The stencil screen is often called a "silk screen," which dates from when silk was used in the printing trade to make the screens; today other screen materials are used, including polyesters and fine stainless-steel wire. This method is simple and inexpensive, but its resolution is limited. It is normally used only for applications in which track widths are greater than about 0.25 mm (0.010 in.).

The second method of transferring the circuit pattern is *photolithography,* in which a light-sensitive resist material is exposed through a mask to transfer the circuit pattern. The procedure is similar to the corresponding process in IC fabrication (Section 35.3.1); some of the details in PCB processing will be described here.

Negative photoresists are used by most circuit fabricators. The resists are available in two forms: liquid or dry film. Liquid photoresists can be applied by roller or spraying. Disadvantages include variability in coating thickness and long exposure times. Dry film resists are more commonly used in PCB fabrication. They consist of three layers: a film of photosensitive polymer sandwiched between a polyester support sheet on one side and a removable plastic cover sheet on the other side. The cover sheet prevents the photosensitive material from sticking during storage and handling. Although more expensive than liquid resists, they can be applied in coatings of uniform thickness, and their processing in photolithography is simpler. To apply, the cover sheet is removed, and the resist film is placed on the copper surface to which it readily adheres. Hot rollers are used to press and smooth the resist onto the surface.

Alignment of the masks relative to the board relies on the use of registration holes in the mask, which are aligned with the tooling holes on the board. Contact printing is used to expose the resist beneath the mask. The resist is then developed, which involves removal of the unexposed regions of the negative resist from the surface. Chemical developing is generally used for both wet and dry resists.

After resist development, certain areas of the copper surface remain covered by resist while other areas are now unprotected. The covered areas correspond to circuit tracks and lands, while uncovered areas correspond to open regions between. *Etching* removes the copper cladding in the unprotected regions from the board surface, usually by means of a chemical etchant (Section 35.4.5). Etching is the step that transforms the solid copper film into the interconnections for an electrical circuit.

Etching is done in an etching chamber in which the etchant is sprayed onto the surface of the board that is now partially coated with resist. Various etchants are used to remove copper, including ammonium persulfate ($(NH_4)_2S_2O_4$), ammonium hydroxide (NH_4OH), cupric chloride ($CuCl_2$), and ferric chloride ($FeCl_3$). Each has its relative advantages and disadvantages. Process parameters (e.g., temperature, etchant concentration, and duration) must be closely controlled to avoid over- or under-etching, as in IC fabrication. After etching, the board must be rinsed and the remaining resist chemically stripped from the surface.

Plating In printed circuit boards, plating is needed on the hole surfaces to provide conductive paths from one side to the other in double-sided boards, or between layers in multilayer boards. Two types of plating process are used in PCB fabrication: electroplating (Section 29.1.1) and electroless plating (Section 29.1.3) Electroplating has a higher deposition rate than electroless plating but requires that the coated surface be metallic (conductive); electroless plating is slower but does not require a conductive surface.

After drilling of the via holes and insertion holes, the walls of the holes consist of epoxy-glass insulation material, which is nonconductive. Accordingly, electroless plating must be used initially to provide a thin coating of copper on the hole walls. Once the thin film of copper has been applied, electrolytic plating is then used to increase coating thickness on the hole surfaces to between 0.025 and 0.05 mm (0.001 and 0.002 in.).

Gold is another metal sometimes plated onto printed circuit boards. It is used as a very thin coating on PCB edge connectors to provide superior electrical contact. Coating thickness is only about 2.5 μm (0.0001 in.).

36.2.4 PCB Fabrication Sequence

In this section we describe the processing sequence for various board types. The sequence is concerned with transforming a copper-clad board of reinforced polymer into a printed circuit board, a procedure called *circuitization.* The desired result, using a double-sided board as an example, is illustrated in Figure 36.4.

Circuitization   Three methods of circuitization can be used to determine which regions of the board will be coated with copper [11]: (1) subtractive, (2) additive, and (3) semiadditive.

In the *subtractive method,* open portions of the copper cladding on the starting board are etched away from the surface, so that the tracks and lands of the desired circuit remain. The process is termed subtractive because copper is removed from the board surface. The steps in the subtractive method are described in Figure 36.5.

The *additive method* starts with a board surface that is not copper clad, such as the uncoated surface of a single-sided board. However, the uncoated surface is treated with a chemical, called a *buttercoat,* which acts as the catalyst for electroless plating. The steps in the method are outlined in Figure 36.6.

The *semiadditive method* uses a combination of additive and subtractive steps. The starting board has a very thin copper film on its surface—5 μm (0.0002 in.) or less. The method proceeds as described in Figure 36.7.

Processing of Different Board Types   Processing methods differ for the three PCB types: single-sided, double-sided, and multilayer. These differences are briefly detailed in the following paragraphs.

A *single-sided board* begins fabrication as a flat sheet of insulating material clad on one side with copper film. The subtractive method is used to define the circuit pattern in the copper cladding. A typical processing sequence is as follows: (1) the board is cut to size, tooling holes are drilled, and the board is cleaned; (2) photoresist is applied to the copper-clad surface; (3) the surface is exposed to ultraviolet light through a circuit mask; (4) the resist is developed, exposing open areas between circuit tracks and lands on the copper; (5) exposed areas are etched, leaving the tracks and lands on the board; (6) the remaining resist is stripped; and (7) lead holes are drilled and deburred.

FIGURE 36.4   A section of a double-sided PCB, showing various features accomplished during fabrication: tracks and lands, and copper-plated insertion and via holes.

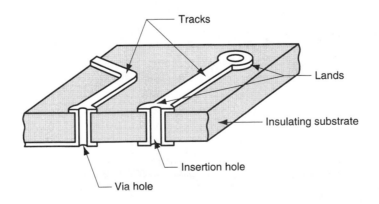

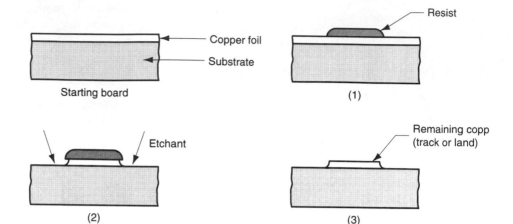

FIGURE 36.5 The subtractive method of circuitization in PCB fabrication: (1) apply resist to areas not to be etched, using photolithography to expose the areas that are to be etched; (2) etch; and (3) strip resist.

A *double-sided board* involves a somewhat more complex processing sequence because it has circuit tracks on both sides that must be electrically connected. The interconnection is accomplished by means of copper-plated via holes that run from lands on one surface of the board to lands on the opposite surface, as shown in Figure 36.4. A typical fabrication sequence for a double-sided board, starting with a board that is copper-clad on both sides and using the semiadditive method, is as follows: (1) the board is cut to size, tooling holes are drilled, and the board is cleaned; (2) via holes, as well as component insertion holes, are drilled; (3) the holes are plated using electroless plating followed by electroplating; (4) resist is applied to surface areas of both sides that are not to be copper plated; (5) a layer of tin is electroplated onto the exposed areas, which will mask copper areas to become tracks, lands, and through holes in a subsequent etching step; (6) resist is stripped to expose areas that have not been tin-plated; and (7) exposed noncircuit copper regions are etched.

FIGURE 36.6 The additive method of circuitization in PCB fabrication: (1) a resist film is applied to the surface using photolithography to expose the areas to be copper plated; (2) the exposed surface is chemically activated to serve as a catalyst for electroless plating; (3) copper is plated on exposed areas; and (4) resist is stripped.

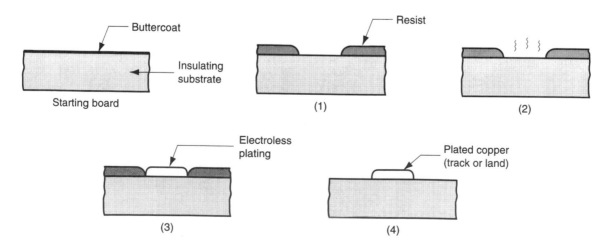

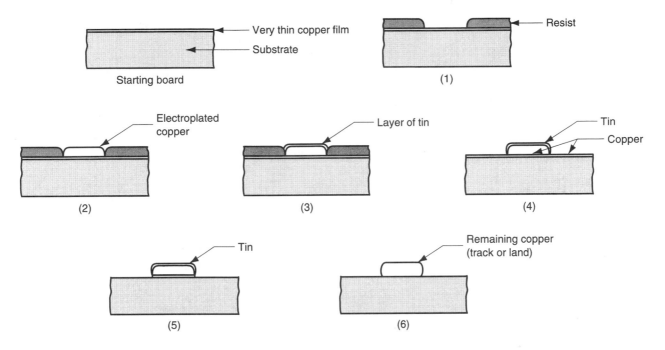

FIGURE 36.7   The semiadditive method of circuitization in PCB fabrication: (1) apply resist to areas that will not be plated; (2) electroplate copper, using the thin copper film for conduction; (3) apply tin on top of plated copper; (4) strip resist; (5) etch remaining thin film of copper on the surface, while the tin serves as a resist for the electroplated copper, and (6) strip tin from copper.

A *multilayer board* is structurally the most complex of the three types, and this complexity is reflected in its manufacturing sequence. The laminated construction can be seen in Figure 36.8, which shows a number of features characteristic of a multilayer PCB. The fabrication steps for the individual layers are basically the same as those used for single-sided and double-sided boards. What makes multilayer board fabrication more complicated is that all of the multiple layers, each with its own circuit design, must first be processed; then the layers must be joined together to form one integral board; and finally the board must itself be put through its own processing sequence. Thus, we can view multilayer PCB fabrication as consisting of three major stages: (1) fabrication of individual layers, (2) joining the layers, and (3) processing of the multilayer board.

A multilayer board consists of *logic layers,* which carry electrical signals between components on the board, and *voltage layers,* which are used to distribute power. Logic

FIGURE 36.8   Typical cross section of a multilayer printed circuit board.

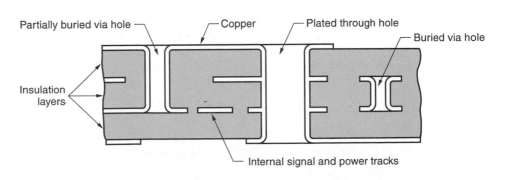

layers are generally fabricated from double-sided boards, while voltage layers are usually made from single-sided boards. Although there are variations in operations and sequence, depending on circuit design, the processing steps for these boards are similar to those already described. Thinner insulating substrates are used for multilayer boards than for their stand-alone single- and double-sided counterparts, so that a suitable thickness of the final board can be achieved.

In the second stage, the individual layers are assembled together. The procedure starts with copper foil on the bottom outside, and then adds the individual layers, separating one from the next by one or more sheets of glass fabric impregnated with partially cured epoxy. After all layers have been sandwiched together, a final copper foil is placed on the stack to form the top outer layer. Registration between layers is extremely important in order to make the correct interconnections. Layers are aligned using tight-fitting pins in the tooling holes of each board. They are then bonded into a single board by heating the assembly under pressure to cure the epoxy. After curing, any excess resin squeezed out of the sandwich around the edges is trimmed away.

At the start of the third stage of fabrication, the board consists of multiple layers bonded together, with copper foil cladded on its outer surfaces. Its construction can therefore be likened to that of a double-sided board; and its processing is likewise similar. The sequence consists of drilling additional through holes, plating the holes to establish conduction paths between the two exterior copper films as well as certain internal copper layers, and the use of photolithography and etching to form the circuit pattern on the outer copper surfaces.

Testing and Finishing Operations   After a circuit has been fabricated on the board surface, it must be inspected and tested to ensure that it functions according to design specifications and contains no quality defects. Two procedures are common: (1) visual inspection and (2) continuity testing. In *visual inspection,* the board is examined visually to detect open and short circuits, errors in drilled hole locations, and other faults that can be observed without applying electrical power to the board. Visual inspections, performed not only after fabrication but also at various critical stages during production, are accomplished by human eye or machine vision (Section 44.5.3).

Continuity testing involves the use of contact probes brought simultaneously into contact with track and land areas on the board surface. The setup consists of an array of probes that are forced under light pressure to make contact with specified points on the board surface. Electrical connections between contact points can be quickly checked in this procedure.

Several additional processing steps must be performed on the bare board to prepare it for assembly. The first of these finishing operations is the application of a thin solder layer on the track and land surfaces. This layer serves to protect the copper from oxidation and contamination. It is carried out either by electroplating or by bringing the copper side into contact with rotating rollers which are partially submersed in molten solder.

A second operation involves application of a coating of solder resist to all areas of the board surface except the lands which are to be subsequently soldered in assembly. The solder resist coating is chemically formulated to resist adhesion of solder; thus, in the subsequent soldering processes (Section 32.2), solder adheres only to land areas. Solder resist is usually applied by screen printing.

Finally, an identification legend is printed onto the surface, again by screen printing. The legend indicates where the different components are to be placed on the board in final assembly. In modern industrial practice, a bar code is also printed on the board for production control purposes.

36.3 PRINTED CIRCUIT BOARD ASSEMBLY

A *printed circuit board assembly* consists of electronic components (e.g., IC packages, resistors, capacitors) as well as mechanical components (e.g., fasteners, heat sinks) mounted on a printed circuit board. This is level two in electronic packaging (Table 36.1). As previously indicated, PCB assembly is based on either pin-in-hole or surface-mount technologies. Some PCB assemblies include both leaded and surface-mounted components. Our discussion here deals exclusively with PCB assemblies that use components with leads. In Section 36.4, we consider surface mount technology and combinations of the two types.

The scope of electronic assembly includes PCB assemblies as well as higher packaging levels such as assemblies of multiple PCBs electrically connected and mechanically contained in a chassis or cabinet. In Section 36.5, we explore the technologies by which electrical connections are made at these higher levels of packaging.

In printed circuit assemblies with leaded components, the lead pins must be inserted into through-holes in the circuit board. We have used the term *pin-in-hole* (PIH) technology to identify this assembly method. Once inserted, the leads are soldered into place in the holes in the board. In double-sided and multilayer boards, the hole surfaces into which the leads are inserted are generally copper plated, giving rise to the name *plated through-hole* (PTH) for these cases. After soldering, the boards are cleaned and tested, and those boards not passing the test are reworked if possible. Thus, we can divide the processing of PCB assemblies with leaded components into the following steps: (1) component insertion, (2) soldering, (3) cleaning, (4) testing, and (5) rework. These steps will be the basis for our discussion of PIH technology.

36.3.1 Component Insertion

In component insertion, the leads of components are inserted into their proper through holes in the PCB. A single board may be populated with hundreds of separate components (DIPs, resistors, etc.), all of which need to be inserted into the board. In modern electronic assembly plants, most component insertions are accomplished by automatic insertion machines. A small proportion (perhaps 5% to 10%) are done by hand for nonstandard components that cannot be accommodated on automatic machines. Industrial robots are sometimes used to substitute for human labor in component insertion tasks.

Automatic Insertion Machines Automatic insertion machines are either semiautomatic or fully automatic. The semiautomatic type involves insertion of the component by a mechanical insertion device whose position relative to the board is controlled by a human operator. Fully automatic insertion machines are the preferred category because they are faster and their need for human attention is limited to loading components and fixing jams when they occur. Our Color Plate 4 shows the working end of a high-speed insertion machine. Automatic insertion machines are controlled by a program that is usually prepared directly from circuit design data. Components are loaded into these machines in the form of reels, magazines, or other carriers that maintain proper orientation of the components until insertion.

The insertion operation involves (1) preforming the leads, (2) insertion of leads into the board holes, and then (3) cropping and clinching the leads on the other side of the board. Preforming is needed only for some component types and involves bending of leads that are initially straight into a U-shape for insertion. Many components come with properly shaped leads and require little or no preforming.

Insertion is accomplished by a workhead designed for the component type. Components inserted by automatic machines are grouped into three basic categories: (a) axial lead, (b) radial lead, and (c) dual-in-line package. The dual-in-line package (Section 35.6.1) is a very common package for integrated circuits. Typical axial and radial lead components are pictured in Figure 36.9. Axial components are shaped like a cylinder, with leads projecting from each end. Typical components of this type include resistors, capacitors, and diodes. Their leads must be bent, as suggested in our figure, in order to be inserted. Radial components have parallel leads and have various body shapes, one of which is shown in Figure 36.9(b). This type of component is exemplified by light-emitting diodes, potentiometers, resistor networks, and fuse holders.

These configurations are sufficiently different that separate insertion machines with the appropriate workhead designs must be used to handle each category. Accurate positioning of the board beneath the workhead prior to each insertion is performed by a high-speed x-y positioning table. For optimum reliability in the insertion operation, the through-hole diameters in the printed circuit board must be larger than the component lead diameters by 0.25 to 0.5 mm (0.01–0.02 in.). This not only facilitates insertion, but it also provides adequate clearance for solder flow in the subsequent soldering operation.

Once the leads have been inserted through the holes in the board, they are clinched and cropped. Clinching involves bending the leads, as in Figure 36.10, to mechanically secure the component to the board until soldering. If this were not done, the component is at risk of being knocked out of its holes during handling. In cropping, the leads are cut to proper length; otherwise, there is a possibility that they might become bent and cause a short circuit with nearby circuit tracks or components. These operations are performed automatically on the underside of the board by the insertion machine.

The three types of insertion machines, corresponding to the three basic component configurations, can be joined to form an integrated circuit board assembly line. The integration is accomplished by means of a conveyor system that transfers boards from one machine type to the next. A computer control system is used to track the progress of each board as it moves through the cell and to download the correct programs to each workstation. One of the problems in managing an integrated assembly line of this type is equalizing the workloads among the stations. Some stations may be assigned a greater number of insertions to perform, causing the other stations to be idle.

Manual and Robotic Insertion Manual insertion is used when the component possesses a nonstandard configuration and therefore cannot be handled by a standard in-

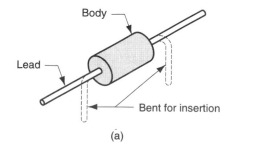
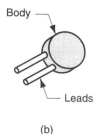

FIGURE 36.9 Two of the three basic component types used with automatic insertion machines: (a) axial lead and (b) radial lead. The third type, dual-in-line package (DIP), is illustrated in Figure 35.19.

Body

Lead

Bent for insertion

(a)

Body

Leads

(b)

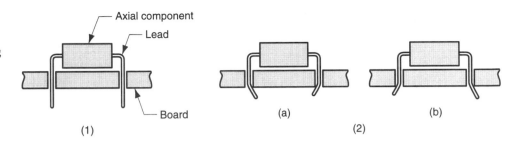

FIGURE 36.10   Clinching and cropping of component leads: (1) as inserted, and (2) after bending and cutting; leads can be bent either (a) inward or (b) outward.

sertion machine. These cases include switches and connectors as well as resistors, capacitors, and certain other components. Although the proportion of component insertions accomplished manually in industry is low, their cost is high due to much lower production rates than automatic insertions.

Manual insertion generally consists of work elements similar to those performed by an automatic insertion machine. The component leads must first be preformed so they properly align with the insertion holes. The component is then inserted into the board, and its leads are clinched and cropped. In the simplest setup, the operator uses the legend that has been printed on the circuit board to determine where each component is to be placed. Human error can be a problem, especially when there are many components to be inserted, each in a different location. In addition, the fact that the board assemblies are often made in low quantities means that the operator cannot fully learn the task, contributing to the error problem. Various schemes have been devised to reduce mistakes. One workstation design presents the components to the operator in a certain order that is coordinated with a computer-controlled light beam directed at the position on the board where the component is to be inserted.

Use of industrial robots (Section 38.2) is another approach to reduce human error in PCB assembly. Two attributes of a robot make such an application feasible: (1) robots can be programmed to perform complicated tasks, and (2) they can be equipped with end-of-arm gripping devices to handle a variety of component styles. Industrial robots cannot be used as substitutes for automatic insertion machines, because robots are too slow. They work at speeds similar to human workers; they are justified based on reducing labor costs and human errors during assembly.

36.3.2 Soldering

The second basic step in PCB assembly is soldering. For inserted components, the most important soldering techniques are hand soldering and wave soldering. These methods as well as other aspects of soldering are discussed in Section 32.2.

Hand Soldering   Hand soldering involves a skilled operator using a soldering iron to make circuit connections. Compared to wave soldering, hand soldering is slow since each solder joint is made one at a time. As a production method, it is generally used only for small lot production and for rework. As with other manual tasks, human error can result in quality problems. Hand soldering is sometimes used following wave soldering to add delicate components that would be damaged in the harsh environment of the wave-soldering chamber. Manual methods have certain advantages in PCB assembly that should be noted: (1) heat is localized and can be directed at a small target area; (2) equipment is inexpensive compared to wave soldering; and (3) energy consumption is considerably less.

Wave Soldering   Wave soldering is a mechanized technique in which printed circuit boards containing inserted components are moved by conveyor over a standing wave of molten solder (Figure 32.9). The position of the conveyor is such that only the underside of the board, with component leads projecting through the holes, is in contact with the solder. The combination of capillary action and the upward force of the wave cause the liquid solder to flow into the clearances between leads and through holes to obtain a good solder joint. The tremendous advantage of wave soldering is that all of the solder joints on a board are made in a single pass through the process.

36.3.3   Cleaning, Testing, and Rework

The final processing steps in PCB assembly are cleaning, testing, and rework. Visual inspections are also performed on the board to detect obvious flaws.

Cleaning   After soldering, contaminants are present on the printed circuit assembly. These foreign substances include flux, oil and grease, salts, and dirt, some of which can cause chemical degradation of the assembly or interfere with its electronic functions. One or more cleaning operations (Section 28.1) must be carried out to remove these undesirable materials.

Traditional cleaning methods for PCB assemblies include hand cleaning with appropriate solvents and vapor degreasing with chlorinated solvents. Concern over environmental hazards in recent years has motivated the search for effective water-based solvents to replace the chlorinated and fluorinated chemicals traditionally used in vapor degreasing.

Testing   Visual inspection is used to detect for board substrate damage, missing or damaged components, soldering faults, and similar quality defects that can be observed by eye. Machine vision systems are being perfected to perform these inspections automatically in a growing number of installations.

Test procedures must be performed on the completed assembly to verify its functionality. The board design must allow for this testing by including test points in the circuit layout. These test points are convenient locations in the circuit where probes can make contact for testing. Various tests can be performed. Individual components in the circuit are tested by contacting the component leads, applying input test signals, and measuring the output signals. More sophisticated procedures include digital function tests, in which the entire circuit or major subcircuits are tested using a programmed sequence of input signals and measuring the corresponding output signals to simulate operating conditions. Equipment for digital function testing is expensive, and much engineering time is required to design and program the appropriate test algorithms.

Another test used for printed circuit board assemblies is the substitution test, in which a production unit is plugged into a mock-up of the working system and energized to perform its functions. If the assembly performs in a satisfactory way, it is deemed as passing the test. It is then unplugged, and the next production unit is substituted in the mock-up.

Finally, a burn-in test is performed on certain types of PCB assemblies that may be subject to "infant mortality." Some boards contain defects that are not revealed in normal functional tests but which are likely to cause failure of the circuit during early service. Burn-in tests operate the assemblies under power for a certain period of time, such as 24 or 72 hr, sometimes at elevated temperatures, such as 40°C (100°F), to force these defects to manifest their failures during the testing period. Boards not subject to infant mortality will survive this test and provide long service life.

Rework   When inspection and testing indicate that one or more components on the board are defective, or that certain solder joints are faulty, it usually makes sense to try to repair the assembly rather than to discard it together with all of the remaining good components. This repair step is called rework and it is an integral part of electronic assembly plant operations. Common rework procedures include touchup (repair of solder faults), replacement of defective or missing components, and repair of copper film that has lifted from the substrate surface. These procedures are manual operations, requiring skilled workers using soldering irons.

36.4 SURFACE MOUNT TECHNOLOGY

One effect of the growing complexity of electronic systems has been the need for greater packing densities in printed circuit assemblies. Conventional PCB assemblies that use leaded components inserted into through-holes have certain inherent limitations in terms of packing density. These limitations are: (1) components can be mounted on only one side of the board, and (2) center-to-center distance between lead pins in leaded components must be a minimum of 1.0 mm (0.04 in.) and is usually 2.5 mm (0.10 in.).

Surface mount technology (SMT) uses an assembly method in which component leads are soldered to lands on the surface of the board rather than into holes running through the board (Historical Note 36.2). By eliminating the need for leads inserted into through-holes in the board, several advantages accrue [6]: (1) smaller components can be made, with leads closer together; (2) packing densities can be increased; (3) components can be mounted on both sides of the board; (4) smaller PCBs can be used for the same electronic system; (5) drilling of the many through-holes during board fabrication is eliminated—via holes to interconnect layers are still required; and (6) undesirable electrical effects are reduced, such as spurious capacitances and inductances. Typical areas on the board surface taken by SMT components range between 20% and 60% compared to through-hole components.

Historical Note 36.2   *Surface mount technology* [4].

Surface mount technology (SMT) traces its origins to the electronic systems in the aerospace and military industries of the 1960s. The first components were small flat ceramic packages with gull-wing leads. The initial reason why these packages were attractive, compared to through-hole technology, was the fact that they could be placed on both sides of a printed circuit board—in effect, doubling the component density. In addition, the SMT package could be made smaller than a comparable through-hole package, further increasing component densities on the printed circuit board.

In the early 1970s, further advances in SMT were made in the form of leadless components—components with ceramic packages that had no discrete leads. This permitted even greater circuit densities in military and aerospace electronics. In the late 1970s, plastic SMT packages became available, motivating the widespread use of surface mount technology. The computer and automotive industries have become important users of SMT and their demand for SMT components has contributed to the significant growth in this technology.

Despite these advantages, the electronics industry has not fully adopted SMT to the exclusion of PIH technology. There are several reasons: (1) owing to their smaller size, surface mount components are more difficult to handle and assemble by humans; (2) SMT components are generally more expensive than leaded components, although this disadvantage may change as SMT production techniques are perfected; (3) inspection, testing, and rework of the circuit assemblies is generally more difficult in SMT because of the smaller scale involved; and (4) certain types of components are not available in surface mount form. This final limitation results in some electronic assemblies which contain both surface mount and leaded components.

The same basic steps are required in the assembly of surface mount components to PCBs as in pin-in-hole technology. The components must be placed on the board and soldered, followed by cleaning, testing, and rework. The methods of placement and soldering the components, as well as certain of the testing and rework procedures, are different in surface mount technology. Component placement in SMT means correctly locating the component on the PCB and affixing it sufficiently to the surface until soldering provides a permanent mechanical and electrical connection. Two alternative placement and soldering methods are available: (1) adhesive bonding of components and wave soldering, and (2) solder paste and reflow soldering. It turns out that certain types of SMT components are more suited to one method while other types are more suited to the other.

36.4.1 Adhesive Bonding and Wave Soldering

The steps in this method are described in Figure 36.11. Various adhesives (Section 32.3) are used for affixing components to the board surface. Most common are epoxies and acrylics. The adhesive is applied by one of three methods: (1) brushing liquid adhesive through a screen stencil; (2) using an automatic dispensing machine with a programmable *x-y* positioning system; or (3) using a pin transfer method, in which a fixture consist-

FIGURE 36.11   Adhesive bonding and wave soldering, shown here for a discrete capacitor or resistor component: (1) adhesive is applied to areas on the board where components are to be placed; (2) components are placed onto adhesive-coated areas; (3) adhesive is cured; and (4) solder joints are made by wave soldering.

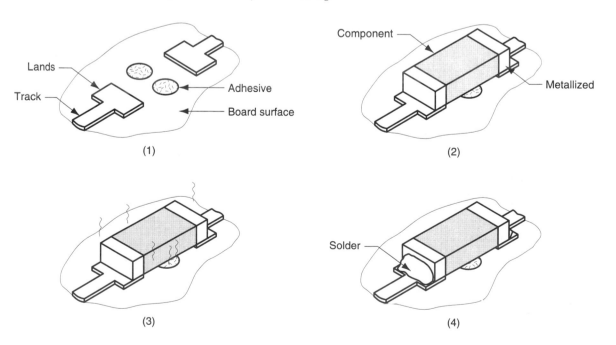

ing of pins arranged according to where adhesive must be applied is dipped into the liquid adhesive and then positioned onto the board surface to deposit adhesive in the required spots.

The components are then placed onto the board surface by automatic placement machines operating under computer control. The term **on**sertion machines is used for these units, to distinguish them from **in**sertion machines used in PIH technology. Onsertion machines operate at cycle rates of up to four components placed per second.

After component placement, the adhesive is cured. Depending on adhesive type, curing is by heat, ultraviolet (UV) light, or a combination of UV and infrared (IR) radiation. With the surface-mount components now bonded to the PCB surface, the board is put through wave soldering. The operation differs from its PIH counterpart in that the components themselves pass through the molten solder wave. Technical problems sometimes encountered in SMT wave soldering include components uprooted from the board, components shifting position, and larger components creating shadows which inhibit proper soldering of neighboring components.

36.4.2 Solder Paste and Reflow Soldering

In this method, which seems more common in industry, a solder paste is used to affix components to the surface of the circuit board. The sequence of steps is depicted in Figure 36.12.

A *solder paste* is a suspension of solder powders in a flux binder. It has three functions: (1) it is the solder—typically 80% to 90% of total paste volume, (2) it is the flux, and (3) it is the adhesive that secures the components to the surface of the board. Methods of applying the solder paste to the board surface include screen printing and syringe dispensing. Properties of the paste must be compatible with these application methods; the paste must flow yet not be so liquid that it spreads beyond the localized area where it is applied.

After solder paste application, components are placed on the board by the same type of onsertion machines used with the adhesive bonding assembly method. A low-temperature baking operation is performed to dry the flux binder; this reduces gas

FIGURE 36.12   Solder paste and reflow method: (1) apply solder paste to desired land areas, (2) place components onto board, (3) bake paste, and (4) solder reflow.

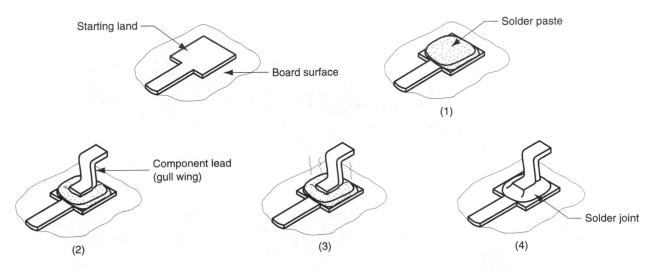

escape during soldering. Finally, the solder reflow process (Section 32.2.3) heats the solder paste sufficiently that the solder particles melt to form a high-quality mechanical and electrical joint between the component leads and the circuit lands on the board.

As in pin-in-hole technology, integrated production lines are used to accomplish the various operations required to assemble SMT printed circuit boards, as in Figure 36.13.

36.4.3 Combined SMT–PIH Assembly

Our discussion of the SMT assembly methods has assumed a relatively simple circuit board with exclusively SMT components on one side only. These cases are unusual because most SMT circuit assemblies combine surface-mounted and pin-in-hole components on the same board. In addition, SMT assemblies can be populated on both sides of the board, whereas PIH components are normally limited to one side only. The assembly sequence must be altered to allow for these additional possibilities, although the basic processing steps described in the two preceding sections are the same.

One possibility is for the SMT and PIH components to be on the same side of the board. For this case, a typical sequence would consist of the steps described in Figure 36.14. More complex PCB assemblies consist of SMT–PIH components as in our figure, but with SMT components on both sides of the board.

36.4.4 Cleaning, Inspection, Testing, and Rework

After the components have been connected to the board, the assembly must be cleaned, inspected for solder faults, circuit tested, and reworked if necessary.

Inspection of soldering quality is somewhat more difficult for surface-mounted circuits (SMCs) because these assemblies are generally more densely packed, the solder joints are smaller, and their geometries are different from joints in through-hole assemblies. One of the problems is the way SMCs are held in place during soldering. In PIH assembly, the components are mechanically fastened in place by clinched leads. In SMT assembly, components are held by adhesive or paste. At soldering temperatures this

FIGURE 36.13   SMT production line; stations include board launching, screen printing of solder paste, several component placement operations, and solder reflow oven (photo courtesy of Universal Instruments Corp.).

FIGURE 36.14   Typical process sequence for combined SMT–PIH assemblies with components on same side of board: (1) apply solder paste on lands for SMT components, (2) place SMT components on the board, (3) bake, (4) reflow solder, (5) insert PIH components, and (6) wave solder PIH components. This would be followed by cleaning, testing, and rework.

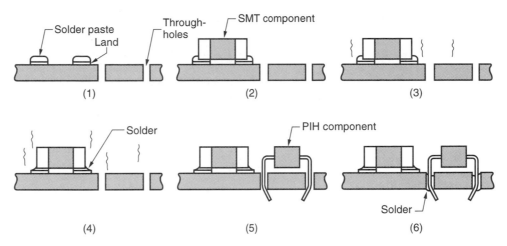

method of attachment is not as secure, and component shifting sometimes occurs. Another problem with the smaller sizes in SMT is a greater likelihood of solder bridges forming between adjacent leads, resulting in short circuits.

The smaller scale also poses problems in SMT circuit testing because less space is available around each component. Contact probes must be physically smaller, and more probes are required because SMT assemblies are more densely populated. One way of dealing with this issue is to design the circuit layout with extra lands whose only purpose is to provide a test probe contact site. Unfortunately, including these test lands runs counter to the goal of achieving higher packing densities on the board.

Manual rework in surface mount assemblies is more difficult than in conventional PIH assemblies, again due to the smaller component sizes. Special tools are required, such as small-bit soldering irons, magnifying devices, and instruments for grasping and manipulating the small parts.

36.5 ELECTRICAL CONNECTOR TECHNOLOGY

PCB assemblies must be connected to back planes, and into racks and cabinets, and these cabinets must be connected to other cabinets and systems by means of cables. The growing use of electronics in so many types of products has made electrical connections an important technology. The performance of any electronic system depends on the reliability of the individual connections linking the elements of the system together. In this section we examine connector technology that is usually applied at the third and higher levels of electronics packaging.

To begin, there are two basic methods of making electrical connections: (1) soldering and (2) pressure connections. Soldering was discussed in Section 32.2 and throughout the current chapter. It is the most widely used technology in electronics. ***Pressure connections*** are electrical connections in which mechanical forces are used to establish electrical continuity between components. Sometimes called ***solderless*** connections, they can be divided into two types: permanent and separable. Our discussion in this section will be organized around these two types of pressure connections.

36.5.1 Permanent Connections

A ***permanent connection*** involves high-pressure contact between two metal surfaces, in which one or both of the parts is mechanically deformed during the assembly process. Permanent connection methods include crimping, press-fit technology, and insulation displacement.

Crimping of Connector Terminals   This connection method is used to assemble wire to electrical terminals. Although assembly of the wire to the terminal forms a permanent joint, the terminal itself is designed to be connected and disconnected to its mating component. There are a variety of terminal styles, some of which are shown in Figure 36.15, and they are available in various sizes. They all must be connected to conductor wire, and crimping is the operation for doing this. ***Crimping*** involves the mechanical deformation of the terminal barrel to form a permanent connection with the stripped end of a wire inserted into it. The crimping operation squeezes and closes the barrel around the bare wire. Crimping is performed by hand tools or by crimping machines. The terminals are supplied either as individual pieces or on long strips that can be fed into a crimping machine. Properly accomplished, the crimped joint will have low electrical resistance and high mechanical strength.

Press Fit Technology   Press fit in electrical connections is similar to that in mechanical assembly, but the part configurations are different. Press fit technology is widely used in the electronics industry to assemble terminal pins into metal-plated through-holes in large PCBs. In that context, a ***press fit*** involves an interference fit between the terminal pin and the plated hole into which it has been inserted. There are two categories of terminal pins: (a) solid and (b) compliant, as in Figure 36.16. Within these categories, pin designs vary among suppliers. The solid pin is rectangular in cross-section and is designed so that its corners press and even cut into the metal of the plated hole to form a good electrical connection. The compliant pin is designed as a spring-loaded device that conforms to the hole contour but presses against the walls of the hole to achieve electrical contact.

Insulation Displacement   ***Insulation displacement*** is a method of making a permanent electrical connection in which a sharp, prong-shaped contact pierces the insulation and squeezes against the wire conductor to form an electrical connection. The method is illustrated in Figure 36.17, and is commonly used to make simultaneous connections between multiple contacts and flat cable. The flat cable, called ***ribbon cable,*** consists of a number of parallel wires held in a fixed arrangement by the insulation surrounding them. It is often terminated in multiple pin connectors that are widely used in electronics to make electrical connections between major subassemblies. In these applications, the in-

FIGURE 36.15   Some of the terminal styles available for making separable electrical connections: (a) slotted tongue, (b) ring tongue, and (c) flanged spade.

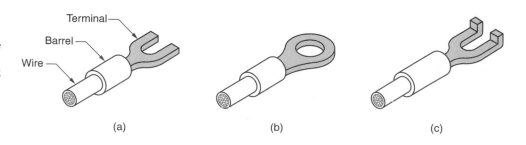

(a)   (b)   (c)

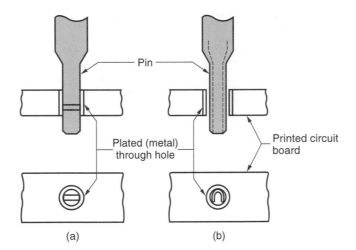

sulation displacement method reduces wiring errors and speeds harness assembly. To make the assembly, the cable is placed in a nest and a press is used to drive the connector contacts through the insulation and against the metal wires.

36.5.2 Separable Connectors

Separable connections are designed to permit disassembly and reassembly; they are meant to be connected and disconnected multiple times. When connected they must provide metal-to-metal contact between mating components with high reliability and low electrical resistance. Separable connection devices are called ***connectors*** and they come in a variety of styles to serve many different applications. Connectors typically consist of multiple contacts, contained in a plastic molded housing, designed to mate with a compatible connector or with individual wires or terminals. They are used for making electrical connections between various combinations of cables, printed circuit boards, components, and individual wires.

A wide selection of connectors is available. The design issues in choosing among them include: (1) power level (e.g., whether the connector is used for power or signal transmission), (2) cost, (3) number of individual conductors involved, (4) types of devices and circuits to be connected, (5) space limitations, (6) ease of joining the connector to its leads, (7) ease of connection with the mating terminal or connector, and (8) frequency of connection and disconnection. Some of the principal connector types include cable connectors, terminal blocks, sockets, and connectors with low or zero insertion force.

Cable Connectors *Cable connectors* are devices that are permanently connected to cables (one or both ends) and are designed to be plugged into and unplugged from a mating connector. A power cord connector that plugs into a wall receptacle is a famil-

FIGURE 36.17 Insulation displacement method of joining a connector contact to flat wire cable: (1) starting position, (2) contacts pierce insulation, and (3) after connection.

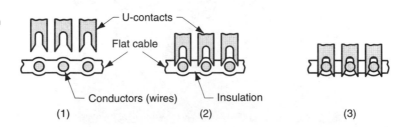

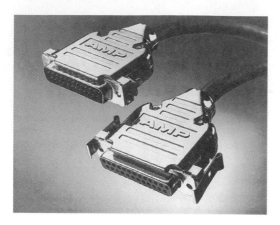

FIGURE 36.18 Multiple pin connector and mating receptacle, both attached to cables (courtesy of AMP Inc.).

iar example. Other styles include the type of multiple pin connector and mating receptacle shown in Figure 36.18, used to provide signal transmission between electronic subassemblies. Other multiple pin connector styles are used to attach printed circuit boards to other subassemblies in the electronic system.

Terminal Blocks *Terminal blocks* consist of a series of evenly spaced receptacles that allow connections between individual terminals or wires. The terminals or wires are often attached to the block by means of screws or other mechanical fastening mechanisms to permit disassembly. A conventional terminal block is illustrated in Figure 36.19.

Sockets A *socket* in electronics refers to a connection device mounted to a PCB, into which IC packages and other components can be inserted. Sockets are permanently attached to the PCB by soldering and/or press-fitting, but they provide a separable connection method for the components, which can be conveniently added, removed, or replaced in the PCB assembly. Sockets are therefore an alternative to soldering in electronics packaging.

Connectors with Low and Zero Insertion Force Insertion and withdrawal forces can be a problem in the use of pin connectors and PCB sockets. These forces increase in proportion to the number of pins involved. Possible damage can result when compo-

FIGURE 36.19 Terminal block that uses screws to attach terminals (photo courtesy AMP Inc.).

nents with many contacts are assembled. This problem has motivated the development of connectors with *low insertion force* (LIF) or *zero insertion force* (ZIF), in which special mechanisms have been devised to reduce or eliminate the forces required to push the positive and negative connectors together and to disconnect them.

REFERENCES

[1] Arabian, J. *Computer Integrated Electronics Manufacturing and Testing.* Marcel Dekker, New York, 1989.

[2] Bakoglu, H. B. *Circuits, Interconnections, and Packaging for VLSI.* Addison-Wesley, Reading Mass., 1990.

[3] Bilotta, A., J. *Connections in Electronic Assemblies,* Marcel Dekker, Inc., New York, 1985.

[4] Capillo, C. *Surface Mount Technology.* McGraw-Hill, New York, 1990.

[5] Coombs, C. F., Jr. (ed.). *Printed Circuits Handbook.* 4th ed. McGraw-Hill, New York, 1995.

[6] Edwards, P. R. *Manufacturing Technology in the Electronics Industry.* Chapman & Hall, London, U.K., 1991.

[7] Kear, F. W. *Printed Circuit Assembly Technology.* Marcel Dekker, Inc., New York, 1987.

[8] Lambert, L. P. *Soldering for Electronic Assemblies.* Marcel Dekker, Inc., New York, 1988.

[9] Marks, L. *Printed Circuit Assembly Design.* McGraw-Hill, New York, 2000.

[10] Prasad, R. P. *Surface Mount Technology.* 2nd ed. Chapman & Hall, New York, 1997.

[11] Seraphim, D. P., Lasky, R., and Li, C-Y. (eds.). *Principles of Electronic Packaging.* McGraw-Hill, New York, 1989.

REVIEW QUESTIONS

36.1. What are the functions of a well-designed electronics package?

36.2. Identify the levels of packaging hierarchy in electronics.

36.3. What is the difference between a *track* and a *land* on a printed circuit board?

36.4. Define what a printed circuit board (PCB) is.

36.5. Name the three principal types of printed circuit board.

36.6. What is a *via hole* in a printed circuit board?

36.7. What are the two basic resist coating methods for printed circuit boards?

36.8. What is etching used for in PCB fabrication?

36.9. What is *continuity testing.* and when is it performed in the PCB fabrication sequence?

36.10. What are the two main categories of PCB assemblies, as distinguished by the method of attaching components to the board?

36.11. What are some of the reasons and defects that make rework an integral step in the PCB fabrication sequence?

36.12. Identify some of the advantages of surface mount technology over conventional through-hole technology.

36.13. Identify some of the limitations and disadvantages of surface mount technology.

36.14. What are the two methods of component placement and soldering in surface mount technology?

36.15. What is a *solder paste*?

36.16. Identify the two basic methods of making electrical connections.

36.17. Define *crimping* in the context of electrical connections.

36.18. What is press-fit technology in electrical connections?

36.19. Define what a *terminal block* is.

36.20. What is a *pin connector*?

MULTIPLE CHOICE QUIZ

There is a total of 17 correct answers in the following multiple choice questions (some questions have multiple answers that are correct). To attain a perfect score on the quiz, all correct answers must be given, since each correct answer is worth 1 point. For each question, each omitted answer or wrong answer reduces the score by 1 point, and each additional answer beyond the number of answers required reduces the score by 1 point. Percentage score on the quiz is based on the total number of correct answers.

36.1. The second level of packaging refers to which one of the following? (a) component to printed circuit board, (b) IC chip to package, (c) intraconnections on the chip, or (d) wiring and cabling connections.

36.2. Surface mount technology is included within which one of the following levels of packaging? (a) zero, (b) first, (c) second, (d) third, or (e) fourth.

36.3. Card-on-board (COB) packaging refers to which one of the following levels in the electronics packaging hierarchy? (a) zero, (b) first, (c) second, (d) third, or (e) fourth.

36.4. Which of the following polymeric materials is commonly used as an ingredient in the insulation layer of a printed circuit board (more than one)? (a) copper, (b) E-glass, (c) epoxy, (d) phenolic, (e) polyethylene, and (f) polypropylene.

36.5. Typical thickness of the copper layer in a printed circuit board is which one of the following? (a) 2.5 mm or 0.100 in., (b) 0.25 mm or 0.010 in., (c) 0.025 mm or 0.001 in., or (d) 0.0025 mm or 0.0001 in.

36.6. Photolithography is widely used in PCB fabrication. Which of the following is the most common resist type used in the processing of PCBs? (a) negative resists or (b) positive resists.

36.7. Which of the following plating processes has the higher deposition rate in PCB fabrication? (a) electroless plating or (b) electroplating.

36.8. In addition to copper, which one of the following is another common metal plated onto a PCB? (a) aluminum, (b) gold, (c) nickel, or (d) tin.

36.9. Which of the following are the soldering processes used to attach components to printed circuit boards in through-hole technology (more than one)? (a) hand soldering, (b) infrared soldering, (c) reflow soldering, (d) torch soldering, and (e) wave soldering.

36.10. In general, which of the following technologies results in greater problems during rework? (a) surface mount technology or (b) through-hole technology.

36.11. Which of the following are methods of forming electrical connections (more than one)? (a) soldering, (b) insulation displacement, (c) retaining rings, and (d) pressure connections.

36.12. Which of the following electrical connection methods produce a separable connection (more than one)? (a) crimping of terminals, (b) terminal blocks, (c) press fitting, and (d) sockets.

MICROFABRICATION TECHNOLOGIES

CHAPTER CONTENTS

37.1 Microsystem Products
    37.1.1 Types of Microsystem Devices
    37.1.2 Industrial Applications
37.2 Microfabrication Processes
    37.2.1 Silicon Layer Processes
    37.2.2 LIGA Process
    37.2.3 Other Microfabrication Processes
37.3 Nanotechnology

An important trend in product design and manufacturing involves products and/or components of products whose features sizes are measured in microns (10^{-3} mm or 10^{-6} m). Several terms have been applied to these miniaturized items. *Microelectromechanical systems* (MEMS) emphasizes the miniaturization of systems consisting of both electronic and mechanical components. The word *micromachines* is sometimes used for these systems. *Microsystem technology* (MST) is a more general term that refers to the products (not necessarily limited to electromechanical products) as well as the fabrication technologies used to produce them. A related term is *nanotechnology,* which refers to even smaller devices whose dimensions are measured in nanometers (10^{-9} m). Figure 37.1 indicates the relative sizes and other factors usually associated with these terms. The figure also provides an overview of the processes described in this chapter.

37.1 MICROSYSTEM PRODUCTS

Designing products that are smaller and are comprised of smaller components and subassemblies means less material usage, lower power requirements, greater functionality per unit space, and accessibility to regions that are forbidden to larger products. In most cases, smaller products should mean lower prices because less material is used; however, the price of a given product is influenced by the costs of research, development, and production, and how these costs can be spread over the number of units sold. The economies of scale that result in lower-priced products have not yet fully been realized in microsystems technology, except for a limited number of cases that we shall examine in this section.

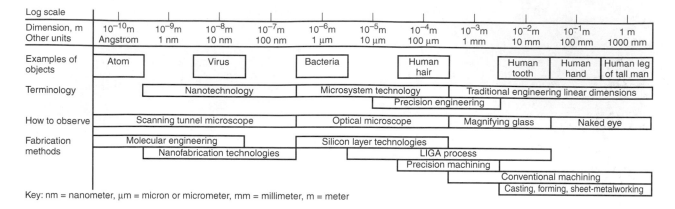

FIGURE 37.1   Terminology and relative sizes for microsystems and related technologies.

37.1.1 Types of Microsystem Devices

Microsystem products can be classified by type of device (e.g., sensor, actuator) or by application area (e.g., medical, automotive). Device types can be classified as follows [3]:

> *Microsensors*—A sensor is a device that detects or measures some physical phenomenon such as heat or pressure. It includes a transducer that converts one form of physical variable into another form (e.g., a piezoelectric device converts mechanical force into electrical current) plus the physical packaging and external connections. Most microsensors are fabricated on a silicon substrate using the same processing technologies as those used for integrated circuits (Chapter 35). Microscopic-sized sensors have been developed for measuring force, pressure, position, speed, acceleration, temperature, flow, and a variety of optical, chemical, environmental, and biological variables. The term *hybrid microsensor* is often used when the sensing element (transducer) is combined with electronics components in the same device. Figure 37.2 shows a micrograph of a micro-accelerometer developed at Motorola Co.

> *Microactuators*—Like a sensor, an actuator converts a physical variable of one type into another type, but the converted variable usually involves some mechanical action (e.g., a piezoelectric device oscillating in response to an alternating electrical field). An actuator causes a change in position or the application of force. Examples of microactuators include [3] valves, positioners, switches, pumps, and rotational and linear motors.

> *Microstructures and microcomponents*—These terms are used to denote a micro-sized part that is not a sensor or actuator. Examples of microstructures and microcomponents include microscopic lenses, mirrors, nozzles, and beams. These items must be combined with other components (microscopic or otherwise) in order to provide a useful function.

> *Microsystems and micro-instruments*—These terms denote the integration of several of the preceding components together with the appropriate electronics package into a miniature system or instrument. Microsystems and micro-instruments tend to be very application specific; for example, microlasers, optical chemical analyzers, and microspectrometers. The economics of manufacturing these kinds of systems have tended to make commercialization difficult.

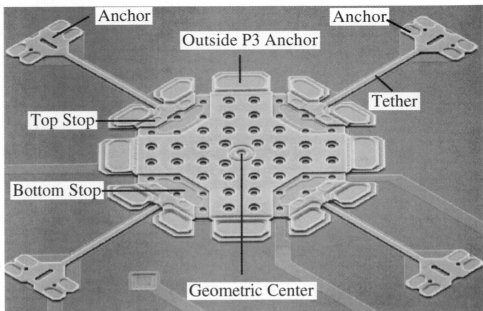

FIGURE 37.2   Micrograph of a micro-accelerometer (photo courtesy of A. A. Tseng, Arizona State University, [5]).

37.1.2 Industrial Applications

The preceding microdevices and systems have been applied in a wide variety of fields. There are many problem areas that can be approached best using very small devices. Some important examples are the following:

Ink-Jet Printing Heads   This is currently one of the largest applications of MST, because a typical ink-jet printer uses up several cartridges each year. The operation of the ink-jet printing head is depicted in Figure 37.3. An array of resistance heating elements is located above a corresponding array of nozzles. Ink flows between the heaters and nozzles. Each resistor can be independently activated under microprocessor control in microseconds. When activated, the liquid ink immediately beneath the heater boils instantly, bursting through the nozzle opening and hitting the paper, where it dries almost immediately to form a dot that is part of an alphanumeric character or other image. Today's ink-jet printers possess resolutions of 1200 dots per inch (dpi), which converts to a nozzle separation of only about 21 μm, certainly in the microsystem range.

FIGURE 37.3   Diagram of an ink-jet printing head.

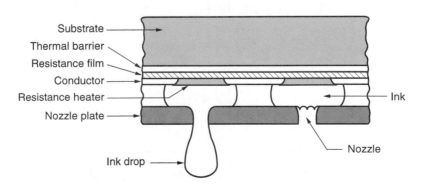

Thin-Film Magnetic Heads   Read–write heads are key components in magnetic storage devices. These heads were previously manufactured from horseshoe magnets that were manually wound with insulated copper wire. Because the reading and writing of magnetic media with higher-bit densities are limited by the size of the read–write head, hand-wound horseshoe magnets were a limitation on the technological trend toward greater storage densities. Development of thin-film magnetic heads at IBM Corporation was an important breakthrough in digital storage technology as well as a significant success story for microfabrication technologies. Thin-film read–write heads are produced annually in hundreds of millions of units, with a market of several billions of dollars per year.

A simplified sketch of the read–write head is presented in Figure 37.4, showing its MST parts. The copper conductor coils are fabricated by electroplating copper through a resist mold. The cross section of the coil is about 2 to 3 μm on a side. The thin-film cover, only a few μm thick, is made of nickel-iron alloy. The miniature size of the read–write head has permitted the significant increases in bit densities of magnetic storage media. The small sizes are made possible by microfabrication technologies.

Compact Discs   Compact discs (CDs) represent important commercial products today, as storage media for audio, video, and computer software storage applications. CDs are mass-produced by plastic molding (Chapter 13) of polycarbonate (Section 8.2). The molds for the process are fabricated using microsystem technology. A master for the mold is made from a smooth, thin layer of photosensitive polymer coated onto a glass plate. The polymer is exposed to a laser beam that writes the data into the surface. When developed, the data are represented in the form of microscopic pits in the surface. The mold is then made by electroforming metal on this polymer master.

Automotive   Microsensors and other microdevices are widely used in modern automotive products. Use of these microsystems is consistent with the increased application of on-board electronics to accomplish control and safety functions for the vehicle. The functions include electronic engine control, cruise control, anti-lock braking systems, air-bag deployment, automatic transmission control, power steering, all-wheel drive, automatic stability control, on-board navigation systems, and remote locking and unlocking, not to mention air conditioning and radio. These control systems and safety features require sensors and actuators, and a growing number of these are microscopic in

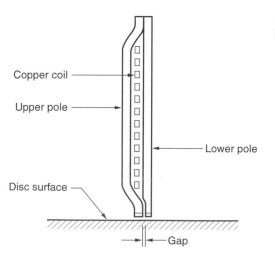

FIGURE 37.4   Thin-film magnetic read–write head (simplified).

size. There are currently 20 to 100 sensors installed in a modern automobile, depending on make and model. In 1970 there were virtually no on-board sensors. A list of some of the specific on-board microsensors are listed in Table 37.1.

Medical Opportunities for using microsystems technology in this area are tremendous. Indeed, significant strides have already been made, and many of the traditional medical and surgical methods have already been transformed by MST. One of the driving forces behind the use of microscopic devices is the principle of minimal-invasive therapy, which involves the use of very small incisions or even available body orifices to access the medical problem of concern. Advantages of this approach over the use of relatively large surgical incisions include less patient discomfort, quicker recovery, fewer and smaller scars, shorter hospital stays, and lower health insurance costs.

Among the techniques based on miniaturization of medical instrumentation is the field of endoscopy,[1] now routinely used for diagnostic purposes and with growing applications in surgery. It is standard medical practice today to use endoscopic examination accompanied by laparoscopic surgery to repair hernias or remove organs such as the gall bladder and appendix. Growing use of similar procedures is expected in brain surgery, operating through one or more small holes drilled through the skull.

Other applications of MST in the medical field now include or are expected to include (1) angioplasty, in which damaged blood vessels and arteries are repaired using surgery, lasers, or miniaturized inflatable balloons at the end of a catheter that is inserted into the vein; (2) telemicrosurgery, in which a surgical operation is performed remotely using a stereo microscope and microscopic surgical tools; (3) artificial prostheses, such as heart pacemakers and hearing aids; (4) implantable sensor systems to monitor physical variables in the human body such as blood pressure and temperature; (5) drug delivery devices that can be swallowed by a patient and then activated by remote control at the exact location intended for treatment, such as the intestine; and (6) artificial eyes.

Chemical and Environmental A principal role of microsystem technology in chemical and environmental applications is the analysis of substances in order to measure trace amounts of chemicals or detect harmful contaminants. A variety of chemical microsensors have been developed. They are capable of analyzing very small samples of

TABLE 37.1 Microsensors installed in a modern automobile.

Micro-device	Application(s)
Accelerometer	Air-bag release
Angular speed sensor	Intelligent navigation systems
Level sensors	Sense oil and gasoline levels
Pressure sensors	Optimize fuel consumption, sense oil pressure, fluid pressures of hydraulic systems (e.g., suspension systems), lumbar seat support pressure, climate control, tire pressure
Proximity and range sensors	Sense distances from front and rear bumpers for parking control and collision prevention
Temperature sensors	Cabin climate control

Compiled from [3] and [6].

---

[1] The use of a small instrument (i.e., an endoscope) to visually examine the inside of a hollow body organ such as the rectum or colon.

the substance of interest. Micropumps are sometimes integrated into these systems so that the proper amounts of the substance can be delivered to the sensor component.

Other Applications There are many other applications of microsystems technology beyond those described above. We list some examples in the following:

> *Scanning probe microscope*—This is one of the latest technologies for observing the microscopic details of surfaces, allowing surface structures to be examined at the sub-nanometer level. In order to operate in this dimensional range, the instruments require probes that are only a few μm in length and that scan the surface at a distance measured in nm. These probes are produced using microfabrication techniques.

> *Biotechnology*—In biotechnology, the specimens of interest are often microscopic in size. In order to study these specimens, manipulators and other tools are needed that are of the same size scale. Microdevices are being developed for holding, moving, sorting, dissecting, and injecting the small samples of biomaterials under a microscope.

> *Electronics*—Printed circuit board (PCB) and connector technologies were discussed in Chapter 36, but they should also be cited here in the context of MST. Miniaturization trends in electronics have forced PCBs, contacts, and connectors to be fabricated with smaller and more complex physical details, and with mechanical structures that are more consistent with the microdevices discussed in this chapter than with the integrated circuits discussed in Chapter 35.

37.2 MICROFABRICATION PROCESSES

Many of the products in microsystem technology are based on silicon, and most of the processing techniques used in the fabrication of microsystems are borrowed from the microelectronics industry. There are several important reasons why silicon is a desirable material in MST: (1) the microdevices in MST often include electronic circuits, so both the circuit and the microdevice can be fabricated in combination on the same substrate; (2) in addition to its desirable electronic properties, silicon also possesses useful mechanical properties, such as high strength and elasticity, good hardness, and relatively low density[2]; (3) the technologies for processing silicon are well-established, owing to their widespread use in microelectronics; and (4) use of single-crystal silicon permits the production of physical features to very close tolerances.

Microsystem technology often requires silicon to be fabricated along with other materials in order to obtain a particular microdevice. For example, microactuators often consist of several components made of different materials. Accordingly, microfabrication techniques consist of more than just silicon processing. Our coverage of the microfabrication processes is organized into three sections: (1) silicon layering processes, (2) the LIGA process, and (3) other processes accomplished on a microscopic scale.

37.2.1 Silicon Layer Processes

The first application of silicon in microsystems technology was in the fabrication of Si piezoresistive sensors for the measurement of stress, strain, and pressure in the early 1960s

---

[2] Silicon is discussed in Section 7.5.2.

[6]. Silicon is now widely used in MST to produce sensors, actuators, and other microdevices. The basic processing technologies are those used to produce integrated circuits (Chapter 35). However, it should be noted that certain differences exist between the processing of ICs and the fabrication of the microdevices covered in this chapter: (1) The aspect ratios in microfabrication are generally much greater than in IC fabrication. The *aspect ratio* is defined as the height-to-width ratio of the features produced, as illustrated in Figure 37.5. Typical aspect ratios in semiconductor processing are about 1.0 or less, whereas in microfabrication the corresponding ratio might be as high as 400 [6]. (2) The sizes of the devices made in microfabrication are often much larger than in IC processing, where the prevailing trend in microelectronics is inexorably toward higher circuit densities and miniaturization. (3) The structures produced in microfabrication often include cantilevers and bridges and other shapes requiring gaps between layers. These kinds of structures are uncommon in IC fabrication. (4) The silicon processing techniques are sometimes supplemented to obtain a three-dimensional structure or other physical feature in the microsystem.

Notwithstanding these differences, let us nevertheless recognize that most of the silicon-processing steps used in microfabrication are the same or very similar to those used to produce ICs. After all, silicon is the same material whether it is used for integrated circuits or microdevices. The processing steps are listed in Table 37.2, together with a brief description and text reference where the reader can obtain a more detailed description. All of these process steps are discussed in previous chapters. As in IC fabrication, the various processes in Table 37.2 used for microfabrication either add, alter, or remove layers of material from a substrate according to geometric data contained in lithographic masks. Lithography is the fundamental technology that determines the shape of the microdevice being fabricated.

Regarding our preceding list of differences between IC fabrication and microdevice fabrication, the issue of aspect ratio should be addressed in more detail. The structures in IC processing are basically planar, whereas three-dimensional structures are more likely to be required in microsystems. The features of microdevices are likely to possess large height-to-width ratios. These 3-D features can be produced in single-crystal silicon by wet etching, provided the crystal structure is oriented to allow the etching process to proceed anisotropically. Chemical wet etching of polycrystalline silicon is isotropic, with the formation of cavities under the edges of the resist, as illustrated in Figure 35.13.

FIGURE 37.5 Aspect ratio (height-to-width ratio) typical in (a) fabrication of integrated circuits and (b) microfabricated components.

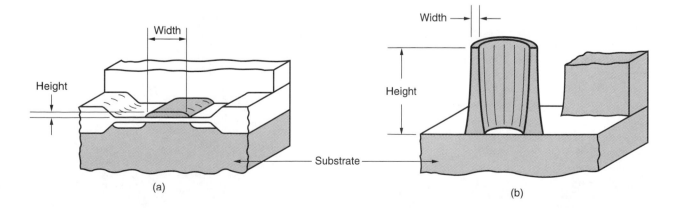

TABLE 37.2   Silicon layering processes used in microfabrication.

Process	Brief Description	Text Reference
Lithography	Printing process used to transfer copies of a mask pattern onto the surface of silicon of other solid material (e.g., silicon dioxide). The usual technique in microfabrication is photolithography.	Section 35.3
Thermal oxidation	(Layer addition) Oxidation of silicon surface to form silicon dioxide layer.	Section 35.4.1
Chemical vapor deposition	(Layer addition) Formation of a thin film on the surface of a substrate by chemical reactions or decomposition of gases.	Sections 29.4 and 35.4.2
Physical vapor deposition	(Layer addition) Family of deposition processes in which a material is converted to vapor phase and condensed onto a substrate surface as a thin film. PVD processes include vacuum evaporation and sputtering.	Section 29.3
Electroplating and Electroforming	(Layer addition) Electrolytic process in which metal ions in solution are deposited onto a cathode work material.	Sections 29.1.1 and 29.1.2
Electroless plating	(Layer addition) Deposition in an aqueous solution containing ions of the plating metal with no external electric current. Work surface acts as catalyst for the reaction.	Section 29.1.3
Thermal diffusion (doping)	(Layer alteration) Physical process in which atoms migrate from regions of high concentration into regions of low concentration.	Sections 28.3.1 and 35.4.3
Ion implantation (doping)	(Layer alteration) Embedding atoms of one or more elements in a substrate using a high-energy beam of ionized particles.	Sections 28.3.2 and 35.4.3
Wet etching	(Layer removal) Application of a chemical etchant in aqueous solution to etch away a target material, usually in conjunction with a mask pattern.	Section 35.4.5
Dry etching	(Layer removal) Dry plasma etching using an ionized gas to etch a target material.	Section 35.4.5

However, in single-crystal Si, the etching rate depends on the orientation of the lattice structure. In Figure 37.6, the three crystal faces of silicon's cubic lattice structure are illustrated. Certain etching solutions, such as potassium hydroxide (KOH) and sodium hydroxide (NaOH), have a very low etching rate in the direction of the (111) crystal face. This permits the formation of distinct geometric structures with sharp edges in a single-crystal Si substrate whose lattice is oriented to favor etch penetration vertically or at sharp angles into the substrate. Structures such as those in Figure 37.7 can be created using this procedure. It should be noted that anisotropic wet etching is also desirable in IC fabrication (Section 35.4.5), but its consequence is greater in microfabrication because of the much larger aspect ratios. The term **bulk micromachining** is used for the relatively deep wet etching process into single-crystal silicon substrate (Si wafer); the term **surface micromachining** refers to the planar structuring of the substrate surface, using much more shallow layering processes.

FIGURE 37.6   Three crystal faces in the silicon cubic lattice structure: (a) (100) crystal face, (b) (110) crystal face, and (c) (111) crystal face.

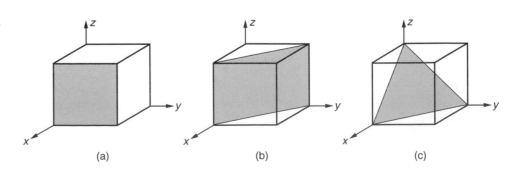

(a)   (b)   (c)

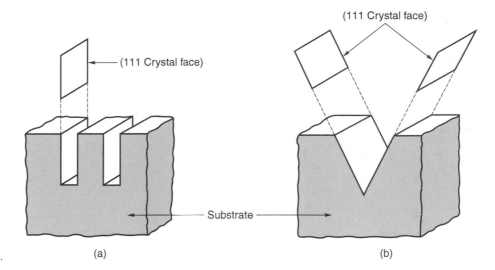

FIGURE 37.7 Several structures that can be formed in single-crystal silicon substrate by bulk micromachining: (a) (110) silicon and (b) (100) silicon.

Bulk micromachining can be used to create thin membranes in a microstructure. However, a method is needed to control the etching penetration into the silicon, so as to leave the membrane layer. A common method used for this purpose is to dope the silicon substrate with boron atoms, which significantly reduce the etching rate of the silicon. The processing sequence is shown in Figure 37.8. In step (2), epitaxial deposition is used to apply the upper layer of silicon so that it will possess the same single-crystal structure and lattice orientation as the substrate (Section 35.4.2). This is a requirement of bulk micromachining that will be used to provide the deeply etched region in subsequent processing. The use of boron doping to establish the etch resistant layer of silicon is called the *p+ etch-stop technique.*

Surface micromachining can be used to construct cantilevers, overhangs, and similar structures on a silicon substrate, as shown in part (5) of Figure 37.9. The cantilevered beams in the figure are parallel to but separated by a gap from the silicon surface. Gap size and beam thickness are in the micron range. The process sequence to fabricate this type of structure is depicted in the earlier parts of Figure 37.9.

Dry etching, which involves material removal through the physical and/or chemical interaction between the ions in an ionized gas (a plasma) and the atoms of a surface

FIGURE 37.8 Formation of a thin membrane in a silicon substrate: (1) silicon substrate is doped with boron, (2) a thick layer of silicon is applied on top of the doped layer by epitaxial deposition, (3) both sides are thermally oxidized to form a SiO_2 resist on the surfaces, (4) the resist is patterned by lithography, and (5) anisotropic etching is used to remove the silicon except in the boron-doped layer.

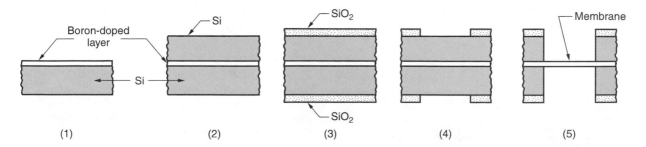

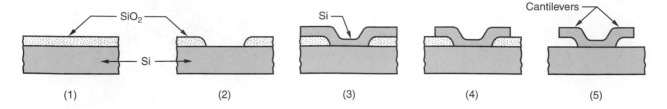

FIGURE 37.9 Surface micromachining to form a cantilever: (1) on the silicon substrate is formed a silicon dioxide layer, whose thickness will determine the gap size for the cantilevered member; (2) portions of the SiO_2 layer are etched using lithography; (3) a polysilicon layer is applied; (4) portions of the polysilicon layer are etched using lithography; and (5) the SiO_2 layer beneath the cantilevers is selectively etched.

that has been exposed to the ionized gas (Section 35.4.5), provides anisotropic etching in almost any material. Its anisotropic penetration characteristic is not limited to a single-crystal silicon substrate. On the other hand, etch selectivity is more of a problem in dry etching; that is, any surfaces exposed to the plasma are attacked.

A procedure called the **lift-off technique** is used in microfabrication to pattern metals such as platinum on a substrate. These structures are used in certain chemical sensors, but are difficult to produce by wet etching. The processing sequence in the lift-off technique is illustrated in Figure 37.10.

37.2.2 LIGA Process

The LIGA process is an important technology of MST. It was developed in Germany in the early 1980s, and the letters **LIGA** stand for the German words **LI**thographie (in particular x-ray lithography), **G**alvanoformung (translated electrodeposition or electroforming), and **A**bformtechnik (molding, in particular, plastic molding). The letters also indicate the LIGA processing sequence. These processing steps have each been described in previous sections of our book: x-ray lithography in Section 35.3.2; electrodeposition and electroforming in Sections 29.1.1 and 29.1.2, respectively; and plastic molding processes in Sections 13.6 and 13.7. Let us examine how they are integrated in LIGA technology.

The LIGA processing steps are illustrated in Figure 37.11. Let us elaborate on the brief description provided in the figure's caption: (1) A thick layer of (x-ray) radiation-sensitive resist is applied to a substrate. Layer thickness can range between several microns to centimeters, depending on the size of the part(s) to be produced. The common resist material used in LIGA is polymethylmethacrylate (PMMA, Section 8.2.2 under "Acrylics"). The substrate must be a conductive material for the subsequent electrodeposition processes per-

FIGURE 37.10 The lift-off technique: (1) resist is applied to substrate and structured by lithography; (2) platinum is deposited onto surfaces; and (3) resist is removed, taking with it the platinum on its surface but leaving the desired platinum microstructure.

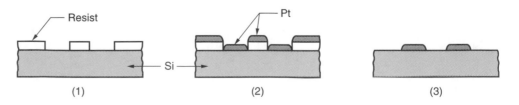

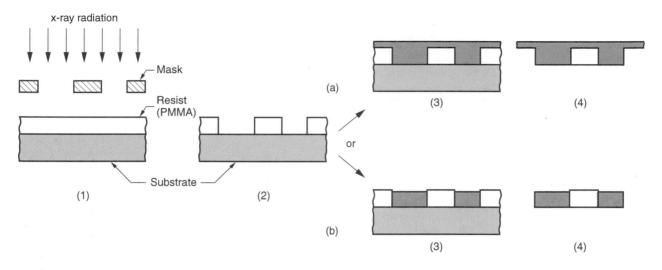

FIGURE 37.11   LIGA processing steps: (1) thick layer of resist applied and x-ray exposure through mask; (2) exposed portions of resist removed; (3) electrodeposition to fill openings in resist; (4) resist stripped to provide (a) a mold or (b) a metal part.

formed. The resist is exposed through a mask to high energy x-ray radiation. (2) The irradiated areas of the positive resist are chemically removed from the substrate surface, leaving the unexposed portions standing as a three-dimensional plastic structure. (3) The regions where the resist has been removed are filled with metal using electrodeposition. Nickel is the common plating metal used in LIGA. (4) The remaining resist structure is stripped (removed), yielding a three-dimensional metal structure. Depending on the geometry created, this metallic structure may be (a) the mold used for producing plastic parts by injection molding, reaction injection molding, or compression molding. In the case of injection molding, in which thermoplastic parts are produced, these parts may be used as "lost molds" in investment casting (Section 11.2.4). Alternatively, (b) the metal part may be a pattern for fabricating plastic molds that will be used to produce more metallic parts by electrodeposition.

As our description indicates, LIGA can produce parts by several different methods. This is one of the greatest advantages of this process in MST: (1) LIGA is a versatile process. Other advantages of LIGA technology include (2) possible high aspect ratios (large height-to-width ratios in the fabricated part), (3) wide range of part sizes, with feasible heights ranging from micrometers to centimeters, and (4) possible close tolerances. A significant disadvantage of LIGA is that it is very expensive, and so large quantities of parts are usually required to justify its application.

37.2.3 Other Microfabrication Processes

Although the principal processes used in microfabrication are those described in the preceding sections, MST research is providing several additional fabrication techniques, most of which are adaptations of full-scale processes. In this section we discuss several of these additional techniques.

Nontraditional and Traditional Processes in Microfabrication   A number of nontraditional machining processes (Chapter 26), as well as conventional manufacturing processes, are important in microfabrication. ***Photochemical machining*** (PCM, Section 26.4.2) is an essential process in IC processing and microfabrication, but we have referred to it in our descriptions here and in Chapter 35 as wet chemical etching (combined with photolithography). PCM is often used with the very conventional processes of *electroplating, electroforming,* and/or *electroless plating* (Section 29.1) to add layers of metallic materials according to microscopic pattern masks.

Other nontraditional processes capable of micro-level processing include [6]: (1) *electric discharge machining,* used to cut holes as small as 0.3 mm in diameter with aspect ratios (depth-to-diameter) as high as 100; (2) *electron-beam machining,* for cutting holes of diameter smaller than 100 μm in hard-to-machine materials; (3) *laser-beam machining,* which can produce complex profiles and holes as small as 10 μm in diameter with aspect ratios (depth-to-width or depth-to-diameter) approaching 50; (4) *ultrasonic machining,* capable of drilling holes in hard and brittle materials as small as 50 μm in diameter; and (5) *wire electric discharge cutting,* or *wire-EDM,* which can cut very narrow swaths with aspect ratios (depth-to-width) greater than 100.

Trends in conventional machining have included its capabilities for taking smaller and smaller cut sizes and associated tolerances. Referred to as *ultra-high-precision machining,* the enabling technologies have included single-crystal diamond-cutting tools and position control systems with resolutions as fine as 0.01 μm [6]. Figure 37.12 depicts one reported application, the milling of grooves in aluminum foil using a single-point diamond fly-cutter. The aluminum foil is 100 μm thick, and the grooves are 85 μm wide and 70 μm deep. Similar ultra-high-precision machining is being applied today to produce products such as computer hard discs, photocopier drums, mold inserts for compact disk reader heads, high-definition TV projection lenses, and VCR scanning heads.

Rapid Prototyping Technologies   Several rapid prototyping (RP) methods (Chapter 34) have been adapted to produce micro-sized parts [7]. RP methods use a layer additive approach to build three-dimensional components, based on a CAD (computer-aided design) geometric model of the component. Each layer is very thin, typically as low as 0.05 mm thick, which approaches the scale of microfabrication technologies. By making the layers even thinner, microcomponents can be fabricated. At time of writing, both of the techniques discussed here are still in research and development.

One approach is called *electrochemical fabrication* (EFAB), which involves the electrochemical deposition of metallic layers in specific areas that are determined by pattern

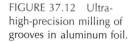

FIGURE 37.12   Ultra-high-precision milling of grooves in aluminum foil.

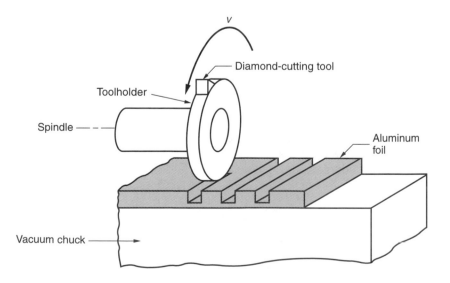

masks created by "slicing" a CAD model of the object to be made (Section 34.1). The deposited layers are generally 5 to 10 μm thick, with feature sizes as small as 20 μm in width. EFAB is carried out at temperatures below 60°C (140°F) and does not require a clean room environment. However, the process is slow, requiring about 40 min to apply each layer, or about 36 layers (a height between 180 and 360 μm) per 24-hour period. To overcome this disadvantage, the mask for each layer can contain multiple copies of the part slice pattern, permitting many parts to be produced simultaneously in a batch process.

Another RP approach, called **microstereolithography,** is based on stereolithography (STL), but the scale of the processing steps is reduced in size. Whereas the layer thickness in conventional stereolithography ranges between 75 μm and 500 μm, microstereolithography (MSTL) uses layer thicknesses between 10 to 20 μm typically, with even thinner layers possible. The laser spot size in STL is typically around 250 μm in diameter, while MSTL uses a spot size as small as 1 or 2 μm. Another difference in MSTL is that the work material is not limited to a photosensitive polymer. The researchers report success in fabricating 3-D microstructures from ceramic and metallic materials. The difference is that the starting material is a powder rather than a liquid.

Photofabrication   This term applies to an industrial process in which ultraviolet exposure through a pattern mask causes a significant modification in the chemical solubility of an optically clear material. The change is manifested in the form of an increase in solubility to certain etchants. For example, hydrofluoric acid etches the UV-exposed photosensitive glass between 15 and 30 times faster than the same glass that has not been exposed. Masking is not required during etching, the difference in solubility being the determining factor in which portions of the glass are removed.

Origination of photofabrication actually preceded the microprocessing of silicon. Now, with the growing interest in microfabrication technologies, there is a renewed interest in the older technology. Examples of modern materials used in photofabrication include Corning Glass Works' Fotoform™ glasses and Fotoceram™ ceramics, and DuPont's Dycril and Templex photosensitive solid polymers. When processing these materials, aspect ratios of around 3 : 1 can be obtained with the polymers and 20 : 1 with the glasses and ceramics.

37.3 NANOTECHNOLOGY

The trend in miniaturization is expected to continue beyond microsystem technology. **Nanotechnology** is the term used for the next generation of even smaller devices and their fabrication processes, which involve the control of feature sizes measured on the nanometer (one nm = 10^{-9} m) scale. Nanostructures consist of physical features whose dimensions are in the range 1 to 100 nm. Structures of this size can almost be thought of as purposely arranged collections of individual atoms and molecules. Two alternative processing technologies will be used to fabricate items of this size: (1) additive molecular processes that build the nanostructure from individual atoms, and (2) nanofabrication technologies similar to microfabrication processes only performed on a smaller scale.

Molecular Engineering   Additive processes that assemble the nanostructure from its molecular components are in the domain of molecular engineering and biotechnology. Nature provides a guide for the kinds of fabrication techniques that might be used. In molecular engineering and in nature, entities at the atomic and molecular level are combined into larger entities, proceeding in a constructive manner toward the creation of some deliberate thing. If the thing is a living organism, the intermediate entities are biological cells, and the organism is grown through an additive process that exhibits massive replication of individual cell formations.

Similar approaches are being explored for fabricating nanostructures other than living organisms. One of these approaches is called *mechanosynthesis*, in which two reactive molecules are brought into contact in a controlled orientation for a specified time in order to achieve some planned synthesis. Scanning tunnel microscopy (STM) is a technology that is currently being used in research laboratories to accomplish a variety of processes that involve the manipulation of atoms and molecules. The processes are classified as parallel and perpendicular. In parallel processes, atoms or molecules are transported in a direction parallel to the "work surface"; whereas in perpendicular processes, the atoms or molecules are lifted and deposited at a different location on the surface. These kinds of manipulations have permitted some fascinating scientific achievements, such as the writing of alphanumeric characters and symbols on an atomic size scale. It is anticipated that engineering products based on this type of nanotechnology will begin to appear around the year 2010 or soon after [4].

Nanofabrication Technologies   Nanofabrication processes are similar to those used in the fabrication of integrated circuits and microsystems, but they are carried out on a scale that is several orders of magnitude smaller than in microfabrication. The processes involve the addition, alteration, and subtraction of thin layers using lithography to determine the shapes in the layers produced. These processes were discussed in Chapter 35 and summarized in Table 37.2. One significant difference is the lithography technologies that must be used at the smaller scales in nanofabrication. Ultraviolet photolithography cannot be used effectively, owing to the relatively long wavelengths of UV radiation. Instead, the preferred technique is high-resolution electron-beam lithography (Section 35.3.2), whose shorter wavelength virtually eliminates diffraction during exposure.

Another process emphasized in nanofabrication is molecular beam epitaxy, a layer addition technology. As described in Section 35.4.2, *molecular-beam epitaxy* (MBE) involves the growth of a single-crystal layer onto an existing crystalline substrate, so that the lattice of the new layer duplicates that of the existing substrate. Physical vapor deposition (Section 29.3) is the process used to accomplish molecular beam epitaxy. The slow rate of growing the new layer in MBE, considered a disadvantage in some applications, is an advantage in nanofabrication, as it permits precise control of the layer thickness.

Nanofabrication techniques have already found applications in transistors for satellite microwave receivers, lasers used in communications systems, and compact disc players.

REFERENCES

[1] Ashley, S. "Getting a Hold on Mechatronics." *Mechanical Engineering.* Vol. 119, No. 5, May 1997, pp. 60–63.

[2] Drexler, K. E., *Nanosystems: Moleculer Machinery, Manufacturing, and Computation,* John Wiley & Sons, Inc., New York, 1992.

[3] Fatikow, S., and Rembold, U. *Microsystem Technology and Microrobotics.* Springer-Verlag, Berlin, Germany, 1997.

[4] Goldin, D, Venneri, S., and Noor, A. "The Great out of the Small." *Mechanical Engineering.* Vol. 122, No. 11, November 2000, pp. 70–79.

[5] Li, G., and Tseng, A. A., "Low Stress Packaging of a Micromachined Accelerometer," *IEEE Transactions on Electronics Packaging Manufacturing,* Vol. 24, No. 1, January 2001, pp. 18–25.

[6] Madou, M. *Fundamentals of Microfabrication.* CRC Press, Boca Raton, Fla. 1997.

[7] O'Connor, L., and Hutchinson, H. "Skyscrapers in a Microworld." *Mechanical Engineering.* Vol. 122, No. 3, March 2000, pp. 64–67.

[8] Paula, G. "An Explosion in Microsystems Technology." *Mechanical Engineering.* Vol. 119, No. 9, September 1997, pp. 71–74.

[9] Tseng, A. A., and Mon, J.–I, "NSF 2001 Workshop on Manufacturing of Micro-Electro Mechanical Systems," in *Proceeding of the 2001 NSF Design, Service, and* *Manufacturing Grantees & Research Conference,* National Science Foundation, 2001.

REVIEW QUESTIONS

37.1. Define microelectromechanical system.

37.2. What is the approximate size scale in microsystem technology?

37.3. Why is it reasonable to believe that microsystem products would be available at lower costs than products of larger, more conventional size?

37.4. What is a hybrid microsensor?

37.5. What are some of the basic types of microsystem devices?

37.6. Why is silicon a desirable work material in microsystem technology?

37.7. What is meant by the term *aspect ratio* in microsystem technology?

37.8. What is the difference between *bulk micromachining* and *surface micromachining*?

37.9. What is meant by the term nanotechnology?

MULTIPLE CHOICE QUIZ

There is a total of 15 correct answers in the following multiple choice questions (some questions have multiple answers that are correct). To attain a perfect score on the quiz, all correct answers must be given, since each correct answer is worth 1 point. For each question, each omitted answer or wrong answer reduces the score by 1 point, and each additional answer beyond the number of answers required reduces the score by 1 point. Percentage score on the quiz is based on the total number of correct answers.

37.1. Microsystem technology includes which of the following (more than one)? (a) LIGA technology, (b) microelectromechanical systems, (c) micromachines, (d) nanotechnology, (e) precision engineering.

37.2. The typical range of feature sizes in microsystem technology is which one of the following? (a) 10^{-3} m to 10^{-2} m, (b) 10^{-6} m to 10^{-3} m, (c) 10^{-9} m to 10^{-6} m.

37.3. Which of the following are current applications of microsystem technology in modern automobiles (more than one)? (a) air-bag release sensors, (b) alcohol-blood-level sensors, (c) driver identification sensors for theft prevention, (d) oil-pressure sensors, and (e) temperature sensors for cabin climate control.

37.4. The most common work material used in microsystem technology is which one of the following? (a) boron, (b) gold, (c) nickel, (d) potassium hydroxide, or (e) silicon.

37.5. The *aspect ratio* in microsystem technology is best defined by which one of the following? (a) degree of anisotropy in etched features, (b) height-to-width ratio of the fabricated features, (c) height-to-width ratio of the MST device, (d) length-to-width ratio of the fabricated features, or (e) thickness-to-length ratio of the MST device.

37.6. Which of the following forms of radiation have wavelengths shorter than the wavelength of ultraviolet light used in photolithography (more than one)? (a) electron beam radiation, (b) natural light, and (c) x-ray radiation.

37.7. *Bulk micromachining* refers to a relatively deep wet etching process into a single-crystal silicon substrate: (a) true or (b) false.

37.8. In the LIGA process, the letters *LIGA* stand for which one of the following? (a) let it go already; (b) little itty-bitty grinding apparatus; (c) lithographic applications; (d) lithography, electrodeposition, and plastic molding; (e) lithography, grinding, and alteration.

37.9. Photofabrication means the same process as photolithography: (a) true or (b) false.

37.10. The typical range of feature sizes in nanotechnology is which one of the following? (a) 10^{-3} m to 10^{-2} m, (b) 10^{-6} m to 10^{-3} m, (c) 10^{-9} m to 10^{-6} m.

Part X
Manufacturing Systems

38 NUMERICAL CONTROL AND INDUSTRIAL ROBOTICS

CHAPTER CONTENTS

38.1 Numerical Control
    38.1.1 The Technology of Numerical Control
    38.1.2 Analysis of NC Positioning Systems
    38.1.3 Precision in Positioning
    38.1.4 NC Part Programming
    38.1.5 Applications of Numerical Control
38.2 Industrial Robotics
    38.2.1 Robot Anatomy
    38.2.2 Control Systems and Robot Programming
    38.2.3 Applications of Industrial Robots
38.3 Programmable Logic Controllers

In this part of the book, we consider several types of manufacturing systems that are commonly associated with the manufacturing and assembly processes discussed in preceding chapters. A *manufacturing system* can be defined as a collection of integrated equipment and human resources that performs one or more processing and/or assembly operations on a starting work material, part, or set of parts. The integrated equipment consists of production machines, material handling and positioning devices, and computer systems. Human resources are required either full-time or part-time to keep the equipment operating. The position of the manufacturing systems in the larger production system is shown in Figure 38.1. As the diagram indicates, the manufacturing systems are located in the factory. They accomplish the value-added work on the part or product.

Manufacturing systems include both automated systems and manually operated equipment. The distinction between the two categories is not always clear, because many manufacturing systems consist of both automated and manual work elements (e.g., a machine tool that operates on a semiautomatic processing cycle but which must be loaded and unloaded each cycle by a human worker). Our coverage includes both categories and is organized into three chapters: Chapter 38 on numerical control, industrial robotics, and programmable logic controllers; Chapter 39 on group technology and flexible manufacturing systems; and Chapter 40 on production lines. A more detailed discussion of automation and manufacturing systems can be found in [5].

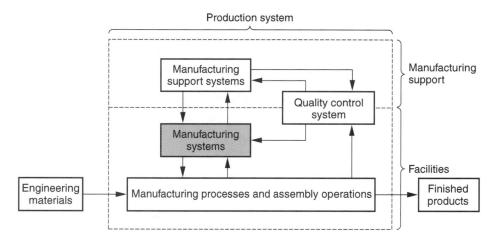

Production system

FIGURE 38.1   The position of the manufacturing systems in the larger production system.

38.1 NUMERICAL CONTROL

Numerical control (NC) is a form of programmable automation in which the mechanical actions of a piece of equipment are controlled by a program containing coded alphanumeric data. The data represent relative positions between a workhead and a workpart. The workhead is a tool or other processing element, and the workpart is the object being processed. The operating principle of NC is to control the motion of the workhead relative to the workpart and to control the sequence in which the motions are carried out. The first application of numerical control was in machining (Historical Note 38.1), and this is still an important application area. A NC machine tool is shown in Color Plate 8 as well as Figures 22.26 and 22.27.

Historical Note 38.1   *Numerical control* [3], [5].

The initial development work on numerical control is credited to John Parsons and Frank Stulen at the Parsons Corporation in Michigan in the late 1940s. Parsons was a machining contractor for the U.S. Air Force and had devised a means of using numerical coordinate data to move the worktable of a milling machine for producing complex parts for aircraft. On the basis of Parson's work, the U.S. Air Force awarded a contract to the company in 1949 to study the feasibility of the new control concept for machine tools. The project was subcontracted to the Servomechanisms Laboratory at the Massachusetts Institute of Technology to develop a prototype machine tool that utilized the new numerical data principle. The M.I.T. lab confirmed that the concept was feasible and proceeded to adapt a three-axis vertical milling machine using combined analog-digital controls. The name *numerical control* (NC) was given to the system by which the machine tool motions were accomplished. The prototype machine was demonstrated in 1952.

The accuracy and repeatability of the NC system was far better than the manual machining methods then available. The potential for reducing nonproduc-

tive time in the machining cycle was also apparent. However, the machine tool builders were unwilling to invest the large sums of money required to develop products based on numerical control. In 1956, the Air Force decided to sponsor the development of NC machine tools at several different companies. These machines were placed in operation at various aircraft companies between 1958 and 1960. The advantages of NC soon became apparent, and the aerospace companies began placing orders for new NC machines. In some cases they even began building their own units.

The importance of part programming was clear from the start. The U.S. Air Force continued to encourage the development and application of NC by sponsoring research at M.I.T. for a part programming language to control NC machines. This research resulted in the development of **APT** in 1958 (APT stands for Automatically Programmed Tooling). APT is a part programming language by which a user can describe the machining instructions in simple English-like statements.

38.1.1 The Technology of Numerical Control

In this section we define the components of a numerical control system, and then proceed to describe the coordinate axis system and motion controls.

Components of an NC System   A numerical control system consists of three basic components: (1) part program, (2) machine control unit, and (3) processing equipment. The *part program* (the term commonly used in machine tool technology) is the detailed set of commands to be followed by the processing equipment. Each command specifies a position or motion that is to be accomplished by the workhead relative to the processed object. A position is defined by its *x-y-z* coordinates. In machine tool applications, additional details in the NC program include spindle rotation speed, spindle direction, feed rate, tool change instructions, and other commands related to the operation. For many years, NC part programs were encoded on 1-in.-wide punched paper tape, using a standard format that could be interpreted by the machine control unit. Today, punched tape has largely been replaced by newer storage technologies in modern machine shops. These technologies include magnetic tape and electronic transfer of NC part programs from a central computer.

The *machine control unit* (MCU) in modern NC technology is a microcomputer that stores the program and executes it by converting each command into actions by the processing equipment, one command at a time. The MCU consists of both hardware and software. The hardware includes the microcomputer, components to interface with the processing equipment, and certain feedback control elements. The MCU may also include a tape reader if the programs are loaded into computer memory from punched tape. The software in the MCU includes control system software, calculation algorithms, and translation software to convert the NC part program into a usable format for the MCU. The MCU also permits the part program to be edited in case the program contains errors, or changes in cutting conditions are required. Because the MCU is a computer, the term *computer numerical control* (CNC) is often used to distinguish this type of NC from its technological predecessors that were based entirely on hard-wired electronics.

The *processing equipment* accomplishes the sequence of processing steps to transform the starting workpart into a completed part. It operates under the control of the machine control unit according to the set of instructions contained in the part program. We survey the variety of applications and processing equipment in Section 38.1.5.

FIGURE 38.2   Coordinate systems used in numerical control: (a) for flat and prismatic work, and (b) for rotational work.

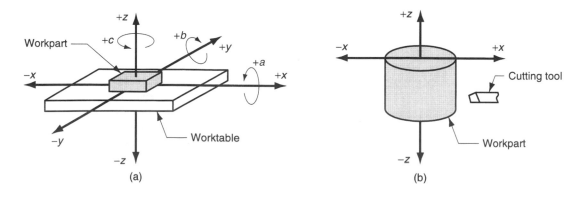

Coordinate System and Motion Control in NC   A standard coordinate axis system is used to specify positions in numerical control. The system consists of the three linear axes (x, y, z) of the Cartesian coordinate system, plus three rotational axes (a, b, c), as shown in Figure 38.2(a). The rotational axes are used to rotate the workpart to present different surfaces for machining, or to orient the tool or workhead at some angle relative to the part. Most NC systems do not require all six axes. The simplest NC systems (e.g., plotters, pressworking machines for flat sheet metal stock, and component insertion machines) are positioning systems whose locations can be defined in an x-y plane. Programming of these machines involves specifying a sequence of x-y coordinates. By contrast, some machine tools have five-axis control to shape complex workpart geometries. These systems typically include three linear axes plus two rotational axes.

The coordinates for a rotational NC system is illustrated in Figure 38.2(b). These systems are associated with turning operations on NC lathes. Although the work rotates, this is not one of the controlled axes. The cutting path of the lathe tool relative to the rotating workpiece is defined in the x-z plane, as shown in our figure.

In many NC systems, the relative movements between the processing element and the workpart are accomplished by fixing the part to a worktable and then controlling the positions and motions of the table relative to a stationary or semi-stationary workhead. Most machine tools and component insertion machines are based on this method of operation. In other systems, the workpart is held stationary and the workhead is moved along two or three axes. Flame cutters, x-y plotters, and coordinate measuring machines operate in this mode.

Motion control systems based on NC can be divided into two types: (1) point-to-point and (2) continuous path. *Point-to-point systems,* also called *positioning systems,* move the workhead (or workpiece) to a programmed location with no regard for the path taken to get to that location. Once the move is completed, some processing action is accomplished by the workhead at the location, such as drilling or punching a hole. Thus, the program consists of a series of point locations at which operations are performed.

Continuous path systems provide continuous simultaneous control of more than one axis, thus controlling the path followed by the tool relative to the part. This permits the tool to perform a process while the axes are moving, enabling the system to generate angular surfaces, two-dimensional curves, or three-dimensional contours in the workpart. This operating scheme is required in drafting machines, certain milling and turning operations, and flame cutting. In machining, continuous path control also goes by the name *contouring.*

Another aspect of motion control is concerned with whether the positions in the coordinate system are defined absolutely or incrementally. In *absolute positioning,* the workhead locations are always defined with respect to the origin of the axis system. In *incremental positioning,* the next workhead position is defined relative to the present location. The difference is illustrated in Figure 38.3.

38.1.2 Analysis of NC Positioning Systems

The function of the positioning system is to convert the coordinates specified in the NC part program into relative positions between the tool and workpart during processing. Let us consider how a simple positioning system, shown in Figure 38.4, might operate. The system consists of a worktable on which a workpart is fixtured. The purpose of the table is to move the part relative to a tool or workhead. To accomplish this purpose, the worktable is moved linearly by means of a rotating leadscrew which is driven by a motor (e.g., stepping motor or servomotor). For simplicity, only one axis is shown in our sketch. To provide x-y capability, the system shown would be piggybacked on top of a second axis perpendicular to the first. The leadscrew has a certain pitch p, mm/thread

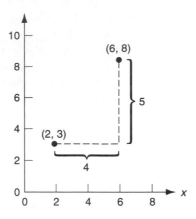

FIGURE 38.3   Absolute vs. incremental positioning. The workhead is at point (2, 3) and is to be moved to point (6, 8). In absolute positioning, the move is specified by $x = 6$, $y = 8$; while in incremental positioning, the move is specified by $x = 4$, $y = 5$.

(in./thread) or mm/rev (in./rev). Thus, the table is moved a distance equal to the leadscrew pitch for each revolution. The velocity at which the worktable moves, which corresponds to the feed rate in a machining operation, is determined by the rotational speed of the leadscrew.

Two basic types of motion control are used in NC systems: (a) open loop and (b) closed loop, as shown in Figure 38.5. The difference is that an open-loop system operates without verifying that the desired position of the worktable has been achieved. A closed-loop control system uses feedback measurement to verify that the position of the worktable is indeed the location specified in the program. Open-loop systems are less expensive than closed-loop systems and are appropriate where the force resisting the actuating motion is minimal. Closed-loop systems are normally specified for machine tools that perform continuous path operations such as milling or turning, in which the resisting forces can be significant.

Open-Loop Positioning Systems   An open-loop positioning system typically uses a stepping motor to rotate the leadscrew. In NC, the stepping motor is driven by a series of electrical pulses generated by the machine control unit. Each pulse causes the motor to rotate a fraction of one revolution, called the step angle. The allowable step angles must conform to the relationship

$$\alpha = \frac{360}{n_s} \tag{38.1}$$

where α = step angle, degrees; and n_s = the number of step angles for the motor, which must be an integer.

FIGURE 38.4   Motor and leadscrew arrangement in a NC positioning system.

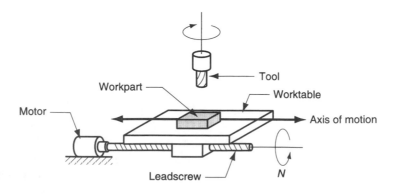

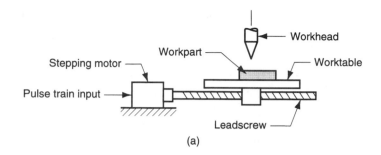

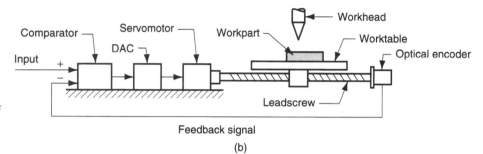

FIGURE 38.5   Two types of motion control in NC: (a) open loop and (b) closed loop.

The angle through which the leadscrew rotates, assuming a one-to-one gear ratio between the motor and the leadscrew, is given by

$$A = \alpha n_p \tag{38.2}$$

where A = angle of leadscrew rotation, degrees; n_p = number of pulses received by the motor; and α = step angle, here defined as degrees/pulse. This equation and others below must be adjusted for the case of a gear ratio different from 1 : 1.

The resulting movement of the table in response to the rotation of the leadscrew can be determined from

$$x = \frac{pA}{360} \tag{38.3}$$

where x = relative x-axis position relative to the starting position, mm (in.); p = pitch of the leadscrew, mm/rev (in./rev); and $A/360$ = the number of revolutions (and partial revolutions) of the leadscrew. By combining the two preceding equations and rearranging, the number of pulses required to achieve a specified x-position increment in a point-to-point system can be found:

$$n_p = \frac{360x}{p\alpha} \tag{38.4}$$

The pulses are transmitted at a certain frequency that drives the worktable at a corresponding velocity or feed rate in the direction of the leadscrew axis. The rotational speed of the leadscrew depends on the frequency of the pulse train as follows:

$$N = \frac{60f_p}{n_s} \tag{38.5}$$

where N = rotational speed, rev/min; f_p = pulse train frequency, Hz (pulses/sec); and n_s = steps/rev, or pulses/rev. For a two-axis table with continuous path control, the relative velocities of the axes are coordinated to achieve the desired travel direction.

The table travel speed in the direction of leadscrew axis is determined by the rotational speed as follows:

$$v_t = f_r = Np \qquad (38.6)$$

where v_t = table travel speed, mm/min (in./min); f_r = table feed rate, mm/min (in./min); N = rotational speed as defined in the previous equation, rev/min; and p = leadscrew pitch, mm/rev (in./rev).

The required pulse train frequency to drive the table at a specified feed rate can be obtained by combining Eqs. (38.5) and (38.6) and rearranging to solve for f_p:

$$f_p = \frac{v_t n_s}{60p} = \frac{f_r n_s}{60p} \qquad (38.7)$$

EXAMPLE 38.1
NC Open-Loop
Positioning

A stepping motor has 150 step angles. Its output shaft is directly coupled to a leadscrew with pitch = 5.0 mm. The worktable of a positioning system is driven by the leadscrew. The table must move a distance of 75.0 mm from its current position at a travel speed of 400 mm/min. Determine (a) how many pulses are required to move the table the specified distance, and (b) what is the required motor speed and pulse rate to achieve the desired table speed?

Solution: (a) Rearranging Eq. (38.3) to find the angle A corresponding to a distance x = 75.0 mm,

$$A = \frac{360x}{p} = \frac{360(75)}{5} = 5400°$$

With 150 step angles, each step angle is $\alpha = \dfrac{360}{150} = 2.4°$

Thus, the number of pulses to move the table 75 mm is $n_p = \dfrac{5400}{2.4} = 2250$ pulses.

(b) Eq. (38.6) can be used to find the motor speed corresponding to the table speed of 400 mm/min

$$N = \frac{v_t}{p} = \frac{400}{5.0} = 80.0 \text{ rev/min}$$

and the pulse rate is given by $f_p = \dfrac{400(150)}{60(5.0)} = 200$ Hz

Closed-Loop Positioning Systems   Closed-loop NC systems, Figure 38.5(b), use servomotors and feedback measurements to ensure that the desired position is achieved. A common feedback sensor used in NC (and also industrial robots) is the optical encoder, illustrated in Figure 38.6. The optical encoder consists of a light source, a photodetector, and a disk containing a series of slots through which the light source can shine to energize the photodetector. The disk is connected, either directly or through a gear train, to a rotating shaft whose angular position and velocity are to be measured. As the shaft rotates, the slots cause the light source to be seen by the photocell as a series of flashes, which are converted into an equivalent series of electrical pulses. By counting the pulses and computing the frequency of the pulse train, worktable position and speed can be determined.

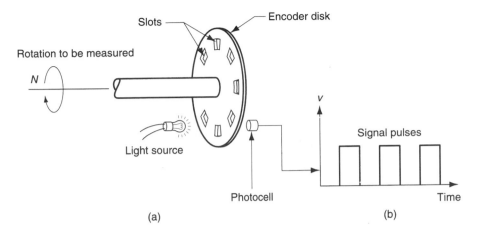

FIGURE 38.6 Optical encoder: (a) apparatus, and (b) series of pulses emitted to measure rotation of disk.

The equations describing the operation of a closed-loop positioning system are similar to those for an open loop system. In the basic optical encoder, the angle between slots in the disk must satisfy the following requirement:

$$\alpha = \frac{360}{n_s} \tag{38.8}$$

where α = angle between slots, degrees/slot; and n_s = the number of slots in the disk, slots/rev; and 360 = degrees/rev. For a certain angular rotation of the shaft, the encoder senses a number of pulses given by

$$n_p = \frac{A}{\alpha} \tag{38.9}$$

where n_p = pulse count; A = angle of rotation, degrees; and α = angle between slots, degrees/pulse. The pulse count can be used to determine the linear x-axis position of the worktable by factoring in the leadscrew pitch. Thus,

$$x = \frac{p n_p}{n_s} \tag{38.10}$$

Similarly, the feed rate at which the worktable moves is obtained from the frequency of the pulse train:

$$f_r = \frac{60 p f_p}{n_s} \tag{38.11}$$

where f_r = feed rate, mm/min (in./min); p = pitch, mm/rev (in./rev); f_p = frequency of the pulse train, Hz (pulses/sec); n_s = number of slots in the encoder disk, pulses/rev; and 60 converts seconds to minutes.

The series of pulses generated by the encoder is compared with the coordinate position and feed rate specified in the part program, and the difference is used by the machine control unit to drive a servomotor which in turn drives the worktable. A digital-to-analog converter (DAC) is used to convert the digital signals used by the MCU into a continuous analog signal to operate the drive motor. Closed-loop NC systems of the type described here are appropriate when there is force resisting the movement of the table. Most metal-cutting machine tool operations fall into this category, particularly those involving continuous path control such as milling and turning.

The equations above assume a gear ratio = 1 : 1. For other gear ratios, adjustments must be made in the relations, as shown in the following example.

EXAMPLE 38.2
NC Closed-Loop
Positioning

A NC worktable is driven by a closed-loop positioning system consisting of a servomotor, leadscrew, and optical encoder. The leadscrew has a pitch = 5.0 mm and is coupled to the motor shaft with a gear ratio of 4 : 1 (four turns of the motor for each turn of the lead-screw). The optical encoder generates 150 pulses/rev of the leadscrew. The table has been programmed to move a distance of 75.0 mm at a feed rate = 400 mm/min. Determine (a) how many pulses are received by the control system to verify that the table has moved ex-actly 75.0 mm, (b) pulse rate, and (c) motor speed that correspond to the specified feed rate.

Solution: (a) Rearranging Eq. (38.10) to find n_p,

$$n_p = \frac{x n_s}{p} = \frac{75(150)}{5} = 2250 \text{ pulses}$$

(b) The pulse rate corresponding to 400 mm/min can be obtained by rearranging Eq. (38.11):

$$f_p = \frac{f_r n_s}{60 p} = \frac{400(150)}{60(5)} = 200 \text{ Hz}$$

(c) Motor speed is the table velocity divided by the pitch, and correcting for the gear reduction:

$$N = \frac{r_g f_r}{p} \tag{38.12}$$

where r_g = gear ratio ($r_g = 4.0$); thus, $N = \dfrac{4.0(400)}{5} = 320 \text{ rev/min}$

Note that pulse count and pulse rate have the same numerical values as in Example 38.1, since the encoder is connected to the leadscrew. However, since the servomotor rotates four times for each rotation of the leadscrew, motor speed is four times the previous step-ping motor value.

38.1.3 Precision in Positioning

Three critical measures of precision in positioning are control resolution, accuracy, and repeatability. These terms are most easily explained by considering a single axis of the position system.

Control resolution refers to the system's ability to divide the total range of the axis movement into closely spaced points that can be distinguished by the control unit. *Control resolution* is defined as the distance separating two adjacent control points in the axis movement. Control points are sometimes called *addressable points* because they are locations along the axis to which the worktable can be specifically directed to go. It is desirable for the control resolution to be as small as possible. This depends on limitations imposed by (1) the electromechanical components of the positioning system, and/or (2) the number of bits used by the controller to define the axis coordinate location.

The electromechanical factors that limit resolution include leadscrew pitch, gear ratio in the drive system, and the step angle in a stepping motor (for an open-loop sys-tem) or the angle between slots in an encoder disk (for a closed-loop system). Together, these factors determine a control resolution, or minimum distance that the worktable can be moved. For example, the control resolution for an open-loop system driven by a stepper motor with a 1 : 1 gear ratio between the motor shaft and the leadscrew is given by

$$CR_1 = \frac{p}{n_s} \tag{38.13}$$

where CR_1 = control resolution of the electromechanical components, mm (in.); p = leadscrew pitch, mm/rev (in./rev); and n_s = number of steps/rev. A similar expression can be developed for a closed-loop positioning system.

Although unusual in modern computer technology, the second possible factor that limits control resolution is the number of bits defining the axis coordinate value. For example, this limitation may be imposed by the bit storage capacity of the controller. If B = the number of bits in the storage register for the axis, then the number of control points into which the axis range can be divided = 2^B. Assuming that the control points are separated equally within the range, then

$$CR_2 = \frac{L}{2^B - 1} \tag{38.14}$$

where CR_2 = control resolution of the computer control system, mm (in.); and L = axis range, mm (in.). The control resolution of the positioning system is the maximum of the two values; that is,

$$CR = \text{Max}\{CR_1, CR_2\} \tag{38.15}$$

It is generally desirable for $CR_2 \leq CR_1$, meaning that the electromechanical system is the limiting factor in control resolution.

When a positioning system is directed to move the worktable to a given control point, the capability of the system to move to that point will be limited by mechanical errors. These errors are due to a variety of inaccuracies and imperfections in the mechanical system, such as play between the leadscrew and the worktable, backlash in the gears, and deflection of machine components. It is convenient to assume that the errors form a statistical distribution about the control point that is an unbiased normal distribution with mean = 0. If we further assume that the standard deviation of the distribution is constant over the range of the axis under consideration, then nearly all of the mechanical errors (99.74%) are contained within ±3 standard deviations of the control point. This is pictured in Figure 38.7 for a portion of the axis range, which includes three control points.

Given these definitions of control resolution and mechanical error distribution, let us now consider accuracy and repeatability. Accuracy is defined in a worst-case scenario in which the desired target point lies exactly between two adjacent control points. Since the system can only move to one or the other of the control points, there will be an error in the final position of the worktable. If the target were closer to one of the control points,

FIGURE 38.7   A portion of a linear positioning system axis, with definition of control resolution, accuracy, and repeatability.

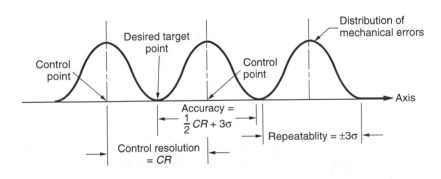

then the table would be moved to the closer control point and the error would be smaller. It is appropriate to define accuracy in the worst case. The *accuracy* of any given axis of a positioning system is the maximum possible error that can occur between the desired target point and the actual position taken by the system; in equation form,

$$\text{Accuracy} = 0.5\,CR + 3\sigma \tag{38.16}$$

where CR = control resolution, mm (in.); and σ = standard deviation of the error distribution, mm (in.).

Repeatability refers to the capability of a positioning system to return to a given control point that has been previously programmed. This capability can be measured in terms of the location errors encountered when the system attempts to position itself at the control point. Location errors are a manifestation of the mechanical errors of the positioning system, which are defined by an assumed normal distribution, as described above. Thus, the *repeatability* of any given axis of a positioning system can be defined as the range of mechanical errors associated with the axis; this reduces to

$$\text{Repeatability} = \pm 3\sigma \tag{38.17}$$

**EXAMPLE 38.3
Control Resolu-
tion, Accuracy, and
Repeatability**

Referring back to Example 38.1, the mechanical inaccuracies in the open-loop positioning system can be described by a normal distribution whose standard deviation = 0.005 mm. The range of the worktable axis is 550 mm, and there are 16 bits in the binary register used by the digital controller to store the programmed position. Determine (a) control resolution, (b) accuracy, and (c) repeatability for the positioning system.

Solution: (a) Control resolution is the greater of CR_1 and CR_2 as defined by Eqs. (38.13) and (38.14):

$$CR_1 = \frac{p}{n_s} = \frac{5.0}{150} = 0.0333 \text{ mm}$$

$$CR_2 = \frac{L}{2^B - 1} = \frac{550}{2^{16} - 1} = \frac{550}{65{,}535} = 0.0084 \text{ mm}$$

$$CR = \text{Max}\{0.0333, 0.0084\} = 0.0333 \text{ mm}$$

(b) Accuracy is given by Eq. (38.16):

$$\text{Accuracy} = 0.5(0.0333) + 3(0.005) = 0.03165 \text{ mm}$$

(c) Repeatability = $\pm 3(0.005) = \pm 0.015$ mm.

38.1.4 NC Part Programming

In machine tool applications, the task of programming the system is called NC part programming because the program is prepared for a given part. It is usually accomplished by someone familiar with the metalworking process who has learned the programming procedure for the particular equipment in the plant. For other processes, other terms may be used for programming, but the principles are similar and a trained individual is needed to prepare the program. Computer systems are used extensively to prepare NC programs.

Part programming requires the programmer to define the points, lines, and surfaces of the workpart in the axis system, and to control the movement of the cutting tool relative to these defined part features. Several part programming techniques are available, the most important of which are (1) manual part programming, (2) computer-assisted part programming, (3) CAD/CAM-assisted part programming, and (4) manual data input.

Manual Part Programming   For simple point-to-point machining jobs, such as drilling operations, manual programming is often the easiest and most economical method. Manual part programming uses basic numerical data and special alphanumeric codes to define the steps in the process. For example, to perform a drilling operation, a command of the following type is entered:

$$n010 \quad x70.0 \quad y85.5 \quad f175 \quad s500$$

Each "word" in the statement specifies a detail in the drilling operation. The n-word ($n010$) is simply a sequence number for the statement. The x- and y-words indicate the x and y coordinate positions ($x = 70.0$ mm and $y = 85.5$ mm). The f-word and s-word specify the feed rate and spindle speed to be used in the drilling operation (feed rate = 175 mm/min and spindle speed = 500 rev/min). The complete NC part program consists of a sequence of statements similar to the above command.

Computer-Assisted Part Programming   Computer-assisted part programming involves the use of a high-level programming language. It is suited to the programming of more complex jobs than manual programming. The first part programming language was Automatically Programmed Tooling (APT), developed as an extension of the original NC machine tool research and first used in production around 1960.

In APT, the part programming task is divided into two steps: (1) definition of part geometry and (2) specification of tool path and operation sequence. In step (1), the part programmer defines the geometry of the workpart by means of basic geometric elements such as points, lines, planes, circles, and cylinders. These elements are defined using APT geometry statements, such as

$$P1 = POINT/25.0, 150.0$$

$$L1 = LINE/P1, P2$$

P1 is a point defined in the x-y plane located at $x = 25$ mm and $y = 150$ mm. L1 is a line that goes through points P1 and P2. Similar statements can be used to define circles, cylinders, and other geometry elements. Most workpart shapes can be described using statements like these to define their surfaces, corners, edges, and hole locations.

Specification of the tool path is accomplished with APT motion statements. A typical statement for point-to-point operation is

$$GOTO/P1$$

This directs the tool to move from its current location to a position defined by P1, where P1 has been defined by a previous APT geometry statement. Continuous path motion commands use geometry elements such as lines, circles, and planes. For example, consider the command

$$GORGT/L3, PAST, L4$$

The statement directs the tool to go right (GORGT) along line L3 until it is positioned just past line L4 (of course, L4 must be a line that intersects L3).

Additional APT statements are used to define operating parameters such as feed rates, spindle speeds, tool sizes, and tolerances. When completed, the part programmer enters the APT program into the computer, where it is processed to generate low-level statements (similar to statements prepared in manual part programming) that can be used by a particular machine tool.

CAD/CAM-Assisted Part Programming   The use of CAD/CAM takes computer-assisted part programming a step further by using a computer graphics system (CAD/CAM system) to interact with the programmer as the part program is being prepared. In the conventional use of APT, a complete program is written and then entered into the computer for processing. Many programming errors are not detected until computer processing. When a CAD/CAM system is used, the programmer receives immediate visual verification when each statement is entered, to determine whether the statement is correct. When part geometry is entered by the programmer, the element is graphically displayed on the monitor. When the tool path is constructed, the programmer can see exactly how the motion commands will move the tool relative to the part. Errors can be corrected immediately rather than after the entire program has been written.

Interaction between programmer and programming system is a significant benefit of CAD/CAM-assisted programming. There are other important benefits of using CAD/CAM in NC part programming. First, the design of the product and its components may have been accomplished on a CAD/CAM system. The resulting design data base, including the geometric definition of each part, can be retrieved by the NC programmer to use as the starting geometry for part programming. This retrieval saves valuable time compared to reconstructing the part from scratch using the APT geometry statements.

Second, special software routines are available in CAD/CAM-assisted part programming to automate portions of the tool path generation, such as profile milling around the outside periphery of a part, milling a pocket into the surface of a part, surface contouring, and certain point-to-point operations. These routines are called by the part programmer as special MACRO commands. Their use results in significant savings in programming time and effort.

Manual Data Input   Manual data input (MDI) is a method in which a machine operator enters the part program in the factory. The method involves use of a CRT display with graphics capability at the machine tool controls. NC part programming statements are entered using a menu-driven procedure that requires minimum training of the machine tool operator. Because part programming is simplified and does not require a special staff of NC part programmers, MDI is a way for small machine shops to economically implement numerical control into their operations.

38.1.5 Applications of Numerical Control

Machining is an important application area for numerical control, but the operating principle of NC can be applied to other operations as well. There are many industrial processes in which the position of a workhead must be controlled relative to the part or product being worked on. We divide the applications into two categories: (1) machine tool applications, and (2) nonmachine tool applications. It should be noted that the applications are not all identified by the name numerical control in their respective industries.

In the machine tool category, NC is widely used for *machining operations* such as turning, drilling, and milling (Sections 22.1, 22.2, and 22.3, respectively). The use of NC in these processes has motivated the development of highly automated machine tools called *machining centers,* which change their own cutting tools to perform a variety of machining operations under NC program control (Section 22.4). In addition to machining, other numerically controlled machine tools include: (1) grinding machines (Section 25.1); (2) sheet metal pressworking machines (Section 20.5.2); (3) tube-bending machines (Section 20.7); and (4) thermal cutting processes (Section 26.3).

In the non-machine tool category, NC applications include (1) tape-laying machines and filament-winding machines for composites (Section 15.2.3 and Section 15.4); (2) welding machines, both arc welding (Section 31.1) and resistance welding (Section 31.2); (3) component insertion machines in electronics assembly (Sections 36.3 and 36.4); (4) drafting machines; and (5) coordinate measuring machines for inspection (Section 44.5.1).

Benefits of NC relative to manually operated equipment in these applications include (1) reduced nonproductive time, which results in shorter cycle times, (2) lower manufacturing lead times, (3) simpler fixturing, (4) greater manufacturing flexibility, (5) improved accuracy, and (6) reduced human error.

38.2 INDUSTRIAL ROBOTICS

An *industrial robot* is a general-purpose programmable machine possessing certain anthropomorphic features. The most apparent anthropomorphic, or human-like, feature of an industrial robot is its mechanical arm, or manipulator. The control unit for a modern industrial robot is a computer that can be programmed to execute rather sophisticated subroutines, thus providing the robot with an intelligence that sometimes seems almost human. The robot's manipulator, combined with a high-level controller, allows an industrial robot to perform a variety of tasks such as loading and unloading machine tools, spot welding automobile bodies, and spray painting. Robots are typically used as substitutes for human workers in these tasks. The first industrial robot was installed in a die-casting operation at Ford Motor Company. The robot's job was to unload die castings from the die-casting machine.

In this section, we consider various aspects of robot technology and applications, including how industrial robots are programmed to perform their tasks.

38.2.1 Robot Anatomy

An industrial robot consists of a mechanical manipulator and a controller to move it and perform other related functions. The mechanical manipulator consists of joints and links that can position and orient the end of the manipulator relative to its base. The controller unit consists of electronic hardware and software to operate the joints in a coordinated fashion to execute the programmed work cycle. *Robot anatomy* is concerned with the mechanical manipulator and its construction. Figure 38.8 shows one of the common industrial robot configurations.

Manipulator Joints and Links   A joint in a robot is similar to a joint in a human body. It provides relative movement between two parts of the body. Connected to each joint are an input link and an output link. Each joint moves its output link relative to its input link. The robot manipulator consists of a series of link–joint–link combinations. The output link of one joint is the input link for the next joint. Typical industrial robots have five or six joints. The coordinated movement of these joints gives the robot its ability to move, position, and orient objects to perform useful work. Manipulator joints can be classified as linear or rotating, indicating the motion of the output link relative to the input link.

Manipulator Design   Using joints of the two basic types, each joint separated from the previous by a link, the manipulator is constructed. Most industrial robots are mounted to the floor. We can identify the base as link 0; this is the input link to joint 1 whose output is link 1, which is the input to joint 2 whose output link is link 2; and so forth, for the number of joints in the manipulator.

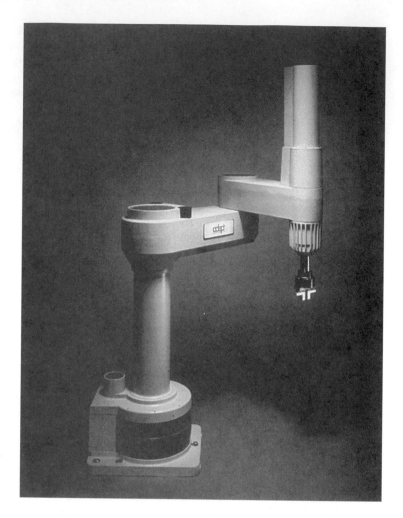

FIGURE 38.8   The manipulator of a modern industrial robot (photo courtesy of Adept Technology, Inc.).

Robot manipulators can usually be divided into two sections: arm-and-body assembly and wrist assembly. There are typically three joints associated with the arm-and-body assembly, and two or three joints associated with the wrist. The function of the arm-and-body is to position an object or tool, and the wrist function is to properly orient the object or tool. Positioning is concerned with moving the part or tool from one location to another. Orientation is concerned with precisely aligning the object relative to some stationary location in the work area.

To accomplish these functions, arm-and-body designs differ from those of the wrist. Positioning requires large spatial movements, while orientation requires twisting and rotating motions to align the part or tool relative to a fixed position in the workplace. The arm-and-body consists of large links and joints, whereas the wrist consists of short links. The arm-and-body joints often consist of both linear and rotating types, while the wrist joints are almost always rotating types.

There are five basic arm-and-body configurations available in commercial robots, identified in Figure 38.9. The design shown in part (e) of the figure and in Figure 38.8 is called a SCARA robot, which stands for "selectively compliant assembly robot arm." It is similar to a jointed arm anatomy, except that the shoulder and elbow joints have vertical axes of rotation, thus providing rigidity in the vertical direction but relative compliance in the horizontal direction.

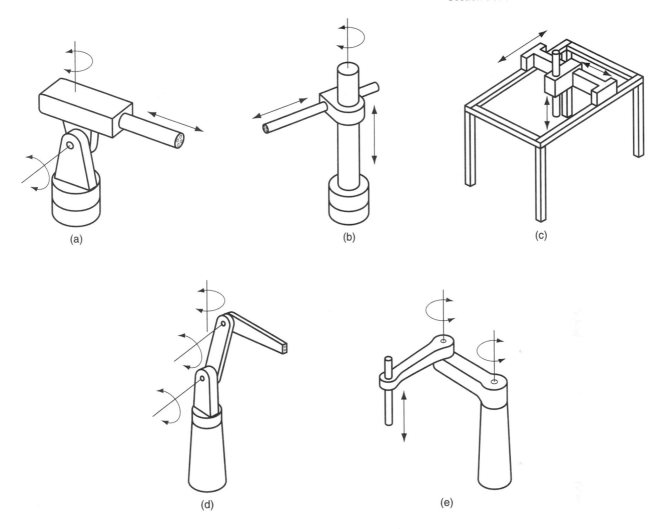

FIGURE 38.9 Five common anatomies of commercial industrial robots: (a) polar, (b) cylindrical, (c) Cartesian coordinate, (d) jointed-arm, and (e) SCARA, or selectively compliant assembly robot arm.

The wrist is assembled to the last link in any of these arm-and-body configurations. The SCARA is sometimes an exception because it is almost always used for simple handling and assembly tasks involving vertical motions. Therefore, a wrist is not usually present at the end of its manipulator. Substituting for the wrist on the SCARA is usually a gripper to grasp components for movement and/or assembly.

Work Volume and Precision of Motion   One of the important technical considerations of an industrial robot is the size of its work volume. *Work volume* is defined as the envelope within which a robot manipulator can position and orient the end of its wrist. This envelope is determined by the number of joints, as well as their types and ranges, and the sizes of the links. Work volume is important because it plays a significant role in determining which applications a robot can perform.

The definitions of control resolution, accuracy, and repeatability developed in Section 38.1.3 for NC positioning systems apply to industrial robots. A robot manipulator is, after all, a positioning system. In general, the links and joints of robots are not nearly as rigid as their machine tool counterparts, and so the accuracy and repeatability of their movements are not as good.

End Effectors An industrial robot is a general-purpose machine. For a robot to be useful in a particular application, it must be equipped with special tooling designed for the application. An *end effector* is the special tooling that connects to the robot's wrist-end to perform the specific task. There are two general types of end effector: tools and grippers. A *tool* is used when the robot must perform a processing operation. The special tools include spot-welding guns, arc-welding tools, spray-painting nozzles, rotating spindles, heating torches, and assembly tools (e.g., automatic screwdriver). The robot is programmed to manipulate the tool relative to the workpart being processed.

Grippers are designed to grasp and move objects during the work cycle. The objects are usually workparts, and the end effector must be designed specifically for the part. Grippers are used for part placement applications, machine loading and unloading, and palletizing. Figure 38.10 shows a typical gripper configuration.

38.2.2 Control Systems and Robot Programming

The robot's controller consists of the electronic hardware and software to control the joints during execution of a programmed work cycle. Most robot control units today are based on a microcomputer system. The control systems in robotics can be classified as follows:

1. *Limited sequence control*—This control system is intended for simple motion cycles, such as "pick-and-place" applications. It does not require a microprocessor and can usually be implemented using limit switches and mechanical stops, together with a sequencer to coordinate and time the actuation of the joints. Robots that use limited sequence control are often pneumatically actuated.

2. *Playback with point-to-point (PTP) control*—As in numerical control, robot motion systems can be divided into point-to-point and continuous path. The program for a point-to-point playback robot consists of a series of point locations and the sequence in which these points must be visited during the work cycle. During programming, these points are recorded into memory, and then subsequently played back during execution of the program. In a point-to-point motion, the path taken to get to the final position is not controlled.

3. *Playback with continuous path (CP) control*—Continuous path control is similar to PTP, except motion paths rather than individual points are stored in memory. In

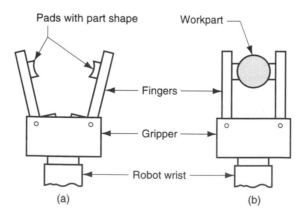

FIGURE 38.10 A robot gripper: (a) open and (b) closed to grasp a workpart.

(a) (b)

certain types of regular CP motions, such as a straight line path between two point locations, the trajectory required by the manipulator is computed by the controller unit for each move. For irregular continuous motions, such as a path followed in spray painting, the path is defined by a series of closely spaced points that approximate the irregular smooth path. Robots capable of continuous path motions can also execute point-to-point movements.

4. *Intelligent control*—Modern industrial robots exhibit characteristics that often make them appear to be acting intelligently. These characteristics include the ability to respond to sophisticated sensors such as machine vision, make decisions when things go wrong during the work cycle, make computations, and communicate with humans. Robot intelligence is implemented using controllers with powerful microprocessors and advanced programming techniques.

Robots execute a stored program of instructions that define the sequence of motions and positions in the work cycle, much like a part program in NC. In addition to motion instructions, the program may include instructions for other functions such as interacting with external equipment, responding to sensors, and processing data.

There are two basic methods used to teach modern robots their programs: (1) leadthrough programming and (2) computer programming languages. *Leadthrough programming* involves a "teach-by-showing" method in which the manipulator is moved by the programmer through the sequence of positions in the work cycle. The controller records each position in memory for subsequent playback. Two procedures for leading the robot through the motion sequence are available: powered leadthrough and manual leadthrough. In *powered leadthrough,* a control box is used to drive the manipulator. The control box, called a teach pendant, has toggle switches or press buttons to control the joints. Using the teach pendant, the programmer moves the manipulator to each location, recording the corresponding joint positions into memory. Powered leadthrough is the common method for programming playback robots with point-to-point control. *Manual leadthrough* is typically used for playback robots with continuous path control. In this method, the programmer physically moves the manipulator wrist through the motion cycle. For spray painting and certain other jobs, this is a more convenient means of programming the robot.

Computer programming languages for programming robots have evolved from the use of microcomputer controllers. The first commercial language was introduced around 1979 by Unimation Inc. Computer languages provide a convenient way to integrate certain nonmotion functions into the work cycle, such as computations and data processing, decision logic, interlocking with other equipment, interfacing with sensors, and interrupts. A more thorough discussion of robot programming is presented in reference [6].

38.2.3 Applications of Industrial Robots

Some industrial work lends itself to robot applications. The following are the important characteristics of a work situation that tend to promote the substitution of a robot in place of a human worker: (1) the work environment is hazardous for humans; (2) the work cycle is repetitive; (3) the work is performed at a stationary location; (4) part or tool handling would be difficult for humans; (5) it is a multishift operation; (6) there are long production runs and infrequent changeovers; and (7) part positioning and orientation are established at the beginning of the work cycle, since most robots cannot see.

Applications of industrial robots that tend to match these characteristics can be divided into three basic categories: (1) material handling, (2) processing operations, and (3) assembly and inspection.

FIGURE 38.11   A portion of an automobile assembly line in which robots perform spot welding operations (photo courtesy of Ford Motor Company).

Material handling applications involve the movement of materials or parts from one location and orientation to another. To accomplish this relocation task, the robot is equipped with a gripper. As noted earlier, the gripper must be custom-designed to grasp the particular part in the application. Material handling applications include material transfer (part placement, palletizing, depalletizing) and machine loading and/or unloading (e.g., machine tools, presses, and plastic molding).

Processing operations require the robot to manipulate a tool as its end effector. The applications include spot welding, continuous arc welding, spray coating, and certain metal cutting and deburring operations in which the robot manipulates a special tool. In each of these operations, the tool (e.g., spot welding gun, spray painting nozzle) is used as the robot's end effector. Two applications of spot welding are illustrated in Color Plate 11 and Figure 38.11. Spot welding is a common application of industrial robots in the automotive industry.

Assembly and inspection applications cannot be classified neatly in either of the previous categories; they sometimes involve part handling and other times manipulation of a tool. *Assembly* applications often involve the stacking of one part onto another part—basically a part handling task. In other assembly operations a tool is manipulated, such as an automatic screwdriver. Similarly, *inspection* operations sometimes require the robot to position a workpart relative to an inspection device, or to load a part into an inspection machine; other applications involve the manipulation of a sensor to perform an inspection.

38.3 PROGRAMMABLE LOGIC CONTROLLERS

Many automated systems operate by turning on and off motors, switches, and other devices to respond to conditions and as a function of time. These control devices use binary variables. They can have either of two possible values, 1 or 0, interpreted as ON or OFF, object present or not present, high or low voltage level, and so on. Binary devices commonly used in industrial control systems include limit switches, photodetectors, timers, control relays, motors, solenoids, valves, clutches, and lights. Some of these devices send a signal in response to a physical stimulus, while others respond to an electrical signal.

A *programmable logic controller* (PLC) is a widely used microcomputer-based device that uses stored instructions in programmable memory to implement logic, sequencing, timing, counting, and arithmetic control functions, through digital or analog input/output modules, for controlling various machines and processes. The PLC was introduced around 1969 in response to specifications proposed by General Motors Corporation. Controls manufacturers saw a commercial opportunity in the PLC, and today it is an important industrial controls technology.

The major components of a PLC, shown in Figure 38.12, are (1) *input and output modules,* which connect the PLC to the industrial equipment to be controlled; (2) *processor*—the central processing unit (CPU), which executes the logic and sequencing functions to control the process by operating on the input signals and determining the proper output signals specified by the control program; (3) *PLC memory,* which is connected to the microprocessor and contains the logic and sequencing instructions; (4) *power supply*—115 V AC is typically used to drive the PLC. In addition, (5) a *programming device* (usually detachable) is used to enter the program into the PLC.

Programming involves entry of the control instructions to the PLC using the programming device. The most common control instructions include logical operations, sequencing, counting, and timing. Many control applications require additional instructions for analog control, data processing, and computations. A variety of PLC programming languages have been developed, ranging from ladder logic diagrams to structured text. A discussion of these languages is beyond the scope of this text, and the reader is referred to our references.

The following advantages are associated with programmable logic controllers: (1) programming a PLC is easier than wiring a relay control panel; (2) PLCs can be

FIGURE 38.12   Major components of a programmable logic controller.

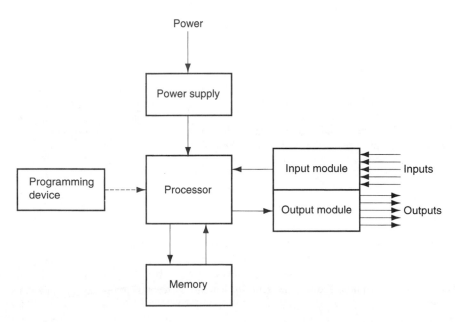

reprogrammed, whereas conventional hard-wired controls must be rewired and are often scrapped instead because of the difficulty in rewiring; (3) a PLC can be interfaced with the plant computer system more readily than conventional controls; (4) PLCs require less floor space than relay controls; and (5) PLCs offer greater reliability and easier maintenance.

REFERENCES

[1] Asfahl, C. R. *Robots and Manufacturing Automation.* Wiley, New York, 1992.

[2] Bollinger, J. G., and Duffie, N. A. *Computer Control of Machines and Processes.* Addison-Wesley Longman, New York, 1989.

[3] Chang, C-H, and Melkanoff, M. A. *NC Machine Programming and Software Design.* Prentice-Hall, Inc., Upper Saddle River, N.J., 1989.

[4] Engelberger, J. F. *Robotics in Practice: Management and Applications of Robotics in Industry.* AMA-COM, New York, 1985.

[5] Groover, M. P. *Automation, Production Systems, and Computer Integrated Manufacturing.* 2nd ed. Prentice-Hall, Upper Saddle River, N.J., 2001.

[6] Groover, M. P., Weiss, M., Nagel, R. N., and Odrey, N. G. *Industrial Robotics: Technology, Programming, and Applications.* McGraw-Hill, New York, 1986.

[7] Hughes, T. A. *Programmable Controllers.* 2nd ed. Instrument Society of America, Research Triangle Park, N.C., 1997.

[8] Jones, C. T., and Bryan, L. A. *Programmable Controllers.* IPC/ASTEC Publications, Atlanta, Georgia, 1983.

[9] Noaker, P. M. "Down the Road with DNC." *Manufacturing Engineering.* November 1992, pp. 35–38.

[10] Pessen, D. W. *Industrial Automation.* Wiley, New York, 1989.

[11] Seames, W. *Computer Numerical Control, Concepts and Programming.* Delmar Publishers Inc., Albany, N.Y., 1995.

[12] Webb, J. W., and Reis, R. A. *Programmable Logic Controllers: Principles and Applications.* 4th ed. Prentice Hall, Upper Saddle River, New Jersey, 1999.

REVIEW QUESTIONS

38.1. Identify and briefly describe the three basic components of a numerical control system.

38.2. What is the difference between point-to-point and continuous path in a motion control system?

38.3. What is the difference between absolute positioning and incremental positioning?

38.4. What is the difference between an open-loop positioning system and a closed-loop positioning system?

38.5. Under what circumstances is a closed-loop positioning system preferable to an open-loop system?

38.6. Explain the operation of an *optical encoder.*

38.7. Why should the electromechanical system be the limiting factor in control resolution rather than the controller storage register?

38.8. What is *manual data input* in NC part programming?

38.9. Identify some of the non–machine tool applications of numerical control.

38.10. What are some of the benefits usually cited for NC compared to using manual alternative methods?

38.11. What is an *industrial robot?*

38.12. How is an industrial robot similar to numerical control?

38.13. What is an *end effector?*

38.14. In robot programming, what is the difference between powered leadthrough and manual leadthrough?

38.15. What is a *programmable logic controller?*

MULTIPLE CHOICE QUIZ

There is a total of 13 correct answers in the following multiple choice questions (some questions have multiple answers that are correct). To attain a perfect score on the quiz, all correct answers must be given, since each correct answer is worth 1 point. For each question, each omitted answer or wrong answer reduces the score by 1 point, and each additional answer beyond the number of answers required reduces the score by 1 point. Percentage score on the quiz is based on the total number of correct answers.

38.1. The standard coordinate system for numerical control machine tools is based on which one of the following? (a) Cartesian coordinates, (b) cylindrical coordinates, or (c) polar coordinates.

38.2. Identify which of the following applications are point-to-point and not continuous path operations (more than one): (a) arc welding, (b) drilling, (c) hole punching in sheet metal, (d) milling, (e) spot welding, and (f) turning.

38.3. The ability of a positioning system to return to a previously defined location is measured by which one of the following terms? (a) accuracy, (b) control resolution, or (c) repeatability.

38.4. The APT command GORGT is which of the following (more than one)? (a) continuous path command, (b) geometry statement involving a volume of revolution about a central axis, (c) name of the monster in a 1960s Japanese science fiction movie, (d) point-to-point command, or (e) tool path command in which the tool must Go Right in the next move.

38.5. The arm and body of a robot manipulator generally perform which one of the following functions in an application? (a) orientation or (b) positioning.

38.6. A SCARA robot is normally associated with which one of the following applications (one answer)? (a) arc welding, (b) assembly, (c) inspection, (d) machine loading and unloading, or (e) resistance welding.

38.7. In robotics, spray-painting applications are which of the following? (a) continuous path, or (b) point-to-point.

38.8. Which of the following are characteristics of work situations that tend to promote the substitution of a robot in place of a human worker (more than one)? (a) frequent job changeovers, (b) hazardous work environment, (c) repetitive work cycle, (d) multiple work shifts, and (e) task requires mobility.

PROBLEMS

Open-Loop Positioning Systems

38.1. A leadscrew with a 7.5 mm pitch drives a worktable in a NC positioning system. The leadscrew is powered by a stepping motor that has 250 step angles. The worktable is programmed to move a distance of 120 mm at a travel speed of 300 mm/min. Determine (a) how many pulses are required to move the table the specified distance; and (b) the required motor speed and pulse rate to achieve the desired table speed.

38.2. Referring to the previous problem, the mechanical inaccuracies in the open-loop positioning system can be described by a normal distribution whose standard deviation = 0.005 mm. The range of the worktable axis is 500 mm, and there are 12 bits in the binary register used by the digital controller to store the programmed position. For the positioning system, determine (a) control resolution, (b) accuracy, and (c) repeatability. (d) What is the minimum number of bits that the binary register should have so that the mechanical drive system becomes the limiting component on control resolution?

38.3. A stepping motor has 200 step angles. Its output shaft is directly coupled to leadscrew with pitch = 0.25 in. A worktable is driven by the leadscrew. The table must move a distance of 5.00 in. at a travel speed of 20.0 in./min. Determine (a) how many pulses are required to move the table the specified distance, and (b) the required motor speed and pulse rate to achieve the desired table speed.

38.4. A stepping motor with 240 step angles is coupled to a leadscrew through a gear reduction of 5 : 1 (5 rotations of the motor for each rotation of the leadscrew).

The leadscrew has 6 threads/in. The worktable driven by the leadscrew must move a distance = 10 in. at a feed rate of 30 in./min. Determine (a) number of pulses required to move the table, and (b) the required motor speed and pulse rate to achieve the desired table speed.

38.5. The drive unit for a positioning table is driven by a leadscrew directly coupled to the output shaft of a stepping motor. The pitch of the leadscrew = 0.18 in. The table must have a linear speed = 35 in./min and a positioning accuracy = 0.001 in. Mechanical errors in the motor, leadscrew, and table connection are characterized by a normal distribution with standard deviation = 0.0002 in. Determine (a) the minimum number of step angles in the stepping motor to achieve the accuracy, (b) the associated step angle, and (c) the frequency of the pulse train required to drive the table at the desired speed.

38.6. The positioning table for a component insertion machine uses a stepping motor and leadscrew mechanism. The design specifications require a table speed of 40 in./min and an accuracy = 0.0008 in. The pitch of the leadscrew = 0.2 in., and the gear ratio = 2 : 1 (2 turns of the motor for each turn of the leadscrew). The mechanical errors in the motor, gear box, leadscrew, and table connection are characterized by a normal distribution with standard deviation = 0.0001 in. Determine (a) the minimum number of step angles in the stepping motor and (b) the frequency of the pulse train required to drive the table at the desired maximum speed.

38.7. The drive unit of a positioning table for a component insertion machine is based on a stepping motor and leadscrew mechanism. The specifications are for the table speed to be 25 mm/s over a 600 mm range and for the accuracy to be 0.025 mm. The pitch of the leadscrew = 4.5 mm, and the gear ratio = 5 : 1 (5 turns of the motor for each turn of the leadscrew). The mechanical errors in the motor, gear box, leadscrew, and table connection are characterized by a normal distribution with standard deviation = 0.005 mm. Determine (a) the minimum number of step angles in the stepping motor and (b) the frequency of the pulse train required to drive the table at the desired maximum speed.

38.8. The two axes of an x-y positioning table are each driven by a stepping motor connected to a leadscrew with a 10 : 1 gear reduction. The number of step angles on each stepping motor is 20. Each leadscrew has a pitch = 5.0 mm and provides an axis range = 300.0 mm. There are 16 bits in each binary register used by the controller to store position data for the two axes. (a) What is the control resolution of each axis? (b) What are the required the rotational speeds and corresponding pulse train frequencies of each stepping motor in order to drive the table at 600 mm/min in a straight line from point (25, 25) to point (100, 150)? Ignore acceleration.

Closed-Loop Positioning Systems

38.9. A NC machine tool table is powered by a servomotor, leadscrew, and optical encoder. The leadscrew has a pitch = 5.0 mm and is connected to the motor shaft with a gear ratio of 16 : 1 (16 turns of the motor for each turn of the leadscrew). The optical encoder is connected directly to the leadscrew and generates 200 pulses/rev of the leadscrew. The table must move a distance of 100 mm at a feed rate of 500 mm/min. Determine (a) the pulse count received by the control system to verify that the table has moved exactly 100 mm, (b) the pulse rate, and (c) motor speed that correspond to the feed rate of 500 mm/min.

38.10. Solve the previous problem, except that the optical encoder is directly coupled to the motor shaft rather than to the leadscrew.

38.11. The worktable of an NC machine tool is driven by a closed-loop positioning system that consists of a servomotor, leadscrew, and optical encoder. The leadscrew has 6 threads/in. and is coupled directly to the motor shaft (gear ratio = 1 : 1). The optical encoder generates 225 pulses per motor revolution. The table has been programmed to move a distance of 7.5 in. at a feed rate = 20.0 in./min. (a) How many pulses are received by the control system to verify that the table has moved the programmed distance? What are (b) the pulse rate and (c) the motor speed that correspond to the specified feed rate?

38.12. A leadscrew coupled directly to a DC servomotor is used to drive one of the table axes of an NC milling machine. The leadscrew has 5 threads/in. The optical encoder attached to the leadscrew emits 100 pulses/rev of the leadscrew. The motor rotates at a maximum speed of 800 rev/min. Determine (a) the control resolution of the system, expressed in linear travel distance of the table axis; (b) the frequency of the pulse train emitted by the optical encoder when the servomotor operates at maximum speed; and (c) the travel speed of the table at the maximum rpm of the motor.

38.13. Solve the previous problem, only the servomotor is connected to the leadscrew through a gear box whose reduction ratio = 12 : 1 (12 revolutions of the motor for each revolution of the leadscrew).

38.14. A leadscrew connected to a DC servomotor is the drive system for a positioning table. The leadscrew pitch = 4 mm. The optical encoder attached to the leadscrew emits 250 pulses/rev of the leadscrew. The motor operates at a speed = 15 rev/s. Determine (a) the control resolution of the system, expressed in linear travel distance of the table axis, (b) the frequency of the pulse train emitted by the optical encoder when the servomotor operates at 14 rev/s, and (c) the travel speed of the table at the operating speed of the motor.

38.15. A milling operation is performed on an NC machining center. Total travel distance = 300 mm in a direction parallel to one of the axes of the worktable. Cutting speed = 1.25 m/s, and chip load = 0.05 mm. The end-milling cutter has four teeth, and its diameter = 20.0 mm. The axis uses a DC servomotor whose output shaft is coupled to a leadscrew with pitch = 6.0 mm. The feedback sensing device is an optical encoder that emits 250 pulses per revolution. Determine (a) feed rate and time to complete the cut and (b) rotational speed of the motor and the pulse rate of the encoder at the feed rate indicated.

38.16. An end-milling operation is carried out along a straight line path that is 325 mm in length. The cut is in a direction parallel to the x-axis on an NC machining center. Cutting speed = 30 m/min, and chip load = 0.06 mm. The end-milling cutter has two teeth, and its diameter = 16.0 mm. The x-axis uses a DC servomotor connected directly to a leadscrew whose pitch = 6.0 mm. The feedback sensing device is an optical encoder that emits 400 pulses per revolution. Determine (a) the feed rate and time to complete the cut and (b) the rotational speed of the motor and the pulse rate of the encoder at the feed rate indicated.

38.17. A DC servomotor is used to drive the x-axis of an NC milling machine table. The motor is coupled directly to the table leadscrew, which has 4 threads/in. An optical encoder is used to provide the feedback measurement. It is connected to the leadscrew using a 1 : 5 gear ratio (one turn of the leadscrew converts to 5 turns of the encoder disk). The optical encoder emits 125 pulses per revolution. To execute a certain programmed instruction, the table must be moved from point (3.5, 1.5) to point (1.0, 7.2) in a straight-line trajectory at a feed rate of 7.5 in./min. Determine (a) the control resolution of the system for the x-axis; (b) the rotational speed of the motor; and (c) the frequency of the pulse train emitted by the optical encoder when the desired feed rate is achieved.

Industrial Robotics

38.18. The largest axis of a Cartesian coordinate robot has a total range of 750 mm. It is driven by a pulley system capable of a mechanical accuracy − 0.25 mm and repeatability = ±0.15 mm. Determine the minimum number of bits required in the binary register for the axis in the robot's control memory.

38.19. A stepper motor serves as the drive unit for the linear joint of an industrial robot. The joint must have an accuracy of 0.25 mm. The motor is attached to a leadscrew through a 2 : 1 gear reduction (2 turns of the motor for 1 turn of the leadscrew). The pitch of the leadscrew is 5.0 mm. The mechanical errors in the system (due to backlash of the leadscrew and the gear reducer) can be represented by a normal distribution with standard deviation = ±0.05 mm. Specify the number of step angles that the motor must have in order to meet the accuracy requirement.

38.20. The designer of a polar configuration robot is considering a portion of the manipulator consisting of a rotational joint connected to its output link. The output link is 25 in. long, and the rotational joint has a range of 75°. The accuracy of the joint–link combination, expressed as a linear measure at the end of the link that results from rotating the joint, is specified as 0.030 in. The mechanical inaccuracies of the joint result in a repeatability error = ±0.030° of rotation. It is assumed that the link is perfectly rigid, so there are no additional errors due to deflection. (a) Show that the specified accuracy can be achieved, given the repeatability error. (b) Determine the minimum number of bits required in the binary register of the robot's control memory to achieve the specified accuracy.

GROUP TECHNOLOGY AND FLEXIBLE MANU-FACTURING SYSTEMS

CHAPTER CONTENTS

39.1　Group Technology
　　　39.1.1　Parts Classification and Coding
　　　39.1.2　Cellular Manufacturing
　　　39.1.3　Benefits and Problems in Group Technology
39.2　Flexible Manufacturing Systems
　　　39.2.1　Flexibility and Automated Manufacturing Systems
　　　39.2.2　Integrating the FMS Components
　　　39.2.3　Applications of Flexible Manufacturing Systems

Group technology is an approach to the production of parts in medium quantities. Parts (and products) in this quantity range are usually made in batches, but batch production requires downtime for changeovers and has high inventory carrying costs. Group technology (GT) minimizes these disadvantages by recognizing that although the parts are different, they also possess similarities. GT exploits the part similarities by utilizing similar processes and tooling to produce them. GT can be implemented by manual or automated techniques. When automation is used, the term flexible manufacturing system is often applied.

39.1　GROUP TECHNOLOGY

Group technology　is an approach to manufacturing in which similar parts are identified and grouped together in order to take advantage of their similarities in design and production. Similarities among parts permit them to be classified into part families. It is not unusual for a factory that produces 10,000 different parts to be able to group most of those parts into 20 to 30 part families. In each part family the processing steps are similar. When these similarities are exploited in production, operating efficiencies are improved. The improvement is typically achieved by organizing the production facilities into

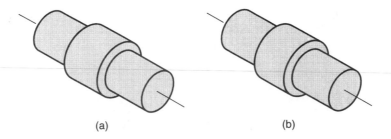

FIGURE 39.1   Two parts that are identical in shape and size but quite different in manufacturing: (a) 1,000,000 units/yr, tolerance = ±0.010 in., 1015 CR steel, nickel plate; and (b) 100/yr, tolerance = ±0.001 in., 18-8 stainless steel.

(a)   (b)

manufacturing cells. Each cell is designed to produce one part family (or a limited number of part families), thereby following the principle of specialization of operations. The cell includes special production equipment and custom-designed tools and fixtures, so that the production of the part families can be optimized. In effect, each cell becomes a factory within the factory.

39.1.1 Parts Classification and Coding

A central feature of group technology is the part family. A *part family* is a group of parts that possess similarities in geometric shape and size, or in the processing steps used in their manufacture. There are always differences among parts in a family, but the similarities are close enough that the parts can be grouped into the same family. Figures 39.1 and 39.2 show two different part families. The parts shown in Figure 39.1 have the same size and shape; however, their processing requirements are quite different because of differences in work material, production quantities, and design tolerances. Figure 39.2 shows several parts with geometries that differ substantially; however, their manufacturing requirements are quite similar.

FIGURE 39.2   Ten parts that are different in size and shape, but quite similar in terms of manufacturing. All parts are machined from cylindrical stock by turning; some parts require drilling and/or milling.

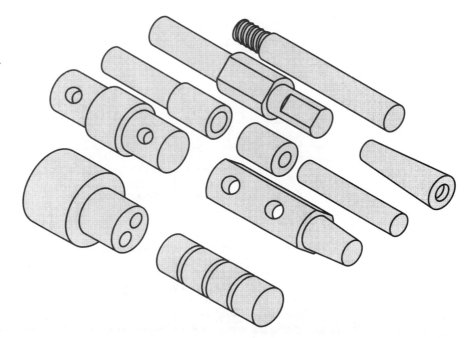

TABLE 39.1 Design and manufacturing attributes typically included in a parts classification and coding system.

Part Design Attributes		Part Manufacturing Attributes	
Major dimensions	Material type	Major process	Major dimensions
Basic external shape	Part function	Operation sequence	Basic external shape
Basic internal shape	Tolerances	Batch size	Length/diameter ratio
Length/diameter ratio	Surface finish	Annual production	Material type
		Machine tools	Tolerances
		Cutting tools	Surface finish

There are several ways by which part families are identified in industry. One method involves visual inspection of all the parts made in the factory (or photos of the parts) and using best judgment to group them into appropriate families. Another approach, called *production flow analysis,* uses information contained on route sheets (Section 41.1.1) to classify parts. In effect, parts with similar manufacturing steps are grouped into the same family. Probably the most widely used method, and also the most expensive, is parts classification and coding.

Parts classification and coding involves the identification of similarities and differences among parts and relating these parts by means of a common coding scheme. Most classification and coding systems are one of the following: (1) systems based on part design attributes, (2) systems based on part manufacturing attributes, and (3) systems based on both design and manufacturing attributes. Common part design and manufacturing attributes used in GT systems are presented in Table 39.1.

Because each company produces a unique set of parts and products, a classification and coding system that may be satisfactory for one company is not necessarily appropriate for another company. Each company must design its own coding scheme. To give the reader an idea of what is involved, we present the basic structure of one of the familiar classification and coding systems in Table 39.2. This system was developed for machined parts by H. Opitz in Germany. The basic code number consists of nine digits containing both design and manufacturing data. Rotational and nonrotational parts are distinguished, as are various part features such as internal bores, threads, and gear teeth. Parts classification and coding systems are described more thoroughly in several of our references [4], [5], [6].

TABLE 39.2 Basic structure of the Opitz parts classification and coding system.

Digit	Description
1	Part shape class: rotation versus nonrotational (Figure 22.1). Rotational parts are classified by length-to-diameter ratio. Nonrotational parts by length, width, and thickness.
2	External shape features; various types are distinguished.
3	Rotational machining. This digit applies to internal shape features (e.g., holes, threads) on rotational parts, and general rotational shape features for nonrotational parts.
4	Plane machined surfaces (e.g., flats, slots).
5	Auxiliary holes, gear teeth, and other features.
6	Dimensions—overall size.
7	Work material (e.g., steel, cast iron, aluminum).
8	Original shape of raw material.
9	Accuracy requirements.

A well-designed classification and coding system (1) facilitates formation of part families, (2) permits quick retrieval of part design drawings, (3) reduces design duplication because similar or identical part designs can be retrieved and reused rather than designed from scratch, (4) promotes design standardization, (5) improves cost estimating and cost accounting, (6) facilitates NC part programming by allowing new parts to use the same part program as existing parts in the same family, (7) allows rationalization and improvement in tool and fixture design, and (8) makes computer-aided process planning (CAPP) feasible (Section 41.1.3); standard process plans can be correlated to part family code numbers, so that process plans for new parts from the same family can be reused or edited.

39.1.2 Cellular Manufacturing

To fully exploit the similarities among parts in a family, production should be organized using machine cells designed to specialize in making those particular parts. One of the principles in designing a group technology machine cell is the composite part concept.

Composite Part Concept Members of a part family possess similar design and/or manufacturing features. There is usually a correlation between part design features and manufacturing operations that produce those features. Round holes are usually made by drilling; cylindrical shapes are made by turning; and so on.

The *composite part* for a given family (not to be confused with a part made of composite material) is a hypothetical part that includes all of the design and manufacturing attributes of the family. In general, an individual part in the family will have some of the features that characterize the family, but not all of them. A production cell designed for the part family would include those machines required to make the composite part. Such a cell would be capable of producing any member of the family, simply by omitting those operations corresponding to features not possessed by the particular part. The cell would also be designed to allow for size variations within the family as well as feature variations.

To illustrate, consider the composite part in Figure 39.3(a). It represents a family of rotational parts with features defined in part (b) of the figure. Associated with each feature is a certain machining operation, as summarized in Table 39.3. A machine cell to produce this part family would be designed with the capability to accomplish all of the operations in the last column of the table.

FIGURE 39.3   Composite part concept: (a) the composite part for a family of machined rotational parts, and (b) the individual features of the composite part.

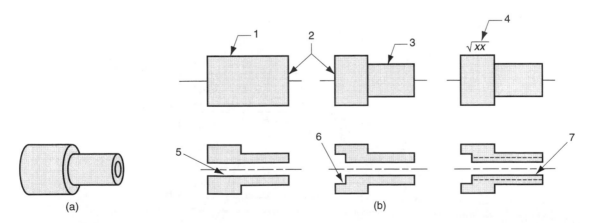

(a)   (b)

TABLE 39.3   Design features of the composite part in Figure 39.3 and the manufacturing operations required to shape those features.

Label	Design Feature	Corresponding Manufacturing Operation
1	External cylinder	Turning
2	Face of cylinder	Facing
3	Cylindrical step	Turning
4	Smooth surface	External cylindrical grinding
5	Axial hole	Drilling
6	Counterbore	Bore, counterbore
7	Internal threads	Tapping

Machine Cell Designs   Machine cells can be classified according to number of machines and level of automation. They can be (a) single machine, (b) multiple machines with manual handling, (c) multiple machines with mechanized handling, (d) flexible manufacturing cell, or (e) flexible manufacturing system. These production cells are depicted schematically in Figure 39.4.

The **single machine cell** has one machine that is manually operated. The cell would also include fixtures and tooling to allow for the feature and size variations within the part family produced by the cell. The machine cell required for the part family of Figure 39.3 would probably be of this type.

FIGURE 39.4   Types of group technology machine cells: (a) single machine, (b) multiple machines with manual handling, (c) multiple machines with mechanized handling, (d) flexible manufacturing cell, and (e) flexible manufacturing system. *Key*: Man = manual operation; Aut = automated station.

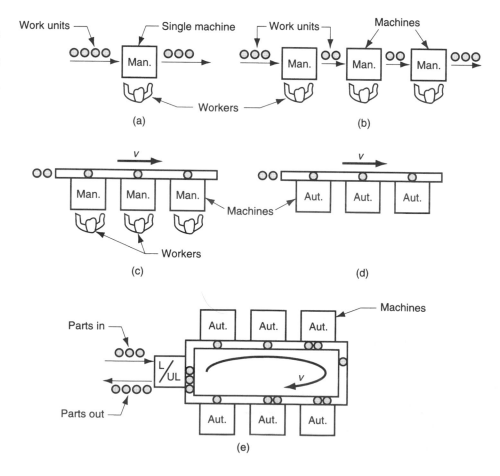

Multiple machine cells have two or more manually operated machines. These cells are distinguished by the method of workpart handling in the cell, manual or mechanized. Manual handling means that parts are moved within the cell by workers, usually the machine operators. Mechanized handling refers to conveyorized transfer of parts from one machine to the next. This may be required by the size and weight of the parts made in the cell, or simply to increase production rate. Our sketch depicts the work flow as being a line; other layouts are also possible, such as U-shaped or loop.

Flexible manufacturing cells and ***flexible manufacturing systems*** consist of automated machines with automated handling. Given the special nature of these production systems and their importance, we devote Section 39.2 to their discussion.

39.1.3 Benefits and Problems in Group Technology

Group technology provides substantial benefits to companies if they have the discipline and perseverance to implement it. The potential benefits include (1) GT promotes standardization of tooling, fixturing, and setups; (2) material handling is reduced because parts are moved within a machine cell rather than within the entire factory; (3) production scheduling is simpler; (4) manufacturing lead time is reduced; (5) work-in-process is reduced; (6) process planning is simplified; (7) worker satisfaction usually improves when working in a GT cell; and (8) higher quality work is accomplished using group technology.

There are several problems in implementing GT, however. One obvious problem is rearranging production machines in the plant into the appropriate machine cells. It takes time to plan and accomplish this rearrangement, and the machines are not producing during the changeover. The biggest problem in starting a GT program is identifying the part families. If the plant makes 10,000 different parts, reviewing all of the part drawings and grouping the parts into families is a substantial task that consumes a significant amount of time.

39.2 FLEXIBLE MANUFACTURING SYSTEMS

A ***flexible manufacturing system*** (FMS) is a highly automated GT machine cell, consisting of a group of processing stations (usually CNC machine tools), interconnected by an automated material handling and storage system, and controlled by an integrated computer system. An FMS is capable of processing a variety of different part styles simultaneously under NC program control at the different workstations.

The FMS relies on the principles of group technology. No manufacturing system can be completely flexible. It cannot produce an infinite range of products. There are limits to the degree of flexibility that can be incorporated in a FMS. Accordingly, a flexible manufacturing system is designed to produce parts (or products) within a range of styles, sizes, and processes. In other words, an FMS is capable of producing a single part family or a limited range of part families.

39.2.1 Flexibility and Automated Manufacturing Systems

Flexible manufacturing systems vary in terms of number of machine tools and level of flexibility. When the system has only a few machines, the term ***flexible manufacturing cell*** (FMC) is sometimes used. Both cell and system are highly automated and computer controlled. The difference between an FMS and an FMC is not always clear, but it is sometimes based on the number of machines (workstations) included. The flexible manufacturing system consists of four or more machines, while a flexible manufacturing cell consists of three or fewer machines [5]. However, this distinction is not universally accepted.

Some highly automated manufacturing systems and cells are not flexible, and this leads to confusion in terminology. For example, a transfer line (Section 40.3) is a highly automated manufacturing system, but it is limited to mass production of one part style, so it is not a flexible system. To develop the concept of flexibility in a manufacturing system, consider a cell consisting of two CNC machine tools that are loaded and unloaded by an industrial robot from a parts carousel, perhaps in the arrangement depicted in Figure 39.5. The cell operates unattended for extended periods of time. Periodically, a worker must unload completed parts from the carousel and replace them with new workparts. This is truly an automated manufacturing cell, but is it a flexible manufacturing cell? One might argue yes, it is flexible since the cell consists of CNC machine tools that can be programmed to machine different part configurations like any other CNC machine. However, if the cell only operates in a batch mode, in which the same part style is produced in lots of several dozen (or several hundred) units, then this does not qualify as flexible manufacturing.

To qualify as being flexible, a manufacturing system should satisfy several criteria. The tests of flexibility in an automated production system are the capability to (1) process different part styles in a nonbatch mode, (2) accept changes in production schedule, (3) respond gracefully to equipment malfunctions and breakdowns in the system, and (4) accommodate the introduction of new part designs. These capabilities are made possible by the use of a central computer that controls and coordinates the components of the system. The most important criteria are (1) and (2); criteria (3) and (4) are softer and can be implemented at various levels of sophistication.

If the automated system does not meet these four tests, it should not be classified as a flexible manufacturing system or cell. Getting back to our illustration, the robotic work cell would satisfy the criteria if it (1) machined different part configurations in a mix rather than in batches; (2) permitted changes in production schedule and part mix; (3) continued operating even though one machine experienced a breakdown (e.g., while repairs are being made on the broken machine, its work is temporarily reassigned to the other machine); and (4) as new part designs are developed, NC part programs are written off-line and then downloaded to the system for execution. This fourth capability also requires that the tooling in the CNC machines as well as the end effector of the robot are suited to the new part design.

FIGURE 39.5   Automated manufacturing cell with two machine tools and robot. Is it a flexible cell?

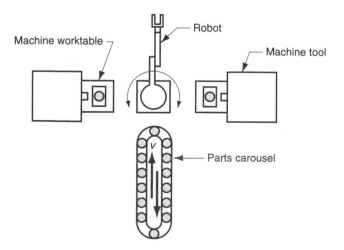

39.2.2 Integrating the FMS Components

A FMS consists of hardware and software that must be integrated into an efficient and reliable unit. It also includes human personnel. In this section we examine these components and how they are integrated.

Hardware Components   FMS hardware includes workstations, material-handling system, and central control computer. The workstations are CNC machines in a machining-type system, plus inspection stations, parts cleaning, and other stations, as needed. For an FMS, a central chip conveyor system is usually included below floor level.

The material-handling system is the means by which parts are moved between stations. The material-handling system usually includes a limited capability to store parts. Handling systems suitable for automated manufacturing include roller conveyors, in-floor towline carts, automated guided vehicles, and industrial robots. The most appropriate type depends on part size and geometry, as well as factors relating to economics and compatibility with other FMS components. Nonrotational parts are often moved in a FMS on pallet fixtures, so the pallets are designed for the particular handling system, and the fixtures are designed to accommodate the various part geometries in the family. Rotational parts are often handled by robots, if weight is not a limiting factor.

The handling system establishes the basic layout of the FMS. Five layout types can be distinguished: (1) in-line, (2) loop, (3) ladder, (4) open field, and (5) robot-centered cell. Types (1), (3), and (4) are shown in Figure 39.6. Types (2) and (5) are shown in Figures 39.4(e) and 39.5, respectively. The *in-line layout* uses a linear transfer system to move parts between processing stations and load/unload station(s). The in-line transfer system is usually capable of two directional movement; if not, then the FMS operates much like a transfer line, and the different part styles made on the system must follow the same basic processing sequence due to the one-direction flow. The *loop layout* consists of a conveyor loop with workstations located around its periphery. This configuration permits any processing sequence, because any station is accessible from any other station. This is also true for the *ladder layout,* in which workstations are located on the rungs of the ladder. The *open field layout* is the most complex FMS configuration, and consists of several loops tied together. Finally, the robot-centered cell consists of a robot whose work volume includes the load/unload positions of the machines in the cell.

The FMS also includes a central computer that is interfaced to the other hardware components. In addition to the central computer, the individual machines and other components generally have microcomputers as their individual control units. The function of the central computer is to coordinate the activities of the components so as to achieve a smooth overall operation of the system. It accomplishes this function by means of software.

FMS Software and Control Functions   FMS software consists of modules associated with the various functions performed by the manufacturing system. For example, one function involves downloading NC part programs to the individual machine tools; another function is concerned with controlling the material handling system; another is concerned with tool management; and so on. Table 39.4 lists the functions included in the operation of a typical FMS. Associated with each function is one or more software modules. Terms other than those in our table may be used in a given installation. The functions and modules are largely application-specific.

Human Labor   An additional component in the operation of a flexible manufacturing system is human labor. Duties performed by human workers include (1) loading and

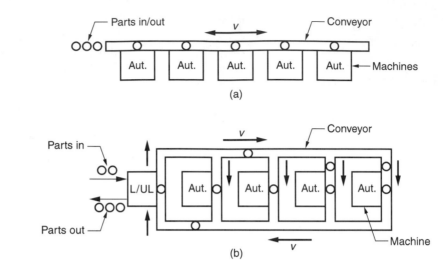

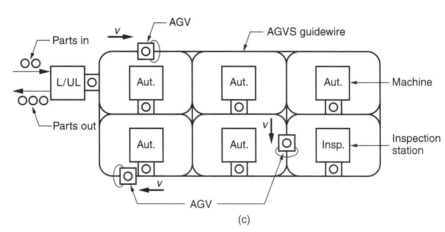

FIGURE 39.6 Three of the five FMS layout types: (a) in-line, (b) ladder, and (c) open field. *Key:* Aut = automated station; L/UL = load/unload station; Insp = inspection station; AGV = automated guided vehicle; AGVS = automated guided vehicle system.

TABLE 39.4 Typical computer functions implemented by application software modules in a flexible manufacturing system.

Function	Description
NC part programming	Development of NC programs for new parts introduced into the system. This includes a language package such as APT.
Production control	Product mix, machine scheduling, and other planning functions.
NC program download	Part program commands must be downloaded to individual stations from the central computer.
Machine control	Individual workstations require controls, usually computer numerical control.
Workpart control	Monitor status of each workpart in the system, status of pallet fixtures, orders on loading/unloading pallet fixtures.
Tool management	Functions include tool inventory control, tool status relative to expected tool life, tool changing and resharpening, and transport to and from tool grinding.
Transport control	Scheduling and control of workpart handling system.
System management	Compiles management reports on performance (utilization, piece counts, production rates, etc.), FMS simulation sometimes included.

unloading parts from the system, (2) changing and setting cutting tools, (3) maintenance and repair of equipment, (4) NC part programming, (5) programming and operating the computer system, and (6) overall management of the system.

39.2.3 Applications of Flexible Manufacturing Systems

Flexible manufacturing systems are typically used for mid-volume, mid-variety production. If the part or product is made in high quantities with no style variations, then a transfer line or similar dedicated production system is most appropriate. If the parts are low volume with high variety, then numerical control, or even manual methods would be more appropriate. These application characteristics are summarized in Figure 39.7.

Flexible machining systems are the most common application of FMS technology. Owing to the inherent flexibilities and capabilities of computer numerical control, it is possible to connect several CNC machine tools to a small central computer, and to devise automated methods for transferring workparts between the machines. Figure 39.8 shows a flexible machining system consisting of five CNC machining centers and an in-line transfer system to pick parts from a central load/unload station and move them to the appropriate machining stations.

In addition to machining systems, other types of flexible manufacturing systems have also been developed, although the state of technology in these other processes has not permitted the rapid implementation that has occurred in machining. The other types of systems include assembly, inspection, sheet-metal processing (punching, shearing, bending, and forming), and forging.

Most of the experience in flexible manufacturing systems has been gained in the machining area. For flexible machining systems, the benefits usually given are: (1) higher machine utilization than a conventional machine shop—relative utilizations are 40 to 50% for conventional batch type operations and about 75% for a FMS due to better work handling, off-line setups, and improved scheduling; (2) reduced work-in-process due to continuous production rather than batch production; (3) lower manufacturing lead times; and (4) greater flexibility in production scheduling.

FIGURE 39.7 Application characteristics of flexible manufacturing systems and cells relative to other types of production systems.

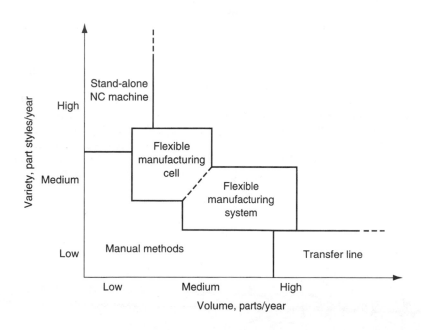

FIGURE 39.8   A five-station flexible manufacturing system (photo courtesy of Cincinnatti Milacron).

REFERENCES

[1]   Black, J. T. *The Design of the Factory with A Future.* McGraw-Hill, New York, 1990.

[2]   Black, J. T. "An Overview of Cellular Manufacturing Systems and Comparison to Conventional Systems." *Industrial Engineering.* November 1983, pp. 36–84.

[3]   Chang, T-C, Wysk, R. A., and Wang, H-P. *Computer-Aided Manufacturing.* 2nd ed. Prentice Hall, Upper Saddle River, N.J., 1997.

[4]   Gallagher, C. C., and Knight, W. A., *Group Technology.* Butterworth & Co. Ltd., London, 1973.

[5]   Groover, M. P. *Automation, Production Systems, and Computer Integrated Manufacturing.* 2nd ed. Prentice-Hall, Upper Saddle River, N.J., 2001.

[6]   Ham, I., Hitomi, K., and Yoshida, T. *Group Technology.* Kluwer Nijhoff Publishers, Hingham, Mass., 1985.

[7]   Houtzeel, A., "The Many Faces of Group Technology." *American Machinist.* January 1979, pp. 115–120.

[8]   Luggen, W. W. *Flexible Manufacturing Cells and Systems.* Prentice Hall, Inc., Englewood Cliffs, N.J., 1991.

[9]   Maleki, R. A. *Flexible Manufacturing Systems: The Technology and Management.* Prentice Hall, Inc., Englewood Cliffs, N.J., 1991.

[10] Moodie, C., Uzsoy, R., and Yih, Y. *Manufacturing Cells: A Systems Engineering View.* Taylor & Francis Ltd., London, 1995.

[11] Snead, C. S. *Group Technology: Foundation for Competitive Manufacturing.* Van Nostrand Reinhold, New York, 1989.

REVIEW QUESTIONS

39.1. Define *group technology.*

39.2. What is a *part family?*

39.3. Define *cellular manufacturing.*

39.4. What is the *composite part concept* in group technology?

39.5. Name some of the possible machine cell designs in group technology.

39.6. What is a *flexible manufacturing system?*

39.7. What makes an automated manufacturing system flexible?

39.8. Name some of the FMS software and control functions.

39.9. Identify some of the applications of FMS technology.

39.10. What are the advantages of FMS technology, compared to conventional batch operations?

MULTIPLE CHOICE QUIZ

There is a total of 12 correct answers in the following multiple choice questions (some questions have multiple answers that are correct). To attain a perfect score on the quiz, all correct answers must be given, since each correct answer is worth 1 point. For each question, each omitted answer or wrong answer reduces the score by 1 point, and each additional answer beyond the number of answers required reduces the score by 1 point. Percentage score on the quiz is based on the total number of correct answers.

39.1. Production flow analysis is a method of identifying part families that uses data from which one of the following sources? (a) bill of materials, (b) engineering drawings, (c) master schedule, (d) production schedule, or (e) route sheets.

39.2. Most parts classification and coding systems are based on which of the following types of part attributes (more than one)? (a) annual production rate, (b) design, (c) manufacturing, and (d) weight.

39.3. Which of the following are part design attributes that are likely to be included in a parts classification and coding system (more than one)? (a) annual production, (b) batch size, (c) length-to-diameter ratio, (d) major process, (e) part dimensions, and (f) tolerances.

39.4. What is the dividing line between a manufacturing cell and a flexible manufacturing system? (a) two machines, (b) four machines, or (c) six machines.

39.5. A machine capable of producing different part styles in a batch mode of operation qualifies as a flexible manufacturing system: (a) true or (b) false.

39.6. The physical layout of a flexible manufacturing system is determined principally by which one of the following? (a) computer system, (b) material handling system, (c) part family, (d) processing equipment, or (e) weight of parts processed.

39.7. Industrial robots can, in general, most easily handle which of the following part types in a flexible machining system (one best answer)? (a) heavy parts, (b) metal parts, (c) nonrotational parts, (d) plastic parts, or (e) rotational parts.

39.8. Flexible manufacturing systems and cells are generally applied in which one of the following areas? (a) high variety, low volume production, (b) low variety, (c) low volume, (d) mass production, (e) medium volume, medium variety production.

39.9. Which of the following technologies is most closely associated with flexible machining systems (one best answer)? (a) lasers, (b) machine vision, (c) manual assembly lines, (d) numerical control, or (f) transfer lines.

40 PRODUCTION LINES

CHAPTER CONTENTS

40.1   Fundamentals of Production Lines
    40.1.1   Product Variations
    40.1.2   Methods of Workpart Transport
    40.1.3   Determining the Number of Workstation Required
40.2   Manual Assembly Lines
    40.2.1   Line Balancing
    40.2.2   Other Factors in Assembly Line Design
40.3   Automated Production Lines
    40.3.1   Types of Automated Lines
    40.3.2   Analysis of Automated Production Lines

Production lines are an important class of manufacturing system when large quantities of identical or similar products are to be made. They are suited to situations where the total work to be performed on the product or part consists of many separate steps. Examples include assembled products (e.g., automobiles and appliances) and mass-produced machined parts on which multiple machining operations are required (e.g., engine blocks and transmission housings). In a production line, the total work is divided into small tasks, and workers or machines perform these tasks with great efficiency. Much credit for the development and refinement of the production line is due to Henry Ford and his engineering staff at the Ford Motor Company in the early 1900s (Historical Note 40.1).

For purposes of organization, we divide production lines into two basic types, manual assembly lines and automated production lines, although hybrid lines consisting of both manual and automated operations are not uncommon. Before examining these particular systems, let us consider some of the general issues involved in production line design and operation.

40.1 FUNDAMENTALS OF PRODUCTION LINES

A *production line* consists of a series of workstations arranged so that the product moves from one station to the next, and at each location a portion of the total work is performed on it. See Figure 40.1. The production rate of the line is determined by its slowest sta-

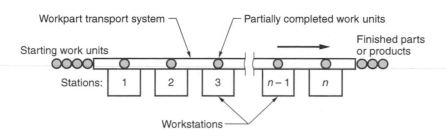

FIGURE 40.1 General configuration of a production line.

tion. Workstations whose pace is faster than the slowest will ultimately be limited by that bottleneck station. Transfer of the product along the line is usually accomplished by a mechanical transfer device or conveyor system, although some manual lines simply pass the product by hand between stations. Production lines are associated with mass production. If product quantities are very high and the work can be divided into separate tasks that can be assigned to individual workstations, then a production line is the most appropriate manufacturing system.

40.1.1 Product Variations

Production lines can be designed to cope with variations in product models so long as the differences between models are not too great (soft product variety, as defined in Section 1.1.2). Three types of line can be distinguished: (1) single model line, (2) batch model line, and (3) mixed model line. A *single model line* is one that produces only one model, and there is no variation in the model. Thus, the tasks performed at each station are the same on all product units.

Batch model and mixed model lines are designed to produce two or more different product models on the same line, but they use different approaches for dealing with the model variations. As its name suggests, a *batch model line* produces each model in batches. The workstations are set up to produce the desired quantity of the first model; then the stations are reconfigured to produce the desired quantity of the next model; and so on. Assembled products are often made using this approach when the demand for each product is medium. The economics in this case favor the use of one production line for several products rather than using many separate lines for each model.

Workstation setup refers to the assignment of tasks to a given station on the line, the special tools needed to perform the tasks, and the physical layout of the station. The models made on the line are usually similar, and the tasks to make them are therefore similar. However, differences exist among models so that a different sequence of tasks is usually required, and tools used at a given workstation for the last model might not be the same as those required for the next model. One model may take more total time than another, requiring the line to be operated at a slower pace. Also, worker retraining or new equipment may be needed for the production of a new model. For these kinds of reasons, changes in the workstation setup are required before production of a new model can begin. These changeovers result in downtime (lost production time) on a batch model line.

A *mixed model line* also produces multiple models; however, the models are intermixed on the same line rather than being produced in batches. While a particular model is being worked on at one station, a different model is being processed at the next station. Each station is equipped with the necessary tools and is capable of performing the variety of tasks needed to produce any model that moves through it. Many consumer products are assembled on mixed model lines. Prime examples are automobiles and major appliances, which are characterized by significant model variations and available options.

Advantages of a mixed model line over a batch model line include: (1) minimized downtime between models; (2) avoidance of high inventories of some models while other models are stocked out; and (3) increased or decreased production rates and quantities of the models in accordance with changes in demand. On the other hand, the problem of assigning tasks to workstations so that they all share an equal workload is more complex on a mixed model line. Scheduling (determining the sequence of models) and logistics (getting the right parts to each workstation for the model currently at that station) are more difficult in this type of line.

40.1.2 Methods of Work Transport

There are various ways of moving work units from one workstation to the next. The two basic categories are manual and mechanized.

Manual Methods of Work Transport   Manual methods involve passing the work units between stations by hand. These methods are associated with manual assembly lines. In some cases, the output of each station is collected in a box or tote pan; when the box is full it is moved to the next station. This can result in a significant amount of in-process inventory, which is undesirable. In other cases, work units are moved individually along a flat table or unpowered conveyor (e.g., a roller conveyor). When the task is finished at each station, the worker simply pushes the unit toward the downstream station. Space is usually allowed for one or more units to collect between stations, thereby relaxing the requirement for all workers to perform their respective tasks in sync.

One problem associated with manual methods of work transport is the difficulty in controlling the production rate on the line. Workers tend to work at a slower pace unless some mechanical means of pacing them is provided.

Mechanized Methods of Work Transport   Powered mechanical systems are commonly used to move work units along a production line. These systems include lift-and-carry devices, pick-and-place mechanisms, powered conveyors (e.g., overhead chain conveyors, belt conveyors, and chain-in-floor conveyors), and other material-handling equipment, sometimes combining several types on the same line. It is not our purpose here to describe the types of material-handling equipment available, but it is appropriate to identify the three major types of workpart transfer systems used on production lines—continuous transfer, synchronous transfer, and asynchronous transfer. These transfer systems can be implemented by various types of hardware.

Continuous transfer systems consist of a continuously moving conveyor that operates at a constant velocity v_c. The continuous transfer system is most common on manual assembly lines. Two cases are distinguished: (1) parts are fixed to the conveyor and (2) parts can be removed from the conveyor. In the first case, the product is usually large and heavy (e.g., automobile, washing machine) and cannot be removed from the line. The worker must therefore walk along with the moving conveyor to complete the assigned task for that unit while it is in the station. In the second case, the product is small enough that it can be removed from the conveyor to facilitate the work at each station. Some of the pacing benefits are lost in this arrangement, since each worker is not required to finish the assigned tasks within a fixed time period. On the other hand, this case allows greater flexibility to each worker to deal with technical problems that may be encountered on a particular work unit.

In *synchronous transfer systems*, work units are simultaneously moved between stations with a quick, discontinuous motion. These systems are also known by the name *intermittent transfer*, which characterizes the type of motion experienced by the work units. Synchronous transfer includes positioning of the work at the stations, which is a requirement for automated lines that use this mode of transfer. Synchronous transfer is not common for manual lines, owing to the rigid pacing involved. The task at each and every station must be finished within the allowed cycle time or the product will leave the station as an incomplete unit. This rigid pacing discipline is stressful to human workers, which is undesirable. By contrast, this type of pacing lends itself to automated operation.

Asynchronous transfer allows each work unit to depart its current station when processing has been completed. Each unit moves independently, rather than synchronously. Thus, at any given moment, some units on the line are moving between stations, while others are positioned at stations. This type of transfer is sometimes called a "power-and-free" system. Associated with the operation of an asynchronous transfer system is the tactical use of queues between stations. Small queues of work units are permitted to form in front of each station, so that variations in worker task times will be averaged and stations will always have work waiting for them. Asynchronous transfer is used for both manual and automated production systems.

40.1.3 Determining the Minimum Number of Workstations Required

Production lines are used for products with high demand. We can develop equations to determine the required number of stations on a production line to meet a given annual demand. Suppose our problem is to design a single model line to satisfy annual demand for a certain product. Management must decide how many shifts per week the line will operate and the number of hours per shift. If we assume 50 weeks per year, then the required hourly production rate of the line will be given by

$$R_p = \frac{D_a}{50SH} \qquad (40.1)$$

where R_p = the actual average production rate, units/hr; D_a = annual demand for the product, units/year; S = number of shifts/wk; and H = hours/shift. If the line operates 52 weeks rather than 50, then $R_p = D_a/52SH$. The corresponding average production time per unit is the reciprocal of R_p:

$$T_p = \frac{60}{R_p} \qquad (40.2)$$

where T_p = actual average production time, converted to minutes.

Unfortunately, the line may not be available for the entire time given by 50SH, because reliability problems result in lost time. These reliability problems include mechanical and electrical failures, tools wearing out, power outages, and similar malfunctions. Accordingly, the line must operate at a faster time than T_p to compensate for these problems. If E = line efficiency, which is the proportion of uptime on the line, then the cycle time of the line T_c is given by

$$T_c = ET_p = \frac{60E}{R_p} \qquad (40.3)$$

Any product contains a certain work content that represents all of the tasks that are to be accomplished on the line. This work content requires an amount of time called the *work content time* T_{wc}. This is the total time required to make the product on the line. If we assume that the work content time can be divided evenly amongst the stations, so that every station has an equal workload whose time to perform $= T_c$, then the minimum possible number of workstations n_{min} in the line can be determined as

$$n_{min} = \text{Minimum Integer} \geq \frac{T_{wc}}{T_c} \qquad (40.4)$$

This minimum number should be interpreted as an ideal value, whose achievement in practice is highly unlikely for the following reasons: (1) *imperfect balancing*—it is very difficult to divide the work content time equally amongst the workstations; some stations will be assigned an amount of work that requires less time than T_c; (2) *task time variability*—there is inherent and unavoidable variability in the time required by a worker to perform a given assembly task; (3) *repositioning time losses*—some time will be lost at each station due to repositioning of the work or the worker; thus, the amount of time available at each station will actually be less than T_c; (4) *manning level*—some stations on manual production lines may have more than one worker; and (5) *quality problems*—defective components and other quality problems will cause delays and rework that will add to the workload. We explore some of these issues in the following sections for the cases of manual and automated lines.

40.2 MANUAL ASSEMBLY LINES

The manual assembly line was an important development in the growth of U.S. industry in the first half of the twentieth century (Historical Note 40.1). It is of global importance today in the manufacture of assembled products including automobiles and trucks, consumer electronic products, appliances, power tools, and other products made in large quantities.

Historical Note 40.1   *Origins of manual assembly lines.*

M anual assembly lines are based largely on two fundamental work principles. The first is division of labor, argued by Adam Smith in England in his book **Wealth of Nations** published in 1776. Smith did not invent division of labor, since there had been instances of its use in Europe for many centuries. He was the first to note its significance in production. The second principle is interchangeable parts, based on the work of Eli Whitney and others at the beginning of the Nineteenth century (Historical Note 1.1). The alternative to interchangeable parts, practiced prior to Whitney's time, was hand filing of individual parts in order to achieve fitting.

Modern production lines can be traced to the meat packing industry in Chicago, Illinois, and Cincinnati, Ohio, where overhead (unpowered) conveyors were used to move carcasses from one worker to the next. These conveyors were later replaced by powered chain conveyors to create "disassembly lines"—predecessor to the assembly line. The organization of work permitted meat cutters to concentrate on single tasks (division of labor).

American automotive industrialist Henry Ford observed the meat-packing industry. Together with colleagues, he designed an assembly line in 1913 in Highland Park, Michigan, for producing magneto flywheels. The result was a fourfold increase in productivity. Stimulated by this success, Ford applied assembly line techniques to chassis fabrication. Using chain-driven conveyors and workstations designed for convenience and comfort (early ergonomics), productivity was increased by a factor of eight, compared to previous single-station assembly methods.

Success of the Ford Motor Company resulted in drastic reductions in the price of the Model T Ford, the principal product of the company at the time. The common American could afford to own an automobile because of Ford's achievement in cost reduction. This forced his competitors and suppliers to imitate his methods, and the manual assembly line became intrinsic to U.S. industry.

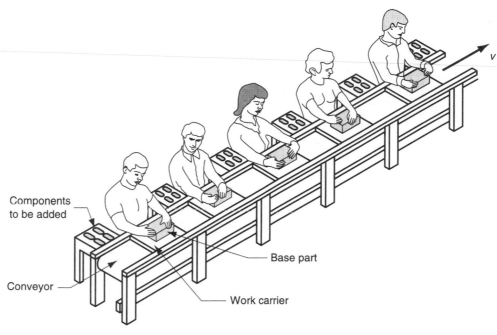

FIGURE 40.2　A portion of a manual assembly line. Each worker performs a task at his or her workstation. A conveyor moves parts on work carriers from one station to the next.

A ***manual assembly line*** consists of multiple workstations arranged sequentially, at which assembly operations are performed by human workers, as in Figure 40.2 (also see Color Plate 12). The usual procedure on a manual line begins with the "launching" of a base part onto the front end of the line. A work carrier is often required to hold the part during its movement along the line. The base part travels through each of the workstations where workers perform tasks that progressively build the product. Components are added to the base part at each station, so that the entire work content has been completed when the product exits the final station. Processes accomplished on manual assembly lines include mechanical fastening operations (Chapter 33), spot welding (Section 31.2), hand soldering (Section 32.2), and adhesive joining (Section 32.3).

40.2.1　Line Balancing

One of the biggest technical problems in designing and operating a manual assembly line is line balancing. This is the problem of assigning tasks to individual workers so that all workers have an equal amount of work. Recall that the entirety of work to be accomplished on the line is given by the work content. This total work content can be divided into ***minimum rational work elements,*** each element concerned with adding a component or joining them or performing some other small portion of the total work content. The notion of a minimum rational work element is that it is the smallest practical amount of work into which the total job can be divided. Different work elements require different times, and when they are grouped into logical tasks and assigned to workers, the task times will not be equal. Thus, simply due to the variable nature of element times, some workers will end up with more work, while other workers will have less. The cycle time of the assembly line is determined by the station with the longest task time.

One might think that although the work element times are different, it should be possible to find groups of elements whose sums (task times) are nearly equal, if not perfectly equal. What makes it difficult to find suitable groups is that there are several constraints on this combinatorial problem. First, the line must be designed to achieve some desired production rate, which establishes in advance the cycle time T_c at which the line must operate, Eq. (40.4). Therefore, the sum of the work element times assigned to each station must be $\leq T_c$.

Second, there are restrictions on the order in which the work elements can be performed. Some elements must be done before others. For example, a hole must be drilled before it can be tapped. A screw that will use the tapped hole to attach a mating component cannot be fastened before the hole has been drilled and tapped. These kinds of requirements on the work sequence are called **precedence constraints.** They complicate the line balancing problem. A certain element that might be allocated to a worker to obtain a task time $= T_c$ cannot be added because it violates a precedence constraint.

These and other limitations make it virtually impossible to achieve perfect balancing of the line, which means that some workers will require more time to complete their tasks than others. Methods of solving the line balancing problem, that is, allocating work elements to stations, are discussed in other references—excellent references indeed, such as [6]. The inability to achieve perfect balancing results in some idle time at most stations. Because of this idle time, the actual number of workers required on the line will be greater than the number of workstations given by Eq. (40.4).

A measure of the total idle time on a manual assembly line is given by the **balancing efficiency** E_b, defined as the total work content time divided by the total available service time on the line. The total work content time has been previously defined; it is the sum of all of the work elements that are to be accomplished on the line. The total available service time on the line can be defined as

$$\text{Total available service time} = wT_s$$

where w = number of workers on the line; and T_s = the longest service time on the line; that is,

$$T_s = \text{Max}\{T_{si}\} \text{ for } i = 1, 2, \ldots n.$$

where T_{si} = the service time (task time) at station i, min. The reader may wonder why we are using a new term T_s rather than the previously defined cycle time T_c. The reason is that there is another time loss in the operation of a production line in addition to idle time from imperfect balancing. Let us call it the **repositioning time** T_r. It is the time required in each cycle to reposition the worker, or the work unit, or both. On a continuous transfer line where work units are attached to the line and move at a constant speed, T_r is the time taken by the worker to walk from the unit just completed to the next unit coming into the station. In all manual assembly lines, there will be some lost time due to repositioning. We assume that T_r is the same for all workers, although in fact repositioning may require different times at different stations. We can relate T_s, T_c, and T_r as follows:

$$T_c = T_s + T_r \qquad (40.5)$$

The definition of balancing efficiency E_b can now be written in equation form as follows:

$$E_b = \frac{T_{wc}}{wT_s} \qquad (40.6)$$

A perfect line balance yields a value of $E_b = 1.00$. Typical line balancing efficiencies in industry range between 0.90 and 0.95.

Eq. (40.6) can be rearranged to obtain the number of workers required on a manual assembly line:

$$w = \text{Minimum Integer} \geq \frac{T_{wc}}{T_s E_b} \qquad (40.7)$$

The utility of this relationship suffers from the fact that the balancing efficiency E_b depends on w, as defined in Eq. (40.6). Unfortunately, we have an equation where the thing to be determined depends on a parameter that, in turn, depends on the thing itself. Notwithstanding this drawback, Eq. (40.7) defines the relationship among the parameters in a manual assembly line. Using a typical value of E_b based on similar previous lines, it can be used to estimate the number of workers required to produce a given assembly.

EXAMPLE 40.1
Manual Assembly Line

A manual assembly line is being planned for a product whose annual demand = 90,000 units. A continuously moving conveyor will be used with work units attached. Work content time = 55 min. The line will run 50 weeks/year, 5 shifts/week, and 8 hours/day. Based on previous experience, assume line efficiency $E = 0.95$, balancing efficiency $E_b = 0.93$, and repositioning time $T_r = 9$ sec. Determine (a) hourly production rate to meet demand; (b) number of workers required; and (c) for comparison, the ideal minimum value as given by n_{min}.

Solution: (a) The hourly production rate required to meet annual demand is given by Eq. (40.1):

$$R_p = \frac{90,000}{50(5)(8)} = \textbf{45 units/hr}$$

(b) With a line efficiency of 0.95, the ideal cycle time is

$$T_c = \frac{60(0.95)}{45} = \textbf{1.2667 min}$$

Given that repositioning time $T_r = 9$ sec = 0.15 min, the service time is

$$T_s = 1.2667 - 0.150 = 1.1167 \text{ min}$$

Workers required to operate the line, by Eq. (40.7) equals

$$w = \text{Minimum Integer} \geq \frac{55}{1.1167(0.93)} = 52.96 \rightarrow \textbf{53 workers}$$

Assuming one worker per station,

$$n = \textbf{53 workstations}$$

(c) This compares with the ideal minimum given by Eq. (40.4):

$$n_{min} = \text{Minimum Integer} \geq \frac{55}{1.2667} = 43.42 \rightarrow \textbf{44 stations.}$$

It is clear that lost time due to repositioning and imperfect line balancing take their toll in the design and operation of a manual assembly line.

40.2.2 Other Factors in Assembly Line Design

The number of workstations on a manual assembly line does not necessarily equal the number of workers. For large products, it may be possible to assign more than one worker to a station. This practice is common in final assembly plants that build cars and trucks. For example, two workers in a station might perform assembly tasks on opposite sides of the vehicle. The number of workers in a given station is called the station *manning level* M_i. Averaging the manning levels over the entire line,

$$M = \frac{w}{n} \tag{40.8}$$

where M = average manning level for the assembly line; w = number of workers on the line; and n = number of stations. Naturally, w and n must be integers. Multiple manning conserves valuable floorspace in the factory because it reduces the number of stations required.

Another factor that affects manning level on an assembly line is the number of **automated stations** on the line, including stations that employ industrial robots (Section 38.2). Automation reduces the required labor force on the line, although it increases the need for technically trained personnel to service and maintain the automated stations. The automobile industry makes extensive use of robotic workstations to perform spot welding and spray painting on sheet-metal car bodies. The robots accomplish these operations with greater repeatability, which translates into higher product quality.

The use of **inventory buffers** in a production line has already been discussed. Inventory buffers help to compensate for service time variability within a given station. They are associated principally with asynchronous transfer systems.

40.3 AUTOMATED PRODUCTION LINES

Manual assembly lines generally use a mechanized transfer system to move parts between workstations, but the stations themselves are operated by human workers. An **automated production line** consists of automated workstations connected by a parts transfer system whose actuation is coordinated with the stations. In the ideal, no human workers are on the line, except to perform auxiliary functions such as tool changing, loading and unloading parts at the beginning and end of the line, and repair and maintenance activities. Modern automated lines are integrated systems, operating under computer control.

Operations performed by automated stations tend to be simpler than those performed by humans on manual lines. The reason is that simpler tasks are easier to automate. Operations that are difficult to automate are those requiring multiple steps, judgment, or human sensory capability. Tasks that are easy to automate consist of single work elements, quick actuating motions, and straight-line feed motions as in machining.

40.3.1 Types of Automated Lines

Automated production lines can be divided into two basic categories: (1) those that perform processing operations such as machining, and (2) those that perform assembly operations. An important type in the processing category is the transfer line.

Transfer Lines and Similar Processing Systems   A **transfer line** consists of a sequence of workstations that perform production operations, with automatic transfer of work units between stations. Machining is the most common processing operation, as depicted in Figure 40.3. Automatic transfer systems for sheet metalworking and assembly are also available. In the case of machining, the workpiece typically starts as a metal casting or forging, and a series of machining operations are performed to accomplish the high precision details (e.g., holes, threads, and finished flat surfaces).

Transfer lines are usually expensive pieces of equipment, sometimes costing millions of dollars; they are designed for jobs requiring high quantities of parts. The amount of machining accomplished on the workpart may be significant, but since the work is di-

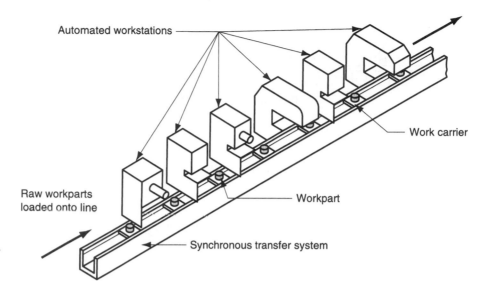

FIGURE 40.3 A machining transfer line, an important type of automated production line.

vided among many stations, production rates are high and unit costs are low compared to alternative production methods. Synchronous transfer is commonly used on automated machining lines.

A variation of the automated transfer line is the ***dial indexing machine,*** Figure 40.4, in which workstations are arranged around a circular worktable, called a dial. The worktable is actuated by a mechanism that provides partial rotation of the table on each work cycle. The number of rotational positions is designed to match the number of workstations around the periphery of the table. The angular rotation of the dial is given by

$$A = \frac{360}{n} \qquad (40.9)$$

where A = rotation angle of the dial, degrees; and n = the number of workstations (more precisely, the number of stop positions) around the outside of the dial. Although the configuration of a dial indexing machine is quite different from a transfer line, its operation and application are quite similar.

Automated Assembly Systems   ***Automated assembly systems*** consist of one or more workstations that perform assembly operations, such as adding components and/or af-

FIGURE 40.4
Configuration of a dial indexing machine.

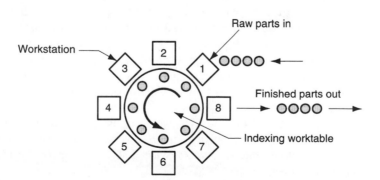

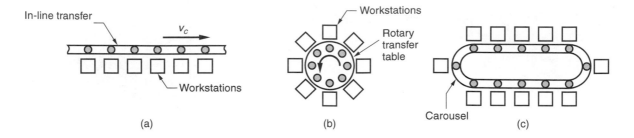

FIGURE 40.5   Three common configurations of multiple station assembly systems: (a) in-line, (b) rotary, and (c) carousel.

fixing them to the work unit. Automated assembly systems can be divided into single station cells and multiple station systems. *Single station assembly cells* are often organized around an industrial robot that has been programmed to perform a sequence of assembly steps. The single robot cannot work as fast as a series of specialized automatic stations, so single station cells are used for jobs in the medium production range.

Multiple station assembly systems are appropriate for high production. They are widely used for mass production of small products such as ball-point pens, cigarette lighters, flashlights, and similar items consisting of a limited number of components. The number of components and assembly steps is limited because system reliability decreases rapidly with increasing complexity.

Multiple station assembly systems are available in several configurations, pictured in Figure 40.5: (a) in-line, (b) rotary, and (c) carousel. The in-line configuration is the conventional transfer line adapted to perform assembly work. These systems are not as massive as their machining counterparts. Rotary systems are usually implemented as dial indexing machines. Carousel assembly systems are arranged as a loop. They can be designed with a greater number of workstations than a rotary system. Owing to the loop configuration, the carousel allows the work carriers to be automatically returned to the starting point for reuse, an advantage shared with rotary systems but not with transfer lines.

40.3.2 Analysis of Automated Production Lines

Line balancing is a problem on an automated line, just as it is on a manual assembly line. The total work content must be allocated to individual workstations. However, since the tasks assigned to automated stations are generally simpler, and the line often contains fewer stations, the problem of defining what work should be done at each station is not as difficult for an automated line as for a manual line.

A more significant problem in automated lines is reliability. The line consists of multiple stations, interconnected by a work transfer system. It operates as an integrated system, and when one component malfunctions, the entire system is adversely affected. To analyze the operation of an automated production line, let us assume a system that performs processing operations and uses synchronous transfer. This model includes transfer lines as well as dial indexing machines. It does not include automated assembly systems, which require an adaptation of the model [6]. Our terminology will borrow symbols from the first two sections: n = number of workstations on the line; T_c = ideal cycle time on the line; T_r = repositioning time, called the transfer time in a transfer line; and T_{si} = the service time at station i. The ideal cycle time T_c is the service time (processing time) for the slowest station on the line plus the transfer time; that is,

$$T_c = T_r + \text{Max}\{T_{si}\} \tag{40.10}$$

In the operation of a transfer line, periodic breakdowns cause downtime on the entire line. Let F = frequency with which breakdowns occur, causing a line stoppage; and T_d = average time the line is down when a breakdown occurs. The downtime includes the time for the repair crew to swing into action, diagnose the cause of the failure, fix it, and restart the line.

Based on these definitions, we can formulate the following expression for the actual average production time T_p:

$$T_p = T_c + FT_d \tag{40.11}$$

where F = downtime frequency, line stops/cycle; and T_d = downtime in minutes per line stop. Thus, FT_d = downtime averaged on a per cycle basis. The actual average production rate is the reciprocal of T_p:

$$R_p = \frac{60}{T_p}$$

as previously given in Eq. (40.2). It is of interest to compare this rate with the ideal production rate given by

$$R_c = \frac{60}{T_c} \tag{40.12}$$

where R_p and R_c are expressed in pc/hour, given that T_p and T_c are expressed in minutes.

Based on this relation, we can define the line efficiency E for a transfer line. In the context of automated production systems, E refers to the proportion of uptime on the line and is really a measure of reliability more than efficiency:

$$E = \frac{T_c}{T_c + FT_d} \tag{40.13}$$

This is the same relationship as earlier Eq. (40.3), since $T_p = T_c + FT_d$. It should be noted that the same definition of line efficiency applies to manual assembly lines, except that technological breakdowns are not as much of a problem on manual lines (human workers are more reliable than electromechanical equipment, at least in the sense we are discussing here).

Line downtime is usually associated with failures at individual workstations. Reasons for downtime include scheduled and unscheduled tool changes, mechanical and electrical malfunctions, hydraulic failures, and normal equipment wear. Let p_i = probability or frequency of a failure at station i, then

$$F = \sum_{i=1}^{n} p_i \tag{40.14}$$

If all p_i are assumed equal, or an average value of p_i is computed, in either case calling it p, then

$$F = np \tag{40.15}$$

Both of these equations clearly indicate that the frequency of line stops increases with the number of stations on the line. Stated another way, reliability of the line decreases as we add more stations.

EXAMPLE 40.2
Automated
Transfer Line

An automated transfer line has 20 stations and an ideal cycle time of 1.0 min. Probability of a station failure is $p = 0.01$ and the average downtime when a breakdown occurs is 10 min. Determine (a) average production rate R_p and (b) line efficiency E.

Solution: The frequency of breakdowns on the line is given by $F = pn = 0.01 \times 20 = 0.20$. The actual average production time is therefore

$$T_p = 1.0 + 0.20(10) = 3.0 \text{ minutes}$$

(a) Production rate is therefore

$$R_p = \frac{60}{T_p} = \frac{60}{3.0} = 20 \text{ pc/hr}$$

Note that this is far lower than the ideal production rate

$$R_c = \frac{60}{T_c} = \frac{60}{1.0} = 60 \text{ pc/hr}$$

(b) Line efficiency is computed as

$$E = \frac{T_c}{T_p} = \frac{1.0}{3.0} = 0.333 \text{ (or 33.3\%)}$$

From this example we see that if a production line operates like this, it spends more time down than up. Achieving high efficiencies is a real problem in automated production lines.

The cost of operating an automated production line is the investment cost of the equipment and installation, plus the cost of maintenance, utilities, and labor assigned to the line. These costs are converted to an equivalent uniform annual cost and divided by the number of hours of operation per year to provide an hourly rate. This hourly cost rate can be used to figure the unit cost of processing a workpart on the line:

$$C_p = \frac{C_o T_p}{60} \tag{40.16}$$

where C_p = unit processing cost, \$/part; C_o = hourly rate of operating the line, as outlined above, \$/hr; T_p = actual average production time per workpart, min/part; and the constant 60 converts the hourly cost rate to \$/min for consistency of units.

REFERENCES

[1] Boothroyd, G., Poli, C., and Murch, L. E. *Automatic Assembly.* Marcel Dekker, Inc., New York, 1982.

[2] Buzacott, J. A. "Prediction of the Efficiency of Production Systems without Internal Storage." *International Journal of Production Research.* Vol. 6, No. 3, 1968, pp. 173–188.

[3] Buzacott, J. A., and Shanthikumar, J. G. *Stochastic Models of Manufacturing Systems.* Prentice-Hall, Upper Saddle River, N.J., 1993.

[4] Chow, W-M. *Assembly Line Design.* Marcel Dekker, Inc., New York, 1990.

[5] Groover, M. P. "Analyzing Automatic Transfer Lines." *Industrial Engineering.* Vol. 7, No. 11, 1975, pp. 26–31.

[6] Groover, M. P. *Automation, Production Systems, and Computer Integrated Manufacturing.* 2nd ed. Prentice-Hall, Upper Saddle River, N.J., 2001.

[7] Riley, F. J. *Assembly Automation, A Management Handbook.* 2nd ed. Industrial Press, New York, 1996.

[8] Wild, R. *Mass-Production Management.* Wiley, London, 1972.

REVIEW QUESTIONS

40.1. What is a *production line*?

40.2. Distinguish between a batch model production line and a mixed model production line.

40.3. What are the advantages of the mixed model line for producing different product styles?

40.4. What are some of the challenges of a mixed model line compared to a batch model line?

40.5. Identify two fundamental principles on which manual assembly lines are based.

40.6. Describe how manual methods are used to move parts between workstations on a production line.

40.7. Briefly define the three types of mechanized workpart transfer systems used in production lines.

40.8. Why are parts sometimes fixed to the conveyor in a continuous transfer system in manual assembly?

40.9. Why must a production line be paced at a rate higher than that required to satisfy the demand for the product?

40.10. What are the reasons why the number of workstations on a manual assembly line cannot be determined simply from the ratio T_{wc}/T_c?

40.11. Why is the line balancing problem different on an automated transfer line than on a manual assembly line?

40.12. Repositioning time on a synchronous transfer line is known by a different name; what is that name?

40.13. Why are single station assembly cells generally not suited to high production jobs?

40.14. What are some of the reasons for downtime on a machining transfer line?

MULTIPLE CHOICE QUIZ

There is a total of 10 correct answers in the following multiple choice questions (some questions have multiple answers that are correct). To attain a perfect score on the quiz, all correct answers must be given, since each correct answer is worth 1 point. For each question, each omitted answer or wrong answer reduces the score by 1 point, and each additional answer beyond the number of answers required reduces the score by 1 point. Percentage score on the quiz is based on the total number of correct answers.

40.1. Batch model lines are most suited to which one of the following production situations? (a) job shop, (b) mass production, or (c) medium production.

40.2. Manual methods of workpart transfer are probably closest to which one of the following mechanized methods of transfer? (a) asynchronous, (b) continuous, or (c) synchronous.

40.3. Precedence constraints are best described by which of the following (one best answer)? (a) launching sequence in a mixed model line, (b) limiting value on the sum of element times that can be assigned to a worker or station, (c) order of work stations along the line, or (d) sequence in which the work elements must be done.

40.4. Which of the following phrases are most appropriate to describe the characteristics of tasks that are performed at automated workstations (more than one)? (a) complex, (b) consists of multiple work elements,

(c) involves a single work element, (d) involves straightline motions, (e) requires sensory capability, and (f) simple.

40.5. The transfer line is most closely associated with which one of the following types of production operations? (a) assembly, (b) automotive chassis fabrication, (c) machining, (d) pressworking, or (e) spot welding.

40.6. A dial-indexing machine uses which one of the following types of workpart transfer? (a) asynchronous, (b) continuous, (c) parts passed by hand, or (d) synchronous.

40.7. The line efficiency (proportion uptime) on an automated line can be increased by which of the following approaches (more than one)? (a) improving the reliability of each workstation on the line, (b) increasing the number of stations n on the line, and (c) reducing the average downtime T_d.

PROBLEMS

Manual Assembly Lines

40.1. A manual assembly line is being designed for a product with annual demand = 100,000 units. The line will operate 50 wks/year, 5 shifts/wk, and 7.5 hr/shift. Work units will be attached to a continuously moving conveyor. Work content time = 42.0 min. Assume line efficiency $E = 0.97$, balancing efficiency $E_b = 0.92$, and repositioning time $T_r = 6$ sec. Determine (a) hourly production rate to meet demand and (b) number of workers required.

40.2. In the previous problem, compute (a) the ideal minimum number of workstations n_{min}; and (b) the number of workstations required if multiple manning can be used and the estimated manning level is $M = 1.4$.

40.3. A manual assembly line produces a small appliance whose work content time = 25.9 min. Desired production rate = 50 units/hr. Repositioning time = 6 sec, line efficiency = 95%, and balancing efficiency is 93%. How many workers are on the line?

40.4. A single model manual assembly line produces a product whose work content time = 48.9 min. The line has 24 workstations with a manning level $M = 1.25$. The product has a work content time = 47.8 min. Available shift time per day = 8 hr, but downtime during the shift reduces actual production time to 7.6 hr on average. This results in an average daily production of 256 units/day. Repositioning time per worker T_r is 8% of cycle time T_c. Determine (a) line efficiency, (b) balancing efficiency, and (c) repositioning time T_r.

40.5. A final assembly plant for a certain automobile model is to have a capacity of 240,000 units annually. The plant will operate 50 weeks/yr, two shifts/day, 5 days/week, and 8.0 hours/shift. It will be divided into three departments: (1) body shop, (2) paint shop, (3) trim-chassis-final department. The body shop welds the car bodies using robots, and the paint shop coats the bodies. Both of these departments are highly automated. Trim–chassis–final has no automation. There are 15.5 hours of direct labor content on each car in this department, where cars are moved by a continuous conveyor. Determine (a) hourly production rate of the plant and (b) number of workers and workstations required in trim–chassis–final if no automated stations are used, the average manning level is 2.5, balancing efficiency = 93%, proportion uptime = 95%, and a repositioning time of 0.15 min is allowed for each worker.

40.6. A product whose total work content time = 50 min is to be assembled on a manual production line. The required production rate is 30 units per hour. From previous experience with similar products, it is estimated that the manning level will be close to 1.5. Assume $E = E_b = 1.0$. If 9 sec will be lost from the cycle time for repositioning, determine (a) the cycle time, and (b) how many workers and (c) stations will be needed on the line?

40.7. A manual assembly line has 17 workstations with one operator per station. Total work content time to assemble the product = 22.2 min. The production rate of the line = 36 units per hour. A synchronous transfer system is used to advance the products from one station to the next, and the transfer time = 6 sec. The workers remain seated along the line. Proportion uptime $E = 0.90$. Determine the balance delay.

40.8. A production line with four automatic workstations (the other stations are manual) produces a certain product whose total assembly work content time = 55.0 min of direct manual labor. The production rate on the line is 45 units/hr. Because of the automated stations, uptime efficiency = 89%. The manual stations each have one worker. It is known that 10% of the cycle time is lost due to repositioning. If the balancing efficiency $E_b = 0.92$ on the manual stations, find (a) cycle time, (b) number of workers, and (c) workstations on the line. (d) What is the average manning level on the line, where the average includes the automatic stations?

40.9. Production rate for a certain assembled product is 47.5 units per hour. The total assembly work content time = 32 minutes of direct manual labor. The line operates at 95% uptime. Ten workstations have two workers on opposite sides of the line so that both sides of the product can be worked on simultaneously. The remaining stations have one worker. Repositioning time lost by each worker is 0.2 min/cycle. It is known that the number of workers on the line is two more than the number required for perfect balance. Determine: (a) number of workers, (b) number of workstations, (c) the balancing efficiency, and (d) average manning level.

40.10. The total work content for a product assembled on a manual production line is 48 min. The work is transported using a continuous overhead conveyor that operates at a speed of 3 ft/min. There are 24 workstations on the line, one-third of which have two workers; the remaining stations each have one worker. Repositioning time per worker is 9 sec, and uptime efficiency of the line is 95%. (a) What is the maximum possible hourly production rate if the line is assumed to be perfectly balanced? (b) If the actual production rate is only 92% of the maximum possible rate determined in part (a), what is the balance delay on the line?

Automated Production Lines

40.11. An automated transfer line has 20 stations and operates with an ideal cycle time of 1.50 min. Probability of a station failure is $p = 0.008$, and average downtime when a breakdown occurs is 10.0 minutes. Determine (a) the average production rate R_p and (b) the line efficiency E.

40.12. A dial-indexing table has six stations. One station is used for loading and unloading, which is accomplished by a human worker. The other five perform processing operations. The longest process takes 25 sec, and the indexing time = 5 sec. Each station has a frequency of failure $p = 0.015$. When a failure occurs, it takes an average of 3.0 min to make repairs and restart. Determine (a) hourly production rate and (b) line efficiency.

40.13. A seven-station transfer line has been observed over a 40-hour period. The processing times at each station are:

Station	1	2	3	4
Process time (min)	0.80	1.10	1.15	0.95

Station	5	6	7
Process time (min)	1.06	0.92	0.80

The transfer time between stations = 6 sec. The number of downtime occurrences = 110, and hours of downtime = 14.5 hours. Determine: (a) the number of parts produced during the week, (b) the average actual production rate in parts/hour, and (c) the line efficiency. (d) If the balancing efficiency were computed for this line, what would be its value?

40.14. A 12-station transfer line was designed to operate with an ideal production rate = 50 parts/hour. However, the line does not achieve this rate, since the line efficiency $E = 0.60$. It costs $75/hour to operate the line, exclusive of materials. The line operates 4,000 hours per year. A computer monitoring system has been proposed that will cost $25,000 (installed) and will reduce downtime on the line by 25%. If the value added per unit produced = $4.00, will the computer system pay for itself within one year of operation? Use expected increase in revenues resulting from the computer system as the criterion. Ignore material costs in your calculations.

40.15. An automated transfer line is to be designed. Based on previous experience, the average downtime per occurrence = 5.0 min, and the probability of a station failure that leads to a downtime occurrence $p = 0.01$. The total work content time = 9.8 min and is to be divided evenly among the workstations, so that the ideal cycle time for each station = $9.8/n$. Determine (a) the optimum number of stations n on the line that will maximize production rate and (b) the production rate R_p and proportion uptime E for your answer to part (a).

41

MANUFACTURING ENGINEERING

CHAPTER CONTENTS

41.1 Process Planning
    41.1.1 Traditional Process Planning
    41.1.2 Make or Buy Decision
    41.1.3 Computer-Aided Process Planning
41.2 Problem Solving and Continuous Improvement
41.3 Concurrent Engineering and Design for Manufacturability
    41.3.1 Design for Manufacturing and Assembly
    41.3.2 Concurrent Engineering

This final part of the book is concerned with the *manufacturing support systems,* which are the set of procedures and systems used by a company to solve the technical and logistics problems encountered in planning the processes, ordering materials, controlling production, and ensuring that the products made by the company meet the required quality specifications. The position of the manufacturing support systems in the overall operations of the company is portrayed in Figure 41.1. As with the manufacturing systems in the factory, the manufacturing support systems include people. People make the systems work. Unlike the manufacturing systems in the factory, most of the support systems do not directly contact the product during its processing and assembly. Instead, they plan and control the activities in the factory to ensure that the products are completed and delivered to the customer on time, in the right quantities, and to the highest quality standards.

The quality control system is one of the manufacturing support systems, but it also consists of facilities that are located in the factory—inspection equipment used to measure and gage the materials being processed and products being assembled. We cover the quality control system in two chapters: Chapter 43, on quality control, and Chapter 44, on measurement and inspection. Other manufacturing support systems covered in this part of the book are production planning and control, Chapter 42, and manufacturing engineering, covered in this chapter.

Manufacturing engineering is a technical staff function that is concerned with planning the manufacturing processes for the economic production of high-quality products. Its principal role is to engineer the transition of the product from design specification to manufacture of a physical product. Its overall goal is to optimize man-

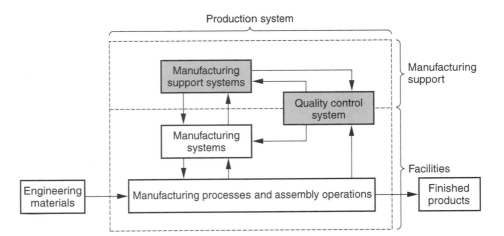

FIGURE 41.1 The position of the manufacturing support systems in the production system.

ufacturing within a particular organization. The scope of manufacturing engineering includes many activities and responsibilities that depend on the type of production operations accomplished by the particular organization. The usual activities include the following:

> *Process planning*—As our definition suggests, this is the principal activity of manufacturing engineering. Process planning includes: (1) deciding what processes and methods should be used and in what sequence; (2) determining tooling requirements; (3) selecting production equipment and systems; and (4) estimating costs of production for the selected processes, tooling, and equipment.

> *Problem solving and continuous improvement*—Manufacturing engineering provides staff support to the operating departments (parts fabrication and product assembly) to solve technical production problems. It should also be engaged in continuous efforts to reduce production costs, increase productivity, and improve product quality.

> *Design for manufacturability*—In this function, which chronologically precedes the other two, manufacturing engineers serve as manufacturability advisors to product designers. The objective is to develop product designs that not only meet functional and performance requirements, but that also can be produced at reasonable cost with minimum technical problems at highest possible quality in the shortest possible time.

Manufacturing engineering must be performed in any industrial organization that is engaged in production. The manufacturing engineering department usually reports to the manager of manufacturing in a company. In some companies the department is known by other names, such as process engineering or production engineering. Often included under manufacturing engineering are tool design, tool fabrication, and various technical support groups.

41.1 PROCESS PLANNING

Process planning involves determining the most appropriate manufacturing processes and the order in which they should be performed to produce a given part or product specified by design engineering. If it is an assembled product, process planning includes deciding the appropriate sequence of assembly steps. The process plan must be developed within the limitations imposed by available processing equipment and productive capacity of the factory. Parts or subassemblies that cannot be made internally must be purchased from external suppliers. In some cases, items that can be produced internally may be purchased from outside vendors for economic or other reasons.

41.1.1 Traditional Process Planning

Traditionally, process planning has been accomplished by manufacturing engineers who are knowledgeable in the particular processes used in the factory and are able to read engineering drawings. Based on their knowledge, skill, and experience, they develop the processing steps in the most logical sequence required to make each part. Table 41.1 lists the many details and decisions usually included within the scope of process planning. Some of these details are often delegated to specialists, such as tool designers; but manufacturing engineering is responsible for them.

Process Planning for Parts   The processes needed to manufacture a given part are determined largely by the material out of which the part is made. The material is selected by the product designer based on functional requirements. Once the material has been selected, the choice of possible processes is narrowed considerably. In our coverage of engineering materials, we provided guides to the processing of the four material groups: metals (Section 6.5), ceramics (Section 7.6), polymers (Section 8.5), and composite materials (Section 9.5).

A typical processing sequence to fabricate a discrete part consists of (1) a basic process, (2) one or more secondary processes, (3) operations to enhance physical properties, and (4) finishing operations, illustrated in Figure 41.2. Basic and secondary

TABLE 41.1   Decisions and details required in process planning.

Processes and sequence—The process plan should briefly describe all processing steps used on the work unit (e.g., part, assembly) in the order in which they are performed.

Equipment selection—In general, manufacturing engineers try to develop process plans that utilize existing equipment. When this is not possible, the component in question must be purchased (Section 41.2.2), or new equipment must be installed in the plant.

Tools, dies, molds, fixtures, and *gages*—The process planner must decide what tooling is needed for each process. Design of these items is usually delegated to the tool design department, and fabrication is accomplished by the tool room.

Cutting tools and *cutting conditions* for machining operations—These are specified by the process planner, industrial engineer, shop foreman, or machine operator, often in accordance with standard handbook recommendations.

Methods—Methods include hand and body motions, workplace layout, small tools, hoists for lifting heavy parts, and so forth. Methods must be specified for manual operations (e.g., assembly) and manual portions of machine cycles (e.g., loading and unloading a production machine). Methods planning has traditionally been the province of industrial engineers. Today's emphasis on self-directed work teams and worker empowerment has shifted much of the responsibilities for methods analysis from IEs to the actual workers who must perform the tasks.

Work standards—Work measurement techniques are applied to establish time standards for each operation.

Estimating production costs—This is often accomplished by cost estimators with help from the process planner.

Material handling—Consideration must be given to the problem of moving materials and work-in-progress in the factory.

Plant layout and *facilities design*—This is usually the responsibility of the plant engineering department working with manufacturing engineering.

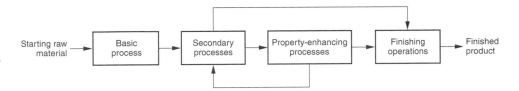

FIGURE 41.2   Typical sequence of processes required in part fabrication.

processes are shaping processes (Section 1.3.1) which alter the geometry of a workpart. A ***basic process*** establishes the initial geometry of the part. Examples include metal casting, forging, and sheet-metal rolling. In most cases, the starting geometry must be refined by a series of ***secondary processes.*** These operations transform the basic shape into the final geometry. There is a correlation between the secondary processes that might be used and the basic process that provides the initial form. For example, when sand casting or forging are the basic processes, machining operations are generally the secondary processes. When a rolling mill produces strips or coils of sheet metal, the secondary processes are stamping operations such as blanking, punching, and bending. Selection of certain basic processes minimizes the need for secondary processes. For example, if plastic injection molding is the basic process, secondary operations are usually not required because molding is capable of providing the detailed geometric features with good dimensional accuracy.

Shaping operations are generally followed by operations to enhance physical properties and/or finish the product. ***Operations to enhance properties*** include heat treating operations on metal components and glassware. In many cases, parts do not require these property-enhancing steps in their processing sequence. This is indicated by the alternate arrow path in our figure. ***Finishing operations*** are the final operations in the sequence; they usually provide a coating on the workpart (or assembly) surface. Examples of these processes are electroplating and painting.

In some cases, property-enhancing processes are followed by additional secondary operations before proceeding to finishing, as suggested by the return loop in Figure 41.2. One example is a machined part that is hardened by heat treatment. Prior to heat treatment, the part is left slightly oversized to allow for distortion. After hardening, it is reduced to final size and tolerance by finish grinding. Another example, again in metal parts fabrication, is when annealing is used to restore ductility to the metal after cold working to permit further deformation of the workpiece.

Table 41.2 presents some of the typical processing sequences for a variety of basic processes.

TABLE 41.2   Some typical process sequences.[a]

Basic Process	Secondary Process(es)	Property-Enhancing Processes	Finishing Operations
Sand casting	Machining	(none)	Painting
Die casting	(none, net shape)	(none)	Painting
Casting of glass	Pressing, blow molding	(none)	(none)
Injection molding	(none, net shape)	(none)	(none)
Rolling of bar stock	Machining	Heat treatment (optional)	Electroplating
Rolling of sheet metal	Blanking, bending, drawing	(none)	Electroplating
Forging	Machining (near net shape)	(none)	Painting
Extrusion of aluminum	Cut to length	(none)	Anodize
Atomize metal powders	Pressing of PM part	Sintering	Painting

[a] Compiled from [6].

The task of the process planner usually begins after the basic process has provided the initial shape of the part. Machined parts begin as bar stock or castings or forgings, and the basic processes for these starting shapes are often external to the fabricating plant. Stampings begin as sheet-metal coils or strips purchased from the mill. These are the raw materials supplied from external suppliers for the secondary processes and subsequent operations to be performed in the factory. Determining the most appropriate processes and the order in which they must be accomplished relies on the skill, experience, and judgment of the process planner. Some of the basic guidelines and considerations used by process planners to make these decisions are outlined in Table 41.3.

The Route Sheet The process plan is prepared on a form called a ***route sheet,*** a typical example of which is shown in Figure 41.3 (some companies use other names for this form). The route sheet is to the process planner what the engineering drawing is to the product designer. It is the official document that specifies the details of the process plan. The route sheet should include all manufacturing operations to be performed on the workpart, listed in the proper order in which they are to be accomplished. For each operation, the following should be listed: (1) a brief description of the operation indicating the work to be done, surfaces to be processed with references to the part drawing, and dimensions (and tolerances, if not specified on part drawing) to be achieved; (2)

TABLE 41.3 Guidelines and considerations in deciding processes and their sequence in process planning.

Design requirements—The sequence of processes must satisfy the dimensions, tolerances, surface finish, and other specifications established by product design.

Quality requirements—Processes must be selected that satisfy quality requirements in terms of tolerances, surface integrity, consistency and repeatability, and other quality measures.

Production volume and *rate*—The process must be capable of meeting the required production volume and rate. (Is the product in the category of low, medium, or high production?) The manufacturing processes and systems are strongly influenced by volume and rate of production.

Available processes—If the product and its components are to be made in-house, the process planner must select processes and equipment available in the factory wherever possible.

Material utilization—It is desirable for the process sequence to make efficient use of materials and minimize waste. When possible, ***net shape*** or ***near net shape*** processes (Section 1.3.1) should be selected.

Precedence constraints—These are technological sequencing requirements that determine or restrict the order in which the processing steps can be performed. Examples: a hole must be drilled before it can be tapped; a powder metal part must be pressed before sintering; a surface must be cleaned before painting; and so on.

Reference surfaces—Certain surfaces of the part must be formed (usually by machining) near the beginning of the sequence so they can serve as locating surfaces for other dimensions that are formed subsequently. Example: If a hole is to be drilled a certain distance from the edge of a given part, that edge must first be machined.

Minimize setups—The number of separate machine setups should be minimized. Wherever possible, operations should be combined at the same workstation. This saves time and reduces material handling. This guideline applies mostly to secondary operations such as machining.

Eliminate unnecessary steps—The process sequence should be planned with the minimum number of processing steps. Unnecessary operations should be avoided. Design changes should be requested to eliminate features not absolutely needed, thereby eliminating the processing steps associated with those features.

Flexibility—Where feasible, the process should be sufficiently flexible to accommodate engineering design changes. This is often a problem when special tooling must be designed to produce the part; if the part design is changed, the special tooling may be rendered obsolete.

Safety—Worker safety must be considered in process selection. This makes good economic sense, and it is the law (Occupational Safety and Health Act).

Minimum cost—The process sequence should be the production method that satisfies all of the above requirements and also achieves the lowest possible product cost.

Part No: 031393	Part Name: Housing, valve				Rev. 2	Page _1_ of _2_	
Matl: 416 Stainless		Size: 2.0 dia × 5. long			Planner: MPG	Date: 3/13/XX	
No.	Operation		Dept.	Machine	Tooling, gages	Setup time	Cycle time
10	Face; rough & finish turn to 1.473 ± 0.003 dia. × 1.250 ± 0.003 length; face shoulder to 0.313 ± 0.002; finish turn to 1.875 ± 0.002 dia.; form 3 grooves at 0.125 width × 0.063 deep.		L	325	G857	1.0 h	8.22 m
20	Reverse; face to 4.750 ± 0.005 length; finish turn to 1.875 ± 0.002 dia.; drill 1.000 + 0.006, −0.002 dia. axial hole.		L	325		0.5 h	3.10 m
30	Drill & ream 3 radial holes at 0.375 ± 0.002 dia.		D	114	F511	0.3 h	2.50 m
40	Mill 0.500 ± 0.004 wide × 0.375 ± 0.003 deep slot.		M	240	F332	0.3 h	1.75 m
50	Mill 0.750 ± 0.004 wide × 0.375 ± 0.003 deep flat.		M	240	F333	0.3 h	1.60 m

FIGURE 41.3　Typical route sheet for specifying the process plan.

the equipment on which the work is to be performed; and (3) any special tooling required, such as dies, molds, cutting tools, jigs or fixtures, and gages. In addition, some companies include cycle time standards, setup times, and other data on the route sheet.

Sometimes a more detailed *operation sheet* is also prepared for each operation listed in the routing. This is retained in the particular department where the operation is performed. It indicates the specific details of the operation, such as cutting speeds, feeds, and tools (if machining), and other instructions useful to the machine operator. Setup sketches are sometimes also included.

In addition to their main purpose, which is to specify the sequence and routing of processes performed on the workpart, route sheets may: (1) provide time standards for each operation, (2) facilitate estimation of production lead times, (3) provide estimates of product costs, (4) aid production scheduling and control, (5) indicate when inspection should be performed, and (6) indicate tools that must be ordered.

Process Planning for Assemblies　For low production, assembly is generally done at individual workstations and a worker or team of workers performs the assembly work elements to complete the product. In medium and high production, assembly is usually performed on production lines (Section 40.2). In either case, there is a precedence order in which the work must be accomplished.

Process planning for assembly involves preparation of the assembly instructions that must be performed. For single stations, the documentation is similar to the processing route sheet in Figure 41.3. It contains a list of the assembly steps in the order in which they must be accomplished. For assembly line production, process planning consists of allocating work elements to particular stations along the line, a procedure called *line balancing* (Section 40.2.1). In effect, the assembly line routes the work units to individual stations, and the line balancing solution determines what assembly steps must be performed at each station. As with process planning for individual parts, any tools and fixtures needed to accomplish a given assembly work element must be decided, and the workplace layout must be designed.

41.1.2 Make or Buy Decision

Inevitably, the question arises whether a given part should be purchased from an outside vendor or made internally. First of all, it should be recognized that virtually all manufacturers purchase their starting materials from suppliers. A machine shop buys bar stock from a metals distributor and castings from a foundry. A plastic molder obtains molding compound from a chemical company. A pressworking company purchases sheet metal from a rolling mill. Very few companies are vertically integrated all the way from raw materials to finished product.

Given that a company purchases some of its starting materials, is it not unreasonable to question whether the company should purchase the parts that would otherwise be made in its own factory. The answer to the question is the ***make or buy decision.*** The make versus buy question is probably appropriate to ask for every component used by the company.

Cost is the most important factor in deciding whether a part should be made in-house or purchased. If the vendor is significantly more proficient in the processes required to make the component, it is likely that the internal production cost will be greater than the purchase price even when a profit is included for the vendor. On the other hand, if purchasing the part results in idle equipment in the factory, then an apparent cost advantage for the vendor may be a disadvantage for the home factory. Consider the following example.

**EXAMPLE 41.1
Make or Buy Cost
Comparison**

Suppose that the quoted price for a certain component from a vendor is $8.00 per unit for 1000 units. The same part made in the home factory would cost $9.00. The cost breakdown on the make alternative is as follows:

$$\text{Unit material cost} = \$2.25 \text{ per unit}$$

$$\text{Direct labor} = \$2.00 \text{ per unit}$$

$$\text{Labor overhead at } 150\% = \$3.00 \text{ per unit}$$

$$\text{Equipment fixed cost} = \$1.75 \text{ per unit}$$

$$\text{Total} = \$9.00 \text{ per unit}$$

Should the component be bought or made in-house?

Solution: Although the vendor's quote seems to favor the buy decision, let us consider the possible effect on the factory if we decide to accept the quote. The equipment fixed cost is an allocated cost based on an investment that has already been made. If it turns out that the equipment is rendered idle by the decision to buy the part, then one might argue that the fixed cost of $1.75 continues even if the equipment is not in use. Similarly, the overhead cost of $3.00 consists of factory floor space, indirect labor, and other costs that will also continue even if the part is bought.

By this reasoning, the decision to purchase might cost the company as much as $8.00 + $1.75 + $3.00 = $12.75 per unit if it results in idle time in the factory on the machine that would have been used to make the part.

TABLE 41.4   Key factors in the make or buy decision.

Factor	Explanation and Effect on Make/Buy Decision
Process available in-house	If a given process is not available internally, then the obvious decision is to purchase. Vendors often develop proficiency in a limited set of processes that makes them cost competitive in external–internal comparisons. There are exceptions to this guideline, in which a company decides that, for its long-term survival, it must develop a proficiency in a manufacturing process technology that it does not currently possess.
Production quantity	Number of units required. High volume tends to favor make decisions. Low quantities tend to favor buy decisions.
Product life	Long product life favors internal production.
Standard items	Standard catalog items, such as bolts, screws, nuts, and many other types of components, are produced economically by suppliers specializing in those products. It is almost always better to purchase these standard items.
Supplier reliability	The reliable supplier gets the business.
Alternative source	In some cases, factories buy parts from vendors as an alternative source to their own production plants. This is an attempt to ensure an uninterrupted supply of parts, or to smooth production in peak demand periods.

On the other hand, if the equipment can be used to produce other components for which the internal prices are less than the corresponding external quotes, then a buy decision makes good economic sense.

Make or buy decisions are rarely as clear as in Example 41.1. Some of the other factors that enter the decision are listed in Table 41.4. Although these factors appear to be subjective, they all have cost implications, either directly or indirectly. In recent years, major companies have placed strong emphasis on building close relationships with parts suppliers. This trend has been especially prevalent in the automobile industry, where long-term agreements have been reached between each car maker and a limited number of vendors who are able to deliver high-quality components reliably on schedule.

41.1.3 Computer-Aided Process Planning

During the last several decades, there has been considerable interest in *computer-aided process planning* (CAPP)—automating the process planning function by means of computer systems. Shop people knowledgeable in manufacturing processes are gradually retiring. An alternative approach to process planning is needed, and CAPP systems provide this alternative. Computer-aided process planning systems are designed around either of two approaches: (1) retrieval systems or (2) generative systems.

Retrieval CAPP Systems   *Retrieval CAPP systems,* also known as *variant CAPP systems,* are based on group technology and parts classification and coding (Section 41.1). In these systems, a standard process plan is stored in computer files for each part code number. The standard plans are based on current part routings in use in the factory, or on an ideal plan that is prepared for each family. Retrieval CAPP systems operate as indicated in Figure 41.4. The user begins by identifying the GT code for the component for which the process plan is to be determined. A search is made of the part family file to determine if a standard route sheet exists for the given part code. If the file contains

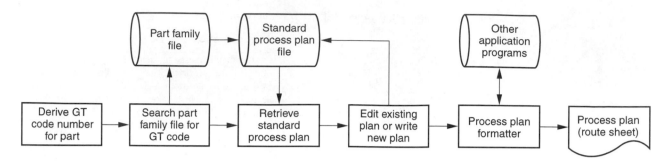

FIGURE 41.4   Operation of a retrieval computer-aided process planning system (source: [6]).

a process plan for the part, it is retrieved and displayed for the user. The standard process plan is examined to determine whether modifications are necessary. Although the new part has the same code number, minor differences in the processes might be required to make the part. The standard plan is edited accordingly. The capacity to alter an existing process plan is why retrieval CAPP systems are also called variant systems.

If the file does not contain a standard process plan for the given code number, the user may search the file for a similar code number for which a standard routing exists. By editing the existing process plan, or by starting from scratch, the user develops the process plan for the new part. This becomes the standard process plan for the new part code number.

The final step is the process plan formatter, which prints the route sheet in the proper format. The formatter may call other application programs: determining cutting conditions for machine tool operations, calculating standard times for machining operations, or computing cost estimates.

Generative CAPP Systems   Generative CAPP systems are an alternative to retrieval systems. Rather than retrieving and editing existing plans from a database, a generative system creates the process plan using systematic procedures that might be applied by a human planner. In a fully generative CAPP system, the process sequence is planned without human assistance and without predefined standard plans.

Designing a generative CAPP system is a problem in the field of expert systems, a branch of artificial intelligence. ***Expert systems*** are computer programs capable of solving complex problems that normally require a human who has years of education and experience. Process planning fits that definition. Several ingredients are required in a fully generative CAPP system:

1. ***Knowledge base***—The technical knowledge of manufacturing and the logic used by successful process planners must be captured and coded into a computer program. An expert system applied to process planning requires the knowledge and logic of human process planners to be incorporated into a knowledge base. Generative CAPP systems then use the knowledge base to solve process planning problems, that is, to create route sheets.

2. ***Computer-compatible part description***—Generative process planning requires a computer-compatible description of the part. The description contains all the pertinent data needed to plan the process sequence. Two possible descriptions are (1) the geometric model of the part developed on a CAD system during product design, or (2) a group technology code number of the part defining its features in significant detail.

3. *Inference engine*—A generative CAPP system requires the capability to apply the planning logic and process knowledge contained in the knowledge base to a given part description. The CAPP system applies its knowledge base to solve a specific problem of planning the process for a new part. This problem-solving procedure is referred to as the "inference engine" in the terminology of expert systems. By using its knowledge base and inference engine, the CAPP system synthesizes a new process plan for each new part presented to it.

Benefits of CAPP   Benefits of computer-automated process planning include the following: (1) process rationalization and standardization—automated process planning leads to more logical and consistent process plans than when traditional process planning is used; (2) increased productivity of process planners—the systematic approach and availability of standard process plans in the data files permits a greater number of process plans to be developed by the user; (3) reduced lead time to prepare process plans; (4) improved legibility compared to manually prepared route sheets; and (5) ability of CAPP programs to be interfaced with other application programs, such as cost estimating, work standards, and others.

41.2 PROBLEM SOLVING AND CONTINUOUS IMPROVEMENT

Problems arise in manufacturing that require technical staff support beyond what is normally available in the line organization of the production departments. Providing this technical support is one of the responsibilities of manufacturing engineering. The problems are usually specific to the particular technologies of the processes performed in the operating department. In machining, the problems may relate to selection of cutting tools, fixtures that do not work properly, parts with out-of-tolerance conditions, or nonoptimal cutting conditions. In plastic molding, the problems may be excessive flash, parts sticking in the mold, or any of several defects that can occur in a molded part. These problems are technical, and engineering expertise is often required to solve them.

In some cases, the solution may require a design change; for example, changing the tolerance on a part dimension to eliminate a finish grinding operation while still achieving the functionality of the part. The manufacturing engineer is responsible for developing the proper solution to the problem and proposing the engineering change to the design department.

One of the areas that is ripe for improvement is setup time. The procedures involved in changing over from one production setup to the next (i.e., in batch production) are time consuming and costly. Manufacturing engineers are responsible for analyzing changeover procedures and finding ways to reduce the time required to perform them. Some of the approaches used in setup reduction are described in Section 42.4.

In addition to solving current technical problems ("fire fighting," as it might be called), the manufacturing engineering department is also responsible for continuous improvement projects. Called *kaizen* by the Japanese, continuous improvement means constantly searching for and implementing ways to reduce cost, improve quality, and increase productivity in manufacturing. It is accomplished one project at a time. Depending on the type of problem area, it may involve a project team whose membership includes not only manufacturing engineering, but other departments such as product design, quality engineering, and production control. The projects deal with: (1) cost reduction, (2) quality improvement, (3) productivity improvement, (4) setup time reduction, (5) cycle time reduction, (6) manufacturing lead time reduction, and (7) improvement of product design to increase performance and customer appeal.

41.3 CONCURRENT ENGINEERING AND DESIGN FOR MANUFACTURABILITY

Much of the process planning function described in Section 41.1 is preempted by decisions made in product design. Decisions on material, part geometry, tolerances, surface finish, grouping of parts into subassemblies, and assembly techniques limit the number of manufacturing processes that can be used to make a given part. If the product engineer designs an aluminum sand casting with features that can be achieved only by machining (e.g., flat surfaces with good finishes, close tolerances, and threaded holes), then the process planner has no alternative but to specify sand casting, followed by the necessary sequence of machining operations. If the product designer specifies a collection of sheet-metal stampings to be assembled by threaded fasteners, then the process planner must lay out the series of blanking, punching, and forming steps to fabricate the stampings and then assemble them. In both of these examples, a plastic molded part may be a superior design, both functionally and economically. It is important for the manufacturing engineer to act as an advisor to the design engineer in matters of manufacturability because manufacturability matters, not only to the production departments but to the design engineer. A product design that is functionally superior and at the same time can be produced at minimum cost holds the greatest promise of success in the marketplace. Successful careers in design engineering are built on successful products.

Terms often associated with this attempt to favorably influence the manufacturability of a product are *design for manufacturing* (DFM) and *design for assembly* (DFA). Of course, DFM and DFA are inextricably coupled, so let us refer to them as DFM/A. The scope of DFM/A is expanded in some companies to include not only manufacturability issues but also marketability, testability, serviceability, maintainability, and so forth. This broader view calls for inputs from many departments in addition to design and manufacturing engineering. The approach is called *concurrent engineering.* Our discussion is organized into two sections: DFM/A and concurrent engineering. DFM/A is a subset of concurrent engineering.

41.3.1 Design for Manufacturing and Assembly

Design for manufacturing and assembly is an approach to product design that systematically includes considerations of manufacturability and assemblability in the design. DFM/A includes organizational changes and design principles and guidelines.

Organizational Changes in DFM/A   To implement DFM/A, a company must change its organizational structure, either formally or informally, to provide closer interaction and better communication between design and manufacturing personnel. This is often accomplished by forming project teams consisting of product designers, manufacturing engineers, and other specialties (e.g., quality engineers, material scientists) to design the product. In some companies, design engineers are required to spend some career time in manufacturing to learn about the problems encountered in making things. Another possibility is to assign manufacturing engineers to the product design department as full-time consultants.

Design Principles and Guidelines   DFM/A also includes principles and guidelines that indicate how to design a given product for maximum manufacturability. Many of these are universal design guidelines, such as those presented in Table 41.5. They are rules of thumb that can be applied to nearly any product design situation. In addition, several

TABLE 41.5   General principles and guidelines in design for manufacturing and assembly.

Guideline	Interpretation and Advantages
Minimize number of components.	Reduce assembly costs. Greater reliability in final product. There is easier disassembly in maintenance and field service. Automation is often easier with reduced part count. Reduced work-in-process and inventory control problems; fewer parts to purchase and reduced ordering costs.
Use standard commercially available components.	Reduced design effort. Avoid design of custom-engineered components. There are fewer part numbers. Better inventory control is possible. Quantity discounts are possible.
Use common parts across product lines.	Group technology (Chapter 39) can be applied. Permits development of manufacturing cells. Quantity discounts are possible.
Design for ease of part fabrication	Use net shape and near net shape processes where possible. Simplify part geometry; avoid unnecessary features. Avoid surface finish that is smoother than necessary because additional processing may be needed.
Design parts with tolerances that are within process capability.	Avoid tolerances less than process capability (Section 43.2). Otherwise, additional processing or sortation are required. Specify bilateral tolerances.
Design the product to be foolproof during assembly.	Assembly should be unambiguous. Design components so they can be assembled only one way. Special geometric features must sometimes be added to components.
Minimize flexible components.	These include components made of rubber, belts, gaskets, electrical cables, etc. Flexible components are generally more difficult to handle.
Design for ease of assembly.	Include part features such as chamfers and tapers on mating parts. Use base part to which other components are added. Design assembly for addition of components from one direction, usually vertically. Avoid threaded fasteners (screws, bolts, nuts) where possible, especially when automated assembly is used; use fast assembly techniques such as snap fits and adhesive bonding. Minimize number of distinct fasteners.
Use modular design.	Each subassembly should consists of five to fifteen parts. Easier maintenance and field service. Facilitates automated (and manual) assembly. Reduces inventory requirements. Reduces final assembly time.
Shape parts and products for ease of packaging.	Can use standard packaging cartons, which are compatible with automated packaging equipment. Shipment to customer is facilitated.
Eliminate or reduce adjustment required.	Product design should minimize the number of adjustments needed, since adjustments are time consuming in assembly.

Compiled from [1], [2], [10].

of our chapters on manufacturing processes include DFM/A principles that are specific to those processes.

The guidelines are sometimes in conflict. For example, one guideline for part design is to make the geometry as simple as possible. Yet, in design for assembly, additional part features are sometimes desirable to avoid incorrect mating of components; and it is also desirable to combine features of several assembled components into a single part to reduce part count and assembly time. In these instances, design for part manufacture conflicts with design for assembly, and a compromise must be found that achieves the best balance between the opposing sides of the conflict.

Other guidelines are specific to a given company because of its particular manufacturing capabilities relative to its competitors (Section 1.1.3). These distinguishing technological capabilities are the sum of the company's available facilities and manufacturing processes, technical competence of its engineering staff, and skill of its labor force. This means that if the company has an excellent design team in a certain product line, this excellence should be exploited in the company's product development strategy. It means that the company should design parts that utilize its factory's available manufacturing processes. It means that if the company's technical people are especially good at devising automation hardware, then this specialty should be exploited in its overall manufacturing strategy. Distinguishing technological competence in manufacturing is often more advantageous than distinctive capability in product design. Competitors can reverse engineer a product just introduced in the marketplace to learn secrets that required much effort to develop. Processing secrets are often more difficult to discover.

Benefits typically cited for DFM/A include [1], [2]: (1) shorter time to bring the product to market, (2) smoother transition into production, (3) fewer components in the final product, (4) easier assembly, (5) lower costs of production, (6) higher product quality, and (7) greater customer satisfaction.

41.3.2 Concurrent Engineering

Concurrent engineering refers to an approach to product design in which companies attempt to reduce the elapsed time required to bring a new product to market by integrating design engineering, manufacturing engineering, and other functions in the company. The traditional approach to launch a new product tends to separate the two functions, as illustrated in Figure 41.5(a). Product design develops the new design, sometimes with small regard for the manufacturing capabilities possessed by the company. There is little interaction between design engineers and manufacturing engineers who might provide advice on these capabilities and how the product design might be altered to accommodate them. It is as if a wall exists between the two functions; when design engineering completes the design, it tosses the drawings and specifications over the wall, so that process planning can commence.

In a company that practices concurrent engineering (also known as *simultaneous engineering*), manufacturing planning begins while the product design is being developed, as pictured in Figure 41.5(b). Manufacturing engineering becomes involved early in the product development cycle. In addition, other functions are also involved, such as field service, quality engineering, the manufacturing departments, vendors supplying critical components, and in some cases customers who will use the product. All of these functions can contribute to a product design that not only performs well functionally, but is also manufacturable, assembleable, inspectable, testable, serviceable, maintainable, free of defects, and safe. All viewpoints have been combined to design a product of high quality that will deliver customer satisfaction. And through early involvement, rather than a procedure of reviewing the final design and suggesting changes after it is too late to conveniently make them, the total product development cycle is substantially reduced.

Concurrent engineering consists of several ingredients: (1) design for manufacturing and assembly, (2) design for quality, (3) design for life cycle, and (4) design for cost. In addition, certain enabling technologies such as rapid prototyping (Chapter 34) are re-

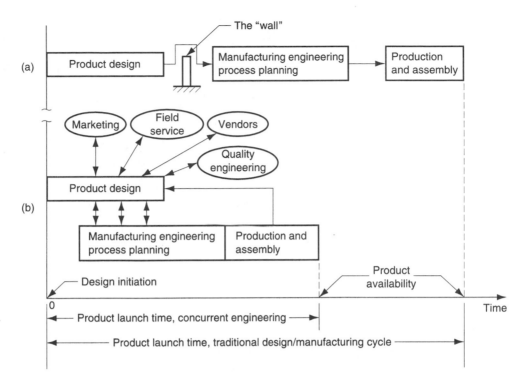

FIGURE 41.5
Comparison of (a) traditional product development cycle, and (b) product development using concurrent engineering.

quired to facilitate these approaches in the company. Also, continuous improvement (Section 41.2) is considered an important component in concurrent engineering.

One might argue that ***design for manufacturing and assembly*** (Section 41.3.1) is the most important aspect of concurrent engineering because it has the greatest impact on production costs and product development time. However, with the growing importance of quality in international competition, and the demonstrated success of those countries and companies that have been able to produce products of high quality, one must conclude that ***design for quality*** (DFQ) is also very important. Chapter 43 is devoted to the topic of quality control and includes a discussion of quality in product design.

Design for life cycle refers to the product after it has been manufactured. In many cases, a product can involve a significant cost to the customer beyond the purchase price. These costs include installation, maintenance and repair, spare parts, future upgrading of the product, safety during operation, and disposition of the product at the end of its useful life. Table 41.6 lists most of the factors associated with life-cycle design of the product.

To the customer, the price paid for the product may be a small portion of its total cost when life-cycle costs are included. Some customers (e.g., federal government) consider life-cycle costs in their purchasing decisions. The manufacturer must often include service contracts that limit customer vulnerability to out-of-control maintenance and service costs. In these cases, accurate estimates of these life-cycle costs must be included in the total product cost.

A product's cost is a major factor in determining its commercial success. Cost affects the price charged for the product and the profit made on it. ***Design for product cost*** refers to the efforts of a company to identify the impact of design decisions on overall product costs and to control those costs through optimal design. Many of the DFM/A guidelines are directed at reducing product cost. It is often useful for a company to develop a product cost model to predict how design alternatives might affect costs of materials, manufacturing, and inspection.

TABLE 41.6 Factors in design for life cycle.

Factor	Typical Issues and Concerns
Delivery	Transport cost, time to deliver, storage and distribution of mass produced items, type of carrier required (truck, railway, air transport).
Installability	Utility requirements (electric power, air pressure, etc.), construction costs, field assembly, support during installation.
Reliability	Service life of product, failure rate, reliability testing requirements, materials used in the product, tolerances.
Maintainability	Design modularity, types of fasteners used in assembly, preventive maintenance requirments, ease of servicing by customer.
Serviceability	Product complexity, diagnostics techniques, training of field service staff, access to internal workings of product, tools required, availability of spare parts.
Human factors	Ease and convenience of use, complexity of controls, potential hazards, risk of injuries during operation.
Upgradeability	Compatibility of current design with future modules and software, cost of upgrades.
Disposability	Materials used in the product, recycling of components, waste hazards.

REFERENCES

[1] Bakerjian, R., and Mitchell, P. *Tool and Manufacturing Engineers Handbook.* 4th. ed. Volume VI: *Design for Manufacturability.* Society of Manufacturing Engineers, Dearborn, Mich., 1992.

[2] Corbett, J., Dooner, M., Meleka, J., and Pym, C. *Design for Manufacture.* Addison-Wesley, Wokingham, England, 1991.

[3] Chang, T-C, Wysk, R. A., and Wang, H-P. *Computer-Aided Manufacturing.* 2nd. ed. Prentice Hall, Upper Saddle River, N.J., 1997.

[4] Eary, D. F., and Johnson, G. E. *Process Engineering for Manufacturing.* Prentice-Hall, Inc., Englewood Cliffs, N.J., 1962.

[5] Groover, M. P., and Zimmers, E. W., Jr. *CAD/CAM: Computer-Aided Design and Manufacturing.* Prentice Hall, Inc., Englewood Cliffs, N.J., 1984.

[6] Groover, M. P. *Automation, Production Systems, and Computer Integrated Manufacturing.* 2nd. ed. Prentice Hall, Upper Saddle River, N.J., 2001.

[7] Kane, G. E. "The Role of the Manufacturing Engineer." *Technical Paper MM70-222.* Society of Manufacturing Engineers, Dearborn, Mich., 1970.

[8] Koenig, D. T. *Manufacturing Engineering.* Hemisphere Publishing Corporation (Harper & Row, Publishers, Inc.), Washington, D.C., 1987.

[9] Kusiak, A. (ed.). *Concurrent Engineering.* Wiley, New York, 1993.

[10] Martin, J. M. "The Final Piece of the Puzzle." *Manufacturing Engineering.* September 1988, pp. 46–51.

[11] Nevins, J. L., and Whitney, D. E. (eds.). *Concurrent Design of Products and Processes.* McGraw-Hill, New York, 1989.

[12] Tanner, J. P. *Manufacturing Engineering.* Marcel Dekker, Inc., New York, 1985.

[13] Tompkins, J. A., White, J. A., Bozer, Y. A., Frazelle, E. H., Tanchoco, J. M. A., and Trevino, J. *Facilities Planning.* 2nd. ed. Wiley, New York, 1996.

[14] Usher, J. M., Roy, U., and Parsaei, H. R. (eds.). *Integrated Product and Process Development.* Wiley, New York, 1998.

[15] Veilleux, R. F., and Petro, L. W. *Tool and Manufacturing Engineers Handbook.* 4th. ed. Vol. V: *Manufacturing Management.* Society of Manufacturing Engineers, Dearborn, Mich., 1988.

REVIEW QUESTIONS

41.1. Define *manufacturing engineering.*

41.2. What are the principal activities in manufacturing engineering?

41.3. Identify some of the details and decisions that are included within the scope of process planning.

41.4. What is a *route sheet?*

41.5. What is the difference between a *basic process* and a *secondary process?*

41.6. What is a *precedence constraint* in process planning?

41.7. In the make or buy decision, why is it that purchasing a component from a vendor may cost more than producing the component internally, even though the quoted price from the vendor is lower than the internal price?

41.8. Identify some of the important factors that should enter into the make or buy decision.

41.9. Name three of the general principles and guidelines in design for manufacturability.

41.10. What is *concurrent engineering,* and what are its important components?

41.11. Identify some of the enabling technologies for concurrent engineering.

41.12. What is meant by the term *design for life cycle?*

MULTIPLE CHOICE QUIZ

There is a total of 18 correct answers in the following multiple choice questions (some questions have multiple answers that are correct). To attain a perfect score on the quiz, all correct answers must be given, since each correct answer is worth 1 point. For each question, each omitted answer or wrong answer reduces the score by 1 point, and each additional answer beyond the number of answers required reduces the score by 1 point. Percentage score on the quiz is based on the total number of correct answers.

41.1. Which of the following are the usual responsibilities of the manufacturing engineering department (more than one)? (a) advising on design for manufacturability, (b) facilities planning, (c) process improvement, (d) process planning, (e) product design, and (f) solving technical problems in the production departments.

41.2. Which of the following would be considered basic processes, as opposed to secondary processes (more than one)? (a) annealing, (b) anodizing, (c) drilling with a twist drill, (d) electroplating, (e) forward hot extrusion to produce aluminum bars, (f) impression die forging, (g) rolling, (h) sand casting, (i) sheet metal stamping, (j) sintering of pressed ceramic powders, (k) spot welding, (l) surface grinding of hardened steel, (m) tempering of martensitic steel, (n) trepanning, (o) turning, and (p) ultrasonic machining.

41.3. Which of the following would be considered secondary processes, as opposed to basic processes (more than one)? (a) annealing, (b) anodizing, (c) drilling with a twist drill, (d) electroplating, (e) forward hot

extrusion to produce aluminum bars, (f) impression die forging, (g) rolling, (h) sand casting, (i) sheet metal stamping, (j) sintering of pressed ceramic powders, (k) spot welding, (l) surface grinding of hardened steel, (m) tempering of martensitic steel, (n) trepanning, (o) turning, and (p) ultrasonic machining.

41.4. Which of the following are operations to enhance physical properties (more than one)? (a) annealing, (b) anodizing, (c) drilling with a twist drill, (d) electroplating, (e) forward hot extrusion to produce aluminum bars, (f) impression die forging, (g) rolling, (h) sand casting, (i) sheet metal stamping, (j) sintering of pressed ceramic powders, (k) spot welding, (l) surface grinding of hardened steel, (m) tempering of martensitic steel, (n) trepanning, (o) turning, and (p) ultrasonic machining.

41.5. Which one of the following types of computer-aided process planning relies on parts classification and coding in group technology? (a) generative CAPP, (b) retrieval CAPP, (c) traditional process planning, or (d) none of the preceding.

PRODUCTION PLANNING AND CONTROL

CHAPTER CONTENTS

42.1   Aggregate Planning and the Master Production Schedule
42.2   Inventory Control
    42.2.1   Types of Inventory
    42.2.2   Order Point Systems
42.3   Material and Capacity Requirements Planning
    42.3.1   Material Requirements Planning
    42.3.2   Capacity Requirements Planning
42.4   Just-In-Time and Lean Production
42.5   Shop Floor Control

Production planning and control are the manufacturing support functions concerned with logistics problems in manufacturing. *Production planning* is concerned with planning what products are to be produced, in what quantities, and when. It also considers the resources required to accomplish the plan. *Production control* determines whether the resources to execute the plan have been provided and, if not, takes the necessary action to correct the deficiency. The scope of production planning and control includes *inventory control,* which deals with the problem of having appropriate stock levels available of raw materials, work-in-process, and finished goods.

Problems in production planning and control differ for different types of manufacturing. One of the important factors is the relationship between product variety and production quantity (Section 1.1.2). At one extreme is *job shop production,* in which many different product types are each produced in low quantities. The products are often complex, consisting of many components, each of which must be processed through multiple operations. Solving the logistics problems in such a plant requires detailed planning—scheduling and coordinating the large numbers of different components and processing steps for so many different products.

At the other extreme is *mass production,* in which a single product (with perhaps some limited model variations) is produced in very large quantities (millions of units). The logistics problems in mass production are simple if the product and process are simple. In more complex cases, the product is an assembly consisting of many components (e.g., automobiles or household appliances), and the facility is organized as a pro-

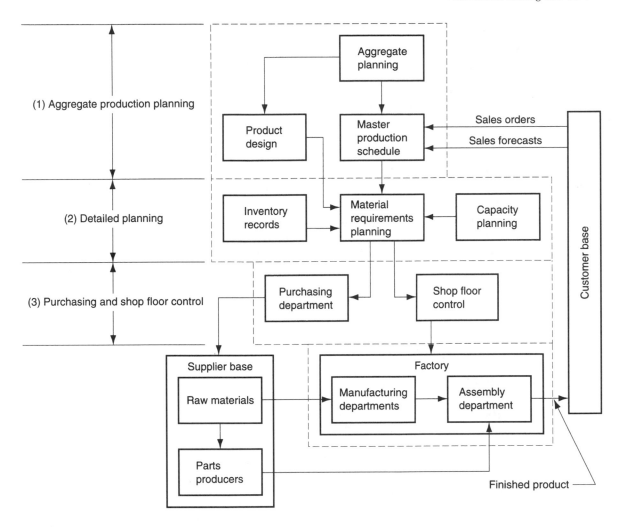

FIGURE 42.1   Activities in a production planning and control system.

duction line (Chapter 40). The logistics problem in operating such a plant is to get each component to the right workstation at the right time so that it can be assembled to the product as it passes through that station. Failure to solve this problem can result in stoppage of the entire production line for lack of a critical part.

To distinguish between these two extremes in terms of the issues in production planning and control, we can say that the planning function is emphasized in a job shop, whereas the control function is emphasized in the mass production of assembled products. There are many variations between these extremes, with accompanying differences in the way production planning and control are implemented.

Figure 42.1 presents a block diagram depicting the activities of a modern production planning and control system and their interrelationships. The activities can be divided into three phases: (1) aggregate production planning, (2) detailed planning of material and capacity requirements, and (3) purchasing and shop floor control. Our discussion of production planning and control in this chapter is organized around this framework.

42.1 AGGREGATE PLANNING AND THE MASTER PRODUCTION SCHEDULE

Any manufacturing firm must have a business plan, and the plan must include what products will be produced, how many, and when. The manufacturing plan should take into account current orders and sales forecasts, inventory levels, and plant capacity considerations. Different types of manufacturing plans are prepared. One difference is in terms of planning horizon; there are (1) *long-range plans* that deal with a time horizon that is more than one year in the future; (2) *medium-range plans* that are concerned with the period six months to one year in the future; and (3) *short-range plans* that consider near future horizons such as days or weeks.

Long-range planning is the responsibility of the highest level executives of the company. It is concerned with corporate goals and strategies, future product lines, financial planning for the future, and obtaining the resources (personnel, facilities, and equipment) necessary so that the company will have a future. As the planning horizon is reduced, the company's long-range plan must be translated into medium-range and short-range plans that become increasingly specific. At the medium-range level are the aggregate production plan and the master production schedule, examined in this section. In the short-range are material and capacity requirements planning, and detailed scheduling of the orders.

The *aggregate production plan* indicates production output levels for major product lines rather than specific products. It must be coordinated with the sales and marketing plan of the company and must consider current inventory levels. Aggregate planning is therefore a high-level corporate planning activity, although details of the planning process are delegated to staff. The aggregate plan must reconcile the marketing plans for current products and new products under development against the capacity resources available to make those products.

The planned output levels of the major product lines listed in the aggregate schedule must be converted into a very specific schedule of individual products. This is called the *master production schedule,* and it lists the products to be manufactured, when they should be completed, and in what quantities. A hypothetical master schedule is presented in Table 42.1(b) for a limited product set, with the corresponding aggregate plan for the product line in Table 42.1(a).

Products listed in the master schedule generally divide into three categories: (1) firm customer orders, (2) forecasted demand, and (3) spare parts. Customer orders for specific products usually obligate the company to a delivery date that has been promised to a customer by the sales department. The second category consists of production output levels based on forecasted demand, in which statistical forecasting techniques are applied to previous demand patterns, estimates by the sales staff, and other sources. The forecast often dominates the master schedule. The third category is demand for individual component parts—repair parts to be stocked in the firm's service department. Some companies exclude this third category from the master schedule because it does not represent end products.

The master production schedule is a medium-range plan because it must consider the lead times required to order raw materials and components, fabricate parts in the factory, and then assemble and test the final products. Depending on type of product, these lead times can run from several months to more than a year. However, although it deals with the midterm horizon, it is a dynamic plan. It is usually considered to be fixed in the near term, meaning that changes are disallowed within about a six-week horizon. However, adjustments in the schedule are possible beyond six weeks to deal with shifts in demand or new product opportunities. It should therefore be noted that the aggregate

TABLE 42.1   (a) Aggregate production plan, and (b) corresponding master production schedule for a hypothetical product line.

(a)	Week									
Product line	1	2	3	4	5	6	7	8	9	10
P models	—	—	—	—	—	—	—	50	150	250
Q models	400	400	400	300	300	300	300	250	250	250
R models	100	100	150	150	200	200	200	250	300	350

(b)	Week									
Product	1	2	3	4	5	6	7	8	9	10
Model P1								50	75	100
Model P2									50	50
Model P3									25	50
Model P4										50
Model Q1	200	200	200	100	100	100	100	50	50	50
Model Q2	200	200	200	200	200	200	200	200	200	200
(etc.)										

production plan is not the only input to the master schedule. Other drivers that may cause it to deviate from the aggregate plan include new customer orders and changes in sales forecast over the near term.

42.2 INVENTORY CONTROL

Inventory control is concerned with achieving a balance between two competing objectives: (1) minimizing the cost of maintaining inventory and (2) maximizing service to customers. Inventory costs include investment costs, storage costs, and cost of possible obsolescence or spoilage. Investment cost is often the dominant factor; a typical case is when the company invests borrowed money at a certain interest rate in materials that have not yet been delivered to the customer. All of these costs are referred to as ***carrying costs.*** The company can minimize carrying cost by maintaining zero inventories. However, customer service is likely to suffer, and customers may decide to take their business elsewhere. This has a cost, referred to as ***stock-out cost.*** A reasonable company wants to minimize stock-out cost and provide a high level of customer service. Customers are of two kinds: (1) external customers, the kind we usually associate with the word; and (2) internal customers, which are the operating departments, final assembly, and other units in the company that depend on the ready availability of materials and parts.

42.2.1 Types of Inventory

Various types of inventory are encountered in manufacturing. The categories of greatest interest in production planning and control are raw materials, purchased components, in-process inventory (work-in-process), and finished products.

Different inventory control procedures are appropriate, according to which type we are attempting to manage. An important distinction is between items subject to independent versus dependent demand. ***Independent demand*** means that the demand or consumption of the item is unrelated to demand for other items. End products and spare parts experience independent demand. Customers purchase end products and spare parts, and their decisions to do so are unrelated to the purchase of other items.

Dependent demand refers to the fact that demand for the item is directly related to demand for something else, usually because the item is a component of an end product subject to independent demand. Consider an automobile—an end product, the demand for which is independent. Each car has four tires (five, if we include the spare), whose demand depends on the demand for the automobile. Thus, the tires used on new automobiles are examples of dependent demand. For every car made in the final assembly plant, four tires must be ordered. The same is true of the thousands of other components used on an automobile. Once the decision is made to produce a car, all of these components must be supplied to build it.

Tires represent an interesting example because they not only experience dependent demand in the new car business, but also independent demand in the replacement tire market.

Different production and inventory control procedures must be used for independent and dependent demand. Forecasting procedures are commonly used to determine future production levels of independent demand products. Production of the components used on these products is then determined directly from the product quantities to be made. Two different inventory control systems are required for the two cases: (1) order point systems, and (2) material requirements planning. Order point systems are covered in the following section. Material requirements planning is covered in Section 42.3.1.

42.2.2 Order Point Systems

Order point systems address two related issues encountered when controlling inventories of independent demand items: how much to order, and when to order. The first issue, determining how many units to order, is often decided by means of economic order quantity formulas. The second issue, determining when to order, can be accomplished using reorder points.

Economic Order Quantity The problem of determining the appropriate quantity that should be ordered or produced arises in cases of independent demand products, in which demand for the item is relatively constant during the period under consideration and the production rate is significantly greater than demand rate. This is the typical ***make-to-stock*** situation. A similar problem is encountered in some dependent demand situations, when usage of the components in the final product is fairly steady over time and it makes sense to pay some inventory carrying costs in order to reduce the frequency of setups. In both of these situations, the inventory level is gradually depleted over time and then quickly replenished to some maximum level determined by the order quantity, as depicted in Figure 42.2.

One can derive a total cost equation for the sum of carrying cost and setup cost for the inventory model in Figure 42.2. The model has a sawtooth appearance, representing the gradual consumption of product down to a zero, followed by immediate replenishment up to a maximum level Q. Based on this behavior, the average inventory level is one-half the maximum level Q. The total annual inventory cost equation is

$$TIC = \frac{C_h Q}{2} + \frac{C_{su} D_a}{Q} \tag{42.1}$$

where TIC = total annual inventory cost (carrying plus ordering costs), \$/yr; Q = order quantity, pc/order; C_h = holding cost (cost of carrying the inventory), \$/pc/yr; C_{su} = cost

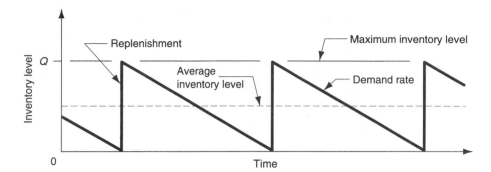

FIGURE 42.2   Model of inventory level over time in the typical make-to-stock situation.

of setting up for an order, \$/setup or \$/order; and D_a = annual demand for the item, pc/yr. In the equation, the ratio D_a/Q = the number of orders (batches of parts produced) per year; it therefore gives the number of setups per year.

The carrying cost C_h is generally taken to be directly proportional to the value of the item; that is,

$$C_h = hC_p \tag{42.2}$$

where C_p = cost per piece, \$/unit; and h = annual holding cost rate, which includes interest and storage charges, $(\text{yr})^{-1}$.

The setup cost C_{su} includes the cost of idle production equipment during the changeover time between batches, as well as whatever labor costs are involved in the setup changes. Thus,

$$C_{su} = T_{su}C_{dt} \tag{42.3}$$

where T_{su} = setup or changeover time between batches, hr; and C_{dt} = cost rate of machine downtime, \$/hr. In cases where parts are ordered from an outside vendor, the price quoted by the vendor usually includes a setup cost, either directly or in the form of quantity discounts. C_{su} should also include the internal costs of placing the order to the vendor.

It should be noted that Eq. (42.1) excludes the actual annual cost of part production, which is D_aC_p. If this cost is included, then annual total cost is given by

$$TC = D_aC_p + \frac{C_hQ}{2} + \frac{C_{su}D_a}{Q} \tag{42.4}$$

By taking the derivative of either Eq. (42.1) or Eq. (42.4), we obtain the economic order quantity (EOQ) formula that minimizes the sum of carrying costs and setup costs:

$$EOQ = \sqrt{\frac{2D_aC_{su}}{C_h}} \tag{42.5}$$

where EOQ = economic order quantity (number of parts that should be produced in the batch), pc; and the other terms are previously defined.

EXAMPLE 42.1
Economic Order Quantity

A certain product is made to stock. Annual demand rate is 12,000 units. One unit of product costs \$10 and the holding cost rate = 24%/yr. Setting up to produce a batch of products requires changeover of equipment, which takes 4 hr. The cost of equipment downtime plus labor = \$100/hr. Determine the economic order quantity and the total inventory costs for this case.

Solution: Setup cost $C_{su} = 4 \times \$100 = \$400$. Holding cost per unit $= 0.24 \times \$10 = \$2.40$. Using these values and the annual demand rate in the *EOQ* formula, we have

$$EOQ = \sqrt{\frac{2(12,000)(400)}{2.40}} = 2,000 \text{ units}$$

Total inventory costs are given by the *TIC* equation:

$$TIC = 0.5(2.40)(2000) + 400(12,000/2000) = \$4800.$$

Including the actual production costs in the annual total costs, by Eq. (42.4) we have

$$TC = 12,000(10) + 4800 = \$124,800.$$

The *EOQ* formula has been a widely used model for deciding "optimum production runs." There are variations of Eqs. (42.1) and (42.4) that take into account additional factors such as rate of production and rate of demand. While the mathematical accuracy of the formula cannot be disputed, it is instructional to note some of the difficulties encountered in its application. One difficulty is concerned with the values of the parameters in the equation, namely setup or ordering cost and inventory carrying costs. These costs are often difficult to evaluate, yet they have an important effect on the calculated *EOQ* value.

A second difficulty is concerned with an erroneous tenet of manufacturing philosophy that has been promulgated by the use of the *EOQ* formula in the United States. It is that long production runs represent an optimum strategy in batch manufacturing. No matter how much it costs to change the setup, the formula gives the optimum production batch size. The higher the setup cost, the longer the production run. In stark contrast with this U.S. approach was the Japanese approach to the problem, which was to develop ways to reduce the setup cost by dramatically reducing the time required to accomplish a changeover. Instead of taking hours to complete a changeover, setup time was reduced to minutes in some factories. Japanese successes in this area have motivated similar efforts to reduce changeover times in U.S. companies. Setup reduction is an important component of just-in-time production, and we consider some of the approaches used to reduce setup time in Section 42.4.

When to Reorder   Determining when to reorder can be accomplished in several ways. Here we describe the reorder point system that is widely used in industry. Refer to Figure 42.3, which provides a more realistic view than Figure 42.2 of the likely variations in demand rate that occur. In a ***reorder point system,*** when the inventory level for a

FIGURE 42.3   Operation of a reorder point inventory system.

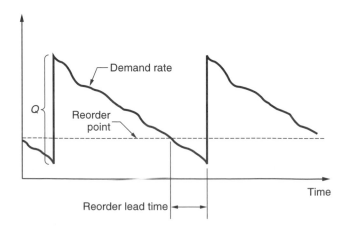

given stock item declines to some point defined as the reorder point, this is the signal to place an order to restock the item. The reorder point is set at a high enough level so as to minimize the probability that a stock-out will occur during the period between when the reorder point is reached and a new batch is received.

Reorder point policies can be implemented using computerized inventory control systems. These systems are programmed to continuously monitor the inventory level as transactions are made, and to automatically generate an order for a new batch when the level falls below the reorder point.

42.3 MATERIAL AND CAPACITY REQUIREMENTS PLANNING

We present two alternative techniques for planning and controlling production and inventory. In this section we cover procedures used for job shop and mid-range production of assembled products. In Section 42.4, we examine procedures more appropriate for high production.

42.3.1 Material Requirements Planning

Material requirements planning (MRP) is a computational procedure used to convert the master production schedule for end products into a detailed schedule for raw materials and components used in the end products. The detailed schedule indicates the quantities of each item, when it must be ordered, and when it must be delivered to achieve the master schedule. *Capacity requirements planning* (Section 42.3.2) coordinates labor and equipment resources with material requirements.

MRP is most appropriate for job shop and batch production of products that have multiple components, each of which must be purchased and/or fabricated. It is the proper technique for determining quantities of dependent demand items that constitute the inventories of manufacturing: raw materials, purchased parts, work-in-process, and so forth.

MRP is relatively straightforward in concept. Its application is complicated by the sheer magnitude of the data that must be processed. The master schedule specifies the production of final products in terms of month-by-month deliveries. Each product may contain hundreds of components. These components are produced from raw materials, some of which are common among the components (e.g., sheet steel for stampings). Some of the components themselves may be common to several different products (these are called *common use items* in MRP). For each product, the components are assembled into simple subassemblies, which are added to form other subassemblies, and so on, until the final products are completed. Each step in the sequence consumes time. All of these factors must be accounted for in material requirements planning. Although each calculation is simple, the large number of calculations and massive amounts of data require that MRP be implemented by computer.

The lead time for a job is the time that must be allowed to complete the job from start to finish. There are two kinds of lead times in MRP: ordering lead times and manufacturing lead times. *Ordering lead time* is the time required from initiation of the purchase requisition to receipt of the item from the vendor. If the item is a raw material stocked by the vendor, the ordering lead time should be relatively short, perhaps a few weeks. If the item is fabricated, the lead time may be substantial, perhaps several months. *Manufacturing lead time* is the time required to produce the item in the company's own plant, from order release to completion.

Inputs to the MRP System   For the MRP processor to function properly, it must receive inputs from several files: (1) master production schedule, (2) product design data, in the form of a bill of materials file, (3) inventory records, and (4) capacity requirements planning. Figure 42.1 shows the data flow into the MRP processor and the recipients of its output reports.

The master production schedule was discussed in Section 42.1. The ***bill-of-materials file*** lists the component parts and subassemblies that make up each product. It is used to compute the requirements for raw materials and components used in the end products listed in the master schedule. Figure 42.4 shows a (simplified) structure of an assembled product. The product consists of two subassemblies, each consisting of three parts. The number of each item in the next level above in the product structure is indicated in parentheses.

The ***inventory record file*** identifies each item (by part number) and provides a time-phased record of its inventory status. This means that not only is the current quantity of the item listed, but also any future changes in inventory status that will occur and when they will occur. These data include gross requirements for the item (how many units will be needed to build products in the master schedule), scheduled receipts, on-hand status, and planned order releases. Each of these data sets indicates the changes by time period in the schedule (e.g., month, week).

How MRP Works   Based on inputs from the master schedule, bill-of-materials file, and inventory record file, the MRP processor computes how many of each component and raw material will be needed in future time periods by "exploding" the end product schedule into successively lower levels in the product structure. The MRP computations must deal with several complicating factors. First, component and subassembly quantities must be adjusted for any inventories on hand or on order. Second, quantities of common use items must be combined during parts explosion to obtain a total requirement for each component and raw material in the schedule. Third, the time phased delivery of end products must be converted into time-phased requirements for components and materials by factoring in the appropriate lead times. For every unit of final product listed in the MPS, the required number of components of each type must be ordered or fabricated, taking into account its ordering or manufacturing lead times. For each component, the raw material must be ordered, accounting for its ordering lead time. And assembly lead times must be considered in the scheduling of subassemblies and final products.

**EXAMPLE 42.2
Material Require-
ments Planning**

Consider the requirements planning procedure for one of the components in product P1 : C4. The required deliveries for P1 are indicated in the master production schedule

FIGURE 42.4   Product structure for assembled product P1 (based on data in [3]).

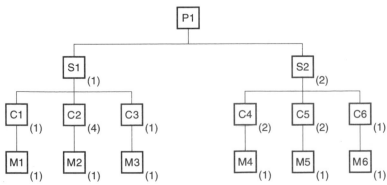

in Table 42.1(b). According to the product structure in Figure 42.4, two units of C4 are required to make subassembly S2, and two S2 units are required to make the final product P1. One unit of raw material M4 is used to make each unit of C4. Ordering, manufacturing, and assembly lead times for these items are known as follows:

Item identification:	P1	S2	C4	M4
Lead time (weeks):	1	1	2	3

The inventory status of M4 is 50 units currently on hand, and zero units of component C4 and S2. There are no scheduled requirements, receipts, or order releases indicated in the inventory record for these items. Neither material M4 nor component C4 are used on any other product—they are not common use items. Determine the time-phased requirements for M4, C4, and S2 to meet the master schedule for product P1. Orders for P1 beyond period 10 are ignored in this problem.

Solution: Table 42.2 presents the solution to this MRP problem. Delivery requirements for P1 must be offset by one week to obtain the planned order releases. S2 must be exploded by 2 units per P1 unit and offset by one week to obtain its order release. C4 is ex-

TABLE 42.2   Material requirements solution to Example 42.2.

Period		1	2	3	4	5	6	7	8	9	10
Item: **Product P1**											
Gross Requirements									50	75	100
Scheduled Receipts											
On hand	**0**										
Net Requirements									50	75	100
Planned Order Releases								50	75	100	
Item: **Subassembly S2**											
Gross Requirements								100	150	200	
Scheduled Receipts											
On hand	**0**										
Net Requirements								100	150	200	
Planned Order Releases							100	150	200		
Item: **Component C4**											
Gross Requirements							200	300	400		
Scheduled Receipts											
On hand	**0**										
Net Requirements							200	300	400		
Planned Order Releases					200	300	400				
Item: **Raw material M4**											
Gross Requirements					200	300	400				
Scheduled Receipts											
On hand	50	50	50	50	50						
Net Requirements				—	150	300	400				
Planned Order Releases		150	300	400							

ploded by 2 units per S2 unit and offset by two weeks to obtain its requirement. And M4 is offset by its three week ordering time to obtain its release date, taking into account current stock of M4 on hand.   ▨

Output Reports and Benefits of MRP   MRP generates various output reports that can be used in planning and managing plant operations. The reports include: (1) order releases, which authorize the placement of orders planned by the MRP system; (2) planned order releases in future periods; (3) rescheduling notices, indicating changes in due dates for open orders; (4) cancellation notices, which indicate that certain open orders have been canceled due to changes in the master schedule; (5) inventory status reports; (6) performance reports; (7) exception reports, showing deviations from schedule, overdue orders, scrap, and so forth; and (8) inventory forecasts, which project inventory levels in future periods.

Many benefits are claimed for a well-designed MRP system, including: (1) inventory reductions, (2) faster response to changes in demand, (3) reduced setup and changeover costs, (4) better machine utilization, (5) improved ability to respond to changes in the master schedule, and (6) helpful in developing the master schedule. Despite these claims, MRP systems have been implemented in industry with varying degrees of success. Some of the reasons for unsuccessful MRP implementations include: (1) inappropriate application, (2) MRP calculations based on inaccurate data, and (3) absence of capacity planning.

42.3.2 Capacity Requirements Planning

Capacity requirements planning is concerned with determining the labor and equipment requirements needed to meet the master production schedule. It is also concerned with identifying the firm's long-term future capacity needs. Capacity planning also serves to identify production resource limitations so that a realistic master production schedule can be planned.

A realistic master schedule must be compatible with the manufacturing capabilities of the plant that is to make the products. The firm must be aware of its production capacity and must plan for changes in capacity to meet changing production requirements specified in the master schedule. The relationship between capacity planning and other functions in production planning and control is shown in Figure 42.1. The master schedule is reduced to material and component requirements using MRP. These requirements provide estimates of the required labor hours and other resources needed to produce the components. The required resources are then compared to plant capacity over the planning horizon. If the master schedule is not compatible with plant capacity, adjustments must be made either in the schedule or in plant capacity.

Plant capacity can be adjusted in the short term and in the long term. Short-term capacity adjustments include: (1) *employment levels*—direct labor in the plant can be increased or decreased according to changes in capacity requirements; (2) *shift hours*—the number of labor hours per shift can be increased or decreased, through the use of overtime or reduced hours; (3) *number of work shifts*—the number of shifts worked per production period can be increased or decreased by authorizing evening and night shifts, and/or weekend shifts; (4) *inventory stockpiling*—this tactic is used to maintain steady employment levels during slow demand periods; (5) *order backlogs*—deliveries to the customer can be delayed during busy periods when production resources are insufficient to keep up with demand; and (6) *subcontracting*, which involves contracting work to outside shops during busy periods, or taking in extra work during slack periods.

Long-term capacity adjustments include possible changes in production capacity that generally require long lead times, including the following types of decisions: (1) *new equipment*—investments in additional machines, more productive machines, or new types of machines to match future changes in product design; (2) *new plants*—construction of new plants or purchase of existing plants from other companies; (3) *plant closings*—closing of plants not needed in the future.

42.4 JUST-IN-TIME AND LEAN PRODUCTION

Just-in-time (JIT) is an approach to production that was developed in Japan to minimize inventories. Work-in-process and other inventories are viewed by thc Japanese as waste that should be eliminated. Inventory ties up investment funds and takes up space (space is much more dear in Japan than in the United States). To reduce this form of waste, the JIT approach includes a number of principles and procedures aimed at reducing inventories, either directly or indirectly. Indeed, the scope of JIT is so broad that it is often referred to as a philosophy. JIT is an important component of "lean production", a principal goal of which is to reduce waste in production operations.[1] *Lean production* can be defined as "an adaptation of mass production in which workers and work cells are made more flexible and efficient by adopting methods that reduce waste in all forms."[2]

In recent years, the JIT philosophy has been embraced by U.S. manufacturing companies. Other terms have sometimes been adopted to give it an American flavor or to indicate slight differences with the Japanese practice of JIT. These terms include *zero inventory* (American Production and Inventory Control Society), *continuous flow manufacturing* (IBM Corporation), and *zero inventory production system* (General Electric Company).

Just-in-time procedures have proven most effective in high-volume repetitive manufacturing, such as the automobile industry [4]. The potential for in-process inventory accumulation in this type of manufacturing is significant because both the quantities of products and the number of components per product are large. A just-in-time system produces exactly the right number of each component required to satisfy the next operation in the manufacturing sequence just when that component is needed—just in time. To the Japanese, the ideal batch size is one part. As a practical matter, more than one part are produced at a time, but the batch size is kept small. Under JIT, producing too many units is to be avoided as much as producing too few units. This is a production discipline that contrasts sharply with traditional U.S. practice, which has promoted use of large in-process inventories to deal with problems such as machine breakdowns, defective components, and other obstacles to smooth production. The U.S. approach might be described as a "just-in-case" philosophy.

Although the principal theme in JIT is inventory reduction, this cannot simply be mandated. Several requisites must be pursued to make it possible: (1) stable production schedules; (2) small batch sizes and short setup times; (3) on-time delivery; (4) defect-free components and materials; (5) reliable production equipment; (6) pull system of

---

[1]The term "lean production" was coined by researchers at Massachusetts Institute of Technology to describe the programs undertaken by Toyota Motors aimed at improving production efficiencies and product quality.

[2] M. P. Groover, *Automation, Production Systems, and Computer-Integrated Manufacturing* [3], p. 834.

production control; (7) a work force that is capable, committed, and cooperative; and (8) a dependable supplier base.

Stable Schedule   For JIT to be successful, work must flow smoothly with minimum perturbations from normal operations. Perturbations require changes in operating procedures—increases and decreases in production rate, unscheduled setups, variations from the regular work routine, and other exceptions. Perturbations in downstream operations (i.e., final assembly) tend to be amplified in upstream operations (i.e., parts feeding). A master production schedule that remains relatively constant over time is one way of achieving smooth work flow and minimizing disturbances and changes in production.

Small Batch Sizes and Setup Reduction   Another requirement for minimizing inventories is small batch sizes and short setup times. We examined the relationship between batch size and setup time in the *EOQ* formula: Eq. (42.5). The Japanese have the *EOQ* formula. They got it from the United States. But instead of using it to compute batch quantities, they focus their efforts on finding ways to reduce setup time, thereby permitting smaller batches and lower work-in-process levels. American manufacturing companies are also adopting setup reduction as a goal. Some of the approaches used to reduce setup time include: (1) performing as much of the setup as possible while the previous job is still running; (2) using quick-acting clamping devices instead of bolts and nuts; (3) eliminating or minimizing adjustments in the setup; and (4) using GT and cellular manufacturing so that similar part styles are produced on the same equipment.

On-Time Delivery, Zero Defects, and Reliable Equipment   Success of JIT production requires near perfection in on-time delivery, parts quality, and equipment reliability. The small lot sizes and parts buffers used in JIT require parts to be delivered before stockouts occur at downstream stations. Otherwise, production must be suspended at these stations for lack of parts. If the delivered parts are defective, they cannot be used in assembly. This tends to promote zero defects in parts fabrication. Workers inspect their own output to make sure it is right before it proceeds to the next operation. Low work-in-process also requires reliable production equipment. Machines that break down cannot be tolerated in a JIT production system. This emphasizes the need for reliable equipment designs and preventive maintenance.

Pull System of Production Control   JIT requires a ***pull system*** of production control, in which the order to produce parts at a given workstation comes from the downstream station that uses those parts. As the supply of parts becomes exhausted at a given station, it "places an order" at the upstream workstation to replenish the supply. This order provides the authorization for the upstream station to produce the needed parts. This procedure, repeated at each workstation throughout the plant, has the effect of pulling parts through the production system. By contrast, a ***push system*** of production operates by supplying parts to each station in the plant, in effect driving the work from upstream stations to downstream stations. MRP is a push system. The risk in a push system is to overload the factory by scheduling more work than it can handle. This results in large queues of parts in front of machines that cannot keep up with arriving work. A poorly implemented MRP system, one that does not include capacity planning, manifests this risk.

One famous pull system is the ***Kanban*** system used by Toyota, the Japanese automobile company. Kanban (pronounced kahn-bahn) is a Japanese word meaning ***card.*** The

Kanban system of production control is based on the use of cards to authorize production and work flow in the plant. There are two types of kanbans: (1) production kanbans, and (2) transport kanbans. A ***production kanban*** authorizes production of a batch of parts. The parts are placed in containers, so the batch must consist of just enough parts to fill the container. Production of additional parts is not permitted. The ***transport kanban*** authorizes movement of the container of parts to the next station in the sequence.

Refer to Figure 42.5 as we explain how two workstations, one that feeds the other, operate in a kanban system. The figure shows four stations, but B and C are the stations we want to focus on here. Station B is the supplier in this pair, and station C is the consumer. Station C supplies downstream station D. B is supplied by upstream station A. When station C starts work on a full container, a worker removes the transport kanban from that container and takes it back to B. The worker finds a full container of parts at B that have just been produced, removes the production kanban from that container, and places it on a rack at B. The worker then places the transport kanban in the full container, which authorizes its movement to station C. The production kanban on the rack at station B authorizes production of a new batch of parts. Station B produces more than one part style, perhaps for several other downstream stations in addition to C. The scheduling of work is determined by the order in which the production kanbans are placed on the rack.

The kanban pull system between stations A and B and between stations C and D operates the same as it does between stations B and C, described here. This system of production control avoids unnecessary paperwork. The cards are used over and over rather than generating new production and transport orders every cycle. An apparent disadvantage is the considerable labor involvement in material handling (moving the cards and containers between stations); however, it is claimed that this promotes teamwork and cooperation among workers.

Work Force and Supplier Base   Another requirement of a JIT production system is workers who are cooperative, committed, and capable of performing multiple tasks. The work force must be flexible to produce a variety of part styles at the feeding stations, to inspect the quality of their work, and to deal with minor technical problems with the production equipment so that major breakdowns do not occur.

Just-in-time extends to the material and component suppliers of the company. Suppliers must be held to the same standards of on-time delivery, zero defects, and other JIT requisites as the company itself. Some of the vendor policies used by companies to implement JIT include: (1) reducing the total number of suppliers, (2) selecting suppliers with proven records for meeting quality and delivery standards, (3) establishing long-term partnerships with suppliers, and (4) selecting suppliers that are located near the company's manufacturing plant.

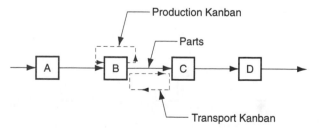

FIGURE 42.5   Operation of a kanban system between workstations.

42.5 SHOP FLOOR CONTROL

The third phase in production planning and control (Figure 42.1) is concerned with releasing production orders, monitoring and controlling progress of the orders, and acquiring up-to-date information on order status. The purchasing department is responsible for these functions among suppliers. The term ***shop floor control*** is used to describe these functions when accomplished in the company's own factories. In basic terms, shop floor control is concerned with managing work-in-progress in the factory. It is most relevant in job shop and batch production, where there are a variety of different orders in the shop that must be scheduled and tracked according to their relative priorities.

A typical shop floor control system consists of three modules: (1) order release, (2) order scheduling, and (3) order progress. The three modules and how they relate to other functions in the factory are depicted in Figure 42.6. They are implemented by a combination of computer systems and human resources.

Order Release   Order release in shop floor control generates the documents needed to process a production order through the factory. The documents are sometimes called the shop packet; it typically consists of (1) a route sheet, (2) material requisitions to draw the starting materials from stores, (3) job cards to report direct labor time used on the order, (4) move tickets to authorize transport of parts to subsequent work cen-

FIGURE 42.6   Three modules in a shop floor control system, and interconnections with other production planning and control functions.

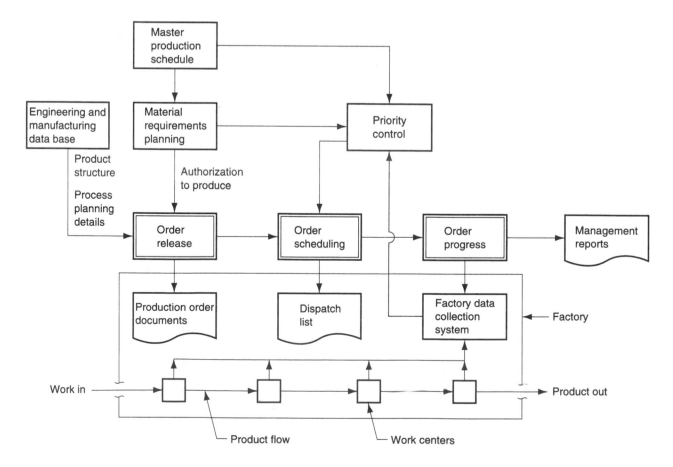

ters in the routing, and (5) parts list—required for assembly jobs. In a traditional factory, these documents move with the production order and are used to track its progress through the shop. In modern factories, automated methods such as bar code technology are used to monitor order status, making some or all of these paper documents unnecessary.

Order release is driven by two principal inputs, as indicated in Figure 42.6: (1) material requirements planning, which provides the authorization to produce; and (2) engineering and manufacturing database that indicates product structure and process planning details required to generate the documents that accompany the order through the shop.

Order Scheduling Order scheduling assigns the production orders to work centers in the factory. It serves as the dispatching function in production planning and control. In order scheduling, a dispatch list is prepared indicating which orders should be accomplished at each work center. It also provides relative priorities for the different jobs (e.g., by showing due dates for each job). The dispatch list helps the department foreman to assign work and allocate resources to achieve the master schedule.

Order scheduling addresses two problems in production planning and control: machine loading and job sequencing. To schedule production orders through the factory, they must first be assigned to work centers. Assigning orders to work centers is called *machine loading.* Loading all of the work centers in the plant is called *shop loading.* Since the number of production orders is likely to exceed the number of work centers, each work center will have a queue of orders waiting to be processed. A given production machine may have 10 to 20 jobs waiting to be processed.

Job sequencing is the problem of deciding the order in which to process jobs through a given machine. The processing sequence is decided by means of priorities among the jobs in the queue. The relative priorities are determined by a function called *priority control.* Some of the rules used to establish priorities for production orders in a plant include (1) *First-come-first-serve*—orders are processed in the sequence in which they arrive at the work center. (2) *Earliest due date*—orders with earlier due dates are given higher priorities. (3) *Shortest processing time*—orders with shorter processing times are given higher priorities. (4) *Least slack time*—orders with the least slack in their schedule are given higher priorities. Slack time is defined as the difference between the time remaining until due date and the process time remaining. (5) *Critical ratio*—orders with the lowest critical ratio are given higher priorities. The critical ratio is defined as the ratio of the time remaining until due date divided by the process time remaining.

The relative priorities of the orders may change over time. Expected demand could be higher or lower for certain products, equipment could break down, orders could be cancelled, or there could be defects in raw materials, among other reasons. Priority control reviews the relative priorities of the production orders and adjusts the dispatch list accordingly. When an order is completed at one work center, it moves to the next machine in its routing. The order becomes part of the machine loading for the next work center, and priority control is again used to determine the sequence among jobs to be processed at that machine.

Order Progress Order progress in shop floor control monitors the status of the orders, work-in-process, and other parameters in the plant that indicate progress and production performance. The objective in order progress is to provide information to manage production based on data collected from the factory.

Various techniques are available to collect data from factory operations. The techniques range from clerical procedures requiring workers to submit paper forms that are later compiled, to fully automated techniques requiring no human participation. The term *factory data collection system* is sometimes used to identify these techniques. More complete coverage of this topic is presented in [3].

Information presented to management is often summarized in the form of reports. The reports include: (1) *work order status reports,* which indicate status of production orders, including current work center where each order is located, processing hours remaining before each order will be completed, whether the job is on time, and priority level; (2) *progress reports* that report shop performance during a certain time period such as a week or month—how many orders were completed during the period, how many orders should have been completed during the period but were not completed, and so forth; and (3) *exception reports* that indicate deviations from the production schedule, such as overdue jobs. These reports are helpful to management in deciding resource allocation issues, authorizing overtime, and identifying problem areas that adversely affect achievement of the master production schedule.

REFERENCES

[1] Bedworth, D. D., and Bailey, J. E. *Integrated Production Control Systems.* 2nd. ed. Wiley, New York, 1987.

[2] Chase, R. B., and Aquilano, N. J. *Production and Operations Management.* 9th. ed. McGraw-Hill, New York, 2000.

[3] Groover, M. P. *Automation, Production Systems, and Computer Integrated Manufacturing.* 2nd. ed. Prentice-Hall, Upper Saddle River, N.J., 2001.

[4] Monden, Y. *Toyota Production System.* 3rd. ed. Engineering and Management Press, Norcross, Ga., 1998.

[5] Orlicky, J. *Material Requirements Planning.* McGraw-Hill, New York, 1975.

[6] Silver, E. A., Pyke, D. F., and Peterson, R. *Inventory Management and Production Planning and Control.* 3rd. ed. Wiley, New York, 1998.

[7] Veilleux, R. F., and Petro, L. W. *Tool and Manufacturing Engineers Handbook.* 4th.. ed. Volume V: *Manufacturing Management.* Society of Manufacturing Engineers, Dearborn, Mich., 1988.

[8] Vollman, T. E., Berry, W. L., and Whybark, D. C. *Manufacturing Planning and Control Systems.* 4th. ed. McGraw-Hill, New York, 1997.

REVIEW QUESTIONS

42.1. What is meant by the term *make-to-stock production?*

42.2. How does aggregate planning differ from the master production scheduling?

42.3. What are the product categories usually listed in the master production schedule?

42.4. What is the difference between dependent and independent demand for products?

42.5. Define *reorder point inventory system.*

42.6. In MRP, what are *common use items?*

42.7. Identify the inputs to the MRP processor in material requirements planning.

42.8. What are some of the resource changes that can be made to increase plant capacity in the short run?

42.9. Identify the principal objective in just-in-time production, as the Japanese view it.

42.10. How is a pull system distinguished from a push system in production and inventory control?

42.11. What are the three phases in shop floor control?

MULTIPLE CHOICE QUIZ

There is a total of 17 correct answers in the following multiple choice questions (some questions have multiple answers that are correct). To attain a perfect score on the quiz, all correct answers must be given, since each correct answer is worth 1 point. For each question, each omitted answer or wrong answer reduces the score by 1 point, and each additional answer beyond the number of answers required reduces the score by 1 point. Percentage score on the quiz is based on the total number of correct answers.

42.1. Which one of the following terms best describes the overall function of production planning and control? (a) inventory control, (b) manufacturing logistics, (c) manufacturing engineering, (d) mass production, or (e) product design.

42.2. Which of the following are the categories usually listed in the master production schedule (more than one)? (a) components used to build the final products, (b) firm customer orders, (c) general product lines, (d) orders for maintenance and spare parts, (e) sales forecasts, and (f) spare tires.

42.3. Inventory carrying costs include which of the following (more than one)? (a) equipment downtime, (b) investment, (c) obsolescence, (d) setup, (e) spoilage, (f) stock-out, and (g) storage.

42.4. Which of the following are the terms in the economic order quantity formula (name three)? (a) annual demand rate, (b) batch size, (c) cost per piece, (d) holding cost, (e) interest rate, and (f) setup cost.

42.5. Order point inventory systems are intended primarily for which of the following more than one)? (a) dependent demand items, (b) independent demand items, (c) low production quantities, (d) mass production quantities, and (e) mid-range production quantities.

42.6. With which of the following manufacturing resources is capacity requirements planning primarily concerned (more than one)? (a) component parts, (b) direct labor, (c) inventory storage space, (d) production equipment, and (e) raw materials.

42.7. The word **kanban** is most closely associated with which one of the following? (a) capacity planning, (b) economic order quantity, (c) just-in-time production, (d) master production schedule, or (e) material requirements planning.

42.8. The term **machine loading** refers most closely to which one of the following? (a) assigning jobs to a work center, (b) floor foundation in the factory, (c) managing work-in-process in the factory, (d) releasing orders to the shop, or (e) sequencing jobs through a machine.

PROBLEMS

Inventory Control

42.1. A product is made to stock. Annual demand is 60,000 units. Each unit costs $4, and the annual holding cost rate = 25%. Setup cost to produce this product is $300. Determine (a) economic order quantity and (b) total inventory costs for this situation.

42.2. Given: annual demand for product X is 20,000 units; cost per unit = $6; holding cost rate = 2.5%/month; changeover (setup) time between products averages 2 hr; downtime cost during changeover = $200/hr. Determine (a) economic order quantity and (b) total inventory costs for this situation.

42.3. A product is produced in batches. Batch size = 2,000 units. Annual demand = 50,000 units, and unit cost of the product = $4. Setup time to run a batch = 2.5 hr, cost of downtime on the affected equipment is figured at $250/hr, and annual holding cost rate = 30%. What would be the annual savings if the product were produced in the economic order quantity?

42.4. A certain piece of production equipment is used to produce various components for an assembled product of the XYZ Company. To keep in-process inventories low, it is desired to produce the components in batch sizes of 150 units (daily requirements for assembly). Demand for each product is 2,500 units per year. Production downtime costs an estimated $200/hr. All of the components made on the equipment are of approximately equal value: C_p = $9/unit. Holding cost rate = 30%/yr. In how many minutes must the changeover (setup) between batches be completed in order for 100 units to be the economic order quantity?

42.5. Current changeover (setup) time on a certain machine = 3 hr. Cost of downtime on this machine is estimated at $200/hr. Annual holding cost per part made on the equipment, C_h = $1. Annual demand for the part is 15,000 units. Determine (a) EOQ and (b) total inventory costs for this data. Also, determine (c) EOQ and (b) total inventory costs, if the changeover time could be reduced to 6 min.

42.6. The two-bin approach is used to control inventory for a particular low-cost component. Each bin holds 1,000 units. The annual usage of the component is 40,000 units. Cost to order the component is around $50. (a) What is the imputed holding cost per unit for this data? (b) If the actual annual holding cost per unit is only 5 cents, what lot size should be ordered? (c) How much more is the current two-bin approach costing the company annually, compared to the economic order quantity?

Material Requirements Planning

42.7. Quantity requirements are to be planned for component C2 in product P1. Required deliveries for P1 are

Item identification:	P1
Lead time (weeks):	1

Given the product structure in Figure 42.4, determine the time-phased requirements for M2, C2, and S1 to meet the master schedule for P1. Assume no common use items and all on-hand inventories and scheduled receipts are zero. Use a format similar to Table 42.2. Ignore demand for P1 beyond period 10.

Item identification:	P1
Lead time (weeks):	1

Given the product structure in Figure 42.4, determine the time-phased requirements for M5, C5, and S2 to meet the master schedule for P1. Assume no common use items. On-hand inventories are: 200 units for M5 and 100 units for C5, zero for S2. Use a format similar to Table 42.2. Ignore demand for P1 beyond period 10.

given in Table 42.1. Ordering, manufacturing, and assembly lead times are as follows:

S1	C2	M2
2	1	2

42.8. Requirements are to be planned for component C5 in product P1. Required deliveries for P1 are given in Table 42.1. Ordering, manufacturing, and assembly lead times are as follows:

S2	C5	M5
1	3	2

42.9. Solve the previous problem except that the following is known in addition to the information given: scheduled receipts of M5 are 250 units in period (week) 3 and 50 units in period (week) 4.

Order Scheduling

42.10. Four products are to be manufactured in Department A, and it is desired to determine how to allocate resources in that department to meet the required demand for these products for a certain week. The demand and other data for the products are given as follows:

Product	1	2	3	4
Weekly Demand	750	900	400	400
Setup Time	6 hr	5 hr	7 hr	6 hr
Operation Time	4.0 min	3.0 min	2.0 min	3.0 min

The plant normally operates one shift (7.0 hours per shift), five days per week and there are currently three work centers in the department. Propose a way of scheduling the machines to meet the weekly demand.

42.11. In the previous problem, propose a way of scheduling to meet the weekly demand if there were four machines instead of three.

42.12. The current date in the production calendar of the XYZ Company is day 15. There are three orders (A, B, and C) to be processed at a particular work center. The orders arrived in the sequence A-B-C at the work center. The following table indicates the remaining process time and production calendar due date for each order:

Order	Remaining Process Time	Due Date
A	5 days	Day 25
B	16 days	Day 34
C	7 days	Day 24

Determine the sequence of the orders that would be scheduled using: (a) first-come-first-serve, (b) earliest due date, (c) shortest processing time, (d) least slack time, and (e) critical ratio.

43

QUALITY CONTROL

CHAPTER CONTENTS

43.1   What Is Quality?
43.2   Process Capability
43.3   Statistical Tolerancing in Product Design
       43.3.1   Natural Tolerance Limits
       43.3.2   Statistical Tolerancing for Assemblies
43.4   Taguchi Methods
       43.4.1   The Loss Function
       43.4.2   Robust Design
       43.4.3   Off-Line and On-Line Quality Control
43.5   Statistical Process Control
       43.5.1   Control Charts for Variables
       43.5.2   Control Charts for Attributes
       43.5.3   Interpreting the Control Charts

Traditionally, *quality control* (QC) has been concerned with detecting poor quality in manufactured products and taking corrective action to eliminate it. Operationally, QC has often been limited to inspecting the product and its components, and deciding whether the measured or gaged dimensions and other features conformed to design specifications. If they did, the product was shipped. The modern view of quality control encompasses a broader scope of activities, including robust design and statistical process control. Our chapter begins by defining product quality.

43.1 WHAT IS QUALITY?

The dictionary defines quality as "the degree of excellence which a thing possesses," or "the features that make something what it is"—its characteristic elements and attributes. The views of the leading experts do not all coincide. Crosby defines quality as "conformance to requirements" [2]. Juran summarizes it as "fitness for use" and "quality is customer satisfaction" [5]. The American Society for Quality Control (ASQC) defines quality as "the totality of features and characteristics of a product or service that bear on its ability to satisfy given needs" [3].

In a manufactured product, quality has two aspects [5]: (1) product features, and (2) freedom from deficiencies. ***Product features*** are the characteristics of the product that result from design; they are the functional and aesthetic features of the item intended to appeal to and provide satisfaction to the customer. In an automobile, these features include the size of the car, the arrangement of the dashboard, the finish on the body, and similar aspects. They also include the available options for the customer to choose. Table 43.1 lists some of the important general product features. The sum of a product's features usually defines its ***grade,*** which relates to the level in the market at which the product is aimed. Cars (and most other products) come in different grades. Some cars provide basic transportation because that is what some customers want, while others are upscale for consumers willing to spend more to own a "better product." The features of a product are decided in design, and they generally determine the inherent cost of the product. Superior features and more of them mean higher cost.

Freedom from deficiencies means that the product does what it is supposed to do (within the limitations of its design features) and that it is absent of defects and out-of-tolerance conditions (Table 43.1). This aspect of quality includes individual components of the product as well as the product itself. Freedom from deficiencies means conforming to design specifications, which is accomplished in manufacturing. Although the inherent cost to make a product is a function of its design, minimizing the product's cost to the lowest possible level within the limits set by its design is largely a matter of avoiding defects, tolerance deviations, and other errors during production. Costs of these deficiencies make a long list indeed: scrapped parts, larger lot sizes for scrap allowances, rework, reinspection, sortation, customer complaints and returns, warranty costs and customer allowances, lost sales, and lost good will in the marketplace.

Thus, product features are the aspect of quality for which the design department is responsible. Product features determine to a large degree the price that a company can charge for its products. Freedom from deficiencies is the quality aspect for which the manufacturing departments are responsible. The ability to minimize these deficiencies has an important influence on the cost of the product. These generalities oversimplify the way

TABLE 43.1   Aspects of quality.

Quality Aspect	Examples
Product features	Design configuration, size, weight
	Distinguishing features of the model
	Ease of use
	Aesthetic appeal
	Function and performance
	Availability of options
	Reliability and dependability
	Durability and long service life
	Serviceability
	Reputation of product and producer.
Freedom from deficiencies	Absence of defects
	Conformance to specifications
	Components within tolerance
	No missing parts
	No early failures

Compiled from [5] and other sources.

things work, because the responsibility for high quality extends well beyond the design and manufacturing functions in an organization.

43.2 PROCESS CAPABILITY

In any manufacturing operation, variability exists in the process output. In a machining operation, which is one of the most accurate processes, the machined parts may appear to be identical, but close inspection reveals dimensional differences from one part to the next. Manufacturing variations can be divided into two types: random and assignable.

Random variations are caused by many factors: human variability with each operation cycle, variations in raw materials, machine vibration, etc. Individually, these factors may not amount to much, but collectively the errors can be significant enough to cause trouble unless they are within the tolerances for the part. Random variations typically form a normal statistical distribution. The output of the process tends to cluster about the mean value, in terms of the product's quality characteristic of interest (e.g., length, diameter). A large proportion of the population of parts is centered around the mean, with fewer parts away from the mean. When the only variations in the process are of this type, the process is said to be in *statistical control.* This kind of variability will continue so long as the process is operating normally. It is when the process deviates from this normal operating condition that variations of the second type appear.

Assignable variations indicate an exception from normal operating conditions. Something has occurred in the process that is not accounted for by random variations. Reasons for assignable variations include operator mistakes, defective raw materials, tool failures, machine malfunctions, and so on. Assignable variations in manufacturing usually betray themselves by causing the output to deviate from the normal distribution. The process is no longer in statistical control.

Process capability relates to the normal variations inherent in the output when the process is in statistical control. By definition, *process capability* equals ± 3 standard deviations about the mean output value (a total of 6 standard deviations),

$$PC = \mu \pm 3\sigma \tag{43.1}$$

where PC = process capability; μ = process mean, which is set at the nominal value of the product characteristic when bilateral tolerancing is used (Section 5.1.1); and σ = standard deviation of the process. Assumptions underlying this definition are (1) the output is normally distributed, and (2) steady state operation has been achieved and the process is in statistical control. Under these assumptions, 99.73% of the parts produced will have output values that fall within $\pm 3.0\sigma$ of the mean.

43.3 STATISTICAL TOLERANCING

The issue of tolerances is critical to product quality. Design engineers tend to assign dimensional tolerances to components and assemblies based on their judgment of how size variations will affect function and performance. Conventional wisdom is that closer tolerances beget better performance. Small regard is given to the cost resulting from tolerances that are unduly tight relative to process capability. The general relationship between tolerance and manufacturing cost is shown in Figure 43.1. As tolerance is reduced, the cost of achieving the tolerance increases at an accelerating rate. This is because additional processing steps may be required to obtain closer tolerances, and/or production machines that are more precise and expensive may be required.

The design engineer should be aware of this relationship. Although primary consideration must be given to function in assigning tolerances, cost is also a factor, and any relief that can be given to the manufacturing departments in the form of wider tolerances without sacrificing product function is worthwhile. Several approaches are available that consider process capability in specifying tolerances. Here we examine two: (1) natural tolerance limits and (2) statistical tolerancing for assemblies. More details on these and other approaches can be found in references [5], [9], and [10].

43.3.1 Natural Tolerance Limits

Design tolerances must be compatible with process capability. It serves no useful purpose to specify a tolerance of ± 0.025 mm (± 0.001 in.) on a dimension if the process capability is significantly wider than ± 0.025 mm (± 0.001 in.). Either the tolerance should be opened further (if design functionality permits) or a different manufacturing operation should be selected. Ideally, the specified tolerance should be greater than the process capability. If function and available processes prevent this, then sorting must be included in the manufacturing sequence to inspect every unit and separate those that meet specification from those that do not.

Design tolerances can be specified as being equal to process capability, defined in Eq. (43.1). The upper and lower boundaries of this range are known as the *natural tolerance limits*. When design tolerances are set equal to the natural tolerance limits, then 99.73% of the parts will be within tolerance and 0.27% will be outside the limits. Any increase in the tolerance range will reduce the percentage of defective parts.

Tolerances are not usually set at their natural limits by product design engineers; tolerances are specified based on the allowable variability that will achieve required func-

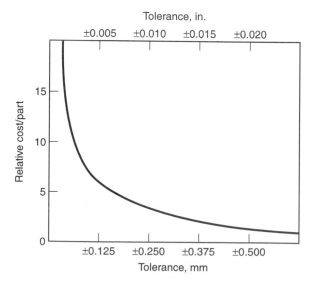

FIGURE 43.1 General relationship between tolerances and manufacturing cost.

tion and performance. It is useful to know the ratio of the specified tolerance relative to the process capability. This is indicated by the *process capability index*:

$$PCI = \frac{T}{6\sigma} \tag{43.2}$$

where PCI = process capability index; T = tolerance range – the difference between upper and lower limits of the specified tolerance; and 6σ = natural tolerance limits. The underlying assumption in this definition is that the process mean is set equal to the nominal design specification, so that the numerator and denominator in Eq. (43.2) are centered about the same value.

Table 43.2 shows the effect of various multiples of standard deviation on defect rate (i.e., proportion of out-of-tolerance parts). The desire to achieve very-low-fraction defect rates has led to the popular notion of "six sigma" limits in quality control (last entry in the table). Achieving six sigma limits virtually eliminates defects in manufactured product, assuming the process is maintained within statistical control.

The process capability of a given manufacturing operation is not always known, and experiments must be conducted to assess it. Methods are available to estimate the natural tolerance limits based on a sampling of the process.

43.3.2 Statistical Tolerancing for Assemblies

Figure 43.2 shows an assembly consisting of three components, in which the overall length must be held to a tolerance of ±0.30 mm (±0.012 in.). To achieve the tolerance on the assembly, what should the tolerance limits of the individual components be? The simple answer is to divide the total tolerance among the components so that the sum of their individual tolerances is equal to the assembly tolerance. If the assembly tolerance is distributed evenly among the parts, then the tolerance of each part is ±0.10 mm (±0.004 in.). This means that if all of the parts are within tolerance, no combination of their dimensions will produce an assembly dimension that is out of tolerance for the assembly. This approach to tolerancing is appropriately called *worst-case design.*

If we assume that the processes making the components are in statistical control, and we are willing to accept a small-fraction defect rate on the overall assembly dimension, then the tolerances on the individual components can be made much wider than under the worst-case design philosophy. A statistical approach to tolerancing can be used for assemblies (and other additive dimensions), which is based on the following relationship between the standard deviation of the assembly dimension and the standard deviations of the component dimensions:

TABLE 43.2 Defect rate when tolerance is defined in terms of number of standard deviations of the process, given that the process is operating in statistical control.

No. of Standard Deviations	Process Capability Index	Defect rate, %	Parts per million
±1.0	0.333	31.74%	317,400
±2.0	0.667	4.56%	45,600
±3.0	1.00	0.27%	2,700
±4.0	1.333	0.0063%	63
±5.0	1.667	0.000057%	0.57
±6.0	2.00	0.0000002%	0.002

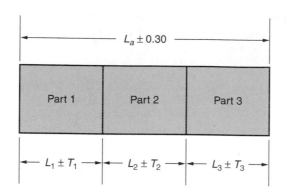

FIGURE 43.2 An assembly consisting of three parts whose overall dimension (L_a) has a tolerance of ±0.30 mm (±0.012 in.).

$$\sigma_a^2 = \sum_{i=1}^{n} \sigma_i^2 \qquad (43.3)$$

where n = number of components.

If tolerances on the individual components are set at some specified multiple of their respective standard deviations (e.g., natural tolerance limits, where $T = 6\sigma$), and it is appropriate to set the tolerance on the assembly using the same multiple, then

$$T_a = \sqrt{\sum_{i=1}^{n} T_i^2} \qquad (43.4)$$

where T_a = tolerance of the assembly dimension; T_i = tolerances of the individual component dimensions; and n = number of components. The summation of individual squared tolerances in Eq. (43.4) is valid whether the component dimensions are added or subtracted to obtain the overall assembly dimension.

Statistical tolerancing equations of this form are based on several assumptions: (1) component dimensions are normally distributed; (2) the distributions are independent; (3) parts making up a given assembly are selected randomly; and (4) processes making the components are in statistical control with process means centered in the tolerance range. Departures from these assumptions in manufacturing will generally result in a higher level of out-of-tolerance assemblies than indicated by the values in Table 43.2.

EXAMPLE 43.1
Statistical Tolerancing Suppose an assembly consists of three components, as in Figure 43.2. The overall dimension of the assembly is $L_a = 75.0 \pm 0.30$ mm, and each part has a dimension of 25.0 mm. If all component tolerances are to be equal, compute the component tolerance using statistical tolerancing.

Solution: In this problem, we solve Eq. (43.4) for the component tolerance, where the number of pieces $n = 3$ and all T_i are equal.

$$T_a = 0.30 = \sqrt{3T_i^2}$$

$$3T_i^2 = (0.30)^2 = 0.09$$

$$T_i^2 = \frac{0.09}{3} = 0.03$$

$$T_i = \sqrt{0.03} = 0.173 \text{ mm}$$

The tolerance on the individual components using statistical tolerancing is ±0.173 mm. This compares with ±0.100 mm under worst-case tolerancing policy. ■

43.4 TAGUCHI METHODS

G. Taguchi has had an important influence on the development of quality engineering, especially in the design area—both product design and process design. In this section we review some of the Taguchi methods. At the risk of oversimplifying his contributions, we boil them down to three topics: (1) loss function, (2) robust design, and (3) off-line and on-line quality control. These topics are briefly discussed in the next sections. More complete coverage can be found among our references [6], [9].

43.4.1 The Loss Function

Taguchi defines quality as "the loss a product costs society from the time the product is released for shipment" [9]. Loss includes costs to operate, failure to function, maintenance and repair costs, customer dissatisfaction, injuries caused by poor design, and similar costs. Some of these losses are difficult to quantify in monetary terms, but they are nevertheless real. Defective products (or their components) that are exposed before shipment are not considered part of this loss. Instead, any expense to the company resulting from scrap or rework of defective product is a manufacturing cost rather than a quality loss.

Loss occurs when a product's functional characteristic differs from its nominal or target value. Although functional characteristics do not translate directly into dimensional features, the loss relationship is most readily understood in terms of dimensions. When the dimension of a component deviates from its nominal value, the component's function is adversely affected. No matter how small the deviation, there is some loss in function. The loss increases at an accelerating rate as the deviation grows, according to Taguchi. If we let x = the quality characteristic of interest and N = its nominal value, then the loss function will be a U-shaped curve, as in Figure 43.3(a). A quadratic equation can be used to describe this curve:

$$L(x) = k(x - N)^2 \tag{43.5}$$

where $L(x)$ = loss function; k = constant of proportionality; and x and N are defined above. At some level of deviation $(x_2 - N) = -(x_1 - N)$, the loss will be prohibitive, and it is necessary to scrap or rework the product. This level identifies one possible way of specifying the tolerance limit for the dimension.

In the traditional approach to quality control, tolerance limits are defined and any product within those limits is acceptable. Whether the quality characteristic (e.g., the

FIGURE 43.3 (a) The quadratic quality loss function. (b) Loss function implicit in traditional tolerance specification.

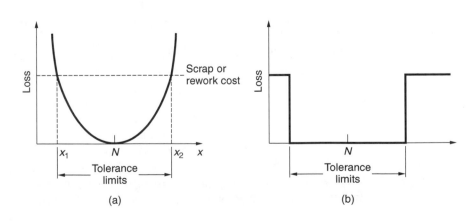

(a)   (b)

dimension) is close to the nominal value or close to one of the tolerance limits, it is acceptable. Trying to visualize this approach in terms analogous to the preceding relation, we obtain the discontinuous loss function in Figure 43.3(b). The reality is that products closer to the nominal specification are of better quality and will provide greater customer satisfaction. In order to improve quality and customer satisfaction, one must attempt to reduce the loss by designing the product and process to be as close as possible to the target value.

43.4.2 Robust Design

A basic purpose of quality control is to minimize variations. Taguchi calls the variations noise factors. A *noise factor* is a source of variation that is impossible or difficult to control and that affects the functional characteristics of the product. Three types of noise factors can be distinguished: (1) unit-to-unit, (2) internal, and (3) external.

Unit-to-unit noise factors consist of inherent random variations in the process or product caused by variability in raw materials, machinery, and human participation. These are noise factors we have previously called random variations in the process. They are associated with a production process that is in statistical control.

Internal noise factors are sources of variation that are internal to the product or process. They include time-dependent factors such as wear of mechanical components, spoilage of raw materials, and fatigue of metal parts; and operational errors, such as improper settings on the product or machine tool. An *external noise factor* is a source of variation that is external to the product or process, such as outside temperature, humidity, raw material supply, and input voltage. Internal and external noise factors constitute what we have previously called assignable variations.

A *robust design* is one in which the product's function and performance are relatively insensitive to variations in design and manufacturing parameters. It involves the design of both the product and process so that the manufactured product will be relatively unaffected by all noise factors.

43.4.3 Off-Line and On-Line Quality Control

Taguchi divides the overall quality system in an organization into two basic functions: off-line quality control and on-line quality control. *Off-line quality control* is concerned with design issues, both product and process design. In the sequence of the two functions, it precedes on-line control. *On-line quality control* is concerned with production operations and relations with the customer after shipment. Its objective is to manufacture products within the specifications defined in product design, utilizing the methods and procedures developed in process design. Traditional QC methods are more closely aligned with this second function, which is to achieve conformance to specification.

Off-Line Quality Control   Off-line quality control consists of two stages: product design and process design. The *product design* stage involves development of a new product or a new model of an existing product. The goals in product design are to properly identify customer needs and to design a product that meets those needs but can also be manufactured consistently and economically. The *process design* stage is what we usually think of as the manufacturing engineering function. It is concerned with specifying the processes and equipment, setting work standards, documenting procedures, and developing clear and workable specifications for manufacturing.

A three-step approach applicable to both of these design stages is outlined: (1) system design, (2) parameter design, and (3) tolerance design. *System design* involves the application of engineering knowledge and analysis to develop a prototype design that will meet customer needs. In the product design stage, system design means the final product configuration, including starting materials, components, and subassemblies. In process design, system design means selecting the most appropriate manufacturing methods, with emphasis on the use of existing technologies rather than development of new ones. Obviously, the product and process design stages overlap, because product design determines the manufacturing process to a great degree. Also, the quality of the product is impacted significantly by decisions made during product design.

Parameter design is concerned with determining optimal parameter settings for the product and process. It is in this stage that a robust design, defined previously, is achieved. This means selecting values of product parameters that result in a product which is insensitive to variations in these parameters. It also means choosing parameter values that minimize the effects of process variations. Taguchi advocates the use of various experimental designs to determine these optimal parameter settings.

In *tolerance design,* the objective is to specify appropriate tolerances about the nominal values established in parameter design. It attempts to achieve a balance between setting wide tolerances to facilitate manufacture and minimizing tolerances to optimize product performance.

On-Line Quality Control   This function of quality control is concerned with production operations and customer relations. In *production,* Taguchi classifies three approaches to quality control:

1. *Process diagnosis and adjustment*—The process is measured periodically and adjustments are made to move parameters of interest toward nominal values.

2. *Process prediction and correction*—Process parameters are measured at periodic intervals so that trends can be projected. If projections indicate deviations from target values, corrective process adjustments are made.

3. *Process measurement and action*—This involves inspection of all units (100%) to detect deficiencies that will be reworked or scrapped. Since this approach occurs after the unit is already made, it is less desirable than the other two forms of control.

The Taguchi on-line QC approach includes *customer relations,* which consists of two elements. First, there is the traditional customer service that deals with repairs, replacements, and complaints. Second, it includes a feedback system in which information on failures, complaints, and related data are communicated back to the relevant departments in the organization for correction. This latter scheme is part of the continuous improvement process advocated by Taguchi.

43.5 STATISTICAL PROCESS CONTROL

Statistical process control (SPC) involves the use of various statistical methods to assess and analyze variations in a process. SPC methods include simply keeping records of the production data, histograms, process capability analysis, and control charts. Control charts are the most widely used SPC method, and this section will focus on them.

The underlying principle in control charts is that the variations in any process divide into two types (Section 43.2): (1) random variations, which are the only variations present if the process is in statistical control, and (2) assignable variations that indicate a departure from statistical control. It is the objective of a control chart to identify when the process has gone out of statistical control, thus signaling that some corrective action should be taken.

A **control chart** is a graphical technique in which statistics computed from measured values of a certain process characteristic are plotted over time to determine if the process remains in statistical control. The general form of the control chart is illustrated in Figure 43.4. The chart consists of three horizontal lines that remain constant over time: a center, a lower control limit (LCL), and an upper control limit (UCL). The center is usually set at the nominal design value. The upper and lower control limits are generally set at ±3 standard deviations of the sample means.

It is highly unlikely that a sample drawn from the process lies outside the upper or lower control limits while the process is in statistical control. Thus, if it happens that a sample value does fall outside these limits, it is interpreted to mean that the process is out of control. Therefore, an investigation is undertaken to determine the reason for the out-of-control condition, with appropriate corrective action to eliminate the condition. By similar reasoning, if the process is found to be in statistical control, and there is no evidence of undesirable trends in the data, then no adjustments should be made since they would introduce an assignable variation to the process. The philosophy, "If it ain't broke, don't fix it," is applicable in control charts.

There are two basic types of control charts: (1) control charts for variables, and (2) control charts for attributes. Control charts for variables require a measurement of the quality characteristic of interest. Control charts for attributes simply require a determination of whether a part is defective or how many defects there are in the sample.

43.5.1 Control Charts for Variables

A process that is out of statistical control manifests this condition in the form of significant changes in process mean and/or process variability. Corresponding to these possibilities, there are two principal types of control charts for variables: $\bar{x}$ chart, and R chart. The $\bar{x}$ **chart** (call it "x bar chart") is used to plot the average measured value of a certain

FIGURE 43.4 Control chart.

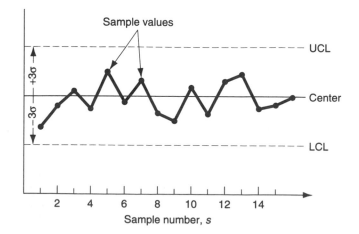

quality characteristic for each of a series of samples taken from the production process. It indicates how the process mean changes over time. The **R chart** plots the range of each sample, thus monitoring the variability of the process and indicating whether it changes over time.

A suitable quality characteristic of the process must be selected as the variable to be monitored on the $\bar{x}$ and R charts. In a mechanical process, this might be a shaft diameter or other critical dimension. Measurements of the process itself must be used to construct the two control charts.

With the process operating smoothly and absent of assignable variations, a series of samples (e.g., $m = 20$ or more is generally recommended) of small size (e.g., $n = 4, 5,$ or 6 parts per sample) are collected and the characteristic of interest is measured for each part. The following procedure is used to construct the center, LCL, and UCL for each chart:

1. Compute the mean $\bar{x}$ and range R for each of the m samples.
2. Compute the grand mean $\bar{\bar{x}}$, which is the mean of the $\bar{x}$ values for the m samples; this will be the center for the $\bar{x}$ chart.
3. Compute $\bar{R}$, which is the mean of the R values for the m samples; this will be the center for the R chart.
4. Determine the upper and lower control limits, UCL and LCL, for the $\bar{x}$ and R charts. Values of standard deviation could be estimated from the sample data and used to compute these control limits. However, an easier approach is based on statistical factors tabulated in Table 43.3 that have been derived specifically for these control charts. Values of the factors depend on sample size n. For the $\bar{x}$ chart:

$$\text{LCL} = \bar{\bar{x}} - A_2\bar{R} \quad \text{and} \quad \text{UCL} = \bar{\bar{x}} + A_2\bar{R} \qquad (43.6)$$

For the R chart:

$$\text{LCL} = D_3\bar{R} \quad \text{and} \quad \text{UCL} = D_4\bar{R} \qquad (43.7)\approx$$

EXAMPLE 43.2
$\bar{x}$ and Bar Charts

Eight samples ($m = 8$) of size 4 ($n = 4$) have been collected from a manufacturing process that is in statistical control, and the dimension of interest has been measured for each part. It is desired to determine the values of the center, LCL, and UCL to construct the $\bar{x}$ and R charts. The calculated values of $\bar{x}$ and R for each sample are given below (measured values are in cm), which is step (1) in our procedure.

Step	1	2	3	4	5	6	7	8
$\bar{x}$	2.008	1.998	1.993	2.002	2.001	1.995	2.004	1.999
R	0.027	0.011	0.017	0.009	0.014	0.020	0.024	0.018

Solution: In step (2), we compute the grand mean of the sample averages.

$$\bar{\bar{x}} = (2.008 + 1.998 + \cdots + 1.999)/8 = 2.000$$

In step (3), the mean value of R is computed.

$$\bar{R} = (0.027 + 0.011 + \cdots + 0.018)/8 = 0.0175$$

In step (4), the values of LCL and UCL are determined based on factors in Table 43.3. First, using Eq. (43.6) for the $\bar{x}$ chart,

$$\text{LCL} = 2.000 - 0.729(0.0175) = 1.9872$$

$$\text{UCL} = 2.000 + 0.729(0.0175) = 2.0128$$

TABLE 43.3 Constants for the $\bar{x}$ and R charts.

Sample Size n	$\bar{x}$ chart A_2	R chart D_3	R chart D_4
3	1.023	0	2.574
4	0.729	0	2.282
5	0.577	0	2.114
6	0.483	0	2.004
7	0.419	0.076	1.924
8	0.373	0.136	1.864
9	0.337	0.184	1.816
10	0.308	0.223	1.777

And for the R chart using Eq. (43.7),

$$\text{LCL} = 0(0.0175) = 0$$

$$\text{UCL} = 2.282(0.0175) = 0.0399$$

The two control charts are constructed in Figure 43.5 with the sample data plotted in the charts.

If the mean and standard deviation for the process were known, an alternative way to calculate the center and upper and lower control limits for the $\bar{x}$ chart would be as follows:

FIGURE 43.5 Control charts for Example 43.2.

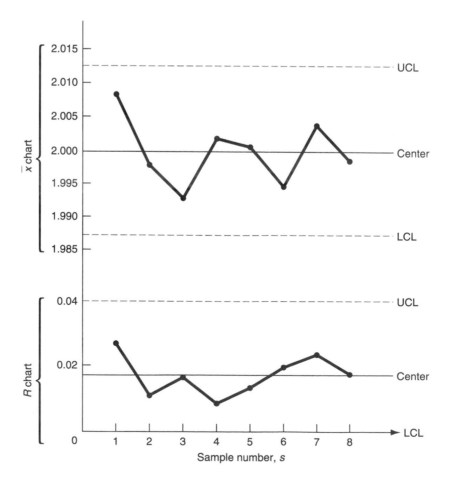

$$\text{LCL} = \mu - \frac{3\sigma}{\sqrt{n}} \quad \text{and} \quad \text{UCL} = \mu + \frac{3\sigma}{\sqrt{n}} \tag{43.8}$$

where μ = process mean; σ = standard deviation of the process; and n = sample size. The LCL and UCL values given by Eq. (43.8) are theoretically the same values as those calculated by Eq. (43.6). However, when first setting up the $\bar{x}$ chart for a process, the mean and standard deviation for the process variable of interest are generally not known. Accordingly, Eq. (43.6) based on measured $\bar{x}$ and $\bar{R}$ values can be conveniently used to compute the control chart parameters. With the control limits set at the values defined in Eqs. (43.6) or (43.8), 99.73% of the random samples drawn from a process that is in statistical control lie inside the control limits.

Readers will note that the standard deviation of the sample means is related to the population standard deviation by the reciprocal of the square root of n, the number of units in the sample:

$$\sigma_{\bar{x}} = \frac{\sigma}{\sqrt{n}} \tag{43.9}$$

where $\sigma_{\bar{x}}$ = standard deviation of the sample mean; and the other terms are defined above.

43.5.2 Control Charts for Attributes

Control charts for attributes do not use a measured quality variable; instead, they monitor the number of defects present in the sample or the fraction defect rate as the plotted statistic. Examples of these kinds of attributes include number of defects per automobile, fraction of bad parts in a sample, existence or absence of flash in a plastic molding, and number of flaws in a roll of sheet steel. The two principal types of control charts for attributes are: (1) *p chart,* which plots the fraction defect rate in successive samples; and (2) *c chart,* which plots the number of defects, flaws, or other nonconformities per sample.

p **Chart**   In the p chart, the quality characteristic of interest is the proportion (p for proportion) of nonconforming or defective units. For each sample, this proportion p_i is the ratio of the number of nonconforming or defective items d_i over the number of units in the sample n (we assume samples of equal size in constructing and using the control chart):

$$p_i = \frac{d_i}{n} \tag{43.10}$$

where i is used to identify the sample. If the p_i values for a sufficient number of samples are averaged, the mean value $\bar{p}$ is a reasonable estimate of the true value of p for the process. The p chart is based on the binomial distribution, where p is the probability of a nonconforming unit. The center in the p chart is the computed value of $\bar{p}$ for m samples of equal size n collected while the process is operating in statistical control:

$$\bar{p} = \frac{\sum_{i=1}^{m} p_i}{m} \tag{43.11}$$

The control limits are computed as three standard deviations on either side of the center. Thus,

$$\text{LCL} = \bar{p} - 3\sqrt{\frac{\bar{p}(1-\bar{p})}{n}} \quad \text{and} \quad \text{UCL} = \bar{p} + 3\sqrt{\frac{\bar{p}(1-\bar{p})}{n}} \tag{43.12}$$

where the standard deviation of $\bar{p}$ in the binomial distribution is given by

$$\sigma_p = \sqrt{\frac{\bar{p}(1 - \bar{p})}{n}}$$

If the value of $\bar{p}$ is relatively low and the sample size n is small, then the lower control limit computed by the first of these equations is likely to be a negative value. In this case, let LCL = 0 (the fraction defect rate cannot be less than zero).

c Chart In the c chart (c for count), the number of defects in the sample are plotted over time. The sample may be a single product such as an automobile, and c = number of quality defects found during final inspection. Or the sample may be a length of carpeting at the factory prior to cutting, and c = number of imperfections discovered in that strip. The c chart is based on the Poisson distribution, where c = parameter representing the number of events occurring within a defined sample space (defects per car, imperfections per specified length of carpet). Our best estimate of the true value of c is the mean value over a large number of samples drawn while the process is in statistical control:

$$\bar{c} = \frac{\sum_{i=1}^{m} c_i}{m} \tag{43.13}$$

This value of $\bar{c}$ is used as the center for the control chart. In the Poisson distribution, the standard deviation is the square root of parameter c. Thus, the control limits are

$$\text{LCL} = \bar{c} - 3\sqrt{\bar{c}} \quad \text{and} \quad \text{UCL} = \bar{c} + 3\sqrt{\bar{c}} \tag{43.14}$$

43.5.3 Interpreting the Charts

When control charts are used to monitor production quality, random samples are drawn from the process of the same size n used to construct the charts. For $\bar{x}$ and R charts, the $\bar{x}$ and R values of the measured characteristic are plotted on the control chart. By convention, the points are usually connected, as in our figures. To interpret the data, one looks for signs that indicate the process is not in statistical control. The most obvious sign is when $\bar{x}$ or R (or both) lie outside the LCL or UCL limits. This indicates an assignable cause such as bad starting materials, new operator, broken tooling, or similar factors. An out-of-limit $\bar{x}$ indicates a shift in the process mean. An out-of-limit R shows that the variability of the process has changed. The usual effect is that R increases, indicating variability has risen. Less obvious conditions may be revealed, even though the sample points lie within the $\pm 3\sigma$ limits. These conditions include: (1) trends or cyclical patterns in the data, which may mean wear or other factors that occur as a function of time; (2) sudden changes in average level of the data; and (3) points consistently near the upper or lower limits.

The same kinds of interpretations that apply to the $\bar{x}$ chart and R chart are also applicable to the p chart and c chart.

REFERENCES

[1] Box, G. E. P., and Draper, N. R. *Evolutionary Operation: A Statistical Method for Process Improvement.* Wiley, New York, 1998.

[2] Crosby, P. B. *Quality is Still Free.* McGraw-Hill, New York, 1999.

[3] Evans, J. R., and Lindsay, W. M. *The Management and*

Control of Quality. 4th. ed. South-Western, Cincinnati, Ohio, 1998.

[4] Groover, M. P. *Automation, Production Systems, and Computer Integrated Manufacturing.* 2nd. ed. Prentice Hall, Upper Saddle River, N.J., 2001.

[5] Juran, J. M., and Gryna, F. M. *Quality Planning and Analysis.* 3rd. ed. McGraw-Hill, New York, 1993.

[6] Lochner, R. H., and Matar, J. E. *Designing for Quality.* ASQC Quality Press, Milwaukee, 1990.

[7] Montgomery, D. C. *Introduction to Statistical Quality Control.* 4th. ed. Wiley, New York, 2000.

[8] Pyzdek, T. *Quality Engineering Handbook.* Marcel Dekker, Inc., New York, 1999.

[9] Taguchi, G., Elsayed, E. A., and Hsiang, T. C. *Quality Engineering in Production Systems,* McGraw-Hill, New York, 1989.

[10] Wick, C., and Veilleux, R. F. *Tool and Manufacturing Engineers Handbook.* 4th. ed. Vol. IV: *Quality Control and Assembly.* Society of Manufacturing Engineers, Dearborn, Mich., 1987.

REVIEW QUESTIONS

43.1. What are the two principal aspects of product quality?

43.2. How is a process operating in statistical control distinguished from one that is not?

43.3. Define *process capability.*

43.4. What is the difference between control charts for variables and control charts for attributes?

43.5. Identify the two types of control charts for variables.

43.6. What are the two basic types of control charts for attributes?

43.7. When interpreting a control chart, what does one look for to identify problems?

MULTIPLE CHOICE QUIZ

There is a total of 12 correct answers in the following multiple choice questions (some questions have multiple answers that are correct). To attain a perfect score on the quiz, all correct answers must be given, since each correct answer is worth 1 point. For each question, each omitted answer or wrong answer reduces the score by 1 point, and each additional answer beyond the number of answers required reduces the score by 1 point. Percentage score on the quiz is based on the total number of correct answers.

43.1. Which of the following would be classified as examples of a product feature, rather than a freedom from deficiency (more than one)? (a) components within tolerance, (b) location of ON/OFF switch, (c) no missing parts, (d) product weight, and (e) reliability.

43.2. If the product tolerance is set so that the process capability index = 1.0, then the percentage of parts that is within tolerance will be closest to which one of the following when the process is operating in statistical control? (a) 35%, (b) 65%, (c) 95%, (d) 99%, or (e) 100%.

43.3. Which of the following principles and/or approaches are generally credited to G. Taguchi (more than one)? (a) acceptance sampling, (b) control charts, (c) loss function, and (d) robust design.

43.4. In a control chart, the upper control limit is set equal to which one of the following? (a) process mean, (b) process mean plus three standard deviations, (c) upper design tolerance limit, or (d) upper value of the maximum range R.

43.5. The R chart is used for which one of the following product or part characteristics? (a) number of rejects in the sample, (b) number of reworked parts in a sample, (c) radius of a cylindrical part, or (d) range of sample values.

43.6. Which one of the following best describes the situations for which the c chart is most suited? (a) control of defective parts, (b) mean value of part characteristic of interest, (c) number of defects in a sample, or (d) proportion of defects in a sample.

43.7. Which of the following identify an out-of-control condition in a control chart (more than one)? (a) consistently increasing value of $\bar{x}$, (b) points near the center, (c) R outside the control limits of the R chart, and (d) $\bar{x}$ outside the control limits of the $\bar{x}$ chart.

PROBLEMS

Note: Problems identified with an asterisk (*) in this set require use of statistical tables not included in this text.

Process Capability and Statistical Tolerancing

43.1. An automatic turning process is set up to produce parts with a mean diameter = 6.255 cm. The process is in statistical control and the output is normally distributed with a standard deviation = 0.004 cm. Determine the process capability.

43.2. *In the previous problem, the design specification on the part is: diameter = 6.250 ± 0.013 cm. (a) What proportion of parts would fall outside the tolerance limits? (b) If the process were adjusted so that its mean diameter = 6.250 cm and the standard deviation remained the same, what proportion of parts would fall outside the tolerance limits?

43.3. A sheet-metal bending operation produces bent parts with an included angle = 92.1°. The process is in statistical control and the values of included angle are normally distributed with a standard deviation = 0.23°. The design specification on the angle = 90 ± 2°. (a) Determine the process capability. (b) If the process could be adjusted so that its mean = 90.0°, determine the value of the process capability index.

43.4. A plastic extrusion process produces extrudate with a critical cross-section dimension = 28.6 mm. The process is in statistical control and the output is normally distributed with standard deviation = 0.53 mm. Determine the process capability.

43.5. *In the previous problem, the design specification on the part is: diameter = 28.0 ± 2.0 mm. (a) What proportion of parts falls outside the tolerance limits? (b) If the process were adjusted so that its mean diameter = 28.0 mm and the standard deviation remained the same, what proportion of parts would fall outside the tolerance limits? (c) With the adjusted mean at 28.0 mm, determine the value of the process capability index.

43.6. An assembly consists of four components stacked to create an overall dimension of 2.500 in., with a bilateral tolerance T_a = 0.020 in. (±0.010 in.). The dimensions of the individual parts are each 0.625 in. All parts will have identical bilateral tolerances. Determine the tolerance (a) under a worst-case design approach, and (b) using a statistical tolerancing approach.

43.7. An assembly is made by stacking 20 flat pieces of sheet-metal to produce a thick laminated structure. The sheet-metal blanks are all cut with the same punch and die to the desired profile, so that the thick assembly has the same profile. All of the parts are cut from the same sheet-metal coil, whose thickness specification is 1/16 in. ± 0.002 in. The thickness of the final assembly is specified as 1.250 ± 0.010 in. Does a statistical tolerancing approach apply in this situation? Why?

43.8. The assembly in Figure P43.8 has a critical assembly dimension C = 5.000 cm. If each part is made from an independent process with process means for part thickness all set to 2.500 cm and standard deviation = 0.005 cm, what is the process capability of the critical dimension C? Assume the opposite sides of each part on the 2.50 cm dimension are parallel.

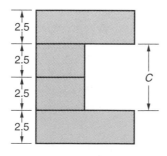

FIGURE P43.8   Assembly for Problem 43.8 (dimensions in cm).

43.9. An assembly consists of three parts stacked to form a final dimension of 30.0 mm with tolerance = ±0.20 mm. The relevant part dimensions making up the 30 mm total are 5 mm, 10 mm, and 15 mm. Parts are produced by independent manufacturing operations, whose process capabilities are proportional to their respective dimensions. Given that the part tolerances are to be a constant proportion of the respective dimensions, determine the tolerance for each part using (a) worst-case design and (b) statistical tolerancing.

43.10. Figure P43.10 shows an assembly in which the critical dimension is C. Each part used in the assembly, including the base part, has a thickness = 10.0 mm, with process capability = ±0.1 mm for the thickness. Given that the process capability index for the parts PCI = 1.0, and the PCI for the assembly will also be 1.0, determine the recommended tolerance for C using: (a) worst-case design and (b) statistical tolerancing.

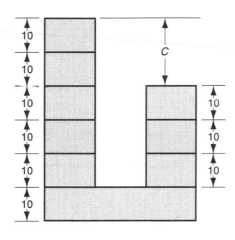

FIGURE P43.10   Assembly for Problem 43.10 (dimensions in mm).

43.11. Solve part (b) of the previous problem, except that the process capability index for the assembly is a more conservative 1.5. The PCI for the individual parts is still 1.0.

Control Charts

43.12. Ten samples of size n = 8 have been collected from a process in statistical control, and the dimension of interest has been measured for each part. (a) Determine the values of the center, LCL, and UCL for the $\bar{x}$ and R charts. The calculated values of $\bar{x}$ and R for each sample are given below (measured values are in mm). (b) Construct the control charts and plot the sample data on the charts.

Step	1	2	3	4	5	6	7	8	9	10
$\bar{x}$	9.22	9.15	9.20	9.28	9.19	9.12	9.20	9.24	9.17	9.23
R	0.24	0.17	0.30	0.26	0.27	0.19	0.21	0.32	0.21	0.23

43.13. Seven samples of five parts each have been collected from an extrusion process that is in statistical control, and the diameter of the extrudate has been measured for each part. (a) Determine the values of the center, LCL, and UCL for $\bar{x}$ and R charts. The calculated values of $\bar{x}$ and R for each sample are given below (measured values are in inches). (b) Construct the control charts and plot the sample data on the charts.

Steps	1	2	3	4	5	6	7
$\bar{x}$	1.002	0.999	0.995	1.004	0.996	0.998	1.006
R	0.010	0.011	0.014	0.020	0.008	0.013	0.017

43.14. In 12 samples of size n = 7, the average value of the sample means is $\bar{\bar{x}}$ = 6.860 cm for the dimension of interest, and the mean of the ranges of the samples is $\bar{R}$ = 0.027 cm. Determine (a) lower and upper control limits for the $\bar{x}$ chart and (b) lower and upper control limits for the R chart. (c) What is your best estimate of the standard deviation of the process?

43.15. In nine samples of size $n = 10$, the grand mean of the samples is $\bar{\bar{x}} = 100$ for the characteristic of interest, and the mean of the ranges of the samples is $\bar{R} = 8.5$. Determine: (a) lower and upper control limits for the $\bar{x}$ chart and (b) lower and upper control limits for the R chart. (c) Based on the data given, estimate the standard deviation of the process.

43.16. A p chart is to be constructed. Six samples of 25 parts each have been collected, and the average number of defects per sample was 2.75. Determine the center, LCL, and UCL for the p chart.

43.17. Ten samples of equal size are taken to prepare a p chart. The total number of parts in these samples was 900, and the total number of defects counted was 117. Determine the center, LCL, and UCL for the p chart.

43.18. The yield of good chips during a certain step in silicon processing of integrated circuits averages 91%. The number of chips per wafer is 200. Determine the center, LCL, and UCL for the p chart that might be used for this process.

43.19. The upper and lower control limits for a p chart are: LCL = 0.19 and UCL = 0.24. Determine the sample size n that is used with this control chart.

43.20. The upper and lower control limits for a p chart are: LCL = 0 and UCL = 0.10. Determine the minimum possible sample size n that is compatible with this control chart.

43.21. Twelve cars were inspected after final assembly. The number of defects found ranged between 87 and 139 defect per car, with an average of 116. Determine the center and upper and lower control limits for the c chart that might be used in this situation.

MEASUREMENT AND INSPECTION

CHAPTER CONTENTS

44.1 Metrology
    44.1.1 Principles of Measurement
    44.1.2 Measurement Standards and Systems
44.2 Inspection Principles
    44.2.1 Testing Versus Inspection
    44.2.2 Manual and Automated Inspection
    44.2.3 Contact Versus Noncontact Inspection
44.3 Conventional Measuring Instruments and Gages
    44.3.1 Precision Gage Blocks
    44.3.2 Measuring Instruments for Linear Dimensions
    44.3.3 Comparative Instruments
    44.3.4 Fixed Gages
    44.3.5 Angular Measurements
44.4 Measurement of Surfaces
    44.4.1 Measurement of Surface Roughness
    44.4.2 Evaluation of Surface Integrity
44.5 Advanced Measurement and Inspection Technologies
    44.5.1 Coordinate Measuring Machines
    44.5.2 Measurements with Lasers
    44.5.3 Machine Vision
    44.5.4 Other Noncontact Inspection Technologies

A basic requirement in manufacturing is that the product and its components meet the specifications established by the design engineer. Design specifications include dimensions, tolerances, and surface finishes of the individual parts comprising the product. These attributes were defined in Chapter 5. Here we consider how to measure and inspect them.

Measurement is a procedure in which an unknown quantity is compared to a known standard, using an accepted and consistent system of units. The measurement may involve a simple linear rule to scale the length of a part, or it may require a sophisticated measurement of force versus deflection during a tension test. Measurement provides a numerical value of the quantity of interest, within certain limits of accuracy and precision.

Inspection is a procedure in which a part or product characteristic, such as a dimension, is examined to determine whether it conforms to the design specification. Many inspection procedures rely on measurement techniques, while others use gaging methods. *Gaging* (also spelled *gauging*) determines simply whether the part characteristic meets or does not meet the design specification—whether the part passes or fails inspection. It is usually faster than measuring, but scant information is provided about the actual value of the characteristic of interest.

Our chapter begins with a discussion of measurement and inspection principles. We then survey the instruments used to measure and inspect part dimensions and surface characteristics—from basic linear rules to computer-automated measuring machines.

44.1 METROLOGY

Metrology is the science of measurement. The science is concerned with six fundamental quantities: length, mass, time, electric current, temperature, and light radiation. From these basic quantities, most other physical quantities are derived, such as area, volume, velocity, acceleration, force, electric voltage, heat energy, and so forth. In manufacturing metrology, our main concern is with measuring the length quantity in the many ways in which it manifests itself in a part or product. These include length, width, depth, diameter, straightness, flatness, and roundness; even surface roughness is defined in terms of length quantities.

44.1.1 Principles of Measurement

Certain concepts and principles apply to virtually all measurements. The most important are accuracy and precision.

Accuracy and Precision *Accuracy* is the degree to which the measured value agrees with the true value of the quantity of interest. A measurement procedure is accurate when it is absent of systematic errors. *Systematic errors* are positive or negative deviations from the true value that are consistent from one measurement to the next.

Precision is the degree of repeatability in the measurement process. Good precision means that random errors in the measurement procedure are minimized. Random errors are usually associated with human participation in the measurement process. Examples include variations in the setup, imprecise reading of the scale, round-off approximations, and so on. Nonhuman contributors to random error include changes in temperature, gradual wear and/or misalignment in the working elements of the device, and other variations. *Random errors* are assumed to obey a normal statistical distribution whose mean is zero and whose standard deviation is given by

$$\sigma = \sqrt{\frac{(x_i - \mu)^2}{n}} \tag{44.1}$$

where σ = standard deviation of the population; x_i = variable of interest; μ = population mean; and n = number of members in the population. The normal distribution has certain well-defined properties, including the fact that 99.73% of the population is included within $\pm 3\sigma$ of the population mean. This is often taken as the indication of a measuring instrument's precision.

The distinction between accuracy and precision is depicted in Figure 44.1. In (a), the random error in the measurement is large, indicating low precision; but the mean value of the measurement coincides with the true value, indicating high accuracy. In (b), the measurement error is small (good precision), but the measured value differs sub-

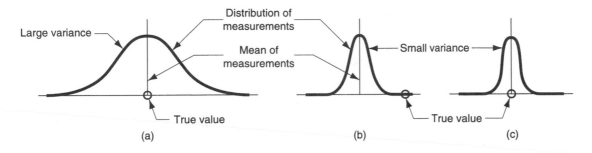

FIGURE 44.1   Accuracy versus precision in measurement: (a) high accuracy but low precision; (b) low accuracy but high precision; and (c) high accuracy and high precision.

stantially from the true value (low accuracy). And in (c), both accuracy and precision are good.

Of course, no measuring instrument can be built that has perfect accuracy (no systematic error) and perfect precision (no random error). Accuracy of the instrument is maintained by proper and regular calibration (explained next). Precision is achieved by selecting the proper instrument technology for the application. A guideline often applied to determine the right level of precision is the ***rule of 10,*** which states that the measuring device must be ten times more precise than the specified tolerance. Thus, if the tolerance to be measured is ±0.25 mm (±0.010 in.), then the measuring device must have a precision of ±0.025 mm (±0.001 in.).

Other Features of Measuring Instruments   Another aspect of a measuring instrument is its capacity to distinguish very small differences in the quantity of interest. The indication of this characteristic is the smallest variation of the quantity that can be detected by the instrument. The terms ***resolution*** and ***sensitivity*** are generally used for this attribute of a measuring device.

Other desirable features of a measuring instrument include ease of calibration, stability, speed of response, wide operating range, high reliability, and low cost. Most measuring devices must be calibrated periodically. ***Calibration*** is a procedure in which the measuring instrument is checked against a known standard. For example, calibrating a thermometer might involve checking its reading in ice (of pure water). For convenience in using the measuring instrument, the calibration procedure should be quick and uncomplicated. Once calibrated, the instrument should be capable of retaining its calibration—continuing to measure the quantity without deviating from the standard. This capability to retain calibration is called ***stability,*** and the tendency of the device to gradually lose its accuracy relative to the standard is called ***drift.***

Some measurements, especially in a manufacturing environment, must be made quickly. The ability of a measuring instrument to indicate the quantity with minimum time lag is called its ***speed of response.*** Ideally, the time lag should be zero; however, this is an impossible ideal. For an automatic measuring device, speed of response is usually taken to be the time lapse between when the quantity of interest changes and the device is able to indicate the change within a certain small percentage of the true value.

The measuring instrument should possess a ***wide operating range,*** which is its capability to measure the physical variable throughout the entire span of practical interest to the user. ***High reliability,*** which can be defined as the absence of frequent failures of the device, and ***low cost*** are, of course, desirable attributes of any engineering equipment.

44.1.2 Measurement Standards and Systems

A common aspect of any measurement procedure is comparison of the unknown value with a known standard. Two aspects of a standard are critical: (1) it must be constant—unchanging over time; and (2) it must be based on a system of units that is consistent and accepted by users. In modern times, standards for length, mass, time, electric current, temperature, and light are defined in terms of physical phenomena that can be relied upon to remain unchanged. For example, the standard for a meter, the basic SI length quantity, is defined as the distance traveled by light in a vacuum in 1/299,792,458 of a second (isn't that helpful?).

Two systems of units have evolved into predominance in the world: (1) the U.S. customary system (U.S.C.S.); and (2) the International System of Units (or SI, for Systeme Internationale d'Unites), more popularly known as the metric system (Historical Note 44.1). Both of these systems are well known. We use both in parallel throughout this book. The metric system is widely accepted in nearly every part of the industrialized world except the United States, which has stubbornly hung on to its U.S.C.S. Gradually, the United States is adopting SI.

Historical Note 44.1   *Measurement systems.*

Measurement systems in ancient civilizations were based on dimensions of the human body. Egyptians developed the cubit as a linear measurement standard around 3000 B.C., which was widely used in the ancient world. The *cubit* was defined as the length of a human arm and hand from elbow to fingertip. Although seemingly Zfraught with difficulties due to variations in arm lengths, the cubit was standardized in the form of master cubit of granite. This standard cubit—524 mm (26.6 in.)—was used to produce other cubit sticks throughout Egypt. The standard cubit was divided into *digits* (a human finger width), with 28 digits per cubit. Four digits equaled a *palm,* and five a *hand.* Thus, a system of measures and standards was developed in the ancient world.

Ultimately, domination of the ancient Mediterranean world passed to the Greeks and then to the Romans. The basic linear measure of the Greeks was the *finger* (about 19 mm or 3/4 in.), and 16 fingers equaled one *foot.* The Romans adopted and adapted the Greek system, specifically the foot, dividing it into 12 parts (called *unciae* by the Romans). The Romans defined five feet as a *pace* and 5,000 feet as a *mile* (such a nice round number—how did we end up with 5,280 feet in a mile?).

During medieval Europe, various national and regional measurement systems developed, many of them based on the Roman standards. Two primary systems emerged in the Western world, the English system and the metric system. The English system defined the yard "as the distance from the thumb-tip to the end of the nose of English King Henry I" [15]. The yard was divide into three feet, and this, in turn, into 12 inches.

Since the American colonies were tied to England, it was natural for the United States to adopt the same system of measurements at the time of its independence. This became the U.S. customary system (U.S.C.S.).

The initial proposal for a metric system is credited to vicar G. Mouton in Lyon, France around 1670. His proposal included three important attributes that were subsequently incorporated into the metric standards: (1) the basic unit was defined in terms of a measurement of the Earth, which was presumed constant—his proposed length measure was based on the length of an arc of one minute of longitude; (2) the units were subdivided decimally; and (3) units had rational prefixes. Mouton's proposal was discussed and debated among scientists in France for the next 125 years. One of the results of the French Revolution was the adoption of the metric system of weights and measures (in 1795). The basic unit of length was the meter, which was then defined as 1/10,000,000 of the length of the meridian between the North Pole and the Equator, and passing through Paris (but of course). Multiples and subdivisions of the meter were based on Greek prefixes.

Dissemination of the metric system throughout Europe during the early 1800s was encouraged by the military successes of French armies under Napoleon. In other parts of the world, adoption of the metric system occurred over many years and was often motivated by significant political changes; this was the case in Japan, China, the Soviet Union, and Latin America.

An act of British Parliament in 1963 redefined the English system of weights and measures in terms of metric units and mandated a changeover to metric two years later, thus aligning Britain with the rest of Europe. This left the United States as the only major industrial nation that was nonmetric. In 1960, an international conference on weights and measures held in Paris reached agreement on new standards based on the metric system. Thus the metric system became the Systeme Internationale (SI).

44.2 INSPECTION PRINCIPLES

Inspection involves the use of measurement and gaging techniques to determine whether a product, its components, subassemblies, or starting materials conform to design specifications. The design specifications are established by the product designer, and for mechanical products they refer to dimensions, surface finish, and similar features. Inspection is performed before, during, and after manufacturing.

Inspections divide into two types: (1) *inspection by variables,* in which the product or part dimensions of interest are measured by the appropriate measuring instruments; and (2) *inspection by attributes,* in which the parts are gaged to determine whether they are within tolerance limits. The advantage of measuring a part dimension is that data are obtained about its actual value. The data might be recorded over time and used to analyze trends in the manufacturing process. Adjustments in the process can be made based on the data so that future parts are produced closer to the nominal design value. When a part dimension is simply gaged, all that is known is whether it is within tolerance or too big or too small. On the other hand, gaging can be done quickly and at low cost.

44.2.1 Testing Versus Inspection

Whereas inspection determines the quality of the product relative to design specifications, testing generally refers to the functional aspects of the product. Does the product operate the way it is supposed to operate? Will it continue to operate for a reasonable period of time? Will it operate in environments of extreme temperature and humidity? In quality control, *testing* is a procedure in which the product, subassembly, part, or material is observed under conditions that might be encountered during service. For example, a product might be tested by operating it for a certain period of time to determine whether it functions properly. If it passes the test, it is approved for shipment to the customer.

Testing of a component or material is sometimes damaging or destructive. In these cases, the items must be tested on a sampling basis. The expense of destructive testing is significant, and great efforts are made to develop methods that do not destroy the item. These methods are referred to as *nondestructive testing* (NDT) or *nondestructive evaluation* (NDE).

44.2.2 Manual and Automated Inspection

Inspection procedures are often performed manually. The work is usually boring and monotonous, yet the need for precision and accuracy is high. Hours are sometimes required to measure the important dimensions of only one part. Because of the time and cost of manual inspection, statistical sampling procedures are generally used to reduce the need to inspect every part.

Sampling Versus 100% Inspection   When sampling inspection is used, the number of parts in the sample is generally small compared to the quantity of parts produced. The sample size may be only 1% of the production run. Because not all of the items in the population are measured, there is a risk in any sampling procedure that defective parts will slip through. One of the goals in statistical sampling is to define the expected risk; that is, to determine the average defect rate that will pass through the sampling procedure. The risk can be reduced by increasing the sample size and the frequency with which samples are collected. But the fact remains that something less than 100% good quality must be tolerated as the price of using a sampling procedure.

Theoretically, the only way to achieve 100% good quality is by 100% inspection; thus, all defects are screened and only good parts pass through the inspection procedure. However, when 100% inspection is done manually, two problems are encountered. The first is the expense involved. Instead of dividing the cost of inspecting the sample over the number of parts in the production run, the unit inspection cost is applied to every part in the batch. Inspection cost sometimes exceeds the cost of making the part. Second, in 100% manual inspection, there are almost always errors associated with the procedure. The error rate depends on the complexity and difficulty of the inspection task and how much judgment is required by the human inspector. These factors are compounded by operator fatigue. Errors mean that a certain number of poor quality parts will be accepted and a certain number of good quality parts will be rejected. Therefore, 100% inspection using manual methods is no guarantee of 100% good quality product.

Automated 100% Inspection   Automation of the inspection process offers a way to overcome the problems associated with 100% manual inspection. *Automated inspection* is defined as automation of one or more steps in the inspection procedure, such as (1) automated presentation of parts by an automated handling system, with a human operator still performing the actual inspection process (e.g., visually examining parts for flaws); (2) manual loading of parts into an automatic inspection machine; and (3) fully automated inspection cell in which parts are both presented and inspected automatically. Inspection automation can also include (4) computerized data collection from electronic measuring instruments.

Automated 100% inspection can be integrated with the manufacturing process to accomplish some action relative to the process. The actions can be one or both of the following: (1) parts sortation, and/or (2) feedback of data to the process. *Parts sortation* means separating parts into two or more quality levels. The basic sortation includes two levels: acceptable and unacceptable. Some situations require more than two levels, such as acceptable, reworkable, and scrap. Sortation and inspection may be combined in the same station. Other installations locate one or more inspections along the processing line, with the sortation station near the end of the line. Inspection data are analyzed and instructions are forwarded to the sortation station indicating what action is required for each part.

Feedback of inspection data to the upstream manufacturing operation allows compensating adjustments to be made in the process to reduce variability and improve quality. If inspection measurements indicate that the output is drifting toward one of the tolerance limits (e.g., due to tool wear), corrections can be made to process parameters to move the output toward the nominal value. The output is thereby maintained within a smaller variability range than possible with sampling inspection methods.

44.2.3 Contact Versus Noncontact Inspection

There are a variety of measurement and gaging technologies available for inspection. The possibilities can be divided into contact and noncontact inspection methods. *Contact inspection* involves the use of a mechanical probe or other device that makes contact with the object being inspected. By its nature, contact inspection is usually concerned with measuring or gaging some physical dimension of the part. It is accomplished manually or automatically. Most of the traditional measuring and gaging devices described in the following section relate to contact inspection. An example of an automated contact measuring system is the coordinate measuring machine (Section 44.5.1).

Noncontact inspection methods utilize a sensor located a certain distance from the object to measure or gage the desired feature(s). Typical advantages of noncontact inspection are (1) faster inspection cycles, and (2) avoidance of damage to the part that might occur due to contact. Noncontact methods can often be accomplished on the production line without any special handling. By contrast, contact inspection usually requires special positioning of the part, necessitating its removal from the production line. Also, noncontact inspection methods are inherently faster because they employ a stationary probe that does not require positioning for every part. By contrast, contact inspection requires positioning of the contact probe against the part, which takes time.

Noncontact inspection technologies can be classified as optical or nonoptical. Prominent among the optical methods are lasers (Section 44.5.2) and machine vision (Section 44.5.3). Nonoptical inspection sensors include electrical field techniques, radiation techniques, and ultrasonics (Section 44.5.4).

44.3 CONVENTIONAL MEASURING INSTRUMENTS AND GAGES

In this section, we consider the variety of manually operated instruments and gages used to measure dimensions such as length, depth, and diameter, as well as features such as angles, straightness, and roundness. This type of equipment is found in metrology labs, inspection departments, and toolrooms. The logical starting topic is precision gage blocks.

44.3.1 Precision Gage Blocks

Precision gage blocks are the standards against which other dimensional measuring instruments and gages are compared. Gage blocks are usually square or rectangular. The measuring surfaces are finished to be dimensionally accurate, parallel to within several millionths of an inch, and polished to a mirror finish. Several grades of precision gage blocks are available, with closer tolerances for higher precision grades. The highest grade—the *master laboratory standard*—is made to a tolerance of ±0.000,03 mm (±0.000,001 in.). Depending on degree of hardness desired and price the user is willing to pay, gage blocks can be made out of any of several hard materials, including tool steel, chrome-plated steel, chromium carbide, or tungsten carbide.

Precision gage blocks are available in certain standard sizes or in sets, the latter containing a variety of different-sized blocks. The gage block sizes in a set are systematically determined so they can be stacked to achieve virtually any dimension desired to within 0.0025 mm (0.0001 in.).

For best results, gage blocks must be used on a flat reference surface, such as a surface plate. A *surface plate* is a large solid block whose top surface is finished to a flat plane. Most surface plates today are made of hard granite. Granite has the advantage of being hard, nonrusting, nonmagnetic, long wearing, thermally stable, and easy to maintain.

Gage blocks and other high-precision measuring instruments must be used under standard conditions of temperature and other factors that might adversely affect the measurement. By international agreement, 20°C (68°F) has been established as the standard temperature. Metrology labs operate at this standard. If gage blocks or other measuring instruments are used in a factory environment where the temperature differs from this standard, corrections for thermal expansion or contraction may be required. Also, working gage blocks used for inspection in the shop are subject to wear and must be calibrated periodically against more precise laboratory gage blocks.

44.3.2 Measuring Instruments for Linear Dimensions

Measuring instruments can be divided into two types: graduated and nongraduated. *Graduated measuring devices* include a set of markings (called *graduations*) on a linear or angular scale to which the object's feature of interest can be compared for measurement. *Nongraduated measuring devices* possess no such scale and are used to make comparisons between dimensions or to transfer a dimension for measurement by a graduated device.

The most basic of the graduated measuring devices is the *rule* (made of steel, and often called a *steel rule*), used to measure linear dimensions. Rules are available in various lengths. Metric rule lengths include 150, 300, 600, and 1000 mm, with graduations of 1 or 0.5 mm. Common U.S. sizes are 6, 12, and 24 in., with graduations of 1/32, 1/64, or 1/100 in.

Calipers are available in either nongraduated or graduated styles. A nongraduated caliper (referred to simply as a *caliper*) consists of two legs joined by a hinge mechanism, as in Figure 44.2. The ends of the legs are made to contact the surfaces of the object being measured, and the hinge is designed to hold the legs in position during use. The contacts point either inward or outward. When they point inward, as in Figure 44.2, the instrument is an *outside caliper* and is used for measuring outside dimensions such as a diameter. When the contacts point outward, it is called an *inside caliper,* which is used to measure the distance between two internal surfaces. An instrument similar in configuration to the caliper is a *divider,* except that both legs are straight and terminate in hard, sharply pointed contacts. Dividers are used for scaling distances between two points or lines on a surface, and for scribing circles or arcs onto a surface.

A variety of graduated calipers are available for various measurement purposes. The simplest is the *slide caliper,* which consists of a steel rule to which two jaws are added, one fixed at the end of the rule and the other movable, shown in Figure 44.3. Slide calipers can be used for inside or outside measurements, depending on whether the inside or outside jaw faces are used. To use, the jaws are forced into contact with the part surfaces to be measured, and the location of the movable jaw indicates the dimension of interest. Slide calipers permit more accurate and precise measurements than simple rules. A refinement of the slide caliper is the *vernier caliper,* shown in Figure 44.4. In this device, the movable jaw includes a vernier scale, named after P. Vernier (1580–1637), a French mathematician who invented it. The vernier provides graduations of 0.01 mm in the SI (and 0.001 in. in the U.S. customary scale), much more precise than the slide caliper.

Variations of the vernier caliper include the *vernier height gage,* used to measure the height of an object relative to a flat surface such as a surface plate, and the *vernier depth gage,* for measuring the depth of a hole, slot, or other recess relative to a top surface.

The *micrometer* is a widely used and very accurate measuring device, the most common form of which consists of a spindle and a C-shaped anvil, as in Figure 44.5. The spindle is moved relative to the fixed anvil by means of an accurate screw thread. On a typical U.S. micrometer, each rotation of the spindle provides 0.025 in. of linear travel. Attached to the spindle is a thimble graduated with 25 marks around its circumference, each mark corresponding to 0.001 in. The micrometer sleeve is usually equipped with a vernier, allowing resolutions as close as 0.0001 in. On a micrometer with metric scale, graduations are 0.01 mm. Modern micrometers (and graduated calipers) are available

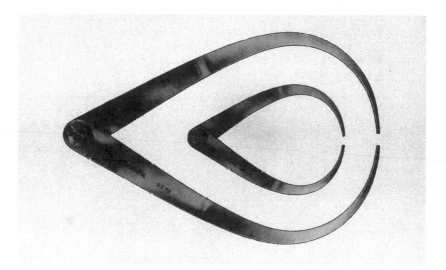

FIGURE 44.2   Two sizes of outside calipers (courtesy of L. S. Starrett Co.).

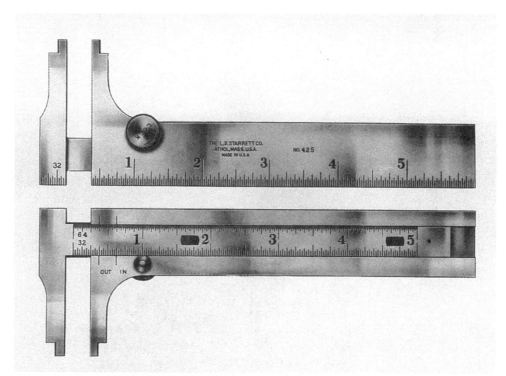

FIGURE 44.3   Slide caliper, opposite sides of instrument shown (courtesy of L. S. Starrett Co.).

with electronic devices that display a digital readout of the measurement (as in our figure). These instruments are easier to read and eliminate much of the human error associated with reading conventional graduated devices.

The most common micrometer types are (1) *external micrometer,* Figure 44.5, also called an *outside micrometer,* which comes in a variety of standard anvil sizes; (2) *internal micrometer,* or *inside micrometer,* which consists of a head assembly and a set of rods of different lengths to measure various inside dimensions that might be encountered; and (3) *depth micrometer,* similar to an inside micrometer but adapted to measure hole depths.

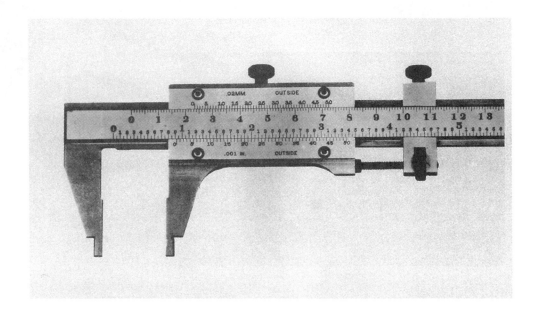

FIGURE 44.4   Vernier caliper (courtesy of L. S. Starrett Co.).

FIGURE 44.5   External micrometer, standard one-inch size with digital read-out (courtesy of L. S. Starret Co.).

44.3.3   Comparative Instruments

Comparative instruments are used to make dimensional comparisons between two objects, such as a workpart and a reference surface. They are usually not capable of providing an absolute measurement of the quantity of interest; instead, they measure the magnitude and direction of the deviation between two objects. Instruments in this category include mechanical and electronic gages.

Mechanical Gages: Dial Indicators   *Mechanical gages* are designed to mechanically magnify the deviation to permit observation. The most common instrument in this cat-

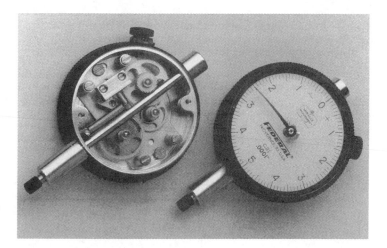

FIGURE 44.6   Dial indicator; top view shows dial and graduated face; bottom view shows rear of instrument with cover plate removed (courtesy of Federal Products Co., Providence, RI).

egory is the ***dial indicator,*** Figure 44.6, which converts and amplifies the linear movement of a contact pointer into rotation of a dial needle. The dial is graduated in small units such as 0.01 mm (or 0.001 in.). Dial indicators are used in many applications to measure straightness, flatness, parallelism, squareness, roundness, and runout. A typical setup for measuring runout is illustrated in Figure 44.7.

Electronic Gages   *Electronic gages* are a family of measuring and gaging instruments based on transducers capable of converting a linear displacement into an electrical signal. The electrical signal is then amplified and transformed into a suitable data format such as a digital readout, as in Figure 44.5. Applications of electronic gages have grown rapidly in recent years, driven by advances in microprocessor technology. They are gradually replacing many of the conventional measuring and gaging devices. Advantages of electronic gages include: (1) good sensitivity, accuracy, precision, repeatability, and speed of response; (2) ability to sense very small dimensions—down to 0.025 μm (1 μ-in.); (3) ease of operation; (4) reduced human error; (5) electrical signal that can be displayed in various formats; and (6) capability to be interfaced with computer systems for data processing.

44.3.4 Fixed Gages

A fixed gage is a physical replica of the part dimension to be inspected or measured. There are two basic categories: master gage and limit gage. A ***master gage*** is fabricated

FIGURE 44.7   Dial indicator setup to measure runout; as part is rotated about its center, variations in outside surface relative to center are indicated on the dial.

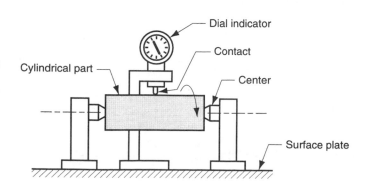

to be a direct replica of the nominal size of the part dimension. It is generally used for setting up a comparative measuring instrument, such as a dial indicator; or for calibrating a measuring device.

A *limit gage* is fabricated to be a reverse replica of the part dimension and is designed to check the dimension at one or more of its tolerance limits. A limit gage often consists of two gages in one piece, the first for checking the lower limit of the tolerance on the part dimension, and the other for checking the upper limit. These gages are popularly known as *GO/NO-GO gages,* because one gage limit allows the part to be inserted while the other limit does not. The *GO limit* is used to check the dimension at its maximum material condition; this is the minimum size for an internal feature such as a hole, and it is the maximum size for an external feature such as an outside diameter. The *NO-GO limit* is used to inspect the minimum material condition of the dimension in question.

Fixed gages must be dimensionally stable and wear resistant. Materials used for these tools are generally alloy steels or tool steels, heat treated and finished to high accuracy. When wear resistance is especially critical, cemented carbide is used. The *rule of 10* is used to determine tolerances for making a fixed gage; that is, the tolerance on the gage dimension is 10% of the tolerance on the part dimension to be checked.

Common limit gages are snap gages and ring gages for checking outside part dimensions, and plug gages for checking inside dimensions. A *snap gage* consists of a C-shaped frame with gaging surfaces located in the jaws of the frame, as in Figure 44.8. It has two gage buttons, the first being the GO gage and the second being the NO-GO gage. Snap gages are used for checking outside dimensions such as diameter, width, thickness, and similar surfaces.

Ring gages are used for checking cylindrical diameters. For a given application, a pair of gages is usually required, one GO and the other NO-GO. Each gage is a ring whose opening is machined to one of the tolerance limits of the part diameter. For ease of handling, the outside of the ring is knurled. The two gages are distinguished by the presence of a groove around the outside of the NO-GO ring.

The most common limit gage for checking hole diameter is the *plug gage.* The typical gage consists of a handle to which are attached two accurately ground cylindrical pieces (plugs) of hardened steel, as in Figure 44.9. The cylindrical plugs serve as the GO and NO-GO gages. Other gages similar to the plug gage include *taper gages,* consisting of a tapered plug for checking tapered holes, and *thread gages,* in which the plug is threaded for checking internal threads on parts.

Fixed gages are easy to use, and the time required to complete an inspection is almost always less than when a measuring instrument is employed. Fixed gages were a fundamental element in the development of interchangeable parts manufacturing (Historical Note 1.1). They provided the means by which parts could be made to tolerances that were

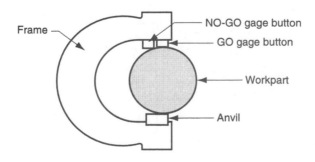

FIGURE 44.8   Snap gage for measuring diameter of a part; difference in height of GO and NO-GO gage buttons is exaggerated.

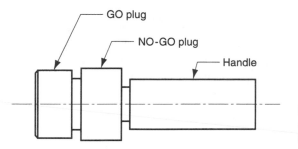

FIGURE 44.9 Plug gage; difference in diameters of GO and NO-GO plugs is exaggerated.

sufficiently close for assembly without filing and fitting. Their disadvantage is that they provide little if any information on the actual part size; they only indicate whether the size is within tolerance. Today, with the availability of high-speed electronic measuring instruments, and with the need for statistical process control of part sizes, use of gages is gradually giving way to instruments that provide actual measurements of the dimension of interest.

44.3.5 Angular Measurements

Angles can be measured using any of several styles of ***protractor.*** A ***simple protractor,*** consists of a blade that pivots relative to a semicircular head that is graduated in angular units (e.g., degrees, radians). To use, the blade is rotated to a position corresponding to some part angle to be measured, and the angle is read off the angular scale. A ***bevel protractor,*** Figure 44.10, consists of two straight blades that pivot relative to each other. The pivot assembly has a protractor scale that permits the angle formed by the blades to be read. When equipped with a vernier, the bevel protractor can be read to about 5 minutes; without a vernier the resolution is only about 1 degree.

High precision in angular measurements can be made using a ***sine bar,*** illustrated in Figure 44.11. One possible setup consists of a flat steel straight edge (the sine bar), and two precision rolls set a known distance apart on the bar. The straight edge is aligned with the part angle to be measured, and gage blocks or other accurate linear measurements are made to determine height. The procedure is carried out on a surface plate to

FIGURE 44.10 Bevel protractor with vernier scale (courtesy of L.S. Starrett Co.).

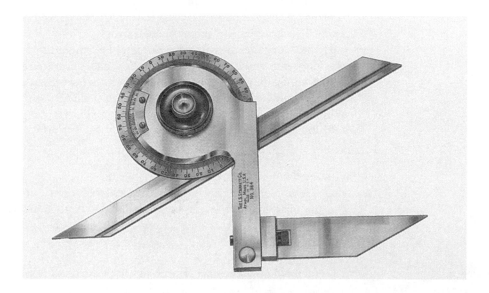

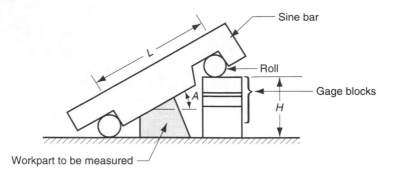

FIGURE 44.11   Setup for using a sine bar.

achieve most accurate results. This height H and the length L of the sine bar between rolls are used to calculate the angle A using

$$\sin A = \frac{H}{L} \tag{44.2}$$

44.4 MEASUREMENT OF SURFACES

In Chapter 5, surfaces are described as consisting of two parameters: (1) surface texture and (2) surface integrity. *Surface texture* refers to the geometry of the surface and is most commonly assessed as *surface roughness* (Section 5.2.2). *Surface integrity* deals with the material characteristics immediately beneath the surface and the changes to this subsurface resulting from the manufacturing processes that created it (Section 5.2.3). In this section, our concern is with the measurement of these two parameters.

44.4.1   Measurement of Surface Roughness

Various methods are used to assess surface roughness. We divide them into three categories: (1) subjective comparison with standard test surfaces, (2) stylus electronic instruments, and (3) optical techniques.

Standard Test Surfaces   Sets of standard surface finish blocks are available, produced to specified roughness values.[1] To estimate the roughness of a given test specimen, the surface is compared to the standard both visually and by the "fingernail test." In this test, the user gently scratches the surfaces of the specimen and the standard set, judging which standard is closest to the specimen. Standard test surfaces are a convenient way for a machine operator to obtain an estimate of surface roughness. They are also useful for design engineers in judging what value of surface roughness to specify on a part drawing.

Stylus Instruments   The disadvantage of the fingernail test is its subjectivity. Several stylus-type instruments are commercially available to measure surface roughness—similar to the fingernail test, but more scientific. An example is the Profilometer, shown in Figure 44.12. In these electronic devices, a cone-shaped diamond stylus with point radius

---

[1]In the U.S.C.S., these blocks have surfaces with roughness values of 2, 4, 8, 16, 32, 64, 128 microinches.

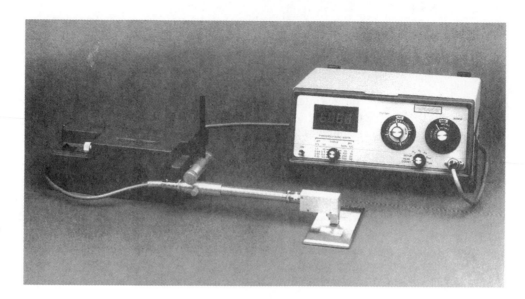

FIGURE 44.12   Stylus-type instrument for measuring surface roughness (courtesy Giddings & Lewis, Measurement Systems Division).

of about 0.005 mm (0.0002 in.) and 90° tip angle is traversed across the test surface at a constant slow speed. The operation is depicted in Figure 44.13. As the stylus head is traversed horizontally, it also moves vertically to follow the surface deviations. The vertical movement is converted into an electronic signal that represents the topography of the surface. This can be displayed as either a profile of the actual surface or an average roughness value. **Profiling devices** use a separate flat plane as the nominal reference against which deviations are measured. The output is a plot of the surface contour along the line traversed by the stylus. This type of system can identify both roughness and waviness in the test surface. **Averaging devices** reduce the roughness deviations to a single value R_a. They use skids riding on the actual surface to establish the nominal reference plane. The skids act as a mechanical filter to reduce the effect of waviness in the surface; in effect, these averaging devices electronically perform the computations in Eq. (5.1).

Optical Techniques   Most other surface-measuring instruments employ optical techniques to assess roughness. These techniques are based on light reflectance from the surface, light scatter or diffusion, and laser technology. They are useful in applications where stylus contact with the surface is undesirable. Some of the techniques permit very high speed operation, thus making 100% inspection feasible. However, the optical

FIGURE 44.13   Sketch illustrating the operation of stylus-type instrument. Stylus head traverses horizontally across surface, while stylus moves vertically to follow surface profile. Vertical movement is converted into either (1) a profile of the surface or (2) the average roughness value.

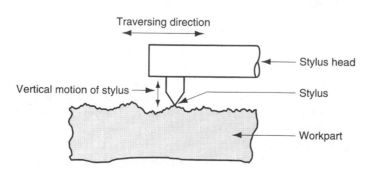

techniques yield values that do not always correlate well with roughness measurements made by stylus-type instruments.

44.4.2 Evaluation of Surface Integrity

Surface integrity is more difficult to assess than surface roughness. Some of the techniques to inspect for subsurface changes are destructive to the material specimen. Evaluation techniques for surface integrity include the following:

> *Surface texture*—Surface roughness, designation of lay, and other measures provide superficial data on surface integrity. This type of testing is relatively simple to perform and is always included in the evaluation of surface integrity.

> *Visual examination*—Visual examination can reveal various surface flaws such as cracks, craters, laps, and seams. This type of assessment is often augmented by fluorescent and photographic techniques.

> *Microstructural examination*—This involves standard metallographic techniques for preparing cross-sections and obtaining photomicrographs for examination of microstructure in the surface layers compared with the substrate.

> *Microhardness profile*—Hardness differences near the surface can be detected using microhardness measurement techniques such as Knoop and Vickers (Section 3.2.1). The part is sectioned, and hardness is plotted against distance below the surface to obtain a hardness profile of the cross section.

> *Residual stress profile*—X-ray diffraction techniques can be employed to measure residual stresses in the surface layers of a part.

44.5 ADVANCED MEASUREMENT AND INSPECTION TECHNOLOGIES

Advanced technologies are substituting for manual measuring and gaging techniques in modern manufacturing plants. They include contact and noncontact sensing methods. In this section we cover: (1) coordinate measuring machines, (2) lasers, (3) machine vision, and (4) other noncontact techniques.

44.5.1 Coordinate Measuring Machines

A *coordinate measuring machine* (CMM) consists of a contact probe and a mechanism to position the probe in three-dimensions relative to surfaces and features of a workpart. See Figure 44.14. The location coordinates of the probe can be accurately recorded as it contacts the part surface to obtain part geometry data.

CMM Construction and Operation In a CMM, the probe is fastened to a structure that allows movement of the probe relative to the part, which is fixtured on a worktable connected to the structure. The structure must be rigid to minimize deflections that contribute to measurement errors. The machine in Figure 44.14 has a bridge structure, one of the most common designs. Special features are used in CMM structures to build high accuracy and precision into the measuring machine, including use of low friction air-bearings and mechanical isolation of the CMM to reduce vibrations. An important aspect in a CMM is the contact probe and its operation. Modern "touch-trigger" probes have a sensitive electrical contact that emits a signal when the probe is deflected from

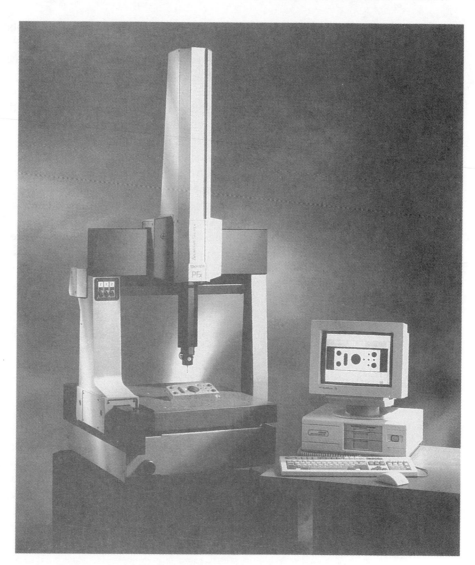

FIGURE 44.14
Coordinate measuring machine (courtesy of Brown & Sharpe Manufacturing Company).

its neutral position in the slightest amount. On contact, the coordinate positions are recorded by the CMM controller, adjusting for overtravel and probe size.

Positioning of the probe relative to the part can be accomplished either manually or under computer control. Methods of operating a CMM are classified as [13] (1) manual control, (2) manual computer-assisted, (3) motorized computer-assisted, and (4) direct computer control.

In ***manual control,*** a human operator physically moves the probe along the axes to contact the part and record the measurements. The probe is free-floating for easy movement. Measurements are indicated by digital read-out, and the operator can record the measurement manually or automatically (paper print-out). Any trigonometric calculations must be made by the operator. The ***manual computer-assisted*** CMM is capable of computer data processing to perform these calculations. Types of computations include simple conversions from U.S.C.S. to SI, determining the angle between two planes, and determining hole-center locations. The probe is still free floating to permit the operator to bring it into contact with part surfaces.

Motorized computer-assisted CMMs power drive the probe along the machine axes under operator guidance. A joystick or similar device is used to control the motion. Low-power stepping motors and friction clutches are used to reduce the effects of collisions between probe and part. The *direct computer-control* CMM operates like a CNC machine tool. It is a computerized inspection machine that operates under program control. The computer also records measurements made during inspection and performs various calculations associated with certain measurements (for example, computing the center of a hole from three or more points on the hole surface).

CMM Measurements and Advantages   The basic capability of a CMM is determination of coordinate values where its probe contacts the surface of a part. Computer control permits the CMM to accomplish more sophisticated measurements and inspections, such as: (1) determining center location of a hole or cylinder, (2) definition of a plane, (3) measurement of flatness of a surface or parallelism between two surfaces, and (4) measurement of an angle between two planes.

Advantages of using coordinate measuring machines over manual inspection methods include [13]: (1) higher productivity—a CMM can perform complex inspection procedures in much less time than traditional manual methods; (2) greater inherent accuracy and precision than conventional methods; and (3) reduced human error through automation of the inspection procedure and associated computations. A CMM is a general-purpose machine that can be used to inspect a variety of part configurations.

44.5.2 Measurements with Lasers

Recall that laser stands for light amplification by stimulated emission of radiation. Applications of lasers include cutting (Section 26.3.3) and welding (Section 31.4.2). These applications involve the use of solid-state lasers capable of focusing sufficient power to melt or sublimate the work material. Lasers for measurement applications are low-power gas lasers such as helium-neon, which emits light in the visible range. The light beam from a laser is: (1) highly monochromatic, which means the light has a single wave length; and (2) highly collimated, which means the light rays are parallel. These properties have motivated a growing list of laser applications in measurement and inspection. We describe two here.

Scanning Laser Systems   The scanning laser uses a laser beam deflected by a rotating mirror to produce a beam of light that sweeps past an object, as in Figure 44.15. A photodetector on the far side of the object senses the light beam during its sweep except for the short time when it is interrupted by the object. This time period can be measured quickly with great accuracy. A microprocessor system measures the time interruption that is related to the size of the object in the path of the laser beam, and converts the time to a linear dimension. Scanning laser beams can be applied in high production on-line inspection and gaging. Signals can be sent to production equipment to make adjustments in the process and/or activate a sortation device on the production line. Applications of scanning laser systems include rolling-mill operations, wire extrusion, machining, and grinding.

Laser Triangulation   Triangulation is used to determine the distance of an object from two known locations by means of trigonometric relationships of a right triangle. The principle can be applied in dimensional measurements using a laser system, as in Figure 44.16. The laser beam is focused on an object to form a light spot on the surface. A position-sensitive optical detector is used to determine the location of the spot. The angle

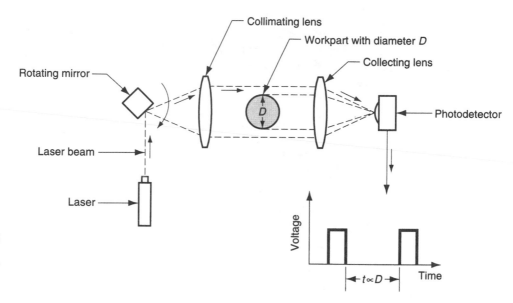

FIGURE 44.15   Scanning laser system for measuring diameter of cylindrical workpart; time of interruption of light beam is proportional to diameter *D*.

A of the beam directed at the object and the distance *H* are fixed and known. Given that the photodetector is located a fixed distance above the worktable, the part depth *D* in the setup of Figure 44.16 is determined from

$$D = H - R = H - L \cot A \tag{44.3}$$

where *L* is determined by the position of the light spot on the workpart.

44.5.3 Machine Vision

Machine vision involves the acquisition, processing, and interpretation of image data by computer for some useful application. Vision systems can be classified as two dimensional or three dimensional. Two-dimensional systems view the scene as a 2-D image, which is quite adequate for applications involving a planar object. Examples include dimensional measuring and gaging, verifying the presence of components, and checking for features on a flat (or almost flat) surface. Three-dimensional vision systems are required for applications requiring a 3-D analysis of the scene, where contours or shapes are involved. The majority of current applications are 2-D, and our discussion will focus (excuse the pun) on this technology.

FIGURE 44.16   Laser triangulation to measure part dimension *D*.

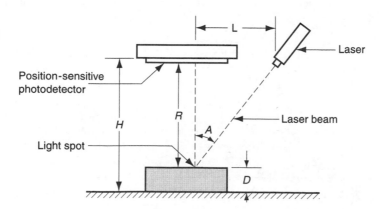

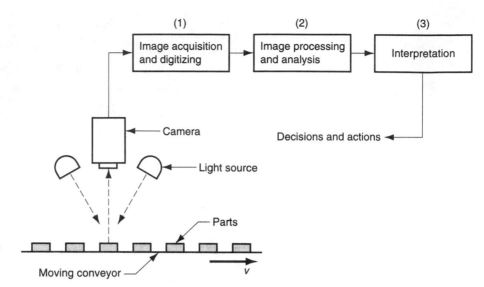

FIGURE 44.17   Operation of a machine vision system.

Operation of Machine Vision Systems   Operation of a machine vision system consists of three steps, depicted in Figure 44.17: (1) image acquisition and digitization, (2) image processing and analysis, and (3) interpretation.

Image acquisition and digitizing is accomplished by a video camera connected to a digitizing system to store the image data for subsequent processing. With the camera focused on the subject, an image is obtained by dividing the viewing area into a matrix of discrete picture elements (called *pixels*), in which each element assumes a value proportional to the light intensity of that portion of the scene. The intensity value for each pixel is converted to its equivalent digital value by analog-to-digital conversion. Image acquisition and digitizing is depicted in Figure 44.18 for a **binary vision** system, in which the light intensity is reduced to either of two values (black or white = 0 or 1), as in Table 44.1. The pixel matrix in our illustration is only 12 × 12; a real vision system would have many more pixels for better resolution. Each set of pixel values is a *frame,* which consists of the set of digitized pixel values. The frame is stored in computer memory. The process of reading all the pixel values in a frame is performed 30 times per second in U.S., 25 cycle/s in European systems.

FIGURE 44.18   Image acquisition and digitizing: (a) the scene consists of a dark-colored part against a light background; (b) a 12 × 12 matrix of pixels imposed on the scene.

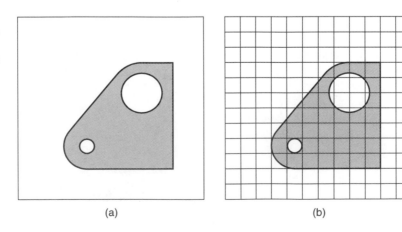

(a)   (b)

TABLE 44.1   Pixel values in a binary vision system for the image in Figure 44.18.

1	1	1	1	1	1	1	1	1	1	1	1
1	1	1	1	1	1	1	1	1	1	1	1
1	1	1	1	1	1	1	1	1	1	1	1
1	1	1	1	1	1	1	0	0	0	1	1
1	1	1	1	1	1	0	1	1	0	1	1
1	1	1	1	1	0	0	1	1	0	1	1
1	1	1	1	0	0	0	0	0	0	1	1
1	1	1	0	0	0	0	0	0	0	1	1
1	1	1	0	1	0	0	0	0	0	1	1
1	1	1	1	0	0	0	0	0	0	1	1
1	1	1	1	1	1	1	1	1	1	1	1
1	1	1	1	1	1	1	1	1	1	1	1

The *resolution* of a vision system is its ability to sense fine details and features in the image. This depends on the number of pixels used. Common pixel arrays include 256×256, 512×512, or 1024×1024 picture elements. The more pixels in the vision system, the higher its resolution. However, system cost increases as pixel count increases. Also, time required to read the picture elements and process the data increase with number of pixels. In addition to binary vision systems, more sophisticated vision systems distinguish various gray levels in the image that permit them to determine surface characteristics such as texture. Called *gray-scale vision,* these systems typically use four, six, or eight bits of memory. Other vision systems can recognize colors.

The second function in machine vision is *image processing and analysis.* The data for each frame must be analyzed within the time required to complete one scan (1/30 or 1/25 s). Several techniques have been developed to analyze image data, including edge detection and feature extraction. *Edge detection* involves determining the locations of boundaries between an object and its surroundings. This is accomplished by identifying contrast in light intensity between adjacent pixels at the borders of the object. *Feature extraction* is concerned with determining feature values of an image. Many machine vision systems identify an object in the image by means of its features. Features of an object include area, length, width, or diameter of the object, perimeter, center of gravity, and aspect ratio. Feature extraction algorithms are designed to determine these features based on the object's area and boundaries. Area of an object can be determined by counting the number of pixels that make up the object. Length can be found by measuring the distance (in pixels) between two opposite edges of the part.

Interpretation of the image is the third function. It is accomplished by extracted features. Interpretation is usually concerned with recognizing the object—identifying the object in the image by comparing it to predefined models or standard values. One common interpretation technique is *template matching,* which refers to methods that compare one or more features of an image with corresponding features of a model (template) stored in computer memory.

Machine Vision Applications   The interpretation function in machine vision is generally related to applications, which divide into four categories: (1) inspection, (2) part identification, (3) visual guidance and control, and (4) safety monitoring.

Inspection is the most important category, accounting for about 90% of all industrial applications. The applications are in mass production, where the time to program and install the system can be divided by many thousands of units. Typical inspection tasks include: (1) *dimensional measurement or gaging,* which involves measuring or gaging certain dimensions of parts or products moving along a conveyor; (2) *verification functions,* which include verifying presence of components in an assembled product, presence of a hole in a workpart, and similar tasks; and (3) *identification of flaws and defects,* such as identifying flaws in a printed label in the form of mislocation, poorly printed text, numbering, or graphics on the label.

Part identification applications include counting different parts flowing past on a conveyor, part sorting, and character recognition. *Visual guidance and control* involve a vision system interfaced with a robot or similar machine to control the movement of the machine. Examples include seam tracking in continuous arc welding, part positioning, part reorientation, and picking parts from a bin (called bin picking). In *safety monitoring* applications, the vision system monitors the production operation to detect irregularities that might indicate a hazardous condition to equipment or humans.

44.5.4 Other Noncontact Inspection Techniques

In addition to optical inspection methods, there are various nonoptical techniques used in inspection. These include sensor techniques based on electrical fields, radiation, and ultrasonics.

Under certain conditions, *electrical fields* created by an electrical probe can be used for inspection. The fields include reluctance, capacitance, and inductance; they are affected by an object in the vicinity of the probe. In a typical application, the workpart is positioned in a fixed relationship to the probe. By measuring the effect of the object on the electrical field, an indirect measurement of certain part characteristics can be made, such as dimensional features, thickness of sheet material, and flaws (cracks and voids below the surface) in the material.

Radiation techniques use x-ray radiation to inspect metals and weldments. The amount of radiation absorbed by the metal object indicates thickness and presence of flaws in the part or welded section. For example, x-ray inspection is used to measure thickness of sheet metal in rolling. Data from the inspection is used to adjust the gap between rolls in the rolling mill.

Ultrasonic techniques use high frequency sound (greater than 20,000 Hz) to perform various inspection tasks. One of the techniques analyses the ultrasonic waves emitted by a probe and reflected off the object. During the setup for the inspection procedure, an ideal test part is positioned in front of the probe to obtain a reflected sound pattern. This sound pattern is used as the standard against which production parts are subsequently compared. If the reflected pattern from a given part matches the standard, the part is accepted. If a match is not obtained, the part is rejected.

REFERENCES

[1] American National Standards Institute, Inc., *Surface Texture.* ANSI B46.1-1978, American Society of Mechanical Engineers, New York, 1978.

[2] American National Standards Institute, Inc. *Surface Integrity.* ANSI B211.1-1986, Society of Manufacturing Engineers, Dearborn, Mich., 1986.

[3] Brown & Sharpe. *Handbook of Metrology.* North Kingston, R.I., 1992.

[4] *Centuries of Measurement.* Sheffield Measurement Division, Cross & Trecker Corporation, Dayton, Ohio, 1984.

[5] DeGarmo, E. P., Black, J. T., and Kohser, R. A. *Materials and Processes in Manufacturing.* 8th ed. Wiley, New York, 1997.

[6] Farago, F. T. *Handbook of Dimensional Measurement.* 3rd ed. Industrial Press Inc., New York, 1994.

[7] *Machining Data Handbook.* 3rd ed. Vol. 2: Machinability Data Center, Cincinnati, Ohio, 1980, Ch 18.

[8] Morris, A. S. *Measurement and Calibration for Quality Assurance.* Wiley, New York, 1998.

[9] Mummery, L. *Surface Texture Analysis—The Handbook.* Hommelwerke Gmbh, Germany, 1990.

[10] Murphy, S. D. *In-Process Measurement and Control.* Marcel Dekker, Inc., New York, 1990.

[11] Nee, J. (ed.). *Fundamentals of Tool Design.* 4th ed. Society of Manufacturing Engineers, Dearborn, Mich., 1998.

[12] Ostwald, P. F., and Munoz, J. *Manufacturing Processes and Systems.* 9th. ed. Wiley, New York, 1997.

[13] Schaffer, G. H. "Taking the Measure of CMMs." Special Report 749, *American Machinist.* October 1982, pp. 145–160.

[14] Schaffer, G. H. "Machine Vision: A Sense for CIM," Special Report 767. *American Machinist.* June 1984, pp. 101–120.

[15] S. Starrett Company. *Tools and Rules.* Athol, Mass., 1992.

[16] Wick, C., and Veilleux, R. F. *Tool and Manufacturing Engineers Handbook.* 4th. ed. Vol. IV: *Quality Control and Assembly.* Society of Manufacturing Engineers, Dearborn, Mich., 1987.

REVIEW QUESTIONS

44.1. How is *measurement* distinguished from *inspection?*

44.2. How does *gaging* differ from *measuring?*

44.3. What are the six fundamental quantities in metrology?

44.4. What is *accuracy* in measurement?

44.5. What is *precision* in measurement?

44.6. What is meant by the term *calibration?*

44.7. Besides good accuracy and precision, what are the desirable attributes and features of a measuring instrument?

44.8. What is the *rule of 10?*

44.9. Automated inspection can be integrated with the manufacturing process to accomplish some action. What are these possible actions?

44.10. Give an example of a noncontact inspection technique.

44.11. What is meant by the term *graduated measuring device?*

44.12. What are the common methods for assessing surface roughness?

44.13. What is a *coordinate measuring machine?*

44.14. Describe a *scanning laser system.*

44.15. What is a *binary vision* system?

44.16. Name some of the nonoptical noncontact sensor technologies available for inspection.

MULTIPLE CHOICE QUIZ

There is a total of 28 correct answers in the following multiple choice questions (some questions have multiple answers that are correct). To attain a perfect score on the quiz, all correct answers must be given, since each correct answer is worth 1 point. For each question, each omitted answer or wrong answer reduces the score by 1 point, and each additional answer beyond the number of answers required reduces the score by 1 point. Percentage score on the quiz is based on the total number of correct answers.

44.1. In measurement and inspection for manufacturing, which one of the following fundamental physical quantities are we most concerned with? (a) electric current, (b) length, (c) light radiation, (d) mass, (e) temperature, or (f) time.

44.2. Which of the following are attributes of the "metric system" of linear measurement (more than one)? (a) based on astronomical distances, (b) defined in terms of the human body, (c) originated in Great Britain, (d) rational prefixes for units, and (e) units are subdivided decimally.

44.3. Which one of the following countries does not embrace the International System of units? (a) China, (b) France, (c) Germany, (d) Japan, (e) Panama, (f) Russia, or (g) United States.

44.4. The two basic types of inspection are inspection by variables and inspection by attributes. The second of these inspections uses which one of the following? (a) destructive testing, (b) gaging, (c) measuring, or (d) nondestructive testing.

44.5. Automated 100% inspection can be integrated with the manufacturing operation to accomplish which of the following actions (more than one)? (a) better design of products, (b) feedback of data to adjust the process, (c) 100% perfect quality, and (d) sortation of good parts from defects.

44.6. Which of the following are examples of contact inspection (more than one)? (a) calipers, (b) coordinate measuring systems, (c) dial indicators, (d) machine vision, (e) micrometers, (f) pneumatic gages, (g) scanning laser systems, (h) snap gages, and (i) ultrasonic techniques.

44.7. A surface plate is most typically made of which one of the following materials? (a) aluminum oxide ceramic, (b) cast iron, (c) granite, (d) hard polymers, or (e) stainless steel.

44.8. Which of the following are graduated measuring instruments (more than one)? (a) bevel protractor, (b) dial indicator, (c) divider, (d) micrometer, (e) outside calipers, (f) sine bar, (g) steel rule, (h) surface plate, and (i) vernier caliper.

44.9. An outside micrometer would be appropriate in the measurement of which of the following (more than one)? (a) hole depth, (b) hole diameter, (c) part length, (d) shaft diameter, and (e) surface roughness.

44.10. In a GO/NO-GO gage, which one of the following best describes the function of the GO gage? (a) checks limit of maximum tolerance, (b) checks maximum material condition, (c) checks maximum size, (d) checks minimum material condition, or (e) checks minimum size.

44.11. Which of the following are likely to be GO/NO-GO gages (more than one)? (a) gage blocks, (b) limit gage, (c) master gage, (d) plug gage, and (e) snap gage.

44.12. Which of the following are contact-sensing methods used in inspection (more than one)? (a) calipers, (b) coordinate measuring machine, (c) laser techniques, (d) machine vision, (e) micrometer, and (f) x-ray radiation.

44.13. Which one of the following is the most important application of vision systems? (a) inspection, (b) object identification, (c) safety monitoring, or (d) visual guidance and control of a robotic manipulator.

INDEX

AA (arithmetic average), 82
Abrasion, 536
Abrasives, 131, 132, 139
Abrasive belt grinding, 602, 603
Abrasive cutoff sawing, 529
Abrasive jet machining (AJM), 614–615
Abrasive machining, 585–606
  buffing, 606
  grinding, *see* Grinding
  history of, 586
  honing, 603–604
  lapping, 604, 605
  polishing, 605, 606
  superfinishing, 605, 606
Abrasive processes, 475
Abrasive water jet cutting (AWJC), 614
Absolute positioning, 863, 864
Accuracy, measurement, 966–967
Acetals, 157
Acid cleaning, 653
Acid core, 743
Acid pickling, 653
Acrylics, 157
Acrylonitrile-butadiene-styrene, 144
Addition polymerization, 146–147
Additives, 154–155, 672
Additive method, 827, 828
Addressable points, 868
Adherends, 745
Adhesion, 537
Adhesives, 745
Adhesive bonding, 11, 16, 744–750,
    836–837
  advantages/disadvantages of, 749, 750
  application technology, 748–750
  and categories of adhesives, 747–748
  definition of, 745
  history of, 745
  and joint design, 746–747
Adjustable bed frame, 459
Advanced composites, 188
Advanced RTM, 326
Aes cyprium, 113
Age hardening, 645
Aggregate production plan, 930, 931
Aging, 645
Agitation, mechanical, 297
Agriculture, 1
Air bending, 447
Air carbon arc cutting, 626
Air-dried sheet, 307
Air vents, 278
AJM, *see* Abrasive jet machining
Alcoa, 110
Alkaline cleaning, 653
Alkyd, 164
Allotropic materials, 30
Allowance, bend, 443, 444
Alloys, 8, 89–93
  aluminum, 109–112
  cast cobalt, 545
  cast irons, 109
  cooling of, 203–205
  copper, 114
  definition of, 89
  eutectic, *see* Eutectic alloys
  low-alloy steels, 103, 104
  melting of, 68

  molybdenum, 119
  nickel, 115
  phase diagrams of, 90–93
  superalloys, 120–122
  tin–lead, 118
  titanium, 116–117
  tungsten, 119
  zinc, 117–118
Alloying, 122
Alloy steels, 94
Altered layer, 80
Alternating copolymers, 151
Alumina, 9, 110, 125, 130, 132, 542
Aluminizing, 656, 663, 664
Aluminum, 15, 109–112
  alloys of, 240
  continuous casting of, 101
  production of, 110
Aluminum Company of America (Alcoa),
  110
Aluminum oxide, 549
Ambient temperature curing, 673
American system, 3
Amino resins, 163
Amorphous (noncrystalline) structures,
  34–36
Analog integrated circuits, 786
Angle of repose, 339
Angular clearance, 439
Angular measurements, 977–978
Anisotropic (etching), 806
Annealing, 253, 640
Anodes, 74
Anodizing, 15, 664, 665
Antioch process, 224
Antioxidants, 155
APCVD, *see* Atmospheric pressure chem-
  ical vapor deposition
Approach (dies), 427
APT (automatically programmed tooling),
  861
Aramids, 159
Arbor (milling), 520
Arc cutting, 625–626
Arc-on time, 695
Arc shielding, 696
Arc strikes, 726
Arc time, 695
Arc welding (AW), 680, 681, 695–705
  and consumable electrodes, 697–703
  definition of, 695
  and nonconsumable electrodes,
    703–705
  technical issues, 695–697
Area defects, 814
Area reduction, 43
Arithmetic average (AA), 82
Arrowhead fracture, 423
Artificial aging, 645
Aspect ratio, 851
Assembly(-ies), 679. *See also* Joining
  design for, *see* Design for assembly
  mechanical, *see* Mechanical assembly
  of metals, 122
  operations, assembly, 11, 16
  process planning for, 917–918
  robotics in, 878
  statistical tolerancing for, 951, 952

Assembly line(s), 4, 19–20, 23, 900–904.
  See also Production lines
  history of, 900
  manual, 901–904
Assignable variations, 949
Assyria, 94
Asynchronous transfer systems, 899
Atactic arrangement, 149–150
Atmospheric pressure chemical vapor
  deposition (APCVD), 670, 671
Atom(s), 24–29
  bonding between, 26–29
  components of, 24–25
  structure of, 25, 26
Atomic number, 24–26
Atomization, 340–341
Attributes, control charts for, 959–960
Attritious wear, 594
Austenite, 94, 95
Austenitizing, 642, 643
Austenitic stainless, 105
Autoclave, 311, 325
Autogenous welding, 681
Automated assembly systems, 905, 906
Automated inspection, 970
Automated production lines, 904–908
  analysis of, 906–908
  automated assembly systems, 905, 906
  multiple station systems, 906
  transfer lines, 904–905
Automated stations, 904
Automated tape-laying machines, 324
Automatically programmed tooling
  (APT), 861
Automatic insertion machines, 831–832
Automatic molds, 286
Automatic tool-changing, 522
Automatic welding, 683
Automatic workpart positioning, 522
Automation, 4
Automotive technologies, 848, 849
Average flow stress, 378
Average roughness, 82
Averaging devices, 979
AW, *see* Arc welding
AWJC, *see* Abrasive water jet cutting

Babylon, ancient, 135
Backer metal, 721
Back plane, 821
Back pressure, 263
Back pressure flow, 263
Back rake angle, 550
Back relief, 427, 428
Backward extrusion, 415
Baekeland, L. H., 144
Bainite, 641
Bakelite, 11, 144, 163
Balancing efficiency, 902
Ball mill, 360–361
Bambooing, 269
Bandsaws, 528
Bardeen, J., 787
Bar drawing, 424
Bar machine, 508
Barrel finishing, 654–655
Barreling, 48
Barrel plating, 662

Basic oxygen furnace (BOF), 22, 94, 98–99
Basic processes, 915
Batch furnaces, 647
Batch model lines, 897
Batch production, 19
Bauxite, 110, 130
Bayer process, 110, 367
BCC (body-centered cubic) crystal structure, 29, 30, 32
Bead coils, 313
Beading, 446, 765, 766
Bearing surface (dies), 427
Bed:
    lathe, 505
    milling, 521
    press, 458
Bell Telephone Laboratories, 787
Bell-type furnaces, 647
Belt(s):
    rubber, 314
    tire, 312
Belt grinding, abrasive, 602, 603
Belt sanding, 603
Benardos, Nikolai, 680
Bench drills, 514
Bend allowance, 443, 444
Bending:
    compression, 470
    draw, 469, 470
    of metal, 377
    of sheet metal, 442–447
    stretch, 469, 470
    of tube stock, 469–470
Bending force, 444, 445
Bending tests, 49–50
Beneficiated sphalerite, 117
Bernoulli's theorem, 199–200
Bessemer, Henry, 94
Bessemer converter, 94, 98, 115
Bethlehem (Pennsylvania), 542
Bevel protractors, 977
Bias tires, belted, 312
Bi-axial winding, 328
Big-end-down mold, 100
Bi-injection molding, 283
Bilateral tolerance, 78
Billet, 388
Bill-of-materials file, 936
Binary phase diagram, 90
Binary vision, 984
Binders:
    in ceramics, 367
    in organic coatings, 671
    in powder metallurgy, 343
Binding, 11
Biotechnology, 850
Blacksmiths, 11, 397
Blades, saw, 556–558
Blank, 437
Blankholder, 453
Blanking:
    chemical blanking, 630–632
    fine, sheet metal, 442
    sheet metal, 437, 457
Blanking clearance, 440
Blanking force, 440
Blank size, 451
Blast finishing, 654
Blast furnaces, 94, 96–97
Blending, of powders, 342–343
Blind holes, 512
Blind risers, 209
Blister copper, 114
Block copolymers, 151
Bloom, 388

Blow-and-blow method, 249
Blow forming, 293
Blowing agents, 297, 298
Blowing process (glass), 249
Blow molding (plastics), 288–291
Blown polymers, 297
BMC, see Bulk-molding compounds
Bobbin, 273
Body-centered cubic (BCC) crystal structure, 29, 30
Boeing 777 (aircraft), 21–22
BOF, see Basic oxygen furnace
Bolster plate, 458
Bolts, 753–755
Bolted joints, 756–758
Boltzmann yield, 815
Bond fracture, 594
Bonding:
    adhesive, 836–837
    die, 812
    molecular, 26–29
    thermocompression, 812
    thermosonic, 812
    ultrasonic, 812
    wire, 812
Bonding materials (for grinding), 588
Borax, 140
Borazon, 134, 542
Boring, 11, 195, 505
Boring machine, 502, 509–511
Boring mills, 509
Boron, 140, 179
Boronizing, 646
Boron nitride, 134
Bose–Einstein statistics, 815
Bottoming, 444
Boule, 789
Box furnaces, 647
Branched polymers, 150–151
Brass, 114, 117
Brattain, W., 787
Braze welding, 740
Brazil, 167
Brazing, 16, 735–740
    definition of, 735
    filler metals/fluxes for use in, 737–739
    joints, brazed, 735–737
    methods of, 738–740
Brazing fluxes, 737, 738
Brazing metal, 735
Breaker plates, 262
Break-in period, 537
Brick, 131
Brinell Hardness Test, 52, 53, 55
Brittle materials, 49–50
Broaches, 556, 557
Broaching, 526–528
Bronze, 113, 114
Bronze Age, 11, 94, 113, 194
Brown & Sharpe, 586
BR (polybutadiene), 169
Brushing, 672
Buffing, 606
Building drum machine, 313
Bulk deformation processes (metal working), 13, 14, 375–376, 386–429
    drawing, 423–429
    extrusion, 413–423
    forging, 397–413
    rolling, 387–397
Bulk micromachining, 852
Bulk-molding compounds (BMC), 321, 325
Buna, 169
Buoyancy (in sand casting), 218, 219
Burnish region, 437

Burrs, 437
Butadiene-acrylonitrile rubber, 170
Butadiene rubber, 169
Butts, 413
Buttercoat, 827
Butt joints (welding), 684
Butyl rubber, 169
Buy or make decision, 918–919

CA, see Cellulose acetate
CAB (cellulose acetate-butyrate), 158
Cable coating, 268, 269
Cable connectors, 841, 842
CAD/CAM-assisted part programming, 872
CAD manufacturing, direct, 774
Calendering, of rubber, 309
Calibration, 967
Calipers, 972–974
Call (molding), 286
Calorizing, 656
Camphor, 144
Cannon, 195, 239
Caoutchouc, 167
Capability, manufacturing, 7–8
Capacity requirements planning, 935, 938–939
Capital goods, 5, 6
CAPP, see Computer-aided process planning
Capscrews, 754
Capstan, 427
Captive threaded fasteners, 755, 756
Carbides, 9
    cemented, 185–186, 369–371, 542, 546–548
    chromium, 186
    coated, 542, 548
    oxide new ceramic, 133
    titanium, 185
    tungsten, 185
"Carboloy," 133
Carbon:
    in ceramics, 138–140
    and fibers, 179
    in stainless steels, 104
Carbon arc welding (CAW), 704
Carbon black, 308
Carbonitriding, 646
Carbon steels, plain, 102
Car-bottom furnaces, 647
Carburizing, 646
Carcass (tires), 312, 313
Card-on-board (COB), 821
Carothers, W., 144
Carriage (lathing), 505
Carrying costs, 931
Case hardening, 645
Cast cobalt alloys, 545
Casting, 11, 13, 69, 193–209, 213–242. See also Bulk deformation processes
    centrifugal, 230–233, 331
    ceramic-mold, 225
    and cleaning, 236
    continuous, 101
    defects in, 237–238
    definition of, 193
    die, 227–230
    drain, 362–363
    expanded polystyrene, 220–222
    and fluidity, 201–202
    furnaces used in, 233–235
    of glass, 249
    heating for, 198–199
    history of, 194–195, 222, 239
    of ingots, 193, 195

inspection methods, 238, 239
investment, 222–224
low-pressure, 226–227
metal, 69, 202–205, 207–208
metals for, 239–240
permanent-mold, 225–233
plaster-mold, 223–225
of polymers, 296–297
and pouring, 199–201, 235, 236
processes involved in, 196–197
and product design, 240–242
of rubbers, 310
sand, 197–198, 214–219
shape, 195
shell molding, 219–220
shrinkage in, 205–209
slip, 362–363
slush, 226
solid, 363
and solidification, 202–205, 207–208
and trimming, 236
vacuum molding, 220, 221
vacuum permanent-mold, 227
wax, 222
Casting alloys, 239–240
Cast iron, 9, 95, 107–109
casting alloys, 239
products, history of, 239
Cast metal, 89
Cast steel, 99–100
Catalyst-activated systems, 162
Catalytic curing, 673
Cathode efficiency, 661
Cathodes, 74
Cavity (of mold), 277
CAW (carbon arc welding), 704
CBN, see Cubic boron nitrides
C chart, 960
CDs (compact discs), 848
Cells, open vs. closed, 298
Cellophane, 144, 158
Cellular layout, 19
Cellular manufacturing, 19, 887–889
Cellular polymers, 297
Celluloid, 144
Cellulose, 158
Cellulose acetate-butyrate (CAB), 158
Cellulose acetate (CA), 144, 158
Cellulose nitrate, 11, 144
Cellulosics, 158
Cemented carbides, 133
in cermet processing, 369–371
as composite materials, 185–186
as cutting tools, 542, 546–548
Cementite, 95
Centers (lathing), 506
Centerburst, 423
Center cracking, 423
Centering, 513
Centerless grinding, 600, 601
Center-type grinding, 598
Centrifugal atomization, 341
Centrifugal casting:
of metal, 230–233
of polymer matrix composites, 331
Centrifugal spraying, 252
Ceramic(s):
as abrasives, 131, 132
boron in, 140
brick, 131
carbide, 133
carbon in, 138–140
as cutting tools, 549
definition of, 125
drying of, 365–366
examples of, 125–126

and fibers, 180
finishing of, 369
hardness tests for, 55
history of, 129–130
importance of, 125
as insulators, 73
mechanical properties of, 127–128
molecular structure of, 36
new, 132–134
nitride, 134
oxide, 132
physical properties of, 128–129
pottery, 131
processing of, 140–141, 358–369
products, 130–132
raw materials, 130
refractory, 131
shaping processes with, 361–365, 367–368
silicon in, 140
sintering of, 366, 368–369
starting materials, preparation of, 359–361, 367
tile, 131
traditional, 129–132
types of, 126
Ceramic coating, 674
Ceramic matrix composites, see under Composite materials
Ceramic-mold casting, 225
Ceramic packaging, 812, 813
Cermets:
as composite materials, 184–185
as cutting tools, 542, 546, 548
definition of, 133
processing of, 369–371
Ceylon, 167
C-frame, 458, 459
Chafee, Edwin, 257
Chain polymerization, 146
Chalcopyrite, 114
Chamfering, 504
Channel bending, 447
Chaplets, 216
Charge carriers, 72, 74
Charge (molding), 285
Charging (of furnace), 247
Chemical blanking, 630–632
Chemical blowing agents, 298
Chemical deposition, 15
Chemical engraving, 632
Chemical fluids, 559
Chemical machining (CHM), 627–633
Chemical milling, 630
Chemical reactions, 537
Chemical reduction, 341–342
Chemical technologies, 849, 850
Chemical vapor deposition (CVD), 668–671, 801–802, 804
Chevron cracking, 423
Chills, 207
Chill-roll extrusion, 271
China, ancient:
and metal forging, 397
pottery in, 129–130
China (ceramic), 130, 131
Chips (integrated circuits), 787, 788
Chip breakers, 550
Chip carriers, 810
Chip formation (metal machining), 481–485
actual chip formation, 483–485
and orthogonal cutting model, 481–483
Chip load, 518
Chip ratio, 482
Chip separation, 812

Chip thickness ratio, 482
Chisel edges, 554
Chloroprene rubber, 170
CHM, see Chemical machining
Chromate coating, 663, 665
Chrome plating, 662
Chromium:
in low-alloy steels, 103
in stainless steels, 104
Chromium carbides, 133, 186
Chromium plating, 662
Chromizing, 646
Chuck, 506, 507
Chucking machine, 508
Chvorinov's Rule, 205
Circuitization, 827
Circuit pattern imaging (PCB production), 825–826
Circular sawing, 529
Civil War, 2, 94
Clamping units, 275, 279–281
Clays, 9, 11, 130
Cleaning. See also Finishing
chemical, 627, 651–654
considerations in, 652
mechanical, 654
in metal casting, 236
of PCBs, 834
in surface mount technology, 838
as surface processing, 15
Clean rooms, 791
Clearance, 438, 439
Cleavage, 49
Climb milling, 516
Closed cells, 298
Closed-die forging, 402
Closed-loop positioning systems, 866–868
Closed molds, 196, 325–327
Closed pores, 338
Cloth, 320
Cluster rolling mills, 394
CMM, see Coordinate measuring machines
CNC (computer numerical control), 508, 509, 862
milling machines, 522
mill-turn center, 524
turning center, 524
turret drill, 514
turret press, 461, 462
winding machines, 329
Coated carbides, 542, 548
Coating processes, 15, 659–676
anodizing, 665
ceramic coatings, 674
chemical vapor deposition, 668–671
conversion coating, 664–665
mechanical plating, 676
organic coatings, 671–674
physical vapor deposition, 665–668
using plastics, 274
using rubbers, 309, 310
Cobalt alloys, cast, 545
Cobalt-based alloys, 121
COB (card-on-board), 821
Coefficient of thermal conductivity, 70
Coefficient of thermal expansion, 68
Coffin, Charles, 680
Cogging, 402
Coiled pins, 762
Coining, 346, 405, 455
Coke, 96–97
Cold-chamber die-casting machines, 228, 229
Cold extrusion, 415, 416
Cold forming (metals), 379

Cold isostatic pressing, 348
Cold roll welding, 720
Cold welding (CW), 720
Cold working, 122, 379
Cold-work tool steels, 107
Collet, 507
Colorants, 154–155
Columns, 525
Combination die, 458
Combined SMT–PIH assembly, 838, 839
Common use items, 935
Compact discs (CDs), 848
Compaction, 343–345, 370
Comparative measurement(s), 974–977
Components, manufactured, 6
Composites (composite materials), 10,
    175–190, 176
  advantages/disadvantages of, 175–176
  ceramic matrix composites (CMCs),
    186–187, 369, 371
  components of, 176–178
  definition of, 175
  metal matrix composites (MMCs),
    184–186, 369
  polymer matrix composites (PMCs),
    187–190
  processing of, 190
  properties of, 181–184
  reinforcing phase for, 178–181
Compound die, 458
Compounding (rubbers), 308
Compression bending, 470
Compression molding, 285–286
Compression section, 261
Compression (sheet metal), 448
Compression tests, 47–49
Computer-aided process planning
    (CAPP), 919–921
  benefits of, 921
  generative CAPP systems, 920–921
  retrieval CAPP systems, 919–920
Computer-assisted part programming, 871
Computer-compatible part description,
    920
Computer numerical control, see CNC
Computer programming languages, 877
Concurrent engineering, 924–926
Condensation polymerization, 148
Conduction, 70, 71
Conductivity, 70, 71, 73
Conductors, 28, 73
Connectors:
  and PCB production, 839–843
  permanent, 840–841
  separable, 841–843
Construction, 1
Consumable electrodes, 696–703
Consumer goods, 5
Contact inspection, 970–971
Contact lamination, 321
Contact molding, 321
Contact printing, 797
Continuity testing, 830
Continuous broaching machines, 527, 528
Continuous casting, 101
Continuous chip, 484
Continuous drawing, 424
Continuous-drive friction welding, 722
Continuous drum cure, 311
Continuous fibers, 178
Continuous furnaces, 647
Continuous laminating, 332
Continuous motion welding, 709
Continuous path control, 876, 877
Continuous path systems, 863
Continuous processes, 6, 416

Continuous tank furnaces, 247
Continuous transfer systems, 898
Contour coating, 274
Contouring, 528, 863
Contour roll forming, 464
Contour turning, 504
Control:
  production, 928–929
  shop floor, 941–944
Control charts, 956–960
Control resolution (in NC), 868, 870
Control systems for industrial robotics,
    876, 877
  intelligent control, 877
  limited sequence control, 876
  playback with continuous path control,
    876, 877
  playback with point-to-point control,
    876
Conventional face milling, 517, 518
Conventional injection molding, 327
Conventional milling, 516
Conventional spinning, 465–466
Conversion coating, 664–665
Coolants, 558
Coolidge, William, 336
Coolidge process, 336
Cooling system, 278
Coordinate measuring machines, 980–982
Coordinate measuring machines (CMM),
    980–982
Coordinate system (in NC), 863
Cope, 197
Cope-and-drag patterns, 216
Copolymers, 151–152
Copper, 11, 113–115
  alloys of, 90–92, 240
  continuous casting of, 101
Copper-nickel alloy system, 90–92
Copper plating, 662
Cores, 216
Corner joint (welding), 684, 685
Corner radii, 408
Corrugating, 447
Corundum, 130
Cost(s), 967
  carrying costs, 931
  minimizing per unit, 575–576
  stock-out costs, 931
Cotter pins, 764
Counterboring, 513
Countersinking, 513
Covalent bonds, 27, 28
CO*SB1*2*SB0* welding, 700
Cracking, 128, 423, 725–726
Crater wear, 535–537
Creep feed grinding, 600, 601
Crete, ancient, 397
Crimping, 765, 840
Crossfeed, 591
Cross-linked polymers, 151
Cross-linking, 155
  in plastics processing, 257
  in vulcanization, 168
Cross-slide, 505
Cross-wire welding, 710
Crown glass, 137, 250
Crucible furnaces, 233–235
Crushing, 359–360
Crystal growing, 793–794
Crystalline ceramics, 9
Crystalline structures, 29–34
  definition of, 29
  deformation in metallic, 32–34
  imperfections in, 30–32
  metallic, 32–34

  types of, 29, 30
Crystallinity (of polymers), 152–153
Crystallites, 152
Crystal-to-slice yield, 814
Crystal yield, 813
Cubic boron nitrides (cBN), 134, 542, 549,
    586
Cubits, 968
Cup drawing, 377, 450
Cupolas (furnaces), 97, 233, 234
Cuprium, 113
Curing:
  of adhesives, 745
  of organic compounds, 673
  solid ground, 777, 778
  of thermosets, 162, 324, 325
Curing time, 745
Curling, 446
Customer relations, 955
Cut, depth of, 479, 480
Cut and peel method, 628
Cutoff, 441, 505
Cutoff length, 83
Cutoff operations (sawing), 528
Cutting:
  abrasive water jet cutting, 614
  arc cutting, 625–626
  conditions for, 479, 480, 502–503,
    511–513, 517–519
  electric discharge wire (EDWC),
    621–623
  forces in metal cutting, 485–490
  and grinding, 593
  interrupted, 516
  orthogonal cutting model, 481–483
  oxyfuel cutting, 626–627
  of polymer matrix composites, 332
  selection of conditions for, 572–578
  of sheet metal, 436–442
  temperature, cutting, 493–495
  tools for, see Cutting tools
  water jet cutting, 613–614
Cutting feed, 479, 480
Cutting fluids, 480, 558–560
  application of, 559–560
  chemical formulation of, 559
  filtration, 560
  functions of, 558–559
Cutting forces, 439, 440, 486, 487
Cutting oils, 559
Cutting speed, 479, 480, 573–578
Cutting tools, 478–479, 534–535
  failure of, 534–535
  fluids, cutting, 558–560
  geometry of, 550–558
  inserts, 552–554
  life of, 535–537
  materials for, 541–549
  wear of, 535–537
CVD, see Chemical vapor deposition
CW (cold welding), 720
Cylindrical grinding, 598–600
Cyprium, 113
Cyprus, 113
Czochralski process, 793

Danner process, 251
Dark Ages, 194
Davy, Humphrey, 110, 680
Day tanks, 247
DDM, see Droplet deposition manufac
    turing
Dead center, 506
Debinding (ceramics), 368
Deburring, electrochemical, 618
Deep drawing, 377

Deep grinding, 601
Defects:
  area, 814
  in casting, 237–238
  in crystals, 30–32
  in extrusion of metals, 422–423
  in extrusion of plastics, 269–270
  point, 30, 31, 814
  and polymer injection molding, 281–282
  in sheet metal drawing, 453–454
  in welding, 725–727
Deficiencies, freedom from, 948
Deflocculants, 343, 367
Deformation processes, *see* Bulk deformation processes
Degree of crystallinity, 152
Degree of polymerization (DP), 149
Demand:
  dependent, 932
  independent, 931–932
Demasking, 628
Dendritic growth, 203
Densification of powders (powder metallurgy), 346
Density:
  of material, 67–68
  of powders, 339
  and temperature, 68
Dependent demand, 932
Deposition processes, *see under* Coating processes
Depth micrometers, 973
Depth of cut, 479, 480
Designation scheme:
  for aluminum alloys, 111, 112
  for copper alloys, 115
  for magnesium, 112, 113
Design (design considerations):
  for casting, 239–240
  for ceramics, 371
  for glassworking, 254
  for integrated circuits packaging, 809–811
  in machining, 578–580
  manipulator (robotics), 873–875
  in manual assembly lines, 903–904
  parameter, 955
  for polymers, 299–301
  for powder metallurgy part designs, 352–354
  process, 954
  product, 954
  for rubbers, 315
  system, 954–955
  tolerance, 955
  in welding, 729
Design for assembly (DFA), 766–768
Design for life cycle, 925, 926
Design for manufacturing and assembly (DFM/A), 922–925
Design for product cost, 925
Design for quality (DFQ), 925
Design of risers, 208–209
Desktop machining, 773
Desktop milling, 773
Destructive testing (welding), 727–728
Devitrification, 136
DFA, *see* Design for assembly
DFM/A, *see* Design for manufacturing and assembly
DFQ (design for quality), 925
DFW, *see* Diffusion welding
Diagonal ply tires, 312
Diagrams, phase, 90–93
Dial indexing machine, 905
Dial indicators, 975
Diamonds, 27, 28

and ceramics, 139–140
  as cutting tools, 542
  synthetic, 549
Die bonding (IC packaging), 812
Die casting, 227–230
Die characteristic, 266
Die configurations, 267–269
Die holders, 457
Dielectric insulators, 73–74
Dielectric strength, 74
Dies:
  draw, 427, 428
  extrusion, 419–421
  forging, 407
  for integrated circuits, 787
  master, 222
  for metal forming, 374, 376
  for sheet metalwork, 457–459
Die sinking, 517
Die swell, 259
Diffusion, 537, 656–657
  mass, 71–72
  thermal, 70, 802–803
Diffusion coating, 656–657
Diffusion welding (DW), 72, 682, 720, 721
Digital integrated circuits, 786, 787
Digits, 968
Dimensional measurement, 985–986
Dimensions, 77, 78
Dimpling, 765, 766
DIP, *see* Dual in-line package
Dip brazing, 740
Dip casting, 310
Dip coating, 673
Dipole forces, 28, 29
Dipping method, 274, 310
Direct CAD manufacturing, 774
Direct computer-control CMM, 982
Direct extrusion, 413–414, 417
Direct-fired furnaces, 647
Direct fuel-fired furnaces, 233
Directional solidification, 207
Direct RTM method, 783
Disc grinders, 602
Discontinuous chip, 484
Discontinuous fibers, 178
Discrete items, 6
Discrete processing, of metal extrusion, 416
Dislocations, 31–33
Displaced ions, 31
Dividers (measuring device), 972
Division of labor, 3
DMC, *see* Dough-molding compounds
DN ratio, 530
Doctor blade method (ceramics), 368
Doctor blade method (plastics), 274
Doehler, H., 228
Dog (lathing), 506, 507
Doping (semiconductors), 657, 802–803
Double-column planers, 526
Double-sided boards, 823, 827, 828
Dough-molding compounds (DMC), 321
Dowel pins, 762
Down milling, 516, 517
Downsprue, 197–198
DP (degree of polymerization), 149
Draft, 389, 408
Drag, 197
Drag flow, 263
Drain casting, 362–363
Draw bench, 427
Draw bending, 469, 470
Draw dies, 427, 428
Drawing:
  of glass, 251, 252
  of metals, 376, 423–429

reverse, 452
  of sheet metal, 447–454
Drawing force, 451
Drawing ratio, 450
Drift, 967
Drill bit, 511
Drilling, 11, 14, 15
  in machining operations, 478, 505, 511–515
  with twist drills, 554–555
Drill jigs, 514
Drill presses, 481, 502, 511, 514–515
Drills, 514–515
Drive-in step (doping), 803
Drive systems, for sheet metal presses, 463
Drop hammers, 406, 407
Droplet deposition manufacturing (DDM), 777–779
Drying (ceramics), 365 366
Drying (organic coatings), 673
Dry machining, 560
Dry plasma etching, 805, 806
Dry pressing (ceramics), 364, 365
Dry-sand molds, 217
Dry spinning, 273
Dry winding, 327
Dual in-line package (DIP), 810, 811
Ductile iron, 108
Ductile solids, 13
Ductility, 36, 42, 43
Dunlop, John, 167
Duplex mills, 521
Duplex stainless, 105
Du Pont Company, 144, 169
Durometer, 54
DW, *see* Diffusion welding
Dyes, 155, 672
Dyeing, 11
Dye-penetrant tests (welding), 727

Earing, 453, 454
Earthenware, 131
EB, *see* Electron beam heating
E-beam lithography, 799
EBM, *see* Electron-beam machining
Ebonite, 144
EBW, *see* Electron-beam welding
ECD (electrochemical deburring), 618
ECEA (end cutting edge angle), 550
ECG, *see* Electrochemical grinding
ECM, *see* Electrochemical machining
Economic order quantity, 932–934
Economics, machining, 573
Edge bending, 443
Edge detection, 985
Edge dislocation, 31
Edge joint (welding), 684
Edge preparation, cutting tool, 553–554
Edging, 402
EDM, *see* Electric discharge machining
EDWC, *see* Electric discharge wire cutting
EDWC (electric discharge wire cutting), 621–623
Effectors, end, 876
Efficiency:
  balancing, 902
  cathode, 661
  welding, 688
EGW, *see* Electrogas welding
Egypt, ancient, 94, 129, 135, 194, 222, 336, 397, 502, 586, 680, 745, 968
Ejection system, 278
Ejector pins, 278
Elastic deformation, 32
Elasticity, 41, 42, 46, 47, 51
Elastic limit, 42

Elastic region, 44
Elastic reservoir molding (ERM), 326
Elastomers, 10, 165–172
    characteristics of, 166–167
    molecular structure of, 37
    as polymer, 61
    and rubber polymers, 144
    thermoplastic, *see* Thermoplastic elastomers
Elastomeric foams, 297
Electrical connector technology (PCB production), 839–843
Electrical fields inspection, 986
Electric-arc furnaces, 99, 235
Electric discharge machining (EDM), 619–623, 856
Electric discharge wire cutting (EDWC), 621–623
Electric furnaces, 94, 247, 647
Electricity, 72–74
Electric power, 4
Electrochemical grinding (ECG), 618, 619
Electrochemical machining (ECM), 75, 615–618
Electrochemical plating, 660
Electrochemistry, 74–75
Electrocleaning, 653
Electrocoating, 673
Electrodes:
    in arc welding, 696–705
    in electrochemistry, 74
Electroforming, 662–663, 856
Electrogas welding (EGW), 701, 702
Electrohydraulic forming, 468
Electroless plating, 663, 856
Electrolysis, 74–75, 342
    for aluminum production, 110
    for copper production, 114
    for nickel production, 115
    of zinc, 117
Electrolytes, 74
Electrolytic cells, 74
Electrolytic cleaning, 653
Electrolytic iron, 94
Electromagnetic forming, 468, 469
Electrons, 24–29
Electron-beam (E-beam) lithography, 799
Electron beam (EB) heating, 648, 649
Electron beam evaporation, 667
Electron-beam machining (EBM), 623–624, 856
Electron-beam welding (EBW), 681, 716–717
Electron cloud, 28
Electronic gages, 975
Electronic grade silicon, 792
Electronics, 12, 16, 850
Electronics assembly and packaging, 819–821. *See also* Integrated circuits; Microfabrication techniques
Electroplating, 15, 75, 660–662, 804, 856
Electroslag welding (ESW), 717–718
Electrostatic spraying, 672–673
Elements, 25
Elemental powders, 351–352
Elevated temperature curing, 673
Elongation, 42, 43
Embossed protrusions, 765, 766
Embossing, 455
Emulsified oils, 559
Emulsion cleaning, 653
Enameling, porcelain, 15
Encapsulation, 297
End cutting edge angle (ECEA), 550
End effectors (robotics), 876
End milling, 517, 518

End milling cutters, 556
End relief angle (ERA), 550
Energy, 490–493
Engineering:
    concurrent, 924–926
    manufacturing, 912–913
    molecular, 857–858
Engineering materials, 36–37
Engineering strain, 41
Engineering stress, 40, 41
Engine lathe technology, 505–506
England, 94, 413, 502, 542, 900, 968
English system of weights and measures, 968–969
Engraving, chemical, 632
Entry region (dies), 427
Environmental technologies, 849, 850
EPDM (ethylene–propylene rubber), 170
Epitaxial deposition, 802
Epitaxial grain growth, 690
Epoxides, 163
Epoxy, 144, 163
Epoxy die bonding, 812
Equipment, production, 16
ERA (end relief angle), 550
ERM (elastic reservoir molding), 326
Errors, random, 966
ESW, *see* Electroslag welding
Etchant, 628, 629
Etching, 628, 629, 804–806, 825–826
Etch selectivity, 805
Eutectic alloys, 93, 204–205
Eutectic composition, 205
Eutectic die bonding, 812
Eutectic temperature, 93, 205
Eutectoid composition, 95
Eutektos, 93
EUV (extreme ultraviolet lithography), 799
Evaporative-foam process, 220
Excessive spatter, 726
Expanded polymers, 297
Expanded polystyrene casting process, 220–222
Expansion fits, 16
Expansion fittings, 762, 763
Expendable mold, 197
Expendable mold casting, 213–225
    ceramic-mold casting, 225
    expanded polystyrene process, 220–222
    investment casting, 222–224
    plaster-mold casting, 223–225
    sand casting, 214–219
    shell molding, 219–220
    vacuum molding, 220, 221
Expert systems, 920
Explosion cladding, 721
Explosion welding (EXW), 721–722
Explosive forming, 467, 468
Extenders (to polymers), 190
External broaching, 527
External centerless grinding, 600
External chills, 207
External cylindrical grinding, 598, 599
External micrometers, 973
External noise factors, 954
Extreme pressure lubrication, 558
Extrudate, 260
Extruded plastics, design considerations for, 300
Extruders, 265, 266
Extrusion, 14
    of cemented carbides, 370
    of ceramics, 364
    of metals, 376, 413–423
    of polymers, 260–272

    of polystyrene foams, 298
    of powders (powder metallurgy), 350
    of rubbers, 309
    slit-die, 270, 271
Extrusion blow molding, 288–290
Extrusion strain formula, Johnson's, 417
EXW, *see* Explosion welding
Eyelets, 760

Fabrication, solid freeform, 774
Face-centered cubic (FCC) crystal structure, 29, 30
Face milling, 478, 517–519
Face milling cutters, 556
Face plate, 507
Facilities, production, 17–20
Facing, 504, 675
Factory system, 3
Failure, tool, 535
Failure region, 537, 538
Fairchild Semiconductor Corp., 787
Faraday, Michael, 75, 169
Faraday's Laws, 75, 660
Fasteners, 11, 752
    integral, 765–766
    threaded, 16, 753–759
Fayin surfaces, 680
FCAW, *see* Flux-cored arc welding
FCC (face-centered cubic) crystal structure, 29, 30, 32
FDM (fused-deposition modeling), 780
Feature extraction, 985
Feed, cutting, 479, 480
Feedback, 970
Feed section, 261
Feldspar, 130
Ferrite, 94, 95
Ferritic stainless, 105
Ferrous casting alloys, 239–240
Ferrous metals, 8, 9, 89, 93–109. *See also* Iron; Steel(s)
Fibers, 178–180
    glass, 251, 252
    and polymer matrix composites, 319
    production of, 272–274
Fiberglass, 179
Fiber-reinforced composites, 182–183
Fiber-reinforced metal matrix composites, 186
Fiber-reinforced polymers (FRPs), 188–190, 317
Fick's first law, 71
Filaments, 272–274. *See also* Fibers
Filament winding, of polymer matrix composites, 327–329
Fillers (in polymers), 154, 190
Filler material, 679
Filler metals (brazing), 737, 738
Fillet, 408
Fillet welds, 684, 685
Film(s):
    polymer, 270
    surface, 340
Filtration, of cutting fluids, 560
Final test yield, 814
Fine blanking, 442
Finger (measurement), 968
Finishes, surface, 82, 568–572
Finishing, 480, 654–656, 915
    buffing, 606
    of ceramics, 369
    of glass, 253–254
    of metals, 122
    of PCBs, 830
    polishing, 605, 606
    of powders (powder metallurgy), 348

superfinishing, 605, 606
surface, 592
Firing, 11, 130, 358, 366, 368–369
Fishtailing, 423
Fitters, 682
Fittings, 762, 763
Fixed gages, 975–977
Fixed mandrel, 429
Fixed-position layout, 18
Fixture, 514
Flakes, 180, 320, 338
Flame cutting, 626
Flame hardening, 648
Flame retardants, 155
Flange welds, 686
Flanging, 446
Flank wear, 535–537
Flash:
    die casting, 230
    forging, 398, 408
Flashing, 282, 711
Flashless forging, 398, 404–406
Flash welding (FW), 711
Flask, 197, 217
Flaskless molding, 217
Flat glass, 250
Flat rolling, 389–393
Flat welding, 729
Flaws (surface texture), 81
Flexible foams, 297
Flexible manufacturing cells, 889
Flexible manufacturing systems (FMS),
    889–894
    applications of, 893–894
    cells in, 889–890
    classification of, 890
    definition of, 889
    hardware components in, 891
    human labor in, 891, 893
    integration of components in, 891–893
    software and control functions in, 891
Flexible overlay process, 675–676
Flexure tests, 49–50
Flint glass, 137
Floating plug, 429
Float process (glass), 251
Flooding application, 560
Flow:
    back pressure, 263
    of metals, 377–378
    of powders, 339
Flow coating, 673
Flow curve, 45
Flow line production, 19
Flow stress, 377–378
Flow turning, 466
Fluids:
    cutting, 480
Fluidity, 57, 201–202
Fluidized bed furnaces, 647–648, 674
Fluids:
    cutting, 558–560
    grinding, 596–597
    properties of, 57–59
Fluorescent-penetrant tests (welding), 727
Fluorine, 26–28
Fluorocarbons, 144
Fluoropolymers, 158–159
Flutes, 554
Flux(es), 696, 712
    brazing, 737, 738
    soldering, 742, 743
Flux-cored arc welding (FCAW),
    700–701
Fly-cutters, 515
Flyer plate, 721

Flywheel, 463
FMS, *see* Flexible manufacturing systems
Foams:
    flexible, 171
    polystyrene, 298–299
    thermoplastic, 283
Foamed material, 184
Foot (measurement), 968
Footwear, rubber, 315
Forces:
    bending, 444, 445
    drawing, 451
    holding, 451
Ford, Henry, 4, 900
Ford Motor Company, 900
Forge hammers, 16
Forge welding, 11, 719
Forging, 14
    equipment for, 405–408
    flashless forging, 404–406
    hobbing, 411, 412
    hot die forging, 412
    impression-die forging, 402–404
    isothermal forging, 412
    of metals, 376, 397–413
    open-die forging, 399–402
    orbital forging, 410, 411
    of powders (powder metallurgy), 350
    radial forging, 409, 410
    roll forging, 410, 411
    spin forging, 466
    swaging, 409, 410
    and trimming, 412–413
    upset forging, 408–409
Forging dies, 407
Forging hammers, 398, 406–407
Forging presses, 398, 407
Forming, 504
    electrohydraulic, 468
    electromagnetic, 468, 469
    explosive, 467, 468
    high-energy-rate, 467–469
    shear, 466
Forming machining operations, 501–502
Form milling cutters, 555
Form turning, 504
Forward extrusion, 413
Forward slip, 390
Foundry, 196
Four-high rolling mills, 394
Fracture, 437
    arrowhead, 423
    bond, 594
    melt, 269
Fractured zone, 437
Fracture failure, 535
Fracture stress, 42
Frame(s), 458, 984
France, 94, 968
Freedom from deficiencies, 948
Free machining steels, 567
Freeze drying, 367
Freezing point, 68
French Revolution, 968
Frenkel defect, 31
Friability, 587
Friction, 48
    in metal forming, 383
    and powders, 339
    and sheet metal drawing, 448
Friction angle, 488–489
Friction force, 485
Friction sawing, 529
Friction welding (FRW), 682, 722–723
Frit, 674
FRPs, *see* Fiber-reinforced polymers

FRW, *see* Friction welding
Fuel-fired furnaces, 233, 647
Full annealing, 640
Fullering, 401, 402
Full-mold process, 220
Furnace(s), 253
    basic oxygen, 94
    blast, 94
    cupola, 97
    electric, 94
    fuel-fired, 233
    for glassmaking, 247
    for heat treatment, 647–648
    open hearth, 94
    used in casting, 233–235
Furnace brazing, 739
Fusion, 35, 68, 202, 726
Fusion-welded joints, 690–691
Fusion welding, 11, 681–682, 694
Fusion zones, 690
FW, *see* Flash welding

Gages, 972–977
Gage blocks, precision, 971–972
Gaging (or gauging), 966, 985–986
Galvanized steel, 117
Galvanizing, 663, 664
Gang drills, 514
Gap frame presses, 458–462
Gases, 25
Gas atomization, 340
Gas carburizing, 646
Gas curing, 311
Gas metal arc welding (GMAW),
    699–700
Gas nitriding, 646
Gas-shielded flux-cored arc welding, 701
Gas tungsten arc welding (GTAW), 703
Gates, 278
Gating system, 197–198
Gear rolling, 396
Gear shapers, 526
General Electric Company, 133, 542, 586
General purpose equipment, 16
Generating machining operations,
    500–501
Generative CAPP systems, 920–921
Germany, 169, 413, 542
GFRP (glass fiber-reinforced plastic), 179
Gilbert, W., 573
Gilbreath, Frank, 3
Gilbreath, Lilian, 3
Glass(es), 134–138
    chemistry/properties of, 135–136
    cooling of, 35
    definition of, 134
    and fibers, 179
    history of, 135
    plate, 250
    products using, 136–137
    recycled, 247
    tempered, 253
    tubular, 251
    viscosity of, 58, 59
Glass blowing, 135
Glass ceramics, 9–10, 137–138
Glass fiber-reinforced plastic (GFRP), 179
Glass former, 136
Glass-transition temperature, 35, 154
Glassworking, 246–254
    finishing operations, 253–254
    and heat treatment, 253
    history of flat glass, 250
    and product design, 254
    raw materials for, 246–247
    shaping processes, 247–252

Glassy phase, 127
Glazing (ceramics), 131, 366
GMAW, *see* Gas metal arc welding
GNP, *see* Gross national product
Gold, 11, 119, 120
GO limit, 976
GO/NO-GO gages, 976
Goodyear, Charles, 11, 144, 167
Goodyear, Nelson, 144
Government Rubber-Styrene (GR-S), 169
Government services, 2
Grades (of quality), 948
Gradual wear, 535
Graduated measuring devices, 972
Graduations, 972
Graft copolymers, 151
Grains, 34
Grain boundary(-ies), 34
Grain size, 34
Graphite, 138–139
Gravity, specific, 67
Gravity drop hammers, 407
Gray, Matthew, 257–258
Gray cast iron, 107, 108
Gray-scale vision (machine vision), 985
Great Britain, *see* England
Greece, ancient, 94, 397
Green (ceramics), 366
Green compact, 344
Green density, 344
Green-sand molds, 217
Green strength, 344
Grinders, tool, 602
Grinding, 11, 14, 360, 586–603
    analysis of, 591–596
    application considerations in, 596–597
    bonding materials for, 588
    centerless, 600, 601
    creep feed, 600, 601
    cylindrical, 598–600
    definition of, 586
    electrochemical, 618, 619
    impact, 361
    physics of, 592–593
    surface, 592, 597–598
    and wheel wear, 594–596
Grinding fluids, 596–597
Grinding ratio, 595
Grinding wheel, 587–591
Grippers, 876
Grooved pins, 762
Groove welds, 684, 685
Gross national product (GNP), 1, 2
Group technology (GT), 19, 884–889
    advantages/disadvantages of, 889
    and cellular manufacturing, 887–889
    definition of, 884
    parts classification/coding in, 885–887
Group technology layout, 19
GR-S (Government Rubber-Styrene), 169
GT, *see* Group technology
GTAW (gas tungsten arc welding), 703
Guerin process, 455–456

Hacksaws, 528
Hacksawing, 528
Hall, Charles, 110
Halogens, 25
Hammers, forging, 398, 406–407
Hammering, 11
Hand lay-up processes, 322–323
Hand (measurement), 968
Hand modeling, 363
Hand molds, 286
Hand molding, 363
Hand soldering, 743, 833

Hand throwing, 363
Hard anodizing, 665
Hard-bake (photolithography), 799
Hardenability, 643, 644
Hardening:
    precipitation, 644–645
    strain, *see* Strain hardening
    surface, 645–646, 648–649
Hard facing, 675
Hardness, 52–55
    of ceramics, 55
    hot, 56
    of metals, 54, 55
    of polymers, 55
    and temperature, 56
    tests for, 52–54
Hard product variety, 7
Hardware, 5
HAZ, *see* Heat-affected zones
HCP (hexagonal close-packed) crystal
    structure, 29, 30, 32, 33
HDPE, *see* High-density polyethylene
Heading processes, 408, 409
Headstock, 505
Heat:
    of fusion, 35, 68, 202
    specific, 69–70
Heat-affected zones (HAZ), 690, 691
Heating:
    of metals, 11, 198–199
    in shaping processes, 13
Heat transfer efficiency, 688
Heat treatment, 11, 15, 122, 123, 639–649
    annealing, 640
    definition of, 639
    electron beam (EB) heating, 648, 649
    flame hardening, 648
    furnaces for, 647–648
    and glassworking, 253
    high-frequency (HF) resistance heating,
        648, 649
    induction heating, 648, 649
    laser beam heating, 649
    martensite transformation, 640–644
    of powders (powder metallurgy), 348
    precipitation hardenability, 644–645
    surface hardenability, 645–646
Helical milling cutters, 555
Helical winding, 328
Helium, 26
Helix angles, 554
Hematite, 96
Hemimorphate, 117
Hemming, 446
HERF, *see* High-energy-rate forming
Heroult, Paul, 110
Hevea brasiliensis (rubber tree), 307
Hexagonal close-packed (HCP) crystal
    structure, 29, 30
HF, *see* High-frequency resistance heating
HFIW, *see* High-frequency induction
    welding
HFRW, *see* High-frequency resistance
    welding
High carbon steels, 102
High-density polyethylene (HDPE), 144,
    160
High-energy-rate forming (HERF),
    467–469
High-frequency (HF) resistance heating,
    648, 649
High-frequency induction welding
    (HFIW), 711
High-frequency resistance welding
    (HFRW), 711, 712
High-impact polystyrene (HIPS), 161

High-pressure steam, 311
High production, 6, 19–20
High-speed machining (HSM), 529–530
High-speed steel (HSS), 542, 544–545
High-speed tool steels, 106, 107
High-strength low-alloy (HSLA) steels,
    103
High-vacuum welding, 716
HIPS (high-impact polystyrene), 161
HMCs (horizontal machining centers), 522
Hob, 411
Hoerni, J., 787
Holding force, 451
Holes, drilled, 512, 825
Hollow profiles (dies), 268
Homopolymer, 151–152
Honeycomb material, 184
Honing, 603–604
Hooke's Law, 41, 42
Hoop winding, 328
Horizontal boring machines, 509, 510
Horizontal broaching machines, 527
Horizontal centrifugal casting, 230
Horizontal machining centers (HMCs),
    522
Horizontal milling machines, 520
Horizontal turning machines, 505
Horsepower, 491, 492
Hoses, rubber, 314, 315
Hot-air tunnel, 311
Hot-chamber die-casting machines, 228,
    229
Hot die forging, 412
Hot dipping, 663–664
Hot extrusion, 415
Hot forming (metals), 379–380
Hot hardness, 56
Hot isostatic pressing, of powders (pow-
    der metallurgy), 349
Hot pressing:
    of cemented carbides, 370
    of ceramics, 367–368
    of powders (powder metallurgy), 350,
        351
Hot pressure welding (HPW), 720
Hot rolling (metal forming), 387
Hot roll welding, 720
Hot-runner molds, 279
Hot working (metals), 57, 379–380
Hot-working tool steels, 107
Hp/rpm ratio, 530
HPW (hot pressure welding), 720
HSLA (high-strength low-alloy) steels,
    103
HSM, *see* High-speed machining
HSS, *see* High-speed steel
Hubbing, 411, 412
Huntsman, B., 542
Hyatt, John, 144, 258
Hybrid composites, 188
Hybrid microsensors, 846
Hydraulic clamps, 280
Hydraulic presses, 407, 463
Hydrodynamic machining, 613
Hydroforming, 456–457
Hydrogen, 25, 26
Hydrogen bonding, 28–29
Hydromechanical clamps, 281
Hydrostatic extrusion, 420, 422
Hypereutectoid steels, 95
Hypoeutectoid steels, 95

ICs, *see* **Integrated circuits**
I.G. Farben, 169
Ilmenite, 116, 133
Image processing and analysis, 985

Immersion, 673
Impact extrusion, 415, 420, 422
Impact grinding, 361
Imperfect balancing (production lines), 900
Imperfections (crystals), 30
Implantation, ion, 657–658, 803
Impregnation, of powders (powder metallurgy), 347, 348
Impression-die forging, 398, 402–404
Improvement, continuous, 921
Impurities (in silicon), 802–803
Incremental forging, 402
Incremental positioning, 863, 864
Independent demand, 931–932
India, 397, 586, 680
Indirect extrusion, 415, 417
Indirect fuel-fired furnaces, 233, 234
Indirect RTM method, 783
Induction brazing, 739
Induction furnaces, 235
Induction heating, 648
Industrial Revolution, 3, 11, 413, 502
Industrial robotics, 873–878
    anatomy of, 873–876
    applications of, 877–878
    control systems for, 876, 877
    definition of, 873
    insertion, robotic, 832, 833
    programming of, 877
    welding, robotic, 683
Industries, 4–6
Inertia friction welding, 722, 723
Infeed, 591
Inference engine, 921
Infiltrated phase (of composites), 180
Infiltration, of powders (powder metallurgy), 348
Infrared brazing, 740
Infrared reflow soldering, 744
Ingots, 99–100
    of aluminum, 110
    of iron, 94
    segregation of, 204
Initiator (chemical catalyst), 146
Injection blow molding, 290
Injection molding, 12
    machines for, 279–281
    of polymer matrix composites, 327
    of polymers, 275–285
    of powders (powder metallurgy), 349–350
Injection units, 275, 279
Ink-jet printing heads, 847
Inorganic adhesives, 747
Inorganic fluxes, 742, 743
Input modules, 879
Insertion holes, 825
Insertion machines, 16
Insertion of components, onto PCBs, 831–832
Inserts:
    cutting tool, 552–554
    in moldings or castings, 765
Inside calipers, 972
Inside micrometers, 973
Inspection, 969–971
    by attributes, 969
    automated, 970
    of casting, 238, 239
    contact vs. noncontact, 970–971
    definition of, 965–966
    electrical fields, 986
    by machine vision, 983–986
    manual, 969
    of PCBs, 830
    radiation techniques for, 986

robotics in, 878
    sampling vs. 100%, 970
    of surface mount technology, 838, 839
    testing vs., 969
    ultrasonic techniques for, 986
    by variables, 969
    of welding, 727
Insulation displacement, 840–841
Insulation materials, in printed circuit boards, 822, 823
Insulators, 73
Integral fasteners, 765–766
Integrated circuits (ICs), 786–815
    analog vs. digital, 786, 787
    clean rooms for processing of, 791
    definition of, 786
    history of, 787
    layer processes used for production of, 800–807
    lithographic technologies used in processing of, 796–800
    packaging of, 808–813
    processing sequence for, 789, 790, 807–808
    and silicon processing, 792–796
    yields in processing of, 813–815
Intelligent control (robotics), 877
Interchangeable parts, 3
Interfaces (of composites), 180
Interference fits, 760–763
Intermediate phase, 90
Intermittent motion welding, 709
Intermittent transfer systems, 899
Intermolecular forces, 28
Internal broaching, 527
Internal centerless grinding, 600, 601
Internal chills, 207
Internal cylindrical grinding, 599, 600
Internal micrometers, 973
Internal mixer, 308
Internal noise factors, 954
International Standard Industrial Classification (ISIC), 5
Interphase (of composites), 180
Interpretation (machine vision), 985
Interrupted cutting, 516
Interstitialcy, 31
Interstitial solid solution, 90
Intramolecular bonds, 27
Inventory, 931–935
    order point systems for controlling, 932–935
    types of, 931–932
    zero, 939
Inventory record file, 936
Inverse lever rule, 92
Investment casting, 222–224
Ion etching, reactive, 806
Ionic bonds, 27
Ion implantation, 657–658, 803
Ion lithography, 799–800
Ion-pair vacancy, 31
Ion plating, 668
Iron, 11. *See also* Steel(s)
    cast, 107–109
    ductile, 108
    electrolytic, 94
    gray cast, 107, 108
    history of, 94
    ingot, 94
    malleable, 109
    ores of, 96
    pig, 97
    production of, 95–97
    white cast, 109
    wrought, 94

Iron Age, 11, 94
Iron-based alloys, 121, 122
Iron–carbon phase diagram, 93–95
Ironing, 454
ISIC (International Standard Industrial Classification), 5
Isocyanate, 164
Isoprene rubber, 170
Isostatic pressing:
    of cemented carbides, 370
    of ceramics, 368
    of powders (powder metallurgy), 348–349
Isotactic arrangement, 149
Isothermal forging, 412
Isothermal forming, 380
Isotropic etching, 805
Israel, ancient, 94

Japan, 397
Jefferson, Thomas, 3
Jig, 514
Jiggering (ceramics), 363
Jig grinders, 602
JIT, *see* Just-in-time
Job shops, 18, 928
Johnson's extrusion strain formula, 417
Joining processes, 16, 679, 734–750. See *also* Welding
    adhesive bonding, 744–750
    brazing, 735–740
    soldering, 740–744
Joints:
    bolted, 756–758
    manipulator, 873
    weld, 683–686
Joint designs (soldering), 741–742, 746–747
Jominy end-quench test, 644
Just-in-time (JIT), 939–941

Kaizen, 921
Kanban system, 940–941
Kaolinite, 126, 130
Keramos, 126
Kernite, 140
Kevlar, 159, 179, 319
Kilby, J., 787
Kiln, 366
Knee-and-column milling machines, 520, 521
Knoop Hardness Test, 52, 53, 55
Knowledge base, 920
Knurled pins, 762
Knurling, 505
Kroll process, 116
Krupp, 133

Laboratory standard, master, 971
Lack of fusion, 726
Lack of penetration, 726
Ladles, 235
Laminar composite structures, 184
Laminated glass, 253
Laminated-object manufacturing (LOM), 779–780
Lamination:
    contact, 321
    continuous, 332
Lanced tabs, 765, 766
Lancing, 455
Land (dies), 427
Lap joint (welding), 684, 685
Lapping, 604, 605
Lasers, 624, 982–983
Laser beam heating (LB), 649

Laser beam machining (LBM), 624–625, 856
Laser-beam welding (LBW), 681, 717
Laser triangulation, 982, 983
Lathes, 481, 502–507
Lattice structures, 29–33
Layer, altered, 80
Layer manufacturing, 774
Layer processes, integrated circuit production, 800–807
  chemical vapor deposition, 801–802
  doping, 802–803
  etching, 804–806
  metallization, 803–804
  thermal oxidation, 800–801
Lay (surface texture), 80, 81
LB (laser beam heating), 649
LBM, *see* Laser beam machining
LBW, *see* Laser-beam welding
LCC (leaded chip carriers), 810
LDPE, *see* Low-density polyethylene
Lead, 92–93, 118
Leaded chip carriers (LCC), 810
Leadless chip carriers (LLCC), 810
Leadthrough programming, 877
Lead time, 935
Leak flow, 264
Lean production, 939
Lehrs (furnaces), 253
LIF, *see* Low insertion force
Life, tool, 537–541
Lift-off technique, 854
Lift-out crucible furnaces, 233
LIGA process, 854–855
Limestone, 96–97
Limits, natural tolerance, 950, 951
Limit dimensioning, 78
Limited sequence control, 876
Limit gages, 976
Limonite, 96
Lines, model, 897–898
Linear dimensions, measurement of, 972–974
Linear polymers, 150
Line balancing, 901–903, 917–918
Line defect, 31
Links, manipulator (robotics), 873
Linotype, 228
Liquids, solidification of, 13
Liquid-based systems, 775–779
  droplet deposition manufacturing, 777–779
  solid ground curing, 777, 778
  stereolithography, 775–777
Liquid carburizing, 646
Liquid nitriding, 646
Liquid-phase epitaxy, 802
Liquid-phase sintering, 351
Liquidus, 68
Lithography (integrated circuit processing), 790, 796–800
  electron-beam lithography, 799
  extreme ultraviolet lithography, 799
  ion lithography, 799–800
  photolithography, 796–799
  x-ray lithography, 799
Live center, 506
LLCC (leadless chip carriers), 810
Loading, wheel, 594
Logic layers, 829–830
Lohmann, H., 133
LOM, *see* Laminated-object manufacturing
London forces, 28, 29
Loom, power, 3
Loss function (Taguchi methods), 953–954

Lost-foam process, 220
Lost pattern process, 220
Lost-wax process, 222
Low-alloy steels, 103, 104
Low alloy tool steels, 107
Low carbon steels, 102
Low-density polyethylene (LDPE), 160
Lower shoe, 457
Low insertion force (LIF), 842, 843
Low-pressure casting, 226
Low-pressure chemical vapor deposition (LPCVD), 670, 671
Low production, 6
Low-quantity production, 18, 19
LPCVD, *see* Low-pressure chemical vapor deposition
Lubricants:
  and ceramics, 138, 139, 367
  and cutting tool technology, 558
  and plastics, 155
  and powder blending, 343
Lubrication, in metal forming, 383

Machinability, 565–567, 566
Machine control unit (MCU), 862
Machine screws, 754
Machine tools, 3, 11, 16–17
  history of, 502
  for metal machining, 480–481
Machine vision inspection, 983–986
Machine welding, 683
Machining, 475–495
  abrasive jet, 614–615
  advantages/disadvantages of, 476, 477
  chemical, 627–633
  chip formation, 481–485
  and cutting temperature, 493–495
  with cutting tools, 478–480
  definition of, 475
  dry, 560
  electrical discharge, 619–623
  electrochemical machining, 75, 615–618
  electron beam, 623–624
  force relationships in, 485–490
  history of, 502
  laser beam, 624–625
  nontraditional, *see* Nontraditional machining
  photochemical, 632–633
  of powders (powder metallurgy), 347
  power and energy relationships in, 490–493
  and product design considerations, 578–580
  surface finish in, 568–572
  tolerances in, 568
  tools for, *see* Machine tools
  ultrasonic, 611–612
Machining centers, 522–524, 872
Machining economics, 573–578
Machining operations, 14, 22, 477–478, 499–529, 872
  abrasive, 585–606
  broaching, 526–528
  cutting, *see* Cutting; Cutting tools
  drilling, 511–515
  grinding, *see* Grinding
  high-speed, 529–530
  milling, 515–522
  planing, 524–526
  sawing, 528–529
  shaping, 524–526
  turning, 502–511
Machining steels, free, 567
Machining times, 573–574
Macromolecules, 145–146

Magnesium, 112–113
Magnesium alloys, 240
Magnetic heads, thin-film, 848
Magnetic particle testing, 727
Magnetic pulse forming, 468
Magnetite, 96
Make or buy decision, 918–919
Make-to-stock, 932, 933
Malaya, 167
Malleable iron, 109
Mandrel, 329, 429
Manganese, 103
Manipulator design, 873–875
Manipulator joints/links (robotics), 873
Mannesmann process, 397
Manning levels, 900, 903
Manual application, 560
Manual assembly lines, 900–904
Manual computer-assisted CMM, 981
Manual control, 981
Manual data input (MDI), 872
Manual insertion, 832, 833
Manual inspection, 969
Manual leadthrough, 877
Manual part programming, 871
Manual spinning, 465
Manufacture (term), 2
Manufacturing:
  cellular, 19
  definitions of, 4, 5
  direct CAD, 774
  droplet deposition, 777–779
  history of, 3–4
  laminated-object, 779–780
  layer, 774
  production vs., 4
Manufacturing engineering, 912–913
Manufacturing engineering department, 20
Manufacturing lead time, 935
Manufacturing support systems, 20, 912
Manufacturing systems, 3, 860–861. *See also* Flexible manufacturing systems; Numerical control
Manufacturing variations, 949
MAPP, 715
Martensite, 640–642
Martensite transformation, 640–644
  and hardenability, 643, 644
  and heat treatment process, 642, 643
  time-temperature-transformation curve, 641–642
Martensitic stainless, 105
Martin, Emile, 94
Martin, Pierre, 94
Masking, 627–628
Mask (photolithography), 796
Massachusetts Institute of Technology, 861
Mass diffusion, 71–72
Mass finishing, 654
Mass production, 19, 928, 929
Masterbatch, 308
Master die, 222
Master gages, 975, 976
Master laboratory standard, 971
Master production schedule, 930–931
Mat, 320
Match-plate patterns, 216
Materials. *See also* Composite materials; Raw materials
  electrical properties of, 72–74
  electrochemical processes in, 74–75
  engineering, 36–37
  manufactured, 6
  manufacturing, 8–10
  mass diffusion in, 71–72

mechanical properties of, *see* Properties, material
melting characteristics of, 68–69
thermal properties of, 69–71
volumetric properties of, 66–68
Material-addition RP, 773
Material handling, 878
Material removal processes, 11, 13–15, 475, 476
Material removal RP, 773
Material requirements planning (MRP), 935–938
Mathematical models, 262–266
Matrix composites, 176–177
ceramic, *see* Composite materials, ceramic matrix composites
metal, *see under* Composite materials
Maudsley, Henry, 502
Maximum distance between centers, 506
Maximum reduction per pass, 426
MBE, *see* Molecular-beam epitaxy
MCU (machine control unit), 862
MDI (manual data input), 872
Mean flow stress, 378
Measurement(s), 965–969, 971–983
accuracy of, 966–967
angular, 977–978
comparative, 974–977
with coordinate measuring machines, 980–982
definition of, 965
history of measurement systems, 968–969
with lasers, 982–983
of linear dimensions, 972–974
precision gage blocks as standards for, 971–972
precision of, 966–967
standards/systems of, 968
of surfaces, 978–980
Measuring devices, 972–974
Mechanical agitation, 297
Mechanical assembly, 16, 752–768
with cotter pins, 764
and design for assembly, 766–768
with eyelets, 760
with inserts in moldings or castings, 765
with integral fasteners, 765–766
interface fits, based on, 760–763
with rivets, 759–760
by sewing, 764
by stapling, 764
by stitching, 763–764
with threaded fasteners, 753–759
Mechanical cleaning, 654
Mechanical control (winding machines), 328
Mechanical energy, 611–615
abrasive jet machining, 614–615
abrasive water jet cutting, 614
ultrasonic machining, 611–612
water jet cutting, 613–614
Mechanical gages, 974–975
Mechanical galvanizing, 676
Mechanical plating, 676
Mechanical presses, 407, 463
Mechanical thermoforming, 295–296
Mechanosynthesis, 858
Media, finishing, 655–656
Medical technologies, 849
Medium carbon steels, 102
Medium production, 6
Medium-quantity production, 19
Medium-vacuum welding, 716
Melamine-formaldehyde, 163
Melt flow, 263–264

Melt fracture, 269
Melting, 35
Melting efficiency, 688
Melting point, 68, 69, 93
Melting process, 68–69
Melt spinning, 273
MEMS (microelectromechanical systems), 845
Mers, 10
Merchant equation, 488–490
Mergenthaler, O., 228
Meros, 143
Mesh count, 337
Mesopotamia, ancient, 135, 194
Metal(s), 11, 88–123
alloys, *see* Alloys
atomic structure of, 36
brazing, 735
casting of, *see* Casting
as conductors, 73
ferrous metals, 93–109
and fibers, 180
grains and grain boundaries in, 34
hardness tests for, 54, 55
heat treatment of, 639–649
iron, 93–98
matrix composites, *see under* Composite materials
nonferrous, *see* Nonferrous metals
phase diagrams of, 90–93
powdered, *see* Powders
precious, 119–120
processing of, 122–123
properties of, 88–89
refractory, 118–119
shaping of, 13. *See also* Metal forming
solidification of, 202–205, 207–208
superalloys, 120–122
viscosity of, 58
Metal bath method, 740
Metal forming, 374–383. *See also* Bulk deformation processes
categories of, 374–377
friction in, 383
lubricants used in, 383
material behavior in, 377–378
and strain rate sensitivity, 380–382
temperature in, 378–380
Metal injection molding (MIM), 349
Metallic bonding, 28
Metallization (integrated circuit production), 803–804
Metallizing (metal spraying), 675
Metalloids, 25
Metallurgy, powder, *see* Powders
Metal machining, *see* Machining
Metal matrix composites, *see under* Composite materials
Metal spraying, 675
Metal tooling, of sheet metal, 454–455
Metalworking, sheet, *see* Sheet metal operations
Metering section, 261
Metric system, 968
Metrology, 966
Micro-accelerometer, 847
Microactuators, 846
Microcomponents, 846
Microelectromechanical systems (MEMS), 845
Microfabrication technologies, 845–858
device types, 846–847
industrial applications of, 847–850
in LIGA processes, 854–855
machining, 855–856
and nanotechnology, 857–858

in photofabrication, 857
and rapid prototyping, 856, 857
in silicon layer processes, 850–854
Micro-instruments, 846
Micromachining, 845, 852
Micrometer, 972–974
Microscopes, scanning probe, 850
Microsensors, 846
Microstereolithography (MSTL), 857
Microstructures, 846
Microsystems, 846
Microsystem technology (MST), 845
Middle Ages, 94, 680
Middle East, 129
Midvale Steel, 542
MIG welding, 700
Mile (measurement), 968
Mills, 360–361
Milling, 11, 14, 15, 478
chemical milling, 630
machining operations, 515–522
Milling cutters, 515, 555–556
Milling machines, 481, 502, 515, 520–522
Mills, rolling, 393–395
MIM (metal injection molding), 349
Minimum rational work elements, 901
Mining, 1
Mist applications, 560
Mixed model lines, 897–898
Mixed-model production line, 20
Mixing, of powders, 342–343
Mixing-activated systems, 162
Modeling, fused-deposition, 780
Model lines, 897–898
Model T Ford, 900
Modulus of elasticity, 42
Moisson, Henri, 133
Molds, 196–198
dry-sand, 217
green-sand, 217
negative, 293–294
for polymers, 277–279
positive, 293–295
for sand casting, 216–218
skin-dried, 217–218
Mold constant, 205
Molded plastic parts, design considerations for, 300–301
Mold erosion, 199
Molding, 13
autoclave, 325
of ceramics, 368
closed mold processes, 325–327
compression, of plastics, 285–286
contact, 321
hand lay-up, 322–323
hand molding, 363
of polymer matrix composites, 320–321, 325–327
of polymers, 275–285
of polystyrene foams, 298
powder injection, 349–350, 368
of rubbers, 310
spray-up, 323–324
Mold steels, 107
Molecular-beam epitaxy (MBE), 802, 858
Molecular engineering, 857–858
Molecular weight (MW), 149
Molecules, bonding between, 26–29
Molybdenum, 103, 119, 544
Monomers, 145
Motion, precision of, 875–876
Motion control (in NC), 863
Motion studies, 4
Motorized computer-assisted CMM, 982
Mouton, G., 968

MRP, *see* Material requirements planning
MSTL (microstereolithography), 857
MST (microsystem technology), 845
Multi-injection molding, 283
Multilayer boards, 823, 829
Multiple cutting edge tools, 479, 554–558
Multiple machine cells, 889
Multiple-spindle drills, 514
Multiple station assembly systems, 906
Multiprobe (wafer testing), 811
Multiprobe yield, 814
Mushet, R., 542
Mushy zone, 203
MW (molecular weight), 149

Nailing, 11
Nanotechnology, 845, 857–858
Nasmyth, James, 502
Natural adhesives, 747
Natural aging, 645
Natural rubber, 306, 307
Natural tolerance limits, 950, 951
NC, *see* Numerical control
NDE, *see* Nondestructive evaluation
NDT, *see* Nondestructive testing
Near net shape processes, 15, 195, 335, 387, 404
Necking, 42–45
Negative molds, 293–294
Negative resist, 796–797
Neon, 26
Neoprene, 169, 170
Net shape processes, 15, 145, 195, 257, 335, 387, 404
Network structure, 151
Neutral flame, 713
Neutral point, 390
New ceramics, 132–134, 366–369
Newton, Isaac, 58
Newtonian fluid, 58, 258
Nickel, 115
    alloys of, 90–92, 121, 240
    in low-alloy steels, 103
    in stainless steels, 104
Nickel plating, 662
Nitrides, 9, 134
Nitriding, 646
Nitrile rubber, 169, 170–171
Nobel Prize, 787
Noble gases, 25
Noble metals, 25, 119
NO-GO limit, 976
Noise factors, 954
Nominal surfaces, 78, 79
Nonconsumable electrodes, 696
Noncontact inspection, 970–971, 971
Noncrystalline (amorphous) structures, 34–36
Nondestructive evaluation (NDE), 969
Nondestructive testing (NDT), 727, 969
Nonferrous casting alloys, 240
Nonferrous metals, 9, 89, 109–120
    aluminum, 109–112
    copper, 113–115
    lead, 118
    magnesium, 109, 112–113
    nickel, 115
    precious metals, 119–120
    refractory metals, 118–119
    tin, 118
    titanium, 116–117
    zinc, 117–118
Nongraduated measuring devices, 972
Nonreversing mills, 393
Nonrotational machined parts, 499, 500

Nonsteel-cutting grades (cemented carbides), 546–547
Nontraditional machining, 610–635
    applications of, 633–635
    chemical machining, 627–633
    and materials, 634, 635
    mechanical energy, processes based on, 611–615
    thermal energy, processes based on, 619–627
Nontraditional processes, 14, 475
Nonvacuum welding, 716
Normal force, 486
Normalizing, 640
Nose radius, 550
Nose radius wear, 536
No-slip point, 390
Notching, 441
Notch wear, 536
Noyce, R., 787
Nucleus (atomic), 24
Numerical control (NC), 514, 861–873
    applications of, 872–873
    components of NC systems, 862
    coordinate systems and motion control in, 863
    definition of, 861
    drill presses, 514
    history of, 861
    part programming with, 870–872
    positioning systems with, 863–870
Nuts, 753
Nylon, 11, 144, 159

OAW, *see* Oxyacetylene welding
OFC, *see* Oxyfuel cutting
Off-line QC, 954–955
Offset bending, 447
OFW, *see* Oxyfuel gas welding
Ohm's law, 73
OHW (oxyhydrogen welding), 715
Oils, 559
On-line QC (Taguchi methods), 955
Opaqueness, 36
Open back inclinable press, 460
Open cells, 298
Open-die forging, 398, 399–402
Open hearth furnace, 94
Open-loop positioning systems, 864–866
Open mold, 196
Open mold processes, 321–325
Open pores, 338
Open risers, 209
Open side planers, 526
Operating point, 266
Operating range, wide, 967
Operation sheets, 917
Operations to enhance properties, 915
Optical lithography, 796
Optical techniques, of surface measurement, 979–980
Orbital forging, 410, 411
Ordering lead time, 935
Order point systems, 932–935
Order progress, 943–944
Order release, 942, 943
Order scheduling, 943
Ores, 96
Organic coatings, 671–674
Organic fluxes, 742
Orsted, Hans, 110
Orthogonal cutting, 481–483, 489, 490
Ostromislensky, I., 144
Output modules, 879
Outside calipers, 972
Outside micrometers, 973

Overaging, 645
Overbending, 444
Overcut, 620
Overlay process, flexible, 675–676
Oxidation, thermal, 800–801
Oxide-based cermets, 186
Oxide ceramics, 132
Oxide film, 80
Oxyacetylene welding (OAW), 712–714
Oxyfuel cutting (OFC), 626–627
Oxyfuel gas welding (OFW), 680, 681, 712–715
Oxygen, in steelmaking, 94
Oxyhydrogen welding (OHW), 715

PAC, *see* Plasma arc cutting
Pace (measurement), 968
Package sealing (IC packaging), 812–813
Packaging hierarchy, 820–821
Packaging (integrated circuits), 808–813
    design of package, 809–811
    processing steps in, 812–813
Pack carburizing, 646
Pack diffusion, 657
Packing characteristics, of powders, 339
Packing factor, 339
PACVD (plasma assisted chemical vapor deposition), 671
Painting, 15
Pale crepe rubber, 307
Pallet shuttles, 522
Palm (measurement), 968
PAN (polyacrylonitrile), 157
Parameter design, 955
Parison, 288
Parkes, Alexander, 144
Parkesine, 144
Parsons, John, 861
Parsons Corporation, 861
Parts:
    in group technology, 885–887
    handling times for, 573–574
    interchangeable, 3
    process planning for, 914–916
Parts sortation, 970
Partial face milling, 517, 518
Particles, 180
    and polymer matrix composites, 320
    powder, 337–338
Particulate processing, 13, 14
Particulates, 180
Parting, 441
Parting line, 197, 277, 407, 408
Parting surfaces, 277
Part programming, 870–872
    CAD/CAM-assisted, 872
    computer-assisted, 871
    manual, 871
    manual data input, 872
Paste, solder, 837–838
Patterns, 197, 215–216
Pattern shrinkage allowance, 206
Pattern tree, 222
PAW, *see* Plasma arc welding
PCBs, *see* Printed circuit boards
PCM, *see* Photochemical machining
PE, *see* Polyethylene
Pearlite, 641, 642
Peening, shot, 654
Pelletized molding compounds, 321
Penetration, 437, 726
Pentlandite, 115
Percussion welding (PEW), 711
Perfect crystals, 30
Perforating, 441

Periodic Table, 25
Peripheral milling, 478, 516–517
Permanent connections (PCB production), 840–841
Permanent mold, 197
Permanent mold casting, 225–233
    centrifugal casting, 230–233
    die casting, 227–230
    low-pressure casting, 226
    semipermanent-mold casting, 225
    slush casting, 226
    vacuum permanent-mold casting, 227
Persia, ancient, 397
P+ etch-stop technique, 853
PET (polyethylene terephthalate), 160
PEW (percussion welding), 711
PGA (pin grid array), 811
PGW (pressure gas welding), 715
Phase(s):
    of composites, 10, 176
    definition of, 89
    natural, 25
Phase diagrams, 90–95
    of copper–nickel alloy system, 90–92
    of iron–carbon alloy system, 93–95
    of tin–lead alloy system, 92–93
Phenol-formaldehyde, 163
Phenolics, 163–164
Phosphate coating, 663, 665
Photochemical machining (PCM), 632–633, 855–856
Photofabrication, 857
Photographic resist method, 628
Photolithography, 796–799, 826
    exposure techniques, 797–798
    photoresists, 796–797
    processing of, 798–799
Photoresist, 628
Physical blowing agents, 297
Physical deposition, 15
Physical production limitations, 8
Physical vapor deposition:
    by ion plating, 668
    by sputtering, 667–668
    by vacuum evaporation, 666–667
Physical vapor deposition (PVD), 665–668
Pickling, acid, 653
Piece rate system, 4
Piece ware, 248–250
Piercing, roll, 396, 397
Pig iron, 94, 97, 98
Pigments, 154–155, 672
PIH, see Pin-in-hole technology
PIM, see Powder-injection molding
Pins, 762, 764
Pin grid array (PGA), 811
Pin-in-hole (PIH) technology, 810, 821, 831
Piping, 423
Pittsburgh Reduction Co., 110
Pixels (machine vision), 984
Plain carbon steels, 102
Plain milling, 516
Plain milling cutters, 555
Plan, aggregate production, 930, 931
Planar coating, 274
Planar process, 789
Plane, shear, 481
Planers, 502, 526
Planer type mills, 521, 522
Planetary model, 25
Planing, 11, 524–526
Planning:
    capacity requirements, 938–939
    material requirements, 935–938
    production, 928–929

Plant capacity, 8
Plant layout, 17
Plasma arc cutting (PAC), 625–626
Plasma arc welding (PAW), 703, 704
Plasma assisted chemical vapor deposition (PACVD), 671
Plasma etching, 805, 806
Plaster-mold casting, 223–225
Plastics, 11–12, 256–301. See also Polymers
    blow molding of, 288–291
    casting of, 296–297
    coating processes using, 274
    compression molding of, 285–286
    extrusion of, see under Polymers
    history of shaping, 257–258
    injection molding of, see under Polymers
    and product design, 299–301
    rotational molding of, 291–292
    sheet and film, production of, 270–272
    thermoforming of, 292–293
    transfer molding of, 286–287
Plastic deformation, 32
    of sheet metal, 437
    in tool wear, 537
Plastic forming (ceramics), 363
Plasticity, 42, 46, 47
Plasticizers, 154, 297
Plastic pressing (ceramics), 364
Plastic region, 42, 44, 45
Plate, surface, 971
Plated through-hole (PTH), 831
Plating, 660–664
    electroforming, 662–663
    electroless, 663
    electroplating, 660–662
    hot dipping, 663–664
    ion, 668
    mechanical, 676
    in PCB production, 826–827
Platinum, 120
Playback with continuous path control, 876, 877
Playback with point-to-point control, 876
PLCs, see Programmable logic controllers
PLC memory, 879
Plies, 312
Plowing, 593
Plug, floating, 429
Plug gages, 976
Plug welds, 684, 685
Plunge-fed, 597
Plunger transfer molding, 286
Plunger-type injection molding machines, 279
PM, see Powders (powder metallurgy)
PMC, see Polymer matrix composites
PMMA (polymethylmethacrylate), 157
Pocket milling, 517, 518
Point angles, 554
Point defects, 30, 31, 814
Pointing process, 428, 429
Point-to-point (PTP) control, 876
Point-to-point systems, 863
Poisson yield, 814–815
Polar winding, 328
Polishing, 11, 605, 606
Poly, 143
Polyacrylonitrile (PAN), 157
Polyamides (PA), 159
Polybutadiene (BR), 169
Polycarbonate, 144, 159–160
Polycrystalline structures, 34
Polydimethylsiloxane, 171
Polyesters, 144, 160, 164
Polyethylene (PE), 11, 144, 160

Polyethylene terephthalate (PET), 160
Polyisoprene, 170
Polymers, 10, 143–173. See also Plastics; Rubber(s)
    additives to, 154–155
    blow molding of, 288–291
    branched, 150–151
    casting of, 296–297
    coating processes using, 274
    compression molding of, 285–286
    copolymers, 151–152
    cross-linked, 151
    crystallinity of, 152–153
    definition of, 143
    elastomers, 165–172
    extrusion of, 260–270
    fibers and filaments, production of, 272–274
    foam processing, 297–299
    foam shaping, 298–299
    hardness tests for, 55
    history of, 144
    importance of, 145
    injection molding of, 275–285
    as insulators, 73
    limitations of, 145
    linear, 150
    molecular structure of, 37, 145–153
    processing of, 172–173
    and product design, 299–301
    rotational molding of, 291–292
    sheet and film, production of, 270–272
    stereoregularity of, 149, 150
    thermal behavior of, 153–154
    thermoforming of, 292–296
    thermoplastic, 155–162
    thermosetting, 162–165
    transfer molding of, 286–287
    types of, 143–145
    viscoelasticity of, 60–61
Polymer foams, 297–299
Polymerization, 146–149
    additions, 146–147
    condensation, 148
    degree of, 149
    step, 147–149
Polymer matrix composites (PMC), 317–332
    centrifugal casting of, 331
    closed mold processes for shaping of, 325–327
    continuous laminating of, 332
    cutting methods with, 332
    definition of, 317
    filament winding of, 327–329
    open mold processes for shaping of, 321–325
    pultrusion/pulforming of, 329–331
    starting materials for, 319–321
    tube rolling of, 331–332
Polymer melts, 258–260
Polymethylmethacrylate (PMMA), 157
Polyol, 164
Polyoxymethylene, 157
Polypropylene (PP), 144, 161
Polystyrene foams, 298–299
Polystyrene (PS), 144, 161
Polytetrafluorethylene (PTFE), 158
Polyurethanes, 144, 164–165, 171, 299
Polyvinylchloride (PVC), 11, 144, 161–162
Porcelain enameling, 15, 131, 674
Pores, 338
Porosity:
    of powders, 340
    welding, 726

Portable spot welding guns, 708
Positioning, automatic workpart, 522
Positioning systems (numerical control), 863–870
    closed-loop positioning systems, 866–68
    open-loop positioning systems, 864–866
    precision of, 868–870
Positive molds, 293–295
Positive resist, 796, 797
Postimpregnation winding, 328
Postmolded packaging, 813
Pot furnaces, 234, 247
Potter's wheel, 363
Pottery, 129–131
Pot transfer molding, 286
Pouring cup, 198
Pouring metal (casting), 199–201, 235, 236
Pouring method, 299
Pouring rate, 199
Pouring temperature, 199
Powders (powder metallurgy) (PM), 13, 334–354
    blending/mixing of, 342–343
    chemistry of, 340
    combined pressing and sintering of, 350, 351
    compaction of, 343–345
    definition of, 336
    densification of, 346
    density of, 339
    elemental, 351–352
    extrusion of, 350
    flow characteristics of, 339
    forging of, 350
    friction characteristics of, 339
    geometry of, 337–338
    history of, 336
    impregnation of, 347, 348
    infiltration of, 348
    injection molding of, 349–350
    isostatic pressing of, 348–349
    materials, PM, 351–352
    packing characteristics of, 339
    porosity of, 340
    pre-alloyed, 352
    processing, PM, 335
    and product design, 352–354
    production methods of, 340–342
    products, PM, 352
    rolling of, 350
    secondary operations involving, 346–348
    sintering of, 345–347, 350–351
    sizing of, 346
    surface films on, 340
Powder-based systems, 781–782
Powder coating systems, 673–674
Powdered metal, 89
Powder injection molding (PIM):
    of ceramics, 368
    of powders (powder metallurgy), 349–350
Power:
    in machining, 490–493
    unit, 491
Power drop hammers, 407
Powered leadthrough, 877
Power hacksaws, 528
Power loom, 3
Power shears, 437
Power spinning, 465
Power supply, 879
PP, *see* Polypropylene
Pre-alloyed powders, 352
Precedence constraints, 902
Precious metals, 119–120
Precipitation, 342, 367

Precipitation hardening, 105, 644–645
Precipitation treatment, 645
Precision, 966–967
Precision forging, 404
Precision gage blocks as standards for, 971–972
Precision of motion, 875–876
Precision of NC positioning systems, 868–870
Predeposition, 803
Prefoaming, 298
Preforms, 320, 325–326
Preload forces, 758
Premolded packaging, 813
Prepegs, 321, 322
Prepeg winding, 327
Press(es), 16
    extrusion, 420
    forging, 407
    for sheet metalwork, 458–463
Press-and-blow method, 249
Press brake, 460, 461
Press fitting, 16, 760, 840
Pressing:
    of cemented carbides, 370
    of ceramics, 364, 365, 367–368
    forging, 398
    glass, 248
    of powders, 343–345, 348–351
Press-type spot welders, 708
Pressure, back, 263
Pressure connections, 839
Pressure gas welding (PGW), 715
Pressure thermoforming, 293–295
Pressworking, 376
Priestley, J., 167
Primary bonds, 27–28
Primary industries, 4, 5
Printed circuit boards (PCBs), 22, 821–843
    board preparation, 824–825
    circuitization, 827
    circuit pattern imaging and etching, 825–826
    cleaning of, 834
    components of, 821
    definition of, 822
    different board types, processing of, 827–830
    electrical connector technology, used in production of, 839–843
    finishing of, 830
    history of, 822
    hole drilling, 825
    insertion of components, 831–832, 832, 833
    insulation materials in, 822, 823
    plating, 826–827
    processes used in fabrication of, 824–827
    processing sequence for, 827–830
    reworking of, 835
    soldering components onto, 833–834
    starting boards, production of, 823–824
    surface mount technology used in production of, 835–839
    testing of, 830, 834
    types of, 823
Printed wiring board (PWB), 822
Printing, three-dimensional, 781–782
Prismatic machined parts, 499
Problem solving, 921
Processes, 10–17, 915
    assembly operations, 16
    continuous processes, 6
    historical development of, 11–12

machinery/tooling for use in, 16–17
    processing operations, 12–15
    secondary, 915
    surfaces, effects on, 85–86
    tolerances, effect on, 85
    types of, 10–11
Process anneal, 640
Process capability, 949
Process design, 954
Processing equipment, 862
Processing of metals, 122–123
Processing operations, 10–15, 878
    classification of, 12
    definition of, 10–11
    property-enhancing processes, 15
    shaping processes, 13–15
    surface processing, 15
Process layout, 18–19
Process planning, 913–921
    for assemblies, 917–918
    computer-aided, 919–921
    definition of, 913
    and make or buy decision, 918–919
    for parts, 914–916
    route sheets in, 916, 917
    traditional, 914–918
Products:
    manufactured, 5–7
    variety of, 6
Product design, 239–240, 954
Product features, 948
Production, 4
    job shop, 928
    lean, 939
    mass, 928, 929
Production capacity, 8
Production control, 928–929
Production equipment, 16
Production facilities, 17–20
Production flow analysis, 886
Production kanban, 941
Production lines, 896–908
    automated, 904–908
    history of, 900
    manual assembly lines, 900–904
    transport methods in, 898–899
    and variations in product models, 897–898
    workstations, determining minimum number of, 899–900
Production planning, 928–931
Production planning and control department, 20
Production quantity, 6
Production rates, maximizing, 573–575
Production schedule, master, 930–931
Product layout, 19
Profile milling, 517, 518
Profiling devices, 979
Profiling mills, 522
Programmable logic controllers (PLCs), 878–880
Programming:
    computer programming languages, 877
    leadthrough programmable, 877
    part, 870–872
Programming device, 879
Progressive dies, 458, 459
Projection printing, 798
Projection welding, 710
Proof strength, 756
Properties, material:
    fluid properties, 57–59
    hardness, 52–55
    mechanical properties, 39–61
    stress–strain relationships, 39–51, 107–109

temperature, effect of, 55–57
viscoelasticity, 60–61
Property-enhancing operations, 13, 15
Prototyping, rapid, *see* Rapid prototyping
Protractors, 977
Protrusions, embossed, 765, 766
Proximity printing, 797, 798
PS, *see* Polystyrene
Pseudoplastic fluid, 59
Pseudoplasticity, 259
PTFE (polytetrafluorethylene), 158
PTH (plated through-hole), 831
PTP (point-to-point) control, 876
Public utilities, 1
Pulforming process, 330, 331
Pull system, 940
Pultrusion process, 330
Punch-and-die sheet metalwork, 436
Punch holder, 457
Punching, 437, 438
Punch (metal tool), 376, 457
Push system, 940
PVC, *see* Polyvinylchloride
PVD, *see* Physical vapor deposition
PWB (printed wiring board), 822
Pyrex, 136

QC, *see* Quality control
Quad flat packages ("quad packs"), 810,
  811
Quality control department, 20
Quality control (QC), 947–960. *See also*
  Statistical process control
  and definition of quality, 947
  and manufacturing variations, 949
  and statistical tolerancing, 949–952
  Taguchi methods for, 953–955
  in welding, 724–728
Quality (definition of), 947
Quality problems, 900
Quantity, production, 6
Quantity production, 19
Quartz, 130, 135
Quartzite, 792
Quenching, 11, 642, 643, 645

Rack (electronics packaging), 821
Rack plating, 662
Radial drills, 514, 515
Radial forging, 409, 410
Radial tires, 312
Radiation curing, 673
Radiation techniques for inspection, 986
Radii, corner, 408
Radiographic testing, 727
Rake angle, 478
Rams, 458, 525
Ram mills, 520
Random copolymers, 151
Random errors, 966
Random variations, 949
Rapid prototyping and manufacturing
  (RPM), 774
Rapid prototyping (RP), 772–783
  applications issues in, 782–783
  categories of, 773
  liquid-based systems, 775–779
  and microfabrication technologies, 856,
    857
  powder-based systems, 781–782
  scope of, 772
  solid-based systems, 779–780
  steps in, 773, 774
Rapid tool making (RTM), 782–783
Rational work elements, minimum, 901

Rayon, 144, 158
R chart, 956–959
Reaction injection molding, 299
Reactive ion etching, 806
Reactive system elastomers, 166
Reamers, 513
Reaming, 505, 513
Reciprocating screw, 275
Reciprocating-screw machines, 279
Reclaimed rubbers, 308
Recovery anneal, 640
Recrystallization, 57, 640
Recycled glass, 247
Recycled rubbers, 308
Redrawing, 452
Reduction, 390
Reflectivity, 36
Reflow soldering, 744, 837–838
Refractory ceramics, 131
Refractory metals, 118–119
Regenerated cellulose, 158
Reinforced composites, fiber-, 182–183,
  186
Reinforced metal matrix composites,
  fiber-, 186
Reinforced plastics, 154
Reinforced reaction injection molding
  (RRIM), 327
Reinforcing agents, 154, 177, 319
Reinforcing fillers (to polymers), 190
Reinforcing phase (for composites),
  178–181
Rejection injection molding (RIM),
  284–285
Relief angle, 478
Reorder point systems, 934–935
Repeatability (in NC), 870
Repositioning time, 902
Repositioning time losses, 900
Repressing, of powders, 346
Resins, amino, 163
Resin injection molding, 326
Resin transfer molding, 326
Resist, 628
Resistance brazing, 740
Resistance heating, 666, 667
Resistance welding (RW), 680, 681,
  705–712
  definition of, 705
  flash welding, 711
  high-frequency resistance welding, 711,
    712
  percussion welding, 711
  power source for, 706–707
  projection welding, 710
  seam welding, 708, 709
  spot welding, 707–708
  upset welding, 711
Resistivity, 73
Resolution, 868, 870, 967, 985
Response, speed of, 967
Retaining rings, 763
Retrieval CAPP systems, 919–920
Reverse drawing, 452
Reverse extrusion, 415
Reversing mills, 393
Reworking:
  of PCBs, 835
  of surface mount technology, 839
Ribbed smoked sheet, 307
Ribbon cable, 840
Ribs, 408
Rigid foams, 297
RIM (rejection injection molding),
  284–285
Rings, 763

Ring gages, 976
Ring rolling, 396
Riser, 198
Riser design, 208–209
Rivets, 16, 759–760
Riveting, 11
RMS (root-mean-square), 82
Road width, 780
Robotics, *see* Industrial robotics
Robust design (Taguchi methods), 954
Rocker-arm spot welders, 708
Rockwell Hardness Test, 52, 53, 55
Roll bending/forming, 464–465
Roller die process, 309
Roller mill, 361
Roll forging, 410, 411
Rolling, 672
  flat, 389–393
  gear, 396
  of glass, 250
  history of, 388
  hot, 387
  of metal, 376, 387–397
  in mills, 393–395
  piercing, roll, 396, 397
  of powders (powder metallurgy), 350
  ring, 396
  shape, 393
  thread, 395–396
  tube, 331–332
Rolling mills, 16, 393–395
Roll method, of coating, 274
Rollover, 437
Roll spot welding, 709
Roll straightening, 464
Roll welding (ROW), 720
Rome, ancient, 113, 135, 745, 968
Root-mean-square (RMS), 82
Rotary tube piercing, 397
Rotating disk method, 341
Rotational machined parts, 499, 500
Rotational molding, 291–292
Rotomolding, 291
Roughing, 480
Roughness, 80, 86
  measurement of surface, 978–980
  surface, 82–83
Route sheets, 916, 917
Roving, 319, 320
ROW, *see* Roll welding
RP, *see* Rapid prototyping
RPM (rapid prototyping and manufacturing), 774
RRIM (reinforced reaction injection
  molding), 327
RTM (rapid tool making), 782–783
Rubber(s), 167–172, 306–332
  for belts, 314
  butadiene, 169
  butyl, 169
  calendering of, 309
  casting of, 310
  chloroprene, 170
  coating with, 309, 310
  compounding of, 308
  ethylene–propylene, 170
  extrusion of, 309
  for footwear, 315
  history of natural, 167
  history of synthetic, 169
  for hoses, 314, 315
  isoprene, 170
  mixing of, 308–309
  molding of, 310
  natural, 167–168, 307
  nitrile, 170–171

Rubber(s) (cont'd)
and product design, 315
production of, 307
reclaimed/recycled, 308
shaping of, 309–310
and shaping plastics, 257–258
styrene-butadiene, 171–172
synthetic, 168–172, 307
thermoplastic, 315
for tires, 312–314
vulcanization of, 311
Rubber forming (sheet metal), 455–457
Rubbing, 593
Rule (measuring device), 972
Rule of mixtures, 181–182
Rule of 10, 967, 976
Runner(s):
molding, 278
sandcasting, 197
Rutile, 116, 133
RW, *see* Resistance welding

Safety monitoring, 986
Salt bath furnaces, 647
Salt bath method, 740
Sampling, 100% inspection vs., 970
Sand blasting, 654
Sand casting, 197–198, 214–219
casting operation, 218
cores, 216
mold characteristics, 217
molds and mold making, 216–218
patterns, 215–216
Sandstone, 130
Sandwich molding, 283
Sandwich structures, 184
Saw blades, 528, 556–558
Sawing, 528–529
Saw milling, 516
SAW (submerged arc welding), 701–703
SBR, *see* Styrene-butadiene rubber
SBS (styrene-butadiene-styrene), 172
Scanning laser systems, 982, 983
SCEA, *see* Side cutting edge angle
Schedule, master schedule, 930–931
Scheelite, 133
Schottky defect, 31
Schroter, K., 133
Scientific management, 3–4
Scleroscope, 54
Scrap, 15
Screening, 825
Screen printing, 825–826
Screen resist, 628
Screw(s), 753–755
characteristics of, 265
reciprocating, 275
Screw-cutting lathes, 502
Screw dislocation, 31
Screw-preplasticizer machines, 279
Screw presses, 407
Screw thread inserts, 755
Sealing, package (IC packaging), 812–813
Seaming, 446, 765, 766
Seam welding, 708, 709
Seam welds, 685–686
Secondary bonds, 28–29
Secondary industries, 5
Secondary processes, 915
Second Industrial Revolution, 3
Segmented chips, 485
Segregation, 92
Selective laser sintering (SLS), 781
Selectivity, etch, 805
Self-centering chuck, 507

Self-shielded flux-cored arc welding, 700, 701
Self-tapping screws, 753–755
Semiadditive method, 827, 829
Semiautomatic molds, 286
Semicentrifugal casting, 232
Semi-chemical fluids, 559
Semiconductors, 74, 657. *See also*
Integrated
circuits
Semi-dry pressing (ceramics), 364
Semiliquids, 13
Semimetals, 25
Seminotching, 441
Semipermanent-mold casting, 225
Sensitivity, 967
Separable connectors, 841–843
Serrated chips, 485
Service sector, 1, 2, 5
Setscrews, 754
Setting:
of eyelets, 760
of thermosets, 162
Setting time, 745
Sewing, 764
SGC, *see* Solid ground curing
Shape casting, 195
Shaper, 502
Shape rolling, 393
Shapers, 525
Shaping, 11, 13–15, 22. *See also* specific
headings, e.g. Casting
of ceramics, *see* Ceramic(s)
of flat glass, 250
in glassworking, 247–252
history of plastic, 257–258
machining operations, 524–526
of metals, 122, *see* Bulk deformation
processes; Casting
of piece ware, 248–250
of polymer foams, 298–299
of polymer matrix composites, 317–332,
321–325, 325–327
of rubbers, 309–310
Sharkskin, 269
Shaving, 442
Shears, 437
Shear angle, 440
Shear force, 486
Shear forming, 466
Shearing (metal), 377, 437
Shear-localized chips, 485
Shear modulus, 51
Shear plane, 481, 484
Shear plane angle, 489
Shear properties, 50–51
Shear spinning, 466–467
Shear strain, 50
Shear strength, 51
Shear stress, 50, 487–488
Shear zone, 484
Sheets:
operation, 917
route, 916, 917
thermoplastic, 270–272
Sheet metal operations, 435–470
bending, 442–447
cutting, 436–442
dies for use in, 457–459
drawing, 447–454
elecrolhydraulic forming, 468
electromagnetic forming, 468, 469
explosive forming, 467, 468
high-energy-rate forming, 467–469
with metal tooling, 454–455
presses for use in, 459–463

roll bending/forming, 464, 465
with rubber forming processes, 455–457
spinning, 465–467
tube stock, 469–470
Sheet metalworking, 376–377
Sheet-molding compounds (SMC), 320–321
Shells, 25, 26
Shell casting, 297
Shell molding, 219–220
Shielded metal arc welding (SMAW), 698–699
Shielding, arc, 696
Shock, thermal, 128
Shockley, W., 787
Shock-resistant tool steels, 107
Shop floor control, 941–944
Shop production, job, 928
Shots, short, 282
Shot (of melt), 276
Shot peening, 654
Shrinkage:
and casting, 205–209
and polymer injection molding, 281–282
Shrinkage voids, 726
Shrink fittings, 762, 763
Shuttles, pallet, 522
Sialon, 134
Side cutting edge angle (SCEA), 550
Side milling, 516
Side rake angle, 550
Side relief angle (SRA), 550
Side risers, 209
Siderite, 96
Siemens process, 792
Signature, tool geometry, 550
Silica, 9, 35, 125, 130, 135
Silicon, 140
Silicon carbide, 9, 130
Silicones, 144, 165, 171
Siliconizing, 657
Silicon layering processes, 850–854
Silicon nitride, 134
Silicon processing, 790, 792–796
crystal growing, 793–794
electronic grade silicon, production of,
792
wafers, shaping of silicon into, 794–796
Silver, 11, 120
Simple die, 458
Simple protractors, 977
Simplex mills, 521
Simultaneous engineering, 924
Sine bars, 977, 978
Single column planers, 526
Single-draft, 424
Single machine cells, 888
Single-model production line, 19–20
Single-phase structure, 89
Single-point tool, 479
Single-point tool geometry, 550–554
Single product lines, 897
Single-screw extrusion machines, 262
Single-sided boards, 823, 827
Single station assembly cells, 906
Sinking, tube, 429
Sink marks, 282
Sintered carbides, 546
Sintered polycrystalline diamonds (SPD),
542, 549
Sintering, 15, 674
of cemented carbides, 370
of ceramics, 366, 368–369
of powders, 345–347, 350–351
and pressing, 350, 351
selective laser (SLS), 781

SI (Systeme Internationale), 969
Size effect, 592
Sizing, of powders (powder metallurgy), 346
Skimming process, 310
Skin-dried mold, 217–218
Slab, 388
Slab milling, 516, 519
Slide, 458
Slide calipers, 972
Slip:
  in casting ceramics, 362
  in porcelain enameling, 674
Slip casting, 362–363
Slip deformation of crystals, 32–34
Slip directions, 32
Slip plane, 32
Slit-die extrusion, 270, 271
Slot milling, 516
Slotting, 441, 516, 528
Slot welds, 684, 685
SLS (selective laser sintering), 781
Slug, 438
Slurry method, 657
Slush casting, 226, 297
SMAW (shielded metal arc welding), 698–699
SMC molding, 325
SMC (sheet-molding compounds), 320–321
Smelting, 11, 114
Smith, Adam, 3, 900
Smithsonite, 117
SMT, see Surface mount technology
Snag grinders, 602
Snap fittings, 763
Snap gages, 976
Snap rings, 763
Soaking, 388, 640
Soaking pits, 388
Sockets, 842
Sodium, 26, 27
Sodium chloride, 27
Soft-bake (photolithography), 798
Soft product variety, 7
Solders, 741–743
Soldering, 11, 16, 740–744
  components onto PCBs, 833–834
  definition of, 740
  hand soldering, 833
  joint designs in, 741–742
  methods of, 743–744
  reflow soldering, 837–838
  solders/fluxes for use in, 742–743
  wave soldering, 834, 836–837
Soldering fluxes, 742, 743
Soldering guns, 743
Solderless connections, 839
Solder paste, 837–838
Solids, 13
Solid-based systems, 779–780
  fused-deposition modeling, 780
  laminated-object manufacturing, 779–780
Solid casting, 363
Solid freeform fabrication, 774
Solid gap frame, 459
Solid ground curing (SGC), 777, 778
Solidification processes, 13. See also specific headings
Solidification time, 202
Solid patterns, 215
Solid-phase sintering, 345
Solid profiles (dies), 267
Solid solution alloys, 89–90
Solid-state electronics, 786

Solid-state sintering, 345
Solid-state welding, 682, 694, 719–723
  cold welding, 720
  diffusion welding, 720, 721
  explosion welding, 721–722
  forge welding, 719
  friction welding, 722–723
  hot pressure welding, 720
  roll welding, 720
  ultrasonic welding, 723
Solidus, 68
Solutions, solid, 89–90
Solution treatment, 645
Solvent cleaning, 653
Solvents, 672
Sonotrode, 723
Sortation, parts, 970
South America, 167
Southeast Asia, 167
Spark sintering, 351
Spatter, excessive, 726
SPC, see Statistical process control
SPD, see Sintered polycrystalline diamond
Special purpose equipment, 16
Specific energy, 491
Specific gravity, 67
Specific heat, 69–70
Speed, cutting, 479, 480
Speed lathe, 508
Speed of response, 967
Sphalerite, 117
Spindle bar machine, 508
Spindle drills, 514
Spin forging, 466
Spinneret, 273
Spinning, 11, 273, 465–467
Spinning jenny, 3
Spiral pins, 762
Split patterns, 216
Spotfacing, 513
Spot welding, 684–686, 707–708
Spraying, 274, 299, 310, 672–674
Spray-up processes, 323–324
Spreading, 390
Springback, 444
Sprues, 197–198, 277
Sputter etching, 806
Sputtering, 667–668, 804
Squaring shears, 437
SRA (side relief angle), 550
Stability, 967
Stainless steels, 103, 104
Stampings, 436
Stamping (metal), 376
Stamping press, 436, 459
Standards, 4, 968, 971
Standard test surfaces, 978
Standoff distance, 613
Stapling, 764
Starting boards, 823–824
Stationary pot furnaces, 234
Statistical control, 949
Statistical process control (SPC), 955–960
  attributes, control charts for, 959–960
  interpretation of control charts, 960
  variables, control charts for, 956–959
Statistical tolerancing, 949–952
  for assemblies, 951, 952
  natural tolerance limits, 950, 951
Steady state wear, 537
Steam engine, 3, 502
Steel(s), 9, 11, 98–107
  casting alloys, 239–240
  continuous casting of, 101
  definition of, 101
  free machining steels, 567

high-speed, see High-speed steel
history of, 94
hypoeutectoid vs. hypereutectoid, 95
low alloy, 103, 104
martensite transformation in, 640–644
plain carbon, 102
production of, 98–101
stainless, 103–105
tool, 105–107
Steel-belted tires, 312, 313
Steel-cutting grades (cemented carbides), 547, 548
Steel rule (measuring device), 972
Step polymerization, 147–149
Stereolithography (STL), 775–777
Stereoregularity of polymers, 149
Sticking, 383, 391
Sticking friction, 383
Stick welding, 698
Stitching, 763–764
STL (stereolithography), 775–777
Stock-out costs, 931
Stoneware, 131
Stool (casting), 100
Stop (dies), 458
Straddle milling, 516
Straightening, 448, 464
Straight pins, 762
Straight-sided frame presses, 460–463
Strain, engineering, 41
Strain hardening, 45–47, 122
  definition of, 34
  temperature effects on, 57
Strain hardening exponent, 45
Strain rate, 381
Strain rate sensitivity, 380–382
Strand casting, 101
Strength:
  dielectric, 74
  proof, 756
  and sand molds, 217
  shear, 51
  tensile, 42, 43
  yield, 42, 43
Strength coefficient, 45
Strength-to-weight ratio, 67
Stress(es), 32–34
  engineering, 40, 41
  fracture, 42
  in welding, 724–725
  yield, 42
Stress-relief annealing, 640
Stress–strain relationships, 39–51
  brittle materials, bending of, 49–50
  compression properties, 47–49
  shear properties, 50–51
  tensile properties, 40–47
  types of, 46–47
Stretch bending, 469, 470
Stretch blow molding, 290
Stretch forming, 463–464
Strip development, 458
Stripper, 457
Stripping (threads), 756
Strip plating, 662
Structural adhesives, 745
Stud welding (SW), 704–705
Stulen, Frank, 861
Stylus instruments, 978–979
Styrene-butadiene rubber (SBR), 169, 171–172
Styrene-butadiene-styrene (SBS), 172
Submerged arc welding (SAW), 701–703
Substitutional solid solution, 89–90
Substrate, 80
Subtractive method, 827, 828

Sulfur, 168
Superalloys, 120–122
Superconductor, 74
Supercooled liquid, 35, 68
Superfinishing, 605, 606
Superheat, 199
Supplies, manufactured, 6
Support systems, manufacturing, 17, 20, 912
Surface(s), 78–84
    characteristics of, 79–80
    integrity, 83–84, 980
    and manufacturing processes, 85–86
    measurement of, 978–980
    nominal, 9, 78
    parting, 277
    roughness, 82–83, 978–980
    texture, 80–83
Surface cleaning, 236
Surface contouring, 517, 518
Surface cracking, 423
Surface defects, 31, 77
Surface films, 340
Surface finishes, 568–572
Surface finishing, 592
Surface grinding, 597–598
Surface hardening, 645–646, 648–649
Surface integrity, 80, 980
Surface micromachining, 852
Surface mount technology (SMT), 810,
    821, 835–839
    adhesive bonding and wave soldering,
        836–837
    cleaning, 838
    combined SMT–PIH assembly, 838
    history of, 835
    inspection of, 838, 839
    rework, 839
    solder paste and reflow soldering,
        837–838
    testing, 839
Surface plate, 971
Surface processing, 15, 651–658. *See also*
        Coating processes
    chemical cleaning, 651–654
    diffusion, 656–657
    ion implantation, 657–658
    mechanical cleaning and finishing,
        654–656
    operations, 13
Surface roughness, 978–980
Surface technology, 79
Surface texture, 80, 978
Surface treatments, 15
Surfacing welds, 686
Swaging, 409, 410
Swell ratio, 260
Swing (lathing), 506
SW (stud welding), 704–705
Synchronous transfer systems, 899
Syndiotactic arrangement, 149–150
Synthetic adhesives, 747–748
Synthetic composites, 176
Synthetic diamonds, 549
Synthetic rubbers, 168–172, 307
Systematic errors, 966
System design, 954–955
Systeme Internationale (SI), 969

Tabs, lanced, 765, 766
Taguchi, G., 953
Taguchi methods, 953–955
    loss function, 953–954
    off-line QC, 954–955
    on-line QC, 955
    robust design, 954
Tailpipe, 423

Tailstock, 505
Tandem rolling mills, 394, 395
Tanks, day, 247
Tank furnaces, continuous, 247
Tantalum carbide, 133
Taps, 513
Tape-laying machines, automated, 324
Taper gages, 976
Taper pins, 762
Taper turning, 504
Tapping, 513
Tapping screws, 754
Task time variability, 900
Taylor, Frederick W., 3, 538, 542
Taylor tool life equation, 538–540
Tearing, 453, 454
Technological processing capability, 7–8
Technology, group, 19
Technology (definition), 1
Tee joint (welding), 684, 685
Teflon, 144, 158
Temperature:
    and alloys, 203–205
    cutting temperature, 493–495
    and density, 68
    and diffusion, 71
    and grinding, 593, 594
    material properties, effect on, 55–57
    and melting characteristics, 68–69
    and metal forming, 378–380
    and phase diagrams, 90–95
    and polymers, 153–154
    recrystallization, 57
    and refractory metals, 118, 119
    and superconductors, 74
    and thermal properties, 69–71
    and viscoelasticity, 60, 61
Temperature-activated systems, 162
Temperature failure, 535
Tempered glass, 253
Tempered martensite, 643
Tempering, 11, 253, 643
Template matching, 985
Tensile strength, 42, 43
Tensile tests, 40–47
Terminal blocks, 842
Ternary polymers, 152
Terneplating, 663, 664
Terpolymers, 152
Tertiary industries, 5
TERTM, *see* Thermal expansion resin
        transfer molding
Testing, 839. *See also* Inspection
    continuity, 830
    final (IC packaging), 813
    machinability, 566
    of PCBs, 830, 834
    of surface mount technology, 839
    vs. inspection, 969
    wafer (IC packaging), 811
    of welding, 727–728
Texas Instruments, 787
Textiles, 11
Texture, surface, 80–83
Thermal behavior of polymers, 153–154
Thermal conductivity, 70, 71
Thermal cracking, 128
Thermal diffusion, 802–803
Thermal diffusivity, 70
Thermal energy, 619–627
    arc cutting processes, 625–626
    electric discharge processes, 619–623
    oxyfuel cutting processes, 626–627
Thermal expansion, 35, 68
Thermal expansion resin transfer molding
        (TERTM), 326, 327

Thermal oxidation, 800–801
Thermal properties, 69–71
Thermal shock, 128
Thermal spraying, 675
Thermal surfacing, 675–676
Thermite, 718
Thermit welding (TW), 718
Thermocompression bonding, 812
Thermoforming:
    of plastics, 292–296
    of polystyrene foams, 298–299
Thermoplastics (TP), 144, 145, 257
Thermoplastic elastomers (TPE), 144, 166,
        172, 315
Thermoplastic foams, 283
Thermoplastic polyester copolymers, 172
Thermoplastic polymers, 10, 37, 50, 59,
        144, 155–162
Thermoplastic polyruethanes, 172
Thermoplastic rubber, 315
Thermosets, *see* Thermosetting polymers
Thermosetting polymers (TS), 10, 37, 61,
        144, 162–165, 257, 283–284
Thermosetting polyurethanes, 171
Thermosonic bonding, 812
Thick molding compounds (TMC), 321
Thickness-to-diameter ratio, 450
Thin film deposition, 15
Thin-film magnetic heads, 848
Thinning (sheet metal), 449
Thompson, E., 680
Threaded fasteners, 16, 753–759
    bolted joints, 756–758
    bolts, 753–755
    methods for, 758–759
    screws, 753–755
    tools for, 758
Thread gages, 976
Threading, 505
Thread rolling, 395–396
Three-dimensional printing, 781–782
Three-high rolling mills, 394
Three-plate molds, 278–279
Through-hole mounting, 810
Through holes, 512
Thrust force, 486, 487
TIG welding, 703
Tile, 131
Tilting-pot furnaces, 234
Times, and production rates, 573–574
Time studies, 4
Time-temperature-transformation curve
        (TTC), 641–642
Tin, 11, 92–93, 118
Tin-based alloys, 240
Tin–lead alloy system, 92–93
Tinning, 663, 664, 741
Tin plating, 662
Tires, rubber, 312–314
Titanium, 116–117
Titanium alloys, 240
Titanium carbides, 133, 185
Titanium nitride, 134
TMC molding, 325
TMC (thick molding compounds), 321
Toggle clamps, 279–280
Tolerances, 77, 78
    in machining, 568
    and manufacturing processes, 85
Tolerance design, 955
Tolerance limits, natural, 950, 951
Tolerancing, statistical, 949–952
Tools, *see* Cutting tools; Machine tools
Tool change times, 573–574
Tool-changing, automatic, 522
Tool-chip thermocouple, 494

Tool geometry, 534, 550–558
    multiple-cutting-edge tools, 554–558
    single-point tool geometry, 550–554
Tool geometry signature, 550
Tool grinders, 602
Tooling, 16, 17, 454–455
Tool life, 537–541
    criteria in production, 540–541
    definition of, 538
    Taylor tool life equation, 538–540
Tool material, 534
Tool post, 505
Toolroom lathe, 508
Tool steels, 105–107
Tool storage drum, 522
Tooth sets, 558
Top risers, 209
Torch brazing, 738–740
Torsion test, 50
TP, *see* Thermoplastics
TPE, *see* Thermoplastic elastomers
Tracer mills, 522
Tracing, 522
Transfer lines, 904–905
Transfer molding, 286–287, 326, 327
Transport kanban, 941
Transport methods in production lines,
    898–899
Transverse rupture strength, 50
The Tribune, 228
Trimming, 236, 412–413, 442
Triplex mills, 521
True centrifugal casting, 230–232
True strain, 44–46
True stress, 44, 45
Truing, 596
TS, *see* Thermosetting polymers
TTC (time-temperature-transformation
    curve), 641–642
Tube drawing, 429
Tube forming, 447
Tube rolling, 331–332
Tube sinking, 429
Tube spinning, 467
Tube stock, 469–470
Tubular glass, 251
Tumbling, 654–655
Tumbling barrel finishing, 654–655
Tundish, 101
Tungsten, 119, 133
Tungsten carbides, 133, 185, 542
Tungsten-type HSS, 544
Turbulence, 199
Turning, 11, 14, 15, 477–478, 502–511
    with bar machine, 508
    with boring machine, 509–511
    with chucking machine, 508
    cutting conditions in, 502–503
    definition of, 502
    with engine lathe, 505–507
    flow, 466
    operations related to, 503–505
    with speed lathe, 508
    with spindle bar machine, 508
    with toolroom lathe, 508
    with turret lathe, 508
Turret drills, 514
Turret lathe, 508
Turret presses, 460–462
Twinning, 33–34
Twin-screw extruders, 262
Twist drills, 554–555
Twisting, 455
Two-high rolling mills, 393
Two-phase structure, 90
Two-plate molds, 277–278

Two-roll miss, 308
Two-stage machines, 279
TW (thermit welding), 718

U-bending, 447
Ultimately reinforced thermoset resin
    injection (URTRI), 327
Ultimate tensile strength, 42
Ultra-high-precision machining, 856
Ultrasonic bonding, 812
Ultrasonic cleaning, 653–654
Ultrasonic machining (USM), 611–612,
    856
Ultrasonic techniques for inspection, 986
Ultrasonic testing, 727
Ultrasonic welding (USW), 682, 723
Ultraviolet light absorbers, 155
Unaffected base metal zones, 691
Undercut, 629
Unilateral tolerance, 78
Unit cell, 29
U.S. Air Force, 861
Unit power, 491
Unit-to-unit noise factors, 954
Universal machining centers, 522, 523
Universal milling machines, 520
Unsaturated polyester, 164
Up milling, 516, 517
Upper shoe, 457
Upright drills, 514
Upset forging, 399, 408–409
Upset welding (UW), 711
Urea-formaldehyde, 163
URTRI (ultimately reinforced thermoset
    resin injection), 327
USM, *see* Ultrasonic machining
USW, *see* Ultrasonic welding
UW (upset welding), 711

Vacancy(-ies), 31
Vacuum evaporation, 666–667, 804
Vacuum forming, 292
Vacuum furnaces, 647
Vacuum molding, 220, 221
Vacuum permanent-mold casting, 227
Vacuum thermoforming, 292–293
Valence electrons, 26
Value, added, 4, 10–11
Vanadium, 103
Van der Waals forces, 28
Vapor degreasing, 653
Vapor-phase epitaxy, 802
Vapor phase reflow soldering, 744
Variables, control charts for, 956–959
Variant CAPP systems, 919
Variations, 949
Variety, product, 6
V-bending, 443
VBM, *see* Vertical boring machines
Vents, air, 278
Vernier calipers, 972
Vernier depth gage, 972
Vernier height gage, 972
Vertical boring machines (VBM), 510, 511
Vertical broaching machines, 527
Vertical centrifugal casting, 231
Vertical machining centers (VMCs), 522
Vertical milling machines, 520
Vertical turning machines, 506
Vertical turret lathe (VTL), 511
Via holes, 825
Vibration, 572
Vibratory finishing, 655
Vickers Hardness Test, 52, 53, 55
"Vicor," 136
Virtual prototype, 773

Viscoelasticity, 60–61, 259–260
Viscosity, 57–59, 258–259
Vise, 514
Vision, machine, 983–986
Vitreous structure, 129
VMCs (vertical machining centers), 522
Voids, 282
Voigtlander, H., 133
Voltage layers, 829–830
Volumetric specific heat, 70
VTL (vertical turret lathe), 511
Vulcanization, 11, 144, 166, 167, 311

Wafers (silicon), 794–796
Wafer testing (IC packaging), 811
Wafer yield, 814
Wagner, E., 228
Warm working (metals), 379
Washers, 755, 757
Waste, 15
Water, 28, 29, 101
Water atomization, 340, 341
Water-hardening tool steels, 107
Water jet cutting (WJC), 613–614
Watt, James, 502
Watt's steam engine, 3
Wave soldering, 743–744, 834, 836–837
Waviness, 80, 83
Ways (lathing), 505
WC (tungsten carbide), 542
Wealth of Nations (Adam Smith), 3, 900
Wear:
    tool, 535–537
    wheel, 594–596
Weaving, 11
Webs, 408, 554
Weldability, 728–729
Welders, 682
Welding, 11, 16, 23, 679–691, 694–729
    advantages/disadvantages of, 679–680
    arc, 695–705
    automation in, 683
    as commercial operation, 682–683
    defects in, 725–727
    definition of, 679
    design considerations in, 729
    diffusion, 72
    electron-beam, 716–717
    electroslag, 717–718
    fusion, 681–682, 694
    fusion-welded joints, 690–691
    history of, 680
    joints, weld, 683–686
    laser-beam, 717
    oxyfuel gas, 712–715
    physics of, 686–689
    quality considerations in, 724–728
    resistance, 705–712
    and safety, 682–683
    solid-state, 682, 694, 719–723
    thermit, 718
    and weldability, 728–729
Welding fixtures, 682
Welding machines, 16
Welding positioners, 682
Weld interface, 690
Weld joints, 683–686
Weld lines, 282–283
Weldment, 11, 679
Weld nuggets, 705
Wet chemical etching, 804, 805
Wet lay-up processes, 322
Wet spinning, 273–274
Wetting agents (ceramics), 367
Wet winding, 327
Wheel, grinding, 587–591

Wheel grade, 589
Wheel loading, 594
Wheel structure, 588–589
Wheel wear, 594–596
Whiskers, 178
White, M., 542
White cast iron, 109
Whitney, Eli, 3, 502, 900
Wickham, Henry, 167
Wide operating range, 967
"Widia," 133
WIG welding, 703
Wilkinson, John, 502
Willert, William, 258
Winding, filament, 327–329
Window glass, 250
Wiping method, 652
Wire bonding (IC packaging), 812
Wire coating, 268, 269
Wire cutting, electric discharge (EDWC),
    621–623

Wire drawing, 424
Wire electric discharge cutting (wire-
    EDM), 856
WJC (water jet cutting), 613–614
Wohler, Friedrich, 110
Wolframite, 133
Wollaston, William, 336
Woodworking, 11
Work content time, 899
Work elements, minimum rational, 901
Work hardening, *see* Strain hardening
Workpart positioning, automatic, 522
Workstations, determining minimum num-
    ber of, 899–900
Work volume, 875
Work volume and precision of motion,
    875–876
World War I, 169, 228
World War II, 2, 144, 169, 745, 787
Worst-case design, 951
Woven roving, 320

Wrinkling, 453, 454
Wrought iron, 94
Wrought metal, 89

X-ray lithography, 799
X-y tracing, 522
X-y-z tracing, 522

Yarn, 319
Yield point, 42
Yield strength, 42, 43
Yield stress, 42

Zero insertion force (ZIF), 842, 843
Zero inventory, 939
Zero level, 820
ZIF, *see* Zero insertion force
Zinc, 117–118
Zinc alloys, 240
Zinc plating, 662